David Hilbert's Lectures on the Foundations of Geometry 1891–1902

Springer
Berlin
Heidelberg
New York
Hong Kong
London
Milan
Paris
Tokyo

David Hilbert's Lectures on the Foundations of Mathematics and Physics, 1891–1933

General Editors
William Ewald, Michael Hallett, Ulrich Majer and Wilfried Sieg

Volume 1
David Hilbert's Lectures on the Foundations of Geometry, 1891–1902

Volume 2
David Hilbert's Lectures on the Foundations of Arithmetic and Logic, 1894–1917

Volume 3
David Hilbert's Lectures on the Foundations of Arithmetic and Logic, 1917–1933

Volume 4
David Hilbert's Lectures on the Foundations of Physics, 1898–1914
Classical, Relativistic and Statistical Mechanics

Volume 5
David Hilbert's Lectures on the Foundations of Physics, 1915–1927
Relativity, Quantum Theory and Epistemology

Volume 6
David Hilbert's Notebooks and General Foundational Lectures

Michael Hallett
Ulrich Majer
Editors

David Hilbert's Lectures on the Foundations of Geometry 1891–1902

Springer

Editors

Michael Hallett
McGill University
Department of Philosophy
855 Sherbrooke St. West
H3A 2T7 Montreal, Canada
e-mail: michael.hallett@mcgill.ca

Ulrich Majer
Philosophisches Seminar
der Universität Göttingen
Humboldtallee 19
37073 Göttingen, Germany
e-mail: umajer@gwdg.de

Library of Congress Control Number: 2004103355

Mathematics Subject Classification (2000):
51-02, 51-03, 03-02, 03-03, 00A30, 01A55, 01A60

ISBN 3-540-64373-7 Springer-Verlag Berlin Heidelberg New York

Springer-Verlag is a part of Springer Science+Business Media
springeronline.com

Printed in Germany

Typeset by the editors.
Cover design: Erich Kirchner, Heidelberg
Printed on acid-free paper 41/3111LK - 5 4 3 2 1 SPIN 11555131

David Hilbert and his wife Käthe, apparently taken in 1892
(Voit Collection, Manuscript Division of the
Niedersächsischen Staats- und Universitätsbibliothek Göttingen)

Preface

The present Volume is the first in a series presenting a selection of the previously unpublished lecture notes of David Hilbert on the foundations of mathematics and natural science spanning, roughly, the period from 1890 to 1933. Hilbert's lecture courses represent an enormous fund of learning and invention, and embrace almost every subject common in the mathematical sciences of his day, including subjects of his own creation like the theory of integral equations (Hilbert Space) and large tracts of mathematical logic. This Volume is devoted to geometry; subsequent volumes will treat logic, arithmetic, proof theory and physics. The lecture notebooks were clearly of importance to Hilbert. Most of them remained in his private possession; he often cites them in his subsequent lectures, and many are heavily annotated in his own hand. In the time immediately after his death a number of plans were explored to publish a selection of these notes, but none of these projects came to fruition. That the present lectures are finally being published is the result of the hard work and support, over some fifteen years, of numerous individuals and institutions.

We wish to express our thanks, first and foremost, to the Deutsche Forschungsgemeinschaft (DFG) for its generous financial support since 1993. To edit the Hilbert *Nachlaß*, even partially, required a considerable institutional apparatus, located in proximity to the archives in Göttingen; without the assistance of the DFG, which enabled us to establish a permanent staff in Göttingen, the present edition could never have been realised. We also acknowledge the indispensable scholarly, editorial and technical contributions to this first volume of Ralf Haubrich, Albert Krayer and Tilman Sauer, all members of that permanent staff. The DFG has also supported over the years a series of workshops and conferences, in addition to numerous journeys and research visits of the General Editors, of Professor Helmut Karzel, our main advisor on geometrical matters, and of the editorial staff.

We furthermore thank the Institut für Wissenschaftsgeschichte at the University of Göttingen (in particular Lorraine Daston, its former director) for giving the project a physical home and for recognising its significance.

Numerous other institutions and individuals have provided significant support, which is here gratefully acknowledged:

- In Göttingen, from the first formative stages of the project onwards, Martin Kneser, Samuel J. Patterson, Günther Patzig, and Helmut Rohlfing gave the enterprise encouragement and advice; the Mathematisches Institut and the Niedersächische Staats- und Universitätsbibliothek, the holders of the original documents, granted the necessary permission for publication.
- The Institute for Advanced Study in Princeton, through the offices of Harry Woolf and Phillip Griffiths, provided the Editors with financial support and a collective working environment in the summer of 1997.

- The Social Sciences and Humanities Research Council of Canada (SSHRCC) has supported this project over many years, and (with help from McGill University) funded a conference on relevant material.
- The Alexander von Humboldt Stiftung supported numerous research visits to Göttingen in both the formative and later years of this project. The University of Pennsylvania Research Foundation likewise over many years provided funding for travel and research.
- Carnegie Mellon University and the University of Pittsburgh twice supported conferences and workshops in which the General Editors and various collaborators played a significant part.
- Catriona Byrne of Springer Verlag has given the Edition abundant support and advice, and has been patient with the inevitable delays; at an earlier stage of the project, similar assistance was provided by both Martin Gilchrist and Elizabeth Johnston of Oxford University Press.

A large number of people have been of assistance in various technical and research capacities. For assistance in Göttingen, we thank: Volker Ahlers, Tobias Brendel, Willem Hagemann, Julia Hartmann, Nina Hehn, Arnim von Helmolt, Pamela Klapproth, Heiko Schilling, Friedericke Schröder-Pander, and many others. All the above were supported by funds from the DFG. For assistance in Montreal, we would like to thank Rebecca Pates, Manuela Ungureanu and, above all, Hans-Jakob Wilhelm. These latter were supported by SSHRCC.

Michael Hallett would like to acknowledge the support of SSHRCC over many years, and the assistance, advice and encouragement he has received from Stephen Menn and, most of all, Emily Carson. That he is still sane is due in no small part to Duffy, Olivia and little Thomas.

Ulrich Majer would like to acknowledge the assistance in editorial questions which he has received from his colleague Herbert Breger, whose advice has been of lasting value. In a project like this, stretching over many years and bringing together many people of different character and temperament, and from different continents, there inevitably come moments when throwing in the towel appears to be the only option. That Ulrich Majer did not succumb to the temptations of an easier life is above all due to the balance and sagacity of his wife Anna-Grethe. The tolerance and patience of his four children (Peter, Dorothea, Christina and Niels) were also crucial factors in his survival as an editor.

It would not be out of place for the other three General Editors to acknowledge here the special contribution of Ulrich Majer. As the official recipient of the DFG funding, and the one among the Editors who was constantly 'vor Ort', he has carried special responsibility, not only towards the DFG but towards the editorial staff as well, and has had to deal with the innumerable problems, large and small, which an edition like this inevitably has to face.

This Volume was originally set up by Ralf Haubrich, who played a central role in the examination and assessment of documents, the preparation and editing of the texts selected, and the design of the editorial apparatus. The latter stages of its compilation were the special responsibility of Albert Krayer, who heroically fulfilled impossible demands with efficiency, grace and good humour, and saved us from countless errors. Oliver Keller, ably assisted by Stefan Krämer, has seen us through the last round of corrections with great skill and diligence. We are deeply indebted to all of them.

Finally, we would like to express our enormous gratitude to Helmut Karzel. His expertise and advice on the geometry lectures in this Volume has been indispensable, and, without his wisdom, the notes and Introductions to the texts would be a great deal poorer.

The Editors

Summer, 2003

Contents

Introduction to the Edition

David Hilbert is broadly and rightly considered one of the greatest mathematicians of the modern era. He made fundamental contributions to a wide range of different subjects, spanning almost the entire breadth of pure and applied mathematics – number theory, geometry, the calculus of variations, mathematical logic, and the theory of linear integral equations, to mention only the best known areas of his research. He opened new paths in well-established disciplines such as elementary geometry, and brought other topics, such as the theory of invariants, to their natural completion. He also worked intensively as a theoretical physicist. For many years the development of a general field theory was at the centre of his attention, and the modern theory of quantum mechanics would be unthinkable without 'Hilbert Spaces'. The breadth and originality of his work brought him into fruitful interaction with some of the greatest mathematicians and scientists of his age. He engaged intellectually with such figures as Wilhelm Ackermann, Paul Bernays, Niels Bohr, Ludwig Boltzmann, Max Born, L. E. J. Brouwer, Georg Cantor, Constantin Carathéodory, Richard Courant, Paul Debye, Albert Einstein, Gottlob Frege, Helmut Hasse, Erich Hecke, Werner Heisenberg, Adolf Hurwitz, Felix Klein, Edmund Landau, Hermann Minkowski, Walter Nernst, Emmy Noether, Henri Poincaré, Bertrand Russell, Karl Schwarzchild, Arnold Sommerfeld, Johann von Neumann, Hermann Weyl, Ernst Zermelo, and many others. A remarkable number of talented young mathematicians and physicists were drawn to the heady world of Göttingen by Hilbert's presence; many went on to become his collaborators or colleagues. Hilbert lectured steadily, first in Königsberg, then in Göttingen, for over forty years; he exercised a profound influence on the development of twentieth century mathematics, not only through his publications, but also through his lectures and through his personal interactions with the 'Hilbert circle'.

Hilbert left behind a rich collection of unpublished writings, including a collection of detailed lecture notes covering virtually the entire span of his teaching career. These notes round out his publications and provide a fuller picture of his thought than does the hitherto published record. They document in detail the evolution of his ideas and the interconnections between his various research interests; they often present the motivation behind his technical innovations, and explore their philosophical implications. Hilbert in the lectures engages in extensive analysis and criticism of other thinkers and situates his own work in the wider context of early twentieth century science.

These discussions, which were often excised from the published record, provide in part a more complex picture of his thinking, and in part an unfamiliar picture. In particular (and decisively for the present Edition) the lecture notes reveal a deep and abiding preoccupation with deepening the foundations of mathematics and natural science – in effect, with what he once called the 'Tieferlegung der Fundamente'.

The importance of Hilbert's unpublished writings was well known to his followers in Göttingen. Already during Hilbert's lifetime, Otto Blumenthal had stressed the importance of Hilbert's unpublished lecture notes on physics. (See *Blumenthal 1922*.) Shortly after Hilbert's death in 1943, plans were set in motion, on the initiative of Arnold Sommerfeld, to publish with Julius Springer excerpts from the Hilbert archive as a fourth volume of the *Gesammelte Abhandlungen*. But despite intensive efforts – first by Erich Hecke, then by Paul Bernays – nothing came of these original plans. The scattering of the Hilbert circle, the Second World War, and the sheer scope of the *Nachlaß* no doubt hindered the design of making accessible to a wider audience a selection of Hilbert's unpublished writings. The present Edition seeks to realise the old plan, at least to the extent of presenting, from the previously unpublished writings, Hilbert's most important contributions to the three chief areas of his foundational research, namely, geometry, logic and arithmetic, and physics. The final volume of the Edition will contain popular lectures and writings of a more philosophical character. We publish here only a fragment of the surviving lecture notes; much work remains to be done on the other aspects of Hilbert's mathematical activity, not to mention on the wider history of science in Göttingen in the early twentieth century. But our hope is to make available at least the core of Hilbert's writings on foundational questions. The following sections explain more concretely the state of the surviving manuscripts, and how they have been prepared for this Edition.

1 The Sources and the Texts

With a small number of exceptions, the texts published in this Edition are based on lecture courses or individual lectures given by Hilbert, and are drawn from two sources:

1. The library (often referred to as the *'Lesesaal'*, or earlier, the *'Lesezimmer'*) of the Mathematisches Institut der Georg-August Universität in Göttingen (present address, Bunsenstraße 3–5, 37073 Göttingen, Germany). In what follows, this Institute will be referred to by the acronym 'MI'. [1]

[1] The *Lesezimmer*, installed by Felix Klein in 1886, was located on the first floor of the *Auditorium* building of Göttingen University until the new building, the *Mathematisches Institut* of today, was completed in 1929. Cf. *Neugebauer 1928* and *Frewer 1979*. The Reading Room in the present library is often referred to as the *Lesesaal*.

2. The *Handschriftenabteilung* (Manuscript Division) of the Niedersächische Staats- und Universitätsbibliothek, Göttingen (present address, Papendiek 14, 37073 Göttingen, Germany). In what follows, this Manuscript Division will be referred to by the acronym 'SUB'.

The notes of Hilbert's lecture courses divide into three categories: (1) lecture notes prepared by Hilbert himself for his own use; (2) polished protocols (here referred to as 'Ausarbeitungen'), typically prepared at his request by his assistants or other collaborators, and worked up after the lectures; and (3) in-class lecture notes (here referred to as 'Mitschriften'). The lecture notes in Hilbert's own hand (or that of his wife, Käthe) are found exclusively in the SUB. The other two classes of notes are mostly found in the MI and the SUB; but some stray copies exist elsewhere.

As a rule, the *Ausarbeitungen* came about as follows. Hilbert would ask one of his assistants or collaborators (generally his own graduate students, but sometimes advanced students, doctoral students of other professors, or other collaborators) to take notes of his lectures and then work them up into a polished finished product.[2] Hilbert himself would generally supervise this process closely, first discussing the lectures in advance with the *Ausarbeiter*, and later correcting the written product, usually before its mimeograph reproduction (if it was reproduced), but occasionally afterwards. This at any rate was the procedure in the first three decades. In later years the assistants or collaborators (the best examples are Paul Bernays and Lothar Nordheim) were sometimes asked to work up lectures in advance, so that in these cases it is sometimes difficult to distinguish between the preparations and the finished protocol.

Somewhere between ten and twenty of these official *Ausarbeitungen* were deposited during Hilbert's lifetime in the *Lesezimmer* of the Mathematical Institute where they were freely accessible to students. Some of the early *Ausarbeitungen* are hand-written, but most of the later ones are typed; a few were mimeographed and available for purchase, and in this way almost attained the status of publications. (The circulation was, however, extremely restricted, and, leaving aside the reproduction of three scripts by the MI in the 1980s,[3] the upper bound for the number of copies of each text could have been no more than a hundred or so.) Some of these texts made their way to the United States (for example, to libraries at Harvard and in Princeton); this fact is not surprising when one considers how many American students came to Göttingen for their doctoral work. But the primary range of distribution would have been in Germany. (The extent of the distribution seems to have

[2]It is instructive to compare the list of the *Ausarbeiter* as noted in the index of lecture courses at the end of this Volume with the list of doctoral students and dissertation subjects contained in Volume III of Hilbert's *Gesammelte Abhandlungen*, i.e., *Hilbert 1935*.

[3]The three scripts are the *Ausarbeitung* of a lecture course 'Zahlentheorie' from 1897/1898 (the *Ausarbeiter* is unknown), and the general lectures 'Natur und mathematisches Erkennen' from 1919 in an *Ausarbeitung* by Paul Bernays, as well as 'Wissen und mathematisches Denken' from 1922/1923 in an *Ausarbeitung* by Wilhelm Ackermann.

been in any case small, and despite widespread advertisement we have been able to locate only a few copies outside of Göttingen.) Other *Ausarbeitungen* remained in Hilbert's private possession until his death; in 1967 some of these protocols were deposited in the MI library, and others were sent to the SUB. (We shall discuss the SUB *Ausarbeitungen* and the division of the collection in a moment.)

The crucial point about all *Ausarbeitungen* – both those that during Hilbert's lifetime were deposited in the *Lesezimmer* and those that he kept in his private possession, but which later found their way into either the *Lesesaal* or the SUB – is that, at the very least, they were inspected by Hilbert, and most must have been formally commissioned by him in advance of the lectures. In short, they thus have something of the status of an officially approved record.

In a few cases we possess copies of *Ausarbeitungen* that were worked up into a polished form, but that do not appear to have been either commissioned or seen by Hilbert or to have had any kind of 'official' status; these notes were presumably intended for the private use of the note taker. (One set of such notes, described below, somehow found its way to the library of the Institute for Advanced Study (IAS) at Princeton.) *Mitschriften*, on the other hand, typically seem to have been in-class notes taken by a member of the audience for private use; but it is possible that some of the *Mitschriften* were recorded at Hilbert's request, or were seen by him before being deposited in the MI or the SUB. The surviving texts typically give no indication either way. The Mathematical Institute now possesses (not counting duplicates) seventy-four 'official' *Ausarbeitungen* for lecture courses, and ten *Mitschriften* for seminars; they cover a span of nearly forty years, and illustrate the enormous variety of Hilbert's engagement as a teacher. (An index of all of Hilbert's recorded activity as a university lecturer is to be found at the end of this Volume.)

The SUB possesses the actual Hilbert archive in the usual sense of the term. The *Nachlaß* was in the possession of the MI until 1967, at which time it was transferred to the SUB, where it has been kept since then under the call-sign '*Cod. Ms. D. Hilbert*'. (We will sometimes refer to documents from the *Nachlaß* by 'SUB xyz', where 'xyz' denotes the manuscript number of the document under *Cod. Ms. D. Hilbert.*) It has been periodically supplemented by contributions from private sources, most recently in 2001 by the addition of some letters from Sommerfeld and Einstein to Hilbert. The extensive *Nachlaß* is divided into six sections: I. Correspondence; II. Materials relating to offers of academic appointments; III. Materials relating to the mathematics department in Göttingen; IV. Materials and manuscripts by Hilbert; V. Manuscripts by other authors; VI. Biographical materials.

For the study of Hilbert's thought, Sections I (Correspondence) and IV (Manuscripts) are of primary interest. However, from the outset we decided not to attempt to edit the correspondence, for a very straightforward reason: the *Nachlaß* consists almost entirely of letters to Hilbert, and contains few letters from Hilbert to others. The search for Hilbert's surviving correspondence would have taken years, and would have delayed the publication of the primary lecture manuscripts. We have therefore consulted (and occasionally cited) those portions of the correspondence known to us; but our editorial efforts have concentrated almost exclusively on Section IV. It is to be hoped that others will undertake the task of editing the correspondence.

The manuscripts in Section IV are in turn subdivided into five groups:

1. Lecture notes taken by Hilbert as a student;
2. Manuscripts of Hilbert's lecture courses (with a supplementary section for his seminars);
3. Lectures and addresses: (a) on mathematicians; (b) on various mathematical problems;
4. Calculations and notes concerning various mathematical problems;
5. A book by Ludwig Kambly (*Die Elementar-Mathematik für den Schulunterricht*, Fifth Edition, Part 4, on stereometry), which Hilbert has annotated.

The texts we have edited, to the extent that they do not come from the MI, come primarily from the subdivisions 2 and 3. Section 2 contains all the extant lecture notes in Hilbert's own hand (or Käthe's); the first of these notes is for lectures on the theory of invariants in the winter semester 1886/87, and the last the introduction to the lectures on mechanics in the summer semester of 1924. Section 2 also contains twelve *Ausarbeitungen*, either in manuscript or typewritten; with two exceptions, these *Ausarbeitungen* are also to be found in the MI. (The two exceptions are the *Ausarbeitungen* by Baule, *Cod. Ms. D. Hilbert 559*, and by Bernays, *Cod. Ms. D. Hilbert 567*, neither of which is complete, and neither of which exists in multiple copies.) But, in contrast to the majority of *Ausarbeitungen* in the MI, those in the SUB contain numerous marginal notes, corrections, and additions in Hilbert's own hand. These protocols seem to have been kept by Hilbert in his private possession, and to have been annotated as he revisited issues in subsequent lectures; indeed, the later notebooks often explicitly cite pages in notebooks from earlier years. One would surmise from these facts that, in the transfer of the *Nachlaß* from the MI to the SUB, all the typescripts with comments in Hilbert's hand were transferred to the SUB; but in fact there are numerous typescripts of this sort in the MI as well. So this principle of dividing the typescripts – if such a principle existed at all – was at best loosely adhered to.

In addition to the lecture notes reproduced here from the MI and the SUB, two further sets of notes have been included in the Edition. The first is a *Mitschrift* by Hellmuth Kneser, which his son Martin Kneser generously placed at our disposal, and which consists of classroom notes of the lectures 'Logische Grundlagen der Mathematik' held by Hilbert and Bernays in the winter of 1922–23. The second consists of excerpts from a manuscript, in an unknown hand, which we located in the Institute for Advanced Study in Princeton, and which is based on the lectures 'Probleme und Prinzipienfragen der Mathematik' held by Hilbert in the winter of 1914–15. The manuscript is in a polished form, and is clearly not a set of in-class notes; but there is no evidence that it represents an official *Aussarbeitung*, or that it was ever seen by Hilbert. We are grateful both to Prof. Kneser and to the IAS for their permission to use this material here.

Finally, we have also included several of the previously unpublished 'lectures and addresses' found in SUB subsection IV.3. Not only is the content of these of great importance, but they also serve to illustrate Hilbert's style in lecturing to his peers.

2 *Criteria for the Selection of Texts*

Because of the sheer extent of the Hilbert *Nachlaß*, even as restricted to foundational matters, it was clear from the outset that the present Edition could include only a selection of the material. Our first guiding principle was to limit the Edition to texts primarily concerned with foundational issues in mathematics and natural science, and more particularly with the broad program of the 'Tieferlegung der Fundamente'. But even under this restrictive condition the volume of material was still too large for every relevant text to be included. We have therefore adopted further criteria for selection.

The primary criterion was to provide for the reader, to as great an extent as the surviving texts allow, a complete and intelligible record of the development of Hilbert's contributions to foundational study, exhibiting both the genesis of his ideas and methods and their subsequent development. All further criteria are subordinate to this one.

Before looking at further criteria in detail, it is important to observe that the state of the texts in the three core areas of (1) geometry, (2) logic and arithmetic, and (3) physics is not uniform. In some cases there are significant gaps; in others, the texts available overlap and duplicate ideas. The precise situation is described in the individual volumes; here we will give only a brief sketch.

The situation for geometry is comparatively straightforward, since we possess a manuscript or typescript for virtually every lecture course Hilbert held on geometry. These lecture notes allow one to trace the development of his geometrical ideas up to the celebrated monograph *Grundlagen der Geometrie* of 1899 and beyond, to the year 1902. In the case of the decisive lectures of the winter semester 1898–99 we are even in the fortunate position of having two scripts to offer: the lecture notes in Hilbert's own hand (*Hilbert 1898c**), and an *Ausarbeitung* of the same lecture course by Hans von Schaper (*Hilbert 1899a**). This is especially interesting, since it allows one to follow Hilbert's thought over a period of several critical months *in statu nascendi*.

In logic and arithmetic the development of Hilbert's ideas is less uniformly documented. For the early period (until roughly 1905), apart from work published by Hilbert, we possess only scattered texts, and only one lecture course directly concerned (in part) with the development of mathematical logic (namely the *Ausarbeitung* by Hellinger of the lectures 'Logische Prinzipien mathematischen Denkens' from 1905). For the later period, especially after Bernays appears on the scene in 1917, the situation is relatively favourable, and every significant step in the rapid development of proof theory can be documented. In order to understand the historical development of proof theory, it is necessary to appreciate the role of Bernays; for this reason, we have also included Bernays's previously unpublished 1918 Habilitation thesis, and the first (1928) edition of the book known as 'Hilbert-Ackermann' (*Hilbert and Ackermann 1928*), which in fact turns out to be substantially identical to Hilbert's 1917–18 logic lectures.

In the case of the natural sciences, when seen against the background of the relatively few published works, the situation is extraordinarily good. Indeed, in certain respects it is rather overwhelming, despite particular gaps in one or another area, for example, the development of the theory of special relativity. In fact, in the case of classical physics and statistical mechanics, there was a serious issue in deciding which lecture courses and which expositions to choose, so rich is the fund of material. In both cases, there exist several different texts with similar content but based on different courses held a number of years apart. The decision we finally took is a compromise among competing criteria (see below). In contrast, there are only two courses on quantum mechanics (we have *Ausarbeitungen* for both, by Nordheim), one from *before* and one from *after* the 'new' quantum theory articulated by Heisenberg and Schrödinger in 1925. Since the first half of the lecture course from 1926 is very similar to that for the earlier course from 1922, we have decided to give the first part of the 1922 course and the second part of the 1926 course. The situation with general relativity theory is similar. Apart from the two 'Mitteilungen' on 'Die Grundlagen der Physik' which Hilbert published in 1915 and 1917 respectively, there exists essentially only one lecture course on 'the foundations of physics' (to use Hilbert's term), though spanning two semesters, the summer semester 1916 and the winter semester 1916/1917. Since the decisive phase in the formulation of the generalised field equations was before the beginning of the summer semester of 1916, we have decided to reproduce the two published 'Mitteilungen' here, in order that the lectures can be seen in their contextual sequence.

With this brief review, we are now in a position to summarise more precisely the selection criteria used. Stated in their order of importance, they are as follows:

1. The most important criterion, as was stated, is that of representing as seamlessly as possible the temporal development of Hilbert's positions on the foundations of mathematics and natural science.
2. The second most important criterion is that overlaps and repetitions in content are to be avoided. In some cases, this was not possible, as the following criterion makes clear.
3. Wherever possible, the texts selected are to be given in full, and lecture courses are not to be cut up. (There are, of course, exceptions; the 1891 lectures on projective geometry are not given in full, and, as just mentioned, the two lecture courses on quantum mechanics are both given only in part.)
4. As a rule, Hilbert's own notes are to be given priority over transcriptions and *Ausarbeitungen*, even when, as in the case of the lectures on mechanics from 1898/1899, this is accompanied by disadvantages, such

as uncertainties in the reading of the text, a poorer ordering of the material, and generally greater difficulty in comprehension. (In the case of the geometry lectures from 1898/1899 we have given *both* Hilbert's own notes and their *Ausarbeitung*. This is mainly because the latter represents Hilbert's work at a riper stage of development.)

5. As a rule, *Ausarbeitungen* 'authorised' by Hilbert are given preference over 'private' transcriptions (for example, the Hellinger *Ausarbeitung* of Hilbert's lectures from 1905 on the 'Logische Principien des mathematischen Denkens' over that by Max Born.)
6. As a completion of criterion 5, unauthorised transcriptions are only adopted or made use of (perhaps in commentary) when they represent a genuine completion or clarification of a part of a text, perhaps when it otherwise would be hard to follow, or when it fills in gaps. The latter is the case with the unauthorised Hellmuth Kneser text mentioned above.

3 The Order and the Reproduction of the Texts

The texts are thematically arranged and will be presented in five volumes, with a sixth planned. The volumes are:

Volume 1: Geometry. Volumes 2 and 3: Logic and Arithmetic. Volumes 4 and 5: Physics. Volume 6: 'Public' lectures and talks, commemorative addresses, diaries and notebooks.

Within each volume, the texts are ordered chronologically; Volume 3 is in this respect the temporal successor of Volume 2. In the case of Volumes 4 and 5, a different procedure was chosen so as to avoid confounding different subjects. Within physics, we have therefore established five different subject areas (see below), and the texts are ordered chronologically *within* these areas.

The individual volumes (or pairs of volumes) cover different time spans that sometimes overlap. To some extent, this reflects what has often been called the shift in Hilbert's central research interests, although the 'periodic' nature of Hilbert's research is less pronounced than has commonly been supposed.

Volume 1 covers the period from 1891, the year of Hilbert's lectures on projective geometry, until 1902, which (in part) deals with the last genuinely new research into meta-geometry which Hilbert published in the year following. The selection of this date, 1902, is by no means meant to indicate that Hilbert ceased to be interested in geometry and its foundations after this point or even ceased to give lectures on this subject. The list of Hilbert's lecture courses quite clearly shows the contrary. Rather, our selection represents the fact that the lectures on geometry subsequent to 1902 do not add anything essential to the basic foundational analysis or method of analysis developed up until that point. Geometry remained for Hilbert the prime example of the application of the axiomatic method in foundational analysis, and it was to serve as the ideal to which the construction and analysis of all other exact sciences should aspire.

Volumes 2 and 3 together cover the period from 1889 to 1933; because of the large number of relevant texts, we were forced to make a cut, and divide the texts into two volumes. The year chosen for this cut, 1917, besides being appropriate for the relative sizes of the two volumes, also reflects a certain change of direction in Hilbert's interests. In the lectures in the winter semester of 1917/1918 entitled 'Prinzipien der Mathematik', with which Volume 3 begins, Hilbert pushes forward his investigations into formal logic which had been set aside after 1905. With the analysis of logic begun in 1917/1918, Hilbert subsequently develops a new way of founding arithmetic, one in which the development of *Beweistheorie* plays a decisive part. These developments are well documented in the succeeding lectures over the next eight years.

Volumes 4 and 5 embrace the period from 1898 to 1927. As stated, the volumes are not themselves divided temporally, but between them cover five different subject areas. This division of Hilbert's interests to a certain degree reflects different periods of Hilbert's involvement in the foundational theories of physics, but only to a limited degree. In the first period, 1898–1910, Hilbert is primarily interested in classical mechanics as the foundation for the whole of physics. In the second phase, which begins in 1911, after his work on the theory of linear integral equations had been completed, Hilbert's interests were directed primarily towards the kinetic theory of gases and the molecular theory of matter. These interests always remained alive, but were interrupted in 1914 by a growing concern with what Hilbert termed the 'Grundlagen der Physik', i.e., the general theory of relativity. The intensive and concentrated work on relativity theory falls in the years 1911–1918. Subsequent to this, Hilbert was interested more in epistemological questions and problems raised by developments in physics. The final phase (1922–1928) embraces Hilbert's work on the mathematical foundations of quantum theory.

Each volume contains a general introduction to the issues and problems raised by the texts gathered in that volume. Each such introduction attempts to analyse briefly the importance of the individual works and their relation to each other, both with respect to their content and historical development.

In addition, the texts carry their own individual introductions. On occasion, two, or sometimes three, texts are grouped together. This occurs either when we present two texts dealing with one and the same lecture course (as is the case with Hilbert's own notes and von Schaper's *Ausarbeitung* for the lectures on Euclidean geometry from 1898/1899), or when the texts stand in such close connection that the provision of different introductions would have made little sense.

The introductions are designed to provide background information for the comprehension of the texts which follow; they draw attention to matters of historical interest, or to noteworthy features of Hilbert's presentation, and also describe whatever significant facts are known to us about the origin and fate of the texts themselves. We have tried to make these introductions useful both to the specialist and to the reader who is not an expert in the field under consideration; but in no respect do the introductions (or the footnotes accompanying the texts themselves) strive for completeness. Our goal was to facilitate the reading of Hilbert's texts, not to impose an interpretation, or to relieve the reader of the burden of working with the texts directly. By the same token, the determination of what features of a text are noteworthy involves an exercise of editorial judgement, and therefore varies somewhat from text to text and from editor to editor; but one can only expect a certain degree of uniformity from an edition encompassing such variegated material.

The texts themselves were edited *critically*; by this is meant basically five things (further details can be found in section 4 below):

1. All deletions, additions, replacements (of one piece of text with another) which are deemed to be (even potentially) of some import are noted in a way which makes it clear what the text was *before* the alteration. In doubtful cases, the judgement was always made in favour of inclusion, i.e., only those alterations which are clearly of no consequence are executed without remark.
2. Longer additions, replacements or remarks by Hilbert are given in a special *first* series of footnotes designed for this purpose. This first series is marked by a short horizontal line under the main text. (For deleted text, see *4.2 Rules of Constitution* below.)
3. Uncertain or ambiguous readings are noted in the first series of footnotes.
4. Obvious mistakes and errors have been corrected, and, whenever it seemed appropriate to do so, commented upon. This commentary, as with other editorial comments on the texts, appears in a *second* series of footnotes, which is marked off from the main text and the first series of footnotes by a horizontal line spanning the whole page. The basic principle is that all the material above the long horizontal line belongs to the text proper, whereas everything beneath the line is either deleted text or commentaries by the editors.
5. Literature mentioned in the text itself is given again in the second series of footnotes. The references are in a standard author/date form, with the precise publication details given in the Bibliography at the end of the volume. Works referred to in the editorial material are rendered in the same way. Quotations have been checked and, if necessary, corrected or completed.

The texts have been adapted to modern conventions, but the changes are very conservative. There are two kinds of such adaptations. (*i*) A short list of *standard corrections* has been carried without any editorial indication whatsoever. These we call 'Silent Emendations' (see section 4). (*ii*) A series of *individual corrections* has been carried out 'quietly', in the sense that the changes were not deemed to be of especial significance, and therefore are not the subject of specific comment in the footnotes, and indeed are not even noted on the page. Nevertheless, they can be recaptured by the reader since they are listed at the end of the text (in the Textual Notes), and coordinated with the corresponding place in the text by means of (our) page numbers and line numbering. These we call 'Quiet Emendations' (again, see section 4).

All other textual corrections are explicitly marked in the second series of footnotes.

Following each text is a short, schematic description of the text, and, where appropriate, a list of the special rules according to which that particular text has been constituted, i.e., special rules applied *in addition* to the general rules in force used to transform the text of the original into the printed text of this Edition.

4 General Rules for the Editing of the Documents

4.1 Presentation of the Documents

The texts are arranged into chapters containing one or more documents. Each chapter has been given an English title; each document or group of documents is preceded by an introductory note discussing the significance, the contents, the historical and scientific background, and, if necessary, any peculiarities of the document(s).

The documents are presented in their original language; for the establishment of the text, see the next section, 'Rules of Constitution'.

Manuscript pagination is given in the outer page margin; page breaks of the original are indicated by vertical lines '|' in the text. If a page begins with a displayed formula or a new paragraph the vertical line is omitted.

Line numbers are printed in the inner page margin.

Footnotes in the source text and Hilbert's marginal remarks are rendered beneath a short, and above a full, horizontal line.

Footnotes by the editors (textual remarks as well as substantial comments) are indicated by superscribed arabic numbers and printed at the bottom of the page beneath a full horizontal line. These notes (a) point out textual features, which are of immediate significance for a proper understanding of the document, including deleted pieces of text, (b) identify places, persons, and literature cited, (c) refer to differences or similarities in other versions of the document (if those exist), (d) indicate errors in the text and (e) elaborate on the scientific and historical context.

At the end of each text a list of Textual Notes is attached, which present less significant differences between the appearance of the text in the original document (including deletions, additions, and substitutions) and the (printed) form of the text in this Edition: The notes are related to the text by page and line numbering; for example '63.7' refers to page 63 (in the pagination of the volume), line 7.

After the Textual Notes a description of the original document is presented. These descriptions are schematised and embrace the following items: (a) provenance of the document; (b) size and appearance of the cover and/or leading page; (c) composition and pagination of the document; (d) the original title; and (e) the genesis of the text of the document.

4.2 Rules of Constitution

The texts published in this Edition are for the most part 'clear texts'; that is, so far as possible they consist of complete and error-free German sentences, exhibiting a minimum of editorial intrusions.

All significant differences between the published text and the source text of the corresponding document are reported either in footnotes to the published text itself, or in the Textual Notes at the end of each published text.

However, some of the original texts contain such a large number of slips of no substantive importance to the understanding of Hilbert's thought that to have reported them all would have overwhelmed the Textual Notes (especially in the case of the manuscripts) and would have distracted the attention of the reader from textual issues of greater significance. Therefore, those slips are executed silently. Examples include failure to capitalize words, failure to insert a full stop at the end of a sentence. The remainder of this section describes the general rules of constitution of the printed text from the source text of a document.

It is important here to distinguish between Hilbert's own *manuscripts*, written either by himself or by his wife Käthe, and the *Ausarbeitungen* by his collaborators, since the two types of document demand different editorial treatment. The following two subsections discuss respectively the rules that apply to Hilbert's manuscripts and those that apply specifically to the *Ausarbeitungen*.

A. Manuscripts

The Layout

The source text has been re-set, and the original physical layout has therefore not been maintained. However, an attempt has been made as far as possible to indicate the placement of marginalia and to preserve the original placement of diagrams relative to the text. The diagrams in the edited text have been reconstructed on the basis of both Hilbert's diagrams and the indications in the accompanying text.

The distinction between paragraphs that commence with an indentation and those that do not has been maintained.

Mathematical formulae are printed in italics, in accordance with standard typesetting practice, regardless of whether the formulae in the source text were underlined or not.

Underlined source text is printed in italics. If an underlining does not extend over a full word or phrase but Hilbert nevertheless clearly intended to emphasize that word or phrase, the full word or phrase is rendered in italics. If a specific feature of underlining is of significance for the text, it may be pointed out by a commentary.

There are exceptional cases in which so much text of a manuscript has been underlined that the editors have decided not to reproduce this in italics. Instead an indication is given in the description of the constitution of the text.

Footnotes, Additional Remarks, and Rendering of Larger Pieces of Text

Authorial footnotes and what appear to be additional remarks are printed immediately below the main body of the text separated off by a short horizontal line. The additional remarks mentioned are often handwritten additions by Hilbert which reflect on the text and cannot for this reason be integrated into the flow of text itself; often they appear to have been made years after the original text was composed. (Of course, the boundaries between these reflections and genuine insertions are imprecise, and in some cases called for the exercise of editorial judgement.)

For the footnotes, the original style of numbering – in most cases $^{*)}$, $^{**)}$, $^{***)}$ or ^{1}, ^{2}, ^{3} – has been preserved. But to avoid confusion some slight modifications are sometimes silently introduced, e.g., '$^{**)}$' instead of '$^{*)}$'. In a few cases the source text contains inconsistencies (e.g., '*' in the text and '$^{*)}$' in the footnote). These inconsistencies have likewise been silently eliminated.

The additional remarks are numbered by superscript uppercase latin letters ('^{A}', '^{B}', '^{C}', etc.). Written mostly in the margins or at the top of the page, and occasionaly even on the blank page verso, they can often be assigned with certainty to a particular point in the text, either because of the positioning of the remark, or because of its content, or because of an insertion sign. If such an assignment is not possible, or not possible without ambiguity, then the additional remark is assigned by the editor to an appropriate location, and the ambiguity recorded. In some cases, where such a procedure seemed necessary, a detailed description of the original appearance of the page has been provided.

Sometimes self-fabricated symbols such as '/⁼','⁼/', '▯', etc. appear in the source text. These symbols indicate the place where long passages, written on an extra sheet of paper or occuring on another page marked by the same symbol, were to be inserted. As a rule the intended insertion is executed without printing the symbol.

Spelling and Punctuation

As a general rule, the spelling and punctuation of the source text is preserved. In particular:

Spelling. Hilbert's lapses from his usual patterns of spelling, slips of the pen and grammatical mistakes are corrected and noted in the Textual Notes.[4] Variations in, and misspelling of, proper names[5] are uniformized resp. corrected and noted in the Textual Notes. Idiosyncrasies and variations of orthography, however, are not corrected.[6]

---

[4]Examples of such mistakes are: 'auf einer Geraden gelegen Punkte' (*Hilbert 1894a**, 9) instead of 'auf einer Geraden gelegene Punkte', 'eine Zweite Weise' instead of 'eine zweite Weise', 'die Nachweis' instead of 'der Nachweis'.

[5]For instance, 'Paskal' for 'Pascal', 'Herz' instead of 'Hertz'.

[6]For instance, 'nicht-Euklidische Geometrie' and 'nichteuklidische Geometrie', 'Variabele' and 'Variable', 'Gerade' and 'Grade', 'anderen' and 'andern', 'gelegene' and 'gelegne'.

Capitalization and Punctuation. When necessary, the first word of a sentence is silently capitalized; the terminal punctuation is silently added. Any editorial capitalization or decapitalization inside a sentence, however, is recorded in the Textual Notes. Commas and semicolons are silently inserted where they are useful in improving readability.[7] Hilbert's inconsistency in his patterns of punctuation in enumerations and similar constructions are silently corrected.[8] Missing full stops are silently added to commonly used abbreviations.[9] Incorrect full stops, commas, and semicolons appearing in the source text are corrected, the correction being noted in the Textual Notes.[10] No attempt, however, has been made to impose a consistent scheme of punctuation, and any change of punctuation which may possibly be queried is explicitly noted in the Textual Notes.

In the case of *mathematical formulae*, any insertion of commas by the editors is noted in the Textual Notes, as opposed to the practice in the case of non-mathematical text;[11] in enumerations like 'A, B, ..., Z' a comma is inserted silently after the ellipsis when missing, and the number of full stops in an ellipsis has been standardized to three.[12]

Common *abbreviations* are not expanded,[13] but those which are not common are, and are noted in the Textual Notes. In the case of references given in the source text, however, abbreviations such as 'Math. Vereinig.', 'Math. Ann.', and 'Math. Zeitschr.' are not expanded. The abbreviations '$\overline{\text{m}}$' und '$\overline{\text{n}}$' are always silently expanded to 'mm' and 'nn' respectively.

Unclear handwriting – which concerns the handwriting in those words where Hilbert was attempting to form all the actual letters but did not form them all completely – is silently corrected. Thus, a letter that could be read as either 'i' or 'e' is silently reported as one or the other depending on what seems to be demanded by the context. Ambiguities are recorded either in the footnotes or in the Textual Notes.

Elided handwriting, which here refers to those places where, instead of forming or half-forming all the actual letters, Hilbert used one or more strokes or an undulating line to stand for some of them, is silently expanded.[14] The same applies where the final letters in a word, or the following punctuation, have been run together.

---

[7] In 'Doch Vorsicht da sie leicht irreleitet', for example, Hilbert has omitted the comma after 'Vorsicht'; this omission, which does not affect readability, has been allowed to stand. Hilbert rarely places a comma after 'd. h.'; this omission, too, has not been corrected here.

[8] For example, if Hilbert gives a list of formulae '1.', '2.' etc. but omits a full stop after one of the numerals, it is silently added to the published text.

[9] For example, 'Bd.', 'z. B.', 'etc.', 'd. h.' instead of the source text reading 'Bd', 'z. B', 'etc', 'd. h'.

[10] For example the comma in 'so giebt es mindestens einen 3ten Punkt, der Geraden' (*Hilbert 1894a**, 13).

[11] For example, Hilbert often writes 'die Punkte A B C liegen', which appears here as 'die Punkte A, B, C liegen'.

[12] Thus Hilbert often writes 'A, B, .. Z', which appears here as 'A, B, ..., Z'.

[13] For instance, 'z. B.', 'etc.', 'd. h.', 'Bd.'.

[14] For example, Hilbert often abbreviates 'sch' with a sign that looks like a combination of 's' and 'h'; and often he replaces the final letters of a word with a dash.

Omissions and Repetitions

Omissions of a word, or a phrase, or of symbols (like parentheses) have been repaired by the editors. Such repairs are signaled in the text by ⟨angled brackets⟩ or, in exceptional cases, recorded in the Textual Notes. These editorial insertions are only added if the text would be grossly incomplete, incorrect or unintelligible without them.

The grammatical errors caused by *Repetitions* of words are corrected in the text, the correction being recorded in the Textual Notes.

Substantial Errors

Substantial errors are not corrected but as a rule are pointed out in the editorial footnotes.

Additions, Deletions, and Substitutions

Hilbert's additions, deletions and substitutions, are executed. They are *always* reported, either overtly on the page or in the Textual Notes, in the following cases:

- False starts that indicate a change of intention.[15]
- Corrections that enhance the precision of the text.
- Corrections which modify the content of a sentence.

Whenever Hilbert's changes are significant for an understanding of the text, they are reported overtly. Additions are marked by means of a ↑textual symbol↓, deletions[1] by means of a footnote, and substitutions ↑by a combination of↓[2] both. (For a full description of textual symbols, see below.)

In less significant cases the differences between the published text and the source text are treated quietly and recorded in the Textual Notes. In these cases the published text itself contains no textual symbol or footnote to indicate an alteration. The citations in the Textual Notes are keyed to the pages and lines of the published text, mostly by means of the traditional 'lemma]' form. Some examples may elucidate this:

XXV.1:	published] publ.
XXV.3:	symbol or footnote] symbol ↑or footnote↓
XXV.4:	an alteration] ⟦a difference⟧⊢↑an alteration↓
XXV.4–6:	The citations ⟨...⟩ form.] ⟨Added.⟩

The word or phrase before the square bracket ']' repeats the reading of the published text; the text after the square bracket reports the source text and/or editorial comments.

[1]Deleted by the author: which are to be recorded overtly are marked

[2]The addition replaces: by means of

[15]For example, 'Die Geometrie unterscheidet sich wesentlich von den rein mathematischen Wissensgebieten wie z. B. ⟦Al⟧ Zahlentheorie, Algebra, Funktiontheorie'. It is highly likely that 'Al' is the beginning of the word 'Algebra'. Therefore one might suppose that Hilbert deleted 'Al' because he wanted to put 'Zahlentheorie' first.

Completely insignificant corrections by Hilbert are silently executed; a few instances are:

Instantaneous corrections so brief that they give no sense of Hilbert's preliminary intention.[16]

Corrections of misspelling or false punctuation that do not indicate a change in the sense of the passage.[17]

Corrections of (non-mathematical) text that seems to have been written heedlessly or carelessly, and which do not indicate any change in the meaning of the sentence.[18] The same holds for any correction which reflects solely a stylistic evolution of the source text.[19]

Corrections of mathematical passages (whether composed of symbols, or of words), which do not affect the sense, i.e., if they are, strictly speaking, not necessary,[20] or if they reflect almost instantaneous and insignificant modifications.[21]

Mistakes of punctuation that arise from the careless placement or execution of an alteration are silently corrected.[22] However, there is an exception to these rules: In the case where the fact of the correction itself can shed light

[16]For instance, 'Um zu dem Gegenstück zu gelangen, bedarf ⟦d⟧ es der elementaren Sätze aus der Geometrie' (*Hilbert 1890**, 4) would be rendered as 'Um zu dem Gegenstück zu gelangen, bedarf es der elementaren Sätze aus der Geometrie'.

[17]For instance, 'Beide Flächen schneiden sich in einer ⟦B⟧↑b↓estimmten Schraubenlinie' (*Hilbert 1890**, 6) would be silently changed to 'Beide Flächen schneiden sich in einer bestimmten Schraubenlinie'.

[18]For example, the following correction does not indicate a change of intention: 'Die Schrauben⟦linie⟧fläche enthält alle Schraubenlinien vom nämlichen h.' In this case the context makes clear that Hilbert could hardly have intended to begin this sentence with 'Die Schraubenlinie'. Hilbert constantly refers here to 'Schraubenlinien' and 'Schraubenflächen' and in this instance exchanged the two words by mistake. The following correction does indicate a change of intention, but a change of minor importance: 'eine Reihe bedeutender Männer, ⟨...⟩ aus deren ⟦Untersuchun⟧ Händen die heutige Geometrie der Lage ⟨...⟩ hervorging' (*Hilbert 1891a**, 11). In this case 'Untersuchun' is silently omitted.

[19]For example, 'Winkel, welcher die Projektion im Kreise ⟦bil⟧ beschreibt' (*Hilbert 1890**, 6) would be changed to 'Winkel, welcher die Projektion im Kreise beschreibt'. Although it is very probable that Hilbert intended to write 'welcher die Projektion im Kreise bildet', his substitution of 'beschreibt' for 'bildet' is considered a purely stylistic change.

[20]For example, 'Winkel ↑t↓' (*Hilbert 1890**, 6) would be rendered by 'Winkel t' because when Hilbert wrote 'Winkel' (without t) he meant that particular 'Winkel' which in this context is denoted by t.

[21]For example, 'folglich schneiden sich AA' und BB' ↑etwa in D↓' (*Hilbert 1891a**, 23) is rendered as 'folglich schneiden sich AA' und BB', etwa in D', since in the very next sentence Hilbert employs the letter 'D' to denote the point of intersection in question.

[22]For example, the source text reading 'Linie. ↑und umgekehrt↓' is printed as 'Linie ↑und umgekehrt↓.' And 'ML schneidet dann alle Strahlen $c, d, \ldots k$. ⟦Ist c der⟧ in C', D', ..., K'' (*Hilbert 1894a**, 20) is printed as 'ML schneidet dann alle Strahlen $c, d, \ldots, k$ in C', D', ..., K''. The source text reading 'P_1, P_2, P_3, ↑und wenn alle Punkte auf einer Seite eines Punktes A liegen↓ .. so giebt' (*Hilbert 1894a**, 38) is printed as 'P_1, P_2, P_3, ... ↑und wenn alle Punkte auf einer Seite eines Punktes A liegen↓ so giebt'.

on the content or development of Hilbert's thought,[23] the correction is not executed silently.

An attempt has been made, wherever possible, to decipher text which has been pasted over. Such passages are described in the Textual Notes (if very short) and in footnotes (if extending over several lines).

B. *Ausarbeitungen of Lectures*

Documents of this type are the result of the work of a collaborator of Hilbert's, in most cases working under his direct supervision. Their aim is to present lecture courses of Hilbert's (in whole or in part) in the form of a polished text based on an accurate record of those lectures. (In the overwhelming number of cases the final text is a typescript but in some exceptional cases, for example the *Ausarbeitung* of the 1898/1899 lectures on Euclidean geometry, the text is hand written.) Many of these documents contain changes, corrections and marginal remarks in Hilbert's own hand. These have been treated with special care. Whereas the main text prepared by the collaborator is treated according to simplified standards and straightforward editorial rules, Hilbert's annotations are always made explicit.

Special Rules of Constitution for Hilbert's Annotations

Drawings, figures and other sketches, sometimes added or supplemented by Hilbert in order to make the text more intelligible, are inserted at the appropriate place. The insertion is annotated in the Textual Notes.

Additions, deletions, and substitutions as well as *underlinings* are executed only if they are obviously intended as an immediate correction of the *Ausarbeitung*, i.e., if they are made by Hilbert in connection with, or shortly after, the establishment of the script. In all other cases, Hilbert's annotations are reproduced in the footnotes or the Textual Notes depending on their significance. Completely trivial annotations (correction of spelling etc.) are executed silently.

Longer remarks and extensive changes, which are not intended as a direct correction of the text, or which are made obviously much later than the establishment of the script, are treated precisely like Hilbert's additional remarks in his own manuscripts. All other changes, like Hilbert's habit of putting parts of the text in brackets, are annotated in the Textual Notes but not executed.

Beside the special rules of constitution just stated, all the rules of constitution in the case of Hilbert's manuscripts remain valid for Hilbert's annotations here.

---

[23] This is the case when, for example, a stylistic correction gives evidence that the original text was copied by Hilbert from a textbook.

Rules of Constitution for Texts Prepared by Hilbert's Collaborators

The rules of constitution for texts, prepared by Hilbert's collaborators are basically the same as those for manuscripts. However, because these texts are relatively 'clean' compared with Hilbert's manuscripts, a number of modifications, mainly simplifications, are possible and were deemed desirable.

Obvious errors in spelling are corrected silently according to the rule of 'overwhelming use', which means that that spelling of a word is adopted, which is the one used in the document in the overwhelming majority of cases.

In typescripts, commas, semicolons etc. occur sometimes too far away from the word to which they belong. They are silently assigned to the correct place.

In the case of *mathematical expressions*, the number of dots '...' continuing a formula is standardised to three. Multiplication points are either omitted altogether or printed in all instances depending on the local use.

Additions, deletions, and substitutions by the collaborators are executed silently, except where they seem substantially relevant.

Substantial mistakes are not corrected but pointed out in footnotes. Occasionally some of the special symbols are missing, and had to be added to a typescript by hand. These and similar changes and mistakes are corrected and annotated in the Textual Notes.

Indentations at the beginning of a paragraph are not reproduced exactly as in the original but are standardised. The same is true for formulae, which are reproduced in a standard form. Furthermore, formulae, when separated from the text, are printed centred.

Beside the emphasising by underlining, typical for manuscripts, other forms of emphasis occur in typescripts: *Sperrschrift* (spaced writing), vertical lines beside the text, etc. Underlining and *Sperrschrift* are reproduced by italics; any combination, like underlining and *Sperrschrift* or double underlining, is not reproduced but noted in the footnotes. The same holds for vertical lines and other features.

C. *Textual Symbols*

The textual symbols used in these volumes are the following:

|	Page break in the source text; the number of the new page is printed in the outer margin.
⟨comment⟩	Editioral insertion.
⟨...⟩	Ellipsis.
⟪dubious⟫	Unsafe reading of words or letters ('dubious' is the unsafe word).
⟪?⟫	Unreadable letter.
⟪??⟫	Several unreadable letters.
⟪???⟫	Unreadable word.
↑addition↓	Authorial addition ('addition' is the word added).
⟦deletion⟧	Authorial deletion ('deletion' is the word deleted).
⟦ori⟧⊣sub↓	Authorial substitution ('ori' is replaced by 'sub').

The Editors

Introduction

1 *The Contents of the Volume*

This Volume contains a selection of David Hilbert's unpublished lectures on the foundations of geometry, representing important stages in his engagement with the subject, and therefore important stages in the formation of his approach to the foundations of the mathematical sciences. More specifically, the Volume contains six Chapters:

Chapter 1: Hilbert's own notes for a course on projective geometry, held in Königsberg in the summer semester of 1891.

Chapter 2: Hilbert's own notes from 1893 and 1894 for a course on the foundations of geometry, given in Königsberg in the summer semester of 1894, although originally announced for 1893.

Chapter 3: Hilbert's notes for two 'Holiday Courses' (*Ferienkurse*) for school teachers, given in Göttingen over the Easter vacations of 1896 and 1898 respectively.

Chapter 4: Two sets of notes for a course on the elements of Euclidean geometry, given in Göttingen in the winter semester of 1898/1899.

Chapter 6: Notes for a course on the foundations of geometry, given in Göttingen in the summer semester of 1902.

In addition, Chapter 5 contains a complete republication of the original edition of the so-called *Festschrift*, the first edition of Hilbert's celebrated work *Grundlagen der Geometrie*, published in 1899.

Chapter 1 presents just less than three-quarters of the 1891 course, and the first part of Chapter 3 gives just over half of the 1896 *Ferienkurs*; the 1898 course is given in full, and the same holds for the material represented in Chapters 2 and 4–6. Chapter 4 contains two major items, first, a set of notes for the lectures from 1898/1899 in Hilbert's own hand, and then an official *Ausarbeitung* made by Hilbert's student Hans von Schaper. This was reproduced in multiple copies, seventy according to its Foreword, firstly, and mainly, for the use of those who attended the course, but it seems to have been widely known. (See the Introduction to Chapter 4.) Chapter 4 also includes all of Hilbert's rather extensive remarks on his own copy of von Schaper's *Ausarbeitung*; these were added by Hilbert (as was his habit) opposite to, and on, the title page, and are given in an Appendix. Chapter 5 not only contains the original edition of the *Festschrift*, but also (in an Appendix) the very important additions to it that Hilbert wrote for inclusion in the first French translation of 1900 and which were retained in the first English translation of 1902. In addition, editorial footnotes to the text of the *Festschrift* contain all the important changes which were made in the second German edition of 1903.

The *Festschrift* has an important place in the development of the foundations of mathematics, and is one of the most celebrated works in the subject's history. Hilbert's *Grundlagen der Geometrie* is widely available in new editions (the most recent edition is the fourteenth), and is also easily obtained in English and French translation. However, the German editions since the eighth of 1956 are based on the seventh edition prepared by Hilbert in 1930, which is much altered from the first edition of 1899, and the most easily accessible translations are also based on this seventh edition. Thus, it is hard to get an impression from these of the original content of the *Festschrift*. Add to this the fact that the first edition itself is hard to obtain, and this would seem to be reason enough to justify its republication. The main reason for its appearance here, though, is that its omission would have left a large gap in the sequence of work on geometry which this Volume represents.

2 *The Editing of the Texts*

The basic editorial principle employed in this Volume is an elementary one, namely that that the texts should be presented in a readable and comprehensible way, but that this must not involve too great a deviation from the manuscripts. Seen in the light of this principle, three of the eight texts used were comparatively straightforward to edit, namely the *Festschrift* itself, and the two *Ausarbeitungen*, those for the 1898/1899 lectures and the 1902 lectures respectively. Hilbert's own notes (in Chapters 1–4) were correspondingly difficult to edit. This holds above all for the lecture notes on the foundations of geometry for 1893/1894 (Chapter 2) and, to a lesser extent, for the 1898 lectures on Euclidean geometry (Chapter 4). Hilbert's notes (and by this is meant also those dictated to Käthe Hilbert) are difficult because they were subject to fairly constant revision, both in the course of writing and later, though how much later is often impossible to judge. The difficulty is especially acute when Hilbert is venturing into new territory. The 1893/1894 lecture notes are an extreme case. Although they were originally written for the course announced in 1893, they were clearly extensively revised for the course actually held in 1894; Hilbert did not write a new manuscript, but reworked the old one. This often produces a text which is not only hard to read, but (more pertinent here) very hard to represent in an edition such as this. In the section of the manuscript entitled 'Einführung der Zahl', Hilbert's method of dealing with the false starts and the new beginnings stretches the confines of the manuscript to breaking point. In addition to strikings out, marginal additions, alterations and renumberings, Hilbert resorts in the end to scissors and paste, and even then has to make additions and cross references, to the point where the manuscript borders on the incomprehensible. It is not clear that any editorial reconstruction could do it full justice. At the most extreme point, we have resorted to a device of little commentary and the presentation of facsimile pages, so as to give the reader at least a raw impression of that with which we were faced. Although sometimes taxing, the difficulties are nowhere near as severe in the case of the notes for the 1891 course on projective geometry or for the 1898/1899 course on Euclidean geometry. (And in this latter case, where the text or a proof or a diagram is hard to reconstruct, there was simple resort to the official *Ausarbeitung* of the lectures, which follows the same path, but is given with exemplary clarity.) Despite these difficulties, we believe we have rendered the content and spirit of all the manuscripts in as comprehensible a way as possible without too much distortion of the texts.

In reproducing Hilbert's own manuscripts, we render Käthe Hilbert's script in a sans serif font to distinguish it from Hilbert's own, thus making the extent of subsequent revision more visible. (This is useful above all in the 1893/1894 lectures given in Chapter 2.) In editing all the lectures, the *Ausarbeitungen* as well as Hilbert's own, we have employed a shorthand device to denote the various authors of emendations found in the manuscripts. For instance, in the case of Hilbert we use '*Hi*' with a small superscript letter, either '*s*' for 'schwarz' (black ink), '*b*' for 'blau' (blue pencil or crayon), '*r*' for 'rot' (red ink), or '*g*' for 'grau' (pencil), thus yielding 'Hi^s' and so on. Käthe Hilbert's hand is denoted by '*HiK*', Hans von Schaper's by '*Sc*', and August Adler's by '*Ad*'. Wherever the hand of the emender is not clear to us, we have used '*Xx*'. All these appear with the appropriate superscripts.

One device we have adopted generally, so as not to overload the text with too many notes detailing alterations and additions, is to execute many of the changes to the text without comment in the text itself, and then to note this in Textual Notes at the end. (The most trivial changes are executed without remark.) Another difficulty concerns Hilbert's substantive additions. In most cases, as mentioned above, there is no certain way of knowing when they were added, and therefore in what relation they stand to the underlying text. (Recall that Hilbert had most of these notes in his possession up to his death in 1943.) To give one or two examples, there are changes in the 1893/1894 lectures which clearly relate to the 1898/1899 lectures, and so may well stem from this later period. In the case of the 1898/1899 lectures in their turn, there are changes executed by Hilbert to his copy of the *Ausarbeitung* which are not in the version given to the *Lesezimmer* of the Mathematical Institute, indicating that they were made at least a little later than 1899. And some of Hilbert's changes to the manuscript of the *Ausarbeitung* of the 1902 lectures clearly date from as late as 1926, when Hilbert might well have been preparing for the lecture course on foundations of geometry he gave in 1927. There is no reason to think that Hilbert's practice generally was any different from that illustrated in these few examples. Hence, the reader is cautioned not to draw from these additions, without further evidence, broad conclusions which relate to the temporal development of Hilbert's ideas.

It goes without saying that this is not an attempt to render Hilbert's manuscripts in their original form. This will certainly not satisfy the reader who puts full weight on 'textual purity', and wishes to see the text in its raw state, complete with flaws. For such a reader, there remains no recourse other than to consult the original manuscripts in the Niedersächsische Staats- und Universitätsbibliothek and the Mathematical Institute in Göttingen. Our task has been to present the content of the lectures in a digestible way.

Each chapter is supplied with commentary of two sorts, that in the Introductions and that in the footnotes to the text. Neither the footnotes nor the Introductions aspire to completeness. The guiding principle was to supply information useful to the reader, firstly, whatever contextual information we have about the origin of the texts, and then specific information about concepts, proofs, constructions, new theoretical developments, sources, allusions and historical connections. We have tried to avoid burdening the reader with assessments and interpretations of the texts. To some extent, assessments and interpretations are inevitable; even the selection of material for a volume such as this already represents a form of interpretation, and the brief description later in this Introduction of the importance of the material in the Volume is by its nature an *assessment* of the lectures. And in the specific commentaries, biases and idiosyncrasies undoubtedly reveal themselves at many turns. Nevertheless, the Editors and commentators have tried to be as unobtrusive as possible.

The Introductions were written by Ralf Haubrich (Chapter 1), Ulrich Majer (Chapter 2), Michael Hallett and Albert Krayer (Chapter 3), and Michael Hallett (Chapters 4–6). The commentary footnotes for Chapter 1 are by Ralf Haubrich, for Chapter 2 by Ralf Haubrich and Ulrich Majer, for Chapter 3, first part, by Ralf Haubrich and Michael Hallett, for Chapter 3, second part, by Michael Hallett, for Chapters 4 and 6 by Michael Hallett.

The Chapters carry rather different degrees of commentary, and the reasons for this are different from case to case.

(*a*) The text of the lectures on projective geometry (Chapter 1) is relatively straightforward, once certain features of Hilbert's orientation are grasped, and thus required rather little by way of detailed exposition and commentary. We only occasionally intervene to explain proofs, diagrams or theoretical developments.

(*b*) The text of the 1893/1894 lectures (Chapter 2) is, as mentioned above, extremely messy, and the text hard to reconstruct. Nevertheless, its overall structure is basically clear, despite the difficulties of detail. The Introduction consequently concentrates on the ways in which the lecture notes do, and the ways in which they do not, point forward to the end of the eighteen-nineties.

(*c*) The 1896 *Ferienkurs* (Chapter 3) alludes to extracts from a lecture which is not given in this Volume (it forms part of Volume 2 of this edition), namely a talk Hilbert gave to the German Society of Mathematicians in 1895 and which was a preparation for his famous *Zahlbericht*. These extracts were clearly meant to be read out, and this necessitated detailed commentary and quotation in the footnotes, not least since much of what is alluded to itself requires elucidation. The geometry part of the 1898 text is sketchy, but nevertheless points forward to much that is set out in detail in the lectures begun later that year. Wherever possible, we have tried to make this clear.

(*d*) The 1898/1899 lectures (Chapter 4) present another case altogether. It was felt that several subjects dealt with in the lectures deserve detailed commentary, each of them for distinct reasons: some because of the importance of theoretical material that was later included in the *Festschrift*; some because the material concerned is dealt with only obliquely in the *Festschrift*; some because of the desire to make explicit Hilbert's sources and his allusions (for instance, with respect to his historical remarks), or the fate of the open problems Hilbert sketches towards the end of the lectures; and some because of the lack of clarity and disconnectedness of the remarks. All this specific commentary has been included in a special section of the Introduction, consequently making it rather long. However, this has at least avoided stretching the footnotes beyond their natural capacity. Most of this commentary applies both to Hilbert's own notes, and to the *Ausarbeitung*. The text underlying the second part of Chapter 4 is Hilbert's own copy of the *Ausarbeitung* of the lectures. The many (disconnected) remarks on geometrical subjects which Hilbert jotted down at the beginning of this manuscript are given in an Appendix to Chapter 4. The remarks are mostly brief, and sometimes appear cryptic; in some cases, they were subsequently stricken out. Nevertheless, it was necessary to give them, and therefore it was thought worthwhile to provide sufficient elucidation to render them comprehensible.

(*e*) Chapter 5 containing the first edition of the *Festschrift* presents another case again. This also has a long Introduction, which, however, consciously avoids commentary on the content of the work itself. Instead, the Introduction contains information about its immediate genesis; about the French and English translations, in whose production Hilbert was involved; about the additions Hilbert made to the French translation, the two most important of which are included in an Appendix to Chapter 5; about changes to the subsequent editions of the *Festschrift*, above all to the second German edition of 1903; and lastly a brief commentary on the origins and aims of Hilbert's *Vollständigkeitsaxiom*, which is the major addition to the French translation and to the second German edition of the *Festschrift*.

(*f*) The commentary to the 1902 lectures in Chapter 6 is again comparatively measured. These lectures in part overlap with the material of the 1898/1899 lectures and the *Festschrift*, and in part go beyond it, in the direction of Hilbert's major publications on geometry in the years 1902 and 1903. Most of the Introduction is concerned with sketching that work.

3 Geometrical Material Omitted from this Volume

None of Hilbert's published writings on geometry appears here, with the single exception of the *Festschrift* in its original version. In addition, there are various unpublished writings of Hilbert on geometrical subjects which are not included. These fall into three main categories, and a few comments about each are in order.

(*A*) Notes for lectures on assorted geometrical subjects. The list of Hilbert's lectures included at the end of this Volume reveals that Hilbert regularly gave lecture courses on geometrical subjects. Apart from those represented in this Volume, and the two major courses to be discussed under (*B*), there are something like another ten lecture courses (depending on exactly what one counts), for which we have five of the sets of notes. (We do not count Hilbert's lectures on 'Zahlbegriff und Quadratur des Kreises', which are discussed under the heading (*C*) below.) But while these notes occasionally contain foundational remarks, they are not primarily concerned with the foundations of geometry, and hence have been omitted here.[24]

To this category of 'non-foundational' work we should perhaps add Hilbert's *Schulheft* from 1879/80. This has in any case recently been published as part of the centenary edition of the *Grundlagen der Geometrie* (*Hilbert 1999a*).

(*B*) There exist two sets of notes for major lecture courses on geometry: (*i*) Notes entitled 'Anschauliche Geometrie', an *Ausarbeitung* by W. Rosemann (*Hilbert 1920**), prepared in connection with lectures in the winter semester of 1920/1921. (Hilbert clearly repeated this course several times in the nineteen-twenties.) (*ii*) Notes entitled 'Grundlagen der Geometrie', an *Ausarbeitung* by Arnold Schmidt (*Hilbert 1927**), prepared in connection with a lecture course in the summer semester of 1927.

The former presents a wide-ranging and profound look at intuitive geometry and its applications in other branches of mathematics. It is, however, not primarily concerned with the foundations of mathematics. More importantly, perhaps, it is well represented in the celebrated book *Hilbert and Cohn-Vossen 1932*, though the book is more detailed than the *Ausarbeitung*. (See Hilbert's Foreword to the book.)[25]

The lecture course from 1927 on the foundations of geometry, judging by the *Ausarbeitung*, represents more than anything a major revision of the material in Hilbert's original *Festschrift* and its descendants up to the sixth edition of 1926, all of which exhibit only minor changes over the second edition of 1903. This revision then served as the basis for the seventh edition of 1930. The *Ausarbeiter*, Arnold Schmidt, is thanked profusely in Hilbert's Foreword

[24]It should be noted that there are many pieces in Hilbert's unpublished work which contain interesting foundational remarks or discussions and which are not being published in this Edition.

[25]The book appeared in English translation as *Hilbert and Cohn-Vossen 1952*. Both the German original and the English translation have experienced several reprints.

to the seventh edition of the *Festschrift*, and Hilbert there ascribes to him much independent work.[26] The revision represented by the seventh edition is important, and any full study of Hilbert's geometrical work should take it, and the 1927 lectures, into account. Nevertheless, three points are to be stressed here: this work is less important for an understanding of the *emergence* of Hilbert's foundational approach to geometry; it is, to a large extent, well represented in the widely available seventh edition of the *Grundlagen der Geometrie*; and finally much of the revision is quite possibly the work of Arnold Schmidt. For these reasons, when taken together, the decision was taken not to include it here.

(*C*) General lectures on foundational issues in which geometry plays a significant part. Throughout his mathematical career, Hilbert gave lecture courses containing wide-ranging discussion of many issues in the foundations of mathematics. The courses (held several times over the decade from 1894) on the concept of number and the problem of squaring the circle are examples, and so are the lectures on the principles of mathematics held regularly over the twenty years from 1905. In the first set just mentioned, the foundations of geometry is obviously a central subject, since one of the aims is to show by analytic means that certain questions are insoluble within elementary geometry. Moreover, some of these lectures contain interesting material on the axiomatisation of complete, ordered fields, and on continuity principles, very much topics of importance in the geometrical lectures. Nevertheless, the main focus of the lectures is on the foundations of analysis, and therefore the foundational discussion fits more closely into Volume 2 of this series. In the second set of lectures mentioned, the axiomatic method is invariably set out as the primary basis for foundational investigation; geometry is then standardly portrayed as providing the best illustration of the efficacy of this method, and this is often followed by a detailed account of what is achieved through these means in geometry. The *Ausarbeitung* for the lectures from 1905 entitled 'Logische Principien des mathematischen Denkens' (*Hilbert 1905a**) provides a good example, since it has a detailed section of ninety pages on the foundations of geometry. Nevertheless, these lecture courses present the axiomatic method as a *general* method of approaching foundational issues. Geometry is invoked as the first and most complete example of how productive the axiomatic method can be, but the courses as a whole are not specifically about geometry. Again it was felt that this kind of foundational discussion, and especially the new insights revealed, fit more appropriately into Volumes 2 and 3 of this series which concentrate on the foundations of logic and arithmetic. This is particularly so in the case of the 1905 lectures and the 1917/1918 lectures, since both contain genuinely new material on the foundations of logic. In short, for good reason, interesting material on the foundations of geometry has been consciously omitted from this Volume.

---

[26] The complete Foreword to the Seventh Edition is reproduced in the Introduction to Chapter 5.

That the decision is, to some extent at least, arbitrary, is underlined by the fact that the 1905 lectures mentioned also contain an interesting section (some seventy pages) on the foundations of physics. The decision to include the *Ferienkurse* in the present Volume has a similar arbitrariness to it, though this time it was felt that the genuinely new insights fit well into the sequence of lectures on geometry which build up to the *Festschrift* of 1899.

Hilbert's more popular lectures (e.g., the lectures on 'Natur und mathematisches Erkennen' from 1919, 'Wissen und mathematisches Denken', and the manuscript entitled 'Über die Einheit in der Naturerkenntnis', probably from 1923/1924) also contain interesting remarks on the foundations of geometry, as do Hilbert's Notebooks. Nevertheless, again geometry is not the main focus, no matter how important it is as an example. It is intended that this Edition be completed with a final Volume 6 containing a selection of these more popular works, and Hilbert's Notebooks.

There is finally one piece in Hilbert's *Nachlaß* which is in part concerned with geometry and which is (again, in part) of foundational interest; this piece consists of notes for a lecture held before the Königliche Gesellschaft der Wissenschaften zu Göttingen on the occasion of its one hundred and fiftieth anniversary.[27] This piece is also omitted from this Volume; instead, we give a brief description here.

First, the notes are short, consisting of less than two sheets in Hilbert's hand. In addition, they are piecemeal, with four separate sections which have no overt connection to one another. They concern: (1) the purpose of carrying through an axiomatisation of Lie's approach to geometry; (2) Hilbert's own axiomatisation of plane Bolyai-Lobatchewsky geometry; (3) (apparently) Dedekind's work on number fields and Galois groups; and lastly (4) Dirichlet's Principle.

Of these, section (2) is a very short summary of the work contained in Hilbert's paper *Hilbert 1903b*, which is briefly discussed in the Introduction to Chapter 6. Section (3) is not concerned with the foundations of geometry, though it does represent Dedekind's work here as a particular example of a general mathematical-philosophical question: how far can one extend facts concerning the finite to the infinite? Section (4) again does not concern the foundations of geometry, and in any case does not contribute anything not already in the record of Hilbert's published work. The remarks of interest to the foundations of geometry are really confined to section (1). This consists of Hilbert's summary of his own treatment of Lie's approach to geometry through groups of transformations, and is therefore a brief summary of the work contained in the paper *Hilbert 1903c*. (See the brief discussion in the Introduction to the 1902 lectures, i.e., Chapter 6.) The remarks begin

[27]The piece is *Cod. Ms. D. Hilbert 582*, entitled 'Vortrag in der Kgl. Gesellschaft der Wissenschaften, d. 8.11.1901 (zum 150jährigen Jubiläum)'. The lecture carries the same date as the paper *Hilbert 1902f*, which subsequently appeared in translation in the first English edition of the *Festschrift* in 1902; see the Introduction to Chapter 5. This memoir presents the first part of *Hilbert 1903c*, mentioned below.

with a brief characterisation of the purpose of an axiomatisation of geometry, which does not differ widely from things Hilbert says elsewhere. But then it continues:

> *Fussend* auf Riemann und Helmholtz hat nun Lie ein Axiomensystem aufgestellt, welches sich von den nach *Euklidischem Muster* entwickelten Systemen wesentlich unterscheidet. Lie's Axiome enthalten *funktionentheoretische Bestandteile*, indem Lie verlangt, dass die Bewegung durch differenzierbare Funktionen vermittelt wird. *Lie musste* diese Voraussetzung machen, wollte er überhaupt seine von ihm ausgebildete Methode der *continuierliche⟨n⟩ Gruppen* anwenden. Es fragt sich, ob die *funktions Bestandteile* nicht bloss wegen des Wunsches diese Methode anzuwenden nöthig waren und nicht vielmehr der Sache selbst fremd ⟨und des⟩wegen *überflüssig* sind. Es zeigt sich, dass sie es in der Tat sind. Wir nähern uns dadurch wieder dem *alten Euklid*, insofern wir dem Begriff Bewegung nicht die weitergehenden infinitesimalen Eigenschaften, die Lie noch nöthig hat, brauchen aufzuerlegen, sondern mit den *elementaren* Postulaten auskommen, die bereits in dem Euklidischen Begriff der Congruenz enthalten sind und die uns allen aus den von der Schule her bekannten Sätzen über die Congruenz von Dreiecken geläufig sind.

4 Hilbert's Geometrical Work Represented in this Volume

The final part of this Introduction contains a brief description of Hilbert's geometrical work as represented in this Volume.

The work contained here forms an integral sequence, in which emerges a particular approach to what can be called 'ordinary' or 'traditional' geometry. Hilbert saw the highpoint of his approach in the various papers that he and his students (for example, Dehn and Hamel) published in the early years of the twentieth century, and which epitomise what Hilbert called the 'axiomatic method'. This became, in Hilbert's eyes, the main tool for examining the foundations of science, and it is explained, or at least alluded to, in virtually all of his mature writings on foundational subjects. Hilbert saw this method as quite distinct from the received view of axiomatisation, according to which an axiom system presents *the* foundation of a science, i.e., the basic concepts and truths on which a theory rests. Hilbert's axiomatic method, on the other hand, was intended as a means for analysing possible foundations, once a certain body of 'facts' is clearly enough delineated to constitute a theoretical domain. It investigates the consistency of the system, its completeness (i.e., how well it represents the 'facts' presented to it), and it investigates the independence of the postulates chosen and the deductive links between central propositions (axioms) and theorems, the main tool in all this being modelling in other mathematical systems, thus interpreting and reinterpreting. Moreover, Hilbert saw nothing canonical in axiom systems; different approaches using a different conceptual apparatus were always possible even if they were not already present. Of the different approaches available, one might be suitable for a given purpose, but not for another. In any case, one aim of axiomatic analysis, with its overridingly meta-theoretic slant, is to investigate the links between different approaches, and also the links to cognate mathematical and physical frameworks. The meta-theoretic point of view demands a certain neutrality towards the basic concepts and propositions, according to which received (or 'everyday') meanings are set aside, and arbitrary reinterpretation is facilitated, this without repudiating entirely those received meanings. This is exactly the approach to traditional geometry which emerges from the sequence of work presented in this Volume.

The 1891 lectures on projective geometry (Chapter 1), are, as far as the mathematics goes, completely conventional, and the theoretical material, although altered to some extent and differently presented, is drawn from largely conventional sources. Projective geometry is seen here as a part of what Hilbert calls the 'geometry of intuition', and his presentation relies explicitly on intuitive arguments. For example, continuity arguments are used without reliance on specific principles, often through a consideration of what happens when a point is allowed to 'wander' along a line. Nevertheless, the lectures are instructive, especially when viewed in the light of Hilbert's *Einleitung*. The 1891 lectures follow a path decisively influenced by von Staudt, namely that of pursuing (projective) geometry synthetically, as free as possible from 'calculation and measuring'. It is therefore concerned with part of that 'body of geometrical facts' which intuition yields. Hilbert's *Einleitung*, however, outlines different general approaches which can be adopted to the mathematical treatment of elementary geometry, for example, the axiomatic approach and that via analytic geometry. The rest of the work presented in this Volume follows the axiomatic approach, and fits in remarkably well with how Hilbert described it in his *Einleitung* to the 1891 lectures (p. 3), namely the investigation of the 'axioms underlying the facts presented by the geometry of intuition' and a 'systematic comparative investigation of those geometries which arise when one or more of the axioms is set aside'. Hilbert later describes the aim of the axiomatic study of the fundamental core of Euclidean geometry as being to 'analyse our intuitional capacity' (see Hilbert's introductions to the 1898/1899 lectures, to the *Festschrift* and to the 1902 lectures); in other words, the body of facts to be analysed is that body which 'ordinary, intuitive, Euclidean geometry' unreflectively yields.

The set of lectures from 1893/1894 (Chapter 2) embarks on the first level of axiomatic analysis, showing the clear influence of Pasch. Incidence and order axioms are introduced as the basic axioms; then in addition there are axioms governing continuity and congruence ('movement'), and finally an analysis of various possible treatments of the theory of parallels. While the influence of Pasch's book is unmistakeable, Hilbert's treatment is essentially different, since Pasch's treatment is really what might best be called an axiomatic presentation of geometry, not an axiomatic analysis. The meta-theoretic approach to geometry is not yet fully in evidence in Hilbert's lectures, although Hilbert prepares it in the sense that he adopts the view that geometry (with respect to its appplications) is just a 'Schema' (p. 60) or 'Fachwerk' of concepts, and that it is up to 'the human understanding' how to apply the schema to appearances, how to 'fill it with material [*Stoff*]' (p. 60). Thus, already present here is the view which Hilbert took as intrinsic to the presentation of a theory according to the axiomatic method, namely that the axiom system should furnish what he calls a 'Fachwerk von Begriffen' abstracted from 'Tatsachen', with the axioms corresponding to 'Grundtatsachen'. The 'completeness' demand (which Hilbert in the 1893/1894 lectures states in association with Hertz's *Bildtheorie*), is then that all the 'Tatsachen' laid before us should be derived in the 'Fachwerk' from the 'Grundtatsachen'.

The 1893/1894 lectures also introduce a completely new kind of question, which eventually leads to a further level of axiomatic analysis. According to Hilbert, it is essential that elementary geometry 'introduces number', and this becomes a central element, not just in these lectures, but also in the 1898 and 1899 treatments of geometry. (For example, in his own notes for the 1898/1899 lectures, on p. 4, Hilbert writes that 'Man kann den Fortschritt einer Naturwissenschaft oder eines Zweiges der Naturwissenschaft geradezu messen an dem Grade, in welchem die Zahl eingeführt ist'.) Hilbert's intention with this injunction is not to resort to analytic geometry, but rather to carry out this 'introduction' purely geometrically. Thus, while the interest is still in trying to see how elementary geometry can be developed synthetically, the new question is (implicitly) introduced of how well synthetic geometry can match full analytic geometry. That is, if geometry can be coordinatised from without, is it possible to show from *within* that the points on a given straight line can be coordinated one-to-one with the real numbers? In the 1893/1894 lectures, Hilbert attempts to show directly that there are 'enough' points to guarantee this. In the 1898/1899 lectures (Chapter 4), however, Hilbert undertakes a finer analysis, namely of what is responsible geometrically for the presence of a complete, ordered field structure. This leads to the '*Streckenrechnung*' which is worked out in the 1898/1899 lectures, and which leads to important new results, for instance that the *planar* Desargues Theorem can be taken as a replacement for *spatial* incidence axioms. The development of the *Streckenrechnung* and the associated work forms a centrepiece of the 1899 *Festschrift* (Chapter 5).

In the 1898/1899 lectures (Chapter 4), two other important elements appear. The first element is a range of questions specific to elementary geometry; the second element is general to Hilbert's axiomatic method.

The questions specific to geometry concern an analysis of the means used to prove basic results in elementary geometry, for instance, in what Hilbert calls 'Schulgeometrie'. For example, how much can be achieved in Euclidean geometry without invoking the full power of the congruence axioms? Or the continuity axioms, even in the weak form of the Archimedean Axiom? These questions are first in evidence in the notes for the second so-called *Ferienkurs* from Easter 1898 (Chapter 3), notes which show Hilbert already deeply engaged with many of the specific problems of this kind which will feature centrally in the 1898/1899 lectures. Many of these questions fall under the heading of what Hilbert in the 1898/1899 lectures calls 'purity of method' questions. The most striking example of this is provided by the examination of Desargues's Theorem. A proof for the planar version of the theorem can be given using just incidence and order axioms, assuming that the spatial incidence axioms are included. But is the use of the *spatial* element in the proof of a *planar* result essential?

The element general to the axiomatic method is the fully developed approach to independence proofs, the construction of a bewildering range of interpretations drawn from the manipulation of whatever mathematical material can be conveniently deployed. It is this which makes it possible to answer many of the questions raised in the finer axiomatic analysis visible in Hilbert's work from 1898 on. Axiomatic investigations since Euclid were partly designed to address injunctions of the form: 'Show that theorem P can be proved using only the concepts and axioms in Σ'. Hilbert's work on elementary geometry in the later 1890s now makes widespread use of a general method which can address in addition the injunction: 'Show that theorem P can*not* be proved using only the concepts and axioms in Σ'. This also involves the final absorption of analytic geometry into Hilbert's study, for the prime providers of interpretations for the independence and/or relative consistency results are models which start from an analytic geometry built over some appropriate number field. In this, the catholicity of Hilbert's approach to geometry is fully revealed.

An important part of this view is stressed in the 1898/1899 lectures, and extends the 'schema' view of the 1893/1894 lectures. It is expressed by saying that basic terms used, thus 'point', 'line', 'plane', 'between', 'congruent', etc., must, for the sake of the analysis, be regarded as devoid of whatever sense one informally attaches to them; it is the axioms, and the axioms alone, which will give properties to the fundamental notions. This view is not directly present in the *Festschrift*, but its traces can be found in the view presented there that geometry concerns itself with 'three systems of things', and that the 'precise and complete description' of the relationships between these 'things' will be given by means of the axioms. The view is philosophically and psychologically important, for it is meant to prepare the reader for the modelling and remodelling which forms a central part of Hilbert's mature work on geometry, and which often represents severe distortion of the intuitive meaning of terms like 'straight line'.

All aspects of this method and this approach to the axiomatic analysis of elementary geometry as developed from 1891 are fully present in the lectures from 1898/1899, and the vast majority of the important results in the *Festschrift* already exist in these lectures (though see the Introduction to Chapter 4). The design of the *Festschrift*, though, is different; it does not follow the slow build-up of the 1898/1899 lectures, influenced by the picture of projective geometry as the most fundamental part of elementary geometry, and it makes earlier use of the simplifying power of the Euclidean Parallel Axiom. Moreover, much of the motivation for, and philosophical discussion of, the results and the investigations found in the 1898/1899 lectures is either lacking in the *Festschrift* or present only in a very cursory way.

The 1902 lectures (Chapter 6) already represent a step away from the *Festschrift*. They are in part a repetition of some of the material now already familiar from the 1898/1899 lectures and the *Festschrift*, in part a presentation of new work along the same lines (the main example being the investigation of the role of the Triangle Congruence Axiom in the proof of the elementary theorem that the base angles in an isosceles triangle are equal), and in part also the investigation of new geometrical questions beyond the 1898/1899 framework, for example, an elementary axiomatisation of the Bolyai-Lobachevsky geometry, or a more elementary treatment of Lie's geometry. In short, the lectures represent in embryo what at the beginning of this section was called the highpoint of Hilbert's work in geometry.

We end this section by pointing out briefly a series of other positions on geometry which are common across many of Hilbert's lectures.

The first is the view stated in the 1894 lectures that geometry is a 'natural science', the 'most perfect, the most complete'. This view is clearly represented later in Hilbert's introductions to the 1898/1899 and 1902 lectures, and also hinted at in the 1898 *Ferienkurs*. It is not stated in the *Festschrift* of 1899, but it can again be found in other lectures, not least where Hilbert presents the project of an axiomatic construction of physics, for example, in his lectures on mechanics from 1905/1906 (*Hilbert 1905b**).

The second common view, again to be found clearly in the 1893/1894 lectures, is the appeal to Hertz's *Bildtheorie* characterization of natural science. This is also referred to in the later lectures from 1898/1899 and 1902, but is not mentioned in the *Festschrift* itself. Not much is said in detail about this view, but it was probably important in Hilbert's formulation of his general conditions on the adequacy of an axiomatisation, the conditions of the mutual independence of the axioms, the consistency of the system, and its completeness 'with respect to the facts before it'. Conditions are first set out explicitly in the *Einleitung* to the *Festschrift*, then in *Hilbert 1900b*, and again in Hilbert's introduction to the 1902 lectures. But note that the *Festschrift* does not state consistency as a condition, although the 1898/1899 lectures quite explicitly concern themselves with the consistency question, and the (relative) consistency of the whole system is actually proved at a relatively early stage in the *Festschrift*. (The syntactic view of relative consistency is also clearly to be seen in the *Festschrift*; i.e., given a model, Hilbert recognises that any proof of inconsistency in the theory being modelled could be transformed into a proof of inconsistency in the theory used for the construction of the model. Nevertheless, the basic idea of consistency is still semantic, since all the consistency proofs are based on the notion of reinterpretation.) Completeness in the sense stated is already raised as a criterion in the 1894 lectures in direct connection with the reference to Hertz's *Bildtheorie*.

The third common view concerns figures. In the 1894 lectures, Hilbert stresses that figures are useful as heuristic devices, and even that their use is like a kind of experimenting (a view also stated in the 1891 lectures on projective geometry), but that nevertheless a proof is only correctly given when it is carried through independently of figures, 'step by step from the axioms'. Only then does the 'Beweisverfahren' make sense. 'Das Figuren machen', i.e., what Hilbert calls 'Experimentalgeometrie', is already over once the axioms are set down. (See the 1893/1894 lectures, p. 11.) The 1898/1899 lectures and the 1902 lectures present a more balanced view. Hilbert admits that the lectures make very frequent use of figures (indeed, very often the proofs given would be incomprehensible if these were not present), but that 'we will never rely on them', that care will be taken that the operations carried out on a figure will 'remain logically correct/valid'. (The wording in the *Ausarbeitungen* for the 1898/1899 and 1902 lectures is close.) In Hilbert's own notes for the 1898/1899 lectures, he says that figures are just a 'Schrift' (i.e., something like a shorthand notation), and a 'mnemonic, not a means of proof'. There is no such statement about the dangers of figures in the *Festschrift* itself, or the need for proofs to be independent of them.

Michael Hallett, Ulrich Majer

Chapter 1

Lectures on Projective Geometry (1891)

Introduction

This Chapter presents manuscript notes (entitled *Projektive Geometrie*) in Hilbert's own hand (*Hilbert 1891a**) for a course of lectures he gave in Königsberg in the summer semester of 1891. There also exists a transcription of these lectures (or a reworking, it is not known which) by Julius Hurwitz entitled *Geometrie der Lage* which resides in the *Nachlaß* of Adolf Hurwitz in Zürich (at the library of the Eidgenössische Technische Hochschule). In a letter to Klein of 30 June, 1891, Hilbert writes:

> Unser Zuhörerkreis besteht jetzt im Wesentlichen aus 2 Studenten, zu denen in meiner Vorlesung über projektive Geometrie noch als dritter Mann der Vorsteher der hiesigen kgl. Kunstschule – ein für Geometrie interessierter Maler – hinzukommt. (See *Hilbert and Klein 1985*, letter 64, p. 74.)

Hilbert's text is divided into an Introduction and eight sections.[1] Hurwitz's *Geometrie der Lage* is structured in the same way, but lacks the introduction; there are no other important divergences, although the text is significantly briefer. What we give here is not the complete text of Hilbert's lecture notes, but rather the Introduction, §§ 1–4 and the last part of § 8.

The Introduction is prefaced (p. 3) by a brief survey of geometry, dividing it into: (1) the geometry of intuition, with a reliance on 'simple facts of intuition'; (2) axiomatic geometry, the investigation of the axioms underlying the geometry of intuition, and a systematic investigation of the system attained when one or more of these axioms is set aside; and (3) analytic geometry, also called later in the Introduction 'Cartesian geometry', the reduction of geometry to analysis via the assumption of a coordination between the points of a line and the real numbers. Hilbert places projective geometry under the geometry of intuition along with what he calls 'school geometry' and *analysis situs.* It is not clear whether this division was a later addition or not (the remarks on p. 2 were certainly added later), but on p. 5 (p. 4 is blank) begins what appears to be the introduction proper. Here Hilbert, echoing Gauß (see *Gauß et al. 1880*, 497, letter from Gauß to Bessel, 9 April, 1830),[2] sets out what he sees as a clear philosophical difference between those branches of mathematics ('pure mathematics', number theory, algebra, function theory) whose results are achieved by 'pure thought' ('reines Denken' or 'rein logisches Denken'), and geometry, which is characterized as the theory of the properties of space, and as such, like physical theories, can never be based on pure thought alone. In addition to this, Hilbert states something like a number-theoretic reductionist position with respect to 'pure' mathematics, namely that a theorem is only in the end considered as proved when it can be

[1]The text seems at some point to have been divided into Parts I and II, with 'I' and 'II' written in crayon in large figures after 'Projektive Geometrie'. 'Projektive Geometrie II' is written on a page (marked p. 83) which is the right-hand side of a sheet which has clearly been inserted, for the sheet splits a sentence that begins on p. 80 and continues on another page, a second page numbered p. 83; pp. 81 and 82 are blank.

[2]The relevant passage is cited below, note to p. 7 of Hilbert's text.

expressed as a relation between numbers.[3] As Hilbert puts it: 'Also die ganze Zahl ist das Element' (p. 6). This Hilbert then gives a brief tour of the history of geometry: Greek geometry, despite its monumental achievements, lacked a unifying *method*; such a method was supplied with the principle of reduction to analysis first suggested by Descartes; this, though, resulted in a separation of geometry from its intuitive roots, a separation which was overcome with the return to synthetic geometry in the nineteenth century, particularly in the pursuit of projective geometry; Hilbert sees the culmination of this in the work of von Staudt, who created a projective geometry in which one 'neither calculates nor measures, but only constructs, in which there is no recourse to the circle or to angle size, but only to the straightedge' (p. 12). There is a further interesting comparison of the relationship between intuitive and analytic geometry in the final paragraph (pp. 107–109), included here.

In § 1 (pp. 13–22), Hilbert introduces some central concepts of projective geometry, including those of points and lines at infinity. Hilbert stresses that these concepts involve nothing difficult, nothing 'supernatural' or 'metaphysical', but constitute just an abbreviated way of referring to 'simple, intuitive facts'. (See pp. 16 and 19.) On p. 19f, he introduces eight 'fundamental laws of intuition' in four pairs dual to each other. Although the term 'axiom' is avoided, in effect these laws correspond to the incidence axioms of projective geometry. On p. 21, it is explained that these laws exemplify the principle of Duality, a Principle which appears here only as an organising 'remark [*Bemerkung*]'; its proof will come 'only later' (p. 22). A proof of the two-dimensional Duality Principle is indeed sketched later, pp. 77–78, in § 5, 'Poles and Polars', not included here. For some further remarks on this proof, see below.

Hilbert begins § 2 (pp. 22–33) with a proof of Desargues's Theorem in both its spatial and planar versions, as well as of their converses (pp. 22–27). He then explains the concept of harmonic position, and shows that the concept of a harmonic set of (four) points on a line is well-defined, i.e., given three points on a line arbitrarily chosen, the fourth point harmonic to these three (the harmonic conjugate) is uniquely determined (pp. 27–29). This is then extended naturally to harmonic lines and harmonic planes, with some important interrelations stressed. Hilbert then points out that if $ABCD$ is a harmonic set (D the harmonic conjugate of A, B and C), then $ADCB$, $CBAD$ and $CDAB$ are also harmonic sets. Thus both A, C and B, D 'belong together', and these pairs are 'coordinated' with each other. Then follows the concept of 'separation', explained by the fact that moving from A to C along the straight line they determine entails passing through either B or D. Then Hilbert adds to p. 31 (on a sheet pasted in) the theorems: (*i*) two pairs of points harmonic to AC are not 'separated', and (*ii*) for any two pairs of points that do not 'separate' each other, then there's always one pair of points harmonic to both pairs. Hilbert's proof-sketches here make informal

[3]This position is also referred to by Hilbert in his 1896 'Ferienkurs' (*Hilbert 1896**, 6–7). See Chapter 3, in particular the note to the relevant page.

use of a continuity assumption (allowing points to 'wander' or 'vary' on a line), although this corresponds to no underlying 'axiom' of continuity.

In § 3 (pp. 33–47), the concepts of *perspective* (pp. 34–36), and *projectivity* (p. 37) are introduced, and the relationship between them is investigated. Using the theorems added to § 2, Hilbert shows that projectivities are ordered correspondences (p. 38f). Then the so-called fundamental theorem of projective geometry (stated on p. 42) is proved, i.e., that a projective relation between two basic configurations is uniquely determined when three elements of the one basic configuration are assigned to three elements of the other. With the fundamental theorem (§ 3), the foundation of projective geometry is complete; § 3 ends with a few of its consequences.

§ 4 (pp. 47–70) is devoted to second-order configurations. In the middle of the section, we find Brianchon's Theorem (p. 54) and its dual, Pascal's Theorem (pp. 62–63). Pascal's Theorem was to become particularly important in Hilbert's later investigations of the foundations of geometry in 1898/1899, beginning in *Hilbert 1898b**, 18–27. (See Chapter 4 in this Volume.)

Sections §§ 5–8 (pp. 71–107) have been omitted in the present edition, with the exception of the final paragraph of § 8 (pp. 107–109). These sections are entitled respectively: § 5, 'Poles and Polars' (pp. 71–84); § 6, 'The 2nd-Order Ruled Surface' (pp. 84–92); § 7, 'Collinearity and Reciprocity' (pp. 92–102); and § 8, 'The 2nd-Order Surface' (pp. 102–109).

The following observations concern largely §§ 1–4, especially the basic text layer.

Hilbert clearly bases his account on the first few sections of the first volume of the third edition of Theodor Reye's textbook *Geometrie der Lage*, i.e., *Reye 1886*. The similarities between this text and Hilbert's notes are certainly very noticeable in §§ 1–3, and extend from the material's arrangement to the adoption, sometimes almost word for word, of comments, definitions, theorems and proofs.[4] The differences between Reye's textbook and Hilbert's notes are therefore all the more instructive. What is striking first of all is that Hilbert's account is shorter and more concise. Several longer passages from Reye's book are not represented at all, and a few deliberations are discussed in less detail. This extends to theorems, to concepts and to formulations. Hilbert's mathematical language tends to be terser, more formal, sometimes less descriptive, and more precise. Hilbert is largely successful in introducing *only* the concepts, definitions and theorems that are necessary for the development of projective geometry. The deductive relations between the theorems are elaborated and clearly demonstrated, and the ordering of theorems is sometimes different from Reye's; moreover, Hilbert's proofs are easy to understand and clearly organized. This makes Hilbert's text distinctively different from Reye's representation of projective geometry. An example of the difference in approach is perhaps afforded by the treatment of Desargues's Theorem. Hilbert argues far more clearly and more briefly than Reye, and the notes that

[4] Some examples are to be found in *Toepell 1986*, § 1.7.

Hilbert added (to pp. 25, 26, and 28) suggest his realisation that he could have gone further in this direction, indicating Hilbert's interest in changes of this kind. (The additions are unfortunately not dated, but were probably made before 1894.)

Also worth noting are differences in interpretation of what Hilbert calls the eight 'fundamental laws of intuition' stated in § 1. Hilbert seems to have taken these as the foundation of projective geometry, as is illustrated by the fact that he gives § 1 the title 'Grundbegriffe', and explains at the beginning of § 2 that now that the fundamental concepts have been determined, the fundamental theorems 'on which the entire science is based' can be developed. Reye uses neither the term 'fundamental concept' nor the term 'fundamental law' at the relevant places. For him, the eight principal propositions are just illustrations of the Principle of Duality (see *Reye 1886*, 24f, 34).

Nevertheless, in its central aspects the structure of Hilbert's text conforms to Reye's book. Reye's work itself goes back to von Staudt's writings, mainly to *von Staudt 1847*. This would help to explain why important steps of the development of the foundation of projective geometry subsequent to von Staudt are not reflected in Hilbert's lectures. For example, the new approach taken in Pasch's *Vorlesungen über neuere Geometrie* (*Pasch 1882*), which for Hilbert was a central source for the lectures in 1893/1894 and later, is not taken into account, at least not noticeably. In particular, Hilbert does not formulate axioms, though, as remarked, the 'Grundgesetze' are proto-axioms, and indeed in the introduction (p. 3), projective geometry is expressly presented as a structure based on intuition. Furthermore, on p. 2 of the introductory remarks, Hilbert states that it might be better to give a course on projective geometry as 'Geometry of Intuition', thereby making use without scruple of 'motion' and 'intuitive relations without resort to calculation'. It seems safe to assume that he is alluding in part to use of continuity assumptions, since difficulties here can be circumvented by appeal to intuition. Aside from this remark, Hilbert devotes no word to the issue of continuity, an issue which had been a central point of contention since the early 1870s. This is somewhat surprising, particularly since in his preface to the third edition, Reye refers to his attempt to free the proof of the fundamental theorem from assumptions of continuity, as is not done by von Staudt, and makes passing reference to further analyses by Klein and Darboux (*Reye 1886*, XII). Moreover, the volume published by Hilbert's Königsberg colleague Lindemann (*Clebsch 1891*) also discusses continuity in great detail. All this changes in Hilbert's later lectures, in large part following the spirit of Pasch. Another illustration of the absence of Pasch's influence is Hilbert's sketch of a proof of the two-dimensional Principle of Duality in § 5, pp. 77–78, not included here. The proof is essentially the same as that given in Reye's book (*Reye 1886*, §§ 136–138). Given a second-order curve in a plane, and using the theory of poles and polars, it is possible to show that to each point (taken as a pole) not on that curve, there is conjugated a line (the polar) in such a way that for any plane figure, we can show the existence of (indeed, construct) a reciprocal (dual) plane figure which has the dual incidence properties. The proof is therefore a straightforward existence proof showing that for each configuration-object, there exists another configuration-object dual to it. The proof of the Duality Principle found in Pasch's book, on the contrary, is, at root, a different kind of proof, namely a proof-theoretic one, for it is a systematic attempt to show how the proof of any theorem in plane projective geometry can be transformed into a proof of the dual statement. The kernel of the idea is that, if the proof of a theorem is deductively correct, then it must work independently of what the terms and concepts in it mean; these can then be systematically replaced by the terms/concepts dual to them without disrupting the form of the proof (and thus the fact that it is a proof) at all. (See *Pasch 1882*, 98.) This is not overtly an existence proof, though it does, of course, have existential consequences, not surprisingly since it depends on there being explicit underlying existence axioms, each of which has its dual. Nevertheless, it means that the Duality Principle could in principle be proved immediately after setting up the axioms, assuming that there are axioms. More importantly, this kind of argument is the core of the method of reinterpretation which is at the centre of Hilbert's later treatment of independence questions, that truth-preserving proofs remain truth-preserving proofs under suitable reinterpretation of their basic mathematical terms.

Hilbert made many corrections and additions to his notes. It is not at all clear when exactly these alterations were effected, but the additions on p. 2 were certainly made after 1900, because they contain a reference to a book by Grünwald, which only appeared in 1900.

The notes to the text which follows are by Ralf Haubrich.

Ralf Haubrich

Projektive Geometrie

⟨Additions on pp. 2–3:⟩ 2

Bezüglich der Axiome und Grundlagen der Geometrie und zwar insbesondere der projektiven Geometrie vergleiche man:
Noth: Arithmetik der Lage.
Pasch: Neuere Geometrie.[1]

Vielleicht besser als „Geometrie der Anschauung" lesen und somit Bewegung etc. (intuitives Verhältnis ohne Benutzung der Rechnung) unbedenklich zu Hülfe nehmen.

Die v. Staudtsche Imaginär-Theorie findet sich am besten bei Grünwald: Lineare Lösung der Aufgaben über das Verbinden und Schneiden imaginärer Punkte etc. (Separatabzug), wovon auch eine Fortsetzung erscheinen soll.[2]

Eintheilung der Geometrie. 3

1.) Geometrie der Anschauung

(führt ihre Behauptung auf einfache Thatsachen der Anschauung zurück, ohne diese selbst, ihre Entstehung und Berechtigung zu untersuchen; sie benutzt unbedenklich die Bewegung, Grenzlage, Parallelismus[3] etc. und ist also euklidische Geometrie). Sie zerfällt in 3 Theile

a. Schulgeometrie (Congruenzsätze, Dreieck, Vieleck, Kreis etc.).
b. Projektive Geometrie (Kegelschnitte, Brennpunkte, Curven im Raume).
c. Analysis Situs.

---

[1] *Noth 1882*, *Pasch 1882*.

[2] *Grünwald 1900*.

[3] The terms 'Grenzlage', 'Bewegung' do not appear in the *Grundschicht* of this text. By 'Grenzlage', Hilbert could be referring to the use of limits. 'Bewegung' almost certainly refers to the treatment of congruence; movement was frequently associated with congruence and congruence proofs, going back to Book I of Euclid's *Elements*, which uses the idea of 'moving' figures so that the one covers the other. Note that the characterization of 'Schulgeometrie' immediately after this refers to 'Congruenzsätze'. Hilbert's later treatment of congruence in the 1898/1899 lectures and the *Festschrift* stresses that its treatment is independent of the notion of movement. See, e.g., *Hilbert 1898c**, 61 or *Hilbert 1899a**, 59–60 (both in Chapter 4 of this Volume).

2.) Axiome der Geometrie
(untersucht, welche Axiome bei den in der Geometrie der Anschauung gewonnenen Thatsachen benutzt werden und stellt systematisch die Geometrien gegenüber, bei welchen einige dieser Axiome weggelassen werden).[4]

3.) Analytische Geometrie
(ordnet den Punkten einer Geraden von vorneherein die Zahl zu und führt so die Geometrie auf die Analysis zurück).

Die Bedeutung von 1.) ist ästhetisch und paedagogisch und praktisch.
2.) erkenntnisstheoretisch.
3.) wissenschaftlich mathematisch.

5 Die *Geometrie* ist die *Lehre von den Eigenschaften* des *Raumes.* Sie unterscheidet sich wesentlich von den *rein mathematischen Wissensgebieten* wie z. B. *Zahlentheorie, Algebra, Funktiontheorie.* Die *Resultate dieser Gebiete* können durch *reines Denken* gewonnen werden, indem man durch klare *logische Schlüsse* die behaupteten Thatsachen auf immer *einfachere* zurückführt, bis man schliesslich nur noch den *Begriff der ganzen Zahl* nöthig hat. Jeder noch so *tief liegende* und *complicirte* Satz der *reinen Mathematik* muss so schliesslich auf *Beziehungen für ganze Zahlen* 1, 2, 3, ... zurückgeführt werden können. Da dieser Weg lang und beschwerlich ist, so hat man Mittel ersonnen, wie man den *Weg ebnet und abkürzt* oder durch *Anlegung von Zwischenstationen sichert* etc. Aber es ist nicht nur diese *Zurückführung möglich*, sondern
6 | auch geboten. *Es gilt heutzutage* ein Satz erst dann als bewiesen, wenn er eine Beziehung zwischen ganzen Zahlen in *letzter Instanz zum Ausdruck bringt.* Also die *ganze Zahl ist das Element.* Zum Begriff der ganzen Zahl können wir auch *durch reines Denken gelangen*, etwa [5] ⟨indem ich⟩ die *Gedanken selber* zähle. *Methoden, Grundlagen der reinen Mathematik* gehören dem reinen Denken an. Ich brauche weiter nichts als *rein logisches* Denken, wenn ich mit Zahlentheorie oder Algebra mich beschäftige.

Ganz anders verhält es sich mit der Geometrie. Ich kann die *Eigenschaften des Raumes* nimmer durch *blosses Nachdenken* ergründen, so wenig wie ich die *Grundgesetze der Mechanik*, das *Gravitationsgesetz* oder irgend ein
7 *anderes physikalisches Gesetz* so erkennen kann. Es ist ja der *Raum* | nicht ein *Produkt meines Nachdenkens*, sondern er ist mir durch *meine Sinne* gegeben. Ich brauche daher zur Ergründung *seiner Eigenschaften* meine Sinne. Ich brauche die *Anschauung und das Experiment*, wie bei der Ergründung physikalischer Gesetze, wo auch noch die *Materie als gegeben durch die Sinne hinzukommt.*

---

[4]This is a fairly accurate, if cursory, description of part of what Hilbert subsequently undertakes. For Hilbert's later characterizations see, e.g., *Hilbert 1894a**, 7–9, *Hilbert 1898c**, 1–7, *Hilbert 1899a**, 1–3, *Hilbert 1899c*, 3, or *Hilbert 1902d**, 1–5. These are all to be found in this Volume.

[5]Deleted: in dem ich etwas denke, dann ein zweites, drittes, etc.

In der That entspringt denn auch die *älteste Geometrie* aus dem *Anschauen der Dinge* im Raume, wie sie das *tägliche Leben bietet*, und, wie alle Wissenschaft zu Anfang, hat sie *Probleme, vom praktischen Bedürfniss* gestellt, und sie beruht auf dem *einfachsten Experiment*, was man machen kann, nämlich auf dem *Zeichnen.*[6]

Die Gegenstände *der Geometrie* der alten *Griechen sind vielseitig* und *mannigfaltig.*[7] Es handelt sich um Sätze *vom Dreieck*, vom *Vieleck*, vom *Kreise* und den dem *Kreise umbeschriebenen Vielecke, Ausmessung von Dreiecken, Kreisen, Kegeln, der Kugel* etc. über *Aehnlichkeit, Gleichheit* und *Congruenz der Figuren* etc. Doch vielseitig wie die *Gegenstände und Resultate* der
griechi|schen Geometrie, so vielseitig sind auch ihre *Methoden*: Es wird *sowohl* 8
gemessen, wie *gerechnet* sowie *construirt*. Eine *systematische Darstellung* von *unübertroffener Klarheit und Schärfe* des ganzen Gebietes gab *Euklid* in seinen *13 Bücher* umfassenden „*Elementen*". Euklid lebte *um 300* vor Chr. in *Alexandrien*. So reich die griechische Geometrie an *Gedanken, an Resultaten* und an *Problemen* war, so hatte sie einen *wesentlichen Mangel*: es fehlte ihr an einer *allgemeinen Methode*, mittelst der allein eine *fruchtbringende Weiterentwickelung* der Wissenschaft möglich ist. In Euklid erscheint die Geometrie *wie fertig, und kein Raum für freie, produktive Betätigung*. In der That, beinahe *2 Jahrtausende* beschränkte man sich darauf, in *höchster Ehrfurcht* und mit
unendlichem Fleisse Euklid zu studieren und zu commentie|ren, ohne dass man 9

---

[6]In the first three paragraphs of this section, Hilbert alludes to a number of views. (*a*) That geometry is the theory of properties of space is Kant's view, among others. (*b*) The view that geometry comes from considering properties of bodies *in* space, i.e., the view that geometry is a form of natural science, although certainly of older provenance, is to be found in *Helmholtz 1876a*, and in *Pasch 1882*, 3. (*c*) The difference between the nature of geometry and that of arithmetic as Hilbert describes it here is an allusion to the famous remark of Gauß to Bessel:

> Nach meiner innigsten Überzeugung hat die Raumlehre in unserm Wissen a priori eine ganz andere Stellung, wie die reine Grössenlehre; es geht unserer Kenntniss von jener durchaus *diejenige* vollständige Überzeugung von ihrer Nothwendigkeit (also auch von ihrer Wahrheit) ab, die der letzteren eigen ist; wir müssen in Demuth zugeben, dass, wenn die Zahl *bloss* unsers Geistes Product ist, der Raum auch ausser unserm Geist eine Realität hat, der wir a priori ihre Gesetze nicht vollständig vorschreiben können. (Gauß to Bessel, 9 April, 1830, in *Gauß 1900*, 201.)

(*d*) Part of the argument for the view that arithmetic at least is 'unsers Geistes Product' is given by Hilbert's allusion to Dedekind's proof (*Dedekind 1888*, Satz 66) for the existence of an actually infinite collection, which relies on the infinity of a conscious subject's 'Gedankenwelt'. (*e*) The view that the results of number theory, algebra and function theory can all ultimately be reduced to theorems about the natural numbers is a view stated in *Dedekind 1888*, Vorwort, VI, and there attributed to Dirichlet. Hilbert also seems to endorse this view here. Note that the view is also stated in *Cantor 1883*, § 4, as a view held, Cantor says, by some of the 'verdienstvollsten Mathematiker der Gegenwart', but rejected by him. In connection with this, Hilbert stresses, as does Dedekind, that the reduction will often be 'lang und beschwerlich'. Dedekind uses the fact that this 'Umschreibung' is often 'mühselig' as an argument for the indispenability of the introduction of new concepts into mathematics. (Op. cit., VI.) See also pp. 6–7 of Hilbert 1896 'Ferienkurs' (this Volume, Chapter 3), and the corresponding note. (*f*) The remark that drawing is a kind of experiment is also found in Hilbert's lectures *Hilbert 1894a**, 11; see Chapter 2 of this Volume.

[7]To a large extent, Hilbert has adapted the remarks in the remainder of the Introduction (excepting the references) from *Hankel 1875*, 1–33. See *Toepell 1986*, 22–25.

weiter kam. Da war es *Descartes* – der Begründer der neueren Philosophie,
welcher ein *neues allgemeines Princip in die Geometrie* (1637)[8] einführte.
Einführung der Coordinaten in die Geometrie. Nach Descartes ist z. B.

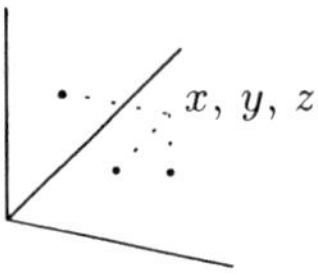

ein Punkt im Raume durch x, y, z gegeben: Da
aus Punkten sich ein *jedes geometrische Gebilde*
zusammensetzen lässt, so kann man somit jedes
geometrische Gebilde durch veränderlich *zu denkende Coordinaten ausdrücken*, und [9] Beziehung
zwischen geometrischen Gebilden drückt sich durch eine *Gleichung aus.* Dieser
Gedanke macht mit *einem Schlage* jedes *geometrische Problem der Analysis
zugänglich.* So wurde *Descartes der Schöpfer der analytischen Geometrie.* Es
wurden zunächst die Sätze der Griechen von *neuem bewiesen* und dann *verallgemeinert.* So trat an Stelle der *Kunstgriffe* und *-mittelchen* durch Cartesius
10 eine *einheitliche Metho|de – die Formel*, das *Rechnen.* So wichtig dieser *Fortschritt* und so *grossartig die Erfolge* waren, so *litt doch schliesslich* die *Geometrie* als solche unter der *einseitigen Ausbildung dieser* Methode. Man *rechnete*
nur noch, *ohne Anschauung von dem Errechneten* zu haben. Man verlor den
Sinn für die geometrische Figur und für *die geometrische Construktion.* Dagegen konnte die *Reaktion* nicht ausbleiben. Die Anregung ging aus von den
praktischen Bedürfnissen der Techniker, welche zu ihrer Arbeit *Zeichnungen* brauchten, und es war der Mathematiker *Monge*, welcher die vereinzelten
Bestrebungen zusammenfasste und die sogenannte „*darstellende Geometrie*"
(Géométrie descriptive um 1800)[10] schuf, d. h. *die Kunst, alle räumlichen
Linien, Flächen, Körper* in einer Ebene als *Aufriss und Grundriss* nach *allgemeinen Regeln* darzustellen und aus solchen Darstellungen die geometrischen
11 Beziehungen *zwischen Gestalt und Lage* jener | Gegenstände abzuleiten. Auf
Monge folgten eine Reihe bedeutender Männer, *Carnot, Poncelet, Chasles,
Steiner, von Staudt etc.*, aus deren Händen die heutige *Geometrie der Lage*
oder besser die *projektive Geometrie* hervorging. Das *bedeutendste Werk* ist
die „systematische Entwickelung der Abhängigkeit geometrischer Gestalten
von einander" von *Steiner (1832).*[11] *Dieses Werk* und *die vorhergehenden bedienen* sich in der Entwickelung und Ableitung der Sätze rein geometrischer
Methoden. Aber dennoch ist die Grundlage auf *Messung und Rechnung* begründet. Andererseits sind die gefundenen Sätze *rein anschaulicher Natur*, sie
beziehen sich auf *Lageverhältnisse.* Es entstand nun die *principielle Frage*, ob
es möglich sei, die *projektive Geometrie ganz vom Messen und Rechnen* frei
12 zu machen. Diese Frage wurde durch | *von Staudt* in seinem *klassischen Werke
Geometrie der Lage (1847)*[12] gelöst. Im *Gegensatze* zu allen seinen Vorgängern, welche immer noch der Rechnung bedurften, machte er – wie er selbst
in *seinem Vorwort sagt* – die *projektive Geometrie* „*zu einer selbständigen*

[8]Hilbert is clearly referring to Descartes's *La Géométrie*, published originally as an Appendix to *Descartes 1637*.

[9]Deleted: jede

[10]*Monge 1811*.

[11]*Steiner 1832*.

[12]*von Staudt 1847*.

Wissenschaft, welche des Messens nicht bedarf."[13] Er schuf eine Geometrie, in der man weder *rechnet* noch *misst*, sondern nur *construirt*, weder *Zirkel* noch *Winkelmaass* benutzt, sondern nur das *Lineal*. Damit war jene *wissenschaftliche Forderung* in *befriedigender Weise erfüllt*. Denn die *Rechnung* muss bei Ableitung von Sätzen über *Lageverhältnisse* als etwas fremdes erscheinen. ↑In dieser Gestalt bildet die projektive Geometrie zwar nur einen *Teil der Geometrie*, aber dieses Teilgebiet ist von einer *wunderbaren Einheitlichkeit und Abgeschlossenheit*.↓ Nach dem *Vorbilde* dieses Werkes gedenke ich meine Vorlesung über projektive Geometrie zu gestalten:

Lehrbücher:

v. Staudt, Geometrie der Lage,
Beiträge zur Geometrie der Lage.
Schröter, Jakob Steiner's Vorlesung über synthetische Geometrie.[14]
Reye, Geometrie der Lage.[15]
Hankel, Elemente der projektiven Geometrie. 13
Thomae, Geometrie der Lage.[16]

Die inhaltsreichsten sind *Schröter und Reye.*

§ 1. Grundbegriffe

Punkt, Gerade, Ebene heissen die Elemente.

Punkte: $A, B, C, \ldots$
Geraden: $a, b, c, \ldots$, auch „Strahlen" genannt.
Ebenen: $\alpha, \beta, \gamma, \ldots$

Die Geraden denken wir uns *immer beliebig verlängert*, die Ebenen *unbegrenzt ausgedehnt*. Indem wir diese Elemente durch *Zusammensetzung* miteinander verbinden, gelangen wir zu *den sogenannten Grundgebilden der projektiven Geometrie*. Die für uns zunächst *wichtigsten Grundgebilde* sind folgende:

1.) Eine Gerade enthält unbegrenzt viele Punkte. Diese Punkte heissen eine (gerade) *Punktreihe*. Die Gerade heisst *Träger der Punktreihe*.
Die Punkte heissen *Elemente der Punktreihe*.

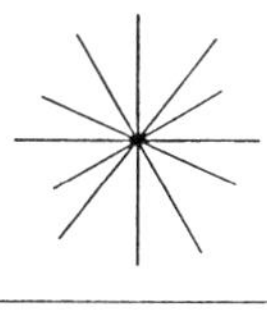

2.) Durch einen *festen Punkt* in einer *festen Ebene* gehen *unendlich viele Geraden*. Diese heissen ein *Strahlbüschel*. | Der *Punkt heisst der Mittelpunkt*, die 14

---

[13]In *von Staudt 1847*, p. III, we find:

Ich habe in dieser Schrift versucht, die Geometrie der Lage zu einer selbstständigen Wissenschaft zu machen, welche des Messens nicht bedarf.

[14]The references are: *von Staudt 1847* (*Geometrie der Lage*); *von Staudt 1856*, *1857*, *1860* (*Beiträge zur Geometrie der Lage*); and *Steiner 1867* (Heinrich Schröter was the editor of this second part of Steiner's lectures).

[15]The third edition of Reye's *Geometrie der Lage* consists of three volumes, *Reye 1886*, *1892b*, *1892a*. In this text, Hilbert appears to have made use only of Volume 1 of this third edition. For example, on p. 70, Hilbert refers to problems in Volume 1 of Reye's third edition which are in neither of the two previous editions.

[16]*Hankel 1875*; *Thomae 1873*.

Ebene heisst die *Ebene des Strahlbüschels.* Die Strahlen heissen *Elemente* des Strahlbüschels.

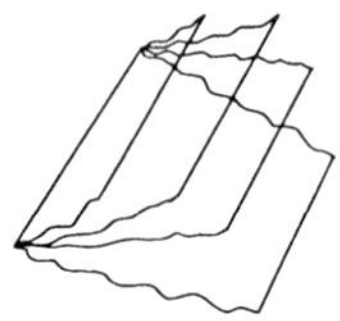

3.) Durch eine *Gerade* gehen unbegrenzt viele Ebenen: *Ebenenbüschel. Achse des Ebenenbüschels. Elemente des Ebenenbüschels.*

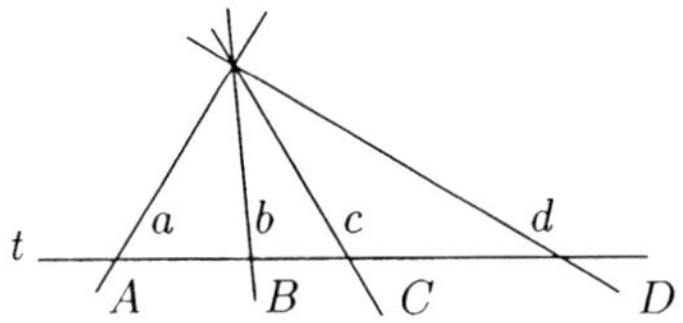

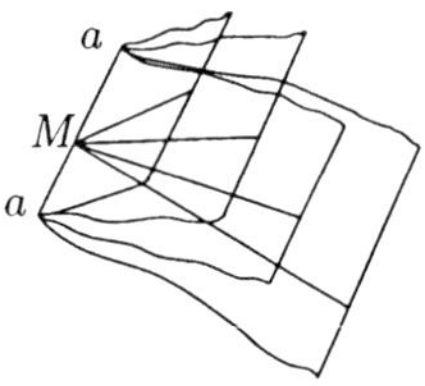

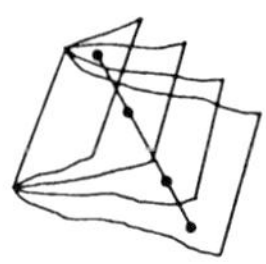

Punktreihe, Strahlenbüschel, Ebenenbüschel heissen die *3 Grundgebilde erster Stufe. Jedes dieser Grundgebilde* kann durch eines der anderen erzeugt werden. *Z. B. die Punktreihe* $ABCD\ldots$ durch das *Strahlenbüschel* $abcd\ldots$. *Durchläuft* ein Punkt den Träger t, so dreht sich der Verbindungsstrahl um M, und *umgekehrt.* Ferner: schneide ich ein Ebenenbüschel mittelst irgend einer Ebene, so entsteht in dieser Ebene ein *Strahlenbüschel*, dessen *Mittelpunkt in der Achse* liegt. *Dreht sich eine Ebene* um die Achse, so dreht sich der
15 Strahl um | den *Mittelpunkt* M. Endlich wird jedes *Ebenenbüschel* von einer Geraden in *einer Punktreihe* geschnitten: Durchläuft der Punkt die Gerade, so dreht sich die Ebene.

Wir fassen den *ersten Fall etwas näher ins Auge*:

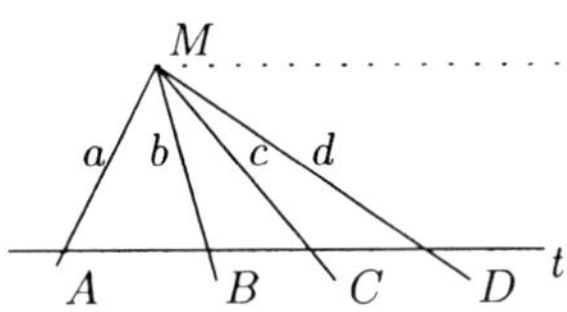

Beschreibt der Strahl etwa von MA *aus* ausgehend ein Büschel, so beschreibt sein Schnitt auf t eine Punktreihe von A über B, C, D ins *Unendliche weiter rückend* und dann von entgegengesetzter Seite her *aus dem Unendlichen* her *wiederkehrend.* Also nur in einer einzigen Lage würde *der Anschauung zufolge* der Strahl nicht die Gerade t schneiden, d. h. *einem einzigen Strahl des Büschels würde* kein Punkt der Punktreihe entsprechen, *nämlich dem Parallelstrahl.* Um diesen *Ausnahme-*
16 *fall zu beseitigen*, sagen wir: dem Parallel|strahl des Büschels entspricht der unendlich ferne *Punkt der Geraden.* Oder *die parallele Linie schneidet t in dem unendlich fernen Punkt.* Denselben müssen wir uns nach *beiden Seiten hin* im Unendlich denken, so dass die gerade Linie durch *Vermittlung dieses unendlich fernen Punktes* als eine *geschlossene Linie erscheint.* Es liegt in dieser Annahme nichts *übernatürliches oder metaphysisches*, sondern nur eine *neue Ausdrucksweise.* Statt zu sagen: *2 Linien sind parallel* oder *2 Geraden haben dieselbe Richtung* sage ich, die beiden Geraden schneiden sich *in einem unendlich fernen Punkt.* Da wir uns nur einen einzigen unendlich fernen Punkt auf der Geraden denken, so ist unsere Annahme gleichbedeutend mit

dem *Parallelaxiome*) des Euklid*, dass durch jeden Punkt eine und *nur eine Parallele möglich ist.* Also auf jeder Geraden liegt ein und nur ein *unendlich ferner Punkt.* Wir wollen | die *Gesammtheit* dieser unendlich fernen Punk- 17 te näher untersuchen. Wir nehmen eine *feste Ebene* und eine Gerade darin. Dann hat diese einen *unendlich fernen Punkt.* Eine Parallele schneidet sie in diesem. Eine 2te Parallele schneidet ebenfalls in dem selben unendlich fernen Punkte, d. h. alle parallelen Geraden haben *einen und denselben unendlich fernen Punkt* gemein. Sie bilden ein *Strahlbüschel.*

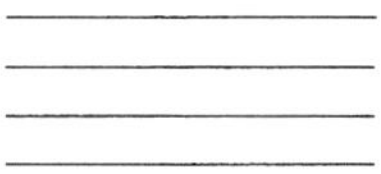

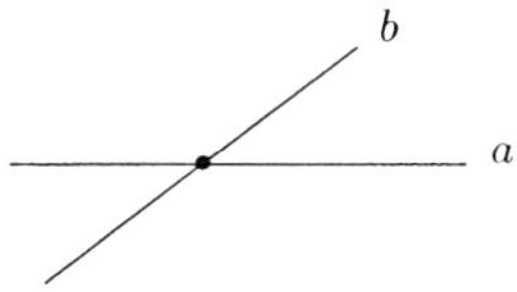

2 sich *schneidende Geraden a, b* haben verschiedene unendlich ferne Punkte. Denn es soll das Gesetz erhalten bleiben, dass *2 Geraden nur einen Punkt gemein haben* oder zusammenfallen. Wir nehmen an, dass alle unendlich fernen Punkte der Ebene auf einer Geraden liegen – *der sogenannten unendlich fernen Geraden Geraden der Ebene.* Wir müssen annehmen, dass dies eine Gerade ist, denn sie wird von jeder Geraden nur in *einem Punkte getroffen.* – Wir nehmen nun eine *Ebene im Raume*, welche also eine un|endlich fer- 18 ne Gerade hat. Nehmen wir nun eine *parallele Ebene*, so muss diese Ebene mit der vorigen alle *unendlich fernen Punkte* gemein haben und umgekehrt.

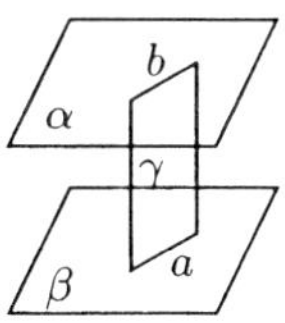

Schneide ich nämlich α und β mit irgend einer Ebene γ, so erhalte ich zwei parallele Linien a und b und diese haben denselben *unendlich fernen Punkt*, und so für alle unendlich fernen Punkte. Wir können also sagen: *2 parallele Ebenen schneiden sich in derselben unendlich fernen Geraden.* Folglich bilden alle *parallelen Ebenen ein Ebenenbüschel.* Wir nehmen nun weiter an, alle unendlich fernen Punkte und daher auch alle unendlich fernen Geraden im Raume liegen auf einer Ebene im Raume. Also alle unendlich fernen Elemente überhaupt denken wir uns in einer unendlich fernen Ebene vereinigt. Als

*) Dieses *Parallelenaxiom* wird geliefert durch unsere Anschauung. *Ob letztere angeboren oder anerzogen ist*, ob *jenes Axiom der Wahrheit entspricht*, ob es durch die *Erfahrung muss bestätigt werden*, oder *ob es dies nicht nöthig hat*, geht uns hier nichts ⟨an⟩. Wir haben mit *der Anschauung* zu thun und *diese erfordert jenes Axiom.*

19 Ebene wird *sie deshalb ge|dacht*, weil sie doch von irgend einer Geraden nur *in einem Punkte geschnitten* wird, nämlich dem unendlich fernen Punkt und von jeder Ebene nur in *einer Geraden*, nämlich der unendlich fernen Geraden.*) Ich hebe noch einmal hervor: Sie müssen hinter *diesen Festsetzungen* nichts besonders *Schwieriges oder Metaphysisches* suchen. Die Einführung der unendlichen Elemente ist weiter nichts als eine *abkürzende Redeweise* für *einfache anschauliche Thatsachen*. Es wird Ihnen der *Vorteil dieser Redeweise* besonders deutlich werden, wenn wir *jetzt dazu übergehen*, die *einfachen Grundgesetze der Anschauung aufzustellen*. Es sind *8 Grundgesetze*, welche sich *paarweise anordnen* und welche wir in folgender Weise gegenüberstellen:

Zwei Punkte A und B bestimmen eine Gerade AB: ihre Verbindungslinie.	Zwei Ebenen α und β bestimmen eine Gerade $\alpha\beta$: ihre Schnittlinie.
Eine Gerade a und ein ausserhalb derselben liegender Punkt B bestimmen eine Ebene aB: ihre Verbindungsebene.	Eine Gerade a und eine nicht durch dieselbe gehende Ebene β bestimmen einen Punkt $a\beta$: ihren Schnittpunkt.
20 Drei Punkte A, B, C, die nicht in einer Geraden liegen, bestimmen eine Ebene ABC: die Verbindungsebene.	Drei Ebenen α, β, γ, die nicht durch eine Gerade gehen, bestimmen einen Punkt: ihren Schnittpunkt.
Zwei Geraden a und b, die einen Punkt gemein haben, liegen in einer Ebene.	Zwei Geraden a und b, die in einer Ebene liegen, haben einen Punkt ab gemein.

Diese Gesetze gelten *allgemein und ausnahmslos*, z. B. das 1.) rechts würde ohne Einführung der unendlich fernen Elemente gelautet haben: *Zwei Ebenen α, β* bestimmen entweder eine Gerade *oder sie sind parallel*; während jetzt auch im letzten Falle eine Gerade bestimmt wird, nämlich die unendlich ferne Gerade. Ebenso der *4te Satz rechts*. Ferner, der 1.) Satz links würde zwar auch ohne unendlich ferne Elemente *allgemein gelten*. Aber jetzt enthält er mehr, nämlich man kann einen Punkt B unendlich fern nehmen, indem man eine Gerade b giebt, d. h. durch einen Punkt A und eine Richtung b ist eine Gerade
21 | bestimmt etc.

*) 2 windschiefe Geraden im Raume haben natürlich keinen, auch keinen unendlich fernen Punkt gemein, weil sie ja nicht parallel sind.

Durch die Art, wie ich die *8 Sätze paarweise* geordnet habe, tritt deutlich das *Princip der Dualität* hervor, welches von *grosser Wichtigkeit und Fruchtbarkeit* ist, weil es den ganzen Stoff in *2 Gruppen von Sätzen* teilt, z. B. Satz 1.) etc. *Punkt und Ebene* entsprechen sich dual. *Die Gerade entspricht sich selbst.* Im letzten Falle ist der dual entsprechende Satz genau die Umkehrung des ersteren. Wir werden später erkennen, dass *allgemein ein jeder Satz* der projektiven Geometrie seine *Ergänzung in einem* anderen findet, der sich sofort durch das Dualitätsprincip ergiebt.

Zur *Uebung möchte ich einige duale Sätze* ableiten.

Sind 4 Punkte A, B, C, D gegeben und schneiden sich die Verbindungslinien AB und CD, so schneiden sich auch AC und BD und ebenso auch AD und BC.	Sind 4 Ebenen α, β, γ, δ gegeben und schneiden sich die Schnittlinien $\alpha\beta$ und $\gamma\delta$, so schneiden sich auch $\alpha\gamma$ und $\beta\delta$ und ferner auch $\alpha\delta$ und $\beta\gamma$.
(Denn die 4 Punkte liegen in einer Ebene.)	(Denn die 4 Ebenen gehen durch einen Punkt.)

Aufgabe: Gegeben 3 Geraden a, b, c, man soll eine Gerade construiren, 22
welche alle drei trifft.

Lösung: Man nehme auf a einen Punkt A, lege die Ebene↑n↓ Ab ↑und AC↓,[17] welche [18]sich dann in der gesuchten Geraden d treffen.[18]

Die Aufgabe entspricht sich *selbst dual.* Es wird daher eine zweite duale Lösung geben:

Man nehme durch a eine Ebene α an, bestimme die Punkte αb und αc, deren Verbindungslinie die gesuchte Gerade d ist.

↑Meine Herren. § 1. *unendlich ferne* Elemente. Keine Richtung, *keine* Stellung, aber sie werden sonst wie endliche Elemente behandelt. *Dualitätsprincip* bisher nur eine *Bemerkung, die den Stoff gruppirt.* Erst *später Beweis.*↓

§ 2. Harmonische Lage.

Haben wir so die *Grundbegriffe festgelegt,* so kommen wir jetzt zu der Entwickelung der *Grundsätze* ↑*Theoreme*↓, auf denen sich die *ganze Wissenschaft aufbaut.*

Satz: *Wenn zwei Dreiecke in verschiedenen Ebenen so gelegen sind, dass je zwei entsprechende Seiten sich schneiden, so laufen die Ver|bindungslinien* 23
entsprechender Ecken durch ein und denselben Punkt.

[17]Erroneously instead of Ac.

[18–18]Substituted for: c in C treffe. AC ist die gesuchte Gerade, die etwa b in B treffen [[⟪???⟫]]

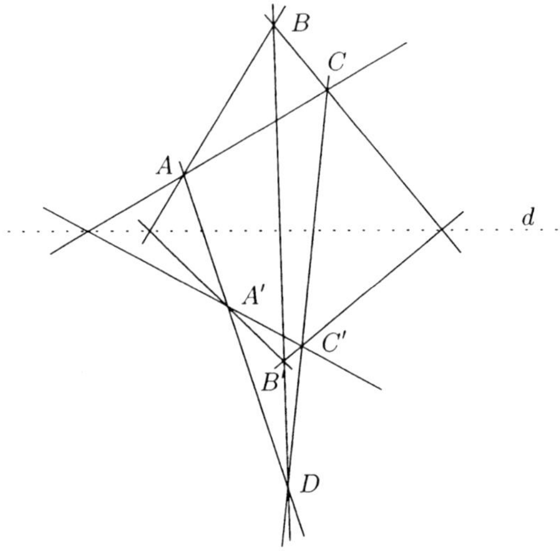

d sei die Schnittlinie der beiden Dreiecksebenen. Auf dieser schneiden sich die 3 Seitenpaare. A, B, A', B' liegen in einer Ebene, folglich schneiden sich AA' und BB', etwa in D. Aus dem entsprechenden Grunde schneiden sich BB' und CC' und AA' und CC', d. h. CC' geht auch durch D.
Umkehrung): Wenn zwei Dreiecke in verschiedenen Ebenen so gelegen sind, dass die Verbindungslinien je 2er entsprechender Ecken durch einen Punkt laufen, so schneiden sich die entsprechenden Seiten.*

Denn da A, B, A', B' in einer Ebene liegen, so schneiden sich AB und $A'B'$ etc.

24 Welcher ist nun *der entsprechende Satz*, wenn die Dreiecke in *derselben Ebene liegen*? Schon die *Zeichnung* verrät es: Das blosse *Schneiden ist dann keine* Bedingung mehr:
Satz: *Wenn 2 Dreiecke in der Ebene so liegen, dass je zwei entsprechende Seiten sich auf einer und der nämlichen Geraden schneiden, so laufen die Verbindungslinien entsprechender Ecken durch den nämlichen Punkt und umgekehrt, wenn die Verbindungslinien entsprechender Ecken durch den nämlichen Punkt laufen, so liegen die Schnittpunkte entsprechender Seiten auf der nämlichen Geraden.*

*) Ist übrigens garnicht nöthig.

25

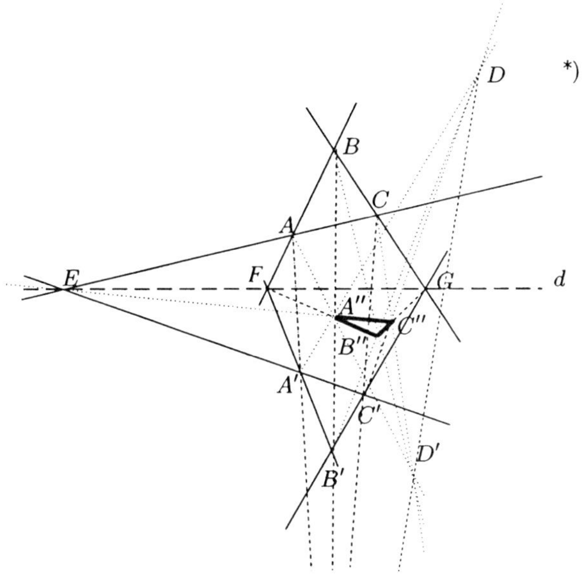

[19]↑ Man lege durch d eine andere Ebene und zeichne in dieser *ein Dreieck* $A''B''C''$, dessen Seiten durch E, F, G gehen. Wir betrachten die *3 Ebenen* $AA'A''$, $BB'B''$, $CC'C''$. Nach dem vorigen Satze schneiden sich *1.)* AA'', BB'', CC'' *in* D' und *2.)* $A'A''$, $B'B''$, $C'C''$ in D. Jene 3 Ebenen gehen also durch *dieselbe Gerade* DD' und daher schneiden sich auch AA', BB', CC', nämlich da, wo DD' die erstere Ebene durchstösst.↓
Somit haben wir den Satz auf *den früheren Raumsatz* zurückgeführt.**) Die Umkehrung kann *indirekt bewiesen* | werden: 26

*) Beachte die Regelmässigkeit dieser Figur und ihr sich selbst dualistisches Entsprechen: Die Figur besteht aus 10 Punkten und 10 Geraden, auf jeder Geraden liegen 3 Punkte und durch jeden Punkt gehen 3 Geraden.
**) Vielleicht ist es kürzer, den Raumsatz überhaupt weg zu lassen und diesen Satz gleich mittels des Raumes zu beweisen.

[19]The paragraph added (which runs up to line 31.7) replaces a previous version. Although its continuation on p. 25 has been pasted over with new text, the whole passage begins on p. 24 as follows:

Ich nehme einen Punkt M an und eine Ebene durch d. MA' schneidet diese Ebene in A'' etc. $B''A''$ schneidet in F. $A''C''$ schneidet in E und $B''C''$ in G. Folglich treffen sich AA'', BB'', CC'' *in einem Punkte* D. MD *schneide die ursprüngliche Ebene in* D'. Ich behaupte, AA', BB', CC' treffen sich in D'. In der That: Es liegen auf einer Geraden *1.)* M, A', A'', *2.)* M, D, D', *3.)* A, A'', D. Folglich liegen alle diese Punkte in einer Ebene und insbesondere schneiden sich auch AA' und MD.

The proof begun here could be continued as follows: Likewise one shows that both the pairs BB', MD and CC', MD intersect. On the other hand, these three points of intersection lie in the original plane, since the lines AA', BB', CC' belong to it. Thus, MD must cut the original plane in D', which was what had to be shown.

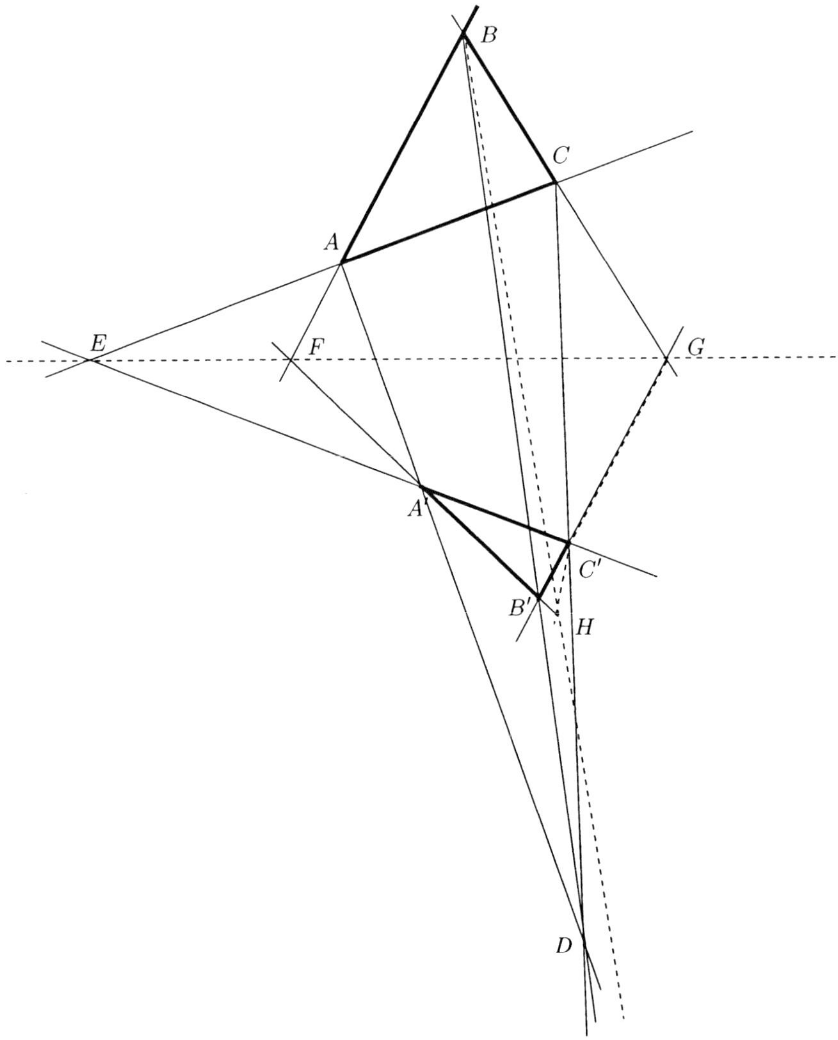

27 BC schneidet EF in G. Ginge nun GC' nicht durch B', sondern schnitte vielmehr GC' die Gerade $A'B'$ in H, so müsste nach dem vorigen Satze BH durch D gehen, was nicht möglich ist, da H nicht auf $BD = BB'$ liegt.*)

Definition:

4 Punkte A, B, C, D einer Geraden heissen harmonische Punkte, wenn sie zu einem Viereck eine solche Lage haben, dass in A und C je 2 Gegenseiten des Vierecks sich schneiden, während durch B und D je eine Diagonale desselben geht.

*) Vielleicht ist es besser, einen einfachen direkten Beweis zu finden.

Ausführung der Construktion.

3 Punkte A, B, C können jedenfalls willkürlich gewählt ⟨werden⟩. *Dann
aber ist durch diese 3 Punkte und ihre Reihenfolge der 4te harmonische
Punkt D völlig bestimmt.* Denn construiren wir etwa nach unten ein $|^A$ ähnli- 28
ches Viereck,

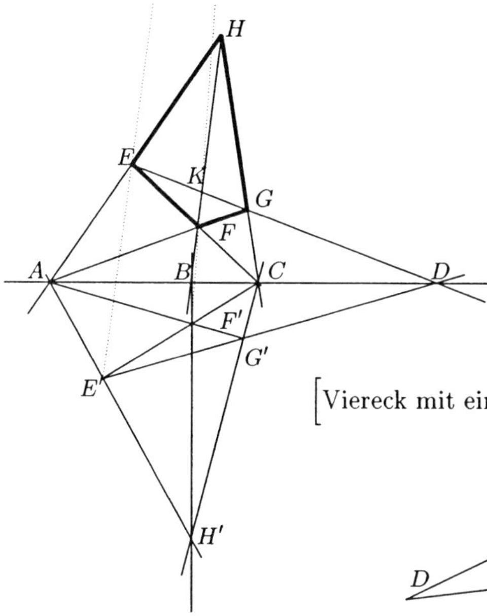

[Viereck mit einspringenden Ecken:

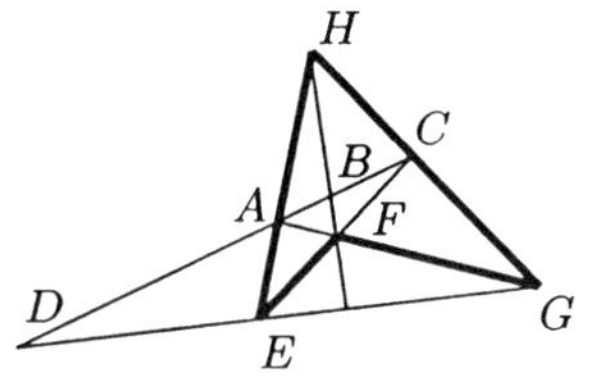

]

so geht $E'G'$ ebenfalls durch D. Denn wir betrachten 1.) die beiden Dreiecke EFH und $E'F'H'$. Nach dem vorigen Satz schneiden sich EE', FF',
HH' in einem Punkte M. 2.) die beiden Dreiecke FHG und $F'H'G'$. Nach
dem vorigen Satze schneiden sich auch FF', HH', GG' in einem Punkte und
daher in demselben Punkte M. 3.) die beiden Dreiecke EHG *und* $E'H'G'$.
Nach der Umkehrung des obigen Satzes | geht daher $E'G'$ durch D. Also bei 29
der Construktion des 4ten harmonischen Punktes kann ich H *ganz willkürlich*
annehmen und dann noch G *oder* E *willkürlich auf* HC *resp.* HA.[20]

4 harmonische Punkte einer Punktreihe.

Wenn man irgend einen Punkt ausserhalb der Punktreihe mit den 4 harmonischen Punkten verbindet, so erhält man

4 harmonische Strahlen eines Strahlbüschels,

---

^{A}Die Unabhängigkeit des 4ten harmonischen ⟨Punktes⟩ von der besonderen Construktion beweise in der Ebene bloss durch Desargues'schen Satz! ⟨Added at the top of p. 28. This remark is not explicitly assigned to any particular point in the text.⟩

---

[20]The point M is not given in the figure, though there are faint dotted lines running up to it. It should be remarked that the *name* 'Desargues's Theorem' appears here for the first, and only, time in the note Hilbert adds, given here as ^{A}. Reye does not use it.

und wenn man eine Gerade ausserhalb beliebig annimmt und durch dieselbe 4 Ebenen legt, so
4 harmonische Ebenen eines Ebenenbüschels;
also *entsprechend den 3 Grundgebilden. 4 harmonische Elemente.*

Wichtigste Eigenschaft der harmonischen Gebilde, welche direkt aus der Definition folgt:

Harmonische Strahlen schneiden auf einer jeden Geraden harmonische Punkte aus.

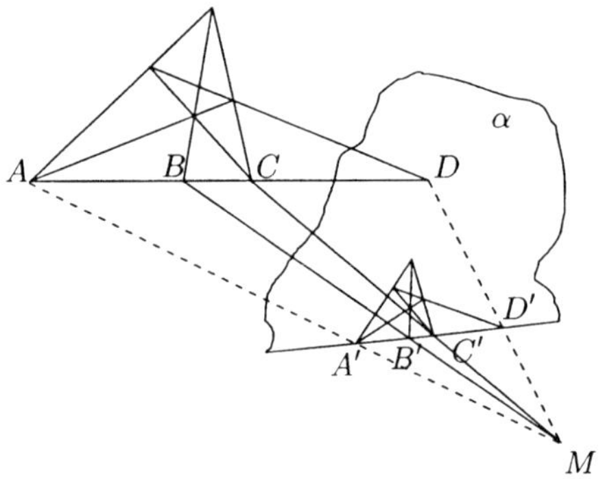

Denn nehme ich einen Punkt ausserhalb der Tafelebene, ziehe die harmonischen Strahlen MA, MB, MC, MD,[21] schneide durch eine beliebige Gerade $A'B'C'D'$, so sind dies
30 har|monische Punkte. Denn denke ich mir durch $A'B'C'D'$ eine Ebene α gelegt, so kann ich die *ganze Viereckscontruktion* auf diese Ebene projiciren.
Folglich schneiden 4 harmonische Strahlen auch aus jeder Ebene 4 harmonische Punkte aus.

4 harmonische Ebenen schneiden aus jeder beliebigen Ebene 4 harmonische Strahlen und aus jeder Geraden 4 harmonische Punkte aus.

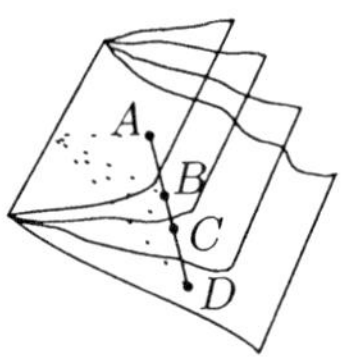

Denn erst lege ich durch $ABCD$ eine Ebene und erhalte so 4 harmonische Strahlen. Irgend eine Ebene schneidet diese 4 harmonische Strahlen in 4 harmonischen Punkten und folglich sind es 4 harmonische Strahlen.
Zusammenfassender Satz:

Aus einem harmonischen Grundgebilde entstehen durch Projiciren und Schneiden immer wieder harmonische Grundgebilde.
Z. B. bei der Viereckscontruktion sind auch $EKGD$ (von H aus) und $HKFB$ (von E aus) harmonische Punkte.[22]

Beweis des Satzes:
↑Vorher den *entsprechenden Satz für die Ebene*, d. h. 3 gegebene Strahlen a, b, c eines Büschels werden von beliebigen Geraden geschnitten und allemal der 4te harmonische Punkt gesucht etc.↓
31 *Werden 3 Ebenen α, β, γ eines Ebenenbüschels* von beliebi|gen Geraden geschnitten und zu den 3 Schnittpunkten allemal[23] *der 4te harmonische Punkt*

[21] A, B, D, C are assumed to be harmonic points, so MA, MB, MC, MD are given as harmonic lines.

[22] For the labelling, refer to the figure on p. 28. The projection from E shows that $HBFK$ must be harmonic; that this is also true of $HKFB$ follows from Hilbert's explanation on p. 31 below.

[23] In the sense of 'jedesmal'.

gesucht, so liegen diese sämmtlich auf ein und derselben Ebene, nämlich der 4ten *harmonischen Ebene.*

Wir wollen nun etwas näher *die Lage von 4 harmonischen Punkten untersuchen.*

Wir haben früher gesehen: *Aus A, B, C* und ihrer Reihenfolge *folgt D eindeutig* bestimmt. *D. h. wenn sich in A und C 2 Paar Gegenseiten schneiden* und durch *B* eine Diagonale ⟨geht⟩, *so durch D die andere. Aus der Definition* sieht man: Ist *ABCD harmonisch,* so auch *ADCB, CBAD, CDAB.* Also *A und C, B und D gehören zusammen* und *heissen zugeordnet.* Die Anschauung zeigt, dass die beiden zugeordneten Paare einer harmonischen Punktreihe *sich immer trennen.* [24]↑Denn wenn ich von *A zu C will,* so muss ich entweder durch *B* oder durch *D*, jenachdem ich den *endlichen* oder *unendlichen Weg* benutze. Ein Punktepaar *AC auf einer Geraden* kann nie durch einen *einzelnen Punkt B*, wohl aber durch ein Punktepaar *BD getrennt werden.* Das Punktepaar *BD kann aber auch* so liegen, dass *es nicht trennt.* Also *zwischen AC giebt es zwei* Strecken-Verbindungen, die endliche und die unendliche. Um über die Lage aller möglichen zu *AC* harmonischen Punktepaare Uebersicht zu erhalten, lassen wir *B auf der endlichen Strecke AC wandern.* Die Punkte *A, C, H, E werden festgehalten* und *AG wird um A gedreht.* Dann wandert *D in entgegengesetzter Richtung* wie *B*. Für $B = A$ wird $D = A$. Dem unendlichfernen Punkt *D* entspricht die Mitte *B*, die so definiert werden kann. $B = C$ wird $D = C$. Zugleich sieht man: *2 zu AC harmonische Paare, etwa BD und B′D′, trennen sich nicht.* Umgekehrt: *Zu 2 sich gegenseitig nicht trennenden Punktepaaren BD und B′D′ giebt es stets ein harmonisches Punktepaar AC.* Denn: lassen wir einen Punkt *C* von *B′* nach *D′* laufen, so entspricht doch *C* einmal der zu *BD* harmonische Punkt *A* und dann der zu *B′D′* harmonisch gelegene ⟨Punkt⟩ *A′*. *A* und *A′* wandern in gleicher Richtung. Aber die von *A* durchwanderte Strecke ist nur ein Teil der von *A′* durchwanderten Strecke, denn *L*

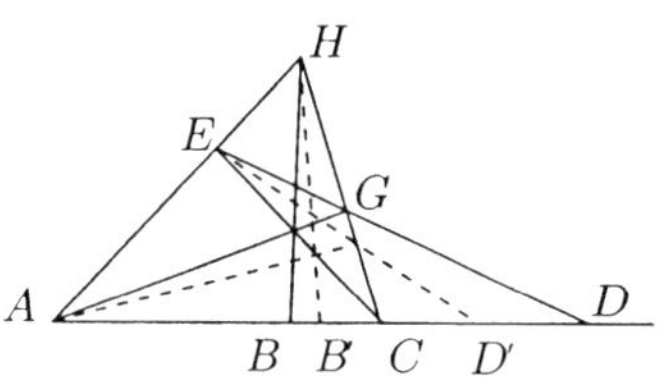

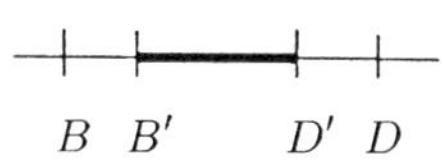

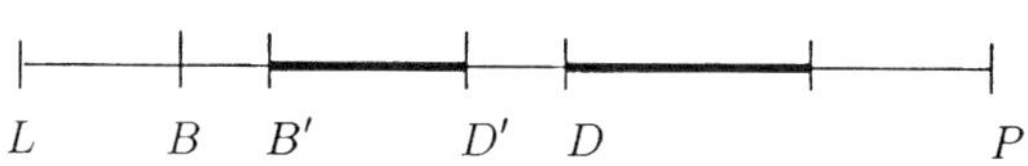

sei zu *BD* und *B′* und *P* der zu *BD* und *D′* harmonische Punkt. Folglich treffen sich *A* und *A′*.↓ Wir wollen nun zeigen, dass auch *die beiden Paare harmonischer* Punkte nicht bevorzugt sind. Diese Bevorzugung liegt in der Definition.

Wir wollen zeigen: Wenn in *A, C* sich die Gegenseiten eines Vierecks schneiden etc., so existirt auch | ein Viereck, dessen Gegenseiten sich in *B, D* 32
schneiden etc.

[24]This addition (up to line 35.35) is written on a sheet pasted to the bottom of p. 31 and directed to this place by an insertion sign.

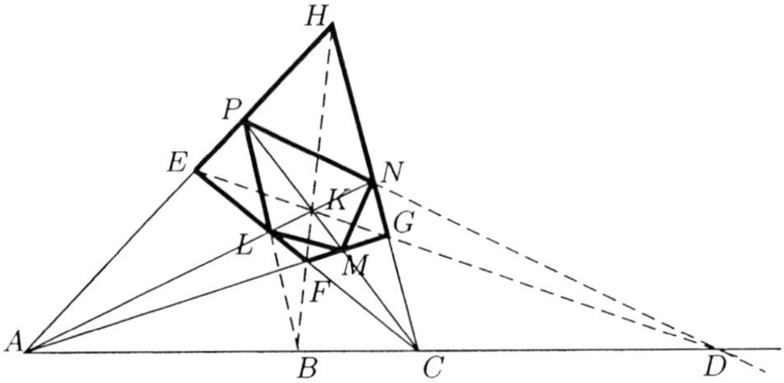

Ich *verbinde A und C mit K* und betrachte das Viereck $LMNP$.

1.) PN *geht durch D*, wegen des Vierecks $PHNK$ und weil A, B, C, D harmonische Punkte sind.

2.) LM *geht durch D* wegen des Vierecks $LKMF$.

3.) NM *geht durch B* wegen des Vierecks $NGMK$.

4.) PL geht durch B wegen " " $ELKP$.

Im Viereck $PLNM$ schneiden sich also die Gegenseiten in B und D, während
33 die Diagonalen durch A und C laufen. | D.h. wenn $ABCD$ harmonische Punkte sind, so sind es auch $DCBA$, $DABC$, $BCDA$, $BADC$.

Aufgaben zur Einübung der Sätze und Constructionen:

Gegeben A, a, b. Man soll durch A nach dem unzugänglichen Schnittpunkte von a und b eine Gerade ziehen:

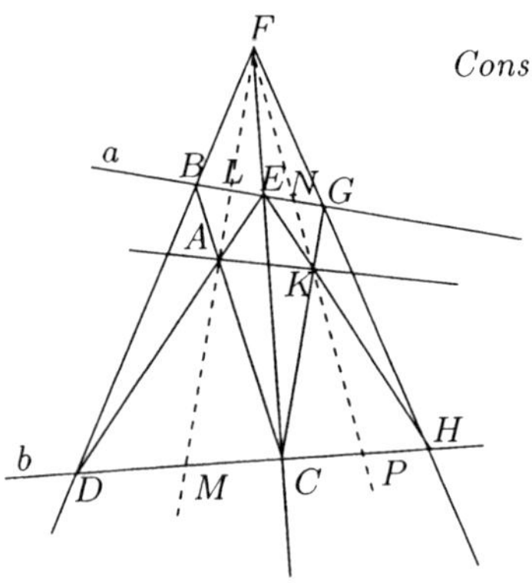

Construktion: A, B, C, D, E, F, G, H, K.

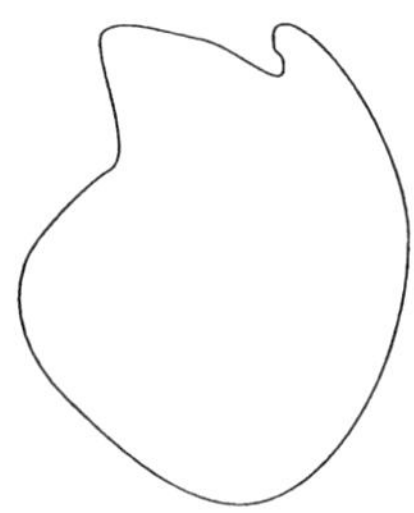

Beweis:[25] *Man verbinde F mit A und K*, so sind $FLAM$ und $FNKP$ harmonisch. Denn bei unserer Viereckskonstruktion waren $HKFB$ harmonisch.

[25]The proof would go somewhat as follows. Let X be the intersection point of a, b whose position is unknown. Choose B and E on a, and draw the straight lines BA and EA. One then obtains C and D respectively as intersection points with b. Then construct DB and CE, and let F be their point of intersection; FA then generates L and M. Choose G, which leads (as the diagram shows) to H and then to K; FK then generates N and P. It is easy to show that $FLAM$ and $FNKP$ are harmonic sets, so the four corresponding lines FX, a, AK and DC must be also harmonic, and will all intersect at X. (Cf. pp. 29–30.) Consequently, it is enough to join A with K to find the required line. It is not clear why Hilbert focuses for a while on $HKFB$.

Ebenso sind $FNKP$ harmonisch, folglich geht nach einem früheren Satze AK *durch ab.*

§ 3. Perspektivität und Projektivität. 34

Schon in § 1 hatten wir 2 Grundgebilde *durch ihre Lage zu einander in Beziehung* gesetzt, so dass *jedem Element des einen Grundgebildes* ein Element *des anderen entspricht.* Dieser Gedanke des eindeutigen Beziehens ist von *grosser Wichtigkeit.* Wir nehmen denselben jetzt auf.

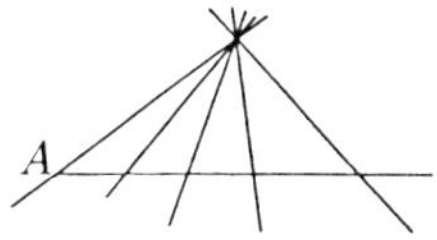

1.) *Strahlbüschel und Punktreihe*

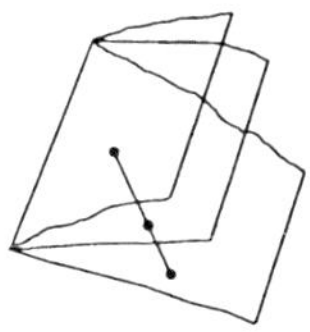

2.) *Ebenenbüschel und Punktreihe*

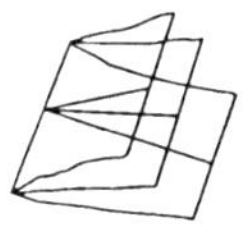

3.) *Ebenenbüschel und Strahlenbüschel*

[26]↑Zwei in dieser Weise durch die Lage auf einander elementweise bezogene Gebilde heissen *perspektiv* oder in *perspektiver Lage* oder auf einander perspektiv bezogen. Also

Definition.

Zwei Grundgebilde heissen perspektiv, wenn das eine der Schnitt des anderen ist.

Aber auch gleichartige Grundgebilde kann man ein-eindeutig elementweise auf einander beziehen, nämlich:↓

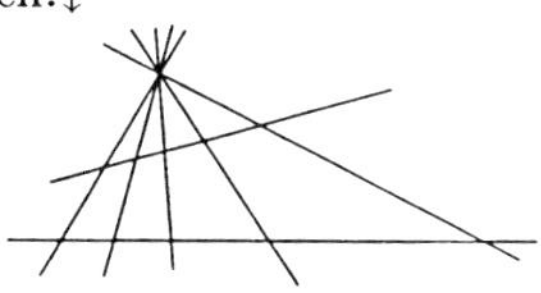

Zwei Punktreihen 35

(Schnitte desselben Strahlen- oder Ebenenbüschels)

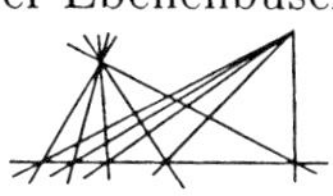

Zwei Strahlenbüschel

wenn sie dieselbe Punktreihe schneiden oder in demselben Ebenenbüschel liegen.

[26] This addition (up to line 37.18) is written on a sheet pasted to the bottom of p. 34. The passage serves as a substitute for three paragraphs deleted on pp. 35–36.

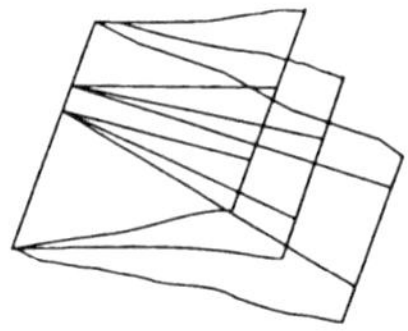

Zwei Ebenenbüschel
wenn sie dieselbe Punktreihe oder dasselbe Strahlenbüschel schneiden.

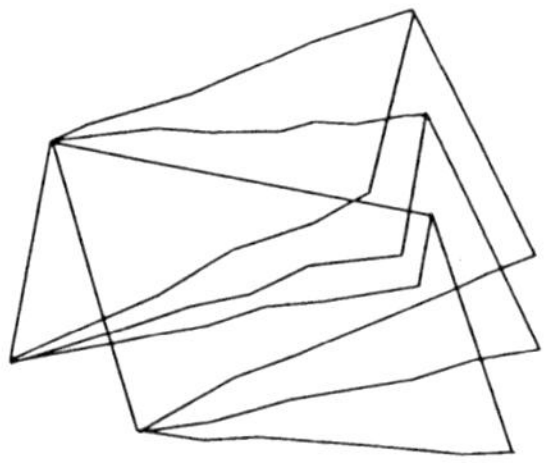

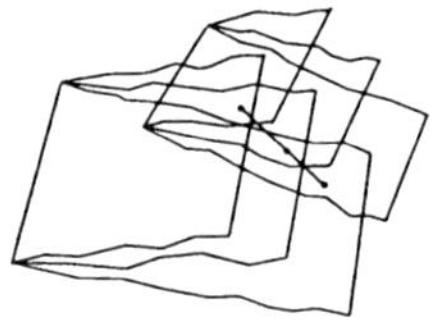

Diese ein-eindeutige Beziehung geschieht also durch Vermittelung *eines 3ten Grundgebildes.*[27]

36

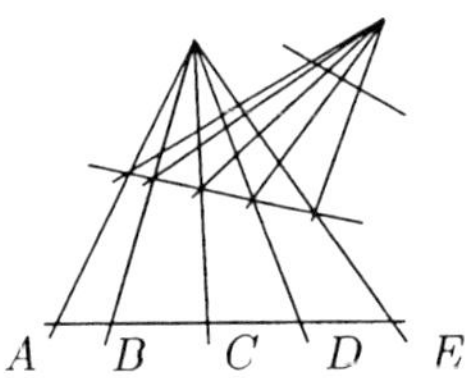

Man kann aber auch auf *sehr viel andere Arten 2 Grundgebilde elementweise und umkehrbar eindeutig auf einander* beziehen, z. B. 2 Punktreihen, welche zu der *nämlichen 3ten Punktreihe* perspektiv sind, werden *im Allgemeinen nicht wieder perspektiv* liegen – ebenso, wenn ich zwei *perspektive Punktreihen oder Strahlenbüschel gegeneinander beliebig verschiebe.*

37 Die *perspektive Beziehung* ↑und die durch Wiederholung oder Verschiebung zusammengesetzten Beziehungen↓ *haben eine ausgezeichnete Eigenschaft,* welche überdies noch bei den letzten angedeuteten Eigenschaften erhalten bleibt, nämlich, wenn man 4 harmonische Elemente herausgreift, so sind die entsprechenden *4 Elemente wieder harmonisch* (§ 2)[28]. Diese Eigenschaft führt nun auf den allgemeinen Begriff der *Projektivität.*

Definition:

Zwei Grundgebilde heissen projektiv, wenn sie umkehrbar eindeutig so auf einander bezogen sind, dass je 4 harmonischen Elementen des einen allemal

---

[27]The following three paragraphs (the first on p. 35, the two others on p. 36) are deleted (see previous note):

Zwei in dieser Weise *durch die Lage* auf einander elementweise bezogene Gebilde heissen perspektiv oder in perspektiver Lage oder auf einander perspektiv bezogen. Also:

⟦Zwei ungleichartige Grundgebilde heissen perspektiv, wenn das entsprechende Element durch Schnitt oder Verbindung, zwei gleichartige Grundgebilde heissen perspektiv, wenn das nämliche durch Vermittelung eines dritten Grundgebildes möglich ist, sie beide zu dem nämlichen dritten Grundgebilde perspektiv sind.⟧

↑*Definition:*↓

Zwei ungleichartige Grundgebilde heissen perspektiv, wenn das eine der Schnitt des anderen ist; zwei gleichartige Grundgebilde heissen perspektiv, wenn sie zu dem nämlichen dritten ungleichartigen Grundgebilde perspektiv liegen.

[28]Hilbert refers to the 'zusammenfassenden Satz' in § 2, p. 30.

4 harmonische Elemente des anderen entsprechen. ↑*projektiv* = $\overline{\wedge}$. *Also Gesetz der Zuordnung der Punkt!*↓

2 perspektive Grundgebilde sind demnach auch projektiv. Ferner: Wenn 2 Gebilde zu einem *dritten projektiv* sind, so sind sie unter *einander projektiv.* Z. B. wenn 2 Gebilde zu einem dritten perspektiv sind, so brauchen sie *nicht perspektiv zu liegen*, wohl aber sind sie *unter einander projektiv.* Ebenso, wenn man eine ganze Reihe von Gebilden hat, von denen immer das folgende zum vorhergehenden perspektiv liegt.

Wenn 2 Grundgebilde projektiv sind, so entspricht jeder Aufeinanderfolge von 38
Elementen des einen Gebildes eine Aufeinanderfolge von Elementen des anderen Gebildes.

Sinn des Satzes: Wenn *z. B. auf einer Punktreihe* die Elemente A, B, C, D, E, $F, \ldots$ aufeinanderfolgen, d. h. so, dass ich von A zu B, B zu C, C zu D, u. s. f. gelangen kann, *ohne einen Punkt der Reihe zu überschreiten*, so haben die entsprechenden Punkte A', B', C', D', E', $F', \ldots$ dieselbe Eigenschaft: oder wenn *ein Punkt* die eine Punktreihe in einer *bestimmten Richtung* durchläuft, so durchläuft der entsprechende Punkt die 2te Punktreihe *ebenfalls so, dass er seine Richtung niemals ändert.*

Wäre die Behauptung unrichtig, so gäbe es also 2 Punkte, etwa E', F', die ich nicht verbinden kann, ohne einen anderen Punkt der Reihe zu überschreiten: auf dem einen (endlichen) Wege etwa P', auf dem anderen (unendlichen) Wege etwa Q'. Dann habe ich entsprechend:

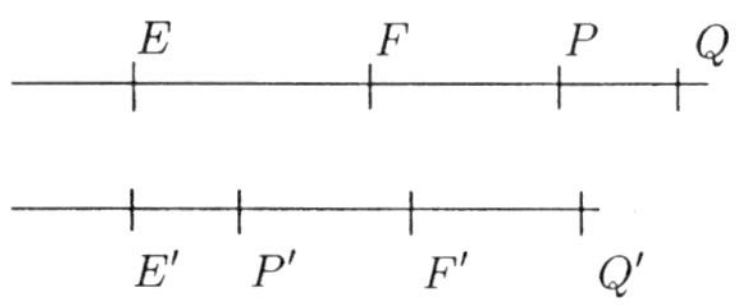

E, F und P, Q trennen sich nicht. Es giebt also 2 Punkte H, K, welche beide 39
Punktepaare harmonisch trennen,

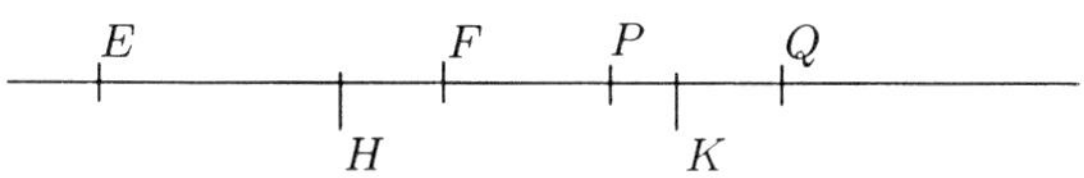

so dass also $HFKE$ *und* $HPKQ$ harmonisch sind. Ich suche die Punkte H', K'. Zu diesen müssen ebenfalls die beiden Punktepaare F', E' und P', Q' harmonisch liegen,[29] was unmöglich ist, da sich diese beiden Paare trennen. Hieraus folgt der Satz durch *Schneiden* mit *einer geraden Linie* auch für *harmonische Strahlen* und *Ebenen.*

Wir können *2 gleichartige Grundgebilde auf einander* legen, z. B. *2 Punktreihen, so dass sie denselben* Träger haben. *Wie viel Elemente können sie entsprechend gemein haben?*

1.) Sind die beiden Punktreihen *entgegengesetzt laufend*, | so haben sie *2 und* 40
nur 2 Elemente entsprechend gemein:

[29] From this follows, according to the theorem given in the addition on p. 31 ('Zwei zu AC harmonische Paare, etwa BD und $B'D'$, trennen sich nicht'), that the pairs F', E' and P', Q' are not 'separated'.

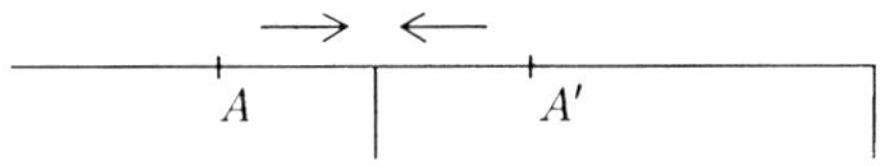

2.) Sind sie *gleichlaufend*, so können auch 2 Elemente entsprechend gemein sein: nämlich P und Q.[30]

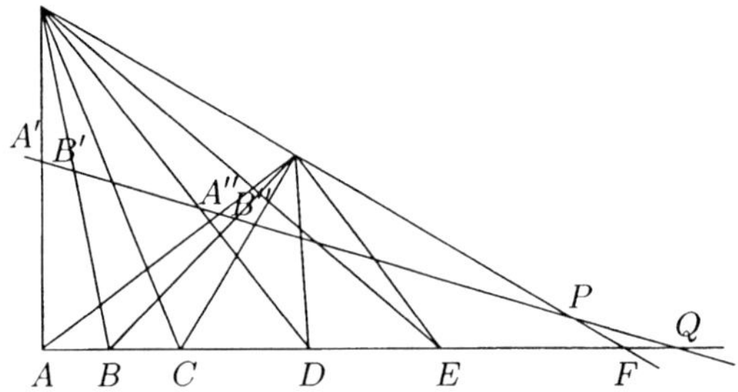

↑Auch *bloss 1 Element*, wenn der Träger $A'B'\ldots\ A''B''\ldots$ durch F geht, und *auch kein Element.*[31] Z. B. wenn man ein Strahlbüschel um seinen Mittelpunkt dreht, so dass *jeder Strahl* mit dem entsprechenden *den nämlichen Winkel einschliesst* und dann mit einem *beliebigen Träger schneidet.*↓

Wir wollen zeigen, dass es auch bei gleichlaufendem Sinne *nie mehr als 2 Punkte* sein können:

Wenn 2 projektive Grundgebilde 3 Elemente entsprechend gemein haben, so haben sie alle Elemente entsprechend gemein.

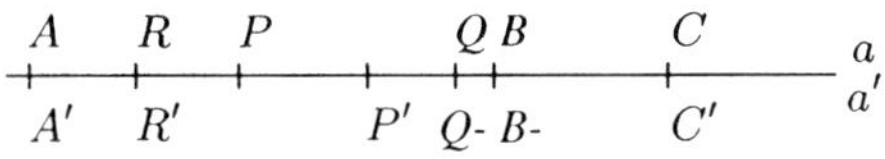

Es möge etwa auf der endlichen Strecke AB einen Punkt P geben, welcher
41 mit dem entsprechenden | Punkte P' (wenn P' nicht auch zwischen A und B fällt, so ist die Unmöglichkeit sofort klar)[32] auf a' nicht zusammenfällt.[33] Ich bewege dann P auf P' zu. Dann bewegt sich auch P' nach rechts, weil entgegengesetzt laufende Punktreihen nur 2 Elemente entsprechend gemein haben. Die entsprechenden Punkte müssen in $B = B'$ oder vor $B = B'$, etwa

[30]The following figure is a little hard to understand. It concerns two sheaves of lines (in the same plane) which are perspective, i.e., the lines of the sheaves intersect in points of a line (call it a), marked here by A, B, C, D, E. Now draw a line (call it c) through the middle points of the sheaves, and let this cut a at F. Now draw another line (call it b) through the two sheaves, cutting the lines of these at, respectively, A', B', C', ... and A'', B'', C'', ... F' and F'' will be the same, namely the point (marked 'P') where b cuts c. Q is the point where b cuts a. Clearly, this is also the point common to the sequences A, B, C, ..., A', B', C', ... and A'', B'', C'', ...

No figure in this form is found in Reye where the corresponding matter is discussed (*Reye 1886*, 53–55), although one can reconstruct Hilbert's figure to some extent from Reye's Fig. 21 on p. 54.

[31]I.e., it could also be the case that the projective point sequences have only one element or none in common.

[32]Because the property of separating points is preserved under projective mappings.

[33]If this held for a point outside AB, then one could construct a fourth harmonic point inside AB for which the same also holds.

in $Q = Q'$ zusammentreffen.[34] Dann bewege ich P' auf P zu. Der entsprechende Punkt P bewegt sich dann ebenfalls nach links. Beide vereinigen sich vor oder in $A = A'$, etwa in $R = R'$. Auf der Strecke RQ fällt kein Punkt mit seinem entsprechenden zusammen. Denn P resp. P' hat stets einen Vorsprung. Dies ist aber unmöglich, weil derjenige Punkt H, welcher von C durch R, Q harmonisch getrennt ist, mit dem entsprechenden Punkte H' zusammenfallen muss. Also der entsprechende Punkt überholt einmal und lässt sich dann überholen.

Dieser Satz ist von *ausserordentlicher Wichtigkeit*, weil er das *Wesen einer projektiven Beziehung in ein helles Licht rückt.*

Für Strahlenbüschel und Ebenenbüschel führen wir den Beweis entweder ebenso oder wir schneiden mit einer Punktreihe.

Wenn man 3 Elemente eines Grundgebilde beliebigen 3 Elementen eines an- 42
deren Grundgebildes zuweist, so ist dadurch stets eine und zwar nur eine projektive Beziehung zwischen den beiden Grundgebilden bestimmt.

Wir führen den Beweis wieder *nur für Punktreihen*:

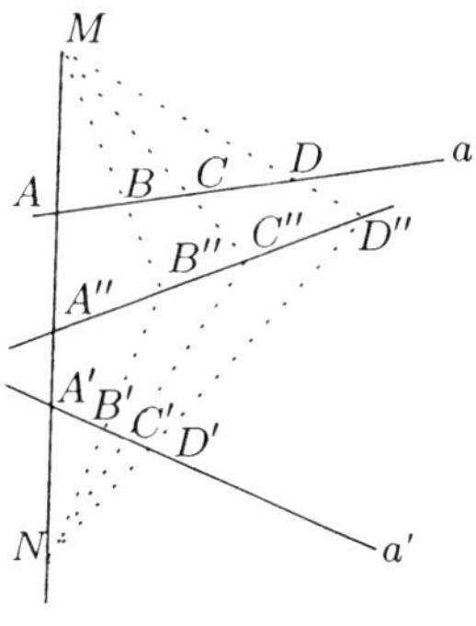

(Wenn a und a' nicht in einer Ebene liegen, so nehme man einen dritten Träger a'', welcher mit beiden a und a' in einer Geraden liegt.) Gegeben sind ABC, $A'B'C'$. Man nehme auf AA' die Punkt M *und* N an. MB, NB' und MC, MC' schneiden sich in B'', C''.[35] M und N mit dem beliebigen Punkte D'' auf $B''C''$ verbunden liefern D, D'. $ABCD$ perspektiv zu MA, MB, MC, MD zu $A''B''C''D''$ zu $N(A''B''C''D'')$[36] zu $A'B'C'D'$, d. h. die beiden Träger a und a' sind
projektiv in der verlangten Weise | auf einander bezogen. Gäbe es noch eine 43
zweite projektive Beziehung von der verlangten Eigenschaft, so wäre etwa der Träger a auf sich selbst mit 3 entsprechend gemeinen Elementen bezogen, was unmöglich ist.

---

[34]That P' cannot lie outside $A'B'$ is shown by the theorem formulated on p. 38, which depends on a continuity assumption. That there must be a point Q' at all can likewise only be shown with a continuity assumption.

[35]It should be 'NC'' here, not 'MC''.

[36]Hilbert means by '$N(A''B''C''D'')$' the lines NA'', NB'', NC'' and ND''.

Zur Uebung noch die dualistisch entsprechende Construktion:

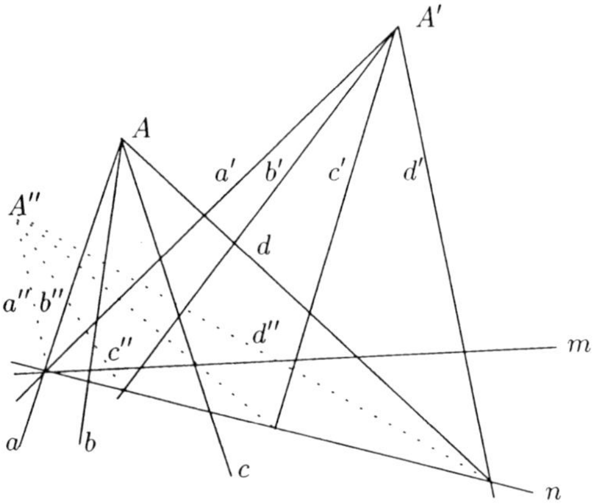

Man sieht *zugleich folgendes: Zwei projektive* Grundgebilde lassen sich stets als Anfangs- und Endglied in einer Reihe von Gebilden betrachten, in welcher jedes folgende Gebilde zum *vorhergehenden perspektiv liegt.* ↑Wichtige Erkenntniss, die zugleich die Auffassung des Begriffes „projektiv" erleichtert. Während früher ganz abstrakt definirt werden musste, wissen wir jetzt ... ↓

Wir können *beliebig viele Paare entsprechender* Punkte allein mit *Hülfe des Lineals construiren.* ↑Z. B. Strahlenbüschel auf Ebenenbüschel projektiv beziehen, ferner auch 2 Gebilde auf demselben Träger.↓

44 Wir beweisen noch leicht folgende 2 Sätze, welche sich dual entsprechen; die wir häufig brauchen werden.

Wenn 2 projektive Punktreihen ihren Schnittpunkt entsprechend gemein haben, so laufen die Verbindungslinien entsprechender Punkte durch einen Punkt.

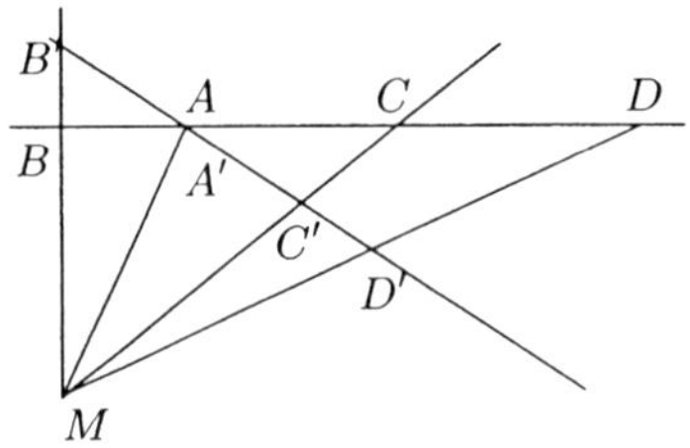

Denn verbinde ich BB' und CC', so erhalte ich von M ein perspektives Strahlenbüschel etc.

Wenn 2 projektive Strahlenbüschel die Verbindungslinie ihrer Mittelpunkte entsprechend gemein haben, so schneiden sich die entsprechenden Strahlen in den Punkten einer Geraden.

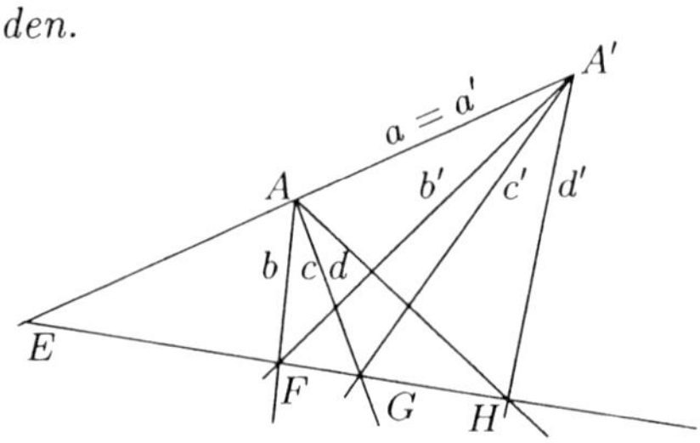

Denn die Punktreihe $EFG\ldots$ ist perspektiv, und durch 3 Paare entsprechender Punkte ist die projektive Zuordnung bestimmt.

Zur Uebung Beweis folgenden Satzes: 45
Liegen *die Eckpunkte eines Sechseckes abwechselnd auf 2 Geraden*, so liegen auch die Schnittpunkte *der 3 Paar Gegenseiten desselben auf einer Geraden.*

46

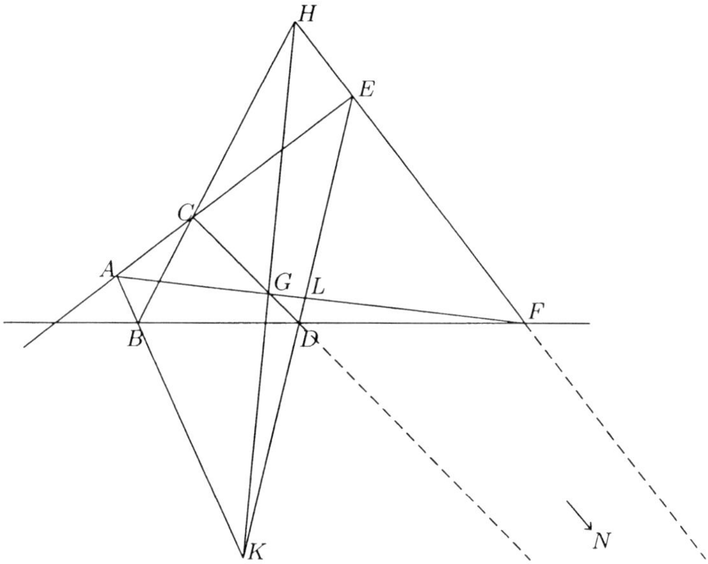

[37]↑ projektiv = $\overline{\wedge}$

Solche Formeln sagen aus, dass die Elemente links und rechts Gebilden angehören, welche projektiv bezogen werden können, so dass die bezüglichen Elemente sich entsprechen. Die Formel sagt also erst etwas aus, wenn mehr als 3 Elemente auf beiden Seiten stehen.↓ [38]

Denn AE, AB, AD, AF projektiv zu CE, CB, CD, CF
projektiv zu $E \quad K \quad D \quad L$ " $E \quad H \quad N \quad F$
(auf ED) (auf EF)
folglich
$E \quad K \quad D \quad L$ projektiv zu $E \quad H \quad N \quad F$
d. h. KH, DN, LF laufen durch einen Punkt G.[39]

---

[37]This addition was originally written to the right of the formulas deleted in the following note, although Hilbert clearly meant it to be relevant to the formulas which follow. That the part 'welche projektiv ⟨...⟩ stehen' is itself an addition, squeezed into the space available under the figure, indicates that the passage was added after the following proof was finished.

[38]The following text, below and to the left of the above figure, was deleted by Hilbert:
Denn AB, AD, AF, AE ⟦projektiv zu ⟪?⟫⟧
projektiv zu CB, CD, CF, CE
Denn AB, AD, AF, A

[39]This latter follows from the first of the two theorems given on p. 44.

47 Beachte die *Regelmässigkeit der Figur*: sie besteht aus 9 Punkten und 9 Geraden. Je *3 Punkte* liegen auf einer Geraden, je *3 Geraden* gehen durch einen Punkt. Die Figur *entspricht sich selbst dualistisch.*[40 41]
↑Meine Herren. Im vorigen § 3 *Begriffe Perspektivität und Projektivität.* Wenn man *2 Grundgebilde* hat und die Elemente. Wichtigster Satz: durch 3 Elemente bestimmte *Beziehung* $A, B, C, D, \ldots \barwedge A', B', C', D', \ldots$ kann *durch lineare Construktion* hergestellt werden. Zur *letzten Figur Bemerkung.*↓

§ 4. Gebilde zweiter Ordnung.

Wir haben bisher die *3 Grundgebilde* behandelt und *die projektive Zuordnung ihrer* Elemente. Nunmehr soll *aus zwei beliebigen Grundgebilden, die projektiv*
48 *sind, eine neues | Gebilde – ein sogenanntes Gebilde 2ter Ordnung – abgeleitet werden.*

Wir nehmen zuerst *2 projektive Punktreihen*: a, a'. Auf a die Punkte A, B, C, auf a': A', B', C'. Dadurch ist die *projektive Beziehung bestimmt,*

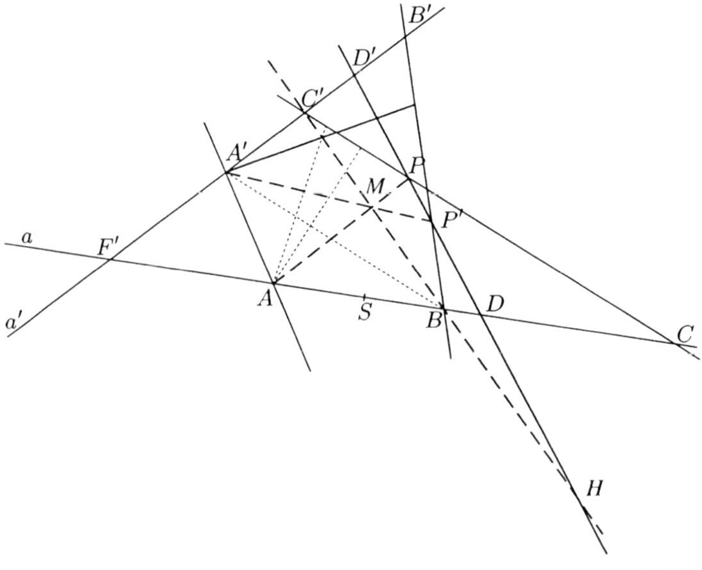

[40] The following text after 'dualistisch' was deleted by Hilbert: und beweist zugleich den dualistischen Satz:

Wenn die Seiten eines Sechsecks (z. B. $AQCDRK$) abwechselnd durch 2 Punkte (B und G) gehen, so laufen die 3 Hauptdiagonalen (Verbindungslinien gegenüberliegender Punkte) durch den nämlichen Punkt:
⟦Sechseck: $ABHGAEH$⟧
Sechsseites: $AB - BH - HG - GA - AE - EH$
Hauptdiagonalen: BA, HE, GP, wo P der Schnittpunkt von EH und AB ist.

[41] Here Hilbert's 'exercise' is the Pappus-Pascal Theorem, and his use of 'dualistisch' also hints at its dual, the Brianchon Theorem; see the next section (§ 4). Hilbert's notation, however, is not particularly clear, having been through several levels of correction, and this (and the lack of explanation) makes the proof sketch difficult to follow. It should be noted that there were even two attempts at figures on p. 45, both deleted, which explains why the page is so truncated. Neither figure was labelled.

und wir können zu jedem *Punkte D den entsprechenden* ⟨Punkt⟩ D' linear construiren. Wir denken uns alle Verbindungslinien AA', BB', CC', ... gezogen: Das System aller dieser Verbindungslinien heisst ein *Strahlenbüschel 2ter Ordnung.* Diese | Strahlen werden leicht so construirt: Man verbinde einen 49
beliebigen Punkt M auf $C'B$ mit A und A': PP' ist ein Strahl. Denn:

$$\underbrace{PC', PM, PB, PH}_{\overline{\wedge}\, CABD} \quad \overline{\wedge} \quad \underbrace{P'C', P'M, P'B, P'H}_{\overline{\wedge}\, C'A'B'D'}$$

q. e. d.

Um alle Strahlen DD' zu erhalten, müssen wir *M auf C'B variiren lassen.* Kommt *M nach B*, so ergiebt sich *der Träger a* und kommt *M nach C'*, so der Träger a'. *Diese Träger gehören also ebenso zu den Strahlen des* Strahlenbüschels 2ter Ordnung. Wir wollen aber *sogar* zeigen, dass diese Träger der ursprünglich gegebenen beiden Punktreihen *überhaupt gar keine* ausgezeichnete Stellung einnehmen. Ich nehme z. B. die Strahlen *CC' und BB' als Träger* und zeige, dass *P und P' projektive Punktreihen* beschreiben. In der That, bewege ich M, so sind die Strahlenbüschel *AM und A'M projektiv*, folglich auch die Punktreihen P und P'.

Das Strahlenbüschel 2ter Ordnung ist daher durch 5 willkürlich herausge- 50
griffene Strahlen völlig ↑eindeutig↓ *bestimmt. Eine 6te Linie* muss eine Bedingung erfüllen, und diese Bedingung folgt auch leicht aus obiger Construktion, wenn wir das Sechseck $ABP'PC'A'$ ins Auge fassen, dessen Hauptdiagonalen sich in einem Punkte M schneiden:

*Brianchonscher Satz: Gehören die 6 Seiten eines Sechseckes einem Strahlenbüschel 2ter Ordnung an, so schneiden sich die Verbindungslinien gegenüberliegender Ecken in einem und demselben Punkte.**)

Durch einen beliebigen Punkt P der Ebene gehen höchstens 2 Strahlen des Büschels 2ter Ordnung. Denn verbinde ich P mit den Punkten $A\,B\,C$ ... des Trägers a und mit $A'\,B'\,C'$ ... des Trägers a', so | erhalte ich 2 projek- 51
tive Strahlenbüschel mit demselben Mittelpunkt. Dieselben haben *höchstens 2 Elemente* entsprechend gemein. *Wir werden so auf die Frage geführt, ob und wann sie nur* 1 *oder* 0 *entsprechend gemein haben.* ↑Solche Punkte, durch welche 2 Strahlen gehen, giebt es natürlich immer:↓

Zu dem Zwecke lassen wir D auf a wandern. Durch D gehen immer 2 Strahlen, nämlich ausser a noch DD'. Es wird aber einen Punkt und nur einen geben, *wenn D' in (aa') fällt.* Diesen Punkt nenne ich den singulären Punkt S des Strahles a. Auf jedem anderen Strahl giebt es ebenfalls einen und nur einen singulären Punkt S. *S ist zugleich die Grenzlage des Schnittpunktes* mit den Nachbarstrahlen. Denn wenn ich mich mit D dem Punkte S nähere, so nähert sich D' dem Schnittpunkt $(aa') = F'$. F' und S sind aber verschiedene Punkte, weil sonst alle Strahlen des Büschels 2ter Ordnung durch einen Punkt laufen würden und daher ein Büschel erster Ordnung bilden würden.

---

*) Das Sechseck kann auch einspringende Ecken haben. Aus 6 Tangenten lassen sich 60 verschiedene Sechsecke bilden; vgl. Schröter, synthetische Geometrie, S. 126.[42]

---

[42] I.e., in *Steiner 1867*.

52 [Das Strahlenbüschel 2ter Ordnung ist auch gegeben 1.) durch 4 Strahlen und einen auf einem derselben gelegenen Berührungspunkt und 2.) durch 3 Strahlen und 2 auf je einem derselben gelegenen Berührungspunkt.]

Die ganze *Anordnung der Strahlen des Büschels* 2ter Ordnung und der singulären Punkte erkennt man anschaulich, wenn man nach dem *Orte der S fragt.*

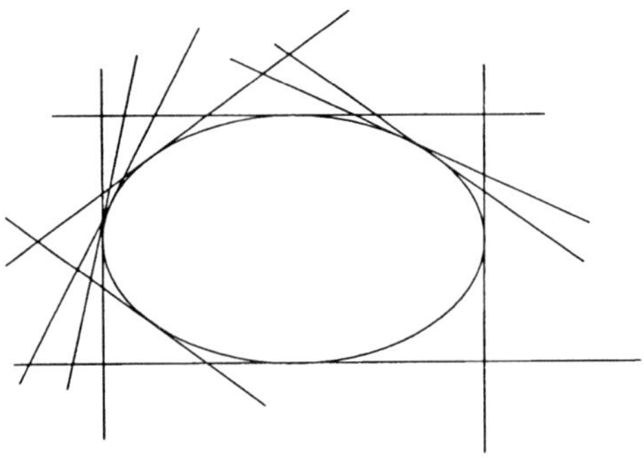

Lässt man D auf dem Träger a wandern, so beschreibt S eine Curve, welche in sich zurückläuft, da D auf seine Anfangsstelle zurückkehrt. Da S die Grenzlage des Schnittpunktes der Nachbarstrahlen ist, so sind die *Strahlen Tangentes.* Die *Curve heisse ein Kegelschnitt.* ↑Man kann etwa die Ebene *schwarz* und die *Strahlen leuchtend* denken.↓[43]

53 Also die *Strahlen eines Strahlenbüschels 2ter Ordnung* sind die *Tangenten eines Kegelschnittes,* sie *umhüllen denselben.* Die *singulären Punkte erfüllen den Kegelschnitt.*

Der Kegelschnitt ist also gegeben *1.) als Tangentengebilde*[44] *2.) als Punktgebilde*[45].

1.)

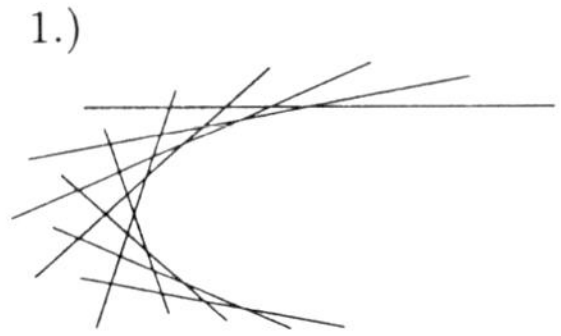

2.)

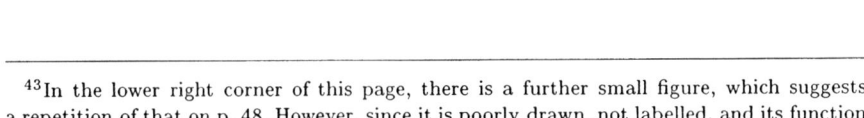

Die Ebene wird durch die Curve in *2 Gebiete* getheilt, *weil dieselbe geschlossen ist.* Die Curve hat *keine Buchtung, da sonst mehr als 2 Tangenten* von einem Punkte möglich wären. [46] Also ein *Aeusseres, von dessen Punkten* 2 Tangenten möglich sind *und ein Inneres, von dessen Punkten keine*

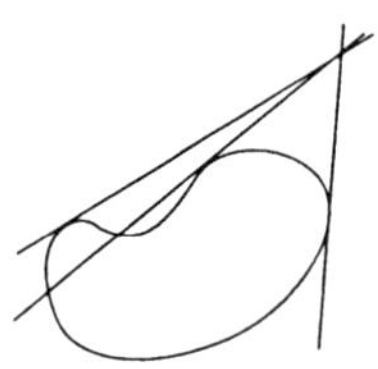

54 *Tan|genten* möglich. In einem Punkte der Curve selbst ist *nur eine Tangente möglich.*

[43] In the lower right corner of this page, there is a further small figure, which suggests a repetition of that on p. 48. However, since it is poorly drawn, not labelled, and its function in the present context is unclear, we have omitted it.

[44] Deleted: : Die Grenzlage des Schnittpunktes der Nachbartangenten liefert den Berührungspunkt und

[45] Deleted: , die Grenzlage der Verbindungslinie mit Nachbarpunkten: die Sehne liefert die Tangente

[46] Deleted: ↑Denn jede Tangente der Curve hat mit derselben immer einen Punkt gemein.↓

Unsere *früheren Sätze* können wir jetzt so aussprechen:

Irgend 2 Tangenten eines Kegelschnittes werden von sämtlichen Tangenten in 2 projektiven Punktreihen geschnitten.

Bildet man aus 6 Tangenten eines Kegelschnittes ein Sechseck, so schneiden sich die Verbindungslinien gegenüberliegender Ecken in einem und demselben Punkte.
(Brianchonscher Satz)

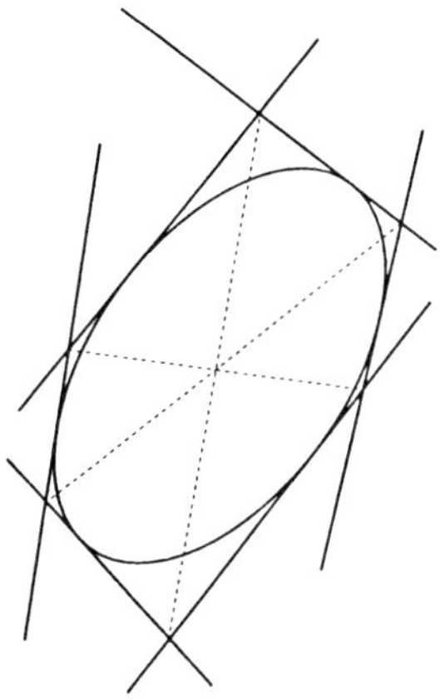

Wir finden nun leicht eine Reihe *weiterer* Sätze vom Kegelschnitt, indem wir diesen *Brianchonschen Satz auf spezielle Fälle anwenden*:

Wenn 2 der 6 Strahlen des Büschels 2ter Ord|nung sich nähern und schlies- 55
slich zusammenfallen, so tritt an Stelle des Schnittpunktes derselben (der Ecke des Sechseckes) der singuläre Punkt des Strahles. So erhalten wir Sätze für 5, 4, 3-Ecke:

In jedem Tangentenfünfeck schneiden sich 2 Diagonale auf der Verbindungslinie des 5ten *Eckpunktes mit dem Berührungspunkt der gegenüberliegenden Seite.*
$\uparrow$(Aufgabe: 5 Tangenten gegeben. Man soll die Berührungspunkte auf denselben mit Lineal construiren.)$\downarrow$

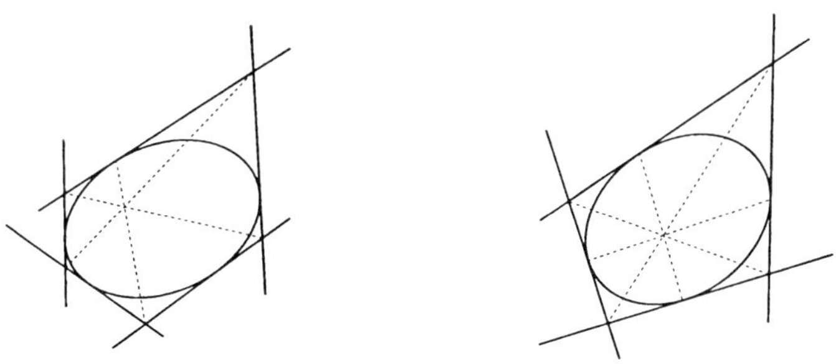

In jedem Tangentenviereck schneiden sich die Diagonalen und die Verbindungslinien der Berührungspunkte von irgend 2 Gegenseiten in einem Punkte.
(Also 4 Linien gehen durch einen Punkt.)

In jedem Tangentendreieck schneiden sich die 3 Verbindungslinien der Ecken mit den Berührungspunkten der gegenüberliegenden Seiten in einem
56 | *Punkte:*

↑*Aufgaben zur Einübung:* Construktion des Kegelschnittes, d. h. beliebig vieler ↑oder 5↓ seiner Tangenten, 1.) wenn 4 Tangenten und auf einer von ihnen der Berührungspunkt 2.) wenn 3 Tangenten und auf 2 von ihnen der Berührungspunkt gegeben sind. (Bei letzterer Aufgabe erst den Dreiecks- und dann den Viereckssatz anwenden!)↓

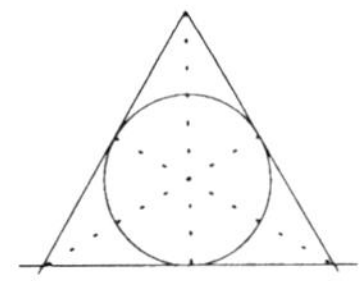

Wir untersuchen *den Satz vom Viereck* etwas genauer. Wir ziehen nämlich 4 Tangenten und betrachten folgende 3 Vierecke:

1.) *ABCD*
2.) *AECF*
3.) *BEDF*

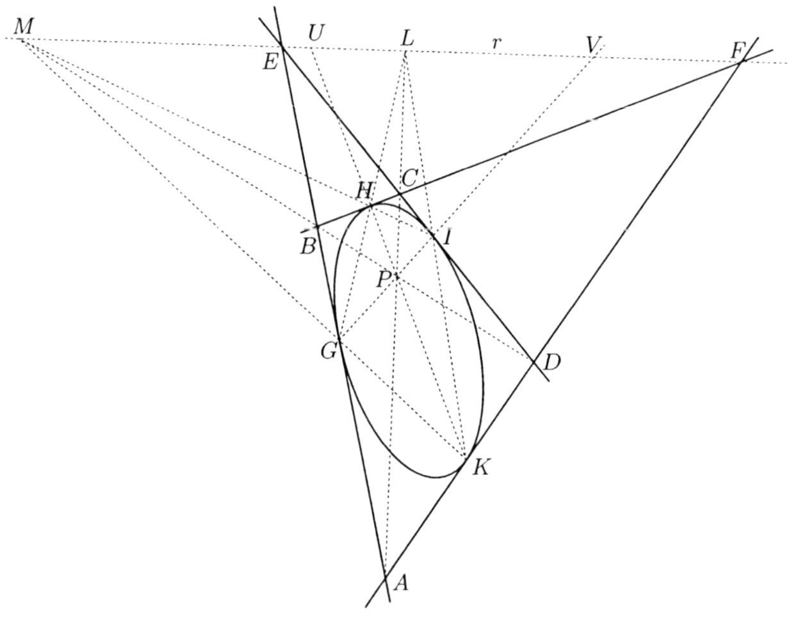

Aus dieser Figur ersieht man zugleich den folgenden Satz:

57 *In jedem Sehnenviereck* (*GHIK*) *liegen die Schnittpunkte der Gegenseiten M, L mit denjenigen der Tangenten von irgend 2 gegenüberliegenden Eckpunkten E, F in einer Geraden.* ↑Dualistisch entsprechend dem vorigen Viereckssatz.↓

Halten wir 3 von den Tangenten, etwa die in H, I, K berührenden, fest und lassen die 4te gleiten, so erhalten wir etwa für eine zweite in G' berührende Tangente die Punkte L', M'. LM und $L'M'$ gehen durch F. Folglich sind die Punktreihen $L, L', \ldots \barwedge M, M', \ldots$, folglich auch $HL, HL', \ldots \barwedge KM, KM', \ldots$ oder $HG, HG', \ldots \barwedge KG, KG', \ldots$, d. h. *der Kegelschnitt ist zu-*

gleich der Ort der Schnittpunkte der entsprechenden Strahlen zweier projektiver Strahlenbüschel.[47]

Also der Kegelschnitt 59 als Punktgebilde zugleich *das Erzeugniss zweier projektiver Strahlenbüschel.*

Zieht man noch die mit Bleifeder ausgezogenen Linien[48] zur Construktion der Punkte E, E', so sieht man, dass die erzeugenden Strahlenbüschel H und K zugleich projektiv sind zu den erzeugenden *Punktreihen* EE', ...

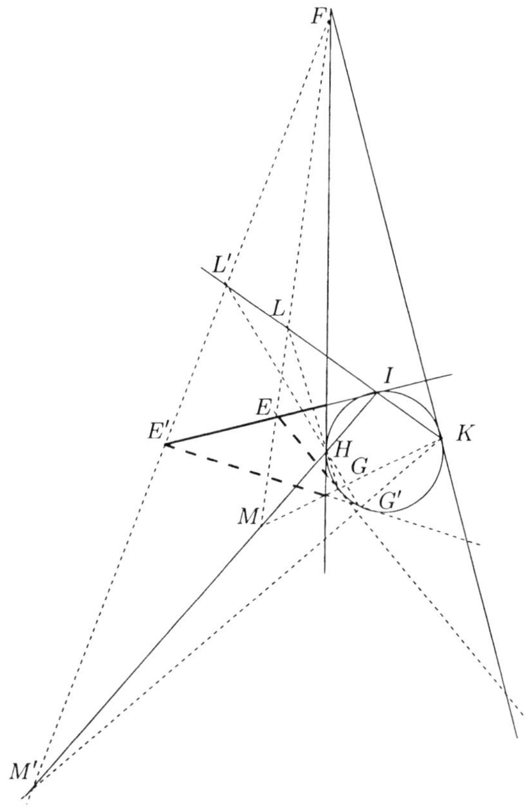

Wir hätten von Anfang an *2 projektive Strahlenbüschel* zu Grunde legen können. 60 Dann wäre nur die *Reihenfolge der Resultate* eine veränderte geworden. Ende wäre Anfang geworden und umgekehrt. Von der ganzen zwischen Anfang und Ende gelegenen Entwickelung überall das dualistische Gegenstück. *Ich deute* diese dualistisch entsprechende Entwickelung nur kurz an:

2 projektive Strahlenbüschel: Durch A die Strahlen a, b, c sollen entsprechen durch A' den Strahlen a', b', c'. Dadurch ist die *projektive Beziehung* völlig bestimmt. Zu jedem d kann der entsprechende d' linear construirt werden. Die Schnittpunkte aller entsprechenden Strahlen heissen eine *Punktreihe* 2ter *Ordnung.* Es werden diese Punkte leicht so construirt: Man ziehe durch bc' die beliebige Gerade m. Die Schnittpunkte von m mit a und a' werden mit $(cc') = I$ und $(bb') = G$ verbunden und pp' schneiden sich in K.

---

[47]In the middle of this page are the beginnings of a figure which appears similar to that on p. 56, but which is stricken through. P. 58 is fully taken up by two other attempts at figures which were not completed, but stricken out. Both again appear to be variations on the current themes.

[48]This refers to the figure on the right, and the three lines in bold ($E'E$, $E'G'$ and EG).

61

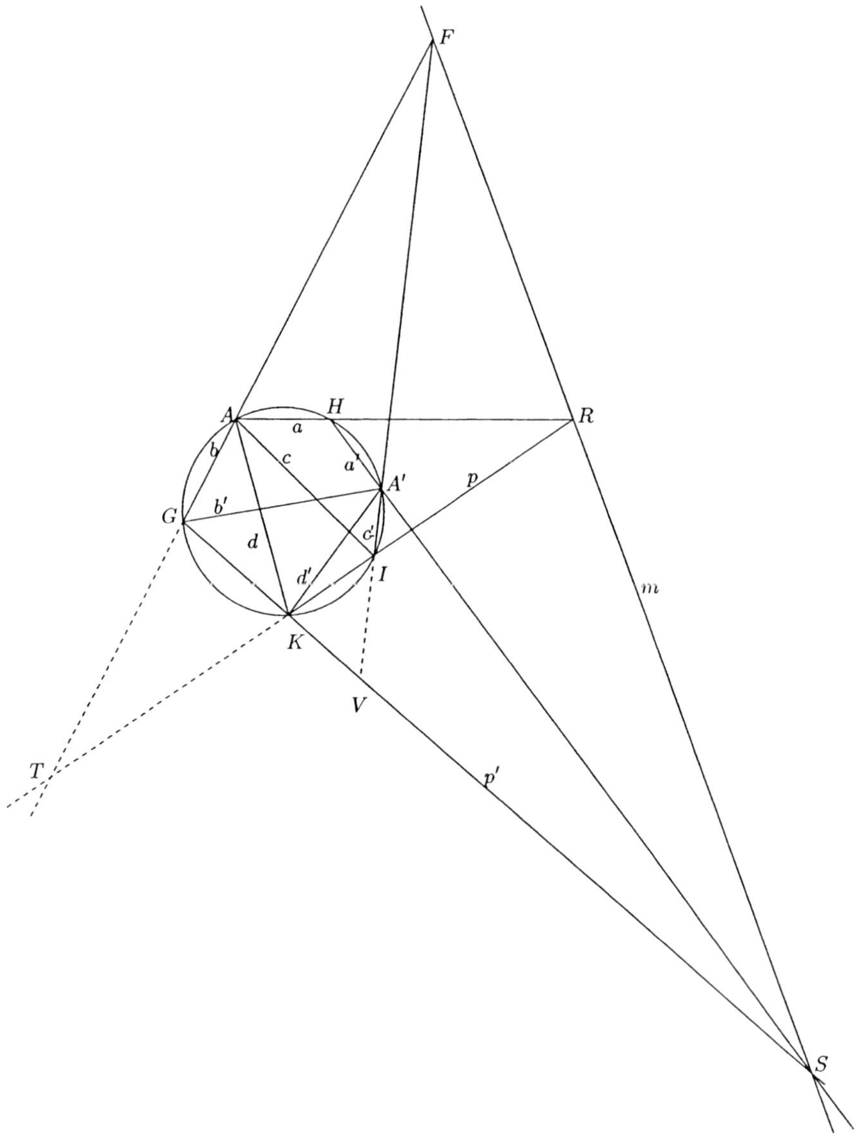

K ist ein verlangter Punkt der Punktreihe. Denn $RSKT \;\overline{\wedge}\; SVKG$, folglich $acdb \;\overline{\wedge}\; a'c'd'b'$ q. e. d.

62 Um alle Punkte dd' zu erhalten, muss man m um F drehen: Man sieht wieder ein, dass A, A' gar keine ausgezeichnete Stellung einnehmen, d. h. durch 5 willkürlich herausgegriffene Punkte ist die Punktreihe 2ter Ordnung bestimmt. Der 6te Punkt muss folgender Bedingung genügen:

Pascalscher Satz:

Gehören die 6 Ecken eines Sechseckes einer Punktreihe 2ter Ordnung an, so liegen die Schnittpunkte gegenüberliegender Seiten auf einer Geraden.

Eine Gerade hat mit dem Kegelschnitt höchstens 2 Punkte gemein. Es giebt aber durch jeden Punkt des Kegelschnittes eine (singuläre) Gerade, nämlich die Tangente, welche mit dem Kegelschnitt nur einen Punkt gemein hat. Er ist zugleich die Grenzlage der Verbindungslinie mit benachbarten Punkten.

Wir erhalten so dasselbe Bild wie oben, nämlich | 1.) Kegelschnitt als 63
Punktgebilde 2.) als Tangentengebilde.

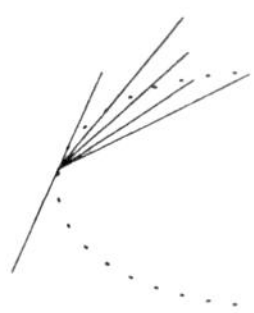

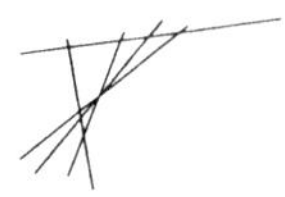

Unsere früheren *Sätze heissen jetzt*:

Wenn irgend 2 Punkte eines Kegelschnittes mit sämmtlichen übrigen verbunden werden, so entstehen projektive Strahlenbüschel. In jedem Sehnensechseck schneiden sich die gegenüberliegenden Seiten in Punkten einer geraden Linie. (*Pascalscher Satz*)

Wenn 2 der 6 Punkte sich nähern und schliesslich zusammenfallen, so tritt an Stelle ihrer Verbindungslinie die Tangente. Auf diese Weise folgt:

In jedem Sehnenfünfeck liegen die Schnittpunkte von 2 Paaren nicht benachbarter Seiten mit demjenigen 3ten Punkte in einer Geraden, in welchem die 5te Seite von der Tangente des gegenüberliegenden Eckpunktes geschnitten wird.

(Aufgabe: *Gegeben sind 5 Punkte. Man soll die Tangenten in denselben construiren.*)(Aufgabe: ⟨...⟩ construiren.)

64

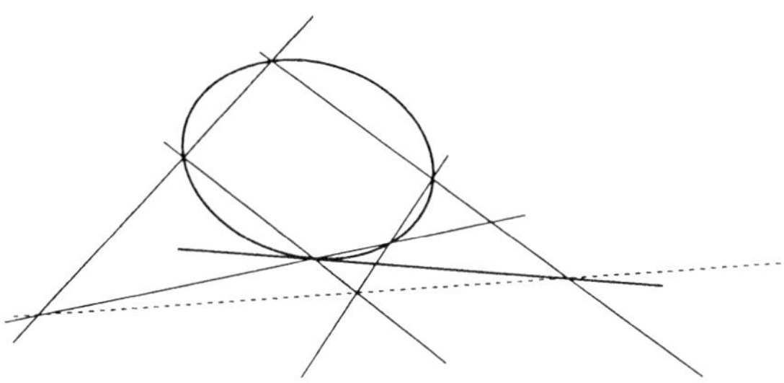

Für das *Sehnenviereck* erhalten wir den uns schon bekannten Satz und endlich:

Die Seiten eines Sehnendreieckes schneiden die in den gegenüberliegenden Punkten gezogenen Tangenten in Punkten einer geraden Linie.

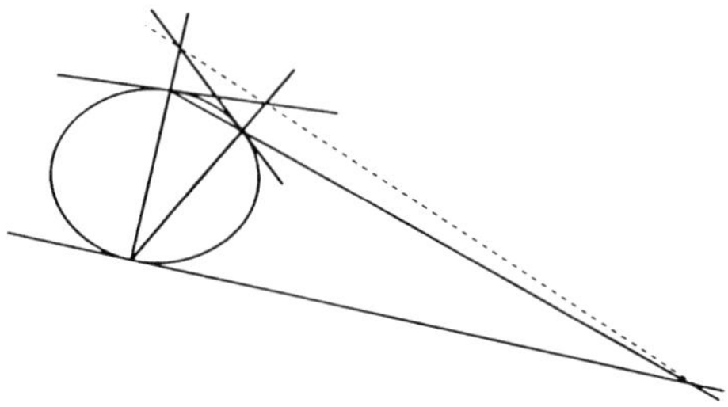

Aufgaben: Construktion beliebig vieler Punkte (auch nur von 5 Punkten) oder der projektiven Strahlenbüschel, welche den Kegelschnitt erzeugen, 1.) wenn *4 Punkte und in einem derselben die Tangente* 2.) ⟨wenn⟩ *3 Punkte und in 2en derselben die Tangenten* gegeben sind.

Durch die entsprechende Entwickelung würden wir den Kreis ganz durchlau-
65 fen und, indem wir die Identität dieser Erzeugungsweise als Punktgebilde | mit der ersteren *als Tangentengebilde* nachweisen, wieder zum Ausgangspunkt zurückkehren. Dabei zeigt sich zugleich, dass aber auch umgekehrt jedes *Paar projektiver Strahlenbüschel einen Kegelschnitt der ursprünglichen Definition nach definiert.*

Ich will zeigen, dass der Ihnen aus der elementaren metrischen Geometrie *bekannte Kreis* ein besonderer Fall des Kegelschnittes ist:

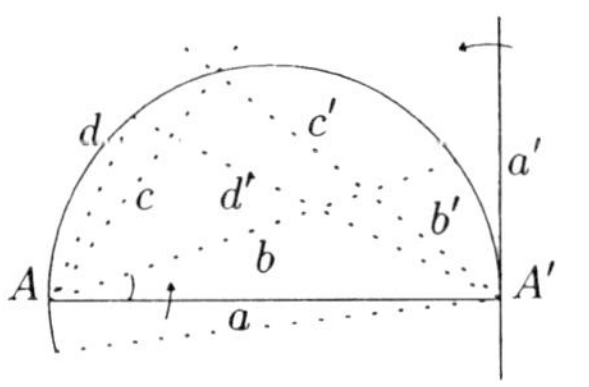

Man nehme 2 gleiche zueinander in einem rechten Winkel gedrehte Strahlenbüschel A, A'; man erhält als Ort der Spitzen rechtwinkliger Dreiecke den Kreis als *Erzeugniss der beiden Strahlenbüschel.* Alle *früheren und alle folgenden Sätze gelten also auch vom Kreise,* z. B. *der Pascalsche, der Brianchonsche Satz und die besonderen Fälle dieser Sätze.* Auch alle übrigen elementaren Sätze vom Kreise erhält man leicht.

Zwei gleiche gleichlaufende Strahlenbüschel

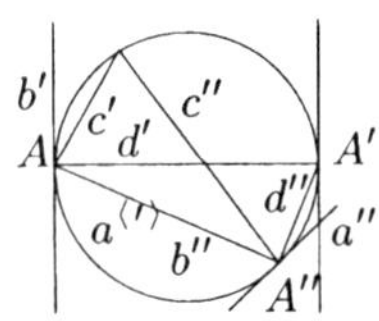

66 *erzeugen stets einen Kreis.*[49] *Denn* nehmen wir auf dem in obiger Weise erzeugten Kreise einen beliebigen Punkt an und verbinden diesen Punkt mit allen Punkten der Kreisperipherie, so sehen wir, dass die beiden Strahlenbüschel nur um einen Winkel gedreht sind.

Wir haben bisher studirt *die Erzeugnisse* von

1.) 2 in einer Ebene liegenden projektiven *Punktreihen,*
2.) " " " *Strahlenbüschel.*

[49]The labelling of the figure is not particularly clear.

Ich füge noch kurz *hinzu auf den Raum bezüglich*:

3.) 2 projektive Strahlenbüschel mit demselben Mittelpunkt und in verschiedenen Ebenen erzeugen einen *Ebenenbüschel 2ter Ordnung.*

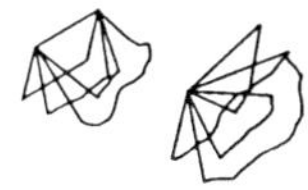

Schneiden wir einen Ebenenbüschel 2ter Ordnung mit einer beliebigen Ebene, so entsteht ein *Strahlenbüschel 2ter Ordnung*, und umgekehrt, wenn wir einen Punkt im Raume mit den Strahlen eines Strahlenbüschels verbinden, so *Ebenenbüschel 2ter Ordnung. Die Ebenen laufen alle durch einen Punkt.*

4.) *2 projektive Ebenenbüschel*, deren Axen sich schneiden, erzeugen einen *Kegel 2ter Ordnung*. Der Schnittpunkt der Axen liegt in der *Spitze des Kegels.*

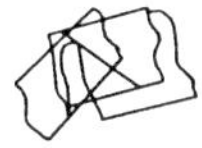

Schneiden wir den Kegel mit einer Ebene, so entsteht eine *Punktreihe 2ter Ordnung* (Kegelschnitt), und umgekehrt, wenn wir einen beliebigen *Punkt mit* den *Punkten einer Punktreihe* | verbinden, so *Kegel 2ter Ordnung.* 67

Es gelten daher entsprechend *alle früheren Sätze*, insbesondere die Identität beider Gebilde *3.) und 4.), insofern* das von den Ebenen eines Ebenenbüschels 2ter Ordnung *umhüllte Gebilde* (d. h. von den Schnittlinien der benachbarten Ebenen erfüllte Gebilde) ein Kegel 2ter Ordnung ist und die *Tangentialebenen eines Kegels 2ter Ordnung* ein *Ebenenbüschel 2ter Ordnung* bilden.

Wichtige Folgerungen über die möglichen Gestalten eines Kegelschnittes. *Der Kegel entsteht*, indem sich eine Gerade bewegt, während sie immer durch einen Punkt geht. Also *von der Spitze aus 2 Hälften. Keine Buchtung*, weil sonst eine Ebene durchgelegt einen *Kegelschnitt mit einer* Buchtung ergeben würde. 1.) Schneide ich mit einer Ebene, so dass nur eine Hälfte getroffen wird, so *ein einziger geschlossener ganz im Endlichen gelegener* Zug: die *Ellipse*. 2.) Ebene, welche beide Hälften trifft, liefert | einen Kegelschnitt, welcher aus *2 Hälften besteht*, welche beide *ins Unendliche gehen.* Denn es existiren dann immer 2 Erzeugende des Kegels, welche 68

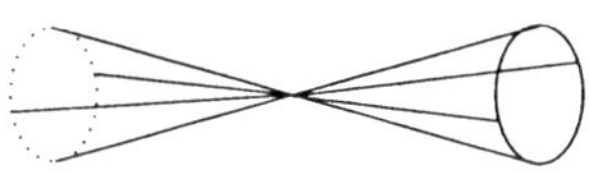

der Ebene parallel sind, wie man erkennt, wenn man die Erzeugende den Kegel beschreiben lässt. Diese beiden unendlich fernen Schnittpunkte gehören dem Kegelschnitt an. Derselbe hat also *2 unendlich ferne Punkte*, durch die hindurch die beiden Hälften miteinander verknüpft sind, wie es *ja sein muss*: *Hyperbel.*

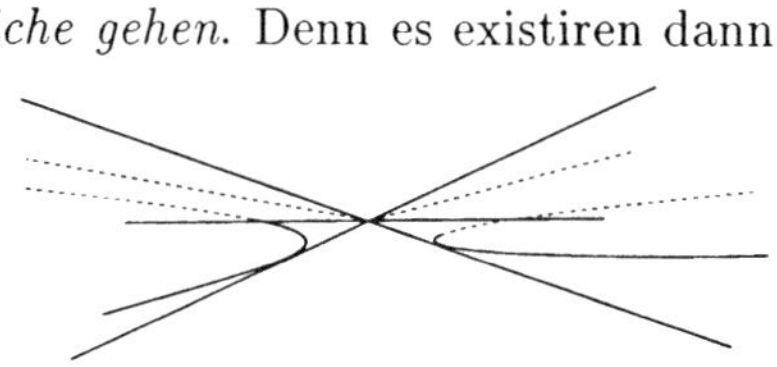

↑Ziehen wir in diesen beiden unendlich fernen Punkten die Tangenten[B], so

[B] Diese Tangenten müssen im Endlichen liegen, da die unendlich ferne Gerade nicht Tangente ist. ⟨Added by Hilbert to the left of the following figure.⟩

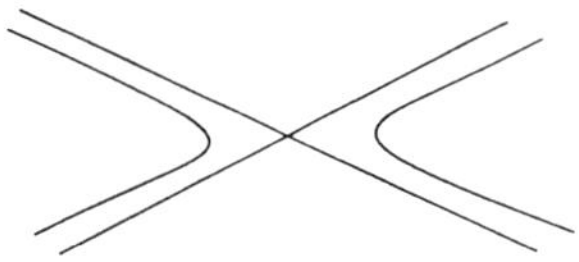

können diese im *Endlichen die Curve nicht mehr schneiden*. Sie *heissen Asymptoten.*↓

3.) Die Ebene ist einer Berührungsebene des Kegels parallel, so dass sie nur
69 einen Mantel trifft, aber doch einen | und *zwar nur einen unendlich fernen Punkt hat.*

Parabel. Es existiren *keine Asymptoten.*

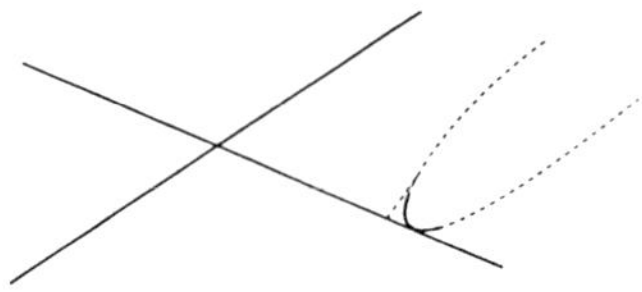

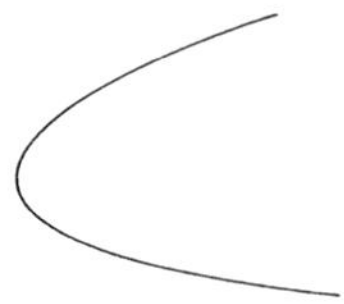

Also Unterscheidung nach der Anzahl der unendlich fernen Punkte: 1.) *Die unendlich ferne Gerade schneidet den Kegelschnitt garnicht,* 2.) *in 2 Punkten,* 3.) *in einem Punkte, d. h. sie berührt.*

Auch bei der Erzeugung durch projektive Strahlenbüschel können wir die 3 Fälle leicht unterscheiden. Es kommt darauf an, ob 0, 2, 1 Paar entsprechender Strahlen parallel laufen, d. h. wenn wir das eine Büschel parallel verschieben, ob dann 0, 2, 1 entsprechend gemein sind.

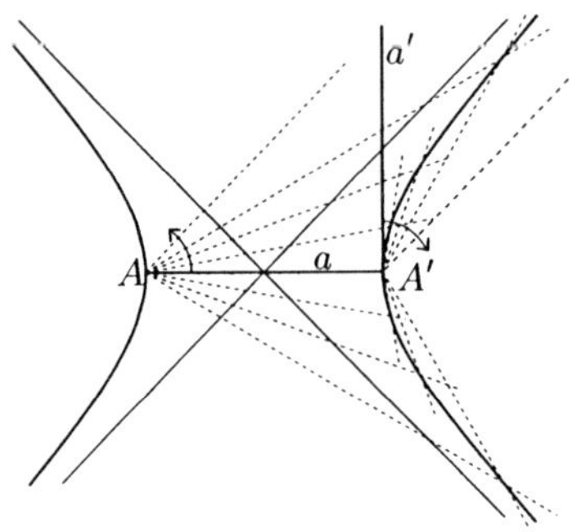

Für *1.) bietet ein Beispiel der Kreis.*

" *2.) drehe ich wieder um einen Rechten*, dann aber ungleichlaufend: Gleiche ungleichlaufende Büschel erzeugen eine gleichseitige Hyperbel:

Die beiden Asymptoten stehen zueinander senkrecht. ↑(Die gleichseitige Hyperbel ebenfalls metrisch.)↓

70 (Eine Hyperbel kann sowohl durch gleichlaufende wie durch ungleichlaufende Strahlenbüschel erzeugt werden, je nachdem die Mittelpunkte auf demselben oder auf verschiedenen Zweigen der Hyperbel liegen.)

Das nähere vgl. Schröter, Synthetische Geometrie der Kegelschnitte[50], § 25 und § 26.

Zeigen, wie die Parabel durch projektive Strahlenbüschel und Punktreihen erzeugt wird.

Modelle zeigen.

Man kann nun *eine Menge von Aufgaben* lösen, z. B.:

Gegeben sind *die beiden Asymptoten einer Hyperbel* und *ein Punkt derselben; man soll* in diesem die Tangente construiren.

[50] *Steiner 1867.*

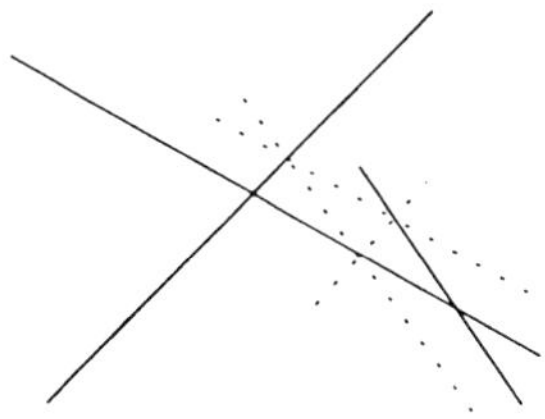

Vgl. Reye, Theil I,[51] Aufgabe 23–46, insbesondere 34 und 46.

⟨...⟩

⟨...⟩ 71–106

⟨...⟩ 107

Wenn wir *das ganze Gebiet der projektiven Geometrie übersehen*, so erkennen wir als die *Grundidee das Princip der umkehrbar eindeutigen Zuordnung*, also im Wesentlichen den Begriff *der Projektivität.* Ordnet man z. B. den Punkten einer Punktreihe die Werte einer *Veränderlichen* zu, so gelangt man
unmittelbar zu | der Einführung *veränderlicher Grössen* in die Geometrie, d. h. 108
der Coordinaten, und in der That, die Einführung *der Coordinaten ist dann die Grundidee der sogenannten analytischen Geometrie*, d. h. diejenige Idee, welche der *Idee der Projektivität* in der eben *dargestellten reinen* Geometrie *entspricht.* An Stelle des *Operirens mit der reinen geometrischen* Anschauung tritt dann die *Rechnung und die Formel – Hülfsmittel* von *einschneidendster Bedeutung.* Die *analytische Geometrie verfährt* so, dass sie von vorneherein den Begriff der *veränderlichen Grösse* einführt und dann für *jede geometrische Anschauung* immer sofort den analytischen *Ausdruck zugleich aufzeigt* und die Beweise durch *letztere liefert.* Auf diese Weise gelingt es rascher, zu *grösster*
Allgemeinheit der Sätze zu gelangen, | als mit der *reinen geometrischen* An- 109
schauung möglich war. Dagegen hat die projektive Geometrie, wie ich *sie vortrug*, *den Vorzug* der *Reinheit, Abgeschlossenheit und Denknothwendigkeit der Methoden.* Mit diesem Hinweis auf die *wichtigste* Schwesterdisciplin, *die analytische Geometrie*, schliesse ich mein Colleg.

[51] *Reye 1886*, 193–197.

Textual Notes

21.7: ⟨Between this line and the previous one, Hilbert leaves approximately a third of a page blank.⟩
21.11: und] u.
21.16–17: benutzt] benutz
21.18: Geometrie). Sie] Geometrie⟦)⟧ ⟦S⟧↑s↓ie
22.1: Geometrie] Geometrie.
22.5: Geometrie] Geometie
22.9: erkenntnisstheoretisch] erkenntniss theoretisch
22.13: *Zahlentheorie*] ⟦Al⟧ *Zahlentheorie*
22.18: *ganze*] *ga*⟪*n*⟫*e*
22.21: *möglich*] ⟦*nöthig*⟧↑*möglich*↓
22.22: *heutzutage*] ⟦ein⟧ *heutzutage*
22.27: ich mit] ich
22.28: Zahlentheorie] Zahlenth.
22.29: *anders*] *anderes*
22.36–1: , wo auch ⟨...⟩ *hinzukommt.*] ⟨Added.⟩
23.2–3: und, wie ⟨...⟩ gestellt,] ⟨Added.⟩
23.3: gestellt,] gestellt.
23.7: *mannigfaltig*] ⟦*M*⟧↑*m*↓*annigfaltig*⟦*keit*⟧
23.7–8: *Kreise* und] *Kreise* ⟪un⟫
23.8: *umbeschriebenen*] *um beschriebriebenen*
23.8: *Ausmessung*] Ausm*essung*
23.10: Doch] Do
23.10: *Gegenstände*] ⟦R⟧ *Gegenstände*
23.12: wie] wie ⟦con⟧
23.12: sowie] so*wie*
23.12–13: von *unübertroffener Klarheit und Schärfe*] ⟨Added.⟩
23.15: griechische] Griechische
23.15: *Gedanken*] ⟦Res⟧ *Gedanken*
23.18: Geometrie] Geometrie,
24.2: (1637)] ↑(1637)↓
24.5: *geometrische*] *geometrisches*
24.8: Beziehung] Beziehg
24.11: *Descartes*] ⟦C⟧ *Descartes*
24.13: und -*mittelchen*] und *mittelchen*
24.13: Cartesius] Car⟪?⟫esius
24.14: wichtig] ⟦wesentlich⟧↑wichtig↓
24.20–21: , welche zu ihrer Arbeit *Zeichnungen* brauchten] ⟨Added.⟩
24.21: der Mathematiker] ⟦Mon⟧ der Mathematiker
24.23: Géométrie] Géometrie
24.23: um 1800] ↑um 1800↓
24.28: *Steiner, von Staudt*] *Steiner,* ⟦und anderer⟧ *von Staudt*
24.31: *Steiner (1832).*] *Steiner. (1832)*
25.2: *construirt,*] *construirt*
25.2–3: weder *Zirkel* ⟨...⟩ *Lineal*] ⟨Added.⟩
25.3: *Lineal*] *Lineal.*
25.11: Staudt,] Staudt;
25.13: Schröter,] Schröter:

25.14:	Reye,] Reye:
25.17:	*Die inhaltsreichsten* sind *Schröter und Reye.*] ⟦⟪*Das*⟫⟧⊣↑*Die*↓ *inhaltsreichste* ⟦ist das von Reye.⟧⊣↑sind *Schröter und Reye.*↓
25.21:	Geraden] Gerade
25.22:	Ebenen] Ebene
25.27:	*1.)*] ↑*1.)*↓
25.29:	Die Punkte ⟨...⟩ *Punktreihe.*] ⟨Added.⟩
25.30:	*2.)*] ↑*2.)*↓
26.3:	3.)] ↑3.)↓
26.4:	*Ebenenbüschels*] ⟨Originally, Hilbert had written 'Achse des ebenen Büschels', which he replaced by 'Ebenenbüschels'.⟩
26.14:	um M, und *umgekehrt.*] um M. ↑und umgekehrt↓
26.19:	Achse] Ache
26.29:	ins *Unendliche*] ins *Unendlich*
26.32:	*zufolge*] *zu folge*
26.35:	sagen wir:] sagen wir
26.36:	*parallele Linie*] *Parallele Linie*
26.39:	*unendlich fernen*] *Unendlichfernen*
26.42–43:	*in einem*] ⟦*im*⟧⊣↑*in ein*↓
26.43:	unendlich fernen] Unendlichfernen
27.1:	eine und] ⟦nur⟧ eine ⟦Para⟧ und
27.2–3:	*unendlich ferner*] *unendlichferner*
27.3:	unendlich fernen] unendlichfernen
27.6:	unendlich fernen] unendl. fer
27.7:	Punkte, d. h. alle] Punkte. d. h. Alle
27.9:	unendlich ferne] unendlichferne
27.10:	*Geraden*] *Gerade*
27.11–12:	unendlich fernen Punkte] unendlichfernen Punkte
27.12–13:	*unendlich fernen Geraden*] *unendlichfernen Geraden*
27.17:	umgekehrt.] umgekehrt. ⟦Dann ziehe ich eine Linie b in β⟧
27.20:	Linien] Linie
27.21:	*unendlich fernen Punkt,*] *unendlichfernen Punkt*
27.22:	unendlich fernen Punkte] unendlichfernen Punkte
27.23:	sagen:] sagen
27.24:	*unendlich fernen Geraden*] *unendlichfernen Geraden*
27.26:	unendlich fernen Punkte] unendlichfernen Punkte
27.26:	unendlich fernen Geraden] unendlichfernen Geraden
27.27–28:	unendlich fernen Elemente] unendlichfernen Elemente
27.28:	unendlich fernen Ebene] unendlichfernen Ebene
27.28:	vereinigt] vereignit
28.1:	*deshalb*] *desshalb*
28.2:	unendlich fernen] unendlichfernen
28.5:	suchen.] suchen,
28.10:	gegenüberstellen:] gegenüberstellen
28.25:	Einführung] Einführug
28.28:	Ferner,] Fer
28.30:	man kann einen] man kan⟪?⟫ ein
28.30:	unendlich fern] unendli⟪??⟫ fern
28.30–31:	eine Gerade] ein Gerade
28.31:	giebt,] giebt.

28n:	Geraden] Gerade
28n:	keinen,] keinen
28n:	unendlich fernen] unendlichfernen
29.3:	*Fruchtbarkeit*] *Fruchbarkeit*
29.17:	Geraden] Gerade
29.19:	nehme] nehm
29.25:	Meine Herren.] M. H.
29.25:	*ferne*] *fernen*
29.28:	*§ 2.*] ↑*§ 2.*↓
29.32:	*Dreiecke*] ⟦Dreiecke⟧ ⟦in verschiedenen Ebenen gelegene⟧ *Dreiecke*
29.32:	*in verschiedenen*] ⟦*ABC und A′B′C′*⟧ *in verschiedenen*
29.33:	*Seiten sich schneiden*] *Seiten* ⟦sich schneiden⟧ ⟦*AB*⟧ ⟦nämlich⟧ *sich schneiden*
29.33:	*laufen*] ⟦*treffen sich*⟧-↑*laufen*↓
29n:	ist] ⟦《sei》⟧ ist
30.1:	Dreiecksebenen.] Dreiecksebenen,
30.2:	*A*, *B*, *A′*, *B′*] *A B A′ B′*
30.5:	*Wenn*] *Wenn* ⟦die Verbindungslinien⟧
30.6:	*entsprechender*] *entsprechenden*
30.8:	*A*, *B*, *A′*, *B′*] *A B A′ B′*
30.11:	Das blosse] Da blosse
30.13:	*dass je zwei*] *dass* ⟦die Schnittpunkte⟧ *je zwei*⟦*er*⟧
30.14:	*sich*] *sich* ⟦in Punkten⟧
30.14:	*auf einer*] *auf* ⟦der nä⟧ *einer* ⟦Gera⟧
31.5:	Ebenen] Ebene
31n:	kürzer] ⟦übersichtlicher⟧-↑kürzer↓
32.3:	durch] durch durch
32.4:	⟨Between this line and the next, another line has been deleted:⟩ ⟦Vier Punkte *A*, *B*, *C*, *D* ↑einer geraden Linie↓ heissen harmonische Punkte, wenn⟧
32.5:	*A*, *B*, *C*, *D*] *A B C D*
32.6:	*in A und C*] ⟦im ersten und dritten von ihnen⟧-↑*in A und C*↓
33.2:	*willkürlich*] *wilkürlich*
33.14:	Construktion] Construcktion
33.16:	*4 harmonische*] *4 Harmonische*
33.17–18:	4 harmonischen] 4 Harmonischen
34.5:	direkt] direckt
34.7:	*Strahlen*] *Strahlen* ⟦《???》⟧ ⟦allemal von einer Gerad⟧
34.20–21:	*Folglich* ⟨...⟩ *aus.*] ⟨Added.⟩
34.20–21:	*harmonische*] *harm.*
34.22:	jeder] ⟦einer⟧-↑jeder↓
34.25:	harmonische Strahlen.] harmonisch Strahlen
34.26–27:	4 harmonische] in 4 Harmonisch
34.27:	4 harmonischen] 4 Harmonischen
34.32:	⟨Between this line and the next, another line has been deleted:⟩ ⟦Etwas näher die Lage von 4 harmonischen Punkten unter⟧
34.33–34:	Z. B. bei ⟨...⟩ Punkte.] ⟨Added.⟩
34.33:	*Z. B.* bei] *z. B.* Bei
34.33:	Viereckskonstruktion] Vierecksconstrucktion

34.36–38: ↑Vorher ⟨...⟩ etc..↓] ⟨The addition is written at the bottom of the page, and directed to this place by an arrow. Hereby, Hilbert seems to have shifted the proposition deleted on p. 31; see note for line 35.2.⟩

34.37: beliebigen] beliebig

34.38: harmonische] harm.

35.2: ⟨Between this line and the next, another line has been deleted:⟩ ⟦↑Entsprechender Satz für die Ebene d. h. für 3 gegeb. Strahlen eines Büschels↓⟧

35.5: gesehen:] gesehen: ⟦$ABCD$ sind⟧

35.10: Anschauung] Anschauuns

35.17: möglichen] möglich

35.20: $B = A$] $B = A$,

35.33: L] ⟦B'' und D''⟧ ⟦E⟧↑L↓

35.34: P] ⟦F⟧↑P↓

35.35: und] u.

35.39: ⟨In the lower margin of p. 31, two lines have been pasted over, apparently a remark. To what it refers and where it was to be placed cannot be deciphered.⟩

36.2–3: und weil ⟨...⟩ sind.] ⟨Possibly added.⟩

36.2: A, B, C, D] $A\ B\ C\ D$

36.3: ⟨Between this line and the next, another line has been deleted:⟩ ⟦2.) NM geht durch B wegen des Vierecks $NGMK$⟧

36.4: *durch*] *dur*

36.10: *Construktionen*] *Construcktionen*

36.11: *unzugänglichen*] *un⟪zu⟫gänglichen*

37.12: *perspektiver*] *perspectiver*

37.15: *Grundgebilde*] ⟦ungleichartive⟧ *Grundgebilde*

37.17: gleichartige] gleichartive

37.19: *Zwei*] ⟦4.)⟧ *Zwei*

37.20: Strahlen-] Strahlen

37.21: *Zwei*] ⟦5.)⟧ *Zwei*

37.22: demselben] dem selben

37.22: Ebenenbüschel] Ebenbüschel

38.1: *Zwei*] ⟦6.)⟧ *Zwei*

38.14: *ausgezeichnete*] *ausgezeichnet*

38.20: *umkehrbar eindeutig*] ↑*umkehrbar eindeutig*↓

38n: perspektiv] perspektiv.

38n: das entsprechende] ⟦man das⟧↑⟦allemal das⟧↓ entsprechende

38n: Verbindung,] Verbindung ⟦gefunden wird;⟧

38n: ist,] ist.

38n: ungleichartige] ungleichartive

38n: gleichartige] gleichartive

38n: ungleichartigen] ungleichartiven

39.5: Z. B. wenn] z. B Wenn

39.9: Wenn ⟨...⟩ so] ⟦In 2 projektiven Grundgebilden⟧↑Wenn 2 Grundgebilde projektiv sind, so↓

39.12: Wenn] Wenn man

39.12–13: $A, B, C, D, E, F, \ldots$] $A\,B\,C\,D, E, F, \ldots$

39.19: E', F'] $E'\ F'$

39.21: (endlichen)] (endlichen

39.21: (unendlichen)] (unendlichen

39.23: H, K] ⟦⟦A, C⟧↑L, M↓⟧↑H, K↓

39.25: $HFKE$] ⟦LEM⟧ $HFKE$

39.25: suche] such

39.26: H', K'] H' K'

39.26: F', E'] F' E'

39.26: P', Q'] P' Q'

39.33: *entgegengesetzt*] *entgegensetzt*

40.1: 2.)] ↑2.)↓ ⟨added by Hi^b⟩

40.3: Auch] aus

40.7–8: Z. B. wenn] z. B; wenn

40.13: *beliebigen*] *belieb.*

40.17: *so*] *so* ⟦haben sie alle Elemente entsprechend gemein⟧ ⟦sind sie identisch.⟧ *so*

40.19–20: (wenn ⟨...⟩ klar)] ⟨Added.⟩

40.19: wenn] Wenn

40.22: entgegengesetzt] entgegengesetz

41.2: links] linkt

41.4: Denn ⟨...⟩ Vorsprung.] ⟨Added.⟩

41.5: unmöglich] umöglich

41.7–8: Also ⟨...⟩ überholen.] ⟨Added.⟩

41.10: ⟨Between this line and the next one there are two others which have been deleted:⟩

⟦Die projektive Beziehung zweier Grundgebilde ist eindeutig bestimmt, wenn 3 Paare entsprechender Elemente gegeben sind.⟧

41.11–12: *Für Strahlenbüschel* ⟨...⟩ Punktreihe.] ⟨Added.⟩

41.14: *und zwar nur eine*] ↑*und zwar nur eine*↓

41.17–21: (Wenn ⟨...⟩ liegt.)] ⟨Possibly added.⟩

41.21: liegt.)] liegt.

41.28: perspektiv] perspek.

41.30: *projektiv* in der verlangten] *projektiv in der ver*langten

42.3: Anfangs-] Anfangs

42.4: *vorhergehenden*] *vorhergehende*

42.8–9: Strahlenbüschel auf Ebenenbüschel projektiv beziehen] Strahlenbüsch auf Ebenbüschel proj. ⟪bezieh⟫

43.1: ⟨Before this line, there is a line which has been deleted:⟩

⟦Lösung vo⟧

43.6–7: angehören,] angehören;

43.7: bezüglichen] bezüglich

43.9: 3 Elemente] 3 Elem.

44.4: Meine Herren.] M. H.

44.4: Im vorigen] im vorig

44.4: § 3] § ↑3↓ ⟨added by Hi^b⟩

44.4: *Perspektivität*] *Persp.*

44.5: *Grundgebilde*] *Grundgeb*

44.5: Satz:] Satz.

44.5: 3 Elemente] 3 Elem.

44.6: *Beziehung*] *Bezei*⟦*g*⟧*chung*

44.6: $A, B, C, D, \ldots \barwedge A', B', C', D', \ldots$] $A, B, C\, D \barwedge A'\, B'\, C'\, D', \ldots$

44.7: *Bemerkung*] *Bemerkg*

44.9:	*3*] ↑*3*↓ ⟨added by Hi^b⟩
44.10:	*Grundgebilden*] *Grundgebilde*
44.11:	*Ordnung* -] *Ordnung*
44.14:	B, C] $B\ C$
44.14:	*bestimmt,*] *bestimmt.*
45.2:	$CC', \ldots$] CC', ⟦D⟧ ⟦con⟧
45.5:	PP'] ⟦D⟧P↑P'↓
45.15:	Strahlenbüschel] Strahlenbüsch
45.16:	Punktreihen] Punktreihe
45.17:	*willkürlich*] *willkürliche*
45.22:	*Brianchonscher Satz:*] *Brianchonscher Satz:* ⟦Bildet man aus 6 Strahlen eines Strahlenbüschels 2ter Ordnung ein Sechseck⟧ ⟦Gehören die Seiten eines Sechsec⟧
45.27:	Trägers] Tragers
45.29:	entsprechend] ⟦gemein⟧ entsprechend
45.34:	nenne] nenn
45.34:	*S*] ↑*S.*↓
45.39:	einen Punkt] ein Punkt
45n:	synthetische] synth.
46.1:	gegeben] geg.
46.14:	ist,] sind
46.19:	*Strahlenbüschels*] *Strahlen büschel*
46.22:	gegeben] geg
46.25–26:	*da sonst* ⟨...⟩ wären.] ⟨Added; this remark replaces most of two lines which have been stricken through.⟩
46.27:	*Aeusseres,*] *Aesseres,*
47.11:	derselben] der selben
47.12:	singuläre Punkt] Singuräle punkt
47.14:	*Diagonale*] *Diagonalen*
47.17:	Aufgabe:] Aufgabe
47.17:	gegeben.] geg.
48.8:	gegeben] geg.
48.9:	Dreiecks-] Dreiecks
48.12:	3 Vierecke:] 3 Vierecke
48.18:	M, L] ↑M, L↓
48.19:	E, F] ↑E, F↓
48.23:	L', M'] $L'\ M'$
48.25:	$KM', \ldots$] KM'
49.19:	Wir] ⟦Wir sehen jetzt nachträglich ein, dass anstatt von⟧ Wir
49.21:	*Strahlenbüschel*] *Strahlbüschel*
49.22:	können] ⟦sollen⟧↑können↓
51.2:	durch] Durch
51.5:	⟨Between this line and the next, another line has been deleted:⟩ ⟦In einem Sehnensechseck sind⟧
51.12:	Bild] Bild,
51.16–17:	*Sehnensechseck*] *Sehnen 6eck*
51.21:	*Sehnenfünfeck*] *Sehnen 5eck*
51.25:	*Gegeben*] *geg.*

52.1–2:	beliebig ⟨...⟩ erzeugen,] ↑oder der projektiven Strahlenbüschel, welche den Kegelschnitt erzeugen↓ beliebig vieler Punkte (auch nur von 5 Punkten)
52.8:	umgekehrt] umgekehrt:
52.9:	*Strahlenbüschel*] *Strahlenbüschen*
52.13–14:	zueinander] zu einander
52.20:	*Kreise,*] *Kreise.*
52.21:	*Brianchonsche*] *Brianschonsche*
52.24–27:	*Denn* ⟨...⟩ gedreht sind.] ⟨Added.⟩
52.25:	an] an,
52.26:	Strahlenbüschel] Strahlbüschel
53.4:	*Ebenenbüschel*] *Ebenbüschel*
53.5:	einen] ein
53.7:	umgekehrt] Umgekehrt
53.8–9:	*Die Ebenen* ⟨...⟩ *Punkt.*] ⟨Added.⟩
53.12:	liegt in der] ⟦ist⟧⊢↑liegt↓ die
53.15:	*Ordnung*] *Ordnug*
53.20:	*Ordnung*] *Ordg*
53.38:	miteinander] mit einander
53n:	unendlich] Unendlich
54.1:	*Endlichen*] *Endlich*
54.6:	Es ⟨...⟩ *Asymptoten.*] ⟨Added.⟩
54.7:	unendlich] Unendlich
54.8:	*Kegelschnitt*] *Kegelsch*
54.8:	*garnicht,*] *garnich*
54.20:	Hyperbel] Hüperbel
54.21–22:	zueinander] zu einander
54.25:	je nachdem] jenachdem
54.27:	Synthetische Geometrie] Synthet. Geometri
54.29–30:	Zeigen ⟨...⟩ wird.] ⟨Added.⟩
54.29:	Punktreihen] Punkreihen
54.32:	z. B.:] z. B.
54.33:	Gegeben] Geg
55.7:	*und*] *un*
55.7:	That,] That.
55.9:	Geometrie] Geometri
55.11:	*Formel*] *Formel.*
55.15–16:	*grösster Allgemeinheit*] *grösseter Allgemeinenheit*
55.18:	*Denknothwendigkeit*] ⟦Nothwendig⟧ *Denknothwendigkeit*
55.19:	Schwesterdisciplin] Schwesterdisdisciplin
55.20:	*Geometrie*] *Geometri*

Description of the Text

Collection: SUB Göttingen, signature *Cod. Ms. D. Hilbert 535.*

Size: Cover size approx. 21.0 × 17.3 cm; page size approx. 21.0 × 16.5 cm.

Cover Annotations: On the front cover is pasted a label with the call number and library stamp, and with the notation, 'Projektive Geometrie // 2st. 1891'.

Composition: Two signatures consisting respectively of twenty and eight double pages, onto which further pages and partial pages have been pasted.

Pagination: Pages are numbered continuously from 1 to 111, although only the right-hand pages are given actual numbers (odd), probably in Käthe Hilbert's hand. Page number 83 is mistakenly given twice, on the title page of the second signature and on the immediately following page of text. The page numbers occasionally (and especially plainly on p. 39) do not appear in their usual place since this is already occupied with text, which suggests that they were added later.

Original Title: On the initial page of the first and of the second signature (pp. 1 and 83 respectively) are two numbered title pages, each with the same title: 'Projektive Geometrie // Sommersemester 1891 2st.'; after the first line, in blue pencil, have been added the Roman numerals 'I' and 'II' respectively. On the first title page, in the upper margin, in addition to the page number, is written 'Hilbert // Rhesastr. 10'; on the second title page, 'Hilbert.' Hilbert most likely originally kept his notes on this lecture course in two separate notebooks, which at some later point were combined into the existing volume.

Text: Pages 3, 5–80, and 83–109 (inclusive) contain continuous text in Hilbert's hand. The verso side of the first title page (p. 2) contains various remarks and annotations, also in Hilbert's hand. Pages 4, 80f., the verso side of the second title page, and pp. 110–112 are blank. The single sheet pasted onto p. 5 and containing pp. 3 and 4 consists of somewhat lighter paper than the rest of the volume. The 'Eintheilung der Geometrie' on those pages is thus probably not a component part of the original lecture course, but was most likely added subsequently. (The fact that these pages lack the blue-colored underlinings that are to be found on almost every other page, and that the handwriting style is clearly different from the surrounding pages, point to the same conclusion.)

Hilbert extensively corrected and annotated his original text, which was written in black ink; for these corrections he employed black ink (Hi^s) as well as blue pencil (Hi^b), the latter being used especially for extensive underlinings. The corrections of Hi^s and Hi^b seem to have been executed at roughly the same period. Passages written by Hi^s are underlined in blue, while corrections by Hi^s are also occasionally underlined by Hi^b. It seems likely that the underlinings Hi^b were added to make the manuscript easier to use in the delivery of the lectures, and moreover Hi^b usually appears as an addition to Hi^s: that is, Hilbert seems first to have executed the corrections in Hi^s, and only then to have partially underlined in blue the corrected text. But the temporal gap is not likely to have been great, as is indicated by the extensive substitution on p. 24, where a text already marked with blue pencil was replaced by Hi^s, and the new text then likewise underlined in the same way. An even more extensive textual revision seems to have occurred on pages 85 to at least 90. Two passages (p. 89 and roughly the lower two thirds of p. 90), clearly belonging to an earlier version, were pasted to the left half of a blank double page. The surrounding text from p. 85 onwards is written with noticeably lighter ink, while the ink of the passages that were pasted in is similar to the ink used up to p. 84. The passages that have been pasted in display the usual blue underlining; not so the remaining text on the pages following p. 87 through to the end of § 6 on p. 92.

Hilbert also executed the drawings in black ink, sometimes adding to them with pencil or (occasionally) blue pencil.

Chapter 2

Lectures on the Foundations of Geometry (1894)

Introduction

Chapter 2 presents Hilbert's own notes for a series of lectures entitled 'Grundlagen der Geometrie' (*Cod. Ms. D. Hilbert 541*), planned for the summer semester of 1893 in Königsberg. The course was not given then because too few registered to attend; it was held instead in the following summer semester, and announced under the title 'Über die Axiome der Geometrie'. This almost certainly occasioned a reworking of the notes.

The course was the sixth on geometrical subjects held by Hilbert up to the end of the academic year in 1894 (see the list of Hilbert's lecture courses at the end of this Volume). It was, however, the first course devoted to the *foundations* of geometry in the full sense, and also the very first lecture course dealing with the foundations of any of the exact sciences. With it, Hilbert turns for the first time to the second approach to geometry stated in the *Einleitung* to the 1891 lectures on projective geometry (Chapter 1 in this Volume), namely 'axiomatic geometry', i.e., the investigation of the axioms underlying 'the geometry of intuition'. Therefore, its significance for the development of Hilbert's work in the foundations of science generally cannot be estimated highly enough.

The basis of his later foundational work was the so-called 'axiomatische Methode', which, although to be employed generally in the analysis of the exact sciences, was first developed in the context of geometry, and geometry always remained Hilbert's primary example of the importance and efficacy of the method. It is true to say that the Axiomatic Method is not exhibited in the 1893/1894 lectures to anything like the degree that became evident in the later lectures of 1898/1899 and 1902, and of course the 1899 *Festschrift*. For one thing, Hilbert does not yet pursue, as he later does, the 'systematic investigation of those geometries which arise when one or more of these axioms [underlying intuition] is set aside' (Hilbert's *Einleitung* to the 1891 lectures on projective geometry, p. 3; Chapter 1 of this Volume) or replaced by its negation. For another, the method of reinterpretation and modelling is not here present to the degree that it was in Hilbert's mature work. Nevertheless, the lecture notes reproduced here mark the beginning of the axiomatic *analysis* (as opposed to axiomatic *presentation*) of geometry, and the scene is clearly set for the later meta-mathematical treatment.

For example, Hilbert expresses here the view that geometry is just a 'schema of concepts' (p. 60) or, to use Hilbert's favourite expression in later years, which here occurs for the first time (p. 7), a 'Fachwerk von Begriffen', which can be applied in different ways, and that it is up to 'the human understanding' how to apply the schema to appearances, how to 'fill it with material [*Stoff*]'(p. 60). There are also some important 'meta' reflections, too, for example, the statement that of the first fourteen axioms, 'none is a consequence of the others' (p. 19). And, of course, the lectures deal, if only implicitly, with the most famous geometrical independence result, namely that of the independence of the Euclidean Parallel Axiom from the other assumptions by presenting a spherical model for hyperbolic geometry (p. 88).

One might helpfully estimate the position of Hilbert's work here by describing it as at once forward looking and traditional.

The basis of the forward looking aspect is, not surprisingly, that Hilbert here chooses for the first time an 'axiomatic' approach for the presentation of geometry, and this structure is the overriding one, despite detours into analytic and projective geometry in the second and fourth parts. The basic axiomatic approach follows Pasch's 'Vorlesungen über die neuere Geometrie' (*Pasch 1882*), an influence not in evidence in the 1891 lectures. For instance, many of the axioms clearly owe their formulation to Pasch, although Hilbert deviates from the structure Pasch gives to his system. But more significant than the mere choice of an axiomatic framework is that Hilbert divides his axioms into *groups* of cognate axioms, and this remained his standard way of analysing axiom systems. Here there are five groups, which remained the case in all Hilbert's subsequent treatments of basic geometry. The notes, therefore, are correspondingly divided into five sections, the primary responsibility of each section being to deal with one of the axiom groups. The basic division is as follows:

'A: Existenzaxiome' (later called 'Axiome der Verknüpfung').

'B: Lagenaxiome' (later called 'Axiome der Anordnung' or 'Axiome der Reihenfolge').
This section has two further sub-divisions, entitled 'Die harmonische Lage' and 'Die Einführung der Zahl'.

'C: Das Stetigkeitsaxiom' (later 'Axiome der Stetigkeit').
This section has two further sub-sections, entitled 'Das Doppelverhältnis' and 'Projektive und analytische Geometrie'.

'D: Die Congruenzaxiome (oder Bewegungsaxiome)' (later called 'Axiome der Kongruenz oder Bewegung'.
There is only one further sub-section here, entitled 'Die Einführung des Maasses'.

'E: Das Parallelenaxiom' (later called the 'Axiom der Parallelen').
This section has four further sub-divisions, entitled 'Hyperbolische Geometrie', 'Elliptische Geometrie', 'Parabolische Geometrie' and 'Das Parallelenaxiom'.

It is important to note several things about this grouping.

Firstly, the groups listed under A–E are exactly the same as those which Hilbert used later in his investigations of Euclidean geometry, i.e., the basic kinds of conceptual building block for Hilbert's construction of geometry are set out here for the first time.

Secondly, as against this, the axioms *within* the groups are not always the same. Indeed, the additions and changes show Hilbert to some extent experimenting with, and reflecting on, the axioms. For instance, Hilbert includes here 'Pasch's Axiom' among the 'Lagenaxiome', whereas he later proves it. Pasch's central 'Kernsatz', or axiom, states that if a straight line in the same plane as the three vertices of a triangle meets one of its sides (other than at its vertices), and does not pass through the third vertex, then it must pass through one of the other two sides. This proposition allows one to speak of a line dividing the plane into two distinct halves. In the 1898/1899 lectures, it was not adopted as an axiom, but is rather proved (straightforwardly) by means of an axiom which states directly the division of any plane into two separate parts by any straight line in it. To take another important example, the congruence axioms deal with segment and angle congruence, but there is no Triangle Congruence Axiom connecting the two notions as in the later lectures and the *Festschrift*. The axiom group C, dealing with continuity, reveals perhaps the biggest difference, however. The continuity axiom Hilbert gives is a fairly standard principle asserting the existence of a limit point for any bounded, infinite sequence of points. This strong principle was used again in *Hilbert 1895b*, but the basic continuity principle Hilbert uses after 1898 is the much weaker Archimedean Axiom.

Thirdly, although the list of the groups of axioms is the same as later, the *order* in which they are deployed is crucially different. In these lectures, continuity is dealt with in the third axiom group, i.e., directly after the axioms of incidence and order, and the Parallel Axiom appears in the last group, group E. In the later lectures and the *Festschrift*, however, continuity is considered last and the Parallel Axiom fourth (third in the *Festschrift*).

It must be stressed, however, that the list of axioms A–E as given above was the result of heavy revision of the manuscript. What Hilbert originally intended under group C was the set of congruence axioms, as the deletion of the heading 'C. Congruenzaxiome' on p. 59 makes clear. Moreover, it seems likely that Hilbert simply left unrevised the paragraph just after this deletion, which begins with the sentence:

> Mit den bisherigen Axiomen, den Existenz- und den Lagenaxiomen, können wir schon eine grosse Menge Thatsachen und Erscheinungen beschreiben.

This, even though the axiom of continuity has already been introduced on p. 38; clearly, when the sentence was first written, the continuity axiom was not among the 'axioms [given] hitherto'. This and other details of the manuscript show clearly that Hilbert saw the necessity of an axiom of continuity only when revising the manuscript for the lectures as given in 1894, and more specifically only when reviewing the difficulties over the 'Einführung der Zahl'. Nevertheless, from this point on, a continuity axiom in some form becomes basic to Hilbert's system of ordinary geometry, and the project becomes very much one of analysing where continuity is really required, and where it is dispensable.

There are four respects in which the difference between the (revised) order of the axiom groups in the 1893/94 lectures and the later treatment is significant.

(*i*) As in later lectures, Hilbert stresses the importance to geometry of introducing number, not by imposition from without, as happens with analytic geometry, but rather by developing an equivalent of the real number structure from within, thus *geometrically*. However, Hilbert seems here to pursue a strategy of introducing real numbers as early as possible, whereas in the later lectures, he is much more circumspect. In these, Hilbert is concerned to analyse exactly which geometrical axioms (theorems) are responsible for the presence of the ordered field structure on a line (see Hilbert's *Einleitung* to his own notes for the 1898/1899 lectures, p. 4; Chapter 4 in this Volume), which requires that there be a detailed analysis of such structure.

(*ii*) In the 1893/1894 lectures, Hilbert tries to demonstrate directly, with the help of just the incidence and order axioms, that there are in fact sufficiently many points for the numbers. His main device is that of the projective construction of the fourth harmonic point. The goal is to show, by means of this construction and a 'lawlike' assignment of numbers to points, that to each point of the line involved exactly one number can be assigned. There are thus two important questions: (1) Are several, indeed infinitely many, points assigned to the same number? (2) Is every real number assigned to a point? Hilbert breaks off his attempt, though not before he has been through a series of attempted improvements. He then introduces (almost certainly in the revision for the 1894 lectures) a powerful continuity axiom asserting the existence of limit points, the axiom appealed to in the new group C. Without some continuity assumption, it cannot be shown either that every point can be assigned a unique real number, or that to every real number there corresponds a unique point. The strong continuity principle Hilbert adduces here will guarantee both; the former (but not the latter) is yielded by the weaker Archimedean Axiom (which is a consequence of the strong continuity principle).

(*iii*) In Hilbert's 1898/1899 lectures, the analysis of the 'Einführung der Zahl' is guided by an important further insight. If there is a coordinatisation, the number coordinates will represent not only points, but also segments, that is, pieces of the line considered as 'directed' (positively or negatively) from a certain arbitrary point (the zero point) on. These segments should then also satisfy certain algebraic conditions once the right algebraic operations are defined for them: they must be subject to addition, multiplication, etc., and the laws they satisfy must be exactly the same as the ordinary laws of calculation satisfied by the real numbers. In other words, a 'Streckenrechnung' should be possible, and Hilbert duly constructs two such. In carrying out this analysis, Hilbert uses *more* assumptions than those initially behind the 'Einführung der Zahl' in the 1893/1894 lectures, for he uses the axioms of congruence (augmented by a Triangle Congruence Axiom), the Axiom of Parallels (essentially as a simplifying tool), and (as a continuity principle) the Archimedean Axiom, in addition to the initial incidence and order axioms. As a result, he achieves a thorough *analysis* of the ordered field structure within geometry without having to resort to the powerful continuity principle of 1893/1894. Indeed, once the analysis of the field structure on segments is carried out, Hilbert rests content with pointing out that the Archimedean Axiom guarantees that every point can be assigned a unique real number; the question as to whether to every real number there can be assigned a point is then left open until the adoption of the *Vollständigkeitsaxiom* after the first edition of the *Festschrift*. (See the Introduction to Chapter 5 in this Volume.) One is tempted to say, therefore, that, far from the 'Einführung der Zahl' being undertaken as early as possible, it is delayed until the very end. (Again, see Hilbert's *Einleitung* to his own notes for the 1898/1899 lectures, p. 5; Chapter 4 in this Volume.) This fits reasonably well with one of the aims of the later work, to reconstruct as much as possible of 'ordinary' geometry without any resort to continuity, even in the weak form of the Archimedean Axiom.

(*iv*) The fact that the Euclidean Parallel Axiom is placed last in the 1893/1894 lectures very likely reflects the influence of Klein on Hilbert's view of geometry, i.e., the view that Euclidean, elliptical and hyperbolic geometry are all special cases of projective geometry. Hilbert is then clear that which of these geometries to choose in applications of geometry to the world is largely an empirical matter. The approach taken in the later lectures, however, is more pragmatic. Hilbert adopts (and analyses) *Euclidean* geometry, and exploits the Euclidean Parallel Axiom as a simplifying assumption, above all in his analysis of the ordered field structure on segments, for the Parallel Axiom enables essential simplifications of the Desargues and Pappus/Pascal Theorems which are so important in Hilbert's analysis, and, crucially, provable without continuity axioms. This pragmatic, direct approach is visible above all in the *Festschrift*. Nevertheless, the initial importance of projective geometry is clear in the later work, not so much in the *Festschrift*, but in the 1898/1899 lectures. (See Chapter 4 of this Volume.) In addition, it might be noted that later (see the 1898 *Ferienkurs*, p. 18; Chapter 3) Hilbert expresses the view that the Parallel Axiom and the Archimedean Axiom do not have the same epistemologically elementary character as the other axioms, and his settled grouping puts the Parallel Axiom as the last group but one, the last being the continuity axioms. (The first edition of the *Festschrift* is the *only* place where the Parallel Axiom appears as group three.) This partially reflects the placement of the Parallel Axiom in the 1893/1894 lectures.

So far we have emphasised the forward looking or 'progressive' elements of the 1893/1894 lectures, though with some caution. But in emphasising this, some conservative or 'traditional' elements have emerged, too. One of these has just been noted, i.e., the treatment of projective geometry (without congruence asssumptions, or 'measure') as the central foundation on which all other synthetic geometries are to be built. As we noted, it is not obvious that Hilbert distanced himself completely from this view. Another aspect which should be stressed is Hilbert's deployment of analytic methods, indeed as early as the second section of the notes dealing with the 'Einführung der Zahl', with its eventual reliance on a powerful continuity axiom. In this, Hilbert leans heavily on the presentation of his teacher Lindemann as given in Lindemann's revision of Clebsch's book (*Clebsch 1891*), in particular with the construction of harmonic points. This reliance is also clearly evident in the analytic presentation of non-Euclidean geometry. Hilbert later distanced himself considerably from this approach, however. Part of the project of his later analysis of geometry is to see how much of standard or traditional geometry (sometimes he calls it 'Schulgeometrie') can be reconstructed without any appeal to analytic methods. Moreover, although analytic and algebraic methods are still very much central to his later investigations, they are confined to the *metatheory*, to the model-theoretic investigation of different geometrical systems.

This text has been the most difficult to edit of all those in this Volume. None of Hilbert's own sets of notes is especially easy to read, but this set in particular has numerous levels of correction involving deletions, additions, both in margins and on later pages, and portions of sheets pasted in, which themselvs have also been corrected. To that is to be added the fact that Hilbert clearly returned to this text at a later time, probably before or during the lectures in 1894, and effected significant structural and mathematical changes. The text with which we were faced, therefore, is anything but a unified one. There is yet another difficulty: the sections on the 'Einführung der Zahl' and the 'Doppelverhältnis' are not only difficult to read, but moreover difficult to understand, and this makes it uncertain how to reconstruct the rather impenetrable text Hilbert has left. The difficulty in understanding has two separate, but intertwined, aspects. Firstly, it is sometimes hard to see exactly what Hilbert is aiming at, and therefore correspondingly hard to assess the extent to which he succeeds or fails. Secondly, this is compounded by a lack of conceptual clarity. In particular, although Hilbert intends to make a sharp distinction between *points* and *numbers*, he does not carry through this distinction consistently, sometimes designating numbers as points and conversely. The construction of harmonic points is a purely geometrical construction, and as such has strictly speaking nothing to do with numbers, even if in the end numbers can be assigned in a canonical way to the harmonic points to effect the 'introduction of numbers'. This confusion has sometimes been noted in footnotes, but not always; we mention it here just to underline the editorial difficulty which the text presents. In order to give the reader some conception of how the text appears, at least in its most impenetrable parts, we have included four pages in facsimile.

As was common practice with him, Hilbert inscribed various additional remarks at the very beginning of the text, starting on the reverse of the title page, and also inserted a long bibliography on pp. 3–5. We have relegated all this material to an Appendix.

The notes on the text are by Ulrich Majer and Ralf Haubrich.

Ulrich Majer

Die Grundlagen der Geometrie

7 Unter den *Erscheinungen oder Erfahrungsthatsachen*, die sich uns ↑bei der Betrachtung der Natur↓ bieten, giebt es eine besonders *ausgezeichnete Gruppe*, nämlich die Gruppe derjenigen *Thatsachen*, welche die äussere Gestalt der Dinge bestimmen. Mit diesen Thatsachen beschäftigt sich die *Geometrie*. Wie jede Wissenschaft dahin zielt, die in ihrem Bereiche liegende Gruppe von Thatsachen zu ordnen, oder die Erscheinungen zu beschreiben, wie Kirchhoff sagt,[1] so thut es die Geometrie mit eben jenen *geometrischen Thatsachen*. Dieses Ordnen oder Beschreiben geschieht vermittelst gewisser *Begriffe*, die durch die Gesetze der Logik unter sich *zu verknüpfen* sind. Die Wissenschaft ist desto fortgeschrittner, d.h. das *Fachwerk der Begriffe* ist desto vollkommner, je leichter jede Erscheinung oder Thatsache *untergebracht* wird. Die Geometrie ist eine Wissenschaft, welche im Wesentlichen so weit fortgeschritten ist, dass alle ihre *Thatsachen* bereits durch *logische Schlüsse* aus früheren abgeleitet werden können. Ganz anders wie
8 z. B. die *Electricitätstheorie* oder Optik, in der noch heute | immer neue Thatsachen entdeckt werden. Dennoch ist sie ↑mit Rücksicht auf ihren Ursprung↓ eine *Naturwissenschaft*, wie ich später deutlich zeigen werde. Da nun nicht alle Begriffe durch *reine Logik* abzuleiten sind, sondern vielmehr aus der *Erfahrung* stammen, so ist die wichtige Frage, die wir in dieser Vorlesung behandeln werden, die nach den *Grundthatsachen*, welche zum Aufbau der *ganzen* Geometrie hinreichen. Diese nicht *beweisbaren* Thatsachen müssen wir von vornherein festsetzen und nennen sie *Axiome*. [A] Dieselben sind bereits *vor mehr als 2000* Jahren mit *bewundernswerthem* Scharfsinn, im Wesentlichen genau, aufgestellt von Euklid. (Um 300 v. Ch. in Alexandrien.) Wenngleich *Euklid* auch erst in neuerer Zeit vollständig verstanden ist. Das Problem unsres Collegs lautet also:

Welches sind die *nothwendigen* und *hinreichenden* und unter sich *unabhängigen* Bedingungen, die man an ein System von Dingen*) stellen muss,[2] damit

[A] Axiom = Hypothese in der Physik. Zähle genau die den Axiomen zu Grunde liegenden Experimente auf.

*) Ding oder Einheit

---

[1] In *Kirchhoff 1877*, Vorrede, p. III, we find: 'Aus diesem Grunde stelle ich es als die Aufgabe der Mechanik hin, die in der Natur vor sich gehenden Bewegungen zu *beschreiben*, und zwar vollständig und auf die einfachste Weise zu beschreiben.'

[2] The terms 'Ding' and 'System' are of fundamental importance in *Dedekind 1888*. The first sentence of § 1 reads: 'Im Folgenden verstehe ich unter einem D i n g jeden Gegenstand unseres Denkens' (p. 1). And some twenty lines later, Dedekind writes: 'Es kommt sehr häufig vor, daß verschiedene Dinge a, b, c ... aus irgend einer Veranlassung unter einem gemeinsamen Gesichtspuncte aufgefaßt, im Geiste zusammengestellt werden, und man sagt dann, daß sie ein S y s t e m S bilden' (pp. 1–2).

jeder Eigenschaft dieser Dinge*) eine geometrische Thatsache *entspricht* und umgekehrt, so dass also mittelst obigen Systems | von Dingen, ein *vollständiges Beschreiben* und *Ordnen* aller geometrischen Thatsachen möglich ist.[B] Im Lauf der Vorlesung wird sich *Gelegenheit* bieten, auf die Bedeutung des eben Gesagten im Einzelnen einzugehn. 9

Wir werden *4 Gruppen von geometrischen* Axiomen unterscheiden.

↑[3] *A. Existenzaxiome* ↑Besser: *Axiome der Verknüpfung*↓

1.) Irgend 2 Punkte A, B bestimmen immer eine und nur eine Gerade a.
A, B heissen auf a gelegen. a heisst die Verbindungsgrade, geht durch A, B.
2.) Irgend 2 auf der Geraden a gelegene Punkte A, B bestimmen die Gerade a.
Oder in Formeln: aus $AC = a$ und $BC = a$, $A \neq B$ folgt $AB = a$.
3.) Irgend 3 nicht auf einer Geraden gelegene Punkte A, B, C bestimmen eine und nur eine Ebene α.

A, B, C heissen in α gelegen oder Punkte von α. α heisst die Verbindungsebene.

4.) Irgend 3 in einer Ebene α ↑aber nicht auf einer Geraden↓ gelegene Punkte A, B, C bestimmen die Ebene α.

D. h. aus $ADE = \alpha$, $BDE = \alpha$, $CDE = \alpha$ folgt $ABC = \alpha$.

Wenn ein Punkt A auf 2 Geraden a, b oder in 2 Ebenen α, β oder in einer Ebene α und auf einer Geraden a liegt, so sagen wir: Die beiden Geraden, Ebenen oder Geraden und Ebenen haben den Punkt A gemein.

5.) *Wenn 2 Punkte A, B einer Geraden a in einer Ebene α liegen, so liegen alle Punkte von a in der Ebene α.*

6.) Wenn 2 Ebenen einen Punkt gemein haben, so haben sie noch wenigstens einen andern Punkt gemein.

↑Zugleich folglich die durch A, B gehende Gerade.↓

[B]so dass also unser System ein vollständiges und einfaches Bild der geometrischen Wirklichkeit werde. ⟨Addition in the upper margin of p. 9, assigned to no definite place in the text.⟩

---

[3]The passages added here (which run to the end of the entry under '8.)') are found on the second sheet pasted to p. 9. They replace the following set of axioms, whose heading however, unlike the axioms, was not deleted:

A. Existenzaxiome.

1.) irgend 2 Punkte bestimmen immer eine und nur eine Gerade.
↑Hier fehlt ein Axiom.↓
2.) irgend 3 nicht auf einer und der nämlichen Geraden gelegne Punkte ↑bestimmen eine und nur eine Ebene.↓
↑Hier ebenfalls.↓
3.) eine Gerade, welche 2 Punkte mit einer Ebene gemein hat, liegt ganz in der Ebene.
↑4.) siehe unten*).
5.) " " ⟨after which some 8–10 words have been erased⟩↓

*) ⟨Here Hilbert refers to Axiom 4 to be found on the first sheet pasted to p. 9, and which itself has been pasted over by the second sheet. The relevant piece pasted over reads:⟩
4. 2 Ebenen, welche einen Punkt gemein haben, haben noch einen anderen Punkt ⟨gemein.⟩

Aber es bleibt unentschieden, ob 2 Geraden in einer Ebene oder 2 Ebenen im Raume überhaupt einen Punkt gemein haben.
7.) Es giebt wenigstens 4 nicht in einer Ebene gelegene Punkte.
8.) In jeder Geraden giebt es wenigstens 2 Punkte, in jeder Ebene wenigstens 3 nicht in einer Geraden gelegene Punkte.
↑Besser 7 mit 8 vertauschen.↓↓

Bemerkung zu Axiomen.[E] Statt „*bestimmen*" sagt man auch durchgehn, verbinden, angehören, liegen auf, gemein haben, enthalten, schneiden, treffen. Punkte werden mit A, $B, \ldots$, Geraden mit a, $b, \ldots$, Ebenen mit α, $\beta, \ldots$ bezeichnet.
⟨Three lines deleted.⟩[5]
10 Das Axiom entspricht einer Beobachtung, wie sich leicht durch *Kugeln, Lineal und Pappdeckel* zeigen lässt.[6] Doch sind diese Erfahrungsthatsachen so *einfach*, von Jedem so *oft beobachtet* und daher so *bekannt*, dass der Physiker sie nicht extra im *Laboratorium* bestätigen darf. ↑Dennoch der Ursprung aus der Erfahrung. Die Axiome sind, wie Hertz sagen würde, Bilder oder Symbole in unserem Geiste, so dass Folgen der Bilder wieder Bilder der Folgen sind, d. h. was wir aus den Bildern logisch ableiten, stimmt wieder in der Natur.↓[7] [8]
↑5 bringt die Thatsache zum strengen Ausdruck,[9] dass der Raum wenigstens 3 Dimensionen hat. Wir werden dieses Axiom später notwendig brauchen.↓
Aus diesen unbeweisbaren Axiomen lassen sich schon eine Reihe von Lehrsätzen logisch ableiten, d. h. beweisen. Z. B.

[E] Eigentlich muss man noch vorher die Existenzsätze ⟦folgenden Inhalts⟧ einschieben: Es giebt ein System von Dingen, die wir Punkte, ein anderes und drittes System von Dingen, die wir Gerade, Ebene nennen wollen. ↑Dies ist jedoch in Folge von Axiom ⟪7⟫ überflüssig.↓[4]

---

[4] The whole addition is written on the first sheet pasted to p. 9 and directed to this place by an insertion sign. The sheet covers a remark in the lower margin of p. 9 in *Hi*[s] which, as far as can be seen, reads the same, aside from the final additional remark.

[5] The following lines are deleted:
Axiom 1. $a = AB$

$$\frac{a > C \ , \ A \neq C}{AC = a} \qquad > \text{ bedeutet: enthält.}$$

[6] On p. 11 of his own notes for the 1898/1899 lectures (see Chapter 4 of this Volume), Hilbert refers to pp. 10–13 of this manuscript.

[7] In *Hertz 1894*, 1, we read: 'Wir machen uns innere Scheinbilder oder Symbole der äußeren Gegenstände, und zwar machen wir sie von solcher Art, daß die denknotwendigen Folgen der Bilder stets wieder die Bilder seien von den naturnotwendigen Folgen der abgebildeten Gegenstände.'

[8] The following lines are deleted:

2.) $\dfrac{\alpha = ABC \quad \alpha > D}{ABD = \alpha}$ $\quad AB \not> C \quad AB \not> D$

3.) $\dfrac{a = AB \quad \alpha = ABC \quad a > D}{\alpha > D}$

[9] Hilbert refers here to Axiom 5 in the first version of the manuscript, which is illegible, but almost certainly identical with Axiom 7 in the revised version.

1. Satz: Durch eine Gerade und einen Punkt welcher nicht auf derselben liegt geht immer eine und nur eine Ebene.

$$\frac{a > A,\ a > B,\ a \not> C \qquad \alpha = ABC}{\alpha > a,\ \alpha > C} \qquad\qquad \frac{\alpha > a,\ a \not> C \qquad \alpha > C}{\alpha = ABC}$$

↑ $>$ heisst „enthält"
$\not>$ " „enthält nicht" ↓

Der Beweis kann auch an der Hand einer geeigneten Figur geführt werden, [F] 11
doch ist die Zuziehung derselben durchaus nichts *nothwendiges*, sie erleichtert die *Auffassung* und ist ein *fruchtbares Mittel* zur Entdeckung neuer Sätze. Doch *Vorsicht* da sie leicht irreleitet. Der Lehrsatz ist erst dann bewiesen, wenn der *Beweis von der Figur vollkommen unabhängig ist.* Der Beweis muss sich Schritt für Schritt auf die *vorangegangnen Axiome* berufen. Das Figuren machen *ist das Experiment* des Physikers, und die *Experimentalgeometrie* ist bereits mit den Axiomen abgeschlossen. Wenn man von dieser Auffassung *im Geringsten* abweicht, so verliert das Beweisverfahren, und damit die ganze Untersuchung *jeden Sinn.*[10] ↑Man schwankt zwischen „überflüssigen" und paradoxen Behauptungen.↓

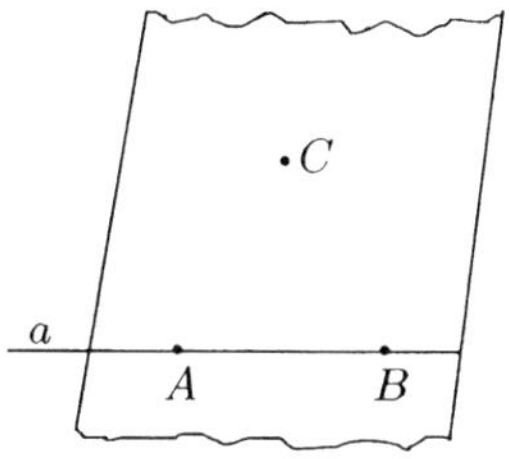

2. Satz: Durch *) 2 Gerade, welche einen Punkt gemein haben, giebt es immer eine und nur eine Ebene.

Man nehme auf der 2ten | Geraden einen Punkt an und führe auf Satz 1 12
zurück. Man kann auch sagen: *2 nicht in* einer *Ebene gelegne Geraden haben keinen* Punkt gemein. Satz 2 ist keineswegs umkehrbar; d. h. man darf nicht schliessen: 2 in einer Ebene gelegene Gerade haben immer einen Punkt gemein. In der That folgt dies nicht aus unsern Axiomen [obwohl eine *Figur leicht dazu verführen* könnte. Genauere Beobachtung zeigt jedoch die *Unmöglichkeit, einen solchen* Satz durch das *Experiment* zu bestätigen. Daher nicht Aufnahme unter die Axiome. Es entzieht sich eben unsrer Erfahrung, ob man für 2 Gerade immer einen *Schnittpunkt finden* kann.] [11] *Wir lassen die Sache daher vorläufig unentschieden, und sagen nur:*

2 Gerade einer Ebene haben entweder einen oder keinen Punkt gemein.

3. Satz: Wenn zwei Ebenen einen Punkt gemein haben, so haben sie eine und nur eine Gerade gemein, diese Gerade heisst die Schnittlinie.

[F] Ein System von Punkten, Geraden, Ebenen *heisst eine Figur.* ⟨Addition in the upper margin of the page, not assigned to any particular place in the text.⟩

*) Definition von „Durch". Wenn A in a oder in α liegt, so geht a, α durch A. Die Wortdefinitionen sind hier nicht immer besonders ausgezeichnet, was künftig vielleicht empfehlenswert ist.

---

[10] A similar formulation can be found in *Pasch 1882*, 43: 'Wenn man von dieser Auffassung im Geringsten abweicht, so verliert der Sinn des Beweisverfahrens überhaupt jede Bestimmtheit.' See also *Pasch 1882*, 99.

[11] The brackets '[' and ']' were very probably added when Hilbert wrote the notes for the 1898/1899 lectures (see Chapter 4 in this Volume), because on p. 11 of that manuscript

13 4. Satz: Wenn 3 Ebenen 2 Punkte gemein haben, so haben sie eine Gerade gemein (nämlich deren Verbindungsgerade). Es ist jedoch nicht gesagt, dass 2 Ebenen immer einen Punkt gemein haben. Daher:

3 Ebenen haben entweder eine Gerade, einen Punkt, oder keinen Punkt gemein.

Eine Gerade hat mit einer Ebene entweder einen oder keinen Punkt gemein. Der Punkt heisse Schnittpunkt oder Treffpunkt.

↑Jedes der 5*) Existenz-Axiome ist von den übrigen 4 unabhängig. In der That wie man leicht sieht, kann man sich je 3 Systeme von Dingen: Punkte, Geraden, Ebenen arithmetisch, d. h. durch Coordinaten construiren, so dass irgend 4 Axiome [12] erfüllt sind, das 5te aber nicht. Doch gehe ich darauf der Kürze wegen nicht ein.↓

⟨The following, set in a smaller font, was in the basic text layer. The first part has been partially pasted over, the rest crossed out, and replaced by the text which follows from line 77.27 to line 78.37.⟩

B. *Lagenaxiome.*

1.) Wenn auf einer Geraden 2 Punkte gegeben sind, so giebt es mindestens einen 3ten Punkt der Geraden, welcher zwischen A und B gelegen heisst, oder zwischen B und A.

Von diesem Begriffe *zwischen* gelten folgende weitere Axiome.

3.) Liegt der Punkt C zwischen A und B, so liegt der Punkt A nicht zwischen B und C.

↑2.) Von 3 Punkten einer Geraden liegt stets einer zwischen den beiden anderen.↓

↑In Formeln schreiben. 2 und 3 zusammen lassen sich aussprechen, wie folgt: Von 3 Punkten einer Geraden liegt stets einer und nur einer zwischen den beiden andern.↓

13a, 13b ⟨Pages 13a, 13b are pasted together; the text cannot be deciphered.⟩

14 Z. B. ↑Axiom 4↓

$$\begin{array}{l} C \sim AB \\ \underline{D \sim AC} \\ D \sim AB \end{array}$$

[13]

Beweis des Satzes 3. Seien A, B irgend 2 Punkte und liege etwa C nicht zwischen A und B, so mag B zwischen A und C liegen (2). Dann liegen die zwischen A und B gelegenen Punkte auch zwischen A und C (4) und so fährt man fort. ↑Läge nämlich etwa D nicht zwischen A und C, so müsste etwa A zwischen C

---

*) Hierbei sind 1, 2 und 3, 4 ↑und 7, 8↓ als je ein Axiom zusammen gerechnet. Man kann diese auch zusammen aussprechen, etwa so: Wenn A, B 2 Punkte sind, so giebt es eine und nur eine Gerade, auf welcher sowohl A, wie B liegt und welche sowohl mit A, wie mit B vereinigte Lage hat. ⟨Addition by Hi^s. The whole addition is on a small strip of paper pasted in for this purpose; this additional note is also on that strip.⟩

Hilbert appears to refer to this very passage. The brackets were probably written by Hi^s and then emphasized by Hi^b.

[12]The numerals '4' and '5' are replacements. Hilbert originally had '6' and '7', respectively.

[13]The sign '$\sim$' means: 'lies between'.

und D liegen und dann liegen auch alle zwischen A und C liegenden Punkte auch zwischen C und D etc.↓

Aus Satz 3 folgt leicht

Satz 4. Eine beliebige Anzahl n von Punkten einer Geraden können so benannt werden $A_1, A_2, \ldots, A_n$, dass A_2 zwischen A_1 und A_3, A_3 zwischen A_2 und A_4, $\ldots, A_{n-1}$ zwischen A_{n-2} und A_n liegen. Dies kann nur noch auf eine zweite Weise $B_1, B_2, \ldots, B_n$ geschehen, wo $B_i = A_{n+1-i}$ ist. ↑Denn liegen etwa alle Punkte zwischen A und B, so setze etwa $A = A_1$, $B = A_n$. Alle Punkte ausser A mögen so zwischen C und D liegen, also auch A_n, d. h. es ist entweder C oder $D = A_n$. Ist $D = A_n$, so setze $C = A_2$. Dann liegt zwischen A_1 und A_2 kein Punkt ↑nach 5↓ und die Reihe ist völlig bestimmt. Die umgekehrte Reihenfolge hätte man erhalten, wenn man B statt $A = A_1$ gesetzt hätte.↓

Mittelst der Definition:

Eine Gerade, welche einen zwischen A und B gelegenen Punkt besitzt, heisst zwischen A und B gelegen

sprechen wir folgendes 7tes und letztes Lagenaxiom aus:

7. Wenn in einer Ebene 3 Punkte A, B, C und eine zwischen A und B gelegene und nicht durch C gehende Gerade gegeben wird, so liegt die Gerade entweder zwischen A und C oder zwischen C und B.[14]

↑ Aus 7. folgt leicht, dass eine Gerade die Punkte einer Ebene in 2 Teile trennt: 2 Punkte desselben Teils haben keinen Punkt der Geraden zwischen sich, 2 Punkte verschiedener Teile haben die Gerade zwischen sich. Man sagt die Punkte liegen auf 19
derselben bez. auf verschiedenen Seiten der Geraden ⟪??⟫ ⟪Liegen A und B⟫ ferner A und C auf derselben Seite der Geraden, so liegen auch B und C auf derselben Seite etc.↓

↑ *B. Lagenaxiome.**) 15

1. Zwischen 2 Punkten A und B giebt es stets wenigstens einen 3ten Punkt der Geraden.[15]
2. Von 3 Punkten einer Geraden liegt stets einer zwischen den beiden anderen.
3. Wenn C zwischen A und B und wenn D zwischen A und C liegt, so liegt D auch zwischen A und B.
4. Wenn C zwischen A und B und D zwischen A und C liegt, so liegt D nicht auch zwischen C und B.
5. Wenn A und B 2 Punkte einer Geraden sind, so giebt es stets einen Punkt C, so dass B zwischen A und C liegt.

↑Es giebt stets ∞ viele Punkte, folgt aus 1.↓

---

*) Diese Lagenaxiome müssen künftig sehr abgekürzt werden und ⟦ausserdem⟧⊣↑vielleicht auch↓ so gefasst werden, dass die elliptische Geometrie auch darunter fällt: Vielleicht so: Wenn irgend n Punkte $n > 3$ auf einer Geraden gegeben sind, so können dieselben auf 2 und nur 2 Arten in eine Reihenfolge gebracht werden. Ist $A, B, C, \ldots, P, Q, R$ die eine, so ist $R, Q, P, \ldots, C, B, A$ die andere. Eine heisst die umgekehrte der anderen. Greift man irgend $m > 3$ Punkte heraus, so erscheinen diese wieder in richtiger Reihenfolge.

---

[14] Here 'C und B' has been pasted over by a small strip of paper, as have a number of words which follow. The function of the strip is to secure an additional double sheet which contains pp. 15–18 in Hilbert's numbering.

[15] The axioms 1, 3, 4 and 5 correspond, in part literally, to *Grundsätze* II, IV, V and VI in § 1 of *Pasch 1882*.

Satz 1. Unter einer beliebigen endlichen Anzahl von Punkten einer Geraden giebt es immer 2 Punkte, zwischen denen alle übrigen liegen.[16]

Beweis: Seien A, B irgend 2 Punkte und liege etwa C nicht zwischen A
16 und B, | so mag etwa B zwischen A und C liegen. (2.) Dann liegen alle zwischen A und B gelegenen Punkte auch zwischen A und C (3.), und so fährt man fort.

Satz 2. Eine beliebige endliche Anzahl von Punkten einer Geraden kann so benannt werden: A, B, $C, \ldots$, P, Q, dass B zwischen A und C, C zwischen B und D und so fort liegen, und dass zwischen A und B, B und $C, \ldots$ keiner der Punkte des Systems liegt. Dies kann nur noch auf eine 2te Weise geschehn und diese ist die umgekehrte: Q, $P, \ldots$, C, B, A.

Beweis: Denn liegen etwa alle Punkte zwischen A und Q und alle Punkte ausser A zwischen M und N, so muss auch Q zwischen M und N liegen. Liegt nun ein Punkt des Systems zwischen A und N, und nehmen wir an, es läge M zwischen A und N, so darf dieser Punkt des Systems, da er zwischen M
17 und N liegen soll, keinen|falls zwischen A und M liegen. (4.) Folglich kann man $M = B$ nehmen und dann so fortfahren, indem man jetzt A und B ausschliesst.

Mittelst der Wortdefinition: eine Gerade, welche einen zwischen A und B gelegenen Punkt besitzt, heisst zwischen A und B gelegen, sprechen wir folgendes 6tes und letztes Lagenaxiom aus.

6. Wenn in einer Ebene 3 nicht in einer Geraden gelegne Punkte und eine zwischen A und B gelegene und durch C nicht gehende Gerade gegeben ist, so liegt diese Gerade entweder zwischen A und C und nicht zwischen B und C oder sie liegt zwischen B und C und nicht zwischen A und C.

Mit anderen Worten: Wenn eine Gerade in ein Dreieck eintritt, so muss sie durch eine der beiden anderen Seiten austreten.[17]

18 Aus 6 folgt leicht, dass eine Gerade die | Punkte einer Ebene in 2 Abteilungen trennt: 2 Punkte desselben Teils haben keinen Punkt der Geraden zwischen sich; 2 Punkte verschiedener Teile haben die Gerade zwischen sich. Man sagt, die Punkte liegen auf derselben bez. auf verschiedenen Seiten der Geraden. Liegen A und B und ferner A und C auf derselben Seite der Geraden, so liegen auch B und C auf derselben Seite.

Liegen A und B auf derselben und A und C auf verschiedenen Seiten, so liegen auch B und C auf verschiedenen Seiten.

Liegen A und B und ferner A und C auf verschiedenen Seiten, so liegen B und C auf derselben Seite der Geraden.[18]↓

19 Definiren wir noch eine Ebene, welche einen zwischen A und B gelegenen Punkt besitzt, als zwischen A und B gelegen,[19] so beweisen wir aus 6[20] leicht: Satz. Eine Ebene, welche zwischen A und B liegt und nicht durch C geht,

[16] *Satz* 1 and its proof correspond to *Lehrsatz* 17 in § 1 of *Pasch 1882*.

[17] *Satz* 6 corresponds to *Grundsatz* IV of § 2 of *Pasch 1882*, which is now known as 'Pasch's Axiom'.

[18] Both propositions are to be found in *Pasch 1882*, 27.

[19] This definition corresponds to that found on p. 30 of *Pasch 1882*.

[20] Substituted for: 7

liegt entweder zwischen A und C oder zwischen C und B. ↑Beweis: Denn die Ebene schneidet die Ebene ABC in einer Geraden etc.↓

↑2 Punkte des Raumes liegen auf derselben oder auf verschiedenen Seiten der Ebene und zwar liegen A und B auf derselben Seite der Ebene und ebenso A und C, so liegen auch B und C auf derselben Seite etc.↓

Die 6 Lagenaxiome sind von einander und von den früheren 5 Existenzaxiomen unabhängig. ↑*Cum grano salis zu verstehen!!!*↓ Oder genauer:[21] Keines der bisher aufgestellen $\overset{14!}{12}$ Axiome ist eine Folge der $\overset{13!}{11}$ übrigen Axiome. Um dies einzusehen, muss man zeigen, dass es ein System von Dingen giebt, welche irgend $\overset{13!}{11}$ Axiome erfüllt, dagegen das $\overset{14!}{12}$^te^ nicht. Ich gehe jedoch hier auf den Beweis nicht ein.

Aus den bisherigen $\overset{14}{12}$ Axiomen ist schon eine grosse Fülle von Sätzen abzuleiten möglich.

Das System aller durch A gehenden Geraden heisst ein Strahlenbüschel.[22]

Satz: Wenn irgend eine endliche Anzahl von Strahlen eines Büschels gegeben ist, so giebt es stets eine nicht dem Büschel angehörige Gerade, welche alle jene Strahlen schneidet.

Beweis: Man nehme auf a einen Punkt A an und wähle dann auf demselben 20
Strahle a einen Punkt B, so dass der Mittelpunkt M des Büschels zwischen A und B liegt. Dann ziehe man durch A irgend eine Gerade l, welche die gegebnen Strahlen in A_1, A_2, ..., A_m schneiden möge. Auf l wähle man C so, dass zwischen A und C keiner der Punkte $A_1, \ldots, A_m$ liegt. Die Verbindungsgerade BC schneidet nach Axiom 6[23] alle Strahlen.

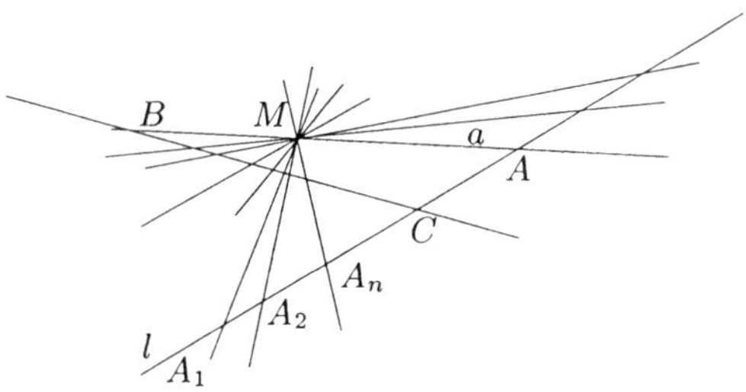

[24]↑Entsprechend gilt für den Raum der folgende Satz:
Wenn im Raume irgend eine endliche Anzahl von Strahlen eines Bündels, d. h., welche durch einen Punkt M gehen, gegeben ist, so giebt es stets eine nicht durch M gehende Ebene, welche diese sämtlichen Geraden schneidet.
Beweis: Man nehme auf 2en der Geraden je einen Punkt A, B an, lege durch A, B, M die Ebene α und wähle auf den übrigen Geraden je einen Punkt

---

[21] In the original text before correction, Hilbert operated with five existence and seven position axioms, whereas in the end he has eight existence and six position axioms. This explains the two different enumerations.

[22] The definition of 'Strahlenbüschel' is incomplete; it requires the condition that all the straight lines are in the same plane.

[23] Substituted for: 7

[24] This addition (which runs up to line 80.19), is assigned by Hi^b to this place in the text by an insertion sign. The whole insertion is on a seperate piece of paper pasted to p. 20.

C, $D, \dots, K$, so dass C, $D, \dots, K$ alle auf derselben Seite von α liegen. Dann lege durch ABC, $ABD, \dots, ABK$ je eine Ebene γ, $\delta, \dots, \kappa$ und ebenso durch AM eine Ebene μ, welche jene Ebenen γ, $\delta, \dots, \kappa$ in den Strahlen c, $d, \dots, k$ schneidet. Auf AM suche einen Punkt M' so, dass A zwischen M und M' liegt; lege durch M' in μ eine Gerade und wähle auf dieser einen Punkt L, welcher auf derselben Seite von α liegt wie die Punkte C, $D, \dots, K$ und so, dass zwischen M' und L keine der Geraden c, $d, \dots, k$ liegt. ML schneidet dann alle Strahlen c, $d, \dots, k$ in C', $D', \dots, K'$. Ist C' der dem Punkt M nächstgelegene Punkt, so ist die Ebene γ die gesuchte. Denn beispielsweise wird DM von γ geschnitten, da L, D, D' auf derselben Seite von α liegen, folglich auch D, D' auf derselben Seite von AB in δ, folglich auch auf derselben Seite von γ und folglich D, M auf verschiedenen Seiten von γ liegen.

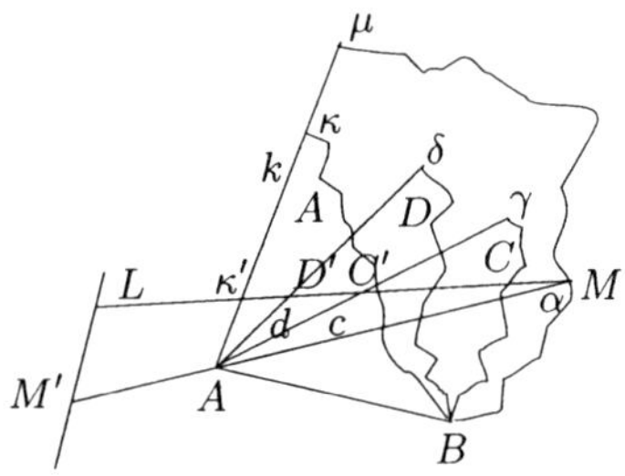

↑Vielleicht ist das folgende bis zum Schluss dieses Abschnittes Ueber das Getrennt-Liegen wenigstens in solcher Ausführlichkeit überflüssig.↓↓

Wenn $C \sim AB$ und wenn $D \not\sim AB$, oder umgekehrt $D \sim AB$ und $C \not\sim AB$, so heissen AB durch CD getrennt.[25] ↑Ist AB durch CD, so ist auch CD durch AB getrennt.↓ Wenn man irgend einen Punkt M mit $ABCD$ verbindet: a, b, c, d, so heissen ab durch cd getrennt.

↑Man kann also z. B., wenn 3 Strahlen a, b, c eines Büschels gegeben sind, stets d so bestimmen, dass ab durch cd getrennt werden.

Wird ab durch cd getrennt, so wird auch cd durch ab getrennt.↓

21

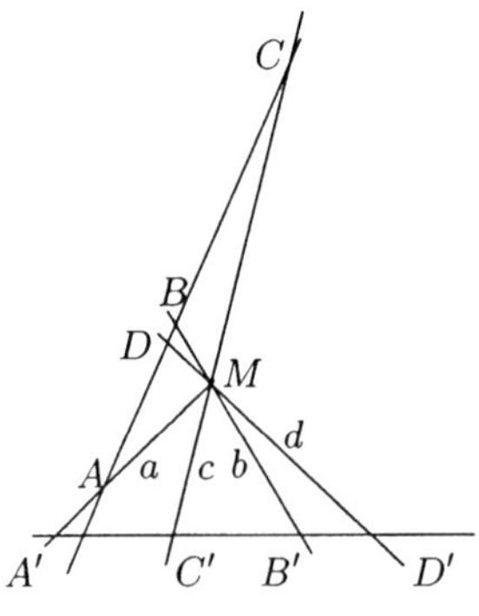

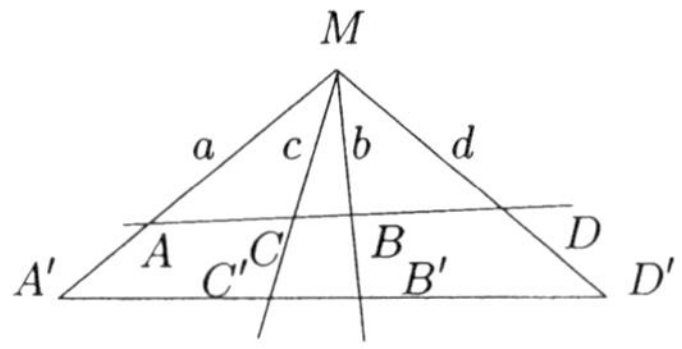

Satz: Wenn die Strahlen ab durch cd getrennt werden, und ein nicht zum Büschel gehöriger Strahl schneidet die 4 Strahlen in A, B, C, D, so wird AB durch CD getrennt.

Beweis:[26] 1.) A, B liegen auf derselben Seite von M wie A' beziehungsweise B'. Dann liegt c auch zwischen A und B und folglich auch C zwischen A und B. Da $D' \not\sim A'B'$, so auch $d \not\sim A'B'$ und folglich auch $\not\sim AB$, d. h. $D \not\sim AB$.

[25] '$\sim$' denotes 'lies between'.

[26] The figure to the right belongs to case 1.) of the proof; that to the left to case 3.).

2.) A bezüglich B liegen beide auf entgegengesetzten Seiten von M, wie A' bezüglich B'. Der Beweis ist ebenso.

3.) A liegt auf derselben, B auf entgegengesetzter Seite von M, wie bezüglich A', B'. $c \not\sim AB$. Dagegen $d \sim AB$.

↑Satz: Wenn a, b, c, d Strahlen eines Strahlenbüschels sind, so werden entweder ab durch cd getrennt, oder ac durch bd oder ad durch bc; und zwar schliesst jede dieser Lagen die beiden anderen aus. ↑Beweis: Man ziehe eine Gerade $ABCD$ durch die 4 Strahlen und ordne dann die Punkte, etwa $ABCD$, nach dem früheren Satze. ↓ ↓

Definition des Ebenen-Büschel ↑Axe des Büschels↓.[27]
Wort dafür, dass die beiden Ebenen $\alpha\beta$ getrennt werden durch $\gamma\delta$.

Satz: Wenn die Ebenen eines Büschels $\alpha\beta$ durch $\gamma\delta$ getrennt werden, und ein nicht durch die Axe des Büschels gehender Strahl schneidet die 4 Ebenen in A, B, C, D, so werden auch AB durch CD getrennt. Beweis wie beim obigen Satz.[G]

Die harmonische Lage. 22

Wir sind jetzt im Stande, die [28]Fundamentalsätze der projektiven[28] Geometrie zu entwickeln.

[H][Es kann dies auf 2 verschiedene Weisen geschehn.

1.) Durch Einführung der idealen Elemente nach Pasch „Vorlesungen über neuere Geometrie" oder nach Schur Separatabzug aus Bd. 39 Math. Ann.[29]
⟨Two paragraphs deleted.⟩[30] 23

[G]Die in diesem Abschnitte aus den Axiomen der Lage abgeleiteten Sätze sagen also im Wesentlichen aus, dass Punkte einer Geraden geordnet werden können und dass die Eigenschaft von 2 Punktepaaren des sich Trennens bei beliebiger Projektion sich nicht ändert. ⟨Addition in the upper margin of p. 22.⟩

[H][ ] Das roth eingeklammerte habe ich in der Vorlesung nicht gesagt und ist auch überflüssig. ⟨Addition by *Hi*ᵀ in the lower margin of p. 22.⟩

[27]In *Pasch 1882*, 31, we find the definition: 'Ebenen, welche (mehr als einen Punkt, mithin) eine Gerade G gemeinsam haben, werden ein Ebenenbüschel genannt; die Gerade G heißt die Axe des Büschels.'

[28–28]Substituted for: Hauptstücke der synthetischen

[29]*Pasch 1882* and *Schur 1891*.

[30]Deleted: 2.) auf dem im folgenden eingeschlagnen Weg:

Satz: Wenn 2 Dreiecke in verschiednen Ebnen so gelegen sind, dass je 2 entsprechende Seiten sich schneiden, so schneiden sich die Verbindungslinien entsprechender Ecken ⟦,wenn überhaupt,⟧ nothwendig in ein und demselben Punkt, oder kein Paar von ihnen hat einen gemeinsamen Punkt.

Beweis: Unter Dreieck versteht man 3 nicht in einer Geraden gelegne Punkte. Die beiden Dreiecke seien ABC und $A'B'C'$. Die Figuren ⟨p. 23⟩ vergl. mein Collegheft projective Geometrie. ⟨*Hilbert 1891a**, 22f.⟩ Wir setzen voraus, es mögen sich AA' und BB' schneiden, etwa in dem Punkte D. Die Schnittlinie der beiden Dreiecks⟨ebenen⟩ sei b. Auf dieser schneiden sich die 3 Seitenpaare. $ABA'B'$, ferner $BB'CC'$ und endlich $AA'CC'$ liegen je in einer Ebene ⟨The proof breaks off here. In *Hilbert 1891a**, 23, it is carried through to the end.⟩

2.) auf dem von Lindemann in seinem Buche, S. 433 Bd. III,[31] eingeschlagenen Wege, welcher wesentlich auf der Construktion des 4ten harmonischen Elementes beruht.

Wir wählen den zweiten Weg, welcher ein arithmetischer ist.]

Wenn man auf einer Geraden 3 Punkte in der Reihenfolge ABC hat, so kann man zwischen B und C einen 4ten Punkt D wie nebenstehend construiren. In der That; nimmt man H beliebig an und dann G zwischen H und C, so liegt AG zwischen H und C und schneidet daher BH in E. Ebenso liegt EC zwischen B und G, weil EC ja zwischen A und G liegt. EC und BG schneiden sich in F, und nun zeige man noch, dass HF die Gerade AC schneidet, nämlich in D. (Denn FH liegt zwischen E und C. E und B liegen aber auf derselben Seite von FH; folglich liegt FH auch zwischen B und C.)

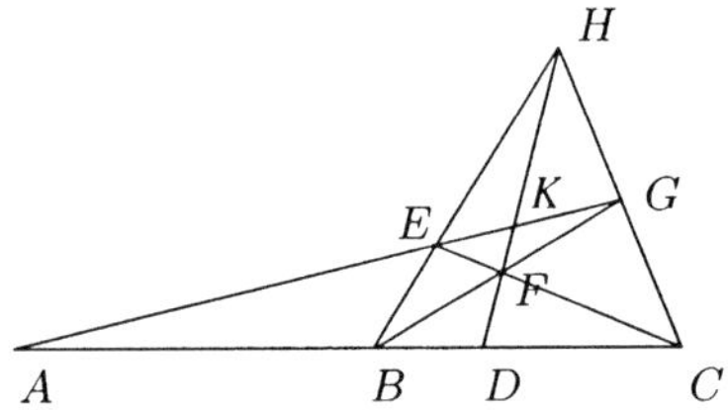

24 Nun wollen wir die wichtige Thatsache einsehen, dass D von der Wahl der Punkte H und G unabhängig, d. h., dass wir bei jeder beliebigen „harmonischen" Construktion immer zu demselben Punkt D gelangen.

Zu dem Zwecke ziehen wir durch C eine Gerade $CG'H'$, so dass G', H' nicht in der alten Construktionsebene ACH liegen[32] und dass ausserdem GG' und HH' sich in einem Punkte M schneiden. Nun construire man E', F', D'. Da HE und $H'E'$ sich schneiden, so liegen H, E, H', E' in

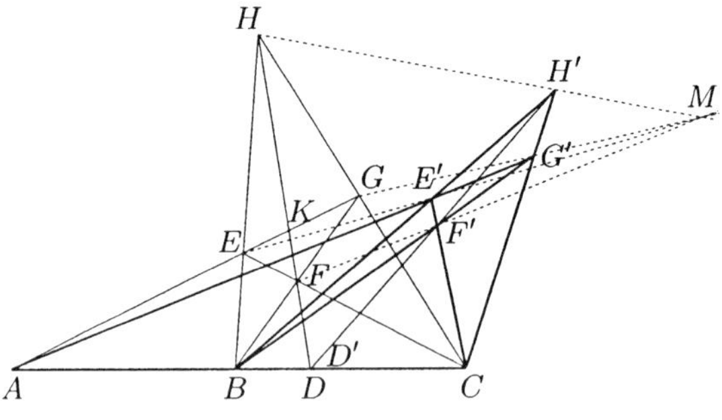

einer Ebene und folglich liegen auch E, E', M in derselben Ebene. Da GE und $G'E'$ sich schneiden, so liegen auch G, E, G', E' in einer Ebene, die von der vorigen verschieden ist. In dieser 2ten Ebene liegen daher auch $EE'M$.
25 E, E', M liegen in 2 verschiedenen Ebenen und daher in einer Geraden. | Da F, E, F', E' mit M in einer Ebene und $FCF'C'$ mit M in einer 2ten Ebene liegen, so liegen auch F, F' und M auf einer Geraden. Also EE' und FF' trifft M.

Folglich liegen $FHF'H'$ in einer Ebene. In derselben Ebene liegen $DF'H'$. In einer anderen Ebene liegen $BF'H'C$ und folglich auch $DF'H'$ d.h. $DF'H'$ liegen in 2 verschiedenen Ebenen und folglich in einer Geraden d. h. $D = D'$.

[31] *Clebsch 1891*. Cf. 173.

[32] This means that the triangle BCH' should not be seen as in the plane of the construction, but inclined to it.

Jetzt wählen wir irgend 2 beliebige Punkte $H'' G''$, so dass jedoch die Gerade $CG''H''$ nicht in die Ebene ACH fällt. Die nebenstehende Construktion liefert $G'H'$. Die harmonische Construktion mittelst $G''H''$ liefere den Punkt D''. Dann ist nach dem Vorigen $D = D'$ und $D' = D''$ und folglich $D = D''$.

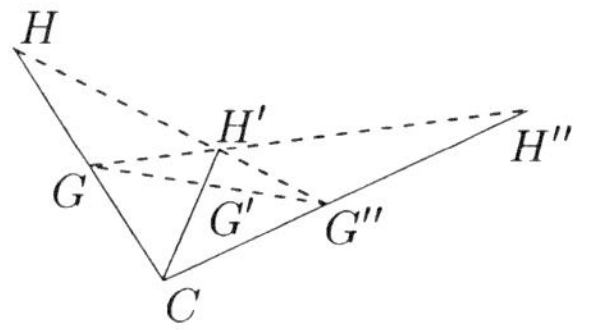

Nun endlich erkennt man die Richtigkeit der Behauptung auch für den Fall, dass der Strahl $CG''H''$ in der alten Construktionsebene ACH liegt. Man wähle nämlich einfach zur Vermittlung einen Strahl | $CG'H'$ in einer neuen Ebene. Dann 26
ist wiederum $D = D' = D''$.

Also wenn A nicht zwischen BC liegt, so lässt sich immer zwischen B und C ein bestimmter 4ter Punkt D construiren, welcher der 4te A zugeordnete harmonische Punkt zu ABC heisst. Aber nicht umgekehrt jeder zwischen B und C gelegne Punkt D muss nothwendig ein 4ter harmonischer Punkt sein. Selbst wenn man A zur freien Verfügung hat. Die Lage von $ABCD$ kann natürlich auch so sein, ↑wie nebenstehende Figur zeigt.↓[33]

↑[34]Es gilt nun der Satz: Wenn A und D die Punkte B und C harmonisch trennen, so trennen auch umgekehrt die Punkte B und C die Punkte A und D harmonisch.

Beweis:[35] Man verbinde B und C mit K, verlängere beide Geraden bis L, P. Dann liegen 1.) A, P, L in einer Geraden; denn würde AP die Gerade HC in L' treffen, so würde der Schnitt von $L'B$ mit PC etwa $K' \neq K$ und folglich auch der Schnitt von HK' mit

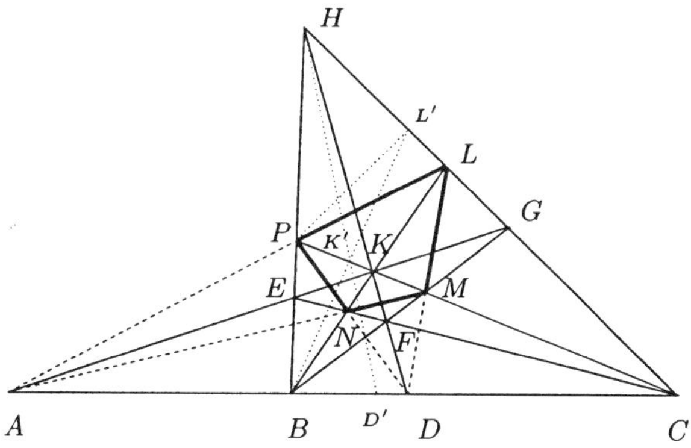

[33]Several words erased and overwritten by the clause inserted.

[34]The paragraphs and the figure on the right replace the following passage: Haupteigenschaft harmonischer Strahlen, wie in Collegheft Projektive Geometrie, nämlich: Wenn man irgend einen Punkt, der nicht auf einer Geraden liegt, ⟪mit 4⟫ harmonischen Punkten dieser Geraden verbindet, so heissen die Verbindungslinien harmonische Strahlen. ⟪Two or three lines pasted over.⟫⟨The 'Collegheft' Hilbert refers to is probably *Hilbert 1891a**; see ibid., p. 29. The 'Haupteigenschaft' mentioned here appears as a definition in the second paragraph added on the sheet pasted to this page.⟩

[35]In what follows it is assumed that the complete quadrilateral $EFGH$ whose diagonals intersect in K, is constructed as on p. 23. It then has to be shown that there is a complete quadrilateral $PNML$, whose diagonals, also intersecting in K, pass through B and C, respectively, while its two pairs of opposite sides intersect in A and D.

AC etwa $D' \neq D$ sein. 2.) liegen L, M, D in einer Geraden.[36] Denn sonst würden wir, wenn wir die Construktion des 4ten Harmonischen mittels CGL ausführten, einen Punkt $D' \neq D$ als 4ten harmonischen erhalten. 3.) P, N, D liegen in einer Geraden,[37] wie wir erkennen, wenn wir die Construktion des 4ten Harmonischen mittels CKP ausführen. 4.) Endlich liegen A, N, M in einer Geraden, denn: Würde AN die Gerade PC in M' schneiden, so würde BM' die Gerade NC in $F' \neq F$ treffen und folglich KF' die Gerade AC nicht, wie es sein soll, in D treffen können. Das Viereck $PNML$ zeigt nun die Richtigkeit der Behauptung.

Nun definire: Wenn man irgend einen Punkt, der nicht auf einer Geraden liegt, mit 4 harmonischen Punkten dieser Geraden verbindet, so heissen die Verbindungslinien 4 harmonische Strahlen eines Strahlenbüschels. Beweise nun die wesentlichste Eigenschaft harmonischer Strahlen:

Satz: Harmonische Strahlen schneiden auf einer jeden sie sämtlich tref-
27 fen↓|den Geraden harmonische Punkte aus. [38]

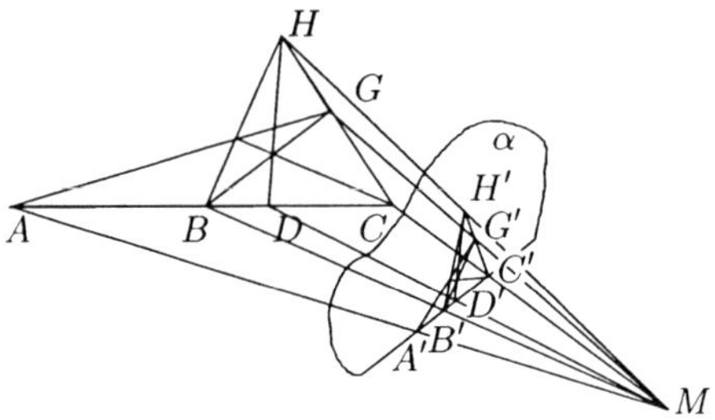

Zum Beweis nehme man zunächst an, dass $A'B'C'D'$ nicht zwischen A und C liegt. Dann wähle man eine Ebene α, welche durch $A'B'C'D'$ geht und daher nicht zwischen A und C, wohl aber zwischen M und $ABCD$ liegt. Dann wähle H auf derselben Seite dieser Ebene α wie AC und mache die harmonische Construktion AHC. Da M auf der anderen Seite liegt, so erhält man durch Aufsuchen der Schnittpunkte M mit AHC die harmonische Con-
28 struktion $A'H'C'$ in der Ebene α. — Liegt $A'B'C'D'$ zwischen | A und C, so kann man leicht eine Gerade $A''B''C''D''$ wählen, welche nicht zwischen A und C liegt und so, dass $A'B'C'D$ nicht zwischen C'', D'' liegt und durch deren Vermittelung unser Satz allgemein bewiesen ist.

[39]↑Harmonische Ebenen eines Büschels schneiden auf jeder Geraden harmonische Punkte aus. [40]↓

---

[36] M denotes the intersection of BG and CP.

[37] N denotes the intersection of BL and CE.

[38] Deleted by Hi^s: Denn in der ursprünglichen Constructionsebene ziehe man eine Gerade l, welche weder zwischen A und H noch zwischen A und C liegt und welche mit der Geraden $A'B'C'D'$ in einer Ebene liegt. ⟨The figure which follows was part of the deleted passage, but remained in the manuscript.⟩

[39] The addition (which runs to 'was zu beweisen ist' on line 85.17) starts in the lower margin, then from 'Beweis' continues on a sheet pasted in for this purpose. The whole addition is assigned to this place in the text by an insertion sign, written in Hi^b.

[40] Deleted by Hi^s: (Vgl. Schluss des Abschnittes „Das Doppelverhältniss".

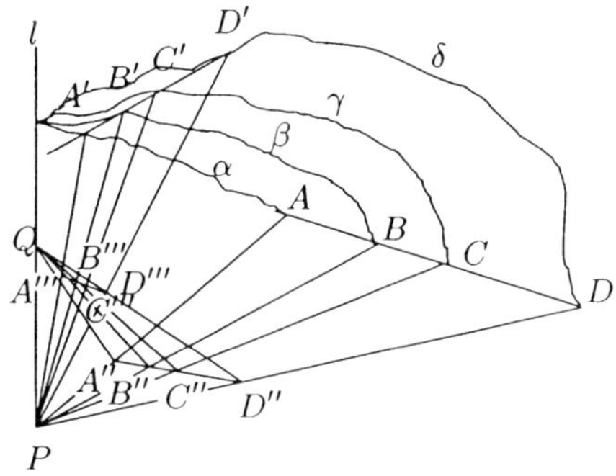

Beweis: Man wähle einen Punkt P auf der Achse l des Ebenenbüschels α, β, γ, δ. A, B, C, D seien harmonische Punkte einer Geraden g; man soll beweisen dass $A'\,B'\,C'\,D'$, die Schnittpunkte mit einer anderen Geraden, ebenfalls harmonische Punkte sind. Construire eine Ebene π, welche alle Geraden l, PA, PB, PC, PD, PA', PB', PC', PD' schneidet, was nach einem früheren Satze stets möglich ist. Die Schnittpunkte seien Q, A'', B'', C'', D'', A''', B''', C''', D'''. Dann liegen Q, A'', A''', ferner Q, B'', B'''; Q, C'', C''', Q, D'', D''' je in einer Geraden a, b, c, d, weil die betreffenden 3 Punkte immer in 2 Ebenen liegen. Nun sind A'', B'', C'', D'' harmonisch, folglich auch A''', B''', C''', D''', folglich auch A', B', C', D', was zu beweisen ist.

$\uparrow$*Die Einführung der Zahl*$\downarrow$[41]

$\uparrow$In allen exakten Wissenschaften gewinnt man erst dann präzise Resultate, wenn die Zahl eingeführt ist. Es ist stets von hoher erkenntnisstheoretischer Bedeutung zu verfolgen, wie dies Messen geschieht.$\downarrow$

Die harmonische Construction setzt uns in den Stand, den Punkten in einer Geraden reelle Zahlen zuzuordnen. Zu dem Zwecke wählen wir auf der Geraden 3 aufeinanderfolgende Punkte und ertheilen diesen die Zahlenwerthe $\uparrow$[42]0, 1, ∞.

---

[41] Hilbert had originally not intended to start a new section at this point, for in the uncorrected version there is simply a new paragraph beginning with 'Die harmonische Construction ...'. That the insertion 'Die Einführung der Zahl' was intended as a new subtitle is made clear by the addition 'Ueberschrift' just before it.

[42] The sentences added replace the following passage (the equation '$\frac{x-\alpha}{x-\beta}/\frac{\gamma-\alpha}{\gamma-\beta} = (\alpha, x, \beta, \gamma)$' belongs to both the deletion and the replacement): α, β, γ. Dann ertheilen wir dem zwischen α und β zu γ harmonisch gelegnen Punkte δ denjenigen Werth $x = \delta$, welcher sich aus der Gleichung

$$\frac{\frac{x-\alpha}{x-\beta}}{\frac{\gamma-\alpha}{\gamma-\beta}} = (\alpha, x, \beta, \gamma) = -1$$

ergiebt. Desgleichen construiren wir den zu α

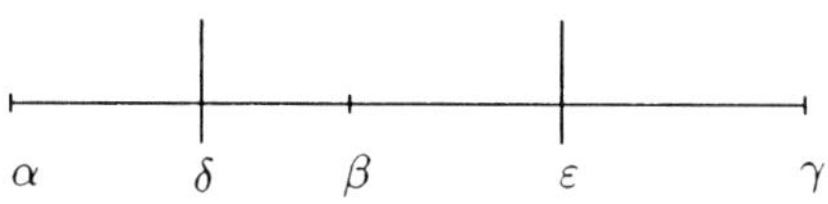

⟨p. 29⟩ zwischen β und γ gelegnen harmonischen Punkt und ertheilen ihm den Werth $x = \varepsilon$, welcher sich aus der Gleichung $(\alpha, \beta, x, \gamma) = -1$ ergiebt. Und so fahren wir fort, immer neue Punkte aus den schon vorhandnen construirend. Der Bruch heisst das Doppelverhältniss der 4 Zahlen. Wir erhalten so eine unendliche Reihe von Punkten, deren jedem eine Zahl zugehört.

Dann nennen wir den Bruch $\dfrac{\frac{x-\alpha}{x-\beta}}{\frac{\gamma-\alpha}{\gamma-\beta}} = (\alpha, x, \beta, \gamma)$ das Doppelverhältniss der 4 Grössen α, x, β, γ. Dabei wird $\frac{\infty+e}{\infty+e'} = 1$ angesehen, wenn e, e' endliche Zahlen bedeuten, und überhaupt geschieht das Rechnen mit ∞ nach den bekannten Regeln $\frac{1}{0} = \infty$, $\frac{1}{\infty} = 0$. Es ist $(\kappa\lambda\mu\nu) = (\nu\mu\lambda\kappa) = \frac{\nu-\mu}{\nu-\kappa}\frac{\lambda-\kappa}{\lambda-\mu}$.↓

29 Jetzt [43] kann man jedenfalls zu jeder positiven rationalen Zahl ↑von der Form $\frac{m}{2^\nu}$↓ wenigstens einen Punkt finden. In der That: Zunächst construiren wir den zwischen 0 und α gelegnen und zu ∞ harmonischen Punkt und nennen denselben $\frac{\alpha}{2}$. ↑Dies heisse die erste harmonische Construktion.↓[I] [44] Es sei ↑ferner↓ α eine positive rationale Zahl, dann ist der zwischen α und ∞ gelegene zu 0 zugeordnete 4te harmonische Punkt mit 2α; ferner der zwischen 2α und ∞ gelegene und zu α zugeordnete 4te harmonische Punkt mit 3α und so fort zu bezeichnen ↑(2[te] harmonische Construktion)↓. [Endlich sei $\frac{1}{n}$ [45] ein zwischen 0
30 und 1 | gelegener Punkt ↑wo n eine ganze positive Zahl bedeutet↓. Dann ist der zu 1 zugeordnete zwischen 0 und $\frac{1}{n}$ gelegne harmonische Punkt zu bezeichnen mit [46] $\frac{1}{2n-1}$ [46] ↑(3te Construktion).↓]

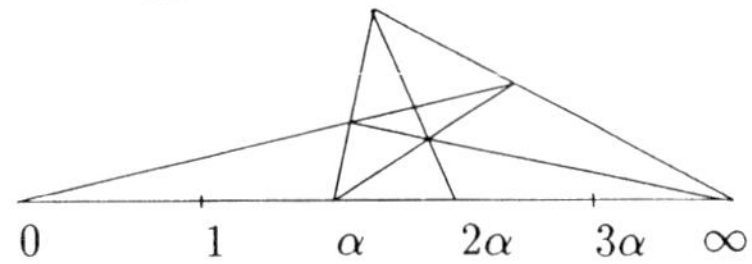

} 2te harmonische Construktion

[↑(Es ist nicht absolut nöthig, die 3te Construktion zu erwähnen.)↓

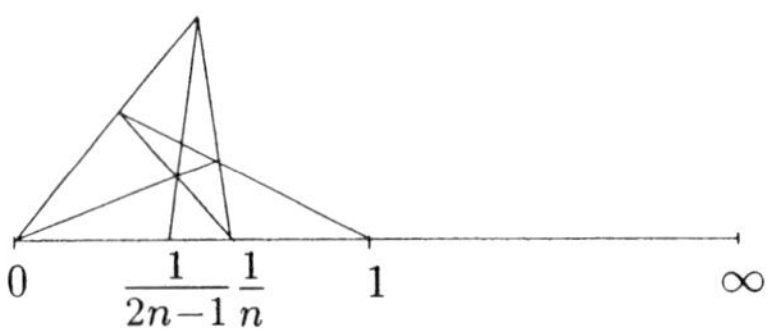

} 3te " "]

[Zu $\frac{1}{2}$ hat man bereits ↑nach der 1[en] Construktion↓ einen Punkt zwischen 0 und 1 ↑daraus nach der 2ten Construktion $\frac{2}{2}$, $\frac{3}{2}$, $\frac{4}{2}$.↓ Für [47] $n = 2$ findet man $\frac{1}{3}$, daraus mittelst der ersten Construktion $\frac{2}{3}$, $\frac{3}{3}$, $\frac{4}{3}$, $\frac{5}{3}$, … [48] Dann vermöge 1 aus $\frac{1}{2}$ findet man $\frac{1}{4}$, daraus

I

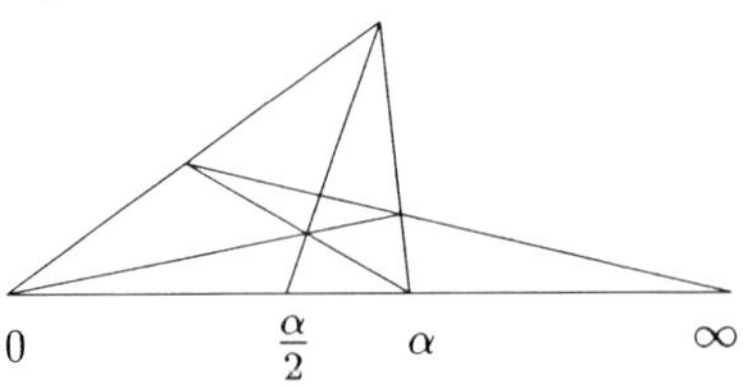

} erste harmonische Construktion

⟨Addition in the lower margin, not specifically assigned to this position.⟩

[43] Deleted: setzen wir $\alpha = 0$, $\beta = 1$, $\gamma = \infty$, dann

[44] Deleted: Dann zeigen wir folgendes allgemein:

[45] Deleted by *Hi*[s]: $\frac{m}{n}$

[46–46] Substituted for: $\frac{m}{2n-m}$

[47] Deleted by *Hi*[s]: $m = 1$,

[48] Deleted: Für $m = 1$, $n = 3$ findet man $\frac{1}{5}$, daraus $\frac{2}{5}$, $\frac{3}{5}$, … Für $m = 2$, $n = 3$

$\frac{2}{4}, \frac{3}{4}, \ldots$ Dann vermöge 3 für $n = 3$ findet man $\frac{1}{5}$. Aus $\frac{1}{3}$ ergiebt sich nach 1 $\frac{1}{6}$. Aus $\frac{1}{4}$ vermöge 3 ergiebt sich $\frac{1}{7}$ etc.]

Aber es ist damit keineswegs gesagt, dass man mit $\frac{2}{4}$ auf denselben Punkt wie $\frac{1}{2}$ oder $\frac{2}{2} = 1$ gelangt. Es könnten also derselben rationalen Zahl mehrere verschie|dene [49]Punkte der Geraden entsprechen und umgekehrt auch demsel- 31
ben Punkt verschiedene Zahlen. Wir haben nun zu zeigen, dass dies nicht der Fall ist, dass vielmehr jedem Punkte, welchen wir überhaupt durch die ↑beiden ersten↓ harmonische⟨n⟩ Construction⟨en⟩ finden, immer nur eine bestimmte rationale Zahl entspricht. Wir zeigen dies [50]wie folgt:[50] Zunächst zeige man die Identität der Punkte $\frac{2}{2}$ und 1. In der That, da $0, \frac{1}{2}, 1, \infty$ harmonisch sind und desgleichen $0, \frac{1}{2}, \frac{2}{2}, \infty$, so ist $\frac{2}{2} = 1$. Ferner zeigen wir, dass $\frac{4}{2} = 2$; denn da $0, \frac{1}{2}, 1, \infty$, ferner $0, 1, 2, \infty$, [51] | harmonische Punkte sind, und wenn man 32

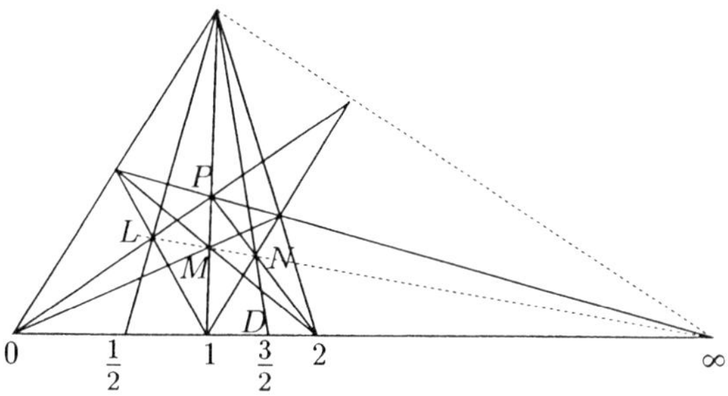

nebenstehende Construction macht, und auch noch den Punkt D construirt, so dass [52] $1, D, 2$↑, ∞↓ harmonisch sind, so liegen L, M, N ↑∞↓[53] auf ein und der nämlichen Geraden, denn die 4 durch ∞ gehenden Strahlen sind harmonisch, weil $1, M, P, Q$ harmonisch sind.[54] Folglich sind auch $P0, P1, P2, P\infty$

[49]Deleted: Punkte der Geraden entsprechen. ↑auch ⟪den Punkten Zahlen⟫↓

↑(Alle die eben construirten sogenannten rationalen Punkte liegen zwischen den Punkten 0 und ∞.)↓⟨Added in the upper margin, but not explicitly assigned to this place.⟩

Wenn wir die Grundpunkte $0, 1, \infty$ anders etwa α, β, γ bezeichnet hätten, so würden wir natürlich

[50–50]Substituted for: zunächst nur für diejenigen Punkte, welche durch die 1te und 2te harmonische Construction gefunden werden. Diese liefern alle und nur die Zahlen, deren Nenner Potenzen von 2 sind.

[51]Deleted by *HiK*ˢ: ferner $\frac{1}{2}, 1, \frac{3}{2}, \infty$

[52]Deleted: 0,

[53]Added by *HiK*ˢ.

[54]It is hard to tell from Hilbert's figure what order the construction follows. However, some sense of it can be made as follows. Assume given Q (the upper point in the figure) and P, and then the harmonic points $0, 1, 2, \infty$. Construct the points $\frac{1}{2}$ and $D = \frac{3}{2}$; the points L, N clearly result from these respective constructions. Now we can construct M: Join the point 1 with L, and take the intersection point with $Q0$; join that point to 2 to get a new line. Now join 1 and N, and take the intersection of this with $Q2$; join this point with 0 to get a second new line. M is then the intersection of these two new lines. That L, M and N all lie on a straight line can be seen as follows. The lines ∞Q and ∞L harmonically separate the lines ∞P and $\infty 0$, as follows from the complete quadrilateral $\infty, P, L, 1$. The same argument will go through with the lines ∞M respectively ∞N instead of ∞L, and

4 harmonische Strahlen. Folglich L, M, N, ∞ 4 harmonische Punkte, und nimmt man Q als Centrum so folgt, dass auch $\frac{1}{2}$, 1, D, ∞ harmonische Punkte
33 sind. Folglich ist $D = \frac{3}{2}$. Da aber 1, D, 2, ∞ harmonisch sind, | so ist $2 = D + \frac{1}{2} = \frac{4}{2}$.[55] Um für rationale Zahlen, deren Nenner höhere Potenzen von 2 sind, das entsprechende nachzuweisen, hat man nur nöthig, in der eben gemachten Entwicklung an Stelle des Punktes 1 der Reihe nach die Punkte $\frac{1}{2}, \frac{1}{4}, \frac{1}{8}, \ldots$ treten zu lassen. Also jedem durch die 1te und 2te harmonischen Construktion gefundenen Punkte entspricht *eine* Zahl $\frac{m}{2^\nu}$ und nur *eine* solche Zahl, und jeder solchen Zahl entspricht auch *ein* und nur *ein* Punkt. Sind α, β, γ 3 solchen Punkten entsprechende Zahlen, und ist $\alpha < \beta < \gamma$ so liegt β zwischen α und γ. In der That, bringt man die 3 Zahlen α, β, γ auf gleiche Benennung, etwa $\frac{m}{2^\rho}$, $\frac{n}{2^\rho}$, $\frac{p}{2^\rho}$, so gehn alle 3 Punkte aus dem Punkte $\frac{1}{2^\rho}$ durch die 2te Construction hervor und zwar bezüglich durch m, n, p-malige Anwendung derselben. Folglich
34 liegt der Punkt α zwischen | β und $\frac{1}{2^\rho}$ und ferner β zwischen α und γ. Wie die Figur zeigt. Da bei der Construction 2 jeder erhaltene Punkt zwischen dem zuvor construirten Punkte und dem gleich darauf construirten Punkte liegt.

Jetzt zeigen wir, dass überhaupt jedem zwischen 0 und ∞ gelegnen Punkte der Geraden eine bestimmte reelle, positive Zahl zugeordnet werden kann. In der That, nehmen wir einen Punkt A zwischen 0 und ∞ an, und bezeichne n_ρ die grösste ganze positive Zahl, so dass $\frac{n_\rho}{2^\rho}$ zwischen 0 und A liegt, so bilden die Zahlen $\frac{n_1}{2}$, $\frac{n_2}{2^2}$, $\frac{n_3}{2^3}$ eine Reihe nicht abnehmender Zahlen, die ausserdem sämmtlich $\uparrow < \beta = \frac{m}{2^\sigma} \downarrow$ bleiben, wenn $\uparrow \beta = \frac{m}{2^\sigma} \downarrow$ ein zwischen A und ∞ gelegner Punkt ist.[56] Die obige Reihe bestimmt eine rationale oder irrationale Zahl a,
35 welche wir dem Punkte A zuordnen wollen. [57] | Aus dieser Bestimmungsweise der Zahl a folgt zunächst, wenn α und β Zahlen von der Form $\frac{n}{2^\rho}$ sind, und $\alpha < a < \beta$ ist, so liegt nothwendig a zwischen α und β. Wir beweisen aber ferner auch leicht folgendes:

Wenn a, b, c irgend welche Zahlen sind, denen Punkte entsprechen und $a < b < c$ ist, so liegt b zwischen a und c. Zum Beweise braucht man nur 2 Zahlen α und β von der Form $\frac{n}{2^\rho}$ zu wählen, so dass $a < \alpha < b < \beta < c$ ist. Denn dann liegt ja a zwischen 0 und α, b zwischen α und β und c zwischen β und ∞.

⟨The following indented passages in smaller font were deleted:⟩

> Wir haben somit bewiesen, dass jedem zwischen 0 und ∞ gelegnen Punkte eine bestimmte[58] positive Zahl entspricht. Aus dem letzteren folgt auch, dass es keinen

---

M respectively N instead of L. The assertion then follows from the uniqueness of the fourth harmonic point. For Hilbert's subsequent argumentation, it still has to be shown that M lies on the line $Q1$.

[55] By means of the second harmonic construction, the points $\frac{2}{2}$, $\frac{3}{2}$, ∞ lead to the point Z, to which then the number $\frac{4}{2}$ is coordinated. (In Hilbert's way of writing: $\frac{3}{2} + \frac{1}{2} = \frac{4}{2}$.) On the other hand, $1 = \frac{2}{2}$, $D = \frac{3}{2}$, and 1, D, 2, ∞ are harmonic. This shows that $Z = 2$, thus that $\frac{4}{2} = 2$.

[56] The existence of $n_\rho < \infty$ is only guaranteed when the Archimedean Axiom is assumed.

[57] Deleted by *HiK*^{s}: Es giebt keinen andern Punkt

[58] Substituted by *HiK*^{s} for: einzige

2ten Punkt giebt, welchem dieselbe Zahl entspricht.[59] Denn sonst würde eine zwischen beiden Punkten gelegne Zahl sich finden lassen, und diese kann nicht zugleich grösser und kleiner sein als ein und dieselbe Zahl. Es ist aber besonders | hervorzu- 36
heben, dass nicht nothwendig für jede positive reelle Zahl ein Punkt auf der Geraden zu existiren braucht. Man kann allerdings von vornherein dem ursprünglich gegebenen System von Punkten noch weitere Punkte so adjungiren, dass allen Zahlen auch Punkte entsprechen. Nöthig ist das jedoch nicht; nöthig ist es nur, jeden Punkt schrittweise, so wie wir ihn brauchen zu adjungiren. Z. B. wenn wir später die Gerade durch einen Kreis oder Kegelschnitt schneiden.[60]

Wir nehmen jetzt statt der Punkte 0, 1, ∞ die 3 Punkte 0, α, ∞ als Grundpunkte, indem wir dem Punkte α eben diese Zahl $\alpha = \frac{n}{2^\rho}$ zuertheilen. Jetzt wenden wir wieder unsre erste und zweite harmonische Construction an. Aber wie gesagt nicht auf die Punkte 0, 1, ∞, sondern auf die Punkte 0, α, ∞. Wir erkennen dann, dass allen Punkten β, γ, ..., die wir überhaupt so erhalten, | dieselben Zahlen wie vorhin, 37
zugeordnet erscheinen. Übrigens werden wir alle und nur diejenigen Zahlen von der Form $\frac{m}{2^\sigma}$ erhalten, wo m durch n theilbar ist.[61] Um zu zeigen, dass wir in der That ↑für dieselben Punkte↓ dieselben Zahlen erhalten[62], [63]nehmen wir an, es seien[63] 0, α, 2α, ∞ harmonisch; ferner α, 2α, 3α, ∞, ferner 2α, 3α, 4α, ∞ und so weiter. Nun[64]

---

[59]This is not correct. Let A and C be distinct points to which the numbers a and c are assigned. Hilbert then argues, as follows, that $a = c$ would lead to a contradiction. Let B be chosen so that it lies between A and C, and let the number b be assigned to it. Then either we would have $a < b < c$ or $c < b < a$, which contradicts $a = c$. But in the previous paragraph, Hilbert has only proved that from $a < b < c$ it follows that B must lie between A and C; what he relies on here is the *converse* assertion.

[60]This passage is similar to a remark of Lindemann in *Clebsch 1891*: '⟨Einer irrationalen Zahl ist⟩ ein bestimmter Punkt unserer Geraden ... zugeordnet, welchem man sich durch eine unendliche Folge von ausführbaren Operationen mehr und mehr nähert. Wir müssen uns nur gewöhnen, einen Punkt, welcher derartig als Grenzlage eines anderen in bestimmter Weise bewegten Punktes erscheint, als wirlich erreicht zu betrachten, wie man es ja auch bei Construction irrationaler (incommensurabler) Strecken in der elementaren Mathematik thun muss. ... die directe Construction irrationaler Zahlen [i.e., die irrationalen Zahlen zugeordneten Punkte] gelingt ... nur (und dann nur in einzelnen Fällen), wenn man andere Hülfsmittel (z.B. den Zirkel, d.h. einen fertig gezeichneten Kreis) zu Hülfe nimmt' (pp. 446–447). The difference between this and Hilbert consists in the fact that Lindemann implicitly assumes a principle of continuity: 'Mit einer geraden Linie verbinden wir den Begriff der stetigen Ausdehnung' (p. 447).

[61]This is only correct if it is assumed that $(n, 2) = 1$.

[62]Here appears an insertion sign, to which corresponds no identifiable text.

[63–63]Substituted for: machen wir beistehende Figur ⟨Here, an illegible figure has been crossed out and replaced by the following⟩

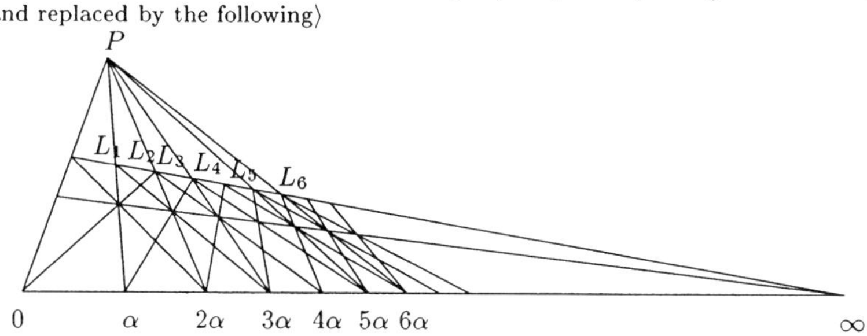

in derselben sind nach Construction

[64]The text breaks off here, having been pasted over with a sheet of paper, which subsequently was largely removed. The first line is still legible: 'Die Thatsache, dass alle Punkte

39 [Unbefriedigend bei dieser ganzen Entwickelung bleibt nur die Bevorzugung der Zahl 2 im Nenner. Um diese Bevorzugung zu beseitigen, müssten wir alle rationalen Zahlen bez. Punkte gleichmässig in Betracht ziehen statt nur die Punkte $\frac{n}{2^\rho}$, d. h. wir müssten auch die 3te harmonische Construktion berücksichtigen und dann den Nachweis führen, dass wir auch dann für ein und denselben Punkt auch nur eine Zahl erhalten. [Dies ist in der That möglich.*)[65] Es ist nämlich nur zu zeigen, dass, wenn ich auf $\alpha = \frac{1}{n}$ die 3te Construktion anwende, der entsprechende Punkt $\beta = \frac{1}{2n-1}$ die Eigenschaft $(2n+1)\beta = 1$[66] hat.]]

Endlich kann man jedem nicht zwischen 0 und ∞ gelegenen Punkt eine (negative) Zahl zuordnen, nämlich die Zahl $-\alpha$, wenn der Punkt zu α zwischen 0 und ∞ der 4te harmonische ist. Natürlich braucht nicht für jede negative Zahl ein Punkt zu existiren, ja nicht einmal für jede negative rationale Zahl
40 von der Form $-\frac{n}{2^\rho}$ braucht es einen | Punkt zu geben. Aber es gelten bezüglich des Begriffes „zwischen" die Sätze: 1. Wenn $\alpha < \beta < 0$ und der Punkt 0 zwischen Punkt α und ∞ liegt, so liegt Punkt β zwischen α und 0↑, falls überhaupt Punkt β existirt↓.

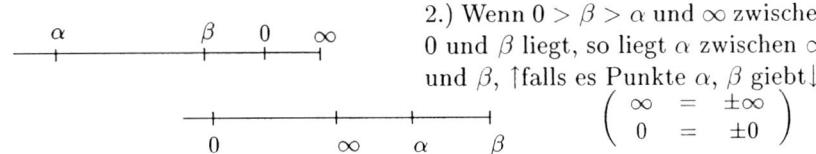

2.) Wenn $0 > \beta > \alpha$ und ∞ zwischen 0 und β liegt, so liegt α zwischen ∞ und β, ↑falls es Punkte α, β giebt↓.

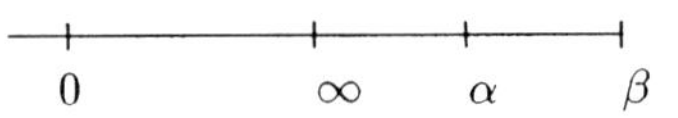

$$\begin{pmatrix} \infty & = & \pm\infty \\ 0 & = & \pm 0 \end{pmatrix}$$

Den Beweis dieser Sätze führt man leicht auf den des vorigen Satzes zurück:

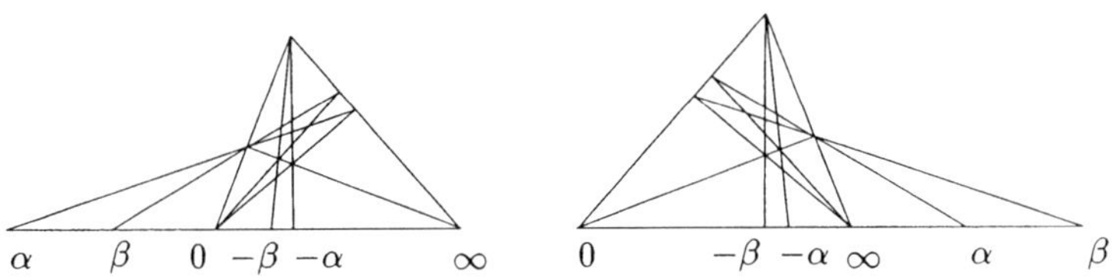

wo $-\alpha$, $-\beta$ nun natürlich positive Zahlen sind.

Ausserdem folgen sofort folgende Existenzsätze:

1. Wenn zu $\alpha < 0$ ein Punkt existirt und 0 zwischen Punkt α und ∞ liegt, so giebt es auch für alle negativen rationalen Zahlen,[67] welche $> \alpha$ sind, Punkte.
41 2. Wenn zu $\alpha < 1$ ein Punkt existirt und ∞ liegt zwischen 0 und diesem Punkt, so giebt es jedenfalls für alle negativen rationalen Zahlen $< \alpha$ Punkte.[68]

---

dieselben'. The next page (38) was completely pasted over and begins now with a new section: 'C. Das Stetigkeitsaxiom'. The original text, however, continues on p. 39.

[65] Hilbert initially had 'wahrscheinlich möglich', but 'wahrscheinlich' was crossed out and a sign for a footnote added in *Hi*[s]. The footnote, reproduced below, is found on a flap pasted to p. 39, which contains, besides the long footnote, six lines of the main text.

[66] This should read: $(2n-1)\beta = 1$

[67] That Hilbert writes explicitly 'rational numbers' may be an indication that he recognises the necessity for an axiom of continuity.

[68] Below this sentence is a special insertion sign with the remark 'siehe vorvorige Seite'. This refers back to section C (beginning on p. 38 and continuing on p. 39) where the same insertion sign is to be found. This explains the sudden switch from p. 41 back to p. 38.

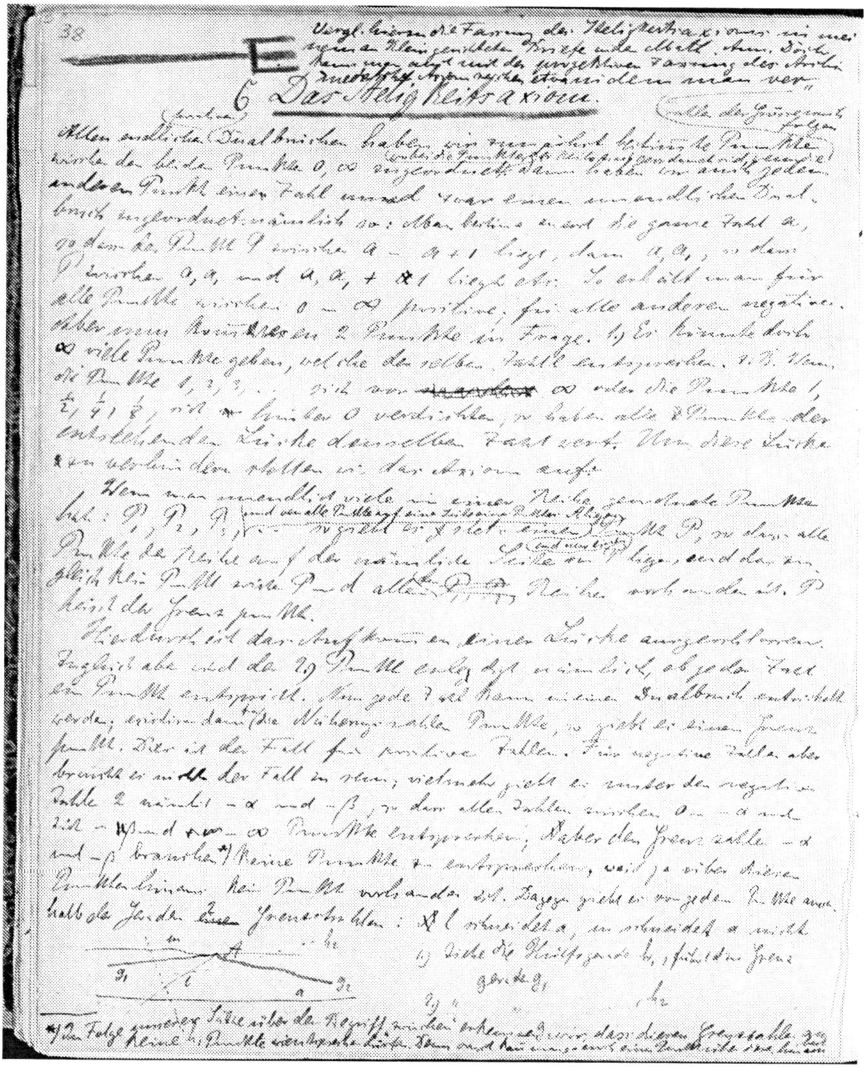

The following four pages are photographic reproductions of pp. 38–41 of the manuscript of Hilbert's notes for his 1893–1894 lectures on the foundations of geometry.

Aus den ersten 3 Axiomen ergeben sich, ähnlich, wenn wir alles zusammenfassen folgende Resultate:

1. Jedem Punkte einer Geraden ist eine bestimmte positive oder negative rationale Zahl zugeordnet. Diese Zuordnung hat folgende Eigenschaften.

2. Wenn die drei Punkte A, B, C ... denen zugeordnet Zahlen α, β, γ ... der Größe nach aufeinanderfolgen, so folgen auch die Punkte der Reihe nach aufeinander.

3. Ergiebt man Grenzwerthe ρ, σ, ... dann keine Punkte zugeordnet sind. Allen zwischen ρ und σ gelegenen Zahlen gehören Punkte zu; allen Zahlen $< \rho$, oder $> \sigma$ gehören keine Punkte an.

4. Das Doppelverhältniß von 4 harmonischen Punkten ist $= -1$ und umgekehrt, wenn das Doppelverhältniß von 4 Punkten $= -1$ ist, so liegen die 4 entsprechenden Punkte harmonisch.

5. Das Doppelverhältniß bleibt bei Projektion erhalten.

Diese 5 Sätze sind (im wesentlichen mit unseren Axiomen A, B, C aequivalent), man könnte daher auch diese 5 Sätze als Axiome nehmen. Die Art der Realität wird im folgenden Abschnitt gezeigt.

[Unbefriedigend bei dieser ganzen Entwicklung bleibt nur die Bevorzugung der Zahl 2 im Nenner. Um diese Bevorzugung zu beseitigen, müßte man alle rationalen Zahlen den Punkten zuordnen. Statt nur die Punkte $\frac{m}{2^n}$ d.h. man müßte auch die 3te harmonische Construktion berücksichtigen und dann den Nachweis führen, daß wir auch dann für ein und denselben Punkt auch nur eine Zahl erhalten. [Dies ist in der That möglich.*) Es ist nämlich nur zu zeigen, daß, wenn ich auf $\alpha = \frac{1}{m}$ die 3te Construktion anwende, der entsprechende Punkt $\beta = \frac{1}{2m+1}$ die Eigenschaft $(2\alpha+\beta) = 1$ hat.]

Endlich kann man jedem nicht zwischen 0 und ∞ gelegenen Punkt eine (negative) Zahl zuordnen nämlich die Zahl $-\alpha$, wenn der Punkt zu α in Bezug auf 0 und ∞ der 4te harmonische ist. Natürlich braucht nicht jeder negativen Zahl ein Punkt zu existiren, ja nicht einmal für jede negative rationale Zahl von der Form $\frac{m}{2^n}$ braucht es einen

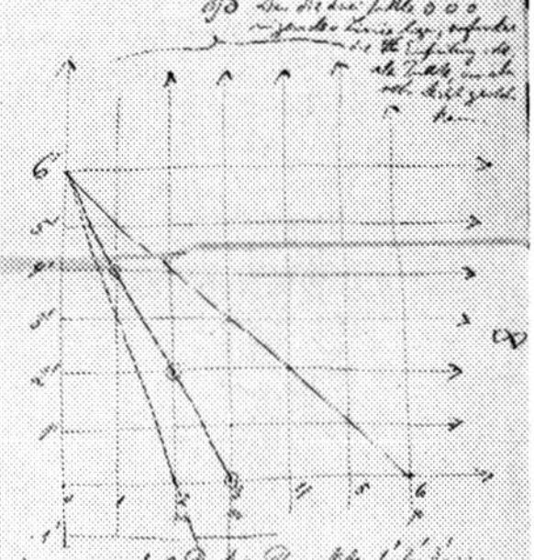

*) Zum Beweise construire man die Punkte $\frac{1}{n(2n-1)}$, von dem aus man dann mit Hülfe der 2ten harmonischen Construktion zu $\frac{1}{2n-1}$ gelangen kann. Der Beweis kann so zu Vereinfachen geführt, wenn man folgende Sätze hat. Es seien $0\,1\,2\,\infty$, $1\,2\,3\,\infty$, $2\,3\,4\,\infty$, ... harmonische Punkte. Sind dann $0\,m\,n\,p$ 3 ganze Zahlen, so dass $\frac{m}{n-m}\frac{p-n}{p} = 1$ ist, so liegen $0, m, n, p$ harmonisch z.B. $m=2$, $n=3$, $p=6$.

Zum Beweise verbindet man irgend einen Punkt P mit $0, 1, 2, \ldots$ und construire auf OP die Punkte $1', 2', 3', \ldots$ so daß $0\,1'\,2'\,P$, $1\,1'\,3'\,P$, $2\,3'\,4'\,P$ harmonisch liegen. Ich verbinde den Punkt ∞ mit $1', 2', 3', \ldots$ Dann zeige, daß die Verbindungslinie von k und l' durch alle diejenigen Gitterpunkte (g, h') geht für welche $\frac{g}{k} + \frac{h'}{l'} = 1$ ist. Dies wird ganz einfach schrittweise gezeigt. z.B. $k=3$ $l'=6$ die Verbindungslinie von $6'$ und $(2, 2')$ muss durch $(1\,4')$ gehen, denn nach dem oben im Text bewiesenen Satze sind einerseits $0P$, $1P$, $2P$, ∞P und andererseits $2'\infty$, $4'\infty$, $6'\infty$, $P\infty$ je 4 harmonische Strahlen. Ferner muss die Verbindungslinie von $(1, 4')$ mit 3 durch $(2\,2')$ gehen, weil einerseits $1P, 2P, 3P, \infty P$ andererseits $0\infty, 2'\infty, 4'\infty, P\infty$ harmonisch liegen etc. Nun verbinde p' mit m, ferner mit n und mit p, wobei wir uns das Gitter fortsetzen, indem wir $-1', -2', (-n+1)'$ construiren und die Punkte mit ∞ verbinden. Die Geraden $p'm$ und $P n$ schneiden sich dann in $(n, (n-p)')$. Man kann P stets so wählen, daß dieser Punkt $(n, (n-p)')$ auch existirt (wegen $\frac{n}{m} + \frac{(n-p)}{p} = 1$). Die Punkte $(n, (n-p)')$, $(n, 0)$, $(n, p-n)'$, P liegen harmonisch, folglich sind die Verbindungslinien mit p' harmonische Strahlen, folglich sind $0, m, n, p$ harmonische Punkte. Will man die Existenz des Gitters nach rechts d.h. nach der

langt dann aber [illegible], wenn man die harmonische Punktkonstruktion ausführt [illegible] kleinen Dreieck [illegible], die [illegible] 0 benachbarten und die Punkte 1, $\frac{1}{2}, \frac{1}{3}, \frac{1}{4}, \frac{1}{5}, \ldots$ [illegible] gelegen ist, von [illegible] harmonisch zu construiren sind.

Punkt zu geben, aber es gelten bezüglich der Begriffe „zwischen" die Sätze: 1) Wenn $\alpha < \beta < 0$ und der Punkt 0 zwischen Punkt α und ∞ liegt, so liegt Punkt β zwischen α und 0, falls überhaupt Punkt β existirt.

~~[illegible]~~

2) Wenn $0 > \beta > \alpha$ und ∞ zwischen 0 und β liegt, so liegt α zwischen ∞ und β, falls α Punkt existirt.

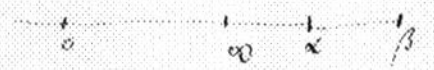

$$\left(\begin{array}{c} \infty = \pm\infty \\ 0 = \pm 0 \end{array}\right)$$

Den Beweis dieser Sätze führt man leicht auf die der vorigen Sätze zurück:

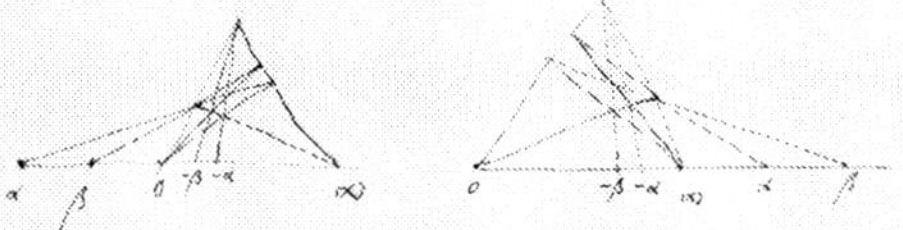

wo $-\alpha, -\beta$ nun natürlich positive Zahlen sind.

Umgekehrt folgen sofort folgende Existenzsätze:

1) Wenn zu $\alpha < 0$ ein Punkt existirt und 0 zwischen Punkt α und ∞ liegt, so gilt er auch für alle negativen rationalen Zahlen, welche $> \alpha$ sind, Punkte

negativen Seite hin vermeiden, so braucht man ja auch nur diejenige Konstruktion von [illegible] nach oben zu verschieben. Diese Gitterkonstruktion scheint jedenfalls der Kernpunkt der Sache zu sein und es empfiehlt sich die Einführung dieses Gitters vielleicht schon früher.

Wie hängt unsere Entwicklung zusammen mit der Frage mit der von Wiener vergl. Hallenser Naturforscherversammlungsberichte 1892 S. 8 und Schoenflies S. 12 behandelten Frage zusammen? Beweise die Wienerschen Behauptungen wenn sie richtig sind, was mir sehr zweifelhaft scheint.]

2. Wenn zu $\alpha < 1$ ein Punkt existirt und ∞ liegt mit 0 und diesem Punkt, so giebt es jedenfalls für alle negativen rationalen Zahlen $< \alpha$, Punkte.

siehe vorvorige Seite

Das Doppelverhältniss.

Für 4 Punkte, denen die Zahlen $\kappa, \lambda, \mu, \nu$ zugeordnet sind ist oben das Doppelverhältniss definirt worden. Bedeuten a, b, c, d beliebige Constante und setzen wir

$$\kappa' = \frac{a\kappa + b}{c\kappa + d}$$

so wird z. B.

$$\kappa' - \lambda' = \frac{a\kappa + b}{c\kappa + d} - \frac{a\lambda + b}{c\lambda + d} = \frac{(ad - bc)(\kappa - \lambda)}{(c\kappa + d)(c\lambda + d)}$$

und hieraus folgt leicht

$$(\kappa, \lambda, \mu, \nu) = (\kappa', \lambda', \mu', \nu').$$

Jetzt beweisen wir das Doppelverhältniss von irgend 4 Punkten bleibt ungeändert wenn wir den Grundpunkten $0, 1, \infty$ nicht

*)[Zum Beweise construire man den Punkt $\frac{1}{n(2n-1)}$, von dem aus man dann mit Hilfe der 2ten harmonischen Construktion zu $\frac{1}{n}$ sowie $\frac{1}{2n-1}$ gelangen kann. Der Beweis ist dann im Wesentlichen geführt, wenn man folgenden Satz hat: Es seien $0\,1\,2\,\infty$, $1\,2\,3\,\infty$, $2\,3\,4\,\infty, \ldots$ harmonische Punkte. Sind dann $0 < m < n < p$ 3 ganze Zahlen, so dass $\frac{m}{n-m}\frac{p-n}{p} = 1$ ist, so liegen $0, m, n, p$ harmonisch, z.B. $m = 2$, $n = 3$, $p = 6$.

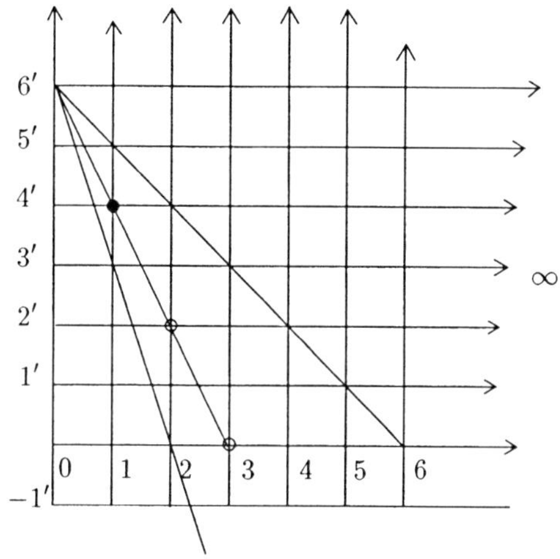

Zum Beweise verbindet man irgend einen Punkt P mit $0, 1, 2, \ldots$ und construire auf $0P$ die Punkte $1', 2', 3', \ldots$ so dass $0\,1'\,2'\,P$, $1'\,2'\,3'\,P$, $2'\,3'\,4'\,P$ harmonisch liegen. Dann verbinde den Punkt ∞ mit $1', 2', 3', \ldots$ Dann zeige, dass die Verbindungslinie von k mit l' durch alle diejenigen Gitterpunkte (g, h') geht, für welche $\frac{g}{k} + \frac{h'}{l'} = 1$ ist. Dies wird ganz einfach schrittweise gezeigt: z. B. $k = 3$, $l' = 6$. Die Verbindungslinie von $6'$ und $(2, 2')$ muss durch $(1, 4')$ gehen, denn nach dem oben im Text bewiesenen Satze sind einerseits $0P$, $1P$, $2P$, ∞P und andererseits $2'\infty$, $4'\infty$, $6'\infty$, $P\infty$ je 4 harmonische Strahlen.[69] Ferner muss die Verbindungslinie von $(1, 4')$ mit 3 durch $(2, 2')$ gehen, weil einerseits $1P$, $2P$, $3P$, ∞P, andererseits 0∞, $2'\infty$, $4'\infty$, $P\infty$ je harmonisch liegen etc. Nun [70] verbinde p' mit m, ferner 3 mit n und mit p [71] . Ausserdem müssen wir noch das Gitter fortsetzen, indem wir $-1'$, $-2'$, $(-p+n)'$ construiren und [72] diese Punkte mit ∞ verbinden. Die Geraden $p'm$ und Pn schneiden sich dann in $(n, (n-p)')$ ↑wegen $\frac{n}{m} + \frac{(n-p)'}{p} = 1$↓. Man kann P stets so wählen, dass dieser Punkt $(n, (n-p)')$ noch existirt. Die Punkte $(n, (n-p)')$, $(n, 0)$, $(n, (p-n)')$, P liegen harmonisch, folglich sind die Verbindungslinien mit p' harmonische Strahlen, folglich sind 0, m, n, p harmonische Punkte. Will man die Erweiterung des Gitters nach unten, d. h. nach der negativen Seite hin vermeiden, so braucht man ja auch nur die ganze Construktion um $p - n$ Sprossen nach oben zu verschieben. Diese Gitterconstruktion scheint jedenfalls der Kernpunkt der Sache zu sein und es empfiehlt sich die Einführung dieses Gitters vielleicht schon früher.

Wie hängt unsere Entwicklung zusammen mit der von Wiener, vgl. Hallenser Naturforschervers.-berichte 1892 S. 8 und Schoenflies S. 12, behandelten Frage zusammen?[73] Beweise die Wienerschen Behauptungen, wenn sie richtig sind, was mir sehr zweifelhaft scheint.]

---

[69] Let g be the straight line through $6'$ and $(2, 2')$. Let R be the point of intersection of ∞P and g, and S the point of intersection of $1P$ and g. The lines $0P$, $1P$, $2P$, ∞P cut the line g in the points $6'$, S, $(2, 2')$, R, which are consequently harmonic points. Then let T be the point of intersection of $4'\infty$ and g. Then the harmonic lines $6'\infty$, $4'\infty$, $2'\infty$, $P\infty$ intersect g in the points $6'$, T, $(2, 2')$, R, which are consequently also harmonic. The assertion follows from the uniqueness of the fourth harmonic point.

[70] Deleted by *Hi*^{s}: ziehe pp' dann

[71] Deleted by *Hi*^{s}: bei den letzteren Geraden

[72] Deleted by *Hi*^{s}: $-1'\infty$, $-2'\infty, \ldots$ construiren

[73] *Wiener 1891*; *Schoenflies 1891*. Hilbert attended Wiener's lecture in Halle in 1891.

38 ↑

C. *Das Stetigkeitsaxiom.*[C]

Allen endlichen positiven Dualbrüchen haben wir zunächst bestimmte Punkte zwischen den beiden Punkten 0, ∞ zugeordnet ↑wobei die Punkte der Reihe nach geordnet sind, wenn die Zahlen der Grösse nach folgen↓. Dann haben wir auch jedem anderen Punkt eine Zahl und zwar einen unendlichen Dualbruch zugeordnet, nämlich so: Man bestimme zuerst die ganze Zahl a, so dass der Punkt P zwischen a und $a+1$ liegt, dann a, a_1, so dass P zwischen a, a_1 und $a, a_1 + 1$ liegt etc. So erhält man für alle Punkte zwischen 0 und ∞ positive; für alle anderen negative. Aber nun kommen 2 Punkte in Frage. 1.) Es könnte doch ∞ viele Punkte geben, welche derselben Zahl l entsprechen. Z.B. wenn die Punkte 1, 2, 3, ... sich vor ∞ oder die Punkte 1, $\frac{1}{2}$, $\frac{1}{4}$, $\frac{1}{8}$, ... sich hinter 0 verdichten, so haben alle Punkte der entstehenden Lücke denselben Zahlwert.[76] Um diese Lücke zu verhindern, stellen wir das Axiom auf:

Wenn man unendlich viele in einer Reihe geordnete Punkte hat: P_1, P_2, P_3, ... ↑und wenn alle Punkte auf einer Seite eines Punktes A liegen↓ so giebt es stets einen ↑und nur einen↓ Punkt P, so dass alle Punkte der Reihe auf der nämlichen Seite von P liegen, und dass zugleich kein Punkt zwischen P und allen Punkten der Reihe vorhanden ist. P heisst der Grenzpunkt.

Hierdurch ist das Aufkommen einer Lücke ausgeschlossen. Zugleich aber wird der 2.) Punkt erledigt, nämlich, ob jeder Zahl ein Punkt entspricht. Nun jede Zahl kann in einen Dualbruch entwickelt werden; existiren dann für die Näherungszahlen Punkte, so giebt es einen Grenzpunkt. Dies ist der Fall für positive Zahlen. Für negative Zahlen aber braucht es nicht der Fall zu sein; vielmehr giebt es unter den negativen Zahlen 2, nämlich $-\alpha$ und $-\beta$, so dass allen Zahlen zwischen 0 und $-\alpha$ und zwischen $-\beta$ und $-\infty$ Punkte entsprechen; aber den Grenzzahlen $-\alpha$ und $-\beta$ brauchen*) keine Punkte zu

[C]Vergl. hierzu die Fassung des Stetigkeitsaxioms in meinem an Klein gerichteten Briefe in den Math. Ann.[74] Doch kann man auch mit der projektiven Fassung des Archimedischen Axioms reichen, etwa in dem man verlangt, dass, wenn man die harmonische Punkteconstruktion ausführt, es keinen Punkt giebt, der zwischen 0 einerseits und den Punkten 1, $\frac{1}{2}$, $\frac{1}{3}$, $\frac{1}{4}$, $\frac{1}{5}$, ... sämtlich gelegen ist, wo $\frac{1}{2}$, $\frac{1}{3}$, $\frac{1}{4}$, ... wie unten harmonisch zu construiren sind.[75] ⟨Addition, begun in the upper margin of p. 38, and then (from '-langt, dass, wenn ⟨...⟩') continued on the reverse of the sheet pasted to the upper margin of p. 39.⟩

*) In Folge unserer Sätze über den Begriff „zwischen“ erkennen wir, dass diesen Grenzzahlen auch keine Punkte entsprechen dürfen. Denn sonst kann man ja auch einen Punkt über diese hinaus bestimmen.

[74] *Hilbert 1895b*. The version given in this letter to Klein corresponds to that given in the second paragraph of this section C.

[75] See also the beginning of the section on the Archimedean Axiom in the *Ausarbeitung* of the 1898/1899 lectures (this Volume, Chapter 4). This version of the Archimedean Axiom is the appropriate projective version of the standard Archimedean Axiom; it is *not* strong enough for Hilbert's purposes here.

[76] Lindemann writes in *Clebsch 1891* that it could be that the points $P_{1/2^\rho}$ 'nicht an der A [$= P_0$], sondern an einer anderen ... verdichten' (p. 446). That this is impossible is of the greatest significance in Lindemann's presentation of the matter.

entsprechen, weil ja über diese Punkte hinaus kein Punkt vorhanden ist. Dagegen giebt es von jedem Punkt ausserhalb der Geraden 2 [77] Grenzstrahlen:[78] l schneidet a, m schneidet a nicht.

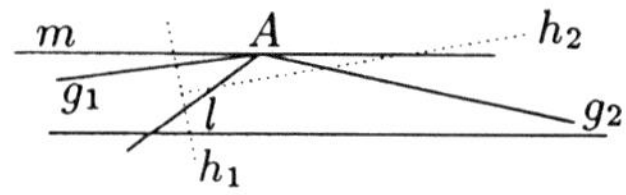

1.) Ziehe die Hülfsgerade h_1, führt zur Grenzgeraden g_1.
2.) " " " h_2, " " " g_2.

[79] Aus unseren 3 Axiomen A, B, C ergeben sich mithin, [80] wenn wir Alles 39
zusammenfassen, folgende Thatsachen:

1. Jedem Punkte einer Geraden ist eine bestimmte positive oder negative reelle Zahl zugeordnet. Diese Zuordnung hat folgende Eigenschaften.

2. Wenn die den Punkten A, B, C,... bez⟪u⟫g zugeordneten Zahlen α, β, γ,... der Grösse nach aufeinanderfolgen, so folgen jene Punkte der Reihe nach aufeinander.

3. Es giebt zwei Grenzzahlen γ_1, γ_2, denen keine Punkte zugeordnet sind. Allen zwischen γ_1 und γ_2 gelegenen Zahlen gehören Punkte zu. Allen Zahlen $< \gamma_1$ oder $> \gamma_2$ gehören keine Punkte zu.

4. Das Doppelverhältnis von 4 harmonischen Punkten ist $= -1$ und umgekehrt, wenn das Doppelverhältnis von 4 Zahlen $= -1$ ist, so liegen die 4 entsprechenden Punkte harmonisch.[81]

5. Das Doppelverhältnis ändert sich nicht bei Projektion.

Diese 5 Sätze sind im wesentlichen mit unseren ↑bisherigen↓ Axiomen A, B, C aequivalent; man könnte daher auch diese 5 Sätze als Axiome nehmen. Dass dies der Fall ist, wird im folgenden Abschnitt gezeigt.↓

Das Doppelverhältniss. 41

Für 4 Punkte, denen die Zahlen κ, λ, μ, ν zugeordnet sind, ist oben das Doppelverhältniss definirt worden. Bedeuten a, b, c, d beliebige Constante und setzen wir $x' = \frac{ax+b}{cx+d}$, so wird z. B.

$$\kappa' - \lambda' = \frac{a\kappa+b}{c\kappa+d} - \frac{a\lambda+b}{c\lambda+d} = \frac{(ad-bc)(\kappa-\lambda)}{(c\kappa+d)(c\lambda+d)}$$

und hieraus folgt leicht[82]

$$(\kappa, \lambda, \mu, \nu) = (\kappa', \lambda', \mu', \nu').$$

---

[77] Substituted for: einen

[78] Strictly speaking, the proposition should read: On the contrary, through each point A outside the straight line a, there is at least one *Grenzstrahl*, and at most two (denoted here by g_1 and g_2), dividing the other straight lines through A into two classes, the lines l which intersect a, and those m which do not.

[79] The following (up to the end of this subsection) is to be found on a sheet pasted to the upper margin of p. 39. Its position here is a little odd, since the *Doppelverhältniss* has not yet been defined.

[80] A, B, C are the three groups of axioms already introduced.

[81] Hilbert proves the second of these assertions on p. 44, Note *). The first follows in essentially the same way.

[82] Here and in the following it is presupposed that $ad - bc \neq 0$.

Jetzt beweisen wir: das Doppelverhältniss von irgend 4 Punkten bleibt unge-
42 ändert, wenn wir den Grundpunkten 0, 1, ∞ nicht | diese Zahlen, sondern beliebige andre Zahlen α, β, γ zuertheilen.[83] Thun wir dies nämlich, so wird im Allgemeinen jedem Punkte der Geraden eine andre Zahl zuzuordnen sein. Um diese zu bestimmen, wählen wir in der obigen linearen Transformation die Constanten a, b, c, d so, dass für $x = 0,\ 1,\ \infty$ das neue $x' = \alpha,\ \beta,\ \gamma$ wird. Ich behaupte dann, dass dem Punkte κ in alter Bezeichnung nunmehr die neue Zahl $\kappa' = \frac{a\kappa+b}{c\kappa+d}$ zuzuordnen ist. In der That, für die rationalen Punkte von der Form $\frac{n}{2^\varrho}$ ist dies an sich klar, weil dieselben durch harmonische Constructionen und durch die Bedingung gefunden waren, dass das Doppelverhältniss $= -1$ ist. Ist also z. B. 0, κ, 1, ∞ harmonisch, so wird in der That $(\alpha, \kappa', \beta, \gamma) = (0, \kappa, 1, \infty) = -1$.

43 Aber auch für die andern Punkte | gilt unsere Behauptung. Wir nehmen einen Punkt κ in alter Bezeichnung, um dies zu zeigen, wo κ nicht von der Form $\frac{n}{2^\varrho}$ ist. Dann wollen wir diesem Punkte in der neuen Bezeichnung diejenige Zahl κ_0 zuordnen, dass wenn in neuer Bezeichnung $\lambda_0 < \kappa_0 < \mu_0$ ist, allemal κ_0 zwischen μ_0 und λ_0 liegt.[84] Wegen der Stetigkeit der linear gebrochnen Function $\frac{ax+b}{cx+d}$ folgt $\kappa' = \kappa_0$. ↑Oder entwickele $\kappa = z,z_1 z_2 \ldots$ in einen Dualbruch, so entsprechen dem Punkt ⟪$\kappa_0 = z$; $\kappa_1 = z,z_1$; $\kappa_2 = z,z_1 z_2$;⟫ in der neuen Bezeichnung die Werte $\kappa_0' = \frac{a\kappa_0+b}{c\kappa_0+d}$, ⟪$\kappa_1'$⟫ .... Wegen der Stetigkeit der linearen Funktion nähert sich diese Reihe einer Zahl und wegen des Stetigkeitsaxioms entspricht dieser Zahl nun eben jener Grenzpunkt.↓

Nun ist ferner das Doppelverhältniss auch unabhängig von der Wahl der Grundpunkte selber.[85] Um dies zu beweisen, nehmen wir statt des Punktes 1 den Punkt β als Grundpunkt. Wenn β eine Zahl von der Form $\frac{n}{2^\varrho}$ ist und wenn wir dann diesem Punkte als Grundpunkt nicht die Zahl 1, sondern die Zahl β ertheilen,
44 so behalten | alle Punkte ihre Zahlen bei, wie oben gezeigt ist, und daher bleibt auch das Doppelverhältniss ungeändert. Um nun zu zeigen, dass dasselbe gilt, auch wenn β nicht von der Form $\frac{m}{2^\varrho}$ ist, zeigen wir, dass bei der Einführung einer neuen Bezeichnungsweise für die Grundpunkte höchstens 2 Punkte ihre Bezeichnungsweise beibehalten können. Dies folgt unmittelbar aus dem oben bewiesenen Satze,[86] wenn man berücksichtigt, dass die Gleichung $x = \frac{ax+b}{cx+d}$ quadratisch in

[83] The introduction of coordinates beginning with $P_0 \mapsto \alpha$, $P_1 \mapsto \beta$, $P_\infty \mapsto \gamma$ is already briefly mentioned in a passage deleted from p. 28. (P_0, P_1, P_∞ denote the *Grundpunkte*, which Hilbert denotes by 0, 1, ∞ respectively.) However, Hilbert does not prove on p. 28 that this introduction of coordinates also leads to an order preserving bijection. He deduces the result in what follows (pp. 42–43).

[84] What is missing here is a proof of the result that there can only be *one* number κ_0. For this the continuity axiom is needed. The insertion after the next sentence shows he is clear about this.

[85] In this paragraph, Hilbert shows that, if β is the number assigned to the *Grundpunkt* P_β (itself β in Hilbert's notation) on the basis of the *Grundpunkte* P_0, P_1, P_∞, then β is still the number assigned on the basis of the new *Grundpunkte* P_0, P_β, P_∞. This result had already been formulated on pp. 36–37, but was then crossed through. For more on the relationship between the maintenance of the *Doppelverhältniss* and the type of coordinate transformation Hilbert deals with here, see *Rédei 1965*.

[86] This refers back to 'Ich behaupte dann, dass ...' on p. 42.

x ist. Wenn wir also ausser den Punkten 0 und ∞ auch noch dem Punkte β die Bezeichnungsweise lassen, so behalten alle Punkte die Bezeichnungsweise.[87] 45

Nun ertheile man den neuen Grundpunkten $0, \beta, \infty$ beziehungsweise die Zahlen $1, 0, \infty$ und verlege nun den ersten Grundpunkt 1 in einen Punkt α. Dann ertheile man den neuen Grundpunkten $\alpha, 0, \infty$ die Werthe $0, \infty, 1$ und verlege so schliesslich den letzten Grundpunkt nach γ. Bei all diesen Prozessen bleibt das Doppelverhältniss ungeändert.*)

Das Doppelverhältniss bleibt auch ungeändert bei Projection; d. h. wenn wir einen Punkt P mit $\kappa, \lambda, \mu, \nu$ verbinden, so sagen wir auch von diesen 4 Strahlen, dass sie das Doppelverhältniss $(\kappa, \lambda, \mu, \nu) = \delta$ haben. Der Satz lautet dann:

Wenn 4 Strahlen eines Strahlbüschels mit dem Doppelverhältniss δ eine Gerade in | 4 Punkten treffen, so besitzen diese 4 Punkte ebenfalls das Doppelverhältniss δ. 46

Erteilen wir 3en von den Punkten A, B, C die Zahlen $0, 1, \infty$,[88] und nehmen wir $\delta = \frac{\varrho}{2^n}$ an, so wird auch D durch eine rationale Zahl α dargestellt sein. Dann führt also eine gewisse Zahl von harmonischen Construktionen von A, B, C auf D. Die[89] entsprechenden harmonischen Construktionen führen

*) Satz: Wenn 4 Punkte ↑$ABCD$ in dieser Reihenfolge↓ das Doppelverhältniss -1 haben, so liegen sie harmonisch.

Beweis: Nimmt man $A = 0$, $B = 1$, $D = \infty$ zu Grundpunkten, so erhält wegen der Voraussetzung C den Wert 2 zuerteilt. Sucht man andererseits zu 0 den 4ten harmonischen zwischen 1 und ∞ gelegenen Punkt, so ist das unsere 2te harmonische Construktion und liefert den Punkt 2, was zu beweisen war.

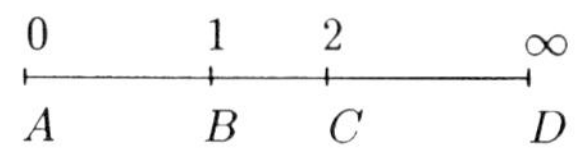

Damit ist vollständig bewiesen, dass irgend welche harmonische Construktionen, wenn man den neu gefundenen Punkten immer solche Zahlen erteilt, dass das Doppelverhältnis $= -1$ ist, immer für den nämlichen Punkt auch die nämliche Zahl ergeben. ⟨Addition on a sheet pasted to the bottom of p. 45.⟩

[87] Deleted: (Wahrscheinlich kann so überhaupt der obige Satz, dass wenn ich einen andern Punkt α an Stelle von 1 als Grundpunkt wähle und ihm die Zahl α ertheile, ⟨p. 45⟩ alle übrigen Punkte ihre Zahlen behalten, leichter und besser bewiesen werden. Vergl. die angeklebte Anmerkung auf einer der früheren Seiten.) ⟨It is possible that, with this last remark, Hilbert is referring to the remark added to p. 37 which no longer survives. See p. 37, note 62.⟩

[88] In the following proof, the *Doppelverhältniss* of the line is defined by means of the points A, B, C, D. The points in question are A', B', C' D', and it is to be shown that their cross ratio is also δ.

[89] Next to the following considerations is a figure which has been crossed out:

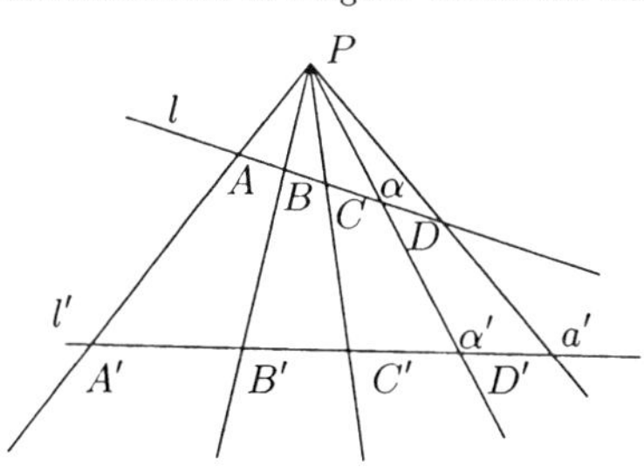

dann nach einem früheren Satze von A', B', C' auf D'. Wenn wir also $A' = 0$, $B' = 1$, $C' = \infty$ nehmen, so wird $D' = \alpha$.

Ist aber δ nicht rational, so wird auch α irrational. [90] Bezeichnen wir die dem
47 Punkte D' auf l' zukommen|de Zahl mit α' und sei $\alpha' \neq \alpha$, so suchen wir eine rationale zwischen α und α' gelegene Zahl a. [91] Es sei z. B.

$$0 < \alpha < a < \alpha'.$$

Verbinden wir P mit a auf l, so muss nach dem eben bewiesenen diese Verbindungslinie auch auf l' den Punkt a ausschneiden. Wenn daher auf l $D = \alpha$ zwischen 0 und a liegt, so muss auch auf l' $D' = \alpha'$ zwischen 0 und a liegen, was einen Widerspruch giebt; daher ist $\alpha = \alpha'$ und daher sind auch die beiden Doppelverhältnisse stets dieselben.

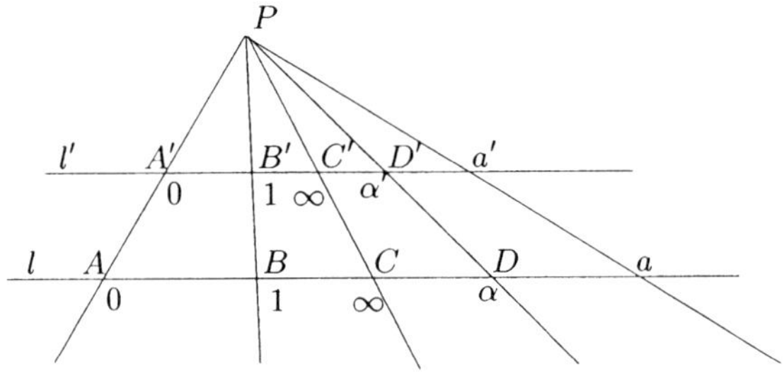

Umgekehrt: Wenn AA', BB', CC' durch ein und den nämlichen Punkt P laufen und die beiden Doppelverhältnisse $(ABCD) = (A'B'C'D')$ gleich sind, so läuft auch DD' durch P.

Denn sonst würde PD etwa durch E' auf l' laufen und dann würde, wegen der Gleichheit der Doppelverhältnisse $(A'B'C'D') = (A'B'C'E')$ den Punkten
48 D' und E' dieselbe | Zahl zukommen, was $D' = E'$ zur Folge hat.*)

Die projektive und die analytische Geometrie.

Die Fundamentalsätze der projektiven Geometrie sind oben vollständig abgeleitet. Man kann nun die weiteren Sätze über Involutionen etc. sowie die Imaginärtheorie von v. Staudt (Theorie der Würfe) in bekannter Weise aufbauen.[92] Aber auch die Fundamentalsätze der analytischen Geometrie (genauer der Geometrie der linearen Transformationen) ergeben sich leicht:

Um in einer Ebene Coordinaten einzuführen, nehme man in derselben ein Dreieck ABC an. Auf der Geraden AB sowohl wie auf AC führe man Zahlen
49 ein und zwar in $A = 0$, $B = \infty$ und $A = 0$, $C = \infty$. Dazwischen | irgendwo

*) Wenn man 4 Ebenen eines Ebenenbüschels durch 2 Gerade schneidet, so sind die Doppelverhältnisse dieser beiden Punktequadrupel auf den Geraden einander gleich.

Beweis: Die Ebenen seien α, β, γ, δ, die Geraden $l = ABCD$ und $l' = A'B'C'D'$. Zunächst Zurückführung auf den Fall $A = A'$. Dann wähle man auf der Axse des Büschels 2 Punkte E und F, so dass DE und $D'F$ sich in G treffen. Dann

[90] Deleted: Sind nun a und b 2 rationale Zahlen $a < \alpha < b$, so verbinde man die beiden Punkte a und b auf l und P; die Verbindungslinien schneiden l' in gewissen Punkten, denen nach dem eben bewiesenen die Zahlen a und b zukommen. Daher liegt auch

[91] Deleted: Der Punkt a auf l und der Punkt a auf l' liegen mit P

[92] The allusion is to *von Staudt 1856* and succeeding volumes.

die Zahl 1. Dann nehme man irgendeinen Punkt P der Ebene und lege durch denselben 2 Strahlen, welche sowohl AB als AC schneiden. Dies ist ja nach einem früheren Satze stets möglich. Die Schnittpunkte mögen für den einen Strahl die Zahlen α, β; für den anderen α', β' haben. Nun bestimme man 2 Zahlen x, y aus den beiden Gleichungen

$$\frac{x}{\alpha}+\frac{y}{\beta}=1 \quad \frac{x}{\alpha'}+\frac{y}{\beta'}=1.$$

Diese beiden Gleichungen werden, da α, β, α', β' nie verschwinden (denn die Strahlen durch P sollen nicht durch A gehen) immer bestimmte endliche Lösungen haben, ausser wenn $\alpha=\infty$ und $\beta=\infty$ ist. D. h. ausser wenn P auf BC liegt und ausser wenn $\begin{vmatrix} \frac{1}{\alpha} & \frac{1}{\beta} \\ \frac{1}{\alpha'} & \frac{1}{\beta'} \end{vmatrix}=0$ ist. Wir schliessen diese beiden Fälle vorläufig aus und zeigen zunächst, dass wir sonst jedenfalls auf dieselben Zahlen x, y geführt werden, welche Strahlen durch P, die AB und AC schneiden, wir auch benutzt hätten. Zu dem Zwecke nehmen wir durch P einen 3ten Strahl an, welcher in α'' und β'' schneiden möge. Dann ist wegen der Gleichheit der Doppelverhältnisse | $\frac{\alpha}{\alpha''-\alpha}\,\frac{\alpha'-\alpha''}{\alpha'}=\frac{\beta}{\beta''-\beta}\,\frac{\beta'-\beta''}{\beta'}$. 50

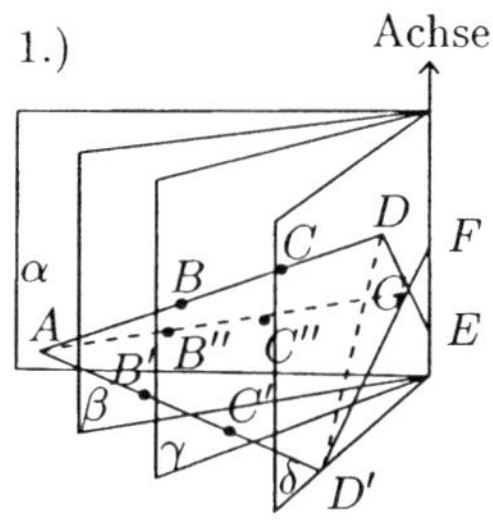

möge AG die beiden übrigen Ebenen in $B''C''$ schneiden. $ABCD$ und $AB'D'C'$ sind beide vom selben Doppelverhältniss wie $AB''C''G$. Man kann immer annehmen, dass die beiden übrigen Ebenen β, γ zwischen $A=A'$ und D, D' liegen. Liegen D, D' auf verschiedenen Seiten der Achse, so wähle man für G irgend einen Punkt der Verbindungslinie DD', so dass GA auch wieder β und γ trifft (2.)). Liegen ⟨p. 49⟩ endlich D und D' beide auf derselben Seite der Achse, aber so, dass β und γ nicht zwischen A und D, D' liegen, so tritt die Figur 3. in Kraft:

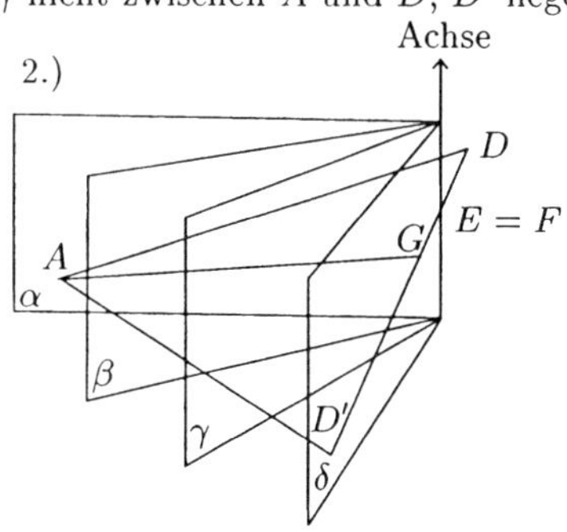

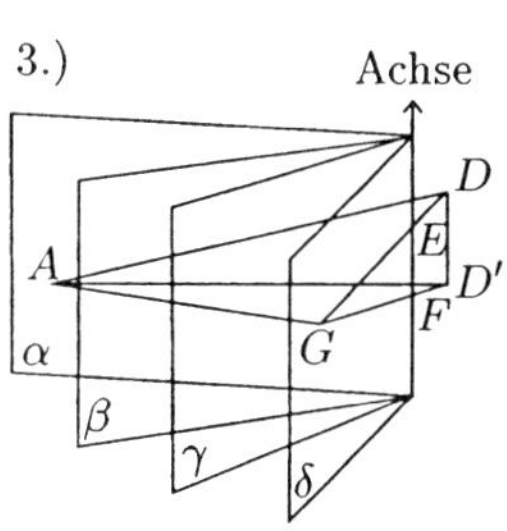

Die Hauptsache ist also, dass G immer auf derjenigen Hälfte der Ebene δ zu liegen kommt, deren Punkte sämtlich sowohl α wie β zwischen sich und A liegen haben. Dann existiren eben die Schnittpunkte B'' und C''. ⟨This footnote was begun on the lower half of p. 48, continued on the lower half of p. 49, and was subsequently fully crossed out.⟩

Diese Gleichung lässt sich in die Gestalt

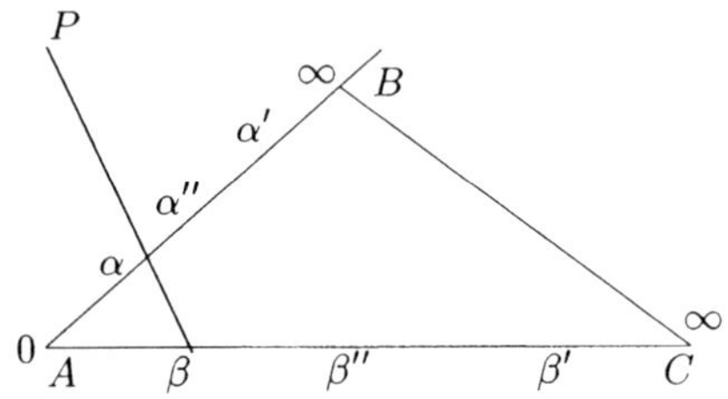

$$\begin{vmatrix} \frac{1}{\alpha} & \frac{1}{\beta} & 1 \\ \frac{1}{\alpha'} & \frac{1}{\beta'} & 1 \\ \frac{1}{\alpha''} & \frac{1}{\beta''} & 1 \end{vmatrix} = 0$$

bringen. Denn letztere Gleichung ist gleichbedeutend mit

$$\begin{vmatrix} \frac{1}{\alpha} - \frac{1}{\alpha''} & \frac{1}{\beta} - \frac{1}{\beta''} \\ \frac{1}{\alpha'} - \frac{1}{\alpha''} & \frac{1}{\beta'} - \frac{1}{\beta''} \end{vmatrix} = 0$$

$$\text{d. h. } \frac{\frac{1}{\alpha} - \frac{1}{\alpha''}}{\frac{1}{\alpha'} - \frac{1}{\alpha''}} = \frac{\alpha''-\alpha}{\alpha}\frac{\alpha'}{\alpha''-\alpha'} = \frac{\beta''-\beta}{\beta}\frac{\beta'}{\beta''-\beta'}$$

d. h. die obigen Werte x, y befriedigen auch die Gleichung $\frac{x}{\alpha''} + \frac{y}{\beta''} = 1$. Bei der obigen Umformung ist $\frac{1}{\alpha'} - \frac{1}{\alpha''} \neq 0$, $\frac{1}{\beta'} - \frac{1}{\beta''} \neq 0$ angenommen, was voraussetzt, dass P nicht auf AB oder AC liegt. Ist dies dennoch der Fall, so ist Richtigkeit der Behauptung selbstverständlich. Die Werte x, y gehören also dem Punkt P allein zu und heissen seine Coordinaten. Jeder Punkt P ist durch seine Coordinaten auch völlig bestimmt; denn aus den beiden Gleichungen folgt, wenn α angenommen wird β, und aus α' folgt β' und somit kann es nur einen Punkt P geben, dem die Coordinaten x, y angehören. Die Coordinaten eines Punktes P sind nur in den beiden oben erwähnten und bisher ausgeschlossenen Fällen nicht bestimmt. Diese beiden Fälle kommen aber auf das nämliche Hinaus. Ist nämlich $\begin{vmatrix} \frac{1}{\alpha} & \frac{1}{\beta} \\ \frac{1}{\alpha'} & \frac{1}{\beta'} \end{vmatrix} = 0$, d. h. $\frac{\alpha}{\alpha'} = \frac{\beta}{\beta'}$, so
51 haben | die Punkte 0, α, α', ∞ dasselbe Doppelverhältniss wie die Punkte 0, β, β', ∞, d. h. P liegt auf BC[93] und umgekehrt, wenn P auf BC liegt, so ist jene Determinante $= 0$. Dem Punkte A erteilen wir die Coordinaten $x = 0$, $y = 0$.

Um dem durch den Ausnahmefall P auf BC bedingten Uebelstande abzuhelfen, führen wir homogene Coordinaten ein, d. h. statt x, y 3 Grössen x_1, x_2, x_3, welche sämmtlich *endlich* sein müssen, und den Gleichungen $\frac{x_1}{\alpha} + \frac{x_2}{\beta} - x_3 = 0$, $\frac{x_1}{\alpha'} + \frac{x_2}{\beta'} - x_3 = 0$ genügen. Ist ϱ eine endliche und von 0 verschiedene Zahl, so stellen ϱx_1, ϱx_2, ϱx_3 denselben Punkt dar. $x_1 = 0$, $x_2 = 0$, $x_3 = 0$ ist stets ausgeschlossen. Ist $\begin{vmatrix} \frac{1}{\alpha} & \frac{1}{\beta} \\ \frac{1}{\alpha'} & \frac{1}{\beta'} \end{vmatrix} \neq 0$ so wird $x = \frac{x_1}{x_3}$, $y = \frac{x_2}{x_3}$. Aber auch falls $\begin{vmatrix} \frac{1}{\alpha} & \frac{1}{\beta} \\ \frac{1}{\alpha'} & \frac{1}{\beta'} \end{vmatrix} = 0$, besitzen unsere beiden homogenen Gleichungen Lösungen, nämlich

[93]For the proof, see above, p. 47.

$$x_1 = \frac{1}{\beta} \qquad x_2 = -\frac{1}{\alpha} \qquad x_3 = 0$$
$$x_1 = \frac{1}{\beta'} \qquad x_2 = -\frac{1}{\alpha'} \qquad x_3 = 0$$

von denen wenigstens eine, etwa die erste, nicht $x_1 = 0$, $x_2 = 0$, $x_3 = 0$ liefert und von denen dann die zweite aus der ersten durch Multiplikation mit $\varrho = \frac{\beta}{\beta'} = \frac{\alpha}{\alpha'}$ hervorgeht. In der Tat ist nunmehr durch die homogenen Coordinaten x_1, x_2, x_3 | auch in dem Ausnahmefall der Punkt P auf BC fest- 52
gelegt. Denn sei etwa $x_2 \neq 0$ also α endlich, so construire auf AB den Punkt $\alpha = -\frac{1}{x_2}$, auf AC den Punkt $\beta = \frac{1}{x_1}$; die Verbindungslinie beider Punkte trifft BC in P und wir haben nur noch zu zeigen, dass die 3 Coordinaten ϱx_1, ϱx_2, 0 auf denselben Punkt führen. Zu dem Zwecke construire auf AB den Punkt $\alpha' = -\frac{1}{\varrho x_2} = \frac{1}{\varrho}\alpha$, auf AC den Punkt $\beta' = \frac{1}{\varrho x_1} = \frac{1}{\varrho}\beta$, wo $\frac{1}{\varrho}$ endlich ist. Dann geht wegen der Gleichheit der Doppelverhältnisse in der Tath $\alpha'\beta'$ ebenfalls durch P.

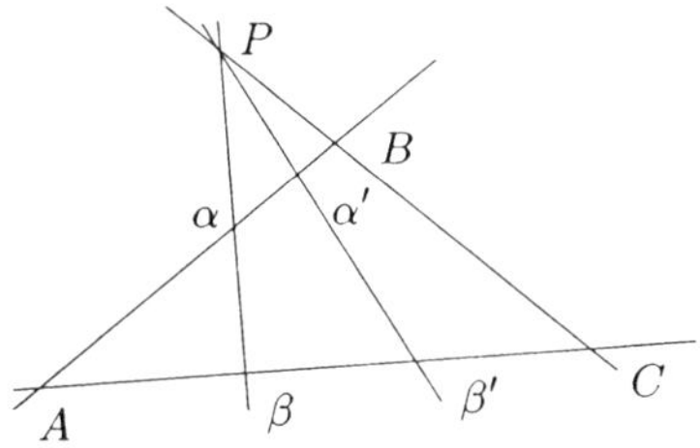

Jedem Punkte P der Ebene ausnahmslos entspricht also ein Wertesystem (x_1, x_2, x_3), wo alle x_1, x_2, x_3 endlich und nicht sämmtlich $= 0$ sind, und allen Systemen $(\varrho x_1, \varrho x_2, \varrho x_3)$ wo ϱ endlich ist, entspricht auch nur jener eine Punkt P der Ebene. Es ist aber keineswegs gesagt, dass | jedem beliebig angenommenen 53
Wertesystem (x_1, x_2, x_3) auch nothwendig überhaupt ein Punkt P der Ebene entsprechen muss. Den Ecken A, B, C entsprechen (0 0 1) (1 0 0) (0 1 0).

Um die Bedingung aufzustellen, dass 3 Punkte P, P', P'' in einer Geraden liegen, unterscheiden wir 2 Fälle: 1.) Sämmtliche 3 Punkte P, P', P'' mögen auf einer Seite von AB liegen, dann können wir leicht auf AC 2 Punkte mit den Zahlen β und b angeben, so dass die Geraden βP, $\beta P'$, $\beta P''$ und bP, bP', bP'' AB schneiden;

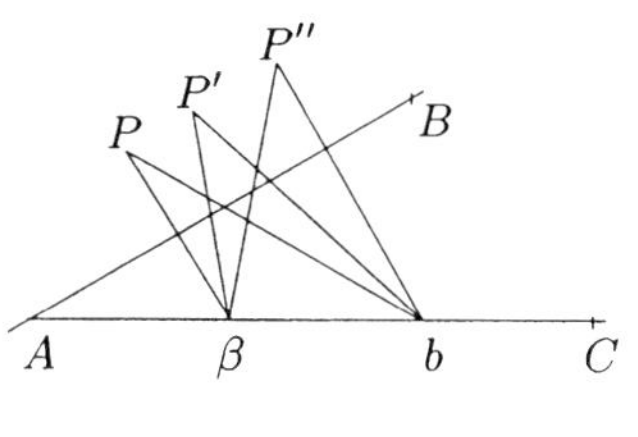

oder 2.) P liegt auf einer, P', P'' liegen auf der anderen Seite von AB. Dann verbinde P mit einem Punkte von AB, welcher auf der anderen Seite von AC liegt. Die Verbindungslinie schneide AC in β. β mit P' und P'' verbunden schneiden AB. Ebenso construire man auf AC einen anderen Punkt b.

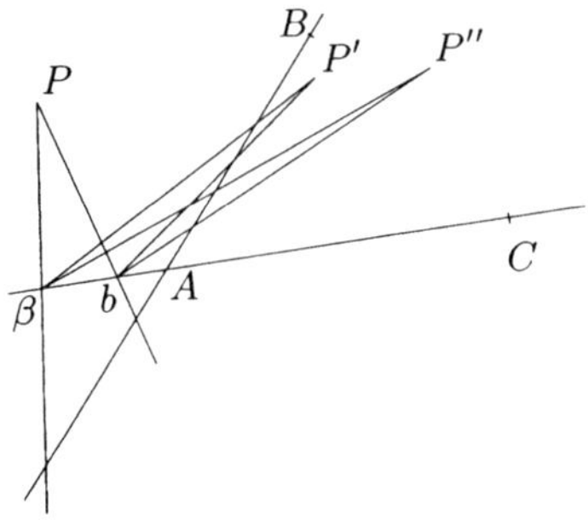

54 In beiden Fällen schneiden die Verbindungslinien von β, b mit P, P', P'' die Gerade AB und zwar in den Punkten α, α', α''; a, a', a''. Bezeichnen wir die Coordinaten von P, P', P'' mit $(x_1\, x_2\, x_3)$, $(x_1'\, x_2'\, x_3')$, $(x_1''\, x_2''\, x_3'')$, so ist

$$\begin{array}{ll}
\frac{x_1}{\alpha} + \frac{x_2}{\beta} - x_3 = 0 & \quad \frac{x_1}{a} + \frac{x_2}{b} - x_3 = 0 \\
\frac{x_1'}{\alpha'} + \frac{x_2'}{\beta} - x_3' = 0 & \quad \frac{x_1'}{a'} + \frac{x_2'}{b} - x_3' = 0 \\
\frac{x_1''}{\alpha''} + \frac{x_2''}{\beta} - x_3'' = 0 & \quad \frac{x_1''}{a''} + \frac{x_2''}{b} - x_3'' = 0
\end{array}$$

d. h.

$$x_1 : x_2 : x_3 = \alpha a(\beta - b) : \beta b(\alpha - a) : \alpha b - a\beta \quad \text{etc.}$$

folglich wird

$$\begin{vmatrix} x_1 & x_2 & x_3 \\ x_1' & x_2' & x_3' \\ x_1'' & x_2'' & x_3'' \end{vmatrix} = \begin{vmatrix} \alpha a(\beta - b) & \beta b(\alpha - a) & \alpha b - a\beta \\ \alpha' a'(\beta - b) & \beta b(\alpha' - a') & \alpha' b - a'\beta \\ \alpha'' a''(\beta - b) & \beta b(\alpha'' - a'') & \alpha'' b - a''\beta \end{vmatrix}$$

$$= (\beta - b)^2 \beta b \begin{vmatrix} \alpha & a & \alpha a \\ \alpha' & a' & \alpha' a' \\ \alpha'' & a'' & \alpha'' a'' \end{vmatrix} = \frac{(\beta - b)^2 \beta b}{\alpha\alpha'\alpha'' a a' a''} \begin{vmatrix} \frac{1}{a} & \frac{1}{\alpha} & 1 \\ \frac{1}{a'} & \frac{1}{\alpha'} & 1 \\ \frac{1}{a''} & \frac{1}{\alpha''} & 1 \end{vmatrix}$$

Nun sind α, α', α'', a, a', $a'' \neq 0$, ferner $\beta - b \neq 0$, $\alpha' - \alpha''$, $a' - a'' \neq 0$. Folglich $\begin{vmatrix} x_1 & x_2 & x_3 \\ x_1' & x_2' & x_3' \\ x_1'' & x_2'' & x_3'' \end{vmatrix}$ dann und nur dann $= 0$, wenn das Doppelverhältniss der 4 Punkte 0, α, α', α'' mit dem der 4 Punkte 0, a, a', a'' übereinstimmt, d. h. wenn P, P', P'' in einer Geraden liegen. Der Fall $\beta = b$ tritt nur ein,
55 wenn einer der 3 Punkte P, P', P'' auf AC liegt, | und $\alpha' = \alpha''$, $a' = a''$ tritt nur ein, wenn einer der 3 Punkte P, P', P'' auf AB liegt. Diese besonderen Fälle erledigen sich leicht direkt. Hieraus folgt, wenn ein Punkt auf der durch $(x_1'\, x_2'\, x_3')$, $(x_1''\, x_2''\, x_3'')$ bestimmten Geraden liegt, so müssen seine Coordinaten x_1, x_2, x_3 der Gleichung

$$u_1x_1 + u_2x_2 + u_3x_3 = 0 \text{ genügen, wo } u_1 = \begin{vmatrix} x_2' & x_3' \\ x_2'' & x_3'' \end{vmatrix}$$

$$u_2 = \begin{vmatrix} x_1' & x_3' \\ x_1'' & x_3'' \end{vmatrix}$$

$$u_3 = \begin{vmatrix} x_1' & x_2' \\ x_1'' & x_2'' \end{vmatrix}$$

ist. Dieses heisst die Gleichung der Geraden; u_1, u_2, u_3 heissen die homogenen Coordinaten der Geraden. Denn es ist klar, dass dieselben bis auf einen Faktor bestimmt sind, wenn man 2 andere Punkte der Geraden wählt. $u_1 = 0$, $u_2 = 0$, $u_3 = 0$ gleichzeitig ist ausgeschlossen ↑ebenso dass eine Coordinate ∞↓. ϱu_1, ϱu_2, ϱu_3 bestimmen die nämliche Gerade. Zu ϱu_1, ϱu_2, ϱu_3 gehört nur eine Gerade. Aber nicht zu jedem Wertsystem u_1, u_2, u_3 braucht eine Gerade zu gehören. Obige Gleichung ist die Bedingung der vereinigten Lage von Punkt $(x_1\,x_2\,x_3)$ und Gerade $(u_1\,u_2\,u_3)$.
Aus der oben benutzen Figur:

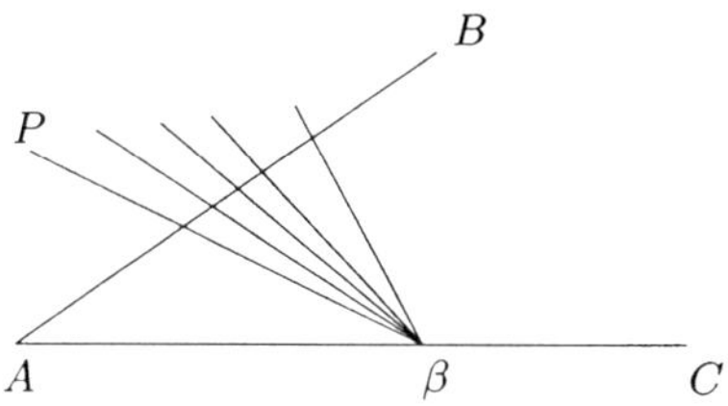

und den bisher gewonnenen Resultaten entnehmen wir folgendes: Wenn 56
$x_1\,x_2\,x_3$, $x_1'\,x_2'\,x_3'$ 2 Punkte sind, so sind $x_1 + \lambda x_1'$, $x_2 + \lambda x_2'$, $x_3 + \lambda x_3'$ lauter und alle Punkte der Verbindungslinie, und zwar bestimmt der Parameter λ eine (projektive) Zahlenzuordnung auf der Linie, wie wir sie oben studirt haben. Denn das λ drückt sich linear gebrochen durch das α aus, und daher giebt $(\lambda_1, \lambda_2, \lambda_3, \lambda_4) = -1$ die Bedingung für die harmonische Lage von den 4 Punkten λ_1, λ_2, λ_3, λ_4. Eben daraus, nämlich aus der linearen Ausdrückbarkeit des λ durch α (die auf AB gelegenen Zahlen) folgt, dass λ_2 zwischen λ_1 und λ_3 liegt, wenn $\lambda_1 < \lambda_2 < \lambda_3$ ist.

Liegen die beiden Punkte $x_1'\,x_2'\,x_3'$ und $x_1''\,x_2''\,x_3''$ auf 2 verschiedenen Seiten einer Geraden $u_1x_1 + u_2x_2 + u_3x_3 = 0$, so wird es ein $\lambda = \lambda_0$ geben, so dass $x_1' + \lambda_0 x_1''$, $x_2' + \lambda_0 x_2''$, $x_3' + \lambda_0 x_3''$ auf der Geraden $u_1x_1 + u_2x_2 + u_3x_3 = 0$ liegt. Daher wird $u_1(x_1' + \lambda x_1'') + u_2(x_2' + \lambda x_2'') + u_3(x_3' + \lambda x_3'')$ für alle $\lambda < \lambda_0$ etwa positiv und dann für alle $\lambda > \lambda_0$ negativ. D. h. $u_1x_1 + u_2x_2 + u_3x_3$ hat für alle auf der einen Seite der Geraden gelegenen Punkte das eine, etwa das positive, und für die auf der anderen Seite gelegenen Punkte dann das negative Vorzeichen. Die Coordinaten aller im Inneren des Dreieckes ABC gelegenen Punkte haben sämmtlich das positive oder sämmtlich das negative Vorzeichen. | Die Seiten AB, BC, CA haben die Gleichungen $x_1 = 0$, $x_2 = 0$, 57

$x_3 = 0$ bezüglich.[94]

Ebenso wie oben die Punktreihe hat man nun auch das Strahlenbüschel $(u_1x_1 + u_2x_2 + u_3x_3) + \lambda(u'_1x_1 + u'_2x_2 + u'_3x_3) = 0$ zu behandeln, wo man wiederum zeigt, dass der Parameter λ, welcher einen einzelnen Strahl eines Büschel bestimmt, projektiv ist etc.

Uebrigens bleibt noch übrig zu zeigen, dass der Punkt A mit Recht die Coordinaten 0 0 1 verdient, d. h. dass auch für diesen der Satz von der vereinigten Lage von Punkt und ↑[95]Gerade gilt. Zu dem Zwecke verbinde P, P' mit b, welche Geraden in a, a' schneiden; $P'a$ trifft in β, $P\beta$ in α. Aus $\frac{x_1}{\alpha} + \frac{x_2}{\beta} = x_3$, $\frac{x_1}{a} + \frac{x_2}{b} = x_3$ folgt

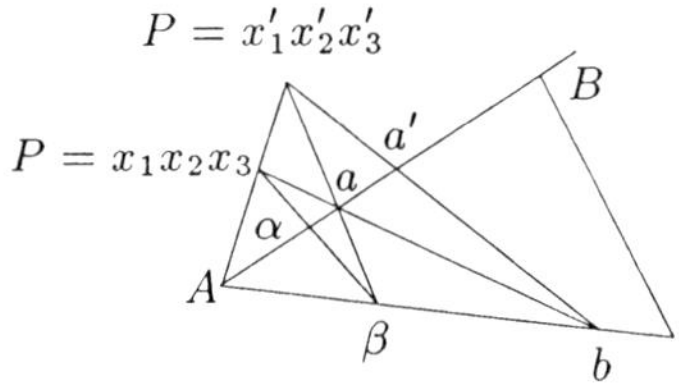

$$x_1 : x_2 : x_3 = \alpha a(\beta - b) : \beta b(\alpha - a) : \alpha b - a\beta$$

und ebenso

$$x'_1 : x'_2 : x'_3 = \alpha a'(\beta - b) : \beta b(\alpha - a') : \alpha b - a'\beta$$

und folglich $x_1x'_2 - x_2x'_1 = \alpha\beta b(\beta - b)(\alpha a - 2\alpha a' + aa')$ und dies ist dann und nur dann $= 0$, wenn A, α, a, a' harmonisch liegen, d. h. wenn $PP'A$ in gerader Linie liegen. Es ist aber $x_1x'_2 - x_2x'_1$ in der That $= \begin{vmatrix} x'_1 & x'_2 & x'_3 \\ x_1 & x_2 & x_3 \\ 0 & 0 & 1 \end{vmatrix}$ was zu beweisen war.↓

↑Durch Adjunktion der Gleichungen als uneigentlicher oder idealer Elemente hat man nun die ausnahmslose Gültigkeit der Sätze: dass irgend 2 Gerade der Ebene je einen und nur einen Punkt gemein haben, siehe unten.↓

Auch sieht man nun leicht, wie eine Coordinatentransformation auszuführen ist. Sind $a_1x_1+a_2x_2+a_3x_3 = 0$, $b_1x_1+b_2x_2+b_3x_3 = 0$, $c_1x_1+c_2x_2+c_3x_3 = 0$ die Gleichungen der Seiten des neuen Dreieckes, so braucht man bloss $y_1 = a_1x_1+\ldots$, $y_2 = b_1x_1+\ldots$, $y_3 = c_1x_1+\ldots$ einzuführen. Denn in der That werden ja die Seiten des neuen Coordinatendreieckes projektiv durchlaufen und auch die Ecken haben die richtigen Anfangswerte.*)

Uebrigens hätte man auch bei dieser ganzen Untersuchung von 2 Strahlenbüscheln in der Ebene ausgehen können, in denen der einzelne Strahl durch

---

*) Man kann hiernach auch solche 3 Geraden als Coordinatenachsen zu Grunde wählen, welche sich unter einander garnicht schneiden.

---

[94]This means: bezüglich A, B, C.

[95]The passage and the figure added here, which run up to line 102.17, are written on a small piece of paper pasted in. They replace the following text, now partially obscured by the addition:

Gerade gilt. Zu dem Zwecke betrachte man die Mannigfaltigkeit ⟪$x_1 + x_2 = 0$⟫ von ⟪Geraden⟫. Aus der Gestalt der linken Seite folgt, dass durch jeden Punkt der Ebene eine bestimmte Gerade ⟪der⟫ Mannigfaltigkeit hindurchgeht, ausser durch den Punkt A mit den Coordinaten ⟪0 0 1⟫ ⟦⟪gäbe⟫ ⟨several words illegible⟩⟧ gehen ⟨three words illegible⟩

je ei|nen Parameter festgelegt ist. Dadurch wird jeder Punkt der Ebene durch 58
2 Parameter festgelegt mit Ausnahme der Punkte, welche in gerader Linie mit den Büschelcentren liegen. Bedingung für das in einer Geraden Liegen von 3 Punkten.

Diese letztere Methode ist auch in der Ebene anwendbar, wenn man 3 Ebenenbüschel zu Hülfe nimmt.

Wenn man die ursprünglichen Punkte, Geraden, Ebenen als eigentliche bezeichnet und nun alle Wertsysteme $(x_1\, x_2\, x_3\, x_4)$ $(u_1\, u_2\, u_3\, u_4)$ $\langle\!\langle(p_{\iota\kappa})\rangle\!\rangle$ auch als Punkte, Gerade, Ebene und zwar zur Unterscheidung als uneigentliche einführt (d. h. ideale Elemente), so kann man unsere Grundsätze der Geometrie sehr viel einfacher aussprechen. Es gelten dann nämlich ausnahmslos die Sätze:

Durch je 2 Punkte geht immer ⟨eine⟩ und nur eine Gerade.

" " 3 " Ebene.

2 Ebenen schneiden sich in einer Geraden.

3 Ebenen, welche nicht die nämliche Gerade gemein haben, schneiden sich stets in einem Punkt.

Die nähere Begründung und Ausführung gehört zum Theil in die projektive, zum Teil in die analytische Geometrie des Raumes. Zugleich erkennt man, dass projektive und analytische Geometrie sich garnicht wesentlich unterscheiden, vielmehr nur die Ausgangspunkte in beiden Zweigen der Geometrie verschiedene sind, während sicherlich die Methoden auf das nämliche hinauslaufen.

Die letzt angeführten Sätze sprechen das so|genannte Reciprocitätsgesetz 59
aus, d. h. in jedem geometrischen Satze für die Ebene können die Worte Gerade und Punkt vertauscht und im Raume Ebene und Punkt vertauscht werden. Aber dies gilt nur, wenn die uneigentlichen Elemente hinzu genommen werden.

Bei Beschränkung auf eigentliche Elemente gilt das Reciprocitätsgesetz nur allgemein im Ebenenbündel, d. h. in allen auf die Strahlen und Ebenen eines Büschels bezüglichen Sätze können wir die Worte „Strahl" und Ebene mit einander vertauschen.

Mit den bisherigen Axiomen,[96] den Existenz- und den Lagenaxiomen, können wir schon eine grosse Menge geometrischer Thatsachen und Erscheinungen beschreiben. Wir brauchen nur für Punkte, Geraden, Ebenen Körper zu nehmen, für hindurch gehen berühren zu nehmen, für bestimmt sein unbeweglich oder fest sein (etwa für den groben Sinn beim Anstossen mit der Hand). Die Körper denken wir uns überhaupt nur in endlicher Zahl und so beschaffen, dass bei der eben genannten Deutung die Axiome erfüllt sind [97]↑(wozu es nötig wird, wie man erkennt, die an Stelle von Punkt, Gerade, Ebene genommenen Körper bezüglich Körnern, Stäben oder gezogenen Fäden, Dräthen

---

[96]Originally, Hilbert intended to open a new section at this point, since he began writing a new title 'C. Congruenzaxi', which he then crossed out. This probably occurred before the insertion of the current section C, 'Das Stetigkeitsaxiom'. For one thing, Hilbert's mention of the 'bisherige Axiome' does not include the continuity axiom; for another, the section actually containing the congruence axioms is designated below as section D.

[97]Addition by *Hi*ˢ in the lower margin, assigned to this place by an insertion sign.

und Pappdeckeln ähnlich zu nehmen)↓ und zwar genau. Dann wissen wir,
60 dass | jedenfalls auch alle bisher aufgestellten Sätze ⟨erfüllt sind⟩ und zwar genau erfüllt sind. Findet man bei der Anwendung die Sätze nicht (nicht genau) erfüllt, so liegt dies an der falschen Anwendung, d. h. die Körper, das Bewegen, Berühren, stimmt nicht mit unserem Schema der Axiome ⟨überein⟩. Es ist dann also nöthig die Dinge: Körper, Berühren, beweglich durch andere zu ersetzen, etwa durch kleinere Körner ↑Klexe, Spitzen↓, schmalere Dräthe, dünnere Pappe, klemmender berühren, Beweglichkeit schon beim Anpusten, zu ersetzen, damit die Axiome erfüllt sind; dann wissen wir stimmen auch die Sätze (genau).[98] Ueberhaupt muss man sagen: Unsere Theorie liefert nur das Schema der Begriffe, die durch die unabänderlichen Gesetze der Logik mit einander verknüpft sind. Es bleibt dem menschlichen Verstande überlassen, wie er dieses Schema auf die Erscheinung anwendet, wie er es mit Stoff anfüllt. Dies kann in der mannigfaltigsten Art geschehen: Aber immer wenn die Axiome erfüllt sind, dann stimmen auch die Sätze. Je leichter und vielseitiger die Anwendung, desto besser*) die Theorie. Ob auf eine Reihe von Erscheinungen
60A das Schema anwendbar ist, ist | natürlich leicht zu entscheiden. Es müssen eben, wenn man Körper, beweglich, berühren geeignet bezieht, die Axiome stimmen.

Eine grosse Reihe von Erscheinungen kann mittelst der bisherigen Sätze nicht gedeutet werden.

Wir haben zwar bei der Anwendung unserer Sätze, d. h. beim Experiment, den Begriff der Beweglichkeit benutzt, aber die Thatsache der Bewegung eines Körpers ↑folgt nicht aus unseren bisherigen Axiomen↓ d. h. die Möglichkeit, die Punkte des Raumes (oder 2er Körper) so auf einander zu beziehen, dass 2 verschiedenen Punkten wieder 2 verschiedene Punkte, 3 in einer Gerade gelegene wieder 3 solche entsprechen, ferner 1 Punkt willkürlich genommen werden kann u. s. f. Daher

*) Jedes System von Einheiten und Axiomen, welches ebenso vollständig die Erscheinungen beschreibt, ist gleichberechtigt. Zeige jedoch, dass das vorliegende angegebene Axiomensystem in gewisser Hinsicht das einzig mögliche ist.[99]

[98] Hilbert's argumentation seems to be coloured by the empiricism of Pasch. The difference, however, is that Hilbert never questions the logically deductive connection between axioms and theorems, and all that is at issue for him is the correct application of the geometrical concepts and axioms to real things. Pasch, however, wants to ground the axioms and theorems of geometry empirically, i.e., ground them through idealisation based on our experience of them.

[99] This might indicate Hilbert's interest in categoricity at this time, but it is quite likely that Hilbert is referring here to the choice of primitive axioms. In a letter to Lindemann of 17 July, 1894, Hilbert writes: 'Etwas Unbefriedigendes scheint mir bei der Aufstellung der Axiome immer noch in dem Umstande zu liegen, dass man bei der Auswahl der Axiome [der Geometrie] eine gewisse Willkür walten läßt und kein rechtes Princip vorhanden ist, warum man nicht lieber gewisse einfache Folgerungen als Axiome nimmt und umgekehrt' (quoted in *Toepell 1986*, 100–101).

D. Die Congruenzaxiome

(oder auch Bewegungsaxiome zu nennen)[100] [D]

⟨The following indented paragraphs were deleted:⟩

> 1.) Wenn 2 Punkte A, A' und 2 durch je einen derselben gehenden Geraden a bez. a' gegeben sind und man nimmt auf a noch einen 2ten Punkt B, auf a' den Punkt S an, so giebt es auf a' auf der durch S bestimmten Seite von A' einen und nur einen Punkte B', so | dass AB congruent $A'B'$ ist: $AB \equiv A'B'$.*) 60B
>
> 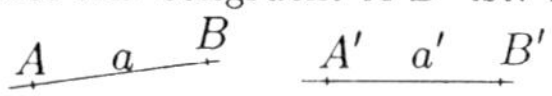
>
>
> Man kann dieses Axiom auch spalten in 2 Axiome, indem man erst $a = a'$ (dies liefert die Verschiebung) und dann $A = A'$ nimmt (dies liefert die Drehung).[101]
>
> *)Mit $AB \equiv A'B'$ ist $BA \equiv B'A'$ gleichbedeutend.

[D]Diese sind durch die beiden folgenden zu ersetzen: Wenn ein Punkt ↑P↓, eine Gerade G durch ihn (auf dieser eine Seite), eine Ebene ↑E↓ durch die Gerade (auf der Ebene eine Seite) und dasselbe noch einmal ↑P' G' E'↓ gegeben sind, so lässt sich für jede Figur, in der P, G, E vorkommen, eine und nur eine congruente Figur angeben, in der P', G', E' vorkommen, so dass PP',... sich entsprechen. Entsprechende Punkte, Gerade, Ebenen ⟨p. 61⟩ schneiden sich wie die entsprechenden. Behandele hintereinander die 3 Verschiebungen und die 3 Drehungen. Rechtwinkeliges Coordinatensystem benutzen. Diese Bewegungen müssen lineare Transformationen sein mit je einem Parameter, weil, wenn ich nach der Verschiebung neue Zahlen einführe, die Bildpunkte gleiche Zahlen erhalten müssen. ⟨Added by *Hi*ˢ in the lower margin of p. 60A and the upper margin of p. 61 and not assigned to any particular place.⟩

---

[100]The genesis of the following pages in the manuscript (up to p. 66) is relatively complicated, and would be scarcely comprehensible solely by means of individual editorial notes concerning deletions, additions and replacements. The main lines of the basic alterations are as follows.

The original version in Hilbert's hand is to be found on the unnumbered pages 60A and 60B, as well as the pages numbered 63–66. The sheet with pages 61 and 62 has been pasted in. The text they contain, originally written in Käthe Hilbert's hand, seems to have been written independently, and certainly before the sheet was pasted in, since it begins on p. 61 with the following heading and opening lines, both later crossed out:

Gedankengang eines über Euklidische
Geometrie zu lesenden Collegs.

> Zunächst Existenzaxiome A und Lagenaxiome B genau wie in meinem Collegheft: Die Grundlagen der Geometrie. Dann die folgenden Gleichheitsaxiome C, von denen die ersten die Gleichheit von Strecken, die letzteren die der Winkel betreffen.

The main portion of the alterations to the original text on pp. 60A and 60B are caused by the introduction of the new section C on continuity. This led not only to a shift from C to D in the numbering of the section 'Congruenzaxiome', but, more importantly, to the replacement of the first set of axioms on pages 60A, 60B and 63, by a new set of axioms on pp. 61, 62 and 63. However, the text written in Käthe's hand ends on p. 62 before Axiom 8 is finished. Hilbert continued Axiom 8 on a strip of paper pasted onto the lower part of page 63, which also includes Axiom 9; the strip presumably covers a previous version of Axiom 4 in the original text.

Lastly, on p. 60A at the bottom and p. 61 in the upper margin, Hilbert adds a further reflection on the congruence axioms, here reproduced as footnote D, considering a 'kinematic' approach to geometry.

[101]The rotation should come before the transport.

Mache mit Maass und Winkel⟪??⟫ die entsprechenden Experimente.
↑Satz: Es giebt 2 und nur 2 Punkte P auf a', so dass $AB \equiv AP$ ist.[102]↓

2.) Wenn $AB \equiv A'B'$ und $AB \equiv A''B''$, so $A'B' \equiv A''B''$ und daraus folgend $AB \equiv AB$.

3.) Wenn auf a ↑der Punkt↓ B zwischen A und C und auf a' der Punkt B' zwischen A' und C' liegt und $AB \equiv A'B'$ ferner $BC \equiv B'C'$, so ist auch $AC \equiv A'C'$.

Satz: Unter der nämlichen Voraussetzung über die Reihenfolge folgt aus $AC \equiv A'C'$ und $AB \equiv A'B'$ auch $BC \equiv B'C'$. Denn wäre BC nicht $\equiv B'C'$ etwa $\equiv B'E'$, so wäre nach dem Axiome 3 $AC \equiv A'E'$, was dem Axiome 1, dass es nur einen Punkt von der betreffenden Beschaffenheit geben soll, widersprechen würde.
↑Satz: AB ist niemals $A'A'$ congruent. Denn sonst wäre $AC \equiv A'C'$ und zugleich $BC \equiv A'C'$, was der Eindeutigkeit (Axiom 1) widerspricht.↓

61 [103]1.) 2 Punkte A, B bestimmen eine Strecke. Wenn A, B auf einer Geraden a gegeben sind, und A', S auf derselben oder auf einer andern a' so kann man auf a' einen und nur einen Punkt B' bestimmen, welcher mit S auf derselben Seite von A' liegt, und so dass $AB = A'B'$ ist. ↑Wenn $AB = A'B'$, so ist auch $A'B' = AB$. AB ist $= BA$.↓

2.) Wenn 2 Strecken einer dritten gleich sind, so sind sie untereinander gleich.

3.) Wenn $AB = A'B'$ ferner $AC = A'C'$ ist und A auf der Geraden $a = ABC$ zwischen B und C liegt, ferner A' auf der Geraden $a' = A'B'C'$ zwischen B'
62 und C' liegt, so ist auch $BC = B'C'$ und heisst | die Summe der beiden Strecken.

[104]↑4.) Wenn $AB = A'B'$, ferner $AC = A'C'$ und B, C auf der Geraden $a = ABC$ auf der nämlichen Seite von A liegen, und B', C' auf der Geraden $a' = A'B'C'$ auf der nämlichen Seite von A', so ist auch $BC = B'C'$ und heisst die Differenz der Strecken AB und AC.

Hier beweise die auf der Seite nebenan stehenden Sätze 1. und 2.↓[105]
Satz 1: Wenn $AB \equiv A'B'$, $AC \equiv A'C'$, $BC \equiv B'C'$ ↑und B zwischen A und C liegt↓, so liegt B' zwischen A' und C'.

Denn im anderen Fall construire $B'E' \equiv BC$, dann muss

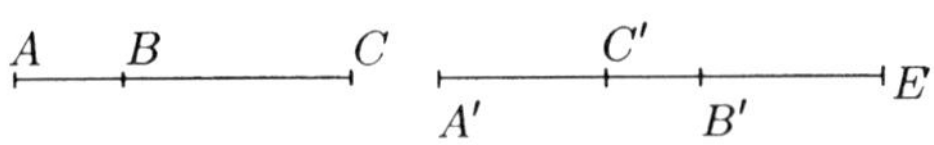

$A'E' \equiv AC$ nach 3 und folglich da es nur einen solchen Punkt geben darf $E' = C'$.

---

[102]This should be: $AB \equiv A'P$.

[103]The following text, up to line 107.25, is to be found largely on the inserted pages 61–62. The exception concerns Theorems 1 and 2, which were added on p. 63, and directed to their place on p. 62 by an arrowed line.

[104]Insertion by *Hi*[s] on a strip of paper pasted to the page and rendering illegible the text pasted over, probably a previous version of Axiom 4 in the present sequence of axioms.

[105]The two following theorems are the ones just mentioned by Hilbert. They appear on p. 63, and followed the deleted passage on pp. 60A/B. Hilbert directed them with an arrowed line to a place between the new Axioms 4 and 5.

↑Definition.↓ Wenn man 2 Punktreihen A, B, C,... auf a und A', B', C',... auf a' hat und alle homologen Punktepaare einander congruent sind

$$AB \equiv A'B', \quad AC \equiv A'C', \quad BC \equiv B'C', \ldots$$

so heissen die beiden Punktereihen congruent

$$[ABC\ldots] \equiv [A'B'C'\ldots].$$

Satz 2: Die Reihenfolgen der homologen Punkte 2er congruenter Punktreihen stimmen überein.

5.) 3 Punkte BAC bestimmen einen Winkel; wenn B' auf AB und zwar auf der nämlichen Seite von A liegt wie B, so ist $\sphericalangle B'AC = \sphericalangle BAC$. ↑Wenn ein Winkel BAC gegeben ist und eine Gerade $A'B'$ und ein Punkt S, welcher nicht auf dieser Geraden liegt, so kann man stets einen Punkt C' bestimmen, welcher mit S auf der nämlichen Seite der Geraden $A'B'$ liegt und so, dass $\sphericalangle B'A'C' = \sphericalangle BAC$ ist. ↓[106]

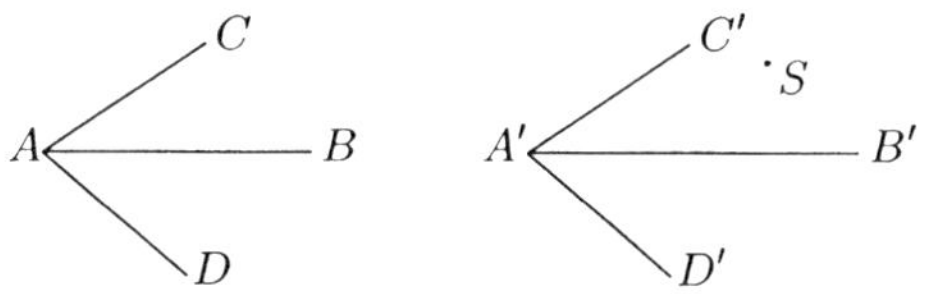

[107]↑Es ist zugleich auch $\sphericalangle BAC = \sphericalangle B'A'C' = \sphericalangle CAB$. Wenn ein Punkt C'' dieselben Bedingungen erfüllt, so liegen $AC'C''$ in einer Geraden.↓[108]

6.) Wenn $AB = A'B'$, $AC = A'C'$, ferner $BC = B'C'$, so ist $\sphericalangle BAC = B'A'C'$ ↑und umgekehrt: Wenn $AB = A'B'$, $AC = A'C'$, $\sphericalangle BAC = B'A'C'$, so ist $BC = B'C'$.↓

7.) Wenn 2 Winkel einem dritten gleich sind, sind sie auch untereinander gleich.

8.) Wenn $\sphericalangle BAC = \sphericalangle B'A'C'$ ist und $\sphericalangle DAB = \sphericalangle D'A'B'$ und einerseits D und C | ↑auf verschiedenen Seiten von AB liegen; andererseits D', C' auf ver- 63
schiedenen Seiten von $A'B'$, so ist $\sphericalangle CAD = \sphericalangle C'A'D'$ und heisst die Summe beider Winkel.

9.) Wenn Alles wie früher, nur einerseits D, C auf derselben Seite von AB und D', C' auf derselben Seite von $A'D'$ liegen, so ist $\sphericalangle CAD = \sphericalangle C'A'D'$ und heisst die Differenz.↓[109]

2 ebene Figuren $ABC\ldots$ und $A'B'C'\ldots$ (d. h. 2 Systeme in je einer 64
Ebene gelegener Punkte) heissen congruent, wenn jedes Paar von Punkten dem homologen Paare $\equiv$ ist: $AB \equiv A'B'$, $AC \equiv A'C'$,... Wir schreiben $[ABC\ldots] \equiv [A'B'C'\ldots]$.

Satz:[110] *Wenn* $[ABC] \equiv [A'B'C']$ *und* ABC *in einer* Geraden liegen, so liegen auch $A'B'C'$ in einer Geraden.

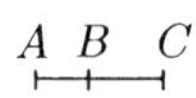

---

[106] Added by HiK^s at the bottom of the page and assigned to Axiom 5 by an insertion sign.

[107] The figure to the left, drawn at the bottom of p. 62, illustrates both Axioms 5 and 8.

[108] Added later by Hi^s to the end of the preceding addition by HiK^s.

[109] Addition by Hi^s on a strip of paper pasted to the lower half of page 63, rendering illegible the text pasted over. The latter probably embodied a version of Axiom 4 of the first set of axioms, which was completely replaced.

[110] Substituted for: *5.*

$\uparrow$Denn $\sphericalangle BAC = \sphericalangle B'A'C'$. Bestimmt man nun B'' zwischen A', C', so dass $AB = A'B''$, so ist auch $\sphericalangle BAC = \sphericalangle B''A'C$ folglich nach Ende von Axiom 5.$\downarrow$ $\uparrow$Satz. Wenn also in einer ebenen Figur 3 Punkte in einer Geraden liegen, so auch die homologen Punkte.$\downarrow$

⟨The following indented paragraphs, beginning with '6. Wenn', and 'Zunächst', were deleted:⟩

> *6. Wenn C, D einerseits auf derselben Seite von AB* liegen und C', D' andererseits auf derselben Seite von $A'B'$ und $[ABC] \equiv [A'B'C']$, $[ABD] \equiv [A'B'D']$, so ist auch $CD \equiv C'D'$.
> Und folglich überhaupt $[ABCD] \equiv [A'B'C'D']$.

Satz. Wenn $[AB\ldots] \equiv [A'B'\ldots]$ und P ist in der ersteren Ebene gegeben, so kann man in der 2ten Ebene P' so bestimmen dass $[AB\ldots P] \equiv [A'B'\ldots P']$. $\uparrow$Diese Bestimmung von P' ist eindeutig, wenn wenigstens 4[111] der Punkte A, B, $C\uparrow$, $D\downarrow$ nicht in einer Ebene[112] liegen.$\downarrow$

> Zunächst: Zeige dass wenn ABC in einer Geraden liegen und P ausserhalb derselben und $[ABC] \equiv [A'B'C']$, $AP \equiv A'P'$, $BP \equiv B'P$ [113] so auch $CP \equiv C'P'$ ⟦(d. h. Axiom 6 gilt auch wenn C auf AB liegt) In der That wäre⟧[114]

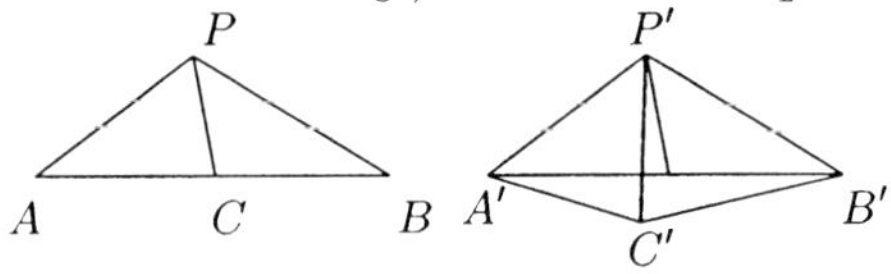

65 PC nicht $\equiv PC'$,[115] so bestimme man E', so dass $A'E' \equiv AC$ und $P'E' \equiv PC$ ist. Dann ist nach 6. $CB \equiv E'B'$ und folglich liegt nach 5. E' auf $A'B'$ und folglich nach 1. $E' = C'$.

Wir nehmen [116]aus den Punkten A, B, $C, \ldots$ derart 2 Punkte heraus, dass alle anderen Punkte mit P zusammen auf der nämlichen Seite von der Verbindungslinie AB (oder auf dieser Verbindungslinie) gelegen sind. (Man sieht leicht ein, dass dies immer möglich ist.) Es seien A, B solche 2 Punkte. Nun construire man P' so, dass $[ABCP] \equiv [A'B'C'P']$ ist, wodurch P' eindeutig bestimmt ist. Wegen $[ABE] \equiv [A'B'E']$ ist folglich $PE \equiv P'E'$ etc.

nach Axiom 6:

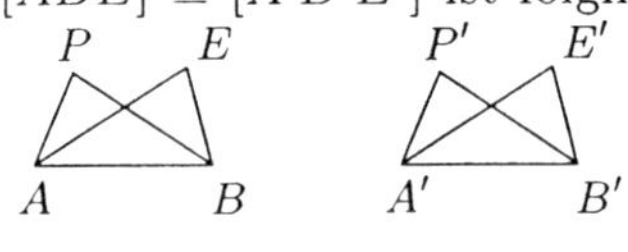

---

[111] Substituted for: 3

[112] It should be 'Gerade' not 'Ebene'.

[113] This should be: $BP \equiv B'P'$

[114] Deleted by *Hi*ˢ before the deletion of the whole passage

[115] This should be: $\equiv P'C'$

[116] Before arriving at the final version of this proof, Hilbert successively deleted two false starts. These are: (*a*) 'Nun bestimme man, um unseren Satz zu beweisen, P' so dass $[ABCP] \equiv [A'B'C'P']$ ist. Wegen $[ABCE] \equiv [A'B'C'E'$'; (*b*) 'Nun nehmen wir an, dass ABC nicht in einer Geraden liegen und bestimmen dann P' so dass $[ABCP] \equiv [A'B'C'P']$ ist, wodurch P' eindeutig bestimmt ist'. Following (*a*) and (*b*) comes 'Nun nehmen wir', which Hilbert has changed to 'Wir nehmen'.

Satz: Wenn in einer Ebene $[ABC\ldots] \equiv [A'B'C'\ldots]$ ist, und man zieht die homologen Verbindungslinien etwa AB und $A'B'$, und C, D [117]und $C'D'$ und AB, CD schneiden sich in F, so schneiden sich auch $A'B'$, $C'D'$ in F', so dass $[AB\ldots F] \equiv [A'B'\ldots F']$ ist. Denn man construire zu dem entsprechen|den 66
Punkt F', so kann F'' sowohl auf $A'B'$, als auf $C'D'$ liegen q. e. d.; d. h.[117]: Homologe Verbindungsgeraden schneiden sich in congruenten Punkten.

[Satz: $AB \equiv BA$.[G] Denn wäre AB etwa $\equiv BC$, so nehme man zwischen A und C einen Punkt E an.

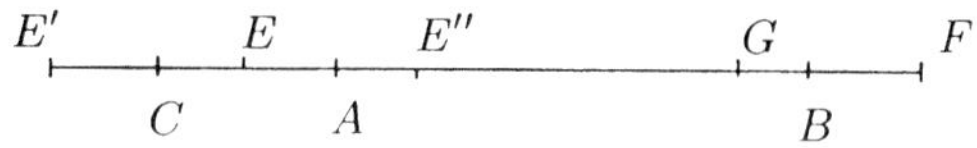

Wegen $AB \equiv BC$ und E ausserhalb AB, so ist $EAB \equiv FBC$, wo F auf der rechten Seite von B liegt. Nun bestimme E', so dass $FC \equiv BE'$ ist, so liegt E' auf der linken Seite von C.

Andererseits wegen $CB \equiv BA$ und E innerhalb CB, so ist $CEB \equiv BGA$, wo G zwischen A und B liegt. Nun bestimme E'', so dass $GA \equiv BE''$ ist, dann liegt E'' zwischen A und B. Nun ist $EB \equiv BE'$ und $EB \equiv BE''$, folglich $BE' \equiv BE''$, was nicht möglich ist wegen Axiom 1. [118]

↑Satz. Wenn auf ein und derselben Geraden $AB \equiv A'B'$, so ist auch $AA' \equiv BB'$.[119] Denn

$$\begin{array}{rl} & AB \equiv B'A' \\ & \underline{BA' \equiv A'B} \\ \text{folglich} & AA' \equiv B'B \equiv BB'. \end{array}$$

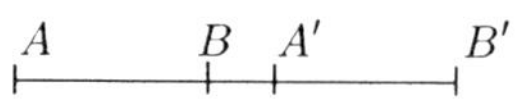

↓]

Aus diesen Sätzen kann ferner der folgende *Hauptsatz* abgeleitet werden:[120] 67

Sind 2 Reihen von je 4 Punkten, welche auf je einer Geraden liegen, einander congruent, so sind ihre Doppelverhältnisse gleich, d. h. aus $[ABCD] \equiv [A'B'C'D']$ folgt $(ABCD) \equiv (A'B'C'D')$.

[121] *Umgekehrt: Wenn* $ABC \equiv A'B'C'$ *und die Doppelverhältnisse sind gleich, d. h.* $(ABCD) = (A'B'C'D')$, *so ist* $ABCD \equiv A'B'C'D'$.[121]

Man verlege nach A, B, C und A', B', C' die Zahlen 0, 1, ∞ und dann sei die D entsprechende Zahl $= \delta$, die D' entsprechende Zahl $= \delta'$.

Erster Fall $\delta =$ rationaler Zahl. Dann gelangt man in einer durch $ABCD$ gelegten Ebene nach einer endlichen Zahl von harmonischen Construktio-

[G]Dieser Satz lässt sich besonders gut durch das Experiment prüfen, ist jedoch oben der Einfachheit wegen unter die Axiome genommen. ⟨Insertion between the proposition and its proof, assigned to this place in *Hi*[s] by an arrow.⟩

[117–117]Substituted for: liegen auf derselben Seite von AB, so liegen auch C', D' auf derselben Seite. Denn im anderen Falle ziehe man $C'D'$, welche Gerade die Gerade $A'B'$ in F' treffen möge. Sucht man den homologen Punkt F, so muss dieser sowohl auf AB als auf CD zwischen C und D liegen, und daher ⟨p. 66⟩ liegen C und D auf verschiedenen Seiten von AB etc. Als Folge dieses Satzes ergiebt sich: Jede Gerade hat zu A, B, C.. genau dieselbe Lage, wie die homologe Gerade zu A', B', C' ... und ferner

[118]Deleted by *Hi*[s]: Aus diesen ⟪6 Congruenzaxiomen⟫ kann ferner ⟨several words erased⟩

[119]This is only correct when AB and $A'B'$ have the same orientation.

[120]Deleted by *Hi*[s]: Hat man auf 2 Geraden je 4 Punkte und sind diese Reihen

[121–121]Insertion added in the lower margin and directed to this place by an insertion sign.

nen $EFGH \dots KL$ zum Punkte D. Man suche in einer durch $A'B'C'D'$ gelegten Ebene die congruente Figur $E'F'G'H'$ auf. Dann trifft $K'L'$ den Punkt D', d. h.

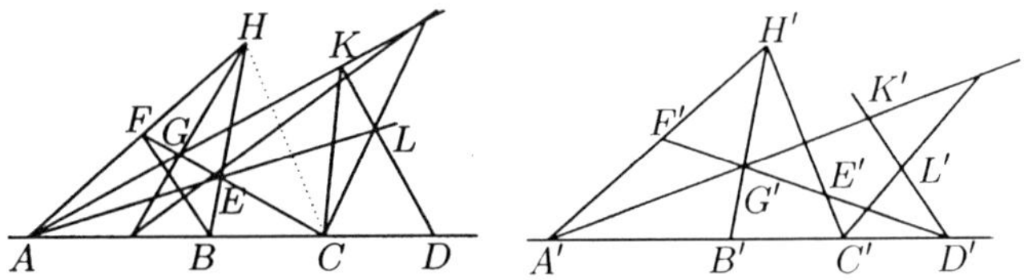

$\delta' = \delta$, d. h. $(ABCD) = (A'B'C'D')$. ↑Daraus folgt die Umkehrung leicht, weil sowohl die Gleichheit der Doppelverhältnisse als das Bestehen der Congruenz $ABCD \equiv A'B'C'D$ den Punkt D' eindeutig bestimmt.↓

Zweiter Fall. δ sei irrational und $\delta' \neq \delta$ und r sei eine zwischen δ und δ' gelegene rationale Zahl. Dann suche man in beiden Punktreihen den Punkt
68 $r = E$ und $r = E'$ auf. Da für den | Fall eines rationalen Doppelverhältnisses die Umkehrung des Satzes richtig ist, so folgt $ABCE \equiv A'B'C'E'$. Ist nun etwa $0 < \delta < r < \delta'$, so folgt D zwischen A und E und E' zwischen A' und D'

und folglich kann nicht $ADE = A'D'E'$ sein.

Der bewiesene Satz lehrt, dass congruente Punktreihen auch projektiv sind. Also wenn man zwei Punktetripel $[ABC] \equiv [A'B'C']$ kennt, so kann man zu jedem Punkt D den congruenten D' durch blosses Ziehen von geraden Linien finden. [122]

Aber auch über die analytische Darstellung der Congruenz erhalten wir Aufschluss. Denken wir nur in der obigen Weise die Punkte der Geraden durch Zahlen dargestellt und seien $[\kappa\lambda\mu\xi] \equiv [\kappa'\lambda'\mu'\xi']$, so $(\kappa\lambda\mu\xi) = (\kappa'\lambda'\mu'\xi')$ d. h. $\xi' = R(\xi)$, wo R eine linear gebrochene Funktion von ξ ist.

Wir wollen jetzt zeigen, dass wenn eine zweite Congruenz durch $\xi' = Q(\xi)$ dargestellt wird, $RQ = QR$ ist. Zum Beweise sei die erstere Congruenz durch den Punkt A und eine Seite, und B und eine Seite von B gegeben; die zweite
69 | Congruenz durch A und C (mit einer Seite, die durch einen Punkt etwa gegeben ist). Nach der 2ten Congruenz bestimme ich den Punkt D:

$$AB \equiv CD$$

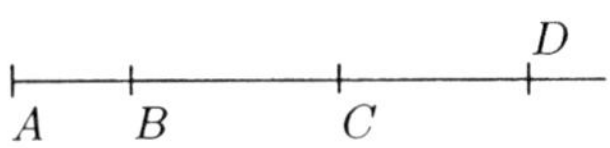

Nun sei ξ beliebig angenommen, so

$$AB \equiv \xi R(\xi) \qquad \text{folglich} \qquad CD \equiv \xi R(\xi)$$

$$AC \equiv R(\xi)\ \boxed{Q(R(\xi))} \qquad\qquad \underline{AC \equiv R(\xi)Q(R)}$$

$$\text{folglich} \qquad AD \equiv \xi Q(R)^{\langle 123\rangle}$$

↑nach Satz 1.) vgl. unten↓

---

[122] Deleted by Hi^s: Wenn $ABCD \equiv A'B'C'D'$, so gehe $B\langle\!\langle B\rangle\!\rangle'$

[123] Initially added, but subsequently deleted: wegen $CD \equiv DC$
$AC \equiv CA$

Andererseits

$$\begin{array}{lll} AC \equiv \xi Q(\xi) & & AC \equiv \xi Q \\ AB \equiv \boxed{R(Q)} & \text{folglich} & \underline{CD \equiv QR(Q)} \\ & & AD \equiv \xi R(Q) \end{array}$$

↑wegen $AC \equiv CA$
$CD \equiv DC$.
↑nach Satz 2.)
vgl. unten↓ ↓

folglich $\xi Q(R) \equiv \xi R(Q)$, d. h. $QR = RQ$.

Wir können nun den oben bewiesenen Hauptsatz über die Gleichheit der Doppelverhältnisse congruenter Punktreihen auf die Strahlen eines Büschels ausdehnen. Zu dem Zwecke erteilen wir 3 Strahlen OA, OB, OC beliebige Zahlen und den congruenten Strahlen $O'A'$, $O'B'$, $O'C'$ die nämlichen. Jedem Strahl durch O, etwa OD, werde dann die Zahl zugeordnet, welche
sein Schnittpunkt mit irgend einer | alle 4 Geraden treffenden Schnittlinie l 70
besitzt.*) Um zu beweisen, dass der zu OD congruente Strahl $O'D'$ dieselbe Zahl hat, construire man die congruente Querlinie l'. Wegen des obigen Hauptsatzes stimmen die Doppelverhältnisse der 4 Treffpunkte auf l und auf l' überein und daher auch die ihnen gleichen Doppelverhältnisse der 4 congruenten Strahlen.

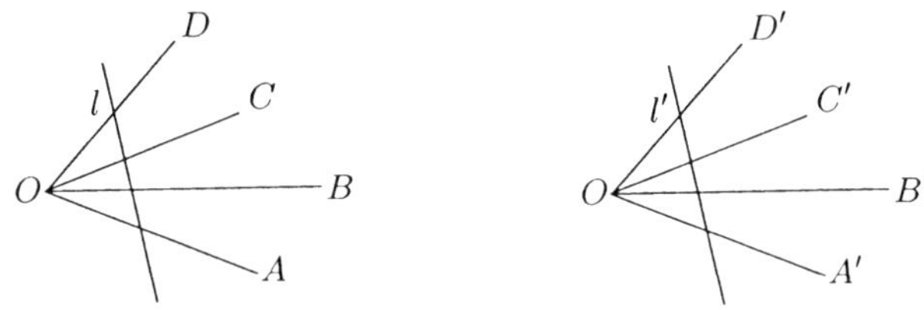

Daraus folgt unmittelbar die Möglichkeit genau wie oben bei der Punktreihe die zum congruenten Strahl gehörige Zahl $\xi' = \mathrm{P}(\xi)$ zu finden, wo P wiederum eine linear gebrochene Function von ξ ist. Stellt $\xi' = \Pi(\xi)$ eine 2te Congruenz im Strahlenbüschel dar, so wird wiederum genau wie oben bei der Punktreihe
geschlossen, dass nothwendig $\mathrm{P}\Pi = \Pi\mathrm{P}$ ist, und zwar wird dies bewiesen | indem 71
man bedenkt, dass die beiden Sätze 1.) und 2.), auf welche sich oben der Beweis für die Punktreihe stützte, auch jetzt für das Strahlenbüschel gilt. Nämlich 2.) der 2te Satz, welcher für die Punktreihe lautete $AB \equiv BA$, lautet für das Strahlenbüschel: $\sphericalangle AOB = \sphericalangle BOA$. Der 1te Satz 1.) lautet für die Punktreihe: Wenn $AB \equiv A'B'$ und $BC \equiv B'C'$, so ist auch $AC \equiv A'C'$, vorausgesetzt, dass die congruenten Stücke an der richtigen Seite angesetzt werden; und lautet für das Strahlenbüschel, wenn $AOB \equiv A'OB'$ und $BOC \equiv B'OC'$, so ist auch $AOC \equiv A'OC'$ wie in der That unsere Axiome aussagen.

*) Diese Darstellung der Strahlen eines Büschels durch Zahlen ist völlig analog der früher behandelten Darstellung der Punkte einer Punktreihe durch Zahlen. Es ist besser diese Zuordnung der Zahlen zu den Büschelstrahlen auch schon früher zu erwähnen und insbesondere in dem Abschnitt „analytische Geometrie" zu benutzen.

⟨The following indented passage was deleted:⟩

Zum Schluss dieses Abschnitts beweisen wir noch einen analytischen Hilfssatz, welcher lautet: Bezeichnen wir mit ω_1, ω_2 die beiden als verschieden angenommenen Wurzeln der Gleichung $\xi = R(\xi)$, so ist, falls in einer Punktreihe $\alpha\beta \equiv \mathrm{AB}$, nothwendig das
72 Doppel|verhältniss $\delta = (\alpha, \beta, \omega_1, \omega_2)$ = dem Doppelverhältniss $(\mathrm{A}, \mathrm{B}, \omega_1, \omega_2)$ oder = dessen reciprokem Werth.

Beweis: Wir vertauschen in δ ω_1 mit ω_2 und nennen den entstehenden Ausdruck ε; dann ist $\delta\varepsilon = 1$, ferner setzt sich $\delta + \varepsilon$ rational aus Doppelverhältnissen der Zahlen ↑κ, λ, μ,↓ κ', λ', μ' zusammen, wie man erkennt, wenn man bedenkt, dass identisch für alle ξ

$$(\xi - \omega_1)(\xi - \omega_2) = \frac{(\lambda-\mu)(\lambda'-\mu')(\xi-\kappa)(\xi-\kappa')}{\uparrow(\lambda'-\mu')(\kappa-\mu)-(\lambda-\mu)(\kappa'-\mu')\downarrow}\uparrow\Big((\kappa\lambda\mu\xi) - (\kappa'\lambda'\mu'\xi)\Big)\downarrow$$

$$\text{wo } (\kappa\lambda\mu\xi) = \frac{\xi-\lambda}{\xi-\kappa}\,\frac{\kappa-\mu}{\lambda-\mu} \text{ genommen ist.}$$

δ genügt also einer Gleichung von der Form $\delta^2 + \Delta\delta + 1 = 0$, wo Δ seinen Werth nicht ändert wenn wir, wie wir das jetzt thun wollen, statt der Punkte α, β, κ, λ, μ, κ', λ', μ' einführen das System congruenter Punkte A, B, K, Λ, M, K$'$, Λ', M$'$. Wir beweisen nun, dass bei diesem Übergang ω_1 und ω_2 ungeändert bleiben. In der That: ω_1, ω_2 sind bestimmt als Wurzeln der Gleichung $(\kappa, \lambda, \mu, \xi) = (\kappa', \lambda', \mu', \xi)$. Nun
77 haben wir |[124] $\kappa' = R(\kappa)$, $\lambda' = R(\lambda)$, $\mu' = R(\mu)$. Ferner sei $\mathrm{K} = Q(\kappa)$, $\Lambda = Q(\lambda)$, $\mathrm{M} = Q(\mu)$, $\mathrm{K}' = Q(\kappa')$, $\Lambda' = Q(\lambda')$, $\mathrm{M}' = Q(\mu')$, $\mathrm{A} = Q(\alpha)$, $\mathrm{B} = Q(\beta)$. Daraus folgt $\mathrm{K}' = Q(R(\kappa)) = R(Q(\kappa)) = R(\mathrm{K})$ und ebenso $\Lambda' = R(\Lambda)$, $\mathrm{M}' = R(\mathrm{M})$, woraus folgt, dass ω_1, ω_2 auch Wurzeln der Gleichung $(\mathrm{K}, \Lambda, \mathrm{M}, \xi) = (\mathrm{K}', \Lambda', \mathrm{M}', \xi)$ sind. Die quadratische Gleichung für δ ändert sich also nicht, wenn ich statt α, β A und B einsetze und damit ist unsere Behauptung bewiesen.

Die entsprechende Behauptung gilt natürlich für die Strahlen eines Büschels, d.h. wenn die beiden Strahlen $\alpha\beta \equiv \mathrm{AB}$ sind, und ω_1, ω_2 die Wurzeln von $\xi = \mathrm{P}(\xi)$, so ist $(\alpha, \beta, \omega_1, \omega_2) = (\mathrm{A}, \mathrm{B}, \omega_1, \omega_2)$ oder dessen reciproken Wert $= (\mathrm{B}, \mathrm{A}, \omega_1, \omega_2) = \frac{1}{(\mathrm{A}, \mathrm{B}, \omega_1, \omega_2)}$, also = einer Invarianten gegenüber der Congruenz.*)

*) Wegen der Gleichung $(\alpha, \beta, \omega_1, \omega_2) = (\mathrm{A}, \mathrm{B}, \omega_1, \omega_2)$ oder $= (\mathrm{B}, \mathrm{A}, \omega_1\,\omega_2)$ ergiebt sich ω_1, ω_2 auch von der besonderen Wahl der $\kappa\,\lambda\,\mu$, $\kappa'\,\lambda'\,\mu'$ unabhängig, was ebenfalls später benutzt wird.

[Um einen genaueren Einblick in das Wesen des Congruenzbegriffes zu erhalten (welcher eben mit dem Begriff der Bewegung übereinstimmt) haben wir die Congruenz-Gleichungen $\xi' = R(\xi)$, $\xi' = \mathrm{P}(\xi)$, d. h. die durch diese Gleichung vermittelte Transformation genauer zu studiren. ⟨]⟩

73 Die Einführung des Maasses

Eine Congruenz auf der Geraden nennen wir auch Verschiebung. Eine Congruenz im Strahlenbüschel: Drehung. Analytisch entspricht beidem eine lineare Transformation: $\xi' = R(\xi)$ bez. $\xi' = \mathrm{P}(\xi)$. Analytische Untersuchung: $R(\xi) = \frac{a\xi+b}{c\xi+d}$, so 2 Fixzahlen α, β. Daher kann diese Transformation auch in Form $\frac{\xi'-\alpha}{\xi'-\beta} = \mu\frac{\xi-\alpha}{\xi-\beta}$ gesetzt werden. 1.) α, β reell, so auch μ reell ↑$\eta' = \mu\eta$ was nur einer anderen Benennung entspricht↓ und 2.) α, β conjugirt imaginär, so $|\mu| = 1$, $\frac{\eta'-i}{\eta'+i} = \mu\frac{\eta-i}{\eta+i}$. 3.) $\alpha = \beta, \eta' = \eta + \nu$, wo nur die Fixzahl ∞ vorhan-

[124] The text ending on p. 72 continues on p. 77. Pp. 73–76 were pasted in before the final page numbering.

den ist. Die aufgezählten Substitutionen werden elliptische, hyperbolische, parabolische[125] genannt.

Was bedeutet die Vertauschbarkeit der Verschiebungen bez. Drehungen:

$$\frac{a\mu\eta+b}{c\mu\eta+d} = \mu\frac{a\eta+b}{c\eta+d} \qquad \langle 126\rangle$$

↑Setze $\eta = 0$, so:↓ $\frac{b}{d} = \mu\frac{b}{d}$ und da $\mu \neq 1$, so $b = 0$ und folglich auch $d = 0$, d. h. die Substitution hat die nämlichen Fixpunkte. Ebenso für die parabolische Substitution

$$\frac{a(\eta+\nu)+b}{c(\eta+\nu)+d} = \frac{a\eta+b}{c\eta+d} + \nu$$

d. h. [127] wenn man nach η differenzirt und dann $\eta = 0$ setzt, ergiebt sich $c = 0$ und dann folgt $a = d$.

Hieraus folgt, dass alle Transformationen, welche einer Verschiebung oder 74
Drehung entsprechen, die nämlichen bestimmten Fixzahlen besitzen und umgekehrt ist mit dieser Eigenschaft auch die geometrische Thatsache $PQ = QP$ der Vertauschbarkeit 2er Congruenzen aequivalent.

Betrachten wir nun zunächst auf einer Geraden eine Verschiebung $\xi' = R(\xi)$ des Punktes $x = a$, so werden die Zahlen $R(a)$, $R^2(a)$, $R^3(a), \ldots$ eine Reihe Punkte definiren, welche immer weiter geordnet nach rechts oder nach links liegen, und daher werden diese Zahlen immer entweder wachsen oder abnehmen und daher werden sie sich einer Grenze γ_2 und die Zahlen $R^{-1}(a)$, $R^{-2}(a)$, $R^{-3}(a), \ldots$ einer Grenze γ_1 nähern. Die Grenzzahlen γ_1, γ_2 sind daher zugleich die Fixzahlen der Transformation; diese ist also nothwendig entweder eine parabolische oder hyperbolische. Da im Strahlenbüschel jeder Zahl ein Strahl entspricht und andererseits durch die Drehung jeder Strahl in einen anderen überführt wird, so ist hier die Transformation nothwendig eine elliptische.

Ist die Verschiebung durch die Formel $\xi' = \xi + M$ gegeben, so nehme man 2 beliebige Punkte und erteile ihnen die Werte 0, 1, dann ist dadurch dass man die Grenzzahl (Fixzahl) $= \pm\infty$ nimmt, allen Punkten eine endliche Zahl und umgekehrt allen endlichen Zahlen ein Punkt zugeordnet. Ist die Strecke $AB = A'B'$ und sind a, b, a', b' die den Punkten A, B, A', B' zugeordneten Zahlen, so folgt aus $a' = a + M$, $b' = b + M$: $a - b = a' - b'$ wird Entfernung von AB genannt: die Entfernung ist eine Invariante bei der Verschiebung. Also Maasseinheit und Anfangspunkt nothwendig, indem man 2 Grundpunkte mit 0, 1 versieht.

Ist die Verschiebung durch die Formel $\xi' = M\xi$ gegeben, erteile man einem 75
Punkte den Wert 1. Dadurch dass man die Grenzpunkte 0, ∞ nimmt, sind dann die Zahlen festgelegt. Strecke $AB = A'B'$ ist mit der Formel $\frac{a}{b} = \frac{a'}{b'}$ aequivalent. $l\left(\frac{a}{b}\right) =$ Entfernung: $la = \alpha$, $lb = \beta$. Allen positiven endlichen Zahlen entsprechen Punkte und umgekehrt. Bei der neuen Bezeichnungsweise mit $l\xi$ entsprechen wieder allen endlichen Zahlen Punkte und umgekehrt. Keine Maasseinheit sondern nur ein Anfangspunkt nothwendig.

[125] The correct order is: hyperbolisch, elliptisch, parabolisch.

[126] Deleted: $bc + ad\mu = ad\mu + bc\mu^2$, d. h. $\mu^2 = 1$ oder $bc = 0$

[127] Deleted: $\eta = 0$: $\frac{a\nu+b}{c\nu+d} = \frac{b}{d} + \nu$

Im Strahlenbüschel ist wiederum die Drehung durch 2 Strahlen OA und OA' gegeben. Ist die entsprechende Transformation

$$\frac{\xi'+i}{\xi'-i} = \mu\frac{\xi+i}{\xi-i}, \text{ so ist } |\mu| = 1, \text{ d. h. } \mu = e^{i\varphi}$$

und φ heisst das Winkelmaass für $\sphericalangle AOA'$.
Hat man eine zweite Drehung um den Winkel $A'OA'' = \varphi$

$$\frac{\xi''+i}{\xi''-i} = \nu\frac{\xi'+i}{\xi'+i} = \mu\nu\frac{\xi+i}{\xi-i}$$

so ist diese der Drehung um den Winkel $\varphi + \varphi'$ aequivalent. Sind die Winkelmaasse φ gleich, so sind die Winkel selber gleich. Denn es ist ja durch φ die Transformation bestimmt. Nur der Anfangspunkt, von dem φ zu zählen ist, muss gegeben sein. Alle gestrekten Winkel $= \pi$ und alle rechten Winkel $= \frac{\pi}{2}$ sind unter einander gleich. Wenn 2 Winkel gleich sind, sind auch die Nebenwinkel gleich. Die Scheitelwinkel sind gleich. Die weitere Ausbildung der Geometrie kann nun mittelst Rechnung geschehen. Wir wenden uns zum letzten Abschnitt: betitelt

Das Parallelenaxiom.[128]

78 *Die hyperbolische Geometrie*[A]

Die fixen Zahlen der Transformation $\xi' = R(\xi)$ d. h. ω_1, ω_2 seien reell und von einander verschieden.*) Die Doppelverhältnisse $(\alpha, \beta, \omega_1, \omega_2)$ werden dann reelle Zahlen. Wir bestimmen nun eine linear gebrochene Funktion $T(\xi)$, so dass $T(\omega_1) = 0$, $T(\omega_2) = \infty$ ↑und ausserdem $T(\alpha)$ positiv ist, etwa $= 1$ oder $= \kappa$↓ und nennen nun die Punkte der Gerade um, nämlich irgend 3 Punkte mit den Zahlen α, β, γ legen wir die Zahlen $T(\alpha)$, $T(\beta)$, $T(\gamma)$ bei. Nach dem am Ende des vorigen Abschnittes bewiesenen Satze wird, wenn $\alpha\beta \equiv \mathrm{AB}$ ist

$$(\alpha, \beta, \omega_1, \omega_2) = (\mathrm{A}, \mathrm{B}, \omega_1, \omega_2) \quad \text{oder} \quad = (\mathrm{B}, \mathrm{A}, \omega_1, \omega_2)$$

d. h. nach der neuen Benennung die vorläufig durch Striche gekennzeichnet werde:

$$(\alpha', \beta', 0, \infty) = (\mathrm{A}', \mathrm{B}', 0, \infty) \quad \text{oder} \quad = (\mathrm{B}', \mathrm{A}, 0, \infty)^{129}$$

[A]Es ist in jeder – pädagogischer und prinzipieller – Hinsicht besser, zuerst die parabolische Geometrie und dann erst die hyperbolische und elliptische zu behandeln. Thue dies künftig. ⟨Added in the upper margin and directed to this section heading by an insertion sign.⟩

*) Auf jeder anderen Geraden gilt dann natürlich das Gleiche für die betreffende Gleichung $\xi = R(\xi)$. Zum Beweise brauche man nur auf der 2ten Geraden die congruenten Punkte gleich benennen.

[128]Pp. 73–76 are contained on a double sheet which has been pasted in. The new text ends on p. 75 with the heading 'Das Parallelenaxiom'. However, p. 76 having been left blank, p. 77 continues the text from p. 72, and was thus reproduced above. Although 'Das Parallelenaxiom' is centred as if signalling the start of a new section, there is no appropriate text immediately following. There is a brief section with the same title right at the end of the manuscript, pp. 93–95. It is likely that Hilbert postponed the subsection on 'Das Parallelenaxiom' to the end of the lecture, and instead continued with the subsection on hyperbolic geometry.

[129]This should be $(B', A', 0, \infty)$.

d. h.

$$\frac{\alpha'}{\beta'} = \frac{\mathrm{A}'}{\mathrm{B}'} \quad \text{oder} \quad = \frac{\mathrm{B}'}{\mathrm{A}'}$$

⟨The following indented passage was deleted:⟩

> Wir beweisen zunächst, dass es nicht Punkte giebt, denen nach neuer Benennung die Zahlen 0, ∞ (also nach alter ω_1, ω_2) entsprechen. In der That sei $\alpha = 0$ [130] ein Punkt, so bestimme man irgend einen anderen Punkt β | etwa $\beta = 1$. Ferner 79
> seien A, B 2 von 0, ∞ verschiedene Zahlen, denen Punkte entsprechen und $\alpha\beta$ sei $\equiv$ AB. Daraus würde wegen $\frac{\alpha}{\beta} = \frac{\mathrm{A}}{\mathrm{B}}$ oder $= \frac{\mathrm{B}}{\mathrm{A}}$ entweder A oder B = 0 folgen, was nicht der Fall ist.
>
> Ferner zeigt sich, dass keiner negativen Zahl ein Punkt entsprechen kann. Denn wäre etwa α negativ und existire ein Punkt α, so construire β, so dass $\alpha 1 \equiv 1\beta$ ist, β aber auf der anderen Seite von 1 als α liegt, so müssen auch allen zwischen α und 1 gelegenen Punkten bei Ausübung dieser Congruenz (Verschiebung) Punkte zwischen 1 und β entsprechen. (Es muss nämlich $\beta = -\alpha$ sein.) Nun nehme man einen Punkt γ, dessen Zahl γ [131] positiv und $< |\alpha|$ und zugleich < 1 ist. [132] Dann liegt der zu γ congruente Punkt nothwendig wieder zwischen α und 1, was ein Widerspruch ist.

und wenn wir wieder bei der neuen Benennung die Striche weglassen:

$$\alpha\beta \equiv \mathrm{AB} \ \text{ hat zur Folge } \ \frac{\alpha}{\beta} = \frac{\mathrm{A}}{\mathrm{B}} \ \text{oder} \ = \frac{\mathrm{B}}{\mathrm{A}}.$$

In der That giebt es ja auch 2 Punkte B, welche bei gegebenen Punkten α, β, A die Congruenz erfüllen. Um sie zu unterscheiden, beweisen wir zunächst, dass es keine Punkte giebt, denen die Zahlen 0, ∞ oder ⟨eine⟩ negative Zahl zukommt. Denn wäre z. B. 0 ein Punkt, so müsste wegen der Relation $\frac{\alpha}{\beta} = \frac{\mathrm{A}}{\mathrm{B}}$
oder $\frac{\mathrm{B}}{\mathrm{A}}$ jede zu 01 congruente Strecke entweder mit dem einen oder | mit 80
dem anderen Ende in den Punkt 0 fallen, was nicht möglich ist. Ferner: die obige durch die Transformation $\xi' = R(\xi)$ vermittelte Congruenz wird jetzt durch die Gleichung $\xi' = c\xi$ vermittelt wo c eine Constante ist, welche der Congruenz (Verschiebung) charakteristisch ist. Da nach den letzten Entwickelungen des vorigen Abschnittes ω_1, ω_2 von der zugrundegelegten Congruenz ganz unabhängig sind, so folgt, dass alle Congruenzen (Verschiebungen) durch Gleichungen von jener Form $\xi' = c\xi$ dargestellt werden, d. h. wenn ξ_1, ξ_2, ξ_3 irgend welche Punkte sind, dann stellt sich die congruente Punktreihe in der Gestalt $c\xi_1$, $c\xi_2$, $c\xi_3$ dar. Gäbe es nun einen Punkt, dem die negative Zahl $-p$ zugehört, so suche den Punkt B, so dass $-p1 \equiv 1\mathrm{B}$ und wo B auf der anderen Seite von 1 liegt als $-p$. Dann müssen auch alle Punkte im Innern der Strecke $-p1$ congruent sein zu Punkten im Innern von 1B. Diese Congruenz (Verschiebung) werde dargestellt durch $\xi' = \gamma\xi$, so kann man jedenfalls ξ absolut so klein bestimmen, dass $\gamma\xi$ nicht im Innern von $-p1$ bleibt und dies ist ein Widerspruch. Wir wissen also: den vorhandenen Punkten entsprechen

---

[130] Deleted by *Hi*^{s} before the whole passage was crossed out; the deleted text is hard to decipher: ein Punkt ⟦, so⟧ und $\beta = 1$, B⟦⟪A⟫⟧ = 2⟪β⟫ = 1 ⟪???⟫, so würde wegen $\frac{\alpha}{\beta} = \frac{\mathrm{A}}{\mathrm{B}}$ A = α folgen und $\beta = 1$ B⟦⟪??⟫⟧ zwei andere, so würde wegen $\frac{\alpha}{\beta} = \frac{\mathrm{A}}{\mathrm{B}}$

[131] Deleted by *Hi*^{s}: absolut

[132] Deleted by *Hi*^{s}: Ist dann $\alpha 1 \equiv \gamma\gamma'$ so $\alpha = \frac{\gamma}{\gamma'}$ oder $\frac{\gamma'}{\gamma}$ also γ' etwa entweder $= \frac{\gamma}{\alpha}$

nur positive Zahlen, und wir haben es noch in der Hand (wegen der obigen $T(\alpha) = \kappa$) einem beliebigen Punkte einen beliebigen positiven endlichen und von $0 \neq$ Wert zu erteilen.

81 Wir können nun auch die obige Unterscheidung treffen. Ist nämlich $\alpha\beta \equiv$ AB, und $\beta > \alpha$ und B > A oder $\beta < \alpha$, B < A, so $\frac{\alpha}{\beta} = \frac{\mathrm{A}}{\mathrm{B}}$ und in den beiden anderen Fällen $\frac{\alpha}{\beta} = \frac{\mathrm{B}}{\mathrm{A}}$, d. h. ist $\alpha\beta \equiv$ AB und folgen in der Reihenfolge $\alpha\beta$AB β auf α und B auf A, oder α auf β und A auf B, so $\frac{\alpha}{\beta} = \frac{\mathrm{A}}{\mathrm{B}}$. Da negative Zahlen nicht vorkommen, so lege ich dem Punkte α nun die Zahl $\mathfrak{a} = k\,\mathrm{l}\,\alpha$ bei, wo k eine positive von $0 \neq$ Zahl ist.[133] Dann ist die zu $\mathfrak{a}_1, \mathfrak{a}_2, \ldots$ congruente Punktreihe einfach durch $\mathfrak{a}_1 + \mathfrak{c}, \mathfrak{a}_2 + \mathfrak{c}, \ldots$ gegeben. Der absolute Wert der Differenz $\mathfrak{a} - \mathfrak{b}$ 2er Punkte $\mathfrak{a}$ und $\mathfrak{b}$ ist dann einer jeden Verschiebung gegenüber invariant und heisst die Entfernung. Wenn $\mathfrak{b}$ zwischen $\mathfrak{a}$ und $\mathfrak{c}$ liegt, so ist die Summe der Entfernungen von $\mathfrak{a}$, $\mathfrak{b}$ und $\mathfrak{b}$, $\mathfrak{c}$ = der Entfernung von $\mathfrak{a}$, $\mathfrak{c}$.

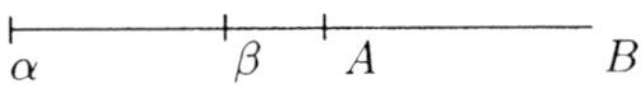

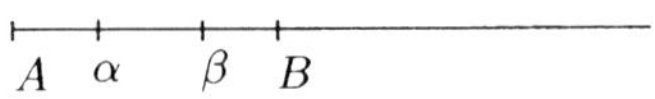

$$\text{Entfernung} = k\ \mathrm{l}(\alpha, \beta, \omega_1, \omega_2).$$

↑2 congruente Strecken sind von gleicher Entfernung, Länge und heissen daher *gleich*.↓

Die vorige Doppeldeutigkeit $\frac{\mathrm{A}}{\mathrm{B}}$ und $\frac{\mathrm{B}}{\mathrm{A}}$ hat für den Logarithmus nur die Doppeldeutigkeit des Vorzeichen zur Folge*).

Bei der neuen Benennung können wieder alle Zahlen ausser $-\infty$ und $+\infty$ durch existirende Punkte vertreten sein. Auch bei der neuen Benennung gilt wieder der Satz: Wenn $\mathfrak{a} < \mathfrak{b} < \mathfrak{c}$, so liegt $\mathfrak{b}$ zwischen $\mathfrak{a}$ und $\mathfrak{c}$.

Zieht man etwa durch den Punkt O nach neuer Benennung eine Gerade und nennt den durch die Congruenz $O\mathfrak{a} \equiv OA$ auf derselben bestimmten
82 Punkt ebenfalls $\mathfrak{a}$, | so hat man die entsprechende Benennung der Punkte auf der zweiten Geraden. Man könnte so jeden Punkt der Ebene bestimmen, falls man noch ein Maass für die Neigung der 2ten Geraden gegen die erste besässe. Dies gelingt, wenn man die Congruenz im Büschel studirt.

Die Gleichung $\xi = \mathrm{P}(\xi)$ hat stets conjugirt imaginäre Wurzeln ω_1, ω_2. Denn wären beide reell und voneinander verschieden, so könnte man alle vorigen Schlüsse von der Punktreihe auf das Strahlenbüschel übertragen und dann dürften allen zwischen ω_1 und ω_2 liegenden Zahlen keine Strahlen entsprechen. Dies kann nicht sein, da ja z. B., wenn man die 3 Grundpunkte 0, 1, ∞ nimmt, allen rationalen ↑positiven und negativen↓ Zahlen Punkte entsprechen. (Die harmonischen Constructionen sind ja im Büschel beliebig ausführbar.) Aber auch zusammenfallende Wurzeln kann jene Gleichung nicht haben. Denn sonst könnte man die eine Congruenz vermittelnde Transformation $\xi' = \mathrm{P}(\xi)$ in die Form $\xi' = \xi + a$ bringen. Die 2malige Anwendung der Congruenz (Drehung) könnte also nie die Identische ergeben. Es giebt

*) d. h. wenn $\mathfrak{ab} \equiv \mathfrak{a}'\mathfrak{b}'$, so ist $\mathfrak{a} - \mathfrak{b} = \pm(\mathfrak{a}' - \mathfrak{b}')$ und umgekehrt.

[133] 'l α' denotes a logarithmic function of α.

aber eine solche Congruenz, welche 2mal angewandt die Identische ausmacht, diejenige Congruenz nämlich, welche durch die Congruenz:

$$[OAC] \equiv [OBD]$$

gegeben ist, wo DOC in gerader Linie und $DO \equiv OC$ angenomen und ferner 83
B und A gemäss der Congruenzen $[DBC] \equiv [CAD]$ bestimmt sind. [Dass B, O, A in gerader Linie liegen, kann wie folgt bewiesen werden: Erweitere ich die Figur DBC auf der linken Seite der obigen Congruenz durch A, so habe ich wegen $[CAD] \equiv [DBC]$ die Congruenz $[ADBC] \equiv [BCAD]$.

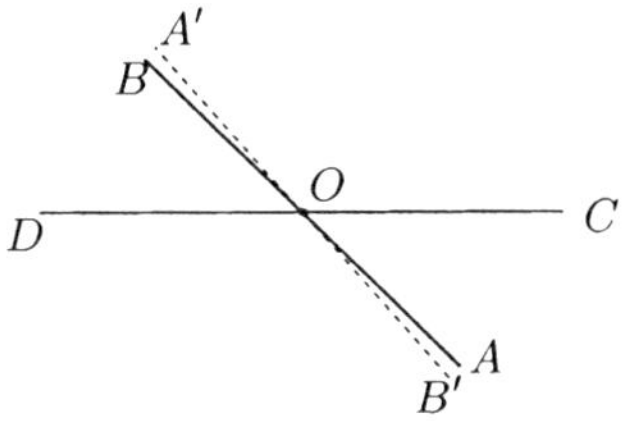

Verlängere ich nun in der Figur links AO um AO über O hinaus bis A', so dass also $OA \equiv OA'$, und suche in der Figur rechts das congruente Stück OB' auf, so muss, da OA' in den Raum OBC fällt, OB' in den congruenten Raum OAD hineinragen, folglich müssten B und B' auf derselben Seite von AA' liegen, wie sich zeigt, wenn ich etwa B mit D und D mit B' verbinde. Dies ist aber andererseits nicht möglich, weil O ein Punkt von AA' ist und zwischen BB' liegt.]

ω_1, ω_2 sind conjugirt imaginär und folglich auch das Doppelverhältniss $\delta = (\alpha, \beta, \omega_1, \omega_2)$ und da der letzte Coeffizient der quadratischen Gleichung für δ der Einheit gleich war, so ist $|\delta| = 1$ und folgt 1δ rein imaginär. Wir definiren

$$ik\,\mathrm{l}(\alpha, \beta, \omega_1, \omega_2) = \text{ Winkel } (\alpha, \beta),$$

derselbe ist nur bis auf additive Vielfache von $2k\pi$ bestimmt.
Wenn ich für das Doppelverhältniss seinen reciproken Wert nehme, so ändert 84
der Winkel nur sein Vorzeichen. Ist $ABC \equiv A'B'C'$, so sind die Winkel absolut und bis auf Vielfache von 2π einander gleich. Umgekehrt ist Winkel $CAB = C'A'B'$ und $AC \equiv A'C'$, $AB \equiv A'B'$, so sind die Figuren $[CAB] \equiv [C'A'B']$.

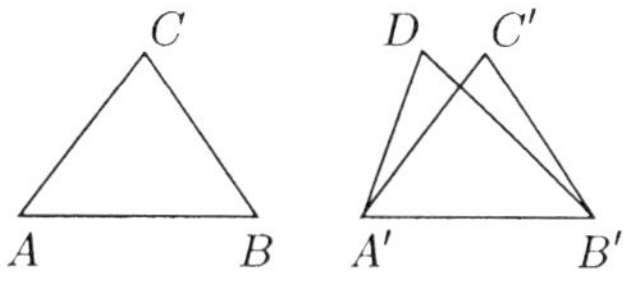

Denn sonst contruire man $[A'D'B'] \equiv [ACB]$. Dann ist nach dem vorigen Winkel $D'A'B' = CAB$ und folglich $= C'A'B'$. Wenn aber diese Winkel gleich sind, so kommt nach der alten projektiven Bezeichnung dem Strahl $A'C'$ dieselbe Zahl zu, wie dem Strahl $A'D'$ und folglich sind es die nämlichen Strahlen (erster Dreieckscongruenzsatz).

Bei Ausübung einer Congruenz (Drehung) ändert sich der Winkel nicht; dies folgt aus der am Ende des vorigen Abschnittes bewiesenen Eigenschaft der Invarianz von $\delta = (\alpha, \beta, \omega_1, \omega_2)$ gegenüber einer Congruenz.

Bei Ausübung zweier Congruenzen nacheinander addiren sich die Winkel, d. h. wenn Winkel $BOA = \varphi_1$, Winkel $BOC = \varphi_2$, so ist Winkel $AOC = \varphi_1 + \varphi_2$, wie oben bei dem Begriff der

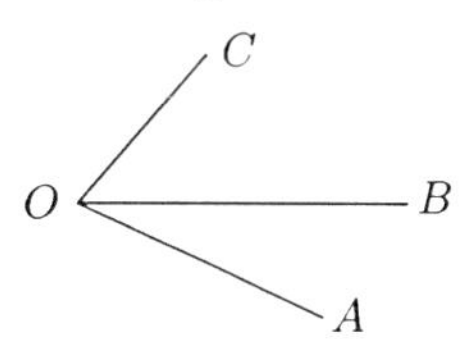

Entfernung folgt, wenn man

$$(\alpha, \beta, \omega_1, \omega_2) = \frac{\alpha - \omega_1}{\beta - \omega_1} \frac{\beta - \omega_2}{\alpha - \omega_2}$$

berücksichtigt.

85 Auch können wir wie vorhin bei der Punktreihe ω_1, ω_2 noch besonders normiren. Wir bestimmen eine linear gebrochene Funktion $T(\delta)$ mit reellen Coefficienten, so dass $T(\omega_1) = +i$, $T(\omega_2) = -i$ ist, und nennen dann

$$T(\alpha) = \alpha', \quad T(\beta) = \beta', \text{ so } \quad (T\alpha, Tb, T\omega_1, T\omega_2) = (\alpha', \beta', i, -i)$$

und wenn wir wieder die Striche weglassen

$$\text{Winkel } = ik\,\mathrm{l}\,\frac{\alpha - i}{\alpha + i}\frac{\beta + i}{\beta - i} = k \operatorname{arcctg} 2\alpha - \operatorname{arcctg} 2\beta,^{134} \quad {}^{*})$$

wodurch also leicht von der alten projektiven Maassbestimmug zu der neuen metrischen übergegangen werden kann. Bei dieser Festsetzung bleibt nur noch die Willkür, wo der Winkel $\varphi = 0$ genommen werden soll, gerade wie oben bei Einführung der Entfernung und wie k gewählt werde. k wollen wir (= 1 nämlich) so wählen dass die Periode $= 2\pi$ werde. Der gestrekte Winkel hat dann nothwendig den Wert $= \pi$ (nach dem oben bewiesenen Additionsgesetz, weil die 2malige Anwendung der Congruenz 2π ergeben muss). Rechter Winkel ist ein solcher der zu sich addirt den gestrekten ergiebt. Sein Wert ist daher $= \frac{\pi}{2}$ und wir haben den Satz: alle rechten Winkel sind einander congruent, wie aus dem oben bewiesenen ersten Dreieckscongruenzsatz folgt. Auch können wir den rechten Winkel leicht construiren, indem wir $BOA \equiv COB$ machen und dann den 4ten harmonischen Strahl BOD suchen,[135] so ist $BOD =$ Rechtem Winkel.

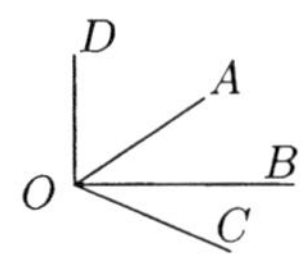

86 Kehren wir nun für den Augenblick zu der alten projektiven Maassbestimmung (Zahlenzuordnung) der Strahlen eines Büschels A und einer Punktreihe a zurück, welche nicht durch A hindurch geht. Da allen negativen ↑(oder allen zwischen ω_1 und ω_2 gelegenen)↓ Zahlen auf der Punktreihe kein Punkt zugehört, wohl aber unendlich vielen negativen, etwa allen rationalen Zahlen, ein Strahl des Büschels (falls man jedem durch einen Punkt von a gehenden Strahl des Büschels die diesem Schnittpunkte zugehörige Zahl zuordnet), so giebt es unendlich viele Strahlen, welche a nicht treffen, und zwar kann man jedenfalls 2 Strahlen l und m↑, n↓

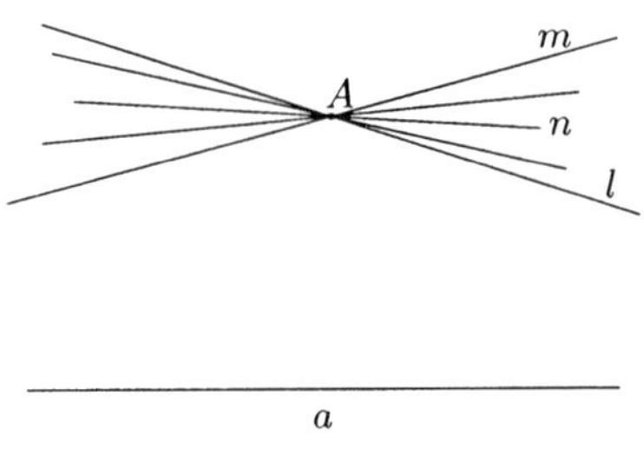

---

*) Nach der Formel: $\mathrm{l}(x + iy) = \frac{1}{2}\mathrm{l}(x^2 + y^2) + i \arctan \frac{y}{x}$ und wenn man links $\mathrm{l}\,\frac{\alpha - i}{\alpha + i}\frac{\beta + i}{\beta - i} = \mathrm{l}\,(i\alpha + 1) - \mathrm{l}\,(-i\alpha + 1) + \mathrm{l}\,(i\beta - 1) - \mathrm{l}\,(-i\beta - 1)$ setzt.

---

[134] The last part of the formula should read: $k(\operatorname{arcctg} 2\alpha - \operatorname{arcctg} 2\beta)$

[135] Presumably, the fourth harmonic ray (*Strahl*) we seek is OD.

durch A construiren ↑welche selbst a nicht treffen↓, so dass alle zwischen l und m gelegenen und von n nicht getrennten Strahlen durch A die Gerade a nicht treffen.

Wir nennen eine jede solche a nicht treffende Gerade zu a parallel und haben daher den Satz: Durch A giebt es zu a unendlich viele Parallelen. Dieser Satz lässt sich durch das Experiment schwer bestätigen oder widerlegen, obwohl das letztere eher als das erstere. Einen experimentell leichter zu prüfenden Satz erhält man aus demselben, nämlich den Satz: In jedem Dreieck ist die Winkelsumme $< \pi$.[136]
Beweis 1.) nach Lindemann, Vorlesungen über Geometrie, Bd. IIa, S. 494,[137] 87
am meisten für diese Vorlesung zu empfehlen, weil analytisch durch Rechnung. 2.) Pasch, neuere Geometrie, S. 162.[138] 3.) Absolute Geometrie nach Bolyai von Frischauf, S. 4 und S. 15.[139]

Das Experiment zeigt nun aber die Winkelsumme nicht von π verschieden. Ein Unterschied von π ist bisher bei noch so grossen Dreiecken in der Astronomie noch nicht zu Tage getreten. D. h. wenn wir kleine Körperteile als Punkte, alle beim Festhalten 2er Punkte unbeweglich bleibenden kleinen Körperteile zu der durch jene Punkte bestimmten Geraden rechnen, oder auch Lichtstrahlen, straff gespannte Fäden u. s. f. als Geraden nehmen, und dann die 3 Winkel eines Dreieckes an einander tragen, so bildet der erste und letzte Schenkel eine Gerade. Infolge dieses Experimentes*) verwerfen wir die hyperbolische Geometrie zum Gebrauch und bewahren sie eventuell für einen später sich nothwendig machenden Gebrauch auf, falls genauere Messungen in grossen Dreiecken einen Unterschied mit π ergeben sollten.[140] Thöricht ist es, wie z. B. Lotze thut, die Möglichkeit der nicht Euklidischen Geometrie von vorneherein zu verwerfen.[141] Die Möglichkeit derselben, d. h. die innere

*) Gauss erklärte – einer Göttinger Tradition zufolge – seine Ueberzeugung von der Richtigkeit der üblichen Geometrie wurzle darin, dass er die Summe der Winkel in dem Dreiecke Inselsberg, Brocken, hoher Hagen gleich 180° gefunden habe.

[136]Deleted by *Hi*^{s}: Beweis: Wäre die Winkelsumme $> \pi$, etwa $= \pi + \delta$, so trage man an einer Seite AB je einen Winkel $< \frac{\delta}{2}$ ab, dann würde der entstehende Winkel $ADB > \pi$ sein ⟨p. 87⟩ müssen, was nicht angeht.
⟨Illegible drawing in the upper margin⟩
↑(Denn die Winkelsumme im Dreieck AEB↓ ⟨added at the bottom of p. 86⟩

[137]*Clebsch 1891*.

[138]*Pasch 1882*.

[139]*Frischauf 1872*.

[140]Killing also emphasizes (*Killing 1893*, 96) that

> ... die Erfahrung ... keines der mitgeteilten Systeme [der euklidischen und nicht-euklidischen Geometrien] mit voller Strenge als das richtige hinstellt. Man kann daher folgende Erwägung anstellen: Nach den Entwicklungen des § 22 (S. 74) sind für unsere Erfahrung, soweit wir sie bis jetzt beurteilen können, unendlich viele Fälle gleich möglich; nur einer von diesen entspricht der euklidischen Geometrie.

[141]*Lotze 1879*. Rudolf Hermann Lotze (1817–1881) was Herbart's successor in the Chair of Philosophy and Psychology in Göttingen, a position he occupied from 1844–1881. Of his central work *System der Philosophie*, planned in three volumes, only the first two, *Logik* (1874) and *Metaphysik* (1879), appeared. Hilbert refers indirectly to the latter.

Widerspruchslosigkeit derselben, kann vielmehr strenge gezeigt werden. Es lässt sich nämlich in der That ein System von Einheiten – Punkte, Geraden, Ebenen – aufstellen (und sogar durch Zahlen rein arithmetisch definiren), für welches alle unsere Axiome – Existenz-, Lagen-, Congruenzaxiome erfüllt sind
88 und | für welches durch einen Punkt zu einer gegebenen Geraden unendlich viele Parallelen (d. h. in derselben Ebene gelegene und nicht schneidende) Geraden existiren. Man findet ein solches System am einfachsten, indem man alle Punkte im Inneren einer Kugel wählt und als Ebenen alle zu der Kugel orthogonal stehenden Kugeln (weil sie im Inneren der Fundamentalkugel liegen) wählt. Natürlich, um auf Lotze's Vorurteil zurückzukommen, kann uns kein Experiment zwingen, die hyperbolische Geometrie anzunehmen. Vielmehr ist es, falls sich experimentell die Winkelsumme $< \pi$ herausstellen sollte, immer noch möglich, mit dem üblichen Euklidschen Raum auszukommen. Wir müssten dann annehmen, dass es eine Kugel giebt, in deren Inneren wir uns befinden, deren Oberfläche durch Verschiebung eines Stabes, wenn wir dieselbe ⟪???⟫ oft ausführen, nie erreicht wird. Jeder Stab in sich gedreht, d. h. jede Achse bildet einen Kreis, ebenso jeder Lichtstrahl, jeder strammgezogene Faden, und zwar solche Kreise, welche auf jener Kugeloberfläche senkrecht stehen. Nun einfacher, durchsichtiger wird es und es bedarf weniger Axiome, wenn wir unter solchen Umständen die hyperbolische Natur unseres Raumes postuliren würden. Und das ist ja doch überhaupt das einzig für uns bestimmende, wenn wir eine Hypothese als Wahrheit annehmen und anerkennen. Auch das Kopernikanische Weltsystem zwingt uns niemand an-
89 zunehmen, wir können vielmehr auch | das Ptolemäische gelten lassen, wenn wir dasselbe richtig ausflicken. Und wie viel umgestaltender, wie viel mehr der ganzen damaligen Auffassung der Erde als ruhender Mittelpunkt widerstrebte es als jetzt etwa die Annahme eines hyperbolischen Raumes. Lotze ist also weit dümmer als die derzeitigen Gegner von Kopernikus und weit schlimmer, da er sich voraus verwahrt, denn es liegen ja noch gar keine dem Euklidischen Raume widersprechende Beobachtungen vor. Es ist als ob der Philosoph instinktiv aus Concurrenzneid dem exacten Forscher jeden Fortschritt hemmen will, daher die Statuirung dieser Vorurteile, in deren Ueberwindung gerade das wahre Wesen der Wissenschaft liegt. Vgl. die Ansichten der übrigen Philosophen: Erdmann, Axiome der Geometrie,[142] Thiele, Grundriss der Logik und Metaphysik[143] und Gino Loria, Theorie der Geometrie, S. 106 Anmerkung.[144]
↑Wundt, 2te Auflage, Logik.[145]↓

[142] *Erdmann 1877.*

[143] *Thiele 1878.*

[144] *Loria 1888.*

[145] *Wundt 1893.* Wilhelm Wundt (1832–1920) was from 1875–1917 Professor of Philosophy in Leipzig, and founded there in 1879 the first institute for experimental psychology. As well as his main work, *Grundzüge der physiologischen Psychologie* (1876), he published in 1880 a two volume *Logik*, to which Hilbert refers. (The second edition of Volume I was published in 1893, Volumes 2 and 3 in subsequent years.)

[146]Die elliptische Geometrie ist durch unsere Lagenaxiome über den Begriff „zwischen" ausgeschlossen.[146] Doch kann man sie dennoch beibehalten, wenn 90 man die Lagenaxiome und die betreffenden Congruenzaxiome abändert, indem man den Begriff zwischen nicht absolut nimmt, nicht davon spricht, dass eine Ebene den Raum in 2 getrennte Theile theilt, sondern nur von 4 Punkten in der Reihenfolge $ABCD$ spricht, dass AC durch BD getrennt werden und so fort. Dann ist die elliptische Geometrie möglich; ihr Aufbau ist in sich selbst widerspruchslos, wie man analytisch und auch geometrisch einsehn kann. Es giebt keine Parallelen, die Winkelsumme im Dreieck ist grösser als π. Die geraden Linien besitzen eine endliche Länge. Betreffs der Anwendung und des Experimentes gilt das Gleiche wie für die hyperbolische | Geometrie. 91

Die parabolische Geometrie.

Die Gleichung $\xi = R(\xi)$ hat nur eine Wurzel. Die entsprechende Transformation kann dann auf die Gestalt $\xi' = \xi + c$ gebracht werden, d. h. man kann 3 Grundpunkten solche Werthe ertheilen, dass congruente Punktsysteme durch Addition einer Constanten erhalten werden. Diese Constante selbst wird die Entfernung genannt, dieselbe ist also eine algebraische Function der projectiven Maassbestimmung. [Die Gleichung der Geraden ist auch in den neuen Coordinaten eine lineare, während sie in den beiden früheren Nicht-Euklidischen Geometrien transcendent wurde. Man bedarf keiner weiteren Axiome zum Aufbau der Euklidischen Geometrie, vergl. die Erklärung des Euklid in Linde|manns 92 Vorlesungen, Bd. 2a, S. 540.[147]

Ueber den Zusammenhang der 3 Geometrien vgl. F. Klein, Nichteuklidische Geometrie, Math. Ann. Bd. 4 u. 6.[148]]

Nach dem Muster der Geometrie sind nun auch alle anderen Wissenschaften, in erster Linie Mechanik, hernach aber auch Optik, Elektrizitätstheorie etc. zu behandeln, wie ja auch Boltzmann[149] (vgl. Catalog der Mathematikerausstellung von W. Dyck[150]) alles auf so einfache Thatsachen (Axiome) will zurückgeführt haben, welche sich durch einen einfachen Mechanismus vor

[146–146]Substituted for:

Die elliptische Geometrie.

Die Gleichung $\xi = R(\xi)$ habe conjugirt imaginäre Wurzeln. Es greift dann dieselbe Betrachtung für die Punktreihe Platz, welche oben bei der hyperbolischen Geometrie auf das Strahlenbüschel angewandt wurde. Man findet, dass es in der Punktreihe 4 Punkte in der Aufeinanderfolge $ACBD$ geben muss, derart dass $ACB \equiv ADB$ ist. Dies aber widerspricht dem ⟨p. 90⟩ Axiom, dass congruente Punkte zwischen congruenten Punkten liegen müssen, und daher ist bei unsrer Wahl der Axiome, insbesondere der den Begriff zwischen betreffenden Axiome, die elliptische Geometrie ausgeschlossen.

[147]*Clebsch 1891*.

[148]*Klein 1871*, *1873*.

[149]Ludwig Boltzmann (1844–1906), physicist and philosopher of science, primarily known for his kinetic theory of gases and his deduction of the Second Law of Thermodynamics. It is interesting that Hilbert mentions him as an ally regarding the axiomatic representation of physics, since he was later in disagreement with him regarding the alleged deduction of the Second Law from the kinetic theory of gases. See Volume 4 of this Edition.

[150]*Dyck 1892*.

unseren Augen schematisch darstellen.[151] Was den Einwand anbetrifft, dass die Sätze der Theorie doch alle nicht genau gelten, so ist derselbe schon oben abgethan: sie gelten wohl genau, d. h. sie gelten, wenn man sie in den richtigen Grenzen anwendet und da müssen sie natürlich auch gelten, d. h. es müssen eventuell ganz neue Axiome hinzutreten, wodurch eine ganze Wissenschaft entsteht. Z. B., wenn wir beobachten, dass der Stab beim Bewegen an einen anderen Ort nicht seine Länge beibehält (Wärmetheorie).

93
Das Parallelenaxiom

Durch einen Punkt giebt es eine und nur eine Gerade, welche eine gegebene Gerade nicht schneidet: Dieselbe heisst parallel.
Parabolische Geometrie.

Zunächst braucht man dies Parallelenaxiom nur für eine Gerade und für einen Punkt anzunehmen; dann folgt es schon für alle.[152]
Dies folgt, wenn man unsere Congruenzsätze und zwar Verschiebung und Drehung verwendet.

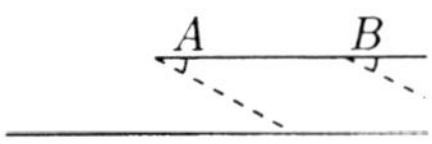

Wenn B ein anderer Punkt der Parallelen ist, so geht zunächst durch B keine zweite Parallele. Denn sonst trage man den gleichen Winkel in A an und verschiebe die Figur nach rechts um AB.[153]
Doch folgt alles schon aus dem Umstande, dass auf der Geraden zusammen fallende Grenzzahlen vorhanden sind. Die Drehung und Abtragung gleicher Strecken und Winkel in neben bezeichneter Weise zeigt das Gewünschte.

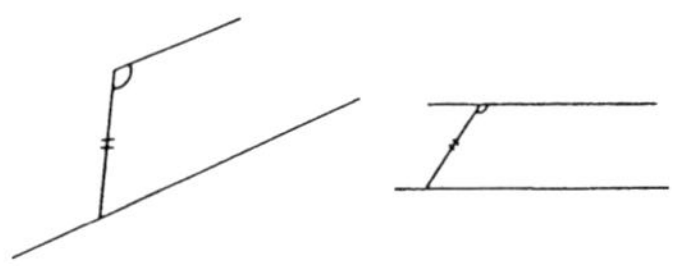

94 Die Gleichheit der Wechselwinkel: Wäre $\alpha' > \alpha$, so erhält man durch Auflegen a' und dann ist ferner $\beta' < \beta$ daher b', a' und b' müssen sich schneiden folglich auch a und b, was nicht sein soll. Daraus

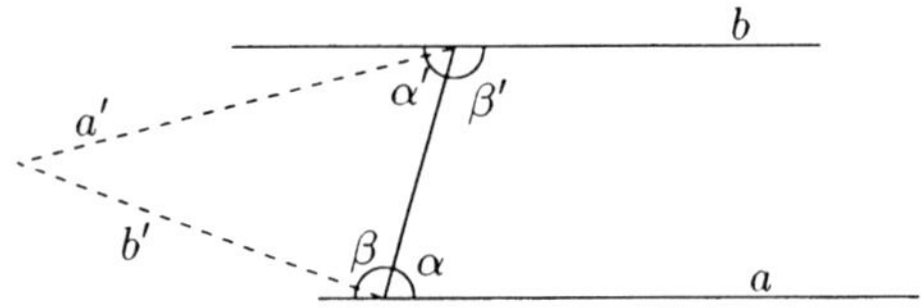

---

[151] *Boltzmann 1892.*

[152] Deleted by Hi^s: Denn Gleichheit der Wechselwinkel folgt durch auf einander Decken

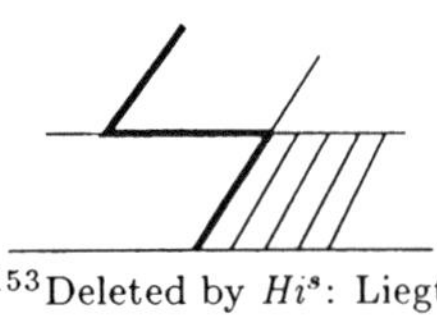

[153] Deleted by Hi^s: Liegt B ausserhalb der Parallelen, so ziehe man durch B eine Parallele, gäbe es noch ein zweite, welche a nicht schneidet, so trage in A den Winkel an

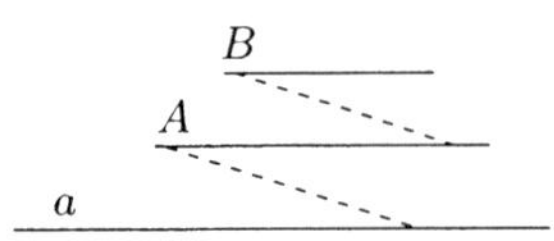

erst folgt Gleichheit der Wechselwinkel. Winkelsumme im Dreiecke = 2 Rechten.

Um recht die Bedeutung des Parallelenaxioms zu erkennen, nehmen wir jetzt das Gegenteil an: a, b seien die Grenzgeraden. [154]

Ich will in dieser Vorlesung den Stoff so ergänzen, dass Ihnen ein möglichst einheitliches Bild vor Augen steht.

In manchen Lehrbücher findet sich der Satz „die Gerade ist die kürzeste Ver- 95
bindung zwischen 2 Punkten.“ Dies hat nur dann einen Sinn, wenn Entfernung definirt ist.[155] Dies ist bei uns der Fall schon auf Grund der ersten 3 Arten von Axiomen: Vgl. meinen Brief an Klein: Ueber die gerade Linie als kürzeste Verbindung zweier Punkte.[156]

Deutung der hyperbolischen Geraden durch Kreise, die die x Achse orthogonal schneiden. Inhalt eines Dreieckes $= 4(\pi - \alpha - \beta - \gamma)$; vgl. Lindemann IIa, S. 494[157].

[154] Deleted:

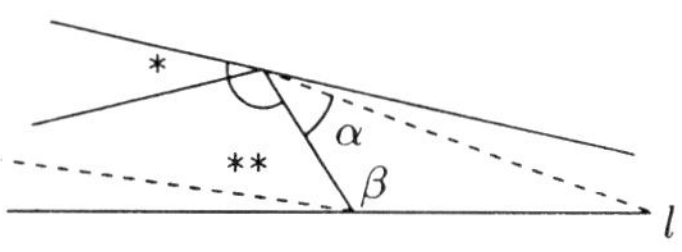

welche nicht schneiden, so zeigen wir zunächst von den Wechselwinkeln $\alpha > \alpha'$. a und b seien die beiden Grenzstrahlen. Wäre nämlich $\alpha \leqq \alpha'$, so trage α ab; es schneidet der erhaltene Schenkel b' die Gerade l

[155] A similar criticism can be found in *Clebsch 1891*, p. 543:

> Dagegen erscheint es unthunlich, eine Gerade als kürzesten Weg zwischen zwei Punkten zu definiren (wie es z. B. «Mehler»$_s$ thut: Hauptsätze der Elementar-Mathematik, 16. Auflage, Berlin 1889, p. 1); denn um eine Entfernung zu messen, muss vorher eine Curve vorhanden sein, längs welcher gemessen werden soll. Das Messen einer Curve aber wird nur durch das Ziehen von Sehnen nach den Begriffen der Integralrechnung verständlich.

[156] *Hilbert 1895b*. The 'kinds of axiom' referred to are: (A) incidence axioms, (B) order axioms, and (C) the continuity axiom.

[157] *Clebsch 1891*.

Appendix: Hilbert's Additions on pp. 2–5

2 Meine Herren. Um nicht durch kritische und literarische Hinweise den Gang der Vorlesung zu unterbrechen, erwähne ich hier schon

1.) Lehrbücher: Killing, Lindemann IIb, S. 433, Pasch, Frischauf (2 Bücher).[158]

2.) Einzelne grundlegende Abhandl.: Riemann, Helmholtz, Lobatschewsky, Klein (2 Abhandl.), Lie, Transformationsgruppen III.[159]

3.) Philosophen: Hoüel, Wundt, Lotze. (Nicht genügend gründlich.)[160]

Vgl. meinen Brief an Klein: Ueber die Gerade als die kürzeste Verbindung 2er Punkte und die am Ende dieses Heftes zu demselben gemachten Erläuterungen.[161]

Es kommt nicht darauf an, im einzeln keine überflüssigen Axiome zu haben. Daher fasse die Axiome über den Begriff zwischen so zusammen: Die Punkte einer Geraden lassen sich ordnen, $AB \ldots KL$ oder $LK \ldots BA$, auf eine und nur eine Weise. Greife ich aus diesen Punkten $A'B' \ldots K'L'$ heraus in dieser Reihenfolge, so sind diese Punkte wieder geordnet. Wenn $ABCD$, so sagt man, AC werde durch BD getrennt. Dann hat man auch den Vorzug, dass man die elliptische Geometrie mitnimmt.

Wenn ich wieder lese, so erst die Euklidische Geometrie.[162]

---

[158]The books referred to here are *Killing 1893*, *Clebsch 1891*, 433–637, *Pasch 1882*, *Frischauf 1872* and *1876*. Because the preface of *Killing 1893* is dated 1 September, 1893, the additions on this p. 2 were not written in the summer semester 1893 when Hilbert first worked on this manuscript. The second volume of Killing's book appeared only in 1898 (*Killing 1898*). Note that 'Lindemann' refers to *Clebsch 1891*, the second half of which was almost completely written by Lindemann who was the editor. It is not clear what Hilbert means with 'IIb'. *Clebsch 1891* is the first part of the second volume, and the pages 433–637 represent the third section (*Dritte Abtheilung*) of this first part. The same uncertainty holds of Hilbert's designation 'Bd. III' used on pp. 4, 23. Later on he writes 'Bd. IIa' (p. 87), 'Bd. 2a' (p. 92), and 'IIa' (p. 95) which more accurately reflects the original division into volumes and parts. Frischauf wrote several books. Hilbert probably refers to *Frischauf 1872*, *1876* since he also mentions these on p. 3 and because they seem to be the most pertinent of Frischauf's works. Later on, Hilbert confined himself to the textbooks of Killing and Pasch (see his *1898c**, 8, *1899a**, 4).

[159]In the light of Hilbert's Bibliography on pp. 3–5, the articles referred to here are probably *Riemann 1876*, 254–269 (first edition: *1866/67*), *Helmholtz 1868* and *1876a*, *Lobachevsky 1837* and *1840*, *Klein 1871* and *1873*, *Lie 1893*, although this is a book, not an article. On p. 4, Hilbert also mentions Klein's *Erlanger Programm*. Nevertheless, Hilbert's references are probably to *Klein 1871*, *1873* since (*i*) Klein's papers from *1871* and *1873* are closer to the subject of Hilbert's lectures than *Klein 1872* (where the *Erlanger Programm* was first set out), and (*ii*) on p. 92 Hilbert again cites *Klein 1871*, *1873*.

[160]On pp. 3, 5 Hilbert refers to *Hoüel 1876*, *Wundt 1893*, and *Lotze 1879*.

[161]The letter to Klein was published as *Hilbert 1895b*, which is briefly discussed on p. 95.

[162]Hilbert's later lectures *did* focus on Euclidean geometry. See in particular Chapter 4 in this Volume.

Lie's Theorie ist zu verwerfen, weil das erste Axiom, dass der Raum eine Zahlenmannigfaltigkeit, zu umfangreich. Will man das aber, so kann man doch lieber sagen, dass 2 Punkte eine Gerade bestimmen, dass ferner bei der Congruenz Geraden in Geraden etc. übergehen. Dann ist Alles sehr einfach.[163]

Bolyai und Lobatschewsky (vgl. Stäckel-Engel und Frischauf, Teubner) haben direkt ohne Durchgang durch die projektive Geometrie, d. h. auf metrische Weise, die Zahl in die Geometrie eingeführt, indem sie den Begriff der Entfernung (also Kugel im Raume) an die Spitze stellen. Denke auch diesen Weg durch.[164]

Axiome der Geometrie. 3

Helmholtz, Ueber die Thatsachen, welche der Geometrie zu Grunde liegen. Göttinger Nach. 1868, und Populär Wissenschaftliche Vorträge, Braunschweig 1871–1876.[165]
Lobatschewsky, ↑Besser Werke Bd. 2↓ Geometrische Untersuchungen über die Theorie der Parallellinien, Berlin 1840 und Crelle Bd. 17.[166]
Frischauf, Absolute Geometrie nach Bolyai.
— Elemente der absoluten Geometrie.
Killing, Die Nicht-Euklidischen Raumformen.
Peano, Die Grundzüge des geometrischen Calculs, deutsch von Schepp.
Schlegel, System der Raumlehre.
Hoüel, Ueber die Rolle der Erfahrung in den exacten Wissenschaften, Grunerts Archiv Bd. 59.[167]
Pasch, Vorlesungen über neuere Geometrie. 4
Beltrami, Interpretation der Nicht-Euklidischen Geometrie, Annal. de l'École Normale Bd. 6, 1869.
Cremona, Ueber Grassmann, Math. Ann. Bd. 14.[168]
Schlegel, Ueber Grassmann, Zeitschrift für Math. Bd. 24.
Klein, Ueber die sogenannte Nicht-Euklidische Geometrie, Math. Ann. Bd. 4 und 6.
Killing, Ueber zwei Raumformen mit constanter Krümmung, Crelle Bd. 86, S. 72.[169]

---

[163] See *Hilbert 1902f*, which discusses Lie's theory.

[164] The references are to *Stäckel and Engel 1895*, *Frischauf 1872*, and/or *Frischauf 1876*.

[165] *Helmholtz 1868*, *1876a*.

[166] The reference to 'Werke Bd. 2' is to *Lobachevsky 1886*. The other references are to *Lobachevsky 1840*, *1837*.

[167] The references to this point are: *Frischauf 1872*, *1876*, *Killing 1885*, *Peano 1891*, *Schlegel 1872*, and *Hoüel 1876*.

[168] This is probably a reference to *Sturm, Schröder and Sohnke 1878*. It is unclear why Hilbert cites Cremona as an author. The reference to Beltrami just before this is clearly to *Beltrami 1869*, which is a French translation of part of *Beltrami 1868*. Hilbert has translated the French title fairly directly into German. The reference to Pasch is to *Pasch 1882*.

[169] These references are *Schlegel 1896*, *Klein 1871*, *1873*, and *Killing 1878*.

Sophus Lie, Ueber die Grundlagen der Geometrie (Erste Abhandlung). Berichte der Gesellschaft der Wissenschaften zu Leipzig. Math. physik. Cl. 42, 1890, S. 284.
Zweite Abhandlung, S. 354.[170]
Riemann, Ueber die Hypothesen, die der Geometrie zu Grunde liegen, ges. Werke.[171]
⟨Two lines deleted.⟩[172]
Klein, Vergleichende Betrachtungen über neuere geometrische Forschungen. Erlangen 1872.
Clebsch-Lindemann, Bd. III, S. 433.[173]
Killing, Meine eingebundenen Separate „Geometrie" grosses Format.[174]
Schur, Vgl. meine ungebundenen Separate. Math. Ann. Bd. 39.[175]
Baltzer, Ueber die Hypothesen der Parallelentheorie, Crelle Bd. 73.
Worpitzky, Ueber die Grundbegriffe der Geometrie, Grunert Arch. 405–421.
Möbius, Der barycentrische Calcül.
v. Staudt, Beiträge zur Geometrie der Lage.
Baltzer, Elemente der Mathematik, 2 Bd.
Thomae, Ebene geometrische Gebilde 1. und 2. Ordnung. Halle Nebert 1873.
Reye, Geometrie der Lage, insbesondere die metrischen Sätze.
Lüroth, Math. Ann. Bd. 8, S. 207.[176]
Wundt, Logik. 2te Auflage 1893.[177]
Auch Lotze, System der Philosophie, Metaphysik, Theil 2.[178]
Lindemann, Separatabzug über die Hypothesen der Geometrie.[179]
Killing, Grundlagen der Geometrie.
Soph. Lie, Theorie der Transformationsgruppen Bd. III.
Hilbert, Gerade Linie als kürzeste Verbindung 2er Punkte, Math. Ann. Bd. 46.
Stolz, allgemeine Arithmetik, 2 Bd. Teubner 1885.
Max Simon, Die Elemente der Geometrie mit Rücksicht auf absolute Geometrie.[180]

---

[170] The two works of Lie are *Lie 1890a*, *1890b*.

[171] *Riemann 1876*, 254–269, i.e., *1866/67*, and originally written in 1854. The title of the memoir reads 'Ueber die Hypothesen, welche der Geometrie zu Grunde liegen.'

[172] Deleted: Meine eingebundenen Separate kleines Format Geometrie und Mechanik.
grosses Format Geometrie (Killing).

[173] What is meant is the third *part* of *Clebsch 1891* which begins on p. 433.

[174] The reference to Killing is not clear. The evidence from *Verzeichnis 1943** is that the offprint of *Killing 1892* is meant.

[175] *Schur 1891*. The article by Schur must have been one among various offprints in Hilbert's possession from volume 39 of the *Mathematische Annalen*.

[176] These references (following that to Schur) are: *Baltzer 1871*, *Worpitzky 1873*, *Möbius 1827*, *von Staudt 1856*, *1857*, *1860*, *Baltzer 1868*, *1870*. *Thomae 1873*, *Reye 1886*, *1892b*, *1892a*, *Lüroth 1875*.

[177] The second edition of Wundt's *Logik* appeared in three volumes (*1893*, *1894*, *1895*) and an index volume (*1902*).

[178] *Lotze 1879*.

[179] This probably refers to an offprint of *Lindemann 1891*.

[180] The references after Lindemann are: *Killing 1892*, *Lie 1893*, *Hilbert 1895b*, *Stolz 1885*, *1886*, and *Simon 1890*.

⟨One line deleted.⟩[181]
F. Klein, Autographirte Vorlesung über Nicht-Euklidische Geometrie.[182]
F. Klein, " " " neuere Geometrie.[183]

[181]Deleted by Hi^s: Cantor, Math. Ann. Bd. 46, 49. ⟨These references are to *Cantor 1895*, *1897*.⟩

[182]*Klein 1890*, *1892*.

[183]Presumably, the references are to *Klein 1893b*, *1893a*.

Textual Notes

72.9: geschieht] geschiet
72.9: gewisser] ⟦der⟧⊣↑gewisser↓ ⟨substituted by Hi^g⟩
72.13: so weit] ⟦bereits⟧ ⟨deleted by Hi^g⟩ so weit
72.15: oder Optik] ⟨'oder' added in pencil, 'oder' then written again, this time in ink; 'Optik' also added in ink.⟩
72.17: *Naturwissenschaft*] ⟦*reine*⟧ ⟨deleted by Hi^s⟩ *Naturwissenschaft*
72.17: Da] ⟦Da⟦n⟧ ⟨deleted by Hi^s⟩⟧⊣↑Es ⟪könn eben⟫↓ ⟨substituted by Hi^g⟩
72.19: so ist] so ist ⟦es eine⟧ ⟨deleted by Hi^s⟩
72.22: *vor mehr als 2000*] *vor* ↑mehr als↓ *2000*
72.23: Scharfsinn,] Scharfsinn
72.24: Wenngleich] ⟦W⟧⊣↑w↓enngleich
72n: Hypothese] Hypothe
73.6: unterscheiden.] unterscheiden. ⟦1.) Existenzaxiome⟧ ⟨deleted by Hi^s⟩
73.8: *A. Existenzaxiome*] *A.) Existenzaxiome*
73.11: *Irgend*] ⟦⟨erased⟩⟧⊣↑*Irgend*↓
73.11: *gelegene*] *gelegen*
73.11: *bestimmen*] *bestimme*
73.12: Formeln:] Formeln
73.13: *3.)*] *3.*
73.13: *gelegene*] *gelegenen*
73.15: heissen] heissen ⟦auf⟧ ⟨deleted by Hi^s⟩
73.15: gelegen] gelegen⟦.⟧ ⟨deleted by Hi^s⟩
73.17: *4.)*] *4.*
73.20: Wenn] ⟦5.⟧ ⟨deleted by Hi^s⟩ Wenn
73.20: einer] eine
73.21: und] u.
73.23: 5.)] 5.
73.25: *6.)*] *6.*
73.25: *einen*] *ein*
73.25: *Punkt*] *Punkt* ⟦*A*⟧ ⟨deleted by Hi^s⟩
73.25: *wenigstens*] *wengsten*
73.27: Zugleich] ⟪zug⟫
73n: *A. Existenzaxiome.*] *A.) Existenzaxiome.*
73n: und] u.
74.3: *7.)*] *7.*
74.4: *8.)*] *8.*
74.5: *Geraden*] *geraden*
74.7–8: verbinden,] ↑verbinden↓ ⟨added by HiK^s⟩
74.9: Geraden] Gerade
74.13: sie] sich
74.15: Hertz] Herz
74.15: Bilder] Bilde
74n: , $A \neq C$] ⟨Added by HiK^s.⟩
74n: $AB \not> C$] ⟨Added by HiK^s.⟩
74n: $AB \not> D$] ⟨Added by HiK^s.⟩
75.1: 1.] 1
75.1: welcher nicht auf derselben liegt] ⟦ausserhalb derselben⟧⊣↑welcher nicht auf derselben liegt↓

75.2: Ebene.] ⟨Added to the right of the two following groups of formulae, and then crossed out: '> heisst „liegt auf" oder „enthält"'; 'enthält' had been added over 'liegt auf'.⟩

75.6: geeigneten] ↑gezeigneten↓

75.18: Untersuchung] ⟦Wissenschaft⟧ ⟨deleted by Hi^s⟩ Untersuchung

75.19: „überflüssigen" und] „überflüssig↑en↓" ⟦und⟧ ⟨deleted by Hi^s⟩ und

75.20: 2.] 2

75.23: sagen:] sagen

75.23: *gelegne*] *gelegnen*

75.25: schliessen:] schliessen

75.31: *nur:*] *nur*

75.33: 3.] 3

75.33–1: und nur eine] ⟨Addition. The text indicates that HiK^s had added 'und nur eine' before 'Schnittlinie' was written.⟩

75n: Wortdefinitionen] ⟦Defini⟧ ⟨deleted by Hi^s⟩ Wortdefinitionen

76.1: 4.] 4

76.2: Verbindungsgerade).] Verbindungsgerade.)

76.3: Daher:] Daher

76.7: oder Treffpunkt] ↑oder Treffpunkt.↓

76.8–12: ↑Jedes ⟨...⟩ ein.↓] ⟨Addition on a small strip of paper pasted in.⟩

76.10: so dass] so dass ⟦jedes der ⟪8 A⟫⟧ ⟨deleted by Hi^s⟩

76.16: *B. Lagenaxiome.*] ⟨Covered by the sheet pasted to p. 13.⟩

76.17–19: 1.) Wenn ⟨...⟩ B und A.] ⟨Reading occasionally unsafe, since this place is covered by the sheet pasted to p. 13.⟩

76.18: Punkt] Punkt,

76.21: so] ⟦↑oder zwischen B und A↓ ⟨added by HiK^s⟩⟧ so

76.24: 2 und 3] 2 u. 3

76.25: und nur] u. nur

76.25: den beiden] d. beiden

76.33: und B,] u. B,

76.33: liegen] liegen ⟦(3)⟧ ⟨deleted by Hi^s⟩

77.1: und C,] u. C,

77.1: zwischen] zwisch

76n: und nur] u. nur

77.1: und D liegen] u. D liegen

77.1: und C] u. C

77.2: und D] u. D

77.5: und A_3,] u. A_3,

77.6: zweite] Zweite

77.8: zwischen A] zwischen A ⟦u. A_n⟧ ⟨deleted by Hi^s⟩

77.8–9: mögen so zwischen] mgen so zwisch

77.9: und D] u. D

77.10: zwischen] zwisch

77.10: und A_2] u. A_2

77.10: kein] kein ⟦⟪?⟫ ⟧ ⟨deleted by Hi^s⟩

77.11: völlig] vollig⟦. B⟧ ⟨deleted by Hi^s⟩

77.11: umgekehrte] Umgekehrte

77.14: Punkt besitzt] Punkt ⟦ent⟧ ⟨deleted by Hi^s⟩ besitzt

77.15: und B] u. B

77.16: Lagenaxiom] ⟦Axio⟧ ⟨deleted by Hi^s⟩ Lagenaxiom

77.17: 3 Punkte A, B, C] 3 Punkt ⟦A⟧ ⟨deleted by Hi^s⟩ A, B, C ⟦g⟧ ⟨deleted by Hi^s⟩
77.20: 2 Teile] 2 ⟪Teile⟫
77.21: Gera] ⟦Eben⟧↑Gera↓
77.22: Gerade] ⟦Ebene⟧↑Gerade↓
77.23: und B] u. B
77.24: A und] A u.
77.24: und C] u. C
77.24: derselben] derselb
77.27: *B. Lagenaxiome.*] *B Lagenaxiome.*
77n: auf 2 und] auf 2 ⟦Arten geordnet werden⟧ ⟨deleted by Hi^s⟩ und
77n: der anderen] der ⟦anderen⟧ ⟨deleted by Hi^s⟩ ⟦ersten⟧ ⟨deleted by Hi^s⟩ anderen
78.3: Beweis: Seien] Beweis seien
78.8: werden:] werden
78.9: so fort] s. f.
78.11: die umgekehrte:] ↑die umgekehrte:↓
78.12: Beweis: Denn] Beweis denn
78.14: und N, und] und ⟦M⟧↑N↓ ⟨substituted by HiK^s⟩, ⟦so⟧ ⟨deleted by Hi^s⟩ und
78.15: darf] ⟦muss⟧↑darf↓
78.16: liegen.] liegen
78.21: Lagenaxiom aus.] Lagenaxiom aus
78.22: gelegne] gelegnen
78.26: eintritt] ein tritt
78.28–29: 2 Abteilungen] 2 ⟦T⟧ ⟨deleted by Hi^s⟩ Abteilungen
78.32: und B] u. B
78.32: und C] u. C
78.34: und B] u. B
78.34: und A] u. A
78.35: und C] u. C
78.36: und B] u. B
78.36: A und] A u.
78.37: und C] u. C
78.38: und B] u. B
78.40: und B] u. B
78.40: und nicht durch C geht] ↑und nicht durch C geht↓
79.1: und C] u. C
79.3: 2 Punkte des Raumes liegen auf] ⟦Die⟧↑2↓ Punkte des Raume ⟦zerfallen in⟧↑liegen auf↓
79.5: und C] u. C
79.6: 6] ⟦⟪erased⟫⟧↑6↓
79.6: einander] einander ⟦unabhängig.⟧ ⟨deleted by Hi^s⟩
79.21: so,] so
79.25: Bündels,] Bündels
79.26: gehen,] gehen
80.1: auf derselben Seite] ⟨With the word 'auf' begins the second piece of paper pasted to p. 20. Through it, text on the lower edge of the first sheet is covered, beginning 'auf derselben Seite von α liegen. Da⟪n⟫ lege ⟪durch⟫ ABC, $ABD, \ldots,$ ABK ⟦⟪?⟫ ⟧ ⟨deleted by Hi^s⟩ je⟪?⟫ eine

Ebene $\gamma, \delta, \ldots, \kappa$. und'. Furthermore, a drawing has been pasted over which appears to correspond to the figure which follows.⟩

80.4: so,] so

80.6: welcher] welche

80.7: liegt] liegt,

80.7–8: so, dass] so dass

80.10: Strahlen $c, d, \ldots, k$ in] Strahlen $c, d, \ldots k$⟦. Ist c der⟧ ⟨deleted by Hi^s⟩ in

80.11: nächstgelegene] nächst gelegene

80.13: beispielsweise] beispielweise

80.18–19: ↑Vielleicht ⟨...⟩ überflüssig.↓] ⟨The addition is to be found on the third sheet pasted in.⟩

80.23: c, d] $c\,d$

80.24: a, b, c] $a\,b\,c$

80.28: wird] ↑wird↓

80.1: A'] A',

81.5–9: ↑Satz: ⟨...⟩ früheren Satze.↓] ⟨Addition, initially by Hi^s in the lower margin of the sheet, then continued by HiK^s on a sheet pasted in. HiK^s had originally written 'bc', 'ad', 'ca', 'ab' and 'cd'. These are replaced by Hi^s with 'ab', 'cd', 'ac', 'ad' and 'bc' respectively.⟩

81.7–8: Beweis: ⟨...⟩ ordne dann] ⟦Beweis man ordne⟧⫣↑Beweis: Man ziehe eine Gerade ⟦ $\underline{A\;\;B\;\;C\;\;D}$⟧ ⟨deleted by Hi^s⟩ $ABCD$ durch die 4 Strahlen und ordne dann↓

81.8: Punkte,] Punkte

81.8: $ABCD$,] $ABCD$

81.9: früheren Satze] ⟪Frühen⟫ Satze

81.10: des] ⟦vom⟧⫣↑des↓

81.11: Wort] ↑Wort↓ Definition

81.12–15: Satz ⟨...⟩ Satz.] ⟨Added by Hi^s.⟩

81.14: C, D] $C\,D$

81.21: Separatabzug] Separat↑abzug↓

81n: dem im folgenden] die im folg.

81n: Dreiecks⟨ebenen⟩] 3eckslinien

81n: sei b. Auf] sei b auf

82.6: und C] u. C

82.7: nimmt] nimt

82.9: und C,] u. C.

82.9: liegt AG] ⟦schneidet AG⟧ ⟨deleted by Hi^s⟩ liegt AG

82.12: und G,] ↑u.↓ G,

82.13: und G] ↑u.↓ G

82.15–16: (Denn ... zwischen B und C.)] ⟨Possibly an addition.⟩

82.18: und G] u. G

82.20: G', H'] $G'\ H'$

82.24: liegen] ↑liegen↓

82.24: E, E', M] E, $E'\ M$

82.24: in derselben Ebene] ↑in derselben Ebene↓

82.28: mit M] ⟨Added by Hi^g, then emphasized by Hi^s.⟩

82.28: mit M] ⟨Added by Hi^g, then emphasized by Hi^s.⟩

83.1:	Jetzt] Setzt
83.2:	$CG''H''$] C⟦H''⟧⟨deleted by Hi^s⟩$G''H''$
83.3:	Die] ⟦↑Die Con↓⟧ ⟨deleted by Hi^s⟩ Die
83.4:	harmonische] Harmonische
83.9:	Strahl $CG''H''$] Strahl CG''⟦G⟧⟨deleted by Hi^s⟩H''
83.27:	] ⟨The text from '-monisch' to the end of the addition (line 84.15) is contained on a sheet pasted to the bottom of p. 26. This has obscured one or more lines of the original text.⟩
83.31:	Geraden] Gerade
83.33:	1.)] ↑1.)↓
83.35:	$L'B$] LB
83.35:	mit PC] mit ⟦H⟧⟨deleted by Hi^s⟩PC
83n:	Wenn man ⟨...⟩ harmonische Strahlen.] ⟨Covered by the sheet pasted in, and two or three lines completely obscured.⟩
84.3:	Punkt] Punkt ⟦LMD'⟧ ⟨deleted by Hi^s⟩
84.5:	Harmonischen] Harmonisch⟨en⟩
84.6:	denn:] denn
84.10:	Nun] ⟦Nun beweise die wesentlichste Eigenschaft harmonischer Strahlen:⟧ ⟨deleted by Hi^s⟩ Nun
84.12:	Verbindungslinien] Verbindungslienien
84.14–15:	sämtlich treffen] sämtlich⟦en⟧ ⟨deleted by Hi^s⟩ treffen⟦de⟧ ⟨deleted by Hi^s⟩
84.22:	und C,] u. C,
84.23:	nicht] ⟦weder⟧⊣↑nicht↓
84.24:	liegt und so, dass $A'B'C'D$ nicht zwischen C'', D'' liegt] ⟦noch zwischen A' und C' liegt⟧⊣↑liegt und so dass $A'B'C'D$ nicht zwischen C'', D'' liegt.↓
85.16:	D''',] D'''
85.17:	D',] D'
85.22:	Punkten] Punkt↑en↓
86.1:	Doppelverhältniss der] Doppelverhältniss ⟦(⟧ ⟨deleted by Hi^s⟩der
86.2:	α, x, β, γ.] $\alpha, x, \beta, \gamma)$
86.5:	positiven] ↑positiven↓ ⟨added by HiK^s⟩
86.8:	harmonische] Harmonische
86.9:	und] und ⟦∞ ↑zu 0↓ ⟨added by HiK^s⟩ gelegne⟧ ⟨deleted by Hi^s⟩
86.13:	positive] posit.
86.15:	Construktion] Construkt
86.16:	2te] 2
86.17:	nöthig,] nothig
86.21:	1^{en} Construktion] 1sten Const
86.21–22:	2ten Construktion] ⟦1⟧⊣↑2↓^{sten} Const.
86.1:	Construktion] Constr.
87.1:	1] 1.
87.3:	denselben] den selben
87.5:	verschie⟨p. 31⟩dene] verschie ⟨p. 31⟩⟦dene⟧
87.5:	Punkte] ⟦Ver⟧schiedene Punkte
87.10:	da] ⟦wie ⟦D.⟧ ⟨deleted by Hi^s⟩ früher gezeigt, wenn⟧⊣↑da↓ ⟨substituted by HiK^s⟩
87.11:	desgleichen] ⟦wenn⟧ ⟨deleted by HiK^s⟩ desgleichen
87.11:	$\frac{4}{2} = 2$;] $\frac{4}{2} = 2$

88.1: $P\infty$] $P\infty$⟦,⟧⟨deleted by HiK^s⟩

88.5: nachzuweisen, hat] nachzuweisen hat,

88.7: harmonischen] harmonische⟦n⟧

88.7–8: Construktion gefundenen] ↑Construktion gefundenen↓

88.9: α, β, γ 3 solchen] α, β, γ, 3 solchen

88.13: p-malige] p malige

88.18: bestimmte] bestimmte,

89.15: Übrigens] Ubrigens

89.18: Nun] ⟦Um nun⟧⊣↑Nun↓ ⟨substituted by HiK^s⟩

90.3: gleichmässig] ↑gleichmässig↓

90.4: müssten] müsste

90.5: den] die

90.11: der 4te harmonische] die 4te harmonische

90.11: für jede negative] ↑für↓ jeder negativen

90.14: 1.] ↑1.↓

90.14: der Punkt] ↑der Punkt↓

90.15: Punkt] ↑Punkt↓

90.17: 2.) Wenn] ⟦2.) Wenn $0 > \alpha$ $\alpha < \beta <$⟧ ⟨deleted by Hi^s⟩ 2.) Wenn

90.21: wo ... sind.] ⟨Possibly an addition.⟩

90.22: Existenzsätze:] Existenzsätze

90.23: 1.] ↑1.↓

90.24: negativen] negativen ⟦$> \alpha$⟧ ⟨deleted by Hi^s⟩

90.25: zwischen] zwisch

90.26: Zahlen $< \alpha$ Punkte] Zahlen $< \alpha$, Punkte

91.1: von dem] vom dem

91.13: Dann] Dan

91.16: diejenigen] diejenige

91.21: $(1, 4')$] $(1\,4')$

91.21: denn] den

91.24: $(2, 2')$] $(2\,2')$

91.25: je] ↑je↓

91.30: $(n, (p-n)'), P$] $(n, (p-n)')\ P$

91.32: Erweiterung] ⟦Construktion⟧ ⟨deleted by Hi^s⟩ Erweiterung

91.32: negativen Seite] ⟨From 'negativen', written on the reverse of the sheet pasted in.⟩

91.36: zusammen] zusammen mit der Frage

91.36: Wiener,] Wiener

91.37: S. 12,] S. 12

92.2: C. *Das Stetigkeitsaxiom.*] C *Das Stetigkeitsaxiom.*

92.3: positiven] ↑positiven↓

92.7: bestimme] bestime

92.8: und] u.

92.9: und $a, a_1 + 1$] und $a, a_1 +$ ⟦a⟧⟨deleted by Hi^s⟩1

92.9: und] u.

92.11–12: Z. B. wenn] z. B. Wenn

92.12: ∞] ⟦unendlich⟧ ⟨deleted by Hi^s⟩ ∞

92.12: $\frac{1}{8}, \ldots$] $\frac{1}{8}$,

92.18: und dass] und das

92.19: Punkten der] ↑der↓ ⟦P_1, P_2⟧ ⟨deleted by Hi^s⟩

92.26:	und] u.
92n:	dass] dass der
93.1:	diese Punkte] diesen Punkten
93.5:	Grenzgeraden] Grenzgerade
93.8:	A, B, C] ↑A, B, C↓ ⟨added by Hi^g⟩
93.12:	zugeordneten] zugeordnet
93.16:	gelegenen] gelegen
93.18:	harmonischen Punkten] harmonische Punkt
93.22:	sind im] sind (im
93.23:	aequivalent] aequivant
93.23:	Dass] Das
94.1:	beweisen wir:] beweisen wir
94.6:	so,] so
94.11:	$(\alpha, \kappa', \beta, \gamma) = (0, \kappa, 1, \infty) = -1$] $(\alpha, \kappa', \beta, \gamma = (0, \kappa, 1, \infty) = -1$
94.18:	Bezeichnung] Bezeichng
94.20:	Reihe] ⟦Zahl⟧ ⟨deleted by Hi^s⟩ Reihe
95.4:	verlege] ⟦verändre⟧ ⟨deleted by Hi^s⟩ verlege
95.8:	⟨In this paragraph and on p. 46 following, the letter 'δ' (with the single exception on line 96.3) is the result of: ⟦⟦d⟧↑D↓⟧↑δ↓.⟩
95.8:	Projection;] Projection
95.13:	⟨In this paragraph, the letters 'C' and 'D' are the result of the substitutions: ⟦D⟧↑C↓, ⟦C⟧↑D↓. This does not apply to the letters 'C'' and 'D''.⟩
95.14:	$\delta = \frac{\varrho}{2^n}$] ⟨$Hi^s$ has replaced 'rational' by '$\delta = \frac{\varrho}{2^n}$'.⟩
95.14:	α] ↑α↓
95.16:	harmonischen] ⟦Construk⟧ ⟨deleted by Hi^s⟩ harmonischen
95n:	C] ⟦D⟧↑C↓
95n:	4ten] 4
96.1:	A', B', C'] A' B' C'
96.3:	δ] ⟦d⟧↑δ↓
96.8:	$D = \alpha$] ↑$D =$ ↓α
96.17–18:	Doppelverhältnisse] Doppelverhältnss
96.19:	Umgekehrt:] Umgekehrt
96.20:	Doppelverhältnisse] Doppelverhältnss
96.25:	*und*] *u.*
96.33:	$A = 0$, $C = \infty$.] $A = 0$ $C = \infty$
96n:	zukommen] zukomen
96n:	und der] ⟦u der Punkt⟧ ⟨deleted by Hi^s⟩ ⟦der⟧ ⟨deleted by Hi^s⟩ und der
97.5:	Gleichungen] Gl.
97.7:	Gleichungen] Gl.
97.7:	denn] Denn
97.8:	endliche] endlich
97.9:	D. h. ausser] d. h. Ausser
97.11:	jedenfalls] jeden falls
97.14:	welcher] welche
97.14:	und] u.
97.14:	möge.] mögen.
97n:	(2.)).] ↑(2.))↓
98.6:	Gleichung] Gl.

98.7: $\frac{1}{\alpha''}$] $\frac{1}{\alpha''}$,

98.10: Gleichung] Gl.

98.12: voraussetzt,] vorausetzt,

98.13: ist ⟨...⟩ selbstverständlich.] ⟦⟪sieht⟫ man die ⟪???⟫ umso leichter ein. ⟦⟪???⟫ dies ist immer der Fall.⟧ ⟨deleted by *Hi*[s]⟩⟧⊢↑ist Richtigkeit der Behauptung selbstverständlich.↓

98.13: Die] ⟦Eine⟧ ⟨deleted by *Hi*[s]⟩ Die

98.14: Jeder] Jede

98.15: Coordinaten] Cordinaten

98.16: Gleichungen] Gl.

98.16: wenn α] wenn α⟦$'$⟧

98.16: angenommen] angenmmen

98.18: Coordinaten] Cordinaten

98.21: wie die] ⟦als⟧ ⟨deleted by *Hi*[s]⟩ wie die

98.25: Uebelstande] Ueberstande

98.26: statt] ⟪Sat⟫

98.26–27: 3 Grössen] 3 ⟪Grossen⟫⟦, wel⟧ ⟨deleted by *Hi*[s]⟩

98.27: und] ↑und↓

98.27–28: den Gleichungen ⟨...⟩ genügen.] ⟦⟨erased⟩⟧⊢↑den Gl. $\frac{x_1}{\alpha} + \frac{x_2}{\beta} - x_3 = 0$, $\frac{x_1}{\alpha'} + \frac{x_2}{\beta'} - x_3 = 0$ genügen.↓

98.30: so wird] ⟦und α⟧ ⟨deleted by *Hi*[s]⟩ so wird

98.30–31: $x = \frac{x_1}{x_3}$, $y = \frac{x_2}{x_3}$.] ⟦$x = \frac{x_1}{x_{⟪?⟫}}$ $y = {}^{x}$⟧ ⟨deleted by *Hi*[s]⟩ $x = \frac{x_1}{x_3}$ $y = \frac{x_2}{x_3}$

98.31–32: homogenen Gleichungen] homogen Gl.

99.7: $x_2 \neq 0$ also] ↑$x_2 \neq 0$ also↓

99.7: construire] ⟦C⟧⊢↑c↓onstruire

99.8: $\alpha = -\frac{1}{x_2}$,] α⟦⟨erased⟩⟧⊢↑ $= -\frac{1}{x_2}$↓,

99.17: ausnahmslos] ausnahmlos

99.20: sind,] sind

99.21: ϱx_2,] ϱx_2

99.22: keineswegs] keines Wegs

99.22: beliebig] beliegig

99.23: (x_1, x_2, x_3)] $(x_1\, x_2\, x_3)$

99.26: 2 Fälle:] 2 Fälle

100.2: a, a', a''] a', a'', a''

100.3: so ist] so so ist

100.8: $x_1 : x_2 : x_3 =$] $x_1 : x_2 : x_3 =$ ⟦$\beta - b : \alpha - a :$ ⟧⟨deleted by *Hi*[s]⟩

100.17: $\beta - b \neq 0$,] $\beta - b \neq 0$

100.17: $\alpha' - \alpha'', a' - a'' \neq 0$.] ↑$\alpha' - \alpha'', a' - a'' \neq 0$↓.

100.18: und] u.

100.25: x_1, x_2, x_3] $x_1\; x_2\; x_3$

100.25: Gleichung] Gl.

101.4: Gleichung] Gl.

101.8: nämliche] nämtliche

101.9: nicht] nich⟨t⟩

101.10: Gleichung] Gl.

101.18: λ_1, λ_2,] $\lambda_1\, \lambda_2$

101.18–19: den 4 Punkten] ↑den↓ 4 Punkten

101.21: und] u.

101.26:	positiv] ⟦> 0 für alle⟧ ⟨deleted by Hi^{s}⟩ positiv
101.28:	gelegenen] gelegen
101.29:	Die Coordinaten aller] ⟦Alle⟧⊣↑Die Coordinaten aller↓
101.1:	Gleichungen] Gleich.
102.2:	Ebenso] ⟦Uebrigens⟧ ⟨deleted by Hi^{s}⟩ Ebenso
102.8:	verbinde] verbinge
102.23:	Gleichungen] Gl.
102.23:	braucht] brauch
102.24:	einzuführen.] ein zu führen.
102.25:	durchlaufen] durch laufen
103.5–6:	Ebenenbüschel] Ebenbüschel
103.9:	als Punkte,] ⟦mit *P*⟧ ⟨deleted by Hi^{s}⟩ als Punkte
103.10:	d. h. ideale] ↑d. h.↓ idealen
103.13:	und] u.
103.15:	2 Ebenen] ⟦⟪?⟫⟧⊣↑2.↓ Ebene
103.15:	einer Geraden.] ein ⟦Punkt, ⟪we⟫⟧ ⟨deleted by Hi^{s}⟩ Geraden
103.16:	3 Ebenen,] ⟦2⟧⊣↑3↓ Eben⟨en⟩,
103.25:	in] In
103.25:	Ebene] Ebene⟦n⟧ ⟨deleted by Hi^{s}⟩
103.26:	und] u.
103.28:	Reciprocitätsgesetz] Reciprocitätsgetzt
103.29:	in] In
103.32:	Existenz-] Existenz
103.40:	Dráhthen] Dräthen
104.6:	beweglich] ⟦B⟧⊣↑b↓eweglich
104.13:	anwendet,] anwenden,
104.14:	mannigfaltigsten] mannigfaltesten
104.18:	bezieht,] bezieht
104.20:	Erscheinungen] Erscheinung
104.23:	die Thatsache] der Thatsache
104.25:	die Punkte] die Punktes
104.26:	verschiedenen] verschieden
104.26:	3 in] 3 ⟦nicht⟧ ⟨deleted by Hi^{s}⟩ in
104.27:	gelegene] gelegen
104.28:	kann] kann⟦. Et⟧ ⟨deleted by Hi^{s}⟩
105.1:	*D*.] ⟦die⟧ ⟨deleted by Hi^{s}⟩ *D*.
105.2:	zu nennen] zunennen
105.4:	1.) Wenn ⟨...⟩ und 2] ⟨Originally underlined by Hi^{b}, but subsequently erased.⟩
105.6:	auf] einen auf
106.1:	Maass] Maas
106.2:	und] u.
106.3:	2.) Wenn ⟨...⟩ so] ⟨Originally underlined by Hi^{b}, but subsequently erased.⟩
106.5:	3.) Wenn ⟨...⟩ auf] ⟨Originally underlined by Hi^{b} (except the addition), but subsequently erased.⟩
106.5:	3.)] 3.
106.5:	*B* zwischen] ⟦⟪*P*⟫⟧⊣↑*B*↓ zwischen
106.8:	der] den
106.8:	Reihenfolge] Reihenfolge,

106.10: 3] ↑3.↓

106.11: widersprechen] wiedersprechen

106.12: Satz:] Satz

106.13: widerspricht.] wiederspricht.

106.15: andern a'] andern ⟦b⟧⊣↑a'↓

106.16: auf a'] auf ⟦b⟧⊣↑a'↓

106.20: und A] und ⟦C⟧ ⟨deleted by Hi^s⟩ A

106.20: $a = ABC$] ⟦AC⟧⊣↑$a = ABC$↓ ⟨substituted by HiK^s⟩

106.21: $a' = A'B'C'$] ⟦$A'C'$⟧⊣↑$a' = A'B'C'$↓ ⟨substituted by HiK^s⟩

106.27: nebenan] neben an

106.27: und] u.

106.28: 1] ↑1↓

106.29: und] u.

106.31: $A'E' \equiv AC$] A'⟦C⟧⟨deleted by Hi^s⟩$E' \equiv AC$

106.31: da es nur einen solchen Punkt geben darf] ⟦nach 1⟧⊣↑da es nur einen solchen Punkt geben darf↓

107.1: A',] A'

107.1: B',] B'

107.2: einander] ↑einander↓

107.6: 2] ↑2↓

107.8: ;] ,

107.9: liegt] liegt,

107.9–10: Wenn ein Winkel BAC] ⟦Wenn ein Winkel BAC⟧⊣↑Wenn ein Winkel BAC↓

107.10: ist] ist,

107.12: so,] so

107.19–20: $\sphericalangle BAC = B'A'C'$] $\sphericalangle BAC = B'A'C'$.

107.21: $BC = B'C'$.] ⟦A⟧⟨deleted by Hi^s⟩$BC = B'C'$.

107.28: 9.)] 9.

107.32: von Punkten] ⟦homologer Punkte⟧⊣↑von Punkten↓

108.1: Bestimmt] Bestimt

108.10: Und] und

108.15: Zunächst:] zunachst

108.18: $P'E' \equiv PC$] ⟦P⟧ ⟨deleted by Hi^s⟩ ⟦$P'C' \equiv PE$ ist⟧ ⟨deleted by Hi^s⟩ $P'E' \equiv PC$

108.24: ist.)] ist.

108.25: so,] so

109.4: dem] den

109.5: q. e. d.;] q. e. d.

109.12: rechten Seite] ⟦anderen⟧⊣↑rechten↓ Seite

109.16–17: folglich $BE' \equiv BE''$, was nicht möglich ist wegen Axiom 1.] ⟦was nicht möglich ist.⟧⊣↑folglich $BE' \equiv BE'$, was nicht möglich ist wegen Axiom 1.↓

109.24: aus] Aus

109.27: A', B', C'] A' B' C'

109.28: sei] ⟦ist⟧⊣↑sei↓

109.28: die D] die D ⟦u D'⟧ ⟨deleted by Hi^s⟩

109n: Hat] ⟦Sin⟧ ⟨deleted by Hi^s⟩ Hat

110.16: E' zwischen] ⟦D' zwischen E' und D'⟧ ⟨deleted by Hi^s⟩ E' zwischen

110.20: geraden] Geraden

110.27: Congruenz] Congurenz

110.34: $R(\xi)$] $R\uparrow(\xi)\downarrow$

110.34: $Q(R(\xi))$] $Q(R\uparrow\xi\downarrow)$

110.34: $R(\xi)Q(R)$] $R\uparrow(\xi)\downarrow Q(R)$

110.35: folglich] ↑folglich↓

111.1: Andererseits] Andereseit

111.2: (ξ)] $\uparrow(\xi)\downarrow$

111.6: den] die

111.8: OA, OB, OC] $\uparrow O\downarrow A$, $\uparrow O\downarrow B$, $[\![A]\!]\uparrow O\downarrow C$, $[\![AD]\!]$⟨deleted by Hi^s⟩

111.9: $O'A'$, $O'B'$, $O'C'$] $\uparrow O'\downarrow A'$, $\uparrow O'\downarrow B'$, $[\![A]\!]\uparrow O'\downarrow C'[\![, A'D']\!]$⟨deleted by Hi^s⟩

111.10: durch O] durch $[\![A]\!]\uparrow O\downarrow$

111.10: etwa OD, werde] ⟦entspricht⟧↑etwa $[\![AC]\!]\uparrow OD\downarrow$ werde↓

111.12: zu OD] zu $[\![AE]\!]\uparrow OD\downarrow$

111.12: Strahl $O'D'$] Strahl $[\![A'E']\!]\uparrow O'D'\downarrow$

111.14: Hauptsatzes] Hauptsatztes

111.20: Strahlenbüschel dar] Strahlenbüschel

111.20: Punktreihe] Punktreihe,

111.21: dass] das

111.22: bedenkt] ⟦sich überzeugt⟧↑bedenkt↓

111.24: 2.)] ⟦⟪1.)⟫⟧↑2.)↓

111.24: 2te] $\uparrow 2^{te}\downarrow$

111.25: $\sphericalangle AOB = \sphericalangle BOA$.] ⟦⟦Wenn man $OA \equiv OB$ ⟨ein oder zwei Wörter unleserlich⟩ ist Figur $[OAB] = [OBA]$.⟧↑$\sphericalangle BAC = \sphericalangle CAB$↓⟧↑$\sphericalangle AOB = \sphericalangle BOA$.↓

111.25: 1] ⟦2⟧↑1↓

111.25: 1.)] ⟦2.)⟧↑1.)↓

111.25: für] ⟦Wenn man auch noch $OC \equiv OA$ macht⟧ ⟨deleted by Hi^s⟩ für

111.26: $AC \equiv A'C'$,] $AC \equiv A'C'$

111.27: werden;] werden.

111.28: $AOB \equiv A'OB'$] $[\![O]\!]A\uparrow O\downarrow B \equiv [\![O]\!]A\uparrow O\downarrow B'$

111.28: $BOC \equiv B'OC'$,] $[\![O]\!]B\uparrow O\downarrow C \equiv [\![O]\!]B\uparrow O\downarrow C'$,

111.29: $AOC \equiv A'OC'$] $[\![O]\!]A\uparrow O\downarrow C \equiv [\![O]\!]A\uparrow O\downarrow C'$

111.29: unsere Axiome aussagen.] aus ↑unsere↓ Axiom ⟦6 leicht folgt.⟧↑aussagen.↓ ⟨substituted by HiK^s⟩

112.3: ω_1, ω_2] $\omega_1\,\omega_2$

112.17: Übergang] Ubergang

112.17: That:] That

112.22: Gleichung] Gl.

112.22: $(\mathrm{K}', \Lambda', \mathrm{M}', \xi)$] $(\mathrm{K}'\,\Lambda'\,\mathrm{M}'\,\xi)$

112.26: wenn] Wenn

112.27: reciproken] reciproquen

112.28: Invarianten] Invarianter

112.29: Gleichung] Gl.

112.29: $(\alpha, \beta, \omega_1, \omega_2)$] $(\alpha, \beta\,\omega_1\,\omega_2)$

112.30: μ,] μ

112.34: Congruenz-Gleichungen] ⟦Transformations- oder⟧ ⟨deleted by Hi^s⟩ Congruenz Gleichungen

112.40: Fixzahlen] Fix⟦punkte⟧↑zahlen↓

112.41: reell] reel

112.42:	Benennung] Benenng
112.42:	entspricht] entsprich
112.43:	β,] β
112.43:	die Fixzahl] der Fix⟦punkt⟧↑zahl↓
113.5:	und da $\mu \neq 1$,] ⟦d. h. $\mu = 1$⟧ ⟨deleted by Hi^s⟩ und da $\mu \neq 1$,
113.9:	$c = 0$] ⟦$d =$⟧ ⟨deleted by Hi^s⟩ $c = 0$
113.12:	Fixzahlen] Fix⟦p⟧ ⟨deleted by Hi^s⟩zahlen
113.13:	geometrische] Geometrische
113.17:	immer] imer
113.18:	immer] imer
113.28:	Grenzzahl] ⟦Unbestimmtheits⟦punkt⟧ ⟨deleted by Hi^s⟩zahl⟧ ⟨deleted by Hi^s⟩ Grenzzahl
113.28:	$= \pm\infty$ nimmt,] $=$ ⟦∞⟧⟨deleted by Hi^s⟩ $\pm\,\infty$ nimmt.
113.28:	allen] ⟦All⟧ ⟨deleted by Hi^s⟩ allen
113.28:	Punkten] Punkte
113.33:	Maasseinheit] Maaseinheit
113.33:	und Anfangspunkt] ↑und Anfangspunkt↓
113.36:	nimmt,] nimt,
113.38:	endlichen] ⟦Zahlen e⟧ ⟨deleted by Hi^s⟩ entdlichen
113.41:	Maasseinheit] Maaseinheit
113n:	$\eta = 0$:] $\eta =$⟦∞: $\frac{a}{c}$⟧ ⟨deleted by Hi^s⟩ 0:
114.1:	Strahlenbüschel] Strahlenbüsche⟪l⟫
114.1:	Drehung] Drehng
114.1:	und] u.
114.3:	$\mu\frac{\xi+i}{\xi-i}$,] ⟦M⟧⟨deleted by Hi^s⟩$\mu\frac{\xi+i}{\xi-i}$,
114.4:	$\sphericalangle AOA'$.] $\sphericalangle$⟦O⟧⟨deleted by Hi^s⟩AOA'.
114.7:	so ist] so ⟦$\mu\nu$⟧ ⟨deleted by Hi^s⟩ ⟦φ⟧ ⟨deleted by Hi^s⟩ ist
114.7–8:	Winkelmaasse] Winkelmaas
114.9:	Anfangspunkt,] Anfangspunkt
114.10:	ist,] ist
114.17:	d. h.] ⟦seien reell⟧ ⟨deleted by Hi^s⟩ d. h.
114.21:	irgend] ⟦⟨erased⟩⟧↑irgend↓
114.25:	Benennung] Benenng,
114.27:	β', 0,] β', 0
114.27:	B$'$, 0,] B$'$, 0
114n:	und] u.
114n:	hyperbolische und elliptische] hyp. u. ellip.
114n:	Gleichung] Gl.
115.1:	d. h.] ⟦oder⟧↑d. h.↓
115.4:	Benennung] Bennennung
115.6:	bestimme] bestime
115.13:	Punkten] Punkte
115.13:	Ausübung] Ausübug
115.15:	γ] ↑γ↓
115.15:	γ] ↑γ↓
115.17:	Widerspruch] Wiederspruch
115.23:	Denn] ⟦Man nehme 3 verschiedene Punkte A, B,⟧ ⟨deleted by Hi^s⟩ Denn
115.25:	Ferner:] Ferner

115.27:	Gleichung] Gl.
115.30:	(Verschiebungen)] Verschiebungen
115.31:	Gleichungen] Gl.
115.31:	d. h. wenn] d. h. Wenn
115.32:	congruente] congruent
115.36:	zu Punkten] Punkte
115.37:	(Verschiebung)] (Verschiebug
115.37:	dargestellt] darstellt
115.39:	Widerspruch.] Wiederspruch.
115n:	etwa] 《etwa》
116.4:	Unterscheidung] ⟦Ent⟧⊣↑Unter↓scheidung
116.6:	d. h. ist] d. h. Ist
116.15:	Entfernungen] Entfernung
116.16:	$\mathfrak{b}$, $\mathfrak{c}$] $\mathfrak{b}$ $\mathfrak{c}$
116.17:	Entfernung] ⟦⟨erased⟩⟧⊣↑Entfernung↓
116.17:	ω_1,] ω_1
116.20:	Logarithmus] Logarith《ng》
116.22:	Benennung] Benenng
116.23:	Benennung] Benenng
116.24:	und] un
116.25:	Benennung] Benenng
116.27:	Benennung] Beneng
116.29:	Maass] Maas
116.31:	Gleichung] Gl.
116.32:	reell] reel
116.35:	Grundpunkte] Gundpunkt
116.36:	positiven] positiv
116.37–38:	(Die ⟨...⟩ ausführbar.)] ⟨Parentheses possibly added.⟩
116.37:	harmonischen] Harmonische
116.38:	ausführbar.)] ausführbar).
116.38:	jene] jede
116.38:	Gleichung] Gl.
116.41:	nie] ⟦nicht⟧ ⟨deleted by *Hi*[s]⟩ nie
117.1:	ausmacht,] ausmacht.
117.2:	welche] welche ⟦⟦durch⟧⊣↑wenn↓ $OA \equiv OB$ und ⟦$OA' \equiv OB'$⟧ $OC \equiv OD$ ist⟧
117.4:	DOC] ⟦DO⟧ ⟨deleted by *Hi*[s]⟩ ⟦AOB⟧ ⟨deleted by *Hi*[s]⟩ DOC
117.5:	der] den
117.12:	congruente] Congruente
117.17:	andererseits] andereseits
117.20:	$\delta =$] ↑$\delta =$↓
117.20:	Gleichung] Gl.
117.21:	imaginär] 《inaginar》
117.24:	additive Vielfache von] ↑additive Vielfache von↓
117.36:	(erster Dreieckscongruenzsatz)] ⟨Possibly an addition.⟩
117.42:	d. h. wenn] d. h. Wenn
118.5–6:	mit reellen Coefficienten] ↑mit reellen Coefficienten↓
118.7:	$(T\alpha, Tb, T\omega_1, T\omega_2)$] $(T\alpha, Tb\,T\omega_1\,T\omega_2)$
118.10:	Maassbestimmug] Maasbestimmug
118.13–14:	k wollen] ⟦$k = 1$⟧ ⟨deleted by *Hi*[s]⟩ k wollen

118.14:	(= 1 nämlich)] ⟨Parentheses possibly added.⟩
118.16:	Additionsgesetz,] Additionsgetz
118.19:	congruent] ⟦⟦congruen⟧ ⟨deleted by Hi^s⟩ gleich (congruent)⟧⊣↑congruent↓
118.22:	$BOA \equiv COB$] $BOA \equiv C$⟦A⟧⟨deleted by Hi^s⟩OB
118.25–26:	Maassbestimmung] Maasbestimmung
118.26:	(Zahlenzuordnung)] (Zahlen⟦benenng⟧ ⟨deleted by Hi^s⟩zuordnung)
118.28:	und ω_2 gelegenen] u. ω_2 gelegen
118.29:	unendlich] ⟦allen⟧ ⟨deleted by Hi^s⟩ unendlich
118.30:	falls] fals
118n:	$\mathrm{l}(i\alpha + 1)$] ⟦⟪$\frac{\mathrm{l}\, i\alpha+1}{-i\alpha+1}$⟫⟧ ⟨deleted by Hi^s⟩ $\mathrm{l}(i\alpha + 1)$
119.3:	treffen.] treffen. ⟦Nun seien B, C 2 Punkte auf a⟧ ⟨deleted by Hi^s⟩
119.5:	Parallelen] Parallele
119.6:	widerlegen] wiederlegen
119.7:	obwohl] obwohl ⟦es nicht⟧ ⟨deleted by Hi^s⟩
119.7:	experimentell] experimentel
119.9:	$< \pi$.] ⟦>⟧ ⟨deleted by Hi^s⟩ $< \pi$.
119.10:	Vorlesungen] Vorles.
119.10:	S. 494,] S. 494.
119.16:	D. h. wenn] d. h. Wenn d. h., we
119.16:	kleine] Kleine
119.17:	bleibenden kleinen] bleibende kleine
119.19:	Geraden] Gerade
119.23:	machenden] machen
119.23:	Messungen] Messung
119.25:	nicht] ⟦hyp⟧ ⟨deleted by Hi^s⟩ nicht
119n:	üblichen] übligen
119n:	Winkelsumme] Winkelsum
120.1:	Widerspruchslosigkeit] Wiederspruchslosigkeit
120.2:	Geraden] Gerade
120.3:	definiren)] definiren
120.5:	Geraden] Gerade
120.5–6:	unendlich viele Parallelen] unendliche Parallele
120.14:	Inneren] inneren
120.15:	deren] ⟦auf deren Oberflächen alle⟧ ⟨deleted by Hi^s⟩ deren
120.15:	Verschiebung] Verschiebng
120.16:	dieselbe] dieselben
120.17:	Achse] Ache
120.18:	Faden,] Faden
120.18:	Kreise,] Kreis
120.19:	einfacher,] einfacher
120.19:	wird es] ⟦und nach unseren gesammten sonstigen physikalischen Aus⟧ ⟨deleted by Hi^s⟩ wird es
120.22:	wenn wir] wenn wir ⟦dies od⟧ ⟨deleted by Hi^s⟩
120.24:	Ptolemäische] Pholemäische
120.26:	widerstrebte] wiederstrebte
120.30:	widersprechende] wiedersprechende
120.32:	Vorurteile] Vorteile
120.34:	Philosophen:] Philosophen

120.35:	Logik und] Logik u.
120.35:	Geometrie,] Geometrie.
121.9:	Dreieck] 3eck
121.18:	Maassbestimmung.] Massbestimmung.
121.19:	früheren] frühreren
121.21:	die] Die
121.23:	Geometrien] Geometrieen
121.23:	F. Klein,] F Klein.
121.27:	behandeln,] behandeln.
121.27:	Boltzmann] Boltzman
121n:	insbesondere] inbesondere
121n:	betreffenden] betreffende
122.1:	schematisch] chematisch
122.4:	gelten,] gelten.
122.9:	giebt es] giebt es ⟦für eine Geraden⟧ ⟨deleted by Hi^s⟩
122.9:	Gerade] Geraden
122.9:	welche] welche ⟦sie nicht schnei⟧ ⟨deleted by Hi^s⟩
122.12–13:	und für einen Punkt] ↑und für einen Punkt↓
122.16:	anderer] ⟦Punkte⟧ ⟨deleted by Hi^s⟩ anderer
122.19:	rechts] rechs
122.19:	AB.] AB. ⟦Da⟧ ⟨deleted by Hi^s⟩
122.21:	Die] ⟦Ebenso durch⟧⊣↑Die↓
122.24:	Gewünschte] Gewünschete
122.30:	und] u.
122n:	Wechselwinkel] Wechselwinkelwinkel
123.4:	Grenzgeraden.] Grenzgeraden
123.5:	ein] als ein
123.5:	möglichst] moglichst
123.8:	2 Punkten."] 2 Punkten.
123.10–11:	kürzeste Verbindung] ⟦die⟧ ⟨deleted by Hi^s⟩ kürzeste Verbindug
123.13:	$= 4(\pi - \alpha - \beta - \gamma)$;] $= 4$⟦π⟧⟨deleted by Hi^s⟩$(\pi - \alpha - \beta - \gamma)$
123n:	zunächst] zunächs
123n:	von den Wechselwinkeln] ⟦wie oben die Gleichheit der Wechselwinkel⟧⊢ ⊣↑von den Wechselwinkeln↓
123n:	$\alpha > \alpha'$.] ⟦$\alpha \neq$ ⟧⟨deleted by Hi^s⟩α⟦ $\leqslant$ ⟧⟨deleted by Hi^s⟩ $> \alpha'$
123n:	a und b] ⟦Denn⟧ ⟨deleted by Hi^s⟩ a u. b
123n:	nämlich $\alpha \leqq \alpha'$,] nämlich $\alpha \leqq \alpha'$
123n:	so trage] ⟦und verlängere ich b nach links, so $\beta' \geqq \beta$⟧ ⟨deleted by Hi^s⟩ so trage
124.1:	Meine Herren.] M. H.
124.2:	hier] hier ⟦das wi⟧ ⟨deleted by Hi^s⟩
124.5:	Abhandl.:] Abhandl.
124.5:	Helmholtz,] Helmholz,
124.5–6:	Lobatschewsky] Lobatschefsky
124.6:	Abhandl.] Abhand
124.6:	Lie,] Lie ⟦III⟧ ⟨deleted by Hi^s⟩
124.6:	III.] III,
124.7:	Philosophen:] Philosophen
124.8:	meinen] Meinen
124.11:	keine] ⟦die ⟪?⟫ ⟪geringen⟫⟧ ⟨deleted by Hi^s⟩ keine

124.12: zusammen:] zusammen⟦,⟧ ⟨deleted by Hi^{s}⟩:
124.13: Geraden] Gerade
124.13: sich ordnen,] sich ordnen
124.13: $LK \ldots BA$,] $LK..BA$
124.14: und nur] u. nur
124.14: eine Weise] ein weise
124.15: Reihenfolge] Reihn《fgl》
124.15: sind] s《in》
124.16: sagt] sgt
124.16: durch] 《du》
125.1: Theorie] Th.
125.1: dass] das
125.2: Zahlenmannigfaltigkeit,] Zahlenmanigfaltikeit
125.3: doch] 《doch》
125.3: sagen] sage
125.3: 2 Punkte ⟨...⟩ ferner] ⟨Addition, under the line to which it is added. The size of the space to the next line ('Dann ist Alles sehr einfach') strongly suggests that the addition was effected before 'Dann ... einfach' was written.⟩
125.3: eine Gerade] ein Gerade
125.3: bestimmen] bestimen
125.4: Geraden in Geraden] Gerade in Gerade
125.6: Bolyai und] Bolay u.
125.6: Lobatschewsky] Lobatschefski
125.9: Kugel] K⟦rei⟧ ⟨deleted by Hi^{s}⟩ugel
125.13: 1868,] 1868.
125.16: Parallellinien,] Parallellinien.
125.1: Bd. 86,] Bd 86.
126.5: der Geometrie zu Grunde] d. Geometrie z. G.
126.10: *Clebsch-Lindemann*,] *Clebsch Lindemann*
126.11: eingebundenen] eingebundene
126.14: Grunert Arch.] ⟨A space is left after 'Grunert Arch.', probably so that the correct Volume number could be added.⟩
126.15: Möbius,] ⟦⟨radiert⟩⟧⊢↑Möbius↓
126.15: Der] der
126.17: Baltzer, Elemente der Mathematik,] ⟨Underlined by Hi^{r}.⟩
126.17: 2 Bd.] ⟨Added by Hi^{r}.⟩
126.18: 1. und 2.] 1 und 2
126.21: Wundt ⟨...⟩ 1893.] ⟨Added by Hi^{g}.⟩
126.25: Soph.] S↑oph.↓
126.25: Theorie der] Theorie d.
126.26: Hilbert,] Hilbert.
126.26: Verbindung] Verbin《dung》
126.28–29: mit Rücksicht ... Geometrie.] ⟨Added.⟩
126n: eingebundenen] eingefundenen
126.–: grosses] gr.
127n: Cantor, Math. Ann. Bd. 46, 49] ⟨Written with a blue coloured crayon.⟩

Description of the Text

Collection: SUB Göttingen, signature *Cod. Ms. D. Hilbert 541.*

Size: Cover size approx. 21 × 16.9 cm; page size approx. 20.9 × 16.4 cm.

Cover Annotations: On the front cover is pasted a label with the call number, library stamp, and the notation, 'Grundlagen der Geometrie // 2st. 1894.', as well as a barely legible numbering ('21'). This numbering is a remnant of one in a catalogue list of Hilbert's lecture notes made in 1943 after Hilbert's death, which is preserved in the library of the Mathematical Institute. The descriptive portion of the label on the spine has been lost.

Composition: A single signature, originally consisting of 22 double pages, onto which further double pages, single pages, partial pages, and strips of paper have been pasted. In particular, between the two outside double pages a single page was bound in such a way as to leave a strip to the left of the fold; an extra double page was pasted to this strip. Moreover, the fifth and sixth sheet of the original signature were pasted together, and a further double page was pasted in behind them. A further page was pasted in following p. 28, and a double page after p. 33.

Pagination: Pages are numbered continuously, with the omission of a single sheet (the right half of the double page 39/40, numbered by the Editors as 60A and 60B). Only the right-hand pages are numbered, with the numeration going from 1 to 99. The fact that the page numbering does not reflect the complex glueing together of the pages suggests that the numbering was supplied only after the manuscript was substantially complete.

Original Title: On p. 1 in Käthe Hilbert's hand, in the middle of the page: 'Die Grundlagen der Geometrie. // Sommer-Semester 1893. nicht gelesen. // Auf 2 Stund. berechnet'; underneath, in Hilbert's hand, 'Sommersemester 1894 gehalten'. At the upper right in Käthe Hilbert's hand: 'Professor Hilbert. // Augustastr. 4.'

Text: Pages 7 to 95 are written with continuous text in Hilbert's or Käthe Hilbert's hand. Page 2, the verso side of the title page, contains various remarks and annotations, in Hilbert's handwriting; the double page pasted in after it contains on pp. 3–5 a bibliography, partly in Käthe Hilbert's hand, and partly in Hilbert's. Pages 6, 76, and 96–100 are blank. Hilbert extensively corrected and annotated his original text, which was written in black ink; entire passages have been either crossed out or made illegible by the pasting-over of substitute text. For these revisions Hilbert used black ink (Hi^s) and, especially for underlinings, blue pencil (Hi^b); he also occasionally used pencil (Hi^g) and red ink (Hi^r). It is not possible to date these revisions and markings with any degree of precision. They possibly belong to the original period of the lectures, but it is more likely that they were revised in preparation for the summer semester of 1894.

Chapter 3

Holiday Courses, 1896 and 1898

Introduction

This Chapter presents two of the *Ferienkurse* which Hilbert held in his Göttingen career, those for 1896 and 1898. The Göttingen *Ferienkurse* were given to *Oberlehrer* (i.e., teachers at *Gymnasien* or comparable institutions) each year from 1892 on; they were held in the Easter break, and were meant as supplementary courses. The courses were initiated at the instigation of the Prussian Education Ministry on the model of those introduced in Berlin shortly before. Mathematics was not one of the subjects given in Berlin; special permission had to be obtained for it to be taught in Göttingen, and for some time it was regarded as temporary and a special case. It was thought that the main need was to supplement teachers' knowledge in the natural sciences, and it seems that permission for mathematics to appear, and remain, on the curriculum was due above all to Klein's interest in the training of teachers. Up to 1899, all *Ferienkurs* subjects were dealt with each year; after this, there were yearly alternations between the mathematical and the descriptive sciences, mathematics falling in the even-numbered years. There were no courses at all in 1911 and 1913, nor during the First World War. A programme was planned for 1922, but was cancelled on financial grounds, and it was only in 1924 that the *Ferienkurse* resumed. After 1926, there was a new organisational structure which placed the responsibility for shaping the courses more firmly in the hands of the teachers themselves. The last year for which the courses seem to have been planned was 1933, but it is not known whether they were held.

We know a little about the organisation of the courses. For one thing, there are files on the *Ferienkurse* in the Universitätsarchiv of the Georg-August Universität, Göttingen (under the signature *KurAVa32*). From 1902 until the outbreak of the War, information about the mathematics courses was published in the *Jahresberichte der deutschen Mathematiker-Vereinigung*, and further information can be found in the relevant volumes of the *Zeitschrift für mathematischen und naturwissenschaftlichen Unterricht*, which contain both announcements of the courses, and detailed reports, for 1906, 1909 and 1912. (See also *Lorey 1916*, especially pp. 296–300.) It seems that, on average, a mathematical *Ferienkurs* in Göttingen lasted in all for two weeks, and about eight *Dozenten* were announced, roughly half from mathematics and half from physics. Thus, such a course was a broad effort, and did not devolve upon one or two leading figures. Giving lectures in the course was not regarded as part of the duties of the *Dozenten*, and they consequently received an extra honorarium for this.

Both Klein and Hilbert regularly took an active part in the courses, certainly up to the Great War. Hilbert himself refers to Klein's *Ferienkurs* lectures on geometry for 1894 (see p. 4 of Hilbert's 1896 course below), and these lectures are the basis of those presented in *Klein 1895* (see Klein's *Vorwort*). We know that Hilbert took part in the courses for 1896, 1898, 1902 and 1912. He was scheduled to give a four-hour series of lectures in 1922 on relativity theory, but this programme was cancelled (see above). It was also announced that he would hold a one-hour lecture in 1933 with the title 'Über neuere logische Forschungen', but, as mentioned, it is not known whether this lecture was held. From these five (or six) *Ferienkurse*, only notes for those from 1896, 1898, 1902 and 1912 are extant. The notes for the first three are bound together into a single item of the Hilbert *Nachlaß* (SUB Göttingen, *Cod. Ms. D. Hilbert 597*); those for the 1912 course are to be found in *Cod. Ms. D. Hilbert 594*. The size of Hilbert's contributions varied. For 1896, Hilbert was announced as giving four, two-hour lectures, and the notes for the 1896 course are indeed marked '$4 \cdot 2 = 8$ stündig'. In the 1898 timetable, Hilbert is alloted five, two-hour lectures, but only three, two-hour lectures in 1902 and 1912. (In all, four hours of lectures were planned for 1922, and just one hour for 1933. The figures allotted for 1896 and 1898 are a trifle misleading; it seems that Hilbert was also responsible for conducting a tour of the Mathematisches Institut!) All the extant notes suggest that the lectures were rich in material, and cannot have been trivial affairs, however breezy.

The notes which survive for the course in 1912 (three pages) are sketchy, referring frequently to material in the lecture courses from 1904/1905 on definite integrals and Fourier series (*Hilbert 1904c**), from 1911/1912 on the kinetic theory of gases (*Hilbert 1911**), and from 1910/1911 on mechanics (*Hilbert 1910**). Hence, it is hard to form a distinct impression of the two lectures given. What is clear is that they dealt with some of the principal difficulties faced by foundational research in physics (the subject announced in the programme is 'Axiomatische Grundlagen der Physik'), and so fall outside the scope of this Volume. Of the three *Ferienkurse* concerning the foundations of mathematics, the 1902 lectures are not represented here. This course was announced in the programme as 'Axiome der Geometrie'. Judging from the five sheets of notes which survive (SUB Göttingen, *Cod. Ms. D. Hilbert 597*; sheet 10 is the first sheet of the notes for the 1902 course), it takes Euclidean geometry as its sole subject, and deals with material under three headings: 'I. Begründung der Geometrie', 'II. Unmöglichkeitsbeweise', 'III. Geometrische Constructionen'. However, these notes, too, like those for *Hilbert 1912**, are also sketchy, and consist in large part of references to other sources, either in the general literature or to Hilbert's own lecture notes, mainly *Hilbert 1899a** and *Hilbert 1894b**, or to *Hilbert 1899c*. Thus, of the extant notes for Hilbert's *Ferienkurse*, we concentrate in this Chapter on those for 1896 and 1898.

There are fifteen pages of notes for the 1896 course, of which we reproduce the first eight and a half pages, and the last passage, from p. 15. In the first three, short pages, Hilbert takes up the accusation that modern mathematics as taught in the universities is abstract and specialised, too far removed from the subjects which schoolteachers have to convey to their pupils. This is then followed by a list of three subjects, all of which, says Hilbert, are pursued at school, '1. Rechnungsoperationen und Zahlbegriff', '2. Euklidische Geometrie', '3. Zahlentheorie und Gleichungstheorie'. Hilbert then argues that all three stand at the centre of mathematical research, and that it is essential for future teachers to be familiar with the achievements of current research in these areas. He states that the first of these was taken as the subject of Klein's *Ferienkurs* in 1894. This might seem a trifle misleading. To judge from *Klein 1895*, the subject of the course was as much geometry as anything, not pure Euclidean geometry to be sure, but rather the three traditional problems of doubling the cube, trisecting the angle, and squaring the circle. Nonetheless, Hilbert is right to say that what Klein deals with falls under the heading of 'Rechnungsoperationen und Zahlbegriff', for he deals among other things with geometrical constructions, constructible, algebraic and transcendental numbers, and the modern algebraic proofs that the traditional problems are not soluble. (*Klein 1895*, 38–42 presents Cantor's proof that the continuum is uncountable as a way of proving the existence of many transcendental numbers.) Klein also sets out (at least in his book) much the same accusation with which Hilbert deals here, and makes it fairly clear that his lectures are meant to constitute a rebuttal. (See *Klein 1895*, *Vorwort* and *Einleitung*.) As the subject of his course, Hilbert chooses the third of the three subjects listed above, which he says is the 'most suitable', since here deep and elementary questions are not easily disentangled (see p. 5). (The programme lists Hilbert as lecturing on 'Elemente der modernen Zahlen- und Gleichungstheorie'.) The rest of the nine pages reproduced here consists of general remarks about the preeminent nature of number theory, and its relations to other branches of contemporary mathematics. This general part contains a number of references to the notes for Hilbert's oral report to the Deutsche Mathematiker-Vereinigung (*Hilbert 1895a**, 'Rede über meinen Bericht'), the greatly expanded, published version of which achieved fame as Hilbert's *Zahlbericht* (*Hilbert 1897a*). These references are for Hilbert himself, and one is to assume that the relevant sections formed part of the lecture. The remaining six pages of the notes (9–15) are devoted to an elementary presentation of the basis of the theory of algebraic number fields, and thus continue to be similar to some of the material found in Hilbert's *Rede*. (*Hilbert 1895a** will be reproduced in full in Volume 2 of this series.)

There are twenty-seven pages of notes for the 1898 *Ferienkurs*, and these are given here in full. The notes begin with a reference to the material of the first three pages of the 1896 course, as, incidentally, does the 1902 course, and one again assumes that this material from 1896 was to be presented in the lectures. In 1898, however, the basis of the rebuttal of the claim that university mathematics is disconnected from 'elementary' mathematics is an examination of the concept of the infinite in contemporary mathematics. (The subject of the course announced in the programme is 'Über den mathematischen Unendlichkeitsbegriff'.) Hilbert proposes to deal first with the infinite in arithmetic, and then with the infinite in geometry. The material Hilbert presents about the former would now be seen as fairly standard, centring on differing infinite cardinalities, the dangers of assuming that properties of finite entities (e.g., finite series) transfer without qualification to infinite entities (e.g., infinite series), etc. But if standard now, one can hardly assume it was standard then, and it is therefore interesting to see that Hilbert stresses to mathematics teachers the importance of Cantor's central discoveries on the powers of some important sets of numbers, although one should note Hilbert's assertion on p. 1 that there is 'admittedly, no systematic theory of the concept of [mathematical] infinity'. Up to this point, therefore, when considered in tandem with the 1896 *Ferienkurs*, it seems that the notes from 1896 and 1898 would have been more appropriately placed alongside *Hilbert 1895a** in the volumes of this edition devoted to Hilbert's writings on the foundations of arithmetic. Indeed, the first half of the 1898 *Ferienkurs*, in broad outline, touches on much of the material standard in Hilbert's various lecture courses on the 'Zahlbegriff und Quadratur des Kreises' (from 1893/1894, 1897/1898 and 1904), material well represented in Volume 2 of this series. Why, then, do these lectures appear in a volume dedicated to Hilbert's work on geometry?

There is a straightforward answer. First, although the 1898 notes are of mixed content, Euclidean geometry is one of the two focuses, occupying the last two-thirds (pp. 9–27) of the lectures. In addition, while nominally about the role of the infinite in geometry, the lectures actually show Hilbert already beginning to grapple in depth with many of the specific problems which feature centrally in the 1898/1899 lectures (*Hilbert 1898c** and *Hilbert 1899a**, given in Chapter 4), as he says in fact on p. 9, even mentioning the forthcoming lecture course. The 1898 *Ferienkurs* thus deserves to be considered as an integral part of the chain in Hilbert's geometric development in the 1890s, a development which leads to the *Festschrift* (*Hilbert 1899c*). Moreover, it presents us with important evidence of Hilbert's occupation with these problems some five or six months before the beginning of the lectures for which the notes *Hilbert 1898c** were written. Further evidence can be found in Hilbert's letter to Hurwitz of 16 March, 1898.[1]

What are these specific problems? Three deserve mention.

The first concerns the import of what Hilbert calls 'Pascal's Theorem', actually a special case of Pascal's Theorem, indeed an affine version of what is usually called Pappus's Theorem. The basic issue here is the connection with a proper treatment of the theory of proportion and the different ways in which one might introduce segment multiplication. The assessments of the place of the Pascal Theorem in the algebra of segments assume central importance in the later lectures, and even more so in the *Festschrift*. The same applies to the Desargues Theorem, also mentioned here.

The second concerns a relatively straightforward theorem about circles: Given three circles, each pair of which intersect, the three chords so generated have a mutual point of intersection. For ease of reference, we will call this the Three Chord Theorem. The result is metamathematically interesting, as Hilbert explicitly shows in the 1898/1899 lectures, for it brings out a fundamental assumption of Euclid's original system, the assumption, roughly, that straight lines and circles which *apparently* intersect do in fact intersect, i.e., that the relevant intersection points exist. The proof of the Three Chord Theorem

---

[1] The letter is to be found in the Hilbert *Nachlaß*, SUB Göttingen, *Cod. Ms. D. Hilbert 272*, and is reprinted in facsimile in *Hilbert 1999a*, 277–280. It is also quoted at length in *Toepell 1986*, 115–118.

turns on just such an existence assumption, which is not satisfied in Hilbert's Euclidean geometry. Moreover, there is more than a hint here of the algebraic result which Hilbert (in *Hilbert 1898c** and *Hilbert 1899a**) associates with the failure of this specific existence assumption, namely, in modern terms, that not all Pythagorean fields are Euclidean. (For more about this theorem, see the Introduction to Chapter 4, Specific Note (2).) Hilbert's treatment of the Three Chord Theorem is an instructive example of the kind of interest he develops in basic geometrical theorems, namely interest in what more basic principles underlie them, and in their proof-theoretical weight. One should note, however, that the Theorem is nowhere discussed in any published work of Hilbert's, neither does it figure in lectures after 1899.

The third problem concerns segment multiplication. The theory of segment products is intimately tied to the Euclidean treatment of the content ('area') of plane rectilinear figures and Euclid's result that the area of a triangle is given by $\frac{1}{2}bh$, where 'b' stands for the length of the base and 'h' for the height. (This is, in effect, Proposition 37 of Book I of the *Elements*: see *Heath 1926a*, 332.) The methodological basis of Hilbert's treatment of these matters is roughly this: one has to define proportion and segment operations (addition, multiplication, etc.) in such a way that the definitions do not *presuppose* that the quantities defined necessarily act like 'pure magnitudes', e.g., that segments automatically possess numerical 'lengths'. I.e., it has to be shown that, when appropriately defined in a geometrical sense, the segment operations form an algebra which satisfy the right 'magnitude' properties. (Showing that the right 'magnitude' properties hold is where the Pascal and Desargues Theorems assume such significance.) 'Surface content' is then defined; it has then to be shown that the classical Euclidean characterisation holds, namely that the surface content of a triangle is indeed $\frac{1}{2}bh$. In this context, Hilbert raises what he calls the Killing-Stolz Postulate, and then proceeds to raise various questions about the status of this 'postulate' (see p. 25 and the accompanying notes), questions which are explicitly dealt with in the 1898/1899 lectures, and whose solutions are expounded in the *Festschrift*.

The examples given here also serve as examples of problems to which Hilbert at this point does not have complete solutions, but for which he does, however, work out solutions over the next year. This text, therefore, shows Hilbert *engaged* in such problems. (In the light of this, as well as of the sketchy nature of large sections of the notes, the reader would be well-advised to read the geometrical part of the 1898 *Ferienkurs* in conjunction with the 1898/1899 lectures.) There are also examples of problems raised here by Hilbert to which he found no solution, for instance, the question raised about the deductive relationship between Pascal's Theorem and the Desargues Theorem. Neither does Hilbert answer his own conjecture about the impossibility of treating volume analogously to his treatment of area or surface content. These unsolved problems reappear explicitly as open problems in the 1898/1899 lectures, and the problem of volume reappears as Problem 7 in *Hilbert 1900e*. (See the Introduction to Chapter 4, Specific Note (5)(c).)

Finally, it should be mentioned that this *Ferienkurs* contains interesting philosophical remarks, for instance a remark (p. 18) that the Axioms of Parallels and Archimedes do not have the same 'empirical' character as the other axioms. And Hilbert here (p. 12) also hints strongly at the view that consistency implies mathematical existence.

Hilbert's treatment of geometry in the 1898 *Ferienkurs* does not have to do directly with the topic announced at the beginning, namely the infinite in geometry, though it would not be hard to adumbrate connections, e.g., via the question of how much of geometry can be developed without introducing continuity axioms which entail the uncountability of the set of points, in particular how much can be carried through of the various classical developments (proportion, etc.) with which Hilbert's audience would have been intimately familiar.[2] Be that as it may, this material is in any case ideal for the audience and for its purpose, i.e., to show that 'higher' mathematics often lurks just behind the 'lower'. It is a thoroughly modern look at a familiar subject, and the metamathematical viewpoint is hard to mistake.

The notes for the 1896 course are by Ralf Haubrich and Michael Hallett, and those for the 1898 course are by Michael Hallett.

Michael Hallett, Albert Krayer

[2]It seems to have been accepted that the actual subject of the geometry part of the course was the foundations of geometry, and not the role of the infinite: see *Liebmann 1898*, 71.

Feriencursus Ostern 1896

↑Anderes Thema für Ferienkurs ist:
Ueber den Zahlbegriff und seine Anwendung in der Funktionentheorie, Zahlentheorie und Geometrie.↓[1]

1 Feriencursus 1896. (Ostern.)

Meine Herren. Es wird bisweilen gegen den Unterricht der Mathematik an den Universitäten ein doppelter Vorwurf erhoben,

1. sagt man uns, die wir den Unterricht leiten: die *Gegenstände*, die Ihr treibt, sind zum Theil so schwierig und abstrakt, dass die Mehrheit der Studirenden sich nicht zu einem *vollen Verständniss* derselben durcharbeiten kann.

2. die Gegenstände stehen *nicht genug im Zusammenhang* mit denjenigen *Gegenständen*, welche der Lehrer in Zukunft auf der Schule selber zu lehren hat und welche er dort braucht.

2 Ich könnte solchen *Vorwürfen* gegenüber hinweisen auf die *Freiheit des* Studiums, die den Einzelnen ja die Auswahl der Universität und der Vorlesungen gestattet und daher ein *Aequivalent* gegen die *Einseitigkeit* eines einzelnen Dozenten bietet, und andrerseits auf die *Souveränität* der Wissenschaft, die uns die Verpflichtung auferlegt, in den Vorlesungen auch die schwierigeren und neueren Auffassungen zur Geltung zu bringen, auch wenn dieselben von dem ausgetretenen Pfade abgelegen erscheinen. Ich könnte *ferner sagen*, dass es vor Allem im Interesse des künftigen Lehrers liegt, seine ganze Wissenschaft kennen zu lernen, damit er sie lieb gewinnt, damit er in den Stand gesetzt wird, später selbst zu ihrer Bereicherung sein Theil beizutragen, denn an der lieb gewonnenen Wissenschaft
3 mitzuarbeiten wird stets zu den edelsten Freuden gehören, die das menschli|che Leben bietet.

Heute aber möchte ich in anderer Weise jenen *Vorwürfen* begegnen. Ich möchte nämlich darauf hinweisen, dass ein solch *scharfer Gegensatz* zwischen der sogenannten Schul- und Universitätsmathematik garnicht vorhanden ist.

Nehmen wir 3 grosse Gebiete der auf der Schule getriebenen Mathematik heraus.

1. Rechnungsoperationen und Zahlbegriff.
2. Euklidische Geometrie.
3. Zahlentheorie und Gleichungstheorie.

[1]Added by Hi^g on the reverse of the title page.

Alle diese 3 Themen stehen, richtig aufgefasst und vertieft, auch in der Litteratur der höheren mathematischen Theorien gerade heute im Mittelpunkt des Interesses, und ich meine, dass die *Beherrschung der neueren Errungenschaften* in diesen *Gebieten* geradezu für | den künftigen Lehrer nothwendig ist. Über 4
das Thema 1 hat Klein hier vor 2 Jahren einen Feriencurs gehalten. *In Punkt 2* handelt es sich um das Euklidische Parallelenaxiom und den Aufbau der Geometrie aus einem kleinsten System unbeweisbarer Axiome. *Entwicklung der nicht Euklidischen Geometrie* auf *Grund der Bemerkungen* von *Riemann und Helmholtz.*

Lehrbücher: Clebsch-Lindemann, Vorlesungen über Geometrie. Bd. 2. Theil I, Abtheil. 3. Leipzig 1891.
Pasch, Neuere Geometrie, Leipzig 1882.
Killing, Grundlagen der Geometrie, Paderborn 1893. Stäckel, Theorie der Parallellinien, Leipzig 1895.[2] Hier findet man auch die Originalabhandlungen citirt, insbesondere die von Klein.
↑*Klein, autographirte Vorlesungen, Nichteuklidische Geometrie.*↓
↑F. Klein: Vorträge über ausgewählte Fragen der Elementargeometrie, Leipzig 1895.↓[3]
Hilbert, Ueber die gerade Linie als die kürzeste,
Lie, Theorie der Transformationsgruppen, Bd. 3. Leipzig 1893. Abtheil. 5.
Math. Ann. 46, wo ich alle Axiome zusammengestellt habe.[4] Dieses Thema 2 5
wäre vielleicht das dankbarste gewesen.[5] Verlockender aber erschien mir das 3te Thema, und ich würde mich sehr freuen, wenn es mir gelänge, Interesse für Zahlentheorie zu wecken, Ihnen *abzugewinnen.*

Das 3te Thema ist überdies das geeignetste, um Ihnen daran die *Grundbegriffe der modernen* Mathematik zu erläutern. Hier kommen wir am raschesten tief in die Sache hinein, wohnen dicht beieinander die leichtesten und die schwierigsten Fragen, besonders in der Zahlentheorie liegt eine Frage, deren Lösung Allgemeinbesitz aller denkenden Menschen geworden ist, dicht neben einem Problem, mit dessen Lösung die grössten Geister sich vergeblich abgemüht haben. ↑Ueberdies ist der principielle und logisch philosophische Wert der Zahlentheorie hoch zu schätzen. Die Fertigkeit der Grundlagen und die Sicherheit der Schlüsse machen es, dass die Zahlentheorie für den künftigen Parlamentsredner, Juristen, Historiker, Zeitungsschreiber vielleicht *wertvoller ist als Erziehungsmittel als andere Dinge in der Mathematik, auch mehr als die Anschauung.*↓ Daher schicke ich zunächst Einiges über die *Zahlentheorie* voraus.

---

[2]The references are to *Clebsch 1891*, *Pasch 1882*, *Killing 1893*, *Stäckel and Engel 1895*.

[3]The references are to *Klein 1890*, *Klein 1892*, *Klein 1895*. Hilbert added these references to Klein in the lower and upper margins respectively.

[4]*Hilbert 1895b*, *Lie 1893*. The reference to Lie interrupts the full reference to Hilbert's paper. Beneath 'kürzeste' there is drawn a line which continues onto p. 5, and leads up the left-hand margin to the first word 'Math.'. This makes it clear where the text after 'kürzeste' continues after p. 4 had been supplemented with the Lie reference.

Comparison of the 1898/1899 lectures and the *Festschrift* with *Hilbert 1895b* shows that Hilbert changed his mind about the latter having 'collected together all the axioms'.

[5]Euclidean geometry forms the major part of the content of the 1898 *Ferienkurs*, reproduced in its entirety later in this Chapter.

6 Die Zahlentheorie ist bekanntlich neben – mündlicher Bericht von mir an die Math. Vereinig., Lübeck 1895, *S. 1–5.* Menschenwerk.[6] Um nun aber Ihnen über den heutigen Stand der Zahlentheorie und ihre Beziehung zur Algebra zu berichten, knüpfe ich an den bekannten Ausspruch von Gauss an, welcher die Arithmetik die Königin der Mathematik nennt.

Sie verdient diesen Namen heute in höherem Maasse noch als zu Gauss Zeiten, und zwar in *3facher Weise*:

1. Seit der von Weierstrass geforderten und durchgeführten Strenge und der von ihm und Cantor präcisirten Begriffe der Irrationalzahlen etc., sind wir gewohnt, in der Analysis und Functionentheorie eine Thatsache dann erst als be-
7 wiesen anzuerkennen, wenn sie sich in letzter Instanz auf die Gesetze der | ganzen rationalen Zahlen zurückführen lässt. Oft freilich *langer* Weg mit *vielen Zwischenstationen* nöthig. Die Königin Arithmetik hat also allein ihre Gesetze in *der übrigen Mathematik zur Anerkennung gebracht.*[7]

---

[6]The reference is to *Hilbert 1895a**, which will appear in Volume 2 of this Edition. Hilbert clearly intends to use here a section from his *Rede* stretching over the first five pages. The second paragraph on p. 1 begins 'Die Zahlentheorie ist bekanntlich neben der Geometrie historisch der älteste Trieb unserer Wissenschaft; ...', and the way Hilbert begins the paragraph here is surely meant to echo this. The single word 'Menschenwerk' then refers to the end of the section Hilbert wishes to use; in the paragraph ending on p. 5 of *Hilbert 1895a**, Hilbert cites the famous remark of Kronecker in the form 'Die ganze Zahl schuf der liebe Gott, alles Übrige ist Menschenwerk'. In the text here, between the words 'Menschenwerk' and 'Um', there is a special sign (in Hi^b) which clearly marks where Hilbert intends to pick up again after citing from his *Rede*. A similar sign (also in Hi^b) appears in *Hilbert 1895a** on p. 5 after 'Menschenwerk', thus very likely marking the end of the quote and signalling a return to the 1896 lecture. The five pages of *Hilbert 1895a** so marked contain general comments about the nature of number theory, which praise 'die Einfachheit ihrer Grundlagen, die Genauigkeit ihrer Begriffe und die Reinheit ihrer Wahrheiten' (op. cit., p. 2). Hilbert also stresses the particular fascination evoked by the fact that many straightforward truths, easily discovered by 'induction', are actually difficult to prove (p. 4). Note that the passages from *Hilbert 1895a** singled out by Hilbert here were later revised. For instance, whereas the section cited begins in the original as cited in this text, after revision it becomes 'Die Zahlentheorie gehört zu dem ältesten Trieb mathematischer Wissenschaft; ...'. This formulation is much closer to the very first sentence of *Hilbert 1897a*, suggesting that the revision of *Hilbert 1895a** was made prior to, or during, its composition.

[7]The four lines beginning 'sind wir gewohnt' and ending with 'zurückführen lässt' are taken word for word from *Hilbert 1895a**, 15, at least in its revised version. Unrevised, there is the beginning of a suggestion that Hilbert wanted to attribute the popularity of the reductionist position to Weierstraß, although the revision indicates that Hilbert changed his mind. The position is not usually associated with Weierstraß. In the *Vorwort* to *Dedekind 1888*, Dedekind attributes the reductionist position to Dirichlet. He then continues:

> Aber ich erblicke keineswegs etwas Verdienstliches darin – und das lag auch Dirichlet gänzlich fern –, diese mühselige Umschreibung wirklich vornehmen und keine anderen als die natürlichen Zahlen benutzen und anerkennen zu wollen. Im Gegenteil, die grössten und fruchtbarsten Fortschritte in der Mathematik und anderen Wissenschaften sind vorzugsweise durch die Schöpfung und Einführung neuer Begriffe gemacht, ... (p. VI).

Perhaps Hilbert's 'langen Weg mit vielen Zwischenstationen' is Dedekind's 'mühselige Umschreibung'.

*Hilbert 1895a**, 10–20, makes it a little clearer what Hilbert might mean by the 'Gesetze der Arithmetik', not the laws of what we call ordinary arithmetic, but rather those which lie behind the theory of algebraic number fields. (See also the following note.)

2. Sie hat sich die Fruchtbarkeit anderer Disciplinen (Functionentheorie, algebraische Functionen) angeeignet, so dass auch in ihr heute der *Riemann'sche Grundsatz* zur Geltung kommt, demzufolge man die Beweise nicht durch Rechnung, sondern lediglich durch Gedanken zwingen soll.[8] Durch diesen Umstand ist die Zahlentheorie *viel methodischer* geworden. Noch Gauss klagt, wie unverhältnissmässig grosse Anstrengung ihn die Bestimmung eines arithmetischen Vorzeichens gekostet hat, mündlicher Bericht, S. 17 getreten ist.[9]

---

[8]In *Minkowski 1905*, 460–461, this 'Grundsatz' is ascribed to Dirichlet. Minkowski speaks of the 'other Dirichlet principle', i.e., as Minkowski puts it, 'mit einem Minimum an blinder Rechnung, einem Maximum an sehenden Gedanken die Probleme zu zwingen'. Minkowski adds that the modern period in the history of mathematics dates from the deployment of this principle. (Page numbers refer to the reprinting in *Minkowski 1911*.) Hilbert sees the beginnings of modern number-theory, as represented, e.g., in the theory of algebraic number fields, in Gauß's *Disquisitiones Arithmeticae* (i.e., *Gauß 1801*). In the Foreword to that work, Gauß states that it is dedicated to 'higher arithmetic', and not to what is ordinarily understood as arithmetic, i.e., the science of 'counting and calculating'. The statements of Gauß, Dirichlet and Hilbert are thus consonant.

[9]The end of this sentence appears not to make sense. However, there is a special insertion sign added in Hi^b between 'gekostet hat,' and 'mündlicher Bericht' which corresponds to an identical sign (also in Hi^b) in *Hilbert 1895a** (the 'mündlichem Bericht'), p. 17, after the words 'gekostet hat,'. These words are in the middle of an anecdote about Gauß, with an extra remark of Hilbert's about the characteristic nature of progress in science. The text spanning pp. 17–18 there reads:

> Noch Gauss klagt, wie unverhältnismässig grosse Anstrengung ihn die Bestimmung eines zahlentheoretischen Vorzeichens gekostet hat, wie ihn jenes Vorzeichen mehr Jahre gekostet habe, als andere grosse Probleme Tage! und plötzlich, wie ein Blitz, sei ihm die Lösung gekommen, ohne dass er sagen könne, woher. Ich glaube, dass an Stelle dieser für das Knabenalter einer Wissenschaft charakteristischen, sprunghaften Entwicklung heute durch den systematischen Ausbau der Körpertheorie ein mehr stetiges Fortschreiten getreten ist.

(See also *Hilbert 1897a*, i.e., *Hilbert 1932*, 65–66. The passage is given here in unaltered form; the second 'gekostet' was later changed to 'aufgehalten'.) In the current text, Hilbert repeats the first part of this passage (from 'Noch Gauss klagt ...' to 'gekostet hat'), and it is after 'gekostet hat' that the insertion signs appear both here and in *Hilbert 1895a**. There is then a vertical line after the reference to 'Bericht, S. 17' followed by the words 'getreten ist'. One is to assume, therefore, that Hilbert intended to read out the remainder of the passage from *Hilbert 1895a**, i.e., from 'wie ihn jenes ...' to 'getreten ist'. The point of this remark in *Hilbert 1895a** is similar to the point made here, namely to underline how number theory has only matured into a 'methodical' theory in recent times. (See also previous note, as well as *Minkowski 1905*, i.e., *Minkowski 1911*, 451, 461.)

Hilbert's source for the anecdote about Gauß is not given, and is not entirely clear, though it is probably a letter from Gauß to Olbers from 3 September, 1805, where Gauß recounts what is almost certainly the same story (see *Olbers 1900*, 268–269). The publication date of the Gauß-Olbers correspondence, 1900, might be a little misleading: according to Schilling's Preface (dated Bremen, November 1899), the correspondence was donated to the Royal Society of Göttingen by Olbers' heirs 'quite some time ago'. It is therefore possible that Hilbert would have known about it through its presence in the Gauß archives in Göttingen. Schilling's footnote to the letter concerned traces the original problem (of whether certain square root signs should be positive or negative) to § 356 of Gauß's *Disquisitiones*. A general method for determining this was given by Gauß's results on certain series ('Gaussian sums'), which yields a generalisation of his famous theorem on quadratic residues. This work was published only in 1811 (see *Gauß 1811*, especially § 33), though already announced in *Gauß 1808*. However, the main result is noted in Gauß's famous *Tagebuch* (*Gauß 1901*) for 30 August, 1805, i.e., just before the letter to Olbers.

8 3. Endlich ist für uns das | Wichtigste: sie hat als siegreiche Königin grosse Provinzen erobert und sich einverleibt. Gauss, mündlicher Bericht, S. 7–8, der modernen Zahlentheorie.[10] Die erfolgreichen Angriffe, die die Zahlentheorie auch auf die Functionentheorie macht, zu schildern, würde hier zu weit führen. Ich bin aber überzeugt, dass in Zukunft die Mathematik noch weit mehr unter den bestimmenden Einfluss der Zahlentheorie treten wird, so dass in der Zahlentheorie die *nothwendige und natürliche* Einheit der Mathematik gegeben ist.

Welches sind nun die Hilfsmittel und Waffen, mit denen die Zahlentheorie diese Erfolge errungen hat?

Serret, Höhere Algebra II, Kap. III, IV, deutsch von Wertheim, Teubner 1878.[11] Dirichlets Vorlesungen über Zahlentheorie, herausgeg. von Dedekind. Braunschweig, 4 Aufl. 1894. Supplement XI.
9 Bachmann, Kreistheilung, Leipzig 1872.
Weber, Lehrbuch der Algebra, Braunschweig 1895. Tannery, Théorie des nombres et de l'algèbre supérieure, Paris 1895.
Hilbert, Theorie der algebraischen Zahlkörper, Bericht der Math. Vereinig.
Minkowski, Geometrie der Zahlen, Leipzig 1896.[12]
⟨...⟩

15 ⟨...⟩

Diese einfachen Begriffe bilden die *sichere Grundlage*, auf der in herrlichem Bau die heutige Algebra ruht. An Fertigkeit und Einheitlichkeit hat dieser Bau in keiner Wissenschaft und auch nicht in der Mathematik seines Gleichen und ich möchte es gern aussprechen, dass ich diesen Bau zu dem grossartigsten rechne, was der menschliche Verstand hervor gebracht hat. Es hat überdies[13]

[10]The reference is again *Hilbert 1895a**. On p. 7 of that text, there is a large insertion sign (in Hi^b and before a paragraph beginning 'Gauss ...') which very probably marks the beginning of the passage Hilbert intended to read. (Again, this reference is to the text *before* revision. After revision the passage begins, not 'Gauss fasst', but 'fasst Gauss...', which is how it is used, with a prefatory clause, in *Hilbert 1897a*, i.e., *Hilbert 1932*, 64.) From this point to close to the end of p. 8, Hilbert deals briefly with Gauß's contribution to higher arithmetic, and the connections between this and general algebraic questions. Consequently, Hilbert sees the particular theory of algebraic number fields as 'der wesentlichste Bestantheil der modernen Zahlentheorie' (*Hilbert 1895a**, 8).

[11]*Serret 1879*. Hilbert's reference to Serret is not completely clear. For one thing, the dating is not consistent with the volume he mentions, there being a first volume of Serret's *Handbuch der höheren Algebra* published in 1878 (*Serret 1878*). However, it is likely that Hilbert is referring to Chs. 3 and 4 in Part 3 of *Serret 1879*. The only parts of the whole work which deal with number-theoretic subjects are these and Part 1, Chs. 1 and 2 of the first volume; the remainder treats of algebra. However, Hilbert almost certainly does not intend a reference to Vol. 1, since these parts deal with continued fractions in connection with the solution of equations. Part 3 of Vol. 2, however, bears the heading 'Die Eigenschaften der ganzen Zahlen', and Chs. 3 and 4 are entitled, respectively, 'Eigenschaften der ganzen Functionen einer Veränderlichen in Beziehung auf einen Primzahlmodul' and 'Bestimmung der nach einem Primzahlmodul irreducibelen [sic] ganzen Functionen in dem Falle, wo der Grad eine Potenz des Moduls ist'. This suggests that the reference is directed to these chapters.

[12]The references are to *Dirichlet 1894*, *Bachmann 1872*, *Weber 1895*, *Tannery 1895*, *Hilbert 1897a* and *Minkowski 1896/1910*, erste Lieferung.

[13]The text breaks off here.

Textual Notes

152.3:	und seine] u. seine
152.4:	und Geometrie] u. Geometrie
152.6:	Meine Herren.] M. H.
152.6:	der Mathematik] d. Math.
152.8:	leiten:] leiten,:
152.10:	derselben] ⟨Added.⟩
152.12–13:	und ⟨...⟩ braucht] ⟨Added.⟩
152.16–17:	und daher ein *Aequivalent* gegen die *Einseitigkeit* eines einzelnen Dozenten bietet,] ⟨Added.⟩
152.20:	abgelegen] ⟦ab⟧⊣↑abge↓legen
152.28:	sogenannten] sogen.
152.28:	Schul-] Schul⟦e⟧
152.28:	Universitätsmathematik] Universitätsmath.
152.29:	Gebiete] Gebiete,
152.29:	Mathematik] Math.
153.2:	Litteratur der höheren mathematischen Theorien] ⟦rein⟧⊣↑Litteratur der höheren↓ math. ⟦Litteratur⟧⊣↑Theorien↓
153.5:	einen Feriencurs] Feriencurs
153.6:	Parallelenaxiom] Parellellenaxiom
153.8:	*Helmholtz*] *Helmholz*
153.9:	Vorlesungen] Vorlesung.
153.10:	Leipzig] Leipzieg
153.13:	Originalabhandlungen] Originalabhandl.
153.14:	insbesondere] insbes.
153.18:	gerade] g⟦e⟧rade
153.23:	, Ihnen *abzugewinnen*] ⟨Added.⟩
153.24:	das] da
153.25:	Mathematik] Math.
153.25–26:	kommen ⟨...⟩ hinein] ⟨Added.⟩
153.27:	besonders] bes.
153.33:	*Erziehungsmittel*] *Erziehungsmittel,*
153.34:	*auch*] *a*⟪*uc*⟫*h*
154.2:	Lübeck] ⟦Brem⟧Lübeck
154.5:	Mathematik] Math.
154.6:	Maasse] Masse
154.14:	*Mathematik*] *Math.*
155.1:	(] ⟨Possibly added by Hi^s.⟩
155.2:	algebraische Functionen)] algebr. Functionen
155.7:	mündlicher] mündl.
156.2:	Gauss,] Gauss.
156.2:	mündlicher] mündl.
156.2:	S. 7–8,] S. 7–8.
156.5:	Mathematik] Math.
156.6:	wird, so] wird so,
156.7:	Mathematik] Math.
156.10:	Algebra II, Kap. III, IV,] Algebra,↑II Kap. III, IV↓ ⟨added by Hi^b⟩
156.10–11:	Wertheim, Teubner 1878] Wertheim ↑Teubner 1878↓
156.11:	Vorlesungen über Zahlentheorie] Vorles. über Zahlenth.

156.12: Supplement XI.] ⟨Added.⟩
156.14: Algebra,] Algebra.
156.16: algebraischen Zahlkörper,] algebr. Zahlkörper.
156.21: Einheitlichkeit] ⟦Sich⟧ Einheitlichkeit
156.24: Es] ↑Es↓ ⟨added by Hi^b⟩

Description of the Text

Collection: SUB Göttingen, signature *Cod. Ms. D. Hilbert 597*, f. 1r–9r. (These numbers refer to the continuous pagination added to the volume by the library; the volume contains three texts, each bearing a separate original pagination by Hilbert or Käthe Hilbert respectively.)

Size: Cover size 20.9 × 16.7 cm; page size 20.9 × 16.4 cm.

Cover Annotations: On the front cover is pasted a label with the notation, in Hilbert's hand, 'Feriencursus: // Zahlentheorie 1896 // Begriff des Unendlichen 1898'.

Composition: The text appears on the first 17 pages of a signature of 7 double pages, which on the following 5 pages contains *Hilbert 1902b**; this signature, in turn, is bound together with and precedes a signature of 8 double pages containing *Hilbert 1898b**.

Pagination: Continuous pagination of the odd-numbered right-hand pages, in Hilbert's hand, from 1 to 15, beginning with p. 1 on f. 2r of the manuscript.

Original Title: On f. 1r (which Hilbert had left unpaginated) 'Feriencursus Ostern 1896. // 4 · 2 = 8 stündig. // Professor Hilbert. // Kreuzbergweg 15.', all in Käthe Hilbert's hand except for the second line, which has been added by Hilbert.

Text: Pages 1 to the upper half of 9 are written with continuous text in Käthe Hilbert's hand (HiK^s), and the remainder, to p. 15, in Hilbert's hand (Hi^s). Hilbert corrected and annotated his original text, which was written in black ink; for these revisions he used black ink (Hi^s) and, especially for underlinings, blue pencil (Hi^b). In addition he made annotations in pencil (Hi^g) on the reverse side of the title page. A precise dating of these annotations is not possible. The blue-pencil underlinings may already have served to organize the manuscript for Hilbert's lectures in 1896, but he also makes reference to parts of the text that were still in draft for the *Ferienkurs* of 1902, and some of the corrections may date from that time.

Feriencursus: Über den Begriff des Unendlichen

⟨Additions on the reverse of the title page:⟩

Beweise, dass es nicht immer möglich ist, 2 Pyramiden mit gleicher Grundfläche und Höhe in congruente Raumteile zu zerlegen. Vgl. Stolz:[1]

Schubert, Mathematische Mussestunden, Göschen in Leipzig:[2] Ein Betrunkener macht einen π mal soweiten Weg, auch wenn die Amplituden noch so klein werden, d. h. auch in der Grenze.

$64 = 65$ [3]

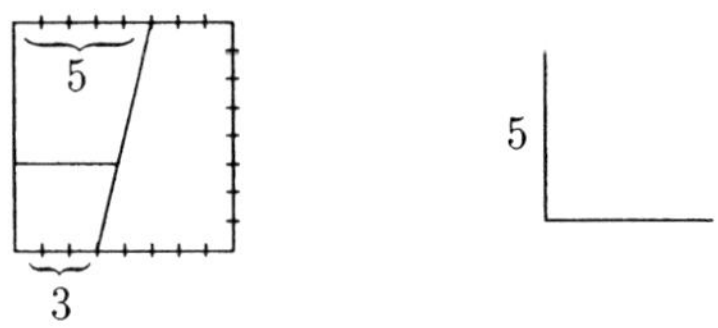

Es ist bisher nicht gelungen, 2 Tetraeder von gleicher Grundfläche und Höhe in eine endliche Anzahl entsprechend congruenter oder symmetrischer Teile zu zerlegen. Ist es unmöglich?[4]

---

1 *Einleitung wie im Feriencursus* Ostern 1896, S. 1–3.

... Ich möchte zeigen, dass die *modernsten Richtungen* und *die äussersten Fortschritte* der Wissenschaft in engster Beziehung zu den *Elementen der Arithmetik und Geometrie* stehen, wie sie ja zuerst auf den *Schulen gelernt und gelehrt* werden. Ich möchte dies zeigen an einem der wichtigsten Begriffe der Mathematik: dem *Begriff des Unendlichen.* Freilich keine systematische

---

[1]The reference is to *Stolz 1885*; see this text, p. 26.

[2]I.e., *Schubert 1898*. The following example of the wandering drunkard could not be found in Schubert's book.

[3]The formula '64 = 65' and the diagrams to its right refer to a pseudo-proof given in *Schubert 1898*, 130–131 as example 3.) of a 'Trugschluß'. The 'Trugschluß' is based on a geometrical illusion that an 8×8 square can be reassembled into a 5×13 rectangle. In this it is crucial that a line that is not in fact straight (it is made up of two straight-line pieces with one slight crook) can be passed off as straight. For the full and correct diagrams, see Schubert, ibid.

[4]This last question about tetrahedra is clearly related to the very first remark above; see also the comments on p. 26 below.

Theorie dieses Begriffes: Eingehende *detaillierte Entwickelungen* und Beweisführungen dürfen wir uns hier gegenseitig *nicht zumuthen.* Schon bei der kurzen Zeit. Ich kann daher nur einzelne Punkte herausgreifen, Ihnen Litteratur nennen, um Sie anzuregen zum späteren eigenen Studium.

I Das *Unendliche in der Arithmetik.*
II " " " " " *Geometrie.*

In II möchte ich besonders die *Elemente der Euklidschen Geometrie* (Congruenzsätze etc.) zur Sprache bringen, und ich würde mich hier sehr freuen, wenn sich eine *Diskussion* darüber an meinen Vortrag anschliessen würde, damit auch ich *Ihre Ansichten über diese Ihnen so geläufigen Dinge erfahre.*

I 2

System von ∞ vielen Dingen, z. B. von Zahlen, Zahlenpaaren, Funktionen. ∞ = *unbegrenztes*, nicht *aufhörendes* Fortzählen möglich ↑– als eine *Einheit logisch* zusammen *aufgefasst.*↓[5]

Beispiele	$1, 2, 3, 4, 5, \ldots$	
	$2, 3, 4, 5, 6, \ldots$	
	$2, 4, 6, 8, 10, \ldots$	
	$1, \frac{1}{2}, \frac{1}{3}, \frac{1}{4}, \frac{1}{5}, \ldots$	
$(r_1, r_2, r_3, \ldots)$	$1, \frac{1}{2}, \frac{1}{3}, \frac{2}{3}, \ldots$	(alle rationalen Zahlen)
$(\alpha_1, \alpha_2, \alpha_3, \ldots)$	$1, \sqrt{2}, \sqrt[3]{3} - \sqrt{2}, \ldots$	(alle algebraischen Zahlen)
(x)	1, 1,011 ...	(alle reellen Zahlen)

(x, y) ... alle Zahlenpaare, z. B. die rationalen Punkte im Inneren eines Quadrates, eines Cubus

(x, y, z) $\quad 0 \leqq x \leqq 1$

$0 \leqq y \leqq 1$

$0 \leqq z \leqq 1$

System aller ganzen rationalen Funktionen mit ganzzahligen Coefficienten.
aller algebraischen Funktionen. $\left(\sqrt{x}, \frac{\sqrt[3]{x-1}}{\sqrt{x}+x^3}, \ldots\right)$
aller stetigen Funktionen. $(\sin x, \sin lx, \ldots)$

Die moderne *Richtung* der Mathematik sieht auf *Genauigkeit der Begriffe* 3
und *Strenge* der Beweise, damit die *sprichwörtliche Zuverlässigkeit* mathematischer Wahrheiten zu *Recht besteht.* Die moderne Mathematik hat eine Reihe von Auffassungen älterer Mathematik *(Funktions-, Zahlbegriff)* als *irrig*

[5]Hilbert uses the term 'system' in Dedekind's sense, i.e., to denote what now would be called sets.

erkannt. Besonders wichtig ist die Auffindung der *Quellen* von Irrthümer. Die vornehmlichste ist: dass man geneigt ist, offenbare Eigenschaften von *endlichen Systemen* auf *unendliche* zu übertragen.
Unter einer bestimmten Zahl von Personen giebt es eine kleinste, grösste, jüngste, älteste .... Aber unter ∞ vielen Zahlen braucht es nicht immer eine kleinste oder grösste zu geben.

Z. B.	$1, 2, 3, \ldots$	keine grösste.
	$1, \frac{1}{2}, \frac{1}{3}, \frac{1}{4}, \ldots$	keine kleinste.
	$1, \frac{1}{2}, 2, \frac{1}{3}, 3, \frac{1}{4}, 4, \ldots$	weder grösste noch kleinste.

Klar bemerkt und in den Consequenzen durchgedacht ist dieser Gedanke von *Weierstrass*, der hauptsächlich 2 Punkte kritisirte.

1.) *Fundamentalsatz der Algebra.*

2.) *Riemannscher Existenzbeweis durch das Dirichletsche Princip.*

Sie sehen, wie tief wir da mit einem Sprunge in die moderne Mathematik hineinkommen.

4 In dem Wertevorrath einer Funktion zwischen zwei Grenzen, z. B. 0 und 1, braucht kein kleinster ↑⟪oder grösster⟫↓ Wert vorhanden zu sein, z. B.

$$f(x) = \frac{\sin \pi x}{1} - \frac{\sin 2\pi x}{2} + \frac{\sin 3\pi x}{3} - \frac{\sin 4\pi x}{4} + \cdots .$$

Dann ist $f(x) = \frac{\uparrow\pi\downarrow x}{2} \quad (0 \leqq x < 1)$
$f(x) = 0 \quad (x = 1)$

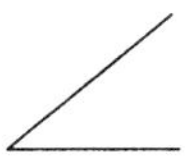

$f(x)$ hat also keinen grössten Wert. $-f(x)$ hat keinen kleinsten Wert in jenem Intervalle.[6]
Lücke bei Cauchy's Beweis: $F(z) = az^n + bz^{n-1} + \cdots = \varphi(x, y) + i\psi(x, y)$, $\varphi^2 + \psi^2 =$ Minimum. Diese Lücke ist durch *Weierstrass* ausgefüllt durch den Satz, dass jede stetige Funktion ein Maximum und Minimum hat.[7]

Schlimmer steht es mit 2.). Hier *stürzten die ganzen Grundlagen* der Riemannschen Theorie der algebraischen Funktionen durch jenen *einfachen Weierstrassischen Einwand.* Es ist zwar gelungen, die Riemannschen Sätze auf *anderem* Wege zu beweisen, aber der so elegante und geniale Weg von Riemann-Dirichlet ist bis jetzt aufgegeben worden. *Weierstrass, Bd. 2, S. 49.*[8]

---

[6]The figure is meant to represent the function in the half-open interval specified.

[7]It is not entirely clear to what result of Cauchy Hilbert is alluding. However, Kline writes:

> Cauchy had used without proof (in one of his proofs of the existence of roots of a polynomial) the existence of a minimum of a continuous function defined over a closed interval. Weierstrass in his Berlin lectures proved for any continuous function of one or more variables defined over a closed bounded domain the existence of a minimum value and a maximum value of the function. (*Kline 1972*, 953.)

In his appreciation of Weierstraß (*Hilbert 1897c*), Hilbert wrote that, with this theorem, Weierstraß created

> ... ein Hilfsmittel, das heute kein Mathematiker bei feineren analytischen oder arithmetischen Untersuchungen entbehren kann. (*Hilbert 1935*, 333.)

[8]The allusion here is to the Dirichlet Principle concerning the existence of minimal

Ich sagte, der menschliche Verstand wäre immer geneigt gewesen, Eigen- 5
schaften des Endlichen auf das Unendliche ohne weiteres zu übertragen. Dieser Punkt ist dann in älterer Zeit eine Hauptquelle von Irrthümern, in neuerer Zeit ein Anlass zu fruchtbarer *Kritik* ↑und *Verschärfung der Begriffe*↓ gewesen.

Z. B. Theorie der Reihen.

Bei endlicher Gliederzahl ist ohne weiteres Vertauschung der Glieder erlaubt; bei unendlichen Gliedern nicht, z. B.

$$1-\tfrac{1}{2}+\tfrac{1}{3}-\tfrac{1}{4}+\tfrac{1}{5}\pm\ldots \quad \text{convergirt} = \log 2,$$

aber

$$1+\tfrac{1}{3}+\tfrac{1}{5}+\ldots = \infty,$$
$$-\tfrac{1}{2}-\tfrac{1}{4}-\tfrac{1}{6}-\ldots = -\infty,$$

also wenn man anders anordnet, kann man alles erhalten.

Z. B.

$$S = \tfrac{1}{1}+\tfrac{1}{3}-\tfrac{1}{2} + \tfrac{1}{5}+\tfrac{1}{7}-\tfrac{1}{4} + \tfrac{1}{9}+\tfrac{1}{11}-\tfrac{1}{6}+\cdots$$
$$S = \left(\tfrac{1}{1}+\tfrac{1}{3}-\tfrac{1}{2}\right) + \left(\tfrac{1}{5}+\tfrac{1}{7}-\tfrac{1}{4}\right) + \left(\tfrac{1}{9}+\tfrac{1}{11}-\tfrac{1}{6}\right)+\cdots$$
$$\log 2 = \left(\tfrac{1}{1}-\tfrac{1}{2}+\tfrac{1}{3}-\tfrac{1}{4}\right)+\left(\tfrac{1}{5}-\tfrac{1}{6}+\tfrac{1}{7}-\tfrac{1}{8}\right)+\left(\tfrac{1}{9}-\tfrac{1}{10}+\tfrac{1}{11}-\tfrac{1}{12}\right)+\cdots$$
$$S-\log 2 = \left(\tfrac{1}{2}-\tfrac{1}{4}\right) + \left(\tfrac{1}{6}-\tfrac{1}{8}\right) + \left(\tfrac{1}{10}-\tfrac{1}{12}\right)+\cdots$$
$$= \tfrac{1}{2}\left(1-\tfrac{1}{2} + \tfrac{1}{3}-\tfrac{1}{4} + \tfrac{1}{5}-\tfrac{1}{6} + \ldots\right) = \tfrac{1}{2}\log 2,$$

d. h. $S = \frac{3}{2}\log 2$.

Begriff der *absoluten Convergenz*. 6

Eine endliche Summe von stetigen Funktionen ist wieder eine stetige Funktion. Nicht so bei ∞ vielen Summanden,

z. B. ↑(Auch die Reihe auf S. 4 oben kann sehr gut als Beispiel gelten.)↓

$$f(x) = x + (x^2 - x) + (x^3 - x^2) + \cdots$$

convergirt für $0 \leqq x \leqq 1$. Denn die Summe der ersten n Glieder $= x^n$,

---

surfaces, and Riemann's use of it (in *Riemann 1851* and *Riemann 1857*). The Principle was so-called by Riemann (see the latter work, § 3), although it seems to go back to Gauß. Weierstraß gave a simple counterexample to it in *Weierstraß 1870*, first published in *Weierstraß 1895*, 49–54, thus the work to which Hilbert refers. The 'other ways of proving Riemann's theorems' refer to, e.g., work of Schwarz and Neumann. In a note to the reprinting of *Klein 1894* in *Klein 1923*, 492, Klein states:

> Ich erinnere mich, daß Weierstrass mir bei Gelegenheit erzählte, Riemann habe auf die Gewinnung seiner Existenzsätze durch das „Dirichletsche Prinzip" keinerlei entscheidenden Wert gelegt. Daher habe ihm auch seine (Weierstrass') Kritik des „Dirichletschen Prinzip" keinen besonderen Eindruck gemacht.

Hilbert clearly believed the Principle to be correct, though, and justifiable, as is clear from his remarks to Problem 23 of *Hilbert 1900e*. Shortly after this, he rehabilitated the Principle in a general form under certain rather weak conditions; see *Hilbert 1900a* and *Hilbert 1904a*. In the former (*Hilbert 1935*, 11), Hilbert writes:

> Aber seit der Weierstraßschen Kritik fand das Dirichletsche Prinzip nur noch historische Würdigung und erschien jedenfalls als Mittel zur Lösung der Randwertaufgabe abgetan. Bedauernd spricht C. Neumann aus, daß das so schöne und dereinst so viel benutzte Dirichletsche Prinzip jetzt wohl für immer dahingesunken sei; ...
>
> Das folgende ist ein Versuch der Wiederbelebung des Dirichletschen Prinzips.

For a general assessment of Hilbert's work on the Principle, and a historical account of the problem, see *Monna 1975*.

also $f(x) = 0, \qquad 0 \leqq x < 1$
$ f(x) = 1, \qquad x = 1,$

d. h. *unstetige Funktion* für $x = 1$.

Begriff der gleichmässigen Convergenz: Maxima der absoluten Glieder müssen eine convergente Reihe bilden.

$$\varphi(x) = x^n - x^{n-1}$$

$$\varphi'(x) = nx^{n-1} - (n-1)x^{n-2} = 0, \qquad x = \frac{n-1}{n}$$

$\varphi(x)$ wird $\left(1 - \frac{1}{n}\right)^n \left(1 - \frac{n}{n-1}\right) = \left(1 - \frac{1}{n}\right)^n \frac{1}{n-1} = M_n$

$M_1 + M_2 + \cdots$ divergirt.[9]

Dann auch gliedweise Integration erlaubt.[10]

Darum die fundamentale Bedeutung der *absoluten gleichmässig convergenten*
7 Reihen, so dass | man in der gesammten Analysis überhaupt gar keine anderen
Reihen gebraucht.

Aber noch viel interessanter ist der positive Gewinn, den die *moderne* Behandlung des Unendlichkeitsbegriffes herbeigeführt hat durch Schaffung der Mächtigkeit.

Siehe die obigen Beispiele! Kann man die verschiedenen ∞ Systeme mit einander vergleichen? Bei *endlichen Systemen* ist die *Definition der Gleichheit* der Anzahl so: gleich, wenn eine Zuordnung möglich ist, so dass umkehrbar eindeutig entspricht. *Damen und Herren* in gleicher Zahl, wenn sie sich paarweise die Hände so reichen können, dass niemand übrig bleibt. Die Systeme nennen wir *Mengen*.

↑Satz.↓ Wenn 2 Mengen M, N einer dritten $P =$ sind, so $M = N$. M heisst $< N$, wenn nicht $= N$, aber $=$ einem Teil von N.

Satz. Wenn $M < N$ und $N < P$, so $M < P$.

8 Aber es ist sehr wichtig, dass der Satz nicht gilt, | wonach der Teil $<$ als
das Ganze ist.

↑Der Begriff der Dimension hält nicht stand: sogar die stetige Abbildung eines Linienstückes auf ein Quadrat ist möglich. Vgl. meine Arbeit Math. Ann. Die

[9] As Hilbert says, the series $f(x) = 1 + \sum_{n=1}^{\infty}(x^n - x^{n-1})$ is convergent in the interval $0 \leq x \leq 1$, but not uniformly so, not even in the interval $0 \leq x < 1$. The example is treated in detail, and thoroughly explained, in *Knopp 1947* (pp. 335–336 of the English edition), as follows. Let the sum of the first n terms of the series, the partial sum function, be $s_n(x)$; the corresponding remainder $f(x) - s_n(x)$ is then denoted by $r_n(x)$. $s_n(x)$ is clearly just x^n, as Hilbert says, and $r_n(x)$ is then $-x^n$. If the series were uniformly convergent in a given interval, then it would follow that the absolute values of the sequence $r_n(x_n)$ would converge to 0 for any sequence x_n of points of the interval. Now choose the sequence of points $x_n = \frac{n-1}{n}$ from $0 \leq x < 1$; the corresponding remainders $r_n(x_n)$ for these points are $[-(1 - \frac{1}{n})]^n$, and the absolute values of this sequence converge to $\frac{1}{e}$, showing that $f(x)$ is not uniformly convergent in the interval in question.

As Hilbert notes, the nth term $(x^n - x^{n-1})$ of the series has a minimum (maximum absolute value) at the point $\frac{n-1}{n}$. Hilbert's M_n appear to be the values of the nth term at those minimum points, although his calculations are somewhat difficult to follow. If this is right, however, it is easy to see that $M_1 + M_2 + \ldots + M_n + \ldots$ would not diverge if the series were uniformly convergent.

[10] Presumably Hilbert means that it is allowed when the series is uniformly convergent.

abbildenden Funktionen sind in eine gleichmässig convergente nach ganzen rationalen Funktionen fortschreitende Reihe entwickelbar.↓[11]

Collegheft: Zahlbegriff und Quadratur des Kreises, 1897/98.
S. 14 – S. 17. ↑(*Abzählbarkeit der rationalen Zahlen* S. 9).↓
↑Die Menge aller reellen Zahlen ist nicht abzählbar.↓
↑Ferner S. 20 –↓S. 21. ↑Beweis, dass die Menge aller stetigen Funktionen nicht grösser ist als die Menge aller reellen Zahlen.↓[12]

Anwendungen: Existenz irrationaler Zahlen $\sqrt{2} = \frac{a}{b}$ etc. Schwieriger, dass e, und noch nicht bekannt, dass $2^{\sqrt{2}}$ irrational ist.[13] ↑Um ein Beispiel zu geben, zeige, dass e irrational.↓

Existenz transcendenten Zahlen.[14]

Collegheft: Quadratur des Kreises, 1894/95, S. 46 – S. 47.[15]

Litteratur: Stolz, Allgemeine Arithmetik.
Lipschitz, Höhere Analysis.
Thomae, Analytische Funktion.
Pringsheim, " " .
Cantor, ↑Beiträge zur Begründung der transfiniten Mengenlehre, Math. Ann. Bd. 46, 49.↓
Hilbert, Math. Ann. Bd. 43, Transcendenz von e, π.
Weber, Algebra, Bd. II, Abschnitt 25.[16]

II 9

Von den *Elementen der Arithmetik* kommen wir zu den *Axiomen der Geometrie*. Ein *altes* Problem, von Euklid mit ausserordentlichem Scharfsinn erfasst. Euklid für uns von besonderer Wichtigkeit, weil er das Fundament aller *geometrischen Wissenschaften* bildet und zugleich das Wesentliche enthält, was noch heute auf unseren *Schulen* gelernt und gelehrt wird. Manches

[11]The work Hilbert refers to is *Hilbert 1891b*, in particular p. 460.

[12]The 'Collegheft' Hilbert refers to is *Cod. Ms. D. Hilbert 549*, which will appear in Volume 2 of this edition. Pp. 14–17 give a general, if sketchy, treatment of power. The countability of the rationals is discussed on pp. 9 and 14–15. On pp. 18–21, Hilbert proves that the set of all continuous functions has the power of the set of all reals numbers, and he calls this '[e]ine sehr überraschende kaum glaubliche Thatsache' (18).

[13]Here appears a favourite example of Hilbert, a special case of the 7th Problem in *Hilbert 1900e*. It was solved by a result of Gel'fond and Schneider, proved independently in 1934, which says: if α, β are algebraic, with $\alpha \neq 0, 1$ and β irrational, then α^{β} is transcendental. (See *Tijdeman 1976*, 242–243.) In 1929, Gel'fond proved a special case which shows that $2^{\sqrt{-2}}$ must be transcendental, and in 1930 Siegel extended the proof to cover $2^{\sqrt{2}}$. Siegel himself recalls having heard Hilbert say in a lecture in 1920 that no one in the room would live to see this problem solved. See *Reid 1970*, 164.

[14]Given the context, Hilbert is probably alluding to Cantor's proof, which follows from the fact that the algebraic numbers are countable, but the reals not.

[15]This is *Hilbert 1894b**, which will appear in Volume 2 of this Edition. Pp. 46–47 deal with the existence of transcendental numbers.

[16]The works referred to here are: *Stolz 1885,1886*, *Lipschitz 1877,1880*, *Thomae 1880*, *Pringsheim 1899*, *Cantor 1895*, *Cantor 1897*, *Hilbert 1893b* and *Weber 1895*.

bei Euklid freilich bedarf der *Ausführung und Ergänzung.* Ich habe mir einige Punkte für den heutigen Zweck überlegt und weil ich im Winter ein Colleg über Euklidische Geometrie zu lesen beabsichtige. Freilich habe ich dabei *viele Schwierigkeiten* gefunden; je mehr ich nachdachte, desto mehr *Lücken habe ich zu* bemerken geglaubt. Was ich heute Ihnen erzähle, sind nur Versuche. Unfertig und sollen hauptsächlich anregende *Gedanken* sein. Leider! Ich möchte Sie aufrufen, sich in diesem Sinne mit den Elementen des Euklid zu beschäftigen als die *berufensten Mitarbeiter.* Ich werde mich glücklich schätzen, wenn ich mit diesem Aufruf Erfolg habe.

10 Die Grundlage bilden die Congruenzsätze. Um diese aber zu verstehen, müssen wir 3 Klassen von Axiomen unterscheiden:

1.) Axiome *der Verknüpfung der Elemente* untereinander. *Elemente: Punkte, Geraden, Ebenen.* Beschränkung auf die ebene Geometrie, daher nur Punkte, Geraden.

Pasch: Vorlesungen über neuere Geometrie, Teubner 1882.
Hilbert, Gerade Linie als kürzeste. Math. Ann. Bd. 46.[17]

Irgend zwei Punkte bestimmen stets eine Gerade.
Auf jeder Geraden giebt es wenigstens 2 Punkte.

Bei Euklid und in den Schulbüchern zur Not wenigstens angedeutet.
2.) *Axiome der gegenseitigen Lage* (über den Begriff „*zwischen*")
Hilbert, Math. Ann. Bd. 46, S. 91.[18]

Fundamental für Geometrie und Arithmetik (d.h. die entsprechenden Sätze, dass von irgend 2 Zahlen α, β immer $\alpha \lesseqgtr \beta$ ist). Wenn man überhaupt von einem *Fundament* der Mathematik sprechen kann, so sind es diese. Sie fehlen bei *Euklid* und in den Schulbüchern, obwohl sehr instruktiv und schöne Aufgaben und Lehrsätze mit ihrer Hülfe bewiesen werden können. *Bei Pasch*[19] *mit grosser Liebe und Breite* behandelt. ↑Die Punkte eines Quadrates können z.B. durch Abbildung auf eine Gerade oder auch nach dem Princip $a + bi < a' + ib'$ (wenn $b < b'$, oder bei $b = b'$, $a < a'$ ist) geordnet werden. Das System aller stetigen Funktionen kann in dieser Weise geordnet werden. „Wohlgeordnet" bei Cantor. Kann auch das System aller unstetigen Funktionen so geordnet werden?[20]↓

11 Durch diese Axiome 2.) ist der Halbstrahl, Halbebene, Halbraum definirt.

3.) *Axiome der Congruenz* (oder *Bewegung*).
Bei Pasch[21] ausführlich, vgl. mein Collegheft Grundlagen der Geometrie, S. 61–62.[22]

---

[17] I.e., *Pasch 1882*, and *Hilbert 1895b*.

[18] I.e., *Hilbert 1895b*, 91.

[19] *Pasch 1882*.

[20] This last question can only be answered positively by using the Axiom of Choice. It is not clear why this passage about ordering is inserted at this point.

[21] *Pasch 1882*.

[22] This is a reference to the 1894 lectures (Chapter 2 in this Volume), where the Axioms of Congruence are to be found, together with a few of their most important consequences.

Nachdem Halbstrahl, Halbebene, Halbraum definirt sind, hat es einen eindeutigen Sinn, wenn wir sprechen von der:
Abtragbarkeit einer jeden Strecke auf einer Geraden.
" " Winkels an eine " " .
Congruenzsätze über Dreiecke.

1.) Wenn $AB \equiv A'B'$, so $A'B' = AB$[23] und 2.) $AB \equiv BA$, gestatten die Schreibweise $a = b$ und die Benennung Strecken statt Punktepaar.[24]
Wenn $a = b$, $a = c$, so $b = c$.
$a + b = b + c$. etc.

Winkel = System von 2 Halbstrahlen.

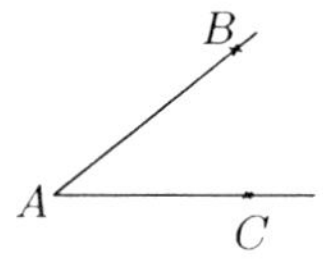

4.) Parallelenaxiom von Euklid. 12
Parallel heissen 2 Geraden, wenn $\alpha = \beta$.[25]
Parallele Linien schneiden sich nicht.
Nichtparallele Linien " " stets.
Unbeweisbar: Orthogonale Halbkreise in der Euklidischen Geometrie.[26] Dass die *Euklidische Geometrie existirt*, folgt aus der *Existenz* der analytischen Geometrie.[27]

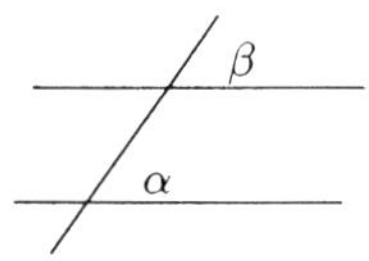

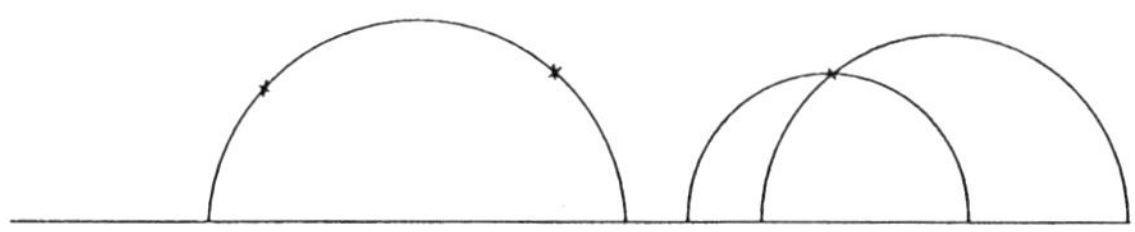

Wir haben Parallele gezogen durch Abtragen eines Winkels. Dass Abtragen eines Winkels möglich ist, folgt aus den Congruenzsätzen. Man kann fragen, ob man nicht mit der geringeren Forderung des Parallelenlinienziehens auskommt. In der That. Zunächst Aufgabe, einen rechten Winkel zu construiren:

---

[23] This should be '≡' instead of '='.

[24] See note to p. 40 in *Hilbert 1898c**, Chapter 4 of this Volume.

[25] This definition differs from that in the 1898/1899 lectures; see pp. 70 and 71 respectively of *Hilbert 1898c** and *Hilbert 1899a** (this Volume, Chapter 4). In the latter, a theorem corresponding to the definition here is proved on p. 79.

[26] The model is given in detail in *Hilbert 1898c**, 78f, and *Hilbert 1899a**, 81f (this Volume, Chapter 4).

[27] A model of Euclidean geometry within analytic geometry will show the relative consistency of the former to the latter. Thus, Hilbert's claim that the *existence* of the latter shows the *existence* of the former means that he here accepts consistency as the decisive characteristic of mathematical existence, and also implicitly accepts the consistency of analytic geometry, that is, of the standard number systems. Cf. the 1898/1899 lectures (this Volume, Chapter 4), i.e., *Hilbert 1898c**, 104, and *Hilbert 1899a**, 167.

Winkel α beliebig, $AB = AC$.[28] $BD \parallel AC$ und $CD \parallel AB$. AD und BC

13

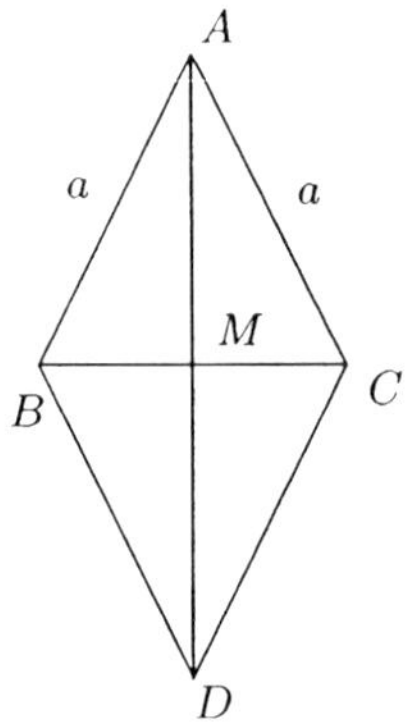

schneiden sich rechtwinklig.

Alle rechten Winkel sind einander gleich, ist dann beweisbar.

Nun soll $\sphericalangle\alpha$ auf l von B' ab abgetragen werden, so mach BMA = rechten Winkel. Dann

$$BM = B'M',$$
$$BA = B'A.$$[29]

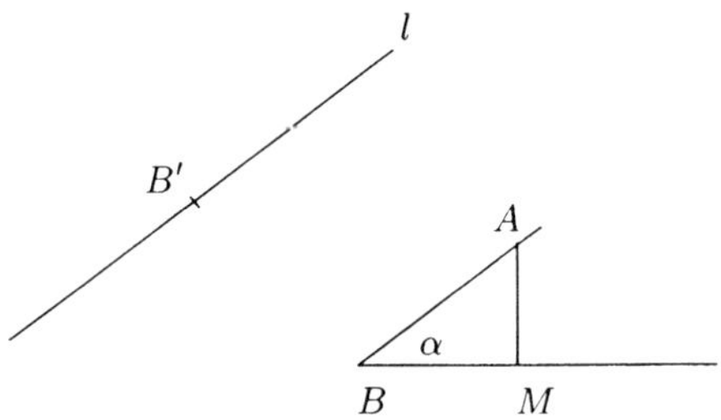

Durch Abtragen eines rechten Winkels allein und durch Abtragen von Strecken kann man alles construiren, was auf Grund der Congruenzsätze construirbar ist.[30] Denn zunächst Construktion eines Rechteckes; mit dessen Hülfe kann man beliebige Parallele ziehen und damit wiederum jeden Winkel abtragen,
14 indem man den rechten Winkel abträgt: Denn man kann | offenbar jeden Winkel abtragen. Denn:[31]

---

[28] In the figure, α is apparently $\sphericalangle BAC$.

[29] This should be '$BA = B'A'$'. To construct a given angle α at any given point B', first construct the segment $B'M'$ congruent to BM; construct a right-angle to l at M; mark off segment $A'M'$ congruent to AM on the perpendicular; join A' and B'. Then $\sphericalangle A'B'M' = \alpha$.

[30] In the 1898/1899 lectures (*Hilbert 1898c**, 47, and *Hilbert 1899a**, 45; this Volume, Chapter 4), Hilbert first constructs a right-angle using segment and angle *Abtragen*, thus the full set of Congruence Axioms, but without the Axiom of Parallels. Here Hilbert asserts that all such constructions are possible using just the Axiom of Parallels and segment *Abtragen* alone. The problem of what is possible using only these means is addressed again at the end of *Hilbert 1899a**, 170–172, and in the *Festschrift* (*Hilbert 1899c*), 79–81.

[31] Just below this, and slightly above and to the right of the two figures below, there is the sketch of a figure which Hilbert has scribbled out in such a way as to render it illegible.

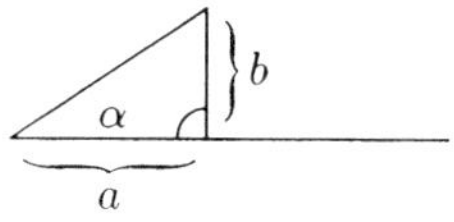

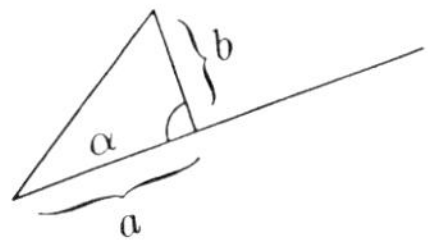

Also Abtragung beliebiger Strecken und Drehung um rechten Winkel sind die geringsten Hülfsmittel, mit denen jede durch die Congruenzsätze construirbaren Figuren construirt werden können. Dazu gehört nicht die Construktion des Schnittpunktes 2er beliebiger Kreise oder eines beliebigen Kreises mit einer beliebigen Geraden, die einen Punkt im Inneren des Kreises enthält ↑oder auch Ziehen einer Tangente an einem Kreis von einem gegebenen Punkte aus?↓ oder auch die Construktion eines Dreieckes | aus 3 Seiten a, b, c, wo $a+b>c$. 15
Dass in jedem Dreiecke stets $a+b>c$ ist, kann bewiesen werden.[32]

Auf Grund der Congruenzsätze kann ein gleichseitiges Dreieck construirt werden.[33]

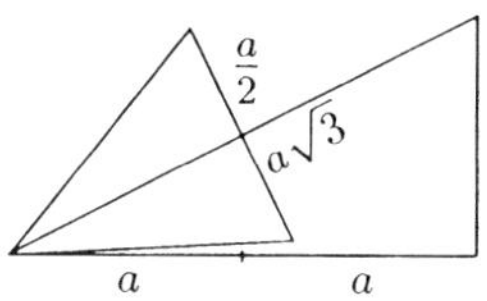

$\frac{a\sqrt{3}}{2}$ = Höhe der gleichseitigen Dreiecke.

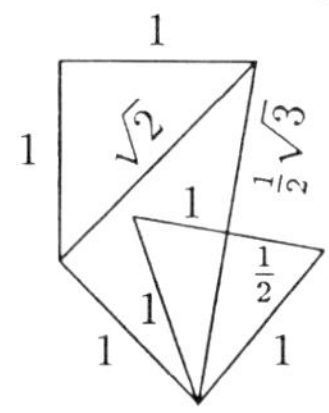

$$\sqrt{1+1+1} = \sqrt{1+\left(\sqrt{1+1}\right)^2}$$

Aber es kann nicht jedes Rechteck in ein Quadrat verwandelt werden.

Es kann nicht an Kreisperipherie (π) die Seiten eines Quadrates $\sqrt{\pi}$ mit gleichem Inhalt construirt werden, obwohl z. B. $\sqrt{\pi^2+1}$ mit Pythagoras construirbar wäre, wenn man den Gebrauch des Zirkels auf Abtragung gegebener Strecken einschränkt.

---

[32]This sentence contains a reference to the problem extensively treated in the 1898/1899 lectures (*Hilbert 1898c**, 62–67, and *Hilbert 1899a**, 61–68; this Volume, Chapter 4), in connection with the Three Chord Theorem; this theorem is given in these lectures on p. 22, and the problem mentioned is alluded to again on p. 23.

[33]The figure below at the right shows how to construct an equilateral triangle of side 1, thus with height $\frac{\sqrt{3}}{2}$, which is the length of half of the hypoteneuse of the second (larger) right-angled triangle. The diagram demonstrates that this can be done with nothing more than 'moving' (and halving) segments and 'moving' right-angles. Hilbert notes that the length of this hypoteneuse is of Pythagorean form (i.e., of the form $\sqrt{1+x^2}$), for $\sqrt{3} = \sqrt{1+(\sqrt{1+1})^2}$.

The right-hand figure was probably intended as a replacement for the figure at the left, which we give here although it was crossed out by Hilbert. It is clearly inappropriate as a guide to constructing $\frac{a\sqrt{3}}{2}$. The right-hand figure itself is very poorly drawn in the text. We have improved it somewhat, based on what is in effect the same drawing found in Hilbert's notes for the 1902 *Ferienkurs*, *Hilbert 1902b**, sheet 11. All the drawings, and the labelling, are in Hilbert's own hand.

Beweise streng, dass z. B. $\sqrt{\pi}$ aus π nicht durch Abtragen von Strecken gefunden werden kann, während dagegen alle Wurzeln aus Zahlen, die Summen von Quadratzahlen sind, durch Abtragen mittelst Pythagoras gefunden werden können. Die genannte Behauptung ist wahrscheinlich gleichbedeutend mit der Behauptung der Transcendenz von π.[34]
↑Dass jedes Rechteck in ein Quadrat verwandelt werden kann, kann auch mit Archimedes nicht bewiesen werden: Vielmehr erst ideale (irrationale) Punkte nöthig, daher besser mein Stetigkeitsaxiom.↓

16 5.) Das Archimedische Axiom

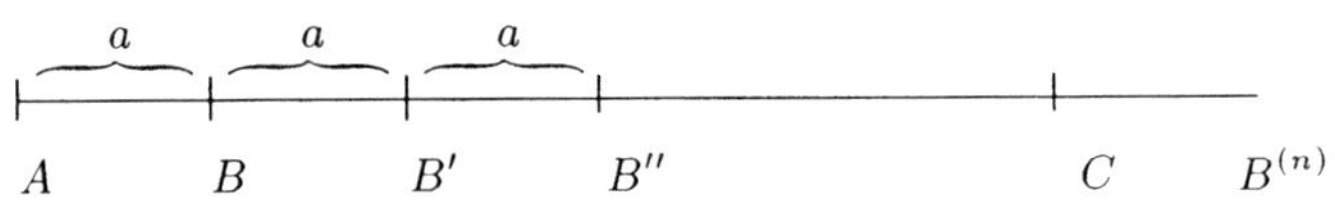

↑im wesentlichen↓ gleichbedeutend mit dem Stetigkeitsaxiom, welches ich in meiner Arbeit über die Gerade als die kürzeste formulirt habe: Jedem Schnitt entspricht eine und nur eine Zahl.[35]
Also Einführung der Zahl in die Geometrie. Gewöhnliche Sätze über die Proportion:

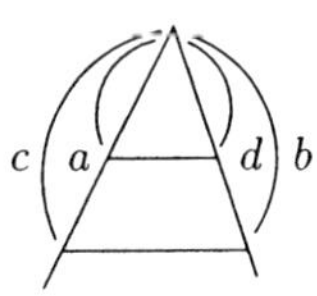

Wenn $a : c = d : b$, so $\|$.
Wenn $\|$, so $a : c = d : b$, oder $ab = cd$.
Das Archimedische Princip bedingt die Eindimensionalität der Geraden. Denn wollte ich die Zahlenpaare (Punkte einer Ebene) (x, y) ordnen nach den Axiomen der gegenseitigen Lage, so etwa

$$x + iy < x' + iy',$$
$$\text{wenn } y < y' \text{ oder bei } y = y', \text{ wenn } x < x'.$$

17 Dann aber würde das Archimedische Princip nicht gelten: denn z. B. 1, 2, 3, 4, ... $< i$.
Auch Geometrie möglich mit den Axiomen 1.) 2.) 3.) 4.) ohne 5.).
Beweis: Construire folgendes System von Funktionen. Zunächst die rationale Funktion $f(t)$ von t mit reellen Coefficienten. Setze $f(t) >$ bez. < 0, wenn $f(t)$ für genug grosse Werte von t $f(t) > 0$ bez. < 0 bleibt, was immer einmal statt hat. Ist $f(t) > 0$, so zähle $\sqrt{f(t)}$ und alle rationalen Funktion von t

---

[34]In the last two paragraph, Hilbert refers to the impossibility of 'squaring the circle' by elementary geometric means, even with use of the compass to mark off given segments. The remarks in the first of these paragraphs, and at the end of the of the paragraph before ('Aber es kann nicht jedes Rechteck ...'), suggest Hilbert's awareness that not every Pythagorean field is Euclidean, a point of great importance in connection with the Three Chord Theorem examined on p. 22f. (See the Introduction to Chapter 4, Specific Note (2).)

[35]Hilbert is referring to the continuity axiom as formulated in *Hilbert 1895b*, and referred to again in *Hilbert 1898c**, 102, and restated in *Hilbert 1899a**, 140–141. (See this Volume, Chapter 4.) This yields Bolzano-Dedekind continuity, but, contrary to what Hilbert says here, it is not 'gleichbedeutend' with the usual form of the Archimedean Axiom, to which he alludes with the diagram. From this weaker form, it follows that to each Dedekind Cut there corresponds a unique real number, but the stronger axiom is required for the converse, i.e., to each real number there corresponds a unique Dedekind Cut.

und $\sqrt{f(t)}$ hinzu. Nun sei $F(t)$ irgendeine Funktion > 0 dieses erweiterten Systems, dann nimm auch $\sqrt{F(t)}$ hinzu etc.[36]

Ein Funktionenpaar x, y dieses Systems sei ein Punkt in der Ebene. Die Verschiebung ist einfach Addition. Die Drehung um den Nullpunkt um $\sphericalangle\alpha$ ⟨berechne⟩, wenn x, y ein beliebiger Punkt ist, wie folgt

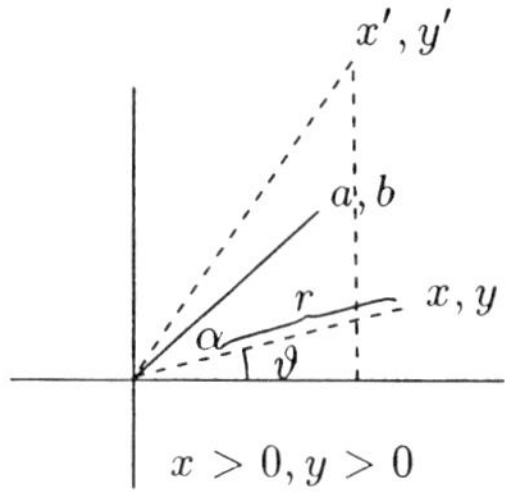

$$\frac{x'}{r} = \cos(\alpha+\vartheta) = \cos\alpha\cos\vartheta - \sin\alpha\sin\vartheta$$
$$\frac{y'}{r} = \sin(\alpha+\vartheta) = \sin\alpha\cos\vartheta + \cos\alpha\sin\vartheta$$

$$\frac{x'}{r} = \frac{a}{\sqrt{a^2+b^2}}\frac{x}{r} - \frac{b}{\sqrt{a^2+b^2}}\frac{y}{r}$$
$$\frac{y'}{r} = \frac{b}{\sqrt{a^2+b^2}}\frac{x}{r} + \frac{a}{\sqrt{a^2+b^2}}\frac{y}{r}$$

↑Vielleicht kann man direkt auch den Winkel ϑ einführen!↓

Ueber Archimedes' Princip Stolz: Allgemeine Arithmetik, Bd. I, Abschnitt V; 18
Paul du Bois Reymond, Allgemeine Funktionentheorie.[37]

↑4.) 5.) haben nicht den *empirischen*, durch eine endliche Anzahl von Versuchen construirbaren Charakter, wie 1.) 2.) 3.).↓

Nachdem wir die 5 Axiome aufgestellt haben, entstehen etwa 3 Fragen:

a.) Sind die 5 Axiome von einander unabhängig?

b.) Wie beweist man die Fundamentalsätze der projektiven oder der analytischen Geometrie?

c.) Welche Axiome kann man hernach entbehren, wenn man die in b.) gemeinten Fundamentalsätze bewiesen hat?

a.) 1.) 2.) 3.) sind offenbar von einander unabhängig.
Dass 4.) keine Folge, ist bereits erwähnt.
" 5.) " " bewiesen.

b.) Wir behandeln hier nur die Sätze über Gleichheit von gradlinig begrenzten Flächen und 2 Schnittpunkt-Sätze, den Desargues und den Pascal.[38]

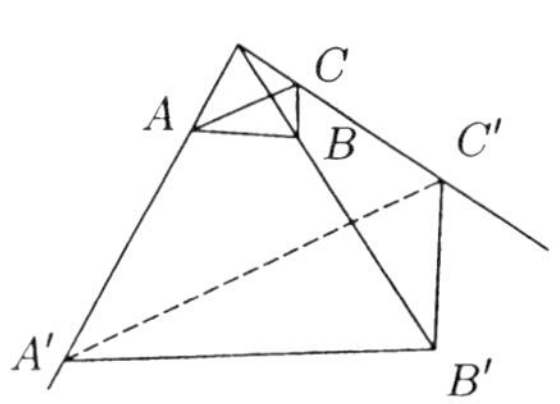

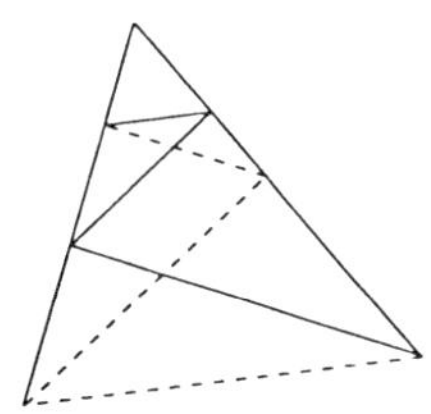

Wir sehen von 5.) ab. 19

Desargues ist im Raume beweisbar. Man braucht bloss obige Figur räumlich zu sehen, so hat man den Beweis aus 1.). Man braucht also gar nicht 3.).

[36]The non-Archimedean model of geometry Hilbert sketches here is probably the one explained more fully in *Hilbert 1899a**, 141f. See Chapter 4 of this Volume.

[37]The references are *Stolz 1885*, 69–70 (and notes), and *du Bois-Reymond 1882*, e.g., p. 46.

[38]What Hilbert calls Pascal's Theorem is not the full projective Pascal Theorem concerning conic sections generally, but rather the affine restriction of Pappus's Theorem, itself a special case.

Es ist nicht bekannt, ob Desargues auch in der Ebene mit Hülfe der Congruenzsätze beweisbar ist. Bei Euklid durch die Proportionen; aber dabei wird angenomen, dass jedem Punkt eine Zahl entspricht, also 5.).
Pascal kann aus den Congruenzsätzen $\uparrow$im Raume$\downarrow$ bewiesen werden. Schur: Fundamentalsatz der projektiven Geometrie, Math. Ann. Bd. 51.[39]
Ich möchte Ihnen einen anderen Beweis mitteilen mittelst einer Methode, welche mir überhaupt geeignet erscheint, Licht in den Zusammenhang von Desargues, Pascal und den Sätzen über Inhaltsgleichheit zu bringen.

Wir können das Produkt ab 2er Strecken a, b auf drei verschiedene Weisen formal einführen:

1.) Wenn $AB \parallel A'B'$, so sei $ab = cd$, falls $\sphericalangle AOB$ = einem beliebigen, aber festen Winkel α ist. Etwa α = einem Rechten. Da die Schenkel vertauschbar, so folgt offenbar:

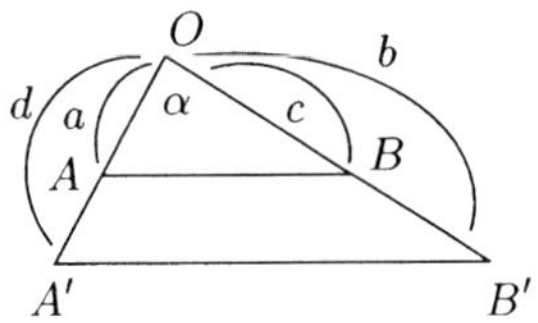

Wenn $ab = cd$, so $cd = ab$.

20 Bedeutung von Desargues: Die Bedeutung der Gleichung

$$ab = cd$$

ist von der Wahl des dabei benutzten Winkels α unabhängig.

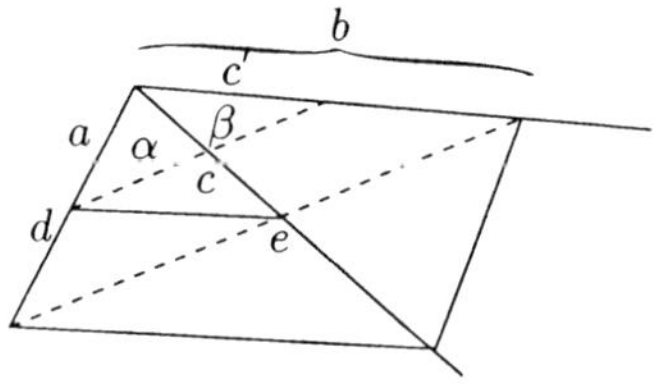

Bedeutung von Pascal: Wir halten a, b fest und suchen e, f, so dass

$$ab = cd \qquad c \neq b \quad (ab = ba)$$

und $ab = ef$,

so $cd = ef$

d. h. wenn $ab = cd$

und $ab = ef$,

so $cd = ef$.

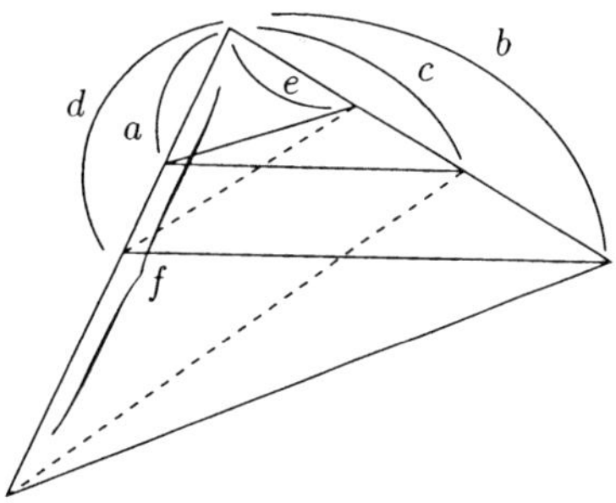

Also wenn 2 Streckenprodukte einem dritten Streckenprodukt gleich sind, so sind sie unter einander gleich.[40]

[39] The reference to Schur is to *Schur 1899* (which appeared in volume 51 of the *Mathematische Annalen*). Although the issue of the *Annalen* containing the former did not appear until 22 December, 1898, Schur's paper is dated 21 March, 1898, and Hilbert, as one of the assistant editors of the journal, would have been aware of it immediately. In the 1898/1899 lectures and the *Festschrift*, Hilbert relies on the fact that both what he calls Pascal's Theorem and the Desargues Theorem can be proved 'in the plane' from the Congruence Axioms, as long as use is made of the Axiom of Parallels. See the 1898/1899 lectures, *Hilbert 1899a**, 100, 108f., and *Hilbert 1899c*, 29–31, 49–50 in Chapters 4 and 5 of this Volume, respectively. In *Hilbert 1898c**, 102 and *Hilbert 1899a**, 169, Hilbert raises the question of whether the Axiom of Parallels can be dispensed with. The answer is yes: see the Introduction to Chapter 4, Specific Note (5)(*b*).

[40] Cf. the statement of Pascal's Theorem in *Hilbert 1899a**, 110, and in *Hilbert 1899c*, 28, this Volume, Chapters 4 and 5 respectively.

(Ich weiss nicht, ob Euklid bei der Anwendung dieses Satzes als ein allgemeines Grössenaxiom eine Lücke aufweist. Der Satz ist natürlich keine allgemeine Grössenthatsache. Denn z. B. definire $a = b$, wenn $(a - b) \leqq 1$,[41] so $2 = 3$, $4 = 3$, aber $2 \neq 4$.)[42]

Zweite Einführung des Streckenproduktes: 21

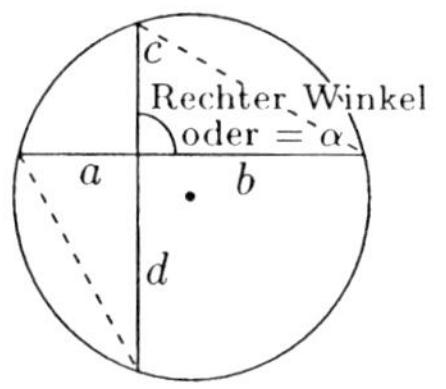

$ab = cd$, wenn die 4 Endpunkte auf einem Kreise liegen, d. h. gleiche Entfernung von einem Centrum.
Aus $ab = cd$ folgt

$$cd = ab = ba = dc.$$[43]

Zurückführung auf die frühere Definition mit Hülfe des Satzes von der Gleichheit der Peripheriewinkel, der ohne weiteres beweisbar.[44]

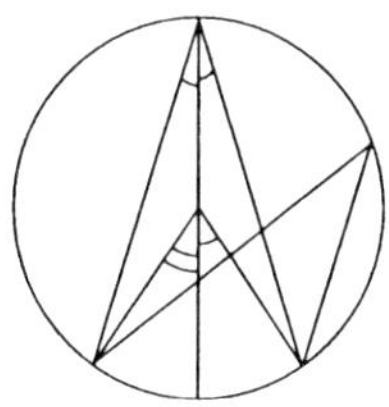

Also wenn $ab = cd$ nach der ersten Definition, so auch nach der zweiten und umgekehrt.

(Der Winkel der beiden Sehnen kann ein beliebiger aber fester sein, z. B. ein Rechter.)

Wir schliessen, dass die Definition von $ab = cd$ wiederum von der Wahl des benutzten Winkels unabhängig ist.

Bedeutung von Pascal:

Wenn die 4 Endpunkte ab und cd auf einem Kreise liegen
und " " ab ef " " " ,
so liegen auch " " cd ef " " .

[41]Very probably Hilbert means $|a - b| \leqq 1$ here.

[42]The reference to Euclid is probably to Common Notion 1, 'Things which are equal to the same thing are also equal to one another' (*Heath 1926a*, 155). What Hilbert means is that one cannot simply *assume* that all 'magnitudes' satisfy the usual properties possessed by the magnitudes with which we are most familiar, say the real numbers or the rationals, since in the example he sets up, transitivity fails. The question implicit in Hilbert's remark is then presumably: was Euclid aware that one should prove that segments have the basic properties we assume 'magnitudes' to have?

[43]In the figure, the horizontal chord splits into two lengths, a and b, the vertical into c and d.

[44]I.e., using the Congruence and Parallel Axioms.

22 D. h. die 3 Sehnen der 3 Kreise schneiden sich in einem Punkt.

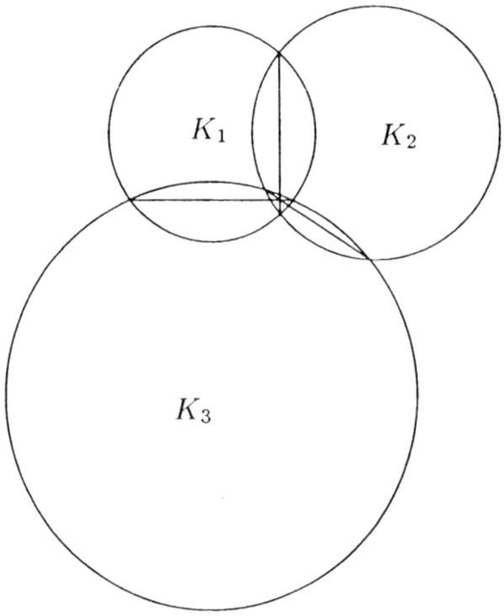

Man errichte nun über K_1, K_2 2 Kugeln K_1', K_2', deren Mittelpunkte in der Ebene liegen und beschreibe durch 3 Punkte auf K_3 und einen Punkt des der Kugeln K_1', K_2' gemeinsamen Kreises eine Kugel K_3'. Die 3 Ebenen durch die 3 Schnittkreise $(K_1'K_2')$, $(K_1'K_3')$, $(K_2'K_3')$ haben dann die beiden allen 3 Kugeln gemeinsamen Punkte gemein und daher deren ganze Verbindungsgerade, d. h. die 3 Sehnen schneiden sich in einem Punkte.[45]

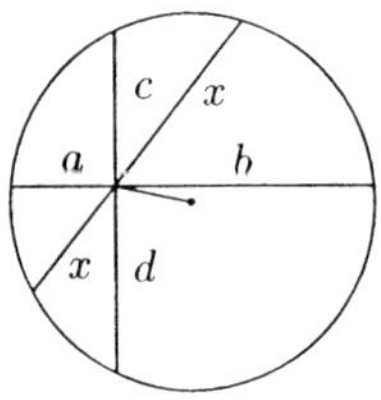

Man könnte auch so schliessen: Bestimme x, so dass $ab = x^2$, $cd = x^2$ und dann y, so dass $ab = y^2$, $ef = y^2$. Aus $ab = x^2$, $ab = y^2$, folgt $x = y$, folgt $cd = x^2$, $ef = x^2$.

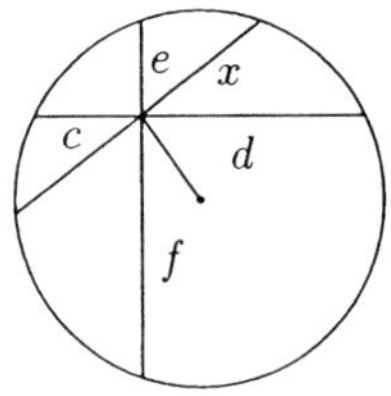

23 Einwand, dass x nicht zu existiren braucht. Ist dieser Beweis streng zu machen?[46] Folgt Pascal in der Ebene aus Desargues und 1.) 2.) 3.)? Folgt Pascal vielleicht auch in der Ebene ohne Desargues aus 1.) 2.) 3.)? Folgt um-

[45]There is a full treatment of the Three Chord Theorem and its metamathematical ramifications in the 1898/1899 lectures, *Hilbert 1898c**, 62–67 and *Hilbert 1899a**, 61–68 respectively (this Volume, Chapter 4). The proof sketch here is given more fully in *Hilbert 1898c**, 63–64, and is easiest to comprehend in *Hilbert 1899a**, 61. The theorem is not dealt with in the *Festschrift*. For a full discussion of the theorem and the connections explored by Hilbert, see the Introduction to Chapter 4, Specific Note (2).

[46]The objection is confirmed by the construction carried out in *Hilbert 1898c**, 66–68 and *Hilbert 1899a**, 66–67. For an explanation, see the Introduction to Chapter 4, Specific Note (2)(*e*). The considerations in the 1898/1899 lectures tie the existence of the square points, and therefore the proof of the Three Chord Theorem, to various other assumptions, e.g., that circles which intersect have two points of intersection or that a straight line with a point in the interior of a circle and a point outside must intersect the circle, and finally to the Euclidean assumption that it is always possible to construct a triangle from three lines, any two of which are greater than the third. See the Introduction to Chapter 4, Specific Note (2)(*c*)–(*f*).

gekehrt Desargues in der Ebene aus Pascal und 1.) 2.) 3.)? Beweise eventuell die Unbeweisbarkeit![47]

Versuch eines Beweises dafür, dass Pascal auch in der Ebene aus den Congruenzsätzen folgt: Man benenne die Punkte der x Achse irgend wie symbolisch durch Parameter u, v, $w,\ldots$ und die gleichweiten Punkte der y Achse ebenso bez. u, v, w, $\ldots$. Ein Punkt in der Ebene werde durch ein Parameterpaar $(u,v),\ldots$ ausgedrückt. Nun definire formal wie auf S. 17 den durch die Drehung entstandenen Punkt

$$x' = \tfrac{a}{\sqrt{a^2+b^2}}x - \tfrac{b}{\sqrt{a^2+b^2}}y$$
$$y' = \tfrac{b}{\sqrt{a^2+b^2}}x + \tfrac{a}{\sqrt{a^2+b^2}}y$$

und beweise aus den Congruenzsätzen, dass

$$\tfrac{a}{\sqrt{a^2+b^2}}(x+\xi) - \tfrac{b}{\sqrt{a^2+b^2}}(y+\eta) = \tfrac{a}{\sqrt{a^2+b^2}}x - \tfrac{b}{\sqrt{a^2+b^2}}y + \tfrac{a}{\sqrt{a^2+b^2}}\xi - \tfrac{b}{\sqrt{a^2+b^2}}\eta$$

etc.

Wegen

$$x'' = \tfrac{a'}{\sqrt{a'^2+b'^2}} \quad \left(\tfrac{a}{\sqrt{a^2+b^2}}x - \tfrac{b}{\sqrt{}}y\right) - \tfrac{b'}{\sqrt{}'}\left(\tfrac{b}{\sqrt{}}x + \tfrac{a}{\sqrt{}}y\right)$$
$$y'' = \tfrac{b'}{\sqrt{}'} \quad \left(\tfrac{a}{\sqrt{}}x - \tfrac{b}{\sqrt{}}y\right) + \tfrac{a'}{\sqrt{}'}\left(\tfrac{b}{\sqrt{}}x + \tfrac{a}{\sqrt{}}y\right)$$

folgt aus der Vertauschbarkeit 2er Drehungen

$$\tfrac{a'}{\sqrt{}'}\tfrac{a}{\sqrt{}} - \tfrac{b'}{\sqrt{}'}\tfrac{b}{\sqrt{}} = \tfrac{a}{\sqrt{}}\tfrac{a'}{\sqrt{}'} - \tfrac{b}{\sqrt{}}\tfrac{b'}{\sqrt{}'}$$

und wenn man für b' : $-b'$ nimmt und addirt, so wird $\tfrac{a'}{\sqrt{}'}\tfrac{a}{\sqrt{}} =$
$\tfrac{a}{\sqrt{}}\tfrac{a'}{\sqrt{}'}$, | d. h. commutatives Gesetz der Multiplikation und damit 24
lineare Gleichung der Geraden und Pascalscher Satz.

Dritte Einführung des Streckenproduktes.

Euklidische Inhaltsgleichheit. Congruente Dreiecke sind gleich. Gleiches $\pm$ Gleichem ist Gleiches.

$ab = cd$ bedeutet die Gleichheit der Rechtecke

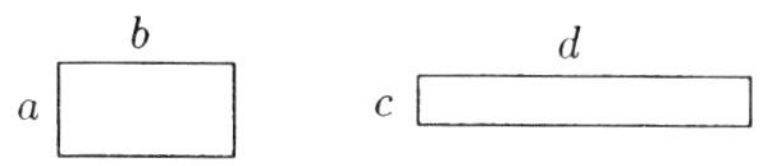

Pythagoras beweisbar z. B. durch den indischen Beweis:

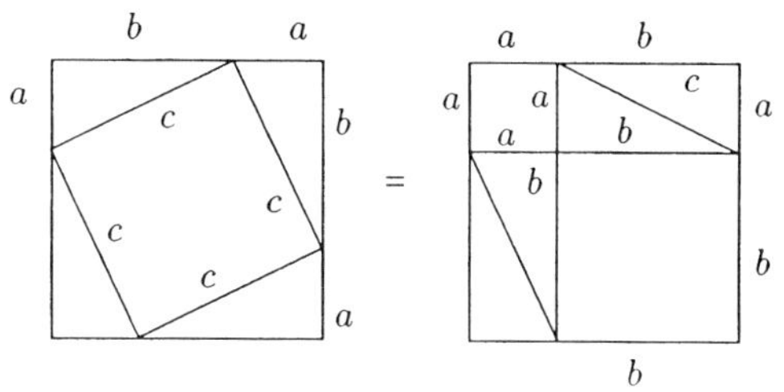

[47]The second question is posed again by Hilbert at the end of the 1898/1899 lectures, *Hilbert 1898c** and *Hilbert 1899a**, pp. 106 and 169 respectively. The answer to it was given positively in *Hjelmslev 1907*. This also answers the first question. The third was answered positively by *Hessenberg 1905b*. See the Introduction to Chapter 4, Specific Note (5)(b).

Dreiecke mit gleicher Höhe und Grundlinie sind inhaltsgleich.
Also alle diese Sätze folgen aus 1.) 2.) 3.) 4.). Wir sind sehr erfreut und hoffen ein neues Mittel zum Beweise eines Desargues oder Pascal zu haben.
25 Denn offenbar $ab = ba = cd = dc$.
Ferner folgt aus $ab = cd$ und $ab = ef$ nothwendig $cd = ef$.
Zum Beweise des Pascal machen wir folgende Ueberlegung:

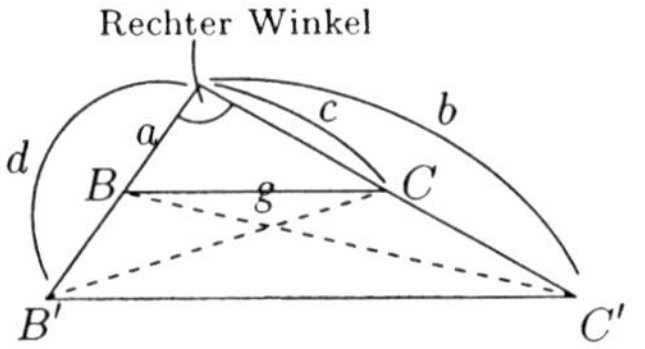

$BC \parallel B'C'$, so $ab = cd$.
Hätten wir auch den umgekehrten Satz, so wäre Pascal bewiesen, d. h. könnten wir aus $\triangle BCB' = BCC'$ die Gleichheit der Höhen schliessen, d. h. könnten wir aus $gh = gh'$ die Gleichung $h = h'$ folgern, so wäre Pascal bewiesen.
Also es kommt auf den Satz an: 2 gleiche Rechtecke mit einer gleichen Seite müssen auch die andere gleich haben oder das Killing-Stolzsche Postulat:

Wie man auch ein Rechteck in n Dreiecke zerlege, man kann nach Fortnahme eines nie durch die $n-1$ übrigen das Rechteck bedecken. (Also Inhalt von Anordnung der Teile unabhängig.)

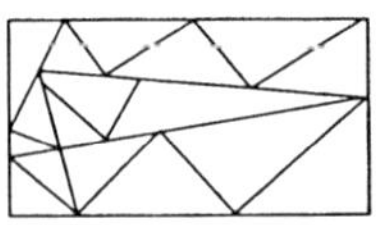

Vergleiche den geometrischen Scherz mit $8 \cdot 8 = 5 \cdot 13$.[48]
Folgt das Killing-Stolzsche Postulat aus Pascal?
Ist das Archimedische Axiom eine Folge des Killing-Stolzschen Postulates?

26 Etwas anderes als die Euklidische Gleichheit ist die Flächengleichheit. Congruente $\triangle$ sind gleich, Gleiches + Gleiches giebt Gleiches.

Euklidisch gleiche Polygone sind auch flächengleich.[49]

Rhéthy, Math. Ann. Bd. 30–40.[50]

---

[48]This jest is mentioned in Hilbert's remarks added to the reverse of the title page above, and explained in the corresponding note.

[49]The key here is the result mentioned by Hilbert two pages before, that triangles with the same base and height have the same surface area or 'content'. In the 1898/1899 lectures, this is proved as a consequence of Hilbert's treatment of the theory of proportion, Euclidean inspired, but more carefully developed, and the theory of surface measurement based on it. These developments begin in the 1898/1899 lectures, i.e., *Hilbert 1898c**, 91, *Hilbert 1899a**, 111 (this Volume, Chapter 4), and lead up to pp. 100 and 137 respectively. *Inhaltsgleichheit* is defined via *Flächengleichheit*. As a central part of this, Hilbert proves what he calls the Killing-Stolz Postulate, established without using the Archimedean Axiom. This answers, negatively, the second question posed here: the Archimedean Axiom does not follow from the Killing-Stolz Postulate. The first question is answered positively: see the *Festschrift* (*Hilbert 1899c*), 48 (this Volume, Chapter 5). In *Hilbert 1899a**, Hilbert uses the Killing-Stolz Postulate to prove that triangles with the same base and height are *inhaltsgleich*, but in the *Festschrift* he proves this first, and then remarks that the Postulate by itself would not be enough, i.e., without Pascal's Theorem or the Archimedean Axiom.

[50]The reference is probably to the three papers of Moritz Réthy, *Réthy 1891*, *Réthy 1893* and *Réthy 1894*.

Gleichheit der Pyramiden von gleicher Grundlinie und Höhe. Euklidisch gleich und flächengleich nicht aus den Congruenzsätzen bisher abgeleitet. ↑Es scheitert daran, dass man nicht zeigen kann, dass Pyramide $= \frac{1}{3}$ Prisma ist. Beweise die Unbeweisbarkeit.↓

Gauss und Gerling!

Nach Stolz*) ist es bisher nicht gelungen, 2 Pyramiden von gleicher Grundlinie und Höhe als inhaltsgleich im engeren Sinne nachzuweisen. (Sind sie überhaupt Euklidisch gleich?)[51]

Aus 1.) 2.) (4.)) kann im Raume (oder mit Desargues in der Ebene) nicht der Pascal bewiesen werden.[52]

Beweis: Es mögen alle Ausdrücke gebildet werden, welche aus t, u hervorgehen durch

1.) Addition $a + b = b + a$

2.) Multiplikation rechtsseitig und linksseitig $a(b+c) = ab + ac$

$(a+b)c = ac + bc$

3.) Division in 1: $\frac{1}{a}$

z. B. 27

$$[\![\langle\!\langle ?? \rangle\!\rangle]\!] + tut^2 \frac{1}{u+t} \quad u^3 \frac{1}{t} + \frac{1}{u}.$$

Eine Grösse dieses Bereiches heisst $= 0$ nur, wenn sie identisch $= 0$ ist; sie heisst > 0, wenn wie folgt: Ist $\alpha > 0$, $\beta > 0$, $\beta > \alpha$, so definire $\frac{1}{\alpha} > \frac{1}{\beta}$, ebenso falls weder α noch β blosse Zahlen sind und $a \cdot \frac{1}{\alpha}$ nicht reducirt werden könnte etwa $= a' \cdot \frac{1}{\alpha'}$

$$a\,\frac{1}{\alpha} > b\,\frac{1}{\beta}\,, \quad \frac{1}{\alpha}\,a > b\,\frac{1}{\beta}, \quad \frac{1}{\alpha}\,a \gtrless \frac{1}{\beta}\,b \quad \text{jenachdem } a \gtrless b$$

etc.**)

*) Allgemeine Arithmetik. Bd. 1. S. 78.

**) Besser und einfacher so: α, β, γ seien Zahlen aus einem System, für welches man schon $\substack{>\\=\\<}$ definirt hat. Dann setze $\alpha \cdot \frac{1}{\beta} + \gamma \gtrless 0$ bez. $\lesseqgtr 0$ jenachdem $\alpha + \gamma \cdot \beta \gtrless 0$, $\beta > 0$ bez. $\alpha + \gamma \cdot \beta \lesseqgtr 0$, $\beta < 0$ ist. Ebenso setze $\frac{1}{\beta}\alpha + \gamma \gtrless 0$ bez. $\lesseqgtr 0$, jenachdem $\alpha + \beta \cdot \gamma \lesseqgtr 0$, $\beta > 0$ bez. $\alpha + \beta \cdot \gamma \gtrless 0$, $\beta < 0$ ausfällt.

[51] The reference to Stolz in Hilbert's footnote is to *Stolz 1885*, S. 78, where we find:

> Bis jetzt ist es nämlich nicht gelungen, zwei Pyramiden von gleichen Grundflächen und Höhen in gleichviele, paarweise congruente Theile zu zerlegen.

The issue reappears as a open problem in the 1898/1899 lectures (*Hilbert 1898c**, 106, *Hilbert 1899a**, 169), and then as Problem 3 in *Hilbert 1900e*. It was solved shortly thereafter by Dehn, negatively. See the Introduction to Chapter 4, Specific Note (5)(c).

[52] See, e.g., *Hilbert 1899c*, 71, 76 (this Volume, Chapter 5). Hilbert alludes here to the use of the planar Desargues's Theorem as a replacement for the spatial axiom in 1.).

Ein Punkt ist immer ein Grössenpaar x, y, eine Gerade das System aller Punkte, für welche

$$ax + by + c = 0$$

etc. Wegen $ab \neq ba$ gilt nicht Pascal; dagegen gilt Desargues.[53]

Aus Pascal und Desargues kann allein durch die Axiome der Verknüpfung jeder Schnittpunktsatz, also auch der Fundamentalsatz der projektiven Geometrie, also auch der Satz, dass jede Ebene durch eine lineare Gleichung und umgekehrt gegeben ist, bewiesen werden.

Beweis. Man verfolge die Streckenrechnung und beweise zunächst die Sätze $a(b + c) = ab + ac$. (Man setze zuvor eine Einheitsstrecke 1 fest und setze $bc = e$, wenn $bc = e \cdot 1 = 1 \cdot e$ ist.)

Ferner $a(bc) = (ab)c$ etc. Dann zeige, dass jeder Satz (d. h. jede Gleichung zwischen Strecken), der unter der Voraussetzung, dass die Strecken Zahlen sind, gilt, auch allgemein gelten muss.[54]

Litteratur: Vgl. Collegheft, Grundlagen der Geometrie, S. 3–5.[55]

---

[53] The reasoning is hard to follow. However, it might be relevant that in *Hilbert 1899c*, 76 (see Chapter 5 of this Volume), Hilbert shows that any model in which Pascal's Theorem fails must be non-Archimedean, since the Archimedean Axiom implies commutativity.

[54] See the 1898/1899 lectures (*Hilbert 1898c**, 104–106, *Hilbert 1899a**, 167–168), and the *Festschrift* (*Hilbert 1899c*), §35, Chapters 4 and 5 of this Volume.

[55] The reference is to *Hilbert 1894a**, 3–5 (see this Volume, Chapter 2, Appendix), which presents a literature list.

Textual Notes

160.2: Feriencursus] Fereiencursus

160.4–5: Grundfläche] Grund⟦linie und⟧fläche

160.6: Leipzig:] Leipzig.:

160.15: *Arithmetik und*] *Arithm. u.*

161.7–8: Congruenzsätze] Conguenzsätze

161.8: etc.] etc..

161.8: bringen,] bringen.

161.20: $(\alpha_1, \alpha_2, \alpha_3, \dots)$] $(\alpha_1, \alpha_2$⟦)⟧$, \alpha_3)$..

161.20: 1] (1

161.21: (x)] ⟦⟪???⟫⟧ (x)

161.21: 1, 1,011] 1, ⟦0⟧ 1,011

161.22: ⟨The variables x, y, z in this line, and also in the next two inequalities '$0 \leqq x \leqq 1$' and '$0 \leqq y \leqq 1$', are replacements. Hilbert originally wrote 'a', 'b', 'c'.⟩

161.27: ganzen rationalen Funktionen] g. r. Funktionen

161.28: algebraischen] algebr.

161.28: $\left(\sqrt{x}, \frac{\sqrt[3]{x-1}}{\sqrt{x}+x^3}, \dots\right)$] $(\sqrt{x}, \frac{\sqrt[3]{x^2-1}}{\sqrt{x}+x^3}$

161.29: $(\sin x, \sin lx, \dots)$] $\sin x, \sin lx, \dots$

161.31: *Richtung* der Mathematik] ⟦Mathematik⟧ *Richtung* der Math.

161.32: *Zuverlässigkeit*] ⟦Nothwendigkeit⟧ *Zuverlässigkeit*

161.33: Mathematik] Math.

162.1: Auffassungen] ⟦Aussagen⟧↑Auffassungen↓

162.1: Mathematik] Math.

162.1: *(Funktions-, Zahlbegriff)*] ↑*(Funktions- Zahlbegriff)*↓

162.2: vornehmlichste] vornemlichste

162.3: *Systemen*] *Systeme*

162.4: Unter] ⟨Before 'Unter' there appears a large 'A', possibly added later. The significance of this sign is not clear, and no corresponding further letters appear subsequently.⟩

162.4: Unter] Unter ⟦∞ vielen⟧

162.4: bestimmten] be⟦li⟧stimten

162.8: $1, \frac{1}{2}, \frac{1}{3}, \frac{1}{4}, \dots$] $1\ \frac{1}{2}\ \frac{1}{3}\ \frac{1}{4} \dots$

162.9: $4, \dots$] 4,

162.14: Mathematik] Math.

162.17: ⟪oder grösster⟫] ⟪od.⟫ g⟪??⟫

162.18: $\frac{\sin \pi x}{1}$] $\frac{\sin [\![\langle\!\langle ?? \rangle\!\rangle]\!] \pi x}{1}$

162.20: 1] ⟦0⟧1

162.24: ausgefüllt] ausfüllt

162.25: Maximum und Minimum] Max. u Min

162.26: *stürzten*] *stürtzten*

162.27: Theorie der algebraischen] Th. der algebr.

162.30: Riemann] Rieman

163.1: sagte] sagt↑e↓ ⟨added by Hi^b⟩

163.4: *Kritik*] *Kriti*⟪*r*⟫

163.7: ohne weiteres] ⟪o⟫⟪?⟫hne weitere

163.9: $1 - \frac{1}{2} + \frac{1}{3} - \frac{1}{4} + \frac{1}{5} \pm \dots$] $1 - \frac{1}{2} + \frac{1}{3} - \frac{1}{4} + \frac{1}{5}$ ..

163.11: $-\frac{1}{2}$] ⟦$1 + \frac{1}{\ }$⟧ $- \frac{1}{2}$

163.16: log 2] $l\,2$

163.17: log 2] $l\,2$

163.18: $\frac{1}{6}$] ⟨The division stroke and '6' are additions by Hi^b.⟩

163.18: $+\ldots)$] ⟨The ellipsis and closing bracket were inserted by the editors.⟩

163.19: log 2.] lg 2

163.24: $f(x) = x$] $\uparrow f(x) = \downarrow [\![(]\!] x [\![-1)]\!]$

163.25: $0 \leqq$] ⟨Added.⟩

163.25: Summe] Sum

164.4: Convergenz:] Convergenz: ↑wenn↓ ⟨addition, which erroneously was not crossed out⟩ ⟦↑Wenn↓⟧

164.4: Maxima] ⟦Die ⟦abso⟧ absoluten Gliedermaxima⟧ ⟦Die⟧ Maxima

164.6: $\varphi(x)$] $\varphi_{[\![n]\!]}(x)$

164.7: $\varphi'(x)$] $\varphi'_{[\![n]\!]}(x)$

164.7: $\frac{n-1}{n}$] $\frac{n-1}{n}$ $[\![= 1 - \frac{1}{n}]\!]$

164.8: $\varphi(x)$ wird] ⟨'$\varphi(x)$ wird' is a replacement for '$\varphi_n(x) =$'. There are several levels of correction here apparently yielding: '$\varphi_n(x) =$'→ '$\varphi\left(\frac{n-1}{n}\right) =$'→ '$\varphi(x)$ wird'. However, the reading is uncertain.⟩

164.8: $\left(1 - \frac{1}{n}\right)^n$] $\left(1 - \frac{1}{n}\right)^{n[\![-1]\!]}$

164.8: $\frac{n}{n-1}$] $[\![\langle\!\langle \frac{1}{x} \rangle\!\rangle]\!]$ $\frac{n}{n-1}$

164.9: $M_1 + M_2 + \cdots$] $[\![\varphi_1 + \varphi_2 + \cdots [\![\varphi]\!]]\!]$ $M_1 + M_2 + \cdot\cdot$

164.11: *gleichmässig*] *gleichmässi*

164.17: Systeme] ⟦Men⟧ Systeme

164.18: *Systemen*] *Syst*emen

164.20: *und*] *un*

165.1–2: ganzen ⟨...⟩ fortschreitende] gan r. Funk⟪??⟫ fort⟪schreit⟫

165.6: aller] ⟦der⟧↑aller↓

165.14: Lipschitz,] Lipschitz.

165.17: Cantor,] Cantor

165.18: Bd. 46, 49.] Bd. ⟦26⟧, ↑46, 49↓ ⟨possibly added⟩

165.22: *Arithmetik*] *Arithm.*

165.23: ausserordentlichem] ausserordentlich

166.6: sollen] ⟦so⟧ ⟦hau⟧ sollen

166.6: anregende *Gedanken*] anreg*ende Gedanken*

166.10–11: Die ⟨...⟩ unterscheiden:] ⟨Possibly added.⟩

166.13: *Geraden,*] *Gerade*

166.14: Geraden] Gerade

166.15: Vorlesungen] Vorles.

166.17–18: Irgend ⟨...⟩ 2 Punkte.] ⟨These two lines are marked at the left by a vertical stroke in Hi^b.⟩

166.19: angedeutet.] ⟨Two empty lines follow in the manuscript.⟩

166.23: immer] ⟪??⟫er

166.24: Mathematik] Math.

166.28: Princip] ⟦Grundsat⟧ Princip

166.29: oder] o⟪?⟫d

166.33: Durch ⟨...⟩ definirt.] ⟨Added.⟩

166.33: Halbebene,] Halbeebene

166.1: S. 61–62.] ⟨Possibly added.⟩

167.1: Nachdem] ⟦Durch die Axiome 2⟧ Nachdem

167.1: Halbebene] Halbeb

167.1: Halbraum] Halbrau
167.7: Benennung] Benug
167.11: 4.)] ⟦3.)⟧↑4.)↓
167.12: Parallel heissen] ⟦Zwei p⟧↑P↓arallel heisse
167.12: Geraden] Gerade
167.13: Parallele] Parallelele
167.13: sich] sich ⟦stets⟧
167.16: Euklidischen] Eukli⟦di⟧schen
167.16–17: *Euklidische*] *Euklische*
167.17: *Existenz* der analytischen] *Existens der ana*lytischen
167.21: auskommt] auskomt
167.21: Zunächst] ⟦Aufgabe⟧ Zunächst
168.1: $AB = AC$] ⟦a = ⟪a⟫⟧ ⟦⟪??⟫⟧ $AB = AC$
168.1: ⟨Both '$\|$' are substituted from '='.⟩
168.2: rechtwinklig] rechtwinklich
168.5: B'] ⟨Substituted from 'A'.⟩
168.12: allein] allein, ⟦kann man⟧
168.16: offenbar] offenbar ⟦um⟧↑offenbar↓
168.17: abtragen.] ⟦drehen⟧↑abtragen:↓
169.2: Congruenzsätze] ⟦Construktion⟧ ⟦au⟧ ⟦vo⟧ die Congruenzsätze
169.4: 2er] 2 er
169.4: beliebiger] beliebg
169.4: beliebigen Kreises] beliebg Kreis
169.5: beliebigen Geraden] beliebg Gerade
169.5: die einen] ⟦deren Ent⟧ die einen
169.6: gegebenen] gegeb
169.11: Höhe] höhe
169.11: gleichseitigen] gleichseitg
169.–: $\sqrt{1+1+1} = \sqrt{1+\left(\sqrt{1+1}\right)^2}$] ⟨Added by Hi^b.⟩
169.16: gegebener] gegeben
170.2: dagegen] dagen
170.9: Archimedische] Archimedeische
170.12: und] u.
170.17: Archimedische] Archimedesche
170.23: Archimedische] Archimedessche
170.23: z. B.] ↑z. B.↓
170.26–171.2: Beweis: ⟨...⟩ etc.] ⟨In this paragraph, the variable 't' is a replacement. Hilbert originally used 'x'.⟩
170.27: Setze] ⟦Dann sei $f(t)$ ei⟧ Setze
170.28: bez.] bz
170.28: immer] imer
170.29–171.1: t und] t u
171.1: irgendeine] irgenein
171.2: nimm] nim
171.3: Funktionenpaar] Funktionenpaar ⟦$F(x$⟧
171.5: ⟨berechne⟩,] nimm,
171.8: $\cos\alpha$] ⟦sin⟧↑cos↓α
171.11: ↑Vielleicht ⟨...⟩ einführen!↓] ⟨Possibly added.⟩
171.12: Archimedes'] Archimedes

171.13:	Reymond,] R⟦ay⟧↑ey↓mond
171.18–19:	analytischen] analytisch
171.21:	bewiesen] ⟦kennt⟧ bewiesen
171.23:	Folge,] Folge
171.26:	2 Schnittpunkt-Sätze] ⟦di⟧ 2 Schnittpunkt Sätze
171.26:	Pascal] Paskal
171.27:	Wir sehen von 5.) ab.] ⟨Probably added.⟩
171.28:	bloss] blos
172.1:	nicht] ↑nicht↓
172.4:	Pascal] Paskal
172.5:	der projektiven Geometrie] d. projektiv Geo
172.8:	Pascal] Paskal
172.9:	Weisen] Weise
172.11:	beliebigen,] ⟦Rechten ist.⟧ beliebigen
172.21:	Bedeutung] Bedeutg
172.21:	Pascal] Paskal
172.21:	halten] hal⟪bt⟫en
172.23:	$(ab = ba)$] ⟨Possibly added.⟩
172.26:	d. h. wenn] d. h. Wenn
173.1:	ob] obb
173.1:	Anwendung] Anwendg
173.2:	Grössenaxiom] ⟦Ax⟧ Grössenaxiom
173.3:	$(a - b) \leqq 1$,] $(a - b) \leqq 1$.
173.7:	Entfernung] Entferng
173.8:	einem Centrum.] ⟪ein Centru⟫
173.14:	Definition] Def.
173.19:	unabhängig ist.] unabhängig.
173.20:	Pascal] Paskal
174.1:	der 3 Kreise] 3 Kreise
174.2–3:	Kugeln] ⟦Halb⟧ Kugeln
174.3:	K_1', K_2',] K_1' K_2',
174.5–6:	und einen] ⟪?⟫d ein
174.13:	Verbindungsgerade] Verbin⟦g⟧↑d↓gsgerade
174.14:	schneiden] ⟦t⟧ schneiden
174.16:	Man] ⟦(⟧Man
174.16:	schliessen:] schliessen: ⟦Unser Satz gilt für $d = c$.⟧
174.17:	Bestimme] Bestime
174.20:	$x = y$,] $x = y$
174.22:	Pascal] Paskal
174.23:	Pascal] Paskal
175.1:	in der Ebene] ↑in der Ebene↓
175.1:	Pascal] Paskal
175.1:	eventuell] e⟦t⟧ventuell
175.5:	$w, \ldots$] $w\,..$
175.6:	bez.] ↑bez.↓
175.7:	S. 17] ⟨Addition in a space left free for this purpose.⟩
175.8:	entstandenen] entstanden
175.23:	Euklidische] Euklische
175.23:	Inhaltsgleichheit] Inhaltgleichheit
175.23:	Gleiches] Gleiches ⟦zu Glei⟧

176.1: inhaltsgleich] Inhaltsgleich

176.3: haben.] haben.
⟦Also Bedeutung des Paskal⟧

176.6: Ueberlegung:] Ueberlegung

176.7: $BC \parallel B'C'$,] ⟦Wenn ab⟧
$BC \parallel B'C'$

176.9: könnten] Könnten

176.11: schliessen,] schliesen

176.12: Gleichung] Gl.

176.13: folgern] fo⟪lglg⟫e⟪rn⟫

176.15: Killing-] ↑Killing-↓

176.16: man kann] mann kan

176.17–18: (Also ⟨...⟩ unabhängig.)] ⟨Added.⟩

176.19–20: Vergleiche ⟨...⟩ $8 \cdot 8 = 5 \cdot 13$.] ⟨Added.⟩

176.19: Vergleiche] Vergl.

176.21: Postulat] ⟦Axiom⟧ Postulat

176.21: Pascal] Paskal

176.22: Archimedische] Archimedesche

176.25: gleich,] gl.

176.25: + Gleiches] + Gl.

176.25: Gleiches.] Gl.

176.26: Euklidisch] Euklische

177.1: von gleicher] vo↑n↓ ⟨added by Hi^g⟩ gle↑icher↓ ⟨added by Hi^g⟩

177.2: Congruenzsätzen] Congruenzesätzen

177.7: als inhaltsgleich] ⟦in gleichvie⟧ als ⟦I⟧⊣↑i↓nhaltsgleich

177.10: Pascal] Paskal

177.11: Ausdrücke] ⟦rati⟧ Ausdrücke ⟦von der Gestalt⟧

177.11–12: hervorgehen] ⟦durch Addition⟧ hervorgehen

177.14: Multiplikation rechtsseitig und linksseitig] M⟪u⟫lipliaktion rechsseitig u linksseitg

177.16: in 1] ⟪in⟫ 1 ⟦rechsseitig⟧

177.19: sie] ⟨·?·⟩ie

177.20: ebenso] ⟨·?·⟩benso

177.21: und $a \cdot \frac{1}{\alpha}$] und ⟦$\frac{a}{\alpha}$⟧ $a \cdot \frac{1}{\alpha}$

177.21: werden] ⟨·?·⟩erden

177.24: etc.] ⟨·?·⟩tc.

177n: Allgemeine Arithmetik.] Allgemein Arithm.

178.1: immer] ⟪ein⟫⟪?⟫

178.4: gilt] ↑⟦nicht⟧⊣↑gilt↓↓

178.5: und] u.

178.7: durch eine] durch

178.7: Gleichung] Gl.

178.9: und] ⟪und⟫

178.10: un] ⟨·?·⟩

178.11: $bc = e$,] ⟦$1 \cdot a = a \cdot 1$⟧ $bc = e$,

178.12: jede Gleichung] jede Gl.

178.13: Voraussetzung] Voraussetz

178.15: Litteratur:] Litteratur

178.15: Collegheft,] Collegheft.

Description of the Text

Collection: SUB Göttingen, signature *Cod. Ms. D. Hilbert 597*, f. 15r–29r. (These numbers refer to the continuous pagination added to the volume by the library; the volume contains three texts, each bearing a separate original pagination by Hilbert or Käthe Hilbert respectively.)

Size: Cover size 20.9 × 16.7 cm; page size 20.9 × 16.4 cm.

Cover Annotations: On the front cover, in Hilbert's hand, is pasted a label with the notation, 'Feriencursus: // Zahlentheorie 1896 // Begriff des Unendlichen 1898'.

Composition: The first 29 pages of a signature of 8 double pages, which is bound together with, and follows, a further signature of 7 double pages, which contains *Hilbert 1896** and *Hilbert 1902b**.

Pagination: Continuous pagination of the odd-numbered right-hand pages, from 1 to 27, beginning on f. 16r of the manuscript with p. 1. Numbers 1 to 25 are in Käthe Hilbert's hand; 27 has been added in an unknown hand.

Original Title: On f. 15r (which Hilbert had left unpaginated) 'Fereiencursus ⟨sic⟩ Ueber den Begriff // des Unendlichen. // Ostern 1898.' all in Käthe Hilbert's hand.

Text: Pages 1 to 27 contain continuous text in Hilbert's hand, written in black ink. There are occasional handwritten corrections and additions, most of which were probably added in the process of writing. Up to p. 12 there are numerous underlinings in blue pencil; only one thereafter, on p. 18. The correction of the diagram on p. 15 is likewise executed in blue pencil. Hilbert added three remarks – presumably at a later time – on the reverse side of the title page.

Chapter 4

Lectures on Euclidean Geometry (1898–99)

Introduction

General Introduction

1. The Lectures and their Ausarbeitung

The first manuscript reproduced in this Chapter (*Hilbert 1898c**) consists of notes, in Hilbert's hand, for a set of lectures on Euclidean geometry in the Wintersemester 1898–1899. In his life of Hilbert, published in *Hilbert 1935*, Otto Blumenthal writes about these lectures:

> Staunen und Bewunderung aber erwachten, als die Vorlesung begann und einen völlig neuartigen Inhalt entwickelte. Eine vorzügliche autographierte Ausarbeitung, hergestellt von dem früh verstorbenen H. v. Schaper, ist noch heute eine empfehlenswerte Einführung in die Axiomatik der Geometrie, denn sie enthält manche Motivierungen und Beispiele, die in der klassisch gefeilten Darstellung des späteren Buches weggefallen sind. (*Blumenthal 1935*, 402.)

The book Blumenthal refers to is, of course, the celebrated *Hilbert 1899c*, otherwise known as the *Festschrift*. The *Festschrift* is reproduced in Chapter 5, and the *Ausarbeitung* Blumenthal mentions is presented later in this Chapter. This latter was prepared, as Blumenthal says, by Hans von Schaper, a doctoral student of Hilbert who submitted his *Doktorarbeit* in 1898 on Hadamard's work on the distribution of the prime numbers. There is a short Foreword by von Schaper dated March 1899, thus, not long before the publication of the *Festschrift*, and which makes explicit reference to this latter:

> Die wichtigsten der in diesem Heft enthaltenen Untersuchungen finden sich, zum Teil in anderer Fassung und ausführlicher, in der Abhandlung von Hilbert in der Festschrift zur Enthüllung des Gauss-Weber-Denkmals (Juni 1899) zu Göttingen. (*Hilbert 1899a**, *Vorwort*.)

Thus we know that the two manuscripts forming the bulk of this Chapter stand in close relationship to the *Festschrift* with respect both to content and the time of their production. Furthermore, we know that Hilbert's part of the Gauß-Weber *Festschrift* was composed rapidly, and that it was Klein's suggestion that Hilbert present the material of these very lectures in celebration of Gauß and Weber (see the Introduction to Chapter 5). When Hilbert actually started working on the lectures is not known. However, the *Ferienkurs* from Easter 1898 (*Hilbert 1898b**, see Chapter 3) already speaks of the 1898/1899 lecture course on Euclidean geometry, and shows that by Easter of 1898 Hilbert was already well on the way to formulating many of the questions which assume a central place in the lectures. (See the Introduction to Chapter 3.) Both the flavour and content of all of these lectures (including the 1898 *Ferienkurs*) is new, certainly when compared, for instance, to *Pasch 1882* or to Hilbert's earlier lectures on the foundations of geometry in 1894 (*Hilbert 1894a**; see Chapter 2). They do indeed contain, as von Schaper unpretentiously says in the Foreword to the *Ausarbeitung*, 'grossenteils neue Entwicklungen und Fragestellungen'.

Von Schaper's *Ausarbeitung* was, as far as we can judge, fairly well-known. Hilbert frequently refers to it as the 'autographierte Vorlesung', a term also used by Blumenthal in the passage quoted earlier, and by von Schaper himself in his Foreword. 'Autographie' very probably refers to a method of reproduction common in Germany at the time (a form of off-set printing).[1] Moreover, the beautifully clear hand in which it is written is not that of von Schaper himself, but more likely that of a professional copyist. (See the description of the manuscript which follows the second text.) In any case, the *Ausarbeitung* was certainly reproduced, as von Schaper says; seventy copies were made (von Schaper, Foreword) mainly to benefit those who attended the lectures. Some of these copies certainly came into the hands of prominent mathematicians and philosophers in the United States as well as in Germany. For example, we know that Frege saw it; his letter to Liebmann of 29 July, 1900 (see *Frege 1976*) speaks of returning, with thanks, Liebmann's copy, though whether Frege saw this before seeing the *Festschrift* itself is not clear. Hausdorff also saw a copy; his letter to Hilbert of 12 October, 1900 mentions not only the *Festschrift*, but also a specific page number of the 'autographierte Ausarbeitung' and an independence proof given there. (See Hilbert's first remark in the Appendix to this Chapter, and the corresponding note.) In the United States, the mathematicians E. J. Townsend (Illinois) and William Osgood (Harvard) clearly knew the text, and very likely others did, too, e.g., E. H. Moore (Chicago), and possibly, through him, his student Veblen. (See the brief remarks on the English translation of the first edition of the *Festschrift* in the Introduction to Chapter 5.)

We know of three copies of the *Autographie* of von Schaper's *Ausarbeitung* which exist outside Göttingen, in libraries in Berlin, Bremen and Oslo, and, in the light of the relatively large number reproduced, there are almost certainly others. The copy in the University Library in Bremen was originally von Schaper's own, since on the flyleaf, and written in ink, it bears in brackets the note 'H. v. Schaper, Göttingen, 8/VI 1899', and below that, in a different hand, stands the name 'Völkel'. The date indicates, perhaps, that the copies were produced only a very short time before the unveiling of the Gauß-Weber Memorial on the 17 June, which the *Festschrift* itself was meant to celebrate, and which seems to have been ready close to the time of the unveiling. (See the Introduction to Chapter 5 of this Volume.) The copy in the library of the Humboldt University in Berlin first belonged to Heinrich Liebmann, and thus in all likelihood is the very copy lent to Frege. The copy in the Mathematics Library at the University of Oslo was previously owned by one C. C. Christensen, an actuary. Beyond that, its provenance is not known.

---

[1] In a footnote to section § 10 of the first edition of the *Festschrift* (see Chapter 5 of this Volume), Hilbert refers to the *Ausarbeitung* by von Schaper and says it was 'autographiert' for the auditors of the lectures. In his English translation (*Hilbert 1902c*), E. J. Townsend, himself one of those auditors, translates the term 'autographiert' as 'manifolded' (p. 30). 'Manifolding' seems to have been the standard term at the time for the reproduction of documents, referring both to methods involving carbon paper or stencils, which is more likely in this case, given the large number of copies made.

Two copies of the text exist in Göttingen. One was deposited in the *Lesezimmer* of the Mathematisches Institut (*Hilbert 1899b**), and the other is Hilbert's own copy (*Hilbert 1899a**), which has been used for this edition. Officially, therefore, page number referals and other textual references are to *Hilbert 1899a**, though they could also apply to any copy of the 'Autographie'. However, *Hilbert 1899a** itself is unique for two reasons.

First, there is the odd remark or alteration which Hilbert has added in the course of the text itself. These are duly noted in a separate series of footnotes, and are clearly marked off from the changes that von Schaper has made to the original text before reproduction. The reader can therefore assume that, in the absence of any note or marking to the contrary, textual oddities, such as square brackets or emphasis, the precise import of which is sometimes hard to discern, stem from work on the text prior to its reproduction and not directly from Hilbert, though they might, of course, have been carried out under Hilbert's direction.

Secondly, there is a rather extensive sequence of remarks which Hilbert wrote in his copy of the *Ausarbeitung* (thus, *Hilbert 1899a**), on the inside front cover, on the title page and on the reverse of the title page. These have little connection with each other and are undated, although what internal evidence there is suggests that they were certainly added after the composition of the *Festschrift*, thus variously in the period 1900–1902. They are therefore placed here in an Appendix following *Hilbert 1899a** itself. The remarks are assigned numbers, mainly to make cross-referencing easier, although it is often a matter of guesswork to say where one remark ends and a separate one begins. Many of the remarks are difficult to comprehend at first reading, and are thus accompanied by expository comments. These immediately follow the corresponding remark, and are set in smaller font.

However, it should be noted that the text *Hilbert 1899b** is also important in itself because of the alteration described in section 2. of this Introduction.

In the remainder of this Introduction, and in the footnotes to both central texts of this Chapter, we will refer to these respectively as 'Hilbert's own notes for the 1898/1899 lectures' (or 'Hilbert's/his own notes'), and 'the *Ausarbeitung*'. The first edition of Hilbert's book on the foundations of geometry will be referred to, as usual, by the nickname 'Festschrift'.

To some extent, the *Ausarbeitung* follows Hilbert's notes in *Hilbert 1898c** closely, but is at the same time more than a faithful reproduction. Furthermore, although it is undoubtedly true, as von Schaper's Foreword states, that the lectures already contain the essence of the work presented later in Hilbert's *Festschrift*, they should not be seen as mere drafts of it, a point made expressly in Blumenthal's remark. Indeed, von Schaper's comment that the *Festschrift* is 'ausführlicher' is not entirely correct, as we will show. There are, consequently, three exegetical questions. What is the relation of the *Ausarbeitung* to Hilbert's original lecture notes? What precisely is the relation of both to the *Festschrift*? And what in turn is the relation of this body of work to Hilbert's previous lectures on the foundations of geometry, mainly the lectures from 1894 (*Hilbert 1894a**, Chapter 2), and work of roughly the same era, like the paper *Hilbert 1895b*?

2. The Relationship of the Ausarbeitung *to Hilbert's Lecture Notes*

There are three senses in which the *Ausarbeitung* is more than a reworking of Hilbert's own notes.

First, after a certain point, the *Ausarbeitung* becomes the more authorative of the two manuscripts, with its composition clearly proceeding faster than that of Hilbert's own notes. From the beginning of the treatment of the Axiom of Parallels, thus from p. 68 onwards in the notes, there appear more and more commonly references to the 'autographierte Vorlesung' or 'Ausarbeitung' of his lectures. Occasionally, these are references to the details of a proof or to the statements of theorems which Hilbert clearly did not want to take the time to write out. (It was Hilbert's standard practice to refer to his other lecture manuscripts; e.g., the 1898/1899 lecture notes often refer early on to *Hilbert 1894a**.) But often the references to the *Ausarbeitung* are to larger and larger sections representing substantial episodes of theoretical development. One example is the reference to the treatment of *Flächeninhalt* on p. 99 of the Hilbert notes, which actually occupies pp. 122–138 of the *Ausarbeitung*. Another example is the treatment of Pascal's Theorem, which is mentioned on p. 93 of the notes, but is only really dealt with in the *Ausarbeitung* (p. 109f), to which Hilbert refers, at the same time leaving gaps to be filled in with the precise page numbers.

The second sense in which the *Ausarbeitung* is not merely a reworking of the lecture notes is that it often contains fully worked out versions of proofs which are only sketched in the notes, or even not clearly given there. Sometimes, Hilbert begins a proof without it being clear from the notes alone how the proof is meant to proceed; very often, if not invariably, the *Ausarbeitung* presents a fully worked-out version along the lines that Hilbert had indicated, suggesting that he did have a certain proof in view. (The footnotes indicate many examples.) In addition, the lecture notes not infrequently present proofs with misleading or inadequate diagrams or with confused notation or which omit cases, and these are very often presented cleanly (and almost always with more careful diagrams) in the *Ausarbeitung*. Occasionally gaps remain (an example where some lacunae are filled and some remain is the proof of Desargues Theorem, first given on pp. 26–27 of the lecture notes, then restated on pp. 24–26 of the *Ausarbeitung*), and sometimes it is by no means clearer in the *Ausarbeitung* what Hilbert intends (e.g., p. 62). In short, the *Ausarbeitung* is not an *ideal* completion of Hilbert's lecture notes, but it is considerably more complete, and without question gives a more complete picture of Hilbert's intentions. At the same time, it must be assumed that Hilbert's own manuscript notes, shorter and pithier, are a fairer representation of his actual lecture style than is the *Ausarbeitung*.

The third sense in which the *Ausarbeitung* is not just a representation of Hilbert's own notes is that both represent work in progress, and the *Ausarbeitung*, being the second working-out of much of the material, often goes further than Hilbert's notes, so presenting a riper picture. The most striking instance of this is the question raised in the notes (p. 33) about Desargues's Theorem. It is asked whether the Planar Desargues Theorem is not only a necessary condition for planes to be embedded in space in such a way that the full set of *spatial* incidence and order axioms hold, but also a *sufficient* condition for this, i.e., whether, if the Planar Desargues Theorem is added as a new postulate to the planar incidence and order axioms, then this will yield *spatial* consequences, so that, in particular, the full set of spatial incidence and order axioms will hold. Hilbert does not settle the question in his notes, and almost certainly had no relevant proof at the time of writing. This is confirmed by the fact that Hilbert raises the question again in the *Ausarbeitung* (p. 32). He adds there:

> Diese Frage ist wahrscheinlich zu bejahen; man könnte im diesem Fall sagen: Der Satz von Desargues ist das Resultat der Elimination der räumlichen Axiome aus I und II.

However, before von Schaper had finished the text, and before the *Festschrift* was published, Hilbert had developed a proof for the positive answer, a proof first given on pp. 159–160 of the *Ausarbeitung* (the build-up from p. 147 is crucial). (The development is repeated in the *Festschrift*, in particular pp. 67–71 and the work surrounding *Satz 35*.) The *Ausarbeitung* makes explicit reference on p. 159 to the passage just quoted. In Hilbert's version (i.e., *Hilbert 1899a**), the whole phrase 'wahrscheinlich zu bejahen; man könnte im diesem Fall sagen' has been replaced by 'zu bejahen', and in the *Lesezimmer* version (i.e., *Hilbert 1899b**) the same result is achieved by the deletion of 'wahrscheinlich' and 'man könnte im diesem Fall sagen', 'zu bejahen' being underlined in rough hand, unlike the usual underlining of the *Ausarbeitung*, which is done carefully with a straightedge. These changes were clearly made because the question had in the meantime been settled. It is not clear whether the changes to *Hilbert 1899b** were made, or authorised, by Hilbert.

3. The Relationship of the 1898/1899 Lectures to the Festschrift

What now of the relationship of these two texts to the *Festschrift*?

As already stated, a great deal of the material presented in the *Festschrift* already appears in the lectures (respectively *Ausarbeitung*) in one form or another. For instance, the axioms are almost the same, although the numbering is different in the *Festschrift*, this latter being the only place where Hilbert presents the Euclidean Axiom of Parallels as III, i.e., before the congruence axioms (IV). (In the lectures, the Axiom of Parallels is put at IV, with the congruence axioms at III, an order to which Hilbert returns in subsequent editions of the *Festschrift*.) However, even here there is a significant difference, for the lectures present, not just the standard version of the Archimedean Axiom, but, unlike the *Festschrift*, alternative versions, too. Nevertheless, the main differences are not to be found among the axioms themselves, but, firstly, in the form in which the (common) material is developed, and, secondly, in the fact that there are important examples of material which is in the lectures but which is not to be found in the *Festschrift*.

First, the difference in form. The *Festschrift* makes essential use of the Parallel Axiom, which is somewhat surprising, given that the titles of the lectures refer specifically to *Euclidean* geometry, whereas the title of the *Festschrift* returns to the unqualified title of the 1894 lectures presented in Chapter 2. There is, however, a fairly natural explanation for the use of the term 'Euclidean' in the title of the lectures. Recall Hilbert's comment on p. 2 of the 1894 lectures, 'Wenn ich wieder lese, so erst die Euklidische Geometrie'. The lectures of 1898/1899 *are* indeed on Euclidean geometry, although these are far from a standard presentation of the basic theoretical lines of that geometry, but rather a profound analysis, metamathematical in spirit, of the theoretical results in the early part of Euclid's *Elements*, of the treatment of congruence, of proportion and area, of continuity, and coordinatisation (the hidden field properties in segment structures), and, of course, of the Parallel Axiom itself, its independence and its place in the system. Many of the results of this analysis find their way into the *Festschrift*, and the Axiom of Parallels is actually made more use of than in the lectures; but the *Festschrift* is nevertheless not shaped and not presented as an analysis of Euclidean geometry in the same way as the lectures obviously are. The *Festschrift* has more the form of a systematic development of geometry with the Axiom of Parallels used more as a simplifying tool, not so much as an object of investigation. This can be seen, for instance, in Hilbert's first treatment of the Desargues Theorem and what he calls Pascal's Theorem (actually what is now known as Pappus's Theorem, a special case of Pascal's Theorem for conics).

Secondly, the treatment of the material in the lectures is more synthetic than that of the *Festschrift*. In the lectures, the axioms are presented group by group, with notable consequences being drawn along the way, and these results and the axioms themselves are often subjected to significant analyses *before* Hilbert moves on to the next group. A proof of the consistency of the whole system is only pointed out in the final stage of the lectures (p. 104 of Hilbert's notes, p. 167 of the *Ausarbeitung*), as if this is the culmination of Hilbert's presentation. The consistency proof is also accompanied by important philosophical remarks on the nature of mathematical existence and the role of relative consistency proofs. In the *Festschrift*, however, the axiom groups are all presented in the first nineteen pages, with only elementary consequences drawn and no meta-investigations, after which the consistency of the whole system is proved immediately (pp. 20–21), using the same algebraic considerations (the countable Pythagorean field construction which occurs several times in the lectures), but now presented without any accompanying philosophical reflections. This architectonic feature of the *Festschrift* enables a far more focused and sharper approach; for instance, as mentioned, it gives Hilbert from the beginning the resources to simplify the Desargues and Pascal Theorems by calling on the Axiom of Parallels in their statements and proofs.

It is worth dwelling on Desargues's Theorem as an example of the different shaping this allows, and of the different intellectual atmosphere of the *Fest-*

schrift. In the lectures, the theorem is stated in a general, projective version, i.e., its statement involves only the axioms I 1, 2 and II (thus, the plane axioms). The *proof* of the theorem, however, employs the full set of axioms from I, i.e., spatial incidence axioms, as well as planar. After proving the theorem, Hilbert raises the question (p. 29 of his lecture notes) as to whether it is possible to find a proof using only the axioms I 1, 2, II, and then a significant metamathematical investigation is undertaken to show that the answer is no. (A planar model is constructed in which I, 1, 2, II hold, but Desargues's Theorem fails. See the lecture notes, pp. 30–33, or the *Ausarbeitung*, pp. 28–31.) All this is done *before* Hilbert moves on to the Congruence Axioms. In the *Festschrift*, the statement of the theorem involves the Parallel Axiom, and the proof from the full spatial (and parallel) axioms is called upon, though not given (p. 49). Hilbert then remarks that one can give a simple *planar* proof using a central result from the theory of proportions (i.e., involving congruence). Thus, congruence can replace the spatial assumptions involved in the usual proof. Hilbert then gives another planar model (thus creating a non-Desarguean affine geometry) in which Desargues's Theorem and the appropriate congruence assumption both fail, i.e., where all the congruence axioms except the triangle congruence axiom (IV 6 of the *Festschrift*, III 6 of the lectures) hold, thus showing (as one might expect) that all the congruence axioms have to hold for the proof of Desargues's Theorem to go through. The investigation is thus different, and certainly adds important information to that of the lectures. As against this, the whole fascinating discussion of the import of Desargues's Theorem, of 'Reinheit der Methode', etc. is quite lacking in the *Festschrift*. (See pp. 29–30, 33 of Hilbert's notes, and pp. 27–28, 31–32 of the *Ausarbeitung*.)

Thirdly, Hilbert's notes are presumably a fair representation of the lectures much as Hilbert gave them. The *Ausarbeitung* is an idealised version of this, something like a publishable, textbook account of the lectures (see the passage from *Blumenthal 1935* quoted above), while nevertheless remaining a representation of Hilbert's *lectures*, with something of the same informality and richness of discussion. On the other hand, the *Festschrift* is more like the kind of mathematics book that we have become used to. It is not quite as laconic and abrupt as modern books tend to be, but it has certain theoretical aims, and sets out to attain these in an efficient way. Conceptual and philosophical discussion is minimised, being confined in the main to the short Conclusion. As mentioned, such discussions are extended and rich in the lectures. Some examples of informal material which is in the lectures, but not in *Festschrift*, or at least, only in a compressed form, have been touched on already: (1) The discussion of what Hilbert calls 'Reinheit der Methode'; (2) the discussion of the meta-problems raised by Desargues's Theorem and its proof, and the discussion of the informal significance of Desargues's Theorem itself; (3) the discussion of the relationship between consistency and existence (Hilbert's notes, p. 104, and the *Ausarbeitung*, p. 167 respectively); (4) the discussion of the theory of proportion and Euclid's way of dealing with this (Hilbert's notes, pp. 92–93, and the *Ausarbeitung*, pp. 112–113). (This theoretical material is dealt with in the *Festschrift*, pp. 35–37, but without discussion.) There are two further examples, both important. Example (5) is the theorem about the three chords generated by three mutually intersecting, co-planar circles, which receives extensive meta-mathematical analysis in the lectures, and which builds on the discussion in the 1898 *Ferienkurs* (see Chapter 3). This work is important, for it can be seen as a further profound analysis of Euclid's system and its assumptions. This analysis makes no direct appearance in any of the editions of the *Festschrift*, nor in subsequent lecture notes, though one of the central results is represented. (See the full discussion in Specific Note (2) appended to this Introduction.) The last major topic to be mentioned here, (6), concerns the two so-called Legendre Theorems, which are not dealt with at all in the *Festschrift*, at least in the first edition, but which are stated and proved in the *Ausarbeitung* (pp. 69–73), and foreshadowed in Hilbert own notes, p. 68. (Hilbert simply refers to the *Ausarbeitung* for the statements and proofs of the theorems, leaving blank spaces for the insertion of page numbers. This is a fairly early example of such references.) The proofs make use of the Archimedean Axiom; Hilbert gave to his student Max Dehn the problem of establishing whether this use is essential. (See Specific Note (5)(*a*).)

The lectures are thus quite distinct in tone and content from the *Festschrift*, despite the large overlap. On the other hand, there are also several respects in which the *Festschrift* represents an important extension of the work of the lectures. One example, already mentioned, is the relation of the Desargues Theorem to congruence; another is the much more extensive section at the end of the book on geometrical constructions, particularly the results on the analytic characterization of constructible points.

4. The Relationship of the 1898/1899 Lectures to Earlier Work

What of the relation of the 1898/1899 lectures to the 1894 lectures presented in Chapter 2? The general character of these latter is described in the Introduction to Chapter 2, but there are some important continuities between the earlier and later lectures which should be stressed here. For instance, both contain a similar description of geometry as a natural science; both appeal in very general terms to Hertz's characterization of natural science (his *Bildtheorie*); both exhibit the same view of geometry as giving what Hilbert calls a 'Fachwerk' or 'Schema' of concepts, in which the primitive terms are taken to have no privileged interpretation; and both make it clear that the axioms of geometry will never hold exactly in nature, only approximately. Moreover, the basic axioms (incidence, order) are not radically different, despite changes in nomenclature and position. The first congruence axioms are much the same, although Hilbert does not deal with the congruence of either 'straight' angles or right-angles in the 1894 lectures, neither does he there have any triangle congruence axiom, although this difference is probably due to the choice of theoretical material dealt with. Nevertheless, there are major differences of form, which can be summarised as follows.

First, the shape of the later lectures is clearer and the focus is sharper. The earlier lectures are projectively (and also analytically) coloured, to the extent that they are designed primarily to show how coordinates could be introduced into geometry before there is any consideration of congruence, and how then hyperbolic and elliptic geometry can be developed on this foundation. The *last* axiom to be introduced is the 'Euclidean' Axiom, i.e., the Axiom of Parallels. It was apparently Hilbert's intention to proceed from the more elementary to the more specialised (Euclidean geometry as a refinement of projective geometry), but this plan is disrupted by the *Einführung der Zahl*, in that Hilbert clearly reaches a point where, to proceed, he requires the use of a powerful continuity assumption. This step is not really analysed. In particular, the question is not dealt with of how much of elementary geometry is possible before use is made of continuity assumptions. (Hilbert only realised quite late that a great deal of basic Euclidean geometry, what he calls 'Schulgeometrie', can be developed without the use of these. See the excerpt from Hilbert's letter to Hurwitz from 31 December, 1899, quoted in *Toepell 1986*, 198–199.) To this extent, the later lectures are surer footed and more clearly directed, i.e., there is a build-up from elementary assumptions to less elementary, to the axioms (parallels, continuity) which Hilbert says (in the 1898 *Ferienkurs*, and again in the *Ausarbeitung*, pp. 145–146) lack the 'immediate' character of the initial axioms. (Cf. p. 106 of Hilbert's own notes.) The impression given by the 1894 lectures is that the earlier number is introduced the better. However, the *overall* goal of the 1898/1899 lectures is just the opposite, so much so, that the question is central there of how far the basis of the geometric 'theory of number' (the *Streckenrechnung*, segment calculus or algebra of segments) can be prepared before one has to resort to continuity assumptions.

The increased sharpness of aim can be seen in Hilbert's own remarks. On the one hand, there is the admiration for von Staudt's goal expressed in the introduction to his lectures from 1891 on projective geometry (see Chapter 1), of creating a projective geometry free from 'measuring and calculating'. On the other hand there is this statement from the 1894 lectures:

> In allen exacten Wissenschaften gewinnt man erst dann präzise Resultate, wenn die Zahl eingeführt ist. Es ist stets von hoher erkenntnisstheoretischer Bedeutung zu verfolgen, wie dies Messen geschieht. (*Hilbert 1894a**, 28.)

But although this sentiment is more or less repeated in Hilbert's notes for the lectures for 1898/1899, there is an important qualification:

> Aber, wenn die Wissenschaft nicht einem unfruchtbaren Formalismus anheimfallen soll, so wird sie auf einem spateren Stadium der Entwickelung sich wieder auf sich selbst besinnen müssen und mindestens die Grundlagen prüfen, auf denen sie zur Einführung der Zahl gekommen ist.
>
> Die Geometrie soll die reichen Mittel der Analysis nicht als Fesseln tragen und die Mittel der Analysis sollen ⟨von⟩ ihr selbst gesuchte und bewusst benutzte Quellen neuer Erkenntnis sein. (*Hilbert 1898c**, 4–5.)

What this amounts to is a refinement of the mathematical task. While Hilbert's *Einführung der Zahl* of 1894 concentrates on establishing and analysing the one-to-one correspondence between points and numbers, it does not examine the algebraic properties of the geometric analogues of the numbers to be introduced, namely the associated field properties. These are the properties which *in fact* allow one to apply the numbers in measuring and for describing the properties of geometric objects (the line, the rectangle, the circle, etc.). This examination is conducted in the 1898/1899 lectures, and is repeated in the *Festschrift*. The question can be posed in this way: If full Euclidean geometry is correctly represented by analytic geometry, then the algebraic structure of the reals must be represented in the line and its segments. Hence, unless synthetic Euclidean geometry falls woefully short of its Cartesian reconstruction, there must be certain theorems of synthetic geometry which are responsible for this, and behind these theorems certain axioms. What are they? And how deeply, if at all, are continuity axioms involved? These are the general questions which appear to lie behind the 1898/1899 lectures. Moreover, these also show a more refined attitude to the way continuity is involved in all this, in that the concentration is really just on the Archimedean Axiom, and on what is and is not possible with it.

As against this, there is still confusion in the 1898/1899 lectures as to the strength of the various continuity axioms. The confusion is visible in both the 1898 *Ferienkurs* (see *Hilbert 1898b**, 16) and the *Ausarbeitung* of the 1898/1899 lectures. For instance, to call the strong continuity axiom of the 1894 lectures and *Hilbert 1895b* a version of the Archimedean Axiom as is done in the *Ausarbeitung* (see p. 141) appears perplexing now. Note also the remark (p. 162) that a system of objects satisfying ordered field axioms whose only continuity axiom is the ordinary Archimedean Axiom must be the usual system of real numbers. The confusion has vanished by the *Festschrift*, with its explicit recognition that Euclidean geometry with only the Archimedean Axiom has a countable model (*Hilbert 1899c*, 21), and this was followed very shortly by the paper*Hilbert 1900b* which adds a completeness axiom to the ordered field axioms, an axiom which guarantees full Dedekind continuity. This addition was mirrored for geometry itself in subsequent editions of the *Festschrift*, indeed from the French translation of the *Festschrift* itself onwards, though no mathematical use is made of this axiom in the geometrical system. (See the Appendix, and the Introduction, to Chapter 5.)

In short, from *Hilbert 1895b* on, Hilbert's work on geometry is increasingly metamathematical in spirit. In this respect, the development of Hilbert's work in the period 1898 to 1903 is foreshadowed in that paper. What characterises the later period is the exploitation of the notion of independence and the exploration of geometries in which key propositions in the Euclidean development fail.

5. Some Brief Remarks on the Editorial Treatment of the Texts

(i) The editorial notes on the texts of Hilbert's lecture notes and their *Ausarbeitung* are mainly direct comments on the specific material and are designed to facilitate its digestion. There are many more for the former than the latter, partly because the *Ausarbeitung* is 'cleaner' than the notes, but partly also because remarks which apply to both have already been made in notes to the former. One important piece of advice to the reader: If difficulties are encountered in reading Hilbert's own notes, e.g., with respect to a passage or even a whole line of development, the corresponding place in the *Ausarbeitung* should be consulted. In particular, the diagrams in the latter are generally much clearer and more informative, and are often, at root, the very same diagrams. Thus, our advice is to read Hilbert's notes first *in conjunction with* the *Ausarbeitung*; only after this should the texts be read separately. The *Ausarbeitung* has, at the end of the text, a very helpful table of contents.

The best 'companion' we know to Hilbert's later work on geometry is the book *Hartshorne 2000*, which came to our attention too late for comprehensive references to be made.

(ii) There are several theoretical developments in these texts which deserve detailed commentary, too detailed to be carried out in footnotes, at least without straining both the patience of the reader and the aesthetics of the page. Since this commentary may be of some service to readers less familiar with the material or with the metamathematical thrust of Hilbert's treatment of it, it was deemed sensible, rather than omit it altogether, to append it to this Introduction. It is hardly necessary to stress that the extended comments, as with those in the footnotes themselves, are not intended to be complete, and that the choice of subjects commented upon is to be seen as idiosyncratic. The detailed notes cover six specific subjects:

(1) The metamathematical questions surrounding the Desargues and Pascal Theorems.

(2) The metamathematical questions raised by the discussion of the 'Three Chord Theorem'.

(3) The treatment of congruence, as compared to that in Euclid.

(4) The historical material concerning the Axiom of Parallels.

(5) The unsolved problems stated by Hilbert at the end of the lectures.

(6) The final paragraphs of Hilbert's lecture notes.

Relevant footnotes in the texts themselves refer the reader back to these notes.

The author wishes to acknowledge his enormous debt to Helmut Karzel in the writing of these Specific Notes and the footnotes themselves.

Specific Notes

(1) The Desargues and Pascal Theorems

The investigation of Desargues's Theorem and Pascal's Theorem emerges through the lectures as one of the central focuses of Hilbert's analysis of traditional geometry (particularly the common core of projective and Euclidean geometry), and it occupies a correspondingly central place in the *Festschrift*, though there in affine rather than projective guise. It is also the area where Hilbert's metamathematical procedures are shown to most dramatic effect. In addition, it represents the core of the mature treatment of the 'Einführung der Zahl', thus forming a strong link to the concern which dominates Hilbert's lectures from 1894. What Hilbert does here is to investigate the algebraic resources available inside the structure of synthetic geometry *independently* of the introduction of specific numerical or continuity assumptions.

(*a*) *The Planar Version of Desargues's Theorem.* Three aspects of this theorem have already been mentioned.

(*i*) While its *statement* only involves planar concepts, the theorem cannot be proved from the planar axioms I 1, 2 and II alone; Hilbert constructs models which show this (pp. 31–33 of his own notes, pp. 28–31 of the *Ausarbeitung*). It is in connection with this that the question of the 'Reinheit der Methode' is raised (see pp. 29–30 of Hilbert's notes, and 27–30 of the *Ausarbeitung*). Thus, to prove the planar version of Desargues's Theorem, some supplementation of I 1, 2 and II is necessary.

(*ii*) Supplementation can be effected either by adding *spatial incidence* assumptions or adding *congruence axioms.* Hilbert shows in the *Festschrift* that all the congruence axioms must be present, in particular, that the triangle congruence axiom must hold, i.e., the axiom which links linear and angle congruence.

(*iii*) Hilbert shows in effect that the Planar Desargues Theorem can itself be treated as an axiom which has spatial consequences: any plane model **M** of I, 1, 2, II *and* Desargues's Planar Theorem can be recoordinatised to provide a model **N** of space (i.e., axioms I and II *in toto*), in which the planes form a part of a 'normal' incidence structure (as described by the axioms of group I). Desargues's Planar Theorem is used to show that **M** can be coordinatised by showing that the collection of segments in **M** form a non-commutative field; this field is then used to construct a 3-dimensional analytic geometry, which is what yields **N**. (See the *Ausarbeitung*, pp. 159–160, and the *Festschrift*, p. 70.) This exhibits a very firm grasp of abstract model construction, and it also reveals that Desargues's *planar* Theorem has hidden *spatial* content, perhaps showing that the spatial proof of the Planar Theorem does not violate 'Reinheit' after all. The key fact is this:

> Die sämtlichen Axiome I, II sind æquivalent den ebenen Axiome I, II und dem Desargues. (*Ausarbeitung*, 159–160.)

This is presumably what Hilbert means when he makes the somewhat cryptic remark:

> ... man könnte im diesem Fall sagen: Der Satz von Desargues ist das Resultat der Elimination der räumlichen Axiome aus I und II. (*Ausarbeitung*, 32. Cf. also the *Festschrift*, 70.)

and what he refers to in the *Festschrift* when he says:

> Wie bereits erwähnt, ist der Satz 32 [Desarguesscher Satz in der Ebene] eine Folge der Axiome I–III; dieser Thatsache gemäss ist die Gültigkeit des Desarguesschen Satzes in der Ebene jedenfalls eine *notwendige* Bedingung dafür, dass die Geometrie dieser Ebene sich als Teil einer räumlichen Geometrie auffassen lässt, in welcher die Axiome der Gruppen I–III sämtlich erfüllt sind. (Ibid., 50.)

This result also shows that, because of its spatial content, the plane version of Desargues's Theorem can be treated as an axiom in place of spatial axioms. Many modern treatments of geometry pursue just this procedure, e.g., *Coxeter 1993*, 7, 14, *Samuel 1988*, 23, and Hilbert himself so treats it, considering for the first time plane Desarguesian geometries. Moreover, the way Hilbert proves his main result (using Desargues's Theorem to coordinatise **M** and **N**) reveals that the theorem is intimately involved in the problem of the internal coordinatisation of synthetic geometry, a project which goes back to von Staudt's 'Wurfrechnung'. For a modern summary of the situation surrounding these two points, including consideration of the various specialisations of the Desargues's Theorem, see *Karzel and Kroll 1988*, 33–45, 59–63.

(*b*) *Hilbert's Pascal Theorem.* As mentioned, what Hilbert calls Pascal's Theorem is actually usually known as Pappus's Theorem. We will continue to call it Pascal's Theorem (sometimes Hilbert's Pascal Theorem) to avoid confusion. (The Pappus theorem concerns only straight lines, whereas the full Pascal Theorem uses the concept of conic section. Pappus's Theorem follows from Pascal's Theorem.) In Hilbert's notes, there are references to the Pascal Theorem on pp. 68 and 92, but no proof, the latter just making open reference to the 'Autographierte Vorlesung'. Pascal's Theorem is first stated on pp. 73–74 of that text; it is then stated differently (p. 107), making essential use of the Axiom of Parallels (introduced after p. 75), and then proved on pp. 109–111. The version of the Pascal Theorem stated and proved in the *Festschrift* (pp. 28–31) is the second, affine version from the *Ausarbeitung*.

Hilbert proves this second version of the theorem from axioms I 1–2, II, the congruence axioms (here III, IV in the *Festschrift*) and the Parallel Axiom (here IV, III in the *Festschrift*), thus, the 'plane' part of I–IV, which is also enough to prove Desargues's Theorem. This result, according to Hilbert in the *Ausarbeitung* (p. 108), had not previously been proved. Schur (in *Schur 1899*) showed that Pascal can be proved without the use of the Parallel Axiom (and without adding any continuity assumption), but he uses the *spatial* incidence axioms. The significance of Hilbert's proof 'in the plane' is then characterized by him as follows:

> Die fundamentale Bedeutung des [Pascal'schen] Satzes liegt daran, daß er uns in den Stand setzt, die Lehre von den Proportionen ohne irgend ein neues Axiom zu begründen. (*Festschrift*, 111–112.)

What Hilbert then shows is that the Pascal Theorem can be used to give a rigorous definition of the notions of proportion and *Flächenmessung* using only plane axioms and without continuity assumptions, thus without reliance on the implicit principle he attributes to Euclid that proportions (or segment products) are kinds of numerical quantity (equivalently, that the end points of segments are always represented by magnitudes), thus without 'ein neues Axiom'. (See Hilbert's notes, pp. 92–93, pp. 111–113 of the *Ausarbeitung*.) Hilbert sets up a *Streckenrechnung*, and shows the commmutativity, associativity and distributivity of the segment multiplication so defined. This is enough to establish the traditional theory of proportion.

The segment calculus on which this rests requires the congruence axioms since it is based on taking products in a right-angled triangle. (See Specific Note (3)(*a*).) Hilbert later builds a second segment calculus with a definition of segment equality which is independent of the congruence axioms, using in effect projective identity under parallel transport. The point here is to reveal the relationships between the Pascal Theorem and the Archimedean Axiom. In short, Hilbert shows that Pascal's Theorem can be proved by relying on spatial axioms, as Schur does, but this time supplemented, not by the congruence axioms, but by one 'numerical assumption', namely the Archimedean Axiom, Hilbert's Axiom V. (For the proof, see pp. 146–164 of the *Ausarbeitung* and pp. 71–76 of the *Festschrift*. The specifically spatial axioms in I can be replaced by the Planar Desargues's Theorem: see the *Ausarbeitung*, p. 146.) The reason is as follows. In his calculus, Hilbert shows that the commutativity property of segment multiplication is in fact *equivalent* to the Pascal Theorem, whereas the other central properties can be established (making essential use of the coordinatisation permitted by the Planar Desargues Theorem) without using congruence assumptions. Given this setting, he shows that the Archimedean Axiom implies the commutativity property, i.e., the Pascal Theorem, and moreover, that if the Archimedean Axiom is *not* added to I, II, IV, then Pascal *cannot* be proved. (The use of the Archimedean Axiom would indeed allow one to make the same 'numerical assumptions' about segments that Euclid did, but while avoiding the congruence axioms.)[2]

(*c*) *Schnittpunktsätze.* On pp. 104–105 of his lecture notes, states that every *Schnittpunktsatz* that can be proved from I–V is already a consequence of Desargues and Pascal. This is addressed in more detail in the *Ausarbeitung* (168), and the *Festschrift* (pp. 76–77), with an even more explicit treatment in the seventh edition (i.e., *Hilbert 1930a*, 111–114). There are two parallel thoughts behind this claim. One is that any pure *Schnittpunktsatz* (i.e., one which is concerned only with statements of incidence and intersection) can be reduced (when the right auxilliary points and lines are added) to a claim about a succession of Desargues or Pascal arrangements. The parallel thought is algebraic. Given the Desargues and the Hilbert Pascal Theorems, then there is a rich underlying analytic geometry over an associative and commutative field. In this geometry, any *Schnittpunktsatz* can be expressed as the solubility of a certain kind of equation built up by using the field operations (the *Ausdruck* of Hilbert's terminology), and the insight is then that the solutions will always exist. Since the fact that the field is commutative is important (it guarantees the commutativity of the right diagrams), the presence of Hilbert's Pascal Theorem is crucial. *Hessenberg 1905b* shows that the Pascal Theorem itself implies Desargues's Theorem in a fairly rudimentary setting (plane incidence and order axioms and the Parallel Axiom), so the result Hilbert states here can be simplified: all *Schnittpunktsätze* can be derived from the Hilbert Pascal Theorem. (See *Hilbert 1930a*, 111.)

---

[2]Euclid's Book V, Definition 4 comes close to the assumption of the Archimedean Axiom for 'magnitudes'. See *Heath 1926b*, 114. See also *Elements* X, 1 (*Heath 1926c*, 14).

(*d*) *Summary.* We are now in a position to provide some elucidation of the passage on p. 68 of Hilbert's lecture notes, where Hilbert makes some general claims about Desargues's and Pascal's Theorems.

(i) 'Ein Rechnen mit Strecken oder eine analytische Geometrie [kann] aufgestellt werden', where the 'Buchstaben' stand for 'Strecken, nicht Zahlen'. This refers to Hilbert's later exhibition of numerical systems developed via the various *Streckenrechnungen* without any specifically numerical assumptions. In effect, what is shown is that the *segments* can form the basis of fields of the right kind suitable for proper coordinatisation, and thus that, to a large extent, analytic geometry is possible without the imposition of number fields from 'outside'. This goes a long way towards explaining how it is that synthetic geometry can match analytic geometry, and to settling Hilbert's long standing occupation with the *Einführung der Zahl.* The Desargues and Pascal Theorems obviously play a central role in this.

(ii) Hilbert's reference to bilinear relations expressing the necessary and sufficent conditions for the 'Vereinigtsein von Punkt und Ebene' is an indirect reference to the result about the Planar Desargues Theorem conjectured on p. 33 of the same text, first proved late in the *Ausarbeitung*, and discussed earlier in this Introduction. If one has Planar Desargues, then one can coordinatise and recoordinatise (see this Note, (*a*)(*iii*)) in such a way that the points and the planes will be correctly related according to the normal incidence relations and will satisfy the right linear equations.

(iii) Desargues is 'the elimination of the spatial axioms'. This refers to the same result; Planar Desargues is not only provable from, but can be used to replace, the spatial incidence axioms.

(iv) Pascal is 'the elimination of the congruence axioms'. This is a reference to the facts established by Hilbert and outlined in (*b*), namely that the Pascal Theorem can be used as a foundation for the theory of proportion without any appeal to the congruence axioms, and that it can be proved without using the congruence axioms, provided that the Archimedean Axiom and the spatial incidence axioms are present.

(2) The Three Chord Theorem
Hilbert's lecture notes, pp. 63f, and the Ausarbeitung, *pp. 61f.*

The theorem states: Given three circles (in the same plane) which pairwise intersect, then the three chords to which the points of intersection give rise have a common intersection point. As in the Introduction to Chapter 3, for ease of reference we will sometimes call this theorem the 'Three Chord Theorem'. Hilbert gives a straightforward proof of the theorem and then subjects it to detailed metamathematical consideration which extends and systematises the discussion in the 1898 *Ferienkurs*, pp. 22–23. Indeed, it forms a fundamental part of Hilbert's analysis of the structure and presuppositions of Euclid's *Elements* and the assumptions behind the constructibility of points. Neither the theorem, nor the metamathematical considerations provoked by it, make an appearance in any edition of Hilbert's *Festschrift*, nor in any later lecture notes. The reason, we suggest, is that the essential results *are* to be found in the *Festschrift*, although stated in more abstract form and without the focus on the interesting example. The Three Chord Theorem thus presents a particularly striking case of evolution from the *Ferienkurs* in 1898 through to the *Festschrift*.

(*a*) *The Theorem.* The Three Chord Theorem is apparently due to Monge. In *Monge 1799*, 97–98, (see also loc. cit., Fig. 1), a similar, but slightly more general theorem, is suggested which is actually reminiscent of the generalization of the Three Chord Theorem originally known as the *Büschelsatz*, a theorem which plays much the same role in circle geometry as Pascal's Theorem does in plane geometry. (For this, see *Karzel and Kroll 1988*, 98–100.) A simple proof of the Three Chord Theorem can be found in *Steiner 1826*, section 4. It can be generalized easily using Steiner's notion of the power of a point with respect to a circle. Consider the line of equal power (the *Potenzgerade*) of any two circles (not necesarily intersecting), that is, the line whose points all have equal power with respect to both circles, lines called *Chordalen* (by Plücker), also known as *radical axes.* This line contains the common chord if there is one, and it is extensively dealt with in *Steiner 1931* originally from 1823–26. The Three Chord Theorem is then a special case of the following: Given three circles, the three *Chordalen* given by the three pairs of circles have a common point of intersection, a theorem which Klein calls the *Hauptsatz der Chordalentheorie.* (See *Klein 1926*, 40, or *Coolidge 1916*, 96.) Another special case of this is the elementary theorem that the perpendicular bisectors of the three sides of a triangle meet in a single point; consider the three vertices as centres of circles of vanishingly small radii and consider their *Chordalen.*

(*b*) *Hilbert's Proof of the Theorem.* Hilbert's proof, which is clearer in the *Ausarbeitung*, pp. 61–64, assumes that the three circles in questions are the equators of three spheres which have just two points P and Q of mutual intersection; it is then shown fairly easily (p. 61) that the three chords generated must all intersect the line PQ. But since neither P nor Q is in the plane of the original circles and their chords, PQ has only one point in common with that plane, and this point must be common to all three chords. Thus, the proof goes through, *given* the assumption that three mutually intersecting spheres intersect in exactly two points, thus, that these points exist. The focus is then on this assumption.

(*c*) *The Euclidean Triangle Property and Its Failure.* In Hilbert's lecture notes, p. 64, it is asserted that this assumption ultimately rests on the Euclidean proposition (*Elements*, I, 22: see *Heath 1926a*, 292–293) that a triangle can be constructed from any three lines which satisfy the 'triangle inequality', i.e., which are such that any two taken together are greater than the third. (See also pp. 63–64 of the *Ausarbeitung.*) The question then is: on what assumption is the proof of this proposition based?

In his lecture notes, Hilbert states this triangle property in the same breath with another (see loc. cit.), namely that, given a straight line and a circle in the same plane, if the line has both a point in the interior of the circle and a point outside it, then it must necessarily intersect it, and indeed in exactly two points. Call this the *line-circle property*. Similar to this is what is sometimes called the *circle-circle property*, namely a circle with points both inside and outside another circle has exactly two points of intersection with it. (For similar properties, see pp. 62, 65 of Hilbert's lecture notes, and pp. 63–64 of the *Ausarbeitung*.) The connection to the triangle property is now unsurprising, for the line-circle property is precisely what Euclid implicitly relies on in the proof of I, 22. Hilbert constructs a model of his geometry (i.e., Axiom Groups I–III) in which the assumption fails, and where Euclid's I, 22 also fails. Thus, the necessary conditions for Hilbert's proof of the Three Chord Theorem are not present in the geometry based on I–III. Moreover, since the Euclidean proof is based on a simple straightedge and compass construction, Hilbert's result is tantamount to saying that his axiom system does not have enough existential 'weight' to justify this particular construction.

(*d*) *The Model.* Hilbert begins to consider this metamathematical problem on p. 65 of his lecture notes, pp. 64f of the *Ausarbeitung*. He constructs what is in effect the smallest Pythagorean sub-field of the reals which contains 1 and π, which yields a countable model of the axioms I–III (indeed of I–V, the whole system of the *Festschrift*) when the usual analytic geometry is constructed from pairs of its elements taken as coordinates. However, the number $\sqrt{1-\left(\frac{\pi}{4}\right)^2}$ will not exist in this field. Since 1, 1, and $\frac{\pi}{2}$ are in the field, the model will possess three lines which satisfy the triangle inequality, but from which no triangle can be constructed. This is well illustrated by the diagram in the *Ausarbeitung*, p. 68: the apex at C (the upper apex of the triangle in the diagram) ought to have coordinates $\left(\frac{\pi}{4}, \sqrt{1-\left(\frac{\pi}{4}\right)^2}\right)$, but this is not a point in the model. The same example shows that there can be a line partially within and partially without a circle but which intersects the circle nowhere: take the vertical line ($x = \frac{\pi}{2}$) in the diagram; this ought to meet the circle $x^2 + y^2 = 1$ at the points with y-coordinates $\pm\sqrt{1-\left(\frac{\pi}{4}\right)^2}$, but these points do not exist in the model.

In his review of Hilbert's *Festschrift*, Sommer remarks that one cannot prove the line-circle property in Hilbert's system (see *Sommer 1899/1900*, 291). What Sommer observes already follows from what Hilbert shows in his lectures, and what he says explicitly (e.g., p. 64 of the *Ausarbeitung*). Sommer, of course, certainly knew that Hilbert had shown this. For one thing, this is the same Julius Sommer, the 'friend', who, along with Minkowski, is thanked by Hilbert at the end of the *Festschrift* for help with proof reading. For another, Sommer refers to Hilbert's lectures course for 1898/1899, both directly, in a different context (p. 292), and indirectly in the context of what he states about the line-circle property. Futhermore, the mathematical core of the result is in fact present in the *Festschrift*, if only implicitly and in a more abstract form. Let us turn to this now, and return to Sommer's remark later.

(*e*) *The Algebraic Significance of the Model.* In modern algebraic terms, what Hilbert has shown is that not every Pythagorean field is Euclidean. An ordered field is said to be *Pythagorean* when the Pythagorean operation holds, that is, when $\sqrt{x^2+y^2}$ is in the field whenever x and y are. The reals form a Pythagorean field, but the rationals do not; nevertheless, there are small (countable) sub-fields of the reals which are Pythagorean. In particular, for each real r, there is a minimal, and countable, Pythagorean sub-field of the reals containing the rationals and r. An ordered field K is said to be *Euclidean* when for any non-negative element $a \in K$, $\sqrt{a} \in K$ holds, too. It is clear that every ordered Euclidean field is Pythagorean, but Hilbert shows here that the converse fails, for his model is formed from a field which is Pythagorean but not Euclidean, since $\frac{\pi}{4}$ is an element, given that π is, and so are $\left(\frac{\pi}{4}\right)^2$ and $1-\left(\frac{\pi}{4}\right)^2$; however, $\sqrt{1-\left(\frac{\pi}{4}\right)^2}$ is not. The key point is then summed up in the following theorem: Given an analytic geometry whose coordinates are given by an ordered Pythagorean field, one can always construct a triangle from three sides satisfying the triangle inequality if and only if the underlying coordinate field is also Euclidean. Indeed, for an analytic geometry based on an ordered field, the Euclidean field property is equivalent to the line-circle property, and this is in turn equivalent to the circle-circle property, a property of obvious direct relevance to Hilbert's proof of the Three Chord Theorem. (See *Hartshorne 2000*, 144–146.)[3] Given this, the connection between the Euclidean field property and the formation of a triangle from any three lines satisfying the triangle inequality is easy to see.

This algebraic result is strongly hinted at in the 1898 *Ferienkurs*, pp. 22–23 (see Chapter 3.). Hilbert poses the question of whether, given a segment product $c \cdot d$, there is a segment x such that $x^2 = c \cdot d$, i.e., a square root for $c \cdot d$. His comment suggests that he thinks this is not always the case, and this is precisely what the counterexample outlined above confirms. For example, in the diagram in the *Ausarbeitung*, 66, the two products concerned would be those formed by the four segments arising from the intersection of the horizontal and vertical chords. The horizontal segment product would be $\left(1+\frac{\pi}{4}\right)\cdot\left(1-\frac{\pi}{4}\right)$, and the vertical product $\sqrt{1-\left(\frac{\pi}{4}\right)^2}\cdot\sqrt{1-\left(\frac{\pi}{4}\right)^2}$, thus $\sqrt{1-\left(\frac{\pi}{4}\right)^2}$ would be the 'x' sought. But the argument Hilbert gives shows that the segment corresponding to this x does not exist in the model. Thus, a question from the 1898 *Ferienkurs* is answered.

Hilbert notes that the problem exhibited here does not just arise because of the involvement of the transcendental number π; he gives an example of an elementary number which will be in any Euclidean field over the rationals, but which is not in the minimal Pythagorean field, namely $\sqrt{1+\sqrt{2}}$. (See Hilbert's own lecture notes, p. 67, and also p. 67 of the *Ausarbeitung*. *Hartshorne 2000*, 147, gives further details of the counterexample.) Another example is given in the *Festschrift*, to which we will come in a moment.

---

[3]Hilbert remarks in the *Ausarbeitung*, p. 64, that the line-circle and circle-circle properties 'hängen eng zusammen'.

(*f*) *The Constructibility of Points.* These algebraic considerations bear closely on the question of the constructibility/existence of points. It is well-known that such questions arise with the first proposition of the *Elements* (I, 1) which shows how to construct an equilateral triangle on a given base AB. The construction takes the two circles whose centres are the endpoints of the base and whose radii are equal to the base; either of the two points of intersection of the circles can be taken as the third apex, C. But do the circles actually intersect? Euclid simply assumes that they will, and a standard objection is that there is no guarantee of this. For instance, Heath notes:

> It is a commonplace that Euclid has no right to assume, without premissing some postulate, that the two circles *will* meet in a point C. To supply what is wanted we must invoke the Principle of Continuity. (*Heath 1926a*, 242.)

And by a Principle of Continuity, Heath means something like Dedekind Continuity (see pp. 237–238). Heath also cites one of Hilbert's contemporaries, Killing, as invoking continuity to show the line-circle property (*Killing 1898*, 43).

In the 1898/1899 lectures, Hilbert himself seems to suggest that the problem might have to do with a continuity assumption. On p. 64 of the *Ausarbeitung*, he says when assuming either the line-circle or circle-circle properties, one is actually assuming that 'the circle is a closed figure'. A related remark is made in the original lecture notes, p. 64, and here Hilbert adds that Euclid has 'a similar sounding axiom'.[4] Moreover, it is just in the context of the failure of continuity in Hilbert's system that Sommer makes his remark about the line-circle property, adding that 'it remains undecided whether or not the circle is a closed figure' (*Sommer 1899/1900*, 291), thus adopting Hilbert's terminology from the lectures.

But continuity is not necessary; just assuming the Euclideanness of the underlying field will do, and results in Hilbert's *Festschrift* make this quite clear. Hilbert takes up the question (not explicitly addressed in the lectures) of which geometrical constructions are performable in his axiom system. He proves two results: (1) Any constructions carried out and justified on the basis of axioms I–V are necessarily straightedge and *Streckenübertrager* (for which a pair of dividers would serve) constructions, the first for drawing straight lines, and the second for measuring off arbitrary segments. (2) The algebraic equivalent to these constructions is the Pythagorean field. In other words, these constructions can be carried out (i.e., the existence of the points constructed justified) in any analytic geometry whose coordinates form a Pythagorean field, even the minimal Pythagorean field built over the rationals. (These are

---

[4] There is, however, no such assumption in the *Elements*, neither in the Postulates nor under the Common Notions. Hilbert may have been referring indirectly to the Euclidean Definitions. For Euclid, a circle is a certain kind of figure, and a figure is 'that which is contained by any boundary or boundaries' (Definition 14; see *Heath 1926a*, 153). Perhaps the somewhat vague 'contained by' and 'boundary' suggest 'closed', and perhaps that the circle has no 'gaps'.

Theorems 40 and 41 of the *Festschrift*, pp. 79–81.)[5] Hilbert then adds to this that:

> Wir können aus diesem Satz sofort erkennen, dass nicht jede mittels Zirkels lösbare Aufgabe auch allein mittelst Lineals und Streckenübertragers gelöst werden kann. (*Hilbert 1899c*, 81.)

Hilbert first gives an example of a real number which cannot be in the minimal Pythagorean field, namely $\sqrt{2|\sqrt{2}| - 2}$. (The example is thus slightly different from that given in the lectures.) Suppose we then pose the problem of constructing a right-angled triangle with hypothenuse of length 1 and the lengths of the two other sides $|\sqrt{2}| - 1$ and $\sqrt{2|\sqrt{2}| - 2}$; this construction cannot be carried out by means of 'Lineal und Streckenübertrager', since the latter length cannot correspond to an element in the minimal Pythagorean field; hence the problem is not soluble in Hilbert's geometry. But, as Hilbert remarks (*Festschrift*, p. 82), the problem is immediately soluble by a compass construction. The number Hilbert specifies is, of course, in any Euclidean ordered field built over the rationals. The central point is now this: If K is the set of all real numbers obtained from the rationals by the operations of addition, multiplication, subtraction and division, and such that K contains square roots for all of its positive elements, then K is Euclidean and is the smallest field over which straightedge and compass constructions can be carried out. (See *Hartshorne 2000*, 147.) The salient point is even clearer in Hilbert's Theorem 44 (p. 86), which deals with the problem of characterising which straightedge and compass constructions *can* be carried out in Hilbert's geometry (i.e., reduced to constructions by straightedge and *Streckenübertrager*). In the statement of the condition, into which we will not go here (see *Festschrift*, Theorem 44, p. 86), Hilbert quite clearly expresses the fact that the algebraic condition corresponding to the compass construction is that each number in the field of coordinates has a square root in the field, i.e., is Euclidean.[6]

Thus, the problem is not to do with a failure of continuity, as Sommer suggests, very possibly leaning on Hilbert's original remark, but rather with the failure of a very much weaker field property, and it seems that Hilbert

[5] Kürschák showed that the *Streckenübertrager* can be dispensed with in favour of an *Eichmaß*, i.e., a device which measures off a single fixed segment. (See *Kürschák 1902*.) Hilbert makes a corresponding adjustment, with acknowledgement to Kürschák's work, in the second edition of the *Festschrift* (see *Hilbert 1903a*, 74, 77.). See also Hilbert's Remark [7] in the Appendix to this Chapter. He also exempts axiom group V from his considerations.

[6] In the 1902 lectures, for the purposes of showing the independence of the *Vollständigkeitsaxiom* Hilbert contructs a (countable) model based on a minimal Pythagorean field. He adds that this geometry (i.e., model) is particularly interesting, for

> ... *sie enthält nur Punkte und Geraden, welche bloß auf Grund des Abtragens von Strecken und Winkeln gefunden werden können.*
>
> Wie wir in unseren *Grundlagen*, pag. 81f. gezeigt haben, kann nicht jede Strecke durch bloßes Abtragen von Strecken construiert werden; z.B. nicht die Strecke $\sqrt{2\sqrt{2}-2}$. (*Hilbert 1902d**, 89–90; see this Volume, Chapter 6.)

would certainly have been fully aware of this, by the time of the writing of the *Festschrift* if not earlier. However, having said this, it is not entirely clear what principle should be added to the axiom system to guarantee Euclideanness; adding the circle-circle property itself as an axiom might appear somewhat *ad hoc*.

Note that, although Euclid's proofs of I, 1 and I, 22 both involve ruler and compass constructions, an adequate construction for the first case *can* be given using Pythagorean operations alone. An equilateral triangle can be constructed just in case $\sqrt{3}$ is in the underlying coordinate field. But $\sqrt{3} = \sqrt{1+\sqrt{2}^2}$, and $\sqrt{2} = \sqrt{1+1^2}$. Hilbert shows this: see the *Ausarbeitung*, p. 173. Hence, the equilateral triangle can be constructed in Hilbert's axiom system. (The actual construction is given in the 1898 *Ferienkurs*, p. 15: see Chapter 3.)

(3) Angle Congruence and the Triangle Congruence Theorems
Hilbert's lecture notes, pp. 36f, and the Ausarbeitung, *pp. 33f.*

Like the analysis of the Three Chord Theorem, this represents a significant part of Hilbert's examination of Euclid's system. It also represents an example of the evolution from the lectures to the *Festschrift*. The notes below concern the role of Euclid's Fourth Postulate ('That all right angles are equal to one another': *Heath 1926a*, 154), the role of triangle congruence, and the role of motion in Euclid's proofs.

(*a*) *Angle Congruence, Axiom 9 and Euclid's Fourth Postulate.* In the original lecture notes (p. 45), Hilbert states his Congruence Axiom 9, 'Alle gestreckten Winkel sind einander ≡', and then poses the question of whether it can be dispensed with on the basis of the other congruence axioms. He has then added another comment on the same page (the date of this is uncertain) which again suggests not postulating Axiom 9, but rather proving it (and the theorem that all 'full' angles are congruent to each other) via Axiom 10 and its immediate consequence, the so-called First Congruence Theorem for triangles, which establishes the side-angle-side criterion for triangle congruence. (The same independence question is also raised in the *Ausarbeitung*, p. 41, but is then consciously set aside.) In the lectures, Hilbert relies on Axiom 9 to prove the four theorems which come immediately after, including the theorems which show the congruence of all right-angles and all 'full angles'. On the same page of his own lecture notes (p. 45), Hilbert remarks that it is 'erroneous' of Euclid to adopt the congruence of all right-angles as an axiom, and Hilbert thus expressly opposes himself to Lindemann (clearly one of the important sources for Hilbert), who endorses Euclid on this point. In *Clebsch 1891*, 547–549, Lindemann asserts that it is correct to adopt this as a postulate since it is more fundamental than any of the principles which imply it. It is presumably this assertion with which Hilbert disagrees, and why he takes the trouble to prove Euclid's axiom. He attributes this 'error', rather enigmatically, to the fact that these 'two authors' (Euclid and Lindemann) do not possess a systematic treatment of the concept 'between'. (See also the *Ausarbeitung*, p. 42.)[7]

Hilbert's procedure in the *Festschrift* shows how to carry out the suggestion made in the comment he adds, namely to use the First Congruence Theorem and avoid Axiom 9 altogether. (This is carried out on pp. 11–16 in the course of developing all the triangle congruence theorems.) There is no direct proof of Axiom 9, but such would be quite straightforward.

(*b*) *Motion and the Triangle Congruence Theorems.* Hilbert remarks (p. 48 of his own lecture notes) that he requires no spatial axioms to prove the congruence theorems, and that

> Auch Euklid scheint dies gemeint und gewusst zu haben, da er nur den ersten Congruenzsatz mittels Decken, d. h. Bewegung, d. h. axiomatisch beweist.

This is an interesting remark, indicating again a close study of Euclid's treatment of congruence. What Hilbert means can be reconstructed as follows. The other triangle congruence theorems are fairly straightforward consequences of the First Congruence Theorem, thus Axiom 10 (IV 6 in the *Festschrift*) without any recourse to spatial axioms. Hilbert then remarks that Euclid appears to have intended to proceed in this way, the difference being that, instead of adopting the First Congruence Theorem as an axiom, he offers a proof of it using 'covering and motion', thus displacement and superposition of figures (*Elements*, I, 4). But this method of proof relies on hidden *spatial* assumptions. Euclid's Common Notion 4 ('Things which coincide with one another equal one another'; see *Heath 1926a*, 155) hints at this, for this would be without substance unless one could *move* objects so as to allow them to 'coincide', and this requires a genuine assumption about the behaviour of planar geometrical objects in space, for instance, an assumption according to which a figure can be 'lifted' out of its plane and set down on a second figure. (Also relevant to this is Hilbert's modification of the Triangle Congruence Axiom in his analysis of the theorem that the base angles in an isoceles triangle are equal. See the Introduction to Chapter 6.) It also assumes that figures retain their properties when so moved, i.e., that space is *homogeneous*. In the light of this, Hilbert suggests that Euclid's proof of I, 4 using displacement and superposition is tantamount to the application of a new axiom about the ambient space, and thus that he does proceed 'axiomatically' even here.[8]

---

[7]The *Festschrift*, p. 16, criticizes only Euclid, as do Hilbert's 1902 lectures (p. 35), not Lindemann. This fact might be relevant in assessing Freudenthal's criticism of the importance Hilbert seems to ascribe to this theorem in the *Festschrift*. For this, and other criticisms, see *Freudenthal 1957*, 121–122. In the seventh edition (*Hilbert 1930a*, 23), Hilbert attributes the idea of the proof to Proclus.

[8]Hilbert is factually incorrect about Euclid in one point, namely Euclid also appeals to displacement and movement in the proof of the Second Triangle Congruence Theorem, I, 8 (the three side criterion of triangle congruence), for this depends on two triangles being 'applied to' each other such that appropriate points correspond. Nevertheless, it does seem that Euclid tries to avoid appeal to such principles wherever possible. For a general treatment of Euclid's attitude, see part of Heath's discussion of Euclid's Common Notion 4,

Hilbert's procedure, of course, avoids both appeal to spatial assumptions, and also to motion. Indeed, he asserts that it is wrong to presuppose the concept of motion in treating the notion of congruence, and that the latter should be used in any proper definition of the former. (See the *Ausarbeitung*, p. 60.) To illustrate that the opposite way of proceding was not uncommon[9], we have only to turn to Lindemann. We find:

> Diese Eigenschaften der Bewegung müssen hervorgehoben, da die ganze Lehre von der Congruenz und Gleichheit der Figuren auf dem Begriffe der Bewegung beruht, wie wir sogleich noch näher ausführen werden (vgl. unten p. 556 f.) (*Clebsch 1891*, 548.)

On p. 556, the general notion of the 'equality of figures' is based directly on that of 'zur Deckung bringen' and thus:

> Hiermit ist deutlich ausgesprochen, dass *der Begriff der Bewegung unerlässlich ist für die Vergleichung verschiedener Figuren hinsichtlich ihrer Grösse, ...* (italics in the original).

For more on how to base congruence on rigid motions, see *Hartshorne 2000*, §17, 148–158.

(4) Hilbert's Historical Remarks on the Parallel Axiom
Hilbert's lecture notes, pp. 69–75, and the Ausarbeitung, *pp. 75–78.*

There are no historical remarks on the Parallel Axiom in any edition of the *Festschrift.*

(*a*) The Parallel Axiom which Hilbert gives in his lectures (p. 70 of his own notes, p. 76 of the *Ausarbeitung*) is not, of course, Euclid's postulate, but what is sometimes called 'Playfair's Axiom', after John Playfair, who proposed it in his English edition of Euclid in 1795 (*Playfair 1795*), although the idea goes back to Proclus. (See *Heath 1926a*, 220, and also 316.)

As Hilbert notes in the *Ausarbeitung*, p. 76, one can derive from the axioms given hitherto that, for every line g and point P outside that line, there is at least one line h through P parallel to g and in the same plane. This follows from the procedure given in Hilbert's own notes (p. 47). Drop a perpendicular k from P onto g. Then, in the plane determined by g and k, draw a line perpendicular to k through P. This is the parallel sought. This theorem appears in the *Elements* as I, 31 (see *Heath 1926a*, 315–316).

(*b*) Hilbert's historical remarks on the Parallel Axiom and its difficulties are based mainly on *Stäckel and Engel 1895*, as well as *Killing 1893* and *Veronese 1894*. All three are referred to in Hilbert's own notes; in the *Ausarbeitung* only *Stäckel and Engel 1895* is mentioned. The first part of this latter work (pp. 6–14) contains a German translation of the first 32 Propositions of Book I of Euclid's *Elements* (taken from *Euclid 1883*), with some proofs,

Heath 1926a, 224–228, and also 249. Thus, the extra 'axioms' are ones that Euclid would want to use only sparingly.

[9]Note that Hilbert's 1903 analysis of the Helmholtz-Lie approach to geometry shows how, using the group concept, one can develop an appropriate axiom system *starting* from the notion of motion. See *Hilbert 1903c*.

Proposition 32 being the one which says that the angle-sum of a triangle is two right-angles, and that the exterior angle is the same as the sum of the two interior and opposite angles. There follow various selections of work on the Parallel Axiom until just after 1830, beginning with that of John Wallis from 1663, and ending with F. A. Taurinus. There are also extensive details given in the authors' Introductions to each selection.

(*c*) On pp. 69–70 of his own notes (p. 76 of the *Ausarbeitung*), Hilbert comments on the genius of Euclid in having recognized that a new axiom was needed in order to guarantee such apparently elementary theorems as the existence of a rectangle. He remarks (p. 74 of his own notes) that Gauß was the first to have grasped why Euclid has the Parallel Axiom as a separate axiom, referring to the 'soon to be published' *Gauß 1900* as evidence of this. In this context, note that in 1816 Gauß commented in print on the fruitlessness of attempting further proofs (which all turn out to be 'Scheinbeweise') of Euclid's Axiom as a way of dealing with the foundation of the theory of parallel lines; he adds that we cannot say ('wenn wir ehrlich und offen reden wollen') that (in 1816) 'we have come any further than Euclid was 2000 years ago' (*Gauß 1816*, i.e., *Gauß 1900*, 170–171). In Schumacher's notes on conversations with Gauß, there is the following entry for 1813 (27 April):

> In der Theorie der Parallellinien sind wir jetzt noch nicht weiter als Euklid war. Diess ist die *partie honteuse* der Mathematik, die früh oder spät eine ganz andere Gestalt bekommen muss. (*Gauß 1900*, 166.)

(*d*) The bulk of Hilbert's historical remarks come on pp. 71–73 of Hilbert's own notes; the remarks are less extensive in the corresponding place in the *Ausarbeitung*, pp. 77–78. Hilbert refers to a large number of figures representing important moments in the history of the the treatment of Euclid's axiom, and many of his comments bear elucidation.

1. On p. 71 of his own notes, Hilbert mentions Christoph Schlüssel, and his twenty-two editions of the *Elements* between 1574 and 1738. Hilbert is referring to *Clavius 1574*, and its various reprints and editions, 'Clavius' being the latinized name taken by Schlüssel. Hilbert's information here was almost certainly taken from *Stäckel and Engel 1895*, 17 (the Introduction to the selection from John Wallis), which gives exactly these details. For further information on Christoph Clavius and the various editions of his Euclid, see *Steck 1981*, 77f.

2. On p. 72 of his own notes (p. 76 of the *Ausarbeitung*), Hilbert mentions Wallis's proof; the relevant work here is *Wallis 1663*, first published in *Wallis 1693*, which explains the date Hilbert gives. Wallis's proof is the first selection in *Stäckel and Engel 1895*, and it is indeed a classic example of the type that Hilbert mentions on p. 71 of his own notes, for it assumes a principle which is equivalent to the Parallel Axiom itself. Indeed, Wallis takes it that, for any figure (e.g., a triangle), there is always an arbitrarily large similar figure. This is equivalent to the Parallel Axiom (given the congruence axioms, of course). According to Wallis, the assumption appears to follow 'from the nature of the comparison of magnitudes' (see *Stäckel and Engel 1895*, 26).

3. As Hilbert remarks (p. 72 of the *Ausarbeitung*), Wallis was an incumbent of the Chair in Oxford founded by Henry Savile, that of the Savilean Professor of Geometry, whose occupant has the duty to lecture on Euclid's *Elements*. (Wallis was one of the first incumbents, as *Stäckel and Engel 1895*, 18, states; the Chair still exists.) Futhermore, Stäckel and Engel quote (pp. 18–19) from Savile's lectures on Euclid, the passage referred to by Hilbert on p. 72 of his own notes, and partially (but inaccurately) quoted by him on p. 77 of the *Ausarbeitung*. The passage is from *Savile 1621*, 140 (*Thirteen Lectures on the Principles of Euclid's Elements*), and reads (in the original):

> In pulcherrimo Geometriæ corpore duo sunt nævi, duæ labes, nec, quod sciam plures, in quibus eluendis & emaculandis, cúm veterum túm recentiorum, ut postea ostendam, vigilavit industria.

Roughly translated, this is:

> On the most beautiful body of Geometry there are two moles, two blemishes, and, so far as I know, no more. As I will later show, the industry, first of the ancients, and then of the moderns, has attempted to wash these blemishes away, to make the body spotless.

Savile's remark is made in the context of the Parallel Postulate, and the problem which this presents is one of the 'blemishes'. Stackel and Engel comment:

> Diese beiden Makel sind die Theorie von Parallellinien und die Lehre von den Proportionen. (*Stäckel and Engel 1895*, 18.)

Hilbert refers to the 'Makel' in his note *) to p. 72 of his own notes just as Stäckel and Engel describe them, and again on pp. 77–78 of the *Ausarbeitung*. The first is finally expunged by the proof of the independence of the Euclidean Parallel Axiom, for this confirms that Euclid was justified in adopting this Axiom as a separate postulate. Hilbert's work on the theory of proportion removes the second blemish, in so far as it shows that the theory of proportion can be developed without the aid of any 'numerical' assumption. (It does not deal with the question of how far the Eudoxan/Euclidean theory of proportion can be reconstructed. See Hilbert's further remark, p. 92 of his own notes.)

4. The references to Saccheri and Lambert (on pp. 72–73 of his own notes, and pp. 77–78 of the *Ausarbeitung*) are to *Saccheri 1733* and *Lambert 1786*. (The title of Saccheri's book contains a direct reference to the remark of Henry Savile recently quoted.) Lambert's results and conjectures on π can be found in *Lambert 1768* and *Lambert 1770b*. Hilbert also mentions Legendre here; the dates (1794–1833) allude to the numerous editions of *Legendre 1794* and to *Legendre 1833*. For what Hilbert calls the Legendre theorems, see Specific Note (5)(a).

5. In the remark added to p. 74 of his own notes, Hilbert refers to Leibniz apparently citing *Veronese 1894*; Leibniz is first discussed there on pp. 635–636. About Leibniz and the Parallel Axiom, Veronese merely says:

> Er glaubt z.B. das Postulat über die Parallelen zu beweisen und bringt einige nicht durchweg genügende Beweise von Proclus.

It is not clear to which places in Leibniz's writings Veronese is referring. However, in *Leibniz 1858*, 209, there is a place where Leibniz briefly discusses the Parallel Postulate, and in particular a proof given by Proclus, which he says he will deal with in another place, although where is not clear.

The pages from Veronese are part of an Appendix entitled 'Historisch-kritische Untersuchungen über die Principien der Geometrie' (pp. 631–684). This explains Hilbert's further reference to 'Sachéri [sic] etc.' and his 'siehe ... vorher u. nachher'.

6. Hilbert mentions (lecture notes, p. 73, the *Ausarbeitung*, p. 78) the work of Gauß, and of various figures who worked on the problem of the Parallel Axiom around's Gauß's time, e.g., Thibaut, Seyffer, Voit and Pfaff. Some remarks are in order.

(*i*) In the note **) on p. 73 of the lecture notes, Hilbert refers to Thibaut and cites *Killing 1893*, 7. Hilbert is actually partially quoting from Killing. At loc. cit. we find:

> Unter den Versuchen, die Parallelentheorie aus den übrigen Axiomen Euklids herzuleiten, gewährt das von Thibaut eingeschlagene Verfahren dadurch besonderes Interesse, daß sein Urheber, der mit Gauß zugleich Professor der Mathematik in Göttingen war, den Versuch zu einer Zeit (1818) veröffentlichte, wo Gauß bereits mehrfach auf das Verfehlte hingewiesen, neue Beweise für das elfte Axiom zu suchen, und erklärt hatte, wir seien nicht weiter gekommen, als Euklid vor 2000 Jahren bereits gewesen sei. Dieser Thibautsche „Beweis" fand anfangs wenig Beachtung; erst später erlangte er teilweise große Beliebtheit und ist dann in manche Lehrbücher übergegangen. (*Killing 1893*, 7.)

Thibaut was Professor in the Philosophische Fakultät in Göttingen from 1802, and titular Professor of Mathematics from 1829. The work referred to here is *Thibaut 1809*. The basis of Thibaut's attempted proof is formed by what he takes to be obvious facts about translation and rotation, which he uses to prove that the exterior angles of a triangle add up to 2π, and thus that the angle sum of the triangle itself is π. For Gauß's reaction to the particular method of proof Thibaut employs, we can look to his correspondence with Schumacher, for something like it was given by Schumacher in a letter to Gauß from 3 May, 1831 (*Gauß 1900*, 210–212 or *Gauß et al. 1860–1865*, Volume 1, 255–258). Gauß's reply to Schumacher from 17 May, 1831 (op. cit., 213, or *Gauß et al. 1860–1865*, Volume 1, 260–261) points out the *petitio principi* involved. The assumption Gauß isolates, call it G, is: Suppose a and b are two intersecting straight lines which are cut by a third line c at angles α and β respectively. Suppose we have a fourth line d which also intersects both a and b, and indeed intersects a at angle α. Then d must intersect b at angle β. As Gauß says:

> Allein dieser Satz [G] ist nicht bloss eines Beweises bedürftig, sondern man kann sagen, dass er im Grunde der zu beweisende Satz selbst ist.

(The relevant parts of the correspondence are also to be found in *Stäckel and Engel 1895*, 227–230.) G is actually equivalent to the assertion that the angle sum of any triangle is π, thus strictly weaker than the Parallel Axiom itself in the form Hilbert gives it, at least if the Archimedean Axiom is not assumed. See Specific Note (5)(*a*).

(*ii*) Hilbert also refers to Seyffer, Voit and Pfaff (p. 73 of the lecture notes; the *Ausarbeitung* refers only to Pfaff, on p. 78). These three mathematicians are mentioned in *Stäckel and Engel 1895*, 214–215 as forerunners of Gauß in the sense they had the view that it is fruitless to try to prove the Parallel Axiom from the other Euclidean axioms. However, they argued for the adoption instead of a 'simpler' postulate to replace it. In the case of Pfaff, this fact is mentioned by Hilbert on p. 78 of the *Ausarbeitung*.

(*iii*) On pp. 74–75 of the lecture notes, and p. 78 of the *Ausarbeitung*, Hilbert mentions the work of Bolyai and Lobatchevsky as being the first published work on non-Euclidean geometry. He gives no references, but he presumably means *Bolyai 1832a* and either *Lobachevsky 1837*, or *Lobachevsky 1840*. But more than anything, he praises Gauß for constructing a non-Euclidean geometry, and thereby taking the 'important step' (the term in the lecture notes, pp. 73–74) and the 'decisive step' (*Ausarbeitung*, p. 78), from 'doubt to action'. Hilbert is here paraphrasing *Stäckel and Engel 1895*, 215. They write:

> Von diesem Skeptizismus zu einem thatkräftigen Handeln überzugehen, sich von der zweitausendjährigen Autorität Euklids zu emanzipieren und eine Geometrie unabhängig vom Parallelenaxiom aufzubauen: das war auch nach den Vorarbeiten von Saccheri und Lambert immer noch ein gewaltiger Schritt. Diesen Schritt gewagt zu haben, ist das Verdienst von Carl Friedrich Gauß.

(*iv*) When Hilbert says (lecture notes, p. 74) that Gauß 'wünschte, dass diese Untersuchungen [nicht] mit ihm untergingen', he is referring to what Gauß wrote in the same letter to Schumacher of 17 May, 1831 cited above in connection with Thibaut:

> Von meinen eignen Meditationen, die zum Theil schon gegen 40 Jahre alt sind, wovon ich aber nie etwas aufgeschrieben habe, und daher manches 3 oder 4mal von neuem auszusinnen genöthigt gewesen bin, habe ich vor einigen Wochen doch einiges aufzuschreiben angefangen. Ich wünschte doch, dass es nicht mit mir unterginge. (*Gauß et al. 1860–1865*, Volume 1, 261.)

The relevant part of this letter is also to be found in *Stäckel and Engel 1895*, 230, and *Gauß 1900*.

(*v*) In the *Ausarbeitung* (p. 78), Hilbert cites the reason Gauß himself gave for not publishing his work on non-Euclidean geometry, namely that he feared the 'Geschrei der Böotier'. Gauß's remark is from a letter to Bessel, 27 January, 1829. (See *Gauß 1900*, 200. The correspondence was first published in *Gauß et al. 1880*; see p. 490. Hilbert actually refers to *Gauß 1900*.) Gauß's allusion is to the people of Boeotia (or Boiotia in German transliteration), a part of Greece to the north-west of Attica. In ancient times, the people of this region were frequently mocked by the Athenians (and the Attic comedians) for their dullwittedness and their lack of intellectual alertness. See *Paulys Realencyclopädie der classischen Altertumswissenschaften*, Volume III, 1 (Metlersche Verlagsbuchhandlung, Stuttgart, 1897), 646. Hence, a colloquial way to translate the passage is to say that Gauß feared 'the clamour of the blockheads'.

(*vi*) Hilbert refers to D'Alembert's having called the problem posed by the Parallel Axiom the 'Ärgernis und die Klippe der elementaren Geometrie' (see the *Ausarbeitung*, p. 78). This is not quite accurate to what d'Alembert actually says. First, Hilbert takes the reference from *Stäckel and Engel 1895*, 211, where we see that in fact d'Alembert's remark was directed against the whole theory of the straight line:

> 'Die Erklärung und die Eigenschaften der geraden Linie, sowie der parallelen Geraden, sind die Klippe und sozusagen das Ärgerniss der Elementargeometrie', hatte d'Alembert in einem bemerkenswerten Aufsatze über die Elemente der Geometrie 1759 ausgerufen und hatte hinzugefügt, man könne allerdings parallele Gerade als solche erklären, die auf einer dritten Geraden senkrecht stehen, dann aber sei erforderlich, zu beweisen, daß der Abstand der beiden Geraden immer gleich dem gemeinsamen Lote sei.

Even Stäckel and Engel's reference is not quite accurate. In 1759, d'Alembert published the *Mélanges de littérature d'histoire et de philosophie*, the fourth volume of which contained his lengthy *Essai sur les Élemens de Philosophie* (*d' Alembert 1759*). In 1767, he added a fifth volume containing the *Éclaircissemens sur différrens endroits des Élemens de Philosophie*. Elucidation XII (in some editions, XI) on the elements of geometry contains the following passage:

> La définition et les propriétés de la ligne droit, ainsi que des lignes parallèles, sont donc l'écueil, et, pour ainsi dire, le scandale des élemens de géometrie. Je ne crains point que les mathématiciens philosophes taxent de puérilité les réflexions que je viens de faire, puis'quelles ont pour objet, non-seulement de porter la plus grande précision dans une science dont la précision est l'âme, mais de montrer par les exemples frappans la nécessité et la rareté des bonnes définitions. (*d' Alembert 1767*, *Éclaircissemen XII.*)

From this, it is clear that d'Alembert's scorn is directed more at what he takes to be poor definitions of the straight line, included under which are the definitions of parallel lines, rather than the parallel problem explicitly. Among the definitions of parallels, he includes (two pages earlier) the one Stäckel and Engel refer to, and the necessity, once adopted, of proving that the lines are the same distance apart wherever this is measured. This alleged definition commits the fallacy, which d'Alembert mentions earlier, of attempting to include 'des idées qu'elle ne doit pas contenir, et qui doivent en être la conséquence'. The definition is also mentioned by Hilbert in the *Ausarbeitung* (p. 77), and he raises a related difficulty; the proposition stated by d'Alembert is proved by Hilbert as an easy consequence of the Parallel Axiom, Proposition 5 in the lectures (p. 80 of both the notes and the *Ausarbeitung*).

It is worth noting that, in this same *Éclaircissemen*, d'Alembert criticises the attempts to base a definition of straight line on how a line might *appear* to us, one of his grounds of criticism being that 'il serait impossible de savoir que la lumière se répand en ligne droit'.

(5) The Unsolved Problems

Remarks on p. 106 in Hilbert's lecture notes, and p. 169 of the Ausarbeitung.

Some brief remarks about each of the three problems posed here. (Cf. also pp. 23 and 26 of the 1898 *Ferienkurs*, i.e., *Hilbert 1898b**; see Chapter 3.) The order in which they are stated is different in the *Ausarbeitung* (p. 169), and they are not stated at all in the *Festschrift*.

(*a*) *The Legendre Theorems.* The first problem concerns the two theorems on the angle sum in a triangle, which Hilbert calls here the Legendre Theorems. The two theorems, provable without the Euclidean Parallel Axiom, are:

(L1) The angle sum in a triangle can never be greater than two right-angles.

(L2) If the angle sum is two right-angles in any *one* triangle, then it is the same for *all* triangles.

The two theorems are stated and proved in the *Ausarbeitung*, pp. 69–73 (Hilbert's own lecture notes refer to the *Ausarbeitung* for both statement and proof). The theorems are not dealt with at all in the original *Festschrift*, although they are stated and proved again in the 1902 lectures, 56–59 (see Chapter 6 of this Volume), with allusions to Dehn's work.[10] The proofs make use of the Archimedean Axiom; the problem is then whether this use is essential. The question is already raised on p. 71 of the *Ausarbeitung* as one about 'Methodenreinheit'.

(L1) appeared in the third edition of Legendre's *Eléments de géométrie* (*Legendre 1800*), and (L2) in *Legendre 1833*, which also contains a summary of Legendre's investigations into the Euclidean Parallel Postulate, of which (L1) and (L2) form a central part. Saccheri had established the two theorems before Legendre in his *Saccheri 1733*, the first explicitly (Proposition IX), and the second implicitly (a consequence of Propositions V, VI, VII and XV). Lambert also comes very close to the two Legendre Theorems without enunciating them exactly. This is not surprising, since *Lambert 1786* pursues in part a method close to Saccheri's. Hilbert notes in the *Ausarbeitung* (p. 78) that Sacherri and Lambert both had the Legendre Theorems before Legendre. Saccheri's memoir (translated into German) and Lambert's are to be found in *Stäckel and Engel 1895*. Lambert's was not published by him, and was found in his Nachlaß. One of Lambert's editors, Johann Bernoulli, states that it was first written in 1766, but this cannot be firmly demonstrated. (See Stäckel and Engel's comments, *Stäckel and Engel 1895*, 141–143.)

The question of the necessity of the Archimedean Axiom in the proofs of the Legendre Theorems (i.e., whether the results can be consistently maintained in a non-Archimedean geometry) was taken up by Hilbert's student

[10]The Legendre theorems are proved again in the seventh edition of the *Festschrift*; the proof of the first theorem is much the same, that of the second is different, apparently avoiding the Archimedean Axiom, as Dehn's results suggest is possible. See *Hilbert 1930a*, § 10.

Max Dehn, at Hilbert's instigation, as Hilbert himself says. For this, see p. 205 of the French translation of the *Festschrift* (here in the Appendix to Chapter 5), and also the excerpt from Hilbert's letter to Hurwitz from 5 November, 1899, quoted in *Toepell 1986*, 257. It is clear from this letter that Dehn had achieved his results by this time, and that Hilbert was very impressed by them. Writing to Hurwitz, he calls them 'so merkwürdig, wie nur möglich'. They are contained in *Dehn 1900a*, published in *Dehn 1900b*, and are summarised in detail by Hilbert himself in the Conclusion to the French and English translations of the *Festschrift*. (The relevant section is included in the Appendix to Chapter 5; see also *Hilbert 1902c*, 127–130.) Hilbert's extended summary appears in no further editions of the *Festschrift*, although from the second edition on there are short remarks on Dehn's work at the end of Chapter II. (For instance, see the relevant footnote to p. 26 of the *Festschrift* in Chapter 5.) Here it will be useful to quote Dehn's own summary. Given the possibility of non-Archimedean geometries (the explicit construction is found in the *Ausarbeitung*, pp. 141–145 and the *Festschrift*, pp. 24–26) Dehn asks:

> *Gelten in einer solchen Geometrie nothwendig die Legendre'schen Sätze?* oder in anderen Worten: *Kann man die Legendre'schen Sätze ohne irgend ein Stetigkeitspostulat beweisen, d. h. ohne vom Archimedischen Axiom Gebrauch zu machen?* Es besteht, wie wir im Folgenden zeigen wollen, in dieser Beziehung ein merkwürdiger Unterschied zwischen den beiden Theoremen: Während das *zweite* sich auf Grund der ersten, zweiten und vierten Axiomgruppe (die dritte Gruppe, das Euklidische Axiom, dürfen wir selbstverständlich nicht benutzen) *beweisen lässt*, ist dasselbe für das *erste unmöglich*. (*Dehn 1900b*, 405–406.)

One of Dehn's models is interesting with respect to the Parallel Axiom Hilbert uses. In the lectures, Hilbert states six propositions, plus that which asserts the existence of a rectangle, which follow easily from Euclid's Parallel Axiom. (See the lecture notes, pp. 75–77; in the *Ausarbeitung*, pp. 78–80 only 1–5 are mentioned.) Many of these are commonly taken to be equivalent to Euclid's Axiom, and Hilbert in his original lecture notes (p. 77) says just this. In the *Ausarbeitung* (p. 80), he is more circumspect, for he says simply:

> Schließlich bemerken wir, daß es den Anschein hat, als wenn jeder dieser fünf Sätze (eventuell in geeignet veränderter Fassung) geradezu als *Ersatz des P.A.* dienen könnte. Doch können wir auf diese interessante Frage nicht eingehen.

(See also Remark [14] in the Appendix to this Chapter.) Hilbert's circumspection was justified; Dehn's work shows that the equivalence sometimes depends on the presence of the Archimedean Axiom. One of Dehn's geometries is 'semi-Euclidean' in the sense that although Euclid's Axiom fails in it (as does the Archimedean Axiom), many characteristically Euclidean theorems hold, e.g., triangles have the angle-sum of two right-angles, similar triangles exist (the proposition which Wallis uses to prove Euclid's Parallel Axiom), all the points on a line parallel to a given line have the same perpendicular distance to that line, and even Saccheri's 'hypothesis of the right-angle' holds, which shows that rectangles can exist (see the lecture notes, p. 77). Thus, without the Archimedean Axiom, all of these propositions are strictly weaker than the Parallel Axiom, and thus taking them as replacements for this axiom is a good deal more problematic. In the 1902 lectures (Chapter 6 of this Volume), Hilbert is quite explicit about this, showing how to prove the Parallel Axiom from the angle-sum theorem *using* the Archimedean Axiom (see pp. 75–83), but also giving a simple model of semi-Euclidean geometry in which the angle-sum theorem holds, but where the Parallel Axiom (like the Archimedean Axiom) fails (see pp. 56–59). His conclusion here is quite unambiguous (p. 83 of the 1902 lectures):

> ... das Parallelenaxiom [kann] durch die Forderung, die Winkelsumme im Dreieck sei 2R, nur mit Zuhilfenahme von V. [the Archimedean Axiom] ersetzt werden.

(*b*) *Plane Proof of Pascal.* The second problem was solved in 1907 by Hjelmslev (*Hjelmslev 1907*), who showed precisely that one can prove the Pappus/Pascal Theorem using the plane part of Axioms I–III alone. Since the plane version of Desargues's Theorem follows from Pascal's Theorem (*Hessenberg 1905b*), then the result of Hjelmslev represents, as Schmidt says in a summary of Hilbert's geometrical work, the final result in the complex of questions regarding the extent of plane geometry:

> Die ebene Geometrie läßt sich ohne axiomatische Einführung eines (nichttrivialen) Schnittpunktsatzes und jede Annahme über Paralellität, Stetigkeit oder Raum algebraisieren. (*Schmidt 1933*, 410.)

(*c*) *Volume.* The third problem reappears as Problem 3 ('Die Volumengleichheit zweier Tetrahedren von gleicher Grundfläche und Höhe') on Hilbert's list of problems given in Paris in 1900 (*Hilbert 1900e*, 266–267). The classical theory of surface content (*Flächeninhalt*) is based on the method of exhaustion, that is, on the division of a surface into an ever larger number of ever smaller sections (e.g., strips), a procedure which is based on some such assumption as the Archimedean Axiom (see *Hilbert 1900e*, 266). One of Hilbert's achievements in the body of work represented in the 1898/1899 lectures, and which carries over to the *Festschrift*, was to structure the theory of proportion in such a way that the comparison of surface content can be carried out without the use of any continuity or numerical assumption, even a weak one like the Archimedean Axiom. This is part of Hilbert's achievement in removing the second of Savile's 'blemishes'. (See the lecture notes, p. 92f, the *Ausarbeitung*, pp. 111–138, and the *Festschrift*, pp. 40–49.) The question at issue here is whether something similar can be done for three-dimensional figures, in other words, for the notion of volume; in the *Ausarbeitung*, p. 169, the problem is in fact stated as one about the measurement of spatial volume.

As Hilbert remarks in his lecture notes (p. 106), the problem was already noticed by Gauß. The reference is to two letters of Gauß to Gerling in 1844 (*Gauß 1900*, 241–244). In the first of these (8 April, 1844), Gauß states that it is regrettable that the equality of the volume of two bodies which are 'symmetric' (i.e., mirror images of each other), although not congruent (presumably in the sense of superposition), is proved only by using the method of exhaustion. Gerling's letter from 15 April, 1844 gives an elementary proof which relies on no such assumption. Gauß, admitting the correctness of Gerling's proof, then specifically mentions Euclid's Proposition 5 from Book XII of the *Elements* (the volumes of tetrahedra with equal height are in the same proportion as the area of their bases: see *Heath 1926c*, 386), saying:

> Mein Bedauern muss ich nun, da jener Satz nicht mehr davon getroffen ist, auf die andern Sätze der Stereometrie beschränken, die annoch [sic] von der der Exhaustionsmethode abhängig sind wie Euklid XII, 5. Vielleicht ist auch hier noch manches zu vebessern; in diesem Augenblick habe ich nicht Zeit, dem Gegendstande weiteres Nachdenken zu widmen ... (*Gauß 1900*, 244.)

Gauß's 'conjecture' might then be interpreted as the conjecture that in these other cases, too, one can give elementary proofs. Hilbert's Problem 3 takes up this challenge. The problem as Hilbert elaborates it is to show that Gauß's conjecture is false, i.e., that it is not possible to treat the concept of volume without the use of some principle like the Archimedean Axiom. To show this, it is enough to describe two tetrahedra with the same height and equal base area such that

> ... die sich auf keiner Weise in kongruente Tetraeder zerlegen lassen, und die sich auch durch Hinzufügung kongruenter Tetraeder nicht zu solchen Polyedern ergänzen lassen, für die ihrerseits eine Zerlegung in kongruente Tetraeder möglich ist. (*Hilbert 1900e*, 267.)

This can be elucidated as follows. In comparing surfaces, the notion which Hilbert takes as the basic one in the theory of surface area is that of division into an equal number of pairwise congruent triangles, i.e., *zerlegungsgleich.* Hilbert also uses the notion of *ergänzungsgleich*, which holds between two figures when they can be supplemented by the addition of congruent triangles so that the results are *zerlegungsgleich.* (See the *Ausarbeitung* of the lectures, pp. 122–138, or the *Festschrift*, pp. 40–49.)[11] Thus, we can see from Hilbert's description of what a counterexample would be that Problem 3 turns on the question of whether the analogues of this will work for the theory of volume, i.e., if no use is made of the Archimedean Axiom; in effect, he conjectures that it will not work.

As Hilbert notes in three short paragraphs added to the second edition of the *Festschrift* (i.e., p. 47 of *Hilbert 1903a*; see the relevant note added to p. 51 of the *Festschrift* in Chapter 5), his conjecture was confirmed by work of Dehn (*Dehn 1902*). Dehn shows how to define polyhedra which have the same volume (defined in a natural way) but which are neither *zerlegungsgleich* or *ergänzungsgleich.* In short, there is no three-dimensional analogue for the reduction of the area of an arbitrary triangle to that of a suitable parallelogram. Thus, these (analogous) notions cannot act as a foundation for the notion of volume. For a rather different proof of Dehn's result, see *Kagan 1903*. For a

[11]The term used in the 1898/1899 lectures and the *Festschrift* is 'flächengleich'. This was systematically changed to 'zerlegungsgleich' in the second edition of the *Festschrift*; see the version given here in Chapter 5, note to p. 40.

very elegant treatment, especially of the comparison between the plane case and the three-dimensional case, see *Boltianskii 1978*. There is a marvellously clear presentation of the whole matter (including a proof of Gerling's result for triangular pyramids, and a clear version of Dehn's solution of Hilbert's problem) in *Hartshorne 2000*, §§ 26–27.

(6) The Final Paragraph in Hilbert's Lecture Notes, p. 109

(*a*) The remark about $AB = BA$ not being provable concerns the congruence of the segments, not their identity, and should thus be that $\overline{AB} \equiv \overline{BA}$ is not provable. See also Hilbert's Remark [8] in the Appendix to this Chapter.

In the lectures, the equivalence in question is given as a separate Axiom, namely III 4 (e.g., see Hilbert's notes, p. 40, or the *Ausarbeitung*, p. 37). It is clear that Hilbert originally wanted to *prove* this axiom on the basis of the other linear congruence axioms, presumably because of its importance in setting up a segment algebra. (See the relevant editorial commentary to p. 40 in Hilbert's lecture notes.) It is a simple consequence of the Triangle Congruence Axiom (III 10 in the lectures, IV 6 in the *Festschrift*), as Hilbert's remark implies, and as his simple proof for the equality of the base angles in an isoceles triangle indicates (e.g., *Festschrift*, p. 17). But such a proof is not 'elementary' in the right sense, since it involves assumptions about the plane; hence Hilbert's stress in his lectures on proving it from the *linear* congruence axioms. Note also what is said in this context in the *Ausarbeitung* concerning the rejection of Helmholtz's 'Monodromieaxiom', which is in effect a very general assumption about the nature of the properties possessed by bodies in virtue of which they are congruent, namely that these properties are possessed *independently* of the bodies' orientation in space. Note also that even this proof of $\overline{AB} \equiv \overline{BA}$ will not go through with the weakened Triangle Congruence Axiom proposed in Hilbert's 1902 lectures, which demands additionally that the triangles involved be in the same orientation. (See Chapter 6, p. 22–25 of the lectures, and *Hilbert 1902/03*. See also the note to Remark [10] in the Appendix to this Chapter.) The statement that $\overline{AB} \equiv \overline{BA}$ is *not* provable is therefore a recognition that the standard linear congruence axioms of the lectures are not enough. Remark [8] in the Appendix suggests that adding the Archimedean Axiom does not alter the matter; in this connection, see the passage from a letter of Schur to Hilbert of 5 January, 1900, quoted in *Toepell 1986*, 164. In the passage of the lecture notes currently under discussion, Hilbert sketches a way of defining length (and therefore segment congruence) which entails that both $\overline{AB} \equiv \overline{BA}$, and the Triangle Congruence Axiom, fail. (The 'usual way of moving angles' is presumably that used in the model of Lobachevskian geometry considered earlier in the Hilbert lecture notes, e.g., pp. 80–81.) An interesting informal model which also entails the failure of segment symmetry is suggested by Hilbert at the beginning of the 1902 lectures, in a remark added on the reverse of the title page. (See Chapter 6.)

In the first edition of the *Festschrift*, Hilbert modifies the statement of the First Congruence Axiom (IV 1) such that symmetry is immediately provable.

IV 1 says that, given a segment $\overline{AB}$, a point A' on any straightline a' and a designated side of A', there is always one and only one point B' on a' on the designated side such that $\overline{A'B'} \equiv \overline{AB}$ *or* (the modification) $\overline{BA}$. $\overline{AB} \equiv \overline{BA}$ is then an immediate consequence of IV 2. From the second to the sixth editions of the *Festschrift*, this modification is dropped, and Congruence Axiom 4 from the lectures is built into the First Congruence Axiom (III 1) explicitly. (E.g., see *Hilbert 1903a*, 7, or the relevant note to p. 10 of the first edition of the *Festschrift* in Chapter 5.) In his review of the *Festschrift* (*Poincaré 1902*, 255), Poincaré, states that Hilbert should have added $\overline{AB} \equiv \overline{BA}$ explicitly alongside the explicit statement that segments are always self-congruent. (He does not note that it is an easy consequence.) Curiously, this explicit statement of $\overline{AB} \equiv \overline{BA}$ *is* in the French translation of the first edition of the *Festschrift*, i.e., *Hilbert 1900c*, 112, which appeared long before Poincaré's review. In any case, the explicit statement is made in the second (German) edition. (The 1902 lectures have *both* the modification *and* the explicit statement; see p. 19 of the lectures, Chapter 6 in this Volume.) In the seventh edition, Hilbert more or less returns to the III 1 of the lectures.

Incidentally, the fact that the 'Dreieckscongruenzaxiom' is mentioned in this passage as IV 6, and not III 10, indicates that these remarks were added after the *Festschrift* was composed, and before the second German edition. (The congruence axioms still form group IV in the French translation.)

(*b*) Hilbert's last sentence mentions both Minkowskian geometry and the concept of the straight line as the shortest distance between two points. In *Hilbert 1895b* it is shown that the axiomatic treatment of this (given the means to define length) rests on the Euclidean theorem that in any triangle, two sides taken together are always greater than the third (*Elements*, I, 20; see *Heath 1926a*, 286). This theorem is a straightforward consequence of the usual notions of segment and angle congruence, assuming that the First Triangle Congruence Theorem (Euclid's I, 4) is present, i.e., in Hilbert's case, the Triangle Congruence Axiom. However, if one drops this axiom and adopts the (weaker) Euclidean I, 20 as an axiom in its stead, then the given characterization will still hold. If one drops the Parallel Axiom as well, then one has the geometry which Hilbert investigated in his *Hilbert 1895b*. If one allows the Parallel Axiom, then one has the geometry which is, in some sense, very close to Euclidean, and which Minkowski made the basis of his important number-theoretic investigations; see *Minkowski 1896/1910*, mentioned by Hilbert in *Hilbert 1895b*. The distance metrics in a Minkowski geometry can be arranged to be either symmetric or asymmetric. The context suggests that Hilbert is attempting to construct an asymmetric Minkowskian geometry.

(*c*) Hilbert's Problem Four on his Paris list of 1900 (see *Hilbert 1900e*) is the general problem of investigating those geometries in which it holds that the straight line is the shortest distance between two points. (He mentions just two such geometries, namely the geometry he set up in 1895, and that of Minkowski.) More generally, one can think of the problem as the investigation of all geometries in which the ordinary lines in an n-dimensional real projective space P^n are geodesics, more specifically the study of the metrics possible on such spaces. Hilbert's problem is, in effect, part of his axiomatic investigation of geometry under the rubric: what happens when one drops or replaces a central Euclidean axiom? The first systematic investigation in the present case was carried out by Hilbert's doctoral student Hamel in two works, *Hamel 1901*, and *Hamel 1903*. For recent accounts, see *Yaglom 1971*, or *Busemann 1976*, which points out the importance of the Desargues Property for the plane case, i.e., the postulate that Desargues's Theorem holds for the plane.

Michael Hallett

Grundlagen der Euklidischen Geometrie

[A]Meine Herren. Collegheft Grundlagen der Geometrie, S. 7–9 „im Einzelnen 1
einzugehen."[3]

Wir werden erkennen, dass die Geometrie eine *Naturwissenschaft* ist, aber eine solche, deren Theorie eine *vollkommene zu nennen* ist, die gleichsam ein *Muster* bildet für die *theoretische Behandlung* anderer Naturwissenschaften.

Bevor wir zum Gegenstande selbst übergehen, möchte ich noch mit wenig Worten die *Stellung* charakterisiren, welche meine Vorlesung zu den von der Schule her mitgebrachten Kenntnissen und überhaupt zu anderen Vorlesungen über Geometrie einnimmt.

Was den *Stoff* anbetrifft, so werden wir uns mit den *Sätzen der elementaren Geometrie* beschäftigen, die wir alle frühzeitig auf der Schule gelernt haben: *Parallelentheorie*, *Congruenzsätze*, *Gleichheit von Polygonen*, Sätze
über den *Kreis etc.* in der | *Ebene* und im *Raume*; also das, was man in den 2
Schulbüchern *Planimetrie* und *Stereometrie* nennt und was wir hier *Euklidische Geometrie nennen.* Diese Geometrie ist gewissermassen die Geometrie des *täglichen Lebens.* Sie liegt aller *Naturbetrachtung* und aller *Naturwissenschaft* zu Grunde. Trotzdem werden die Elemente dieser Geometrie in den Universitätsvorlesungen sehr *stiefmütterlich* behandelt – was umso weniger zu rechtfertigen ist, als ja gerade der angehende *Gymnasiallehrer* künftig einen Hauptteil seiner Thätigkeit darin findet, die Elemente der Euklidischen Geometrie selbst zu lehren. Also schon dieser praktische Grund sollte die Abhaltung einer Vorlesung über Euklidische Geometrie an den Universitäten recht-
fertigen. Wie steht es nun thatsächlich gegenwärtig mit den *geometri|schen* 3
Vorlesungen in dieser Hinsicht? Zunächst die Vorlesungen über *analytische*

[A]Es ist doch besser und einfacher, ideale Punkte, Geraden, Ebenen einzuführen (nach Pasch oder Schur, Math. Ann., Bd. 39, vgl. meinen Separatabzug, Geometrie II). Der Desargues ermöglicht die Einführung der idealen Elemente.[1]

W. M. Kutta, Gesch. der Geometrie mit konstanter Zirkelöffnung, Nova Acta, Abh. der Kais. Leop. Carol. Akad., Bd. 71, Halle 1897.[2]

⟨Added on reverse of the title page.⟩

[1]The references are to *Pasch 1882* and *Schur 1891* respectively. It is not clear to what the 'Separatabzug, Geometrie II' refers.

[2]The reference is to *Kutta 1897*.

[3]The reference is to the introductory paragraphs of Hilbert's 1894 lecture notes, pp. 7–9 (Chapter 2 in this Volume).

Geometrie, welche für den weitesten Zuhörerkreis bereichernd sind, und die daran anschliessenden auf der nämlichen *analytischen Methode* beruhenden Vorlesung über *algebraische Flächen*, über *krumme Linien* und Flächen etc. Da wird am Anfange gesagt: Jeder Punkt z.B. in der Ebene ist durch 2 Coordinaten a, b bestimmt, das sind die *senkrechten Abstände* von den beiden *festen Achsen*. Alle Punkte einer Geraden erfüllen die Gleichung $Ax + By + C = 0$ und umgekehrt. Wenn also ein Punkt a, b auf jener Geraden liegen soll, so

$$Aa + Bb + C = 0 \quad \text{etc.}$$

Mit diesen Prämissen ist dann sofort aus der *Geometrie eine Rechenkunst* geworden.[4] Es ist klar, dass man bei Benutzung des *rechten Winkels, der*
4 *Parallelen*, der *Längen und Abstände* alles Principielle | aus der elementaren Geometrie *voraussetzt*. Man schlägt da also den Weg ein, auf dem man so *rasch als möglich um jeden Preis* vorwärts zur *Einführung der Zahl* in die Geometrie gelangt. Nun ist in der That in jeder exakten Wissenschaft die Einführung der Zahl ein vornehmstes Ziel. Man kann den Fortschritt einer Naturwissenschaft oder eines Zweiges der Naturwissenschaft geradezu messen an dem Grade, in welchem die Zahl eingeführt ist. Aber, wenn die Wissenschaft nicht einem *unfruchtbaren Formalismus* anheimfallen soll, so wird sie auf einem späteren Stadium der Entwickelung sich wieder auf sich *selbst besinnen müssen* und mindestens die Grundlagen prüfen, auf denen sie zur *Einführung der Zahl* gekommen ist.[5]

5 Die Geometrie soll die reichen | Mittel der Analysis *nicht als Fesseln* tragen und die Mittel der Analysis sollen von ihr *selbst gesuchte und bewusst benutzte Quellen neuer Erkenntniss* sein.

Dementsprechend wird in unserer Vorlesung die Einführung der Zahl in die Geometrie gerade zum Schluss als *Endziel* erscheinen, welches das ganze bis dahin aufgeführte Gebäude der Geometrie *krönt*.

Nicht anders verhält es sich mit den Vorlesungen über projektive Geometrie und Geometrie der Lage. Hier wird am Anfange an die *Anschauung* appellirt, d. h. an die Kenntnisse und die *tägliche Erfahrung* des Zuhörers, ohne dass da etwa principiell auseinander gehalten wird, welche Elemente der *Erfahrung* angehören und welche Thatsachen dann logische Folgen sind.

Mitunter werden freilich an einigen Universitäten von einigen Dozenten
6 Vorlesungen über Nicht-|Euklidische Geometrie gehalten, wobei dann einige *Brosamen* für die Euklidische gewöhnlich abfallen. Natürlich ist eine solche Vorlesung auch kein hinreichender Ersatz. Wir werden gerade umgekehrt an geeigneter Stelle das wesentliche der Nicht-Euklidischen Geometrie auseinandersetzen, um damit die Unabhängigkeit des berühmten *Parallelenaxioms* zu erweisen.

---

[4] With respect to the connection between *Rechenkunst* and analytic geometry, see Hilbert's 1891 lectures on projective geometry, pp. 9–10 (this Volume, Chapter 1).

[5] The 'Einführung der Zahl' is a common theme in Hilbert's geometry texts in the 1890s. See, e.g., Hilbert's lectures from 1894, pp. 7, 28–40 passim, the present lectures, p. 103, and the *Ausarbeitung*, pp. 166–167. In connection with the reference to 'unfruitful formalism', cf. Frege's letter to Hilbert 1 October, 1895 (*Frege 1976*, 58).

⟨Deletion⟩[6]

Aber es ist nicht nur ein *praktisches* und ein *wissenschaftliches* Bedürfniss, die Elemente der Euklidischen Geometrie erneut zu untersuchen, sondern die Resultate, die wir erhalten werden, lohnen, hoffe ich, reich die aufgewandte Mühe. Wir werden auf eine Reihe scheinbar sehr *einfacher* und doch sehr *tiefliegender und schwieriger Probleme* geführt werden. Wir werden zu ganz *neuen* und wie ich | glaube *fruchtbaren* Fragestellungen Anregung erhalten 7 und werden nahe und bemerkenswerte Zusammenhänge erkennen zwischen den *Elementen der Arithmetik und der Geometrie* und damit wieder einen Grund gewahr werden für die *Einheit* der mathematischen Wissenschaft.

Da Vorkenntnisse nur in *geringem* und *leicht anzueignendem* Maasse erforderlich sind, so glaube ich, dass für jedes Semester die Vorlesung verständlich bleiben wird. Die Hauptschwierigkeit, die sich dem Verständniss bietet, möchte ich kurz hervorheben:

Es erfordert einige Mühe und Wachsamkeit, sich beharrlich Dinge, Vorstellungen, Anschauungen hinweg zu denken, mit denen man vertraut ist, und sich in einen Standpunkt des Nichtwissens zurückzuversetzen. Dieser Mühe sich zu unterziehen, wird aber leichter, wenn man den Zweck klar erkennt;[7] ich will im Einzelnen oft darauf hinweisen.

[8]↑Ja wir werden uns geeigneter Figuren sogar bedienen, aber immer nur so, dass jeder Schluss rein logisch richtig und von der Figur unabhängig ist. Die Figur ist gewissermassen nur eine Schrift, nur eine Mnemotechnik, nicht ein Beweismittel. ↑Bei unseren subtilen Schlüssen, die den strengsten Forderungen der Logik Stich halten sollen, würde eine Berufung auf die Figur fehl schlagen.↓[9]

Wie sehr man ganz grob fehlgehen kann beim blossen sich Berufen auf die Figur, zeigt folgender mathematischer Scherz:

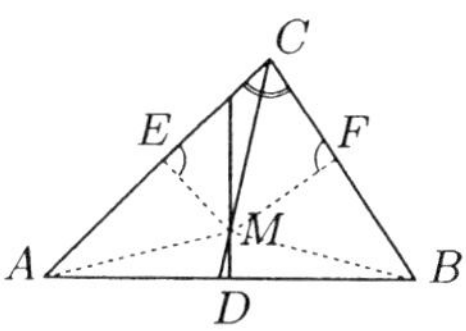

Es soll bewiesen werden, dass jedes Dreieck gleichschenklig und daher auch gleichseitig ist: *Halbire* $\sphericalangle ACB$ und errichte die *Mittensenkrechte auf* AB. Beide Geraden schneiden sich in M. Ziehe MA, MB und fälle die Senkrechten ME, MF auf AC und BC. Dann wegen $\triangle MEC \equiv MFC$ ist $ME = MF$ und $EC = FC$. Wegen $\triangle DAM \equiv DBM$ ist $AM = BM$ und

[6] Deleted: Wenn ich aber dies Semester einmal die Grundlagen der Euklidischen Geometrie Ihnen entwickeln will – und falls ich sehe, dass diese Vorlesung Anklang hat, so will ich sie in regelmässigen Zeitläufen wiederholen.

[7] This passage is very close to one in *Pasch 1882*:

> Es erfordert einige Mühe und Wachsamkeit, sich beharrlich Dinge hinwegzudenken, mit denen man vertraut ist, und auf einen Standpunkt zurückzugehen, von dem man sich weit entfernt hat. Diese Mühe ist aber bei der Prüfung der folgenden Darstellung unerlässlich, wenn aber der Zweck derselben erreicht werden soll. (p. 3)

[8] Addition (up to line 224.5), pasted in on a separate sheet.

[9] One finds a similar warning against reliance on figures in *Pasch 1882*, 43, and a similar declaration that a theorem is only to be counted as proved when the proof can be carried through without the use of a figure (op. cit., 99). Pasch also stresses (p. 43) that the role of figures (in proofs) is really a heuristic one. In the 1894 lectures, p. 11, Hilbert stresses this too, but also calls (loc. cit.) the drawing of figures a kind of 'experimenting'.

folglich $\triangle AEM \equiv BFM$, folglich $AE = BF$ und somit $AE + EC = BF + FC$ q. e. d.
Die Lösung des Paradoxons beruht auf der *unrichtigen Zeichnung der Figur*, wie sie leicht finden werden. (M liegt stets unterhalb AB ausserhalb $\triangle ABC$.)↓

8 Von Litteratur will ich Ihnen den ganzen Strom von Lehrbüchern und Schriften über *Nicht-Euklidische* Geometrie nicht nennen, da uns diese doch erst *mittelbar* angehen, sondern nur 2 Lehrbücher erwähnen, in denen Sie am ehesten das uns Interessirende finden:

Pasch, Neuere Geometrie, 1882. ↑(klar)↓
Killing, Grundlagen der Geometrie, 2 Bd. ↑(inhaltsreich)↓
Stäckel, Theorie der Parallellinien, Leipzig 1895.[10]

Einzelne Schriften, die ich benutze, werde ich im Laufe der Vorlesung nennen.

Im Ganzen 5 Gruppen von Axiomen bei unserer Betrachtung als die Hauptpfeiler beim Aufbau der Geometrie:

I Axiome der [11]Verbindung[11]
II " " [12]der Reihenfolge und Anordnung[12]
IV Axiom des Euklid über die Parallelen
III Axiome der Congruenz oder Bewegung
V Axiom des Archimedes.[13]

9 Die *Punkte*, *Geraden* und *Ebenen* nehmen wir als Elemente. Also es giebt ein System von Dingen, die wir *Punkte* nennen und mit A, B, $C, \ldots$ bezeichnen, ein *anderes* und bez. *drittes* System von Dingen, die wir *Geraden* (a, b, $c, \ldots$) und *Ebenen* (α, β, $\gamma, \ldots$) nennen. Punkt, Gerade, Ebene sind weiter nichts wie Benennungen von Dingen; wir verknüpfen keine Anschauung und keine weiteren Eigenschaften damit. ↑System geg., d.h. man kann sie von einander unterscheiden $A \neq B$.↓

I *Axiome der* [14] *Verbindung.*[14]

1. Irgend 2 von einander verschiedene Punkte A, B bestimmen stets eine Gerade a.
A, B heissen *auf a gelegen*, a heisst *durch A, B gehend* oder *Verbindungsgerade*, A, B heissen *Punkte von a*. $AB = a$.

[10]These references are to *Pasch 1882*, *Killing 1893*, *Killing 1898* and *Stäckel and Engel 1895*. Hilbert leaves several lines free immediately after the literature list.

[11–11]Substituted for: Verknüpfung

[12–12]Substituted for: Lage

[13]The axioms have been renumbered here; Hilbert crosses out the 'III' next to the Axiom of Parallels and writes 'IV' instead. Since the numbering next to '*Axiome der Kongruenz*' is correct, even though it is in fourth position, it seems that Hilbert did the renumbering immediately after writing down the Axiom of Parallels. This is then put into fourth place by means of an arrow inserting it between III and V. In the *Festschrift*, Hilbert restored the Parallel Axiom to third place, but in later editions, it again moves back to fourth.

[14–14]Substituted for: Verknüpfung

↑2. **Irgend 2 von einander verschiedene Punkte einer Geraden** a **bestimmen diese Gerade** a.↓
Zwei von einander verschiedene Geraden haben nie mehr als einen Punkt gemein.[15]

Wenn $AB = a$, $AC = a$, $B \neq C$, so auch $BC = a$.

3. **Irgend 3 nicht auf derselben Geraden gelegene Punkte** A, B, C **bestimmen eine Ebene** α.

A, B, C heissen *auf* α *gelegen* oder *Punkte von* α. α heisst die | *Verbin-* 10
dungsebene. $ABC = \alpha$.

4. **Irgend 3 Punkte** A, B, C **einer Ebene** α, **welche nicht auf derselben Geraden liegen, bestimmen diese Ebene** α.
Aus $ADE = \alpha$, $BDE = \alpha$, $CDE = \alpha$, folgt $ABC = \alpha$.[16]

Wenn ein Punkt A auf zwei Geraden a, b oder in 2 Ebenen α, β oder in einer Ebene α und auf einer Geraden a liegt, so sagen wir: die beiden Geraden, Ebenen oder Gerade und Ebene haben den *Punkt* A *gemein.*

5. Wenn 2 Punkte A, B einer Geraden a in einer Ebene α liegen, so liegt die Gerade a vollständig in der Ebene α.
6. Wenn 2 Ebenen α, β einen Punkt A gemein haben, so haben sie wenigstens noch einen weiteren Punkt B gemein.[17]

Aber es bleibt unentschieden, ob 2 Geraden in einer Ebene oder 2 Ebenen im Raume überhaupt einen Punkt gemein haben.

7. Auf jeder Geraden giebt es wenigstens 2 Punkte, in jeder Ebene wenig- 11
stens 3 nicht auf einer Geraden gelegene Punkte, und im Raume giebt es wenigstens 4 nicht in einer Ebene gelegene Punkte.

↑Da es später erheblich sein wird, Geometrie der Ebene und des Raumes zu trennen, so bemerke ich hier, dass in der ebenen Geometrie nur die beiden ersten Axiome in Frage kommen würden.↓

Collegheft Grundlagen der Geometrie, S. 10–13.[18]

(Die klassische Darlegung von Hertz über die Erfordernisse eines guten Bildes vgl. Mechanik S. 1–4.)[19]

S. 12 [ ] lasse weg.[20]

Nun müsste gezeigt werden, dass sich die 7 Axiome dieser Gruppe nicht widersprechen. Wir werden später zeigen, dass sich sogar die sämtlichen Axiome der 3 Gruppen I, II, III nicht widersprechen.[21]

---

[15] This sentence was originally meant to be Axiom 2, and was initially underlined and marked '2.' (probably in *Hi*^{s}). Hilbert then added what is now Axiom 2 at the bottom of p. 9, placing it before the old Axiom 2. by means of an arrow. The original marking '2.' and the underlining were then erased.

[16] Only correct when A, B and C are not colinear.

[17] Deleted: und folglich die durch A gehende Gerade a.

[18] I.e., *Hilbert 1894a**, 10–13.

[19] The reference is *Hertz 1894*, 1–4.

[20] The brackets were added by Hilbert in blue pencil. This is probably a reference to the material marked in square brackets on p. 12 of Hilbert's 1894 lecture notes. See Chapter 2.

[21] Hilbert delays this until he shows the consistency of the whole axiom system. See this text, p. 104, and the *Ausarbeitung*, pp. 8–9.

Ferner sind sie auch im Wesentlichen von einander unabhängig, was im Einzeln zu zeigen wäre, z. B. Axiom 1 und 2 sind unabhängig von einander. Denn sei jeder Punkt durch eine ganze rationale positive Zahl p und jede Gerade ebenfalls durch eine ganze rationale positive Zahl g definirt, so möge die
12 Zuordnung nach Axiom 1 durch die | Gleichung

Grösster Ganzer ⟨22⟩ $\left[\frac{p_1\cdot p_2}{2}\right] = g$ ⟨22⟩[23]

vermittelt werden. Dann ist Axiom 2 nicht erfüllt. Denn zum Beispiel[24]

$$\left[\frac{1\cdot 2}{2}\right] = 1, \left[\frac{1\cdot 3}{2}\right] = 1;$$

dagegen ist

$$\left[\frac{2\cdot 3}{2}\right] = 3, \quad \text{also} \neq 1.$$ [25]

Oder einfacher, wir repräsentiren den Punkt durch eine positive oder negative Zahl und die Gerade durch eine positive Zahl und vermitteln Axiom 1 durch die Gleichung

$${p_1}^2 {p_2}^2 = g,$$

so ist

$$1^2 \cdot 2^2 = 4\,, \text{ und } 1^2(-2)^2 = 4$$
$$\text{aber } 2^2 \cdot (-2)^2 = 16 \neq 4\,.$$

13 Man sieht, dass es auf jeder Geraden auch wirklich ∞ viele Punkte giebt.[26]

Axiom 5 ist von allen übrigen unabhängig. Beweis kurz so, wenn man die Möglichkeit der euklidischen Geometrie anticipirend zugiebt, oder man muss Alles analytisch umsetzen: Die Punkte seien die Punkte des Euklidischen Raumes ausgenommen ein bestimmter Punkt O. Die Geraden seien die durch O gehenden Kreise[27] und die Ebenen seien die Ebenen.

Wie kann man zeigen, dass Axiom 6 von den übrigen Axiomen unabhängig ist? Man nehme in dem gewöhnlichen Raume von einer bestimmten Geraden

---

[22–22] Substituted for: $\left[\frac{p}{p}\right] = g$

[23] That is, $[x]$ denotes the largest integer $\leqslant x$.

[24] Deleted:

$\left[\frac{4}{3}\right] \frac{=}{6} 1, \left[\frac{5}{3}\right]$

$\frac{⟪3⟫\cdot⟪4⟫}{2} = 4 \quad \frac{⟪3⟫\cdot 5}{2}$

$\left[\frac{2\cdot 6}{3}\right] = 4, \left[\frac{2\cdot 7}{3}\right] = 4$

Dagegen ist $\left[\frac{6\cdot 7}{3}\right] = 14$

⟨These formulas have been crossed out so radically that they are almost illegible.⟩

[25] This independence proof is presented more clearly (with the necessary modification separating points and lines) in the *Ausarbeitung*, p. 9. Much the same holds for the proof which follows.

[26] If Hilbert's remark here is meant to refer to the two models just given, then it is clearly wrong; if not, then the first sentence of the next section concerning order axioms makes the statement particularly odd. Note that the remark is missing from the corresponding place in the *Ausarbeitung*.

[27] We should add: 'as well as the straight lines of the Euclidean space passing through O'. See the *Ausarbeitung*, p. 10.

a nur einen ihrer Punkte, etwa P, als wirklich, während alle anderen Punkte von a unnahbar seien. Geraden seien alle gewöhnlichen Geraden ausser a. Ebenen seien alle gewöhnlichen. Dann gelten alle Axiome ausser 6, indem 2 durch a gehende Ebenen nur den einen Punkt P miteinander gemein haben.

↑Diejenigen Fragen der Geometrie, die bloss die Axiome dieser Gruppe I benutzen, kann man als die Geometrie der Configurationen bezeichnen. Also man nimmt noch hinzu, dass auch 2 Geraden stets einen Punkt bestimmen und fragt dann nach den einfachsten Configurationen C_{nm}, wo also jede Gerade n Punkte enthält und durch jeden Punkt m Geraden gehen z. B: Litteratur:↓

II Axiome der Reihenfolge und Anordnung (oder Lage) 14

Ueber die Punkte einer Geraden selber wissen wir aus I nichts ausser, dass es wenigstens 3 geben muss. Eine Geometrie der Geraden wäre bisher nicht möglich. Also jetzt Axiome über die gegenseitige Lage. Begriff „zwischen".

1. Wenn A, B, C Punkte einer Geraden sind und C zwischen A und B liegt, so liegt auch C zwischen B und A.
2. Wenn A, B 2 Punkte einer Geraden sind, so giebt es stets wenigstens einen Punkt C, der zwischen A, B liegt, und ausserdem wenigstens einen Punkt D, so dass B zwischen A und D liegt. (Natürlich auch einen Punkt E, so dass A zwischen E und B liegt.)
   ↑(Eine Strecke kann verlängert werden.)↓
3. Unter irgend 3 Punkten einer Geraden giebt es stets einen und nur einen 15
   Punkt, welcher zwischen den beiden anderen liegt.
   Also einer immer ausgezeichnet.
4. Irgend 4 Punkte A, B, C, D einer Geraden können stets so angeordnet werden, dass B zwischen A, C und zwischen A, D; ferner C zwischen A, D und zwischen B, D liegt. [28]

Wir wollen den Inbegriff aller Punkte, welche zwischen A, B liegen, eine Strecke nennen, und mit $\overline{AB}$ bezeichnen, so $\overline{AB} = \overline{BA}$, und von jedem zwischengelegenen Punkte sagen, er liege auf der Strecke. A, B heissen Endpunkte der Strecke. Ein Punkt liegt „auf" oder „ausserhalb" einer Strecke.[29]

Alle Sätze über die Lage von Punkten auf einer Geraden sind mit Axiom 1–4 zu beweisen, z. B.

[28] Deleted: Ausser dieser Anordnung und der umgekehrten $DCBA$ giebt es keine von der selben Beschaffenheit.

[29] Hilbert uses the concept 'Inbegriff' (i.e., totality, collection) to characterize a segment (as in the *Ausarbeitung*, p. 12), and if *Inbegriffe* are taken extensionally, then certainly $\overline{AB} = \overline{BA}$. However, it does mean that Hilbert's system is not elementary, even at this early stage. The *Festschrift* employs Dedekind's concept 'System' at this point. 'System' has already been used in this text (p. 9), as well as in the *Ausarbeitung*, pp. 8–9. In the 1898 *Ferienkurs*, Hilbert uses both 'System' and 'Menge' (see, e.g., pp. 2, 7).

16 Satz:

Wenn A, B, C, D 4 Punkte einer Geraden sind, so dass C zwischen A, B und D zwischen A, C liegt, so liegt D auch zwischen A, B, aber nicht zugleich auch zwischen C, B.

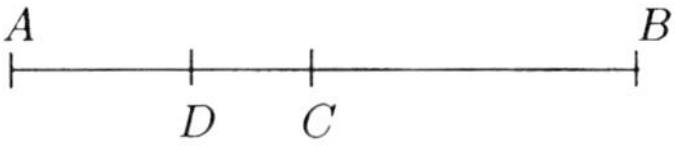

Denn in der nach Axiom 4 möglichen Anordnung muss es heissen

$$\begin{array}{ll} & A \ldots C \ldots B \\ \text{und} & A \ldots D \ldots C \end{array}$$

und daher

$$A\ D\ C\ B.$$

Satz. Auf jeder Strecke $\overline{AB}$ giebt es ∞ viele Punkte.

Beweis. Nach 2 giebt es auf $\overline{AB}$ zunächst einen Punkt C, dann auf $\overline{AC}$ einen Punkt C', welcher nach 3 auch auf $\overline{AB}$ liegen muss etc.

$\uparrow$Satz. Ist $ABCD$ eine Anordnung gemäss 4, so giebt es ausser dieser und der umgekehrten Reihenfolge $DCBA$ keine von der nämlichen Beschaffenheit. Beweis auf Grund von 3.$\downarrow$

17 Am wichtigsten sind die beiden folgenden Sätze: **Sind** A, B, C, D, $E, \ldots, K$ [30] **eine endliche Anzahl von Punkten einer Geraden**, so **lassen sich dieselben stets in eine Reihe bringen, so dass** B **zwischen** A **einerseits und** C, D, $E, \ldots$ **andererseits, ferner** C **zwischen** A, B **einerseits und** D, E, $F, \ldots$ **andererseits liegt etc.**

Beweis. Man ordne erst irgend 4 Punkte so

$$X_1 \ \ldots\ X_2\ \ldots\ X_3\ \ldots\ X_4,$$

so nimm einen 5ten hinzu und ordne 1, 2, 3, 5. Es sei

$$\begin{array}{llllll} 1 & 2 & 5 & 3, & \text{so schreibe} & \\ 1 & 2 & 5 & & 3 & 4 \end{array}$$

Hätte man 1 2 3 5, so müsste man 2 3 4 5 ordnen, etwa 2 3 5 4, so:

$$1\ 2\ 3\ 5\ 4$$

etc.[31]

Die Punkte in dieser Reihenfolge heissen kurz „*geordnet*".

18 Der zweite wichtige allgemeine Satz ist folgender:

Jeder Punkt O, welcher auf der Geraden a liegt, trennt die übrigen Punkte dieser Geraden in 2 Abtheilungen von folgender Beschaffenheit: Ist A ein Punkt der einen und B irgend ein Punkt der anderen Abtheilung, so liegt O stets zwischen A, B; sind dagegen A, C Punkte derselben Abtheilung, so liegt O nicht zwischen A, C.

---

[30] Deleted: so giebt es stets

[31] The proof is much more clearly structured as a proof by induction in the *Ausarbeitung*, pp. 14–15.

Beweis. Man bestimme A, B, so dass O zwischen A, B liegt. Ist dann X irgendein Punkt, so sind nach 4 nur folgende Fälle möglich.

$$AOXB \quad \text{oder} \quad AOBX,$$
$$XAOB \quad \text{oder} \quad XAOB,$$ [32]

und je nachdem wird X der Abtheilung mit B bez. mit A ⟨zugeordnet⟩ etc. ↑Man sage, O teilt die Gerade in zwei *Halbstrahlen*, oder man spricht auch von *den beiden Seiten* des Punktes O auf der Geraden.↓

Wir werden an die Definition des Begriffes der Irrationalzahl erinnert, wo die Abtheilung der kleineren und der grösseren. Das Grösser kleiner sein von Zahlen ist | das arithmetische Analogon zu dem geometrischen Begriffe 19
zwischen.

Dass die 4 Axiome der Gruppe II gewiss unter einander nicht in Widerspruch stehen, zeigt sich, wenn wir als Punkte der Geraden die reellen positiven oder negativen Zahlen nehmen und sagen, dass Punkt β zwischen α, γ liegt, wenn entweder

$$\alpha < \beta < \gamma \quad \text{oder} \quad \alpha > \beta > \gamma$$

ist. Dann sind alle Axiome 1–4 erfüllt. ↑Doch ist die natürliche Reihenfolge nicht die einzige, z. B. $\alpha < \beta$, wenn α, β entweder beide rational oder beide irrational sind, ist dagegen α rational und β irrational, so stets $\alpha < \beta$.↓

Man denke nicht etwa, dass die Axiome 1–4 für die Gerade sich mit der Eindimensionalität der gewöhnlichen Anschauung decken. Man kann vielmehr sehr wohl alle Punkte der gewöhnlichen Euklidischen Ebene, oder sogar des Raumes so anordnen, dass alle Axiome 1–4 für sie gelten. Man denke die Punkte durch ein rechtwinkliges Coordinatensystem festgelegt. Dann setze

$$(a,b,c) < (a',b',c')$$

wenn $a < a'$ oder, falls $a = a'$, $b < b'$ oder, falls $a = a'$, $b = b'$, $c < c'$ ausfällt. Nun liege Punkt P' | $= (a',b',c')$ zwischen $P = (a,b,c)$ und $P'' = (a'',b'',c'')$, 20
wenn entweder

$$P < P' < P''$$
$$\text{oder} \quad P > P' > P''.$$

↑Also von irgend 3 Punkten der Ebene liegt immer einer zwischen den beiden anderen, z. B.

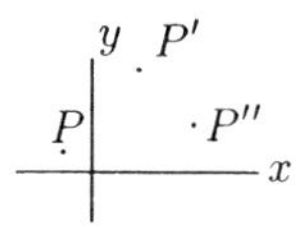

oder

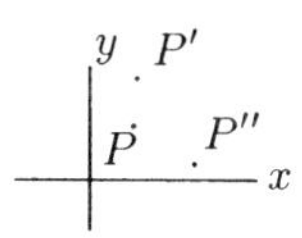

[32] One of the two '$XAOB$' should be '$AXOB$'.

oder

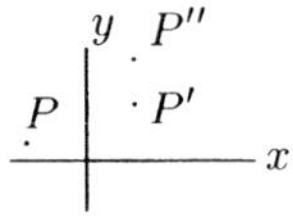

oder

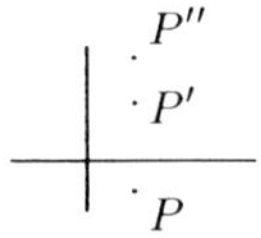

Immer liegt P' zwischen P und P''.$\downarrow$[33]
Das liegt eben daran, dass man für die complexen Zahlen $c = a + bi$ sehr wohl $<, >$ definiren kann, so dass den gewöhnlichen Gesetzen genügt wird, nämlich eben

$$a + bi < a' + b'i$$

$$\text{wenn } a < a' \text{ oder für } a = a'\,,\; b < b' \text{ ausfällt.}$$

Die Axiome 1–4 sind natürlich von den Axiomen der Gruppe I unabhängig, weil die Axiome der Gruppe I nichts über die Punkte einer Geraden unter sich aussagen.

21 Axiom 4 ist keine Folge von 1–3. Denn definire die Punkte durch Zahlen und definire C $(= \gamma)$ zwischen A $(= \alpha)$ und B $(= \beta)$, wenn

$$\gamma > \alpha \text{ und } \gamma > \beta$$

ist,[B] so sind 1–3 erfüllt, aber nicht 4. Denn seien α, β, γ, δ 4 Punkte und sei α, β, γ, δ die Reihenfolge, welche 4 vorschreibt, so müsste β zwischen α und γ liegen, also $\beta > \gamma$. Nun soll aber auch γ zwischen β und δ liegen, also $\gamma > \beta$ ausfallen, was nicht möglich ist.

$\uparrow$Definirt man unser zwischen = „ausserhalb" in gewöhnlichem Sinne, so sind alle Axiome 1–2, 4, aber nicht 3 erfüllt.[34]

Zwischen ist ja logisch zunächst eine blosse Beziehung von einem Punkt zu irgend 2 anderen; sie bekommt erst Inhalt durch die Axiome. Wenn Sie wollen, dürfen wir erst dann das Wort „zwischen" anwenden. Aber desshalb dürfen Sie nicht denken, dass unsere Untersuchungen überflüssig sind. Sie sind eben die logische Analyse unseres Anschauungsvermögens. Ueber Wesen und Bedeutung dieser Untersuchungen vgl. die sehr zutreffenden Ausführungen in

[B](d. h. wenn unser „zwischen" = „vor" in gewöhnlichem Sinne verstanden wird.) ⟨Added to the right of the preceding displayed equation.⟩

[33]This insertion replaces 'Z. B.' followed by the beginnings of a misleading diagram.

[34]What Hilbert means here is something like the following. Suppose we have a model G with a relation Z ('zwischen') satisfying Axioms 1–4. We can then define a new three-place relation A ('ausserhalb') by $A(a, b, c)$ if and only if $Z(a, b, c)$ or $Z(a, c, b)$. G together with A is then a model of axioms 1, 2 and 4, but not 3.

Pasch, Neuere Geometrie, S. 99–100.↓[35]

5. Jede Gerade a, welche in der Ebene α liegt, trennt die Punkte dieser Ebene α in 2 Gebiete von folgender Beschaffenheit: ein jeder Punkt A des einen Gebietes bestimmt mit jedem Punkt B des anderen Gebietes zusammen eine Strecke $\overline{AB}$, innerhalb welcher ein Punkt der Geraden a liegt; dagegen bestimmen irgend 2 Punkte A und B des nämlichen Gebietes eine Strecke $\overline{AB}$, welche keinen Punkt der Geraden a enthält.

Man nennt die Punkte des einen Gebietes auf einer Seite, die Punkte des 22
anderen Gebietes auf der anderen Seite der Geraden gelegen oder man nennt beide Seiten auch Halbebenen. Wenn 2 Punkte auf verschiedenen Seiten einer Geraden a liegen, so sagt man auch, die Gerade geht zwischen den beiden Punkten.

Satz: Es sei $ABC = \alpha$ eine Ebene und a eine durch keinen Punkt A, B, C gehende Gerade. Trifft dann die Gerade a das Dreieck ABC in einer Seite (Strecken AB, BC, AC) desselben, so muss sie nothwendig noch eine der anderen Seiten treffen.[C]

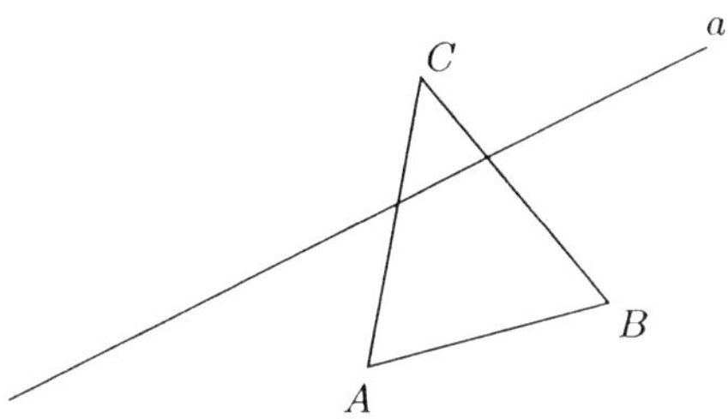

[C]Besser das Paschsche Axiom (vgl. Festschrift II 5) und gieb den schönen Beweis des Lehrsatzes 11, Pasch, S. 25.[36] ⟨Added in the upper margin.⟩

[35]The part of this addition beginning with 'dass unsere' (and ending 'Pasch, Neuere Geometrie, S. 99-100') is written on a strip of paper pasted to the bottom of p. 21. The passage contains the first appearance of the expression 'logische Analyse unseres Anschauungsvermögens', which is used again in the *Ausarbeitung*, p. 2, and then in the Introduction to the *Festschrift*. The reference is to *Pasch 1882*, 99–100. These two pages in Pasch's work contain extensive remarks on the method and purpose of proof in geometry. Pasch asserts that the main dangers to fully rigorous proof are the reliance on figures and the formulation of concepts and propositions in the 'Sprache des täglichen Lebens'.

[36]The references here are to *Pasch 1882*, 25, *Lehrsatz* 11, and the *Festschrift*, pp. 6–7. The *Satz* proved here is standardly known as Pasch's Axiom, and Hilbert's note (clearly added later) points out the possibility of taking the *Satz* as the axiom instead of Axiom 5, the procedure actually followed in the *Festschrift*, as the note also makes clear. (The two propositions are equivalent in the presence of the other order axioms.) The position of Pasch's Axiom in *Pasch 1882* is a little complicated. Pasch adopts as an axiom (*Kernsatz*) a version of 'Pasch's Axiom' restricted to the case of segments. He then proves the full Pasch Axiom as *Lehrsatz* 10, p. 25. *Lehrsatz* 11, to which Hilbert refers, states (p. 25):

> Sind A, B, C drei nicht in gerader Linie gelegenen Punkte, a ein Punkt der Geraden BC zwischen B und C, b ein Punkt der Geraden AC zwischen A und C, c ein Punkt der Geraden AB zwischen A und B, so liegen die Punkte a, b, c nicht in gerader Linie.

This follows easily from Pasch's Axiom. *Lehrsatz* 11, though, is much weaker than the Pasch Axiom; the latter asserts the *existence* of a point (given four others with certain properties), whereas 11 is not an existence assertion at all.

Beweis. a gehe zwischen A, C durch, so liegen A, C auf verschiedenen Seiten. Träfe nun a die Strecke $\overline{AB}$ nicht, so liegen A, B auf der nämlichen Seite und folglich liegen dann B, C auf verschiedenen Seiten, d. h. a trifft $\overline{BC}$.

23 5. ist von 1–4 und von I unabhängig.

Beweis: Als Punkte, Geraden, Ebenen nehme man die Punkte, Geraden, Ebenen des gewöhnlichen Euklidischen Raumes bezogen auf ein rechtwinkliges Coordinatensystem, so dass die Axiome I erfüllt sind.

Dann setze man für jede Gerade eine Richtung fest, etwa: für die die xy-Ebene durchstossenden in der Richtung der positiven z-Achse, für die in der zur xy-Ebene parallelen Geraden in der Richtung der y-Achse und für die zur x-Achse parallelen in der Richtung der positiven x-Achse. Nun sei ein rationaler Punkt ein solcher, dessen sämtliche Coordinaten x, y, z rationale Zahlen sind. Um die Axiome II 1–4 zu erfüllen, setzen wir nun fest, dass die rationalen für sich sowie die irrationalen für sich auf jeder Geraden wie gewöhnlich folgen, dass aber jeder irrationale Punkt hinter allen rationalen folgen soll. Dann ist II 5 nicht erfüllt. Denn nimmt man etwa in der xy-Ebene ein Dreieck mit rationalen Ecken A, B, C und auf $\overline{AC}$ einen rationalen Punkt D, ferner auf CB einen irrationalen Punkt E. Die Gerade $a = DE$ trifft dann BC im Punkte E, welcher nach unserer

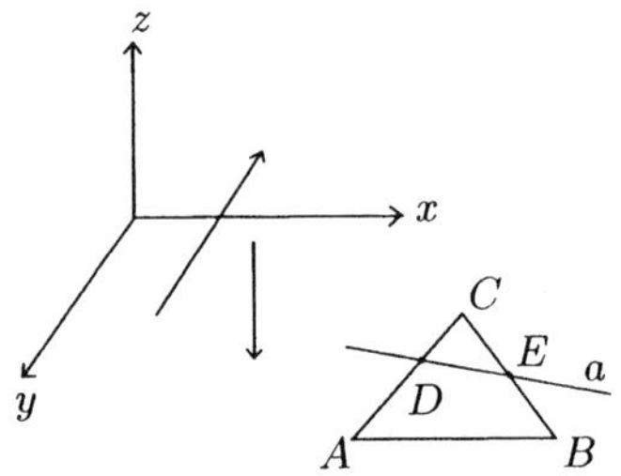

24 Festsetzung ausserhalb der Strecke BC[37] liegt | oder: A, C liegen auf verschiedenen Seiten von a, A, B und B, C aber auf der nämlichen Seite, was nicht sein sollte.[D]

[D]Will man dies bloss für die Ebene (also indem man von der Gruppe I nur die ebenen Axiome 1, 2 nimmt) zeigen, so nehme man als Punkte bez. Geraden die Punkte bez. Geraden des gewöhnlichen Euklidischen Raumes und die Punkte der Geraden in der natürlichen Folge, so gilt II 5 nicht.[38] ⟨Added in the upper margin of p. 23.⟩

[37]I.e., $\overline{BC}$. Hilbert often omits the overlining when referring to segments, and from around pp. 25–26 on, the omission is more or less standard. This will not be explicitly marked, except where the consequent ambiguity threatens to cause serious confusion.

[38]Two comments are called for here. (1) The proof of the independence of Axiom 5 from the others is not clearly set out. E is irrational, so, although on BC, it cannot lie between B and C. Now join D and E to form the line a. a passes through $\overline{AC}$ by assumption (at D), but not through $\overline{BC}$. Hence, by Pasch's Axiom, a must pass through $\overline{AB}$, say at point F. Since F is between A and B, it must be a rational point; but it is also not hard to see that F must be irrational. This shows that Pasch's Axiom is false, and therefore that Axiom 5 is also false. The proof given in the *Ausarbeitung*, pp. 20–21 is clearer, though the diagram is equally misleading. It is simply pointed out there that in this model, although DE meets $\overline{AC}$ at D, it cannot intersect either $\overline{AB}$ or $\overline{BC}$, and this violates the Pasch Axiom. (2) With respect to the note, the model Hilbert intends to specify here is the model given in the *Ausarbeitung*, pp. 19–20.

Für den Raum wird aber ein entsprechendes Axiom nicht mehr nöthig sein. Vielmehr beweisen wir leicht:

Satz: Jede Ebene α trennt die Punkte des Raumes in 2 Gebiete von folgender Beschaffenheit: Jeder Punkt A des einen Gebietes bestimmt mit jedem Punkt B des anderen Gebietes eine Strecke $\overline{AB}$, innerhalb welcher ein Punkt von α liegt. Dagegen bestimmen irgend 2 Punkte A, C des nämlichen Gebietes eine Strecke $\overline{AC}$, welche keinen Punkt von α enthält.

Beweis. Es sei A ein Punkt, der nicht auf α liegt, so rechne zu dem nämlichen Gebiet alle Punkte P des Raumes, so dass $\overline{AP}$ keinen Punkt von α enthält. Alle andern Punkte Q rechne zur anderen Seite.

1.) $\overline{PP'}$ enthält keinen Punkt von α, weil die Ebene α sonst das Dreieck APP' in einer Geraden a treffen würde.

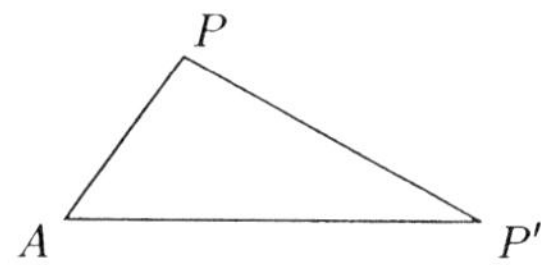

2.) $\overline{QQ'}$ enthält keinen Punkt von α, weil die Ebene | α das Dreieck AQQ' in der Geraden a schneidet und die Strecken $\overline{AQ}$ und $\overline{AQ'}$ von a geschnitten werden. 25

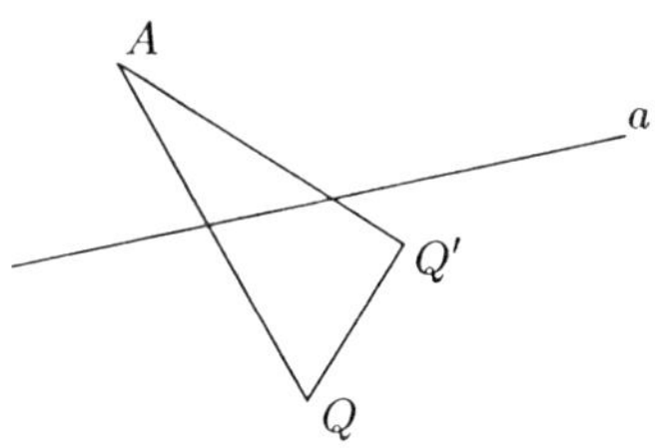

3.) $\overline{PQ}$ enthält stets einen Punkt von α, wegen des Dreieckes APQ.

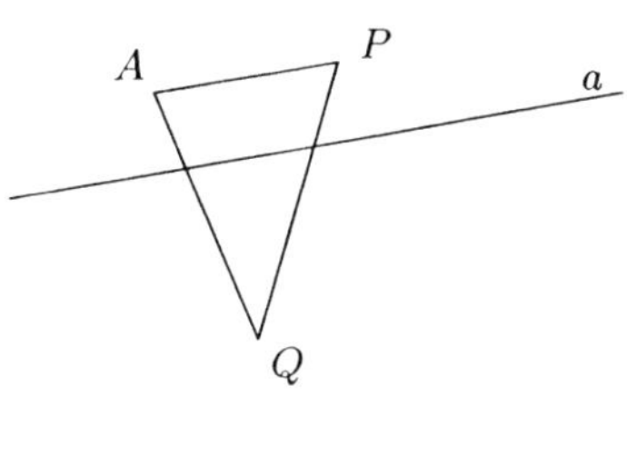

Halbraum auf der einen, *Halbraum* auf der anderen Seite. ↑Der Reichtum der Sprache ist der Darstellung einer logischen Entwickelung immer hinderlich. Vorzug der französischen Sprache.↓ Als Uebung beweise die Sätze: Mein Collegheft Die Grundlagen der Geometrie, 1894, S. 19 unten und S. 20, angeheftetes Blatt.[39]

Mit I und II können wir schon mehrere sehr wichtige Sätze beweisen, vor Allem den folgenden, welcher später für uns eine fundamentale Rolle spielen wird:

Satz von Desargues. Wenn 2 Dreiecke ABC und $A'B'C'$ in einer Ebene so lie-
26 gen, dass | je 2 entsprechende Seiten, nämlich AB und $A'B'$, AC und $A'C'$, BC
und $B'C'$, sich auf einer Geraden schneiden, so treffen sich die Verbindungslinien entsprechender Ecken, also die Geraden AA', BB', CC', überhaupt garnicht oder laufen sämmtlich durch den nämlichen Punkt.[40]

Beweis. ↑Wir nehmen an, es liege Dreieck ABC auf der einen Seite und Dreieck $A'B'C'$ auf der anderen Seite von der Geraden DEF.↓ AA' und BB' mögen sich in M schneiden. Dann lege durch M eine Gerade m, welche nicht in der Ebene der Dreiecke verläuft. Wähle auf m 2 Punkte P, Q, so dass M, A', A und M, P, Q die gleiche Reihenfolge bilden.[41] QA' und PA schneiden sich nothwendig in A'', PB und QB' in B''. P, A'', A, P, B'', B, E liegen in einer Ebene; ↑Q↓, A'', A', ↑Q, B'',↓ B', E liegen in einer
⟨27⟩ anderen Ebene. Daher liegen A'', B'', E in gerader Linie. ↑[42]Nun verbinde
B'' mit F, dann muss diese Verbindungslinie von DA'' in C'' geschnitten werden.[43] Jetzt betrachten wir die Lage von Dreieck ABC zu $A''B''C''$. Da die Voraussetzungen des Satzes gelten, so muss CC'' ebenfalls durch P gehen, wenn wir annehmen, dass der Satz bereits bewiesen ist, wenn die Dreiecke nicht in einer Ebene liegen. Ebenso liegen C'', C', Q auf einer Geraden↓. Die Punkte P, C, C'', Q, C'', C', M liegen also in einer Ebene und da C, C', M überdies auch in der Ebene der Dreiecke liegen, so liegen C, C', M in einer Geraden, was bewiesen werden sollte.[44]

---

[39] The reference is to the 1894 lecture notes, pp. 19–20. The theorems about sheaves of lines (rays) there dealt with are stated and (in part) proved in the *Ausarbeitung*, pp. 22–23.

[40] At the beginning of this statement of Desargues's Theorem, after 'dass', Hilbert has deleted the phrase 'die Verbindungsgeraden AA', BB' und CC' durch den ⟨p. 26⟩ nämlichen Punkt laufen'. This phrase indicates that Hilbert actually began to formulate Desargues's Theorem in its (equivalent) converse version, also known as Desargues's Theorem. The converse is not mentioned here, although it is mentioned in the *Ausarbeitung*, p. 24, and in the 1891 lecture notes, pp. 22-23.

[41] Hilbert is apparently assuming here that M does not lie between A and A'.

[42] Text replaces: Wir bezeichnen die Ebene $DEFA''B''$ mit α. Dann trifft PC die Ebene α in C''. P, B'', C'', B, C, F liegen in einer Ebene, ausserdem liegen B'', C'', F in der Ebene α; folglich liegen B'', C'', F in einer Geraden. Ebenso folgt, dass A'', C'', D in einer Geraden liegen. Verbinden wir nun Q mit C', so schneidet ⟨p. 27⟩ diese Gerade die Ebene α ebenfalls in ⟨C''⟩ d.h., C'' ist der Schnitt von DA'' und FB''. Wir müssen daher denselben Punkt C'' auch als Schnittpunkt von QC' mit der Ebene α erhalten

[43] Here we have to know that E is between D and F, and A'' between E and B''.

[44] The proof given here is more like a proof sketch than a proof proper, since it does not take all cases into account. The *Ausarbeitung*, pp. 24–26 presents it more clearly with more appropriate, and less confusing, figures.

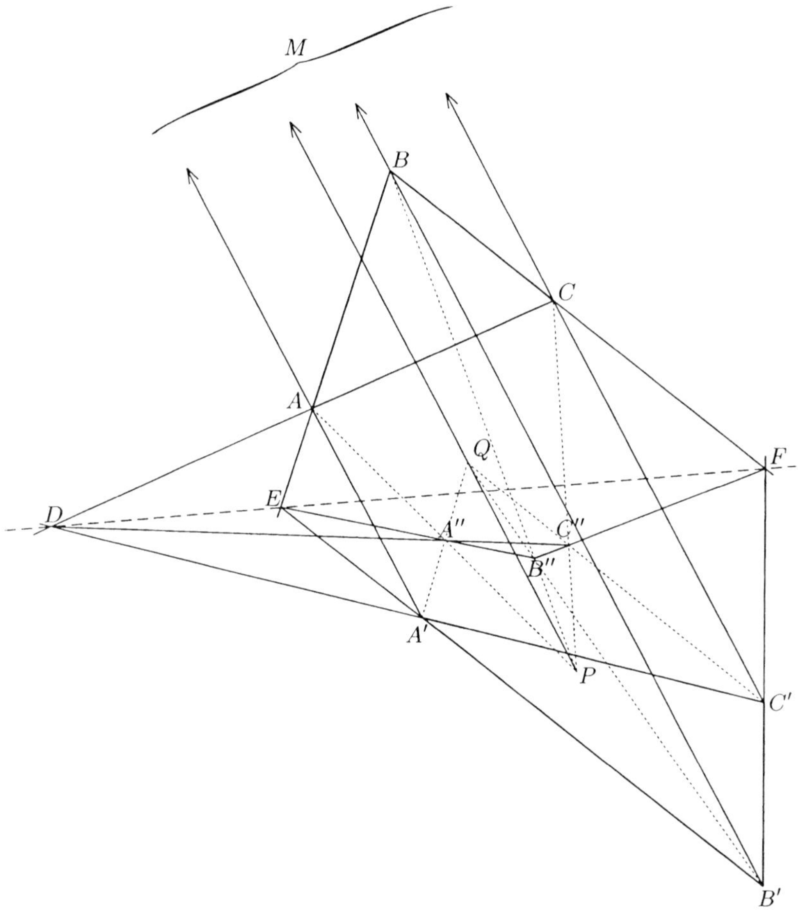
M
B
C
A
Q
F
E
D
A''
C''
B''
A'
P
C'
B'

28 Es ist also nur noch zu zeigen, dass unser Satz gilt, | wenn die beiden Dreiecke nicht in der nämlichen Ebene liegen:

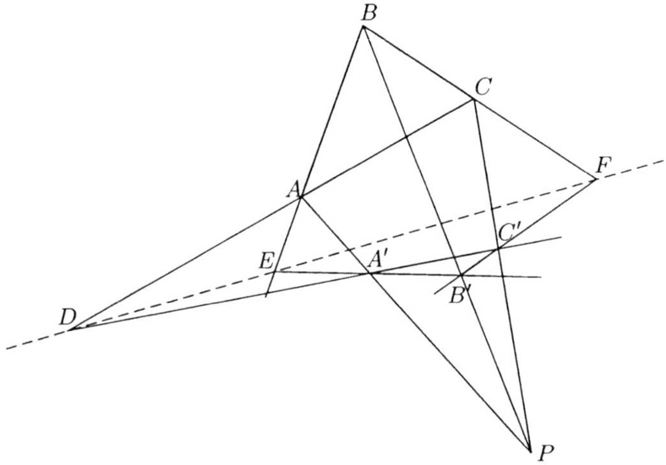

AA' und BB' mögen sich in P schneiden.

B, A, E	B', A', E	liegen in einer Ebene γ
B, C, F	B', C', F	" " " α
C, A, D	C', A', D	" " " β

α, β haben die Punkte C, C', folglich auch ihre Verbindungslinie gemein; andererseits haben sie den Punkt P gemein. Folglich P auf CC', d. h. P, C, C' liegen auf einer Geraden, q. e. d.

29 Ich sagte, dass der Desargues'sche Satz dem *Inhalte* nach wichtig ist. Für jetzt ist es aber noch mehr sein *Beweis*, weil wir eine wichtige *Ueberlegung* oder vielmehr *Fragestellung* daran knüpfen wollen. Der Satz ist ein ebener; doch der Beweis operirt im Raume. Es entsteht die Frage: giebt es etwa einen Beweis, der bloss die *linearen und ebenen Axiome*, also I 1–2, II 1–5, benutzt. Also sind hier zum ersten Mal die *Hülfsmittel der Beweisführung* einer *Kritik* unterworfen. Es ist modern, überall die *Reinheit* der Methode zu garantieren. In der That ist dies auch in Ordnung: vielfach befriedigt es unseren Verstand nicht, wenn zum Beweise eines arithmetischen Satzes *Geometrie* oder einer *geometrischen Wahrheit Funktionentheorie* herangezogen wird. Häufig freilich hat das Heranziehen verschiedengearteter Hülfsmittel einen *tieferen, berechtigten* Grund und schöne und *fruchtbare Beziehungen* werden aufgedeckt, z. B. Primzahlproblem und $\zeta(x)$ Funktion, Potentialtheorie und analytische Funktionen etc. Jedenfalls soll man aber an einem solchen *Vorkommnis* des Zusammengreifens verschiedener Gebiete nie achtlos vorübergehen.

30 Wir wollen vielmehr zeigen, dass der Desargues'sche Satz vermöge I 1–2, II 1–5 *unbeweisbar* ist. Man wird sich also die *Mühe sparen*, nach einem Beweise in der Ebene zu suchen. Für uns das erste einfachste Beispiel für den Beweis der *Unbeweisbarkeit.* Also: Zu unserer Befriedigung ist es nöthig, entweder einen Beweis, der nur in der Ebene operirt, *zu finden* oder *zu zeigen, dass es einen solchen nicht giebt.* Beweis so, dass wir ein System von Dingen =

Punkten und Dingen = Geraden angeben, für welche Axiome I 1–2, II 1–5 gelten aber nicht der Desarguessche Satz, d. h. dass eine ebene Geometrie mit den Axiomen I, II möglich ist ohne Desarguesscher Satz.

Beweis. Punkte = *Punkte* einer gewöhnlichen *Euklidischen* Ebene ausser einer Geraden von O bis ins ∞. Als *Geraden* nehmen wir folgende Curven. Jede Gerade der unteren Halbebene, wenn sie auf $O\,\infty$ *stösst*, soll sie dort ihr *Ende* haben, wenn sie auf den anderen Teil O bis $-\infty$ stösst, soll sie sich dort in einem *Kreise* fortsetzen, der die *gleiche Richtung*

31

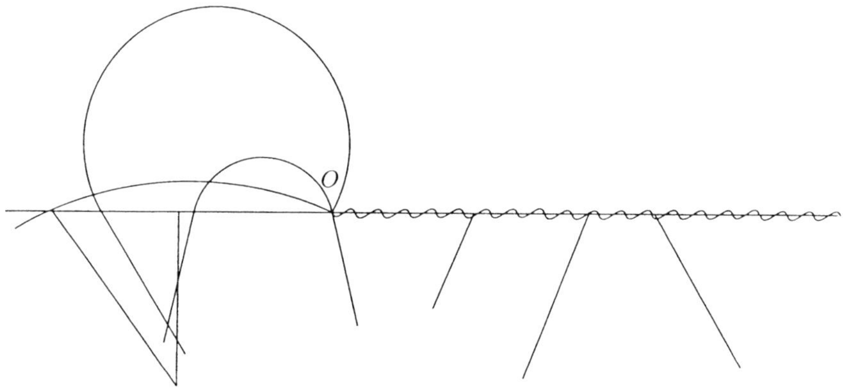

hat, wie die Gerade und durch O geht und in O ihr *Ende* hat.[45] Die *Punkte* auf der Geraden und in der Ebene in Bezug auf jede Gerade sollen gewöhnlich, d. h. natürlich geordnet sein. In der That gelten alle Axiome I 1–2, II 1–5. Denn wenn sich die *Kreishälften* in der oberen Halbebene *schneiden*, so divergiren die Geradenhälften unten *auseinander*. Denn Winkel $\alpha > \beta$ und $\alpha = \alpha$, $\beta = \beta$ nach dem Satz vom Winkel zwischen Sehne und Tangente.[46]

[45] The specification of the model is clearer in the *Ausarbeitung*, pp. 28–29.

[46] In considering Axiom I,1, that any two points determine a line, Hilbert fails to treat the case where one point has positive y-coordinate, and one point negative; in other words, he fails to show that in this case, there is always a circle passing through 0 and the upper point and which cuts the line $-\infty$, 0 in a point below 0 such that the tangent to the circle at that point is a straight-line which passes through the given point in the lower half-plane. This, however, is correct. Let $A = (a_1, a_2)$ and $B = (b_1, b_2)$ be the two given points in the upper and lower half-planes respectively, thus $a_2 > 0$, and $b_2 < 0$. Assume $M = (m_1, m_2)$ is the centre of the circle we are seeking. Simple calculation tells us that the existence of (m_1, m_2) amounts to the solubility of the following quadratic equation:

$$4a_2m_1^2 - 2(a_1b_2 + a_2b_1)m_1 + b_2(a_1^2 + a_2^2) = 0$$

This is indeed soluble, since $(a_1b_2+a_2b_1)^2 - 4a_2b_2(a_1^2+a_2^2) > 0$, given that $a_2 > 0$, and $b_2 < 0$. It remains to show the uniqueness of the solution. Put $d = (a_1b_2+a_2b_1)^2 - 4a_2b_2(a_1^2+a_2^2)$, and $c = (a_1b_2 + a_2b_1)$. Then $m_1 = \frac{c\pm\sqrt{d}}{4a_2}$. But since clearly $\sqrt{d} > |c|$, and $a_2 > 0$, we get two distinct values for m_1, one < 0, and one > 0. But it is obvious that no circle with $m_1 > 0$ would yield a connecting 'straight line' between the upper and lower half-planes, since it would actually cross the x-axis only at 0 and a point > 0, and thus would never 'pass through' the x-axis. Hence, the straight line connecting A and B is uniquely determined by taking the solution with $m_1 < 0$.

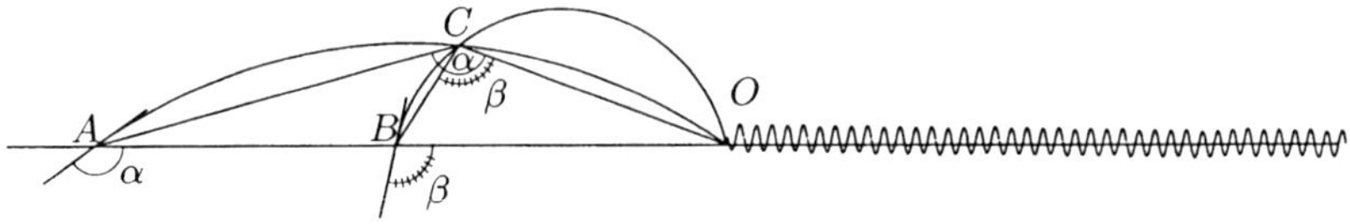

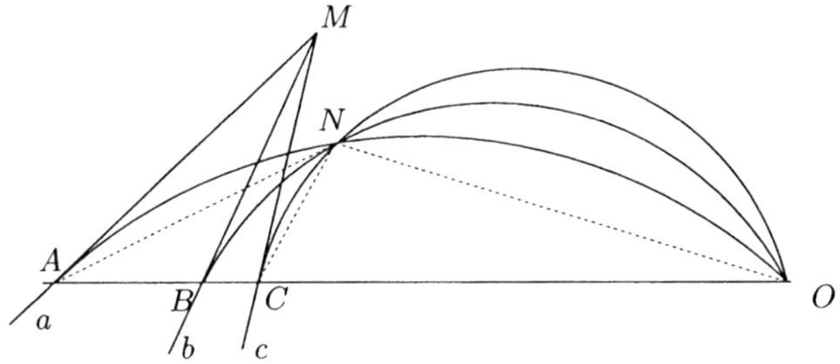

32 Es seien a, b, c 3 Geraden, die sich in der oberen Halbebene sowohl als gewöhnliche Euklidische Geraden in M, als auch in unserer Geometrie (als Kreise fortgesetzt) in N schneiden. Dann müssten

$$\begin{array}{rl} & \sphericalangle MAO = \pi - \sphericalangle ANO \\ & \underline{\sphericalangle MBO = \pi - \sphericalangle BNO} \\ \uparrow\text{folglich bei Substitution}\downarrow & \sphericalangle AMB = \sphericalangle ANB \end{array}$$ [47]

d. h., die 4 Punkte A, M, N, B *müssten auf einem* Kreise liegen. Ebenso würde folgen, dass A, M, N, C auf einem Kreise liegen müssten, was beides nicht *vereinbar* ist, d. h. 3 Geraden a, b, c, die sich als gewöhnliche Geraden verlängert in einem Punkte M schneiden, schneiden sich in *unserer* Geome-
33 trie niemals in einem Punkte, | sondern[48] stets in 3 Punkten.[49] Denn wenn 2 *divergirende Geraden* auf O, $-\infty$ auftreten, so schneiden sich die beiden Kreise, weil dann

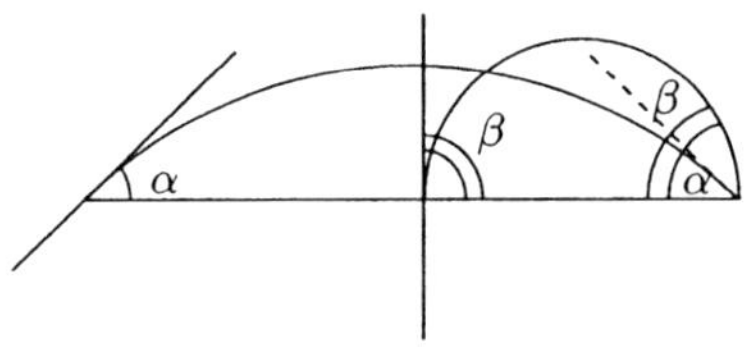

$\sphericalangle\beta > \sphericalangle\alpha$ ist.[50]

[47] Because $\sphericalangle AMB \equiv \sphericalangle MBO - \sphericalangle MAO \equiv \pi - \sphericalangle BNO - (\pi - \sphericalangle ANO) \equiv \sphericalangle ANO - \sphericalangle BNO \equiv \sphericalangle ANB$.

[48] There is a figure in the text just above this which has been crossed out and is illegible. In crude outline, it appears similar to that on p. 32.

[49] That is, one point for each pair of lines.

[50] This only covers the case where the two lines make angles α and β with the x-axis, with $\alpha < \beta$ and α further to the left. In the case where α is to the left of β and *larger*, then it has to be shown that, while they might intersect at 0 and below the x-axis, they do not intersect above it.

Nun construire wie folgt:

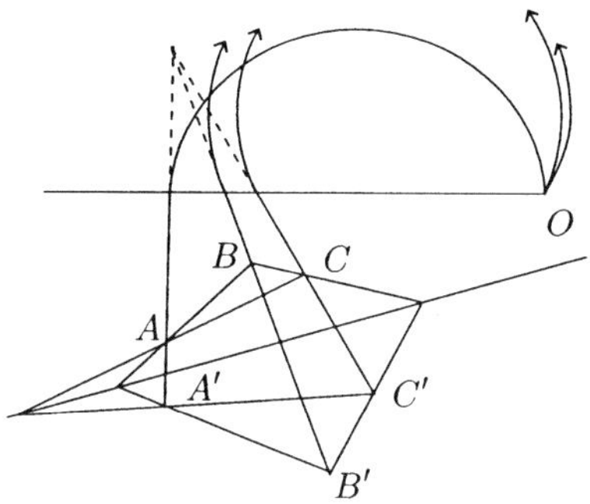

Die 3 punktirten Geraden schneiden sich, und daher können sich die Geraden unserer Geometrie nicht ⟨in einem Punkt⟩ schneiden und zwar in 3 verschiedenen Punkten q. e. d. Uebrigens könnte man auch O dahin legen, dann würde CC' überhaupt keine der beiden Geraden AA', BB' treffen.[51]

[52]↑Beispiel für die Gültigkeit des Desarguesschen Satzes: Punkte = Punkte eines Euklidischen Halbraumes exc. der Grenzebene, Gerade = orthogonale Halbkreise, Ebene = orthogonale Halbkugeln. Also in jeder orthogonalen Ebene gilt der Desarguesche Satz.[53]

---

[51]With respect to Hilbert's 'man könnte auch O dahin legen', Hilbert has an arrow leading from the word 'dahin' to a point in the diagram just above, a point which is on the horizontal axis to the left of O, and between the points of intersection with $\overline{BB'}$ and $\overline{CC'}$. Thus, although $\overline{AA'}$ and $\overline{BB'}$ would both continue into the upper half-plane, $\overline{CC'}$ would not, and would thus not intersect them.

The construction of this plane model in which Desargues's Theorem fails answers the 'purity of method' question raised by Hilbert on p. 29. In exhibiting this model, Hilbert constructs for the first time planar non-Desarguesian geometries. In the *Ausarbeitung*, pp. 138–139, Hilbert gives a model (a simple modification of the model just given here) which shows that adding the Parallel Axiom to I 1–2, II does not enable the proof of Desargues's Theorem to go through. The plane model showing the failure of Desargues's Theorem given in the *Festschrift*, pp. 51–55 is different; this also has to be a model in which the standard triangle congruence results fail. The model there is the first example of a plane affine non-Desarguesian geometry.

[52]The following addition (up to line 240.11) is actually written on p. 34 of the manuscript. It is marked at this place on p. 33 with an insertion sign, before which is written 'Vgl. folgende Seite'.

[53]The model is illustrated in the figure which follows, an example of the so-called *Poincaré model* for hyperbolic geometry.

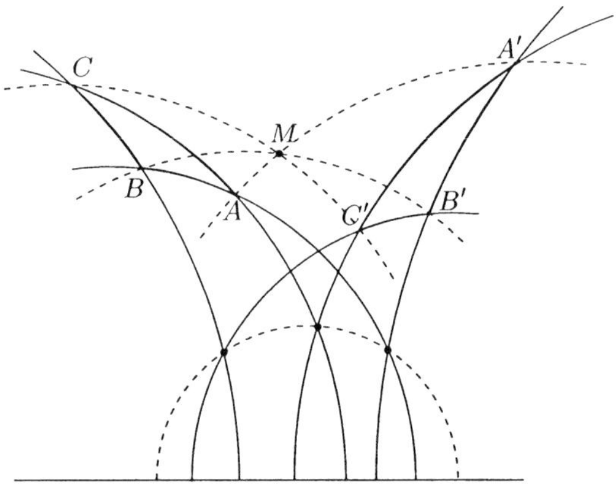

Der Desarguessche Satz ist jedenfalls eine *nothwendige* Bedingung dafür, dass zum System von Punkten und Geraden, die Axiome I 1–2, II 1–5 erfüllen, noch ein solches System von Ebenen zugefügt werden kann, damit auch I 3–7 gilt, d. h. Desargues ist *nothwendiges* Kriterium dafür, dass die Ebene als Ebene im Raume aufgefasst werden kann. Ist der Desarguessche Satz auch ein *hinreichendes* Kriterium dafür? D. h. lässt sich dann stets ein System von Dingen (Ebenen) hinzufügen, dass alle Axiome I, II erfüllt sind und das vorgelegte System als Teilsystem des ganzen Systems aufgefasst werden kann? Dann würde der Desarguessche Satz das alleinige sein, was die Ebene für sich vom Raume profilirt und wir könnten sagen, dass nun alles mit Desargues in der Ebene beweisbar ist, was überhaupt im Raume beweisbar ist.[54] ↓

34 Möglichkeit der Construktion des 4ten *harmonischen Punktes* D zu A, B, C zwischen B und C und Beweis, dass D von der Construktion unabhängig ist. Collegheft Grundlagen der Geometrie, Sommer 1894, S. 23–28. Damit wird wieder der Raum benutzt und man beweist wie früher, dass die Benutzung der *räumlichen Axiome nothwendig ist.*[55]

35 Irgend *ein System von Geraden, die durch den nämlichen* Punkt laufen, können geordnet werden, indem man mit einer Geraden schneidet und die Anordnung der Schnittpunkte auf die der Geraden überträgt. Aber man kann mit

---

[54]This result is one of those entirely new to Hilbert, although he does not prove it here, and it seems highly likely that he did not have the proof when he wrote this section of the lecture notes. The proof is given first in the *Ausarbeitung*; see the Introduction to this Chapter and Specific Note (1)(a).

[55]The text referred to is that of the lecture notes for 1894, pp. 23–28. This construction is carried out in the *Ausarbeitung*, pp. 32–33; there, Hilbert makes it clear that it is the proof that the construction is independent of the particular choice of the point D which demands the use of spatial axioms, thus all of I and II. (For a proof, see the 1894 notes, pp. 24–26.) One way to do this is to use Desargues's Theorem (see e.g., *Coxeter 1993*, 19–20), though something weaker than Desargues's Axiom will do, namely what is sometimes called the *Lesser Desarguesian Axiom* (*das kleine Desargues*). This adds a further condition to the statement of Desargues's Theorem: If two triangles in a plane are such that the lines joining the corresponding vertices go through a common point, and that point is colinear with two of the intersection points of corresponding sides, then the third such point is also colinear with these.

jeder Geraden beginnen. Von irgend 3 Geraden ist keine ausgezeichnet, von 4 Geraden kann man sagen, dass 2 durch die beiden anderen *getrennt* werden etc. Entsprechend für Ebenen, die durch die nämlichen Geraden laufen.

Die Axiome I und II sind die wesentlichen Axiome, mit denen die *projektive Geometrie* operirt. Fügt man noch ein geeignet auszusprechendes Stetigkeitsaxiom hinzu (Math. Ann., Bd 46, Hilbert: Ueber die Gerade als die kürzeste Verbindung zwischen 2 Punkten), so kann man die projektive Geometrie folgerecht aufbauen.[56]

III Axiome der Congruenz.[E] 36

Die wichtigste und schwierigste Gruppe.
Zunächst die *linearen Axiome*, wo wir bereits einige recht merkwürdige Dinge kennenlernen werden.
1.) Wenn 2 Punkte A, B auf einer Geraden a und ferner zwei Punkte A', S auf derselben oder einer anderen Geraden a' gegeben sind, so kann man stets auf a' einen und nur einen Punkt B' finden, so dass A' nicht zwischen SB' liegt und dass

$$AB \equiv A'B' \quad \langle 57 \rangle$$

wird. Jede Strecke ist sich selbst $\equiv$.

Also Möglichkeit des *Abtragens einer Strecke* auf einem durch S gegebenen Halbstrahl. Möglichkeit des *Verschiebens* aller Punkte auf allen Geraden um eine gegebene Strecke.
2.) Wenn $AB \equiv A'B'$ und $AB \equiv A''B''$, so ist auch $A'B' \equiv A''B''$.
Folgerung. Da $AB \equiv AB$, so folgt aus $AB \equiv A'B'$ nothwendig $A'B' \equiv AB$, ↑was durchaus nicht denknothwendig ist, wenn man z. B. annimmt, dass der Maasstab (Schnur) bei jedem Abtragen sich um die Längeneinheit vergrössert!![58] Nun kann man sagen „2 Strecken sind einander congruent"↓ und jetzt können

[E] Diese Axiome der Congruenz erscheinen etwas *trocken* und *umständlich*. Auch mir erschienen sie es früher und manchmal ging mir es wie ein *Mühlrad* im Kopfe herum, wenn ich *andere Mathematiker* davon sprechen hörte. In der That was man liest ist meist sehr confus. Ich bitte Sie nur noch diese Sätze *auszuhalten*. Wir werden dann schon einige *Früchte* ernten. Es ist nöthig, wenn man die Sätze der Geometrie nicht *bloss technisch handhaben*, sondern *ihren Geist verstehen* will, wenn man über den *2000 Jahre alten* Euklid auch nur ein Urteil haben will. ⟨Addition in the upper margin.⟩

[56] This refers to *Hilbert 1895b*. The continuity axiom stated there already appears in the 1894 lecture notes, and is the third version of the Archimedean Axiom stated in the *Ausarbeitung*, pp. 140–141, in effect a Dedekind–Bolzano principle asserting the existence of unique limits, i.e., stronger than the standard Archimedean Axiom. See the corresponding notes to the *Ausarbeitung*.

[57] Deleted: $A'B' \equiv AB$

[58] Hilbert makes a similar remark in the 1894 lecture notes, p. 92 which concerns moving a metal rod into zones with different temperatures.

37 wir 2.) auch | so aussprechen: *Wenn 2 Strecken einer dritten* $\equiv$, so sind *sie unter einander* $\equiv$.

Man kann nicht sagen, dass solch' ein Axiom ein *allgemeines Grössenaxiom* ist und daher *für alles* gilt. Nein, wenn ein solches Axiom auf Geometrie angewandt wird, so ist es eben ein *geometrisches* Axiom, ein *wesentliches Axiom unserer Anschauung.* ↑Congruent ist ja zunächst nur eine Beziehung. Setze ich z. B. fest: 2 Zahlen a, b heissen $\equiv$, wenn $|a-b| \leqq 1$, so $2 \equiv 3$, $3 \equiv 4$, aber $3 \not\equiv 4$.[59] ↑Wenn also, wie in der Physik 2 Grössen = heissen, wenn sie unterhalb eines Beobachtungsfehlers liegen, so kann man, bei dieser Definition der Gleichheit, nicht einmal obigen Satz anwenden.↓↓Wohl kann man sagen, es liegt im Begriff der Congruenz: man darf das Wort congruent (bez. gleich) nur anwenden, wenn solche Axiome gelten, schön, dann liegt eben mein Axiom darin ausgesprochen, dass die Anwendung des Wortes congruent hier erlaubt ist.[60]

38 ⟨Deletion⟩[61]

3. Es seien A, B, C Punkte einer Geraden, so dass B zwischen A, C liegt und A', B', C' Punkte derselben oder einer anderen Geraden, so dass B' zwischen A', C' liegt: wenn dann

$$AB \equiv A'B', \quad BC \equiv B'C'$$

so ist stets auch

$$AC \equiv A'C'.$$

3 ist von 1 und 2 unabhängig. Um dies zu sehen bestimme man, dass das Abtragen einer Länge l von a auf a' so geschehe, dass die Länge $e^l - 1$ aufgetragen
39 werde, | aber umgekehrt das Abtragen von L auf a'[62] als Länge $\log(L+1)$.[63] Dann ist 3 nicht erfüllt, weil

$$e^{l_1} - 1 + e^{l_2} - 1 \neq e^{l_1+l_2} - 1.$$

Nun wollen wir die Bedeutung von Axiom 1–3 uns klar machen, indem wir Folgerungen ziehen.

Satz. ↑Es seien A, B, C Punkte einer Geraden, so dass B zwischen A und C liegt; es seien ferner A', B', C' Punkte einer anderen oder derselben Geraden, so dass B', C' auf der nämlichen Seite von A' liegen: wenn dann

$$AB \equiv A'B', \quad AC \equiv A'C'$$

---

[59] '$3 \not\equiv 4$', should read '$2 \not\equiv 4$'.

[60] Here Hilbert formulates a proposition, which he subsequently adds (by means of insertion signs) to p. 39. See lines 242.29–243.1.

[61] Deleted: 4. Es seien A, B, C Punkte einer Geraden, so dass B zwischen A, C liegt; es seien ferner A', B', C' Punkte einer anderen oder derselben Geraden, so dass B' zwischen A', C' liegt: wenn dann

$$AB \equiv A'B', \quad AC \equiv A'C'$$

so ist stets auch

$$BC \equiv B'C'.$$

Bevor wir diese Sätze einer Kritik unterwerfen, sie auf ihre Unabhängigkeit prüfen, wollen wir ihre Bedeutung uns klarmachen und Folgerungen ziehen.

[62] This should read 'von a' auf a'.

[63] In other words, given a segment S of length L on a', then it is stipulated that any segment of length $log(L+1)$ on a will be congruent to S. See the *Ausarbeitung*, p. 35.

so liegt B' zwischen A', C', und es ist stets $BC \equiv B'C'$.↓[64]

Beweis. Wäre B' nicht zwischen A', C', sondern läge C' zwischen A', B', so suche

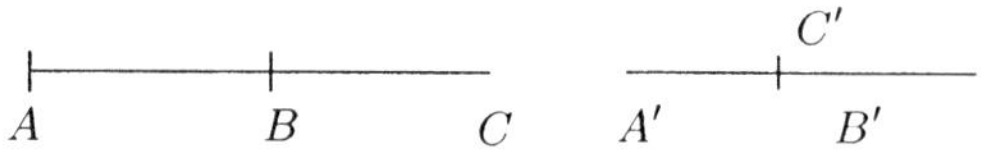

C'', so dass $BC \equiv B'C''$ ist und B' zwischen $A'C''$, dann ist nach 3 auch $AC \equiv A'C''$ und nach Voraussetzung $AC \equiv A'C'$, so $C' = C''$, q. e. d.

Satz. Wenn $AB \equiv A'B'$, $AC \equiv A'C'$, $BC \equiv B'C'$ und B zwischen A, C liegt, so liegt auch B' zwischen A', C'.

Beweis. Denn liege etwa C' zwischen A', B', 40

so liegen B', C' auf derselben Seite von A' und daher sind die Voraussetzungen des vorigen Satzes erfüllt.

Nun Definition und Satz 2 auf S. 63 meines Collegheftes Grundlagen der Geometrie, 1894, S. 63.

[65]Axiom 4.[65] Es ist stets $AB = BA$.[66]

Sehr wichtig weil es zeigt, dass Endpunkte nicht voreinander ausgezeichnet sind und es daher erlaubt ist, sie mit einem Buchstaben a, b, c, ... zu bezeichnen.[67]

Ebensowichtig ist die Folgerung

Satz. Es seien A, B, C, D 4 Punkte einer Geraden, so dass B, C zwischen A und D liegen: wenn dann | $AB \equiv CD$ ist, so ist auch $AC \equiv DB$. 41

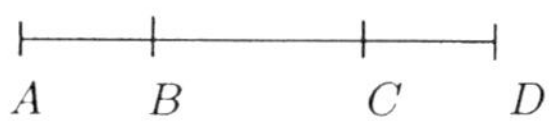

[64] The passage added here (the whole sentence beginning with 'Es seien ... ' is from p. 37. There (in Hi^b) it is circled, marked with an insertion sign and the words 'verg. die folgen Seite'; here (after 'Satz') there is (also in Hi^b) an insertion sign followed by 'Vgl. auf vorvoriger Seite'.

[65-65] Substituted for: Satz

[66] The following line is deleted: Beweis. Collegheft Grundlagen d. Geom., 1894, S. 66.

[67] The '*Collegheft*' referred to here twice is the notebook for the 1894 lectures. The first reference is to the *Definition* and *Satz 2* on p. 62 of the text, and the second (deleted) is to p. 66.

Axiom 4 was originally intended to be a theorem, and the deleted reference to p. 66 of the 1894 lecture notes is, of course, to its proof. (The restructuring here might indicate that the rather confused revision of the axioms carried out in the 1894 notes, pp. 60f. stems from much later than 1894.) In the *Ausarbeitung* Hilbert seems again to have changed his mind and wants to prove $\overline{AB} \equiv \overline{BA}$ on the basis of the *linear* congruence axioms. However, he breaks the proof off, as he says, 'in order not to be held up too much', and adds the proposition there, too, as Axiom 4. (See the *Ausarbeitung*, pp. 35–36. See also Hilbert's Note E on p. 66 of the 1894 notes, set immediately after the enunciation of the proposition.) That '$\equiv$' is an equivalence relation follows from Axioms 1 and 2; as Hilbert's remark here implies, Axiom 4, or the corresponding theorem, is required to guarantee in addition that we can always assign a unique absolute 'length' to a segment $\overline{AB}$ independently of the order in which we take the end-points. See the Introduction to this Chapter, Specific Note (6)(a).

Beweis: $AB \equiv DC$

$\underline{BC \equiv CB}$

nach [68]Axiom 3[68] $AC \equiv DB.$

Definition. Wenn $AB = a$, $BC = b$ und B zwischen A, C liegt, so heisse AC die Summe der Strecken a, b und werde mit $a + b$ bezeichnet.

Setzen wir $AB = a$, $BD = b$, so haben wir mit Rücksicht auf den vorigen Satz

$$AD = a + b \equiv AC + CD \equiv b + a$$

d. h.

$$a + b \equiv b + a \qquad \text{oder} \qquad a + b = b + a$$

↑denn wir nennen congruente Strecken auch gleich.↓ Also das commutative Gesetz der Addition gilt für unsere Summendefinition.

Definition. Liegt C zwischen A, B, so nennen wir AC die Differenz und schreiben $a - b$. Wir können die Strecke 0 und negative Strecken für die Rechnung einführen, in der üblichen Weise, ohne auf Widersprüche zu gelangen. Die Strecke Null wird nur von einem Punkt gebildet. Alle Strecken Null sind einander congruent.

Gleiche Strecken zu gleichen Strecken addirt geben gleiche Strecken ist
42 direkte Folge oder nur andere Aus|drucksweise von Axiom 3. Aehnlich folgt: Gleiches von Gleichem subtrahiert giebt Gleiches.

Definition. Wenn B zwischen A, C liegt, so[69] $AB = a$, $AC = b$, so heisse $a < b$. Diese Definition hätte von uns schon früher aufgestellt werden können, aber jetzt wissen wir, dass dem Congruenzbegriff gegenüber das grösser und kleiner sein eine Invariante ist.

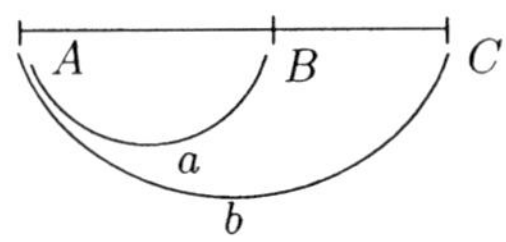

Wenn $a < b$, $b < c$, so $a < c$; ist ein Satz über den Begriff zwischen.

Nun die ebenen Congruenzsätze:

Definition. 2 von einem Punkt A (Scheitel) ausgehende Halbstrahlen a, b (Schenkel) zusammen mit einem nicht auf a, b gelegenen Punkt W bestimmen einen Winkel. Dabei kann W durch irgend einen Punkt W_1 oder W_2 ersetzt werden, wenn auf WW_1 bez. auf WW_1 und W_1W_2 kein Punkt von a, b liegt. Die Punkte W, W_1, W_2 heissen die Winkelpunkte oder Punkte des Winkels.

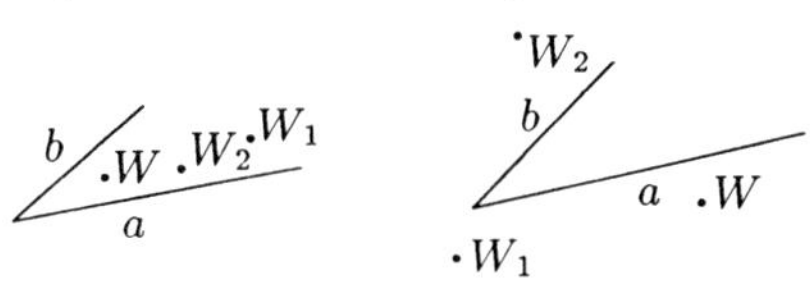

[68–68] Substituted for: Satz

[69] This clearly should be 'und'.

Axiom 5. 43
Wenn ein Winkel abW und ferner ein Halbstrahl a' gegeben ist und ferner eine Seite S der durch a' bestimmten Geraden, so giebt es einen und nur einen Halbstrahl b', der mit a' den nämlichen Anfangspunkt hat, und einen nicht auf a', b' gelegenen Punkt W', so dass

$$\sphericalangle abW \equiv \sphericalangle a'b'W' \quad \text{und} \quad \sphericalangle a'b'W' \equiv \sphericalangle abW$$

ist und dass alle Winkelpunkte von $\sphericalangle a'b'W'$ auf der gegebenen Seite S liegen oder alle Punkte der gegebenen Seite S Punkte des Winkels $\sphericalangle a'b'W'$ sind.[70]

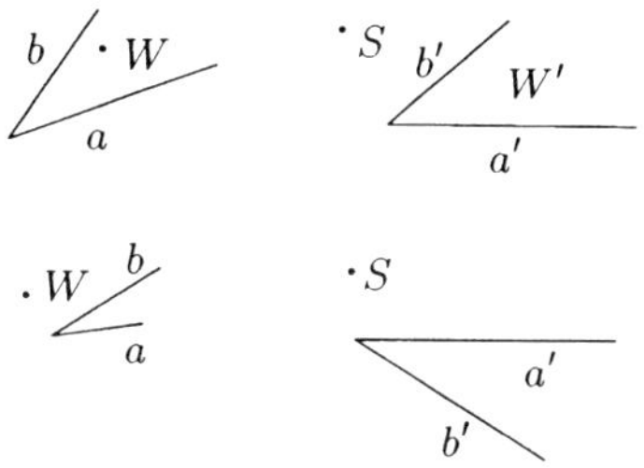

Jeder Winkel ist sich selbst congruent

$$\sphericalangle abW \equiv \sphericalangle abW' .$$

Also Möglichkeit, einen Winkel abzutragen oder die Ebene um einen Winkel zu drehen.

Axiom 6. Wenn 2 Winkel einem dritten $\equiv$ sind, so sind sie unter einander $\equiv$. 44

Axiom 7. Es seien in einer Ebene α $\sphericalangle abW, \sphericalangle bcU$ 2 Winkel ohne gemeinsame Winkelpunkte und desgleichen in der Ebene β $\sphericalangle a'b'W$, $\sphericalangle b'c'U'$ 2 Winkel ohne gemeinsame Winkelpunkte: wenn dann

$$\sphericalangle abW \equiv \sphericalangle a'b'W' \quad \text{und} \quad \sphericalangle bcU \equiv \sphericalangle b'c'U'$$

ist, so ist stets

$$\sphericalangle acW \equiv \sphericalangle a'c'W' .$$

Hieraus auch wieder die entsprechenden Sätze wie oben aus Axiom 3 auf der Geraden.

Congruenz von Halbstrahlen von Büscheln.[71]

Axiom 8. $\sphericalangle abW \equiv \sphericalangle baW$.

Bezeichnung des Winkels durch **einen** griechischen Buchstaben. Summe 2er Winkel $\alpha+\beta$, wobei man eventuell bloss den doppelt überdeckten Teil rechnet. $\alpha + \beta = \beta + \alpha$ etc. Wie auf S. 41 für die Gerade. Wenn $\alpha > \beta, \beta > \gamma$, so $\alpha > \gamma$.[72]

---

[70]The formulation in the *Ausarbeitung*, p. 39 is cleaner. One essential thing specified there, which Hilbert omits to mention here, is that a plane α has to be specified along with a' and the side S.

[71]It is not clear what Hilbert means here.

[72]The way of taking sums and differences for angles is importantly different from that for segments in that, as Hilbert indirectly indicates, one must calculate modulo a 'complete' angle, i.e., 2π. This means that angle transitivity as stated here fails. In addition, Hilbert does not show that '$<$' for angles is a total odering, i.e., of any two angles α, β one of $\alpha < \beta$,

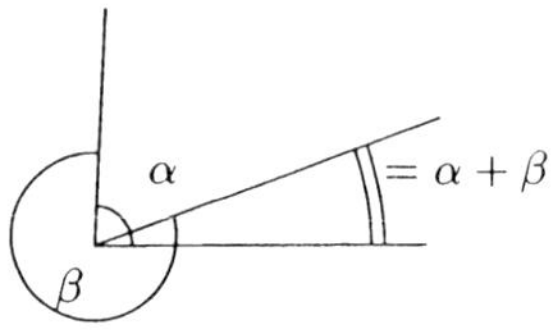

Definition: Wenn $a = b$, so heisse $\sphericalangle aaW$ ein Vollwinkel. Wenn a, b zusammen die verschiedenen Halbstrahlen | derselben Geraden sind, so heisst $\sphericalangle abW$ ein gestreckter Winkel. Auch wird der $\sphericalangle aa$ ohne Winkelpunkte ein Winkel 0 genannt.

Axiom 9. Alle gestreckten Winkel sind einander $\equiv$. $\uparrow$(Ist dies Axiom 9 nicht vielleicht doch eine Folge von III 1–8 und 10??)[73] $\downarrow$

Folgerung: Alle Vollwinkel und alle Winkel 0 sind einander $\equiv$. Das Addiren, Subtrahieren geschieht in üblicher Weise, nur Vollwinkel wird $= 0$ gerechnet und in Ungleichungen hat man daher zu vermeiden, dass auf der einen Seite Winkel vorkommen, die $>$ einem Vollwinkel werden.

[74]$\uparrow$Es folgen sofort die Sätze:

Wenn 2 Winkel gleich sind, so sind auch ihre Nebenwinkel gleich.

Die Scheitelwinkel sind einander gleich.[75]

Definition. Wenn ein Winkel seinem Nebenwinkel = ist, so heisst er ein rechter Winkel.

Satz. Alle rechten Winkel sind einander gleich. (Ist mit Unrecht wie ich im Gegensatz zu Lindeman glaube bei Euklid ein Axiom*).)

Beweis. Ist a und b je ein Rechter, und $a > b$, so

$$a + a = b + b, \quad (a - b) + a = b, \quad \text{d. h.} \quad a < b,$$

was ein Widerspruch. Euklid und Lindemann haben nicht unsere scharfen Definitionen und Axiome über den Begriff zwischen; dies erklärt den Unterschied.$\downarrow$[76]

Die Vereinigung oder vielmehr das Band zwischen den Axiomen über Strecken und Winkelcongruenz bildet der folgende Congruenzsatz.

---

*) Vielleicht denkt sich Euklid die Gleichheit aller gestreckten Winkel nicht als Axiom! Vgl. Lindemann, S. 547. Suche die Gleichheit aller Vollwinkel und daraus die Gleichheit aller gestreckten eventuell mittelst Axiom 10 (erster Congruenzsatz) zu beweisen.

$\alpha = \beta$, $\alpha > \beta$ holds. In the *Ausarbeitung*, p. 42, Hilbert expressly states that this will follow from Axiom 7. This is hinted at in this text (p. 45), but see the *Ausarbeitung*, pp. 40–41. Note that neither angle arithmetic nor angle ordering are introduced in the *Festschrift*.

[73]The question of the independence of Axiom 9 resurfaces in the *Ausarbeitung*, pp. 41–42, but is put to one side. It is dealt with in the *Festschrift*: see the Introduction to this Chapter, Specific Note (3)(a).

[74]The following addition, which extends to line 246.25 and includes Hilbert's footnote *) is on a separate piece of paper pasted to p. 45.

[75]Hilbert has defined neither 'Nebenwinkel' nor 'Scheitelwinkel'. These are not defined in the *Ausarbeitung* either, and Hilbert consciously omits the definitions: see ibid., p. 41. However, they *are* defined in the *Festschrift*, p. 13, after which the theorems Hilbert mentions here are proved with the help of the Second Congruence Theorem (op. cit., 14–15).

[76]In this paragraph, and the note to it, Hilbert refers to Euclid's Fourth Postulate, and then states his disagreement with Lindemann's support for this Postulate in *Clebsch 1891*, 547–549. See the Introduction to this Chapter, Specific Note, (3)(a).

Axiom 10. Wenn in 2 Dreiecken ABC und $A'B'C'$

$$AB \equiv A'B', \quad AC \equiv A'C', \quad \sphericalangle BAC \equiv \sphericalangle B'A'C',$$

so ist auch stets

$$BC \equiv B'C', \quad \sphericalangle ACB \equiv \sphericalangle A'C'B', \quad \sphericalangle ABC = \sphericalangle A'B'C'$$

↑d.h. die beiden Dreiecke heissen congruent↓ (erster Congruenzsatz). (2 Seiten und der eingeschlossene Winkel).

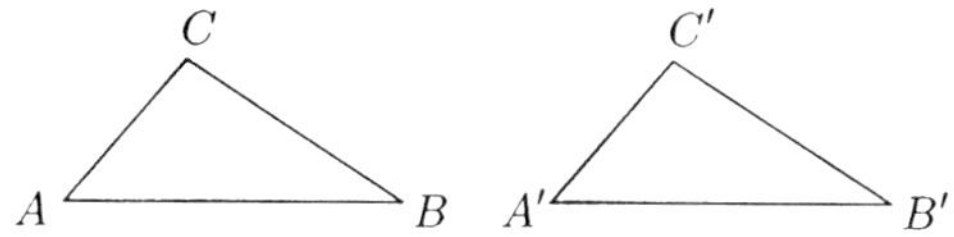

Dieser Congruenzsatz ist sicher ein von 1–9 unabhängiges Axiom. Denn 46
nimm im Euklidischen Raume das Abtragen von Strecken und Winkeln überall in gewöhnlichem Euklidischen Sinne abgesehen von einer Ebene α', welche eine bestimmte Ebene α in irgend einem Winkel schneiden möge. In der Ebene ist das Abtragen von Winkeln, wie gewöhnlich. Dagegen sei $\sphericalangle abW$ in $\alpha = \sphericalangle a'b'W'$ in α', wenn AA' senkrecht auf α, Ebene $AA'aa'$ senkrecht auf α und ebenso bb' senkrecht auf α und $WW' \perp$ auf α ist, d. h. wenn $\sphericalangle ab$ die senkrechte Projektion von $\sphericalangle a'b'$ auf α ist. In α' kann man also einen Winkel abtragen erst durch zurückgehen in die Ebene α. Alle Axiome sind erfüllt, aber nicht 10, weil ja die congruenten Winkel im gewöhnlichen Sinne von einander verschieden und daher auch die dritten Dreiecksseiten von einander verschieden sind.

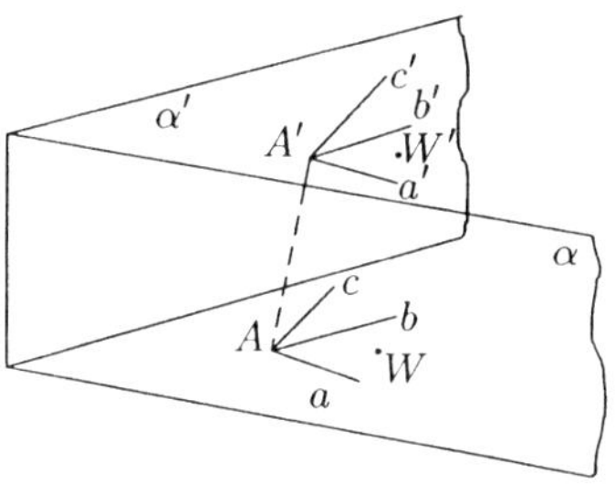

Will man also $\sphericalangle a'b'$ in α' um den Scheitel drehen, so dass etwa a' in den gegebenen Halbstrahl c' fällt, so muss man $a'b' \perp$ projiciren und in α ab so drehen, dass a in die senkrechte Projektion c von c' fällt und dann den erhaltenen Winkel wieder zurückprojiciren.

⟨Deletion⟩[77] 47

⟨Deletion⟩[78]

[79]↑Es wird sich nun das wichtige Resultat herausstellen, dass hiermit alle Congruenzaxiome erschöpft sind und insbesondere keine räumlichen Axiome nöthig sind. Auch Euklid scheint dies gemeint und gewusst zu haben, da er nur

[77] Deleted: Definition. 2 Dreiecke heissen congruent, wenn alle homologen Winkel und Seiten gleich sind.

[78] Deleted: Satz. Wenn in 2 Dreiecken ABC und $A'B'C'$

$$AB = A'B', \quad AC = A'C', \quad \sphericalangle BAC = \sphericalangle B'A'C',$$

so sind die beiden Dreiecke congruent.
Beweis

[79] This addition appears in the upper, and upper right, margins.

den ersten Congruenzsatz mittels Decken, d. h. Bewegung, d. h. axiomatisch beweist.[80] ↓

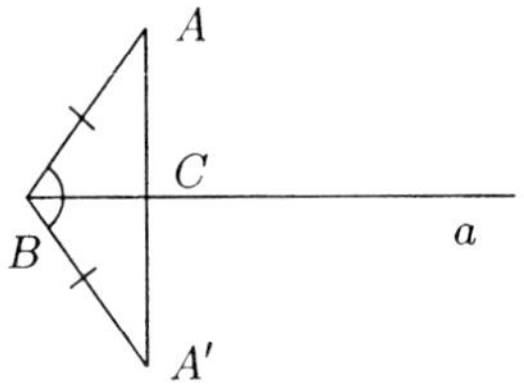

Nun können wir die *Existenz rechter Winkel* nachweisen: Will man von A ein Loth auf a fällen, so verbinde A mit einem Punkt B von a; trage nach unten an BC in B den $\sphericalangle ABC$ ab und mache auf dem neuen Schenkel $BA' = BA$. Verbinde A mit A', so ist AA' das verlangte Loth, da $\sphericalangle ACB = \sphericalangle BCA'$ wird.

Die übrigen Dreieckscongruenzsätze können bewiesen werden. ↑Beim weiteren Aufbau Vorsicht und oftmals von den üblichen Schulbüchern abweichen.↓

Satz. Wenn 2 Dreiecke in einer Seite und den beiden anliegenden Winkeln übereinstimmen, so sind sie untereinander congruent. (*2*[ter] *Congruenzsatz*)

Denn wäre $AB \neq A'B'$, etwa $AB = AB''$, so wäre $\triangle ABC \equiv \triangle A'B''C'$, folglich $\sphericalangle ACB = \sphericalangle A'C'B'' = \sphericalangle A'C'B'$, was unmöglich.[81]

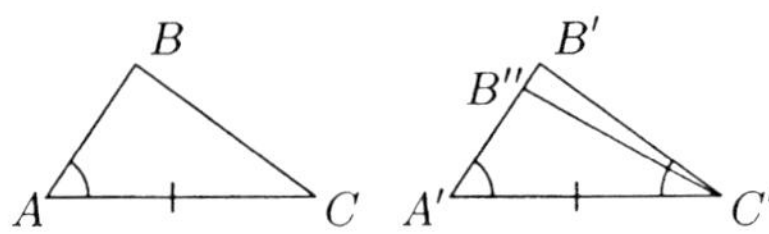

48 *Satz.* Wenn in Dreieck ABC gegeben ist, so ist der Nebenwinkel von $\sphericalangle ABC$ (d. h. der sogenannte Aussenwinkel) stets $>$ als $\sphericalangle CAB$.

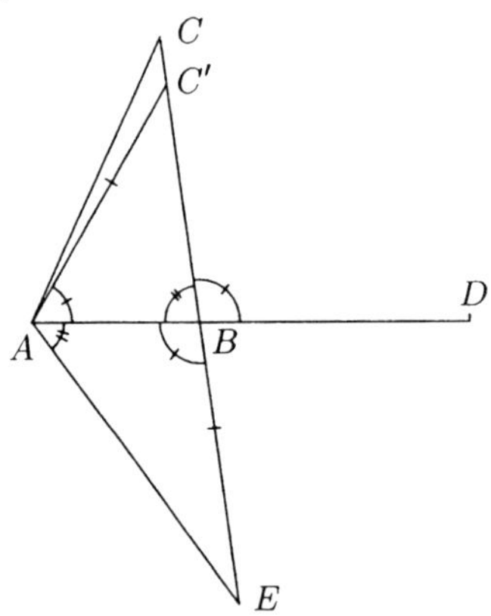

Beweis.[F] Wäre $\sphericalangle DBC \leqq BAC$, so mache $\sphericalangle C'AB = \sphericalangle DBC$, wo eventuell $C' = C$ sein darf. Nun verlängere CB über B hinaus bis E, so dass $BE = AC'$ wird, so $\triangle ABC' \equiv \triangle BAE$, wegen 2 Seiten und eingeschlossener Winkel $\sphericalangle ABE = \sphericalangle DBC = \sphericalangle C'AB$. Folglich $\sphericalangle BAE = \sphericalangle ABC$, d. h.

$$\sphericalangle C'AB + \sphericalangle BAE = \sphericalangle CBA + \sphericalangle BAE\,^{83}$$

= einem gestreckten, d. h. C', A, E *liegen auf einer Geraden*, was nicht möglich, weil C', B, E auf einer Geraden liegen.

[F](Den Euklidischen auch sehr einfachen Beweis vgl. Stäckel, Theorie der Parallellinien, S. 11.[82] Vielleicht ist dieselbe vorzuziehen.) ⟨This remark is set to the right of the figure.⟩

[80]For elucidation of this remark, see the Introduction to this Chapter, Specific Note (3)(*b*).

[81]Here and in what follows Hilbert frequently uses the ordinary equality symbol for congruence, instead of the special symbol '≡'.

[82]The reference is to *Stäckel and Engel 1895*, 11. There we find Euclid's Proposition 16 from Book I, which says: 'In any triangle, if one of the sides be produced, the exterior angle is greater than either of the interior and opposite angles' (quoted from *Heath 1926a*, 279). This is more general than the theorem Hilbert proves, and Hilbert's proof is different.

[83]The second '$\sphericalangle BAE$' here should be '$\sphericalangle ABE$'.

Folgerung. In einem Dreieck können nicht 2 | Winkel vorkommen, von denen 49 jeder $\geqq$ ein Rechter ist ↑d. h. es sind mindestens 2 Winkel $<$ ein Rechter (*spitz* im Gegensatz zu *stumpf* $>$ Rechter)↓.

Satz. Wenn 2 Dreiecke in einer Seite, einem anliegenden und dem gegenüberliegenden Winkel übereinstimmen, so sind sie unter einander congruent.

Lässt sich *in der üblichen Weise nicht* beweisen, da wir ja nicht den Satz von der Winkelsumme im Dreieck = gestrecktem Winkel haben.[84] Vielmehr so:

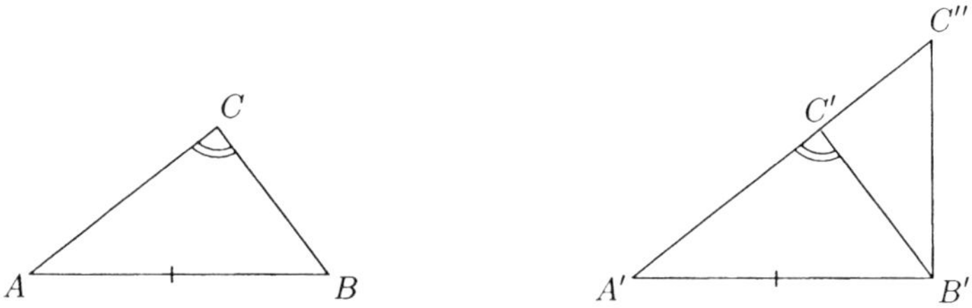

Sei $AC \neq A'C'$ etwa $AC = A'C''$, so wäre Dreieck $ABC \equiv A'B'C''$, d. h. $\sphericalangle ACB \equiv \sphericalangle AC''B' \equiv \sphericalangle A'C'B'$, was wegen des oben bewiesenen Satzes vom Aussenwinkel nicht möglich ist.

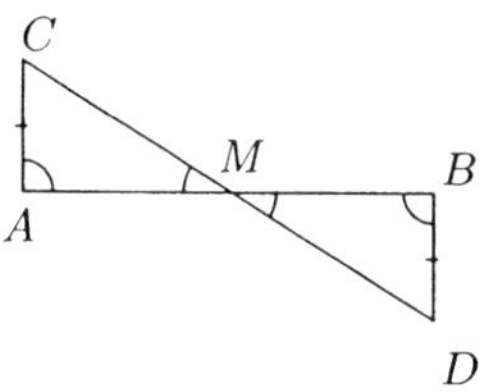

Man kann eine Strecke AB stets *halbiren.* Man 50 errichte in A eine Senkrechte[G] $= AC$, in B eine Senkrechte $BD = AC$ nach der anderen Seite, so trifft CD in der Mitte M.

↑Wenn 2 Strecken *gleich sind*, so sind auch ihre Hälften =, wird wie die Gleichheit rechter Winkel, S. 45, bewiesen.↓

Satz: Im gleichschenkligen Dreiecke sind die Basiswinkel =. Wenn in einem Dreiecke die Basiswinkel = sind, so ist dasselbe gleichschenklig.

⟨Deletion⟩ [85] ⟨51⟩

Ehe ich Ihnen den sehr einfachen Beweis mitteile, erwähne ich meine vergeblichen Versuche.

[G]braucht nicht $\perp$ zu sein, sondern unter irgend einem Winkel $\sphericalangle CAB = \sphericalangle ABD$. ⟨This remark is added just to the right of the figure, and is assigned to the word 'Senkrechte' by an insertion sign.⟩

---

[84]It is not clear to what Hilbert's 'the usual proof' refers. In Euclid's *Elements*, the theorem is Proposition 26 from Book I, and Euclid certainly does *not* rely on the theorem that the angle-sum in any triangle is two right-angles; this is only proved in Proposition 32 of Book I. Moreover, the proof Hilbert gives here is very close to that Euclid gives for one case.

[85]Deleted:

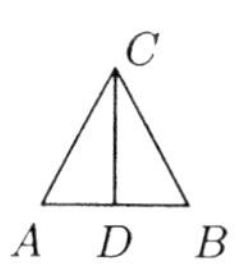

Beweis. Man ⟦könnte versucht sein von der⟧⟊↑fälle von der↓ Spitze C ↑auf↓ eine $\perp$ ⟦zu fällen⟧ auf die Basis AB, ⟦aber man würde dann nicht wissen dass die Senkrechte in das Innere des Dreieckes fällt. Daher ⟦lieber⟧ halbiere AB in M und verbinde ⟨p. 51⟩ C mit M, so $\triangle AMC \equiv BMC$ etc.⟧ Trifft dieselbe zwischen AB in D etwa auf, so

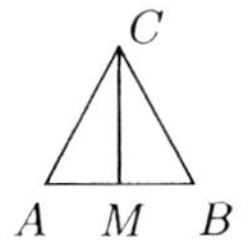

1.) Halbire AB in M, so in den $\triangle AMC$, BMC 3 Seiten gleich. Dieser sogenannte 3te Congruenzsatz erfordert aber gerade den Satz vom gleichschenkligen Dreieck. 2.) Man fälle von C eine Senkrechte auf AB, aber dann sogenannter 4ter Congruenzsatz, der beim Beweis wieder erfordert den Satz, dass im $\triangle$ der grösseren Seite der grössere Winkel gegenüberliegt und zu diesem Beweise brauchen wir wieder gerade den Satz vom gleichschenkligen $\triangle$. 3.) Halbire den Winkel ACB. Dies ist aber eine schwierige Aufgabe ohne den Satz vom gleichschenkligen $\triangle$, aber ausführbar. [86]

52 Der einfache Beweis:

$$\triangle ABC \equiv \triangle BAC \quad \text{etc.}$$

Ebenso die Umkehrung.
Der Euklidische Beweis ist umständlicher.[87]

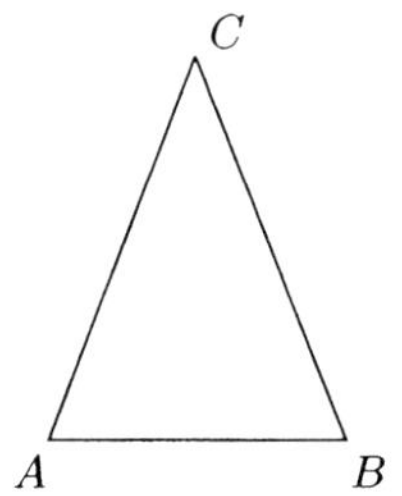

Satz. Wenn 2 Dreiecke in den 3 Seiten übereinstimmen, so sind sie einander $\equiv$.
↑(3ter Congruenzsatz)↓
Beweis.

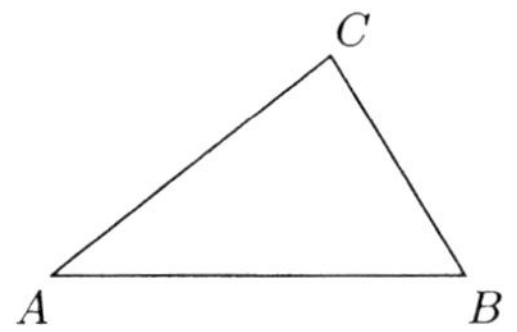

[86]Deleted: (

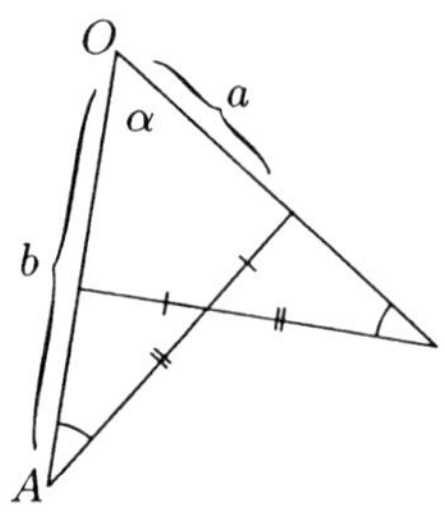

Von A fälle $\perp$ auf den anderen Schenkel. Ist $a < b$, so $a' = a$, $b' = b$ etc. Ist $a > b$, so errichte in A eine Senkrechte auf dem ersten Schenkel OA. Diese muss den 2ten Schenkel treffen etwa in C und es muss dann $OA < OC$ sein etc., wie oben.)

[87]The theorem in Euclid is Book I, Proposition 5 (though this proposition claims a bit more than Hilbert's); the converse is Proposition 6. (See *Heath 1926a*, 251–256.) For the simple proof Hilbert mentions here, see the *Ausarbeitung*, p. 50. The proof is due to Pappus: see Heath, loc. cit., 254. Heath notes of this proof:

> This will no doubt be recognised as the foundation of the alternative proof frequently given by modern editors, though they do not refer to Pappus.

Heath goes on to say that the proofs often given, unlike Pappus's original, and unlike that employed by Hilbert here, assumed a second copy of the triangle lifted off, turned, and placed on the original, and thus, like Euclid's proof, again depend on displacement and superposition.

Trage das $\triangle ABC$ ab nach $A'B'C''$.

Dann $\sphericalangle A'C'C'' = \sphericalangle A'C''C'$

$\underline{\sphericalangle B'C'C'' = \sphericalangle B'C''C'}$

$\sphericalangle A'C'B' = \sphericalangle A'C''B' = \sphericalangle ACB$ etc.[88]

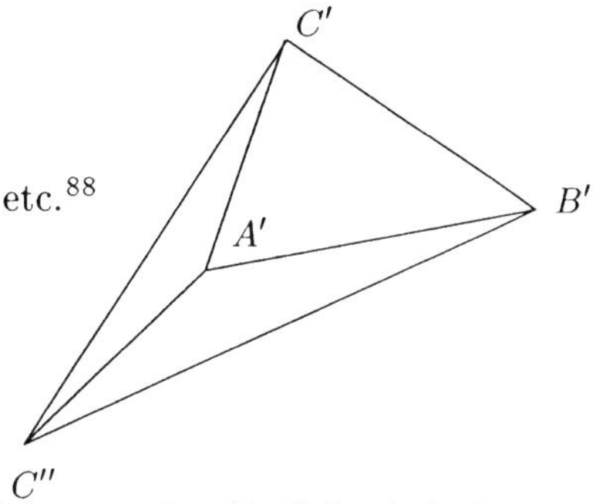

Jeder Winkel kann halbirt werden: Man mache die Schenkel gleich und halbire die Strecke zwischen ihren Endpunkten.

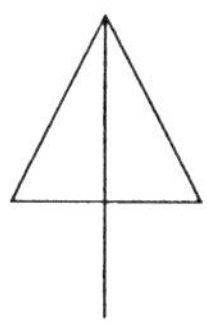

Dem grösseren Winkel in einem Dreieck liegt stets die grössere Seite gegenüber. 53

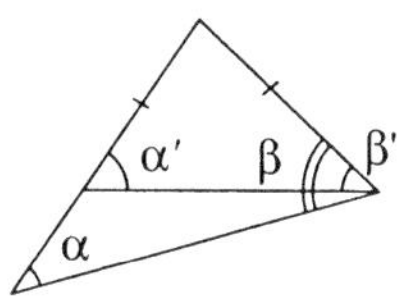

Beweis.
$\alpha' > \alpha$, $\beta > \beta'$, also $\beta > \alpha$.[89]

Die Summe zweier Seiten im Dreieck ist stets grösser als die dritte. ↑(Auch bei Euklid erst an späterer Stelle bewiesen!)↓

Beweis.

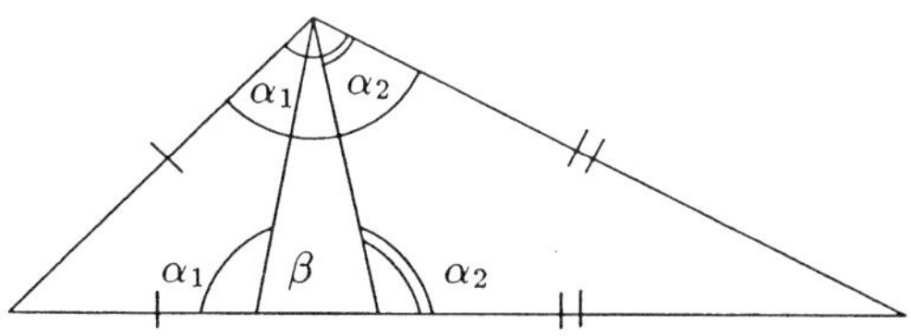

$\alpha > \alpha_1 + \alpha_2$, $\alpha_2 > \beta$, also $\alpha > \alpha_1 + \beta$, also $\alpha >$ ein gestreckter, was nicht der Fall ist.

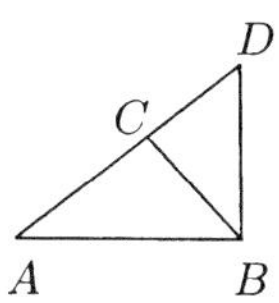

↑[Besser ist der Euklidische Beweis: mache $CD = CB$, $\sphericalangle CDB = \sphericalangle CBD$, folglich $\sphericalangle ABD > ADB$, folglich $AC + CD = AC + CB > AB$.]↓[90]

[88]This proof seems more appropriate to the diagram constructed in the *Ausarbeitung*, p. 50.

[89]For Hilbert's proof, see the *Ausarbeitung*, p. 51.

[90]Hilbert adds this passage in square brackets, together with the small figure, to the right of the larger figure. The proof sketched in the addition is indeed Euclid's proof, and the diagram is Euclid's: see Euclid, *Elements*, Book I, Proposition 20, *Heath 1926a*, 286–287.

↑Die Definition der Geraden als der kürzesten ist ohne Sinn, bevor Entfernung definirt und dies ist bei unserer Anordnung erst sehr spät allgemein möglich. Vgl. Stäckel-Engel, S. 29.↓[91]

54 Satz. Wenn 2 Dreiecke in 2 Seiten und dem der grösseren Seite gegenüberliegenden Winkel übereinstimmen, so sind sie einander $\equiv$.[92]

Jetzt kann der übliche Beweis geführt werden:

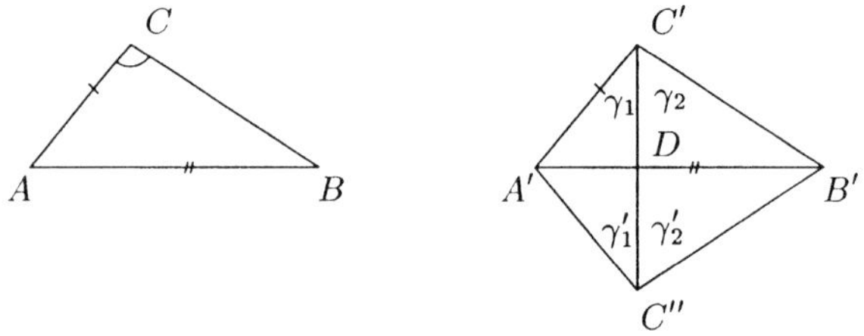

$AB > AC$, daher $\sphericalangle C > \sphericalangle B$, $\sphericalangle C' > \sphericalangle B'$, daher $\sphericalangle B$, $\sphericalangle B'$ beide < ein Rechter und daher ihre Summe < als ein Gestreckter. Nun mach $\sphericalangle C'A'C'' = \sphericalangle A$[93] und $AC = A'C''$, so dass $\triangle A'C''B' \equiv \triangle ACB$ wird und insbesondere $\sphericalangle B = \sphericalangle A'B'C''$ ist. $C'C''$ muss AB oder deren Verlängerung nothwendig in einem von B' verschiedenen Punkte D treffen, da $\sphericalangle C'B'C''$ nicht = einem

---

Hilbert repeats the proof in the *Ausarbeitung*, pp. 51–52. The first proof, with the sparsely labelled diagram, is a little hard to follow. The idea is this. Assume that the third side (the base in this diagram) is greater than the two other sides. Then it should be possible to cut off on the base two segments, the first to the left, marked here with a single stroke, equal to the left-hand side of the original triangle, also marked with a single stroke, and the second to the right, marked with a double stroke, equal to the right-hand side of the original triangle, also marked with a double stroke. This will leave a third segment in the middle, unmarked here. Then we have two isoceles triangles, one at the left and one at the right, and whose base angles α_1 and α_2 resp. are equal to each other. Assuming that β is the angle complementary to α_1 on the base, then Hilbert's calculations (given the theorem on p. 48) make sense.

[91]The reference is to *Stäckel and Engel 1895*, 29. This cites a place in *Wallis 1663*, where Wallis says (in rough translation from the Latin):

> In truth, I would have no objection if *Euclid* had set up yet more unprovable postulates, for example, if he had postulated (after *Archimedes*), that *the straight line is the shortest line of all the lines between the same points* (thus, he would not have required to set up nineteen propositions before he proved [Book I, Proposition 20] that two sides of a triangle taken together are always greater than the third), and other things, which are clear in and of themselves. (*Wallis 1693*, 677.)

Archimedes's assertion about straight lines appears in *On the Sphere and the Cylinder*. For discussion, see *Heath 1926a*, 166–167. For further commentary, see *Veronese 1894*, 635. *Hilbert 1895b* is an axiomatic investigation of the proposition that the straight line is the shortest between two points.

[92]This is called the 'Fourth Congruence Theorem' by Hilbert in the *Ausarbeitung*, p. 52. In Euclid's *Elements*, Proposition 4 of Book I is Hilbert's First Congruence Theorem (a direct consequence of Axiom 10); Proposition 8 (Book I) is what Hilbert here calls the Third Congruence Theorem, and Hilbert's Second and Fourth Congruence Theorems are both parts of Euclid's Proposition 26 (Book I). For discussion, see the Introduction to this Chapter, Specific Note (3)(*b*).

[93]This should be $\sphericalangle B'A'C'' = \sphericalangle A$. We must assume here that C'' lies in the same plane as A', B', C' and on the opposite side of $A'B'$ from C'.

gestreckten ist. Wegen $A'C' = AC = A'C''$ ist $\sphericalangle A'C'C'' = \sphericalangle A'C''C'$ etc. $\sphericalangle\gamma_1 = \gamma_1'$, $\sphericalangle\gamma_2 = \sphericalangle\gamma_2'$ und daher $\sphericalangle A'C'B' = \sphericalangle A'C''B'$.[94]

Irgend ein System von endlich vielen Punkten heisst eine Figur – insbe- 55
sondere ebene Figur.

Definition. 2 ebene Figuren heissen einander congruent, wenn sich ihre Punkte so zuordnen lassen, dass die auf diese Weise einander zugeordneten (homologen) Strecken und Winkel gleich (congruent) sind.

Wenn 3 Punkte einer Figur auf einer Geraden liegen, so liegen die homologen Punkte der congruenten Figur auch auf einer Geraden.

Die Reihenfolge der Punkte auf einer Geraden in einer Figur und die Anordnung der Punkte in bezug auf irgend eine Gerade in einer Figur stimmt mit der Reihenfolge und Anordnung der homologen Punkte bez. der homologen Geraden einer congruenten Figur überein.

Allgemeinster Congruenzsatz in der Ebene:

Satz. Wenn $(A, B, C, \ldots)$ und $(A', B', C', \ldots)$ congruente ebene Figuren sind und P ein Punkt in der Ebene der ersteren, so lässt sich stets in der Ebene der zweiten ein Punkt P' finden, so dass $(A, B, C, \ldots, P)$ und $(A', B', C', \ldots, P')$
congruente ebene Figuren sind. | Enthalten die ursprünglichen Figuren min- 56
destens 3 nicht in einer Geraden gelegene Punkte, so ist die Construktion des Punktes P' nur auf eine Weise möglich.

Beweis. Mache $\sphericalangle PAB = \sphericalangle P'A'B'$, $PA = P'A'$.

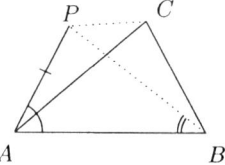

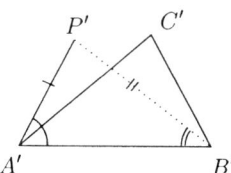

Dann ist $PB = P'B'$, $\sphericalangle PBA = \sphericalangle P'B'A'$, $\sphericalangle CBP = \sphericalangle C'B'P'$, folglich auch $PC = P'C'$ etc.

Es ist also durch die beiden congruenten Dreiecke eine bestimmte Transformation der Ebene in eine andere oder sich gegeben – eine Transformation, bei welcher jede Figur in eine congruente Figur übergeführt wird – wie wir sagen „zur Deckung gebracht" oder „bewegt" wird. Damit ist gerade umgekehrt durch unsere Congruenzsätze definiert das Wort zur „**Deckung** bringen" oder „**bewegen**", während man sonst fälschlicher Weise letzteres Wort als selbstverständlich definirt annimmt.

Die Transformation oder die | Bewegung ist festgelegt, wenn ein Punkt A, eine 57
Gerade a durch ihn, eine Seite auf a von A und eine Seite S von der Geraden a – ferner A', a', S' gegeben ist:

---

[94] Hilbert clearly runs into a confusion here. The proof is correctly given in the *Ausarbeitung*, pp. 52–53.

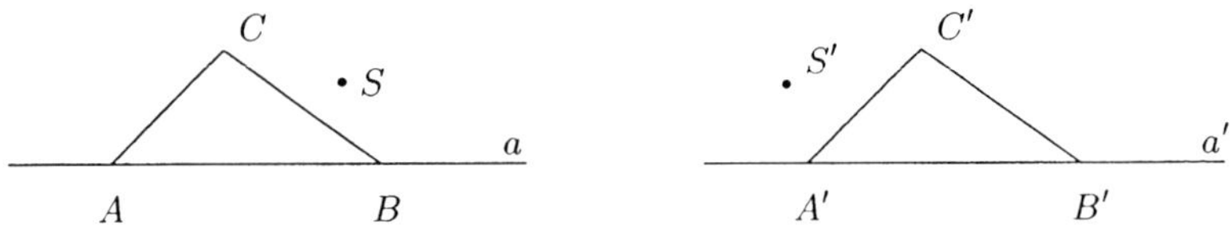

Denn dann construire $\triangle ABC \equiv \triangle A'B'C'$.

Damit sind die Congruenzsätze in der Ebene ⟨bewiesen⟩. Es ist sehr merkwürdig, dass kein **räumliches** Congruenzaxiom nöthig ist. Um dies zu sehen, ↑müssen wir den Satz auf S. 55 unten auf den Raum übertragen↓, vor Allem ist die Abtragung eines Ebenen-Winkels zu zeigen.[95]

Zunächst das Loth erichten und fällen. ↑Es kommt überall darauf an, vom Parallelenziehen unabhängig zu schliessen.↓

Satz. Wenn eine Gerade auf 2 in einer Ebene gezogenen Geraden gleichzeitig $\perp$ steht, so steht sich auf allen Geraden in der Ebene, die sie treffen $\perp$.[96]

Beweis.

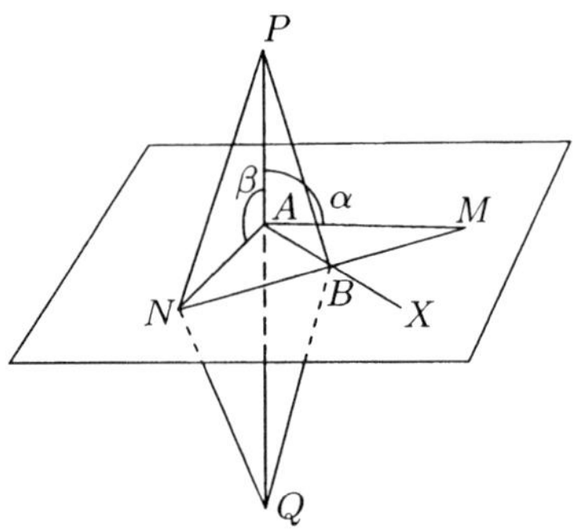

α, β seien rechte Winkel, verlängere PA um sich selbst bis Q. Man soll zeigen, dass PAX [97] ein rechter ist. Verbinde M mit N. MN schneide AX in B. $\triangle PNM \equiv \triangle QNM$. Folglich $\sphericalangle PNM = \sphericalangle QNM$, folglich $\triangle PNB \equiv \triangle QNB$, folglich $PB = QB$, folglich $\triangle PAB \equiv QAB$, d. h. $\sphericalangle PAB \equiv \sphericalangle QAB =$ einem Rechten.

[95] After making the insertion, Hilbert neglects to transfer 'ist' from after 'sehen' to its present position.

[96] In the following diagram, AN and AM are the two lines in the plane, P is a point outside this plane, such that both $\sphericalangle PAN$ and $\sphericalangle PAM$ are right angles. X is any point in the plane determined by A, N and M. The diagram is much clearer in the *Ausarbeitung*, pp. 54–55. Cf. also note to Remark [3] in the Appendix to the *Ausarbeitung*, where Gauß's various considerations on the definition of a plane are discussed.

[97] 'PA' should be '$\sphericalangle PAX$', X being any point in the plane not on the two given lines.

Das Errichten eines Lothes in A ist nun leicht: Man ziehe irgend zwei Geraden durch A, $= a, b$. Errichte in A eine zu a und eine zu b senkrechte Ebene, die sich im gesuchten Loth schneiden.[98] 58

Um ein Loth zu fällen von P, ziehe eine Gerade a in der Ebene, fälle $PB \perp$ auf a. M sei ein Punkt auf a. Auf a in B errichte eine Senkrechte b in der Ebene und fälle auf b eine Senkrechte von $P = PA$; diese ist das verlangte Loth. Denn verlängere PA um sich selbst bis Q. Dann a steht auf der Ebene $BPAQ \perp$. Daher $\sphericalangle QBM =$ einem Rechten. Daher $\triangle PBM \equiv \triangle QBM$ (wegen $PB = QB$), folglich $PM = QM$, daher $\triangle PAM = \triangle QAM$, also $\sphericalangle PAM = \sphericalangle MAQ =$ Rechter q. e. d.

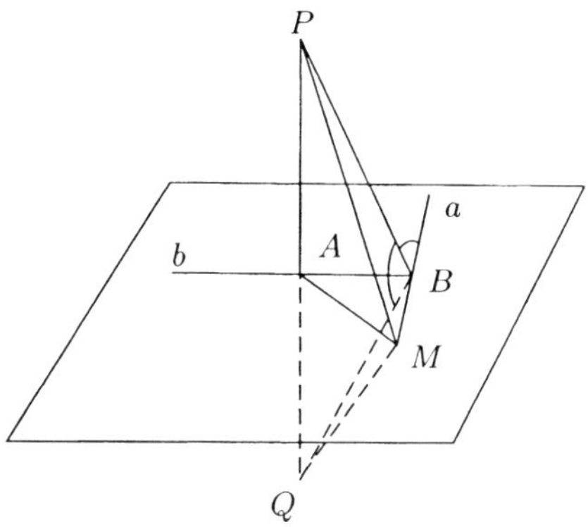

Satz: 2 Lothe auf derselben Ebene liegen stets in einer Ebene.

Beweis.

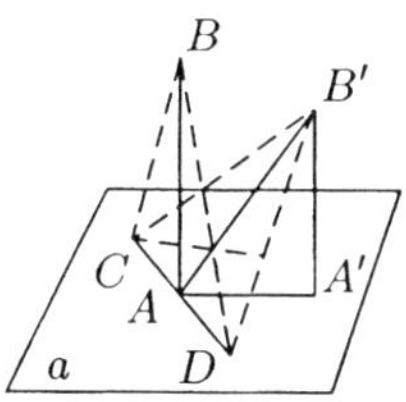

AB und $A'B'$ seien die beiden Lothe auf α. Es werde durch A in α die Gerade $CD \perp AA'$ gezogen und $AC = AD$ gemacht. Dann ist $B'C = B'D$, folglich $B'AD = \sphericalangle B'AC =$ einem Rechten. Daher muss $B'A$ in der Ebene ABA' liegen. Denn diese Ebene ist $\perp$ zu AD und daher muss ja auch die Schnittgerade von $DB'C$ mit $A'AB \perp$ auf AD stehen. Es giebt aber nur eine Senkrechte.

Wenn zwei Ebenen sich in der Geraden a schneiden, so errichte in A auf a in beiden Ebenen Lothe, so heisst der Winkel zwischen diesen der Winkel der Ebenen. Denn dasselbe ist von der Wahl des Punktes A auf a unabhängig: 59

Zum Beweise fälle von M (einem Punkte des Lothes in β ein Loth MF auf das Loth in α, so ist MF Loth in α nach dem vorigen. Ebenso Alles in A'. Ferner fälle von N (Schnitt von AM' mit $A'M$) ein Loth auf α: NG. Es ist $AM = A'M'$ gemacht und daher $AN = A'N$. Folglich $\triangle NAG \equiv \triangle NA'G$, folglich $\sphericalangle NAG = \sphericalangle NA'G$. Nun trifft AG den Punkt F' und $A'G$ den Punkt F – weil 2 Lothe auf derselben Ebene in

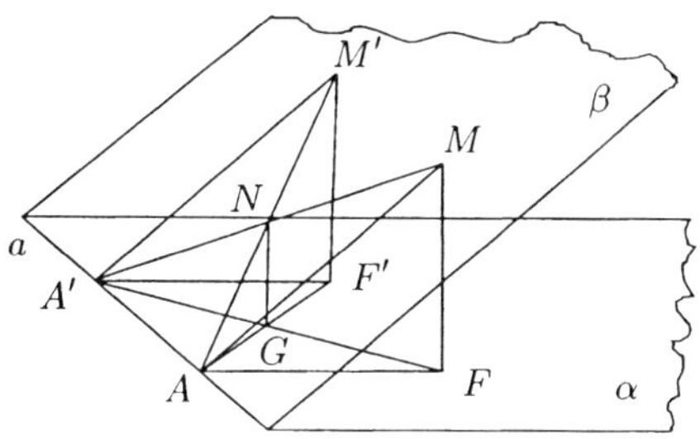

[98]It is easy to show that such perpendicular planes exist. See the *Ausarbeitung*, p. 55.

einer Ebene liegen. Daher $\triangle AF'M' \equiv \triangle A'FM$ und daher $FM = F'M'$, d.h. $\triangle AFM \equiv \triangle A'F'M'$ und $\sphericalangle MAF = \sphericalangle M'A'F'$.[99]

60 Damit können wir einen räumlichen Satz beweisen, der dem linearen Axiom 1 und ebenen Axiom 5 entspricht, nämlich:

Satz. Wenn eine Ebene α und eine Seite S derselben und eine in ihr gelegene Gerade a gegeben sind, so kann man stets eine Ebene durch a legen, die mit α einen gegebenen Winkel einschliesst.

Also Drehung einer Ebene um eine Gerade ist dadurch möglich. Aber noch mehr:

Allgemeinster Congruenzsatz im Raume:

Wenn $(A, B, C, \ldots)$ und $(A', B', C', \ldots)$ congruente Figuren sind und P ein beliebiger Punkt im Raume, so lässt sich stets ein Punkt P' finden, so dass $(A, B, C, \ldots, P)$ und $(A', B', C', \ldots, P')$ congruente Figuren sind. Enthalten die ursprünglichen Figuren mindestens 4 nicht in einer Ebene gelegene Punkte, so ist die Construktion des Punktes P' nur auf eine Weise möglich.[100]

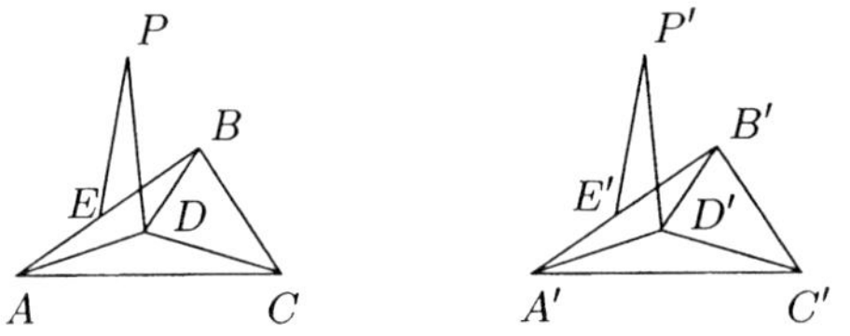

Beweis. Fälle von P ein Loth auf $ABC = PD$, suche D', so dass $(A\,B\,C\,D) \equiv (A', B', C', D')$, errichte in D' ein Loth $= DP = D'P'$, so $(A\,B\,C\,D\,P) \equiv (A'\,B'\,C'\,D'\,P')$.

Beweise zuerst $AP = A'P'$, $BP = B'P'$. Dann fälle von P auf AB und von P'
61 auf $A'B'$ | ein Loth $PE = P'E'$ etc.

Die Congruenzsätze über Ecken im Raume sind hierin enthalten und also auch sämtlich beweisbar (z. B. 2 Ecken, die in 2 Seiten und dem eingeschlossenen Winkel und in der Reihenfolge übereinstimmen, sind einander congruent). Zur Deckung bringen, Bewegung, Transformation in sich, und zwar ist die Bewegung festgelegt, wenn ein Punkt A, eine Gerade a durch A, eine Seite von A, eine Seite von a und eine Ebene durch α und eine Seite von α und A', a', α' gegeben sind.[101]

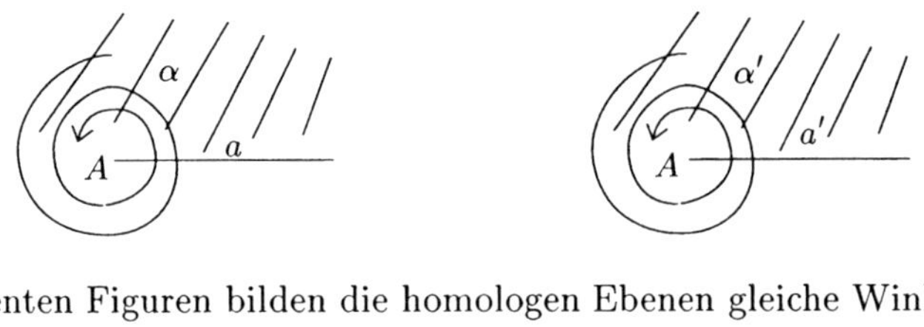

In congruenten Figuren bilden die homologen Ebenen gleiche Winkel.

[99] The whole proof is explained more carefully, with a better diagram, in the *Ausarbeitung*, p. 58.

[100] The definition of the congruence of two spatial figures is missing here; see the *Ausarbeitung*, pp. 53, 59.

[101] This is more carefully explained in the *Ausarbeitung*, pp. 59–60; see in particular the remarks in about the proper relationship between the concept of congruence and that of movement. For elucidation of the relationship between the notions of congruence and transformation and the (geometrical) notion of (rigid) motion defined with their help, see *Hartshorne 2000*, section 17.

In congruenten Figuren liegen homologe 4 Punkte in einer Ebene, weil die betreffenden Ebenen den gleichen Winkel 0 bilden.[102]
Beweis: Man erweitere die congruenten Figuren durch Fällen von Lothen, so dass die Winkel der Ebenen Winkel zwischen Geraden werden:

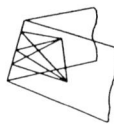

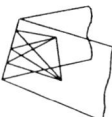

Also in 2 congruenten Figuren sind alle Eigenschaften der Verbindung (Axi- 62
om I) und der Anordnung (Axiom II) die gleichen.

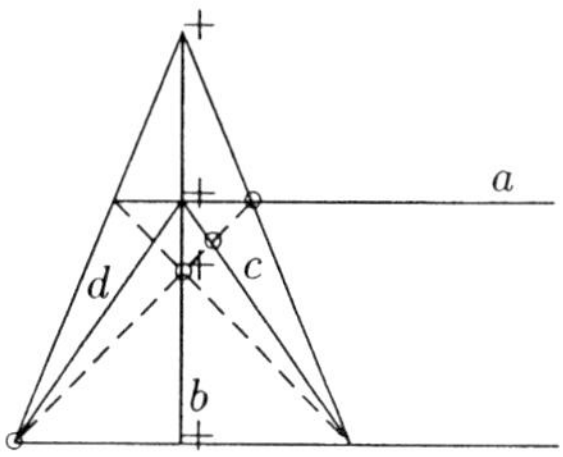

Construktion harmonischer Strahlen durch rechte Winkel ist möglich: Wenn $\sphericalangle bc = \sphericalangle bd$ und $\sphericalangle ab =$ Rechter, so sind $a\,b\,c\,d$ harmonische Strahlen, wie nebenstehende Figur lehrt. Denn ∘ ∘ ∘ ∘ sind harmonische Punkte, weil + + + + es sind. Ueber Kreise, Kugeln gelten die gewöhnlichen Schnittsätze. *Eine Gerade* hat mit einem Kreise ↑und der Kugel↓ keinen, einen *oder 2 Punkte* gemein.
Desgleichen ein Kreis mit Kreis oder Kugel.
Eine *Ebene oder Kugel* hat mit einer Kugel keinen, einen Punkt oder einen Kreis gemein. Eine Gerade, die durch einen Peripheriepunkt in einen Kreis oder Kugel eintritt, tritt durch einen andern heraus: *Inneres und Äusseres eines Kreises, einer Kugel.*

[H] Wir stossen aber bereits auf eine *eigentümliche Schwierigkeit*, wenn wir 63
folgenden Schnittpunktsatz ableiten wollen.

Wenn K_1, K_2 2 sich schneidende Kreise und K_3 ein Kreis ist, der die gemeinsame *Sehne von* K_1, K_2 *einmal passirt*, wie in nebenstehender Figur, so schneiden die 3 gemeinsamen Sehnen sich in einem Punkt.
Beweis. Man lege durch die Mittelpunkte M_1, M_2, M_3 Kugeln. Dann kommt alles darauf an, zu zeigen, dass diese 3 Kugeln sich in 2 Punkten schneiden. Man ziehe den K_1, K_2 gemeinsamen Kreis K_1K_2,[103] fälle von M_3 auf die gemeinsame Sehne ein Loth M_3F, beschreibe um F mit FG einen Kreis in

[H] Sätze über Gleichheit der Centri- und Peripheriewinkel auf gleichem Bogen können nicht bewiesen werden, da sie nicht gelten. ⟨This remark is added in the upper margin of p. 63 and not directed to any particular place.⟩

[102] It is not clear what Hilbert means here; in particular, it is not clear what the diagram just given is meant to illustrate. Neither the discussion nor the diagram reappears in the *Ausarbeitung*.

[103] This has to be K_{12}.

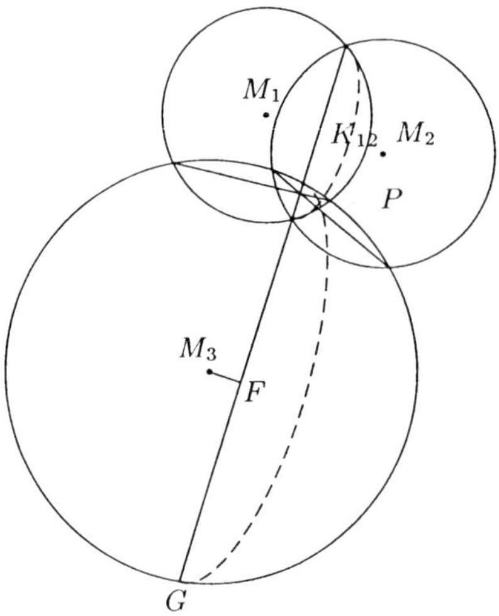

der Ebene von K_{12}. Trifft dieser in P und in Q, so sind Q die gemeinsamen Punk-
64 te und man schliesst | nun so: Die *3 Ebenen* durch die *drei Schnittkreise* K_{12}, K_{13}, K_{23} haben die Punkte P, Q und daher deren Verbindungsgrade PQ gemein. PQ möge die Ebene der Kreise in dem Punkt C *durchstossen.* Dann liegt C *auf allen 3 gemeinsamen* Sehnen q. e. d.

Die Existenz von P, Q hängt davon ab, ob man stets aus 3 Seiten, von denen die Summen 2er > als die 3te ist, ein Dreieck construiren kann, oder ob ein Kreis, der in dem Inneren eines Kreises teils und teils ausserhalb desselben Punkte hat, die Peripherie durchschneiden muss, ob also in *diesem Sinne der Kreis* eine geschlossene Figur ist (bei Euklid ein ähnlichklingendes Axiom: der Kreis ist eine geschlossene Figur).

Man muss dies leider *verneinen* und es muss daher *dahingestellt* bleiben, ob obiger Satz von den gemeinsamen Sehnen 3er Kreise beweisbar ist.[104]

65 Es ist *nicht möglich*, auf Grund der bisherigen Axiome I, II, III *zu beweisen*:

1.) dass eine Gerade, welche im Inneren eines Kreises verläuft, die Peripherie dieses Kreises trifft.

2.) " ein Kreis, welcher zum Teil im Inneren, zum Teil ausserhalb eines Kreises verläuft, die Peripherie dieses Kreises trifft, d. h. dass aus irgend 3 Seiten, von denen 2 zusammen > als die dritte sind, stets ein Dreieck sich construiren lässt.[105]

Beweis: Wir legen im *gewöhnlichen Euklidischen* Raume ein *rechtwinkliges* Coordinatenkreuz zugrunde und denken alle *Punkte als existirend*, deren Coordinaten Zahlen des Bereiches sind, die durch die 4 Species und durch die 5te Rechnungsart: $\sqrt{a^2+b^2}$ $\uparrow$besser $\sqrt{1+x^2}$, da $\sqrt{a^2+b^2} = a\sqrt{1+\left(\frac{b}{a}\right)^2}$ $\downarrow$, wo

---

[104]The whole treatment of this theorem, not least the accompanying diagram, is clearer in the *Ausarbeitung* (see pp. 61–64). Here, Hilbert's proof is based on the clear assumption that, if the three circles in question are the equators of three spheres, these spheres will have just two points (P and Q) of mutual intersection. The viability of the proof thus depends on justifying the existence of these two points. This then gives rise to the metamathematical question concerning triangle existence stated in the previous paragraph here. See the Introduction to this Chapter, Specific Note (2).

[105]Hilbert here begins to consider the metamathematical problem mentioned in the previous note. He constructs a model of the axioms in groups I–III containing three segments of which any two are greater than the third but which do not form a triangle. See the Introduction to this Chapter, Specific Note (2). In this model, as Hilbert points out, the other assumptions, 1.) and 2.) here, also fail. See also the *Ausarbeitung*, pp. 64–68.

a, b Zahlen des Bereiches sind, hervorgehen, also Quadratwurzel ziehen aus Summen von Quadraten wird den 4 Spezies adjungirt.[106] Als Gerade, Ebene nimmt man Geraden bez. Ebenen, die durch irgend 2 bez. 3 jener Punkte gehen. Hier gelten die Axiome I, II, III. Denn Verschiebung auf einer Geraden wird durch Addition, Substraktion gegeben und Drehung in der Ebene um den Winkel α zwischen der x-Achse und der Verbindungsgeraden des Punktes a, b mit dem 0-Punkt des Coordinatensystems wird durch die Formel

66

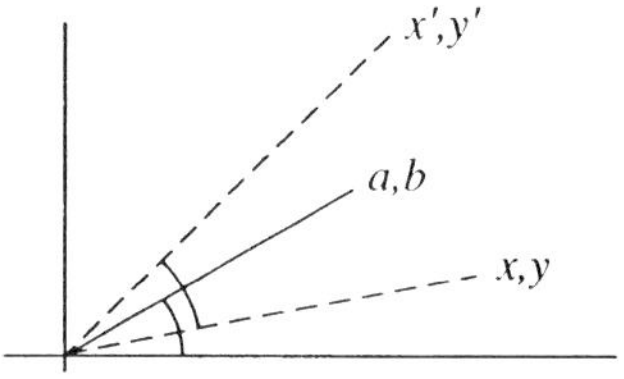

$$x' = \frac{a}{\sqrt{a^2+b^2}}x - \frac{b}{\sqrt{a^2+b^2}}y$$
$$y' = \frac{b}{\sqrt{a^2+b^2}}x + \frac{a}{\sqrt{a^2+b^2}}y$$

gegeben und folglich ist auch Drehung möglich. Nehmen wir uns die Gerade $x = \frac{\pi}{4}$ und den Kreis mit dem Radius 1 um den Anfangspunkt, so schneidet dieser Kreis die Gerade nicht. Denn sonst müsste die y-Coordinate des Schnittpunktes

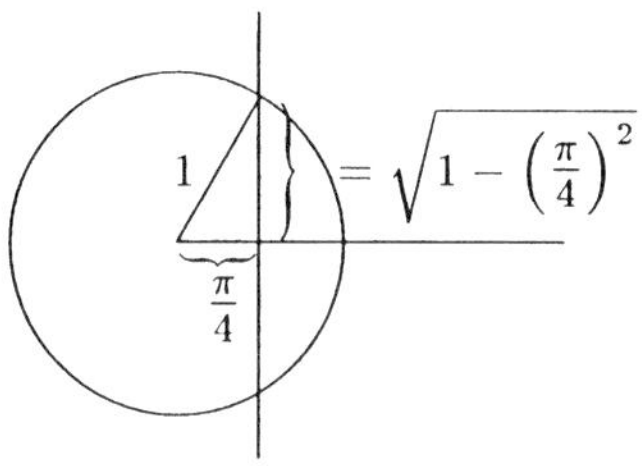

$$\sqrt{1-\left(\frac{\pi}{4}\right)^2} = \left(\sqrt{1+\pi^2}, \sqrt{3+(\pi+2)^2}\right) = A(1,\pi)$$

sein, wo rechts irgendein aus Quadratwurzeln zusammengesetzter Ausdruck ⟨steht⟩, unter denen eine Quadratsumme steht. Dies ist aber nicht möglich. Denn aus algebraischen Gründen könnte eine solche Beziehung für die transcendente Zahl π nur gelten, wenn sie identisch gilt, d. h. für jedes t:

$$\sqrt{1-\left(\frac{t}{4}\right)^2} = \left(\sqrt{1+t^2}, \sqrt{3+(t+2)^2}\right) = A(1,t).$$

Die rechte Seite $A(1,t)$ ist für solche t stets selber reell; dagegen wird $\sqrt{1-\left(\frac{t}{4}\right)^2}$

[106]Starting from the space of ordinary Euclidean analytic geometry, Hilbert isolates the field based on the four standard arithmetic field operations, and the 'Pythagorean operation', that is, $\sqrt{x^2+y^2}$ for any x, y in the field, or, as he adds, the operation $\sqrt{1+x^2}$. The field obtained from 1 and any real number t by closure under these five operations yields the smallest Pythagorean sub-field of the reals containing 1 and t. In Hilbert's notation, $A(1,t)$ stands for any member of this field; more precisely, it stands for any expression composed of a finite combination of the operations permitted starting from 1 and t. The model he constructs here is the analytic geometry taken over such a sub-field where t is π.

These sub-fields are countable. Early in the *Festschrift* (pp. 19–20), just after setting out all the axioms, Hilbert shows that a field of this kind yields a countable model for all the axioms of the *Festschrift*. In the *Ausarbeitung*, p. 142, Hilbert also uses such a field as the basis for a construction of a non-Archimedean geometry.

67 für $t > 4$ imaginär. Es | gibt also keinen Schnittpunkt von Kreis und Gerade.[107]

Die Heranziehung der transcendenten Zahl π ist hier nicht nöthig, z. B. $\sqrt{1+\sqrt{2}}$ ist durch unsere 5 Operationen nicht darstellbar. Denn die im Bereiche α liegenden algebraischen Zahlen haben die Eigenschaft, dass alle zu α conjugirten Zahlen reell sind. Die zu $\sqrt{1+\sqrt{2}}$ conjugirte Zahl $\sqrt{1-\sqrt{2}}$ ist aber imaginär. Oder

$$\begin{aligned} & 1+\sqrt{2} = \alpha^2 + \beta^2 + \dots \\ \text{hätte}\quad & 1-\sqrt{2} = \alpha'^2 + \beta'^2 + \dots \end{aligned}$$

zur Folge, was nicht möglich ist.

Nun folgt, dass aus den Seiten 1, 1, $\frac{\pi}{2}$ kein Dreieck sich construiren lässt. Denn sonst könnte ja auch dessen Höhe $= \sqrt{1-\left(\frac{\pi}{4}\right)^2}$ construirt werden.

Auf diese Betrachtungen werden wir später noch einmal zurückkommen, wenn wir nach dem vollständigen Aufbau der Geometrie die Hülfsmittel untersuchen, die zur Construktion dienen können. Wir werden dann den feinen Unterschied kennen lernen, der darin beruht, ob der Zirkel unbeschränkt oder nur zum Abtragen von Strecken und Winkeln (es genügt des rechten Winkels) gebraucht wird.[108]

---

[107] Much is left unclear in the preceeding paragraphs.

(1) The figure is misleading, wrongly labelled and rather unenlightening. Hilbert considers an isoceles triangle, two of whose sides are of length 1 and the third, the base, of length $\pi/2$. The base and the side of length 1 on the left in the diagram meet at the origin; the rest of the triangle is not represented. The base runs along the positive x-axis from $(0,0)$ and ends outside the unit circle drawn; the third side joins this end-point with the intersection of the line marked with '1' and the circle. The vertical line drawn is meant to cut the triangle in two, and therefore to cut the base at the point $(\pi/4, 0)$, which explains Hilberts '$\pi/4$' in the diagram. The coordinates of the three vertices of the triangle are thus $(0,0), (\pi/2, 0)$, and $(\pi/4, \sqrt{1-(\frac{\pi}{4})^2})$. The corresponding figures in the *Ausarbeitung*, pp. 66, 68 are much better.

(2) Hilbert argues that $\sqrt{1-\left(\frac{\pi}{4}\right)^2}$ cannot be in the field. For, if it were represented by an expression $A(1,\pi)$, then $\sqrt{1-\left(\frac{t}{4}\right)^2}$ would always be representable by the expression $A(1,t)$. But if we take any $t > 4$, $\sqrt{1-\left(\frac{t}{4}\right)^2}$ is the square root of a negative number, whereas, since t is real, $A(1,t)$ must always represent a real number. (The argumentation is more comprehensible in the *Ausarbeitung*, pp. 66–67.)

(3) The natural conclusion is that the formula Hilbert writes here directly after the diagram on p. 66 is supposed to represent the *assumption* that $\sqrt{1-\left(\frac{\pi}{4}\right)^2}$ is a number in the field $A(1,\pi)$, and the *fact* that the two numbers given in the middle term in round brackets *are* in the field.

[108] Hilbert does not return to these matters in these lecture notes, although there is a section in the *Ausarbeitung*, pp. 170–173 which takes up directly the question of geometrical constructions. (See also Hilbert's 1898 *Ferienkurs*, pp. 12–14, Chapter 3 in this Volume.) This is generalized in the *Festschrift*, Chapter VII. For comments on the algebraic import of Hilbert's result, see the Introduction to this Chapter, Specific Note (2)(e).

Die beiden Legendre'schen Sätze[I] über Winkelsumme im Dreieck und der 68
Beweis des Pascalschen Satzes nach Schur, vgl. die Autographie dieser Vorlesung S. – S. .[109]

So wie der Desargues gewissermassen die Elimination der räumlichen Axiome ist, so entsteht der Pascal durch die Elimination der Congruenzaxiome und zwar ist der Pascal auch die hinreichende Bedingung dafür, dass eine Congruenzdefinition möglich ist. Denn wie z. B. Schur zeigt, gilt der Fundamentalsatz der projektiven Geometrie, d. h. es kann ein Rechnen mit Strecken oder eine analytische Geometrie aufgestellt werden, wo die Buchstaben freilich Strecken, nicht Zahlen bedeuten und wo die bilineare Relation die nothwendige und hinreichende Bedingung für das Vereinigtsein von Punkt und Ebene ausdrückt.[110]

Die auf den Axiomengruppen I, II, III sich aufbauende Geometrie heisst die (nicht-Euklidische) Gauss'sche Geometrie (auch von Lobatscheffsky und Bolyai) (auch absolute Geometrie).

IV Das Parallelenaxiom. 69

Gruppe I S. 9, II S. 14, III S. 36.

↑Dies ist die *Grundlage*, ohne die wir die *feinen und schönen* Fragen im Folgenden gar nicht verstehen!↓

Die in diesen 3 Gruppen niedergelegten Axiome sind nicht anderes als eine *gründliche Ausführung der Axiome, die Euklid* zum Beweise der ersten 28 Sätze gebraucht und die hernach von allen Geometern als *nothwendige Thatsachen* der Geometrie betrachtet worden sind, oder unbewusst als *definirende Eigenschaften der Anschauung* von Punkten, Geraden, Ebenen betrachtet sind. Wenn auch Euklid diese Axiome nicht so vollständig ausgesprochen hat, so entsprechen sie doch dem, was er und seine Nachfolger bis in die neuere Zeit gemeint haben. Als Euklid aber nun weitere Sätze beweisen wollte, die die Anschauung unmittelbar liefert, z. B. Vorhandensein eines *Viereckes mit 4 rechten* Winkeln, so erkannte er, dass er mit den bisherigen Axiomen nicht
| reichte und stellte sein *berühmtes Parallelenaxiom* auf, das wir als einziges 70
Axiom der Gruppe IV wie folgt formuliren wollen:

In einer Ebene lässt sich durch einen Punkt ausserhalb einer Geraden nur eine Gerade ziehen, welche die erstere Gerade nicht schneidet: diese Gerade heisst eine Parallele zu jener Geraden.[111]

[I] (auch von Lambert und Saccheri, vgl. Stäckel-Engel, Parallellinien.) ⟨Added in the upper margin.⟩

---

[109] Here Hilbert has left spaces for page numbers of the 'Autographie' (i.e., *Ausarbeitung*) to be inserted. (The reference is pp. 69–72.) This occurs relatively frequently from this point on. For discussion of the two Legendre Theorems on the angle sum of a triangle, see the Introduction to this Chapter, Specific Note (5)(*a*).

[110] This paragraph deserves detailed comment; see the Introduction to this Chapter, Specific Note (1), especially (*a*) and (*d*).

[111] For comment, see the Introduction to this Chapter, Specific Note (4)(*d*).

Welchen *Scharfsinn* es erforderte, diesen Satz als Axiom hinzustellen, werden Sie am besten aus dem kurzen *historischen Abriss* erkennen: Stäckel-Engel, Parallellinien, Teubner 1895.[112]
Euklid 300 v. Chr.
Schon im Alterthum Versuche, ein folgerichtiges System ohne ein solches Par-
71 allelenaxiom aufzubauen, indem man die ↑*beständige*↓ *Gleichheit* | des *Abstandes* 2er Geraden als Merkmal ihres Parallelseins ansieht. Der Fehler beruht darin, dass es solche Geraden *nicht zu geben braucht.* Auch konnte die Euklidische Darstellung trotz ihrer *Anfeindungen* nicht verdrängt werden. Der Euklid herrschte im Alterthum, im Mittelalter bis in die Neuzeit. Erwähnt sei die Euklidausgabe des deutschen Mathematikers Ch. *Schlüssel* von *1574*, die bis *1738* nicht weniger als 22 mal gedruckt worden ist.[113] Mit dem 17ten Jahrhundert beginnen die Veröffentlichungen, die ausschliesslich der Parallelentheorie gewidmet sind und alle im Grunde den *Zweck haben*, das Euklidische Parallelenaxiom zu beweisen: Meist laufen sie darauf hinaus, dass auf eine mehr oder weniger verborgene Weise eine Thatsache der
72 Anschauung entlehnt oder eine Forderung eingeführt wird, die mit | dem Parallelenaxiom *aequivalent* ist. *Analogie* mit der Aufgabe, die *Quadratur des Kreises* zu finden. Aber durchaus nicht immer nutzlose Spekulationen, sondern oftmals bedeutende Fortschritte. Interessant, welche feine und bestimmte Empfindung für die Schwierigkeit des Problems man hatte, ist es, wenn man den Ausspruch von *Henry Savile 1621* liest, vgl. Stäckel-Engel S. 18.*) Auf dem durch Savile errichteten Lehrstuhl lehrte J. *Wallis* mit seinem neuen Beweisversuche (1693). Dann *Sacchéri*, Euklid, von jedem Makel befreit (1733), *Lambert* (1766), derselbe, der 1767 die Irrationalität von π bewies
73 und die *Transcendenz von* π behauptete, die beide schon im | Wesentlichen die Sätze von Legendre (1794–1833) beweisen.[115] Alle drei Mathematiker *glaubten fest an die Beweisbarkeit des Parallelenaxioms.* Sacchèri und Legendre glaubten auch, den Beweis vollständig erbracht zu haben. Von den Arbeiten Lambert's und Legendre's an nimmt die Frage nach dem Beweise

*) Diese beiden Makel sind es gerade, die unsere Axiome IV und V betreffen; ihre *Tilgung* wird in der That unsere *vornehmste und interessanteste* Aufgabe sein.[114]

[112] Meant is *Stäckel and Engel 1895*. See the Introduction to this Chapter, Specific Note (4)(*c*).

[113] What Hilbert refers to is *Clavius 1574*, and its various reprints and editions. See the Introduction to this Chapter, Specific Note (4)(*d*), Remark 1.

[114] For an explanation of 'die beiden Makel', see the Introduction to this Chapter, Specific Note 4(*d*), Remark 3.

[115] The works referred to up to this point are: *Wallis 1663* (published in *Wallis 1693*, hence Hilbert's date), *Saccheri 1733* and *Lambert 1786*. Lambert's results and conjectures on π can be found in *Lambert 1768* and *Lambert 1770b*. See the Introduction to this Chapter, Specific Note (4)(*d*), Remarks 2–4.

des Parallelenaxioms die Aufmerksamkeit der Mathematiker *immer mehr* in Anspruch, vgl. Stäckel-Engel, S. 211–212.[116] Dagegen werden die vermeintlichen Beweise auch immer wieder von anderen Mathematikern verworfen, so von *Seyffer, Voit und Pfaff*. Vgl. Stäckel-Engel S. 215.*)

Vom Zweifeln bis zum Handeln war noch ein grosser wichtiger Schritt, den Gauss that, | [A] indem er eine vom Parallelenaxiom unabhängige Geome- 74
trie aufbaute und dadurch die *Unabhängigkeit des* Parallelenaxioms zeigte.[120] Man kann so sagen, dass Gauss der *erste war* nach etwa *2100 Jahren*, der den Gedanken von Euklid *verstand und klar begriff*, warum Euklid**) das Parallelenaxiom unter die Axiome gestellt hat. Gauss selbst hat nichts über seine schwerwiegenden und tief gehenden Untersuchungen veröffentlicht. Wie weit Gauss gesehen hat, wird der jetzt bald zur Veröffentlichung gelangende *Bd. 8* seiner Werke zeigen. Er hat seine Hauptergebnisse aufgeschrieben, weil er nicht wünschte, dass „*diese Untersuchungen mit ihm untergingen*". Die ersten Veröffentlichungen erfolgten von anderer Seite, nämlich | von *Lobatscheffsky* 75
und Bolyai, die ich schon früher nannte.[121]

---

*) Erwähnt sei noch, dass *Thibaut 1818* noch einen auch in manche Lehrbücher übergegangenen Beweis veröffentlichte zu einer Zeit, wo Gauss in Briefen auf das Verfehlte einer solchen Beweismöglichkeit schon hinwies; Thibaut war mit Gauss gleichzeitig hier Professor. *Killing I, S. 7.*[117]

↑Auch *vor einigen Tagen las* ich die Ankündigung eines völlig strengen Beweises von einem unbekannten Franzosen.↓[118]

[A]Interessante Bemerkungen über Leibniz, der ebenfalls das Parallelenaxiom zu beweisen suchte, Sachéri etc. siehe in Veronese, Grundzüge der Geometrie, Teubner 1894, S. 635–637 und vorher und nachher.[119]

**) So nimmt er *unter allen Gelehrten* des Alterthums eine ganz einzige Stellung ein: er ist noch heute modern, er wird heute ungeändert auf den Schulen gelehrt, während sonst die Mathematik und alle exakten Wissenschaften sich so völlig geändert haben.

---

[116]I.e., *Stäckel and Engel 1895*, 211-212. These are the first two pages of Stäckel and Engel's Introduction to the section on Gauß.

[117]Hilbert is referring to *Killing 1893*, 7, which describes the work of Thibaut (1775–1832); see the Introduction to this Chapter, Specific Note (4)(*d*), Remark 6(i).

[118]This sentence is added by Hilbert in space to the right of 'Stäckel-Engel S. 215. *)'. It is not clear to what Hilbert is referring, although it is quite possibly the work referred to in the *Ausarbeitung*, p. 77, footnote *).

[119]The note is added in the upper margin of p. 74, and is possibly meant as a supplement to the previous note. The work cited is *Veronese 1894*, and Leibniz is indeed discussed on pp. 635–636. See the Introduction to this Chapter, Specific Note (4)(*d*), Remark 5.

[120]Hilbert is here paraphrasing *Stäckel and Engel 1895*, 215. See the Introduction to this Chapter, Specific Note (4)(*d*), Remark 6(iii).

[121]The works Hilbert refers to are *Gauß 1900*, and probably *Lobachevsky 1840* and *Bolyai 1832a*. For the precise quote of Gauß, see the Introduction to this Chapter, Specific Note (4)(*d*), Remark 6(iv).

Vor dem Unabhängigkeitsbeweise empfehlen sich noch einige *sachliche Bemerkungen.*
Satz 1: Die Summe der Winkel eines Dreieckes ist stets = 2 Rechten. Der Aussenwinkel eines Dreieckes ist = der Summe der gegenüberliegenden Innenwinkel.

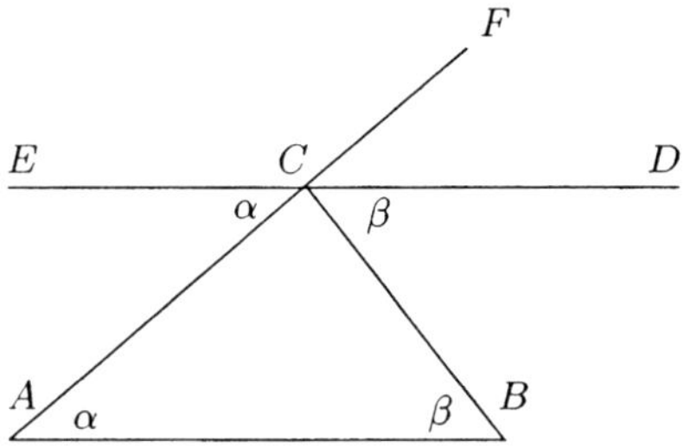

Mache $ECA = \alpha$ und $DCB = \beta$, so können sich die Geraden CE und AB nicht schneiden, wie schon früher bewiesen. Denn sonst würde ein Dreieck entstehen, wo die Summe 2er Winkel = 2 Rechten ist. Ebenso können CD und AB sich nicht schneiden, d. h. ECD ist eine Gerade etc.
Verlängert man $AC = ACF$, so $\sphericalangle FCD = ECA = \alpha$, d. h. $FCB = \alpha + \beta$, q. e. d.

76 ⟨Deletion⟩[122]

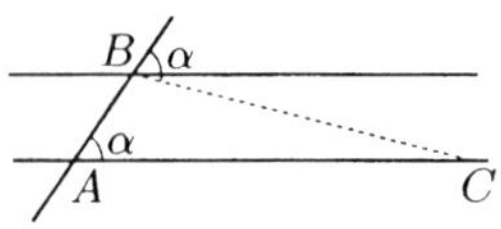

Satz 2. Wenn 2 Parallele von einer beliebigen dritten Geraden geschnitten werden, so sind die Gegenwinkel und die Wechselwinkel gleich. Beweis. $\triangle ABC$ hätte zwei Winkel, deren Summe = 2 Rechten, was nicht möglich.[123]

Satz 3. 2 Gerade, welche einer dritten parallel sind, sind untereinander parallel.

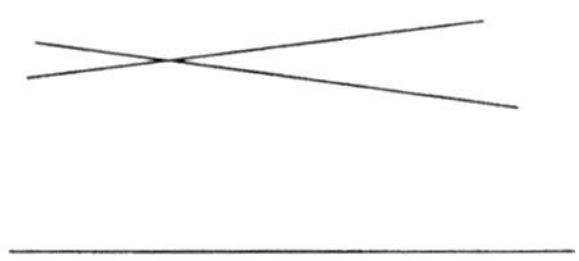

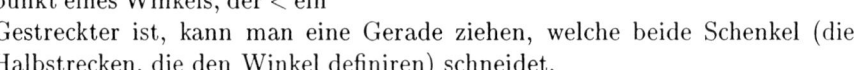

Denn wenn sie sich schnitten, gäbe es durch den Schnittpunkt 2 Parallelen.
Satz 4. Durch jeden Winkelpunkt eines Winkels, der < ein Gestreckter ist, kann man eine Gerade ziehen, welche beide Schenkel (die Halbstrecken, die den Winkel definiren) schneidet.

[122] Deleted: Umgekehrt: der letztere Satz ⟨i. e., Satz 1; Satz 2 was added after the passage here deleted⟩ ersetzt vollständig das Parallelenaxiom. Denn ziehe die Gerade ECD, so dass $\sphericalangle ACE = \sphericalangle A$ ist, $\sphericalangle DCB = \sphericalangle B$ und folglich schneidet ECB ⟨this should be ECD⟩ die Basis AB nicht. Jede andere Gerade durch C schneidet.

[123] The proof Hilbert sketches in the *Ausarbeitung*, p. 79, is simpler.

Denn man ziehe durch P die Parallelen a', b' und dann | eine 77

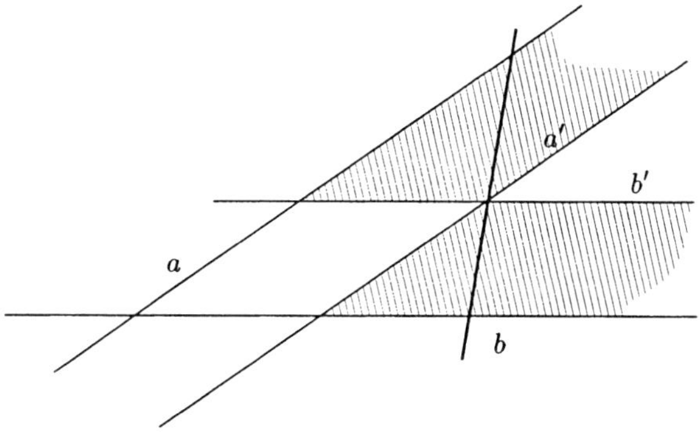

in den schraffirten Winkelräumen verlaufende Gerade.
Satz 5. Die sämtlichen Punkte einer Geraden haben von einer Parallelen gleichen Abstand.

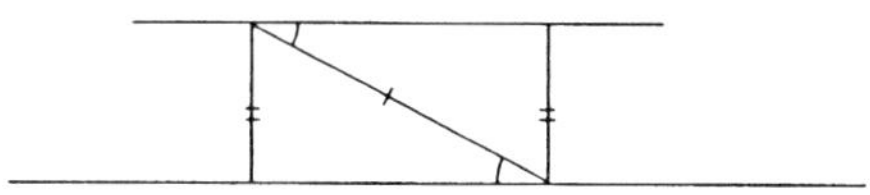

↑Man kann noch andere Sätze hinzufügen, die, wie man sieht, ebenfalls dem Parallelenaxiom gleichwertig sind, z. B.
Satz 6: Durch irgend 3 nicht in einer Geraden liegende Punkte giebt es stets einen Kreis.[124]↓
Es scheint, dass *jeder dieser Sätze* eventuell nach geeigneter umgeänderter Fassung als Ersatz für das Parallelenaxiom dienen kann. Auf diese interessante Frage können wir hier noch nicht eingehen.
↑(Existenz des *Viereckes mit 4 Rechten*)
Bei *Benutzung* des *Parallelenaxioms* ist sie zu *bejahen.*[125]↓

[B] Von den *vergeblichen Beweisversuchen* erwähne ich des *principiellen* 78
Interesses wegen nur den folgenden:

[B]Man kann 6^{tens} auch sagen: durch irgend 3 nicht in einer Geraden liegende Punkte giebt es stets einen Kreis. ⟨Added in the upper margin and not directed to any particular place.⟩

---

[124]This proposition is not mentioned in the *Ausarbeitung*. It is due to Wolfgang Bolyai: see *Bonola 1912* (or *Bonola 1908*), Ch.II, § 29.

[125]With respect to whether any of propositions 1–6 or the existence of rectangles can replace the Parallel Axiom, Hilbert's remark (repeated in the *Ausarbeitung*, p. 80) that 'we cannot go into this interesting question here' suggests that he did not see the situation as quite so straightforward. Indeed, it is not; see the Introduction to this Chapter, Specific Note (5)(a) which discusses briefly the Legendre Theorems and Dehn's work.

Vgl. autographirtes Heft, S.

(Euklid's Grössensatz ist: *Totum parte majus est.*[C])[126]

↑Man braucht das Parallelenaxiom nur für eine *bestimmte Ebene* und eine *bestimmte Gerade* in ihr und einen *beliebigen Punkt* in ihr, da wegen des allgemeinen Congruenzsatzes dann allgemein.*)↓

Nun die Hauptsache, nämlich der Beweis der
Unabhängigkeit des Parallelenaxioms.
Zunächst für die Ebene, d. h. Es soll ein System von Dingen: Punkte und ein System von Dingen: Geraden aufgewiesen werden, für welche alle unsere linearen und ebenen Axiome I–III, aber nicht IV gilt.

Man nehme die *Punkte einer Halbebene* und die zur *Grenzgeraden* ⊥ *Kreise.*[127] Auszuschliessen sind die *Punkte der Grenzgeraden.* Erfüllt sind Axiome I–II.

79 Bez. III zuvor folgende Ueberlegungen:

Vgl. die Autographie dieser Vorlesung S. –S. und mein Collegheft Eindeutige Funktionen mit Transformationen in sich:[128] α, β, γ, δ seien reell. $z = x + iy \ldots$
Es sei $Sz = z' = \left(z : \frac{\alpha z+\beta}{\gamma z+\delta}\right)$, d. h. z ersetze durch $\frac{\alpha z+\beta}{\gamma z+\delta}$ $(= z')$. $\alpha\delta - \beta\gamma \neq 0$. Geraden und Kreise gehen in Geraden und Kreise über. Winkel bleiben erhalten. Beweis Collegheft Eindeutige Funktion mit Transformationen in sich, S. 8–10.[129] Um die Winkeltreue für die Transformation der reciproken

[C]Höchstens könnte man das Parallelenaxiom durch die Forderung ersetzen: der Teil (eines Winkelraumes) ist nicht kleiner als das Ganze (des Winkelraums). ⟨Added just above and to the right of the citation from Euclid, with a line apparently leading to 'Totum'.⟩

*) Ob auch ein *bestimmter Punkt* genügt, ist vielleicht richtig, aber mir unbekannt; folgt wiederum bei Heranziehung des Archimedischen Axioms.

[126]The reference to the 'Autographirtes Heft' is to the *Ausarbeitung*, and indeed pp. 80–81. The mention of Euclid's 'Grössensatz' here partially explains Hilbert's cryptic *Ergänzung*. In the text, he mentions the 'following' example of a failed proof of the Axiom of Parallels, but gives none, only this reference to the *Ausarbeitung*. However, the example he gives there appeals to just this Euclidean 'Grössensatz' applied to the collection of *Winkelpunkte* of an angle. This explains the context of the *Ergänzung*, if not the content. See the *Ausarbeitung*, p. 81 and notes.

[127]The lines should also include the half-lines of the upper half-plane which are perpendicular to the x-axis.

[128]The open reference is to the *Ausarbeitung*, pp. 81–94. The *Collegheft* mentioned by Hilbert is *Hilbert 1892**. The whole independence proof is much better organised in the *Ausarbeitung*, where Hilbert gives all the relevant details of the transformations.

[129]This is *Hilbert 1892**, 8–10, where Hilbert sketches exactly these results for the transformations at issue here. He also refers to *Thomae 1880*, 11–12.

Radien $\left(z : \frac{1}{z}\right)$ einzusehen, nehmen wir irgend 2 durch $x = a$, $y = b$ gehende Kreise[D]

$$\begin{aligned}(x-u_1)^2+(y-v_1)^2 &= \overbrace{(a-u_1)^2+(b-v_1)^2}^{r_1}\\ (x-u_2)^2+(y-v_2)^2 &= \underbrace{(a-u_2)^2+(b-v_2)^2}_{r_2}.\end{aligned}$$ [130]

In ihren Gleichungen sind die Coefficienten von x, y bez.[E]

$$\begin{array}{cc|cc} -2u_1, & -2v_1 & -2u_1, & +2v_1\\ -2u_2, & -2v_2 & -2u_2, & +2v_2. \end{array}$$

Der *Cosinus des Winkels ihrer Tangenten* ist daher

$$\frac{u_1u_2+v_1v_2}{\sqrt{u_1{}^2+u_2{}^2}\sqrt{v_1{}^2+v_2{}^2}},$$

woraus durch Rechnung die Winkeltreue leicht folgt.[131]
↑(Uebrigens auch leicht aus der *conformen Abbildung*, die durch die lineare Funktion, die eine analytische Funktion ist, vermittelt wird.)↓

Nun nehmen wir $\alpha\delta - \beta\gamma = 1$, so gehen die Punkte der oberen ↑positiven↓ 80
Halbebene in Punkte der oberen Halbebene über (Collegheft Automorphe Funktionen S. 78)[F] und es bleiben invariant die beiden Integrale

$$1.) \int \frac{|dz|}{y} = \textit{nicht-Euklidische Länge}, \quad 2.) \iint \frac{dxdy}{y^2} = \textit{nicht-Euklidischer Inhalt}.$$

D

$$\frac{1}{x+iy} = \frac{x-iy}{x^2+y^2}$$
$$x' = \frac{x}{x^2+y^2}$$
$$+y' = \frac{-y}{x^2+y^2}$$

⟨Addition in the margin, written to the right of the two following lines. Hilbert did not assign the addition to any particular place.⟩

E
$$\begin{aligned}&\big(x-u_1(x^2+y^2)\big)^2+\big(-y-v_1(x^2+y^2)\big)^2\\ &\qquad= \big\{(a-u_1)^2+(b-v_1)^2\big\}(x^2+y^2)^2\end{aligned}$$

oder $1-2u_1x+u_1{}^2(x^2+y^2)+2v_1y+v_1{}^2(x^2+y^2)$
$$= \big\{(a-u_1)^2+(b-v_1)^2\big\}(x^2+y^2)$$

⟨Addition in the margin, written to the right of the two following lines. Hilbert did not assign the addition to any particular place.⟩

F Insbesondere die *Orthogonal-Kreise zur Grenzgeraden* bleiben orthogonal. ⟨Addition in the upper margin.⟩

---

130 These should be respectively:

$$(x-u_1)^2+(y-v_1)^2 = (a-u_1)^2+(b-v_1)^2 = r_1^2$$
$$(x-u_2)^2+(y-v_2)^2 = (a-u_2)^2+(b-v_2)^2 = r_2^2$$

where u_1, u_2 are arbitrary, and $v_1 = v_2 = 0$, since the circles are perpendicular to the x-axis.

131 The expression for the cosine is wrong. It is correctly given in the *Ausarbeitung*, p. 86.

Vgl. Collegheft Automorphe Funktionen, S. 82–83.[132]
Damit nun auch die Axiome III gelten, treffen wir die Festsetzung, dass das *Abtragen der* Winkel die gewöhnliche Bedeutung direkt habe, da es

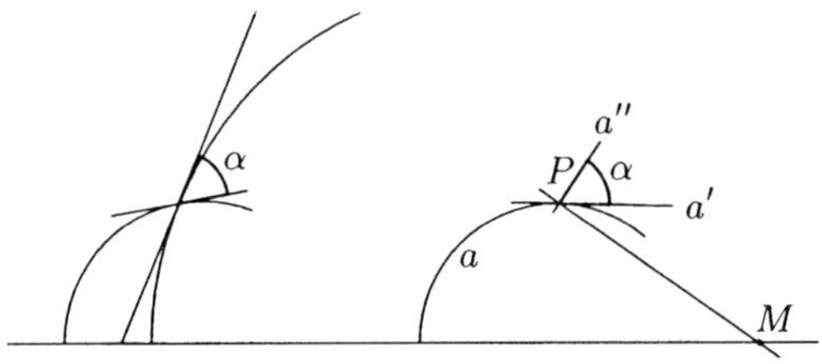

immer einen Kreis ⟨gibt⟩, der durch einen gegebenen Punkt geht, dort eine
81 gegebene Richtung hat und auf | der Grenzgeraden orthogonal steht.[133]

Das Abtragen der Strecke AB auf $A'S'$[134] von A' aus geschieht mittelst der nicht-Euklidischen Länge, indem wir

$$\int_A^B \frac{|dz|}{y} = \int_{A'}^{B'} \frac{|dz|}{y}$$

machen. Dies ist immer möglich. Denn $\int_A^B$ ist, da B nicht auf der Grenzgeraden

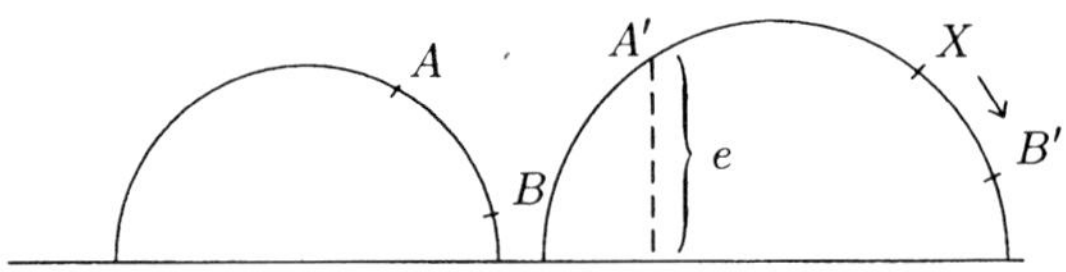

liegt, gewiss endlich. B' existirt. Denn wenn X ein variabeler Punkt x, y auf

---

[132]The two references in this paragraph are to *Hilbert 1892**, 77–78, 82–83. See the *Ausarbeitung*, pp. 84–87.

[133]Hilbert defines angle congruence for the new model as just congruence in the Euclidean sense. There are now two related problems, first to show that transformations of the kind Hilbert is considering leave a given angle unchanged, and second to show that a given angle α can be 'moved' to a given point P on a given line a (in the new sense) with a given direction. The former is dealt with expressly in the *Ausarbeitung*, pp. 85–87; the second is fleetingly addressed in this paragraph.

This latter can be explained as follows. Suppose two semi-circles standing orthogonal to the *Grenzgerade* (thus, lines in the new sense) intersect with angle α, i.e., their tangents at the point of intersection have ordinary Euclidean angle α. Suppose now we wish to form an angle α at P on a 'new line' a (i.e., a is a semi-circle standing orthogonal to the *Grenzgerade*) in the given direction (along the semi-circle a to the right of P in the diagram). First draw the Euclidean tangent to a at P, and call this straight line a'. Now move the Euclidean angle α to P on the line a' in the given direction, so that the two Euclidean lines forming α are a' and a''. Take a perpendicular to a'' through P, and let M be the point of intersection of this perpendicular with the *Grenzgerade*. Draw a semi-circle in the upper half-plane with centre M through P. This semi-circle will be orthogonal to the *Grenzgerade*, and will thus be the new line sought. The labelling of the right-hand part of the figure (apart from 'α') has been added by the editors.

[134]This should be $A'B'$.

$A'B'$ ist, so wird $\int\limits_{A'}^{X}$ unendlich, wenn X sich dem Grenzpunkte nähert (indem man x als Funktion von y betrachtet und $\left(\frac{dx}{dy}\right)^2 > 0$ bedenkt).

$$\text{Denn} \quad \int\limits_{A'}^{X} \qquad > \int\limits_{A'}^{X} \frac{dy}{y} = le - ly = le + l\frac{1}{y}.$$

Nun sind alle Axiome III 1–9 erfüllt. Aber nun auch III 10 (S. 45).[135]

Wir beweisen dies erst für den Fall, dass ein Dreieck eine Seite = einer 82
wirklichen Geraden hat.

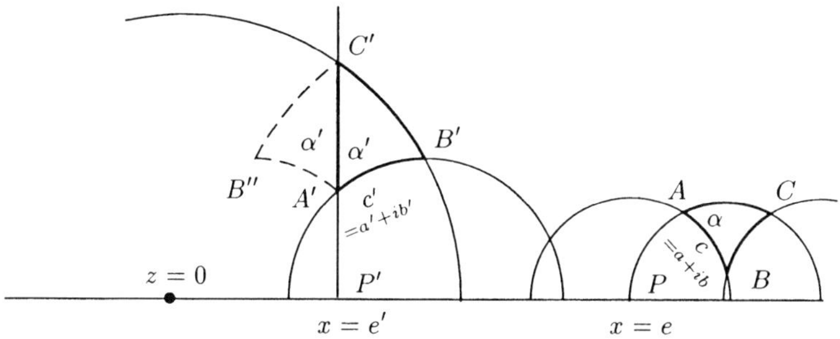

Voraussetzung $AC = A'C'$, $AB = A'B'$, $\alpha = \alpha$,[136] Behauptung: die übrigen homologen Stücke sind gleich.

Beweis. Wir bestimmen α, β, γ, δ reell und nicht sämtlich = 0, so dass, wenn

$$S = \left(z : \frac{\alpha z+\beta}{\gamma z+\delta}\right)$$

gesetzt wird,

$$c'(\gamma c + \delta) = \alpha c + \beta \quad \text{und} \quad e'(\gamma e + \delta) = \alpha e + \beta$$

wird. Es kann nicht $\gamma c+\delta$ null sein, da sonst auch $\alpha c+\beta = 0$ folgen würde. Es kann auch nicht $\gamma e + \delta = 0$ sein, da sonst $\alpha e + \beta = 0$ und folglich $\alpha\delta - \beta\gamma = 0$ wäre. Dann aber wäre $\frac{\alpha c+\beta}{\gamma c+\delta} = \frac{(\alpha c+\beta)(\gamma \bar{c}+\delta)}{|\gamma c+\delta|}$ = reell = c',[137] was nicht der Fall ist. Daher haben wir

$$c' = Sc, \quad e' = Se.$$

Zugleich sehen wir $\alpha\delta - \beta\gamma \neq 0$. [138] 83

---

[135]The missing integrands in this paragraph are each $\frac{|dz|}{y}$. Here 'l' denotes the logarithmic function. See the *Ausarbeitung*, pp. 87–89.

[136]This should be: $\alpha = \alpha'$.

[137]This should be: $\ldots \frac{(\alpha c+\beta)(\gamma\bar{c}+\delta)}{|\gamma c+\delta|^2} \ldots$. Hilbert uses $\bar{c}$ to denote the complex conjugate of c. His calculations are not especially perspicuous here and in what immediately follows.

[138]Deleted: Denn sonst würde $c' = rc$ folgen, wo r reell ist d. h. Die Verbindungsgerade von A

[139] [140] Nun sehen wir zugleich

$$c' = \frac{(\alpha c+\beta)(\gamma \bar{c}+\delta)}{|\gamma c+\delta|} \quad \text{und daher}$$

$$b' = \frac{(\alpha\delta+\gamma\beta)b}{|\gamma c+\delta|} \quad \text{und daher}$$

$d = \alpha\delta - \beta\gamma > 1$. Wir können also, indem wir

$$\alpha\,,\ \beta\,,\ \gamma\,,\ \delta \quad \text{durch} \quad \frac{\alpha}{\sqrt{d}}\,,\ \frac{\beta}{\sqrt{d}}\,,\ \frac{\gamma}{\sqrt{d}}\,,\ \frac{\delta}{\sqrt{d}}$$

ersetzen, annehmen, dass es eine Substitution S mit der Determinante 1 giebt und

$$c' = Sc, \quad e' = Se.$$

Wegen $AC = A'C'$ und weil die Reihenfolge der Punkte P, A, C mit P',
84 A', C' auf den Kreisbögen stimmt, muss

$$C' = SC$$

sein. Wo liegt SB? Entweder in B' oder in B'', wenn $\sphericalangle C'A'B' = \alpha$ und $A'B'' = A'B'$ im gewöhnlichen Sinne der Gleichheit gemacht wird. Dann ist natürlich, weil $A'B'$ und $A'B''$ gegen die Grenzgerade symmetrisch liegen,[141] im Nicht-Euklidischen Sinne

$$A'B'' = A'B' = AB.$$

Die Dreiecke $A'B'C'$ und $A'B''C'$ sind im gewöhnlichen und, da sie zur Grenzgeraden die gleiche Lage haben, auch im Nichteuklidischen Sinne congruent. Die homologen Stücke von $\triangle ABC$ sind also den homologen Stücken von $\triangle A'B'C'$ oder $\triangle A'B''C'$ congruent, je nachdem

$$S(ABC) = A'B'C' \quad \text{oder} \quad A'B''C' \quad \text{q. e. d.}$$[142]

86 Das Parallelenaxiom gilt nicht. Denn

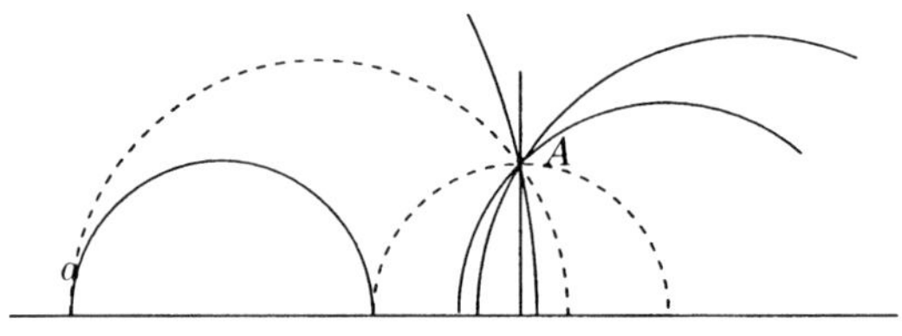

Ebenso gelten unsere 5 Sätze, S. 75–77, *nicht.*
Zu Satz 1. Dass wir ein Dreieck mit beliebig kleiner Winkelsumme construiren können. Erkennen wir, wenn wir zwischen 2 parallelen Geraden im eigentlichen Sinne den Berührungskreis ein wenig vergrössern und dann für die eine Gerade einen sehr grossen Halbkreis nehmen, der die andere Gerade schneidet:

---

[139] Deleted: $\dfrac{\alpha c+\beta}{\gamma c+\delta} = \alpha e+\beta$

[140] Deleted: Nun liegen aber A A' auf derselben Seite der Grenzgeraden. Daher ist die Determinante

[141] What Hilbert presumably means is: because $A'B'$ and $A'B''$ lie symmetrically with respect to the line $A'C'$ which is perpendicular to the *Grenzgerade*.

[142] This proof is carried out much more perspicuously in the *Ausarbeitung*, pp. 89–91, although, exceptionally, the figure there is less informative.

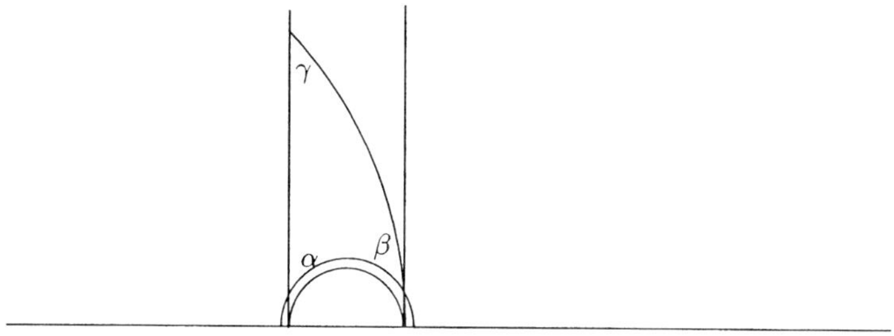

Entsprechend den Legendreschen Sätzen hat jedes Dreieck eine Winkelsumme $< 2R$. Denn der nicht-Euklidische Inhalt eines Dreieckes ist 87

$$I = 4(\pi - \alpha - \beta - \gamma) > 0.$$

Zu Satz 3.[143] Zwei sich schneidende Geraden brauchen beide eine dritte nicht zu schneiden.

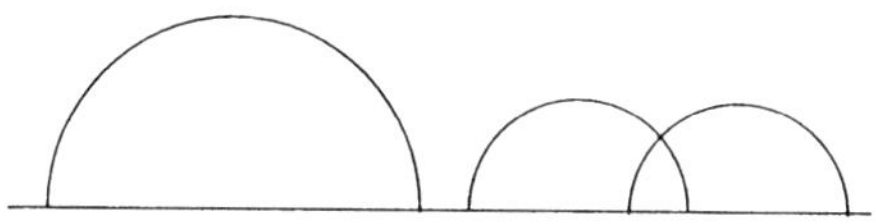

Zu Satz 4. In dem schraffirten Winkelraum giebt es den Punkt A, durch den keine Gerade beide Schenkel trifft.

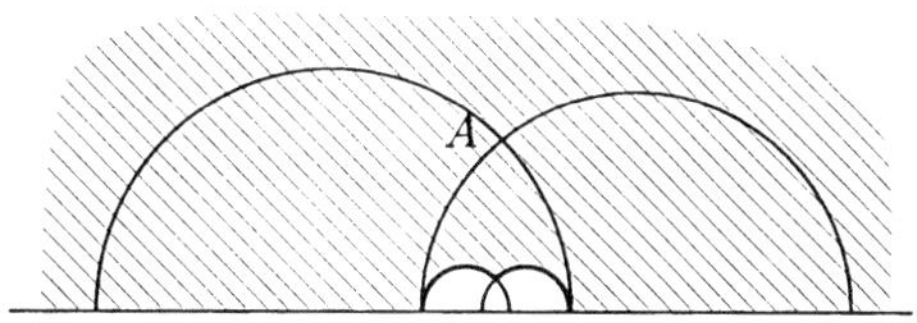

Zu Satz 5. Es giebt kein Rechteck. Wenn

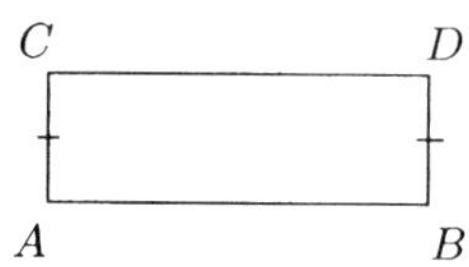

$AC = BD$ und $A = B =$ einem Rechten, so sind die Winkel bei C und D einander gleich, aber $<$ als ein Rechter.[144] 88

Sie haben damit *im Grunde eigentlich* die ganze *Nichteuklidische* ↑*Gausssche Lobatschefskysche*↓ Geometrie in der Ebene und alle ihre *Geheimnisse* sind enthüllt und auch im *wesentlichen erschöpft.*

Nun ist aber das Parallelenaxiom nicht etwa eine Folge der räumlichen Axiome. Zu dem Zwecke denke man sich in der gewöhnlichen Geometrie eine Ebene als Grenzebene $z = 0$ und die orthogonalen Halbkreise bez. Halbkugeln als Ebenen, so gilt Alles wie vorhin, wobei es uns wesentlich zu statten kommt, dass wir nur ebene Congruenzaxiome haben und *alle räumlichen Congruenz-*

[143] For Hilbert's comment on the second of the five theorems, see the *Ausarbeitung*, p. 93.

[144] *Satz* 5 does not actually assert the existence of the rectangle, but rather the equidistance of two parallels to each other. The theorem on rectangles is unnumbered and follows *Satz* 6 in this text (p. 77). However, the fact that this existence theorem follows from the Parallel Axiom is used explicitly in the proof of *Satz* 5 in the *Ausarbeitung*, p. 80.

sätze beweisen konnten. Nicht-Euklidischer Inhalt ist jetzt

$$\int \sqrt{\left(\frac{dx}{dt}\right)^2 + \left(\frac{dy}{dt}\right)^2 + \left(\frac{dz}{dt}\right)^2}\ \frac{dt}{z}.$$

89 Der *Desarguessatz* gilt, wie schon früher hervorgehoben. Aber auch der *Pascal* muss wegen der geltenden Congruenzsätze bestehen.[145]
Noch *andere Deutung* der Punkte, Geraden, Ebenen der Gaussschengeometrie durch die Mittel der Euklidischen Geometrie ist möglich und sei hier kurz erwähnt.
Transformation der Kreisfläche in sich. Killing, Bd. 1, S. 13–17. Vgl. autographierte Ausarbeitung.
Transformation T, die $x^2 + y^2 = 1$ in sich überführt.
Formeln Killing, S. 16.

$$\begin{vmatrix} a & b & c \\ a' & b' & c' \\ a'' & b'' & c'' \end{vmatrix} \begin{vmatrix} a & b & -c \\ a' & b' & -c' \\ a'' & b'' & -c'' \end{vmatrix} = -1 \quad \text{d. h.} \begin{vmatrix} a & b & c \\ a' & b' & c' \\ a'' & b'' & c'' \end{vmatrix} = \pm 1 \quad \text{also} \neq 0$$

d. h. die umgekehrte Transformation T^{-1} in sich existirt. Der Mittelpunkt M geht in den Punkt $x' = \frac{c'}{c}$, $y' = \frac{c''}{c}$, c stets $\neq 0$ und $\left(\frac{c'}{c}\right)^2 + \left(\frac{c''}{c}\right)^2 - 1 = -\frac{1}{c^2}$, d. h. < 0, d. h. der Mittelpunkt geht in einen inneren Punkt über. Umgekehrt. Für jeden inneren Punkt P giebt es zunächst eine Transformation T wenig-
90 stens, so dass $TP = M$. Trans|formation T, so dass $TM = M$ giebt es nur die Drehung. Denn $c' = 0$, $c'' = 0$, $c = 1$; daraus $a = 0$, $b = 0$ etc.
Definition des Abtragens *eines Winkels.* Definition *der Strecke*: Killing S. 15.
Beweis des Congruenzaxioms 10.[146]

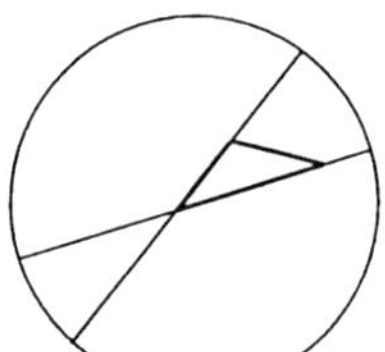

Analog im Raume die Transformation der Kugel in sich.
(Cayley'sche Maassbestimmung)

[145]Hilbert is referring to the model of Desargues's Theorem given on p. 33 of this text. Pascal's Theorem follows without the Parallel Axiom in a space which satisfies the congruence axioms: see *Schur 1899*, mentioned by Hilbert several times in the 1898/1899 lectures and the *Festschrift.*

[146]The work referred to over the last page is *Killing 1893*, 13–17, i.e., §7. Hilbert's treatment is sketchy and would not be readily comprehensible without the treatment in the *Ausarbeitung* to which Hilbert also refers; see pp. 93–99. (Hilbert does not refer to Killing's book there.) *Killing 1893*, 14–15, deals with those projective transformations of a disc onto itself which leave the centre fixed. On op. cit., p. 15 Killing gives an argument as to why any angle at the centre M of the given unit circle remains unaltered by one of these transformations. He also asserts that one can then prove Euclid's triangle congruence theorems, among which is Hilbert's III, 10. This model is basically Klein's model of hyperbolic geometry.

↑Die Existenz der gewöhnlichen Euklidischen Analytischen Geometrie werden wir später nachweisen.↓[147]
Sie haben damit zugleich im Wesentlichen die Principien derjenigen Geometrie, die man die nicht-Euklidische nennt und in der alle Axiome I–III gelten, aber nicht IV.
↑Das räumliche Analogon des Parallelogramms des Parallelenaxioms kann bewiesen werden:↓[148]
Es folgt *sofort*:

Satz: Durch einen Punkt P ausserhalb einer Ebene ↑E↓ lässt sich stets *nur eine Ebene* ziehen, welche die erstere Ebene nicht schneidet; diese Ebene heisst eine Parallel-Ebene zu jener Ebene.

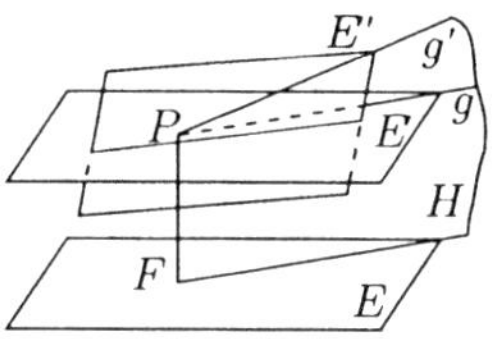

Beweis: Man fälle von P auf E das Loth PF und ziehe zu PF eine $\perp$ Ebene durch P. Diese Ebene E' schneidet E nicht. Dagegen jede andere Ebene E'' durch P schneidet E ↑wie↓ man erkennt, wenn man durch PF eine Ebene H legt, welche E', E'' in 2 verschiedenen Geraden g, g' schneidet.

Ideale Punkte, Geraden, Ebenen: Vgl. die autographirte Vorlesung S. [149]. 91
Von den idealen Punkten, Geraden und der idealen Ebene gelten die Axiome I, aber keineswegs die Axiome II, III. Einführung idealer Begriffe ist in der Mathematik ein wichtiges, fruchtbares und häufig vorkommendes Hülfsmittel: Negative, irrationale, imaginäre, ideale Zahlen etc., transfinite Zahlen etc.[150]
Jeder Schnittpunktsatz der Ebene gilt auch für ideale Elemente, wenn er für die wirklichen richtig ist. Beweis

z. B. Desargues:

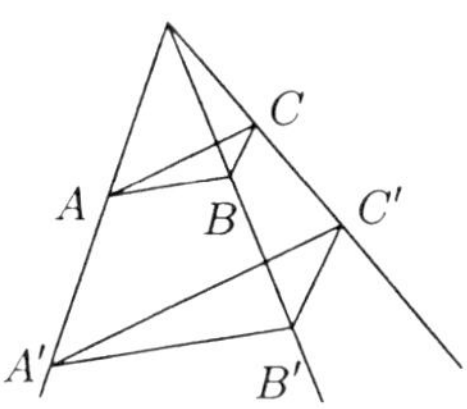

Pascal:

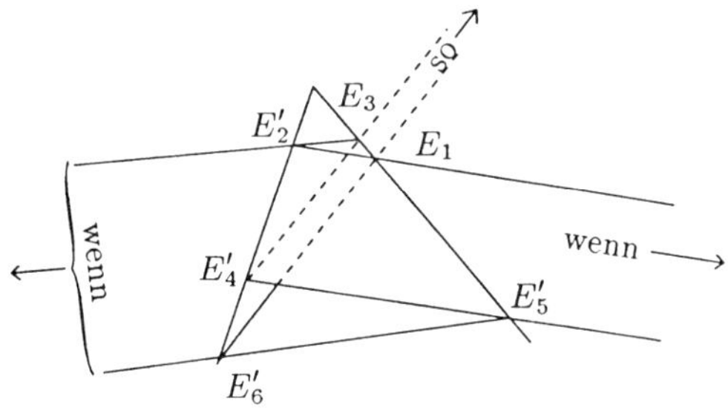

Vgl. autographirte Vorlesung:[151]

[147] See p. 104 of this text, and the *Ausarbeitung*, p. 167.

[148] I. e., from the Parallel Axiom. Hilbert has presumably neglected to delete 'Parallelogramms' here.

[149] The relevant pages are 101–106 of the *Ausarbeitung*.

[150] This is Hilbert's first mention of the use of ideal elements as a general method in mathematics. See also p. 103 of this text, and then the *Ausarbeitung*, p. 101 and the important remarks on 166–167.

[151] The reference is to the *Ausarbeitung*, pp. 105–107, where Hilbert proves this assertion. The Parallel Axiom enables Hilbert to adopt the simplified versions of the Desargues and Pascal Theorems which he uses here, in the *Ausarbeitung* and in the *Festschrift* for the construction of various coordinate fields.

↑Indem man andere Geraden ins Unendliche schiebt, kann man noch andere spezielle Fälle finden, die nützlich sind, z. B.

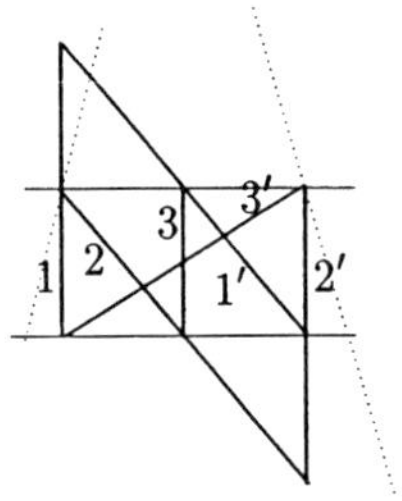

Dieser Fall kann durch Congruenzsätze leicht bewiesen werden. Suche den allgemeinen durch Projektion zu erhalten.[152] ↓

Projektive Bemerkungen: Desargues und Pascal entsprechen sich selbst dual. Desargues = Zwei sich selbst um- und einbeschriebenen Fünf-Ecken. Pascal = 3 sich cyklisch einbeschriebenen Dreiecken. Vgl. Schoenflies, Math. Ann. Bd. 31.[153]

92 Die auf den Axiomgruppen I–IV beruhende Geometrie nennen wir Euklidische Geometrie. Wie weit reicht ihre Gewalt?

Zunächst Beweis von Pascal in der Ebene, also mit Hülfe der Schulgeometrie. Schur hatte ihn im Raume bewiesen.[154] Der folgende Beweis ist sehr merkwürdig und für das folgende von fundamentaler Bedeutung:

Autographirte Vorlesung: Beweis des Pascal.[155]

Merkwürdig ein wie feines Verständniss Henry Savile 1621 an den Tag legte, als er von den beiden Makeln sprach (Stäckel-Engel, Parallelentheorie, S. 18), die dem herrlichen Körper der Geometrie anhaften. Der zweite – die Lehre von den Proportionen lässt sich nun in der That vollständig tilgen, so dass Euklid in soweit recht hatte, kein neues Axiom zu brauchen.[156]

Die Euklidische und in alle Schulbücher übernommene Lehre von den Proportionen in der Elementargeometrie beruht auf folgenden 2 Sätzen:

1.) Wenn $AB \parallel A'B'$, so $a : b = a' : b'$.
2.) Wenn $a : b = a' : b'$, so $AB \parallel A'B'$.

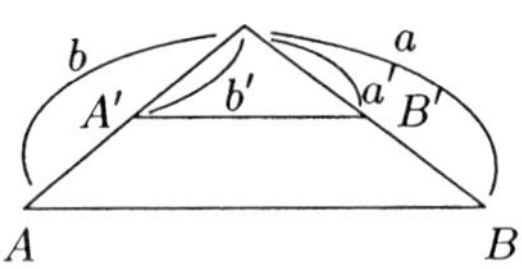

Die Beweise sind richtig, wenn a, a' durch Abtragen der nämlichen Strecke entstanden sind, z. B. für den Fall $a = 2a'$. Wenn dies aber nicht geht (in-
93 commensurabler Fall) so werden 2 Irrthümer | auf einmal begangen 1.) dass man jedem Punkt überhaupt eine Zahl zuordnen kann, setzt ein neues Axiom voraus.[157] 2.) dass für die neu einzuführenden Zahlen wieder die alten Rech-

[152] It is not clear to what Hilbert is referring here, nor is the point of the figure clear.

[153] The reference is to *Schoenflies 1888*, 58–59.

[154] The reference is to *Schur 1899*.

[155] See the *Ausarbeitung*, pp. 108–111.

[156] The reference is to *Stäckel and Engel 1895*, 18. See the Introduction to this Chapter, Specific Note (4)(d), Remark 3.

[157] Hilbert adds the Archimedean Axiom partly for the purpose of guaranteeing this.

nungsregeln wie für die rationalen Zahlen ⟨gelten,⟩ insbesondere commutatives Gesetz der Multiplikation gilt. ↑Also dass aus $a : b = c : d$, immer $a : c = b : d$ folgt etc.↓

Es entsteht die Frage: lässt sich dennoch die Lehre von den Proportionen schon hier in der Euklidischen Geometrie begründen. In der That:
Es sei ein rechter Winkel gegeben, und eine beliebige Strecke $\neq 0$ als Strecke 1 bezeichnet.

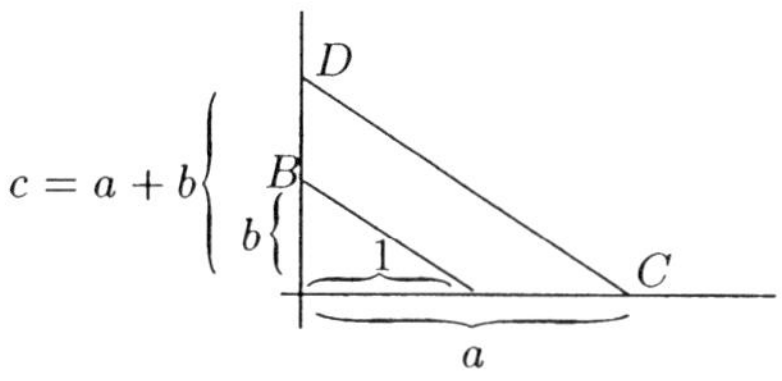

Wenn $AB \parallel CD$, so heisst $OD = ab$, oder wenn wir $OD = c$ nennen, so gilt die Streckengleichung[158]

$c = ab$. ↑(Es ist die Proportion $1 : b = a : ab$, nur anders geschrieben.)↓

Das Produkt ab, wo a der erste, b der 2te Faktor heisst, ist eindeutig definirt und wieder eine Strecke.

Von hier an bis S. dieses Collegheftes siehe die autographirte Vorlesung S. – S. .[159]

Wenn $ab = 0$, so ist entweder $a = 0$ oder $b = 0$. Wenn $a_1 > a_2$, so $a_1 b > a_2 b$: 94
Grösseres mit gleichem multiplicirt giebt Grösseres.

1.) $ab = ba$ (Commutatives Gesetz) nach Pascal.

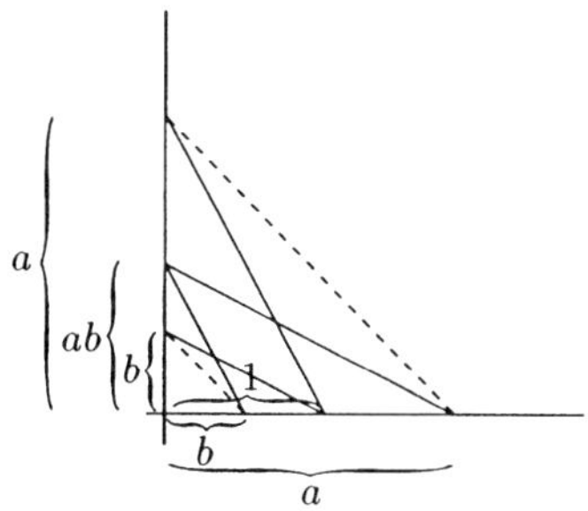

[158] The figure is confusing as it stands; see the corresponding figure in the *Ausarbeitung*, p. 113.

[159] It is not clear to which later page of the present notes Hilbert refers, or thus to which pages of the *Ausarbeitung* he refers. The sections on the three laws $ab = ba$, $a(bc) = (ab)c$ and $a(b + c) = ab + ac$, on negative quantities, and on analytic geometry over the resultant field, are properly proved in the *Ausarbeitung*, pp. 114–121, whereas they are only indicated sketchily here. The correct page in the present text would thus be p. 99.

2.) $a(bc) = (ab)c$ Associatives Gesetz.

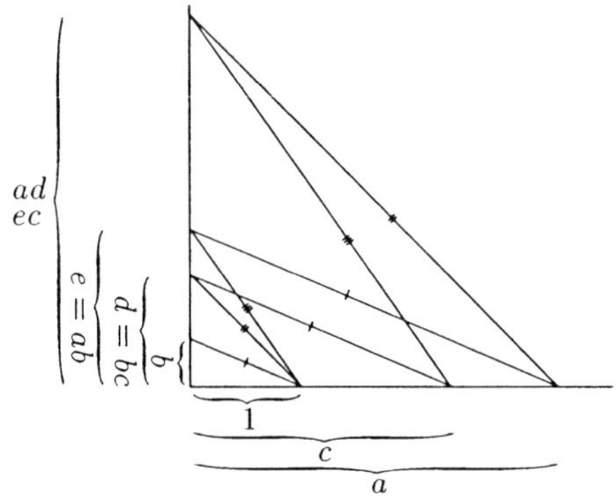

Also $ad = ec$ nach Pascal.

3.) $a(b + c) = ab + ac$ Distributives Gesetz.

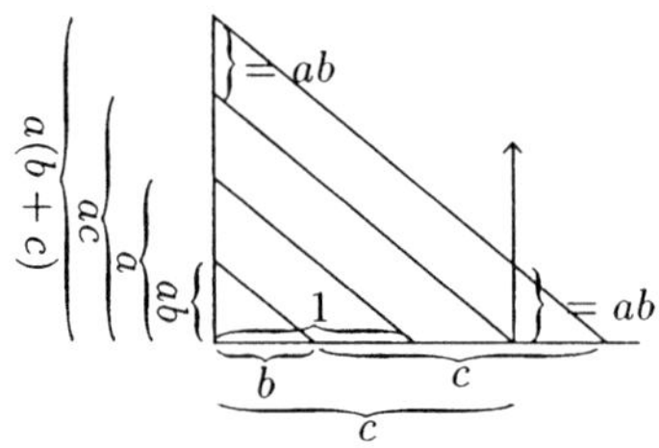

95 Wenn man negative Strecken wie üblich definirt, so gelten alle Rechnungsgesetze auch für diese und die gewöhnliche Vorzeichenregel.

Wenn $b \neq 0$, so giebt es immer eine Strecke a, so dass $c = ab$ wird; man schreibt $a = \frac{c}{b}$. Also Division durch jede von $0 \neq$ Strecke möglich und jede Gleichung darf durch eine von $0 \neq$ Strecke auf beiden Seiten dividirt werden. Denn:

$$bA = bB, \quad \text{so } b(A - B) = 0, \quad A = B.$$

Die Proportion $a : b = a' : b'$ bedeutet $ab' = ba'$, also folgt

$$a : a' = b : b \quad \text{etc.}$$

Zwei Dreiecke heissen ähnlich, wenn die homologen Winkel = sind.

Satz. In 2 rechtwinkligen ähnlichen Dreiecken mit den Katheten a, b und a', b' gilt die Proportion $a : b = a' : b'$.

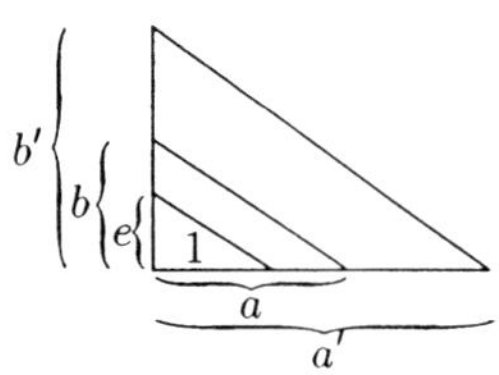

Beweis. Es ist nach der Definition des Streckenproduktes

$$b = ea, \; b' = ea', \quad \text{folglich}$$

$$ab' = ba' \quad \text{q. e. d.}$$

Die Umkehrung dieses Satzes folgt leicht.

Aus diesem Satz kann auch der entsprechende Satz für beliebige ähnliche Dreiecke $a : b : c = a' : b' : c'$ abgeleitet werden, woraus die beiden obigen Fundamentalsätze der Lehre von den Proportionen folgen. Und hieraus der Satz von Desargues.

Wir nehmen in unserer Ebene 2 zueinander $\perp$ Geraden als x-, y-Achse. 96
Von dem Schnittpunkt an rechnen die Strecken, so dass jeder Punkt auf der x-Achse durch eine Strecke bestimmt ist und zwar positive Strecke bez. negative, jenachdem der Punkt auf der einen Seite oder auf der anderen vom Nullpunkt aus liegt. Diesem kommt die Strecke 0 zu. Ebenso auf der y-Achse. Bestimmung eines beliebigen Punktes P durch die senkrechten Abstände, Coordinaten x, y. Jetzt ziehe durch den Nullpunkt eine beliebige Gerade, nenne einen bestimmten und einen beliebigen Punkt C, P, fälle von C, P die Lothe, so sind die Verbindungslinien den Fixpunkten parallel, daher

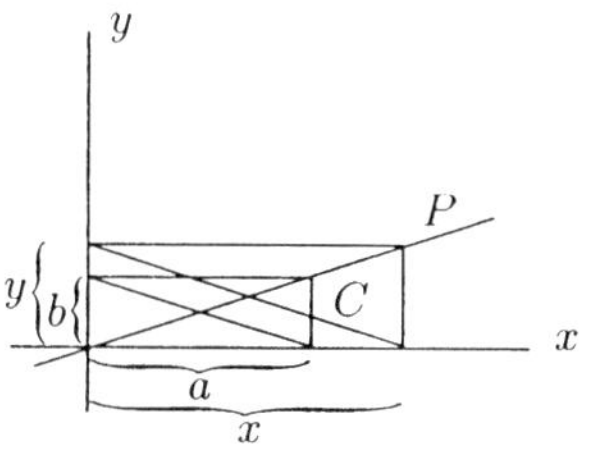

$$a : b = x : y, \quad \text{d. h.} \quad bx - ay = 0,$$

Gleichung der Gerade.[160] Ist nun eine beliebige Gerade gegeben, die die x-Achse an der Stelle $x = c$ schneidet,

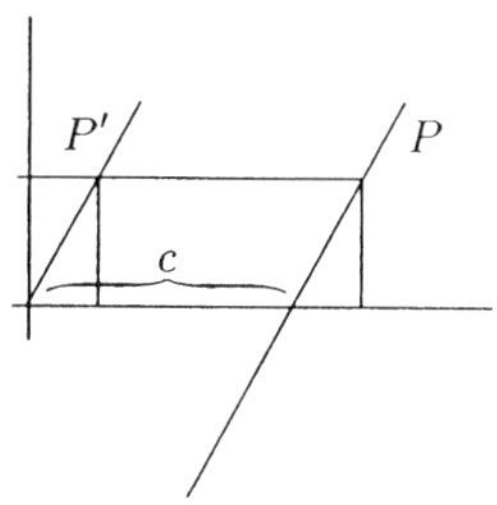

so ist die Gleichung der Parallelen durch den Nullpunkt $bx - ay = 0$, folglich die Gleichung der ursprünglichen Geraden

$$b(x - c) - ay = 0$$
$$bx - ay - bc = 0.$$

Umgekehrt jede Gleichung von der Form 97

(1) $\quad ux + vy + w = 0 \quad (u : v : w)$

stellt eine Gerade dar. Denn $u \neq 0$, so obige Form etc. Alle Geraden schneiden sich. Nur Parallele nicht, d. h. wenn

$$u_1 : u_2 = v_1 : v_2.$$

Wenn man auf einer Geraden (1) 3 Punkte

$$x_1, y_1 ; \quad x_2, y_2 ; \quad x_3, y_3$$

hat, so liegt 2 zwischen 1 und 3, wenn

$$x_1 < x_2 < x_3 \quad \text{oder} \quad x_1 > x_2 > x_3$$

[160]That is, drop perpendiculars from C (the fixed point) and P (the arbitrary point) to the two axes, and assign coordinates (a, b) to C and (x, y) to P. The diagram suggests that by the 'Verbindungslinien' Hilbert means the lines joining the coordinate values for C and P respectively.

ist; dann ist zugleich

$$y_1 < y_2 < y_3 \quad \text{oder} \quad y_1 > y_2 > y_3\,,$$

wie durch Multiplikation mit u und Subtraktion von w leicht gezeigt wird.

Alle Punkte x, y, für welche

$$ux + vy + w < 0 \quad \text{bez.} \quad > 0$$

liegen auf einer bez. der anderen Seite der Geraden (1). Denn sei z. B

$$ux_1 + uy_1 + w < 0 \quad \text{und} \quad ux_2 + uy_2 + w > 0$$

so construire eine Gerade durch 1, 2,[161] die $ux+vy+w = 0$ in x_3, y_3 schneidet, so $ux_3 + vy_3 + w = 0$

$$\begin{array}{r|l} b & ux_1 + vy_1 + w \quad < \quad ux_3 + vy_3 + w \quad < \quad ux_2 + vy_2 + w \\ -v & ax_1 + by_1 + c = 0, \quad ax_3 + by_3 + c = 0, \quad ax_2 + by_2 + c = 0 \end{array}$$

98 folglich

$$(bu - av)x_1 + (bw - cv) \lesseqgtr (bu - av)x_3 + (\quad) \lesseqgtr (\quad)x_2 + (\quad)$$

$$x_1 \quad \lesseqgtr \quad x_3 \quad \lesseqgtr \quad x_2 .$$

Nun kann man durch Rechnung nach der Methode der analytischen Geometrie den Satz zeigen, dass das Doppelverhältniss von 2 vier Punkte, die perspektiv liegen, gleich ist. Ferner die Sätze von Menelaus und Ceva (?).[162] Folglich Satz, dass die Höhen, Transversalen etc. sich in einem Punkte schneiden.

Satz, dass die Abschnitte der Sehnen im Kreise gleiches Produkt haben. $ab = cd$ wegen der Aehnlichkeit der Dreiecke. Theorie des Feuerbachschen Kreises und der merkwürdigen Punkte des Dreieckes, Ptolemäischer Satz etc.[163]

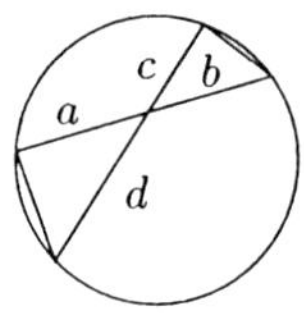

Analog im Raume

$$ux + vy + wz + t = 0.$$

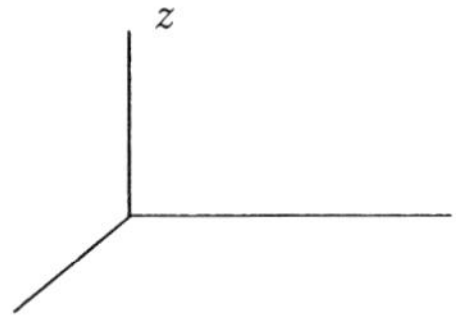

99 ^{G} Nun eine wichtige und interessante Anwendung auf die Flächeninhalte: Vgl. autographirte Vorlesung S. –S. .[164]

↑Wenn 2 Polygone einem dritten flächengleich sind, so sind sie untereinander flächengleich, erfordert im Beweis schon einen Gedanken, *der bei der Defini-*

^{G}Idealer Punkt $= \infty$. Rechnung mit ∞: $\infty + a = \infty$ ist Festsetzung, weil Abtragen nicht möglich. $a\infty = b\infty$, $\frac{a}{0} = \infty$. ⟨Added in the upper margin of p. 99 and not directed to any specific place.⟩

[161] Hilbert means (x_1, y_1) and (x_2, y_2). In general, the notation here is confusing, and the whole development of this paragraph is given more clearly in the *Ausarbeitung*, pp. 120–21.

[162] Hilbert's question mark in brackets is probably to signal doubt about his spellings of the names. They are correct.

[163] In the *Ausarbeitung*, p. 122, Hilbert also mentions all these theorems, and then adds the remark that they can all be proved 'without any consideration of continuity'.

[164] The treatment of *Flächeninhalt* is in the *Ausarbeitung*, pp. 122–138.

tion des Integrals benutzt wird, nämlich die Ueberlagerung der beiden Einteilungen.↓

Unsere Freude über die Leichtigkeit der Beweisführungen nach Euklidischem Beispiele wird aber sehr getrübt, wenn wir näher hinsehen und erkennen, dass es in Wahrheit mit diesen Euklidischen Schlüssen sehr schlimm bestellt ist. Zwar Alles richtig, aber sämtliche Behauptungen sind leer und bedeutungslos, so lange nicht vor Allem gezeigt ist, dass es Polygone verschiedenen Inhaltes giebt und ferner, dass wenn 2 Rechtecke mit gleicher einer und verschiedener anderer Seite nicht inhaltsgleich sind.

$$\boxed{I_1 \,|\, I_2} \qquad I_1 + I_2 \neq I_1$$

Es könnte $I_1 = 2I_1$ apriori sein oder $I_1 + I_2 = I_1$, d. h. es handelt sich um den Beweis eines von Killing aufgestellten Satzes.

Es ist nicht möglich, ein Rechteck so zu zerschneiden in n Dreiecke, dass nach Wegnahme eines von ihnen bei geeigneter Anordnung die übrigen $n-1$ Dreiecke ebenfalls noch bedecken.

Die Fassung von Stolz: Killing, Bd II, S. 23.

[H]Oder auch: Wenn 2 Dreiecke mit gleicher Grundlinie gleichen Inhalt haben, 100
so haben sie auch gleiche Höhe. Giebt es überhaupt Dreiecke, die nicht inhaltsgleich sind? Totum parte majus est ist anwendbar? Apriori natürlich nicht, da eben dieser allgemeine Grössensatz sich in einen geometrischen Satz verwandelt, sobald er auf unsere geometrischen Begriffe angewandt wird. Stolz glaubt den Satz entweder als Axiom nehmen zu müssen, und Killing beweist ihn mit Hülfe des Archimedischen Axioms. Beides trifft nicht das Wesentliche, in dem der Satz ohne Archimedes beweisbar ist.

Beweis: Autographirte Vorlesung S.   – S.   .[165]

Aus dem Killing-Stolzschen Satze könnte umgekehrt eine Begründung der Lehre von den Proportionen entnommen werden, indem man festsetzt, es solle

$$a : c = d : b$$

bedeuten, dass das aus den Seiten a, b gebildete Rechteck dem aus c, d gebildeten Rechteck gleich sei, und man beweist, dass dann nothwendig $AB \parallel DC$ ausfällt und umgekehrt etc.[166]

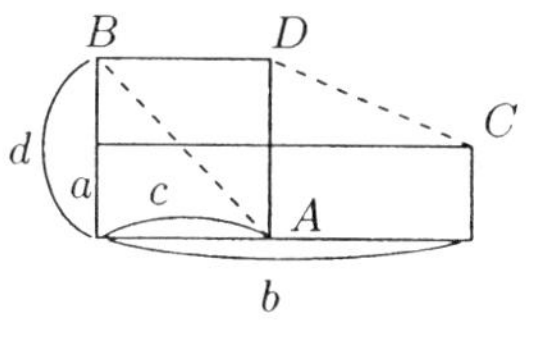

[H] *Man muss flächengleich und inhaltsgleich* beides haben. Vgl. autographirte Vorlesung. ⟨Added in the upper margin of p. 100 and not directed to any specific place.⟩

---

[165] That is, the *Ausarbeitung*, pp. 122–38. The reference to Killing's book at the end of p. 99 is to *Killing 1898*, 22–31. There is also reference to the 'Killing-Stolz Theorem' in the *Ausarbeitung*, pp. 129-130, and also in the *Festschrift*, p. 49. The latter adds a little more information, and specific references to Schur and Stolz.

[166] The figure to the right is not correct; there are several important points unlabelled and lines undrawn. The appropriate diagram can be found (with an appropriate proof) in the *Ausarbeitung*, pp. 130–131.

101 V Archimedisches Axiom.

In 3 Fassungen:

1.) Liegt B auf einer Geraden zwischen A und C und construirt man die Punkte A, B, B', B'', B''', ... in dieser Reihenfolge, so dass

$$AB = BB' = B'B'' = \dots$$

wird, so treffen wir in der Reihe der Punkte B, B', B'', ... sicher einen solchen Punkte $B^{(n)}$, so dass C zwischen A und $B^{(n)}$ liegt.
Wir können auch sagen: Wenn a, b irgend 2 Strecken, so giebt es eine ganze Zahl n, so dass $na > b$.

Bei dieser Fassung sind die Congruenzsätze nöthig, d. h. Axiome I, II, III.

2.) Projektive Fassung: Liegt B auf einer Geraden zwischen A und D und ist C irgend ein Punkt zwischen B und D und construirt man die Punkte A, B, B', B'', B''' in dieser Reihenfolge, so dass

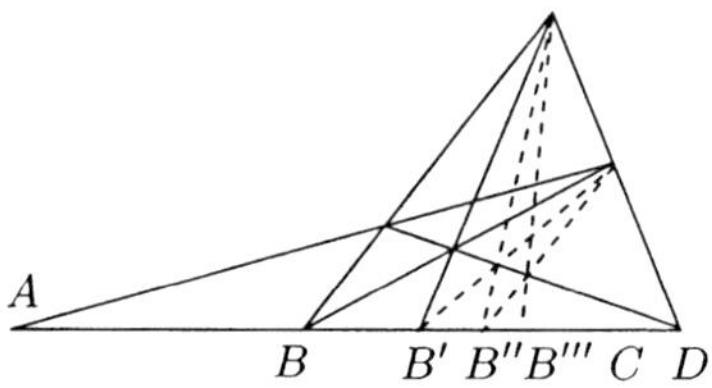

$$ABB'C\,,\ BB'B''C\,,\ B'B''B'''C\,,$$

harmonische Punkte sind: so treffen wir in jener Punktreihe sicher einen Punkt $B^{(n)}$ an, so dass C zwischen A und $B^{(n)}$ liegt.

102 Bei dieser Fassung brauchen wir nur die Axiome I und II.

3.) Vgl. meine Note über kürzeste.

Vergl. autographirte Vorlesung S. ⟨....⟩–S. ⟨....⟩ *)[167]

Desargues ist nicht mittelst der ebenen Axiome I, II, V beweisbar, wohl aber aus Desargues der Pascal. Da Alles auf Pascal beruht, so ist diese Erkenntniss sehr wichtig und wir wollen daher genauer das Verhältniss von Desargues und Pascal erörtern.

*) Natürlich kommt man durch fortgesetztes Halbiren einer Strecke nicht in jede Strecke hinein, z. B. $\frac{1}{2^n} > \frac{1}{t}$. Bei gewöhnlichen Zahlen kann dies nicht passiren, obwohl auch für unsere complexen Zahlen $(1, t)$ der Satz gilt, dass eine Zahl $= 0$ ist, wenn sie absolut kleiner ist als jede beliebige positive Zahl. In dem rechtwinkligen $\triangle$ mit den Katheten 1 und $\frac{1}{t}$ ist der kleinere Winkel kleiner als jeder Winkel, der aus einem Rechten durch fortgesetzte Halbirung entstehen kann.

[167] Hilbert's '3.)' refers to the third version of the Archimedean axiom, and the further reference is to his paper *Hilbert 1895b*, 92. This version is stated, essentially unaltered, in the *Ausarbeitung*, 140–141. Neither 2.) nor 3.) appears in the *Festschrift*.

Concerning the footnote, various remarks are in order.

(1) Although a footnote mark appears in the text, the footnote itself is unmarked, and so there is no certain way of telling to which place in this text (or the *Ausarbeitung*) it refers. (It appears at the foot of the page under a line across the sheet; it was probably an addition.) The content concerns non-Archimedean geometries, i.e., geometries in which there are segments which can never be surpassed by integral multiples of smaller

Also wir nehmen nun im Raume I, II an oder nehmen Desargues und die ebenen Axiome I, II und fragen, was weiter folgt. Zu dem Zwecke suchen wir eine Rechnung mit Strecken unabhängig von III zu begründen.[168]

Obwohl nicht nothwendig, so setzen wir noch das Parallelenaxiom IV voraus. Desarges nimmt dann die Form an: Wenn in 2 Dreiecken entsprechende Seiten paarweise ||, so liegen die Dreiecke perspektiv und wenn 2 Dreiecke perspektive liegen und 2 Paar Seiten paarweise parallel sind, so sind auch die 3ten Seiten parallel.

$$
\begin{aligned}
a + b &= b + a \\
a + (b + c) &= (a + b) + c \\
a(bc) &= (ab)c \\
a(b + c) &= ab + ac
\end{aligned}
$$
[169]

---

segments. t here yields one such; since, if $\forall n[n < t]$ in the non-Archimedean ordering, then $\forall n[\frac{1}{t} < \frac{1}{2^n}]$. Now suppose that $\frac{1}{t}$ stands for a segment (i.e., the segment from 0 to $\frac{1}{t}$). Consider the segment from 0 to 1; in continued halving of this, the right end-points of the resulting segments beginning with 0 will be marked by the decreasing numbers $\frac{1}{2^n}$. But these segments will never fall wholly inside the segment $(0, \frac{1}{t})$, which explains Hilbert's reference to *'fortgesetztes Halbieren'*. (See also the *Ausarbeitung*, 145.) Hilbert then hints at an analogous point about angles. The example is surely taken from Hilbert's first model of a non-Archimedean geometry presented in the *Ausarbeitung*, 141–145. The same model is given in the *Festschrift*, 24–25. A more complicated model, this time involving a non-commutative, non-Archimedean field is given in the *Ausarbeitung*, 164–166, and the *Festschrift*, 73–75.

(2) 'Our complex numbers' is a (rather misleading) term introduced by Hilbert in the *Ausarbeitung*, 142, to describe the elements which form the basis of the non-Archimedean (analytic) geometry defined there. Hilbert starts from the smallest (countable) Pythagorean field containing 1 and an arbitrarily given real number t. The new 'complex' numbers are then in effect the functional expressions defining the elements of this field, on which Hilbert defines a total ordering which is non-Archimedean. Later in the *Ausarbeitung*, 160–162, he gives axioms for 'complex numbers in the most general sense' (160). These are, in effect, the axioms for Archimedean ordered fields. (On p. 162, Hilbert mistakenly says that these axioms give the system of 'ordinary real numbers'.) He also then refers to numbers satisfying these axioms minus the Archimedean Axiom and the axiom expressing the commutativity of multiplication as 'complex numbers'. The axioms for 'complex numbers' reappear in the *Festschrift*, 26–28 ('§13. Complexe Zahlensysteme'), then further as the core of the axiom system for the theory of real numbers (complete ordered fields) in *Hilbert 1900b*, 182–183 (Axiom groups I–III). There, to the Archimedean Axiom is added Hilbert's *Vollständigkeitsaxiom*.

[168]For comments on the developments alluded to here and the next paragraph, see the Introduction to this Chapter, Specific Note (1).

[169]The connection between the revised Desargues Theorem which Hilbert gives here and the addition and multiplication principles stated is that these latter are just the rules governing addition and multiplication (together with $(a + b)c = ac + bc$) which can be proved in a *Streckenrechnung* using Desargues Planar Theorem but neither the congruence axioms nor Pascal's Theorem. This is shown in the *Ausarbeitung*. Pascal's Theorem is required to prove in addition the commutativity of multiplication: see the Introduction to this Chapter, Specific Note (1)(b).

103 Nathan der Weise: – Der Wunder höchstes ist,
Dass uns die wahren, echten Wunder so
Alltäglich werden können, werden sollen.
Ohn' dieses allgemeine Wunder etc.
(Nathan der Weise, meine Ausgabe, S. 17)[170]

Ich denke immer, dass die Einführung der Zahl, die Thatsache, dass die gewöhnliche Zahl in den Naturwissenschaften eine solche Rolle spielt und diese erst zu Wissenschaften macht, ein solches grösstes Wunder ist, und naturgemäss wird die Mathematik sich damit beschäftigen, genau die Stelle zu bezeichnen, wo diese Einführung und wie sie stattfindet.[171]

Wenn alle Axiome I–V gelten, so kann man jeder Strecke eine bestimmte reelle Zahl zuordnen:

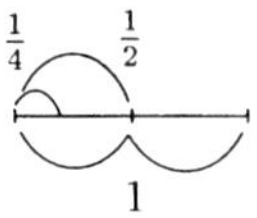

Entwickelung im duadischen Zahlensystem. Gleichen Zahlen ist auch dieselbe Strecke zugeordnet. Denn sonst kann man immer $\frac{1}{2^n}$ so finden, dass diese Strecke < als die Differenz der beiden vorgelegten Strecken etc. Man kann jeder reellen Zahl einen idealen Punkt entsprechen lassen, so dass für das System alle wirklichen und idealen Punkte sämtliche Axiome I–V gelten und es mithin überflüssig ist, zwischen den wirklichen und reellen Punkten einen Unterschied zu machen.[172]

104 Gewöhnliche analytische Geometrie.

Diese ist möglich, weil die Eigenschaften der reellen Zahlen sich nicht einander widersprechen, sondern alle miteinander verträglich sind. Also die Axiome I–V sind sämtlich miteinander verträglich. Folglich widersprechen sich auch erst recht nicht einzelne dieser Axiome und daher Existenz der nicht-Euklidischen Geometrie.

Existiren heisst die den Begriff definirenden Merkmale (Axiome) widersprechen sich nicht, d. h. es ist nicht möglich, aus allen ⟨Axiomen⟩ mit Ausnahme eines durch rein logische Schlüsse einen Satz zu beweisen, welcher dem letzten Axiom widerspricht.[173]

Jeder überhaupt richtige, d. h. mittelst Axiome I–V zu beweisende Schnittpunktsatz ist eine logische Folge aus Desargues und Pascal. Denn ein Schnittpunktsatz muss doch stets folgende Form haben: Gegeben sind irgend welche Punkte und Geraden eventuell durch jene Punkte, die durch eine bestimmte
105 Anzahl von Parametern $p_1, \ldots, p_r$ ausgedrückt werden. | Wenn man nun

[170]The reference is to Gotthold Ephraim Lessing's *Nathan der Weise* (1779), Act I, Scene 2. The passage quoted is spoken by Nathan to his adopted daughter Recha.

[171]Cf. the 1893/1894 lectures, 28. See Chapter 2 and its Editorial Introduction.

[172]See also the corresponding passage in the *Ausarbeitung*, 166–167.

[173]Hilbert gives a syntactic characterization of consistency, equivalent to the usual one, assuming the right tautologies can be proved. According to Hilbert's remark here, the 'possibility' of the geometrical systems rests on the fact that the 'Eigenschaften' of the real numbers are not contradictory, so in fact we have a relative 'existence' proof. (See also the 1898 *Ferienkurs*, 12, and notes.) This is made more explicit in the *Ausarbeitung*, 167. In the *Festschrift*, 21, Hilbert makes it clear that what we have is a relative syntactic consistency

Geraden zieht, zum Schnitt bringt etc., so sollen schliesslich 3 Punkte auf einer Geraden bleiben. Wir verfolgen nun die Construktion, indem wir die betreffende Rechnungsoperation ausführen, so soll ein gewisser Ausdruck

$$A(p_1, \ldots, p_r) = 0$$

für alle besonderen Werte von $p_1, \ldots, p_r$ ⟨werden,⟩ folglich auch identisch für alle $p_1, \ldots, p_r$ ⟨verschwinden,⟩ d. h. mittelst der Beweise, d. h. auch mittelst der formalen Rechnungsregeln, d. h. mittels Desargues und Pascal.[174]
[175]↑ 106

Geometrie Muster einer Naturwissenschaft: Fortgeschrittenste etc. Collegheft Grundlagen der Geometrie, S. 92.[176]

Möglichkeit der Euklidischen Geometrie. Mit Axiom V ist die Analyse unserer Anschauung beendet und die analytische Behandlung ermöglicht.

Es liegt im Wesen einer jeden Theorie, dass sie in der Erfahrung nicht genau stimmt. Denn sonst könnten wir nur die einzelnen Erscheinungsthatsachen beschreiben. Man sagt dann oft, die Theorie wäre falsch. Dies nur, wenn untereinander in Widerspruch. Wir müssen uns doch schlüssig machen, was wir in der Welt der Erscheinung als Punkte, Geraden etc. nehmen wollen und dann, selbst Axiom I–III zugegeben, so würden doch Euklid und Archimedes immer nur aus der Erfahrung genommen werden können. Altes Collegheft S. 92, S. 87–89. Vgl. ferner Pasch, S. 3 Mitte und S. 4 oben, S. 18.[177]

---

proof: any proof of an inconsistency from the axioms of Euclidean geometry would give rise to an identifiable inconsistency in the countable Pythagorean sub-field of the real numbers there defined to 'model' the axioms.

The relation between existence (or 'possibility') and consistency is stated in the 1893/1894 lectures, 87 (see Chapter 2) and alluded to in the 1898 *Ferienkurs*, 12 (see Chapter 3); it is reiterated here and in the *Ausarbeitung*, 167, and then in Hilbert's letter to Frege of 29 December, 1899 (see *Frege 1976*, letter XV/4, also in *Frege 1980*). It is not stated in *Hilbert 1899c*. In fact, consistency is not even explicitly listed in the *Einleitung* to the *Festschrift* as a goal of the investigation, unlike, e.g., the *Ausarbeitung*, 167. Nevertheless, in the *Festschrift*, Hilbert gives (p. 21) for his axioms the model mentioned above as soon as these have been set out in full and their major consequences drawn. Showing the consistency of axiom systems (like that for geometry) *is* stated as a goal of the 'axiomatic method' at the beginning of *Hilbert 1900b*.

The passage here also states that the axioms give the defining characteristics of the concept being axiomatized; on this, see the correspondence between Hilbert and Frege in *Frege 1976*, or *Frege 1980*.

[174]For elucidation of these two paragraphs, see the Introduction to this Chapter, Specific Note (1)(c).

[175]The text from here to the beginning of p. 108a is on a separate sheet pasted to p. 106.

[176]The reference is to the 1893/1894 lectures (see Chapter 2), where Hilbert states the view that all the main physical sciences should follow the example and method of geometry.

[177]Concerning the inexactness of theories when applied to the world of experience, see the

So alt die Gegenstände die wir behandeln und so alt ihr Urheber: Euklid: das Princip des Beweises der Unbeweisbarkeit ⟨–⟩ ↑Also Lösung einer Aufgabe unmöglich oder mit gewissen Mitteln unmöglich. Damit hängt auch die Forderung der Reinheit der Methode zusammen.↓ ⟨–⟩ modern zuerst an den beiden Problemen: Quadratur des Kreises und Parallelenaxiom. Aber wir wollen dieses als modernes Princip erheben: Man darf nicht abstehen, wenn eine mathematische Sache misslingt, man darf erst zufrieden sein, wenn wir ihre Unbeweisbarkeit eingesehen haben. Fruchtbarstes und tiefstgreifendes Princip in der Mathematik.[178]

Beweis von Legendre-Sätzen ohne V.

Beweise Pascal in der Ebene durch I–III ohne IV.

Problem der Gleichheit körperlicher Volumina, z. B. zweier Pyramiden mit gleicher Höhe und Grundlinie (Gauss' Vermuthung).[179] ↓

108a ↑Berechnung des äussersten, d. h. am weitesten nach rechts auf der positiven x-Achse gelegenen Punktes, der noch von einem Kreise getroffen wird, welcher die gegebene Ellipse $\frac{x^2}{a^2} + \frac{y^2}{b^2} = 1$ in 4 Punkten schneidet.↓[180]

Die Ellipse $\frac{x^2}{a^2} + \frac{y^2}{b^2} = 1$ werde in 4 getrennten Punkten von einem Kreise geschnitten, welcher die x-Achse in den Punkten mit den Abszissen ξ_1 und $\xi > \xi_1$ treffen möge. Dann lässt sich dieser Kreis um den Punkt ξ solange

---

1893/1894 lectures, 92 (Chapter 2), and the *Ausarbeitung*, 168–169. The statement that a theory is only false when it is inconsistent is connected to the view of existence mentioned in the note before last. 'Euklid' und 'Archimedes' refer to the Parallel Axiom and the Archimedean Axiom, respectively. The reference to 'Altes Collegheft' is to the 1893/1894 lectures, 87–89, 92, the former pages being explicitly concerned with the empirical interpretation of the Axiom of Parallels. The reference to Pasch is to *Pasch 1882*, 3–4 and 18. On p. 3, Pasch states his view of geometry as the study of a certain cluster of concepts used to describe things in the external world, in principle no different from any other basic physical science. On p. 4, Pasch states:

> Die Anwendung dieser Begriffe bleibt mit einer gewissen Unsicherheit verbunden, wie dies bei fast allen Begriffen, die wir zur Auffassung der Erscheinungen geschaffen haben, der Fall ist.

On p. 18, Pasch confronts the limits to an empirical justification of the repeated sub-division of a segment.

[178]More specific problems are mentioned in the *Ausarbeitung*, 169, and in the *Schlußwort* to the *Festschrift*, 89. Note the connection between what is stated here and a passage in *Hilbert 1900e*, 261 (*Hilbert 1935*, 297) which stresses

> ... die Überzeugung, daß ein jedes bestimmte mathematische Problem einer strengen Erledigung notwendig fähig sein müsse, sei es, daß es gelingt, die Beantwortung der gestellten Frage zu geben, sei es, daß die Unmöglichkeit seiner Lösung und damit die Notwendigkeit des Mißlingens aller Versuche dargetan wird ...

It is clear that, when Hilbert declares (op. cit., 262, *Hilbert 1935*, 298):

> Da ist das Problem, suche die Lösung. Du kannst sie durch reines Denken finden; denn in der Mathematik gibt es kein Ignoramibus.

he means 'solution' in this general sense. He also stresses, as he does here, the fruitfulness of such 'impossibility' investigations, even in physics.

[179]Some brief remarks about each of the problems posed here can be found in the Introduction to this Chapter, Specific Note (5).

[180]The pages here marked 108a–108c (and pasted in) are largely in von Schaper's hand,

drehen, bis wenigstens 2 der Schnittpunkte zusammenfallen, sodass wir einen Kreis erhalten, der die Ellipse etwa in P berührt und in P_1 und P_2 schneidet. Erweitern wir diesen Kreis derart, dass er immer noch durch P geht und dort die Ellipsen berührt, wobei ξ wächst, so müssen entweder P_1 und P_2 einmal zusammenrücken; oder es rückt etwa P_1 nach P, d. h. der Kreis wird zum Krümmungskreis; in diesem Falle erweitern wir den Kreis noch mehr, und müssen dann auch zu einem doppelt berührenden Kreise gelangen; denn wir wissen, dass es durch jeden Punkt der Ellipse einen ganz ausserhalb von ihr verlaufenden doppelt berührenden Kreis giebt. Jedenfalls erhalten wir eine obere Grenze von ξ, wenn wir sämtliche doppelt | berührende (natürlich 108b
ausserhalb der Ellipse verlaufende) Kreise in betracht ziehen. Wie gesagt, giebt es durch jeden Punkt der Ellipse einen solchen Kreis, und zwar liegt dieser symmetrisch zur y-Achse. Es giebt aber nur einen K_1, denn jeder andere K_2 müsste, wie leicht zu sehen, die Ellipse in denselben beiden Punkten berühren wie der erste. (Da nämlich K_2 ganz innerhalb K_1 verlaufen müsste, andrerseits ausserhalb der Ellipse liegen soll, so muss K_2 auch durch den zweiten Berührungspunkt von K_1 gehen etc.) Wir brauchen also nur die zur y-Achse symmetrischen doppelt berührenden Kreise zu betrachten. Sei x_1, y_1 einer der Berührungspunkte, so ist die Gleichung der Normalen:

$$\frac{a^2}{x_1}(x - x_1) = \frac{b^2}{y_1}(y - y_1)\,.$$

Sie schneidet die y-Achse im Punkte:

$$0\,,\ y_1\frac{b^2-a^2}{b^2} = -\varepsilon y_1\,.$$

Das ist der Mittelpunkt des doppelt berührenden Kreises. Sein Radius R ergiebt sich aus:

$$\begin{aligned} R^2 &= (x_1 - 0)^2 + (y_1 - (-\varepsilon y_1))^2 \\ &= {x_1}^2 + {y_1}^2(1+\varepsilon)^2\,. \end{aligned}$$

Die Gleichung dieses Kreises ist also: 108c

$$x^2 + (y + \varepsilon y_1)^2 = R^2\,,$$

seine Schnittpunkte mit der Abszissenachse haben die Abszissen:

$$\begin{aligned} \xi &= \pm\sqrt{{x_1}^2 + {y_1}^2(1+\varepsilon)^2 - (\varepsilon y_1)^2} \\ &= \pm\sqrt{{x_1}^2 + {y_1}^2(1+2\varepsilon)} \end{aligned}$$

oder, da ${x_1}^2 = a^2 - \frac{a^2}{b^2}{y_1}^2$ ist:

$$\begin{aligned} \xi &= \pm\sqrt{a^2 + {y_1}^2\left(1 + 2\frac{a^2-b^2}{b^2} - \frac{a^2}{b^2}\right)} \\ &= \pm\sqrt{a^2 + {y_1}^2\frac{a^2-b^2}{b^2}}\,. \end{aligned}$$

---

except for a few calculations at the end of 108c, and the addition in the first paragraph ('Berechnung' to 'schneidet').

The consideration of ellipses and tangent circles is clearly the same calculation as that executed in § 23 of the *Festschrift*. This is done in the course of setting up the plane model in which both the planar Desargues theorem *and* the Triangle Congruence Axiom fail, showing that the latter is crucial in the proof of the former using congruence. (See the *Festschrift*, pp. 51-55.)

$|\xi|$ wird also am grössten, wenn $|y_1|$ am grössten wird, d. h. für $|y_1| = b_1$; nämlich

$$|\xi|_{\max} = \sqrt{2a^2 - b^2}\,.$$

In unserem Beispiel ist $a = 1$, $b = \frac{1}{2}$, also wird

$$|\xi|_{\max} = \tfrac{1}{2}\sqrt{7}$$

$$\text{also:} \qquad < \tfrac{3}{2}. \quad -$$

q. e. d.

⟨Addition⟩[181]

109 $AB = BA$ ist jedenfalls nicht beweisbar, wenn man von dem Dreieckscongruenzaxiom (Axiom IV, 6) absieht; auch wenn man alle übrigen Axiome als erfüllt annimmt. Um dies einzusehen, denke man sich im gewöhnlichen Euklidischen Raume eine feste Kugel und innerhalb derselben einen Punkt P, welcher nicht der Mittelpunkt ist. Dann sei eine beliebige Strecke $\underrightarrow{AB}$ mit Richtung gegeben, so ziehe durch P eine Parallele in der gegebenen Richtung zu $\underrightarrow{AB}$ und diese Parallele durchstosse die Kugeln im Punkte K, so nenne $\frac{AB}{PK}$ die Länge der Strecke $\underrightarrow{AB}$ wozu AB, PK gewöhnliche Euklidische Strecken sind. Wenn man dann Abtragen von Winkeln in gewöhnlicher Weise definirt, so hat man die gewünschte Geometrie (in der übrigens nach Minkowski die Gerade die kürzeste ist).[182]

---

[181] Added in Hi^s:
⟨Added in the bottom margin at the left:⟩

$$a = 1$$
$$b = \frac{1}{2}$$

⟨Added in the bottom margin, beginning to the right of ‘q. e. d.’, partly unclear:⟩

$$a = 1$$
$$b = \frac{1}{2}$$

$$a^2 + 4b^2 = 1$$
$$1 - b^2 \qquad a^2 = 1 - 4b^2$$
$$= 1 - b^2$$

[182] It is not entirely clear to what Hilbert’s remarks here relate; for some discussion, see the Introduction to this Chapter, Specific Note (6)(a). The date of remark is closely tied to the appearance of the *Festschrift*, for it is only in the *Festschrift* that the Triangle Congruence Axiom is numbered IV 6.

Textual Notes

221.2:	Meine Herren.] M. H.
221.2:	der Geometrie,] d. Geometrie
221.13:	*Parallelentheorie,*] ↑*Parallelentheorie*↓
221.13:	*Polygonen*] *Poly*⟦⟪??⟫⟧*gonen*
221.14–15:	man in den Schulbüchern] ⟦wir mit⟧↑mann↓ ↑in den Schulbüchern↓
221n:	doch] doch ⟦richtig⟧
221n:	Punkte, Geraden, Ebenen] Punkte Geraden Ebenen
222.3:	*algebraische*] *algebr.*
222.4:	Jeder Punkt] ⟦Alle Pun⟧ Jeder Punkt
222.6:	Gleichung] Gl.
222.10:	Benutzung] Benutung
222.11:	*Parallelen,*] *Paralleten*
222.18:	späteren] ⟦höheren⟧↑späteren↓
222.20:	*Einführung*] *Einführg*
222.21:	ist.] ist. ⟦So werden wir in unserer Vorlesung⟧
222.22:	tragen] tragen ⟦u. soll so⟧ tragen
222.23:	sollen von] sollen
222.23:	*gesuchte*] ⟦⟪*gesu*⟫⟧ ⟪*gesuchte*⟫
222.30:	*tägliche*] ↑*tägliche*↓
222.30:	Zuhörers] ⟦Schülers ⟪oder⟫⟧ Zuhörers
222.31:	dass] das
222.36:	Vorlesung] Vorles.
222.38:	*Parallelenaxioms*] *Parallelaxioms*
223.3:	Euklidischen Geometrie] Euklishen Geometri
223.6:	*tiefliegender*] *tiefliegengender*
223.6:	*schwieriger*] *schwierigen*
223.6:	*Probleme*] ⟦*Fragen*⟧ *Probleme*
223.8:	nahe] nahe ⟦⟪??⟫⟧
223.11:	Maasse] Masse
223.17:	Nichtwissens] Nichtwissens sich
223.22:	Figur ist] Figur
223.26:	blossen] Blossen
223.27:	mathematischer] Mathematischer
223.29:	gleichschenklig] gleichschencklig
223.31:	Geraden] Gerade
224.4:	wie] die
224.11:	inhaltsreich] inhaltreich
224.12:	Stäckel, ⟨...⟩ 1895.] ⟨Possibly added.⟩
224.16:	Geometrie:] Geometrie
224.24:	bezeichnen] bezeichen
224.24:	*Geraden*] *Gerade*
224.25–26:	sind weiter] weiter
224.30:	**von einander verschiedene**] ↑**von einander verschiedene**↓
225.1:	**einander**] **einan**
225.3:	Geraden] Gerade
225.6:	**derselben**] ⟦**einer u.**⟧ **derselben**
225.6:	**gelegene**] **gelegenen**
225.6:	A, B, C] ↑A, B, C↓

225.10: **Irgend**] **Irdend**

225.10: A, B, C] ↑A, B, C↓

225.11: **Geraden**] **Gerade**

225.12: folgt] folg

225.16–17: Wenn ⟨...⟩ Ebene] ⟨Printed; appears on a strip taken from a copy of *Hilbert 1895b* and pasted in here.⟩

225.18–19: Wenn ⟨...⟩ gemein.] ⟨Printed; appears on a strip taken from a copy of *Hilbert 1895b* and pasted in here.⟩

225.18: Wenn] ⟦α. –⟧ Wenn

225.19: gemein.] gemein. –

225.22–24: Auf ⟨...⟩ Punkte.] ⟨Printed; appears on a strip taken from a copy of *Hilbert 1895b* and pasted in here.⟩

225.22: Auf] – Auf

225.23: 3] 3,

225.24: Punkte.] Punkte. –

225.28: Grundlagen der Geometrie,] Grundl. d. Geom.

225.28: S. 10–13.] ⟨Possibly added.⟩

226.3: ganze rationale positive] ↑ganze rationale positive↓

226.4: ganze rationale positive] ↑ganze rationale positive↓

226.4: definirt] ⟦bestimmt⟧ definirt

226.6: Grösster Ganzer] ↑Grösster Ganzer↓

226.11: Oder] ⟦Oder: wir vermitteln Axiom 1 durch die Gl.⟧ Oder

226.11: repräsentiren] representiren

226.13: Gleichung] Gl.

226.23: und] u.

226.24: Axiom 6] 6

226.25: Geraden] Gerade

227.1: Punkte] Punkt

227.2: unnahbar] ⟦nicht erreichbar⟧⊣↑unnahbar↓

227.2: Geraden] Gerade

227.8: C_{nm}] ↑C_{nm}↓

227.10: Litteratur:] ⟨Possibly added.⟩

227.11: und] u.

227.13: ausser,] ausser

227.16–22: 1. ⟨...⟩ werden.)] ⟨Written on a separate strip of paper and pasted in here. What it covers cannot be determined.⟩

227.16: und] un

227.17: und] u.

227.24: anderen] Anderen

227.29–32: Wir ⟨...⟩ Strecke.] ⟨Written on a separate strip of paper and pasted in here, probably covering nothing.⟩

227.32: Ein ⟨...⟩ Strecke.] ↑Ein Punkt liegt „auf“ oder „ausserhalb“ einer Strecke.↓

227.32: „auf“ oder „ausserhalb“] ⟨Quotation marks in Hi^b.⟩

228.2–4: Wenn ⟨...⟩ B.] ⟨Written on a separate piece of paper and pasted in here, probably covering nothing. The addition includes the diagram.⟩

228.2: A, B, C, D] $ABCD$

228.10–12: Satz ⟨...⟩ etc.] ⟨Written on a separate strip of paper and pasted in here, probably covering nothing.⟩

228.10: Satz.] Satz ⟦1.⟧

228.12:	einen] ein
228.12:	nach 3] nach 3.
228.13:	Anordnung] Anordung
228.14:	umgekehrten] Umgekehrten
228.15:	Beweis ⟨...⟩ von 3.] ⟨Probably added.⟩
228.18:	**so dass]** **dass**
228.18–19:	**zwischen** A] **zwischen** A ⟦und C und A, D, A, E ⟪??⟫ C und⟧
228.19:	**andererseits,]** **anderseits**
228.20:	**einerseits und]** **und**
228.20:	$F, \ldots$] F.
228.25:	5] 5 ⟦4⟧
228.26:	ordnen,] ordnen.
228.32:	in 2] in 2 ⟦Gebi⟧
228.32:	Abtheilungen] Ab⟦schnitte von⟧teilungen
228.32:	Beschaffenheit:] Beschaffenheit: ⟦ein jeder Punkt A der einen Abtheilung bestimmt mit jedem Punkt B⟧
228.33:	und] u.
228.33:	Abtheilung,] Abth.
228.34:	dagegen] dagen
228.1:	zwischen] zwisch
229.2:	irgendein] irgenein
229.2:	so sind nach] so nach
229.5:	wird ⟨...⟩ etc.] wirf X ⟦zu de⟧ der Abth. mit B bez. mit A etc.
229.8:	erinnert] definirt
229.9:	Abtheilung] Abt.
229.17:	die natürliche] natürliche
229.20:	die Axiome] Axiome
229.21:	Anschauung] Anschauung sich
229.24:	rechtwinkliges] rechtwinkliches
229.27:	$= (a', b', c')$] $= (a', b', c)$
229.27:	$P'' = (a'', b'', c'')$,] $P'' = (a''\, b''\, c'')$
229.30:	P''.] P''
230.24:	dass] das
230n:	verstanden wird] verstanden
231.1:	Geometrie] Geom.
231.2–7:	Jede ⟨...⟩ enthält.] ⟨Printed; appears on a strip taken from a copy of *Hilbert 1895b* and pasted in here.⟩
231.2:	liegt,] liegt
231.4:	$\overline{B}$] ⟦A⟧⊦↑B↓
231.5:	$\overline{AB}$] $\overline{AB}$
231.9–12:	oder ⟨...⟩ Punkten.] ⟦⟪???⟫ damit die beiden Halbebenen.⟧⊦↑oder man nennt beide Seiten auch Halbebenen. Wenn 2 Punkte auf verschiedenen Seiten einer Geraden a liegen, so sagt man auch, die Gerade geht zwischen den beiden Punkten.↓
231.13–15:	**Satz:** ⟨...⟩ nothwendig] ⟦Satz Tritt eine Gerade a einer Ebenen in ein Dreieck ABC durch eine Seite desselben ein, so muss sie nothwendig durch ei⟧⊦↑⟦Satz Schneidet die Gerade a einer Ebenen ABC⟧⊦↑**Satz:** Es sei $ABC = \alpha$ eine Ebene und a eine durch keinen Punkt A, B, C gehende Gerade. Trifft dann die Gerade a↓ das Dreieck ABC in einer Seite (Strecken AB, BC, AC) desselben, so muss sie nothwendig↓

232.3:	B, C] $B\ C$
232.5:	Geraden] Gerade
232.6:	auf] auch
232.9–10:	xy-Ebene] x, y Ebene
232.12:	xy-Ebene] $x\,y$ Ebenen
232.12:	Geraden] Gerade
232.13:	y-Achse] y Achse
232.14:	x-Achse] x Achse
232.15:	x-Achse] x Achse
232.17:	Coordinaten] Cordinaten
232.18:	für sich] für,
232.20–21:	nimmt man] nimm
232.21:	xy-Ebene] $x\,y$ Ebene
232.21:	A, B, C] ABC
232n:	natürlichen] natürlich
233.1:	wird] wir
233.11:	$\overline{PP'}$] $\overline{PP}$
233.14:	$\overline{QQ'}$] QQ'
233.17:	$\overline{PQ}$] PQ
233.17:	stets] ⟦jetzt⟧⊣↑stets↓
234.4:	Die] die
234.10:	dass] dass ⟦die Verbindungsgeraden AA', BB' und CC' durch den ⟨p. 26⟩ nämlichen Punkt laufen⟧
234.11:	sich auf] auf
234.18:	M, A', A] $MA'A$
234.18:	M, P, Q] MPQ
234.19–20:	P, A'', A, P, B'', B, E] ⟦PA''⟦B''⟧⊣↑APQ↓⟦ liegen mit E⟧⊣↑AB↓⟧⊢⊣↑$PA''APB''BE$↓
234.20:	↑Q↓, A'', A', ↑Q, B'',↓ B', E] ↑Q↓$A''A'$↑QB''↓$B'E$
234.21:	A'', B'', E] $A''B''E$
234.22–23:	geschnitten werden] geschnitten
234.26:	C'', C', Q] $CC'Q$
234.26–27:	Die Punkte] d. h. $QC''C'$ liegen in einer Geraden. Die Punkte
234.27:	P, C, C'', ⟨…⟩ Ebene] ⟦$CC'C$⟧ ⟦$P, QCC'C''$ liegen in einer Ebene.⟧⊢⊣↑⟦C, C', C'', P, Q, M⟧⊣↑PCC'', $QC''C'M$↓ liegen also in einer Ebene↓
234.28:	C, C', M] $CC'M$
234n:	P, B'', C'', B, C, F] $PB''C''BCF$
234n:	ausserdem liegen B'', C'', F] ausserdem liegen $B''C''F$
234n:	folglich liegen B'', C'', F] folglich liegen $B''C''F$
234n:	A'', C'', D] $A''C''D$
236.7:	C, C',] CC'
236.8–9:	P, C, C'] PCC'
236.10:	wichtig ist] wichtig
236.15:	Also sind] Also
236.18:	Satzes] Satz
236.18:	*Geometrie*] ⟦Funktionentheorie⟧⊣↑*Geometri*↓
236.19:	*Wahrheit*] *Wah*⟦*h*⟧*rheit*
236.21:	*berechtigten*] *berechtigtigten*
236.28:	unserer] unsere
237.3:	Axiomen] Axiom.

237.5: *Geraden*] *Gerade*
237.10: hat] haben
238.1: seien] sei
238.1: Geraden,] Gerade
238.2: Geraden] Gerade
238.7: A, M, N, B] $AMNB$
238.8: A, M, N, C] $AMNC$
238.9: Geraden] Gerade
238.9: Geraden] Gerade
238.11: niemals in einem Punkte,] ⟦in 3 verschiedenen Punkten⟧⊢↑⟦nicht⟧⊢↑niemals↓ in einem Punkte↓
238.11: ⟨p. 33⟩ sondern] ⟦und man kann sie so wählen, dass sie sich in 3 verschieden Punkten ⟨p. 33⟩ schneiden.⟧ sondern
238.12: $O, -\infty$] $O\ -\infty$
238.1: $\sphericalangle\beta > \sphericalangle\alpha$] $\uparrow\sphericalangle\downarrow\beta > \uparrow\sphericalangle\downarrow\alpha$
239.3: schneiden] schneiden ⟦q. e. d.⟧
240.1: Desarguessche] Desargues
240.1: dass] dass man
240.2: Axiome] Axiom
240.5: Desarguessche] Desargues
240.8: werden kann?] werden.
240.12: Construktion des 4ten] ↑Construktion des 4ten↓
240.12: *Punktes*] *Punkte*
240.14: der Geometrie,] d. Geom.
240.14: Sommer] S.
240.15: der Raum] Raum
240.17: laufen,] ⟨Comma added by Hi^b.⟩
241.1: Geraden] Gerade
241.1: beginnen] beginnen ⟦dann aller⟦《??》⟧dings bis auf die umgekehrte Folge bestimmt⟧
241.6: Stetigkeitsaxiom hinzu] Stetigkeitsaxiom
241.15: Punkt] Punkte
241.15: so] ⟦welcher mit S auf⟧ so
241.17: $A'B'$] $A'B'$,
241.24: denknothwendig ist] denknothwendig
241.24: annimmt] annimt
241.25: Maasstab] Maastab
241.25: Abtragen] abtragen
241n: *Mathematiker*] *Math*
242.5: angewandt wird] angewandt
242.5: *wesentliches*] *wesentlich*
242.7: Zahlen] Zahl
242.9: Beobachtungsfehlers] Beobachtungsfehler
242.11: gleich)] gleich
242.17–18: Geraden ⟨...⟩ liegt] Geraden ↑so dass B' zwischen A', C' liegt↓
242.23: einer Länge l] ↑einer Länge l↓
242.23: geschehe] gesche
242.23: $e^l - 1$] ⟦$+L$《?》⟧⊢↑$e^l - 1$↓
242.29: Es] Satz Es
242.29: zwischen] zwisch

242.30: und] u.
242.30: A', B', C'] $A' B' C'$
242.31: Geraden,] Gerade
242.31: B', C'] $B' C'$
242n: B', C'] $B' C'$
243.1: zwischen] zwisch
243.1: A', C',] A', C'.
243.1: und ⟨...⟩ $BC \equiv B'C'$.] ⟨Possibly added.⟩
243.2: läge] liege
243.2: A', B'] $A'B'$
243.5: $C' = C''$,] $C = C''$.
243.14: es] e⟦r⟧↑s↓
244.–: $b + a$] $b + a$.
244.12: das] dass
244.12: commutative] comutative
244.17–18: Die ⟨...⟩ congruent.] ⟨Added.⟩
244.20–21: Aehnlich ⟨...⟩ Gleiches.] ⟨Added.⟩
244.20: folgt] folglt
244.22: Definition.] ↑Definition.↓
244.30: $a < c$;] $a < c$
244.33: gelegenen] gelegen
244.35: WW_1] WW'
244.36: Die] ⟦Das System von allen P⟧ Die
244.36: W_2] W_2,
245.2: abW] ab⟦S⟧↑W↓
245.2: Halbstrahl] ⟦Halbstrahl a' und⟧ ⟦Gerade a'⟧ Halbstrahl
245.3: Seite S] Seite ↑S↓
245.3: bestimmten] bestimten
245.3: einen] eine
245.4: der mit] der
245.4: einen] ein
245.5: a', b'] $a'b'$
245.5: Punkt W'] Punkt ⟦S'⟧↑W'↓
245.7: Seite S] Seite ⟦von der durch a' bestimmten Geraden⟧↑S↓
245.8: gegebenen] gegeb.
245.14: gemeinsame] gemeine
245.17: $\equiv$] $=$
245.20: Axiom 3] Axiom 3 ⟦in der Ebene⟧
245.24: 2er] 2
245.26: etc. Wie auf S. 41 für die Gerade.] ↑etc. Wie auf S. 41 für die Gerade.↓
245.26–27: Wenn $\alpha > \beta, \beta > \gamma$, so $\alpha > \gamma$.] ↑Wenn $\alpha > \beta, \beta > \gamma$, so $\alpha > \gamma$.↓
246.2: $\sphericalangle aaW$] $\sphericalangle a$⟦b⟧aW
246.5: gestreckter] gestrekter
246.9: III 1–8] III 1–8,
246.10: Folgerung] Folgerng
246.13: einem Vollwinkel] ein Vollwinkel
246.23: unsere] unseren
246.24: und] u.
246.27: folgende] folgen

246n:	gestreckten] gestrekten
247.5:	(erster Congruenzsatz)] ⟨Probably added.⟩
247.8:	nimm] Nimm
247.16:	senkrecht] senk⟪tech⟫
247.25:	$a'b'$] ab
247.25:	projiciren] projijiren
247.26:	die] ⟦die⟧
247n:	Definition] ⟦Satz⟧↑Definition↓
247n:	Dreiecke] Dreiecke ⟦sind cong⟧
248.1:	Bewegung] Bewegng
248.3:	Nun können] ⟦Beweis der übrigen Congruenz⟧ Nun können
248.5:	einem Punkt B] ⟦2⟧↑einem↓ Punkt⟦en⟧ B ⟦C⟧
248.9–10:	$\sphericalangle ACB = \sphericalangle BCA'$] $\sphericalangle ACB = BCA'$
248.14:	(2ter *Congruenzsatz*)] ⟨Probably added.⟩
248.19:	in] ⟦in⟧↑ein↓
248.19:	Dreieck] Dreick
248.19:	gegeben] ⟦die Seite⟧ gegeben
248.20:	(d. h. der sogenannte Aussenwinkel)] ↑(d. h. der sogenannte Aussenwinkel↓
248.21:	Beweis] Beweise
248.22:	$\sphericalangle C'AB = \sphericalangle DBC$] $\sphericalangle C'AB = DBC$
248.22:	eventuell] ev.
248.23:	verlängere] verlänge
248.28:	$\sphericalangle C'AB + \sphericalangle BAE$] $\sphericalangle C'AB + BAE$
248.31–249.3:	auf ⟨...⟩ Rechter)] ⟨Replaces '⟨p. 49⟩ auf einer Geraden liegen'.⟩
249.3:	Rechter] Rech
249.10:	$\sphericalangle ACB \equiv \sphericalangle AC''B' \equiv \sphericalangle A'C'B'$] $\sphericalangle ACB \equiv AC''B' \equiv A'C'B'$
249.10:	oben] Oben
249.11:	nicht möglich ist.] nicht möglich ist. ⟦Satz ⟦Wenn 2 Dreiecke⟧ Im gleichschenkligen Dreieck sind die Basiswinkel gleich. Beweis. Man verfahre als wolle man eine Senkrechte von der Spitze ↑C↓ auf die Basis ↑AB↓ fällen:⟧
249.12:	stets] ↑stets↓
249.12:	*halbiren.*] *halbiren* ⟦indem man irgend einen Punkt⟧
249.19:	gleichschenkligen] gleichschenklichen
249.20:	gleichschenklig] gleichschenklich
250.13:	Ebenso die Umkehrung.] ⟨Possibly added.⟩
250.14:	Der ⟨...⟩ umständlicher.] ⟨Possibly added.⟩
250.14:	Euklidische] Euklische
251.13:	späterer] später
251.15:	gestreckter] gestrekter
252.1:	Sinn] Sinne
252.8:	Gestreckter] Gestrekter
252.8:	Nun] Nun ⟦mache $\sphericalangle A'B'C'' = \sphericalangle B$ und⟧
253.2:	$\sphericalangle \gamma_1 = \gamma_1'$] $\sphericalangle$ $\sphericalangle \gamma_1 = \gamma_1'$
253.10:	Die Reihenfolge] ⟦Wenn von 3 Punkten ↑A, B, C↓ einer Geraden in einer Figur⟧ Die Reihenfolge

253.10:	in einer Figur] ↑in einer Figur↓
253.11:	Gerade] Geraden
253.12:	Anordnung] Anordng
253.14:	Allgemeinster ⟨...⟩ Ebene:] ⟨Added.⟩
253.17:	$(A', B', C', \ldots, P')$] $(A',, B', C', .. P')$
253.18:	ursprünglichen] ⟨Added.⟩
253.19:	Geraden] Gerade
253.19:	gelegene] gelegenen
253.22:	$\sphericalangle PBA = \sphericalangle P'B'A'$] $\sphericalangle PBA = P'B'A'$
253.24:	durch ⟨...⟩ Dreiecke] ⟦da⟧durch ↑die beiden congruenten Dreiecke↓
253.25:	eine andere oder] ↑eine andere oder↓
253.26:	Figur in] Figur der
253.27:	gebracht"] gebracht
253.27:	oder] ober
253.27:	„bewegt"] bewegt
253.29–30:	, während ⟨...⟩ animmt.] ⟨Added.⟩
253.32:	ihn ⟨...⟩ A] ihn ↑eine Seite auf a von A↓
253.32:	von] ⟦auf⟧⊣↑von↓
254.16:	$\sphericalangle PAB \equiv \sphericalangle QAB$] $\sphericalangle PAB \equiv QAB$
255.2:	Geraden] Gerade
255.2:	A,] A.
255.16–17:	verlängere ⟨...⟩ Dann] ⟨Added.⟩
255.16:	verlängere] verlänge
255.19:	$\triangle PAM = \triangle QAM$] $\triangle PAM = QAM$
255.19:	$\sphericalangle PAM = \sphericalangle MAQ$] $\sphericalangle PAM = MAQ$
255.21–30:	Satz ⟨...⟩ Senkrechte.] ⟨Added on a separate piece of paper pasted onto the bottom of p. 58.⟩
255.25:	Dann ist] Dann ist ⟦$BC = BD$, $A'C = A'D$⟧
255.31:	errichte] errichten
255.43:	$\triangle NAG \equiv \triangle NA'G$] $\triangle NAG \equiv NA'G$
256.1:	$\triangle AF'M' \equiv \triangle A'FM$] $\triangle AF'M' \equiv A'FM$
256.3:	räumlichen] ⟨Added.⟩
256.5:	Satz.] ↑Satz↓
256.10:	Allgemeinster ⟨...⟩ Raume:] ⟨Possibly added.⟩
256.14:	ursprünglichen] ⟨Added.⟩
256.16–23:	Beweis ⟨...⟩ etc.] ⟨Possibly added, including the figures.⟩
256.28–29:	eine Gerade a ⟨...⟩ von A,] eine Gerade a ↑eine Seite von A↓ durch ⟦ihn⟧⊣↑A↓
256.29:	eine Seite von a] ⟨Added, probably before 'von α und A', a', α' gegeben sind.' was written.⟩
257.1:	Figuren] Figur
257.4:	Ebenen] Ebene
257.7:	Construktion] Construcktion
257.20:	durch] durch durch
257.20:	*und*] *u.*
257.24:	Kreis ist, der] ⟦durch das⟧⊣↑Kreis ist, der↓ ⟨Replacement, carried out before '*passirt*, wie in nebenstehender Figur' was written.⟩
257.25:	*Sehne*] ⟦Stück⟧⊣↑*Sehne*↓
257.25:	*einmal*] ⟨Added.⟩
257.27:	M_2, M_3] $M_2\, M_3$

257n: Centri-] Centri
258.1: Trifft] Triff
258.9: Verbindungsgrade PQ] Verbindungsgrade ↑PQ↓
258.15: Existenz] Existens
258.17: Dreieck] Dreiecke
258.18: Inneren] Innere
258.19: Peripherie] Periepherie
258.19: durchschneiden] durch schneiden
258.24: *beweisen*:] *beweisen*
258.25–26: Peripherie] Periepherie
258.27: "] – –
258.27: Kreis] Kreise
258.31: *rechtwinkliges*] *rechtwinkliches*
258.34: Rechnungsart] Rechngsart
259.5: gegeben] geg.
259.6: x-Achse] x Achse
259.7: 0-Punkt] 0 Punkt
259.15: $= A(1, \pi)$] ⟨Added.⟩
259.15: $\pi + 2$] $\pi +$ ⟦《??》⟧↑2↓
259.20: $= A(1, t)$] ⟨Probably added.⟩
259.1: selber] selbser
260.4: α] ↑α↓
260.4: algebraischen] ↑algebraisch↓
260.4: zu α] ↑zu α↓
260.17: gebraucht wird] gebraucht
261.5: Pascal] Paskal
261.11: und] u.
261.21: *Ausführung*] *Ausführung:*
261.29: *rechten*] *Rechten*
261.29: Winkeln] winkeln
261.29: Axiomen] ⟦Sätzen⟧↑Axiomen↓
261.31: folgt] folgt,
261.33: **welche**] **welche** ⟦**jene**⟧
261.34: **eine**] ⟦⟦die⟧↑eine↓ Parallele zu j⟧ **eine**
261n: und] u.
261n: Parallellinien.] Paralllinien
262.7: Parallelseins] Parallelesi《m》es
262.9: *Anfeindungen*] ⟦*Anfenidungen*⟧↑*Anfeindungen*↓ ⟨Correction by Hi^b.⟩
262.10: Alterthum,] Altherth.
262.11: Euklidausgabe] Euklidausgabe von
262.11: des deutschen Mathematikers] ↑des deutschen Mathematikers↓
262.14: Parallelentheorie] Parallelenth.
262.21: Empfindung] Empfindg
262.23: Lehrstuhl] Lehrstuht
262.25–26: , derselbe ⟨...⟩ behauptete] ⟨Added.⟩
262.27: Mathematiker] Math.
262.29: auch,] auch
262n: unsere] unsere ⟦Gru⟧
262n: und] u.
263.9: warum] 《warum》 ⟦Axiom⟧

263.9–10:	das Parallelenaxiom unter] unter
263n:	hinwies;] hinge⟪wies⟫
263n:	Thibaut] Tibaut
263n:	Franzosen] Francosen
263n:	beweisen] beweisen,
263n:	und nachher] u. nacher
263n:	nimmt er] nimmt
263n:	Mathematik] Math.
263n:	exakten] exackten
264.3:	*Satz 1*] *Satz* ↑*1*↓
264.6:	Geraden] Gerade
264.8:	entstehen] enstehen
264.8:	Summe] Sume
264.13–17:	*Satz 2* ⟨...⟩ möglich.] ⟨Added.⟩
264.18:	*Satz 3*] *Satz* ⟦*2*⟧↑*3*↓
264.23:	*Satz 4*] *Satz* ⟦*3*⟧↑*4*↓
264n:	Umgekehrt:] Umgekehrt
264n:	$\triangleleft ACE = \triangleleft A$] $\triangleleft ACE =$ ⟦α,⟧$\triangleleft A$
265.1:	Parallelen] Paralelen
265.1:	⟨p. 77⟩ eine] eine ⟦in den schrafflrten Winkel⟨p. 77⟩räumen verlaufende Gerade⟧
265.2:	Gerade.] Gerade. ⟦Es scheint, dass umgekehrt⟧
265.3:	*Satz 5*] *Satz* ⟦*4*⟧↑*5*↓
265.14:	eventuell] ev.
265.15:	interessante] ⟨Added.⟩
265.17–18:	↑(Existenz ⟨...⟩ *bejahen.*↓] ⟨In the text, these two lines appear in the reverse order.⟩
266.1:	autographirtes] Autographirtes
266.2:	*est*] *ist*
266.6:	Hauptsache,] Hauptsache
266.11:	*Grenzgeraden*] *Grenzgerade*
266.15:	die] Die
266.16:	Transformationen] transformationen
266.18:	Es] ⟦Beweis. ⟪α⟫⟧ Es
266.19:	Geraden] Gerade
266.19:	in Geraden] in Gerade
266.20:	mit Transformationen] m. Transf.
266.21:	reciproken] reciprocen
266n:	Archimedischen] Archimedeschen
267.5:	Gleichungen] Gl.
267.11:	lineare] linearen
267.14:	(Collegheft] Collegheft
267.15:	S. 78)] S. 78
267.16:	*nicht-Euklidische*] *nicht Euklische*
267.16:	*nicht-Euklidischer*] *nicht Euklidischer*
267n:	*Orthogonal-Kreise*] *Orthogonal Kreise*
268.1:	Automorphe] Automorp⟦f⟧↑⟪h⟫↓e
268.2:	Damit] ⟦Orthogonale Krei⟧ Damit

268.2: dass das] das

268.4: gegebenen] gegeben

268.6: $A'S'$] $A'⟦F⟧S'$

268.7: Euklidischen] Euklischen

269.1: nähert] näher

269.1–2: (indem ⟨...⟩ bedenkt)] ⟨Added.⟩

269.5: erst] erst,

269.6: wirklichen] ⟨Added.⟩

269.7: Behauptung:] Beh.

269.8: gleich.] gl.

269.9: Beweis.] Beweis,

269.13: kann nicht $\gamma c + \delta$] ⟦können nicht γ, δ beide⟧⊣↑kann nicht $\gamma c + \delta$↓

269.13–16: Es ⟨...⟩ ist.] ⟨Added in the lower margin, and directed to this place by an insertion sign.⟩

270.6: annehmen] annehm

270.13: im ⟨...⟩ wird.] ⟦$= AB$⟧⊣↑im gewöhnlichen Sinne der Gleichheit↓ gemacht wird. gemacht wird.

270.14: liegen,] liegen

270.18: Nichteuklidischen] Nichteuklischen

271.2: nicht-Euklidische] nicht Euklidische

271.4: Geraden] Gerade

271.16: *Lobatschefskysche*] *Lobatschefsky*

271.17: enthüllt] enthüllt.

271.20: $z = 0$] ⟨Added.⟩

271.21: Alles] aAlles

272.3: *Pascal*] *Palcal*

272.6: Euklidischen] Euklischen

272.8–9: autographierte] Autographierte

272.17: der] Der

272.17: Umgekehrt] Umgeht

272.18: P] ↑P↓

272.20: Denn ⟨...⟩ etc.] ⟨Added.⟩

272.20: $c'' = 0$, $c = 1$] $c'' = 0$ ↑$c = 1$↓

272.23: Analog ⟨...⟩ sich.] ⟨Possibly added.⟩

272.24: Maassbestimmung] Maasbestimmung

273.14: andere] ande

273.15: schneidet E] schneidet E ⟦da sie jedenfalls eine Gerade ↑durch P↓ enthält, welche mit PF in P einen Winkel $<$ wie⟧

273.18: Geraden, Ebenen] Gerade, Ebene

273.18: Vgl. die autographirte Vorlesung S.] ⟨Added.⟩

273.19: Von] ⟦Erweiterung der Axiome⟧ Von

273.25: Pascal] Paskal

273.26: autographirte] Autographirte

274.1: Geraden] Gerade

274.8: Pascal] Paskal

274.9: um-] um

274.9: Pascal] Paskal

274.10: Schoenflies] Schönfliess

274.12: Euklidische] Euklische

274.14: Pascal] Paskal

274.17: Pascal] Paskal
274.19: Parallelentheorie,] Parallelenth.
274.21: Proportionen] Porportionen
274.24: Sätzen:] Sätzen
275.8: $OD = ab$,] $OD = ab$.
275.11: definirt] defenirt
275.13–14: Von ⟨…⟩ S. .] ⟨Probably added.⟩
275.13: autographirte] autographierte
275.15–16: Wenn $a_1 > a_2$ ⟨…⟩ Grösseres.] ⟨Probably added.⟩
276.8: dividirt] divirt
276.20–1: Die ⟨…⟩ Desargues.] ⟨Added.⟩
276.22: werden,] werden
276.23–1: woraus die beiden obigen Fundamentalsätze der Lehre von den Proportionen folgen] ↑woraus die beiden obigen Fundamentalsätze der Lehre von den Proportionen folgt↓
276.1: hieraus] hieraus aus
277.2: Von dem] ⟦Dem⟧↑Von dem↓
277.2: Schnittpunkt] Schnittpunkt ⟦ordnen wir die Zahl a zu, jedem⟧
277.12: C, P] CP
277.16: gegeben] geg
277.18: Gleichung] Gl.
277.19: Gleichung] Gl.
277.22: Gleichung] Gl.
278.8: schneidet,] schneidet.
278.17: (?).] (?)
278.27: autographirte] Autographirte
278.29: Gedanken,] Gedanken
278.29–279.2: *der bei* ⟨…⟩ *Einteilungen.*] ⟨This part of the sentence is written in the bottom margin, and is connected to the first part with a '*)'.⟩
278n: Idealer] ⟦Unendlichferner⟧↑Idealer↓
278n: Rechnung] Rechg
278n: Festsetzung] Festsetzng
279.6: Behauptungen] Behauptung
279.8: mit gleicher] ⟪mit⟫ geleicher
279.8–9: verschiedener anderer] verschieden ⟪ander⟫
279.11: sein] sein.
279.11: es] Es
279.14: Anordnung] Anordng
279.16: Stolz:] Stolz
279.19: majus est] majus ist
279.23: Axioms] Axiom
279.26: Killing-Stolzschen] Killingstolzschen
279.31: gebildete] gebilde
279.31: aus] auch
279.32: sei,] sein
279n: autographirte] Autographirte
280.2: In] in
280.3: einer] ⟦der⟧↑einer↓
280.3: Geraden] Geraden ⟦g⟧
280.3: und C] und ⟦B⟧↑C↓

280.3:	construirt] ⟦B' zwischen B u. C, D'' zwisch B' und C, etc. und ⟪??⟫⟧ construirt
280.4:	A, B, B', B'', B''', ...] ↑A B↓⟦A B⟧ $B'B''B'''$ ...
280.4:	dass] das
280.6:	treffen wir] ⟦⟦giebt⟧↑⟪gelangen⟫↓⟧↑treffen wir↓ es
280.6:	B, B', B'', ...] B B' B'' ..
280.6:	sicher einen solchen] ⟦⟦⟦ein⟧↑zwei↓⟧↑sol↓⟧ Punkt ⟦$B^{(n)}$, $B^{(n+1)}$, so das C⟧ ↑sicher↓ zu einen solchen
280.7:	dass] ⟪dass⟫
280.7:	zwischen] zwisch.
280.7:	und] u.
280.13–14:	und ist C irgend ein Punkt zwischen B und D] ↑und ist C irgend ein Punkt zwischen B und D↓
280.15:	A, B,] ↑A B↓
280.25:	Pascal] Paskal
280.25:	Pascal] Paskal
280.26:	und] u.
280.27:	Pascal] Paskal
280n:	fortgesetztes] Fortgesetztes
280n:	positive] posit.
281.5–8:	Desargues ⟨...⟩ parallel.] ⟨Possibly added, including the equations.⟩
281.5:	entsprechende] entsprechen
281.6:	die Dreiecke] die Dreieck
281.6:	Dreiecke] Dreieck
282.10:	Einführung] Einführg
282.18:	einen] eine
282.20:	und] u.
282.21:	machen.] machen:
282.24:	widersprechen,] widersprechen.
282.24:	Axiome] Axiom
282.25:	widersprechen] wiedersprechen
282.32:	mittelst] mittest
282.32:	Axiome] ↑Axiom↓
282.33:	und] u.
282.34:	Gegeben] Geg
282.35:	eventuell] ev.
282.35:	eine] ein
283.1:	Geraden] Gerade
283.3:	Rechnungsoperation] Rechgoperation
283.3:	ausführen,] ausführen
283.3:	Ausdruck] ⟪Ausdukt⟫
283.5:	besonderen] besondere
283.7:	und] u.
283.10:	Grundlagen der Geometrie,] Grundlg d. Geo.
283.11:	Euklidischen] Euklid.
283.12:	Anschauung] Anschaug
283.12:	Behandlung ermöglicht] Behandlg ⟪ermö⟫glicht
283.13:	Erfahrung] Erfahrg
283.14–15:	Erscheinungsthatsachen] Erscheingsthatsachen
283.15:	Theorie wäre] Th. wahre

283.17: Erscheinung] Erscheing
283.17: Geraden] Gerade
283.1: und] u.
283.1: oben,] oben.
284.1–2: So ⟨...⟩ Mathematik.] ⟨Added on a separate strip of paper which has been pasted in.⟩
284.1: behandeln] behandel
284.2: Lösung] Lösg
284.5: und] u.
284.5: Parallelenaxiom] Prallelenaxiom
284.7: mathematische] math.
284.7: erst] ⟦《??》⟧⥉erst↓
284.9: Mathematik.] Math.
284.10: Legendre-Sätzen] Legendre Sätzen
284.12: körperlicher] Körperlicher
284.13: und] u.
284.13: Gauss'] Gauss
285.11: verlaufende] verlaufenden
286.10: IV, 6)] IV, 6
286.14: gegebenen] geg.
286.14: Parallele] Paralle
286.14: gegebenen] geg.
286n: $4b^2 = 1$] ⟦3⟧⥉4↓b^2 = ⟦0⟧⥉1↓
286n: $1 - b^2$] $1 -$ ⟦4⟧b^2 ⟨This concerns the formula to the left of the main group of formulas.⟩

Description of the Text

Collection: SUB Göttingen, signature *Cod. Ms. D. Hilbert 551.*

Size: Cover size approx. 21 × 17 cm; page size approx. 20.9 × 16.3 cm.

Cover Annotations: On the front cover is pasted a label with the call number, library stamp, and numbering ('36'), and with the notation, 'Grundlagen // der // Euklidischen Geometrie // 2st. 1898/99'. The label on the spine bears the notation, 'Euklidische Geometrie 1898/'; the remainder of the label has been lost.

Composition: Three signatures consisting of 21, 1, and 7 double pages respectively, onto which further double pages, single pages, partial pages, and strips of paper have been pasted. The second signature seems to have been created by folding back the outside page of the third signature, which originally was the second and consisted of 8 double pages, in order to obtain a blank page on which Hilbert could repeat the title. The point of this procedure was probably to provide the binder with clear directions.

Pagination: Continuous pagination, normally of the right-hand pages. The right-hand pages of the first signature are numbered with the odd numbers from 1 to 83; those of the second and third signatures are numbered on the right-hand side with the even numbers from 86 to 116. The number 112 is absent, since the sheet containing pp. 94–97 and 110–113 was left uncut between pp. 111 and 112. The fact that, contrary to the usual practice, number 23 does not appear, manifestly because the margin was already full, and that the opposite page instead bears the number 22, suggest that the numeration was supplied after the manuscript had been at least partially completed.

Original Title: Pasted onto p. 1, the first page of the first signature, and the first page of the second signature are two unnumbered title pages (in our pagination, pages 0 and 85) which bear the identical titles 'Grundlagen der Euklidischen // Geometrie // 2st. Winter 1898/99'. Underneath these, the Roman numerals 'I' and 'II' respectively, which have been added in blue pencil. On the first title page stands in addition 'Prof. Hilbert, Göttingen.' Hilbert very probably had his lecture notes in two separate notebooks, which were subsequently bound together into the present form.

Text: Pages 1 to 84 and 86 to 105 are written with continuous text in Hilbert's hand; on pp. 10, 11, and 21 printed passages cut out of copies of *Hilbert 1895b* have been pasted into the text. The verso side of the first title page, a page pasted onto p. 106, and p. 109 also contain various remarks and elucidations added by Hilbert. A double page pasted onto the blank p. 108 (in our numeration, pp. 108a–c) is in the handwriting of Hans von Schaper, who was responsible for writing up the lectures; there is also a superscription in Hilbert's hand (presumably added later), and two further remarks by Hilbert on p. 108c. Pages 110–117 are blank.

Hilbert extensively corrected and annotated his original text, which was written in black ink; for these revisions Hilbert also used black ink (Hi^s) and especially for underlinings in the first part of the manuscript, blue pencil (Hi^b). These revisions and markings cannot be dated with any degree of precision. They possibly belong to the original period of composition, and probably came to an end after the von Schaper *Ausarbeitung* had been prepared. In contrast to the *Ausarbeitung*, in Hilbert's own manuscript of the lectures there occur no annotations that can be dated with certainty after the time of the completion of the *Festschrift* (cf. *Hilbert 1898c**, 109, in particular the footnote).

Elemente der Euklidischen Geometrie

0 *Vorwort.*

Das vorliegende Heft enthält die von mir unter Aufsicht von Herrn Prof. Hilbert angefertigte Ausarbeitung seiner Vorlesung über die Elemente der Euklidischen Geometrie (Göttingen, W.S. 98/99).

Da die Vorlesung grossenteils neue Entwicklungen und Fragestellungen enthält, erschien es zweckmässig und entsprach den Wünschen vieler Zuhörer, die Ausarbeitung zu vervielfältigen, zunächst und hauptsächlich nur für den Gebrauch der Zuhörer selbst. So entstand die vorliegende Autographie, von der nur 70 Exemplare hergestellt wurden. Die wichtigsten der in diesem Heft enthaltenen Untersuchungen finden sich, zum Teil in anderer Fassung und ausführlicher, in der Abhandlung von Hilbert in der Festschrift zur Enthüllung des Gauss-Weber-Denkmals (Juni 1899) zu Göttingen.

Göttingen, im März 1899.

H. v. Schaper.

1 *Einleitung.*

Die elementare (Euklidische) Geometrie hat zum Gegenstande die Thatsachen und Gesetze, die uns das räumliche Verhalten der Dinge darbietet. Ihrer Struktur nach ist sie ein System von Sätzen, die – im grossen und ganzen wenigstens – auf rein logischem Wege aus gewissen selbst unbeweisbaren Sätzen, den Axiomen, hergeleitet werden. [1]Dieses Verhalten, wie wir es in geringerer Vollkommenheit z. B. auch bei der mathematischen Physik finden, drückt sich am kürzesten in dem Satz aus: *Geometrie ist die vollkommenste Naturwissenschaft.*

Je mehr nun eine Naturwissenschaft ihrem Ziele: „logische Herleitung aller zu ihrem Gebiet gehöriger Thatsachen aus gewissen Fundamentalsätzen" sich nähert, desto notwendiger wird es, diese Axiome selbst genau zu untersuchen, ihre gegenseitigen Beziehungen zu erforschen, ihre Anzahl möglichst zu vermindern u. dergl.

[1]From this point to the end of the paragraph, the text is encircled by Hi^g and directed by an arrow to the remark 'Verhalten ⟨...⟩ Raume', here marked as '^{A}' in line 303.2.

Diese Untersuchung soll in dieser Vorlesung für die [2]Euklidische[2] Geometrie [A] angestellt werden; wir werden dabei nicht nur einem wissenschaftlichen Bedürfnis genügen, sondern auch wichtige Resultate und fruchtbare Gesichtspunkte erhalten.

Es ist von Wichtigkeit, den Ausgangspunkt unserer Untersuchung genau 2
zu fixiren: *Als gegeben betrachten wir die Gesetze der reinen Logik und speciell die ganze Arithmetik.* (Ueber das Verhältnis zwischen Logik und Arithmetik vgl. *Dedekind*, Was sind und was sollen die Zahlen?) Unsere Frage wird dann sein: *Welche Sätze müssen wir zu dem eben definierten Bereich „adjungieren", um die Euklidische Geometrie zu erhalten?*[3]

Mit Benutzung eines Ausdrucks von *Hertz* (in der Einleitung zu den „Principien der Mechanik") können wir unsere Hauptfrage so formulieren: *Welches sind die notwendigen und hinreichenden und unter sich unabhängigen Bedingungen, denen man ein System von Dingen unterwerfen muss, damit jeder Eigenschaft dieser Dinge eine geometrische Thatsache entspreche und umgekehrt, damit also diese Dinge ein vollständiges und einfaches „Bild" der geometrischen Wirklichkeit seien?*

Endlich können wir unsere Aufgabe als eine *logische Analyse unseres Anschauungsvermögens* bezeichnen; die Frage, ob unsere Raumanschauung apriorischen oder empirischen Ursprung habe, bleibt dabei unerörtert. –

Durch das gesagte bestimmt sich das *Verhältnis dieser Vorlesung zu denen über analytische und [4]projektive[4] Geometrie.* | In beiden Disciplinen werden 3
die principiellen Fragen nicht behandelt; in der analytischen Geometrie beginnt man mit der Einführung der Zahl, wir dagegen werden die Berechtigung hierzu genau zu untersuchen haben, sodaß bei uns die Einführung der Zahl geradezu den Schluß bilden wird; in der projectiven Geometrie appelliert man von vorneherein an die Anschauung, wogegen wir ja die Anschauung analysieren wollen, um sie dann sozusagen aus ihren einzelnen Bestandteilen wieder aufzubauen.[5]

Wir werden im folgenden häufig Gebrauch von Figuren machen, wir werden uns aber niemals auf sie verlassen. Stets müssen wir dafür sorgen, daß die an einer Figur vorgenommenen Operationen auch rein logisch gültig bleiben. Es kann dies gar nicht genug betont werden; im richtigen Gebrauch der Figuren liegt eine Hauptschwierigkeit unserer Untersuchung. (Vgl. darüber *Pasch*, Neuere Geometrie, pag. 3.)[6]

Welches Unheil der unüberlegte Gebrauch von Figuren stiften kann, mag das folgende *Beispiel* zeigen. *Wir beweisen den absurden Satz*:

[A]= Verhalten der Gegenstände ⟪nach⟫ Gestalt = Lage im Raume

---

[2–2]Deleted by *Hi*^{g}

[3]This paragraph constitutes the major difference in substance between the *Einleitung* to this text and that to Hilbert's own lecture notes.

[4–4]Replaced (in *Hi*^{g}) by: synthetische

[5]'Analyzing intuition' is not mentioned in the *Einleitung* to Hilbert's own lecture notes, but only first on p. 21 of that text.

[6]The reference is to *Pasch 1882*, 3.

Jedes Dreieck ist gleichschenklig.

Halbieren wir den Winkel C, errichten wir die Mittelsenkrechte auf AB, und
4 fällen wir vom Schnittpunkt M | der genannten Graden die Lote ME und MF.

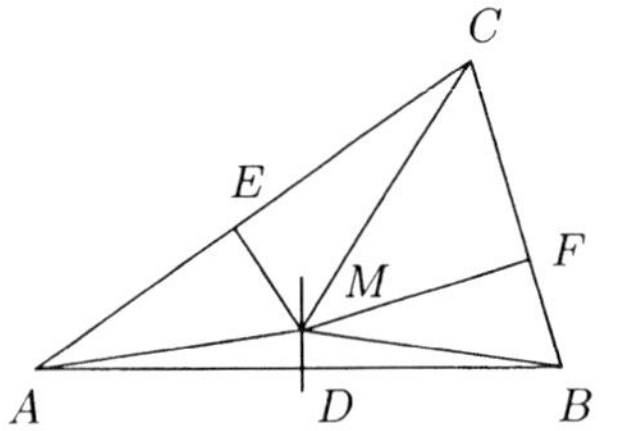

Dann ist:

1) $\triangle CEM \cong \triangle CFM$,

2) $\triangle AMD \cong \triangle BMD$,

3) $\triangle AEM \cong \triangle BFM$.

Also hat man: $CE = CF$, $AE = BF$, woraus durch Addition folgt:

$$AC = BC. \quad \text{q. e. d.}$$

Der Fehler liegt lediglich in der falsch gezeichneten Figur. –

Litteratur. Wir nennen nur 2 Werke:

1) *Pasch*, Vorlesungen über neuere Geometrie 1882, ein sehr klares, aber auch sehr weitläufiges Buch.
2) *Killing*, Grundlagen der Geometrie, 2 Bände, weniger klar, aber sehr inhaltlich.[7] [B]

Gehen wir nun an unsere Aufgabe. *Zum Aufbau der Euklidischen Geometrie denken wir uns drei Systeme von Dingen, die wir Punkte, Grade und Ebenen nennen und mit A, B, C, ...; a, b, c, ...; α, β, γ, ... bezeichnen wollen.*[9] Durch die gewählten Namen dürfen wir uns nicht verleiten lassen, diesen Dingen etwa geometrische Eigenschaften beizulegen, wie wir sie ge-
5 wöhnlich mit | diesen Bezeichnungen verbinden. *Bis jetzt wissen wir nur, daß jedes Ding des einen Systems von jedem Dinge der beiden andern Systeme verschieden ist. Alle weiteren Eigenschaften erhalten diese Dinge erst durch die Axiome, die wir in fünf Gruppen zusammenfassen:*

I. Axiome der Verknüpfung.
II. Axiome der Anordnung oder Reihenfolge.
III. Axiome der Kongruenz oder Bewegung.
IV. Euklid's Parallelenaxiom.
V. Axiom des Archimedes.

Diese Axiome werden wir nun im einzelnen untersuchen.

[B] Vgl. Feriencurs S. 16.[8] ⟨Added by Hi^9.⟩

[7] The works referred to here are: *Pasch 1882*, *Killing 1893* and *Killing 1898*.

[8] Hilbert is presumably referring to the 1898 *Ferienkurs* (see Chapter 3 of this Volume). The page cited appears to have no connection to Killing.

[9] Note the different formulation from Hilbert's own lecture notes, p. 9, which has 'wir nehmen ...', and 'es gibt ein System ...'. The formulation 'wir denken ⟨uns⟩ ...' is also used in the *Festschrift*, p. 4 (see Chapter 5 of this Volume, p. 437).

I. Axiome der Verknüpfung.

Wir unterscheiden sieben.

1. *Irgend zwei von einander verschiedene Punkte A, B bestimmen stets eine Grade a; wir setzen:*

$$\underline{AB = a.}$$

Statt „bestimmen" werden wir auch andere Ausdrücke gebrauchen, z. B. A liegt auf a, A ist ein Punkt von a, a geht durch A und B, a verbindet A und B; wir meinen dabei aber nichts anderes, als was das Axiom 1. aussagt.

2. *Irgend zwei von einander verschiedene Punkte einer* | Graden bestimmen 6
diese Grade; d. h. wenn $AB = a$ und $AC = a$ und $B \neq C$, so ist auch $BC = a$.

Wollten wir nur ebene Geometrie treiben, so wären wir jetzt mit der ersten Gruppe fertig; eine Folgerung aus 1. und 2. ist z. B. daß zwei verschiedene Graden nicht zwei Punkte gemeinsam haben können.

3. *Irgend drei nicht auf derselben Graden liegende Punkte A, B, C bestimmen stets eine Ebene:*

$$\underline{ABC = \alpha.}$$

Auch hier werden wir sagen: A, B, C liegen in α u. s. w.

4. *Irgend drei Punkte A, B, C einer Ebene α, die nicht auf derselben Graden liegen, bestimmen diese Ebene.*

5. *Wenn zwei Punkte A, B einer Graden a in einer Ebene liegen, so liegt die Grade a vollständig in der Ebene, d. h. jeder Punkt von a liegt in α.*

6. *Wenn zwei Ebenen α, β einen Punkt A gemein haben, so haben sie wenigstens noch einen weiteren Punkt B gemein.*

Der Ausdruck: „α und β haben den Punkt A gemein" heisst natürlich nur: „A liegt in α und in β."

Wir betonen, daß 6. die Frage, ob zwei Ebenen stets oder auch nur jemals einen Punkt gemein haben, vollständig offen lässt.

7. *Auf jeder Graden giebt es wenigstens zwei Punkte, in jeder* | *Ebene we-* 7
nigstens drei nicht auf einer Graden gelegene Punkte, und im Raum giebt es wenigstens vier nicht in einer Ebene gelegene Punkte.

Jedes dieser Axiome entspricht einer Thatsache der Beobachtung, die man etwa mit Kugeln, Drähten und Pappdeckeln anstellen kann. Über das Verhältnis zwischen Axiomen und Thatsachen vgl. man die schönen Ausführungen bei *Hertz*, Prinzipien der Mechanik.[10]

Wie man aus den Axiomen folgert, zeigen wir, indem wir folgende *zwei Sätze* beweisen.

a) *Durch eine Grade a und einen nicht auf ihr liegenden Punkt C geht stets eine und nur eine Ebene.*

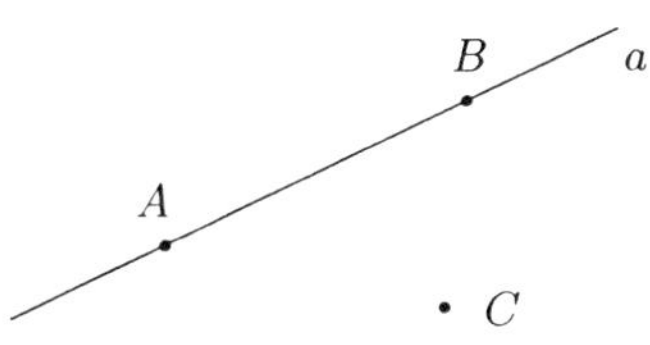

Beweis: Nach 7. giebt es auf a wenigstens zwei Punkte A, B; da C nicht auf a liegt, so ist $A \neq C$, $B \neq C$, und A, B, C liegen nicht auf einer Graden. Also bestimmen A, B, C nach 3. eine Ebene α, in welcher

[10]The reference is to *Hertz 1894*, 1–4.

nach 5. die Grade a vollständig liegt. Aus 4. folgt, daß es nur eine Ebene von der verlangten Eigenschaft giebt.

b) *Durch zwei Grade a, b die einen Punkt C gemein haben, giebt es stets eine und nur eine Ebene.*

8 *Beweis:* Nach 7. giebt es auf a und b mindestens noch je einen Punkt A und B, sodass $A \neq C$, $B \neq C$. A, B, C liegen nicht auf einer Graden, bestimmen also, und zwar eindeutig, eine Ebene α, in der a und b *vollständig* liegen.

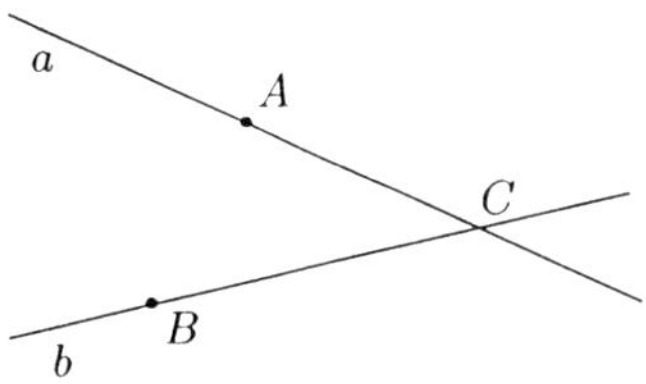

Diesem Satz kann man auch die Form geben:

Zwei nicht in einer Ebene liegende Graden haben keinen Punkt gemein. Dieser Satz ist aber keineswegs umkehrbar; alles, was wir bis jetzt über gemeinsame Elemente wissen, ist dies, wie man leicht aus 1.–7. beweist:

Zwei Graden einer Ebene haben einen oder keinen Punkt gemein; zwei Ebenen haben keinen Punkt oder eine Grade gemein; drei Ebenen haben keinen Punkt oder einen Punkt oder eine Grade gemein; eine Ebene und eine Grade haben keinen oder einen Punkt gemein.

Nachdem wir die Axiome der ersten Gruppe aufgestellt haben, drängen sich zwei Fragen auf:

1.) *Sind diese Axiome unter einander verträglich?*

2.) *Sind sie von einander unabhängig, d. h. ist etwa ein Axiom eine Folge der übrigen?*

Die Beantwortung der ersten Frage schieben wir noch hinaus, da wir später
9 ja zu zeigen haben werden, daß die Axiome aller 5 Gruppen | miteinander verträglich sind. Dagegen wollen wir hier auf die *Frage der Unabhängigkeit* näher eingehen, jedoch nicht etwa beweisen, daß jedes unsrer sieben Axiome von allen andern unabhängig ist, sondern nur *an einzelnen Beispielen das Verfahren* erläutern, welches in folgendem besteht:

Um zu zeigen, daß ein Axiom $\mathfrak{A}$ keine logische Folge der Axiome $\mathfrak{B}$, $\mathfrak{C}$, $\mathfrak{D}$, ... ist, geben wir ein System von Dingen an, bei welchem $\mathfrak{B}$, $\mathfrak{C}$, $\mathfrak{D}$ ... gelten, $\mathfrak{A}$ aber nicht. Solche Systeme wird uns nach dem pag. 2 gesagten am einfachsten die Arithmetik liefern.

Wir wollen drei Beispiele geben.

a.) *2. ist keine Folge von 1.*

Beweis: Als Punkte nehmen wir die ganzen rationalen positiven Zahlen, als Graden die ganzen rationalen negativen Zahlen (damit kein Punkt mit einer Graden identisch werden kann, pag. 5). Zwei Punkte $A = p_1$, $B = p_2$ mögen eine Grade g bestimmen nach dem Gesetz:

$$\left[\frac{p_1 p_2}{2}\right] = -g,$$

wo $[x]$ in üblicher Weise die grösste unterhalb x liegende ganze Zahl bezeichnet. 1. ist selbstverständlich erfüllt, denn g berechnet sich eindeutig aus p_1, p_2; hingegen ist 2. nicht erfüllt; setzen wir etwa $A = 1$, $B = 2$, $C = 3$, so wird $AB = -1$, $AC = -1$, dagegen $BC = -3$.

Oder man nimmt als Punkte die positiven und negativen ganzen rationalen Zahlen, als Graden die mit $i = \sqrt{-1}$ multiplizierten positiven ganzen rationalen Zahlen, und als Zuordnungsgesetz:

$$i \cdot p_1^2 p_2^2 = g.$$

1. gilt selbstverständlich, 2. aber nicht; denn setzen wir $A = 1$, $B = 2$, $C = -2$, 10
so wird $AB = 4i$, ebenso $AC = 4i$, dagegen $BC = 16i$.[11]

b.) *5. ist keine Folge der übrigen 6 Axiome.*

Beweis: Unsere Systeme von Dingen werden wir hier dem Euklidischen Raum entnehmen und damit scheinbar eine grobe Inkonsequenz begehen, da doch die Euklidische Geometrie das Endziel aller unsrer Betrachtungen sein soll. *Wenn schon hier Eigenschaften des Euklidischen Raumes benutzt werden, so ist das lediglich als eine abkürzende Bezeichnung gewisser arithmetischer Beziehungen aufzufaßen*; so steht z. B. „Punkt" (des Euklidischen Raumes) für „Tripel reeller Zahlen", „Ebene des R_3" für „Gesamtheit der einer linearen Gleichung zwischen x, y, z genügenden Zahlentripel" u. s. w.

Nach dieser Bemerkung wird das folgende kein Missverständnis erzeugen.

Als Punkte nehmen wir die Punkte des Euklidischen Raumes, mit Ausnahme eines einzigen Punktes O; als Graden nehmen wir die durch O gehenden Kreise; als Ebenen die gewöhnlichen Ebenen. Es braucht nicht näher ausgeführt zu werden, daß alle Axiome der ersten Gruppe, mit Ausnahme von 5. gültig sind.[12]

c.) *6. ist keine Folge der übrigen sechs Axiome.*

Beweis: Als Punkte bezeichnen wir die Punkte des Euklidischen Raumes; dabei scheiden wir aber sämtliche Punkte einer Geraden a aus, bis auf einen einzigen Punkt P, der wieder zu unsern Punkten gehören soll; als Graden
bezeichnen wir sämtliche Graden des gewöhnlichen | Raumes, a ausgenommen; 11
als Ebenen endlich nehmen wir die Ebenen des R_3. Dann gelten alle Axiome I, nur das 6. nicht: denn zwei Ebenen, die im R_3 die Grade a gemein haben, haben in unserer soeben definierten Geometrie nur den Punkt P gemein.

Zum Schluß bemerken wir, daß die Unabhängigkeitsbeweise hier noch sehr einfach sind; später hingegen werden sie viele Schwierigkeiten darbieten.

II. Die Axiome der Anordnung oder Reihenfolge.

Bisher wissen wir über die Punkte einer und derselben Graden so gut wie nichts; hierüber geben uns die fünf Axiome der zweiten Gruppe Aufschluß, die man auch als *Axiome über den Begriff „zwischen"* bezeichnen könnte. Sie sind zuerst von *Pasch* aufgestellt worden, dessen Darstellung wir hier aber bedeutend vereinfachen werden.

[11] Note that multiplying by i removes the ambiguity found in the same proof in Hilbert's own lecture notes.

[12] Straight lines passing through 0 are to be taken as straight lines in the new model, too. However, the model does not serve Hilbert's purpose here, since both Axioms 3 and 4 also clearly fail. Three points may be colinear in the usual sense, although not in the new sense, thus will lie in infinitely many (standard) planes. The failure of 3 and 4 was pointed out to Hilbert by Hausdorff in a letter: see Appendix, Remark [1], and the note thereto.

Die *vier ersten Axiome dieser Gruppe* beziehen sich auf die *Geometrie auf der Graden* und lauten:

1. *Wenn A, B, C Punkte einer Graden sind, und C zwischen A und B liegt, so liegt C auch zwischen B und A.*

Damit ist die erste Eigenschaft des Begriffs „zwischen“ gegeben, die weiteren Eigenschaften dieses wichtigen Begriffs kommen in den folgenden Sätzen zum Ausdruck.

2. *Wenn A, B zwei Punkte einer Graden sind, so giebt es stets wenigstens*
12 *einen Punkt C, der zwischen A und B liegt, und wenigstens ei|nen Punkt D, sodaß B zwischen A und D liegt.*

3. *Unter irgend drei Punkten einer Graden giebt es stets einen und nur einen, der zwischen den beiden andern liegt.*

4. *Irgend vier Punkte A, B, C, D einer Graden können stets so angeordnet werden, daß B zwischen A und C und zwischen A und D, und ferner C zwischen A und D und zwischen B und D liegt.*

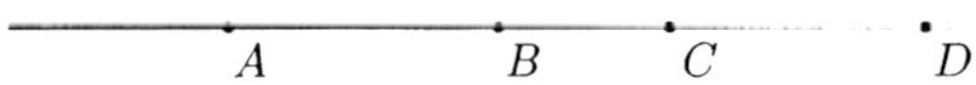

Definition: Den Inbegriff aller zwischen zwei Punkten A und B einer Graden liegenden Punkte nennen wir die „*Strecke* $\overline{AB}$“. Von den Punkten zwischen A und B sagen wir, daß sie „*innerhalb*“ der Strecke liegen, alle übrigen Punkte der Graden liegen „*ausserhalb*“ der Strecke. –

Mit den bisher aufgestellten Axiomen lassen sich bereits viele geometrische Sätze beweisen; wir geben einige Beispiele dafür und beweisen zuerst:

a) *Sind A, B, C, D Punkte einer Graden, sodaß C zwischen A und B, und D zwischen A und C liegt, dann liegt auch D zwischen A und B, aber nicht zwischen C und B.*

Beweis: Wir ordnen die Punkte A, B, C, D so an, wie es dem Axiom 4. entspricht; dann müssen offenbar C und D die zweite oder dritte Stelle erhalten. Wir kommen so auf folgende Anordnungen: A, C, D, B, B, D, C, A, A, D, C, B, B, C, D, A. Von diesen Anordnungen genügen aber die beiden
13 ersten den | Voraussetzungen nicht. Denn bei beiden Anordnungen läge C zwischen A und D, was nach 3. der Voraussetzung „D zwischen A und C“ widerspricht; bei der dritten und vierten Anordnung ergiebt sich nach 4., dass

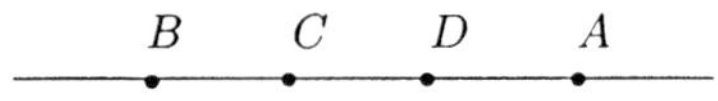

C zwischen B und D liegt, also nach 3. nicht D zwischen C und B liegen kann.

b) *Auf jeder Strecke giebt es unendlich viele Punkte.*

Beweis: Nach 2. giebt es auf $\overline{AB}$ wenigstens einen Punkt C; ebenso giebt es auf $\overline{AC}$ wenigstens einen Punkt C'; nach dem Satz a) liegt dann auch C' auf $\overline{AB}$, aber nicht auf $\overline{CB}$. Weiter giebt es auf $\overline{AC'}$ wenigstens einen Punkt C'', der ebenfalls auf $\overline{AB}$, aber nicht auf $\overline{C'B}$ liegt, also, da C auf $\overline{C'B}$ liegt, nicht mit C identisch sein kann. Auf diese Weise erhalten wir immer neue Punkte

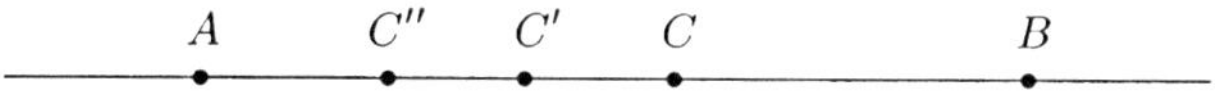

von $\overline{AB}$, ohne jemals zu Ende zu kommen.

c) *Ist* A, B, C, D *eine dem Axiom 4. entsprechende Anordnung von vier Punkten, so giebt es ausser dieser Anordnung nur noch die umgekehrte* D, C, B, A, *welche dem Axiom 4. entspricht.*

Der *Beweis* ist im wesentlichen schon bei a) gegeben.

d) *Sind* A, B, C, $D, \ldots, K$ *eine endliche Anzahl von Punkten einer Graden, so lassen sich dieselben stets in eine Reihe bringen, sodaß* B *zwischen* A *einerseits und* C, $D, \ldots, K$ *andrerseits,* C *zwischen* A, B *einerseits und* $D, \ldots, K$ *andrerseits u. s. w. liegt.* | *Diese Anordnung ist eindeutig bestimmt,* 14
bis auf ihre Umkehrung: $K, \ldots, D, C, B, A$. (Dieser Satz ist eine Erweiterung des 4. Axioms.) Wir nennen die Punkte A, B, $C, \ldots, K$ schlechtweg „*geordnet*".

Beweis: Unser Satz gilt für 4 Punkte A, B, C, D nach Axiom 4 und nach Satz c). Wir zeigen, daß der Satz gültig bleibt für $n+1$ Punkte, wenn er für n Punkte bewiesen ist. Der Beweis soll nur skizziert werden; die Ausführung im einzelnen bietet keine Schwierigkeiten, ist aber etwas schwerfällig. Sei also $A_1 A_2 \ldots A_n$ die verlangte Anordnung bei n Punkten. Nehmen wir einen weiteren Punkt X hinzu, so sind nach 3. zunächst drei Fälle möglich:

1.) A_1 liegt zwischen X und A_n,
2.) A_n liegt zwischen X und A_1,
3.) X liegt zwischen A_1 und A_n.

Im Fall 3. beweist man weiter, daß es eine und nur eine Zahl m giebt, sodaß X zwischen A_m und A_{m+1} liegt. Schließlich zeigt man, daß folgende Anordnungen in den drei Fällen die verlangten Eigenschaften haben:

1.) $X A_1 A_2 \ldots A_n$
2.) $A_1 A_2 \ldots A_n X$
3.) $A_1 A_2 \ldots A_m X A_{m+1} \ldots A_n$

und daß sie nebst ihren Umkehrungen die einzigen sind, die jene Eigenschaften 15
besitzen.

e) *Jeder Punkt* O, *welcher auf einer Graden* a *liegt, trennt die übrigen Punkte dieser Graden in zwei Abteilungen von folgender Beschaffenheit: Ist* A *ein Punkt der einen,* B *ein Punkt der andern Abteilung, so liegt stets* O *zwischen* A *und* B. *Sind dagegen* $A_1 A_2$ *Punkte derselben Abteilung, so liegt* O *nicht zwischen* A_1 *und* A_2.

Beweis: Seien A, B zwei Punkte, die nicht ein und derselben Abteilung angehören, sei ferner X irgend ein von O, A, B verschiedener Punkt der Graden; wenn wir dann die 4 Punkte gemäss dem Axiom 4. ordnen, so sind je nach der Lage von X folgende Fälle möglich: 1) $XAOB$, 2) $AXOB$, 3) $AOXB$, 4) $AOBX$. In den beiden ersten Fällen nehmen wir X zu der durch A, in den beiden anderen Fällen zu der durch B vertretenen Abteilung. Es ist leicht zu zeigen, daß diese Abteilungen die verlangte Beschaffenheit haben.

Man sagt, daß O die Grade a in zwei „*Halbstrahlen*" teilt; oder daß jeder Punkt von a entweder *rechts* oder *links* von O liegt, u. s. w.

16 Der eben bewiesene Satz zeigt, daß *zwischen der Geometrie* | *und den Grundsätzen der Arithmetik eine weitgehende Analogie* herrscht. In der That lassen sich bekanntlich die rationalen Zahlen „ordnen" im Sinne des Satzes d). Jede rationale Zahl α trennt alle übrigen in zwei Abteilungen, so daß jede Zahl der einen Abteilung grösser ist als jede Zahl der andern Abteilung. Indem man weiter diese Eigenschaft, einen „Schnitt" in der Reihe der rationalen Zahlen zu machen, zur wesentlichen Eigenschaft des Zahlbegriffs erhebt, gelangt man zu den Irrationalzahlen. Dem geometrischen Begriff „zwischen" entspricht in der Arithmetik der Begriff des grösser und kleiner seins. —

Ehe wir näher auf die Beziehungen zwischen unsern vier Axiomen eingehen, haben wir noch *zwei wichtige Bemerkungen* zu machen:

1.) *Durch die Axiome 1.–4. ist keineswegs die Eindimensionalität der Graden gegeben.* In der That lassen sich z. B. alle Punkte des dreidimensionalen Raumes so anordnen, daß die Axiome 1.–4. gelten. Wir brauchen nur jeden Punkt durch seine rechtwinkligen Koordinaten xyz zu charakterisieren und dann folgende Festsetzungen zu machen: Jeder Punkt mit grösserer x-Koordinate soll in der Reihe hinter jedem Punkt mit kleinerer x-Koordinate
17 stehen; bei Punkten | mit gleichem x soll das y, bei gleichem x und y soll z den Ausschlag geben.

2.) *Die gewöhnliche Anordnung der Punkte auf einer Graden und entsprechend die Anordnung der reellen Zahlen ihrer Grösse nach ist nicht die einzige, bei welcher der Begriff „zwischen" mit seinen Merkmalen 1.–4. gültig ist.* Eine andere Anordnung enthält man z. B, indem man zuerst die rationalen Zahlen ihrer Grösse nach ordnet, ebenso die irrationalen, und ausserdem festsetzt, daß jede irrationale Zahl grösser heissen soll als jede rationale.

Wir behaupten jetzt, daß *die Axiome 1.–4. von einander unabhängig* sind, und beweisen 2 Beispiele:

1) *4. folgt nicht aus 1., 2., 3.*

Beweis: Wir repräsentieren die Punkte A, B, $C, \ldots$ durch reelle Zahlen α, β, $\gamma, \ldots$ und setzen fest: C liege zwischen A und B, sobald $\gamma > \alpha$ und zugleich $\gamma > \beta$ ist (wenn also der Punkt C nach gewöhnlichem Sprachgebrauch „*hinter*" A und B liegt). Daß die Axiome 1.–3. gelten, ist evident, 4. aber gilt nicht, denn nehmen wir an, $ABCD$ sei eine Anordnung von 4 Punkten im Sinne des Axioms 4. Dann müsste sein:

$$\beta > \alpha, \quad \beta > \gamma, \quad \beta > \delta,$$

und gleichzeitig:

$$\gamma > \alpha, \quad \gamma > \beta, \quad \gamma > \delta,$$

was nicht möglich ist.

18 2.) *3. folgt nicht aus 1., 2., 4.*

Beweis: Wir setzen fest, C liege zwischen A und B, wo etwa $\alpha < \beta$, sobald entweder $\gamma < \alpha$ oder $\gamma > \beta$ ist (d. h. sobald C „*ausserhalb*" $\overline{AB}$ liegt.)
1., 2., 4., gelten, 3. aber nicht; denn von 3 gegebenen Punkten A, B, C liegt nicht einer, sondern es liegen stets zwei ausserhalb der übrigen. —

Nach den „linearen" Axiomen 1.–4. kommen wir nun zu dem *einzigen „ebenen" Axiom in II.*

5. *Jede Grade a, welche in einer Ebene α liegt, trennt die Punkte dieser Ebene α in zwei Gebiete von folgender Beschaffenheit: Ein jeder Punkt A des einen Gebietes bestimmt mit jedem Punkt B des andern Gebietes eine*[13] *Strecke $\overline{AB}$, innerhalb welcher ein Punkt der Graden a liegt; dagegen bestimmen irgend zwei Punkte A und C eines und desselben Gebietes eine Strecke $\overline{AC}$, welche keinen Punkt von a enthält.*

Durch dieses Axiom wird der Begriff „*Halbebene*" eingeführt.

Wir beweisen mit Hülfe von 5. den Satz:

Sei $ABC = \alpha$ eine Ebene und a eine durch keinen der drei Punkte A, B, C gehende Grade in ihr, welche die Strecke $\overline{AB}$ schneidet, dann schneidet a notwendig eine und nur eine | der Strecken $\overline{AC}$ und $\overline{BC}$. 19

Beweis: Nach Voraussetzung liegen A, B auf verschiedenen Seiten von a.

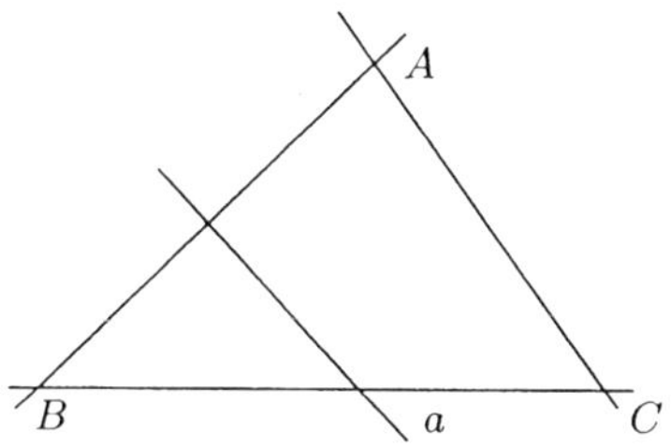

Nehmen wir nun an, $\overline{AC}$ werde von a nicht geschnitten; dann liegen A und C auf der nämlichen Seite von a, also liegen B und C auf verschiedenen Seiten von a, d.h. $\overline{BC}$ wird von a geschnitten.

Nehmen wir ferner an, $\overline{AC}$ und $\overline{BC}$ werden beide von a geschnitten; dann müssten A und B auf der nämlichen Seite von a liegen, was nicht möglich ist, da nach Voraussetzung $\overline{AB}$ von a geschnitten wird.[14]

Wir zeigen nun, daß das Axiom 5. wirklich notwendig ist, um den vorstehenden Satz zu beweisen, oder:

5. ist keine Folge von 1.–4. und von den Axiomen der Gruppe I.

Beweis: Wir zeigen, daß der obige Satz nicht mehr richtig ist, sobald wir von allen uns bis jetzt bekannten Axiomen einzig das Axiom II, 5. fortlassen.

a) Denken wir uns zunächst einmal, daß wir dies nur für ebene Geometrie beweisen wollten, sodaß von der Gruppe I nur die | beiden ersten Axiome in 20
Betracht kommen. Als die Ebene unserer neuen Geometrie bezeichnen wir den Inbegriff sämtlicher Punkte des gewöhnlichen Raumes. Dann gelten I, 1., 2. und II, 1.–4. Ferner ist klar, daß ich für jedes Dreieck ABC Grade ziehen kann, die nur eine einzige Seite des Dreiecks schneiden. Dann kann aber 5. nicht gültig sein.

b) Für den Raum ist der Beweis ein wenig länger. Wir bezeichnen auf jeder Graden eine bestimmte Richtung als positiv und ordnen nun die Punkte

---

[13]The following was added by Hilbert (in *Hi*^{g}) in the left-hand margin, but later crossed out and partly cut off: 'eine Gerade, ⟨d⟩ie ⟦die Gerade ⟨a⟩ ⟪in⟫ einem ⟨zw⟩ischen A u. B ⟨ge⟩legenen Punkte ⟨s⟩chneidet⟧↑mit der Gerade ⟨a⟩ einen ⟨zw⟩ischen A u. B ⟨ge⟩legenen Punkt ⟨g⟩emein hat↓ ⟨*Hi*^{g}⟩'. Two lines later, in the last clause of axiom 5, Hilbert added: '... keinen ⟨Punkt⟩ ... gemein hat.'

[14]The *Satz* is 'Pasch's Axiom'. See p. 22 of Hilbert's own lecture notes, and the corresponding editorial note.

auf jeder Graden in der schon auf pag. 17 geschilderten Weise, d. h. wir setzen fest, daß jeder irrationale Punkt jedem rationalen folgen soll, während im übrigen die gewöhnliche Reihenfolge beibehalten wird. Als einen rationalen Punkt bezeichnen wir einen solchen, dessen drei gewöhnliche Koordinaten x y z sämtlich rational sind.

Sei nun ABC ein Dreieck mit lauter rationalen Ecken, D sei ein rationaler Punkt auf $\overline{AC}$, E ein irrationaler Punkt auf $\overline{BC}$.[15] Dann schneidet die Grade $DE = a$ nicht die Strecke $\overline{BC}$, denn E liegt ja nach der Definition nicht zwischen B und C. a tritt also in das $\triangle ABC$ ein, ohne wieder herauszutreten. Der

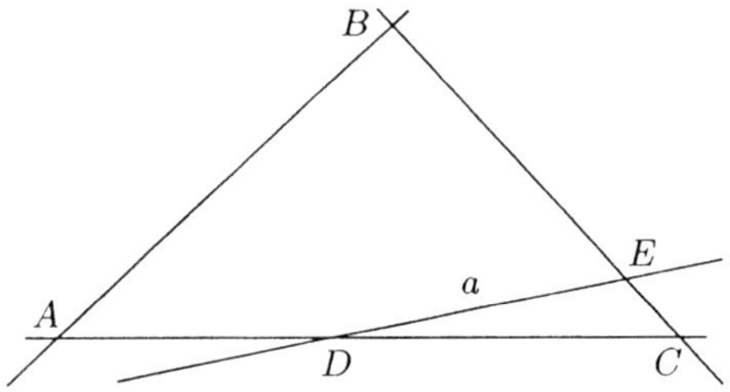

21 Satz pag. 19 gilt also hier nicht, und | folglich auch nicht das Axiom 5. –

Ein „räumliches"Axiom brauchen wir in der II. Gruppe nicht, vielmehr werden wir den zu 5. analogen Satz der räumlichen Geometrie nun *beweisen.* Er lautet:

Jede Ebene α trennt die Punkte des Raumes in zwei Gebiete von folgender Beschaffenheit: Jeder Punkt A des einen Gebiets bestimmt mit jedem Punkt B des andern Gebiets eine[16] *Strecke $\overline{AB}$, innerhalb derer ein Punkt von α liegt; dagegen bestimmen irgend zwei Punkte A und C eines und desselben Gebietes eine Strecke $\overline{AC}$ die keinen Punkt*[17] *von α enthält.*

Beweis: Es sei A ein Punkt, der nicht auf α liegt. Dann rechne ich zu dem einen Gebiet alle Punkte P von der Eigenschaft, daß zwischen A und P, also auf $\overline{AP}$ kein Punkt von α liegt; zu dem andern Gebiet alle Punkte Q, sodass auf $\overline{AQ}$ ein Punkt von α liegt. Nun ist zu zeigen:

1) Auf $\overline{PP'}$ liegt kein Punkt von α.
2) Auf $\overline{QQ'}$ liegt kein Punkt von α.
3) Auf $\overline{PQ}$ liegt stets ein Punkt von α.

1) Nach Voraussetzung liegt weder auf $\overline{AP}$ noch auf $\overline{AP'}$ ein Punkt von α. Nehmen wir nun an, auf $\overline{PP'}$ läge ein Punkt von α, dann hätten die Ebene α und die Ebene APP' diesen Punkt und folglich eine Grade a gemein. Diese Grade geht durch keinen der Punkte A, P, P'; sie schneidet $\overline{PP'}$, sie muss also
22 nach pag. 19 entweder $\overline{AP}$ oder $\overline{AP'}$ | schneiden, was gegen die Voraussetzung ist.

---

[15] This should be the line BC, not the segment $\overline{BC}$.

[16] Addition by Hilbert in the right-hand margin (in Hi^g), and assigned to this position by a correction character. As with the previous addition, it was later crossed out by Hilbert (in Hi^g): Gerade, die mit der Ebe⟨ne⟩ α ein⟨en⟩ zwi⟨schen⟩ A ⟨u. ⟩ B geleg⟨enen⟩ P⟨un⟩k⟨t⟩ ⟨⟨gemein⟩⟩ hat.

[17] To the right of 'keinen Punkt', Hi^g has added in the margin, and later on deleted, 'keinen', which obviously was meant to indicate an alternative formulation corresponding to the one reported in the note to line 312.22.

2) Nach Voraussetzung liegt auf $\overline{AQ}$ ein Punkt von α, ebenso auf $\overline{AQ'}$. Die Schnittgrade der Ebenen α und AQQ' trifft also zwei Seiten des $\triangle AQQ'$; folglich kann sie nach pag. 19 nicht auch noch die dritte Seite $\overline{QQ'}$ treffen.

3) $\overline{AP}$ enthält nach Voraussetzung keinen Punkt von α, $\overline{AQ}$ dagegen enthält einen solchen. Die Schnittgrade der Ebenen α und APQ trifft also die Seite $\overline{AQ}$ und trifft nicht die Seite $\overline{AP}$ im $\triangle APQ$; also trifft sie die Seite $\overline{PQ}$.

Dieser Satz erlaubt uns, von einem „*Halbraum*“ und von den „*beiden Seiten der Ebene*“ α zu sprechen. –

Von weiteren Sätzen, die sich auf Grund der bisherigen Axiome beweisen lassen, heben wir folgende hervor:[18]

a) *Sei in einer Ebene ein Punkt M und eine endliche Anzahl durch ihn hindurch laufender Graden a, b, $c\dots$ gegeben; dann läßt sich stets eine Grade finden, welche diese sämtlichen durch M gehenden Graden schneidet.*

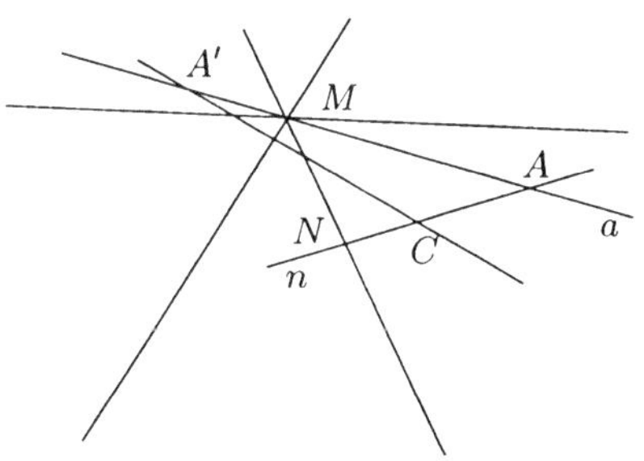

Beweis: Ich nehme auf a 2 Punkte A und A' an, sodaß M zwischen A und A' liegt; ferner lege ich durch A eine Grade $n \neq a$ von der Art, daß sie wenigstens noch eine | der Graden 23
b, $c,\dots$ trifft. Unter allen Schnittpunkten N', $N'',\dots$ (deren Anzahl also wenigstens $= 1$ ist) der Graden n mit Strahlen des Büschels M giebt es nun stets einen N (und höchstens zwei), sodaß zwischen A und N kein weiterer solcher Schnittpunkt liegt. Jetzt wähle ich auf $\overline{AN}$ irgend einen Punkt C und verbinde ihn mit A'; dann ist $A'C$ eine Grade von der verlangten Beschaffenheit. In der That sei x irgend einer der Strahlen durch M. Dann liegen A und A' auf verschiedenen Seiten von x, weil ja $\overline{AA'}$ in M von x geschnitten wird. A und C liegen auf derselben Seite von x, weil ja kein Strahl die Strecke $\overline{AN}$ und a fortiori die Strecke $\overline{AC}$ trifft. Also liegen A' und C auf verschiedenen Seiten von x, d. h. x wird von $A'C$ geschnitten, was zu beweisen war.

b) Der analoge Satz im Raum lautet:

Wenn durch einen Punkt im Raum eine endliche Anzahl von Graden geht, so giebt es stets eine Ebene, die alle diese Graden schneidet.

Bei dem Beweis wollen wir uns nicht aufhalten, sondern gleich übergehen zu dem

Satz von Desargues.

Wenn zwei Dreiecke ABC und $A'B'C'$ in einer Ebene so liegen, daß je zwei entsprechende Seiten AB und $A'B'$, AC und $A'C'$, BC und $B'C'$ sich in drei Punkten E, D, F einer Graden schneiden, so treffen sich die Verbindungslinien entsprechender Ecken, | AA', BB', CC' entweder gar nicht, oder 24
sie laufen sämmtlich durch denselben Punkt M.

(Dieser Satz ist übrigens umkehrbar.)

[18] These two theorems are not in Hilbert's own lecture notes.

Den *Beweis* führen wir so, daß wir zuerst die Gültigkeit des Satzes unter der Annahme beweisen, daß der Satz für Dreiecke ABC und $A'B'C'$ gelte, *die nicht in einer Ebene liegen*; hinterher zeigen wir dann, daß er für diesen Fall wirklich gilt.

Nehmen wir an, daß sich AA' und BB' in M schneiden; dann ist zu zeigen, daß auch CC' durch M geht.

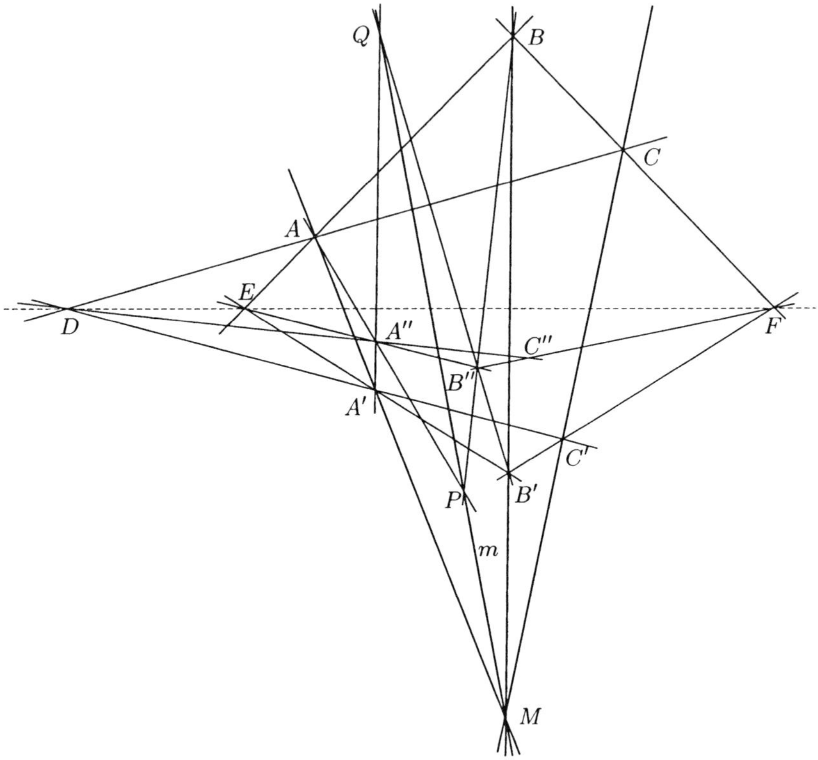

Wir wollen dies beweisen *für den Fall, daß* A, B, C *auf der einen,* A', B', C' *auf der andern Seite der Graden* DEF *liegen*; in den übrigen Fällen modifiziert sich der Beweis etwas.

Wir legen nun eine Grade m durch M, *die nicht in der Ebene der beiden Dreiecke liegt*, und nehmen auf m zwei Punkte P, Q an derart, daß die Reihenfolge MPQ mit der Reihenfolge $MA'A$ übereinstimmt. Die Graden $\overline{AP}$ und $\overline{A'Q}$ müssen sich nun notwendig schneiden, etwa in A''; denn nach Wahl von P, Q liegt gleichzeitig P zwischen M und Q, und A' zwischen M und A. Ebenso müssen sich $\overline{BP}$ und $\overline{B'Q}$ schneiden, etwa in B''; denn nach Voraussetzung ist $MB'B$ dieselbe Reihenfolge wie $MA'A$, also auch wie MPQ, und es liegt demnach gleichzeitig P zwischen M und Q, und B' zwischen M und B.

Da A'' auf $\overline{AP}$, B'' auf $\overline{BP}$ liegt, so liegen die 5 Punkte P, A'', B'', A, B in einer Ebene; in dieser Ebene liegt auch E als Punkt von AB; also liegen P, A'', B'', A, B, E in einer Ebene. Genau so findet man, daß auch Q, A'',

B'', A', B', E in einer Ebene liegen. Diese beiden Ebenen sind von einander verschieden; wären sie nämlich gleich, so müssten sie mit der Dreiecksebene zusammenfallen, was unmöglich ist, da z. B. P nicht in der Dreiecksebene liegt. Daraus folgt: *die Punkte* A'', B'', E (die beiden Ebenen gemeinsam sind) *liegen auf einer Graden* oder: $A''B''$ *geht durch den Punkt* E.

Jetzt verbinde ich A'' mit D und B'' mit F; dann liegen $A''D$ und $B''F$ in einer Ebene $A''B''DEF$ und schneiden sich notwendig in einem Punkte C''. Denn erstlich kann ohne Beschränkung der Allgemeinheit angenommen werden, daß E zwischen D und F liegt; ferner kann man zeigen, | daß die ganze 26
Figur auf einer Seite der Dreiecksebene liegt, sodaß entweder A'' zwischen E und B'' oder B'' zwischen E und A'' liegt. Im ersten Falle betrachten wir $\triangle EFB''$ und die Grade $A''D$ im zweiten aber $\triangle EDA''$ und die Grade $B''F$. Aus dem Satz p. 19 folgt beidemale *die Existenz von* C''.

Jetzt betrachten wir die Dreiecke $A''B''C''$ und ABC. Sie haben die Eigenschaft, daß die entsprechenden Seiten sich in drei Punkten einer Graden schneiden; und sie liegen ausserdem in zwei verschiedenen Ebenen. Für solche Dreiecke wollten wir unsern Satz vorläufig als bewiesen betrachten, p. 24. Dann ergiebt sich, daß CC'' *durch* den Schnittpunkt P von AA'' und BB'' hindurchlaufen muss. Genau so findet man durch Betrachtung der Dreiecke $A''B''C''$ und $A'B'C'$, daß $C'C''$ *durch* Q gehen muss. Die beiden Graden $CC''P$ und $C'C''Q$ bestimmen eine Ebene, in der auch M als Punkt von PQ liegt. Ausser in dieser Ebene liegen C, C', M auch noch in der von ihr verschiedenen Dreiecksebene; und hieraus folgt endlich, daß C, C', M auf einer Graden liegen.

Jetzt ist der *Beweis* nachzuholen, *daß unser Satz gilt, wenn* ABC *und* $A'B'C'$ *nicht in einer Ebene liegen.* Mögen sich AA' und BB' in M schneiden [vgl. die Figur pag. 24], dann ist zu zeigen, daß auch CC' durch M geht. Nun sieht man:

$$\begin{array}{lccc} A,\, B,\, E,\, B',\, A' & \text{bestimmen} & \text{eine Ebene} & \gamma, \\ B,\, C,\, F,\, C',\, B' & " & " & \alpha, \\ C,\, A,\, D,\, A',\, C' & " & " & \beta. \end{array}$$

Die Ebenen α und β haben die Punkte C und C' gemein; sie haben aber auch 27
M gemein, weil M sowohl auf AA', also in β, als auch auf BB', also in α liegt. Hieraus folgt dann, daß C, C', M auf einer Graden liegen, womit der Satz von Desargues vollständig bewiesen ist. —[19]

Dieser Satz giebt uns nun Gelegenheit zur Erörterung einer wichtigen Frage. *Der Inhalt nämlich des Desargues'schen Satzes gehört durchaus der ebenen Geometrie an; zu seinem Beweise aber haben wir den Raum gebraucht.* Wir sind daher hier zum ersten Mal in der Lage, eine *Kritik der Hülfsmittel eines Beweises* zu üben. In der modernen Mathematik wird solche Kritik sehr häufig geübt, wobei das Bestreben ist, *die Reinheit der Methode* zu wahren, d. h. beim Beweise eines Satzes wo möglich nur solche Hülfsmittel zu benutzen,

[19]The proof given here is a fuller version of that sketched in Hilbert's own lecture notes on pp. 26–28. Nevertheless, it is not, as Hilbert asserts, complete. For instance, it operates under the assumption that M is not between A and A', and the further assumption that E is between D and F.

die durch den Inhalt des Satzes nahe gelegt sind. Dieses Bestreben ist oft erfolgreich und für den Fortschritt der Wissenschaft fruchtbar gewesen.

Untersuchen wir nun von diesem Gesichtspunkte aus unsern Satz, so werden wir uns die Aufgabe stellen: *entweder den Desargues lediglich mit Hülfe der ebenen Axiome zu beweisen, oder zu zeigen, daß ein solcher Beweis unmöglich ist.* Wie kann man nun zeigen, daß ein an sich richtiger Satz mit gewissen Hülfsmitteln unbeweisbar ist? Wir werden offenbar dasselbe Princip anwenden müssen wie bei den Unabhängigkeitsbeweisen [p. 8, 9], d. h. um es
28 gleich | für den vorliegenden Fall zu spezialisieren:

Wir werden eine ebene Geometrie konstruieren müssen, in der die sämtlichen ebenen Axiome (I, 1., 2., II, 1.–5.) gelten, während der Satz von Desargues nicht gilt.

Eine solche Geometrie wollen wir nun in der That konstruieren.

Als *Punkte unserer Geometrie* nehmen wir die Punkte einer Ebene, scheiden jedoch sämtliche Punkte eines Halbstrahls $0,+\infty$ aus. *Die Graden definieren wir so*, daß zwei beliebige Punkte stets eine (u. nur eine) Grade bestimmen, und zwar folgendermassen: Zwei Punkte der unteren Halbebene bestimmen eine Grade im gewöhnlichen Sinne. Diese Graden sollen auch in unserer Geometrie Grade sein; schneiden sie $0,+\infty$, so sollen sie dort endigen, schneiden sie $0,-\infty$, so sollen sie sich in die obere Halbebene als Kreisbögen fortsetzen, die in 0 endigen und im Schnittpunkt mit $0,-\infty$ die Grade aus der

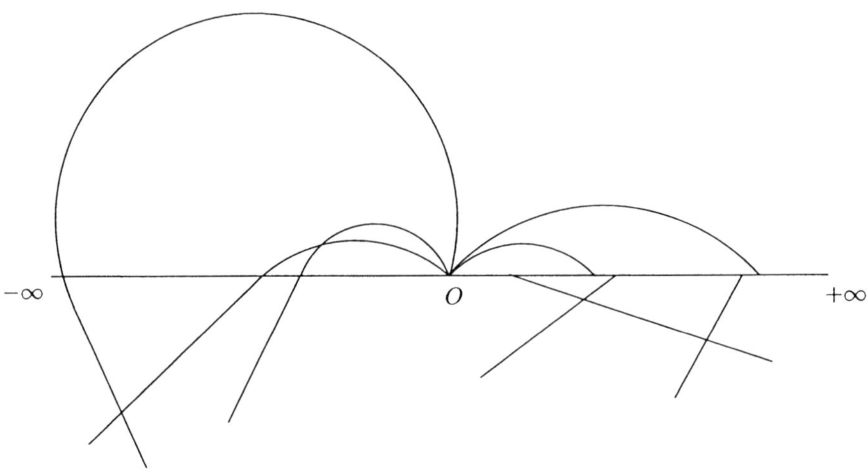

untern Halbebene zur Tangente haben. Zwei Punkte der oberen Halbebene
29 sollen als Verbindungsgrade | einen in 0 endigenden Kreisbogen bestimmen, der entweder auch in $0,+\infty$ endigt, oder sich über $0,-\infty$ gradlinig in die untere Halbebene fortsetzt, und zwar in Richtung der Tangente. Haben wir endlich einen Punkt der oberen und einen der unteren Halbebene, so soll auch hier die Verbindungsgrade aus einem (im gewöhnlichen Sinne) gradlinigen Stück

und einem Kreisbogen bestehen, in derselben Weise wie in den beiden ersten Fällen.[20]

In dieser Geometrie gilt nun I, 1. nach Voraussetzung; ferner gilt I, 2.: Zwei Grade schneiden sich höchstens in einem Punkt. Es folgt nämlich aus dem Satz vom Sehnen- und Peripheriewinkel, daß wenn sich zwei von 0 auslaufende Kreisbögen schneiden, ihre gradlinigen Fortsetzungen in der unteren Halbebene divergieren, sich also nicht schneiden, und umgekehrt. Weiter gelten alle Axiome II, wenn ich *die Anordnung der Punkte auf einer Graden im Sinne der gewöhnlichen Geometrie* nehme; insbesondere kann ich auf jeder Graden, wenn irgend zwei Punkte A, B auf ihr gegeben sind, einen Punkt C angeben, sodaß B zwischen A und C liegt; diese Eigenschaft bleibt für unsere Geometrie dadurch gewahrt, daß die Punkte der Halbgeraden $0,+\infty$ nicht Punkte unserer Geometrie sind. Aber *der Satz von Desargues gilt in unserer Geometrie nicht.* Um das einzusehen, legen wir die beiden Dreiecke der Figur p. 24 so, daß sie selbst sowie auch die drei Schnittpunkte der Seitenpaare in der unteren Halbebene liegen, daß dagegen der Schnittpunkt | M der die entsprechenden Ecken verbindenden Graden a, b, c in 30
der oberen Halbebene liegt, und die Graden a, b, c den Halbstrahl $0,-\infty$ in A, B, C schneiden. (Eine Verwechselung mit der Bezeichnung p. 24 ist nicht zu befürchten.) Jetzt setzen wir in A, B, C die Kreisbögen $\overset{\frown}{0A}$, $\overset{\frown}{0B}$, $\overset{\frown}{0C}$ an, die in unserer Geometrie die Graden a, b, c in die obere Halbebene fortsetzen. Und nun ist zu zeigen, daß $\overset{\frown}{0A}$, $\overset{\frown}{0B}$, $\overset{\frown}{0C}$ sich in drei verschiedenen Punkten schneiden. Nehmen wir an, sie hätten einen gemeinsamen

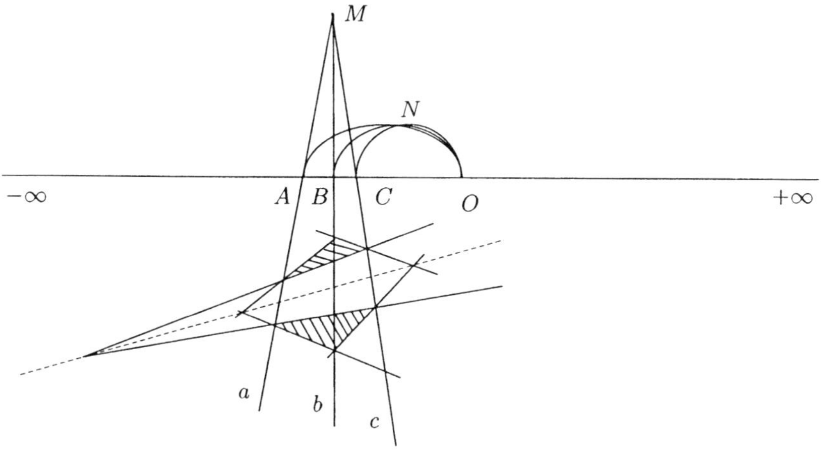

Schnittpunkt N, so wäre offenbar:

$$\left.\begin{array}{l}\sphericalangle MA0 = 2R - \sphericalangle AN0\\ \sphericalangle MB0 = 2R - \sphericalangle BN0\end{array}\right\}\ \text{folglich: } \sphericalangle AMB = \sphericalangle ANB;$$

[20]To complete the specification of the model, we should also add the following case: If A, B are two points from the upper half-plane which are collinear with 0, then the half-line of AB in the upper half-plane is also a line in the model.

folglich müssen A, B, M, N auf einem Kreise liegen; ebenso müssen A, C, M, N auf einem Kreise liegen. Und das wäre nur möglich, wenn M und N
31 auf der Graden ABC lägen, was nicht der Fall ist. Somit ist | gezeigt, daß a, b, c sich in drei verschiedenen Punkten schneiden, der Satz von Desargues also nicht gilt. — Wenn man nur das letztere zeigen wollte, so würde es genügen, etwa einen der Punkte A, B, C auf den Halbstrahl $0,+\infty$ zu legen, wo dann a, b, c überhaupt keinen gemeinsamen Schnittpunkt hätten, während etwa a, b sich schneiden. —

Der Satz von Desargues gilt nach dem vorangehenden überall da, wo wir drei Systeme von Dingen haben, für die die Axiome I und II gelten. Nehmen wir beispielsweise als Punkte die sämtlichen Punkte eines von einer Ebene begrenzten Halbraums, ausgenommen die Punkte der Grenzebene selbst; als Ebenen nehmen wir die Halbkugeln, deren Mittelpunkte auf der Grenzebene liegen und als Graden die Schnitthalbkreise der Halbkugeln. Hier gelten die Axiome I u. II, woraus sich ergiebt: *Der Satz von Desargues gilt in der Halbebene für die zur Grenzgraden orthogonalen Halbkreise.* —

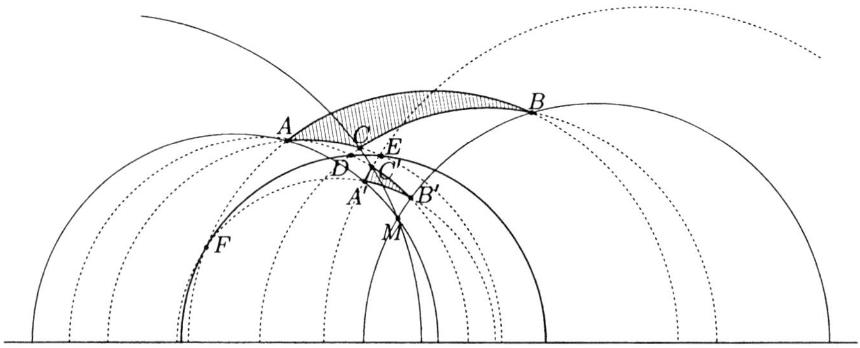

Wir können unsere Betrachtung so zusammenfassen: *Der Satz von Desar-*
32 *gues ist die notwendige Bedingung dafür, daß die Ebene als ein | Stück des Raumes angesehen werden kann. Es fragt sich, ob er auch die hinreichende Bedingung dafür ist*; d. h.: ob man zu zwei Systemen von Dingen, für welche die Axiome I, 1., 2. und II und ausserdem der Desargues gelten, stets ein drittes System von Dingen so finden kann, daß *alle* Axiome I und II gelten. Diese Frage ist wahrscheinlich zu bejahen; man könnte in diesem Fall sagen:[21] Der Satz von Desargues ist das Resultat der Elimination der räumlichen Axiome aus I und II. —

Auf den Axiomen I und II beruht auch *die Lehre von den harmonischen Punkten.* Sind auf einer Graden drei Punkte A, B, C gegeben,[22] so kann man einen Punkt D zwischen B und C so konstruieren:

[21] In the copy of this *Ausarbeitung* in the *Lesezimmer* of the Mathematical Institute (i.e., *Hilbert 1899b**), 'wahrscheinlich' and 'man könnte in diesem Fall sagen' are crossed out, and 'zu bejahen' is underlined by Xx^g; see the Introduction to this Chapter.

[22] With B between A, C.

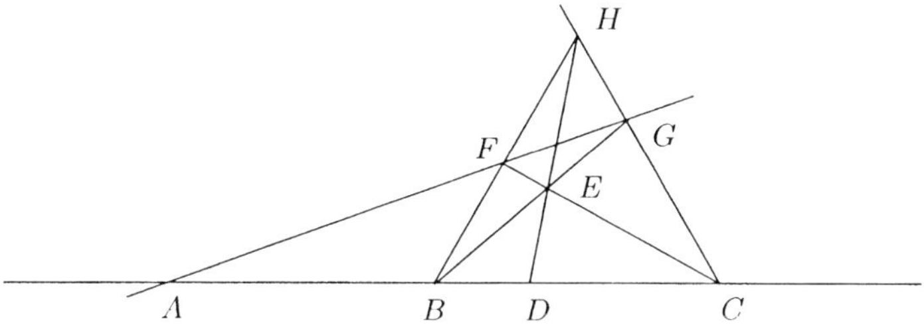

Man legt durch C eine Grade und nimmt auf ihr einen Punkt H an. Auf $\overline{HC}$ nimmt man G und verbindet G mit A; GA schneidet dann BH zwischen B und H in F. CF schneidet BG zwischen B und G in E. Endlich trifft HE ABC zwischen B und C in D. Wir sagen dann: A, B, D, C sind harmonische Punkte. Die Hauptaufgabe ist nun, zu zeigen daß die Lage von D lediglich von der Lage von A, B, C abhängt, nicht aber von der Wahl der Punkte H, G. Dieser Beweis kann wieder nur mit Hülfe der räumlichen | Axiome geführt 33
werden. —

Zum Schluss bemerken wir, *daß die projektive Geometrie sich im wesentlichen auf den Axiomen I, II aufbaut; es bedarf nur noch der Hinzufügung eines Stetigkeitsaxioms.* Wir verweisen auf eine Arbeit von Hilbert, Math. Ann. 46. —[23]

III. Die Axiome der Kongruenz.

A. *Lineare Axiome* stellen wir drei auf.

1. *Wenn zwei Punkte A, B auf einer Graden a, und ferner zwei Punkte A', S auf derselben oder auf einer andern Graden a' gegeben sind, so kann man auf $A'S$ stets einen und nur einen Punkt B' finden, sodaß A' nicht zwischen B' und S liegt und daß $\overline{AB}$ kongruent $\overline{A'B'}$ ist, $\overline{AB} \equiv \overline{A'B'}$. Es ist stets $\overline{AB} \equiv \overline{AB}$.*

Dies Axiom soll aequivalent sein folgendem: Ich kann eine gegebene Strecke auf einer gegebenen Graden von einem gegebenen Punkte nach einer gegebenen Seite (S) hin stets auf eine und nur eine Weise „*abtragen*."

2. *Wenn $\overline{AB} \equiv \overline{A'B'}$ und $\overline{AB} \equiv \overline{A''B''}$ ist, so ist auch stets $\overline{A'B'} \equiv \overline{A''B''}$.*

Eine Folge dieser beiden Axiome ist der *Satz*:

Wenn $\overline{AB} \equiv \overline{A'B'}$ ist, so ist auch $\overline{A'B'} \equiv \overline{AB}$.

Beweis: Nach 1. ist $\overline{AB} \equiv \overline{AB}$; aus den beiden Kongruenzen $\overline{AB} \equiv \overline{A'B'}$ und $\overline{AB} \equiv \overline{AB}$ folgt nach 2.: $\overline{A'B'} \equiv \overline{AB}$. q. e. d.

[23] The work referred to here is *Hilbert 1895b*. For the remark about the continuity axiom, see p. 35 of Hilbert's lecture notes, and the note there.

34 Dieser Satz erlaubt uns zu sagen: Zwei Strecken sind *untereinander* kongruent. Das Axiom 2. enthält dann etwas weniger als der Satz: Sind zwei Strecken einer dritten kongruent, so sind sie auch untereinander kongruent.

Man findet häufig die Auffassung (möglicherweise hatte auch Euklid sie), daß Sätze wie der eben bewiesene reine Grössenbeziehungen seien und daher gar keines besonderen Beweises bedürfen und nicht als besondere geometrische Axiome aufzuführen seien. Das ist aber unrichtig; es kommt ganz darauf an, wie ich Beziehungen „gleich", „kongruent" u. s. w. definiere. Setze ich z. B. zwei reelle Zahlen a, b gleich, sobald $|a-b| \leqq 1$ ist, so folgt aus $a=b$, $b=c$ nicht mehr, daß $a=c$ ist; z. B. ist dann $2=3$, $3=4$, aber $2 \neq 4$.

So ist auch der Satz $\overline{AB} \equiv \overline{AB}$ aus Axiom 1. keineswegs selbstverständlich; denn es steht nichts im Wege, das Abtragen so zu definieren, dass dabei jedesmal die abgetragene Strecke sich (im gewöhnlichen Sinne) um ein bestimmtes Stück verlängert. Man kann nun sagen, daß in solchen Fällen die Ausdrücke „gleich", „kongruent" u. s. w. nicht gebräuchlich sind. Das ist zwar richtig, ändert aber an der Sache nichts; denn dann nehmen unsere Axiome bloss eine andere Form an, z. B.: „die Anwendung des Wortes „gleich" soll in dem und dem Falle erlaubt sein."

3. *Es seien A, B, C Punkte einer Graden a, sodass B zwischen A und C liegt, und A', B', C' Punkte derselben oder einer andern Graden a', sodaß B'*
35 *zwischen A' | und C' liegt; wenn dann $\overline{AB} \equiv \overline{A'B'}$ und $\overline{BC} \equiv \overline{B'C'}$ ist, so ist auch $\overline{AC} \equiv \overline{A'C'}$.*

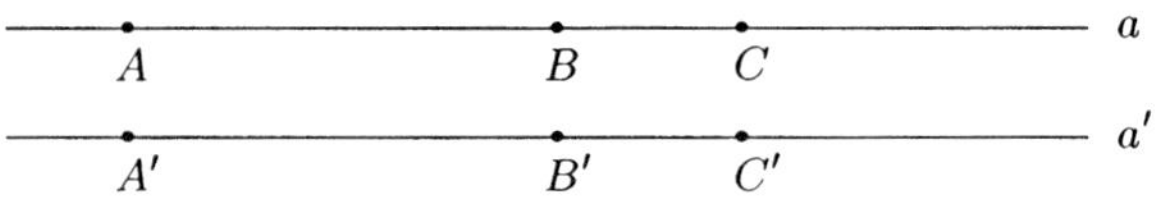

Verweilen wir einen Moment bei der Frage nach der *Unabhängigkeit unserer drei Axiome von einander*. Wir begnügen uns damit, den Satz zu beweisen:

3. ist keine Folge von 1. und 2.

Beweis: Es habe $\overline{AB}$ auf a die Länge l; dann soll jede kongruente Strecke $\overline{A'B'}$ auf a' die Länge $l' = e^l - 1$ haben; umgekehrt soll der Länge l' auf a' die Länge $l = \log(l'+1)$ auf a kongruent heissen. Ist nun $\overline{AB} = l_1$, $\overline{BC} = l_2$, so wird $\overline{AC} = l_1 + l_2$, ferner $\overline{A'B'} = e^{l_1} - 1$, $\overline{B'C'} = e^{l_2} - 1$, $\overline{A'C'} = e^{l_1} + e^{l_2} - 2$, wogegen das Axiom 3. verlangt, dass $\overline{A'C'} = e^{l_1+l_2} - 1$ sein soll.[24] _

Unser Ziel bei den linearen Kongruenzaxiomen ist das Rechnen mit Strekken; dazu wird es vor allem nötig sein, zu zeigen, daß wir eine Strecke $\overline{AB}$ im Sinne der Kongruenz mit einem einzigen Zeichen a bezeichnen dürfen. Wir wollen daher folgenden sehr wichtigen Satz beweisen:

Es ist stets $\overline{AB} \equiv \overline{BA}$.

[24] In the right-hand margin, as a second example by which the independence proof could be accomplished, *Hi*^{g} has added two equations: '$l' = l^2$' and '$l^2 + l'^2 = (l+l')^2$'. In order to adapt the new definition of congruence to the proof in the main text, the second equation should read '$l_1'^2 + l_2'^2 = (l_1' + l_2')^2$'.

Diesen Satz findet man meist unter den Axiomen (vgl. Helmholtz' Monodromieaxiom, welches einen ähnlichen Inhalt wie unser Satz hat);[25] wir aber werden ihn auf Grund der linearen Kongruenzaxiome *beweisen.**) Dazu brauchen wir einige *Hülfssätze*:

a) *Seien A, B, C Punkte einer Graden, sodass B zwischen A und C liegt, | ferner A', B', C' Punkte derselben oder einer andern Graden, sodaß B', C'* 36
auf einer Seite von A' liegen; wenn dann $\overline{AB} \equiv \overline{A'B'}$ und $\overline{AC} \equiv \overline{A'C'}$ ist, so liegt stets B' zwischen A' und C' und es ist $\overline{BC} \equiv \overline{B'C'}$.

Beweis: Nehmen wir an, es liege nicht B' zwischen A' und C', dann liegt nach Voraussetzung C' zwischen A' und B' etwa bei (B'). Nun bestimme ich

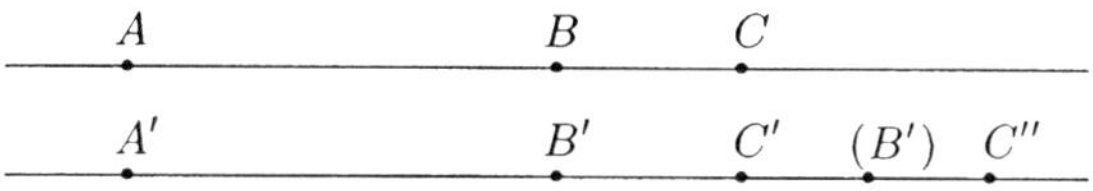

(was nach 1. stets möglich ist) C'' so, dass $\overline{B'C''} \equiv \overline{BC}$ wird, und daß B' zwischen A' und C'' liegt. Aus $\overline{AB} \equiv \overline{A'B'}$ und $\overline{B'C''} \equiv \overline{BC}$ folgt nach 3.: $\overline{AC} \equiv \overline{A'C''}$, also, da $\overline{AC} \equiv \overline{A'C'}$ ist, nach 2.: $\overline{A'C'} \equiv \overline{A'C''}$, also nach 1.: $C' = C''$. Dann läge also gleichzeitig $C' = C''$ zwischen A' und B', und B' zwischen A' und $C' = C''$; das ist nach II, 3. unmöglich, folglich liegt B' zwischen A' und C'. Nehmen wir weiter an, es sei $\overline{BC} \not\equiv \overline{B'C'}$; dann lässt sich stets C'' so bestimmen, daß $\overline{BC} \equiv \overline{B'C''}$ ist, und B' zwischen A' und C'' liegt. Dann folgt aus 3.: $\overline{AC} \equiv \overline{A'C''}$, und nach 2. wegen $\overline{AC} \equiv \overline{A'C'}$: $\overline{A'C'} \equiv \overline{A'C''}$, also nach 1.: $C' = C''$, d. h.: $\overline{BC} \equiv \overline{B'C'}$, q. e. d.

b) *Eine Umkehrung des Satzes a) ist folgender Satz:*

Wenn $\overline{AB} \equiv \overline{A'B'}$, $\overline{AC} \equiv \overline{A'C'}$ und $\overline{BC} \equiv \overline{B'C'}$ ist, und wenn noch B zwischen A und C liegt, so liegt auch B' zwischen A' und C'.

Beweis: Läge etwa C' zwischen A' und B', so könnte ich C'' so bestimmen, dass $\overline{B'C''} \equiv \overline{BC}$ wird, und B' zwischen A' und C'' liegt. Dann folgt wie oben $\overline{AC} \equiv \overline{A'C''}$, weiter $C' = C''$, was nicht möglich ist, da C' zwischen A' und B', B' zwischen A' und C'' liegt

Axiom 3. sowohl wie die Hülfssätze a) und b) verknüpfen die Begriffe „kon- 37
gruent" und „zwischen".

Wie schon ⟨auf⟩ pag. 35 angedeutet, lässt sich auf Grund der vorangehenden Axiome die Beziehung $\overline{AB} \equiv \overline{BA}$ nach Anbringung einer kleinen Modifikation beweisen. Um uns aber nicht zu sehr aufzuhalten, wollen wir lieber jenen Satz selbst einführen als

*) vgl. dagegen pag. 37.

[25] The reference is to one of the hypotheses on which Helmholtz's geometrical system is based. In *Helmholtz 1868*, we find:

> IV. Endlich müssen wir dem Raume noch eine Eigenschaft beilegen, die der *Monodromie* der Functionen einer complexen Grösse analog ist, und die sich darin ausspricht, dass zwei congruente Körper auch noch congruent sind, nachdem der eine eine Umdrehung um irgend eine Rotationsaxe erlitten hat. (Op. cit., § 1.)

Helmholtz then goes on to make a more precise, analytic statement of the principle. The extended passage is quoted in full in *Lie 1893*, 439, where Lie consciously assigns the term 'Axiome' to what Helmholtz calls 'Hypothesen' (438). This is probably the source of Hilbert's reference to Helmholtz's 'Monodromieaxiom'.

4. lineares Kongruenzaxiom: Es ist stets $\overline{AB} \equiv \overline{BA}$.[26]
Für das weitere brauchen wir noch folgende Definition: Sind A, B, C, $D, \ldots$ auf a und A', B', C', $D', \ldots$ auf a' zwei Punktreihen, und ist $\overline{AB} \equiv \overline{A'B'}$, $\overline{AC} \equiv \overline{A'C'}$, $\overline{BC} \equiv \overline{B'C'}$, ... so heissen die *Punktreihen kongruent*. A und A', B und B', C und C', ... heissen homologe Punkte. Dann gilt der *Satz:*

Die Reihenfolgen der homologen Punkte zweier kongruenter Punktreihen stimmen überein. –

Auf Grund unseres 4. Axioms können wir nun, worauf bereits p. 35 hingewiesen wurde, eine *Rechnung mit Strecken* einführen. *Wir dürfen und wollen eine Strecke* $\overline{AB}$ *mit einem einzigen Buchstaben* a *und alle kongruenten Strecken mit demselben Buchstaben bezeichnen. Ferner wollen wir statt des Kongruenzzeichens* $\equiv$ *bei Strecken das Gleichheitszeichen* $=$ *einführen.* Dann gestalten sich die Sätze über Streckenkongruenzen sehr einfach, wie folgendes Beispiel zeigt.

Satz: *Seien vier Punkte A, B, C, D einer Graden in dieser Reihenfolge gegeben; wenn dann* $\overline{AB} \equiv \overline{CD}$ *ist, so ist auch* $\overline{AC} \equiv \overline{DB}$.

38 *Beweis*: Nach Voraussetzung und nach 2. und 4. ist:

$$\overline{AB} = \overline{DC}, \quad \overline{BC} = \overline{CB},$$

also ist nach 3:

$$\overline{AC} \equiv \overline{DB} \quad q.e.d.$$

Wie gestaltet sich nun dieser Satz bei Einführung der Bezeichnungen $a, b, \ldots$ für Strecken? Um das zu sehen, geben wir erst folgende *Definition:* Wenn $\overline{AB} = a$, $\overline{BC} = b$ ist, und B zwischen A und C liegt, so nennen wir $\overline{AC}$ *die Summe der Strecken* a *und* b und schreiben dafür $\overline{AC} = a + b$.

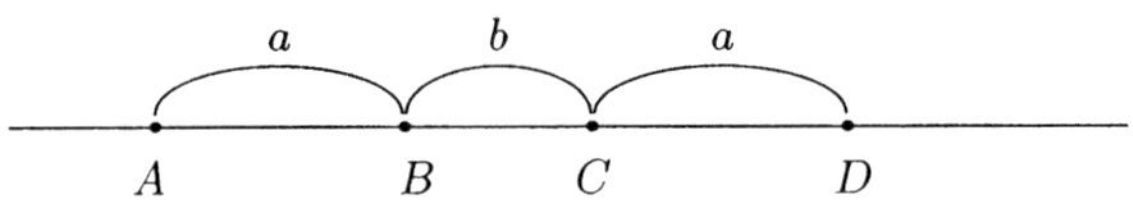

Unser Satz sagt dann einfach:

$$a + b = b + a,$$

d. h. *die Addition von Strecken ist eine kommutative Operation.*

Dieser Satz ist keineswegs selbstverständlich, er sagt keine allgemeine Grössenbeziehung aus, sondern eine ganz bestimmte geometrische Thatsache; denn die a, b sind durchaus keine Zahlen, sondern eben nur Symbole für gewisse geometrische Gebilde.

Genau so wie die Summe führen wir die *Differenz zweier Strecken* ein, weiter *die Strecke null, negative Strecken*; endlich die Begriffe „*grösser*" „*kleiner*" etwa folgendermassen: Ist $\overline{AB} = a$, $\overline{AC} = b$, und liegt B zwischen A und C, so sei $a < b$ oder $b > a$.

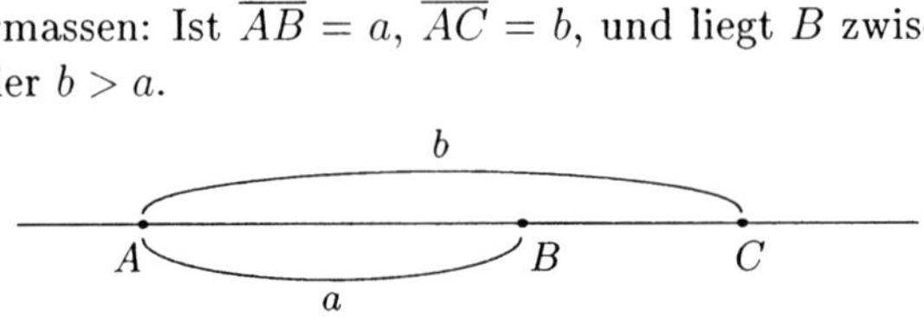

[26] See Hilbert's own lecture notes, p. 40.

Wir haben dann den allgemeinen Satz:

Die Begriffe $>$, $<$ *bei Strecken sind invariant gegenüber der Kongruenz | oder was dasselbe sagt: der Verschiebung.* — 39

Wir kommen nun zu den ebenen Axiomen und haben da zunächst:

B. Winkelkongruenzen.

Definition des Winkels: Zwei von einem Punkt A ausgehende Halbstrahlen a, b mit einem nicht auf a oder b gelegenen Punkte W bestimmen einen Winkel.

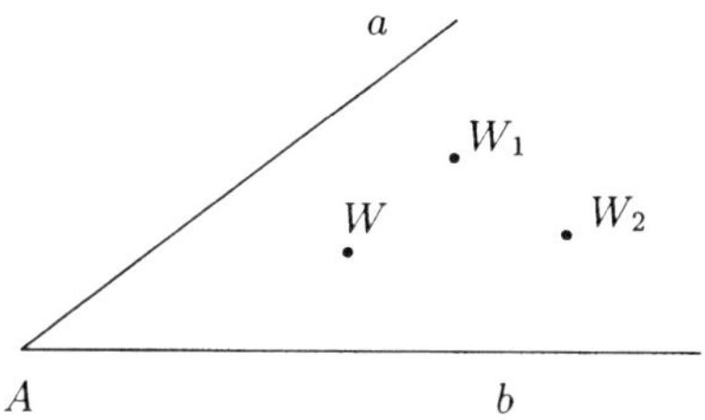

A heisst der *Scheitel*, a, b die *Schenkel* des Winkels. Dabei kann W durch irgend einen Punkt W_1 oder W_2, u. s. w. ersetzt werden sobald auf $\overline{WW_1}$ bezüglich auf $\overline{WW_1}$ und $\overline{W_1W_2}$ kein Punkt von a und kein Punkt von b liegt. $W, W_1, \ldots$ u. s. w. heissen *Winkelpunkte*.

5. *Es seien gegeben ein Winkel $\sphericalangle abW$, ein Halbstrahl a', eine Ebene α durch a', und in dieser Ebene eine Seite S der durch a' bestimmten Graden; dann giebt es einen und nur einen Halbstrahl b', der denselben Anfangspunkt wie a' hat, und einen in α, aber nicht auf a' oder b' liegenden Punkt W', sodass $\sphericalangle abW \equiv \sphericalangle a'b'W'$ ist, und daß entweder alle Winkelpunkte von $\sphericalangle a'b'W'$ auf der gegebene Seite S liegen, oder alle Punkte der gegebenen Seite S Punkte des Winkels $\sphericalangle a'b'W'$ sind. — Es ist stets $\sphericalangle abW \equiv \sphericalangle abW$.*

Dies Axiom sagt also die Möglichkeit aus, *einen gegebenen Winkel* in einem gegebenen Punkt an eine gegebene Grade nach gegebener Seite hin *abzutragen*.

Dies Axiom ist analog dem 1. pag. 33, ebenso werden die folgenden Axiome 40 analog sein den Axiomen 2.–4.

Wir fügen zu 5. gleich hinzu: Aus $abW \equiv a'b'W'$ folgt $a'b'W' \equiv abW$; wir dürfen also sagen: Zwei Winkel sind *unter einander kongruent*.

Die folgenden Axiome lauten:

6. *Sind zwei Winkel einem dritten kongruent, so sind sie unter einander kongruent.*

7. *Es seien in einer Ebene α die Winkel abW und bcU ohne gemeinsame Winkelpunkte, und ebenso in β $a'b'W'$ und $b'c'U'$ ohne gemeinsame Winkelpunkte. Ist dann $abW \equiv a'b'W'$ und $bcU \equiv b'c'U'$, so ist auch stets $acW \equiv a'c'W'$,* oder was dasselbe ist: $acU \equiv a'c'U'$.

8. *Es ist stets $abW \equiv baW$.*

Auf Grund von 8. können wir nun wieder die *Abkürzung* $\sphericalangle abW \equiv \alpha$ einführen, und nun grade wie bei den Strecken, *Summe und Differenz zweier Winkel* definieren, wie nicht näher ausgeführt zu werden braucht. Es folgt dann auch hier der *Satz*:

$$\underline{\alpha + \beta = \beta + \alpha.}$$ [27]

[27] This use of Greek letters is not unambiguous, since Hilbert also uses them to denote planes.

Weiter lassen sich die *Beziehungen* $>$; $<$ definieren, *die gegenüber der Kongruenz (Drehung in der Ebene) invariant sind*, wie sich aus 7. beweisen lässt.

Bei diesem Rechnen mit Winkeln tritt aber eine Besonderheit auf, die bei Strecken nicht vorkam, und die damit zusammenhängt, daß zur Definition
41 des Winkels (p. 39) zwei Halbstrahlen a, b nicht ausreichen, vielmehr | noch ein Punkt W hinzugenommen werden muss. Jene Besonderheit besteht in folgendem: Wir können zwei Winkel zunächst nur dann addieren, wenn bei der Abtragung keine gemeinsamen Winkelpunkte auftreten. Wir können aber die Addition retten*), wenn wir verabreden, daß im Fall gemeinsamer Winkelpunkte nur der beiden Winkeln gemeinsame Teil der Ebene als Summe der Winkel gelten soll; oder: *wir nennen zwei Winkel gleich (kongruent), wenn sie sich nur um Vollwinkel* unterscheiden. Wir definieren nämlich: Jeder Winkel aaW heisst ein *Vollwinkel*, der Winkel aa, der gar keinen Winkelpunkt enthält, heisst der *Winkel null*, ist a die Verlängerung von b, so heisst abW ein *gestreckter Winkel.*

Ueber diese Begriffe, die bei Strecken kein Analogon haben, müssen wir nun ein weiteres Axiom aufstellen.

9. *Alle gestreckten Winkel sind einander kongruent.*

Es wäre hier eine interessante Aufgabe, die Unabhängigkeit des Axioms 9. von den vorangehenden Axiomen nachzuweisen.[28] Wir wollen aber nicht dabei verweilen, sondern gleich *Folgerungen* aus unsern Axiomen ziehen; es gelten die Sätze:

1.) *Alle Vollwinkel sind einander kongruent.*

2.) *Sind zwei Winkel gleich**), so sind auch ihre Nebenwinkel gleich.*

3.) *Scheitelwinkel sind einander gleich.*

Die Begriffe „Nebenwinkel" und „Scheitelwinkel" sind leicht zu definieren. Der Beweis der Sätze 1.)–3.) beruht vor allem auf den Axiomen 7. und 9.

42 Wir kommen zu einer neuen wichtigen *Definition*:

Wenn ein Winkel seinem Nebenwinkel gleich ist, so heisst er ein rechter Winkel.

Wir heben hervor, daß wir die *Existenz* rechter Winkel erst später beweisen werden;[29] jedenfalls aber gilt der *Satz:*

4.) *Alle rechten Winkel sind einander gleich.*

Beweis: Seien α, β rechte Winkel, dann sind $\alpha+\alpha$ und $\beta+\beta$ gestreckte Winkel, also nach 9.:

$$\alpha+\alpha=\beta+\beta$$

Nehmen wir nun an, es sei $\alpha>\beta$ (NB.: Wir haben nicht ausdrücklich bewiesen, es folgt aber, besonders aus Axiom 7., daß zwischen zwei beliebigen Winkeln stets eine und nur eine der Beziehungen $=$, $>$, $<$ besteht); dann

---

*) Ungleichungen dagegen lassen sich auf diesen Fall nicht übertragen.

**) *wir brauchen „gleich" in derselben Bedeutung wie „kongruent".*

---

[28] See Hilbert's own lecture notes, p. 45.

[29] See p. 45 of this text.

schreiben wir:

$$(\alpha - \beta) + \alpha = \beta.$$

Da $\alpha - \beta > 0$ ist, so müsste $\beta > \alpha$ sein; das ist aber gegen die Annahme. Also ist $\alpha = \beta$, q. e. d.

Der Satz 4.) tritt bei Euklid wie auch bei Lindemann meiner Meinung nach irrtümlicherweise als Axiom auf; das rührt daher, daß diese Autoren nicht die Sätze über den Begriff „zwischen" systematisch aufgebaut haben, wie wir es thaten.[30]

C. Das Band zwischen den Sätzen über Streckenkongruenz und denen über Winkelkongruenz bildet das folgende Axiom (der sog. 1. Kongruenzsatz.)

Wenn in zwei Dreiecken ABC und $A'B'C'$ $\overline{AB} \equiv \overline{A'B'}$, $\overline{AC} \equiv \overline{A'C'}$, 43
$\sphericalangle BAC \equiv \sphericalangle B'A'C'$ ist, dann ist auch stets $\overline{BC} \equiv \overline{B'C'}$, $\sphericalangle ACB \equiv \sphericalangle A'C'B'$, $\sphericalangle ABC \equiv \sphericalangle A'B'C'$; wir sagen dafür kurz: die Dreiecke sind kongruent.

Bei diesem Axiom wollen wir einmal wieder etwas länger verweilen, und den Satz beweisen:

10. ist keine Folge der vorhergehenden Axiome.

Beweis: Im gewöhnlichen, Euklidischen Raum denken wir uns durch eine Grade zwei Ebenen α und α' gelegt. Das Abtragen von Strecken soll ausnahmslos im gewöhnlichen Sinne gelten, das Abtragen von Winkeln im allgemeinen auch; in Bezug auf die Ebenen α und α' aber sollen folgende Festsetzungen gelten:

(also AA' senkrecht in A zur Ebene α)

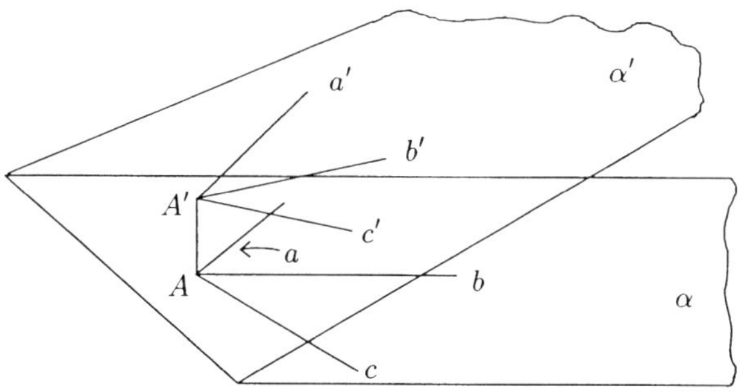

In α soll Parallelverschiebung und Drehung von Winkeln im gewöhnlichen Sinne gelten, ebenso in α' die Parallelverschiebung; ein Winkel $a'b'W'$ in α' soll seiner orthogonalen Projektion abW in α gleich heissen; die Drehung eines Winkels $a'b'W'$ in α' um seinen Scheitel, sodass etwa a' die gegebene Lage c' erhält muss demnach so ausgeführt werden, daß zuerst $a'b'W'$ orthogonal | auf α projiziert, dann in der Ebene α im gewöhnlichen Sinne gedreht 44
wird, sodaß der Schenkel a in die Projektion c von c' fällt, und endlich der entstandene Winkel nach α' zurückprojiziert wird.

[30]For the reference to Lindemann, and Hilbert's procedure in the *Festschrift*, see the Introduction to this Chapter, Specific Note (3)(a).

In der so definierten Geometrie gelten thatsächlich alle Axiome, z. B. das von der Gleichheit aller gestreckten Winkel; denn ein gestreckter Winkel ändert sich ja nicht bei der Projektion. Axiom 10. aber gilt nicht. Liegen nämlich die Dreiecke ABC und $A'B'C'$ in den Ebenen α und α', und ist $\overline{AB} \equiv \overline{A'B'}$, $\overline{AC} \equiv \overline{A'C'}$, $\sphericalangle BAC \equiv B'A'C'$ im Sinne unserer Geometrie, so ist nach der gewöhnlichen Auffassung $\sphericalangle BAC \not\equiv \sphericalangle B'A'C'$, und folglich $\overline{BC} \not\equiv \overline{B'C'}$. Da nun in Bezug auf Strecken die Auffassungen unserer Geometrie und die der Euklidischen übereinstimmen, so ist auch im Sinne der ersteren $\overline{BC} \not\equiv \overline{B'C'}$, womit die Ungültigkeit von 10. gezeigt ist. —

Wir geben nachträglich (zu 10.) eine *genaue Definition des Dreieckswinkels*: Unter einem Dreieckswinkel verstehen wir einen Winkel, dessen Schenkel zwei Dreiecksseiten sind, *während ein Punkt der dritten Seite als Winkelpunkt genommen werden soll.*

Wir werden später zeigen, daß wir weitere Kongruenzaxiome nicht nötig haben; vorerst aber wollen wir aus unsern Axiomen eine ganze Reihe zum Teil sehr wichtiger Sätze herleiten. Wir werden sehen, daß es dabei sehr auf die Reihenfolge ankommt, in der wir die Sätze vornehmen. Diese Reihenfolge wird von der in den Lehrbüchern der Elementargeometrie üblichen stark abweichen;
45 sie wird dagegen vielfach über|einstimmen mit der Reihenfolge bei Euklid. So führen uns diese ganz modernen Untersuchungen dazu, den Scharfsinn dieses alten Mathematikers recht zu würdigen und aufs höchste zu bewundern. —

Wir beweisen zuerst: *Es giebt rechte Winkel*, indem wir gleich zeigen, *wie man von einem Punkt A auf eine Grade a ein Lot fällen kann.*

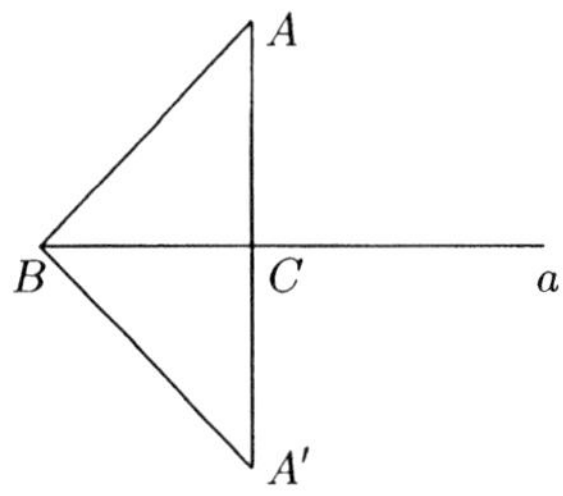

Man verbinde A mit irgend einem Punkte B auf a und trage den Winkel AB, a in B an a nach der andern Seite hin ab. Den neu entstehenden Schenkel $A'B$ mache man $= AB$ und verbinde A mit A'. Da A und A' auf verschiedenen Seiten von a liegen, muss $\overline{AA'}$ die Grade a in einem Punkte C schneiden. ACA' ist dann das verlangte Lot; denn nach 10. ist:

$$\triangle A'BC \cong \triangle ABC$$

d. h. $\sphericalangle BCA' = \sphericalangle BCA$. Nun ist $\sphericalangle A'CA$ ein gestreckter, also ist nach Definition $\sphericalangle ACB$ ein rechter, q. e. d. —

Der sog. *II. Kongruenzsatz* lautet:

Wenn zwei Dreiecke in einer Seite und den beiden anliegenden Winkeln übereinstimmen, so sind sie kongruent.

Beweis: Wir nehmen an, es sei $\overline{AC} = \overline{A'C'}$, $\sphericalangle A = \sphericalangle A'$, $\sphericalangle C = \sphericalangle C'$. Sei nun, entgegen der Behauptung, $\overline{AB} \neq \overline{A'B'}$, sei also etwa $\overline{AB} < \overline{A'B'}$. Tragen
46 wir dann $\overline{AB}$ auf $\overline{A'B'}$ von A' aus nach der |Seite von B' ab, so muss der

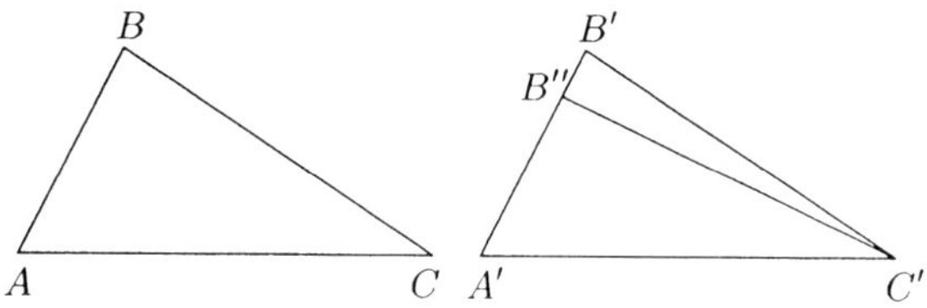

Endpunkt B'' dieser Strecke zwischen A' und B' fallen (nach den linearen Kongruenzaxiomen). Wir verbinden dann B'' mit C' und sehen leicht nach 10., daß $\triangle ABC \cong \triangle A'B''C'$, d. h. auch $\sphericalangle ACB = \sphericalangle A'C'B''$ ist. Nun ist nach Voraussetzung $\sphericalangle ACB = \sphericalangle A'C'B'$, sodass also $\sphericalangle A'C'B' = \sphericalangle A'C'B''$ sein müsste. Das ist aber nicht möglich, denn $\sphericalangle A'C'B''$ ist ja ein Teil von $\sphericalangle A'C'B'$. Demnach ist unsere Annahme $AB \neq A'B'$ falsch, es ist $AB = A'B'$, und nun folgt aus 10. sofort, daß $\triangle ABC \cong \triangle A'B'C'$ ist, q. e. d.

Von grosser Bedeutung ist der

Satz vom Aussenwinkel.

In jedem Dreieck ABC ist der Nebenwinkel CBD des Winkels CBA (der „Aussenwinkel") grösser als jeder der beiden andern Dreieckswinkel, z. B. als $\sphericalangle CAB$.

Beweis: Nehmen wir erstens an, es sei $\sphericalangle CBD < \sphericalangle CAB$; dann können wir $\sphericalangle CBD$ in A an AB so abtragen, daß der neu entstehende Schenkel die Seite BC in C' schneidet. Vgl. die Axiome über Winkelkongruenzen und den Satz pag. [31]. Ist zweitens $\sphericalangle CBD = \sphericalangle CAB$, so liegt die Sache einfacher; wir nehmen dann C selbst als Punkt C' an. In beiden | Fällen verlängern wir nun 47
BC über B hinaus um $\overline{AC'}$, bis zum Punkt E, und verbinden A mit E. Dann ist, wegen $AB = BA$, $AC' = BE$,

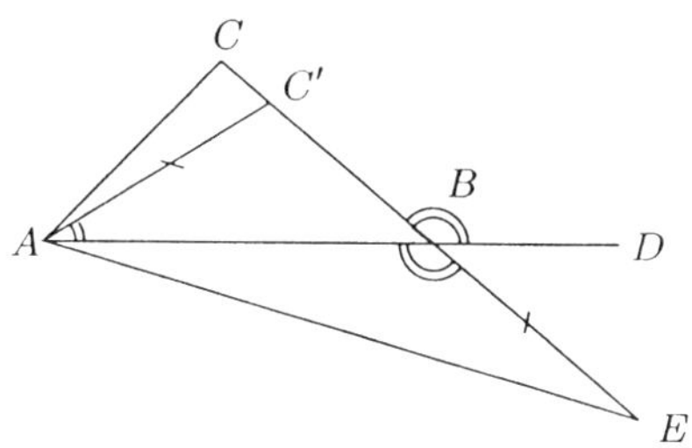

$\sphericalangle C'AB = \sphericalangle C'BD = \sphericalangle EBA$:

$$\triangle ABC' \cong \triangle BAE$$

folglich ist:

$$\sphericalangle ABC' = \sphericalangle BAE.$$

Nun war $\sphericalangle ABE = \sphericalangle BAC'$, also sind auch die Summen der Winkel links und rechts gleich, d. h. $\sphericalangle C'BE = \sphericalangle C'AE$. Nun ist $\sphericalangle C'BE$ ein gestreckter, also ist auch $\sphericalangle C'AE$ ein gestreckter Winkel, d. h. C', A, E liegen auf einer Graden; aber auch C', B, E liegen auf einer Graden; diese beiden Graden hätten also zwei Punkte C', E gemein, ohne zusammenzufallen (denn A liegt nicht auf $C'E = BC$); das ist aber unmöglich (cf. die Axiome I) und somit ist der Satz vom Aussenwinkel bewiesen.

Hieran knüpft sich die *Folgerung, daß in jedem Dreieck wenigstens zwei Winkel „spitz", d. h. kleiner als ein Rechter sein müssen.* Wären nämlich etwa α und β beide stumpf, so wäre der Aussenwinkel bei α einerseits spitz, als Nebenwinkel eines stumpfen Winkels, andererseits stumpf, denn er müsste ja

[31]The page reference is blank here; it is not entirely clear to which 'Satz' Hilbert wishes to refer.

nach dem Satz vom Aussenwinkel $> \beta$ sein. So kommen wir also auf einen Widerspruch. –

Weiter: *Zwei Dreiecke sind kongruent, wenn sie in einer Seite, einem*
48 | *anliegenden und dem gegenüberliegenden Winkel übereinstimmen.*

Beweis: Sei $AB = A'B'$, $\sphericalangle A = \sphericalangle A'$, $\sphericalangle C = \sphericalangle C'$.

Wir nehmen an, es sei $AC \neq A'C'$, und tragen AC auf $A'C'$ von A' aus nach C' hin ab, was auf den Punkt C'' führen mag. Dann fällt entweder C'' zwischen A' und C', oder C' zwischen A' und C''. In jedem Fall verbinden wir C'' mit B' und haben nach 10:

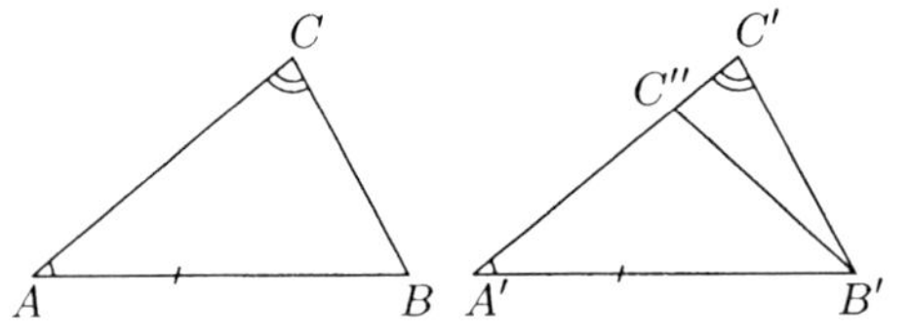

$$\triangle ABC \cong \triangle A'B'C'',$$

d. h. es muss sein $\sphericalangle C = \sphericalangle A'C''B' = \sphericalangle A'C'B'$. Nach dem Satz vom Aussenwinkel ist diese Gleichheit nicht möglich. Wir haben also: $AC = A'C'$ und nach 10: $\triangle ABC \cong \triangle A'B'C'$. –

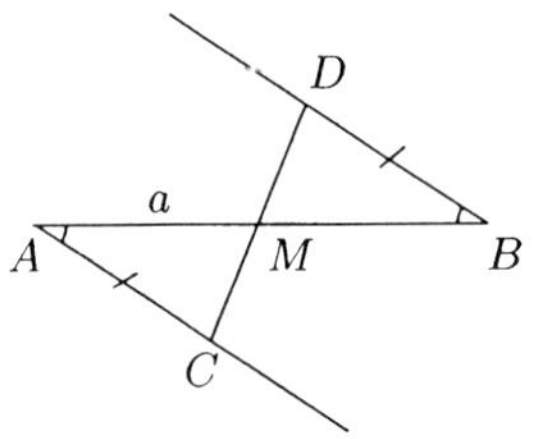

Wir können jetzt die *Aufgabe lösen: Eine beliebige Strecke zu halbieren.*

Sei $\overline{AB}$ die gegebene Strecke; wir tragen in A und B gleiche Winkel an, aber nach verschiedenen Seiten der Graden a und machen die freien Schenkel einander gleich: $AC = BD$. Die Strecke CD muss die Grade a schneiden, in M etwa, da C und D auf verschiedenen Seiten von a liegen. Dann ist M der gesuchte Halbierungspunkt. Denn es ist nach dem eben bewiesenen Kongruenzsatz $\triangle AMC \cong \triangle BMD$,
49 | d. h. es ist $AM = BM$, wie verlangt wurde. –

Der Satz vom gleichschenkligen Dreieck.

Im gleichschenkligen Dreieck sind die Basiswinkel einander gleich. Wenn in einem Dreieck die Basiswinkel gleich sind, so ist es gleichschenklig. Wir beweisen nur den ersten Teil; die Umkehrung wird ganz analog bewiesen.

Der *Beweis* dieses Satzes kann, wie wir unten sehen werden, überaus einfach geführt werden. Aber alle Beweise, die man in den üblichen Lehrbüchern findet, sind meist unbrauchbar. Bei dieser eigentümlichen Thatsache müssen wir einen Moment verweilen. Es scheinen sich zunächst drei Weisen darzubieten, um den Satz herzuleiten:

1.) Man könnte AB in M halbieren und dann zeigen wollen, daß $\triangle ACM \cong \triangle BCM$ ist. Dazu braucht man aber den sog. III. Kongruenzsatz, der seinerseits erst mit Hülfe des Satzes vom gleichschenkligen Dreieck beweisbar ist.

2.) Man könnte von C auf AB ein Lot CM fällen; aber das führt auf den sog. IV. Kongruenzsatz, über den dieselbe Bemerkung wie bei 1) zu machen ist.

3.) Durch Halbirung des Winkels C kommt man allerdings zum Ziel, aber nur auf einem sehr langen Wege; vor allem muss man dazu erst die Halbierung eines Winkels einwandfrei einführen, was bis jetzt noch nicht geschehen war.

Ein wirklich strenger und sehr einfacher Beweis ist folgender: Es ist $CA = CB$, $BC = AC$ nach Voraussetzung, $\sphericalangle C = \sphericalangle C$, also $\triangle BCA \cong \triangle ACB$, also $\sphericalangle ABC = \sphericalangle BAC$, q. e. d. 50

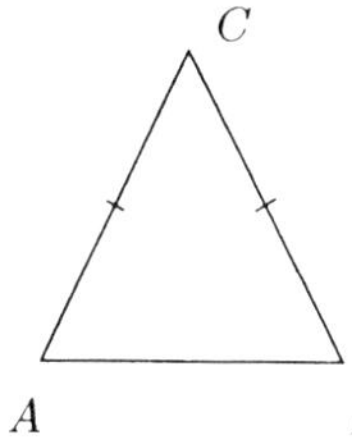

Nun beweisen wir den *III. Kongruenzsatz:*

Zwei Dreiecke sind kongruent, wenn sie in allen drei Seiten übereinstimmen.

Beweis: Wir tragen $\triangle ABC$ an $\triangle A'B'C'$ in der Weise ab, daß erstens A auf A', B auf B' fällt und daß C' und C'' auf verschiedenen Seiten von $A'B'$ liegen; ferner machen wir $\sphericalangle B'A'C'' = \sphericalangle BAC$ und $A'C'' = AC$. Dann ist nach 10: $\triangle A'B'C'' \cong \triangle ABC$, d. h. auch: $CB = C''B'$. Nun verbinden wir C' mit C''; dann ist $A'C'C''$ ein gleichschenkliges Dreieck, weil $A'C' = AC = A'C''$ ist; also ist nach dem vorigen Satz $\sphericalangle A'C''C' = \sphericalangle A'C'C''$. Ebenso ist $\triangle B'C'C''$ gleichschenklig, denn es ist $B'C' = BC = B'C''$. Also ist auch: $\sphericalangle B'C''C' = \sphericalangle B'C'C''$. Addirt man nun die beiden Winkelgleichheiten, so kommt:

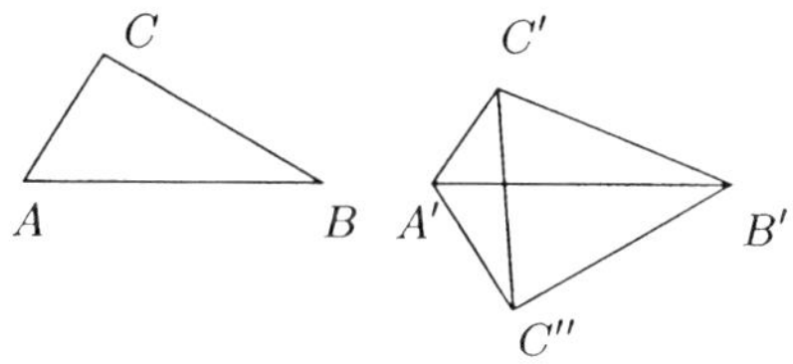

$$\sphericalangle A'C''B' = \sphericalangle A'C'B'.$$

Der Winkel links ist aber $= \sphericalangle ACB$, also wird $\sphericalangle C = \sphericalangle C'$. Nun können wir 10. anwenden und haben $\triangle ABC \cong \triangle A'B'C'$, q. e. d.

Aufgabe: Einen gegebenen Winkel zu halbieren. 51

Wir machen $AB = AC$, halbieren BC in M und ziehen AM. Dann ist nach dem III. Kongruenzsatz $\triangle ABM \cong \triangle ACM$, also $\sphericalangle BAM = \sphericalangle CAM$, wie verlangt wurde.

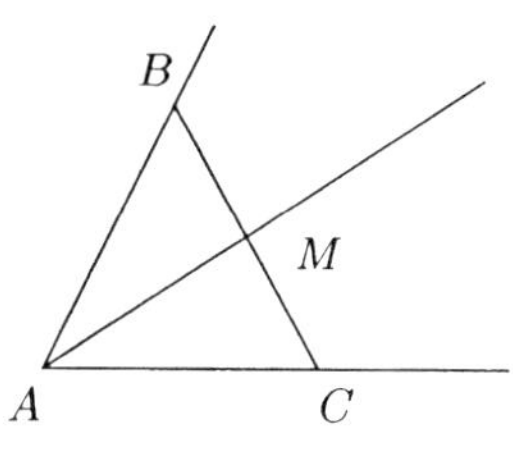

In jedem Dreieck liegt der grösseren Seite der grössere Winkel gegenüber, und umgekehrt.

Beweis: Sei $AC > BC$; dann ist zu zeigen, daß $\beta > \alpha$[32] ist. Wir tragen BC auf AC in C nach A hin ab. Der Endpunkt D dieser Strecke fällt dann wegen $AC > BC$ zwischen A und C. Dann ist $\triangle BCD$ gleichschenklig, also ist $\sphericalangle CDB = \sphericalangle CBD$. Nun ist einerseits $\sphericalangle ABC > \sphericalangle CBD$, andrerseits ist $\sphericalangle CDB$ als Aussenwinkel $> \sphericalangle DAB$. Also ist schließlich $\sphericalangle ABC > \sphericalangle BAC$, q. e. d.

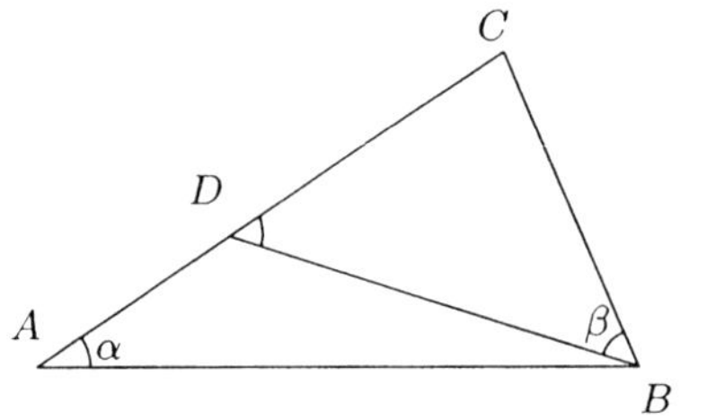

Die Umkehrung beweist man ganz analog.

Der nächste Satz soll sein:

In jedem Dreieck ist die Summe zweier Seiten größer als die dritte.

Beweis: Wir verlängern AC über C hinaus um CB bis zum Punkte D, und verbinden D mit B. Dann ist $\sphericalangle CDB = \sphericalangle CBD$ und $\sphericalangle CBD < \sphericalangle ABD$, also auch $\sphericalangle ABD > \sphericalangle CDB$. Folglich ist, nach dem vorigen Satz:
52 $AD > AB$, und da $AD = AC + CB$, | so ist auch $AC + CB > AB$, q. e. d. —[33]

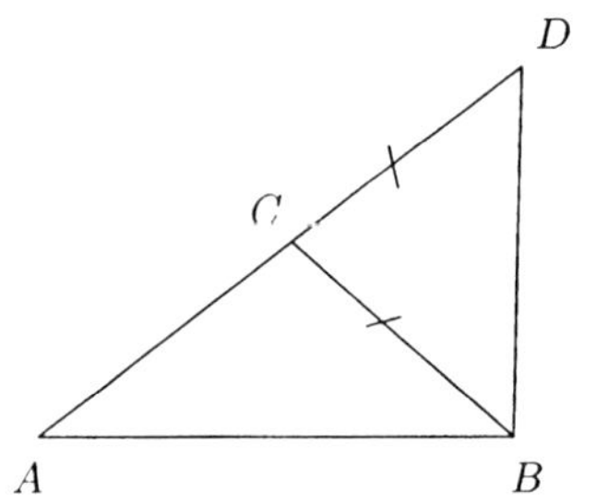

Bei manchen tritt dieser Satz als Folge der Definition auf: Die grade Linie ist die kürzeste Verbindungslinie zwischen irgend zwei Punkten. Eine solche Definition ist aber sinnlos, wenn man den Begriff der Länge nicht definiert.[34]

Der IV. Kongruenzsatz: Zwei Dreiecke sind kongruent, wenn sie in zwei Seiten und dem der größeren Seite gegenüberliegenden Winkel übereinstimmen.

Beweis: Sei $AB = A'B'$, $AC = A'C'$, $AB > AC$, $\sphericalangle C = \sphericalangle C'$. Dann ist nach p. 51 auch $\sphericalangle C > \sphericalangle B$, $\sphericalangle C' > \sphericalangle B'$ und folglich sind B und B' beide spitz [p. 47]. Wir tragen nun $\sphericalangle A$ in A' an $A'B'$ an nach der entgegengesetzten Seite von C' und machen den freien Schenkel $A'C'' = AC$. Dann ist nach 10: $\triangle A'B'C'' \cong \triangle ABC$, also $\sphericalangle A'B'C'' = \sphericalangle B$. Da nun B und B' beide spitz sind, so ist $B + B' = \sphericalangle C'B'C''$ kleiner als ein gestreckter, d. h. C', C'', B' liegen nicht auf einer Graden. Also bilden C', C'', B' ein wirkliches Dreieck, welches offenbar gleichschenklig ist. Denn es ist zunächst $A'C' = A'C''$ also $\sphericalangle A'C'C'' = \sphericalangle A'C''C'$; nach Voraussetzung ist $\sphericalangle A'C''B' = \sphericalangle ACB = \sphericalangle A'C'B'$, ⟪al⟫so ist auch $\sphericalangle B'C''C' = \sphericalangle B'C'C''$ und also $\triangle C'C''B'$ gleichschenklig, d. h. $B'C' = B'C''$.

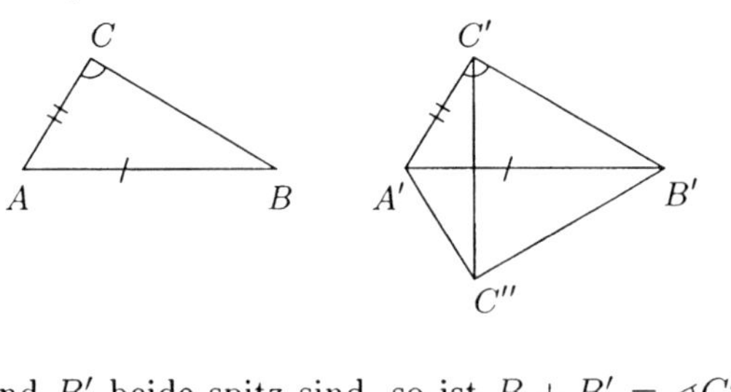

[32] Angles α, β have been added to the figure.

[33] See Hilbert's own lecture notes, p. 53.

[34] See Hilbert's own lecture notes, p. 53, and the corresponding note.

Nach dem III. Kongruenzsatz ist daher $\triangle A'B'C'' \cong \triangle A'B'C'$ und folglich 53
$\triangle ABC \cong \triangle A'B'C'$. –

Um jetzt die bisherigen Kongruenzsätze zusammenzufassen, *definieren* wir: *Ein System von irgend welchen endlich vielen Punkten heisst eine Figur; liegen alle Punkte der Figur in einer Ebene, so heisst sie eine ebene Figur.*

Zwei Figuren heissen: kongruent, wenn ihre Punkte sich paarweise einander so zuordnen lassen, daß die auf diese Weise einander zugeordneten Strecken und Winkel sämtlich einander kongruent sind.

Kongruente Figuren haben auf Grund der Kongruenzaxiome folgende Eigenschaften: 3 Punkte einer Graden liegen auch in jeder kongruenten Figur auf einer Graden; die Reihenfolge der Punkte auf einer Graden, sowie die Anordnung der Punkte in entsprechenden Ebenen in Bezug auf entsprechende Ebenen ist in kongruenten Figuren dieselbe.

Wir fassen nun unsere ebenen Konguenzsätze zusammen in den *allgemeinen Kongruenzsatz: Wenn $(A, B, C, \ldots)$ und $(A', B', C', \ldots)$ kongruente ebene Figuren sind, und P ein Punkt in der Ebene der ersten, so lässt sich in der Ebene der zweiten Figur stets ein Punkt P' finden derart, dass $(A, B, C, \ldots, P)$ und $(A', B', C', \ldots, P')$ wieder kongruente Figuren sind. Enthalten die beiden Figuren wenigstens drei nicht auf einer Graden liegenden Punkte, so ist die Konstruktion von P' nur auf eine Weise möglich.*

Wir erkennen hieraus, daß wir zwei Ebenen einander (bezw. eine | Ebene 54
sich selbst) punktweise eineindeutig zuordnen können, wenn wir in jeder Ebene einen Punkt, eine Grade durch diesen Punkt, eine Seite auf dieser Graden u. eine Seite der Ebene inbezug auf die Grade festlegen. Wir sagen, daß *bei dieser Transformation die beiden Ebenen zur Deckung gebracht werden, bezw. die eine Ebene in sich bewegt wird. Damit haben wir den Begriff der Bewegung auf den Kongruenzbegriff gegründet.* –

Wir schließen hier die Betrachtung der ebenen Kongruenzbeziehungen ab und zeigen nun, *daß wir für den Raum keine Kongruenzaxiome mehr nötig haben.* Wir müssen zu diesem Zweck die wichtigsten stereometrischen Sätze durchgehen; im allgemeinen können wir dabei die üblichen Beweise benutzen, nur haben wir genau darauf zu achten, daß wir nirgends parallele Linien anwenden, da wir von diesen erst im nächsten Abschnitt sprechen werden.

Zuerst beweisen wir:

Wenn eine Grade auf zwei Graden einer Ebene gleichzeitig senkrecht steht, so steht sie auf allen Graden dieser Ebene, die sie überhaupt treffen, senkrecht.

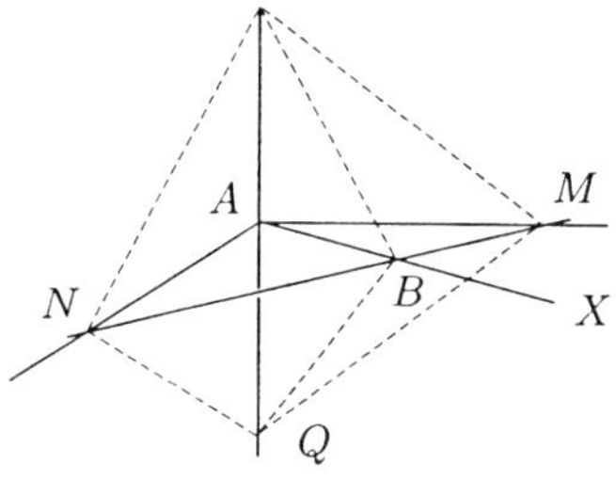

Beweis: Wir verlängern die auf AM und AN senkrecht stehende Strecke AP über A hinaus um sich selbst, so|daß $\overline{AQ} = \overline{AP}$ 55
wird. Nehmen wir nun irgend einen Halbstrahl AX durch A in der Ebene AMN. Die Punkte M, N lassen sich offenbar stets so wählen, daß die Grade MN entweder AX oder die Verlängerung von AX über A hinaus schneidet, etwa in B. Nun ist nach 10: $\triangle PAM \cong \triangle QAM$, $\triangle PAN \cong \triangle QAN$, folglich ist $PM = QM$, $PN = QN$,

und weiter $\triangle PMN \cong \triangle QMN$. Also ist $\sphericalangle PMN = \sphericalangle QMN$, daraus folgt wieder, daß $\triangle PBM \cong \triangle QBM$ ist, woraus sich $PB = QB$ ergiebt.
Dann ist endlich, nach dem III. Kongruenzsatz, $\triangle PAB \cong \triangle QAB$, also $\sphericalangle PAB = \sphericalangle QAB$. Die Summe dieser Winkel ist ein gestreckter, also jeder ein rechter. q. e. d.

Wir sagen: *AP steht senkrecht auf der Ebene AMN.*

Nun beweisen wir die *Existenz von Senkrechten auf einer Ebene, indem wir gleich die Aufgabe lösen, in einem Punkt A einer Ebene α auf dieser eine Senkrechte zu errichten.*

Wir ziehen durch A zwei Grade a, b, in der Ebene α. Dann können wir mit Hülfe des vorigen Satzes eine Ebene finden, die in A auf a senkrecht steht, [wir brauchen nur durch a zwei Ebenen zu legen, in jeder von ihnen auf a in A ein Lot zu errichten und durch diese Lote eine Ebene zu legen] und ebenso eine, die in A auf b senkrecht steht. Beide Ebenen schneiden sich notwendig in einer Graden, und diese ist die verlangte senkrechte.

56 Nun wollen wir umgekehrt *von einem Punkte P auf eine Ebene α ein Lot fällen.*

Wir ziehen in α irgend eine Grade a und fällen von P[35] das Lot auf a [pag. 45]; es möge a in B schneiden. Nun ziehen wir in α eine Grade b, die in B auf a senkrecht steht, und fällen endlich von P[36] auf b ein Lot PA. Wir behaupten, daß AP das verlangte Lot auf α ist. Das ist gezeigt, sobald wir wissen, daß AP ausser auf b noch auf irgend einer Graden in α durch A senkrecht steht. Verbinden wir also A mit einem Punkt M auf a, verlängern wir ferner AP über A hinaus um sich selbst bis Q. Dann ist zunächst $\triangle PAB \cong \triangle QAB$ nach 10. Also ist $PB = QB$. Weiter war $a \perp BP$, und $a \perp AB$, also nach dem vorigen Satze auch $a \perp BQ$, d. h. es ist $\sphericalangle PBM = \sphericalangle QBM =$ einem rechten. Also ist nach 10. $\triangle PBM \cong \triangle QBM$, also $PM = QM$. Dann ist nach dem III. K. S. $\triangle AMP \cong \triangle AMQ$, woraus wie pag. 55, folgt: $\sphericalangle PAM =$ einem rechten Winkel. q. e. d.

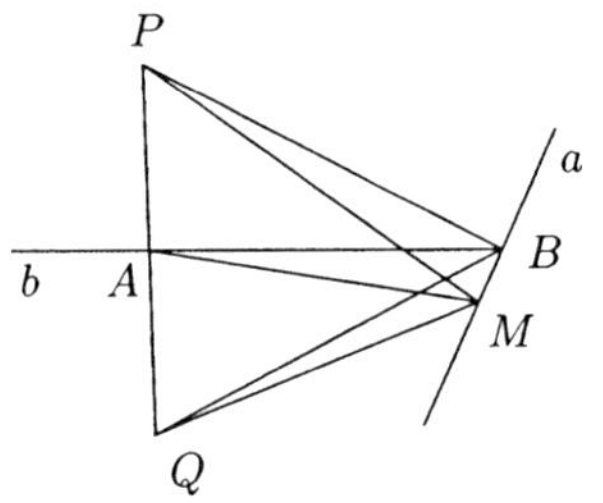

Wir weisen schon hier darauf hin, daß die eben bewiesenen Sätze uns später bei der Paralleltheorie sehr wesentliche Dienste leisten werden. —

Zwei Lote auf einer Ebene liegen stets in einer Ebene.

57 *Beweis*: Sei also $AB \perp \alpha$, $A'B' \perp \alpha$; wir haben zu zeigen, daß AB, $A'B'$ in einer Ebene liegen. Wir ziehen dazu in α senkrecht zu AA' durch A eine Grade CAD und machen $AC = AD$. Dann ist nach 10: $\triangle CA'A \cong \triangle DA'A$,

---

[35] Originally, both copies of the *Ausarbeitung* erroneously read 'A'. In the copy in the *Lesezimmer* of the Mathematical Institute, this has been changed to 'P' by Xx^g, while in Hilbert's own copy it remained unchanged.

[36] Originally both copies had 'A', and only the copy in the *Lesezimmer* has been corrected to 'P', this time by Xx^s.

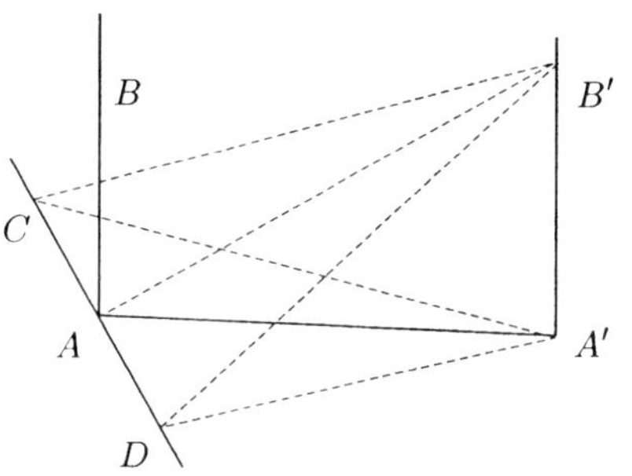

folglich $CA' = DA'$, also nach 10: $\triangle CA'B' \cong \triangle DA'B'$, folglich $CB' = DB'$, also nach dem III. Kongr. S.: $\triangle AB'C \cong \triangle AB'D$, also $\sphericalangle CAB' = \sphericalangle DAB'$, d. h. $CD \perp AB'$. Nun ist auch $CD \perp AA'$ und $CD \perp AB$; daraus folgt aber, daß AA', AB, AB' und also auch AB und $A'B'$ in einer Ebene liegen, wenn man die Umkehrung des Satzes pag. 54 benutzt, die wir der Kürze halber hier ohne Beweis anführen:

Steht eine Grade auf drei durch einen Punkt gehenden Graden gleichzeitig senkrecht, so liegen diese drei Graden in einer Ebene.[37]

Um zu dem allgemeinsten Kongruenzsatz im Raum zu gelangen, haben wir nur noch folgenden *Satz* zu beweisen: *Zwei Ebenen α, β mögen sich in der Graden a schneiden, A und A' seien Punkte auf a. Errichten wir nun in A und A' Lote auf a, und zwar sowohl in α als auch in β, die wir der Kürze halber mit $A\alpha$, $A\beta$, $A'\alpha$, $A'\beta$ bezeichnen, so ist der Winkel zwischen den Loten in A gleich dem Winkel zwischen den Loten in A'.*[38] Wir nennen diesen Winkel den *Neigungswinkel der beiden Ebenen α und β.*

Beweis: Auf dem Lot $A\beta$ nehmen wir einen Punkt M und fällen von M ein Lot auf $A\alpha$, nämlich MF. Dann steht MF senkrecht auf der Ebene α, | nach der 58
Konstruktion p. 55, 56. Jetzt tragen wir die Strecke $\overline{AM}$ von A' aus auf $A'\beta$ ab, und fällen von dem entstehenden Punkt M'[39] aus ein Lot $M'F'$ auf $A'\alpha$;

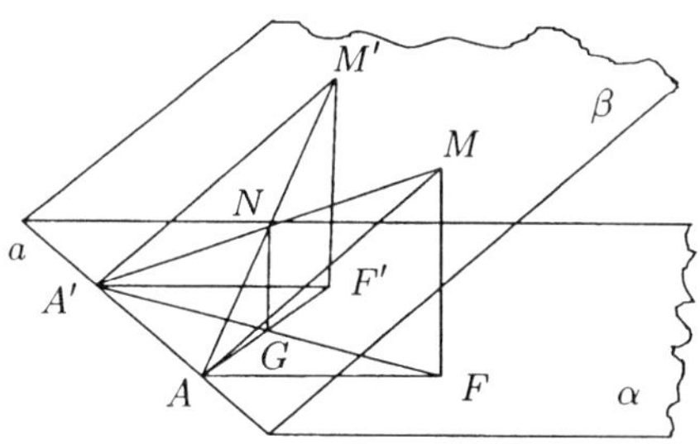

dann steht auch $M'F'$ senkrecht auf α. Dann schneiden sich die Graden $A'M$ und AM' in einem Punkte N der Ebene β, wie man leicht erkennt. Von N aus fällen wir endlich ein Lot NG auf α. Dann haben wir folgende Reihe von Kongruenzen:

[37]This 'converse' can be easily proved as follows. Suppose that line d is simultaneously perpendicular to the lines a, b, c, and let S be the point where all four of these lines meet. Consider the planes α determined by a and d, and β determined by b and c. S belongs to both α and β. Hence, by Axiom I, 6 there is at least one other point P which is in both planes, and hence (by I, 5) the line SP belongs to both planes. Call this line g. We show that g must be the same as a. d is perpendicular to two distinct lines b and c in β, and since g is in β, and has a common point S with b, c and d, then d must also be perpendicular to g by the Theorem proved on p. 54. Thus, a and g are pependicular to d and also have a common point S. But a and g lie in the same plane α, and there can only be one perpendicular to d in this plane at the same point S on d. Hence g must be identical to a; hence a is in the same plane as b and c.

[38]Originally both copies of the *Ausarbeitung* erroneously read 'B'. In the *Lesezimmer* copy, this has been changed to 'A'' by Xx^s, while in Hilbert's copy it remained unchanged.

[39]The construction only works if the point M' lies on the same side of a as M, as is shown in the diagram.

1) $\triangle AA'M \cong \triangle AA'M'$ nach dem Axiom 10. Also ist $\sphericalangle AMA' = \sphericalangle A'M'A$, also 2) $\triangle A'NM' \cong \triangle ANM$, nach dem II. Kongruenzsatz. Daraus folgt wieder $A'N = AN$, und hieraus: 3) $\triangle A'NG \cong \triangle ANG$, nach dem IV. K. S. Also ist auch $\sphericalangle NA'G = \sphericalangle NAG$.

Nun verlängere ich AG über G hinaus; diese Verlängerung muss notwendig durch F' gehen. Denn der Fußpunkt F' liegt erstens in der Ebene α, zweitens nach dem Satz p. 56 in der Ebene $ANM'G$,[40] also liegt er auf der Schnittgraden beider Ebenen, das ist aber AG. Ebenso liegt F auf der Verlängerung von $A'G$ über G hinaus.

Nun ist 4) $\triangle AM'F' \cong \triangle A'MF$, nach dem IV. K. S.[41] also ist $M'F' = MF$. Daraus folgt endlich, wieder nach dem IV. K. S.,[42] daß 5) $\triangle A'M'F' \cong \triangle AMF$ ist, woraus sich ergiebt $\sphericalangle M'A'F' = \sphericalangle MAF$. q. e. d.

Mit Hülfe dieses Satzes können wir nun an eine Ebene α in einer gegebenen
59 Grade a *eine zweite Ebene unter gegebenem Winkel* | *antragen. Es ergiebt sich also die Drehung einer Ebene um eine Grade rein logisch aus den linearen und ebenen Kongruenzaxiomen, während die Verschiebung auf einer Graden und die Drehung einer Graden um einen Punkt in einer Ebene axiomatisch festgelegt werden musste* [Axiome 1. und 5.].

Nun ist es leicht, den *allgemeinsten Kongruenzsatz für den Raum,* ein Analogon zu dem Satz pag. 53, zu beweisen, den wir mit Vorausschickung einer Definition so aussprechen:

Zwei Figuren im Raum heissen kongruent, wenn ihre Punkte sich einander so zuordnen lassen, daß alle homologen Strecken und Winkel einander gleich sind. — Sind $(A, B, C, \ldots)$ und $(A', B', C', \ldots)$ kongruente Figuren und ist P ein beliebig gegebener Punkt, so lässt sich stets ein Punkt P' finden, sodaß die Figuren $(A, B, C, \ldots, P)$ und $(A', B', C', \ldots, P')$ kongruent sind. Enthält die Figur $(A, B, C, \ldots)$ mindestens vier nicht in einer Ebene liegende Punkte, so ist die Konstruktion von P' nur auf eine einzige Weise möglich.

Dieser Satz sagt die Existenz einer gewissen *umkehrbar eindeutigen Transformation des Raumes in sich aus, die wir als Kongruenz oder Bewegung*

[40] That is, the theorem stated on the last line of p. 56, the perpendiculars to α in question here being NG and $M'F'$.

[41] It seems this should be 'according to the Second Congruence Theorem', since it is not clear how the Fourth would be applied here. Even the use of the Second seems unclear. It follows from 1) on p. 58 that AM' is congruent to $A'M$, and also from 3) that $\sphericalangle MA'F \equiv \sphericalangle M'AF'$, since these angles are just $\sphericalangle NA'G$ and $\sphericalangle NAG$ respectively. But then to apply the Second Congruence Theorem, it has to be shown that $\sphericalangle A'MF \equiv \sphericalangle AM'F'$. Of course, $\sphericalangle MFA' \equiv \sphericalangle M'F'A$, since MF and $M'F'$ are perpendicular to α, and thus to FA' and $F'A$ respectively. But to conclude that $\sphericalangle A'MF \equiv \sphericalangle AM'F'$, one either has to use the fact that NG and MF resp. $M'F'$ are parallel, and therefore that MF (resp. $M'F'$) cuts $A'M$ (resp. AM') in the same angle as NG, or that the angle-sum in any two triangles is the same, both of which require some form of the Axiom of Parallels.

[42] This appeal to the Fourth Congruence Theorem is also a little open to question. AM and $A'M'$ are congruent by construction, and it has just been shown that MF and $M'F'$ are also congruent. We now have to show that the angles opposite the greater of each of these pairs are also congruent. It can easily be argued that $\sphericalangle MFA \equiv \sphericalangle M'F'A'$, since both MF and $M'F'$ are perpendicular to AF and $A'F'$ respectively. But how do we know that MA resp. $M'A'$ are the greater of each of the two pairs of sides?

bezeichnen können. Die Bewegung hat nach den früheren Sätzen die Eigenschaft, Punkte in Punkte, Grade in Grade, Ebenen in Ebenen zu transformieren; sie lässt ferner sämtliche Beziehungen der Verbindung und der Reihenfolge ungeändert. Diese Transformation ist vollständig be|stimmt, wenn in den 60
beiden Räumen je eine Ebene α, eine Grade a in α, ein Punkt A auf a, ferner eine Seite des Raumes in bezug auf die Ebene α, eine Seite der Ebene α in bezug auf a und endlich eine Seite von a in bezug auf den Punkt A gegeben sind.

Den Begriff der Bewegung des Raumes haben wir hier, genau so wie früher den der Bewegung der Ebene, auf die Kongruenzaxiome gestützt. Der umgekehrte Weg, die Kongruenzaxiome und -sätze mit Hülfe des Bewegungsbegriffs zu beweisen ist falsch, da sich die Bewegung ohne den Kongruenzbegriff gar nicht definieren lässt.[43]

Ehe wir zu den Axiomen IV und V übergehen, die den interessantesten Teil unserer Untersuchung bilden werden, wollen wir sehen, welche Sätze sich etwa noch mit Hülfe der drei ersten Axiomgruppen beweisen lassen.

Wie steht es zunächst mit der *Geometrie des Kreises und der Kugel?*

Kreis und Kugel lassen sich auf Grund der bisherigen Axiome natürlich sehr einfach definieren durch Strahlbüschel bezw. Strahlbündel und Abtragung von Strecken. Ferner gelten unverändert die *Sätze über den Schnitt zweier Kreise einer Ebene und über den Schnitt einer Ebene mit einer Kugel*; alles dies braucht nicht näher erörtert zu werden. Interesse bietet erst der *Satz*:

Wenn sich drei Kreise einer Ebene in je zwei Punkten schneiden, so laufen 61
die drei gemeinsamen Sehnen alle durch einen Punkt.

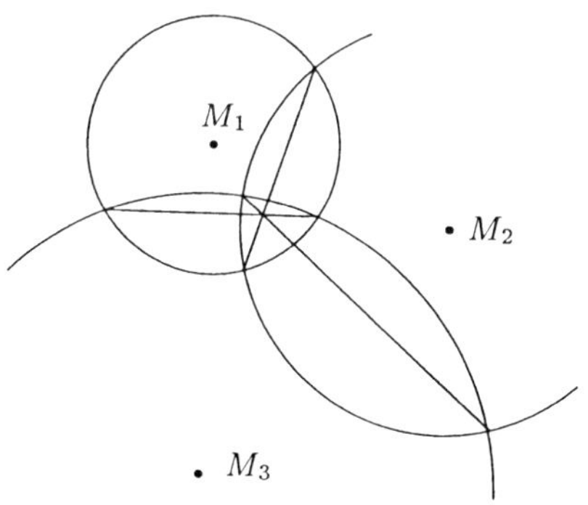

Der Beweis dieses Satzes wird gewöhnlich folgendermassen geführt. Um die Mittelpunkte M_1, M_2, M_3 der drei Kreise denken wir uns Kugeln gelegt, welche aus der Zeichenebene grade unsere drei Kreise ausschneiden. Diese drei Kugeln schneiden sich zu je zweien in Kreisen K_{12}, K_{23}, K_{31}, und diese Kreise definieren drei Ebenen α_{12}, α_{23}, α_{31}, in denen natürlich die drei Sehnen, um die es sich handelt, liegen. Sind nun P und Q die beiden Punkte, die allen drei Kugeln gemeinsam sind, so erkennen wir, daß die drei Ebenen α_{12}, α_{23}, α_{31} sich in der Graden PQ schneiden, also müssen auch die drei Sehnen durch PQ hindurchgehen, da P und Q nach Symmetrie auf verschiedenen Seiten der Zeichenebene liegen. Nun liegen die

[43] Hilbert states here that the proper conceptual dependence is that of (presumably rigid) motion on congruence and not the other way around, this latter being not uncommon: see the Introduction to this Chapter, Specific Note (3)(*b*). Hilbert of course founds the theory of the equality of plane figures quite independently of any concept of motion; see, for instance, this text, 122–138.

drei Sehnen alle in einer Ebene; diese Ebene hat mit PQ einen und nur einen Punkt gemein, und durch diesen Punkt müssen also die drei Sehnen laufen. —

An diesem Beweis ist aber auszusetzen, daß *die Existenz der Punkte P*
62 *und Q auf eine unbewiesene Annahme gegründet* | *ist*; wir wissen gar nicht, ob drei Kugeln, die sich paarweise schneiden, gemeinsame Punkte haben.

Ehe wir diesen Gegenstand näher untersuchen, schalten wir die Bemerkung ein, daß auf Grund der bisherigen Axiome die *Konstruktion harmonischer Strahlen und Punkte durch rechte Winkel* möglich ist. [Cf. pag. 32]

Es mögen die Strahlen 1, 2 einen rechten Winkel bilden, ferner soll 2 den Winkel zwischen 3 und 4 halbieren. Dann ist zu zeigen, daß 1, 2, 3, 4 harmonische Strahlen sind. Zum Beweis ziehen wir die in der Figur angegebenen Hülfslinien (auf 3 und 4 sind gleiche Strecken abgetragen). Dann sind nach der früheren Konstruktion pag. 32 die mit Kreuzen bezeichneten Punkte harmonisch, also auch die durch dieselben Strahlen ausgeschnittenen, mit Kreisen versehenen Punkte. Diese werden aber andrerseits durch die Strahlen 1, 2, 3, 4 ausgeschnitten, also sind diese Strahlen harmonisch. q. e. d.[44]

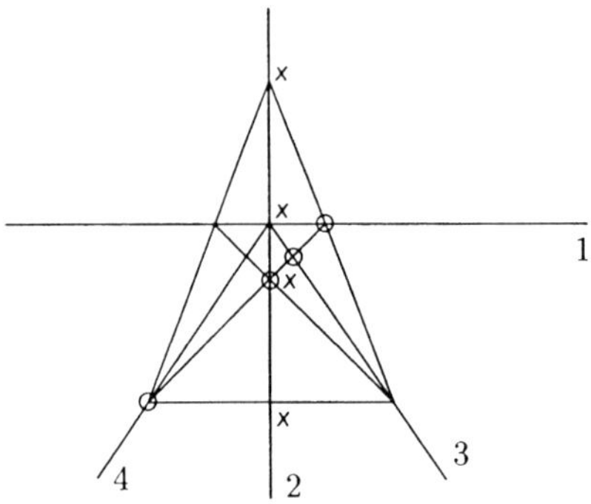

Kehren wir nun zur Geometrie der *Kreise und Kugeln* zurück. Wir geben zunächst die *Schnittpunktsätze* ausführlich an, auf die schon pag. 60 hingewiesen wurde.

63 1) *Eine Grade hat mit einem Kreis oder einer Kugel entweder keinen oder einen oder zwei Punkte gemein.*

2) *Ein Kreis hat mit einer Kugel keinen oder einen oder zwei Punkte gemein, oder er liegt ganz auf der Kugel.*

3) *Eine Ebene oder eine Kugel hat mit einer Kugel entweder keinen oder einen Punkt oder einen Kreis gemein.*

4) *Eine Grade, die durch einen Peripheriepunkt in einen Kreis oder eine Kugel eintritt, muss durch einen andern Peripheriepunkt wieder heraustreten. Dasselbe gilt von einem Kreise, der in einen andern Kreis derselben Ebene oder eine Kugel eintritt.* —

Jetzt fahren wir fort in der Betrachtung des Sehnensatzes. Wir haben ihn bewiesen p. 61 unter der Annahme, daß die drei Kugeln um M_1, M_2, M_3 zwei Punkte P, Q gemein haben. Wir wollen diese Annahme nun genauer prüfen. Sei wieder K_{12} der den Kugeln 1 und 2 gemeinsame Kreis, dessen Ebene auf $M_1M_2M_3$ senkrecht steht, und dessen Mittelpunkt C die Sehne s_{12}

[44]This construction is something of an oddity here. In Hilbert's own notes (p. 62), it comes before the treatment of the Three Chord Theorem, whereas here it interrupts its treatment.

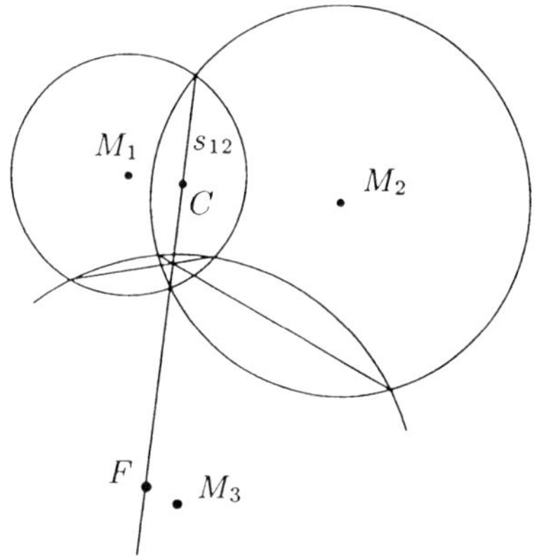

halbiert. Die Ebene von K_{12} schneidet die Kugel 3 in einem Kreise, dessen Mittelpunkt F auf der Verlängerung von s_{12} liegt. Es kommt also darauf an zu zeigen, daß die Kreise um C und F sich schneiden. Die Punkte P und Q, | die wir so suchen, 64
haben von C und F die Entfernungen CP, CQ, FP, FQ und es ist offenbar $FP + CP > CF$, d. h. P und Q wären die Ecken von Dreiecken, von denen wir wissen, daß die Summe zweier Seiten grösser ist als die dritte. Wir haben also zunächst das Ergebnis: *Der Sehnensatz gilt sicher dann, wenn es stets möglich ist, aus drei gegebenen Seiten, wenn nur die Summe zweier grösser ist als die dritte, ein Dreieck zu konstruieren.*

Daß dies thatsächlich *nicht* möglich ist, werden wir bald sehen. Fürs erste schliessen wir die Betrachtung des Sehnensatzes mit der Bemerkung ab: *Der Sehnensatz gilt möglicherweise auch dann, wenn die genannte Konstruktion nicht stets möglich ist; wir müssen hier die Frage nach seiner Gültigkeit offen lassen.* —

Nun kommen wir auf die obige Konstruktion zurück u. zeigen:

Auf Grund der bisherigen Axiome sind folgende Sätze nicht beweisbar:

1) *Eine Grade, die im innern eines Kreises verläuft, trifft notwendig deßen Peripherie.*

2) *Wenn ein Kreis teils im innern, teils im äussern eines andern Kreises derselben Ebene verläuft, so trifft er notwendig dessen Peripherie*, und zwar nach p. 53 in 2 Punkten.

Beide Sätze hängen eng zusammen; [man könnte sie etwas allgemein so aussprechen: *Der Kreis ist eine geschlossene Figur;*] insbeson|dere kommt 2) 65
grade auf die oben verlangte Dreieckskonstruktion hinaus, wie man leicht sieht.

Beweis: Wir stellen uns eine Geometrie her, in der sämtliche Axiome I–III gelten, die Sätze 1), 2) aber nicht. Zu diesem Zwecke legen wir im R_3 ein gewöhnliches Koordinatensystem xyz zugrunde. Von den Punkten des R_3 sollen aber nur diejenigen unserer Geometrie angehören, deren sämtliche Koordinaten dem Zahlbereich angehören, der folgendermassen definiert ist:

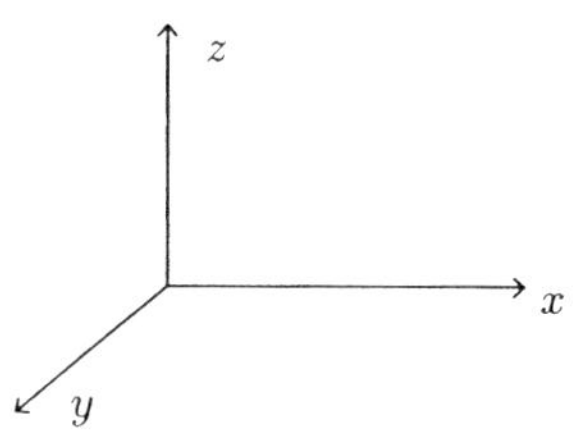

Wir bilden aus der Zahl 1 und einer transcendenten Zahl, z.B π, durch die vier Species alle möglichen Kombinationen, deren Gesamtheit den „Körper" $k(\pi)$ bildet. Dieser ganze Körper soll unserem Zahlbereich angehören, ausserdem aber auch alle Zahlen der Form $\sqrt{1+x^2}$, wo x irgend eine schon bekannte Zahl ist; man sieht sofort, daß auch

$$\sqrt{x^2+y^2} = x\cdot\sqrt{1+\left(\tfrac{y}{x}\right)^2},\ \sqrt{x^2+y^2+z^2} = \sqrt{\left(\sqrt{x^2+y^2}\right)^2+z^2}\ \text{u. s. w.}$$

unserm Bereich angehören, d. h. wir dürfen aus Quadratsummen Quadratwurzeln ziehen. Wir bemerken, *daß an dieser Stelle zum ersten Mal der Pythagoräische Lehrsatz in unsere Betrachtungen hereinspielt.*[45]

Nachdem wir so die Punkte unserer Geometrie definiert haben, werden wir als Grade und Ebenen einfach die Verbindungsgraden und Verbindungsebenen irgend welcher von diesen Punkten betrachten.

66 In der so definierten Geometrie gelten offenbar die Axiome | I und II; aber auch die Kongruenzaxiome III, d. h. es ist Verschiebung und Drehung möglich. Denn die Verschiebung drückt sich analytisch durch Addition aus, die ja unter den fünf erlaubten Operationen sich befindet; und ebenso wird die Drehung um einen Punkt durch erlaubte Operationen dargestellt; z. B. hat man für die Drehung der xy-Ebene in sich um den Anfangspunkt folgende Formeln:

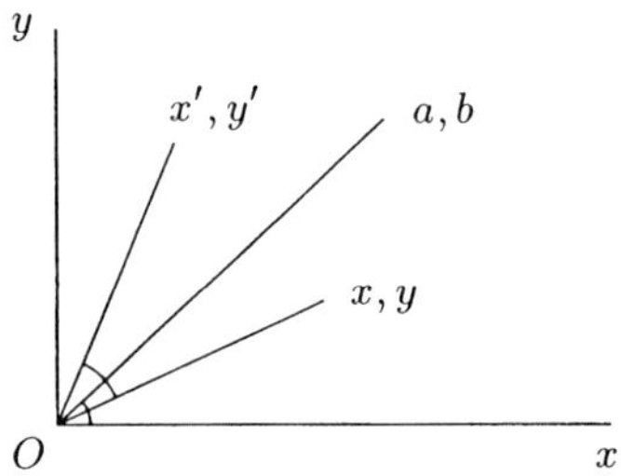

$$x' = \frac{a}{\sqrt{a^2+b^2}}x - \frac{b}{\sqrt{a^2+b^2}}y$$

$$y' = \frac{b}{\sqrt{a^2+b^2}}x + \frac{a}{\sqrt{a^2+b^2}}y;$$

die Bedeutung der Buchstaben ist aus der Figur ersichtlich.[46]

Nun aber wollen wir zeigen, *daß eine Grade im innern und im äussern eines Kreises verlaufen kann, ohne ihn zu schneiden.* Sei die Gleichung des Kreises $x^2+y^2=1$, die der Graden $x=\frac{\pi}{4}$. Diese hat Punkte im innern des Kreises, z. B. $x=\frac{\pi}{4}$, $y=0$, und auch im äussern, z. B. $x=\frac{\pi}{4}$, $y=1$. In der gewöhnlichen Geometrie schneiden sich Grade und Kreis in den Punkten $x=\frac{\pi}{4}$, $y=\pm\sqrt{1-\left(\frac{\pi}{4}\right)^2}$. Wir wollen zeigen, daß diese Punkte in unserer Geometrie nicht existieren,

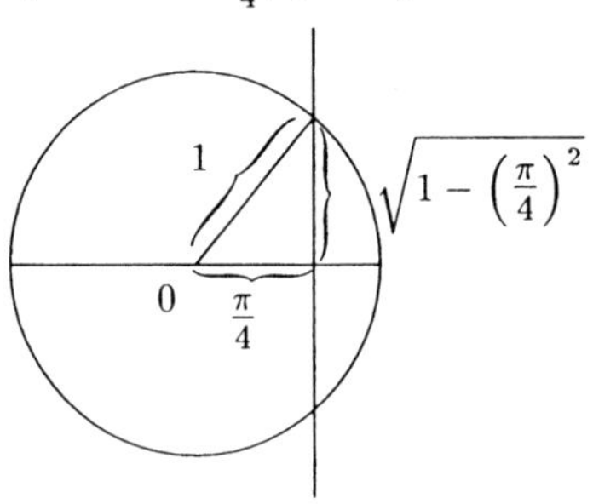

67 mit andern Worten, | daß $\sqrt{1-\left(\frac{\pi}{4}\right)^2}$ sich nicht durch die fünf erlaubten Operationen aus 1 und π erzeugen lässt. Nehmen wir an, es gelte eine Beziehung

$$\sqrt{1-\left(\frac{\pi}{4}\right)^2} = A(1,\pi),^{47}$$

[45]Hilbert here constructs the smallest Pythagorean sub-field of the reals containing π. See the relevant note to Hilbert's own notes, p. 65.

[46]I.e., one rotates about O through an angle α the line joining O and (x,y) to take it to the line joining O and (x',y'), α being the angle between the line joining O and (a,b) and the x-axis.

[47]$A(1,\pi)$ stands for an expression denoting an arbitrary element of the smallest Pythagorean sub-field just constructed.

wo A einen aus den fünf erlaubten Operationen gebildeten Ausdruck bezeichnet. Eine solche algebraische Beziehung für eine transzendente Zahl wie π kann aber nur dann richtig sein, wenn sie *identisch* gilt, d. h. wenn *für jedes* t

$$\sqrt{1-\left(\frac{t}{4}\right)^2} = A(1,t)$$

ist.

Die rechte Seite ist für reelle t stets selber reell, wie sich aus der Definition der 5 Operationen ergiebt; dagegen kann die linke Seite für reelle t imaginär werden. Also ist eine Gleichung der vorausgesetzten Art nicht möglich, und es giebt also keinen Schnittpunkt von Kreis und Grade.

Anmerkung: Die Heranziehung transzendenter Zahlen ist gar nicht notwendig; z. B. wenn man allein auf die Zahl 1 die 5 Operationen anwendet, so lassen sich schon sehr einfache Zahlen angeben, zu denen man nicht gelangt, z. B. die Zahl $\sqrt{1+\sqrt{2}}$. Wäre nämlich $\sqrt{1+\sqrt{2}}$ durch unsere 5 Operationen darstellbar, so müsste sich $1+\sqrt{2}$ als Summe von Quadraten darstellen lassen:

$$1+\sqrt{2} = \alpha^2 + \beta^2 + \gamma^2 + \cdots + \lambda^2,$$

wo α, β, $\gamma, \ldots, \lambda$ algebraische Zahlen sind, die durch unsere 5 Operationen erhalten werden können. Algebraische Betrachtungen führen | dann dazu, daß 68
diese Gleichung auch richtig bleiben muss, wenn man $\sqrt{2}$ durch die konjugierte Zahl $-\sqrt{2}$, also α, β, $\gamma, \ldots, \lambda$ durch geeignete zu α, $\beta, \ldots, \lambda$ konjugierte Zahlen $\overline{\alpha}$, $\overline{\beta}$, $\overline{\gamma}, \ldots, \overline{\lambda}$ ersetzt, also:

$$+1-\sqrt{2} = \overline{\alpha}^2 + \overline{\beta}^2 + \overline{\gamma}^2 + \cdots + \overline{\lambda}^2$$

$\overline{\alpha}$, $\overline{\beta}$, $\overline{\gamma}, \ldots, \overline{\lambda}$ sind reell, also ist die rechte Seite positiv, die linke ist aber negativ. Unsere Annahme ist also falsch, $\sqrt{1+\sqrt{2}}$ mit unsern 5 Operationen nicht konstruierbar.

Jetzt ist leicht zu zeigen, daß sich z. B. *aus den Seiten* $a=1$, $b=1$, $c=\frac{\pi}{2}$ *kein Dreieck konstruieren lässt.* Gäbe es nämlich ein solches Dreieck, so könnte man es zunächst so verschieben, daß A in den Punkt $x=0$, $y=0$, und B in den Punkt $x=\frac{\pi}{2}$,[48] $y=0$ fiele. Dann aber fiele C auf den Punkt $x=\frac{\pi}{4}$, $y=\sqrt{1-\left(\frac{\pi}{4}\right)^2}$, deßen Nichtexistenz wir soeben bewiesen haben. —[49]

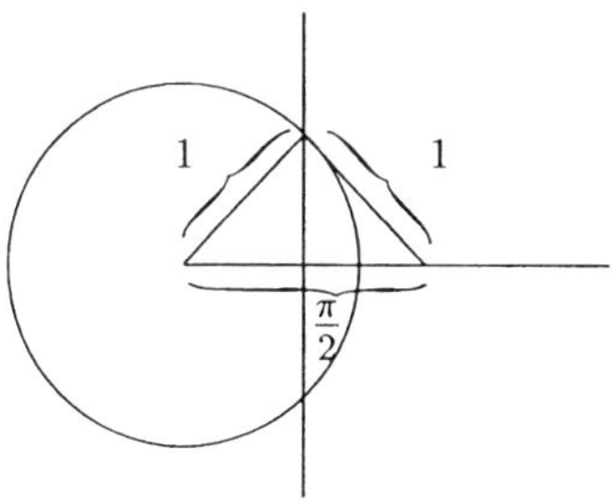

An diese Betrachtungen werden wir später noch einmal anknüpfen, wenn wir untersuchen, mit welchen Hülfsmitteln gegebene Zahlen sich konstruieren lassen. Wir werden dabei u. a. finden, daß es einen wesentlichen Unterschied ausmacht, ob man den

[48] Originally both copies of the *Ausarbeitung* erroneously read '1'. In the copy in the *Lesezimmer*, this has been changed to '$\frac{\pi}{2}$' by Xx^g, while in Hilbert's own copy it remained unchanged.

[49] For a general consequence of this result, see the Introduction to this Chapter, Specific Note (2)(e).

Zirkel unbeschränkt oder nur zum Abtragen von Strecken und von Winkeln benutzt. —[50]

69 Bevor wir die Betrachtung der vierten Gruppe von Axiomen begin|nen, wollen wir noch über die *Beweisbarkeit oder Unbeweisbarkeit einiger wichtiger Sätze mit Hülfe der Axiome I–III* einiges bemerken.

Betrachten wir zunächst die

Legendre'schen Sätze über die Winkelsumme im Dreiecke.

Zum besseren Verständnis bemerken wir: Der Satz, daß die Winkelsumme im Dreieck zwei Rechte beträgt, ist auf Grund der Axiome I–III durchaus unbeweisbar, wie unter IV ausführlich gezeigt werden wird; einen bestimmten *Inhalt* dagegen hat der Satz schon an dieser Stelle.

Die beiden Legendre'schen Sätze lauten:

1.) *In einem Dreieck kann die Summe der drei Winkel niemals grösser als 2 Rechte sein.*

2.) *Wenn in irgend einem Dreieck die Summe der drei Winkel gleich 2 Rechten ist, so ist sie es in jedem Dreieck.*[51]

Den ersten dieser Sätze beweist Legendre so:

Sei im $\triangle ABC$ $\alpha \leqq \beta$, $\alpha \leqq \gamma$, und $AC \leqq AB$.

Halbieren wir nun BC in D und verlängern wir AD um sich selbst über D hinaus bis E. Nun betrachten wir $\triangle ABE$. Da wegen der Kongruenz der Dreiecke ADC und EDB[52] $\sphericalangle BED = \sphericalangle CAD$ und $\sphericalangle EBD = \sphericalangle ACD$ ist, so ist $\sphericalangle EAB + \sphericalangle AEB = \alpha$, $\sphericalangle ABE = \beta + \gamma$, also die Winkelsumme im $\triangle ABE$ gleich $\alpha + \beta + \gamma$, d. h. gleich der Winkelsumme im $\triangle ABC$. — Nun wollen wir

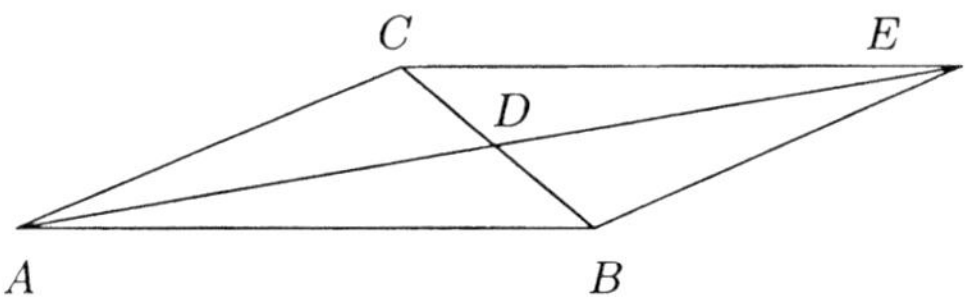

zeigen, daß $\sphericalangle BAE$ der klein-
70 ste Winkel | im $\triangle ABE$ und zugleich $\leqq \frac{\alpha}{2}$ ist.

Zunächst ist $\sphericalangle BAE < \alpha < \beta + \gamma = \sphericalangle ABE$; weiter ist auch $\sphericalangle BAE \leqq \sphericalangle BEA$; wäre nämlich $\sphericalangle BAE > \sphericalangle BEA$, so müsste nach pag. 51 auch $BE > BA$ sein, oder da $BE = AC$ ist, auch $AC > BA$, gegen die Voraussetzung. Also ist $\sphericalangle BAE$[53] der kleinste Winkel im $\triangle ABE$. Daß außerdem $\sphericalangle BAE \leqq \frac{\alpha}{2}$ ist, folgt nun leicht daraus, daß $\sphericalangle BAE + \sphericalangle BEA = \alpha$ und $\sphericalangle BAE \leqq \sphericalangle BEA$ ist.

Wir können uns also aus jedem Dreieck ABC ein anderes Dreieck konstruieren mit derselben Winkelsumme, indem der kleinste Winkel höchstens halb so gross ist wie der kleinste Winkel von ABC. Nehmen wir nun an, die Winkelsumme im $\triangle ABC$ sei $2R + \delta$. Dann führe ich die angegebene Kon-

[50] See p. 170 of this text.

[51] For discussion of the Legendre Theorems, see the Introduction to this Chapter, Specific Note (5)(a).

[52] By the First Congruence Theorem, i.e., Axiom 10.

[53] Originally both copies of the *Ausarbeitung* erroneously read '$\sphericalangle ABE$'. In the copy in the *Lesezimmer* this has been changed to '$\sphericalangle BAE$' by Xx^s, while in Hilbert's own copy it remained unchanged.

struktion n mal aus und wähle n so gross, daß $\frac{\alpha}{2^n} < \delta$ wird. Ich erhalte dann ein Dreieck, in welchem ein Winkel $< \delta$ ist, während die Winkelsumme $2R+\delta$ beträgt. Hieraus folgt, daß in diesem Dreieck die Summe zweier Winkel grösser als 2 Rechte wäre. Nun ist früher bewiesen, daß der Aussenwinkel zu $\sphericalangle A$ im Dreieck ABC stets grösser als der Winkel B ist. Es wäre dann also der $\sphericalangle A$ im Dreieck + dem Nebenwinkel zu $\sphericalangle A$ grösser als $\sphericalangle A + \sphericalangle B$, d. h. grösser als 2 Rechte, was nicht möglich ist.

In diesem Legendre'schen Beweis steckt nun aber eine Annahme, die nicht 71
aus den Axiomen I–III bewiesen werden kann (wie wir später zeigen werden). *Daß man nämlich stets ein n so finden kann, dass $\frac{\alpha}{2^n} < \delta$ wird bei gegebenem α und δ, ergiebt sich erst als Folge des Archimedischen Axioms, das wir* später unter V behandeln.

Also: *Der Legendre'sche Beweis ist für uns nicht zwingend; ob der Legendre'sche Satz nicht dennoch aus den Axiomen I–III folgt, lassen wir dahingestellt;* fast ist zu vermuten, daß er sich in der That so beweisen lässt. —[54]

Den *Beweis des 2. Legendre'schen Satzes* führen wir einfacher, als es wohl sonst zu geschehen pflegt, wobei wir übrigens den 1. Satz als richtig annehmen.

Zuerst zeigen wir: Wenn die Winkelsumme im $\triangle ABC$ zwei Rechte ist, und wenn $\triangle A'B'C'$ ganz im innern von ABC liegt, so ist auch in $A'B'C'$ die Winkelsumme 2 Rechte. Wir verbinden die Ecken der beiden Dreiecke, wie die Figur zeigt, derart daß ABC in 7 Dreiecke zerfällt. (Wir haben der Einfachheit halber angenommen, daß die Verbindungslinien der Ecken sich nirgendwo schneiden; wie der Beweis sich im andern Fall gestaltet, ist leicht zu sehen.)

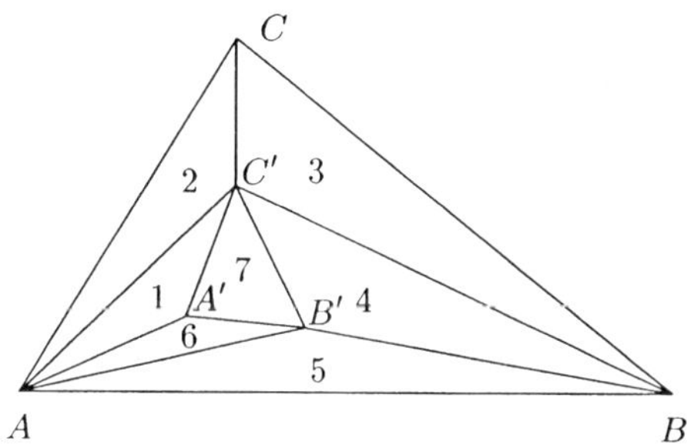

| Seien $w_1, \ldots, w_7$ die Winkelsummen der 7 Dreiecke; dann ist $w_1 + \cdots + w_7 = \alpha + \beta + \gamma + 3 \cdot 4R$, also nach Voraussetzung: 72

$$= (2 + 12)R = 14R.$$

Andrerseits ist nach dem 1. Legendre'schen Satz:

$$w_1 \leqq 2R, \ldots, w_7 \leqq 2R.$$

Hieraus folgt, daß notwendig $w_1 = 2R$, $w_2 = 2R, \ldots$ und speziell auch die Winkelsumme in $A'B'C'$ $w_7 = 2R$ sein muss.

Nun zeigen wir, *daß man jedes gegebene Dreieck in das innere eines Dreiecks mit der Winkelsumme* $2R$ oder, was gleichwertig ist, eines Vierecks mit der Winkelsumme $4R$ *bringen kann.* Sei ABC ein Dreieck mit der Winkelsumme $2R$. Fällen wir etwa von C ein Lot CD auf AB, dann ist wie eben bewiesen, auch im $\triangle BCD$ die Winkelsumme $2R$, also wenn wir $\triangle BCE \cong \triangle BDC$

[54] For the work of Hilbert's student Dehn on the Legendre Theorems, see the Introduction to this Chapter, Specific Note (5)(a).

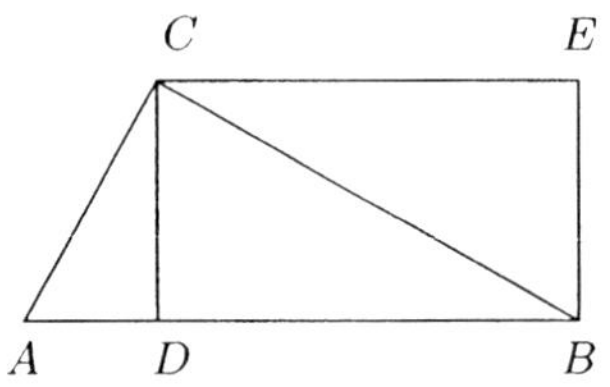

machen, die Winkelsumme des Vierecks $BDCE$ $4R$, $\sphericalangle B = \sphericalangle C = \sphericalangle D = \sphericalangle E = R$. Durch Aneinanderlegen solcher Rechtecke kann man, *unter Voraussetzung des Archimedischen Axioms,* jeden beliebigen Bereich der Ebene überdecken, d. h. auch das vorgegebene $\triangle A'B'C'$ einschliessen. Dann ist, wie oben bewiesen, auch im $\triangle A'B'C'$ $w = 2R$. q. e. d.

73 *Nicht beweisbar* auf Grund der Axiome I–III ist auch *z. B. der Satz von der Gleichheit zweier Peripheriewinkel über demselben Bogen*; denn der Beweis dieses Satzes erfordert den Satz, daß jeder Aussenwinkel eines Dreiecks gleich der Summe der gegenüberliegenden Winkel ist. Und dieser Satz lässt sich hier noch nicht beweisen. —

Weiter wollen wir noch einige Bemerkungen *über Schnittpunktsätze* anfügen.

Wir haben bisher einen solchen, nämlich den *Satz von Desargues* kennengelernt, und gefunden, dass er nicht beweisbar ist, wenn man nur die *ebenen* Axiome aus den Gruppen I und II zugrunde legt. Jetzt erhebt sich die Frage, ob der Satz von Desargues etwa beweisbar ist, wenn man zu den ebenen Axiomen I und II noch die Kongruenzaxiome III hinzunimmt.

Weiter bietet sich die Frage, ob aus den Axiomen I, II, III nicht noch weitere Schnittpunktsätze folgen. Das ist thatsächlich der Fall; wir können nämlich auf Grund der Axiome I–III den *Pascal'schen Satz für einen aus zwei Graden bestehenden Kegelschnitt* beweisen, der so lautet: *Liegen die Ecken $E_1, E_2, \ldots, E_6$ eines Sechsecks abwechselnd auf zwei sich schneidenden Graden, und schneiden sich die drei Paare gegenüberliegender Seiten in den Punkten $D_1 = (E_1E_2, E_4E_5)$, $D_2 = (E_2E_3, E_5E_6)$, $D_3 = (E_3E_4, E_6E_1)$, so*
74 *liegen diese drei Punkte | auf einer Graden.*

Der Beweis dieses Satzes (vgl. *Schur, Math. Ann. 51.*)[55] kann geführt

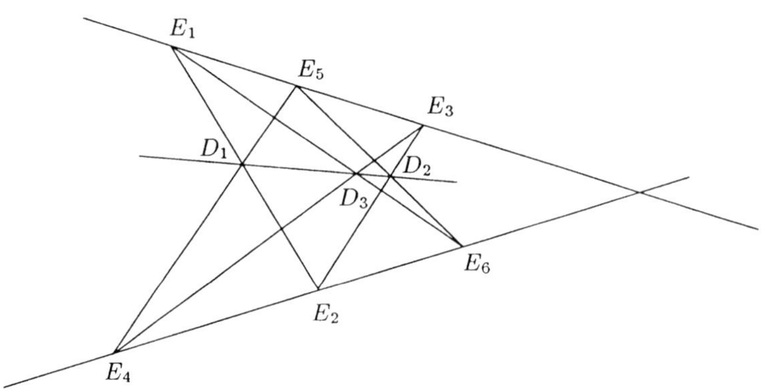

[55]The reference is to *Schur 1899*.

werden mit Hülfe aller Axiome I–III, nicht etwa nur der ebenen. Er beruht im Grunde auf dem einschaligen Hyperboloid und benutzt wesentlich die Spiegelung des Raumes an einer Ebene.[56]

Wir werden später zeigen, daß man den Pascal ohne Kongruenzsätze, allein aus I und II, nicht beweisen kann; dagegen lässt er sich beweisen, wenn man zu I und II die Stetigkeit hinzunimmt (V). — Eine weitere Frage ist, ob man den Pascal beweisen kann mit den ebenen Axiomen I–III, wenn man noch den Desargues hinzunimmt.[57]

Schliesslich bemerken wir, daß *Desargues und Pascal die beiden einzigen Schnittpunktsätze* sind, *die wir jetzt beweisen können*; wir werden später hierauf zurückkommen. —

Werfen wir einen kurzen *Rückblick* auf die bisher betrachteten Axiome.

Die Axiome I, II bildeten im wesentlichen die Grundlage der *projektiven* 75
Geometrie.

Die auf den Axiomen I–III ruhende Geometrie, vielfach Bolyai-Lobatschewsky'sche genannt, wollen wir als *Gaussische Geometrie* bezeichnen. Sie bildet einen wesentlichen Teil dessen was man im engeren Sinne Nicht-Euklidische Geometrie nennt. Eine Zusammenstellung der *Litteratur* über Gaussische Geometrie findet man im *2. Bande der Werke von Lobatschewsky.*[58]

Die Axiome I–IV werden uns die *Euklidische Geometrie* liefern.

Endlich wollen wir die auf den sämtlichen Axiomen I–V aufgebaute Geometrie die *analytische Geometrie* nennen. —[59]

IV. Das Parallelenaxiom.

Die bisher betrachteten Axiome sind im wesentlichen diejenigen, die den ersten 28 Sätzen des Euklides zugrunde liegen. Euklides erkannte, dass so einfache geometrische Thatsachen wie *z. B. die Existenz eines Rechtecks* auf Grund dieser Axiome *nicht beweisbar* sind; er musste also, um weiter zu kommen, ein neues Axiom aufstellen, das be|rühmte sogen. *Parallelenaxiom*, dem 76
wir folgende Form geben:

In einer Ebene lässt sich durch einen Punkt ausserhalb einer Graden nur eine einzige Grade ziehen, welche die erste Grade nicht schneidet. Die zweite Grade heisst eine Parallele zur ersten.

Um den Inhalt dieses Axioms ganz scharf hervorzuheben, bemerken wir: Dass es durch einen Punkt einer Ebene zu einer gegebenen Graden *wenigstens* eine Parallele, d. h. die gegebene Grade nicht schneidende giebt, folgt aus den Axiomen I–III; IV aber behauptet, daß es immer *nur* Parallele giebt.

[56]This is an indirect reference to the method of Schur's proof.

[57]The various questions about Pascal's Theorem raised in this paragraph are settled later in this text. See respectively pp. 147–160 (especially the summary on p. 160) and 107–111, and then also the *Festschrift*, p. 71–76. See also the Introduction to this Chapter, Specific Note (1).

[58]This is a reference to *Lobachevsky 1886*.

[59]The third version of the 'Archimedean Axiom' given below is a strong enough continuity axiom to yield full analytic geometry. The other two versions are not; see this text, 140–141. The ordinary Archimedean Axiom is the only continuity axiom used in the first edition of the *Festschrift*.

Welchen Scharfsinn die Aufstellung dieses Axioms erforderte, erkennen wir am besten, wenn wir einen Blick auf die *Geschichte des Parallelenaxioms* werfen; wir dürfen uns dabei kurz fassen und übrigens auf *Engel u. Stäckel, die Theorie der Parallellinien von Euklid bis Gauss* verweisen.[60]

Was *Euklides* (ca. 300 v. Chr.) selbst angeht, so sei nur noch bemerkt, daß er z. B. den Satz vom Aussenwinkel vor Einführung des Parallelaxioms beweist, ein Zeichen, wie tief er in den Zusammenhang der geometrischen Sätze eingedrungen war.[61]

77 Schon im *Altertum* begannen die Versuche, das Pa|rallelenaxiom auf Grund der übrigen zu beweisen; man wollte z. B. die Parallele als Grade gleichen Abstandes definieren, in dieser Definition steckt aber die Forderung, daß die Punkte, die von einer Graden gleichen Abstand haben, selbst eine Grade bilden, und diese Forderung ist grade das Parallelenaxiom.[62]

Seit dem *17. Jahrhundert* finden wir von neuem Beweisversuche*),[63] die aber alle erfolglos waren: immer schleicht sich in den Beweis unbemerkt eine Thatsache der Anschauung ein, die dem Parallelaxiom gleichwertig ist. Dennoch sind diese Untersuchungen nicht nutzlos gewesen; denn es wurden dabei auch manche richtige Sätze gefunden, und man erkannte wenigstens, daß das Problem nicht so einfach ist, wie es zuerst scheint.

So sagt der Engländer *Henry Savile* 1621: es seien duo macula in pulcherimo geometriae corpore vorhanden: die Theorie der Parallelen und die Lehre von den Proportionen, (also genau die Probleme, die wir unter IV und V behandeln werden.)

Wallis gab 1693 einen neuen „Beweis“ des P. A.

Einen wesentlichen Fortschritt bezeichnet das Buch von *Saccheri* 1733: Euclides ab omni naevo vindicatus, weiter auch *Lambert* 1766, *Legendre* 1794–
78 1833. Dieser Fortschritt besteht | im wesentlichen darin, daß diese Mathematiker die oben besprochenen sog. Legendre'schen Sätze über die Winkelsumme im Dreieck bewiesen; andererseits glaubten auch sie, wie alle ihre Vorgänger, das Parallelenaxiom wirklich bewiesen zu haben.

Unter den Mathematikern, die sich mit unserm Problem beschäftigten, nennen wir noch *d'Alembert* (der es das Ärgernis und die Klippe der elementaren Geometrie nannte), *Fourier*, *Lagrange*.

Pfaff hielt das Axiom für unbeweisbar, wollte es aber durch ein einfacheres ersetzen.[64]

---

*) sie sind bis heut nicht ausgestorben; 1898 erschien in Paris: La Théorie des Parallèles démontrée rigoureusement, par M. Frolov.

---

[60] For comments on Hilbert's historical remarks, see the Introduction to this Chapter, Specific Note (4).

[61] This theorem is Book I, Proposition 16. See the comments in *Heath 1926a*, 280–281.

[62] This version of the Axiom of Parallels goes back to Posidonius (1st century BC), Geminus (probably 1st century BC), and Proclus. See *Heath 1926a*, 220.

[63] This is probably the work referred to by Hilbert in his own notes, p. 73.

[64] For the reference to d'Alembert, see the Introduction to this Chapter, Specific Note (4)(*d*)6(*vi*). Hilbert clearly takes his references to Fourier, Lagrange and Pfaff from *Stäckel and Engel 1895*, pp. 211–212 and 215 respectively. These are mostly secondhand accounts;

Den entscheidenden Schritt that *Gauss.* Er stellte, wie wir aus seinen Briefen und dem Nachlass (Werke Bd. 8) wissen, eine Geometrie auf, in welcher das Parallelenaxiom nicht gilt. Er selbst veröffentlichte nichts darüber, weil er sich vor dem „Geschrei der Böotier" scheute. Doch erschienen bereits zu Gauss' Lebzeiten die grundlegenden Arbeiten von *Bolyai* und *Lobatschewsky.*[65]

Hiermit schliessen wir den historischen Überblick und geben zunächst

Folgerungen aus dem Parallelenaxiom.

1. Die Summe der Winkel eines Dreiecks ist stets gleich 2R. (Jeder Aussenwinkel ist gleich der Summe der gegenüberliegenden Innenwinkel.)

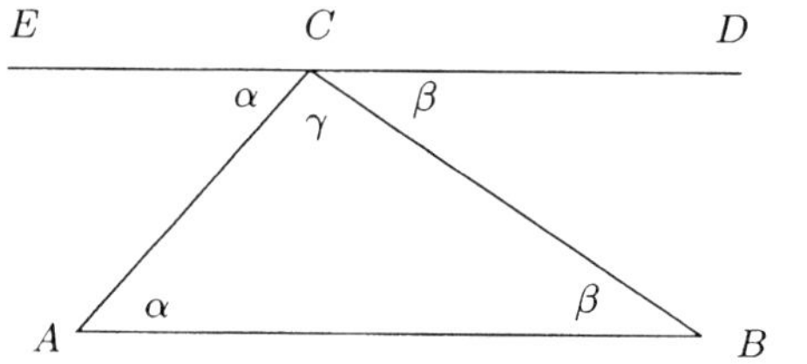

Beweis: Machen wir $\sphericalangle ECA = \alpha$, $\sphericalangle DCB = \beta$, so kann weder CE noch CD AB schneiden.[66] Nach dem P. A. ist dann ECD eine Grade, also $\sphericalangle ECD = \alpha + \beta + \gamma = 2R$. 79

2. Wenn zwei Parallelen von einer dritten Graden geschnitten werden, so sind die Gegenwinkel gleich.

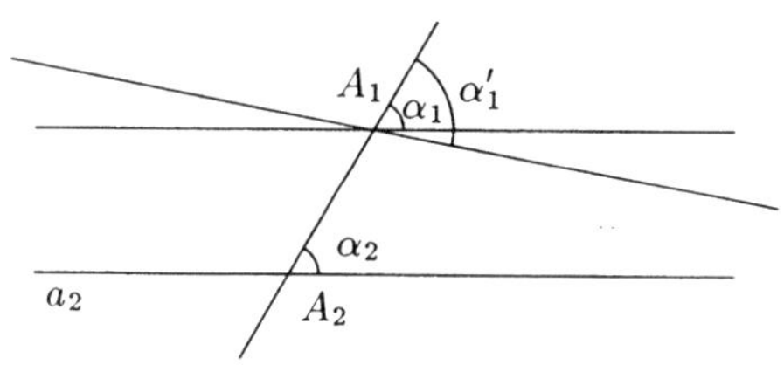

Beweis: Wäre etwa $\alpha_1 \neq \alpha_2$, so könnte man durch A_1 eine Grade unter dem Winkel $\alpha_1' = \alpha_2$ legen und bekäme also durch A_1 zwei Parallelen zu a_2, gegen das P. A.[67]

3. Zwei Graden, die einer dritten parallel sind, sind unter sich parallel.

---

for the references, see *Stäckel and Engel 1895*, 211–215, especially the bibliographical index. See also the very comprehensive *Stäckel 1923*.

[65]For Hilbert's references, and more on Gauß's comment, see the Introduction to this Chapter, Specific Note (4)(*d*), 6(iii) and 6(v).

[66]If, e.g., CE were to intersect BA, say in point P, then PAC would form a triangle, to which $\sphericalangle CAB$ would be an exterior angle. But $\sphericalangle CAB$ would *equal* one of the interior and opposite angles, $\sphericalangle PCA$, both being α, contradicting the exterior angle theorem (p. 46). The same holds for CD.

[67]Hilbert's proof here is not complete. To complete it, we have to combine the argument here with that hinted at in Hilbert's own notes, p. 76. Call the two given parallel lines a_1 through A_1 and a_2 through A_2. Construct the new line through A_1 as Hilbert says, and call it a_1'. There are now two possibilities: either a_1' intersects a_2, or it does not. In the first case, we get a triangle which violates the exterior angle theorem (p. 46), and in the second case, we have two distinct lines, a_1 and a_1' through A_1 which do not intersect a_2, which violates the Parallel Axiom. Hilbert passes over the first possibility.

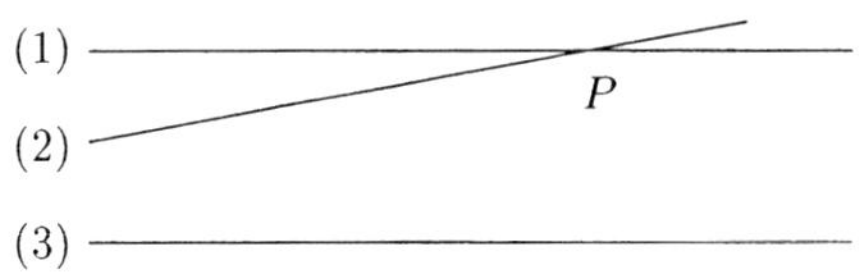

Beweis: Wären (1) und (2) nicht parallel, so gäbe es ja durch ihren Schnitt-punkt P zwei Parallelen zu (3), was nicht möglich ist.

4. Durch jeden Winkelpunkt eines Winkels, der kleiner als ein gestreckter ist, kann ich stets Graden ziehen, die beide Schenkel [nicht etwa ihre Verlängerungen] schneiden.
80 *Beweis:* Wir ziehen durch den Winkelpunkt P die Parallelen zu den Schenkeln; diese Parallelen sind von einander verschieden, da $\alpha < 2R$ ist, und bilden also in P zwei Paare von Scheitelwinkeln. Jede Grade durch P, die innerhalb der schraffierten Winkel verläuft, hat dann die verlangte Eigenschaft, wie sich aus den Axiomen II ergiebt.[68]

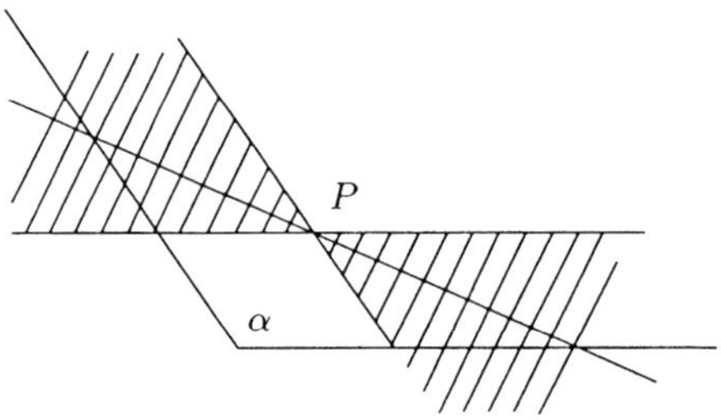

5. Sämtliche Punkte einer Graden haben von einer Parallelen gleichen Abstand.[69]

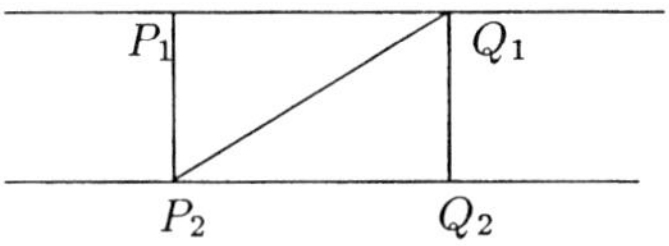

Beweis. Aus dem P. A. folgt zunächst, dass die Figur $P_1Q_1Q_2P_2$ ein Rechteck ist.[70] Dann ergiebt sich sogleich, daß: $\triangle P_2Q_1P_1 \cong \triangle Q_1P_2Q_2$ ist;[71] also ist $P_1P_2 = Q_1Q_2$, q. e. d. —

Schließlich bemerken wir, daß es den Anschein hat, als wenn jeder dieser fünf Sätze (eventuell in geeignet veränderter Fassung) geradezu als *Ersatz des P. A.* dienen könnte. Doch können wir auf diese interessante Frage nicht eingehen.[72]

Unter den vielen angeblichen Beweisen des P. A. wollen wir nur einen ganz kurz besprechen. Angenommen, die von einander verschiedenen Graden PM und PN seien beide parallel zur Graden ACB. Nun ist einerseits
81 $\sphericalangle MPN < \sphericalangle ACB$, andrerseits | enthält $\sphericalangle MPN$, als Inbegriff seiner Winkel-

---

[68]Better to say, from Axiom Group II *together* with the Parallel Axiom.

[69]'Same distance' is to be understood here in the following sense. Suppose two lines a_1 and a_2 are parallel, and that P_2 and Q_2 are any two points on a_2. Erect perpendiculars at P_2 and Q_2 respectively, and let P_1 and Q_1 be the respective points of intersection with a_1. Then P_1P_2 and Q_1Q_2 are congruent.

[70]Let a_1 and a_2 be parallel, and let P_1, Q_1, Q_2, P_2 be four points as described in the previous note. It is easy to see, using Theorem 2 on the previous page, that all the angles of the figure $P_1Q_1Q_2P_2$ are right angles.

[71]This follows from the Second Congruence Theorem (p. 45), using the facts that $\overline{Q_1P_2} \equiv \overline{P_2Q_1}$, that $\sphericalangle Q_1P_2Q_2 \equiv \sphericalangle P_2Q_1P_1$, according to Theorem 2 from the previous page, and that $\sphericalangle P_1P_2Q_1 \equiv \sphericalangle P_2Q_1Q_2$, again using Theorem 2.

[72]See Hilbert's own notes, p. 77, and the second editorial note on that page.

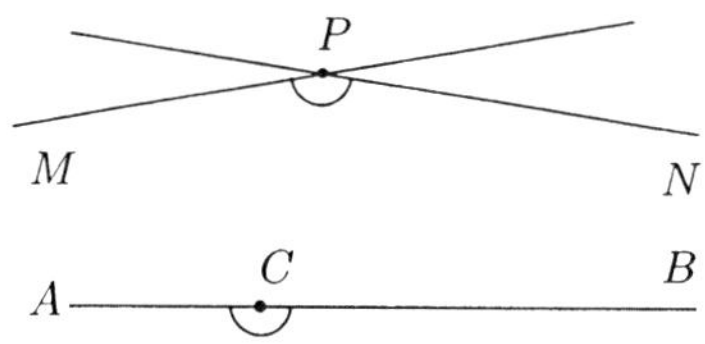

punkte aufgefasst, den $\sphericalangle ACB$ vollständig, und hierin liegt ein Widerspruch gegen den Satz: Das ganze ist grösser als ein Teil.

Dieser Beweis ist nun völlig verfehlt; denn:

1.) Es ist unzulässig, allgemeine Grössenbeziehungen wie den obigen Satz ohne weiteres auf geometrische Dinge zu übertragen (wie schon früher betont wurde).[73] Dies erkennt man hier sehr deutlich; mit Hülfe jenes Satzes kann man nämlich auch beweisen, daß bei Parallelen die Gegenwinkel *ungleich* sein müssen (also das grade Gegenteil des Parallelenaxioms!).

2.) Der Satz: „totum parte maius" wird hier auf Systeme von unendlich vielen Dingen angewendet; und das ist sogar schon in der reinen Grössenlehre nicht erlaubt, wie in den Elementen der Mengenlehre gezeigt wird. —[74]

Wir kommen nun zur Hauptaufgabe, nämlich zum

Beweis der Unabhängigkeit des Parallelenaxioms von den Axiomen I–III.

Wir stellen uns die Aufgabe, eine „Nichteuklidische Geometrie", d. h. eine Geometrie ohne P. A. zu konstruieren | und zwar in möglichst einfacher Weise. 82

Zunächst geben wir den *Beweis für die Ebene*; die Erweiterung auf den Raum wird später sehr einfach sein.

Wir denken uns in einer gewöhnlichen Euklidischen Ebene ein Koordinatenkreuz xy. *Die Punkte unserer Geometrie seien die Punkte der „positiven" Halbebene* ($y > 0$), *mit Ausschluß der Punkte der x-Achse* ($y = 0$).

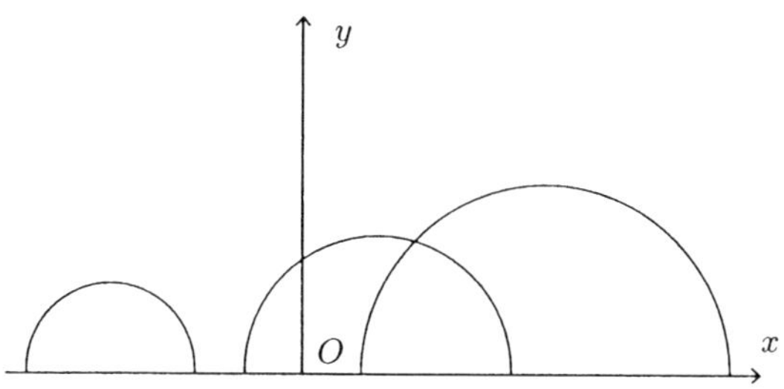

Als Graden nehmen wir die *zur x-Achse orthogonalen Halbkreise*.[75] In dieser Geo-

---

[73] This is a reference to the discussion on p. 34.

[74] What Hilbert cites in Latin here is actually Common Notion V ('The whole is greater than the part') from Euclid's *Elements* (see *Heath 1926a*, 155). The source of the Latin rendering is not clear, although *Clebsch 1891* gives first the Greek versions of all the Postulates and Common Notions ('basing themselves' on Heiberg's text from 1883, *Euclid 1883*), and then, alongside these, Latin translations, one assumes by one of the authors. (Heiberg's text does not give Latin versions.) Common Notion V is translated in *Clebsch 1891*, 554 more or less as Hilbert gives it ('Et totum parte maius est'). Hilbert cites it in Greek in the *Festschrift* (43), and he has also just given it here in German at the end of the proof.

The reference to the 'elements of set theory' is simply explained. Once comparison of sets by size is based on the notion of one-to-one correspondence, it follows that there are infinite sets where one is a proper part (subset) of another, and yet of the same size.

[75] The lines should also include the half-lines of the upper half-plane which are perpendicular to the x-axis.

metrie gelten die sämtlichen ebenen Axiome I und II, wie früher schon gelegentlich benutzt wurde.[76]
Es kommt nur noch darauf an, die Axiome III gültig zu machen durch geeignete Definition der Kongruenzbegriffe (Abtragen von Strecken und Winkeln).

Zu diesem Zwecke brauchen wir einen kurzen *Exkurs über lineare Transformationen der Halbebene in sich.* Wir wollen die xy-Ebene, wie es in der Funktionentheorie gebräuchlich ist, als *Ebene der komplexen Variablen* $z = x + iy$ auffassen; was im Grunde nur eine, allerdings sehr bedeutende, Vereinfachung im Ausdruck ist.

83 Betrachten wir jetzt folgende Transformation:

$$z' = \frac{\alpha z + \beta}{\gamma z + \delta}, \quad \alpha, \beta, \gamma, \delta \text{ reell.}$$

Sie definiert ein eindeutiges Entsprechen zwischen z und z', ausser wenn $\alpha\delta - \beta\gamma = 0$ ist; wir wollen daher *ein für alle mal* $\alpha\delta - \beta\gamma \neq 0$ annehmen.

Die obige Transformation wollen wir auch durch S, oder durch $\left(z : \frac{\alpha z+\beta}{\gamma z+\delta}\right)$ bezeichnen.

Aus der Realität der α, β, γ, δ *folgt, dass bei unserer Transformation die* x*-Achse in sich übergeht. Denn für reelle* z *wird eben auch* z' *reell.*

Für die weitere Untersuchung sind nun zwei Formeln sehr nützlich, die wir erst hinschreiben und dann erklären wollen. Sie lauten:

$$1.) \quad \left(z : \frac{\alpha z + \beta}{\delta}\right) = \left(z : \frac{\alpha}{\delta} z\right)\left(z : z + \frac{\beta}{\delta}\right).$$

$$2.) \quad \left(z : \frac{\alpha z + \beta}{\gamma z + \delta}\right) = \left(z : z + \frac{\delta}{\gamma}\right)\left(z : \frac{1}{z}\right)\left(z : \frac{\beta\gamma - \alpha\delta}{\gamma^2} z\right)\left(z : z + \frac{\alpha}{\gamma}\right).$$

Rechterhand stehen „Produkte" von Transformationen; sie sind so zu verstehen, daß man die als Faktoren auftretenden Transformationen hintereinander ausführt, indem man von rechts her beginnt.

Die Richtigkeit der Formeln zeigt die Ausrechnung. Ihr Nutzen besteht in folgendem:

84 *Die Formeln 1) und 2) zeigen uns, daß die allgemeinste | reelle lineare Transformation sich zerlegen lässt in Transformationen der einfachen Gestalten:* $(z : z + \lambda)$, $(z : \mu z)$, $\left(z : \frac{1}{z}\right)$, λ, μ *reell.*

Unsere Aufgabe wird aber sein, *diese einfachen Transformationen näher zu untersuchen.*

Zwei Eigenschaften der linearen Transformationen kommen für uns in Betracht, die wir, obwohl sie sehr bekannt sind, doch der Vollständigkeit halber nun beweisen wollen.

1. Bei der linearen Transformation gehen Kreise in Kreise über; grade Linien gelten dabei als Kreise, die durch den unendlich fernen Punkt der z*-Ebene gehen.*

Wir beweisen diese Eigenschaft, indem wir zeigen, dass sie für jede der drei einfachen Transformationen gilt, aus denen wir, wie gezeigt ist, die allgemeine Transformation zusammensetzen können. $z' = z + \lambda$ $(x' = x + \lambda,\ y' = y)$

[76] Desargues's Theorem holds for the model Hilbert gives here: see p. 31.

ist eine Verschiebung der Ebene entlang der x-Achse, $z' = \mu z$ ($x' = \mu x$, $y' = \mu y$) ist eine Ähnlichkeitstransformation, d. h. jede Figur wird in eine ihr ähnliche transformiert. Diese beiden Transformationen haben also offenbar die behauptete Eigenschaft. Aber auch $z' = \frac{1}{z}$ ($x' = \frac{x}{x^2+y^2}$, $y' = \frac{-y}{x^2+y^2}$) ist eine „Kreisverwandtschaft" wie wir durch direktes Ausrechnen bestätigen. Ist nämlich $(x-u)^2 + (y-v)^2 = r^2$ die Gleichung eines Kreises, dann ist die Gleichung der transformierten Kurve:

$$\left(\frac{x}{x^2+y^2} - u\right)^2 + \left(\frac{-y}{x^2+y^2} - v\right)^2 = r^2$$ 85

oder

$$(u^2 + v^2 - r^2)(x^2 + y^2) - 2ux + 2vy + 1 = 0,$$

d. h. wieder die Gleichung eines Kreises wie zu beweisen war.

2. Bei der linearen Transformation bleiben die Winkel erhalten, (sie ist „winkeltreu").

Diese Eigenschaft haben, wie bekannt, alle durch analytische Funktionen vermittelte Transformationen.

Wir brauchen sie aber hier nur für lineare Transformationen, und zwar nur für den Winkel zwischen zwei Orthogonal-Kreisen. Auch hier benutzen wir die Zerlegung von pag. 83.[77] Die Transformationen $z' = z + \lambda$ und $z' = \mu z$ sind selbstverständlich winkeltreu, es bleibt nur $z'' = \frac{1}{z}$ zu untersuchen übrig. Diese Transformation ist bis auf eine Spiegelung an der x-Achse mit der Transformation durch reziproke Radien identisch und letztere lässt die Winkel, speciell die Winkel zwischen zwei Orthogonalkreisen ungeändert.

Wir beweisen das folgendermassen. Die Figur zweier sich schneidender Orthogonalkreise ist vollständig bestimmt, wenn ihre Schnittpunkte A, B, C, D mit der x-Achse gegeben sind; ihre Abszissen seien a, b, c, d. Nun drücken wir den Winkel φ der beiden Kreise durch a, b, c, d aus.

Die Figur zeigt, dass $\cos\varphi = -\cos M_1SM_2$ ist. 86

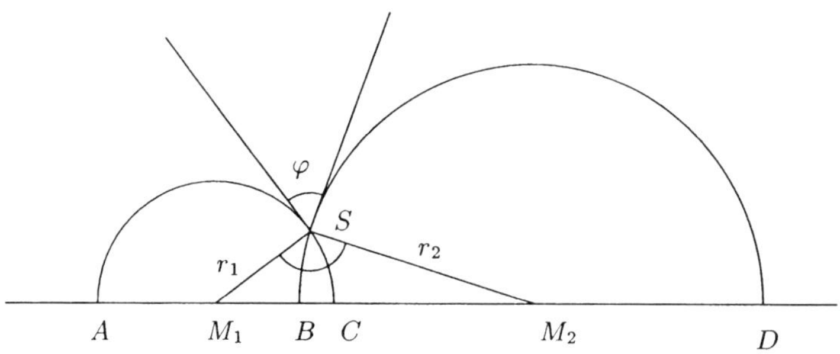

Nun folgt aus dem Kosinussatz

$$\cos M_1SM_2 = \frac{r_1^2 + r_2^2 - \overline{M_1M_2}^2}{2r_1r_2}.$$

[77]And p. 84.

Hierin setzen wir $r_1 = \frac{c-a}{2}$, $r_2 = \frac{d-b}{2}$, $M_1 M_2 = \frac{d+b-(c+a)}{2}$. Dann wird

$$\cos\varphi = \frac{2(ac+bd)-(ab+cd)-(ad+bc)}{ab+cd-(ad+bc)}.$$

Nun transformieren wir unsere Figur durch reziproke Radien; wir brauchen dazu nur a, b, c, d durch $\frac{1}{a}$, $\frac{1}{b}$, $\frac{1}{c}$, $\frac{1}{d}$ zu ersetzen. Machen wir diese Substitution in dem Ausdruck für $\cos\varphi$, so sehen wir, daß dieser ungeändert bleibt, und das war zu beweisen. —

Aus der ersten Eigenschaft folgt nun, sobald man $v = 0$ setzt, *daß bei einer linearen Transformation mit reellen Koeffizienten α, β, γ, δ die zur x-Achse orthogonalen Kreise in einander übergehen.* Und zwar werden dabei *die oberen*
87 *Halbkreise in obere Halbkreise übergehen*, sobald die Transformation | die obere Halbebene in sich überführt. Dies ist nun, wie man aus den Zerlegungen auf p. 83[78] sieht, dann und nur dann der Fall, *wenn $\alpha\delta - \beta\gamma > 0$ ist*; wir wollen daher von nun an *annehmen, es sei $\alpha\delta - \beta\gamma = +1$.* Dann gilt, wie man durch Ausrechnen findet, die Beziehung: $y' = \frac{y}{|\gamma z+\delta|^2}$, wo $|\gamma z + \delta|$ den absoluten Wert von $\gamma z + \delta$ bezeichnet.

Die bisher gewonnenen Resultate fassen wir nun so zusammen: *Bei einer linearen Transformation $z' = \dfrac{\alpha z+\beta}{\gamma z+\delta}$ (α, β, γ, δ reell; $\alpha\delta - \beta\gamma = +1$) gehen die Punkte unserer Nichteuklidischen Geometrie in Punkte über, die Graden in Graden, und der Winkel zwischen zwei Graden ändert sich nicht.*

Um nun die Kongruenz zu definieren, haben wir nur noch den Begriff der *Länge* nötig. *Wir wollen unter der nichteuklidischen Länge eines die Punkte $x_1 y_1$ und $x_2 y_2$ verbindenden Kurvenstückes das Integral verstehen: $\int_{x_1 y_1}^{x_2 y_2} \frac{|dz|}{y}$ hinerstreckt an der Kurve.* Das Integral hat einen Sinn, da y niemals null wird. Es lautet, ausführlicher geschrieben:

$$\int_{t_0}^{t_1} \sqrt{\left(\frac{dx}{dt}\right)^2 + \left(\frac{dy}{dt}\right)^2}\,\frac{dt}{y}$$

wenn $x = x(t)$, $y = y(t)$ die Integrationskurve darstellen.

Wählen wir als Integrationskurve die durch $x_1 y_1$ und $x_2 y_2$ hindurchlau-
88 fende nichteuklidische Grade, so nennen | wir das obige Integral die *nichteuklidische Entfernung* der beiden Punkte.

Wir beweisen nun eine Reihe von Sätzen, die obiges Integral betreffen. Vor allem gilt:

Die *nichteuklidische Länge eines Kurvenstückes bleibt bei linearer Transformation ungeändert.* (Dasselbe gilt von dem Integral $I = \iint \frac{dx dy}{y^2}$, dem „nichteuklidischen Flächeninhalt".)

Es ist nämlich $\frac{dz'}{dz} = \frac{1}{(\gamma z+\delta)^2}$, also $|dz'| = \frac{|dz|}{|\gamma z+\delta|^2}$ und $y' = \frac{y}{|\gamma z+\delta|^2}$, also $\frac{|dz'|}{y'} = \frac{|dz|}{y}$, wie zu beweisen war.

Weiter gilt: Die *nichteuklidische Entfernung eines Punktes von sich selbst ist null*; denn es fallen dann obere und untere Grenze des Integrals zusammen.

[78] And p. 84.

Ist P ein fester Punkt, und lässt man den Punkt Z auf einem durch P gehenden Orthogonalhalbkreis nach der einen oder andern Seite von P weg bis zur x-Achse laufen, so wächst die nichteuklidische Entfernung PZ beständig, und zwar über alle Grenzen.

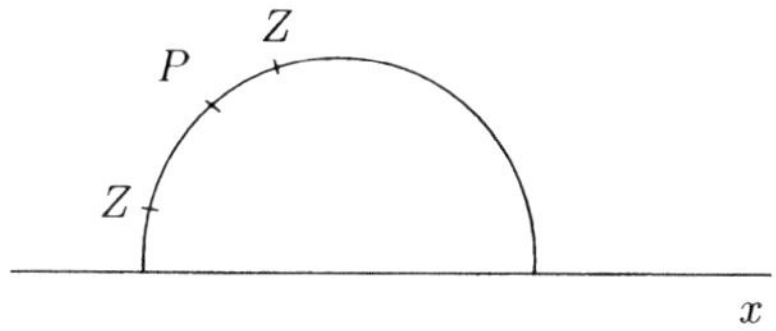

Dass PZ beständig wächst, folgt daraus, daß der Integrand $\frac{|dz|}{y}$ stets positiv ist. Aber PZ wächst auch über alle Grenzen; denn es ist:

$$PZ = \int_P^Z \frac{|dz|}{y} = \int_P^Z \frac{\left|\sqrt{1+\left(\frac{dy}{dx}\right)^2}\right|}{y}|dx| > \int_P^Z \frac{|dy|}{y}.$$ 89

Nun ist von einem gewissen Punkt an stets $|dy| = -dy$, folglich ist $PZ > \log\frac{1}{y} - \text{const.}$, wo y die Ordinate im Punkte Z ist. Da nun $\log\frac{1}{y}$ für hinreichend kleine y beliebig gross wird, so ist unser Satz bewiesen.

Nun sind wir imstande, *die Kongruenz von Strecken und Winkeln* zu definieren:

Zwei Winkel unserer Geometrie sollen kongruent heissen, wenn sie im Euklidischen Sinne gleich sind.

Zwei Strecken Z_1Z_2 und $Z_1'Z_2'$ sollen kongruent heissen, sobald die Nichteuklidischen Entfernungen Z_1Z_2 und $Z_1'Z_2'$ gleich sind.

Damit haben wir erreicht, daß in unserer Geometrie die ersten neun Axiome der III. Gruppe gelten.

Wir wollen nun *beweisen, dass auf Grund unserer Festsetzungen auch das 10. Axiom, nämlich der erste Kongruenzsatz gilt*: Zwei Dreiecke sind kongruent, wenn sie übereinstimmen in zwei Seiten und dem eingeschlossenen Winkel.

Der Einfachheit halber wählen wir zunächst das eine Dreieck $A'B'C'$ so, daß eine Seite, etwa $A'C'$ auch im Euklidischen Sinne grade ist.

Sei nun $AC = A'C'$, $AB = A'B'$, $\sphericalangle BAC = \sphericalangle B'A'C'$ (alle Gleichheiten nichteuklidisch aufgefasst) dann ist zu | zeigen, daß $\sphericalangle B = \sphericalangle B'$, $\sphericalangle C = \sphericalangle C'$, 90
$BC = B'C'$ ist.

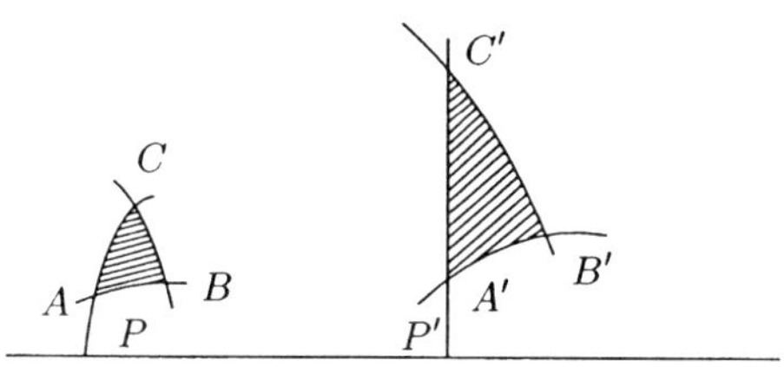

Die Grade AC, über A hinaus verlängert, treffe die x-Achse in P; $A'C'$, über A' hinaus verlängert, treffe die x-Achse in P'. Den Punkten A, A', P, P' mögen die Zahlen $c = a + ib$, $c' = a' + ib'$, e, e' entsprechen.

Wir beweisen nun, daß es *eine lineare Transformation der Halbebene in sich giebt, die P in P', A in A' überführt.* Seien α, β, γ, δ die (reellen)

Koeffizienten dieser Transformation, so müssen die Gleichungen gelten:

$$c'(\gamma c + \delta) = \alpha c + \beta$$
$$e'(\gamma e + \delta) = \alpha e + \beta.$$

Das sind thatsächlich drei von einander unabhängige lineare homogene Gleichungen für die 4 Unbekannten α, β, γ, δ. Demnach lassen sich α, β, γ, δ (bis auf einen konstanten Faktor) hieraus bestimmen, derart, dass diese Grössen nicht sämtlich null sind.

Nun kann weder $\gamma c + \delta$ noch $\gamma e + \delta$ null sein. Denn aus $\gamma c + \delta = 0$ folgt, wenn $\gamma \neq 0$ ist: $c = -\frac{\delta}{\gamma}$, was unmöglich ist; denn c ist nicht reell; ist aber $\gamma = 0$, so folgt $\delta = 0$, also auch $\alpha c + \beta = 0$, also entweder $c = -\frac{\beta}{\alpha}$, was nicht möglich ist, oder $\alpha = \beta = 0$, was auch nicht möglich ist. — Aus $\gamma e + \delta = 0$
91 folgte: | $\alpha e + \beta = 0$, also $\alpha\delta - \beta\gamma = 0$. Dann wäre aber

$$\frac{\alpha c + \beta}{\gamma c + \delta} = \frac{(\alpha c + \beta)(\gamma \bar{c} + \delta)}{|\gamma c + \delta|^2},$$

was unmöglich ist, da $\frac{\alpha c+\beta}{\gamma c+\delta} = c'$ komplex ist.

Also ist $\gamma c + \delta \neq 0$, $\gamma e + \delta \neq 0$, d. h. wir dürfen schreiben: $c' = \frac{\alpha c+\beta}{\gamma c+\delta}$, $e' = \frac{\alpha e+\beta}{\gamma e+\delta}$, womit die Existenz der verlangten Substitution bewiesen ist:

$$S = \left(z : \frac{\alpha z + \beta}{\gamma z + \delta}\right), \quad \alpha, \beta, \gamma, \delta \text{ reell.}$$

Thatsächlich ist $\alpha\delta - \beta\gamma > 0$, da C und C' beide in derselben (oberen) Halbebene liegen, und wir können, indem wir Zähler und Nenner durch $\sqrt{\alpha\delta - \beta\gamma}$ dividieren, es erreichen, daß $\alpha\delta - \beta\gamma = +1$ wird.

Die Eigenschaften der linearen Transformation und das Verhalten der nichteuklidischen Entfernung lassen nun ohne weiteres erkennen, dass bei unserer Transformation S das $\triangle ABC$ genau in das $\triangle A'B'C'$ übergeht, woraus dann die Kongruenz beider Dreiecke folgt.

Endlich können wir uns von der Annahme, daß $A'C'$ eine wirkliche Grade im gewöhnlichen Sinne sei, frei machen. Sind nämlich zwei beliebige Dreiecke I, II vorgelegt, so haben wir nur ein drittes III mit einer euklidisch graden Seite dazu zu nehmen, so dass $I \cong III$, $II \cong III$ ist; dann folgt: $I \cong II$.

Demnach haben wir das *Schlussresultat*: *In der von uns konstruierten Geometrie gelten sämtliche Axiome der Gruppen I–III.*

92 *Das Parallelenaxiom gilt aber in dieser Geometrie keineswegs*; vielmehr lassen sich durch einen Punkt A stets unendlich viele Graden ziehen, die eine gegebene Grade a nicht schneiden, wovon uns der Anblick der Figur überzeugt.

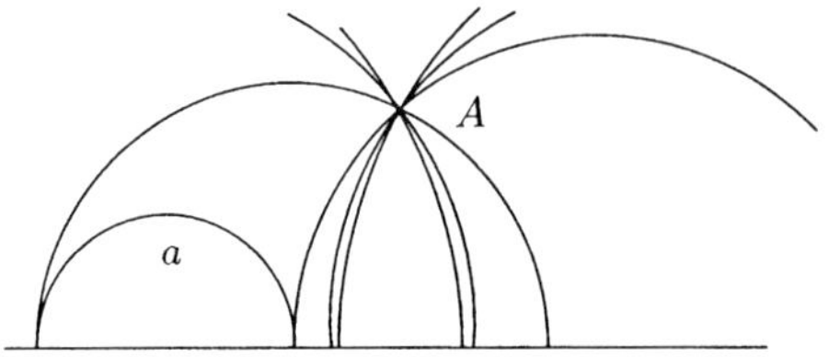

Damit ist der Beweis der Unabhängigkeit des Parallelenaxioms von den *ebenen* Axiomen I–III geführt.

Wir wollen jetzt zeigen, dass in der oben definierten Geometrie die sämtlichen 5 Folgerungen, die wir seiner Zeit aus dem Parallenaxiom ableiteten, ungültig sind, daß also diese 5 Sätze nicht schon aus den Axiomen I–III folgen.

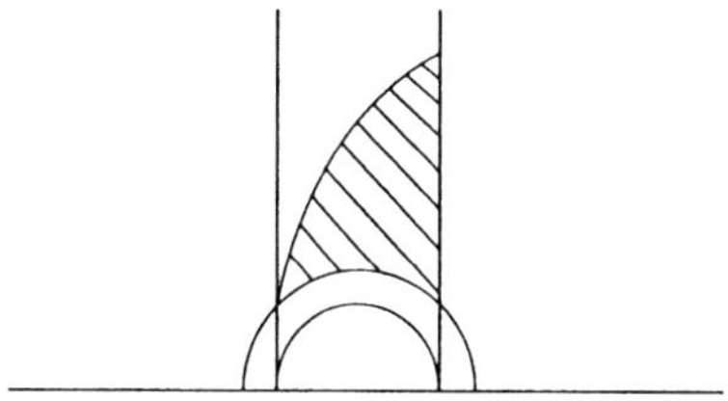

1. *Es giebt Dreiecke, deren Winkelsumme kleiner ist als jede vorgegebene Grösse.* [Cf. das schraffierte Dreieck in der Figur.] Die genauere Untersuchung liefert für die Winkelsumme eines Dreiecks:

$$\alpha + \beta + \gamma = \pi - \frac{I}{4}.$$

Hier ist I, der nichteuklidische Inhalt [cf. p. 88] stets positiv, sodaß wir sogar sagen dürfen: *Es giebt in unserer Geometrie kein Dreieck, dessen Winkelsumme w zwei Rechte beträgt, stets ist $w < \pi$.*

2. Gegenwinkel bei Parallelen sind durchaus nicht stets gleich, wie die Figur 93
der vorigen Seite erkennen lässt.[79]

Und ebenso zeigt sie:

3. Zwei Grade können einer dritten parallel sein, ohne untereinander parallel zu sein.[80]

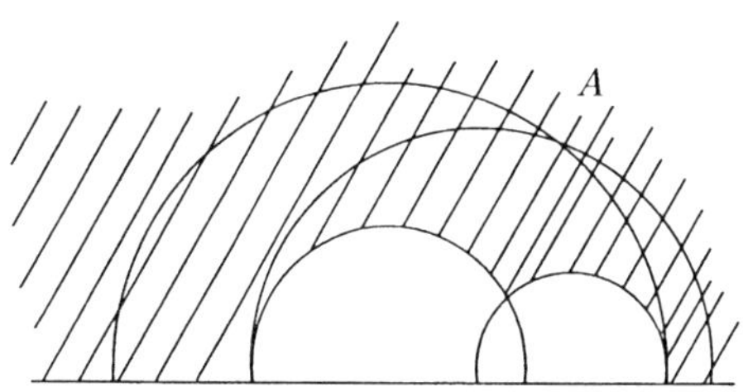

4. Man kann in jedem Winkel, der kleiner ist als ein gestreckter, Punkte A angeben derart, dass keine durch A gehende Grade beide Schenkel des Winkels schneidet. In der Figur ist der Winkelraum schraffiert, A ist offenbar ein Punkt von der gewünschten Beschaffenheit; eine durch A gehende Grade schneidet entweder keinen oder einen Schenkel des Winkels, aber niemals beide.

5. Es giebt in unserer Geometrie kein Rechteck, weil es keine Dreiecke mit der Winkelsumme $2R$ giebt.

Es ist jetzt sehr leicht, auch eine *räumliche Geometrie ohne das IV. Axiom* zu konstruieren. Wir nehmen als Punkte alle Punkte eines Euklidischen Halbraums, der von der xy-Ebene begrenzt sein mag, die Punkte dieser Grenzebene selbst ausgenommen. Die nichteuklidischen Ebenen seien die zur xy-Ebene orthogonalen Halbkugeln, die Graden seien die orthogonalen Halbkreise. Dann gelten zunächst alle Axiome I, II. Um auch die Axiome III, die sich ja nur auf die Ebene beziehen, gültig zu machen, haben | wir nur die Kongruenz geeig- 94

[79]The figure is presumably the second on p. 92. In the upper half-plane, draw a perpendicular to the x-axis (*Grenzgerade*) which cuts both circles and is tangent to neither. In general, such a line will cut the two circles at different angles.

[80]The figure below ilustrates this. The two intersecting and unshaded semi-circles are not parallel to each other, since they intersect in the upper half-plane, but both are parallel to both the other semi-circles drawn, since semi-circles which stand perpendicular to the *Grenzgerade* and are tangent there are parallel lines in the model. Hilbert develops a more precise treatment of parallels in non-Euclidean geometry in *Hilbert 1903b*.

net zu definieren. Als Winkel nehmen wir die euklidischen Winkel, als nichteuklidische Länge bezeichnen wir das Integral $\int\limits_{t_0}^{t_1} \sqrt{\left(\frac{dx}{dt}\right)^2 + \left(\frac{dy}{dt}\right)^2 + \left(\frac{dz}{dt}\right)^2} \cdot \frac{dt}{z}$. Dann gelten auch alle Axiome III; IV gilt natürlich nicht.

Und damit ist die Unabhängigkeit des Axioms IV von sämtlichen Axiomen I–III vollständig bewiesen.

Dass in unserer Geometrie der Satz von Desargues gilt, haben wir schon früher gezeigt, aber auch der Pascal gilt, da er ja eine Folge der Axiome I–III ist. _[81]

Wir geben noch einen *zweiten Beweis der Unabhängigkeit des Parallelenaxioms*, indem wir noch *eine andere nichteuklidische Geometrie* angeben, die in mancher Hinsicht einfacher ist als die oben besprochene.

Wir betrachten *in der xy-Ebene den Kreis* $x^2 + y^2 - 1 = 0$. Die *Punkte im inneren dieses Kreises* sollen *die Punkte unserer nichteuklidischen Geometrie* sein, dagegen sollen die Punkte des Kreises selbst nicht dazu gehören. *Die Graden der Nichteuklidischen Geometrie sollen die gewöhnlichen Euklidischen Graden sein, aber nur soweit sie im innern des Kreises* $x^2 + y^2 - 1 = 0$ *verlaufen.* In der so definierten Geometrie gelten die Axiome I und II.

Um die *Kongruenz* zu definieren, betrachten wir diejenigen *Kollineationen der Ebene, bei welchen unser Einheitskreis in sich übergeht.* Eine beliebige
95 Kollineation der Ebene ist ge|geben durch die Gleichungen:

$$x' = \frac{a'x + b'y + c'}{ax + by + c}, \quad y' = \frac{a''x + b''y + c''}{ax + by + c},$$

wo die $a, \ldots, c''$ reelle Konstanten sind. Die Gesamtheit dieser Operationen bildet eine (9–parametrige) *Gruppe*, d. h. wenn man zwei Kollineationen zusammensetzt, so entsteht wieder eine Kollineation.

In dieser Gruppe ist als Untergruppe enthalten die Gesammtheit derjenigen Kollineationen, die den Einheitskreis in sich überführen. Eine solche Kollineation muss so beschaffen sein, daß identisch giebt:

$$x'^2 + y'^2 - 1 = x^2 + y^2 - 1$$

(Genauer: $x'^2 + y'^2 - 1 = \mathrm{const}(x^2 + y^2 - 1)$; doch darf const. $= 1$ genommen werden, weil die $a, \ldots, c''$ nur bis auf einen Faktor bestimmt sind.) Aus dieser Bedingung ergeben sich *6 Relationen für die 9 Koeffizienten:*

$$\begin{array}{ll} a'^2 + a''^2 - a^2 = 1 & a'b' + a''b'' - ab = 0 \\ b'^2 + b''^2 - b^2 = 1 & b'c' + b''c'' - bc = 0 \\ c'^2 + c''^2 - c^2 = -1 & c'a' + c''a'' - ca = 0. \_ \end{array}$$

Diese Kollineationen sind vor allen Dingen umkehrbar, denn ihre Determinante ist $= \pm 1$; es ist nämlich:

$$\begin{vmatrix} a' & a'' & a \\ b' & b'' & b \\ c' & c'' & c \end{vmatrix}^2 = - \begin{vmatrix} a' & a'' & -a \\ b' & b'' & -b \\ c' & c'' & -c \end{vmatrix} \cdot \begin{vmatrix} a' & a'' & a \\ b' & b'' & b \\ c' & c'' & c \end{vmatrix} = - \begin{vmatrix} 1 & 0 & 0 \\ 0 & 1 & 0 \\ 0 & 0 & -1 \end{vmatrix} = 1.$$

Der Mittelpunkt M des Einheitskreises geht stets in einen ganz bestimmten
96 *Punkt im innern des Kreises über.* Denn für | $x = 0$, $y = 0$ wird $x' = \frac{c'}{c}$, $y' = \frac{c''}{c}$, und es ist $x'^2 + y'^2 = \frac{c'^2+c''^2}{c^2} = 1 - \frac{1}{c^2} < 1$. —

Umgekehrt kann man stets Kollineationen, bei denen der Einheitskreis in sich übergeht, so angeben, daß ein gegebener Punkt im innern des Kreises in den Mittelpunkt übergeht, wie man leicht sieht.

Endlich gilt der Satz: *Jede Kollineation, die den Einheitskreis in sich überführt und zugleich den Mittelpunkt festlässt, ist eine Drehung der Ebene um M.*

In diesem Fall ist nämlich $\frac{c'}{c} = 0$, $\frac{c''}{c} = 0$, folglich $c' = 0$, $c'' = 0$. Dann ergiebt sich mit Hülfe der 6 Koeffizientenrelationen: $c = \pm 1$, $a = 0$, $b = 0$, sodass die genannten Kollineationen die Form annehmen:

$$\begin{array}{l} x' = a'x + b'y \\ y' = a''x + b''y \end{array} \qquad \left(\begin{array}{l} a'^2 + a''^2 = 1 \\ b'^2 + b''^2 = 1 \end{array} \quad a'b' + a''b'' = 0 \right)$$

und das sind die Formeln für die Drehung. —

Nun definieren wir zunächst den *Winkel zwischen zwei Graden, die sich in einem Punkte P schneiden*, und zwar folgendermassen: Wir suchen eine unserer Kollineationen, die P in M transformiert. Dabei gehen die Graden durch P in zwei Durchmesser des Kreises über. Diese Collineation ist, wie wir oben bewiesen haben, bestimmt bis auf eine Drehung um M. Daraus folgt aber, daß der *Winkel zwischen den | beiden Durchmessern völlig bestimmt* 97
ist, wenn P und die durch P gehenden beiden Graden gegeben sind. Diesen Winkel zwischen den beiden Durchmessern will ich als nichteuklidischen Winkel der beiden Graden bei P bezeichnen. — Um die *Länge einer Strecke AB* zu definieren, verfahre ich so: Ich verlängere AB nach beiden Seiten bis zum Schnitt mit dem Einheitskreis in S und T. Unter den 6 Werten, welche das *Doppelverhältnis der 4 Punkte A, B, S, T* besitzt, wähle ich denjenigen aus, der *positiv und* >1 *ist*, und bezeichne ihn mit δ. (In der Figur wäre $\delta = \frac{AS}{AT} : \frac{BS}{BT}$.)[82] Dann nenne ich $\log\delta$ *die Länge der Strecke AB* (siehe folgende Seite). Näher können wir hier auf diese Massbestimmung (Cayley) nicht eingehen, wir begnügen uns mit einem Hinweis auf: *Cayley*, Mathematical papers, II, p. 561ff. *Klein*, Vorlesung über Nicht-Euklidische Geometrie I. p. 65ff. *Fricke-Klein*, Automorphe Funktionen, I. p. 6, 7. —[83] *Auf Grund der Definitionen von Winkel und Strecke gelten*

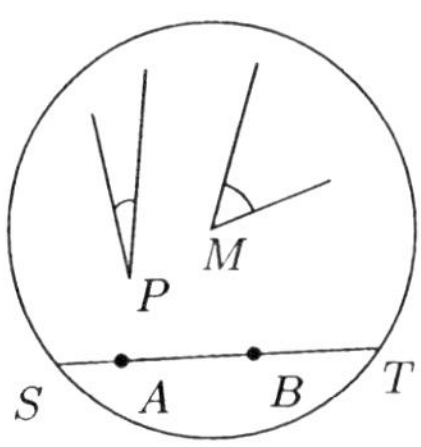

[81]See pp. 31 and 73–74.

[82]AS, AT, etc. stand for the Euclidean distances between the various points. The figure meant is the one on this page.

[83]The works referred to are: (i) *Cayley 1889*, 561ff, which is the first page of the reprinting of Cayley's 'A sixth memoir on quantics' (561–606), i.e., *Cayley 1859*; (ii) *Klein 1892*, 65ff; and (iii) *Fricke and Klein 1897*, 6–7. The work Hilbert refers to derives from Cayley's memoir, but the modification using the logarithm of the anharmonic ratio is due to

nun die Axiome III, 1–9. Wieder ist die *Aufgabe, das 10. Axiom, den ersten Kongruenzsatz, zu beweisen.*

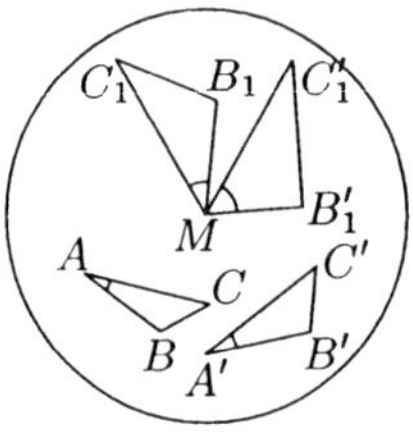

Wir setzen voraus, es sei $AB = A'B'$, $AC = A'C'$, $\sphericalangle A = \sphericalangle A'$.[84] Nun transformieren wir so, daß einmal A, dann A'
98 nach M fällt, und erhalten so die | beiden
Dreiecke MB_1C_1 und $MB_1'C_1'$. Diese beiden Dreiecke sind, wie man sofort sieht, auch im gewöhnlichen euklidischen Sinne kongruent, also ist, wenn wir den 1. Kongruenzsatz aus der Euklidischen Geometrie anwenden $B_1C_1 = B_1'C_1'$, $\sphericalangle B_1 = \sphericalangle B_1'$, $\sphericalangle C_1 = \sphericalangle C_2'$; Diese Gleichheiten gelten aber auch im nichteuklidischen Sinne; und hieraus folgt, wenn man die Invarianz der Winkel und Strecken bei unsern Kollineationen berücksichtigt, daß $BC = B'C'$, $\sphericalangle B = \sphericalangle B'$, $\sphericalangle C = \sphericalangle C'$ ist. — q. e. d.[85]

Wir haben nunmehr eine ebene Geometrie mit allen Axiomen I–III. Dass

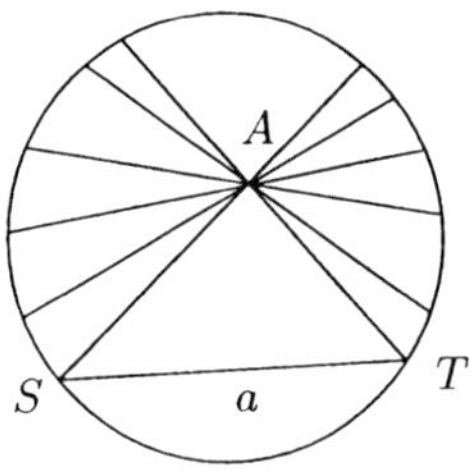

das Parallelenaxiom nicht gilt, lehrt ein Blick auf die Figur: Durch jeden Punkt A giebt es ein ganzes Büschel von Strahlen, die eine gegebene Grade a nicht schneiden. Sämtliche Strahlen des Büschels liegen zwischen zwei bestimmten Strahlen AS und AT, den „Grenzstrahlen", die ebenfalls a nicht schneiden, denn die Punkte S und T gehören unserer Geometrie nicht an.

Auch diese Geometrie kann leicht auf den Raum ausgedehnt werden; man wird dann diejenigen Kollineationen des Raumes betrachten, welche die Kugel $x^2 + y^2 + z^2 - 1 = 0$ in sich überführen. —

Wir wollen nachträglich (siehe vorige Seite) ausdrücklich bemerken: *Bei*
99 *unsern Kollineationen bleiben Strecken und | Winkel erhalten.* Das erstere
ergiebt sich aus den bekannten Eigenschaften des Doppelverhältnisses; das zweite erkennt man folgendermassen. Es möge $\sphericalangle\alpha$ bei P durch die Kollineation S in $\sphericalangle\alpha'$ bei P' übergehen. Nun können wir zwei Kollineationen T und T'finden, welche P bezw. P' in M, α in α_1, α' in α_1' transformieren. Bezeichnen wir die noch unbekannte Operation, bei der α_1 in α_1' übergeht, mit U, so gilt die Gleichung $T'^{-1}UT = S$ (das Produkt ist von rechts nach

Klein (*Klein 1871*), which is what is reproduced in the section of the Klein lectures cited (*Klein 1892*, 65f). Hilbert has probably just used the specific pages of the Fricke-Klein book mentioned for the reference to Cayley. However, the whole *Einleitung* to this book (entitled 'Entwicklungen über projective Maassbestimmungen', pp. 3–58) is devoted to the theoretical development Hilbert makes use of here (pp. 82–99), for which previously he makes reference to his own *Hilbert 1892** (see Hilbert's own notes, p. 80).

[84] Here AB, $A'B'$ etc. stand for the non-Euclidean distances just defined.

[85] It would be better to say, both here and on the next page, 'Winkelgrössen' and 'Streckenlängen'. Clearly Hilbert feels that not enough has been said by way of explanation, for he inserts another paragraph at the beginning of p. 99.

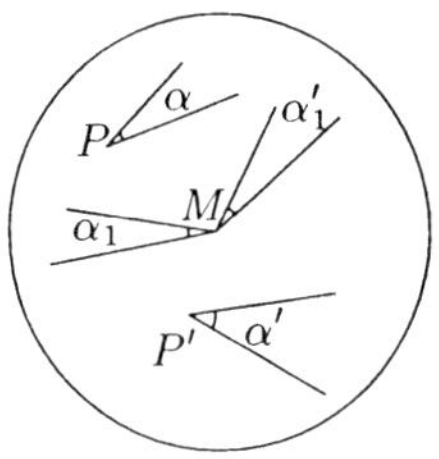

links zu lesen), aus der folgt:

$$U = T'ST^{-1}.$$

Hieraus folgt, daß U eine Kollineation ist, die den Einheitskreis in sich überführt; andererseits wissen wir, daß U den Punkt M fest lässt. Demnach ist U eine einfache Drehung um M, also ist, im euklidischen und im nichteuklidischen Sinne $\sphericalangle \alpha_1 = \sphericalangle \alpha_1'$. Vermöge der Definition des nichteuklidischen Winkels gelten im nichteuklidischen Sinne die Gleichungen: $\sphericalangle \alpha = \sphericalangle \alpha_1$, $\sphericalangle \alpha' = \sphericalangle \alpha_1'$, also folgt endlich $\sphericalangle \alpha = \sphericalangle \alpha'$, wie zu beweisen war. —

Wir ergänzen nun das Parallelenaxiom durch folgenden Satz: *Durch einen beliebigen Punkt P ausserhalb einer Ebene ε lässt sich nur eine Ebene ε' legen, welche die erste Ebene nicht schneidet. ε' heisst eine Parallelebene zu ε.*

Beweis. Ich fälle von P auf ε das Lot PF, und lege durch P | eine zu PF 100
senkrechte Ebene ε'. Dann folgt aus den Kongruenzsätzen, daß ε' und ε sich nicht schneiden. Nehmen wir nun an, es gebe eine von ε' verschiedene Ebene ε'' durch P, die ε nicht schneidet. Dann legen wir durch PF eine Ebene ζ, die nicht grade durch die Schnittgrade von ε' und ε'' geht. ζ schneide die Ebenen ε, ε', ε'' in den Graden FE, PE', PE''. Dann hätten wir in der Ebene ζ durch P zwei verschiedene Parallelen zu FE, was dem Axiom IV widerspricht. —

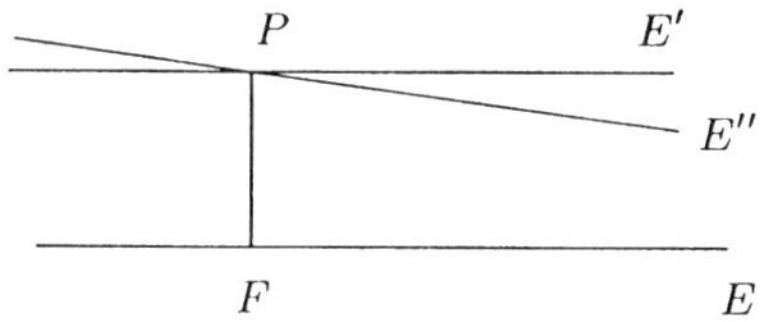

Nachdem wir das Parallelenaxiom eingeführt und seine Unabhängigkeit von den vorhergehenden Axiomen bewiesen haben, sind zwei Fragen zu erledigen:

1. *Können früher bewiesene Sätze vielleicht auch bewiesen werden wenn man das Parallelenaxiom zu Hülfe nimmt, dafür aber einige der Axiome I–III fortlässt?* Lässt sich beispielsweise der Satz von Desargues beweisen aus dem Parallelenaxiom und den ebenen Axiomen I–III?

2. *Wieweit können wir die Geometrie mit Hülfe des Parallelenaxioms aufbauen, ehe wir ein neues Axiom nötig haben?* Es wird sich zeigen, daß wir die ganze „*Schulgeometrie*" (Pythagoras, Sätze über Flächengleichheit etc.) allein mit unseren Axiomen I–IV aufbauen können; und das ist ein neues Resultat.

Zur Einleitung bemerken wir, dass wir jetzt mit Hülfe | des Axioms IV 101
imstande sind, die Axiome I zu erweitern und ihnen gewisse andere Sätze dualistisch gegenüber zu stellen. Das wesentliche hierbei ist, dass wir ideale Punkte, Graden, Ebenen einführen. (Analog verfährt man in der Mathematik häufig, z. B. wenn man in der höheren Zahlentheorie neben den „wirklichen" auch ideale Zahlen einführt.)

Bleiben wir zunächst in der *Ebene*. Den Begriff „idealer Punkt" führen wir durch folgende Festsetzung ein: *Statt zu sagen, zwei Graden a und b sind parallel wollen wir sagen, sie schneiden sich in einem idealen Punkt.* Demgegenüber nennen wir die bisher allein betrachteten Punkte *wirkliche Punkte*.

Eine solche Erweiterung des Begriffes „Punkt" ist nun aber nur dann zweckmässig, wenn die idealen Punkte wenigstens einige Eigenschaften mit den wirklichen Punkten gemeinsam haben. *Wir werden also zunächst die Axiome der I. Gruppe daraufhin untersuchen, ob sie auch für ideale Punkte gültig bleiben.* Wir wollen diese Untersuchung für die Ebene ausführlich anstellen, um uns später beim Raume kürzer zu fassen. Das Axiom I, 1 besagt, daß irgend zwei Punkte stets eine Grade bestimmmen. Sei nun einer der beiden Punkte ideal; dann verlangt das Axiom, daß durch zwei parallele Grade a, b und einen wirklichen Punkt P eine Grade g bestimmt werde.

102 Diese Grade darf offenbar weder a noch b in einem wirklichen Punkte A, B schneiden, denn sonst hätten die Graden g und a, ohne zusammenzufallen, sowohl A als auch den durch a, b bestimmten idealen Punkt gemein, und das ist im Widerspruch mit Axiom I, 2, nach welchem zwei Grade höchstens einen Punkt gemein haben können; und auch dies Axiom soll für ideale Punkte gelten.

Nach dem Parallelenaxiom giebt es aber nur eine Grade durch P, die weder a noch b schneidet, nämlich die zu a und b durch P gezogene Parallele p. Diese Parallele aber ist bereits durch die Grade a allein und den Punkt P bestimmt. Das führt nun zu folgendem Satze: *Jeder Graden ist ein einziger idealer Punkt zuzuschreiben; alle unter einander parallelen Graden laufen durch ein und denselben idealen Punkt.* Jeder ideale Punkt der Ebene ist also gewissermassen repräsentiert durch eine bestimmte *Richtung.* Wenn wir diese Festsetzungen gemacht haben, so gelten die Axiome I, 1. und I, 2. falls einer der beiden Punkte ideal ist.[86] Und sie werden auch für zwei ideale Punkte gelten, wenn wir noch festsetzen: *Die Gesamtheit aller idealen Punkte einer Ebene wollen wir die ideale Grade der Ebene nennen.* So haben wir das Resultat:

A. *Irgend zwei Punkte bestimmen stets eine und nur eine Grade.*

103 Damit haben wir die Axiome I, 1., 2. auf ideale Punkte erweitert. Der Nutzen der idealen Punkte tritt aber erst hervor, wenn wir diesem Satze folgenden zur Seite stellen:

B. *Irgend zwei Graden einer Ebene bestimmen stets einen und nur einen Punkt, nämlich ihren Schnittpunkt.*

Dieser Satz gilt offenbar nur dann, wenn wir ideale Punkte zulassen. Dass er wirklich gilt, folgt aus den Festsetzungen über ideale Punkte. Die beiden Sätze A und B unterscheiden sich nur dadurch, daß die Worte „Punkt" und „Grade" in ihnen vertauscht sind; wir sagen: *Die beiden Sätze A, B entsprechen sich (in der Ebene) in dualistischer Weise. Jedem Satz, der aus A und B gefolgert ist, entspricht also ein anderer, der ihm dualistisch gegenübersteht.* —

Nun wollen wir analoge Überlegungen für den Raum anstellen; dann gelangen wir zu folgenden Resultaten: *Jede Grade im Raum hat einen einzigen idealen Punkt; parallele Grade laufen durch denselben idealen Punkt. Jede Ebene besitzt eine einzige ideale Grade; parallele Ebenen haben dieselbe ideale Grade. Die Gesamtheit aller idealen Punkte (oder Graden) nennen wir die ideale Ebene des Raumes.* — Alle Axiome der Gruppe I gelten dann auch für

[86] In case at most one of the two points is ideal.

ideale Elemente; wir erhalten auch im Raume eine *Dualität derart, dass jedes Axiom und jeder aus ihnen folgende Satz richtig bleibt, wenn man die Worte „Punkt" und „Ebene" vertauscht, während die Grade | sich selbst dualistisch* 104 *entspricht.* Beispielsweise wollen wir einige solche Sätze mit ihren dualen Gegenstücken anführen.

1.) Zwei Punkte bestimmen stets eine einzige Grade.	1.) Zwei Ebenen schneiden sich stets in einer einzigen Graden.
2.) Drei nicht auf einer Graden liegenden Punkte bestimmen stets eine einzige Ebene.	2.) Drei Ebenen, die nicht durch eine Grade gehen, schneiden sich in einem einzigen Punkt.
3.) Durch eine Grade und einen nicht auf ihr liegenden Punkt ist eine einzige Ebene bestimmt.	3.) Eine Grade und eine nicht durch sie hindurchgehende Ebene schneiden sich in einem einzigen Punkt.
4.) Zwei durch einen Punkt gehende Graden bestimmen eine einzige Ebene.	4.) Zwei in einer Ebene liegende Grade schneiden sich in einem einzigen Punkt.

Links stehen Axiome bezw. Sätze, die wir von früher her kennen, die rechts stehenden dualistisch entsprechenden Sätze auszusprechen, erlaubt uns erst das Parallelenaxiom. *Wir erkennen, welche wichtige Rolle dies Axiom für den symmetrischen Aufbau der Geometrie spielt.*

Noch *einige Bemerkungen* zum Begriff der idealen Elemente. Man findet häufig die Namen: *„unendlich ferner Punkt"* etc. Dieser Name erscheint uns aber *wenig passend*, weil er störende Nebenvorstellungen erweckt: — *Auch in der Gaussischen Geometrie*[87] *kann man die idealen Punkte einführen*; | doch 105 gestalten sich dann die Verhältnisse *erheblich komplicierter*, wie man allein daraus ersehen kann, dass dort jeder Graden unendlich viele ideale Punkte zukommen. —

Endlich weisen wir darauf hin, *daß nun nicht etwa auch die Axiome II oder gar III für ideale Elemente gelten, wie übrigens leicht erkannt wird.* —

Wenden wir uns nun zu erneuter Betrachtung der *Schnittpunktsätze*. Wir behaupten:

Ist ein Schnittpunktsatz bewiesen, für den Fall dass alle in ihm vorkommenden Punkte, Graden, Ebenen wirklich sind, so gilt er auch, wenn ideale Elemente in ihm vorkommen.

Der *Beweis* wird mit Hülfe einer *Projektion im Raume* geführt. Wir nehmen außerhalb der Zeichenebene einen Punkt P an und verbinden P mit allen in Betracht kommenden (wirklichen und idealen) Punkten der Figur in der Zeichenebene. Wir erhalten so eine endliche Anzahl von Graden durch P. Nun können wir, wie leicht zu zeigen, eine zweite Ebene so annehmen, dass sie von sämtlichen durch P laufenden Strahlen in wirklichen Punkten geschnitten wird. Hierdurch wird die Figur der Zeichenebene in der Weise projiziert, daß das Schneiden von Graden erhalten bleibt. Nach Voraussetzung gilt der

[87] It is not precisely clear what Hilbert means by 'Gaussian Geometrie'.

betreffende Schnittpunktsatz in der zweiten Ebene, also gilt er auch in der
106 | Zeichenebene. –

Wir wollen nun, indem wir den eben bewiesenen Satz anwenden, die beiden früher besprochenen Schnittpunktsätze, den von Desargues und den von Pascal, durch Einführung idealer Elemente specialisieren, wodurch wir zu Sätzen gelangen, die weiterhin von grosser Bedeutung sind.[88]

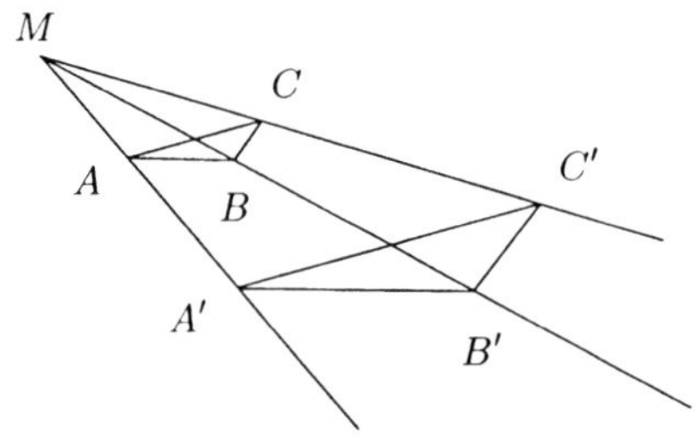

Beim Satz von Desargues wollen wir annehmen, daß die Grade, auf welcher die drei Schnittpunkte entsprechender Seiten liegen, zur idealen Graden wird, während der Schnittpunkt der Verbindungslinien entsprechender Ecken wirklich bleiben soll. Dann können wir den *Satz von Desargues* so fassen: *Wenn sich die drei Graden AA', BB', CC' in einem Punkte M schneiden, und wenn $AB \parallel A'B'$, $BC \parallel B'C'$ ist, so ist auch $AC \parallel A'C'$.*

Ebenso wollen wir beim Pascalsatz die sogenannte Pascalgrade zur idealen Graden machen. Dabei geht die folgende Figur:

107

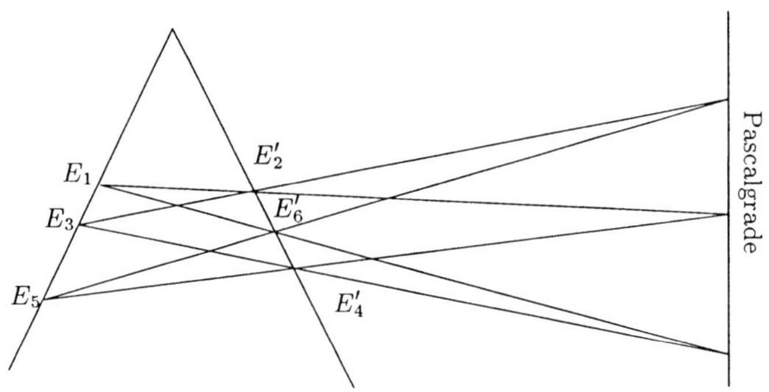

über in diese:

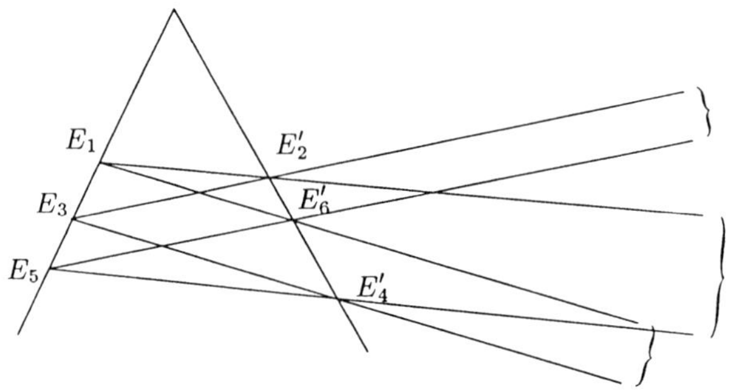

[88]These are the versions of Desargues's and Pascal's Theorems which Hilbert gives in the *Festschrift* (*Hilbert 1899c*), the latter on p. 28, the former on pp. 49–50.

und der Pascalsatz erhält folgende Form: *Liegen die sechs Punkte* E_1, E_2', E_3, E_4', E_5, E_6' *abwechselnd auf zwei sich in einem (wirklichen) Punkte schneidenden Graden, und ist* $E_1E_2' \parallel E_4'E_5$, $E_2'E_3 \parallel E_5E_6'$, *so ist auch* $E_3E_4' \parallel E_6'E_1$.

Die fundamentale Bedeutung dieses Satzes wird sich bald herausstellen. –

Wir beantworten jetzt die erste der pag. 100 aufgeworfenen Fragen, indem wir den Satz beweisen: *Die Sätze von Pascal | und Desargues lassen sich* 108
ohne Zuhilfenahme des Raumes beweisen, sobald man nur das Parallelenaxiom hinzunimmt, d. h. also, auf Grund der ebenen Axiome I–IV. Dieser Satz ist bisher noch niemals bewiesen worden.

Wir beginnen mit der Bemerkung, dass wir *mit Hülfe der ebenen Axiome I–IV die gewöhnlichen Sätze über den Kreis* beweisen können, vor allen den Satz von den *Peripheriewinkeln* über gleichen Bögen, sowie den Satz vom *Sehnenwinkel*; [die ohne das Axiom IV nicht galten.] Hieraus ergiebt sich u. a. der Satz, den wir bald brauchen werden: *Ist in einem Viereck A, B, C, D die Summe zweier gegenüberliegender Winkel 2 Rechte, so liegen die vier Punkte A, B, C, D auf einem Kreise.* –

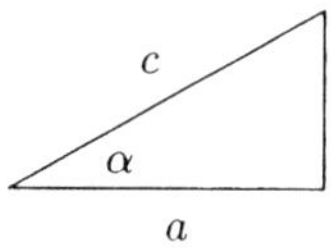

Wir beweisen nun zuerst einen Hülfssatz, den wir mit Hülfe einer sogleich einzuführenden *Bezeichnung* formulieren wollen. *Sei in einem rechtwinkligen Dreieck die Hypotenuse c, eine Kathete a, der Winkel zwischen a und c sei* α, dann wollen wir schreiben: $a = \alpha c$. In der That ist a durch α und c vollständig bestimmt.

Unser *Hülfssatz* lautet dann einfach:

Es ist stets: $\alpha\beta c = \beta\alpha c$.

Zum *Beweise* machen wir $AB = c$ und tragen in A | an AB zu beiden Seiten 109
die Winkel α und β ab. Dann fällen wir von B auf die freien Schenkel dieser Winkel die Lote BC und BD, verbinden C mit D und fällen schliesslich das Lot AE auf CD.

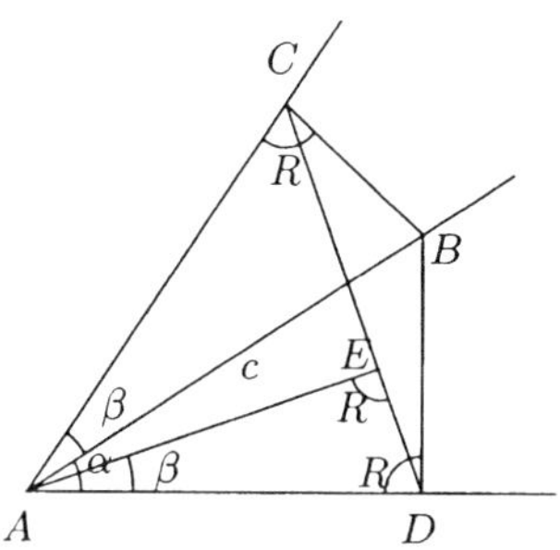

Nun zeigen wir zuerst, dass $\sphericalangle EAD = \beta$ ist.

Wir haben nämlich:

$$\sphericalangle EAD + \sphericalangle ADE = R$$
$$\sphericalangle\beta + \sphericalangle ABC = R;$$

nun liegen, wegen $\sphericalangle ACB + \sphericalangle ADB = 2R$, die 4 Punkte A, C, B, D auf einem Kreise,[89] demnach sind $\sphericalangle ADE$ und $\sphericalangle ABC$ Peripheriewinkel über demselben Bogen AC, also ist $\sphericalangle ADE = \sphericalangle ABC$, also ist auch $\sphericalangle EAD = \sphericalangle\beta$, wie wir beweisen wollten. Es folgt daraus: $\sphericalangle CAE = \sphericalangle\alpha$. Nun ist, wie man sofort sieht:

$$\beta c = AC, \quad \alpha\beta c = \alpha AC = AE,$$
$$\alpha c = AD, \quad \beta\alpha c = \beta AD = AE,$$

[89] It would be simpler to say that, because $\sphericalangle ACB$ and $\sphericalangle ADB$ are right-angles, A, B, C, D must lie on a circle.

womit bewiesen ist: $\alpha\beta c = \beta\alpha c$. —

Mit Hülfe dieses Satzes wollen wir nun den *Pascal* in der Ebene beweisen, und zwar *zunächst für den Spezialfall, daß die Pascalgrade die ideale Grade ist, und daß die Graden* $E_1E_3E_5$ *und* $E_2'E_4'E_6'$ *sich unter einem rechten Winkel schneiden*(siehe die Anmerkung).

110

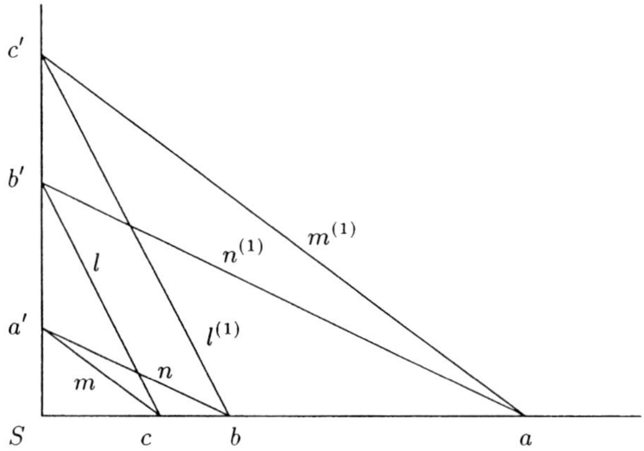

Der Pascal'sche Satz kann für diesen Fall so ausgesprochen werden. *Auf den Schenkeln eines rechten Winkels seien je drei Strecken abgetragen von den Längen* a, b, c, a', b', c'; *durch geeignete Verbindung der 6 Endpunkte entstehen 6 rechtwinklige Dreiecke mit den Hypotenusen* l, $l^{(1)}$, m, $m^{(1)}$, n, $n^{(1)}$. *Wenn dann* $l \parallel l^{(1)}$, $m \parallel m^{(1)}$ *ist, so ist auch* $n \parallel n^{(1)}$.

Zum Beweise nehmen wir an, daß l; m; n mit den beiden Schenkeln die Winkel λ, λ'; μ, μ'; ν, ν' bilden.

Dann bilden nach Voraussetzung auch $l^{(1)}$; $m^{(1)}$ mit den Schenkeln die Winkel λ, λ'; μ, μ'. Zu beweisen ist, dass $n^{(1)}$ mit den Schenkeln die Winkel ν, ν' bildet.

Denken wir uns von S aus auf jede der sechs Hypotenusen ein Lot gefällt, so können wir die Länge jedes dieser Lote mit Hülfe der Katheten und spitzen Winkel des betreffenden Dreiecks auf doppelte Weise ausdrücken, z. B. das
111 auf l gefällte Lot ist $= \lambda b' = \lambda' c$. So erhalten wir sechs | Gleichungen und der Pascal lässt sich nunmehr so aussprechen: *Von den sechs Gleichungen:*

$$1)\ \lambda b' = \lambda' c \qquad 3)\ \mu c' = \mu' a \qquad 5)\ \nu a' = \nu' b$$
$$2)\ \lambda c' = \lambda' b \qquad 4)\ \mu a' = \mu' c \qquad 6)\ \nu b' = \nu' a$$

ist jede eine Folge der fünf übrigen. Wir wollen beispielsweise zeigen, daß 6) aus 1)–5) folgt. Wenn wir zuerst 5) mit $\mu\lambda'$ „multiplizieren", so kommt wegen der Kommutativität der Operationen $\lambda, \mu, \ldots, \nu'$:

$$\nu\lambda'\mu a' = \nu'\mu\lambda' b,$$

oder, wenn wir links 4), rechts 2) anwenden:

$$\nu\lambda'\mu' c = \nu'\mu\lambda c'$$

oder wegen der Kommutativität:

$$\nu\mu'\lambda' c = \nu'\lambda\mu c';$$

nun wenden wir links 1), rechts 3) an:

$$\nu\mu'\lambda b' = \nu'\lambda\mu' a,$$

oder:

$$\lambda\mu'\nu b' = \lambda\mu'\nu' a.$$

Hieraus folgt aber offenbar:

$$\nu b' = \nu' a, \quad \text{q. e. d.} \_$$

Anmerkung: Dieser Beweis überträgt sich übrigens sofort auf den Fall, daß die Graden $E_1E_3E_5$ und $E_2'E_4'E_6'$ bei S einen *beliebigen* Winkel einschliessen. Nur muss man dann nicht die oben eingeführten Winkel $\lambda, \ldots, \nu'$, sondern die Winkel benutzen, welche die von S gefällten Lote bei S mit den Schenkeln Sa und Sa' bilden. Ist $\sphericalangle S = R$, so kommt beides auf eins hinaus.

Die fundamentale Bedeutung des soeben bewiesenen Satzes liegt darin, daß er uns in den Stand setzt, die | Lehre von den Proportionen ohne irgend ein 112 *neues Axiom zu begründen.* Wir sehen also, dass auch hier *Euklid* schliesslich Recht behält: auch er führt die Lehre von den Proportionen ohne neues Axiom ein. Allerdings müssen wir hinzufügen: *Die Art dieser Einführung bei Euklid ist gänzlich verfehlt.*

Euklid basiert nämlich die Lehre von den Proportionen auf folgende zwei Sätze:

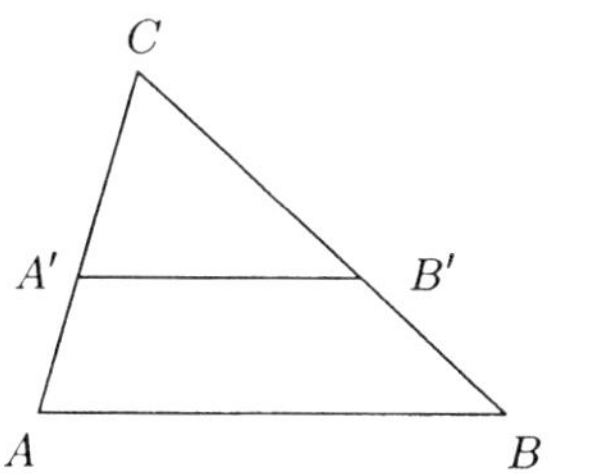

1) *Wenn in einem Dreiecke ABC zu AB die Parallele $A'B'$ gezogen wird, so ist $AC : BC = A'C : B'C$.*

2) *die Umkehrung: Wenn in einem Dreieck $AC : BC = A'C : B'C$ ist, so ist $AB \parallel A'B'$.*

Die Beweise dieser Sätze bei Euklid sind in dem Falle durchaus streng, wo AC und BC beide durch wiederholtes Abtragen einer und derselben Strecke entstanden sind. Nun aber beruft sich Euklid auf allgemeine Grössenbeziehungen, indem er die obige Proportion als eine *Zahlen*gleichung auffasst, und schliesst so, dass der Satz bei beliebiger Lage von A und A' gültig bleibt. Hiergegen ist einzuwenden: 1) Dass man eine Proportion zwischen *Strecken* stets als eine *Zahlen*relation auffassen darf, ist ein neues Axiom (welches wir unter V besprechen werden). 2) Selbst wenn | man dies neue Axiom eingeführt hat, 113 muss man ausdrücklich beweisen, daß die dadurch neu eingeführten Zahlen (cf. später) denselben Rechnungsgesetzen folgen wie die bereits bekannten. _

Statt im folgenden von Proportionen zu sprechen, wollen wir, weil es einfacher ist, von Gleichungen zwischen Produkten reden. An die Spitze stellen wir demgemäss die

Definition des Streckenprodukts.

Gegeben seien zwei Strecken a, b. *Auf dem einen Schenkel eines rechten Winkels tragen wir die Strecke b, auf dem andern die Strecke a und ausserdem eine Strecke, die während der ganzen Untersuchung dieselbe bleiben soll, und die wir mit 1 bezeichnen wollen. Nun ziehen wir die Grade $1b$ und ziehen an ihr die Parallele durch a. Diese Parallele trifft den andern Schenkel des rechten Winkels in einem Punkte P. Die Strecke SP wollen wir das Produkt der Strecken a und b nennen und mit ab bezeichnen.* — Wir betonen, daß diese Definition *rein geometrisch* ist; ab ist keineswegs ein Produkt zweier *Zahlen*.

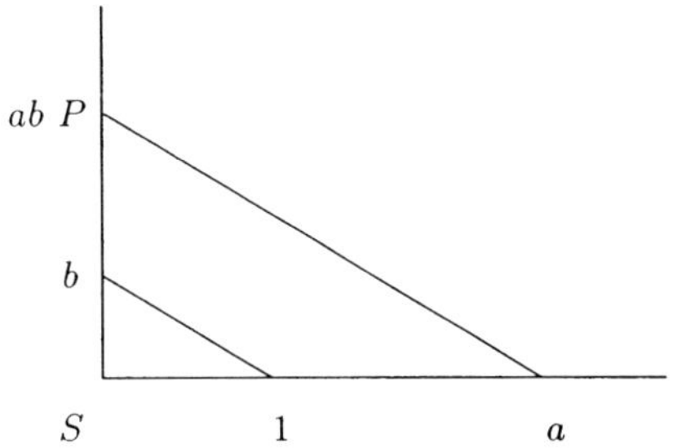

114 Aus dieser Definition ergeben sich zunächst die folgenden Sätze:

Es ist stets $a \cdot 1 = a, \quad 1 \cdot a = a.$

Wenn $a \cdot b = 0$ *ist, so ist entweder* $a = 0$ *oder* $b = 0$.

Wenn $a_1 > a_2$, $b_1 > b_2$ *ist,* (das Zeichen $>$ ist rein geometrisch aufzufassen!) *so ist* $a_1 b > a_2 b$ *und* $ab_1 > ab_2$.

Dieser Satz sagt nichts anderes aus wie: Die Anordnung der Punkte auf einer Graden bleibt erhalten, wenn man diese Punkte durch parallele Grade auf eine andere Grade überträgt. —

Wir werden jetzt, mit Hülfe des Pascal'schen Satzes, beweisen, *dass das soeben definierte Streckenprodukt ab die fundamentalen Gesetze der Zahlenmultiplikation befolgt, nämlich das kommutative, assoziative und distributive Gesetz, die sich in den Formeln ausdrücken:*

$$1)\ \underline{ab = ba}. \quad 2)\ \underline{a(bc) = (ab)c}. \quad 3)\ \underline{a(b+c) = ab + ac}.$$

1.) Um das *kommutative Gesetz zu beweisen,* konstruieren wir zuerst die Strecke ab nach der Definition, indem wir durch den Endpunkt von a die +Parallele zu $1b$ ziehen; der Endpunkt von ab heisse E. Die Strecke ba erhalten wir, wenn wir durch den
115 Endpunkt von b | die Parallele zu $1a$ ziehen; der Endpunkt von ba heisse E'. Dann ist zu zeigen, daß E und E' zusammenfallen, oder anders ausgedrückt, daß die Verbindungslinie $Eb \parallel 1a$ ist. Das ist aber genau der Pascal'sche Satz wie wir ihn bewiesen haben; es kommt dabei in Betracht, daß die Graden aa und bb parallel sind.

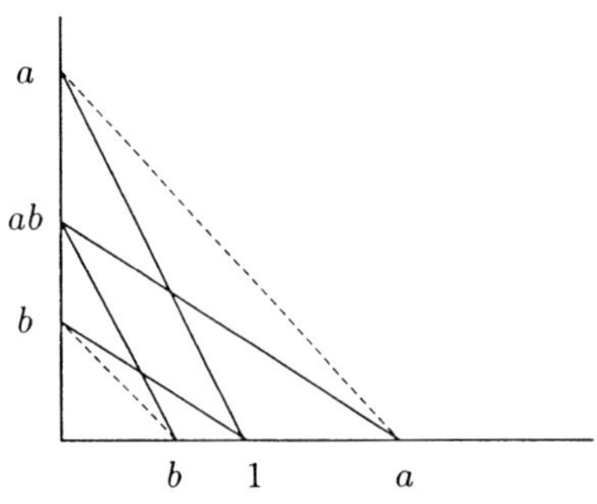

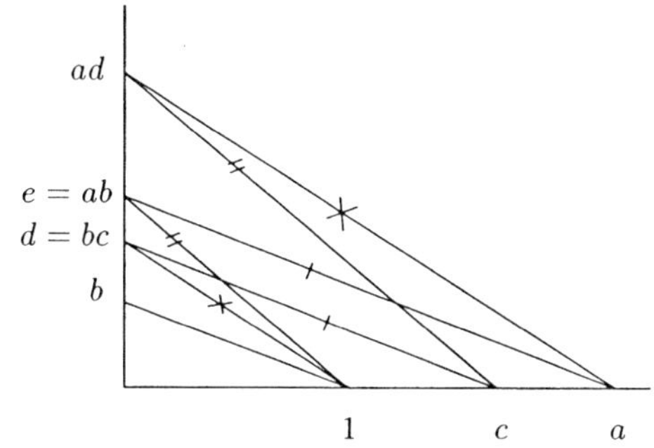

2.) *Beim Beweis des assoziativen Gesetzes* (wobei wir $ab = ba$ benutzen dürfen) verfahren wir ganz analog; wir konstruiren erst $bc = d$, dann ad, dann $ab = e$, endlich ec. Dass die Endpunkte von ad und ec zusammenfallen, ist wiederum nichts wie der Pascalsche Satz, wie die Figur zeigt.[90]

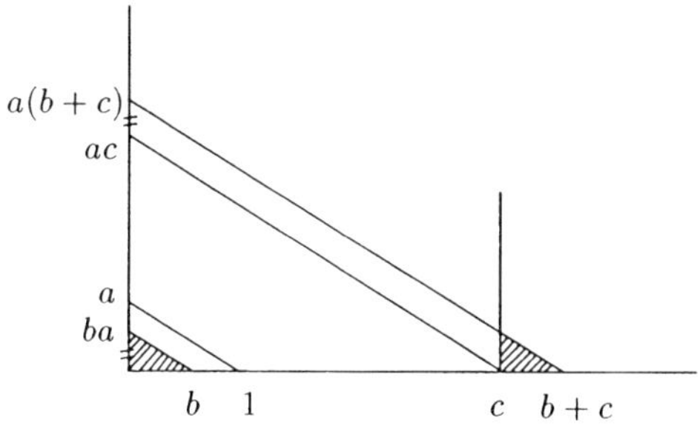

3.) *Der Beweis des distributiven Gesetzes* kommt darauf hinaus, die Gleichheit der in der Figur doppelt durchstrichenen Strecken zu zeigen.

Diese Gleichheit folgt aber, sobald man durch c eine Vertikale zieht, aus der Kongruenz der schraffierten Dreiecke, sowie aus dem Satz von | der Gleichheit der Gegenseiten im Par- 116
allelogramm.

Wir können nun auch *negative Strecken* einführen und für ihre Multiplikation die bekannten Vorzeichenregeln festsetzen. Wir können ferner durch jede Strecke $b \neq 0$ *dividieren*, indem wir definieren: $a = \frac{c}{b}$ $(b \neq 0)$ soll bedeuten $ab = c$. In der That, aus c und $b \neq 0$ ist a eindeutig bestimmt. —

Ferner führen wir die *Schreibweise* ein

$$a : b = c : d,$$

und verstehen darunter nichts anderes wie $ad = bc$. —

Nun sind wir im stande, *die beiden Euklidischen Fundamental-Sätze der Lehre von den Proportionen streng zu beweisen.*

Wir beginnen mit der *Definition*: Zwei Dreiecke sollen *ähnlich* heissen, wenn ihre Winkel paarweise übereinstimmen. Dann gilt der Satz:

In zwei ähnlichen rechtwinkligen Dreicken mit den Katheten a, b bezw. a', b' gilt die Proportion:

$$a : b = a' : b', \quad \text{d.h.}: \quad ab' = a'b.$$

[90]Here, by means of commutativity of multiplication, Hilbert proves the associativity of multiplication in the algebra he has set up. (The same result is proved in the *Festschrift*, p. 34–35.) This foreshadows Hessenberg's result in *Hessenberg 1905b* that Desargues's Theorem follows from the Pascal Theorem, although Hilbert is operating with his Axioms I–IV, from which Desargues's theorem already follows.

117

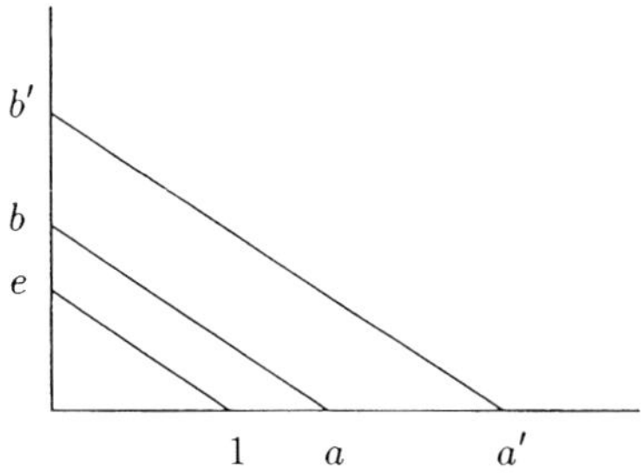

Zum *Beweis* tragen wir beide Dreiecke in einen rechten Winkel ein, und ziehen durch den Punkt 1 des einen Schenkels die Parallele zu den beiden Hypotenusen.

Dadurch erhalten wir die Strecke e auf dem andern Schenkel, und es ist nach Definition $b = ea$, $b' = ea'$, folglich $eab' = ea'b$, also $ab' = a'b$. q. e. d.

Ebenso leicht lässt sich die *Umkehrung* beweisen.

Nun ist es nicht schwer, *diesen Satz auf beliebige ähnliche Dreiecke zu übertragen*, d. h. zu beweisen: *In ähnlichen Dreiecken sind die Seiten proportional, d. h. es ist:*

$$\underline{a : b : c = a' : b' : c'.}$$

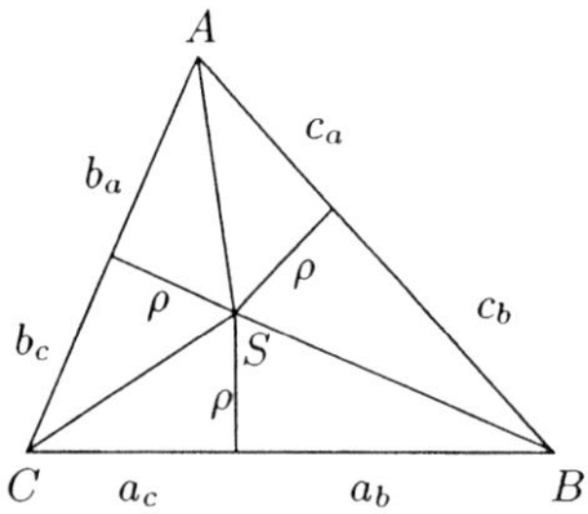

Zum *Beweise* teilen wir jedes der beiden Dreiecke ABC und $A'B'C'$ in sechs paarweise ähnliche rechtwinklige Dreiecke, indem wir den Mittelpunkt S des eingeschriebenen Kreises
118 (dessen Radius ϱ | bezw. ϱ' sei) mit den Ecken verbinden und von S Lote auf die drei Seiten fällen. Dann liefert der vorige Satz:

$$a_b : \varrho = a'_b : \varrho', \quad a_c : \varrho = a'_c : \varrho', \quad \text{also } a : \varrho = a' : \varrho'$$
$$b_c : \varrho = b'_c : \varrho', \quad b_a : \varrho = b'_a : \varrho', \quad \text{also } b : \varrho = b' : \varrho';$$

hieraus folgt: $a : b = a' : b'$, und weiter: $a : b : c = a' : b' : c'$, q. e. d.

Damit haben wir die Lehre von den Proportionen vollständig streng auf Grund der Axiomgruppen I–IV aufgebaut.

Unser Ziel ist jetzt, zu beweisen, daß die bisher betrachteten Schnittpunktsätze (Desargues und Pascal) die einzigen sind, die wir jetzt beweisen können.

Zunächst zeigen wir, *daß wir jetzt die analytische Geometrie der Graden und Ebene aufbauen können.*

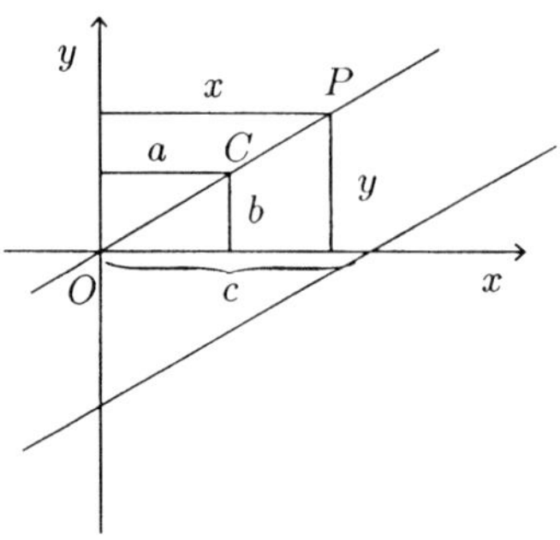

Wir zeichnen uns ein rechtwinkliges Achsenkreuz in der Ebene. Die eine Achse nennen wir x-Achse, die andere y-Achse. Ihr Schnittpunkt heisse der Nullpunkt. Dann können wir jeden Punkt P der Ebene durch zwei *Strecken* x und y, charakterisieren,
119 die | positiv oder negativ sein können, (von *Zahlen* ist hier noch gar keine Rede, die werden erst durch das V. Axiom eingeführt werden). x sei das Lot von P auf die y-Achse, y das auf die x-Achse. x und y nennen wir die *Koordinaten von P*.

Wir ziehen nun durch den Nullpunkt irgend eine Grade, die durch den Punkt C mit den Koordinaten a, b gehen möge. Durch a, b ist offenbar diese Grade völlig bestimmt. Sei ferner $P(x, y)$ *irgend ein* Punkt der Graden OC. Dann ist die Frage: *welcher Bedingung müssen x, y genügen, wenn eben x, y auf OC liegen soll?*

Nun gilt nach den vorangehenden Sätzen:

$$a : b = x : y \quad \text{oder} \quad bx - ay = 0.$$

Das ist die gesuchte Bedingung, denn auch umgekehrt sieht man, daß jeder Punkt, dessen Koordinaten diese Bedingung erfüllen, auf der Graden OC liegt. Wir wollen daher die obige Gleichung als die *Gleichung der Graden OC bezeichnen.*

Nun ist es leicht, die Gleichung einer beliebigen Graden zu finden, die etwa die x-Achse bei $x = c$ schneiden mag. Ziehen wir nämlich zu dieser Graden die Parallele durch O, so ist deren Gleichung $bx - ay = 0$; hieraus ergiebt sich offenbar die gesuchte Gleichung, wenn wir x durch $x - c$ ersetzen, was zu der Gleichung führt:

$$bx - ay - bc = 0.$$ 120

Also haben wir den Satz: *Jede Grade wird durch eine lineare Gleichung in x und y dargestellt. Umgekehrt stellt jede lineare Gleichung in x und y eine Grade dar.*

Wir wollen nun *die fundamentalen Eigenschaften der Graden aus ihrer Gleichung herleiten.*

Zwei Grade schneiden sich in einem Punkte, oder sie sind parallel. In der That haben die Gleichungen

$$ux + vy + w = 0 \quad \text{und} \quad u_1x + v_1y + w_1 = 0$$

stets ein Paar gemeinsamer Wurzeln x, y, die beide endlich sind, ausser wenn $uv_1 - u_1v = 0$ ist. (Parallelität.)

Von drei Punkten x_1y_1, x_2y_2, x_3y_3 einer Graden liegt der zweite zwischen den beiden andern, wenn $x_1 \gtrless x_2 \gtrless x_3$ und $y_1 \gtrless y_2 \gtrless y_3$ ist, wo in jeder von beiden Ungleichungen entweder beide obere oder beide untere Zeichen gelten sollen. (Übrigens kann auch Gleichheit vorkommen, was nichts wesentliches ändert).

Sollen nun unsere Axiome über den Begriff „zwischen" gelten, so müssen wir zeigen, daß die eine der beiden Ungleichungen die andere nach sich zieht. Nehmen wir also an, es sei $ux + vy + w = 0$ die Gleichung der Graden und $x_1 \gtrless x_2 \gtrless x_3$. Dann folgt, da wir ja mit Strecken wie mit Zahlen rechnen dürfen:

$$ux_1 + w \gtrless ux_2 + w \gtrless ux_3 + w$$ 121

oder:

$$-vy_1 \gtrless -vy_2 \gtrless -vy_3,$$

also:

$$y_1 \gtrless y_2 \gtrless y_3, \qquad \text{q. e. d.}$$

Jede Grade teilt die Ebene in zwei Teile von bekannter Eigenschaft. Wir wollen sagen, daß xy auf der einen bezw. andern Seite der Graden liegt,

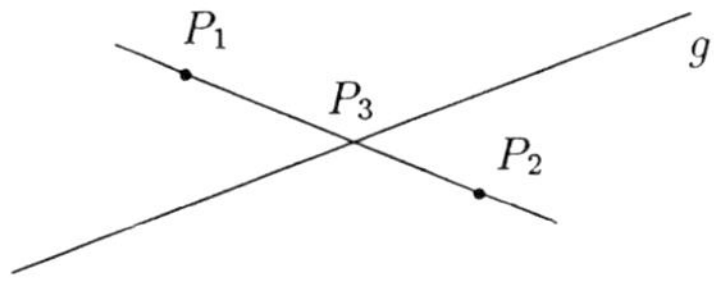

wenn $ux + vy + w > 0$ bezw. < 0 ist. Wir zeigen nun, daß, wenn $P_1(x_1 y_1)$ und $P_2(x_2 y_2)$ auf verschiedenen Seiten der Graden g liegen, notwendig auf der der Strecke $P_1 P_2$ ein Punkt von g liegt. Zum Beweis nehmen wir an, die Verbindungsgrade $P_1 P_2$ habe die Gleichung $ax + by + c = 0$. Ihr Schnittpunkt mit g sei P_3 $(x_3 y_3)$. Dann haben wir nach Vorraussetzung:

$$ux_1 + vy_1 + w \gtrless 0 = ux_3 + vy_3 + w = 0 \gtrless ux_2 + vy_2 + w$$

und

$$ax_1 + by_1 + c = 0, \qquad ax_3 + by_3 + c = 0, \qquad ax_2 + by_2 + c = 0.$$

Multiplizieren wir die obere Zeile mit b, die untere mit v, so ergiebt sich durch Subtraktion:

$$(bu - va)x_1 + bw - vc \gtrless (bu - va)x_3 + bw - vc \gtrless (bu - va)x_2 + bw - vc$$

Hieraus folgt aber sogleich: $x_1 \gtrless x_3 \gtrless x_2$, d. h. P_3 liegt zwischen P_1 und P_2, q. e. d.

Diese Betrachtungen werden uns später nützlich sein.

122 Wir können jetzt, was wir aber nicht ausführen wollen, *den Satz von der Erhaltung des Doppelverhältnisses beweisen* (in der Figur ist: $\frac{1\,2}{2\,3} : \frac{1\,4}{4\,3} = \frac{1'\,2'}{2'\,3'} : \frac{1'\,4'}{4'\,3'}$) und damit *die projektive Geometrie* begründen; im Anschluß daran können wir die Sätze von Menelaus und Ceva fernerhin den *Sehnen- und Sekantensatz* in der Theorie des Kreises, die Theorie des Feuerbach'schen Kreises, *den Satz von Ptolemaeus beweisen*, alles *ohne jede Stetigkeitsbetrachtung.*

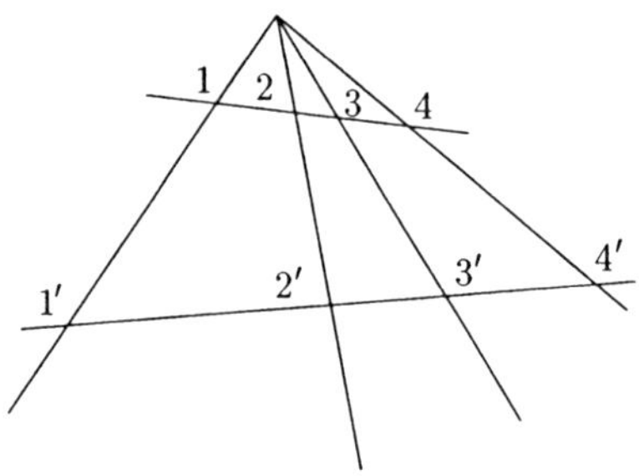

Auch die *analytische Geometrie des Raumes* lässt sich für Graden und Ebenen jetzt aufbauen. Wir finden als *Gleichung einer Ebene* $ux + vy + wz + t = 0$, und könnten hieran ähnliche Betrachtungen knüpfen wie an die Gleichung der Graden in der Ebene.

[Beiläufig bemerken wir, dass wir für die idealen Punkte auch das Zeichen ∞ einführen können, für welches dann z. B. folgende Regeln gelten:

$$\infty \pm a = \infty, \quad a\infty = b\infty, \quad \frac{a}{0} = \infty.] \text{ --}$$

Von unseren Axiomen I–IV machen wir jetzt eine letzte und wohl die
123 interessanteste Anwendung, indem | wir die *Theorie der Flächenmessung* auf sie basieren.

Wir beginnen mit der *Definition*: Unter einem *Polygon* verstehen wir einen gebrochenen Linienzug, bei dem Anfangs- und Endpunkt zusammenfallen, und der (in leicht verständlicher Ausdrucksweise) keine Doppelpunkte enthält.

Ein solches Polygon teilt die Ebene in zwei Gebiete von der Art, dass sich zwei in demselben Gebiet liegende Punkte stets durch einen gebrochenen Linienzug verbinden lassen, der das Polygon nicht schneidet. Diesen Satz kann man übrigens bereits auf Grund der Axiome I, II beweisen. Dasjenige der beiden Gebiete, welches keinen idealen Punkt enthält, nennen wir *das Innere*, das andere *das Äussere des Polygons*.

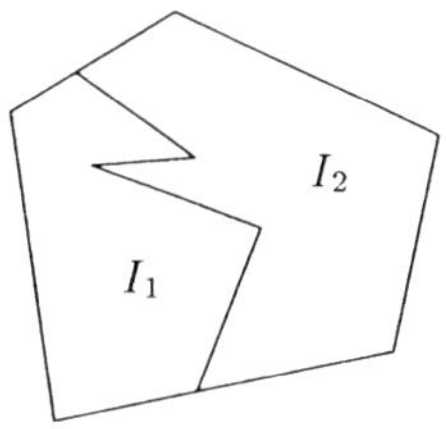

Verbindet man zwei Punkte des Polygons durch einen gradlinigen Zug, der ganz im Innern verläuft, so sagen wir, *das Polygon I sei zerlegt in die beiden Polygone I_1 und I_2* und schreiben:

$$I = I_1 + I_2.$$

(Das ist wieder eine *rein geometrische* Beziehung!)

Durch Schluss von n auf $n+1$ kann man zeigen, daß *jedes Polygon sich in Dreiecke zerlegen lässt.* –

Die *Definition der Flächengleichheit* fassen wir nun so: *Zwei Polygone* 124 *heissen Flächengleich, wenn sie in gleich viele paarweise kongruente Dreiecke zerlegt werden können.* Etwas weiter ist der Begriff der *Inhaltsgleichheit*, den wir so definieren:

Zwei Polygone heissen inhaltsgleich, wenn es möglich ist, flächengleiche Stücke so hinzuzufügen, dass die entstehenden Polygone flächengleich sind.

Das ist im wesentlichen die Euklidische Definition. Er gründet seine Theorie der Inhaltsgleichheit auf die Lehre von den Proportionen; wir werden zeigen, dass eine solche Begründung thatsächlich möglich ist, daß aber in Euklid's Beweisen erhebliche Lücken bestehen.

Wir beweisen nun zuerst den *Satz*: *Sind zwei Polygone einem dritten flächengleich, so sind sie untereinander flächengleich.* (Diesen Satz hat Euklid auch; doch beweist er ihn durch Berufung auf einen allgemeinen Satz über Grössen – ein Irrthum den wir bereits mehrmals erwähnten.)

Sei also $I_1 = I_3$, $I_2 = I_3$; wir wollen zeigen, dass $I_1 = I_2$ ist. Nach Voraussetzung lässt sich sowohl für I_1 als auch fur I_2 eine Dreiecksteilung angeben, sodass | jeder dieser beiden Teilungen eine Dreiecksteilung von I_3 „entspricht" 125 wie wir kurz sagen wollen. Bei Superposition dieser beiden Teilungen wird im allgemeinen jedes Dreieck der einen Teilung durch Grade, die der andern Teilung angehören, zerlegt in irgendwelche Polygone. Nun füge ich noch soviele (punktierte) Graden hinzu, daß jedes Dreieck jeder der beiden Teilungen in lauter kleinere *Dreiecke* zerlegt wird. Die entsprechende Teilung in kleinere Dreiecke kann ich dann in jedem Dreieck von I_1 und I_2 anbringen. Dann sind offenbar I_1 und I_2 in gleich viele paarweis kongruente Dreiecke zerlegt, d. h. es ist $I_1 = I_2$. –

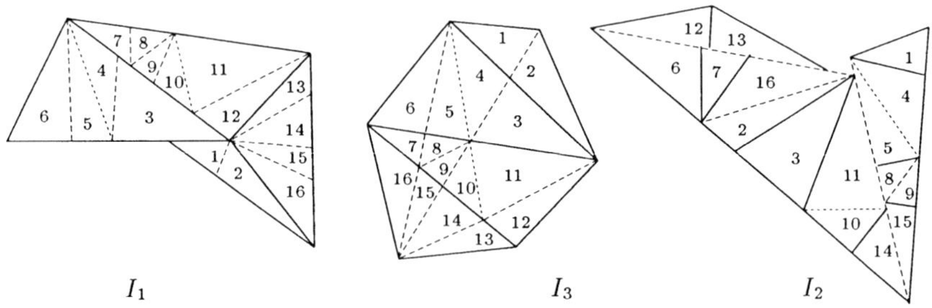

Wir weisen auf die *Analogie* hin, die *zwischen diesem Beweis und dem*
126 *Existenzbeweis des bestimmten Integrals* | besteht.

Aus der Definition der Flächengleichheit folgt sofort: *Flächengleiches zu flächengleichem addiert giebt flächengleiches. Dagegen gilt für die Subtraktion*, was wir besonders betonen, nur der Satz: *Flächengleiches von flächengleichem subtrahirt giebt inhaltsgleiches.*

Auf Grund unserer Definitionen beweisen wir jetzt den Satz: *Zwei Parallelogramme sind inhaltsgleich, wenn sie gleiche Grundlinie und gleiche Höhe haben.*

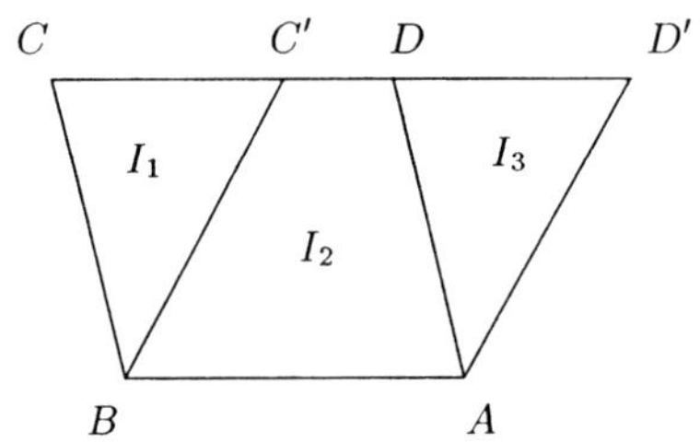

Zum *Beweise* legen wir die Parallelogramme so aufeinander, dass die gleichen Grundlinien zusammenfallen. Dann unterscheiden wir folgende *drei Fälle*: 1) C' zwischen C und D; 2) 3) D zwischen C und C', und zwar 2) $AE > DE$, 3) $AE < DE$. Wir bezeichnen $ABCD$ mit P, $ABC'D'$ mit P'. Für die Fälle 1) 2) werden wir sogar *Flächen*gleichheit beweisen. *Im Fall 1)* haben wir sofort: $I_1 = I_3$, $I_2 = I_2$ folglich: $I_1 + I_2 = I_3 + I_2$, d. h.

$$P = P'.$$

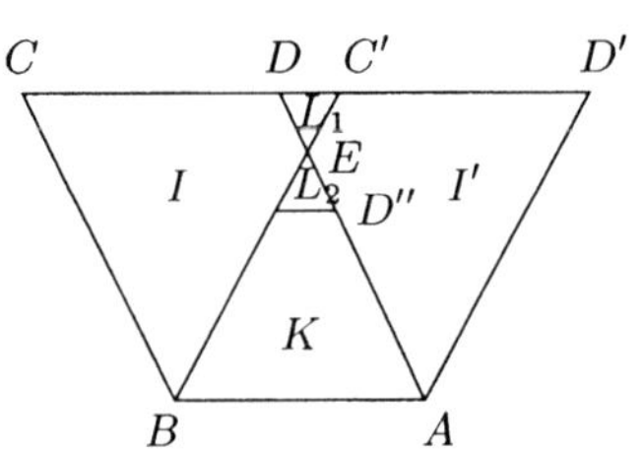

Im Falle 2) tragen wir DE in E ab, sodass $ED'' = ED$ wird, und ziehen | durch
127
D'' die Parallele zu AB. Dann ist $L_1 = L_2$, folglich $I + L_1 = I + L_2$, und ebenso:

$$I' + L_1 = I' + L_2.$$

Nun ist $I + L_2 = I' + L_2$, also auch $I + L_1 = I' + L_1$ und: $I + L_1 + K = I' + L_1 + K$, d. h. $P = P'$. — Man kann auch leicht direkt (ohne Benutzung von L_2) eine Zerlegung von P und P' in kongruente Dreiecke angeben, indem man $FEG \parallel AB$ und $FM \parallel AD'$, $GN \parallel BC$ zieht.

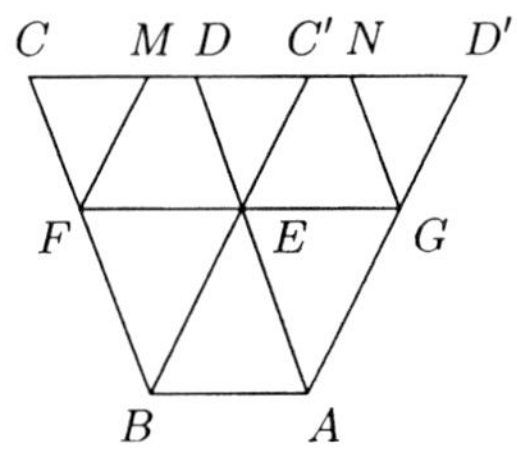

Im Fall 3) können wir nur *Inhalts*gleichheit beweisen. [Der Beweis der Flächengleichheit erfordert hier das Archimedische Axiom.] Die Dreiecke $BC'C$ und $AD'D$ sind kongruent, also um so mehr flächengleich. Subtrahieren wir von jedem Dreieck das $\triangle C'DE$ und fügen wir $\triangle ABE$ hinzu, so

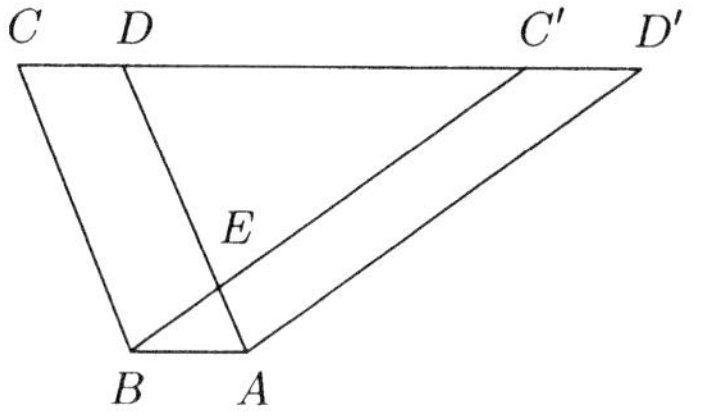

sehen wir, dass | P und P' inhaltsgleich sind. — (Das Gleichheitszeichen soll 128
weiterhin im allgemeinen Inhaltsgleichheit bedeuten.)

Hieraus folgt nun der Satz: *Zwei Dreiecke von gleicher Grundlinie und Höhe sind inhaltsgleich.*

Sei ABC das eine Dreieck, $A'B'C'$ das andere, $AB = A'B'$. Halbieren wir AC in D, BC in E und machen wir $EF = DE$, so ist das Parallelogramm $ABFD$ inhaltsgleich dem $\triangle ABC$, ferner hat es dieselbe Grundlinie AB und die halbe Höhe. Führe ich nun am $\triangle A'B'C'$ dieselbe Konstruktion aus, so wird nach dem vorigen Satze $ABFD = A'B'F'D'$, also auch $\triangle ABC = \triangle A'B'C'$. —

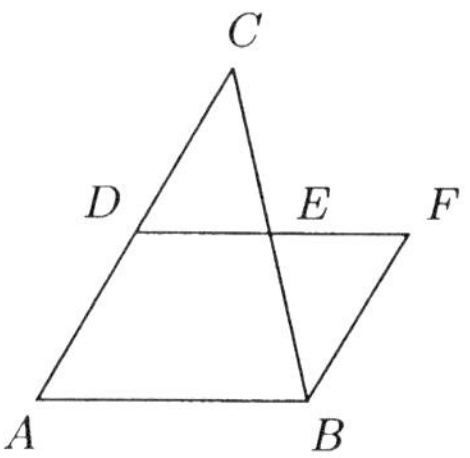

Wir bemerken, *daß im Raume ein analoger Satz wahrscheinlich nicht gilt*, wie schon *Gauss* vermutete.[91]

Endlich können wir mit Hülfe des letzten Satzes in der gewöhnlichen Art den *Pythagoräischen Lehrsatz* beweisen: *Das Quadrat über der Hypotenuse im rechtwinkligen Dreieck ist inhaltsgleich der Summe der Kathetenquadrate.* —

Die bewiesenen Sätze über Inhaltsgleichheit sind vollkom|men streng; in- 129
dessen erkennt man bei näherer Untersuchung, daß sie sämtlich *vorläufig gar keinen Inhalt* haben. Wir wissen ja noch gar nicht, *ob es überhaupt Polygone giebt, die verschiedene Inhalte haben.* Und nicht nur dies müssen wir wissen, wenn wir mit unseren Sätzen etwas anfangen wollen, sondern *wir brauchen noch folgenden Satz: Sind A, B, C, D, C', D' sechs verschiedene Punkte, derart, daß $ABCD$ und $ABC'D'$ Rechtecke sind, und dass C zwischen B und C' liegt, so sind diese Rechtecke nicht inhaltsgleich.*

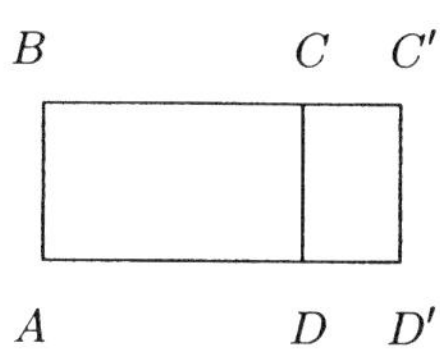

Wir werden sehen, dass wir *nur* diesen Satz brauchen, den wir auch so fassen können: *Wenn zwei inhaltsgleiche Dreiecke gleiche Grundlinien haben, so haben sie auch gleiche Höhe.*

Euklid hat die Schwierigkeit, die hier vorliegt, bei Seite gelassen. Wir werden den genannten Satz, der offenbar die Umkehrung eines vorhergehenden ist, auf Grund der Lehre von den Proportionen beweisen. Es darf uns

[91] See this text, p. 169, Problem 3, or equivalently the third problem mentioned in Hilbert's own notes, p. 107. See also the Introduction to this Chapter, Specific Note (5)(c).

nicht wundern, wenn dieser Beweis nicht ganz einfach ist. Denn die Inhaltsgleichheit zweier Dreiecke sagt doch nach Definition nur aus, dass gewisse
„entsprechende" Dreiecksteilungen existieren; dabei kann die Anzahl der Drei-
130 ecke sehr gross sein, und man | sieht nicht unmittelbar, wie man von da aus,
bei Gleichheit der Grundlinien, auf Gleichheit der Höhen schliessen soll.

Der zu beweisende Satz ist zuerst von *Killing* und *Stolz* besonders formuliert worden; letzterer giebt ihm die Fassung: *Zerlegt man ein Rechteck durch Grade in Dreiecke, und lässt man auch nur eines dieser Dreiecke fort, so kann man das Rechteck mit den übrigen auf keine Weise mehr bedecken.*

Unsere Darstellung ist übrigens von der bei Stolz und Killing *durchaus verschieden.* Denn der erste stellt den in Rede stehenden Satz als Axiom auf, der zweite dagegen beweist ihn mit Hülfe des Archimedischen Axioms, was kein besonderes Interesse hat.[92]

Dem Beweise schicken wir zwei Hülfssätze voraus:

1. Hülfssatz. Wenn für die vier Strecken a, b, c, d die Gleichung $ab = cd$ besteht, so sind die aus a, b einerseits und c, d andrerseits gebildeten Rechtecke

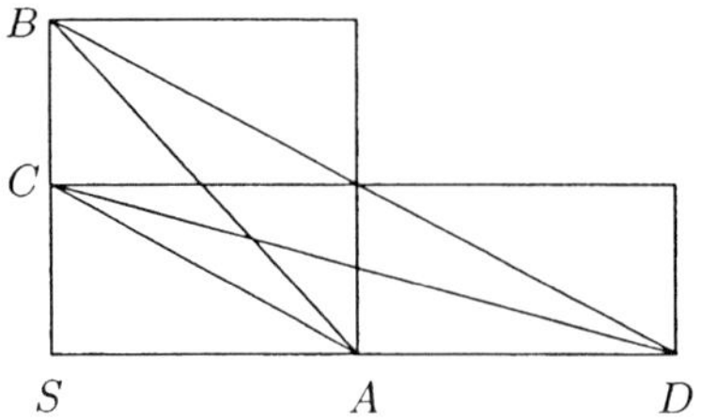

inhaltsgleich. Der *Beweis* ist ein-
fach. Aus der Voraussetzung $a : c = d : b$ folgt, dass die Verbindungsgraden $\overline{AC}$ und $\overline{BD}$ parallel sind.[93]
Hieraus folgt, dass $\triangle ACB = $
131 $\triangle ACD$ ist, also, wenn wir auf | bei-
den Seiten $\triangle ACS$ addieren, dass auch

$$\triangle ABS = \triangle CDS$$

ist, also auch:

$$\triangle ABS + \triangle ABS = \triangle CDS + \triangle CDS. \quad \text{q. e. d.}$$

2. Schwieriger ist der *2. Hülfssatz* zu beweisen: *Wenn die aus a, b bezw. c, d gebildeten Rechtecke inhaltsgleich sind, so ist $ab = cd$.*

Wir beginnen damit, den Begriff „*Flächenmass*", und zwar zunächst beim Dreieck, zu definieren.

Def. I. Wir wollen unter *dem Flächenmass eines Dreiecks das halbe Produkt aus einer Seite und der auf ihr senkrecht stehenden Höhe verstehen.* Die

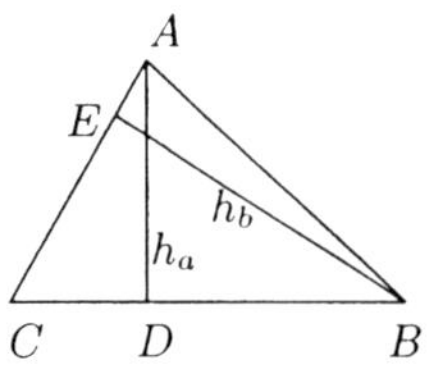

Berechtigung dieser Bezeichnungsweise ergiebt sich erst daraus, *dass das genannte Produkt unabhängig ist von der Wahl der Dreiecksseite.* In der That folgt aus der Ähnlichkeit der Dreiecke ACD und BCE:

$$AC : AD = BC : BE,$$

oder:

$$b : h_a = a : h_b,$$

[92] For the references to Killing and Stolz, see p. 100 of Hilbert's own notes.

[93] In the diagram given, a represents $\overline{SA}$, b represents $\overline{SB}$, c represents $\overline{SC}$ and d represents $\overline{SD}$.

also

$$a \cdot h_a = b \cdot h_b \quad \text{q. e. d.}$$

Def. II. Nun definieren wir weiter: *Denken wir uns ein beliebiges Polygon irgendwie in Dreiecke zerlegt, so wollen wir die Summe der Flächenmasse aller dieser Dreiecke als das Flächenmass des Polygons bezeichnen.*

Hierin steckt implizite die Behauptung, dass *die genannte Summe unab-* 132
hängig ist von der Art der Einteilung des Polygons, und diese Behauptung wollen wir beweisen. (Dieser Beweis ist der Kern dieser ganzen Untersuchung.) Man kann leicht zeigen, dass alles darauf ankommt, folgenden Satz zu beweisen:

Wenn man ein Dreieck ABC auf irgend eine Weise in Dreiecke zerlegt, so ist die Summe der Flächenmasse dieser Dreiecke (also das Flächenmass von ABC im Sinne der Definition II) gleich dem Flächenmass von ABC im Sinne der Def. I.

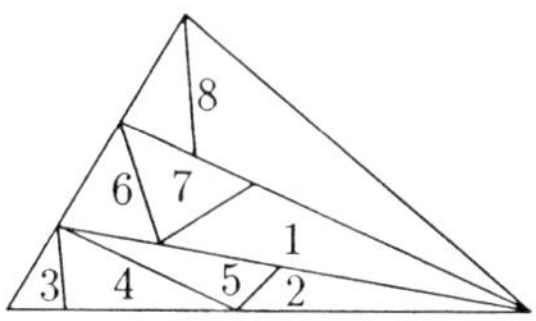

Der Beweis dieses Satzes wird erleichtert, wenn wir den Begriff der *kanonischen Zerschneidung eines Dreiecks* einführen. So wollen wir nämlich eine Dreiecks-Einteilung [94] dann nennen, wenn sie *durch fortgesetztes Ziehen von Transversalen* hergestellt werden kann. (Vgl. die Figur, in der die Teilungslinien so numeriert sind, wie man sie successive zu ziehen hat.)

Nun zeigen wir *erstens, dass für eine kanonische | Zerschneidung der zu* 133
beweisende Satz gilt, und zweitens, dass jede beliebige Dreiecks-Teilung sich auf kanonische zurückführen lässt.

Erstens. Denken wir uns ein Dreieck $\triangle$ dadurch in Dreiecke δ_i zerlegt,

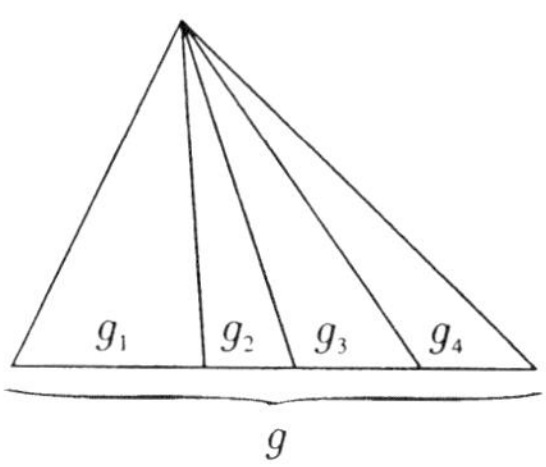

dass von einer Ecke aus Transversalen gezogen sind, wobei die Grundlinie g von $\triangle$ in Stücke g_i zerfallen mag; diese Art der Zerlegung wollen wir andeuten durch $\triangle = S\delta_i$. Bezeichnen wir nun das Flächenmass eines Dreiecks $\triangle$ im Sinne der Definition I mit $F(\triangle)$, so haben wir $F(\delta_i) = \frac{1}{2} g_i h$, also:

$$\sum F(\delta_i) = \sum \frac{1}{2} g_i h = \frac{1}{2} h \sum g_i = \frac{1}{2} g h = F(\triangle),$$

d. h. der zu beweisende Satz gilt für eine Zerschneidung S. Folglich gilt er auch für eine Zerschneidung $S^{(1)} S^{(2)} \ldots S^{(k)}$; in dieser Form ist aber jede kanonische Zerschneidung darstellbar, also gilt unser Satz für jede solche.

Zweitens. Um zu zeigen, daß jede beliebige Dreieckszerschneidung eines Dreiecks ABC sich auf kanonische zurückführen lässt, verfahren wir so. Sei das Dreieck $\triangle$ in beliebiger Weise in Dreiecke $\triangle_i$ zerlegt: $\triangle = \sum \triangle_i$.

[94]Deleted by Sc^s: eines Dreiecks

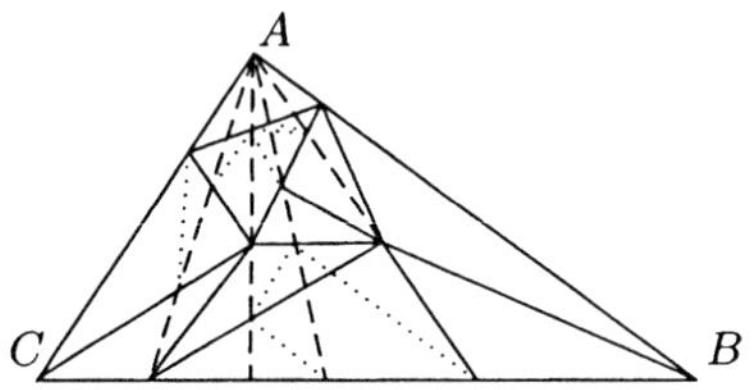

134 Nun ziehen wir etwa von der Ecke A durch jeden Eckpunkt der $\triangle$-Teilung eine Transversale; durch diese Transversalen zerfällt $\triangle$ in gewisse Dreiecke D_k und zwar in kanonischer Weise: $\triangle = \sum D_k$. Durch die Linien der $\triangle$- und der D-Teilung zusammen zerfällt $\triangle$ in Dreiecke und Vierecke. Ziehen wir noch in jedem dieser Vierecke eine (punktierte) Diagonale (wodurch keine neuen Endpunkte entstehen) so zerfällt sowohl $\triangle$ als auch jedes $\triangle_i$ und D_k in Dreiecke δ. Wir wollen zeigen, dass die δ-Teilung jedes $\triangle_i$ und jedes D_k kanonisch ist.

In der That, die δ-Teilung jedes D_k ist so beschaffen, dass sowohl das innere von D_k als auch eine Seite, nämlich die A gegenüberliegende, von Teilpunkten frei ist; daraus folgt aber, dass die Zerschneidung kanonisch ist. Hieraus folgt weiter, dass auch die δ-Teilung des Gesamtdreiecks $\triangle$ kanonisch ist, sodass wir haben:

$$F(\triangle) = \sum F(\delta),$$

wo über alle δ zu summieren ist.

135 Aber auch die δ-Teilung jedes $\triangle_i$ ist kanonisch. Bemerken wir zunächst,

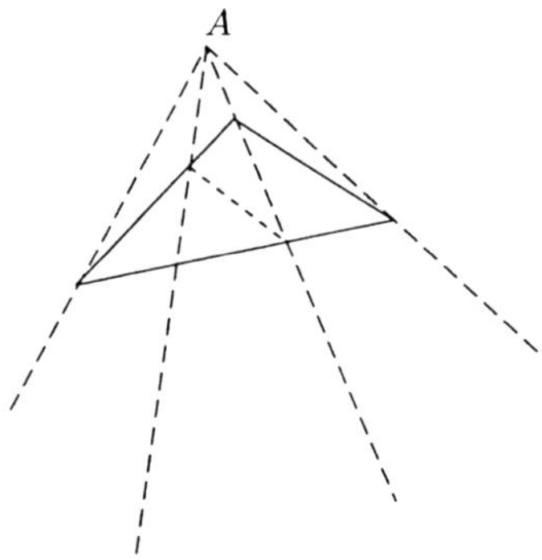

dass im innern keines $\triangle_i$ Teilpunkte liegen. Nach Konstruktion geht durch jede Ecke von $\triangle_i$ eine Transversale durch A; und zwar giebt es stets eine Ecke von $\triangle_i$, für welche diese Transversale nicht im äusseren von $\triangle_i$ verläuft; sie geht also entweder durch das innere von $\triangle_i$, oder sie fällt mit einer Seite von $\triangle_i$ zusammen. Im ersten Fall haben wir eine Dreiecksteilung ohne innere Punkte mit einer Transversale; eine solche aber ist kanonisch. Im zweiten Fall haben wir eine Dreiecksteilung ohne innere Punkte, bei der überdies noch eine Seite

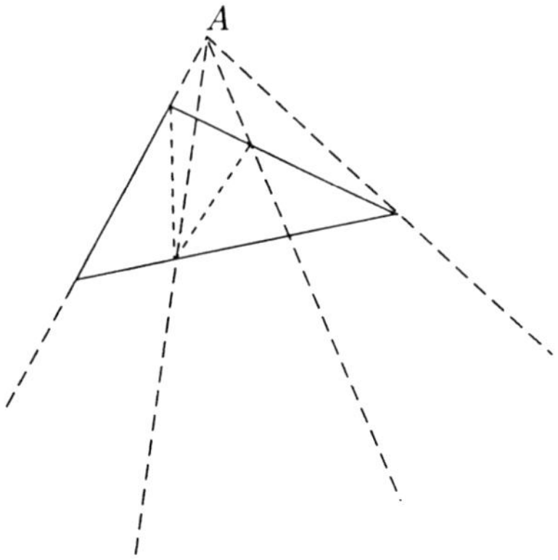

frei von Teilpunkten ist. Läge nämlich auf der mit der Transversale durch A zusammenfallenden Seite von $\triangle_i$ ein Eckpunkt der δ-Teilung von $\triangle_i$, so könnte das nur ein Schnittpunkt dieser Seite mit einer andern Transversale durch A, d. h. der Punkt A selber sein, was unmög-
136 lich ist. Also liegt | auf der betreffenden Seite von $\triangle_i$ kein Teilpunkt, und daraus schliessen wir weiter, dass die δ-Teilung jedes $\triangle_i$ kanonisch ist.

Daraus folgt aber:

$$F(\triangle_i) = \sum\nolimits_{\triangle_i} F(\delta)$$

und folglich:

$$\sum_i F(\triangle_i) = \sum F(\delta)$$

wo rechts über alle δ zu summieren ist. Kombinieren wir diese Gleichung mit der oben erhaltenen:

$$F(\triangle) = \sum F(\delta),$$

so bekommen wir endlich die zu beweisende Gleichung:

$$F(\triangle) = \sum F(\triangle_i). -$$

Hiermit ist denn auch bewiesen, *daß jedem Polygon P im Sinne der Definition II ein ganz bestimmtes Flächenmass zukommt, welches wir von nun an mit $F(P)$ bezeichnen wollen.*

Der *Beweis des 2. Hülfssatzes* erledigt sich jetzt sehr rasch. Zunächst folgt aus den Definitionen:

Flächengleiche Polygone haben gleiches Flächenmass.

Bezeichnen wir nun die aus a, b, bezw. c, d gebildeten Rechtecke mit R_1 und R_2, so lassen sich nach Voraussetzung zwei flächengleiche Polygone E_1 und E_2 finden, sodass $R_1 + E_1$ und $R_2 + E_2$ flächengleich sind, dass also gilt:

$$F(R_1 + E_1) = F(R_2 + E_2)$$

oder

$$F(R_1) + F(E_1) = F(R_2) + F(E_2).$$

Von dieser Gleichung zwischen Streckenprodukten subtrahieren wir $F(E_1)$ 137
$= F(E_2)$ und erhalten:

$$F(R_1) = F(R_2).$$

Nun ist offenbar $F(R_1) = ab$, $F(R_2) = cd$, also folgt:

$ab = cd$, was zu beweisen war. –

Aus diesem Beweise ergiebt sich: *Inhaltsgleiche Polygone haben gleiches Flächenmass.*

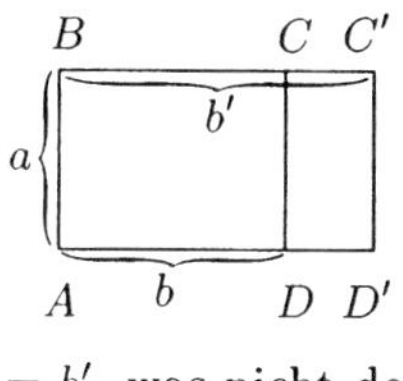

Nun können wir leicht beweisen, dass in der nebenstehenden Figur *die Rechtecke $ABCD$ und $ABC'D'$ nicht inhaltsgleich sein können.* Wären sie es nämlich, so hätten sie gleiches Flächenmass, d. h. es wäre $ab = ab'$, also $b = b'$, was nicht der Fall ist.

Ebenso ergiebt sich der Satz: *Inhaltsgleiche Dreiecke von gleicher Grundlinie haben gleiche Höhe.* Denn aus:

$$\frac{1}{2}gh = \frac{1}{2}gh'$$

folgt ohne weiteres:

$$h = h'.$$

Endlich gilt auch *der Killing-Stolz'sche Satz.*
Im entgegengesetzten Fall nämlich würde man, entgegen unserm Theorem, für ein Polygon zwei verschiedene Flächenmasse finden können. –

138 Jetzt lassen sich auch die Begriffe „*inhaltsgrösser*" | und „*inhaltskleiner*" so definieren, dass für zwei Polygone P und Q stets eine und nur eine der Beziehungen gilt:

$$P < Q, \quad P = Q, \quad P > Q.$$

Es soll nämlich $P < Q$ heissen, wenn $F(P) < F(Q)$ ist.

Damit haben wir denn die Theorie der Flächenmessung vollständig begründet. –

Wir wollen den Abschnitt IV abschliessen mit einer *Bemerkung über den Satz von Desargues.*

Wir haben seinerzeit gezeigt, daß *der Satz von Desargues nicht bewiesen werden kann mit Hülfe der ebenen Axiome I und II; wir wollen jetzt zeigen, dass er auch dann noch unbeweisbar ist, wenn man das Parallelaxiom hinzunimmt.* Wir erinnern uns zu diesem Zwecke an die früher (p. 28) erörterte Geometrie ohne Desargues. In dieser Geometrie gilt freilich auch das Parallelenaxiom nicht; so giebt es durch P in der Figur ein Parallelen*büschel* zu g. Jene Geometrie ist also für den vorliegenden Zweck unbrauchbar. Indessen können wir aus ihr eine brauchbare folgendermassen herstellen. Wir legen um irgend einen Punkt der negativen

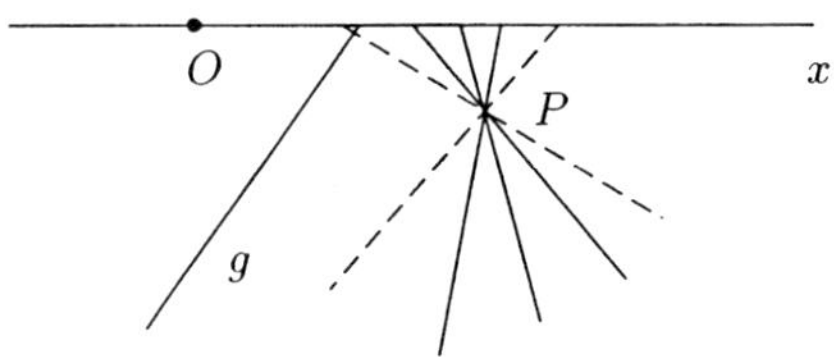

139 x-Achse einen Kreis, welcher nicht die positive x-Achse schneidet. | Ausserhalb dieses Kreises soll überall die gewöhnliche Euklidische Geometrie gelten; innerhalb des Kreises dagegen sollen die graden Linien im Sinne der Geometrie p. 28 definiert sein. In der so definierten Geometrie gilt der Desargues nicht; man erkennt dies sofort, wenn man die Figur p. 30 ganz ins innere des genannten Kreises bringt. Wohl aber gilt das Parallelenaxiom. Man kann nämlich zeigen, daß zwei Grade unserer Geometrie, die durch den Kreis hindurchgehen, sich im innern des Kreises (in S) dann und

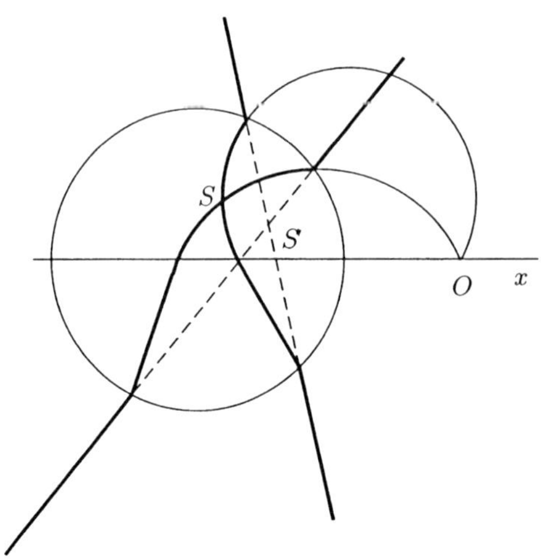

nur dann schneiden, wenn sie sich als Euklidische Grade im innern des Kreises (in S') schneiden würden. _ [95]

V. Das Axiom des Archimedes.

Wir wollen dieses letzte unserer Axiome zuerst *in drei verschiedenen Formen* aussprechen.

1. Archimedes selbst giebt es in folgender Form:[96]

Liegt der Punkt B auf einer Graden zwischen A und C, und konstruieren wir die Punkte B', B'', B''',... derart, daß die Reihenfolge $ABB'B''B'''$... gilt, und dass ferner | $AB = BB' = B'B'' = \ldots$ ist, dann giebt es in der 140
Reihe der Punkte B', B'', B''',... einen Punkt $B^{(n)}$ derart, dass C zwischen A und $B^{(n)}$ liegt.

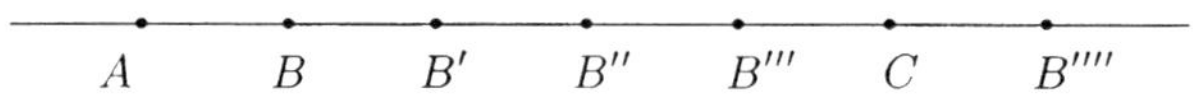

In der Sprache der Streckenrechnung heisst das: *Sind a, b irgend zwei Strecken, so kann man die ganze positive Zahl n stets so wählen, daß $na > b$ wird.*

2. Etwas allgemeiner ist die *projektive Fassung des Axioms: Es liege auf einer Graden B zwischen A und D, und C zwischen B und D; man konstruiere die Punkte B', B'',... derart, dass die Reihenfolge $ABB'B''$... gilt, und daß $ABB'D$, $BB'B''D$, $B'B''B'''D$,... je vier harmonische Punkte sind. Dann giebt es in der Reihe der Punkte B einen Punkt $B^{(n)}$, welcher zwischen C und D liegt.*

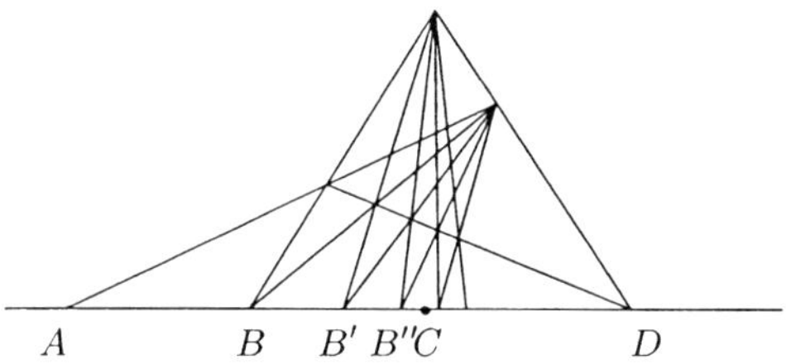

3. In gewisser Hinsicht noch allgemeiner ist die folgende Fassung: *Wenn B, B', B'',... eine unendliche Reihe von Punkten einer Graden ist, und D ein weiterer Punkt dieser Graden, sodass die Reihenfolge $BB'B''\ldots D$ gilt, so giebt es stets einen Punkt P, welcher folgende Eigenschaften besitzt: Sämtliche Punkte B', B'', B''',... | liegen zwischen B und P;* 141

[95] This plane model in which the Axiom of Parallels holds but the planar Desargues's Theorem does not is similar to the one used in the *Festschrift* (*Hilbert 1899c*), pp. 52–55. The major differences are that there Hilbert uses an ellipse to 'distort' the Euclidean lines, whereas here he uses a circle, and that inside the ellipse of the *Festschrift* model, the new straight lines run solely along arcs of Euclidean circles, whereas here they run along arcs of certain Euclidean circles glued below the x-axis to segments of Euclidean straight lines. The Axiom of Parallels holds in the *Festschrift* model for a reason similar to that explained for the model here: two Euclidean lines intersect inside the ellipse if and only if the two corresponding new straight lines intersect just once inside the ellipse. With the *Festschrift* model, however, Hilbert is concerned to show in detail that, while the congruence axioms 1–5 hold in his model, 6 fails, and there is no discussion of that here.

[96] This is not Archimedes's formulation, and neither is the axiom new with Archimedes, as Archimedes himself states, attributing it to Eudoxus. For Archimedes's statement of the principle, its use by Euclid, and historical remarks, see *Heath 1926c*, 15–16.

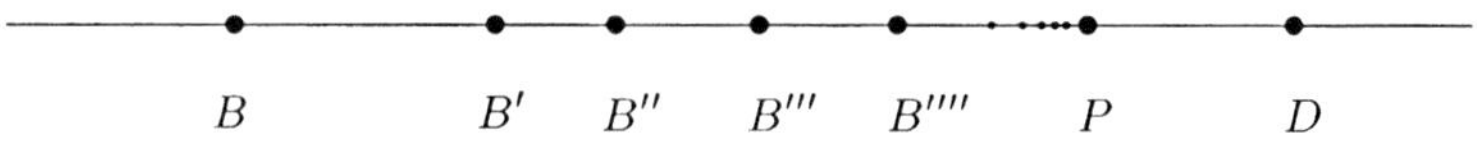

jeder andere Punkt P', für welchen dasselbe gilt, liegt zwischen P und D.[97]

In der Ausdrucksweise der Mengenlehre sagt dieser Satz die Existenz eines *Grenzpunktes* der unendlichen Punktmenge aus. Es ist kaum nötig, auf die Analogie dieses Satzes mit der Theorie der Dedekind'schen Schnitte hinzuweisen.

Die drei gegebenen Fassungen des Axioms sind nicht völlig gleichwertig.[98]

Wir bemerken noch, dass man das Axiom für den ganzen Raum hat, sobald es auf einer einzigen Graden gilt. –

Wir haben nun wieder die Aufgabe, zu beweisen, *daß unser neues Axiom von allen früheren unabhängig ist.* Wir geben den Beweis ausführlich nur für die Ebene; die Ausdehnung auf den Raum ergiebt sich dann ganz von selbst.

Wir werden uns eine *Geometrie* konstruieren, *in welcher alle Axiome I–IV gelten, das V. aber nicht.* Zu diesem Zwecke betrachten wir den Funktionenkörper, der entsteht, wenn man zum Bereich der rationalen Zahlen eine reelle Veränderliche t adjungiert, mit andern Worten die Gesamtheit der Grössen, die sich aus 1 und t durch endlich oft angewandte Addition, Subtraktion,
142 | Multiplikation, Division erzeugen lassen. Den so gewonnenen Funktionenbereich erweitern wir noch, indem wir jede Funktion von der Form $\sqrt{1+u^2}$ aufnehmen, wo u irgend eine bereits vorkommende Funktion ist. Die so definierte Gesamtheit (die übrigens eine abzählbare Menge bildet) wollen wir als eine Zahlenmannigfaltigkeit auffassen und demgemäss die einzelne Funktion $F(t)$ als *komplexe Zahl* bezeichnen. Dieser Zahlbereich hat offenbar die Eigenschaft, dass man durch die vier Species und durch die Operation $\sqrt{1+u^2}$ nicht aus ihm hinauskommt.[99]

---

[97]It is implied by Hilbert's formulation that P must lie between B and D, but it should be stated that P could coincide with D; moreover, P' could coincide with D or even lie to the right of it (in the diagram).

[98]Only Version 1 appears in Hilbert's *Festschrift* (*Hilbert 1899c*). Version 3, which is first deployed by Hilbert in the 1893/1894 lectures (p. 38), and then in *Hilbert 1895b*, is in fact much stronger than the other two. It imposes a version of Bolzano-Dedekind continuity, and guarantees that any coordinate field generated by the geometry (e.g., by a *Streckenrechnung*) will be isomorphic to the real number field. Versions 1 and 2 guarantee only that such a coordinate field will be an Archimedean field (and thus that to each point on the line, a real number can be uniquely assigned), of which the usual rational field is an example. (Every sub-field of the reals is an Archimedean ordered field, and every such field is monotone embeddable in the real field.) From the French translation of the *Festschrift* on, and also in *Hilbert 1900b*, Hilbert uses a *Vollständigkeitsaxiom* to guarantee Dedekind continuity, and not an axiom like Version 3 here. The *Vollständigkeitsaxiom*, when combined with a standard version of the Archimedean Axiom (e.g., Version 1 here) will imply Dedekind continuity, although the presence of the Archimedean Axiom is essential: the *Vollständigkeitsaxiom* on its own, unlike Version 3 here, does not imply Dedekind continuity. See the Introduction to Chapter 5, section 6.

[99]The smallest Pythagorean field over t has been used by Hilbert before in this text; see pp. 65, 67, and also pp. 65, 67 of Hilbert's own notes, and the relevant notes. Here, however, Hilbert does something rather different. Each element of the field can be considered as a function of t, represented here by $F(t)$, and earlier by $A(1,t)$. Hilbert now takes the

Wir wollen zusehen, *wie man mit diesen Zahlen rechnen kann.* Seien α und β zwei solche komplexen Zahlen. Der Sinn einer Gleichung $\alpha = \beta$ ist dann klar; dagegen müssen wir nun verabreden, was eine Ungleichung $\alpha > \beta$ bedeuten soll. Die α, β sind nach Definition algebraische Funktionen von t, und zwar reell für alle reellen t. Hieraus folgt, dass von einem gewissen t ab z. B. α sein Vorzeichen nicht mehr wechselt, d. h. entweder beständig positiv oder beständig negativ bleibt. Je nachdem für die Funktion $\alpha - \beta$ das erste oder das letzte der Fall ist, wollen wir $\alpha - \beta > 0$ ($\alpha > \beta$) oder $\alpha - \beta < 0$ ($\alpha < \beta$) setzen.

Es ist dann klar, dass für beliebiges α, β stets eine und nur eine der 143
Beziehungen $\alpha = \beta$, $\alpha > \beta$, $\alpha < \beta$ gelten muss (hierbei müssen wir noch eine Festsetzung über das Vorzeichen der Quadratwurzeln machen; wir wollen verabreden, dass alle Wurzeln positiv gerechnet werden sollen), ferner dass aus $\alpha > \beta$, $\beta > \gamma$ folgt $\alpha > \gamma$; dass $\alpha\beta$ nur dann $= 0$ sein kann, wenn mindestens ein Faktor verschwindet, etc. Kurz, unsere komplexen Zahlen verhalten sich so wie reelle Zahlen im gewöhnlichen Sinne.

Hingegen gilt bei diesen Zahlen nicht das Archimedische Axiom. Betrachten wir z. B. die Zahlen 1 und t, so erkennen wir, dass es keine noch so grosse Zahl n giebt, sodass $n > t$ würde; denn nach Definition ist $0 < t - n$, da ja t eben selber unbegrenzt wächst, während n fest ist.

Um nun aus diesen Zahlen eine Geometrie aufzubauen, verfahren wir genau nach dem Vorbilde der analytischen Geometrie. [Vgl. für das folgende pag. 65ff., 118ff.]

Sind x, y zwei komplexe Zahlen, so nennen wir das Zahlenpaar x, y einen *Punkt*. Sind ferner u, v, w drei komplexe Zahlen, so soll $u : v : w$ *eine Grade* definieren. Die *vereinigte Lage von Punkt und Grade* sei charakterisiert | durch 144
die Gleichung $ux + vy + w = 0$.

Sind x_1y_1, x_2y_2, x_3y_3 Punkte einer Graden, so sagen wir, dass der zweite *zwischen* den beiden andern liegt, sobald $x_1 \gtrless x_2 \gtrless x_3$ ist; hieraus ergiebt sich denn durch Rechnung grade wie früher, dass auch $y_1 \gtrless y_2 \gtrless y_3$ ist. Ist x_1y_1 irgend ein Punkt, so kann ich die beiden Seiten der Graden $u : v : w$ durch das Vorzeichen von $ux_1 + vy_1 + w$ charakterisieren. Kurz, *alle Axiome I, II, und ebenso IV gelten.*

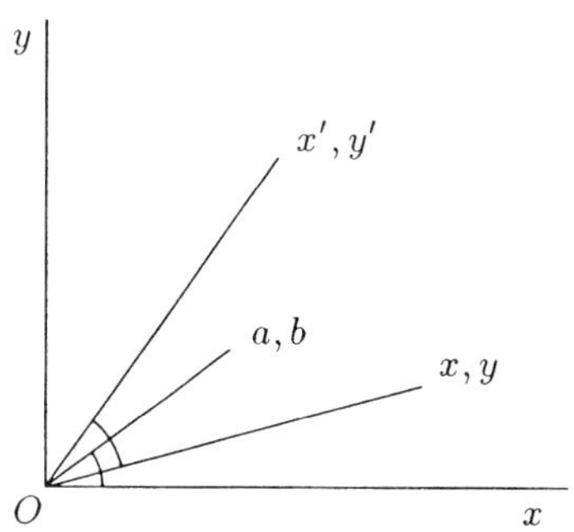

Aber auch die *Kongruenzaxiome* gelten, wenn man folgende Festsetzungen macht (hier tritt hervor, warum wir die Operation $\sqrt{1+u^2}$ einführen mussten).

functions themselves as the elements of the field, and not the corresponding numbers with their natural ordering in the real field, and defines new field operations making essential use of properties of the functional expressions. Once again, Hilbert is referring to 'complex numbers' in his special sense, the axioms for which (i.e., for ordered fields) are first given explicitly later in this text, 160–162, then in §13 ('Complexe Zahlensysteme') of the *Festschrift* (*Hilbert 1899c*), pp. 26–28, and also in *Hilbert 1900b*.

Die *Parallelverschiebung* (das Streckenabtragen) sei definiert durch $x' = x+a$, $y' = y+b$. Unter der *Entfernung zweier Punkte* $x_1 y_1$ und $x_2 y_2$ verstehen wir den Ausdruck $\sqrt{(x_1 - x_2)^2 + (y_1 - y_2)^2}$. *Die Drehung um einen Punkt* (das Abtragen von Winkeln) definieren wir durch die Formeln (die durch die Figur motiviert werden):

$$x' = \frac{a}{\sqrt{a^2+b^2}}x - \frac{b}{\sqrt{a^2+b^2}}y$$
$$y' = \frac{b}{\sqrt{a^2+b^2}}x + \frac{a}{\sqrt{a^2+b^2}}y.$$

145 Die Ausführbarkeit der Operationen rechts folgt aus unsern Festsetzungen. Durch blosse Rechnung folgt dann, dass alle Kongruenzaxiome, insbesondere auch der 1. Kongruenzsatz für Dreiecke, gültig sind.

Das Archimedische Axiom gilt natürlich nicht. Betrachten wir z. B. die auf der Graden $y = 0$ liegenden Punkte 1,0 und t,0; wie schon oben bemerkt, kann man dann durch Abtragen der Strecke 1 den Punkt t,0 niemals erreichen, da für jedes n gilt: $t - n > n$. Auch der Satz, dass man durch fortgesetztes Halbieren einer Strecke einem Endpunkte beliebig nahe kommen kann, gilt hier nicht, wie die Ungleichung $\frac{1}{t} < \frac{1}{2^n}$ zeigt.

Hiermit ist die Unabhängigkeit des Axioms V von allen vorangehenden bewiesen, zunächst für die Ebene; die Übertragung auf den Raum liegt auf der Hand. –

Wir bemerken, dass *auch nicht etwa das Parallelenaxiom eine Folge von I, II, III, V ist*; das ergiebt sich einfach daraus, dass in den Beispielen von Geometrieen ohne Parallelenaxiom [p. 82–99] das Axiom V gilt. –

Es möge hier noch eine *allgemeine Bemerkung über den Charakter unserer Axiome I–V* Platz finden. Die Axiome I–III sprechen sehr einfache, man
146 könnte sagen ursprüngliche, Thatsachen aus; ihre Gültigkeit in der | Natur lässt sich durch das Experiment leicht nachweisen. Hingegen ist die Gültigkeit der Axiome IV, V nicht so unmittelbar einleuchtend; ihre experimentelle Bestätigung erfordert eine grössere Zahl von Versuchen. –

Zum Schluss wollen wir noch einen Einblick gewinnen in die *Beziehungen, die zwischen den Sätzen von Desargues, von Pascal und dem Axiom des Archimedes bestehen.* Zunächst bemerken wir, dass wir *im folgenden die Kongruenzaxiome III ein für alle mal ausschalten* wollen; denn mit ihrer Hülfe ist der Desargues wie der Pascal auf Grund der ebenen Axiome I, II beweisbar. *Dagegen wollen wir das IV. Axiom hinzunehmen*; das wäre allerdings nicht nötig, doch vereinfacht es die Untersuchungen sehr bedeutend.

Wir wissen, daß *der Desargues* unbeweisbar ist aus den ebenen Axiomen I, II und aus IV [p. 138];[100] er *bleibt unbeweisbar, wenn man V hinzunimmt*; denn in der pag. 138ff. konstruierten Geometrie gilt ersichtlich das V. Axiom, und zwar in der 3. Fassung. Der Pascal verhält sich ganz anders; wir behaupten:

1.) Der Pascal ist beweisbar auf Grund der Axiome I, II, IV, V (wo man die räumlichen Axiome I, II durch den Desargues ersetzen kann).

[100] Pp. 138–139.

2.) Der Pascal ist nicht beweisbar, wenn man V weglässt, d. h. allein auf 147
Grund der Axiome I, II, IV.

Diese Sätze werden wir beweisen und dabei ihren inneren Grund deutlich erkennen.

Dass der Pascal sich aus I, II, IV, V beweisen lassen muss, folgt schon daraus, dass man ihn in der projektiven Geometrie beweist, zu deren Voraussetzungen die Stetigkeit, d. h. das V. Axiom, gehört.

Unsere erste Aufgabe soll sein, eine *neue von den Kongruenzaxiomen unabhängige Streckenrechnung* einzuführen. Wir denken uns durch einen Punkt O zwei Graden OX und OY gezogen und nehmen auf jeder von ihnen einen Punkt an, E und E'. Jede der Strecken OE und E' nennen wir 1 und setzen überhaupt fest, dass jede zu EE' parallele Grade auf OX und OY gleiche Strecken abschneiden soll. Um nun die Summe $a + b$ zweier Strecken $a = OA$, $b = OB$ zu definieren, ziehe ich AA' parallel zu EE', ferner durch B eine Parallele zu OY, durch A' eine Parallele zu OX. Diese beiden mögen sich in S schneiden. | Dann ziehe ich durch S eine Parallele DD' 148
zu EE'; jede der Strecken OD, OD' nenne ich die *Summe* $a + b$. Durch diese Festsetzung ist das Streckenabtragen definiert, natürlich nur auf den Graden OX und OY; mehr brauchen wir aber auch nicht, wie sich zeigen wird.

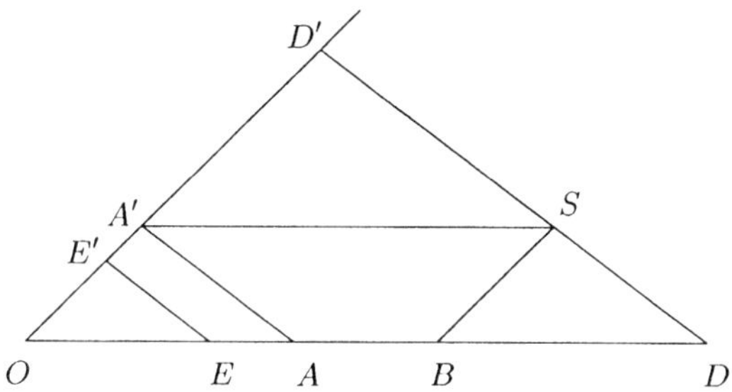

Wir wollen jetzt zeigen, daß die soeben definierte Addition sowohl das kommutative als auch das assoziative Gesetz befolgt; bei diesem Beweise müssen wir uns lediglich auf den Satz von Desargues stützen, von dem übrigens nur der spezielle Fall hier vorkommt, in welchem die Seiten der Dreiecke paarweise parallel sind.

1.) Das *kommutative Gesetz:* $a + b = b + a$.

Es sei $a = OA = OA'$, $b = OB = OB'$. Wir konstruieren die Punkte A'', B'', indem wir $A'A'' \parallel B'B'' \parallel OX$ und $AB'' \parallel A''B$[101] $\parallel OY$ ziehen; dann verlangt unser Satz, dass $A''B'' \parallel AA'$ ist. Oder, anders gewandt: Wenn im $\triangle ODD'$ ist: $AA' \parallel BB' \parallel DD'$, $AB'' \parallel A''B \parallel OD'$, $A'A'' \parallel OD$, dann ist auch $B'B'' \parallel A'A''$. Diesen Satz beweisen wir so: Wir verlängern AA' und $A''B$ bis zum Schnitt in E. Nach Voraussetzung | sind in den Dreiecken $A'A''D'$ und ABS die Seiten paarweise 149
parallel. Daraus ergiebt sich mit Hülfe des Desargues, dass D', S, E auf einer

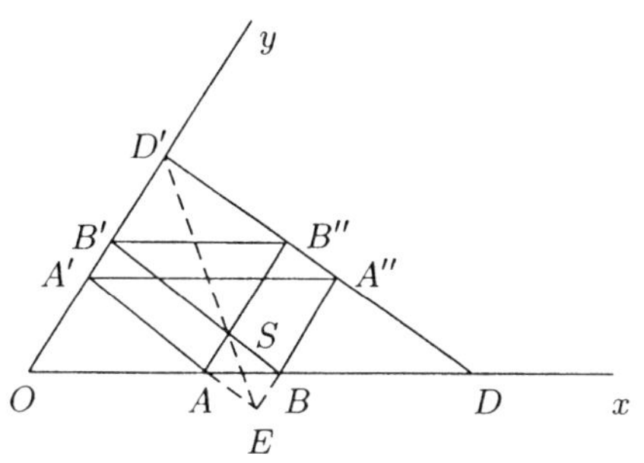

[101] Here Hilbert (in Hi^g) switches 'A''' und 'B'.

Graden liegen. Demnach sind die Dreiecke $A'A''E$ und $B'B''S$ perspektiv, woraus sich die Parallelität von $A'A''$ und $B'B''$ ergiebt.

2.) Das assoziative Gesetz $(a+b)+c = a+(b+c)$ führt auf die folgende ohne weiteres verständliche Figur. Zu beweisen ist die Parallelität von $A'A''$ und $B'B''$. Dieser Beweis reduziert sich sofort auf den vorigen, denn der schraffierte Teil der Figur ist identisch mit der zu $a+b = b+a$ gehörigen Figur. —

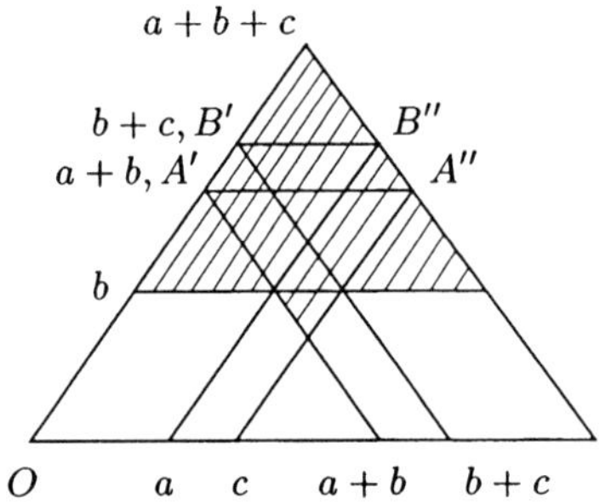

Das Produkt ab zweier Strecken definieren wir jetzt, analog der früheren Definition [p. 113] folgendermassen: Man trage auf OX die Strecken 1 und b, auf OY die Strecke a ab, verbinde 1 mit a und ziehe durch b die Parallele zu $1a$;
150 diese | schneidet auf OY die Strecke ab ab.

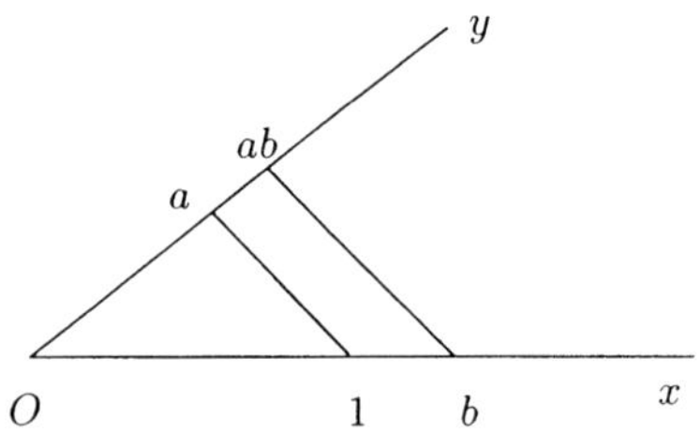

Dies Produkt folgt dem assoziativen Gesetz:

$$\underline{a(bc) = (ab)c.}$$

Zum Beweise dient die nebenstehende Figur, die sich so beschreiben lässt: Zwei Vierecke 1234 und $1'2'3'4'$ liegen mit ihren Ecken abwechselnd auf den Schenkeln eines Winkels XOY, und es ist: $1'2' \parallel 12$, $2'3' \parallel 23$, $3'4' \parallel 34$; zu beweisen ist, dass auch $4'1' \parallel 41$ ist. Den Beweis führen wir wieder mit Hülfe des Desargues. In den Dreiecken $23S$ und $2'3'S'$ sind die Seiten paarweise parallel; dann ergiebt sich, dass O, S, S' auf einer Graden liegen. Folglich liegen die Dreiecke $14S$ und $1'4'S'$ perspektiv woraus denn $4'1' \parallel 41$ folgt.

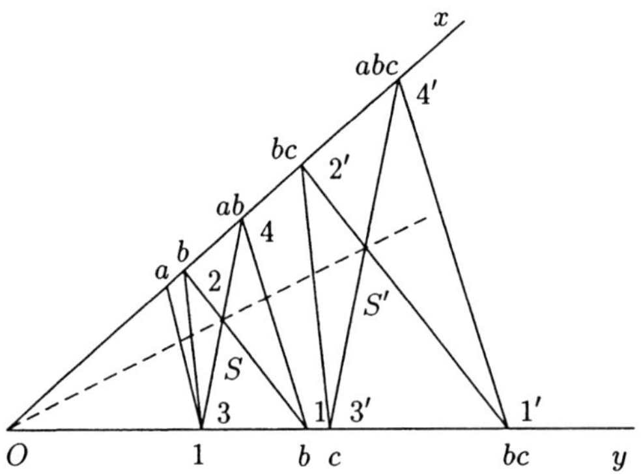

Das kommutative Gesetz ab = ba lassen wir bei Seite; denn von früher her wissen wir, dass es mit dem Pascal'schen Satz identisch ist, und es ist ja grade unser Zweck, die Unbeweisbarkeit des Pascal auf Grund von I, II, IV nachzuweisen. —

Hingegen wollen wir *das distributive Gesetz* beweisen und zwar in den beiden Formen:

$$\alpha)\quad \underline{a(b+c) = ab + ac} \quad \text{und} \quad \beta)\quad \underline{(a+b)c = ac + bc.}$$ 151

α) Zum Beweise der Formel $a(b+c) = ab+ac$ brauchen wir die folgende Figur, die wir zunächst beschreiben wollen. Es ist $A'A'' \parallel C'C''$, $A'B'' \parallel A''B'$, $B'D_1 \parallel C'D_2 \parallel OY$, $B''D_2 \parallel C''D_1 \parallel OX$, $D_2E' \parallel D_1E'' \parallel A'B''$. Zu beweisen ist, dass $E'E'' \parallel A'A''$ ist. Zum Beweise brauchen wir folgende (in der Figur punktierte) Hülfslinien: $E'H \parallel OY$, $E''H \parallel OX$, D_2FG_2 als Verlängerung von $C'D_2$, also $\parallel OY$, D_1FG_1 als Verlängerung von $C''D_1$, also $\parallel OX$; endlich die Verbindungslinien $A'C''$, $A'E''$, $A''C'$, $A''E'$; $E''F$, $E''G_1$, $E'F$, $E'G_2$, G_1G_2.

Der Beweis läuft darauf hinaus zunächst zu zeigen, daß die schraffierten Dreiecke zu je zweien perspektiv liegen, nämlich $OA'E''$ und HG_2E' einerseits, $OA''E'$ und HG_1E'' andrerseits.

1.) Da $A'E' \parallel B''D_2 \parallel C''F$ ist, so liegen die Dreiecke $A'B''C''$ und $E'D_2F$ perspektiv; also ist $A'C'' \parallel E'F$. Also ist, da auch die Dreiecke $A'C''E''$ und $E'FG_2$ perspektiv liegen, $A'E'' \parallel E'G_2$. Also sind in den Dreiecken $OA'E''$ und HG_2E' die Seiten paarweise parallel; also schneiden sich OH, $A'G_2$, $E''E'$ in einem Punkte S.

152

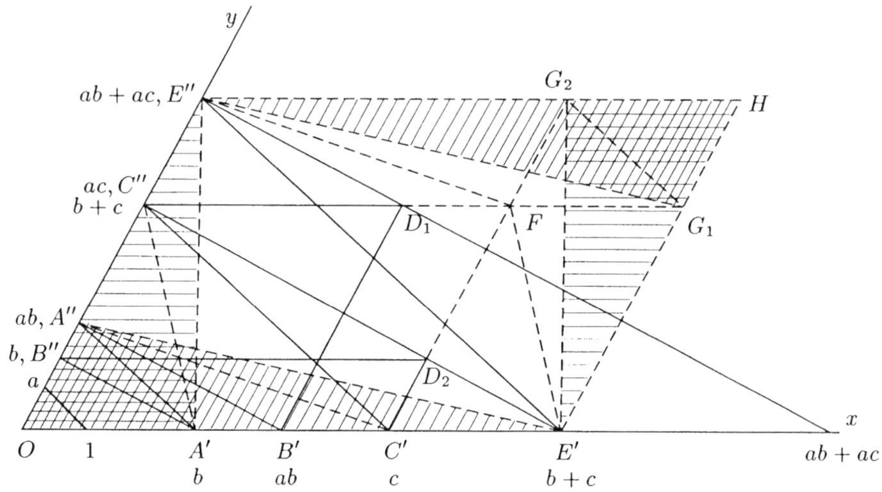

2.) Ebenso findet man, dass die Dreiecke $OA''E'$ und HG_1E'' die Seiten paarweise parallel haben, dass also auch OH, $A''G_1$, $E'E''$ sich in einem Punkte, also auch in S schneiden müssen.

3.) Hieraus folgt, dass die Dreiecke $OA'A''$ und HG_2G_1 perspektiv liegen; also ist $G_1G_2 \parallel A'A''$.

4.) Endlich betrachten wir die Figur $C'C''G_1\ E'E''G_2C'$; aus ihr ergiebt sich [cf. den Beweis von $a + b = b + a$], dass $E'E'' \parallel C'C''$, also auch $\parallel A'A''$ ist, q. e. d.

β) Beim Beweise von $(a+b)c = ac+bc$ kommt man auf die folgende Figur. In ihr ist $AB \parallel HI \parallel A'B' \parallel H'I'$, | $DF \parallel D'F'$, $BD \parallel B'D'$, $AG \parallel A'G' \parallel OY$, 153

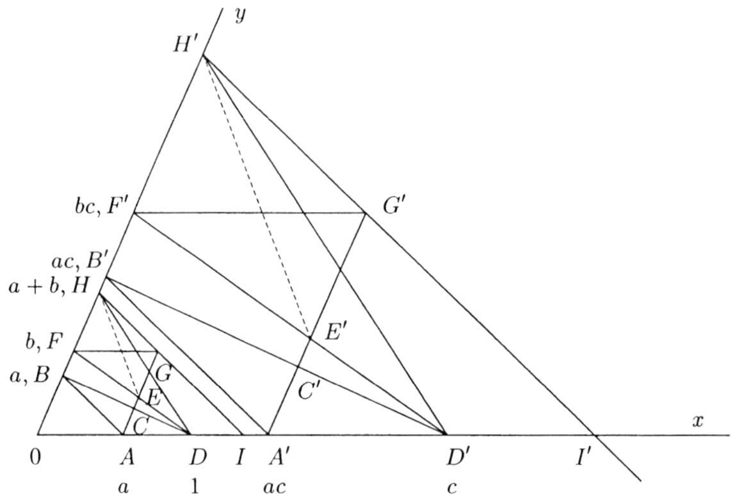

$FG \parallel F'G' \parallel OX$; zu beweisen ist: $DH \parallel D'H'$. Wir ziehen die Hülfslinien EH und $E'H'$ und finden: 1) $\triangle ABC$ und $\triangle A'B'C'$ haben paarweis parallele Seiten, also ist OCC' eine Grade. 2) Dann folgt aus $\triangle CDE$ und $\triangle C'D'E'$, dass OEE' eine Grade ist. 3) Aus $\triangle EFG$ und $\triangle E'F'G'$ folgt,[102] dass OGG' eine Grade ist. 4) Aus der perspektiven Lage von $\triangle EGH$ und $\triangle E'G'H'$ folgt, dass $EH \parallel E'H'$ ist. 5) Endlich folgt aus der perspektiven Lage von $\triangle DEH$ und $\triangle D'E'H'$, dass $DH \parallel D'H'$ ist, q. e. d..

In der von uns allein auf Grund der Axiome I, II, IV eingeführten Streckenrechnung gelten also *die gewöhnlichen formalen Rechnungsregeln, ausgenommen das kommutative Gesetz der Multiplikation.* Ferner existieren die *Strek-*
154 *ken* 0 *und* 1; | unter 0 verstehen wir naturgemäss eine Strecke mit zusammenfallenden Endpunkten. Ist a eine beliebige Strecke, so gelten die folgenden Regeln:

$$a + 0 = a; \quad 1 \cdot a = a \cdot 1 = a, \quad 0 \cdot a = a \cdot 0 = 0.$$

Aus $ab = 0$ folgt entweder $a = 0$ oder $b = 0$.

Ferner gelten die *Gesetze der Anordnung*: Sind a, b irgend zwei Strecken, so ist entweder $a > b$, oder $a = b$, oder $a < b$. Aus $a > b$, $b > c$ folgt $a > c$ etc. Aus $a > b$ folgt $a + c > b + c$, $ac > bc$, $ca > cb$.

Weiter ist es leicht, *„negative“ Strecken* einzuführen, d. h. solche, die < 0 sind. Dann giebt es stets eine und nur eine Strecke x, sodass $a + x = b$ wird; wir setzen $x = b - a$, und es ist also die *Subtraktion* von Strecken „unbeschränkt“ und eindeutig ausführbar. Ähnliches gilt von der *Division.* Man zeigt zunächst, daß die Gleichungen $ax = 1$ und $xa = 1$ stets eine einzige Lösung haben (ausser wenn $a = 0$ ist) und zwar beide dieselbe, die wir mit a^{-1} bezeichnen wollen. Dann sind ersichtlich die Gleichungen $ax = b$ und $xa = b$ stets auf eine einzige Weise lösbar; ihre Wurzeln sind $\alpha = a^{-1}b$ und $\alpha = ba^{-1}$.

[102]In 2.) and 3.), Hilbert means 'From the fact that $\triangle CDE$ and $\triangle C'D'E'$ resp. $\triangle EFG$ and $\triangle E'F'G'$ have pairwise parallel sides, it follows ...'

Wir wollen noch die *Gleichung der Graden* aufstellen. Jeden Punkt P der Ebene denken wir uns bestimmt durch die Strecken x, y, welche die zu $0X$ und $0Y$ durch P gezogenen | Graden auf den Achsen abschneiden. Es 155
handelt sich darum, eine Beziehung zwischen x und y aufzustellen, die für jeden Punkt x, y einer bestimmten Graden gilt. Betrachten wir zunächst eine durch 0 gehende Grade. Sei $P_1(x_1y_1)$ ein Punkt auf ihr; wir ziehen durch P_1 die Parallelen P_1X_1 und P_1Y_1 zu den Achsen und verbinden X_1 mit Y_1. Endlich ziehen wir durch 1 auf $0X$ die Parallele zu X_1Y_1, die auf $0Y$ die Strecke e abschneiden möge. Dann ist $y_1 = ex_1$. Ist nun $P_2(x_2y_2)$ irgend ein anderer Punkt unserer Graden, so ist nach dem Desargues $X_2Y_2 \parallel X_1Y_1$, folglich $y_2 = ex_2$.[103]

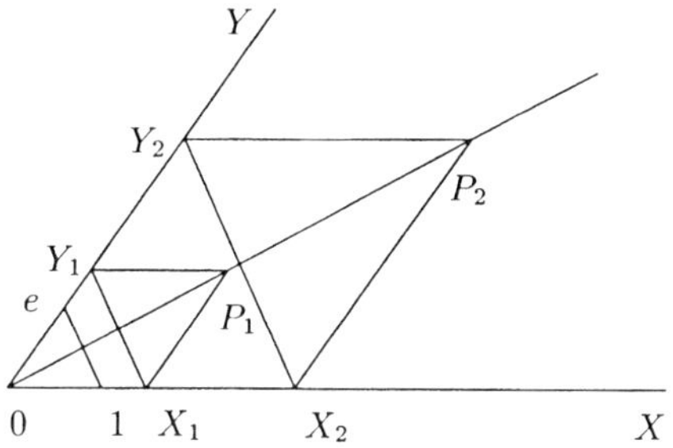

Demnach ist $y = ex$ die Gleichung unserer Graden, wofür wir allgemeiner $ax + by = 0$ schreiben können.

Wir kommen nun zu dem allgemeineren Fall, dass unsere Grade g' nicht durch O geht, sondern etwa durch einen Punkt O' der x-Achse; es sei $OO' = c$. Wir ziehen dann durch O die Parallele g zu g', und beweisen nun, dass, im Sinne unserer Streckenrechnung, g' aus g durch Parallelverschiebung hervorgeht, d. h. dass die zu gleichen y gehö|renden x beider Graden sich um c 156
unterscheiden. Dazu brauchen wir die folgende Figur, in welcher $A'D \parallel O'C$ zu beweisen ist. Wir ziehen als Hülfslinien DD' als Verlängerung von CD, $E'D'$ als Verlängerung von $A'E'$, $O'C' \parallel OY$, AC, $A'C'$, CE, $C'E'$.

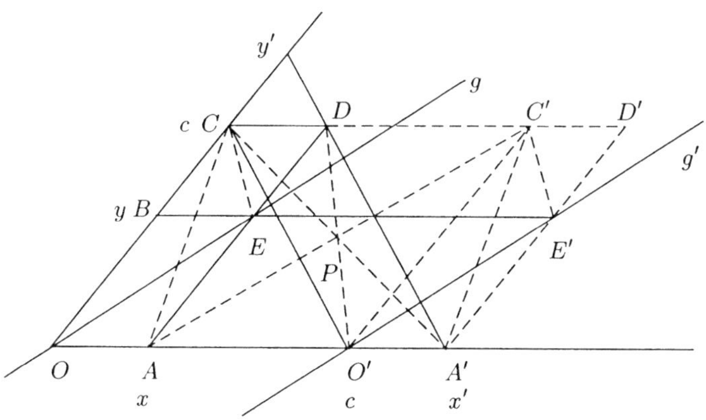

[103]The handwriting in the text and the diagram does not make it uniformly clear which instance of the letters 'x' and 'y' are in upper-case and which in lower-case, or whether we should read 'O' or '0'. However, it seems clear that 'x_i', 'y_i' etc. are generally meant to denote the segments ('lengths') themselves, whereas 'X_i', 'Y_i' etc. denote the corresponding end-points of the segments. The reading offered in the printed text follows this principle; hence '0' stands ambiguously for the null-segment and the point of origin of the coordinate system.

Dann liegen 1) $\triangle OCE$ und $\triangle O'C'E'$ perspektiv, also ist $CE \parallel C'E'$; da 2) $\triangle ACE$ und $\triangle A'C'E'$ perspektiv liegen, ist $AC \parallel A'C'$. Folglich haben 3) $\triangle ACD$ und $\triangle C'A'O'$ die Seiten paarweise parallel, sie liegen also perspektiv, d. h. AC', CA', DO' gehen durch einen Punkt P. Also liegen 4) $\triangle ACO'$ und $\triangle C'A'D$ perspektiv, also ist $CO' \parallel A'D$, q. e. d.

Nun können wir leicht die Gleichung der Graden g' finden. Sind x', y' die
157 Koordinaten eines ihrer Punkte, x, y | die Koordinaten des entsprechenden Punktes von g, so ist nach dem eben bewiesenen Satze stets $x' = x + c$. Die Gleichung von g lautet $ax + by = 0$; demnach ist die Gleichung von g' $a(x - c) + by = 0$, oder wenn $-ac$ mit c bezeichnet wird: $ax + by + c = 0$.

Wir haben also den Satz: *Bei unserer Streckenrechnung wird jede Grade durch eine lineare Gleichung in x und y, bei welcher die Koeffizienten vor den Variablen stehen, dargestellt.* Und man sieht leicht: *Umgekehrt stellt jede Gleichung von solcher Beschaffenheit eine Grade dar.*

Die Angabe, dass die Koeffizienten vor den Variabeln stehen, ist notwendig, weil wir ja das kommutative Gesetz der Multiplikation nicht haben. Es lässt sich leicht zeigen, dass eine Gleichung von der Form $xa' + yb' + c' = 0$ im allgemeinen keine Grade darstellt. Im entgegengesetzten Falle nämlich wäre diese Grade wie eben bewiesen auch durch eine Gleichung $ax + by + c = 0$ darstellbar, beide Gleichungen müssten also genau dieselben Streckenpaare x, y liefern. Bringen wir also (durch rechtsseitige bezw. linksseitige Multiplikation mit a'^{-1} bezw. a^{-1}) die Gleichungen auf die Form $x = yB' + C'$, $x = By + C$, so müsste für jedes y gelten: $yB' + C' = By + C$. Für $y = 0$ kommt $C = C'$,
158 für $y = 1$ kommt dann $B = B'$; d. h. es wäre für | jedes y: $yB = By$. Das ist aber nicht der Fall, da wir eben das kommutative Gesetz der Multiplikation nicht haben. –

Wir können nun umgekehrt zeigen: *Wenn eine Streckenrechnung von der vorher gekennzeichneten Art gegeben ist, so lassen sich aus ihr die sämtlichen ebenen Axiome I und II, sowie der Satz von Desargues ableiten.*

Wir wollen den *Beweis nur für das Axiom II, 5* ausführen, welches von der *Teilung der Ebene durch eine Grade* handelt; wir verfahren analog pag. 121.

Es sei $ax + by + c = 0$ die Gleichung einer Graden g. Wir rechnen einen Punkt x,y der Ebene zur einen oder andern Seite der Graden g, je nachdem $ax + by + c > 0$ oder < 0 ist. Seien x_1y_1 und x_2y_2 Punkte derart, dass:

$$ax_1 + by_1 + c < 0, \quad ax_2 + by_2 + c > 0$$

ist.

Die Gleichung ihrer Verbindungsgraden G sei $Ax + By + C = 0$, sodass also: $Ax_1 + By_1 + C = 0$, $Ax_2 + By_2 + C = 0$ ist.

Wir wollen zeigen, dass G und g sich in einem Punkte x_3y_3 schneiden, der auf G zwischen x_1y_1 und x_2y_2 liegt, d. h. dass $x_1 \gtrless x_3 \gtrless x_2$ ist. Durch linksseitige Multiplikation mit b^{-1} bezw. B^{-1} erhalten wir:

$$b^{-1}ax_1 + y_1 < \; -b^{-1}c \; < b^{-1}ax_2 + y_2$$

und:

$$B^{-1}Ax_1 + y_1 = -B^{-1}C = B^{-1}Ax_2 + y_2$$

also

$$(B^{-1}A - b^{-1}a)x_3 = b^{-1}c - B^{-1}C.^{104}$$ 159

Also wird die obige Ungleichung:

$$(B^{-1}A - b^{-1}a)x_1 > (B^{-1}A - b^{-1}a)x_3 > (B^{-1}A - b^{-1}a)x_2,$$

also: $x_1 \gtrless x_3 \gtrless x_2$. –

NB. Wir haben beim Beweis stillschweigend angenommen, dass gewisse Strecken $\neq 0$ sind; das ist natürlich für den Satz selbst unwesentlich.

Mit Hülfe dieser Resultate können wir nun vorerst eine Frage erledigen, die wir früher (pag. 32)[105] offen lassen mussten; wir wollen nämlich zeigen: *Der Satz von Desargues ist die hinreichende Bedingung dafür, dass die Ebene als ein Stück des Raumes angesehen werden kann.* In der That, wir haben eben gesehen, daß auf Grund des Desargues jede Grade sich darstellen lässt durch $ax + by + c = 0$, und daß umgekehrt aus dieser Darstellung der Desargues folgt. Nun ist es leicht, einen Raum, d. h. ein System von Punkten, Graden, Ebenen, die sämmtliche Axiome I, II erfüllen, anzugeben, von welchem die xy-Ebene ein Stück ist. Als Punkte dieses Raumes nehmen wir alle Streckentripel xyz, als Ebenen die Inbegriffe von Punkten, die einer linearen Gleichung $ax + by + cz + d = 0$ genügen, als Graden die Punktmannigfaltigkeiten, die in zwei Ebenen zugleich liegen, die also den Gleichungen $ax + by + cz + d = 0$, $a'x+b'y+c'z+d' = 0$ gleichzeitig genügen. Hiermit ist denn, bei Heranziehung des Satzes pag. 28 gezeigt: *Die sämtlichen Axiome I, II sind | aequivalent den* 160 *ebenen Axiomen I, II und dem Desargues.*[106]

Nach all diesen Vorbereitungen können wir jetzt an die Frage der Beweisbarkeit des Pascal'schen Satzes gehen. Wir haben auf Grund der Axiome I und II eine Streckenrechnung aufgebaut, in welcher die Gesetze der gewöhnlichen reellen Zahlen gelten, bis auf das kommutative Gesetz der Multiplikation und das Axiom des Archimedes. Wir wissen ferner, dass der Pascal aus $ab = ba$ folgt und umgekehrt. Wir sehen also, dass der Pascal allein auf Grund der Axiome I und II dann und nur dann beweisbar ist, wenn bei Ausschluß des Archimedischen Axioms, das kommutative Gesetz der Multiplikation eine Folge der übrigen Rechnungsgesetze ist; dass ferner der Pascal aus I, II und V dann beweisbar ist, wenn $ab = ba$ eine Folge aller übrigen Rechnungsgesetze, einschliesslich des Archimedischen Axioms ist. Wir werden zeigen, dass der zweite Satz richtig ist, der erste nicht.

Unsere Untersuchung hat uns auf die *Grundgesetze der Arithmetik* geführt. Wir wollen daher jetzt zunächst definieren, wann wir ein System von Dingen $a, b, c, \ldots$ als *komplexe Zahlen im allgemeinsten Sinne* bezeichnen wollen; wir wollen folgendes verlangen:

I a) Aus irgend zwei Dingen a, b soll sich eindeutig ein drittes c finden lassen, was wir durch $a + b = c$ ausdrücken | wollen. Diese Operation heisse *Addition*; 161

[104] This equation comes from subtracting the two equations $B^{-1}Ax_3 + y_3 = -B^{-1}C$ and $b^{-1}ax_3 + y_3 = -b^{-1}c$, since (x_3, y_3) lies on both g and G.

[105] Pp. 31–32.

[106] Here Hilbert means Desargues's Theorem restricted to the plane. The theorem on p. 28 says that Desargues's theorem is independent of the plane incidence and order axioms. Note also that Hilbert has used the existence of unique parallels to prove this result.

gleiches zu gleichem addiert soll gleiches ergeben; aus $a = b$, $b = c$ soll $a = c$ folgen.
I b) Es soll stets möglich sein, auf eine einzige Weise x, y so zu bestimmen, dass $x + a = b$, $a + y = b$ wird. Hierzu fügen wir die Forderung der *Existenz der Null*, d. h. einer Zahl 0, die bei beliebigem a die Gleichungen erfüllt:

$$a + 0 = a, \quad 0 + a = a.$$

II a) Neben der Addition soll es noch eine zweite Weise geben, aus zwei Zahlen a, b eindeutig eine dritte c zu finden: $ab = c$. Auch für diese Operation (*Multiplikation*) soll gelten: aus $a = b$ folgt $ac = bc$, $ca = cb$.
II b) Es soll stets möglich sein, x und y eindeutig zu bestimmen aus $ax = b$, $ya = b$, ausser wenn $a = 0$ ist. Dazu kommt die Forderung der *Existenz der eins*, für welche bei beliebigem a gelten soll $a \cdot 1 = 1 \cdot a = a$. Ferner sei $a \cdot 0 = 0 \cdot a = 0$.

Die genannten Eigenschaften dienen wie gesagt zur Definition des Begriffs „komplexe Zahl". In speziellen Fällen können bei komplexen Zahlen noch *weitere Gesetze* gelten, die wir kurz aufzählen wollen.

1) *Gesetze der Anordnung.*

Zwischen irgend zwei Zahlen a, b besteht eine und nur eine der Beziehungen: $a > b$, $a = b$, $a < b$. Aus $a > b$, $b > c$ folgt $a > c$; Aus $a > b$ folgt
162 $a + c > b + c$, $c + a > c + b$, | $ac > bc$, $ca > cb$[107] u. s. w. Hier lassen sich die Begriffe „positive" und „negative Zahl" einführen; wir nennen a positiv oder negativ, je nachdem $a > 0$ oder $a < 0$ ist.

2) *Die formalen Rechnungsregeln.*

α) Addition: $a + b = b + a$, $a + (b + c) = (a + b) + c$.
β) Multiplikation: $ab = ba$, $a(bc) = (ab)c$.
γ) Distributives Gesetz für Addition und Multiplikation:
$a(b + c) = ab + ac$, $(a + b)c = ac + bc$.

Auf Grund der Existenz der 1 und der formalen Rechnungsregeln ergiebt sich die *Existenz der ganzen Zahlen*: $1 + 1 = 2$, $1 + 1 + 1 = 3$, ... Wir bemerken, dass für ganze Zahlen m, n und beliebiges a die kommutativen Gesetze $mn = nm$, $ma = am$ bewiesen werden können, sobald die übrigen formalen Rechnungsregeln gelten. Näher können wir hier auf diese Fragen nicht eingehen.

3) *Das Archimedische Axiom.*

Ist $0 < a < b$, so lässt sich stets mit Hülfe einer endlichen Anzahl von Summanden erreichen, dass $a + a + a + a + \ldots > b$ wird. –

Wenn alle Eigenschaften 1)–3) gelten, haben wir das System der gewöhnlichen reellen Zahlen vor uns.[108]

---

[107] Assuming $c > 0$. This condition is stated explicitly in the *Festschrift* (*Hilbert 1899c*), 27.

[108] This is not right, for the rational field satisfies conditions 1.) – 3.). It would have been true if Hilbert had meant the Archimedean Axiom in the third version given on pp. 140–141, but the version he states here for numbers is really like the (weaker) first or second. This mistake is not made in the *Festschrift*.

Man kann sich ganz allgemein die Frage vorlegen, wie die genannten Gesetze unter einander zusammenhängen. In unserm Falle – wir wollen ja die beiden | Sätze pag. 146/147 beweisen – haben wir nur folgendes zu zeigen: 163
1.) *$ab = ba$ ist eine Folge aller übrigen Gesetze mit Einschluss des Archimedischen Axioms.*
2.) *$ab = ba$ folgt nicht aus den übrigen Gesetzen, wenn man das Archimedische Axiom fortlässt.*
1.) Bemerken wir zunächst noch, dass die Forderung, n stets so bestimmen zu können, dass $na > b$ wird, im Sinne unserer Streckenrechnung genommen, genau die projektive Fassung des Archimedischen Axioms ist; es lässt sich nämlich leicht zeigen, dass 0, a, $2a$, ∞ etc. vier harmonische Punkte sind. –

Nehmen wir nun an, es sei $a > 0$, $b > 0$ und $ab - ba = d > 0$. Wir wählen eine positive Zahl $\varepsilon < 1$, die ausserdem $< \frac{d}{a+b+1}$ ist. Nach dem Archimedischen Axiom kann man zwei ganze Zahlen m, n so finden, dass $m\varepsilon < a \leqq (m+1)\varepsilon$, $n\varepsilon < b \leqq (n+1)\varepsilon$ ist.

Durch Multiplikation folgt:

$$ab \leqq (m+1)\varepsilon(n+1)\varepsilon, \quad ba > n\varepsilon m\varepsilon,$$

oder, da für die ganzen Zahlen das kommutative Gesetz gilt:

$$ab \leqq mn\varepsilon^2 + (m+n+1)\varepsilon^2$$
$$ba > mn\varepsilon^2.$$

Also:

$$d \leqq (m+n+1)\varepsilon^2 = (m\varepsilon + n\varepsilon + \varepsilon)\varepsilon,$$

also auch

$$< (a+b+1)\varepsilon,$$

d. h.

$$\varepsilon > \frac{d}{a+b+1},$$

was unserer Annahme widerspricht. Damit ist 1.) bewiesen. 164

2.) Um die zweite Behauptung zu beweisen, haben wir ein *System komplexer Zahlen anzugeben, in welchem alle vorher aufgezählten Gesetze gelten, ausgenommen das Archimedische Axiom und das kommutative Gesetz der Multiplikation.* Ein solches System bilden wir uns folgendermassen: Es seien s, t zwei Parameter, für welche das Gesetz $ts = 2st$ gilt, mit denen aber im übrigen nach den gewöhnlichen Regeln gerechnet werden soll. Als komplexe Zahl bezeichnen wir jeden Ausdruck der Form $Z = \sum_{\sigma,\tau}^{\infty} r_{\sigma\tau} s^\sigma t^\tau$ (wo die $r_{\sigma\tau}$ irgend welche reelle, die σ, τ positive oder negative ganze Zahlen bedeuten), wenn folgende Bedingungen erfüllt sind: Unter allen vorkommenden σ giebt es ein kleinstes, σ_0; unter allen mit einem gegebenen s^σ multiplizierten t^τ giebt es eine niedrigste Potenz; diese sei für s^{σ_0} speziell $= t^{\tau_0}$, der zugehörige Zahlenkoeffizient sei r_0. Dann lässt sich, wie leicht zu sehen, Z auf die Form bringen:

$$Z = r_0 s^{\sigma_0} t^{\tau_0} \left\{1 + t \cdot \mathfrak{P}(t) + s t^{-\tau_1} \cdot \mathfrak{P}_1(t) + s^2 t^{-\tau_2} \cdot \mathfrak{P}_2(t) + \dots\right\}$$

Hierin sind $\mathfrak{P}(t)$, $\mathfrak{P}_1(t)$, $\mathfrak{P}_2(t), \dots$ Reihen, die nach ganzen positiven Potenzen von t fortschreiten, die τ_i sind $\geqq 0$, und endlich für jedes endliche i; σ_0, τ_0 sind positiv oder negativ, alle Zahlkoeffizienten reell. Auf die Konvergenz der unendlichen Reihen kommt es gar nicht an. —

Um die *Gesetze der Anordnung* gültig zu machen, definieren wir die Be-
165 ziehungen „grösser", „kleiner" wie folgt. | Sind zwei Zahlen $A = rs^\sigma t^\tau$, $A' = r's^{\sigma'}t^{\tau'}$ gegeben, so setzen wir $A > A'$, wenn entweder $\sigma < \sigma'$, oder $\sigma = \sigma'$, $\tau < \tau'$, oder $\sigma = \sigma'$, $\tau = \tau'$, $r > r'$ ist. Dann besitzt, wenn Z, Z' irgend zwei verschiedene Zahlen sind, die Differenz $Z - Z'$ ein nach dieser Definition grösstes Glied A; wir setzen $Z > Z'$, sobald $A > 0$ ist. Analog wird der Begriff „kleiner" definiert.

Man erkennt, dass auf Grund dieser Definitionen die Gesetze der Anordnung gelten.

Dass Addition, Subtraktion, Multiplikation stets ausführbar sind, ist klar; nur für die *Division* muss dies näher erörtert werden. Es genügt offenbar zu zeigen, dass der Ausdruck $(1 + t\mathfrak{P}(t) + st^{-\tau_1}\mathfrak{P}_1(t) + \dots)^{-1} = (1 - R)^{-1} = 1 + R + R^2 + R^3$ wieder eine Zahl des Bereiches ist. 1.) Da in R nur positive Potenzen von s vorkommen, so gilt dasselbe von $1 + R + R^2 + \dots$ 2.) Für die mit s^m multiplizierten Glieder kommen von R nur die ersten Glieder bis $s^m t^{-\tau_m}\mathfrak{P}_m(t)$ in Betracht. Ist $-\mathrm{T}_m$ die kleinste der Zahlen $-\tau_1$, $-\tau_2, \dots, -\tau_m$, so ist der niedrigste Exponent von t der, von R^k herrührend, im Koeffizienten von s^m auftritt, jedenfalls $> k - m(\mathrm{T}_m + 1)$, also positiv, sobald $k > m(\mathrm{T}_m + 1)$ ist. Also, da $m(\mathrm{T}_m + 1)$ eine endliche Anzahl ist, giebt es unter den mit s^m multiplizierten Potenzen von t stets eine mit kleinsten Exponenten. 3.) Es ist noch zu zeigen, dass zu gegebenem $s^m t^n$ ein endlicher
166 Zahlkoeffizient r_{mn} gehört. Nun kommt für | die Berechnung dieses Koeffizienten nur eine endliche Anzahl von Potenzen von R in Betracht, die nämlich, deren Exponent $< m(\mathrm{T}_m + 1) + n$ ist; aus jeder Potenz von R nur eine endliche Anzahl von Gliedern $s^i t^{-\tau_i}\mathfrak{P}_i(t)$, aus jedem $\mathfrak{P}_i(t)$ höchstens die $n + \tau_i$ ersten Glieder. Also ist r_{mn} endlich. — Damit sind für unser Zahlensystem alle erforderlichen Eigenschaften bewiesen; $Z \cdot Z' = Z' \cdot Z$ gilt natürlich nicht, ebensowenig das Archimedische Axiom. q. e. d. —[109]

Auf Grund des Archimedischen Axioms kann nun die *Einführung der Zahl in die Geometrie* erfolgen.

Seien etwa auf einer Graden die Punkte O, 1 und ein beliebiger Punkt P gegeben. Dann lässt sich nach dem Archimedischen Axiom n so angeben, dass P zwischen n und $n+1$ liegt. Durch fortgesetztes Halbieren der Strecke $n, n+1$ lässt sich dann ein im allgemeinen ins unendliche verlaufender Dualbruch angeben, der als Repräsentant der Strecke OP dienen kann. *Auf diese Weise wird jedem Punkt P der Graden eine ganz bestimmte reelle Zahl zugeordnet; jeder Zahl entspricht höchstens ein Punkt der Graden.* —

[109] Hilbert's presentaton is difficult to follow. For a clear account of this non-Archimedean field, see *Artin 1957*, 43.

Dass auch wirklich jeder reellen Zahl ein Punkt der Graden entspricht, folgt aus unsern Axiomen nicht. Man kann es aber erreichen durch *Einführung von idealen (irrationalen) Punkten (Cantor'sches Axiom.)*. Es lässt | sich zeigen, 167
dass diese idealen Punkte den sämtlichen Axiomen I–V genügen; es ist daher gleichgültig, ob wir sie erst hier oder schon an einer früheren Stelle einführen wollen. Die Frage, ob diese idealen Punkte wirklich „existieren", ist aus dem genannten Grunde völlig müssig; für unsere erfahrungsmässige Kenntniss von den räumlichen Eigenschaften der Dinge sind die irrationalen Punkte nicht nothwendig. Ihr Nutzen ist lediglich ein methodischer; *erst mit ihrer Hülfe ist es möglich, die analytische Geometrie in ihrer vollen Ausdehnung zu entwickeln.* —

Hiermit sind wir am Ziel unserer Untersuchung angelangt. Wir können jetzt mit wenigen Worten die *Frage nach der Verträglichkeit der Axiome I–V* erledigen, anders ausgedrückt, die Frage nach der Existenz der Euklidischen Geometrie. Nach Einführung der analytischen Geometrie ist ja diese Frage der Arithmetik zugewiesen, und wir können sagen: *Die Euklidische Geometrie existiert, sofern wir aus der Arithmetik den Satz herübernehmen, daß die Gesetze der gewöhnlichen reellen Zahlen auf keinen Widerspruch führen.* Damit ist zugleich die Existenz aller derjenigen Geometrieen nachgewiesen, die wir im Laufe der Untersuchung betrachtet haben. — [110]

Jetzt können wir auch übersehen, daß *jeder Schnittpunktsatz eine Folge aus Desargues und Pascal ist.*

Jeder Schnittpunktsatz hat doch folgende Form: Man nimmt gewisse 168
Punkte und Graden willkürlich an, nur dass etwa bestimmte Punkte auf bestimmten Graden liegen, und konstruiert dann neue Schnittpunkte. Die Behauptung ist dann, dass gewisse drei Punkte auf einer Graden liegen, bezw. gewisse drei Graden durch einen Punkt gehen. In die Sprache der analytischen Geometrie übersetzt heisst dies: *Ein Ausdruck* $A(p_1, p_2, p_3, \ldots)$*, der aus den Parametern* p_1, p_2, $p_3, \ldots$ *vermittelst der vier Spezies zusammengesetzt ist, soll identisch, d. h. für alle Parameterwerte verschwinden.* Dieser Ausdruck muss dann verschwinden auf Grund der formalen Rechnungsgesetze, d. h. aber wie wir früher sahen, auf Grund der Pascal und Desargues. Man erkennt hieraus, dass man selbst durch Projektion aus beliebig hohen Räumen keine neuen Schnittpunktsätze erhalten kann. — [111]

Wir haben in dieser Vorlesung gewissermassen eine *Theorie* der Geometrie gegeben; wir wollen nun noch eine Bemerkung über die *Anwendung* dieser Theorie *auf die Wirklichkeit* machen. *Die geometrischen Sätze gelten in der Natur niemals mit voller Genauigkeit, weil die Axiome von den Objekten niemals genau erfüllt werden.* Dieser Mangel an Uebereinstimmung liegt aber im Wesen | jeder Theorie; denn eine Theorie, die bis ins einzelne mit der 169
Wirklichkeit übereinstimmte, wäre nur mehr eine genaue Beschreibung der Dinge. — [112]

---

[110] See Hilbert's own lecture notes, and the note there.

[111] For elucidation of these two paragraphs, see the Introduction to this Chapter, Specific Note (1)(*c*).

[112] See Hilbert's lecture notes, p. 106.

Ein wesentliches Stück unserer Untersuchung waren die *Beweise für die Unbeweisbarkeit* gewisser Sätze; wir erinnern hier zum Schluss daran, dass derartige Beweise in der modernen Mathematik überhaupt eine grosse Rolle spielen und sich als fruchtbar erwiesen haben; man denke nur an die Quadratur des Zirkels, an die Auflösung der Gleichungen 5. Grades durch Wurzelziehen, an Poincaré's Satz, dass es beim Dreikörperproblem eindeutige Integrale ausser den bekannten nicht giebt, etc.[113]

Weiter seien noch einige interessante *Probleme* genannt, *die bis jetzt noch unerledigt sind.*

1.) Lassen sich Pascal und Desargues vielleicht beweisen aus den ebenen Axiomen I, II, III, ohne IV?

2.) Beweis der Legendre'schen Sätze über die Winkelsumme im Dreieck aus I, II, III, ohne V.

3.) Begründung der Lehre von der Messung von Raumvolumina. —[114]

Wir schliessen diese Vorlesung mit Bemerkungen über:

170 *Elementare geometrische Konstruktionen.*[115]

Eine erste Frage ist: *Kann man die mit Zirkel und Lineal ausführbaren Konstruktionen nicht auch mit geringeren Hülfsmitteln ausführen?*

Es zeigt sich hier, dass der Zirkel das Lineal überflüssig macht; dass ein einziger fester Kreis den Zirkel überflüssig macht.

Weiter kann man fragen: *welche Konstruktionen können allein auf Grund der Kongruenzsätze ausgeführt werden?*

Aus früheren Betrachtungen [pag. 60ff.] ergiebt sich, dass es hier nicht möglich ist, z. B. ein Dreieck aus seinen drei Seiten zu konstruieren. Dagegen lassen sich alle diejenigen *Konstruktionen* ausführen, *bei denen man mit Strecken- und Winkel-abtragen auskommt.* Es zeigt sich also, dass es einen Unterschied macht, ob man den Zirkel unbeschränkt oder nur zum Abtragen von Strecken und Winkeln gebraucht. Analytisch sind die letzteren Konstruktionen dadurch charakterisiert, dass ausser den vier Spezies nur noch die Operation $\sqrt{1+x^2}$ vorkommt.

171 Hieran schliesst sich die weitere *Frage nach den* | *geringsten Hülfsmitteln, mit denen sich die letztgenannten Konstruktionen ausführen lassen.* Es ergiebt sich, dass *das Abtragen von Strecken allein bereits das Winkelabtragen enthält.* Wir wollen dies, da sich dabei allerlei einfache Konstruktionen ergeben, kurz zeigen, indem wir der Reihe nach folgende Aufgaben lösen:

1.) *Durch einen gegebenen Punkt A zu einer gegebenen Graden eine Parallele zu ziehen.*

---

[113]See also Hilbert's own notes (p. 106), and the *Festschrift* (*Hilbert 1899c*, 89), *Schlusswort.*

[114]These are the same problems as are mentioned in Hilbert's own notes, p. 106; for comments, see the Introduction to this Chapter, Specific Note (5).

[115]See Hilbert's remarks on constructions in his notes to this text (in the following Appendix), and Ch. VIII of the *Festschrift* (*Hilbert 1899c*), which is a generalization and completion of this section.

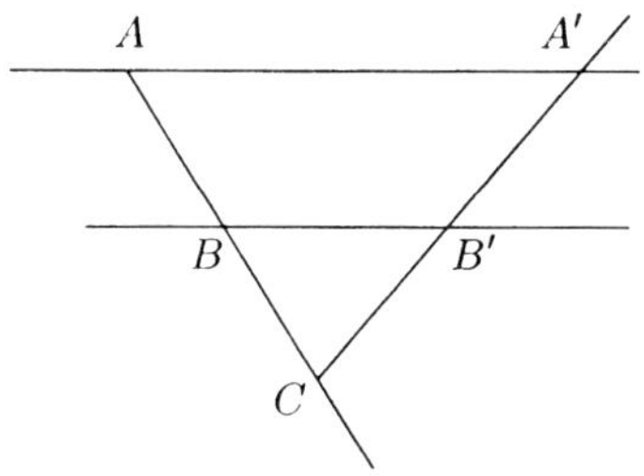

Man ziehe AB beliebig, mache $BC = AB$, ziehe CB' beliebig, mache $B'A' = CB'$; dann ist $AA' \parallel BB'$.

2.) *In einem gegebenen Punkt A einen rechten Winkel zu konstruieren.*

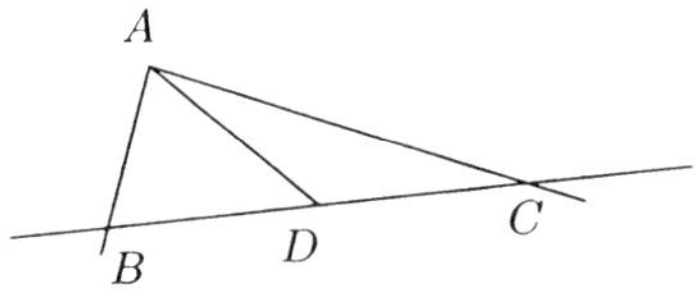

Man ziehe durch A eine beliebige Grade AD, durch D eine beliebige Grade BC und mache $AD = BD = CD$. Dann ist $\sphericalangle BAC$ ein Rechter.

3.) *Von einem gegebenen Punkt A auf eine gegebene | Grade eine Senkrechte zu fällen.* 172

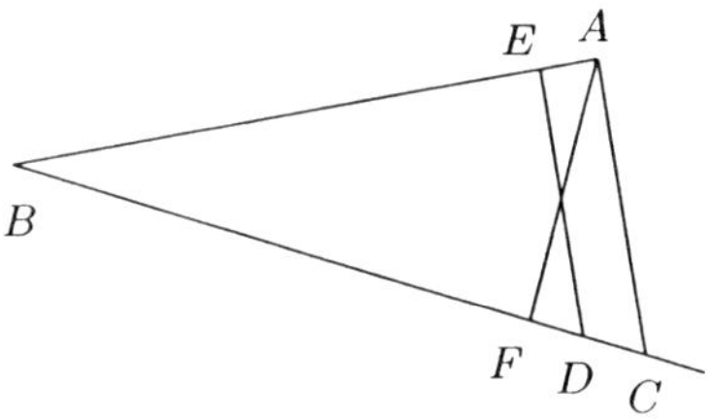

Man konstruiere nach 2.) in A einen rechten Winkel BAC, mache $BD = BA$, ziehe nach 1.) $DE \parallel AC$ und mache $BF = BE$. Dann ist $AF \perp BC$.

Analog kann man in F auf BC die Senkrechte errichten.

4.) *Eine gegebene Strecke AB zu halbieren.*

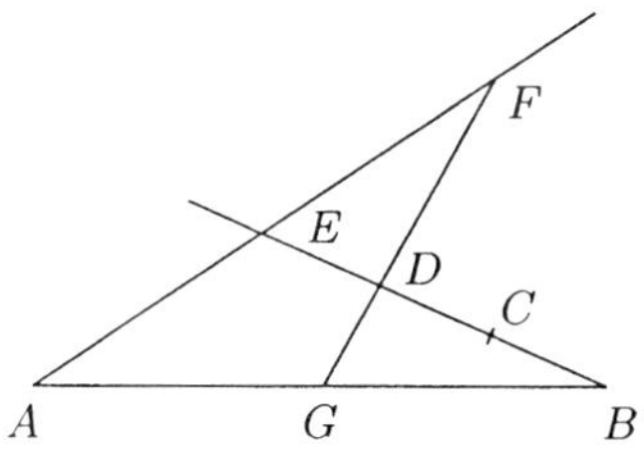

Man ziehe BC beliebig, mache $BC = CD = DE$, $AE = EF$, und ziehe FDG; dann ist $AG = GB$.

5.) *Einen gegebenen Winkel α an eine gegebene Grade BC anzutragen.*[116]

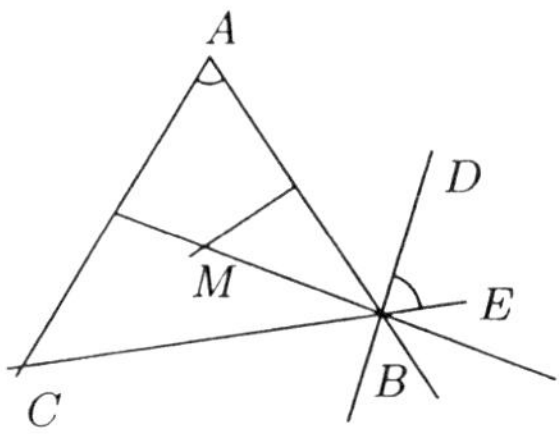

Man konstruiere nach 4.), 3.) den Mittelpunkt M des dem $\triangle ABC$ umbeschriebenen Kreises, und ziehe $BD \perp MB$. Dann ist $DBE = \alpha$.[117] Durch Parallelenziehen kann man den Scheitel dieses Winkels | an jeden vor- 173 gegebenen Punkt auf BC bringen. –

[116] In the Fig., α is the angle $\sphericalangle CAB$.

[117] E is a point on the line BC so chosen that B lies between C and E.

Weiter zeigt sich, *dass das Streckenabtragen durch Winkelhalbieren ersetzt werden kann.* —[C]

Endlich kann man fragen, welche Aufgaben nun auf Grund der fünf oben charakterisierten Spezies lösbar sind. Es ergiebt sich das interessante Resultat, dass *alle überhaupt mit Zirkel und Lineal konstruierbaren regulären Polygone durch Streckenabtragen allein konstruierbar* sind.

Beim regulären Dreieck z. B. kommt es auf die Konstruktion von $\sqrt{3}$ an; und diese ist ausführbar, denn es ist $\sqrt{3} = \sqrt{1 + (\sqrt{2})^2}$, $\sqrt{2} = \sqrt{1+1}$.[118]

[C] Um eine Senkrechte auf einer gegebenen Geraden a zu ziehen, kann auch folgende Construktion dienen, wo man $FG = FC = FD$ und $EG = DG$ und $AG = CG$ macht.

⟨Added by *Hi*[s]⟩

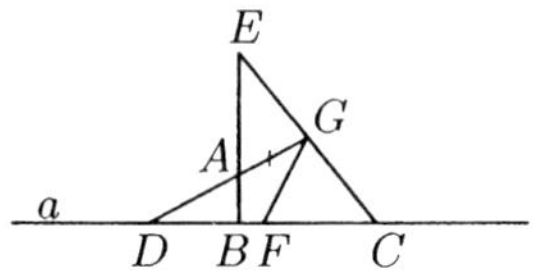

[118] See Hilbert's 1898 *Ferienkurs* (*Hilbert 1898b**, 15), this Volume, Chapter 3.

Inhalt.

Appendix: Hilbert's Notes at the Beginning of 'Elemente der Euklidischen Geometrie'

[1] Vgl. die an mich gerichteten Briefe von Hausdorff u. Liebmann.

In a letter to Hilbert dated 12 October, 1900 (*Cod. Ms. D. Hilbert 136*), Hausdorff discusses various points arising from the *Festschrift* and von Schaper's *Ausarbeitung* of the 1898/1899 lectures. The first point concerns Hilbert's Axiom I 2 (*Festschrift*, 5), and asserts (with a model) that Hilbert's paraphrase of it which follows immediately is not in fact equivalent. (The paraphrase is in the 1898/1899 lectures as well as the *Festschrift*. There is no such paraphrase in the second edition; see *Hilbert 1903a*, 3.) The second concerns the proof sketched in the *Ausarbeitung*, p. 10, for the independence of I 5 from the other incidence axioms; Hausdorff points out that Axioms I 3 and I 4 also both fail in the model Hilbert gives. (See note to the *Ausarbeitung*, p. 10.) Hausdorff then gives a model which does show the required independence. This independence is not explicitly dealt with in the *Festschrift*. The third main point concerns Axiom I 7, which Hausdorff considers unneccessary. The Axiom asserts that every line contains at least two points, every plane at least three non-colinear points, and that there are at least four non-coplanar points. In the second edition (*Hilbert 1903a*, 3), Hilbert splits this Axiom into two parts; the first, concerning the lines and the planes, is inserted as an addition to the plane axioms, as Axiom I 3, and the second, Axiom I 8, is added to the spatial axioms. The last point concerns the theorem that in any triangle, two sides taken together are always greater than the third. In the *Ausarbeitung*, this is proved (pp. 51–52) on the basis of the Triangle Congruence Axiom, whereas Hausdorff points out that Hilbert's own *Hilbert 1895b* establishes this theorem without any such 'plane congruence axiom'. It is to be noted that this work of Hilbert operates within a nowhere concave solid, where the notion of *length* can be defined, and as Hausdorff says the theorem is not a 'mere consequence' of the other axioms. The theorem is not mentioned in the *Festschrift*, nor in subsequent editions of the *Grundlagen*.

The corresponding letter from Liebmann is not in the Hilbert *Nachlaß*.

---

[2] Notwendigkeit einer Definition des Begriffes „zwischen" hat schon Gauss erkannt. Vgl. Werke Bd. 8 S. 222.

Hilbert is referring to *Gauß 1900*, 222. This concerns the mathematical parts of a letter from Gauß to Wolfgang Bolyai from 6 March, 1832 in which Gauß gives his reactions to some of the results of Johann Bolyai presented in *Bolyai 1832a*. (The complete letter is in *Gauß et al. 1899*, 107–113.) In particular, Gauß gives a 'purely geometrical proof' for the theorem that, in hyperbolic geometry, the difference of the sum of the angles of a triangle from two right-angles is proportional to the area of the triangle. The first lemma concerns a curvilinear, asymptotic triangle (thus one whose vertices all lie at infinity) which has a definite finite area. The area included by the three lines he calls the area 'between' the lines. In a footnote he adds:

> Bei einer vollständigen Durchführung müssen solche Worte, wie »zwischen«, auch erst auf klare *Begriffe* gebracht werden, was sehr gut angeht, was ich aber nirgends geleistet finde.

Gauß's proof sketch is not really 'purely geometrical' by modern standards, nor is the 'between' relation mentioned here Hilbert's. Nevertheless, Gauß's remark might be seen as a general plea for the use of precise definitions and concepts.

---

[3] „Possibilitatem plani demonstravi" Gauss.

This remark, attributed to Gauß, is actually to be found in Gauß's *Tagebuch*, entry 72 (dated Göttingen, 28 July, 1797), which reads simply: 'Plani possibilitatem demonstravi', or 'I have proved the possibility of the plane'. (See *Gauß 1917*, 521.) This was clearly of great interest to Gauß. In the letter to Bolyai from 6 March, 1832, and cited in the note to Remark [2], Gauß writes:

> Um die Geometrie vom Anfange an ordentlich zu behandeln, ist es unerlässlich, die Möglichkeit eines Planums zu beweisen; die gewöhnliche Definition enthält zu viel, und implicirt eigentlich subreptive schon ein Theorem. Man muss sich wundern, dass alle Schriftsteller von Euklid bis auf die neuesten Zeiten so nachlässig dabei zu Werk gegangen sind; allein diese Schwierigkeit ist von durchaus verschiedener [Natur] mit der Schwierigkeit zwischen Σ und S zu entscheiden, und jene ist nicht gar schwer zu heben. Wahrscheinlich finde ich mich auch schon durch Dein Buch hierüber befriedigt. (*Gauß 1900*, 224, or *Gauß et al. 1899*, 112. The brackets around 'Natur' are in the published versions of the letter.)

(Σ and S refer to the systems of Euclidean geometry and Johann Bolyai's absolute geometry respectively. A similar passage can be found in the letter from Gauß to Bessel, 27 January, 1829; see *Gauß 1900*, 200, or *Gauß et al. 1880*, 487–490, for the complete letter.) The 'usual definition' of a plane which Gauß refers to is: a plane is a surface such that any straight line joining two points of the surface lies wholly within the surface. Gauß goes into this further in two short pieces in the *Nachlaß* marked [2] and [3] respectively (see *Gauß 1900*, 194–195). Stäckel, the editor of the geometrical part of Gauß's *Nachlaß*, dates [2] to between 1820 and 1830, and [3] to March 1832 since 'sie bezieht sich auf Gauss's Brief an Bolyai vom 6. März 1832'. (See *Gauß 1900*, 199, and also Stackel's note to entry 72 in the Gauß *Tagebuch*, *Gauß 1917*, 521.) In [2], Gauß notes that the definition of a plane involves a theorem which effectively asserts that when ABD, AFG, ACE are three lines coming out from a point A and when the straight line joining B with C goes through the point F, then the straight line joining D and E cannot go 'above' or 'below' the line AFG but must intersect it. He does not give the proof of this theorem but claims that it is not too difficult. In [3], Gauß characterizes a plane differently. Suppose we have a line AB and consider the point A; all the straight lines through A which stand at right angles to AB are in (or form) a plane. In other words, a plane is generated at the point A when we take a line perpendicular to AB and rotate it with AB as axis. Gauß then proves that the straight line through any two points of a plane characterized in this sense lies wholly in the plane, thus that the plane so characterized is a plane in the old sense. Presumably Stäckel's reason for tying [3], but not [2], to Gauß's letter to Bolyai of 6 March 1832 is that the former could equally well be tied to the letter to Bessel of 27 January, 1829, and that the latter, unlike the former, attempts to give a more satisfactory definition. Gauß's result in [3] is what Hilbert, in Remark [9] below, sets himself to prove. Note Hilbert's further mention of Gauß's 'Plani possibilitatem demonstravi' in the last of these remarks, [19].

Note also that, in the geometrical axiom system adopted in Hilbert's 1898/1899 lectures and the *Festschrift*, the 'old' characterization of a plane is built into one of the axioms, I, 5 (*Existenzaxiom* 5 in the 1893/1894 lectures).

---

[4] Parallelen Axiom ist auch ersetzbar (nach Gauss) durch die Forderung eines ∞ grossen Dreieckes mit ∞ grossen Seiten.

It is not clear exactly what Hilbert's statement means. One possibility is that it refers to the following comment of Gauß in a letter to Wolfgang Bolyai of 16 December, 1799:

> Zwar bin ich auf manches gekommen, was bei den meisten schon für ein Beweis gelten würde, aber was in meinen Augen so gut wie NICHTS beweist, z.B. wenn man beweisen könnte, dass ein geradliniges Dreieck möglich sei, dessen Inhalt grösser wäre als eine jede gegebene Fläche, so bin ich imstande die ganze Geometrie völlig

streng zu beweisen. Die meisten würden nun wohl jenes als ein Axiom gelten lassen; ich nicht; es wäre ja wohl möglich, dass, so entfernt man auch die drei Endpunkte des Dreiecks im Raume von einander annähme, doch der Inhalt immer unter (infra) einer gegebenen Grenze wäre. Dergleichen Sätze habe ich mehrere, aber in keinem finde ich etwas Befriedigendes. (*Gauß 1900*, 159, or *Gauß et al. 1899*, 36–37. The complete letter can be found in the second source on pp. 34–38.)

---

[5] In der Lobatschefkyschen Geometrie kann man ohne Archimedes die Flächengleichheit von Dreiecken mit gleicher Winkelsumme in der Ebene durch Zerschneiden nachweisen. Vgl. Bolyai-Frischauf S. 46 etc. u. Frischauf S. 38 durch Benutzung der Linien gleichen Abstandes.

The works referred to in this remark are Frischauf's treatment of Bolyai's geometry, *Frischauf 1872*, and Frischauf's own textbook on absolute geometry, *Frischauf 1876*. The theoretical construction Hilbert refers to here is much the same in both treatments, thus pp. 46–51 in the first, and pp. 38–42 in the second. Frischauf states (*Frischauf 1872*, 46, or *Frischauf 1876*, 38):

> Die Linien gleichen Abstandes gestatten eine Auffindung der Beziehung zwischen der Winkelsumme und der Fläche eines geradlinigen Dreiecks.

An 'equidistant line' to a given line PQ is one such that the perpendiculars to PQ from any two points on the equidistant line are of equal length. In Bolyai's 'absolute geometry' one can use 'equidistant lines' to prove an analogue of Euclid's theorem that triangles on the same base and between the same parallels have the same area. (See *Elements*, Book I, Proposition 37, i.e., *Heath 1926a*, 332.) Take two lines which are both equidistant to a given line and a base NR formed from the straight line connecting two points on one of the equidistant lines. The result in absolute geometry is then: any two triangles with the straight line base NR and whose third apex is on the other equidistant line have the same angle-sum and the same area. Using this result, one can then show quite generally that $\triangle ABC$ has the same angle sum as $\triangle A'B'C'$ if and only if they have the same surface area, and further that the surface areas of the two triangles are in the same ratio as the difference in their angle-sums from two right-angles. As Hilbert in effect points out, there is no use of the Archimedean Axiom in these proofs.

---

[6] A. Adler hat Wiener Akad. 1891 gezeigt, dass durch Verschiebung eines rechten Winkels die allgemeinste cubische Gl⟨eichung⟩ aufgelöst werden kann.

Hilbert here is referring to the paper *Adler 1890*, which deals with constructions executable by a draughtsman whose only means of construction is a wooden right-angle. He shows that, assuming one has several right-angled triangles at one's disposal, one can solve in principle any cubic equation, and with perfect accuracy, assuming the means used to draw the right-angles are perfect. Hilbert's remark implicitly suggests replacing the use of the draughtsman's right-angle tool with the 'movement' of an angle, that is, with a congruence assumption which allows one to construct right-angles at any point and on any line. The Adler referred to here was very probably the *Ausarbeiter* of the 1902 lectures, given here in Chapter 6.

---

[7] Kürschak hat gezeigt, dass man beim Streckenabtragen nur nöthig hat eine bestimmte Strecke abtragen.

The reference is to Josef Kürschák, and to his paper *Kürschák 1902*. Kürschák here shows that all constructions which can be carried out by means of a straightedge and some device for moving segments of arbitrary length (a *Streckenübertrager*) can be carried out instead by using a straightedge and what he calls an 'Aichmass', a device that allows one to measure off just a single given segment a. This is an effective simplification of the treatment of constructions in Hilbert's *Festschrift* (*Hilbert 1899c*), §36, where Hilbert uses straightedge and a *Streckenübertrager*. From the second edition of the *Grundlagen* on (*Hilbert 1903a*, 74), Hilbert refers to Kürschák's simplification using only an 'Eichmass'.

---

[8] Ohne Dreiecks-Congruenzsätze ist $AB \equiv BA$ auch mit Archimedes nicht beweisbar; denn nimm zum Beweise $AB = \frac{1}{2}BA$ etc.

See the Introduction to this Chapter, Specific Note (6)(a).

---

[9] Beweise den Satz, dass eine Grade, die mit der Ebene 2 Punkte gemein hat, ganz in der Ebene liegt, aus den Congruenzsätzen durch die bekannte geometrische Construktion, mit deren Hülfe man nach den Schulbüchern beweist, dass 3 auf einer Geraden in einem Punkte $\perp$ Geraden in einer Ebene liegen (Gauss: possibilitatem plani demonstravi).

See note to Remark [3].

---

[10] Wahrscheinlich genügt es – wenigstens wenn Parallelenaxiom und Archimedes angenommen wird – statt des Dreieckscongruenzsatzes den Satz von der Gleichheit der Basiswinkel im gleichschenkligen Dreieck als Axiom zu nehmen. – Suche den Desargues aus Pascal zu beweisen[[, wobei es nöthig sein wird den Sommerschen Beweis für den Pascalschen Satz in meinen Grundlagen so zusammenzuziehen, dass dabei nur die Thatsache: wenn $\sphericalangle\alpha = \sphericalangle\beta$ so $\sphericalangle\alpha + \delta = \sphericalangle\beta + \delta$ benutzt wird. Es ist leicht den Satz vom Sehnenviereck bloss mit Hülfe dieser Thatsache und Parallelverschieben, ohne Congruenzsätze zu beweisen:

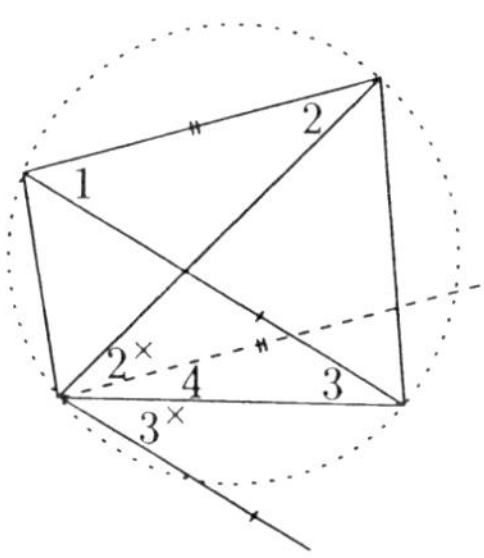

Voraussetzg $\sphericalangle 2 = \sphericalangle 3$. Man ziehe die punktirten Parallelen I u. II. Dann ist $\sphericalangle 2^{\times} = \sphericalangle 2$, $\sphericalangle 3^{\times} = \sphericalangle 3$. $\sphericalangle 4 + 3^{\times} = 1$ folglich $4 + 2^{\times} = 1$ q. e. d.]]H.↓[1]

This sequence of remarks deserves four comments.

(1) The axiomatic investigation of the theorem showing the equality of the base angles in an isoceles triangle is the subject of Hilbert's *Hilbert 1902/03*. (It is also dealt with in Hilbert's 1902 lectures, pp. 107–125a, here Chapter 6.) See the discussion in the Introduction to Chapter 6.

(2) Hessenberg showed in *Hessenberg 1905b* that Desargues's Theorem *can* be proved from Pascal's Theorem. (Hessenberg gives two proofs, one for the affine case and one for the projective.) Hilbert's remark confirms that this and the previous notes were added before 1905.

(3) In the remark about proving Desargues from the Pascal Theorem, note that all the material referring to the 'Sommerschen Beweis' has been crossed out by Hilbert. The reference to Sommer must be to his *Sommer 1899/1900*, where (p. 297) Sommer sketches a somewhat different proof of Hilbert's Pascal Theorem from Desargues's Theorem based on

---

[1] The material enclosed in double square brackets was crossed out by *Hi*^{s}.

the assumption of commutativity. There seems no relation between Sommer's proof sketch, or Hilbert's further suggestion, and Hessenberg's proof.

(4) In the last part of the remark, the proof Hilbert gives concerns four points lying on a circle. Hilbert's diagram, however, omits the circle; we have included it in our rendering of the diagram.

---

[11] Wenn man einen Kreis mit Sinne durch die unendlichferne Gerade hindurch führt, so kommt derselbe mit umgekehrtem Sinne zurück. Wenn man also von 2 Paaren Handschuhen die rechten verloren hat, so braucht man nur einen der übrig gebliebenen durch die ∞ ferne Ebene hindurch zu werfen, um ein richtiges Paar zu erhalten.

⟨III⟩

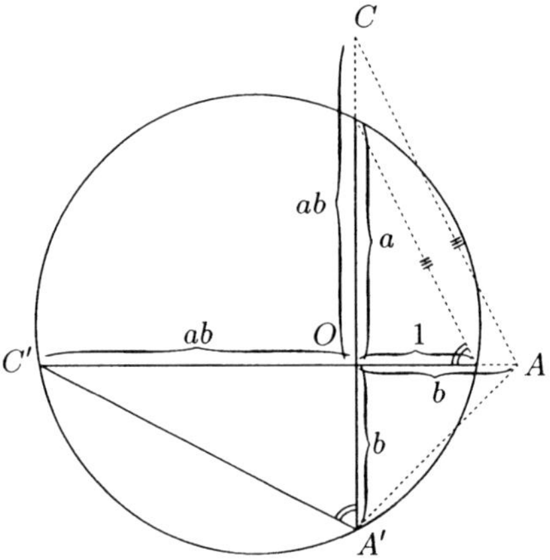

$\triangle OAC \equiv \triangle OA'C'$

[12] ⟦Der von mir modificirte Mollerup'sche Beweis für $ab = ba$ beruht darauf, dass man nachweist, dass meine Construktion von ab auch so gemacht werden kann, dass man a, b entgegengesetzt nach oben u. unten abträgt und dann durch die Endpunkte von a, b, 1 einen Kreis schlägt, welcher dann entgegengesetzt zu 1 die Strecke ab abschneidet. Da ist ja ab symmetrisch in a, b.⟧[2]

Note that this whole remark, with the diagram, has been stricken through by Hilbert. The reference is clearly to *Mollerup 1903a* (see also *Mollerup 1904*), and this is referred to again in Hilbert's 1902 lectures, 96. In this article, Mollerup gives a simple and ingenious proof for the commutativity of segment multiplication, different from Hilbert's. It involves first taking two lengths/segments OA, OB, emanating from a central point O, and then a third OD which goes in the opposite direction from OA. The points A, B and D determine a circle. The proof then centres on the properties of the various similar triangles determined by the relevant chords. At the end of the paper, Mollerup states (p. 280):

> Herr Hilbert macht mich darauf aufmerksam, dass meine Vereinfachung des Beweises des commutativen Gesetzes auch in seiner Darstellung erreicht wird, wenn er das Streckenproduct ab mittlest eines Kreises definirt. Auf die Schenkel eines rechten Winkels O werden die Strecken $OA = a$ und $OC = 1$, endlich $OB = b$ in entgegengesetzter Richtung von OA abgetragen. Die Punkte A, B und C bestimmen einen Kreis, der zum zweiten Male von CO in D geschnitten wird. OD wird dann dem Streckenproduct ab gleich gesetzt; dieses Product ist dann in den Factoren symmetrisch. Natürlicherweise ist der Beweis des Satzes dem Hilbert'schen ganz analog.

Thus, Mollerup describes Hilbert's construction given here. If we let c stand for what Hilbert defines as ab, then we have both the ratios $\frac{a}{c} = \frac{1}{b}$ and $\frac{c}{a} = \frac{b}{1}$, from which it follows that c equals both ab and ba.

---

[2]Deleted by *Hi*^{s}.

[13] Die gewöhnliche analytische Geometrie nenne Cartesische Geometrie.

---

[14] ⟦Wie beweist man das Parallelenaxiom in der Ebene ohne Archimedes allein auf Grund meiner Axiome I–II, III wenn man weiss dass es ein Dreieck gibt dessen Winkelsumme = 2R ist.⟧[3]

Note that this passage is stricken through here. According to Dehn's results (see *Dehn 1900b*), without the Archimedean Axiom, it cannot be shown that, if the angle-sum in all triangles (indeed, just one) is two right-angles, then the Parallel Axiom holds. Thus, without the Archimedean Axiom, the angle-sum theorem is not a replacement for the Parallel Axiom. Hilbert first states Dehn's results himself in the French translation of the *Festschrift* (*Hilbert 1900c*, 207), and then in subsequent editions of the *Festschrift*. This suggests that this particular note was written before Dehn results were established. See the Appendix and Introduction to Chapter 5.

---

[15] In der Physik und in der Natur überhaupt auch in der praktischen Geometrie gelten alle Axiome (vielleicht auch sogar das Archimedische Axiom) nur angenähert. Man muss aber die Axiome genau annehmen und dann die genauen logischen Consequenzen ziehen, weil man sonst gar keinen logischen Überblick erhielte. Nothwendig endliche Anzahl von Axiomen, wegen der Endlichkeit unseres Denkens.

Compare the remark about the approximate truth of axioms with that in Hilbert's own notes for the 1898/1899 lectures, 106. Note Hilbert's insistence on the finitude of axiomatisation. Hilbert also adumbrates this condition in *Hilbert 1900b* and *Hilbert 1900e*, but is not explicit about the reason, although the strong implication there is that the finitude of the axiomatization of the theory of real numbers liberates us from any doubt concerning the 'actual' existence of an uncountable infinity. Here, Hilbert gives a more general reason why axiomatizations must be finite.

---

[16] Wenn man die Nicht-Euklidische Geometrie durch die Geraden innerhalb eines Kreises veranschaulicht und die Entfernungen durch den log. des Doppelverhältnisses, so kann man als Winkel, den wirklichen Winkel nehmen, den die zur Ebene durch die beiden Geraden gelegten Ebenen auf der durch den Kreis gelegten Halbkugel bestimmen.

Hilbert here appears to refer to the model of hyperbolic geometry known as the (plane) Poincaré model. This can be achieved as follows. Start from the model of hyperbolic geometry in the interior of a disc in the plane, where the straight lines are the Euclidean chords of the disc (minus the end-points); then set a sphere on the disc touching the plane at its centre, and with radius the same as that of the disc. Consider now the larger disc P on the plane formed by the stereographic projection of the equator of the sphere. Take a straight line in the original model and then the plane it determines perpendicular to the given plane. This plane cuts out a semi-circle on the lower half-sphere; project this down to the plane by the stereographic projection, and we get an arc of a circle which is perpendicular to P at its end-points. These arcs (minus end-points) are the straight-lines of the new model, and Hilbert is referring to distance and angles with respect to these. For Hilbert's very clear presentation of this, see *Hilbert and Cohn-Vossen 1932*, §36.

---

[3]Deleted by *Hi*^{s}.

[17] Complexe Geometrie = Geometrie, in der nicht alle Axiome gültig sind nach Analogie der complexen Zahlen. Auch die Analysis des Unendlich z. B. wann gelten die Gesetze des gliedweisen Integrirens, Stetigkeit unendlicher Reihen etc., die sonst nur für endliche Systeme gelten, ist in diesem Sinne eine complexe Algebra.
Wie ist es mit einer complexen Elektricitätstheorie?

Note that in the *Ausarbeitung* of the 1898/1899 lectures (pp. 142, 160–162), and in the *Festschrift* (*Hilbert 1899c*, §13), Hilbert calls the axioms for ordered fields (rather misleadingly) axioms for 'complex number systems'. The allusion is thus to the examination of the geometries which arise when various key axioms are not satisfied, i.e., non-Euclidean, semi-Euclidean, non-Archimedean, non-Pythagorean geometries. (Cf. Hilbert's 1891 lectures, p. 4, in Chapter 1.) What Hilbert implicitly expresses then is a programme along similar lines which applies potentially to all axiomatic investigations, particularly in physics. The reference to the treatment of infinity is presumably to the fact that some principles which hold of finites (sets or numbers) apply to infinites, too, while others are not satisfied.

⟨IV⟩ [18] Vgl. Ueber den Ursprung u. die Bedeutung der geometrischen Axiome. Vorträge u. Reden von Helmholtz Bd II S. 1, S. 391, S. 394

The page references are to *von Helmholtz 1896*; on p. 1 begins Helmholtz's essay 'Ueber den Ursprung und die Bedeutung der geometrischen Axiome' (1870), which Hilbert mentions, and then on pp. 391 and 394 begin respectively two Appendices to a later essay by Helmholtz, 'Die Thatsachen in der Wahrnehmung' (1878), which is also contained in the same volume (pp. 213–247). One of these Appendices is entitled 'Der Raum kann transcendental sein, ohne dass es die Axiome sind' (pp. 391–393), and the other 'Die Anwendbarkeit der Axiome auf die physiche Welt' (pp. 394–406), and both refer to specific places in the lecture they are meant to elucidate. It is possible that Hilbert is referring to Volume 2 of the fifth edition of Helmholtz's *Vorträge und Reden* which appeared in 1903, since the pagination is the same.

---

[19] Vor Allem arbeite genau durch Gauss Werke Bd 8 S. 159–268 insbesondere Begründung des Planum S. 194.

The reference is to the section in *Gauß 1900* devoted to geometry (157–268). Hilbert also refers expressly to the two notes on the theory of the plane which are on pp. 194–195 of loc. cit. See the note to the Hilbert's remark [3], above.

Textual Notes

303.11–12: Mit ⟨...⟩ Mechanik“)] ⟦Mit Benutzung eines Ausdrucks von *Hertz* (in der Einleitung zu den „Prinzipien der Mechanik“)⟧ ⟨deleted by Hi^g⟩

303.19–20: die ⟨...⟩ unerörtert. —] ⟦die Frage, ob unsere Raumanschauung apriorischen oder empirischen Ursprung habe, bleibt dabei unerörtert. —⟧ ⟨deleted by Hi^g⟩

303n: = Verhalten ⟨...⟩ Raume] Verhaltn der ⟦⟪raum⟫⟧ Gegenstnde ⟪nach⟫ Gestalt = Lage im Rame

304.33: untersuchen.] untersuchen. ↑Zunächst Alles in der Ebene↓ ⟨added by Hi^g⟩

304n: 16] ⟦17⟧⊢↑16↓ ⟨Hi^g⟩

305.3–8: 1. ⟨...⟩ verbindet A und B] ⟨Enclosed in two corresponding markings by Hi^g.⟩

305.6: Ausdrücke] ⟦Ausdrücke⟧⊢↑Wendungen↓ ⟨Hi^s⟩

305.7: und B] und ↑durch↓ ⟨added by Hi^s⟩ B

305.9–10: 2. ⟨...⟩ $BC = a$.] ⟨Enclosed in two corresponding markings by Hi^g.⟩

305.14–30: 3. ⟨...⟩ Punkte.] ⟨Enclosed in squared brackets by Hi^g.⟩

305.14: liegende] liegenden

305.17: Auch hier werden wir sagen] ⟦Auch hier werden wir sagen⟧⊢↑Wir gebrauchen auch die Wendungen↓ ⟨Hi^g⟩

305.17: in α] in α↑; A, B C sind Punkte von α↓ ⟨added by Hi^s⟩

305.20–21: die ⟨...⟩ Ebene α.] ⟦die Grade a vollständig in der Ebene, d. h. jeder Punkt von a liegt in α.⟧⊢↑jeder Punkt von a in α, wir sagen dann die Gerade a liegt in der Ebene α↓ ⟨Hi^g⟩

305.24–27: Der ⟨...⟩ lässt.] ⟨Enclosed in parentheses by Hi^g.⟩

305.30: gelegene] gelegenen

306.3: es stets] es ↑stets↓ ⟨added by Hi^s⟩

307.13: Euklidischen] Euklid.

307.16: Missverständnis] Missverständniss

307.32: oder] od.

308.3–15: 1. ⟨...⟩ liegt.] ⟨Enclosed in two corresponding markings by Hi^g, including both diagrams.⟩

308.3–4: und C ⟨...⟩ zwischen B] ⟨Hi^s has replaced 'C' by 'B' and vice versa.⟩

308.5–7: Damit ⟨...⟩ Ausdruck.] ⟨Enclosed in parentheses by Hi^g.⟩

308.8: 2. *Wenn* A, B ⟨...⟩] ⟨To the right of the following two manuscript lines Hi^g has added:⟩ B, C miteinander vertauschen!

308.8: wenigstens] ⟨Enclosed in parentheses by Hi^g.⟩

308.9: wenigstens] ⟨Enclosed in parentheses by Hi^g.⟩

309.6–9: $D, \dots, K$ ⟨...⟩ andrerseits] ⟦$D \dots K$ ⟨...⟩ C, $D \dots K$ andrerseits, C ⟨...⟩ $D, \dots K$ andrerseits⟧⊢↑D, $E \dots K$ ⟨...⟩ C, D, $E \dots K$ andrerseits, ferner C ⟨...⟩ D, $E \dots K$ andrerseits, sodann D zwi↓ ⟨unfinished revision by Hi^g ⟩

309.31: Abteilungen] ⟦Abteilungen⟧⊢↑Teile↓ ⟨Hi^g⟩

309.31: Ist] ⟨In the manuscript, 'Ist' begins a new paragraph (without indentation), the continuation with 'Beschaffenheit' was marked by Hi^g.⟩

310.13: dreidimensionalen] dreiminensionalen

310.35: $\beta > \gamma$] ⟨Underlined by Hi^s.⟩

310.35: δ] δ ↑weil B zwischen A, C und A, D liegt↓ ⟨added by Hi^s⟩

310.37: $\gamma > \beta$] ⟨Underlined by Hi^s.⟩

310.37: δ] δ ↑weil C zwischen A, B und B, D liegt↓ ⟨added by Hi^{s}; the first 'B' should be 'D'⟩

311.7: eines und desselben] ⟦des nämlichen⟧↑eines und desselben↓ ⟨Hi^{s}⟩

314.10: Wir] ⟨Indentation inserted by editors.⟩

318.17: zusammenfassen] zusammenfaßen

326.10–11: Dreieckswinkels] Dreieckwinkels

326.12: Dreiecksseiten] Dreieckseiten

326.19: über|einstimmen] ueber|einstimmen

331.4: irgend welchen] ↑irgend welchen↓ ⟨added by Hi^{g}⟩

331.5: der Figur] ↑der Figur↓ ⟨added by Hi^{g}⟩

331.7: auf diese Weise] ↑auf diese Weise↓ ⟨added by Hi^{g}⟩

331.8: einander] ↑einander↓ ⟨added by Hi^{g}⟩

331.12: in entsprechenden Ebenen] ↑in entspr⟪ec⟫hd⟪en⟫ Eben.↓ ⟨added by Hi^{g}⟩

331.12–13: entsprechende Ebenen] ⟦homologe Grade⟧↑entsprechende Graden↓ ⟨Hi^{g}⟩

331.13: dieselbe] ⟦die nämliche⟧↑dieselbe↓ ⟨Hi^{g}⟩

331.41: AMN] A, M. N

334.23: einander gleich] einandergleich

334.24: Sind] ⟦Sind⟧↑Wenn↓ ⟨Hi^{g}⟩

336.17: abgetragen).] abgetragen.

336.42: dessen] deßen

337.46: schon bekannte Zahl] ⟦Zahl von $k(\pi)$⟧↑schon bekannte Zahl↓ ⟨Sc^{s}⟩

338.1: $x \cdot \sqrt{1 + \left(\frac{y}{x}\right)^2}$] $x\,.\,\sqrt{1 + \left(\frac{y}{x}\right)^2}$

340.8: besseren] ⟦besonderen⟧↑besseren↓ ⟨Sc^{s}⟩

341.13: *Beweis*] ⟨Additionally emphasized by an undulating line drawn under the word.⟩

341.14: *Satz*] ⟨Additionally emphasized by an undulating line drawn under the word.⟩

341.14: lassen] laßen

345.1: *Gauss*] ⟨Underlined twice.⟩

345.2: Nachlass] Nächlass

346.1: Wären (1)] ⟨In the manuscript, the underlining of 'Beweis' is prolonged under these two words. The purpose of this prolongation, however, seems to be just to separate the text from the diagram.⟩

350.8: linearen Transformation] lin. Transf.

353.35: nichteuklidischen] Nichteuklidischen

354.13: nichteuklidischen] Nichteuklidischen

354.15: Nichteuklidischen] N. E.

354.23: (9–parametrige)] ⟨Parentheses presumably added by Hilbert before the text was copied.⟩

354.27: Kollineation] K.

356.11: Euklidischen Geometrie] Euklid. G.

357.42: *Ebene*] ⟨Underlined twice.⟩

358.10: Diese] ⟨Indentation inserted by editors.⟩

359.1: *Dualität*] ⟨Underlined twice.⟩

359.9: liegende] liegenden

365.22: der] von der

366.1: Zum] ⟨Indentation inserted by editors.⟩

366.23: fällen] fallen
368.14: $\gtrless$ ⟨...⟩ $\gtrless$] ⟪?⟫ $(bu - va)x_3 + bw - vc \lessgtr$
369.5: Innere] innere
369.6: Äussere] äussere
369.8: Innern] ⟦innern⟧⊣↑Innern↓ ⟨Xx^g⟩
371.33: C] B
372.20: $\overline{AC}$] $\overline{ac}$
372.20: $\overline{BD}$] $\overline{bd}$
373.12–14: (also ⟨...⟩ Def. I.] ⟦also das Flächenmass von ABC im Sinne der Definition II.⟧⊣↑(also das Flächenmass von ABC im Sinne der Definition II) gleich dem Flächenmass von ABC im Sinne der Def. I.↓ ⟨Sc^s⟩
374.1: Nun] ⟨Indentation inserted by editors.⟩
374.6: $\triangle = \sum D_k$] $\triangle = SD_k$
375.2: $\sum_{\triangle_i}$] $\sum_{\uparrow\triangle_i\downarrow\langle\text{added by } Sc^s\rangle}$
375.4: $\sum_i$] $\sum i$
377.26: *3.*] ⟨Indentation inserted by editors instead of vertical space in manuscript.⟩
378.6: gleichwertig.] gleichwertig; ⟦wir werden darauf zurückkommen.⟧ ⟨deleted before the manuscript was duplicated⟩
379.7: Je nachdem] Jenachdem
380.8: Die] ⟨Indentation inserted by editors.⟩
381.45: D'] D
382.38: $1'4'S'$] $1'4'S$
385.7: $P_1(x_1y_1)$] $P_1(xy_1)$ ⟨The absence of the index for x is probably due to the copying process.⟩
398.8: abtragen.] abtragen↑.↓ ⟨added by Hi^s⟩ ⟦⟪zu k⟫⟧ ⟨deleted by Hi^s⟩
399.1: Dreiecks-Congruenzsätze] ↑Dreiecks↓ ⟨added by Hi^s⟩ ⟦Congruenzsatz⟧⊣↑Congruenzsätze↓ ⟨Hi^s⟩
399.15: die Thatsache] der ⟦Satz⟧⊣↑Thatsache↓ ⟨Hi^s⟩
399.17: Es ist] ⟦Bekanntlich ist es⟧⊣↑Es ist↓ ⟨Hi^s⟩
400.1: unendlichferne] unendlifferne
401.5: Physik ⟨...⟩ überhaupt] ⟦Natur ⟦g⟧ d.h. der Physik⟧⊣↑Physik und in der Natur überhaupt↓ ⟨Hi^s⟩
401.7: angenähert ⟨...⟩ aber] angenähert ⟦d.h. man muss⟧⊣↑Man muss aber↓ ⟨Hi^s⟩
401.13: wirklichen] ↑wirklichen↓ ⟨added by Hi^s⟩
401.13: nehmen] ↑nehmen↓ ⟨added by Hi^s⟩

Description of the Text

Collection: SUB Göttingen, signature *Cod. Ms. D. Hilbert 552.*
Size: Cover size approx. 21.1 × 17 cm; page size approx. 21.1 × 16.8 cm.
Cover Annotations: On the front cover is pasted a label of recent provenance with the call number and library stamp. On the spine is a label with the notation, 'Euklidis⟨che Ge⟩ometrie 1898' and the library numbering '35' on the part of the label that overlaps onto the front cover; this corresponds to the numbering of the catalogue list in the Mathematisches Institut.
Composition: 22 signatures of two double pages each; on the first page of the first signature is pasted a further double page. The double page and the signatures 1, 3, 7, 10, 13, 14, 15, and 16 bear on the upper right-hand corner of the first page the pencilled letter 'H', probably as a sign that they belong to the copy assembled for Hilbert.
Pagination: The first double page, bearing the title and the Preface, is not numbered. The rest of the volume is continuously numbered from 1 to 175; the number 1 has been omitted from the first page. The last page is unnumbered and blank.
Original Title: 'D Hilbert, // Elemente der Euklidischen Geometrie. // Göttingen, Wintersemester 1898/99'.
Text: Autographic reproduction in black ink, with corrections and marginalia added by Hilbert in pencil (Hi^g) and black ink (Hi^s). On the verso side of the front cover, and the recto and verso side of the title page are numerous remarks by Hi^s. The handwriting of the document is not that of von Schaper, the *Ausarbeiter*, but possibly that of a professional copyist.

A second copy of the reproduction is in the possession of the Mathematisches Institut of the Georg-August Universität, Göttingen (Inv. Nr. 6808). It contains scattered traces of having been worked over at a later date, in all likelihood by somebody other than Hilbert.

J.K. Whittemore M. Feldblum E. Townsend V. Snyder M. Dehn W.F. Osgood
A. Walter A. Guldberg A. Viterbi A. Ziwet J. Sommer E. Zermelo F. Diestel
L.W. Reid M.J. Hatzidakis H. v. Schaper
M. Brendel D. Hilbert F. Klein W. Schur F. Schilling G. Bohlmann

The Göttingen Mathematical Society in 1899

Chapter 5

The Foundations of Geometry: The Festschrift *of 1899*

Introduction

1. The origin of the Festschrift.

This Chapter presents Hilbert's so-called *Festschrift* (*Hilbert 1899c*), perhaps one of the most celebrated works in the history of mathematics. Although from its second edition (1903) on, it became a self-standing monograph, it first appeared as the first half (92 pages) of a volume published in 1899 by B.G. Teubner of Leipzig, the 'Festschrift zur Feier der Enthullung des Gauss-Weber-Denkmals in Gottingen' (*Festschrift 1899*); hence its nickname. Apparently the decision to erect a monument had been reached in 1892[1], and it was unveiled with due ceremony on 17 June, 1899. (Following the report in the *Jahresbericht der deutschen Mathematiker-Vereinigung*, **8**, p. 10, Hilbert laid a wreath.) According to Hilbert's biographer, Constance Reid, Klein saw the unveiling as an opportunity, not just to honour Gauß and Weber, but to

> ... emphasize once again the organic unity of mathematics and the physical sciences. ... Carrying on and extending the tradition of mathematical abstraction combined with deep interest in physical problems was central to Klein's dream for Göttingen. (*Reid 1970*, 61.)

The continuity was to be represented above all in the *Festschrift*, with Hilbert's contribution and that of Emil Wiechert, whose 'Grundlagen der Elektrodynamik' (*Wiechert 1899*) forms the second part. Wiechert was an Extraordinary Professor of Geophysics and Director of the Geophysikalisches Institut in Göttingen, working mainly on electrodynamics and seismography. (He became an Ordinarius in 1905.)[2] The volume was edited by a 'Fest-Comitee', but, according to Reid again, it was Klein who asked Hilbert and Wiechert to represent the two streams of mathematics and mathematical physics:

> So now he [Klein] asked Emil Wiechert to edit his recent lectures on the foundations of electrodynamics for a celebratory volume, and asked Hilbert to do the same for his lectures on the foundations of geometry. (*Reid 1970*, 61.)

The lectures of Hilbert referred to are, of course, those from 1898/1899, reproduced in Chapter 4. Hilbert's and Wiechert's texts were given independent pagination.

Hilbert's contribution must have been written relatively quickly. There is mention of it in von Schaper's Foreword to his *Ausarbeitung* of Hilbert's 1898/1899 lectures (see Chapter 4), and which bears the date March 1899. The nature of von Schaper's remark suggests that the structure of the *Festschrift* contribution was already fairly clear, and von Schaper even gives the month of appearance as the June approaching, presumably chosen to coincide with the unveiling of the monument. There is also mention of the *Festschrift*

---

[1] See the *Göttinger Tageblatt* for 18 June, 1899, cited in *Toepell 1986*, 253.

[2] Coincidentally, Wiechert was also one of Hilbert's two 'Gegner', fellow students of mathematics, at the public defence for his *Promotion* in Königsberg in 1885. See *Reid 1970*, 16, 61–62.

in a letter from Hilbert to Wiechert dated Saturday, 8 April, 1899 and sent from Nervi, where Hilbert was clearly on holiday. He writes:

> Lieber Wiechert.
> Während wir hier am Strande des mittelländischen Meeres zwischen Palmen und Orangen beladenen Bäumen spazieren gehen, besinne ich mich, dass es nützlich und nothwendig ist, dass von der Gaussfestschrift, so weit sie auf mein Teil fällt, wenigstens die Figuren noch im April gedruckt werden. Ich habe zu dem Zweck die Figuren meiner Arbeit über die Grundlagen der Geometrie in meinem Schreibtische deponiert und meiner Schwiergermutter überantwortet. Wenn also die Herstellung des Druckes der Festschrift bereits beginnen kann, so würde ich Dich bitten, die Figuren Dir von meiner Schwiegermutter geben zu lassen und sie der Druckerei zu übermitteln. Den besten Dank dafür im Voraus.[3]

Thus, already in April, the publication is clearly pressing.

Hilbert sent the proof sheets to Minkowski. (He is one of those thanked at the end of the *Festschrift* for reading the proofs.) In a letter from Zürich dated 11 May, 1899, Minkowski writes: 'Morgen sende ich die Correctur zurück ...' (*Minkowski 1973*, 115). This probably refers to the *Festschrift* proofs. In a letter from the following day, Minkowski adds:

> Deine „Grundlagen der Eukl. Geometrie" haben mir äußerst gut, namentlich auch durch die leichte Darstellung, gefallen und werden sicher allgemein viel Anklang finden. (Minkowski to Hilbert, 12 May, 1899.)[4]

The title given by Minkowski suggests either von Schaper's *Ausarbeitung* of the 1898/99 lectures or the *Festschrift*, though it is very likely that the same manuscript is at issue in both of the letters. It seems possible that the von Schaper *Ausarbeitung* appeared only in late May or early June (see the Introduction to Chapter 4), despite the fact that its Preface bears the date March 1899. Perhaps Minkowski had been sent 'proofs' of this manuscript. Or perhaps the *Festschrift* originally bore a title mentioning the foundations of *Euclidean* geometry. In any case, in a letter from 5 June, there is no doubt about what is being mentioned. Minkowski writes:

> Soeben habe ich den letzten Bogen der Correctur Deiner Festschrift Dir zurückgesandt. Dein Aufsatz hat mir wirklich sehr gefallen und wird gewiß auch allgemein bei den Mathematikern Anklang finden. Man merkt ihm auch in keiner Weise an, daß Du daran zuletzt so schnell arbeiten mußtest, und nach mehrjährigem Durcharbeiten wäre er gewiß nicht so frisch herausgekommen. Daß das Euklidische Axiom über die rechten Winkel beseitigt ist, wird besonders auffallen, doch glaube ich noch, daß dies im

[3]See the Wiechert Nachlaß in the Niedersächsische Staats- und Universitätsbibliothek, Göttingen, i.e., *Cod. Ms. Wiechert, 40*.

[4]This letter is not to be found in the volume *Minkowski 1973*, but rather among Hilbert's correspondence with Léonce Laugel. The reference is thus to the Hilbert *Nachlaß*, in the Niedersächsische Staats- und Universitätsabibliothek, Göttingen, i.e., *Cod. Ms. Hilbert, 211*, Beilage 2. In what follows, references to the Hilbert *Nachlaß* will simply take the form '*Cod. Ms. Hilbert*'.

> Grunde mit einer etwas anderen Einführung der Winkel zusammenhängen wird. (*Minkowski 1973*, 116.)[5]

Another letter from Minkowski, dated 24 June, 1899, states that he had not yet received the copy promised to him, so the actual publication must have followed fairly quickly on receipt of the proofs. By the beginning of July, however, Minkowski apparently had received his copy of the *Festschrift* and had reported about it to his colleagues in Zürich. Hurwitz wrote to Hilbert on 5 July, 1899:

> Im letzten Kolloquium hat uns Minkowski auch über Ihre wunderschöne Festschrift vorgetragen; wir sind alle auf die Fortsetzung des Vortrages am nächsten Mittwoch gespannt. Daß ich bisher noch kein Exemplar der Festschrift erhalten habe, beruht wohl auf einem Versehen. Ich würde Ihnen sehr dankbar sein, wenn Sie mir möglichst bald ein Exemplar zuschicken wollten, da ich mich sehr gern in die Lektüre Ihrer Arbeit vertiefen möchte. (*Cod. Ms. Hilbert, 160*, letter 48.)

Apart from his friends Minkowski and Hurwitz, Hilbert sent copies of the *Festschrift* to several colleagues, e.g., to Otto Hölder (Hölder to Hilbert, 29 June, 1899, *Cod. Ms. Hilbert, 156*, letter 2), Léonce Laugel (author of the first translation of the *Festschrift* into French, of whom more below), Paul Volkmann (Volkmann to Hilbert, 15 May, 1899, *Cod. Ms. Hilbert, 416*, letter 11, and 2 January 1900, ibid., letter 12).

Both Hilbert's and Wiechert's texts were also soon available separately. In issue number 4/5 of the 'Mitteilungen der Verlagsbuchhandlung B.G. Teubner in Leipzig' from October 1900, it was announced:

> Vielfach geaüßerstem Wunsche zufolge sind D. Hilberts: Grundlagen der Geometrie nunmehr auch einzeln zu haben ebenso wie E. Wiecherts: Grundlagen der Elektrodynamik. (*Mitteilungen 1900*, 138.)

No-one nowadays speaks of Wiechert's *Grundlagen* (except in contexts like these). But Hilbert's text and its descendants became famous. It was translated into French almost immediately by Léonce Laugel, a translation which appeared first as a journal article (*Hilbert 1900c*), and then in the same year as a monograph (*Hilbert 1900d*), which marks the first appearance of the *Festschrift* as a separate book. It also appeared in an English translation by E. J. Townsend in 1902 (*Hilbert 1902c*).[6] A second, revised German edition was published (as a self-standing book) by Teubner in 1903. On 24 November, 1902, Alfred Ackermann wrote to Hilbert on behalf of the Teubner publishing house:

---

[5] Minkowski is here referring to Euclid's Postulate 4, 'That all right-angles are equal to one another' (*Heath 1926a*, 154), which Hilbert avoids. See the Introduction to Chapter 4, Specific Note 3(a).

[6] A second English translation by Leo Unger, this time of the tenth edition (*Hilbert 1968*), including the Supplements by Bernays, appeared in 1971; see *Hilbert 1971a*. The same year also saw a second French translation, by Paul Rossier, also of the tenth edition, with substantial commentary by Rossier; see *Hilbert 1971b*.

> Von Ihren Grundlagen der Geometrie möchte ich eine neue Auflage haben und zwar eine, die den Gegenstand etwas breiter behandelt, damit auch der Lehrer das Werk lesen und verstehen kann. Es ist mir nämlich verschiedentlich aus den betreffenden Kreisen dieser Wunsch geäussert worden. Über die äusseren Bedingungen werden wir uns im Falle einer Zusage, die mich mit lebhaftem Danke erfüllen würde, unschwer einigen. Mit der Drucklegung möchte ich dann noch im Laufe des nachsten Jahres beginnen, wenn Ihnen das möglich ist. (Ackermann to Hilbert, 24 November, 1902; see *Cod. Ms. Hilbert, 403*, item 19.)

To the title of the second edition is added:

> Zweite, durch Zusätze vermehrte und mit fünf Anhängen versehene Auflage. (*Hilbert 1903a*, title page.)

Its origin as part of the *Festschrift* is mentioned only in the table of contents, in a single line of very small print inserted between the listing of the title and Introduction; the line reads: 'Aus der Festschrift zur Feier der Enthüllung des Gauß-Weber-Denkmals in Göttingen, Leipzig, 1899'. About the five Appendices added to the second edition, more later.

During Hilbert's lifetime, five further editions were published, in 1909[3], 1913[4], 1922[5], 1923[6], and 1930[7]. After Hilbert's death, to date seven more editions have appeared, in 1956[8], 1962[9], 1968[10], 1972[11], 1977[12], 1987[13] and the centenary edition 1999[14]. This surely makes Hilbert's *Festschrift*, not just a classic, but also a mathematical 'best seller'.[7] Here we present the text of the first edition, the original *Festschrift*, partly because this is the least accessible, but mainly because it represents a crucial element of the sequence of work which begins with the *Ferienkurs* of 1898 (Chapter 3) and the lectures of 1898/1899 (Chapter 4), and which ends with the lectures of 1902 (Chapter 6). We have refrained from adding footnotes commenting on the text, as well as from giving substantial introductory commentary on the text itself. In part, this is because there is, in effect, already substantial commentary on the *Festschrift* in the introductions and notes to Chapters 3 and 6 and particularly Chapter 4. It is also in part because the *Festschrift* has been much discussed, and in addition because the focus of this Volume is more on Hilbert's unpublished writings on geometry than on his published work. Of the numerous commentaries which exist, we mention only a few by way of introduction: (1) The review of the first edition of the *Festschrift* by Henri Poincaré which appeared in 1902 (*Poincaré 1902*). (2) For the immediate development of the *Festschrift*, see the book by Toepell (*Toepell 1986*), and the two articles by him in the centenary edition *Hilbert 1999a*, namely *Toepell 1999a* and *Toepell 1999b*. (3) For some remarks on the development of the *Festschrift* over its various editions, and (very) critical commentary on it, see the article by Freudenthal (*Freudenthal 1957*). (4) For the development

[7]Perhaps not surprisingly, three of the most celebrated books in the history of mathematics were geometry books: Euclid's *Elements*, which has appeared in countless editions; Legendre's *Éléments de géometrie*, which experienced twelve editions; and then Hilbert's *Festschrift*, with fourteen editions.

of geometry after Hilbert's work, see the book *Karzel and Kroll 1988* and the article *Kiechle et al. 1999*. The best detailed, mathematical commentary on the project represented in the *Festschrift* (and in effect the 1898/1899 lectures, too) is *Hartshorne 2000*, which (as the title suggests) also contains many interesting historical remarks.

The remaining remarks in this Introduction are devoted to the differences between the first edition and the second, and in particular those major items which appeared first in the French and English translations, these being then incorporated into the second German edition of 1903. The footnotes we *have* added to the text of the *Festschrift* are of two kinds, those documenting the textual changes in the second German edition (referred to by the abbreviation 'E2'), and those translating Hilbert's bibliographical references into the bibliographical style of the present Volume.

2. The French and English Translations.

(*a*) *The French Translation of 1900.* A French translation of the *Festschrift* was published in Volume 17 (3rd Series) of the *Annales scientifiques de l'école normale supérieure*, pp. 103–209 (*Hilbert 1900c*). The translation was by Léonce Laugel, and the article was issued in May 1900. It was also issued in that year as a separate monograph by the same publisher, Gauthier-Villars (*Hilbert 1900d*). The page setting in the monograph has exactly the same appearance as in the article, although naturally the pagination itself is different. In addition, the monograph has four pages of title apparatus, in which incidentally there is no reference to the fact that the original *Festschrift* was produced in connection with the unveiling of the Gauß-Weber Monument, and also includes a table of contents at the end, just as in the original German edition. There appears no way to date the appearance of the monograph exactly, but there is circumstantial evidence that it appeared subsequent to the article. In a footnote on p. 204 of the article, there is a reference to p. 208 of the same article. This reference remains in the monograph, although 'p. 208' makes no sense in that context. This tends to indicate that the setting for the article was used later for the monograph, and thus that the article preceded the monograph. There is also some evidence that the publication as an article was initially to be *instead* of a monograph publication; see below. Consequently, our primary reference will be to the article (which in any case is more easily accessible), and the excerpts from the translation given in the Appendix to this Chapter are also drawn from this source.

There must have been a substantial correspondence between Laugel and Hilbert, since the Hilbert *Nachlaß* contains twenty-five letters and cards from Laugel, though, unfortunately, no letters (and no copies of letters) from Hilbert. The correspondence started before the *Festschrift* was written; for instance, in a letter from November, 1898, Laugel mentions that he had translated the first 100 pages of Hilbert's *Zahlbericht* for Hermite.[8] There are various *Beilage* included with the correspondence, among which is a letter from Laugel to Minkowski from 10 May, 1899, to which Laugel attaches a brief CV of himself. According to this, Laugel (born in 1855) was in the French diplomatic service from 1874 to 1882 (after which he describes himself as being 'on leave of absence'), first in Paris, and then in Washington; was a member of various mathematical societies; and had translated various works of Gauß, Riemann, Weierstraß and Klein, with others planned. He says nothing about his mathematical education. Minkowski mentions Laugel in two letters to Hilbert. In the first, from 11 May, 1900, he says that he has enquired about 'L.', perhaps suggesting that Hilbert is seeking advice before engaging 'L.' for some purpose or other, possibly to translate the *Festschrift*. (Minkowski also notes that Laugel was a diplomat. See *Minkowski 1973*, 115.) In the second letter, from the following day, Minkowski mentions Laugel by name, and that he is sending on to Hilbert the letter to him from Laugel, including the brief autobiographical sketch.[9]

Whatever the reason it was first sent, Laugel received the *Festschrift* in June, 1899 as witnessed by Laugel's letter to Hilbert, dated 29 June, 1899:

> Mille remerciements pour l'envoi si amiable de votre magnifique Festschrift dont je viens de commencer la lecture. (*Cod. Ms. Hilbert, 211*, item 2.)

He seems to have started translating the work fairly quickly, since in a letter dated 4 September, 1899, he wrote to Hilbert:

> Je suis en train de traduire votre merveilleuse Festschrift. Cela va lentement car il faut être précis et il ne faut pas que cela soit trop indigne de l'admirable original. Je ne pense pas que vous ayiez d'objection à la publication en France. (Ibid., item 3.)

At some point, Hilbert must have authorised the translation, since Laugel writes (in an undated letter; ibid., item 25) thanking Hilbert for this, and saying that he will be in touch with both Teubner and Gauthier-Villars, the proposed publisher.

The correspondence, without Hilbert's side, is not all that interesting. Laugel makes a few elementary suggestions about formulation, some of which were taken up (and which also appear in the second German edition), but some not, and some of which were special to the French edition. There is also a remark or two pointing out misprints which were corrected both in the translation and the second German edition. He also explains (in an un-

[8]See Laugel to Hilbert, 15 November, 1898, *Cod. Ms. Hilbert, 211*, item 1. Laugel also later translated *Hilbert 1900a* on the Dirichlet principle (*Hilbert 1900f*), and, for the official conference proceedings, Hilbert's celebrated paper on mathematical problems delivered at the 1900 International Congress of Mathematicians in Paris: see *Hilbert 1902e*.

[9]This letter is not in *Minkowski 1973*, but is found along with Hilbert's correspondence from Laugel (*Cod. Ms. Hilbert, 211*, Beilage 2). Hilbert had also clearly enquired about Painlevé and Hadamard in connection with some award or other, for Minkowski's first letter discusses 'P.' and 'H.', the second one mentioning Painlevé and Hadamard by name. It is highly unlikely that Laugel is being mentioned in the same connection.

dated letter: see ibid., item 23) why he translates 'flächengleich' as 'equal by addition' and 'inhaltsgleich' as 'equal by subtraction'.[10]

The most interesting thing to emerge from the correspondence concerns the plan for the translation to appear with Gauthiers-Villars as a separate monograph. In an undated letter[11], Laugel writes to Hilbert that

> J'ai fini Grundlagen en ce moment; j'attends une réponse de Gauthier Villars. Je crois qu'il aimerait à le publier. (*Cod. Ms. Hilbert, 211*, item 24.)

And from Laugel's letter of 8 January, 1900, it appears that Picard, partly at the request of Hermite, had impressed upon Gauthier-Villars the 'extreme importance' of the work and the 'great honour' it would be to publish it. However, Laugel writes to Hilbert on 17 January (presumably 1900) saying that Gauthier-Villars 'no longer wishes to publish the thing [presumably the translation of the *Festschrift*] separately', giving as a reason that 'these matters are addressed to a restricted audience'. He adds that Gauthier-Villars 'is certainly at fault, but we can do nothing'. (See ibid., item 20.) The letter also includes a note to Laugel from Picard, saying that Darboux will 'take the Hilbert memoir you have translated and his notes' either in the 'Annales or instead in his Bulletin'. This latter is very likely a reference to the *Bulletin des sciences mathématiques et astronomiques*, of which Darboux was the founder. As we know, the *Annales* was chosen. A letter dated 3 February, 1900 states a plan whereby the offprints from the *Annales* 'puissent être publiés sous forme de volume' (ibid., item 6). No other letters from Laugel throw light on this matter, however, though they do suggest that the journal publication preceded that of the monograph. There is certainly no light thrown on the change of heart on the part of Gauthier-Villars which saw the eventual publication of the translation as a separate monograph under his imprint.

Laugel sent Hilbert the manuscript for review and comments around the middle of January, 1900, and incorporated his suggestions. (See below.) It went to the printer at the beginning of March (see the letter to Hilbert from Laugel on the 28 February, 1900; *Cod. Ms. Hilbert, 211*, item 7). It appears that the first proofs were not sent to Hilbert, but were read by Laugel himself, a geometer called Gérard (a Professor at the Lycée Charlemagne), and by Stäckel in Kiel. Subsequent proofs were sent to Hilbert at the end of April. (See Laugel to Hilbert, 28 April, ibid., item 8.) Laugel thanks both Gérard and Stäckel for help with the proofs in the published translation itself; see p 111, n. 1. (By chance, this note is reproduced in the selections included in the Appendix to this Chapter.)

The alterations Laugel made in the course of translating the text are minor, consisting mainly in the changes of formulation and the correction of

---

[10]In the second French translation, the translator, Rossier, coins the neologisms 'équidécomposable' and 'équicomplémentaire' for these: see *Hilbert 1971b*, p. 100.

[11]Some of Laugel's missives are dated, but some, including some which clearly have to do with the *Festschrift*, are not, or not fully. This makes it impossible to pin the relevant discussions unequivocally to the *Festschrift*, though it is hard to see what else could have been at issue.

misprints mentioned previously. In addition, there are three footnotes added by Laugel, two of which simply note Hilbert's supplementation of the text, one of these also expressing the debt of gratitude to Gérard and Stäckel just mentioned. However, there are significant additions by Hilbert. We list them in the order in which they appear.

(1) By far the most important addition is at the end of § 8 (pp. 123–124). This section states the Archimedean Axiom, which is the sole axiom of continuity, and the addition presents the *Vollständigkeitsaxiom*, which is given in the French translation as 'Axiom d'intégrité'. This was not the axiom's first appearance in print; it also figures in Hilbert's axiomatisation of the real numbers (complete, ordered fields) published as *Hilbert 1900b*, which is referred to in the French translation, and which carries the date 12 October, 1899. Nevertheless, this is the first appearance of the axiom in the context of the *geometrical* system. Hence, this supplement to the French translation is included in an Appendix to this Chapter. When incorporated into the second German edition of the *Festschrift*, the *Vollständigkeitsaxiom* is built into Group V, the continuity axioms, where it also appears in the axiomatisation of complete, ordered fields in *Hilbert 1900b* (Group IV there). In Hilbert's addition to the French translation, though, the axiom is seemingly given, less as an extension of Group V, but (more appropriately perhaps) as a supplement to the whole axiom system. Hilbert does point out that the axiom allows one to show that there is a one-to-one onto correspondence between the points on a line and the real numbers, and that therefore it allows one to show that any convergent sequence actually converges to a point. He also stresses that no use is actually made of the axiom in the subsequent development, and this remained the case even after the axiom had been more thoroughly incorporated into Axiom Group V as V 2. There is a fuller discussion of the form, content, and purpose of the axiom in section 5 of this Introduction below.

(2) The second addition concerns a footnote added by Hilbert to the beginning of § 18, pointing out the similarity between his own work on plane areas and that of the mathematician Gérard already mentioned above. The footnote reads:

> En ce qui concerne la théorie des aires dans le plan, nous appelons avant tout l'attention sur les Travaux suivant de M. Gérard: *Thèse de Doctorat sur la Géometrie non euclidienne* (1892) et *Géometrie plane* (Paris, 1898). M. Gérard a exposé une théorie tout à fait analogue à celle du § 20 du présent Travail relativement à la mesure des polygones. La difference est que M. Gérard emploie des transversales parallèles, tandis que moi je me sers de tranversales issues d'un sommet. En outre, le lecteur pourra comparer les Travaux suivants de F. Schur, où l'on trouve aussi une exposition analogue: *Sitzungsberichte der Dorpater Naturf. Ges.*, 1892, et *Lehrbuch der analytischen Geometrie*, Leipzig, 1898, Introduction. Enfin, je renverrai encore à un Travail de O. Stolz: *Monatshefte für Math. und Phys.*, 5e année, 1894.

To this Laugel then adds:

En outre, M. Gérard a encore traité la question des aires, par diverses méthodes, dans le *Bulletin de Mathématiques spéciales* (mai 1895), dans le *Bulletin de la Société mathématique de France*, dans le *Bulletin de Mathématiques élementaires* (janvier 1896, juin 1897, juin 1898). (*Hilbert 1900c*, 149, n. 1)

Neither part of this note is to be found in the second, German edition of the *Festschrift*.

Laugel had brought Hilbert's attention to the work of Gérard in a letter from 28 February, 1900 (*Cod. Ms. Hilbert, 211*, item 7). Moreover, one of the *Beilagen* (number 2) to the correspondence from Laugel is a note (presumably in Laugel's hand) listing all the publications given here, and implying that a mathematician of Hilbert's stature (Hilbert is referred to neutrally, as if the list was not initially addressed to Hilbert himself) could do a great deal for the reputation of the unknown Gérard by mentioning his work. This is precisely what Hilbert does in his note.

The works of Gérard referrred to by Hilbert are *Gérard 1892* (the doctoral dissertation), and *Niewenglowski and Gérard 1898* (the monograph). In addition, Hilbert refers to the articles *Schur 1893* and *Stolz 1894*, as well as Schur's book *Schur 1898*. Laugel's footnote adds references to *Gérard 1895b*, *Gérard 1895a*, *Gérard 1896*, *Gérard 1897* and *Gérard 1898*.

(3) The third addition is to the Conclusion. Apart from a few opening remarks about the Helmholtz-Lie construction of geometry using groups of transformations, Hilbert concentrates on a detailed description of Dehn's work on the Legendre Theorems. The two Legendre theorems on the angle-sum in a triangle were proved by Hilbert in the 1898/1899 lectures (Chapter 4 of this Volume) making use of the Archimedean Axiom in their proof, as was standard. The theorems make no appearance in the first edition of the *Festschrift*, or (named as such) in the second edition either. Towards the end of the 1898/1899 lectures (p. 106 of Hilbert's lecture notes, p. 169 of the *Ausarbeitung* by von Schaper), Hilbert states as a problem to show whether the Legendre Theorems are provable without the use of any continuity principle. This was the problem Dehn tackled and solved in work completed in 1900. (See the Introduction to Chapter 4, Specific Note (5)(a).) The account added to the French translation amounts to over three pages of text, and although a brief description of Dehn's work on the Legendre Theorems was added to the second German edition of the *Festschrift* (not in the Conclusion, but at the end of the section § 12), this extensive discussion of Dehn's work did not appear again, aside, that is, from its reproduction in the English translation of 1902. (Perhaps one reason is that, at the time of publication of the French translation, Dehn's work had not yet appeared in print, though it came out shortly afterwards.) Hence, we have included this as the second selection from the French translation given in the Appendix to this Chapter.[12]

---

[12]Laugel's card to Hilbert from 28 April, 1900 mentions that he is returning Hilbert's 'manuscript' on Dehn's work. See *Cod. Ms. Hilbert, 211*, item 8. There is no mention of a manuscript on the *Vollständigkeitsaxiom*.

(*b*) *The English Translation of 1902.* An authorised English translation of the *Festschrift* appeared with the Open Court Publishing Company of Chicago in 1902. (The London agents were Kegan Paul, Trench, Trübner and Co.) The translator was E. J. Townsend. From 1900, Townsend was Professor of Mathematics at the University of Illinois, Urbana-Champaign, but from 1898 to 1900, he was a doctoral student of Hilbert's working primarily on analysis. As with Laugel, there is a substantial collection of letters from Townsend to Hilbert in the *Nachlaß*, although again no (copies of) letters from Hilbert. Townsend must have attended the lecture course on Euclidean geometry held by Hilbert in 1898/1899. In a letter dated 20 November, 1900 from Champaign, Townsend actually offers to translate Hilbert's lectures, not, interestingly, the *Festschrift*. (See *Cod. Ms. Hilbert, 409*, item 2.) Hilbert clearly demurred. In a later letter (3 June, 1901; ibid., item 4), it clearly having been agreed in the meantime that Townsend would translate the *Festschrift* itself, Townsend mentions that Osgood at Harvard had proposed that the translation should also include, either in the text or in an Appendix, selections from the 1898/1899 lectures covering material which makes no appearance in the *Festschrift* itself, and he requests Hilbert's opinion about this. Once again, Hilbert clearly demurred. Townsend writes on the 9 October, 1901 (see ibid., item 5) that the translation is finished, and will be sent to the publishers after a read through by Osgood.

The translation has a brief Preface by Townsend, which opens by calling the reader's attention to the fact that the material in the *Festschrift* originated in the 1898/1899 lectures. Townsend also states:

> In the French edition, which appeared soon after [the publication of the *Festschrift*], Professor Hilbert made some additions, particularly in the concluding remarks, where he gave an account of the results of a recent investigation made by Dr. Dehn. These additions have been incorporated in the following translation. (*Hilbert 1902c*, iii.)

The *Vollständigkeitsaxiom* is called (ibid., p. 25) the 'Axiom of Completeness', as is now standard in English discussions. The rest of the Preface then consists of a sketch of the main results of Hilbert's work. Townsend ends by thanking 'Professor Osgood of Harvard, Professor Moore of Chicago, and Professor Halsted of Texas' for valuable suggestions (ibid., p. iv). Apart from the inclusion of the additions to the French translation (which, incidentally, include the footnote on Gérard's work mentioned above; see ibid., p. 57), there are several other modifications.

Firstly, there are three footnotes added by Townsend to the text itself all of them pointing out further literature, including Moore's paper, *Moore 1902* which shows the redundancy of Axiom II 4 (see Hilbert's lectures from 1902, Chapter 6 in this Volume, note to p. 14), and Moulton's paper on non-Desarguesian geometries (*Moulton 1902*). Townsend also twice points out the paper *Schur 1902*. Townsend had written to Hilbert on 21 June, 1902 (ibid., item 8), pointing out papers by Moore and Moulton relevant to the work of the *Festschrift* which were about to appear in the *Transactions of the American Mathematical Society*. He then asks Hilbert whether he should cite these in his translation, together with Schur's paper in the *Mathematische Annalen*, presumably *Schur 1902*. Hilbert clearly agreed.[13] Secondly, the translation contains an Appendix, which bears the following footnote by Townsend:

> The following is a summary of a paper by Professor Hilbert which is soon to appear in full in the *Math. Annalen*. (*Hilbert 1902c*, 133.)

The forthcoming paper referred to is clearly *Hilbert 1903c*. Townsend does not state that the summary is in fact by Hilbert himself, although this is evident from the text itself; indeed, a quick comparison reveals that it is in fact almost exactly the introductory part, corresponding to pp. 381–388, of the paper referred to, translated into English. The remainder of the paper (pp. 388–422, §§1–42) consists of the detailed working out of the proofs. The paper carries the date 10 May, 1902. On the 12 March, 1902 (ibid., item 6) Townsend had written asking of Hilbert what has happened to his new 'Grundlagen der Geometrie', clearly the paper *Hilbert 1903c*, and he also asks whether he should wait for it or not. The next letter from 18 June, 1902 (ibid., item 7) reveals that the Appendix is in fact the paper *Hilbert 1902f*, although *this* just constitutes the first part of the later *Hilbert 1903c*. Townsend says that the note 'from Nov. 1901' came too late to be bound together with the translation. (The paper *Hilbert 1902f*, from the *Göttinger Nachrichten* is indeed dated 8 November, 1901.) However, Townsend goes on:

> Ich habe es übersetzt aber und es wird als Anhang gedruckt und *nachher* wird es alle gemeinsam gebunden werden.

Townsend's translation was rather brutally criticised by G. B. Halsted in *Halsted 1902a*. (Another paper of Halsted's from the same year, *Halsted 1902b*, is a balanced review of the place of Hilbert's *Festschrift* in the development of geometry, and, although Halsted mentions the new English edition, he makes no reference to the translation or the translator.) The criticism seems interesting in the light of a passage from one of Townsend's letters to Hilbert. Townsend writes:

> Herr Professor Halsted aus Texas hat mir geschrieben und Ihren Brief an ihm mit geschickt. Leider war er zu spät angekommen, weil alles schon bestimmt war und die Übersetzung schon begonnen. Sonst hätte ich alles übergeben, weil es scheint, dass Sie es lieber haben, wenn Herr Halsted sie macht. Wenn Sie ein starkes Gefühl darüber haben, schreiben Sie mir, bitte, und ich will [sic] noch es überhanden. (3 June, 1901: *Cod. Ms. Hilbert, 409*, item 4.)

According to Hedrick in his review of both the French and English translations (*Hedrick 1903*, 162), Halsted's somewhat harsh criticism is in large measure

[13] One slight inconsistency here is that, in this letter, Townsend states that the matter is urgent since the translation is already 'in press'. However, three days earlier, on the 18 June, 1902 (409/07), he had written to Hilbert stating that the translation 'ist schliesslich erschienen', somewhat contradicting the later letter, unless 'erschienen' refers to the final proofs and not the finished book itself.

unfair.[14] However, although Hedrick dismisses Halsted's attacks, he does offer some serious criticisms of his own, and he does not offer the positive praise he gives to Laugel's French translation. (See *Hedrick 1903*, in particular pp. 162–164.)

3. The Second Edition of the Festschrift.

The changes in the second edition are not really the changes in scope and style that Ackermann of Teubner's appears to have been calling for. The actual alterations fall into four groups.

(1) *Appendices.* As announced on the title page, there are five Appendices, all reprintings of papers, some incorporating minor revisions such as the specification of more recent references, etc. The papers they reproduce/correspond to are: I, *Hilbert 1895b*; II, *Hilbert 1902/03*; III, *Hilbert 1903b*; IV, *Hilbert 1903c*; V, *Hilbert 1901*. These all present further results in metageometry in the style of investigation established by the work of 1898/1899, and deal respectively with: the characterisation of the straight line as the shortest distance between two points; the relationship of the Triangle Congruence Axiom to the elementary theorem showing the equality of the base angles in an isoceles triangle; the axiomatisation of Bolyai-Lobachevsky geometry in the style of the *Festschrift*; the Helmholtz-Lie approach to geometry via groups of transformations, but using much weaker assumptions than Lie does; and lastly surfaces with constant Gaussian curvature. In between the first and second editions of the *Festschrift*, Hilbert held another series of lectures on the foundations of geometry. (See Chapter 6.) This was not, as might be thought, a preparation for the second edition of the *Festschrift*, but it does contain preliminary work on the material which appears in Appendices II–IV.

(2) *Minor Revisions.* Axiom I 7 was expanded into two axioms, I 3 and I 8; Axiom II 4 was dropped and replaced by Satz 4 (p. 5), following Moore's observation (*Moore 1902*) that II 4 can be proved on the basis of the other axioms of groups I and II. Revisions of the text were made throughout the book, ranging from minor deletions, substitions, or insertions of single words to substantial insertions and revisions of full passages, and minor reorganisation, especially of the axiom groups. We have noted these changes in footnotes throughout our presentation of the first edition of the *Festschrift.* One significant change was to put the third axiom group, which consists of just the Axiom of Parallels, *after* the fourth group of axioms, i.e. the Axioms of Congruence, in other words, to return to the ordering of the 1898/1899 lectures, an ordering which then remained fixed in all subsequent editions. The revisions occasioned a renumbering of axioms, paragraphs, and theorems summarised in the following synopsis:

---

[14] E. R. Hedrick was also a doctoral student of Hilbert's; his thesis, on differential equations, was orally examined in February, 1901.

First Edition	Second Edition
I 3	I 4
I 4	I 5
I 5	I 6
I 6	I 7
II 4	Satz 4
II 5	II 4
Satz 4	Satz 5
Satz 5	Satz 6
Satz 6	Satz 7
Satz 7	Satz 8
§ 5	§ 7
III	IV
§ 6	§ 5
IV 1	III 1
IV 2	III 2
IV 3	III 3
IV 4	III 4
IV 5	III 5
IV 6	III 6
§ 7	§ 6
V	V 1
Satz 28	Satz 29
Satz 29	Satz 30
Satz 30	Satz 31
Satz 31	Satz 32
Satz 32	Satz 33
Satz 33	Satz 34
Satz 34	Satz 35

In addition, the second edition differs from the first in a number of changes in orthography and layout:

1. Orthographical revisions are carried out as a consequence of the *Rechtschreibreform* ratified in 1902; for example, 'dass' is changed to 'daß', and 'congruent' is changed to 'kongruent'.

2. Figures which, in the first edition, were numbered consecutively as 'Fig. 1', 'Fig. 2' are no longer numbered at all in the second edition.

3. Some formulae which, in the first edition, were printed in-line are now set in a separate line.

There were also some important terminological and conceptual changes which deserve brief comment.

(*a*) 'Definition' → 'Erklärung'. In the first twenty pages of the first edition of the *Festschrift*, Hilbert uses two designations when introducing definitions, 'Erklärung' and 'Definition'. The primary meaning of the former word is elucidation, clarification or explanation, but it can also mean definition; the meaning of the latter is just definition. (Both words are latinate at root, but the latter more recognisably so.) Sometimes when Hilbert uses *Erklärung*, it seems to mark something like a very brief elucidation of a primitive notion, as in the very first paragraph of Chapter 1 (p. 4) where we find

> Erklärung. Wir denken ⟨uns⟩ drei verschiedene Systeme von Dingen: ...

The same is true of the *Erklärungen* of the primitive terms 'zwischen' (p. 6) and 'congruent' (p. 10). Sometimes, however, the use marks the introduction, or simplification, of new terminology, as with the terms 'on the same side as' and 'on the opposite side to' on p. 9, terminology which is justified by Theorem 7 proved just before this. We are not given a *definition* of the new terminology, but Hilbert's reference to Theorem 7 shows that it could easily be turned into one. Sometimes Hilbert does give definitions of new notions, as with the *Definitionen* of *Streckenzug* (a line joining two points and composed of a finite number of segments) and *Polygon* on p. 8. Another important example is the definition of 'angle' given on p. 11.[15] There seems, thus, a less than sharp separation between *Erklärungen* and *Definitionen*. This is shown by the example of the notion of 'half-line', which falls under an *Erklärung* (p. 8), but which in turn is important in the *definition* of angle. In any case, while all the *Erklärungen* remain in the second edition, the *Definitionen* are systematically changed into *Erklärungen*. With this, whatever sense the first edition might give of a distinction between 'elucidation' and 'definition' is lost. The uniform method of designation remains in all subsequent editions.

There is another use Hilbert makes of the idea of defining, a non-standard one, at least at that time. In the first edition, Hilbert describes Axiom Groups II and IV as 'defining' the central concepts they govern, 'between' and 'congruent' (pp. 6 and 10) respectively, a usage of 'define' which makes no appearance in the 1898/1899 lectures. Hilbert was sharply criticised by Frege for using the notion of definition in this sense. (See the letter from Frege to Hilbert of 27 December, 1899 in *Frege 1976*, 60–64, or *Frege 1980*, 34–38.) Despite Frege's strong letter[16], Hilbert did not abandon this use of 'define'. He defends his view in two letters replying to Frege (29 December, 1899, and 22 September, 1900; see *Frege 1976*, 65–69, 79 and *Frege 1980*, 38–43, 50–51 respectively), and Axiom Groups II and (now) III are introduced in exactly the same way in the second and all subsequent editions of the *Festschrift*. Hilbert's procedure came to be known as that of 'implicit definition'.

---

[15]Note that some definitions, like that of 'segment' on p. 6, use a notion which is *not* officially a primitive, namely that of a system of points.

[16]The point was also pursued by Frege in a further letter from 6 January, 1900 (see *Frege 1976*, 70–76, or *Frege 1980*, 43–48), and then in a series of papers, *Frege 1903a*, *Frege 1903b* and *Frege 1906*.

(*b*) 'Streckenübertrager' $\rightarrow$ 'Eichmass' (p. 74f. in the second edition). This change was occasioned by work of Josef Kürschák (*Kürschák 1902*). Kürschák shows that all constructions which can be carried out by means of a straightedge and some device for moving segments of arbitrary length (a *Streckenübertrager*, e.g., a pair of dividers) can be carried out instead by using a straightedge and what he calls an 'Aichmass', a device that allows one to measure off just a single given segment a. This is an essential simplification of the treatment of constructions in section §36 of the first edition of the *Festschrift* where Hilbert uses straightedge and a *Streckenübertrager*. From the second edition on, Hilbert refers to Kürschák's simplification and drops 'Streckenübertrager' in favour of 'Eichmass'. This simplification was noted by Hilbert in his Remark [7] on his copy of the *Ausarbeitung* of the 1898/1899 lectures; see the Appendix to Chapter 4.

(*c*) 'Flächengleich' $\rightarrow$ 'zerlegungsgleich'. This seems to be merely a change in terminology which more nearly accords with how the term is defined (using a decomposition into a finite number of pieces).

(*d*) 'Flächenmass eines Dreiecks' ($F(\Delta)$ for short) $\rightarrow$ 'Inhaltsmaß' ($J(\Delta)$ for short). This again seems to be just a uniformisation in terminology which makes the theorem finally proved on p. 45 of the second edition read 'Inhaltsgleiche Polygone haben gleiches Inhaltsmaß', in place of the first edition's 'Inhaltsgleiche Polygone haben gleiches Flächenmass' (p. 43).

(3) *Legendre Theorems.* As we have remarked, in the French translation of 1900 Hilbert adds to the Conclusion an extensive note on Dehn's work on the Legendre Theorems and the Archimedean Axiom, a note which is included in the English translation of 1902 as well. This addition is not taken up in the second edition. Instead, Hilbert adds a much more truncated discussion of Dehn's construction (and some of its consequences) at the end of § 12 (pp. 23–24) where non-Archimedean geometries are first discussed. (The 1902 lectures are also important with respect to the Legendre Theorems; see the Introduction to Chapter 6 in this Volume.)

(4) The Continuity Axioms, Axiom Group V. The addition in the second edition of the *Vollständigkeitsaxiom* to Axiom Group V (see § 8, pp. 16–17) is the most significant conceptual change. (This new axiom takes its place as V 2, while the Archimedean Axiom now becomes V 1.) This axiom was not first introduced in the second edition of the *Festschrift*, but in the French translation published in 1900, or, even earlier, in Hilbert's paper on the concept of real number (*Hilbert 1900b*). For more discussion of this, see the discussion of the French translation above and section 5 of this Introduction.

(5) In section § 13 of the *Festschrift*, Hilbert gives seventeen ordered field axioms (axioms for what he calls 'Complexe Zalhensysteme'), the Archimedean Axiom being the sole continuity axiom. The system given in *Hilbert 1900b* is the same, except for the addition of a second continuity axiom, the *Vollständigkeitsaxiom.* Despite the fact that the *Vollständigkeitsaxiom* is stated in the French translation of 1900, it is not given as an addition to the continuity axioms, and no corresponding change is effected in the axioms for ordered fields. As already mentioned, however, by the time of the second German

edition of the *Festschrift*, the axiom is incorporated into the geometrical system as a second continuity axiom (mirroring the system of *Hilbert 1900b*), and Hilbert does therefore effect a corresponding change to the field axioms as stated in § 13. (See *Hilbert 1903a*, 25–26.)

(6) At the end of § 21 of the second edition, after having developed the central theory governing the measures of surface content, Hilbert (p. 47) raises the problem of whether an analogous development (thus, one avoiding continuity assumptions) is also possible for polyhedra. The question, which goes back to an observation of Gauß, makes no appearance in the first edition of the *Festschrift*. However, it had already been raised in the 1898/1899 lectures (see the Introduction to Chapter 4, Specific Note (5)(*c*)), and it was stated again by Hilbert as an unsolved problem in his celebrated lecture from 1900 on mathematical problems (*Hilbert 1900e*, Problem 3), where he puts forward the conjecture that the problem *cannot* be solved positively. This conjecture was confirmed shortly thereafter by Dehn, and Hilbert presumably adds this note here because he can point to a solution of the problem previously raised. Hilbert refers to *Dehn 1900c*, *Dehn 1902* and *Kagan 1903*.

(7) In the last chapter VII, on construction by straightedge and *Eichmaß* (formerly *Streckenübertrager*) Hilbert systematically changes the reference to reliance on Axioms I–V to reliance on I–IV. The reason is clear: in constructions, no use is (or can be) made of the Archimedean Axiom.

4. Variations in Later Editions.

We refrain from any serious attempt to describe the alterations in subsequent editions. What we give here is a very short survey of what the various editions contain.

Third Edition, 1909 (*Hilbert 1909*). The title page reads:

> *Grundlagen der Geometrie*, dritte, durch Zusätze und Literaturhinweise von neuem vermehrte und mit sieben Anhängen versehen Auflage. Mit zahlreichen in den Text gedruckten Figuren. Leipzig und Berlin. Druck und Verlag von B. G. Teubner.

The book (with VI + 279 pages) appeared as Volume VII in the Teubner series *Wissenschaft und Hypothese*. The two appendices new to this edition were reprintings of the papers *Hilbert 1900b* and *Hilbert 1904b*. The footnotes throughout the book, including those to the Appendices, represent substantial additions to the citations of the work of others; indeed, this applies to all the editions.

Fourth Edition, 1913 (*Hilbert 1913*). The details are the same as for the third edition, except the book now has VI + 258 pages.

Fifth Edition, 1922 (*Hilbert 1922*). Again, the details on the title page are the same as for third edition. The book has VII + 265 pages, and has the same appendices as the two previous editions. This edition is the first with a *Vorwort* by Hilbert, which states:

> Die vorliegende 5. Auflage ist ein anastatischer Nachdruck der 4. Auflage. Ich habe mich darauf beschränkt, zum Schluß S. 259–265 einige wenige

> Zusätze und notwendige Berichtigungen hinzuzufügen, für deren Ausarbeitung ich ins besondere P. Bernays zu herzlichstem Dank verpflichtet bin.

The notes are added at the end to Appendix II (p. 259; for a description, see the Introduction to Chapter 6), Appendix III (p. 263), and Appendix VIII [sic] (p. 265). There is, of course, no Appendix VIII, and the last note concerns rather § 21 in Chapter IV, which is to do with the work in *Süß 1921* on the theory of the measure of content of polyhedra and its relation to the Archimedean Axiom.

Sixth Edition, 1923 (*Hilbert 1923*). This is an exact reproduction of the fifth edition, the *Vorwort* being identical, except that 'fünfte' is changed to 'sechste'. The third note added is no longer marked as 'Zu Anhang VIII', but just 'Zu Kapitel IV, § 21'.

Seventh Edition, 1930 (*Hilbert 1930a*). This was the last edition to be published in Hilbert's lifetime, and represents the most substantial revision both to the axioms and the shape of the theoretical development. The page count is now VII + 326, with ten appendices in all. The title page reads:

> *Grundlagen der Geometrie*, siebente umgearbeitete und vermehrte Auflage, mit 100 in den Text gedruckten Figuren. Leipzig und Berlin. Druck und Verlag von B. G. Teubner.

The *Vorwort* by Hilbert makes it clear that Hilbert's student Arnold Schmidt had much to do with the revision. Schmidt was the *Ausarbeiter* of Hilbert's lectures on the foundations of geometry in 1927 (*Hilbert 1927**, not included in this Volume), which was an important step in the development of the radically revised seventh edition. The *Vorwort* is as follows:

> Die vorliegende siebente Auflage meines Buches „Grundlagen der Geometrie" bringt gegenüber den früheren Auflagen erhebliche Verbesserungen und Ergänzungen, und zwar teils nach meinen späteren Vorlesungen über diesen Gegenstand, teils wie sie durch die inzwischen von anderen Autoren erzielten Fortschritte bedingt worden sind. Dementsprechend ist der Text des Hauptstückes des Buches umgearbeitet worden. Hierbei hat mich meiner Schüler H. Arnold Schmidt aufs tatkräftigste unterstützt. Er hat für mich nicht nur die ins einzelne gehende Arbeit geleistet, sondern von ihm rührren auch zahlreiche selbständige Bemerkungen und Zusätze her: insbesondere ist die neue Fassung des Anhanges II von ihm selbstständig hergestellt worden. Ich spreche ihm hiermit für seine Hilfe meinen herzlichsten Dank aus. Zu den Anhängen der früheren Auflagen habe ich noch als Anhang VIII meinen in Münster i. W. gehaltenen Vortrag „Über das Unendliche", als Anhang IX meinen Hamburger Vortrag „Über die Grundlagen der Mathematik" und als Anhang X meinen auf dem Internationalem Mathematiker-Kongreß in Bologna gehaltenen Vortrag „Probleme der Grundlegung der Mathematik" hinzugefügt.
>
> Göttingen, Januar 1930.

The new appendices referred to by Hilbert are the papers *Hilbert 1926*, *Hilbert 1928a* and *Hilbert 1930b*. However, the first of these is only a *partial* reprinting of the original paper, and leaves out the original attempt at a proof of the Continuum Hypothesis, and the third, which Hilbert refers to as the lecture held in Bologna in 1928, differs significantly from the version published in 1928 in the proceedings of the International Congress (*Hilbert 1928b*). Moreover, the paper *Hilbert 1900b*, which appeared unaltered in the third to the sixth editions, is here changed, a key sentence from p. 184 (in the original pagination of the paper) being omitted. The sentence in question reads:

> Um die Widerspruchslosigkeit der aufgestellten Axiome zu beweisen, bedarf es nur einer geeigneten Modifikation bekannter Schlußmethoden.

The axiom system in question, of course, is that for complete, ordered fields, i.e., for the system of real numbers. Regardless of these alterations, the seventh edition does present five of Hilbert's important papers on the foundations of mathematics beyond the particular concern with the foundations of geometry (Appendices VI-X). This was no doubt the reason that these papers were omitted from the *Gesammelte Abhandlungen* published not long afterwards (*Hilbert 1932*, *Hilbert 1933*, *Hilbert 1935*). The omission of all five from subsequent editions of the *Festschrift* is therefore to be seen as unfortunate, and a complete collection of Hilbert's published papers on foundational subjects is sorely needed.

In addition to the various changes to the general foundational papers, there were changes to the other appendices on the foundations of geometry, most notably to Appendix II concerning the theorem of the equality of the base angles in an isoceles triangle. This alteration is referred to in Hilbert's *Vorwort*, and is briefly described in the Introduction to Chapter 6.

Editions After Hilbert's Death. The seven editions to have appeared after Hilbert's death (1956[8], 1962[9], 1968[10], 1972[11], 1977[12], 1987[13] and 1999[14]) are based on Hilbert's text from the seventh edition, with only minor alterations, a text which differs significantly from that of the first edition. Moreover, these subsequent editions drop those appendices which do not directly have to do with Hilbert's work on geometry, thus Appendices VI and VII from the third edition and the further VII-X from the seventh. Editions eight to eleven were edited and prepared by Paul Bernays, and they contain various Supplements written by Bernays himself, Supplements which attempt to elucidate, or elaborate on, various subjects from both the text and the remaining (geometrical) appendices. The twelfth and thirteenth editions are just reprints of the eleventh. The fourteenth edition, the centenary edition, edited by Toepell (*Hilbert 1999a*), is a reprint of the previous three editions supplemented further by historical material, the papers *Toepell 1999a*, *Toepell 1999b*, *Kiechle et al. 1999*, and by Hilbert's *Schulheft* (*Hilbert 1999b*), as well as by very useful bibliographies. None of these editions appears in the *Wissenschaft und Hypothese* series, unlike the third to the seventh. The fourteenth (centenary) edition is marked as Supplement 6 in the series *Teubner-Archiv zur Mathematik*.

5. The Vollständigkeitsaxiom.

Let us turn finally to a brief discussion of the most important change in the French translation and the second edition, the addition of the *Vollständigkeitsaxiom.*

The *Vollständigkeitsaxiom* given in the second edition of the *Festschrift* is as follows:

> V 2 (Axiom der Vollständigkeit). *Die Elemente (Punkte, Geraden, Ebenen) der Geometrie bilden ein System von Dingen, welches bei Aufrechterhaltung sämtlicher genannten Axiome keiner Erweiterung mehr fähig ist,* d.h. : zu dem System der Punkte, Geraden, Ebenen ist es nicht möglich, ein anderes System von Dingen hinzuzufügen, so daß in dem durch Zusammensetzung entstehenden System sämtliche aufgeführten Axiome I – IV, V 1 erfüllt sind. (*Hilbert 1903a*, 16.)

This axiom remains in all subsequent editions. Hilbert explains 'Erweiterung' as follows:

> Die Aufrechterhaltung sämtlicher Axiome, von der in diesem Axiom die Rede ist, hat man so zu verstehen, daß nach der Erweiterung sämtliche früheren Axiome in der früheren Weise gültig bleiben sollen, d. h. sofern man die vorhandenen Beziehungen der Elemente, nämlich die vorhandene Anordnung und Kongruenz der Strecken und Winkel nirgends stört, also z.B. ein Punkt, der vor der Erweiterung zwischen zwei Punkten liegt, dies auch nach der Erweiterung tut, Strecken und Winkel, die vorher einander kongruent sind, dies auch nach der Erweiterung bleiben. (*Hilbert 1903a*, 17.)

Even with this, it is not entirely clear what Hilbert means. However, for a first attempt at clarification, the language and concepts of modern model theory are not out of place.[17] Thus consider as given some relational structure which is a model of the axioms I–V 1. A putative extension of this model would then be another relational structure which includes the domain of the original model, possibly with more objects, but which is such that the basic relations (e.g., 'between' or 'congruent') are the same when restricted to the original domain. In other words, the original model must be a *substructure* of the extensions Hilbert is considering. Given this, we can state what the *Vollständigkeitsaxiom* demands thus: No model of the axioms I–V 1 can be a proper substructure of any other model of these axioms. One could read Hilbert's formulation slightly more liberally and consider, not substructures,

[17] Note that, in his earlier lectures, particularly *Hilbert 1894a**, 60 (see Chapter 2), and to some extent in the 1898/1899 lectures, too, and quite unequivocally in his correspondence with Frege at the end of 1899 (see letter of 29 December, 1899, in *Frege 1976* or *Frege 1980*), Hilbert describes an axiom system as a schema awaiting interpretation (in *Hilbert 1894a**, 59, he uses the word 'Deutung'). There is no talk of schemata in the *Festschrift* in any of its editions, but Hilbert's practice (constantly shifting between various interpretations as a way of investigating the logical relations between propositions) is certainly of a piece with just this way of describing matters.

but rather *embeddings*, where an extension is a model which has a substructure isomorphic to the given model. The *Vollständigkeitsaxiom* would then demand that no model of I–V 1 is properly embeddable in an extension.

From the seventh edition on, the *Vollständigkeitsaxiom* is replaced by what Hilbert calls the *Axiom der linearen Vollständigkeit*, which states:

> *Die Punkte einer Geraden bilden ein System, welches bei Aufrechterhaltung linearen Anordnung (Satz 6), des ersten Kongruenzaxioms und des Archimedischen Axioms (d. h. der Axiome I 1–2, II, III 1, V 1) keiner Erweiterung mehr fähig ist*, d. h. es ist nicht möglich, zu dem System von Punkten Punkte auf *a* hinzuzufügen, so daß in dem durch Zusammensetzung entstehenden System sämtliche aufgeführten Axiome erfüllt sind. (*Hilbert 1930a*, 30.)

This version, which, Hilbert states in a footnote, stems from an observation of Paul Bernays, appears essentially weaker, for it is a linear axiom and only refers to some of the axioms, not all. But Hilbert proves immediately (pp. 31–32) the so-called *Satz der Vollständigkeit*, which in effect is the old (apparently stronger) version of the axiom. (The proof is straightforward: if one could extend the system in any way, then it can be shown that there must be a *new* point on an *old* line, contradicting the linear *Vollständigkeitsaxiom*.)[18] It is unclear in Hilbert's formulation whether the axiom is meant to refer to all lines or just one given line; Baldus makes it clear that, since the congruence axioms must be satisfied, then it would be enough to demand the completeness of a single line. (See *Baldus 1928*, 327.)

The axiom has a slightly odd appearance, not least because it both refers to the other axioms (or some of them in the seventh edition version), and to what would now be called *models* of those axioms. One worry, addressed by Carnap and Bachmann in *Carnap and Bachmann 1936*, is whether the axiom is properly formulable in a logical language. The answer, of course, is that it is, and this is what they begin to develop. However, one does have to formalise the requisite model-theoretic notions, and be able to quantify over all the models which come into question, and this certainly will make the axiom far from geometrically elementary. Another apparent peculiarity is that it refers to other axioms. This has become, if not commonplace, then not unknown. For instance, some formulations of the axioms for the predicate calculus include a final axiom (schema) which says that the universal quantification of any axiom will also be an axiom. Moreover, it should be noted that Bernays (in *Bernays 1955*) restates the mathematical content of the axiom (at least when used in the context of the complete, ordered field of the real numbers) in several mathematically unexceptional ways. (One

[18]The linear *Vollständigkeitsaxiom* does not appear in the 1927 lectures (i.e., *Hilbert 1927**) in the *Ausarbeitung* by Schmidt. It is, however, reported in Baldus's 1928 paper on the *Vollständigkeitsaxiom*, together with the proof hinted at here. Baldus says that he was made aware of it by Bernays, 'auf grund einer Vorlesungsbemerkung von Herrn D. Hilbert'. See *Baldus 1928*, 325–326, especially p 325.

such formulation is: the system of real numbers is an ordered field such that any addition of an element to the field will either disrupt the denseness of the ordering, or give it a first or last element. See op. cit., p. 221.) These matters will not be discussed further here; we confine ourselves instead to a few remarks about the purpose and provenance of the axiom, and what it achieves.

1. The Purpose of the Axiom. In the *Ausarbeitung* of the 1898/1899 lectures (pp. 139–141), Hilbert gives three versions of what he calls 'the Archimedean Axiom'. The first is the usual Archimedean Axiom, and the second a projective version of this. But the third is a much stronger principle concerning the existence of limit points for increasing, but bounded, infinite sequences of points on a line, in effect guaranteeing Bolzano-Dedekind continuity. As Hilbert says:

> Es ist kaum nötig, auf die Analogie dieses Satzes mit der Theorie von Dedekind'schen Schnitte hinzuweisen. (*Hilbert 1899a**, 141.)

This Limit Point Principle, as we may call it, is first stated publicly by Hilbert in *Hilbert 1895b*, although a version of it is used in the 1894 lectures on the foundations of geometry (see Chapter 2); it implies the usual Archimedean Axiom. However, in the 1898/1899 lectures, as opposed to those of 1894, Hilbert makes no use whatsoever of this third version of 'the Archimedean Axiom'. Indeed, it cannot have been adopted officially as an axiom. Speaking of the ordinary Archimedean Axiom, Hilbert says:

> Auf Grund des Archimedischen Axioms kann nun die *Einführung der Zahl in die Geometrie* erfolgen. (*Hilbert 1899a**, 166.)

Then he continues, a little later on the same page:

> Dass auch wirklich jeder reellen Zahl ein Punkt der Graden entspricht, folgt aus unseren Axiomen nicht. Man kann es aber erreichen durch *Einführung von idealen (irrationalen) Punkten (Cantor'sches Axiom).* (*Hilbert 1899a**, 166.)[19]

Thus, 'our axioms' do not guarantee this, although, if the stronger version of the Archimedean Axiom were amongst them, they would. Moreover, part of Hilbert's project, it seems, is to investigate how much of basic geometry can (or cannot) be developed without appealing even to the ordinary Archimedean Axiom. In the first edition of the *Festschrift*, there is no room for ambiguity, since the usual Archimedean Axiom is the only continuity axiom even mentioned.[20]

In short, the axiom system of the 1898/1899 lectures and the *Festschrift* exhibits two, related, features. Firstly, with the usual Archimedean Axiom alone we cannot show that to each real number there corresponds a point,

[19]See also the 1898 *Ferienkurs* (*Hilbert 1898b**, in Chapter 4), pp. 15–16.

[20]The reference to 'Cantor's Axiom' in the passage quoted is to Cantor's adoption of the 'axiom' that to each real number there corresponds a unique point in a (geometrical) line. See *Cantor 1872*, § 2. Cantor's characterisation of the reals using Cauchy sequences is first to be found in § 1 of this article. There is a similar 'axiom' in *Dedekind 1872*.

which means that the project undertaken in the 1894 lectures (complete co-ordination of points on a line with numbers) cannot be carried out with this axiom alone. The fairly elementary examples given by Hilbert in the discussion surrounding the Three Chord Theorem and the failure of the Euclidean field property illustrate the extent of this failure. (See the Introduction to Chapter 4, Specific Note (2), especially (*e*) and (*f*).) Secondly, the absence of Bolzano-Dedekind continuity means that the system of axioms has countable models, as Hilbert himself points out (p. 21 of the *Festschrift*; Hilbert uses the minimal Pythagorean field construction). Both ways of putting the matter point to a gap between analytic (Cartesian) geometry and Hilbert's system, and these difficulties raise the issue of 'completeness' and the role of the *Vollständigkeitsaxiom* in this regard.

In the Introduction to the *Festschrift*, Hilbert states the desire to give a 'simple and complete' system of axioms. However, the irrational 'gaps' in the line mean that the system of the *Festschrift* is *incomplete* with respect to an analytic geometry based on the usual real number system, since the latter can prove the existence of points which the former cannot. The *Vollständigkeitsaxiom* ensures completeness in this respect when added to the other axioms, including the Archimedean Axiom. This is the primary sense of completeness with which Hilbert was concerned. After presenting the *Vollständigkeitsaxiom* in the French translation, Hilbert says:

> ... cet axiome rend possible la correspondance univoque et reversible des points d'une droite et de tous les nombres réels. D'ailleurs, dans le cours des présents recherches, nous ne nous sommes servi nulle part de cet «axiome d'intégrité» [*Vollständigkeitsaxiom*]. (*Hilbert 1900c*, 124.)

I. e. 'rendering possible this correspondence' is the sole purpose of the axiom. This function of the axiom is stressed in the second edition of the *Festschrift* as well, where the new axiom is integrated into Group V as V 2:

> In der Tat reicht das Archimedische Axiom allein nicht aus, um mit Benutzung der Axiome I–IV unsere Geometrie als identisch mit der gewöhnlichen analytischen „Cartesischen" Geometrie nachzuweisen (vgl. § 9 und § 12). (*Hilbert 1903a*, 17.)

(The references to §§ 9 and 12 are clearly to the existence of countable models of the system with just the Archimedean Axiom.) Hilbert continues:

> Dagegen gelingt es unter Hinzunahme des Vollständigkeitsaxiom – obwohl dieses Axiom unmittelbar keine Aussage über den Begriff der Konvergenz enthält –, die Existenz der einem Dedekindschen Schnitte entsprechenden Grenze und den Bolzanoschen Satz vom Vorhandensein der Verdichtungsstellen nachzuweisen, womit dann unsere Geometrie sich als identisch mit der Cartesischen Geometrie erweist. (*Hilbert 1903a*, 17.)

In other words, Hilbert's system with the *Vollständigkeitsaxiom* is complete with respect to 'Cartesian' geometry. As Hilbert puts it three pages later:

> Wie man erkennt, gibt es unendlich viele Geometrien, die den Axiome I–IV, V 1 genügen, dagegen nur e i n e, nämlich die Cartesische Geometrie, in

> der auch zugleich das Vollständigkeitsaxiom V 2 gültig ist. (*Hilbert 1903a*, 20.)

This is also an unmistakeable sign that Hilbert acknowledges what we would now call the *categoricity* of the axiom system with the *Vollständigkeitsaxiom*. We will come back to this role of the axiom below, after looking at the motivation behind the axiom. Here we just note that, once the equivalence to Cartesian geometry has been pointed out, no further use is to be made of the axiom:

> In den nachfolgenden Untersuchungen stützen wir uns wesentlich nur auf das Archimedische Axiom und setzen im allgemeinen das Vollständigkeitsaxiom nicht voraus. (*Hilbert 1903a*, 17.)

This is the same point as is made in the French translation.

2. The Provenance of the Axiom. Hilbert originally contemplated adding a Bolzano-Dedekind continuity principle to the system of axioms for Euclidean geometry, a principle such as that used in the 1894 lectures and *Hilbert 1895b*, and stated in the 1898/1899 lectures. Moreover, that this was also seen as a possible replacement for the Archimedean Axiom is suggested by the formulation of the axiom as a third 'version' of the latter. (See the *Ausarbeitung* of the 1898/1899 lectures, pp. 140–141, or the 1898 *Ferienkurs*, pp. 15–16, respectively Chapters 3 and 4 of this Volume.) On the other hand, such an axiom would seem to be primarily a numerical principle borrowed directly from analysis, and when formulated correctly would involve quantification over all subsets of rationals or all Cauchy sequences. One of Hilbert's purposes was to frame his geometry as far as possible synthetically, and independently of the concept of number, and then to show that there is no shortfall between this system and that of analytic geometry. Viewed from this perspective, the Bolzano-Dedekind continuity principle might seem inappropriate, as perhaps excessively 'numerical'. But this objection is not especially cogent. For one thing, the axiom as framed in the 1898/1899 lectures is not particularly ungeometrical. It does make use of the notion of a bounded, infinite sequence of points on a given line, but this just relies on the application to that line of the order relation for its points established through the axioms of Group II. Secondly, the encroachment of a 'numerical' element already occurs through the Archimedean Axiom, for this makes essential use of existential quantification over the natural numbers.[21] Moreover, any implicit involvement of strong set-

[21]The non-elementary nature of the Archimedean Axiom was noted by Hilbert in his 1905 lectures 'Die logischen Principien des mathematischen Denkens'. After the statement of the standard Archimedean Axiom as one of the axioms for complete, ordered fields, Hilbert remarks:

> Hier tritt als neues logisches Element hinzu der Begriff der beliebigen endlichen Anzahl, wie wir ihn für die hier notwendige Menge von Additionsprocessen brauchen; doch nehmen wir das vorläufig wieder als von vornherein gegeben an. (*Hilbert 1905a**, 16–17.)

The same 'logical element', of course, appears in the Archimedean Axiom in its geometrical setting. Note the statement at the beginning of the *Ausarbeitung* of the 1898/1899 lectures (pp. 2–3) that 'the laws of logic and the whole of arithmetic' are assumed as given. See Chapter 4 of this Volume. Hilbert sometimes called the Archimedean Axiom 'Das Axiom

theoretic elements in Bolzano-Dedekind continuity principles would not have presented a barrier to Hilbert. For one thing, the principle when used in analysis involves an equally powerful encroachment of set theory, and it might be argued that this compromises the *analytic* character of these axioms. More to the point, the *Vollständigkeitsaxiom* eventually adopted apparently involves an even stronger set-theoretic quantification, this time over all models of the axioms.

A much more plausible objection to a standard Bolzano-Dedekind continuity axiom is that such an axiom would render the Archimedean Axiom superfluous. This would be highly unsatisfactory. One of Hilbert's purposes was to investigate to what extent the standard involvement of the Archimedean Axiom is essential in various developments, e.g., that surrounding the Pascal Theorem or the proof of the Legendre Theorems (see the Introduction to Chapter 4), or in proving the theorem that the base angles of an isoceles triangle are equal when only a weaker version of the Triangle Congruence Axiom is present (see the Introduction to Chapter 6). This investigation would no longer be precise or even informative if the Archimedean Axiom were not stated as a separate principle.

Hilbert's solution is not to replace the Archimedean Axiom, but to adopt an axiom which *complements* it, and which, when the two are taken in tandem, implies that the system is as strong as analytic geometry. This is what the *Vollständigkeitsaxiom* is designed to do. The fact that no essential use is made of it after its statement means that the original investigations, particularly those surrounding the use of continuity in the form of the Archimedean Axiom, can be taken over without modification and why the *Vollständigkeitsaxiom* 'forms the keystone of the whole axiom system'. (See p. 17 of *Hilbert 1903a*.)

The conceptual origin of the axiom is easy to uncover, taking the 1898/99 lectures as a guide. On the one hand, we know that there is at least one uncountable model of the axioms I–V (i.e., V 1), since real Euclidean (analytic) 3-space is a model. On the other hand, it is consistent to assume that there are only countably many points, since the axioms have as a model the number triples built over a countable sub-field of the reals, i.e., a substructure of the whole field. It follows that it is consistent to assume that more points can be added to this substructure than just those demanded by the axioms, provided, of course, that such points continue to satisfy the axioms I–V. The presence of the Archimedean Axiom, however, guarantees that each point, whether 'old' or newly added, can be assigned a real number triple; hence, any model of Axioms I–V will be (isomorphic to) a sub-model of analytic Euclidean 3-space. Thus, while one can add what Hilbert calls 'ideal points', these points will in fact always correspond to real numbers (real number triples), and so will be what Hilbert calls extra 'irrational' points. (See the 1898/1899 lecture notes, p. 103, the *Ausarbeitung* of these notes, pp. 166–167, both in Chapter 4 of this

des Messens'. See, e.g., *Hilbert 1905a**, 34, where this term is explained by the fact that the axiom guarantees that any 'number' can be 'measured' by any other; see also *Hilbert 1917**, 18. (These lectures will appear in Volumes 2 and 3 respectively of this edition.)

Volume or the *Festschrift*, p. 39 in Chapter 5.) The Archimedean Axiom thus guarantees that there can be no model which is a genuine *extension* of the full real number field, in other words, that the full field of real numbers will be the 'limit of completion' to which additions will tend. What the *Vollständigkeitsaxiom* does is then to demand that no model of I–V falls short of the real number field either, i.e., to put it loosely, that all additions of (irrational) points by piecemeal procedures which could have been made have already been made.[22] This makes it immediately obvious why the axiom guarantees Bolzano-Dedekind completeness, and why Hilbert thinks of the Archimedean Axiom as a preparation for the *Vollständigkeitsaxiom* (*Hilbert 1903a*, 17). It is only in the presence of the Archimedean Axiom that the *Vollständigkeitsaxiom* 'completes' the arithmetisation of the geometrical line in a unique way.

There is a fairly direct connection to Dedekind in the consideration. In his 1872 essay on continuity and the irrational numbers (*Dedekind 1872*), Dedekind asserts that the completion of the spatial line is a conceptual completion which is not forced on us by anything we know about the nature of space. He goes on:

> Die Annahme dieser Eigenschaft [the Cut Property] der Linie ist nichts als ein Axiom, durch welches wir erst der Linie ihre Stetigkeit zuerkennen, durch welches wir die Stetigkeit in die Linie hineindenken. Hat überhaupt der Raum eine reale Existenz, so braucht er doch nicht notwendig stetig zu sein; unzählige seiner Eigenschaften würden dieselben bleiben, wenn er auch unstetig wäre, so könnte uns doch wieder nichts hindern, falls es uns beliebte, ihn durch Ausfüllung seiner Lücken in Gedanken zu einem stetigen zu machen; diese Ausfüllung würde aber in einer Schöpfung von neuen Punktindividuen bestehen und dem obigen Prinzip gemäß auszuführen sein. (*Dedekind 1872*, 11.)

Hilbert adopts a related view concerning the addition of 'ideal (irrational) points' in his remarks on pp. 166–167 of the *Ausarbeitung* of the 1898/1899 lectures. Adding these points is not necessary 'für unsere erfahrungsmässige Kenntniss von den räumlichen Eigenschaften der Dinge'. They are, as he calls them, 'ideal points', and the question whether they 'genuinely exist' is 'completely idle'. Hilbert goes on to say this:

> ... Ihr Nutzen ist lediglich ein methodischer; *erst mit ihrer Hilfe ist es möglich, die analytische Geometrie in ihrer vollen Ausdehnung zu entwickeln.* (*Hilbert 1899a**, 167; see Chapter 4.)

(Note again the stress on matching analytic geometry.)

When Hilbert first applied the *Vollständigkeitsaxiom* to his system of Euclidean geometry is not clear. Since it appeared in the French translation, it

[22] In his 1894 lectures, Hilbert speaks of adjoining points in a 'stepwise' fashion as and where the adjunctions are deemed important. See *Hilbert 1894a** (in Chapter 2), pp. 35–36. This comes before Hilbert adds a full continuity principle (p. 38). 'Adding points' by piecemeal procedures would make sense in some contexts, e.g., adding points to guarantee that all straightedge and compass constructions can be carried out, i.e., that the Euclidean field property is satisfied.

must have been formulated for geometry before about March of 1900. We can, however, say a little more. Given that the purpose of the axiom when used in geometry is to ensure that the system matches analytic geometry, it seems clear that Hilbert adopted the *Vollständigkeitsaxiom* from his axiomatiation of the theory of complete, ordered fields. As mentioned above, the axiom appears as the second continuity axiom for ordered fields in *Hilbert 1900b*, dated October 1899. It does not, however, appear among the ordered field axioms given in either the 1898/1899 lectures or the first edition of the *Festschrift*. Hence, it seems that the *Vollständigkeitsaxiom* was developed for ordered fields in the summer or autumn of 1899. Some evidence that it was actually rather early in the summer is given in a letter from Minkowski to Hilbert from 24 June, 1899, written shortly after his return from the festivities in Göttingen surrounding the unveiling of the Gauß-Weber Monument:

> Die schönen Tage in Göttingen kommen mir, nachdem ich in die Züricher Wirklichkeit zurückgekehrt bin, heute wie ein Traum vor; doch ist an ihrer Existenz wohl ebenso wenig zu zweifeln wie an der Deiner 18 = 17 + 1 Axiome der Arithmetik. (*Minkowski 1973*, 116–117.)

The *Festschrift* (§ 13) contains seventeen axioms for arithmetic (ordered fields), which are exactly the first seventeen axioms of the system in *Hilbert 1900b*. (The axioms are numbered 1–17 in the *Festschrift*, but *not* in the *Hilbert 1900b* system, where they are grouped, and the groups labelled, rather as the geometrical axioms are in the *Festschrift*.) Perhaps the additional eighteenth axiom Minkowski mentions here is the eighteenth axiom of *Hilbert 1900b*, the second continuity axiom, i.e., the *Vollständigkeitsaxiom*.

A few remarks on the nature of the relationship between the *Vollständigkeitsaxiom* and the aim of providing a 'simple and complete' system of axioms for geometry stated by Hilbert in the *Einleitung* to the *Festschrift*. The *Vollständigkeitsaxiom* helps to achieve completeness in the sense of permitting parity with analytic geometry. This is a form of relative deductive completeness, a claim about provability within synthetic geometry relative to what is provable within another theory, analytic geometry, for, once the *Vollständigkeitsaxiom* is added, there can be no theorems about the existence of points, lines and planes provable in analytic geometry which are not also provable in Hilbert's synthetic geometry. It is clear that this was Hilbert's immediate aim with the *Vollständigkeitsaxiom*. But does this conform to his goal in seeking a 'simple and complete' axiom system for geometry?

The statement in the *Einleitung* to the *Festschrift* itself does not make it clear exactly what is to be aimed at in seeking a complete axiom system. One guide is that the term 'completeness' figures in Hilbert's various allusions to Hertz's *Bildtheorie* (in the 1894 lectures, pp. 8–9, in the *Ausarbeitung* of the 1898/1899 lectures, p. 2, and in the 1902 lectures, p. 2; see Chapters 2, 4 and 6 of this Volume respectively). For instance, in the 1898 *Ausarbeitung* he says that a 'complete and simple' *Bild* will be one in which each property of things of the system will correspond to 'geometrical facts' and conversely. In the 1894 lectures, it is stressed (pp. 7–8) that what is sought is a set of *Grundtatsachen* sufficient for the construction (i.e., logical derivation) of the whole of geometry. Thus, completeness appears to mean 'deductive completeness with respect to the geometrical facts'.

By the time of Hilbert's 1905 lectures on the logical principles of mathematical thought, the criterion of completeness had indeed crystalised somewhat along these lines. Here, the third adequacy condition on axiom systems (the first two are the consistency of the system and the independence of the axioms from one another) is formulated as follows:

> Ist das Axiomsystem *vollständig*, d. h. hat es wirklich alle vorgelegten Thatsachen zur logischen Consequenz. (*Hilbert 1905a**, 12.)[23]

In the case of Euclidean geometry, however, there are various ways in which 'the facts before us' can be interpreted. If interpreted as 'the facts presented in school geometry' (or the initial stages of Euclid's *Elements*), then arguably the system of the original *Festschrift* is adequate. If, however, the facts are those given by geometrical intuition, then matters are less clear. (Witness the problem with the 'line-circle property' examined in the Introduction to Chapter 4, Specific Note (*2*).) Certainly the *Vollständigkeitsaxiom* addresses some of these problems, and, as mentioned previously, achieves completeness in a more precise sense, since it guarantees deductive parity with analytic geometry. Thus, another way to interpret 'the facts before us' is as what is provable by invoking analytic geometry.

Hilbert later moved towards characterisations of completeness which appears closer to deductive completeness as we now understand it. For instance, in his 1917/1918 lectures on the principles of mathematics, explaining what is meant by saying that an axiom system is to be examined as to its completeness, Hilbert says:

> ... d. h. daraufhin, ob alle Fragen, welche die betrachteten Gegenstände und die in den Axiomen vorkommenden Beziehungen zwischen ihnen betreffen, mit Hilfe der Axiome prinzipiell logisch entscheidbar sind. (*Hilbert 1917**, 3.)[24]

Moreover, Hilbert even uses the notion of Post-Completenesss in proving the completeness of propositional logic in these lectures. Nevertheless, even here, there is less clarity than the passage above suggests, and Hilbert still uses the term 'completeness' in several, quite distinct senses. In particular, in his analysis of logic, which is what is most at issue in the 1917/1918 lectures, Hilbert still speaks of completeness in the two senses 'being adequate to the analysis of a particular subject matter' and 'capturing all traditional logical inferences', and both notions have a good deal in common with being adequate to 'the facts before us'.

Nevertheless, Hilbert's axiom system in the second edition of the *Festschrift* appears to achieve more than this, since it is *categorical*, i.e., any two models of it must be isomorphic. The reason for this appears to be simple, namely that the associated system of axioms for the real number system is

---

[23] See Volume 2 of this edition.

[24] See Volume 3 of this edition.

itself categorical, and the geometrical *Vollständigkeitsaxiom*, acting with the Archimedean Axiom, forces a one-to-one mapping of the points onto the system of real numbers. And in an Archimedean ordered field, it is surely the addition of Hilbert's *Vollständigkeitsaxiom* for *fields* which imposes categoricity. In short, the *Vollständigkeitsaxiom* seems indissolubly bound up with categoricity.[25] Moreover, we now tend to think automatically that categoricity implies full deductive completeness.[26] But two points should be made here, the first concerning categoricity and deductive completeness, and the second concerning the *Vollständigkeitsaxiom* and the categoricity of the geometrical axiom system.

Firstly, establishing a precise relationship between categoricity and deductive completeness only makes sense once one is presented with a precise deductive system. Moreover, the argument which shows that categoricity implies deductive completeness relies on the *completeness* of the logical system being used. Hilbert had at this stage no precise notion of a logical system, and even if he had had such a system at his disposal, the formulation of the *Vollständigkeitsaxiom* would have required use of an incomplete second-order system.

Secondly, the stress on the role of the *Vollständigkeitsaxiom* in the proof of the categoricity of the geometrical axiom system is misleading. While the *Vollständigkeitsaxiom* certainly plays the key role in establishing the 'point completeness' of the lines, Baldus points out that, if categoricity is the aim, it is not just the Archimedean Axiom which is important as 'preparation'. (See *Baldus 1928*.) The point is simply made. If the Parallel Axiom is left out of Hilbert's list of axioms I–V 1, then a reasonable axiomatisation of what was called by Bolyai 'Absolute Geometry' results. If the *Vollständigkeitsaxiom* is now added, the axiom will have the same effect of maximising the number of points on any line. However, the system so constructed will *not* be categorical. One could certainly add to it the Euclidean Parallel Axiom, but one could also add to it an axiom allowing many parallels to a given line through a given point outside it. Hence, Absolute Geometry is not categorical even though it will be linearly complete in the sense the *Vollständigkeitsaxiom* (in tandem with the Archimedean Axiom) demands. Moreover, it is precisely this sense of *linear* completeness which is underlined with the reformulation of the axiom in the seventh edition of 1930, i.e., shortly after Baldus's paper was published.

Michael Hallett, with the assistance of Ralf Haubrich and Tilman Sauer

[25]The concept of categoricity itself was not unknown to Hilbert. It is pointed out for the theory of complete, ordered fields in both the lectures for 1899/1900 on 'Zahlbegriff und Quadratur des Kreises' (see *Hilbert 1897b**, 42), and in the 1905 lectures (see *Hilbert 1905a**, 21). Both sets of lectures will figure in Volume 2 of this edition. For a discussion of some of the senses of the term 'completeness of an axiom system' as understood in the 1920s, of which categoricity is one, see *Fraenkel 1928*, 347–354.

[26]Hilbert later proposed deductive completeness as a replacement for the categoricity condition in the special cases of number theory and analysis, as being one which meets the 'Anforderungen finiter Strenge'. See *Hilbert 1928b*, 139.

Grundlagen der Geometrie

3

> So fängt denn alle menschliche Erkenntnis mit Anschauungen an, geht von da zu Begriffen und endigt mit Ideen.
>
> Kant, Kritik der reinen Vernunft, Elementarlehre 2. T. 2. Abt.

Einleitung.

Die Geometrie bedarf – ebenso wie die Arithmetik – zu ihrem folgerichtigen Aufbau nur weniger und einfacher Grundthatsachen. Diese Grundthatsachen heissen A x i o m e der Geometrie. Die Aufstellung der Axiome der Geometrie und die Erforschung ihres Zusammenhanges ist eine Aufgabe, die seit *Euklid* in zahlreichen vortrefflichen Abhandlungen der mathematischen Litteratur[1]) sich erörtert findet. Die bezeichnete Aufgabe läuft auf die logische Analyse unserer räumlichen Anschauung hinaus.

Die vorliegende Untersuchung ist ein neuer Versuch, für die Geometrie ein e i n f a c h e s und v o l l s t ä n d i g e s System von einander u n a b h ä n g i g e r Axiome aufzustellen und aus denselben die wichtigsten geometrischen Sätze in der Weise abzuleiten, dass dabei die Bedeutung der verschiedenen Axiomgruppen und die Tragweite der aus den einzelnen Axiomen zu ziehenden Folgerungen möglichst klar zu Tage tritt.

---

1) Man vergleiche die zusammenfassenden und erläuternden Berichte von *G. Veronese*, „Grundzüge der Geometrie", deutsch von *A. Schepp*, Leipzig 1894 (Anhang), und *F. Klein*, „Zur ersten Verteilung des *Lobatschefskiy*-Preises", Math. Ann. Bd. 50.[1]

---

9 Grundthatsachen. Diese Grundthatsachen] Grundsätze. Diese Grundsätze

Note 1 *Lobatschefskiy*] *Lobatschefsky*

16 e i n f a c h e s und v o l l s t ä n d i g e s] v o l l s t ä n d i g e s und m ö g l i c h s t e i n f a c h e s

16–17 einander u n a b h ä n g i g e r Axiome] Axiomen

[1] *Veronese 1894*, *Klein 1898*.

Kapitel I. 4

Die fünf Axiomgruppen.

§ 1.

Die Elemente der Geometrie und die fünf Axiomgruppen.

Erklärung. Wir denken ⟨uns⟩ drei verschiedene Systeme von Dingen: die Dinge des ersten Systems nennen wir *Punkte* und bezeichnen sie mit A, B, C, ...; die Dinge des zweiten Systems nennen wir *Gerade* und bezeichnen sie mit a, b, c, ...; die Dinge des dritten Systems nennen wir *Ebenen* und bezeichnen sie mit α, β, γ, ...; die Punkte heissen auch die *Elemente der linearen Geometrie*, die Punkte und Geraden heissen die *Elemente der ebenen Geometrie* und die Punkte, Geraden und Ebenen heissen die *Elemente der räumlichen Geometrie* oder *des Raumes*.

Wir denken die Punkte, Geraden, Ebenen in gewissen gegenseitigen Beziehungen und bezeichnen diese Beziehungen durch Worte wie „liegen", „zwischen", „parallel", „congruent", „stetig"; die genaue und vollständige Beschreibung dieser Beziehungen erfolgt durch die *Axiome der Geometrie*.

Die Axiome der Geometrie gliedern sich in fünf Gruppen; jede einzelne dieser Gruppen drückt gewisse zusammengehörige Grundthatsachen unserer Anschauung aus. Wir benennen diese Gruppen von Axiomen in folgender Weise:

I 1–7. Axiome der *Verknüpfung*,
II 1–5. Axiome der *Anordnung*,
III. Axiom der *Parallelen* (*Euklidisches* Axiom),
IV 1–6. Axiome der *Congruenz*,
V. Axiom der *Stetigkeit* (*Archimedisches* Axiom).

§ 2. 5

Die Axiomgruppe I: Axiome der Verknüpfung.

Die Axiome dieser Gruppe stellen zwischen den oben erklärten Begriffen Punkte, Geraden und Ebenen eine *Verknüpfung* her und lauten wie folgt:

I 1. *Zwei von einander verschiedene Punkte A, B bestimmen stets eine Gerade a; wir setzen $AB = a$ oder $BA = a$.*

Statt „bestimmen" werden wir auch andere Wendungen brauchen, z. B. A „liegt auf" a, A „ist ein Punkt von" a, a „geht durch" A

22–26 ⟨In E2 the table reads:⟩
I 1–8. Axiome der *Verknüpfung*,
II 1–4. Axiome der *Anordnung*,
III 1–6. Axiome der *Kongruenz*,
IV. Axiom der *Parallelen*,
V 1–2. Axiome der *Stetigkeit*.

32 *Gerade a; wir setzen $AB = a$ oder $BA = a$.*] *Gerade a.*

33 brauchen] gebrauchen

34 A „liegt auf" a, A „ist ein Punkt von" a,] ⟨deleted⟩

„und durch" B, a „verbindet" A „und" oder „mit" B u. s. w. Wenn A auf a und ausserdem auf einer anderen Geraden b liegt, so gebrauchen wir auch die Wendung: „die Geraden" a „und" b „haben den Punkt A gemein" u. s. w.

I 2. *Irgend zwei von einander verschiedene Punkte einer Geraden bestimmen diese Gerade; d. h. wenn $AB = a$ und $AC = a$, und $B \neq C$, so ist auch $BC = a$.*

I 3. *Drei nicht auf ein und derselben Geraden liegende Punkte A, B, C bestimmen stets eine Ebene α; wir setzen $ABC = \alpha$.*

Wir gebrauchen auch die Wendungen: A, B, C „liegen in" α; A, B, C „sind Punkte von" α u. s. w.

I 4. *Irgend drei Punkte A, B, C einer Ebene α, die nicht auf ein und derselben Geraden liegen, bestimmen diese Ebene α.*

I 5. *Wenn zwei Punkte A, B einer Geraden a in einer Ebene α liegen, so liegt jeder Punkt von a in α.*

In diesem Falle sagen wir: die Gerade a liegt in der Ebene α u. s. w.

I 6. *Wenn zwei Ebenen α, β einen Punkt A gemein haben, so haben sie wenigstens noch einen weiteren Punkt B gemein.*

I 7. *Auf jeder Geraden giebt es wenigstens zwei Punkte, in jeder Ebene wenigstens drei nicht auf einer Geraden gelegene Punkte und im Raum giebt es wenigstens vier nicht in einer Ebene gelegene Punkte.*

Die Axiome I 1–2 enthalten nur Aussagen über die Punkte und Geraden, d. h. über die Elemente der ebenen Geometrie und mögen daher die *ebenen Axiome der Gruppe* I heissen, zum Unterschied von den Axiomen I 3–7, die ich kurz als die *räumlichen Axiome* bezeichne.

Von den Sätzen, die aus den Axiomen I 1–7 folgen, erwähne ich nur diese beiden:

---

1 „mit" B u. s. w.] „mit" B. Wenn A ein Punkt ist, der mit einem anderen Punkte zusammen die Gerade a bestimmt, so gebrauchen wir auch die Wendungen: A „liegt auf" a, A „ist ein Punkt von" a, „es gibt den Punkt" A „auf" a u. s. w.

2 auf a] auf der Geraden a

6–7 *Gerade;* ⟨...⟩ *$BC = a$.*] *Gerade.*

7 ⟨After this line the following axiom is inserted; cf. I 7 of E1:⟩

I 3. *Auf einer Geraden gibt es stets wenigstens zwei Punkte, in einer Ebene gibt es stets wenigstens drei nicht auf einer Geraden gelegene Punkte.*[2]

9 *Ebene α; wir setzen $ABC = \alpha$.*] *Ebene α.*

12 *Punkte A, B, C einer Ebene α,*] *Punkte einer Ebene,*

13 *diese*] *die*

15 *in α.*] *in der Ebene α.*

20–22 I 7. ⟨...⟩ Punkte.] I 8. *Es gibt wenigstens vier nicht in einer Ebene gelegene Punkte.*

23–24 enthalten ⟨...⟩ mögen daher] mögen

26 kurz als die *räumlichen Axiome*] als die *räumlichen Axiome der Gruppe I*

[2] For the resulting renumbering of axioms, see the synopsis p. 420.

S a t z 1. Zwei Geraden einer Ebene haben einen oder keinen Punkt gemein; zwei Ebenen haben keinen Punkt oder eine Gerade gemein; eine Ebene und eine nicht in ihr liegende Gerade haben keinen oder einen Punkt gemein. 6

S a t z 2. Durch eine Gerade und einen nicht auf ihr liegenden Punkt, so wie auch durch zwei verschiedene Geraden mit einem gemeinsamen Punkt giebt es stets eine und nur eine Ebene.

§ 3.

Die Axiomgruppe II: Axiome der Anordnung[1]).

Die Axiome dieser Gruppe definiren den Begriff „zwischen" und ermöglichen auf Grund dieses Begriffes die *Anordnung* der Punkte auf einer Geraden, in einer Ebene und im Raume.

E r k l ä r u n g. Die Punkte einer Geraden stehen in gewissen Beziehungen zu einander, zu deren Beschreibung uns insbesondere das Wort „*zwischen*" dient.

Fig. 1.

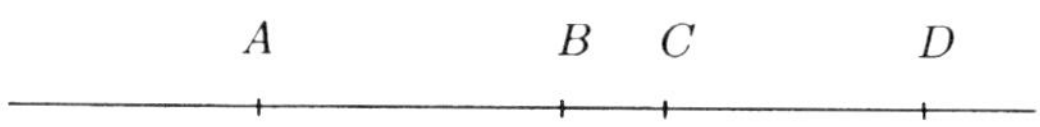

Fig. 2.

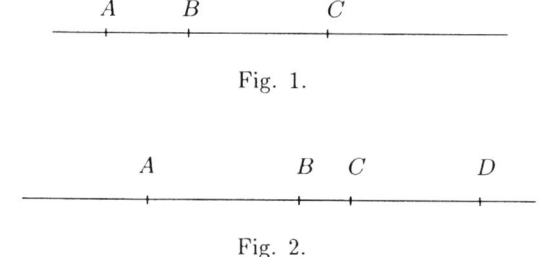

II 1. *Wenn A, B, C Punkte einer Geraden sind, und B zwischen A und C liegt, so liegt B auch zwischen C und A.*

II 2. *Wenn A und C zwei Punkte einer Geraden sind, so giebt es stets wenigstens einen Punkt B, der zwischen A und C liegt und wenigstens einen Punkt D, so dass C zwischen A und D liegt.*

II 3. *Unter irgend drei Punkten einer Geraden giebt es stets einen und nur einen, der zwischen den beiden andern liegt.*

II 4. *Irgend vier Punkte A, B, C, D einer Geraden können stets so angeordnet werden, dass B zwischen A und C und auch zwischen A und D und ferner C zwischen A und D und auch zwischen B und D liegt.*

D e f i n i t i o n. Das System zweier Punkte A und B, die auf einer Geraden a liegen, nennen wir eine *Strecke* und bezeichnen dieselbe mit AB oder BA. Die Punkte zwischen A und B heissen Punkte der Strecke AB oder auch *innerhalb* der Strecke AB ge|legen; alle übrigen Punkte der Geraden a heissen 7

1) Diese Axiome hat zuerst *M. Pasch* in seinen Vorlesungen über neuere Geometrie, Leipzig 1882,[3] ausführlich untersucht. Insbesondere rührt das Axiom II 5 von *M. Pasch* her.

Note 1 von *M. Pasch*] inhaltlich von *M. Pasch*

28–30 II 4. ⟨...⟩ liegt.] ⟨This Axiom corresponds to S a t z 4. in § 4 of E2⟩

31–32 D e f i n i t i o n. ⟨...⟩ nennen wir] E r k l ä r u n g. Wir betrachten auf einer Geraden a zwei Punkte A und B; wir nennen das System der beiden Punkte A und B

32 oder] oder mit

34–440.2 alle ⟨...⟩ Strecke AB.] die Punkte A, B heissen *Endpunkte* der Strecke AB. Alle übrigen Punkte der Geraden a heißen *außerhalb* der Strecke AB gelegen.

ausserhalb der Strecke AB gelegen. Die Punkte A, B heissen *Endpunkte* der Strecke AB.

II 5. *Es seien A, B, C drei nicht in gerader Linie gelegene Punkte und a eine Gerade in der Ebene ABC, die keinen der Punkte A, B, C trifft: wenn dann die Gerade a durch einen Punkt innerhalb der Strecke AB geht, so geht sie stets entweder durch einen Punkt der Strekke BC oder durch einen Punkt der Strecke AC.*

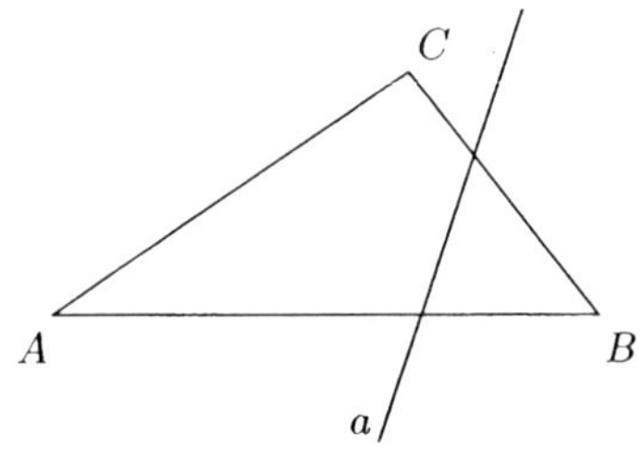

Fig. 3.

Die Axiome II 1–4 enthalten nur Aussagen über die Punkte auf einer Geraden und mögen daher die *linearen Axiome der Gruppe* II heissen; das Axiom II 5 enthält eine Aussage über die Elemente der ebenen Geometrie und heisse daher das *ebene Axiom der Gruppe* II.

§ 4.

Folgerungen aus den Axiomen der Verknüpfung und der Anordnung.

Zunächst leiten wir aus den linearen Axiomen II 1–4 ohne Mühe folgende Sätze ab:

S a t z 3. Zwischen irgend zwei Punkten einer Geraden giebt es stets unbegrenzt viele Punkte.

S a t z 4. Sind irgend eine endliche Anzahl von Punkten einer Geraden gegeben, so lassen sich dieselben stets in einer Reihe A, B, C, D, E, ..., K anordnen, sodass B zwischen A einerseits und C, D, E, ..., K andererseits, ferner C zwischen A, B einerseits und D, E, ..., K andererseits, sodann D

---

8 *Punkt innerhalb*] *Punkt*

9 *stets*] *gewiß auch*

19–20 Zunächst ⟨...⟩ ab.] Aus den Axiomen I und II folgen die nachstehenden Sätze:

22 ⟨After this line the following theorem is added; cf. II 4 of E1:⟩

S a t z 4. Sind irgend vier Punkte einer Geraden gegeben, so lassen sich dieselben stets in der Weise mit A, B, C, D bezeichnen, daß der mit B bezeichnete Punkt zwischen A und C und auch zwischen A und D und ferner der mit C bezeichnete Punkt zwischen A und D und auch zwischen B und D liegt[1]).

1) Dieser in der ersten Auflage als Axiom bezeichnete Satz ist von *E. H. Moore*, Transactions of the American Mathematical Society 1902,[4] als eine Folge der aufgestellten ebenen Axiome der Verknüpfung und der Anordnung erkannt worden.

23 S a t z 4.] S a t z 5 (Verallgemeinerung von Satz 4).[5]

24–25 in einer Reihe A, B, C, D, E, ..., K anordnen, sodass B] in der Weise mit A, B, C, D, E, ..., K bezeichnen, daß der mit B bezeichnete Punkt

[3] *Pasch 1882.*

[4] *Moore 1902.*

[5] Regarding the renumbering in E2, cf. p. 420.

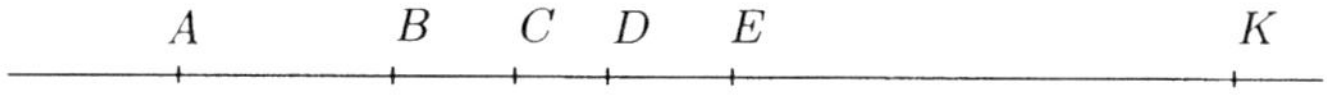

Fig. 4.

zwischen A, B, C einerseits und $E, \ldots, K$ andererseits u. s. w. liegt. Ausser dieser Anordnung giebt es nur noch die umgekehrte Anordnung $K, \ldots, E, D, C, B, A$, die von der nämlichen Beschaffenheit ist.

S a t z 5. Jede Gerade a, welche in einer Ebene α liegt, trennt die übrigen Punkte dieser Ebene α in zwei Gebiete, von folgender | Beschaffenheit: ein 8

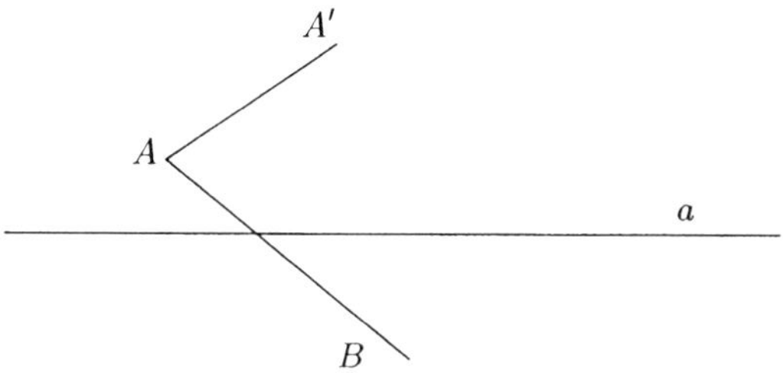

Fig. 5.

jeder Punkt A des einen Gebietes bestimmt mit jedem Punkt B des anderen Gebietes eine Strecke AB, innerhalb derer ein Punkt der Geraden a liegt; dagegen bestimmen irgend zwei Punkte A und A' ein und desselben Gebietes eine Strecke AA', welche keinen Punkt von a enthält.

E r k l ä r u n g. Es seien A, A', O, B vier Punkte einer Geraden a, so dass O zwischen A und B, aber nicht zwischen A und A' liegt; dann sagen wir: die Punkte A, A' liegen *in der Geraden a auf ein und derselben Seite vom Punkte O*, und die Punkte A, B liegen *in der Geraden a auf verschiedenen Seiten vom Punkte O.* Die sämtlichen auf ein und derselben Seite von O gelegenen Punkte der Geraden a heissen auch ein von O ausgehender *Halbstrahl*; somit trennt jeder Punkt einer Geraden diese in zwei Halbstrahlen.

Fig. 6.

Indem wir die Bezeichnungen des Satzes 5 benutzen, sagen wir: die Punkte A, A' liegen *in der Ebene α auf ein und derselben Seite von der Geraden a* und die Punkte A, B liegen *in der Ebene α auf verschiedenen Seiten von der Geraden a.*

D e f i n i t i o n. Ein System von Strecken $AB, BC, CD, \ldots, KL$ heisst ein *Streckenzug*, der die Punkte A und L miteinander verbindet; dieser Streckenzug wird auch kurz mit $ABCD \ldots KL$ bezeichnet. Die Punkte innerhalb

2 dieser Anordnung] dieser Bezeichnungsweise

2 umgekehrte Anordnung] umgekehrte Bezeichnungsweise

4 übrigen] nicht auf ihr liegenden

16 enthält.] enthält. ⟨Footnote added in E2:⟩ Vgl. den Beweis bei *M. Pasch* l. c. S. 25.[6]

25 trennt] teilt

27 Indem] E r k l ä r u n g. Indem

31 D e f i n i t i o n.] E r k l ä r u n g.

[6] *Pasch 1882.*

der Strecken AB, BC, CD, ..., KL, sowie die Punkte A, B, C, D, ..., K, L heissen insgesamt die *Punkte des Streckenzuges.* Fällt insbesondere der Punkt L mit dem Punkt A zusammen, so wird der Streckenzug ein *Polygon* genannt und als Polygon $ABCD \dots K$ bezeichnet. Die Strecken AB, BC, CD, ..., KA heissen auch die *Seiten des Polygons.* Die Punkte A, B, C, D, ..., K heissen die *Ecken des Polygons.* Polygone mit 3, 4, ..., n Ecken heissen bez. *Dreiecke, Vierecke, ..., n-Ecke.*

Wenn die Ecken eines Polygons sämtlich von einander verschieden sind und keine Ecke des Polygons in eine Seite fällt und endlich irgend zwei Seiten eines Polygons keinen Punkt mit einander gemein haben, so heisst das Polygon *einfach.*

9 Mit Zuhülfenahme des Satzes 5 gelangen wir jetzt ohne erhebliche Schwierigkeit zu folgenden Sätzen:

S a t z 6. Ein jedes einfache Polygon, dessen Ecken sämtlich in einer Ebene α liegen, trennt die Punkte dieser Ebene α, die nicht dem Streckenzuge des Polygons angehören, in zwei Gebiete, ein Inneres und ein Aeusseres, von folgender Beschaffenheit: ist A ein Punkt des Inneren (i n n e r e r P u n k t) und B ein Punkt des Auesseren (ä u s s e r e r P u n k t), so hat jeder Streckenzug, der A mit B verbindet, mindestens einen Punkt mit dem Polygon gemein; sind dagegen A, A' zwei Punkte des Inneren und B, B' zwei Punkte des Aeusseren, so giebt es stets Streckenzüge, die A mit A' und B mit B' verbinden und keinen Punkt mit dem Polygon gemein haben. Es giebt Gerade in α, die ganz im Aeusseren des Polygons verlaufen, dagegen keine solche Gerade, die ganz im Inneren des Polygons verläuft.

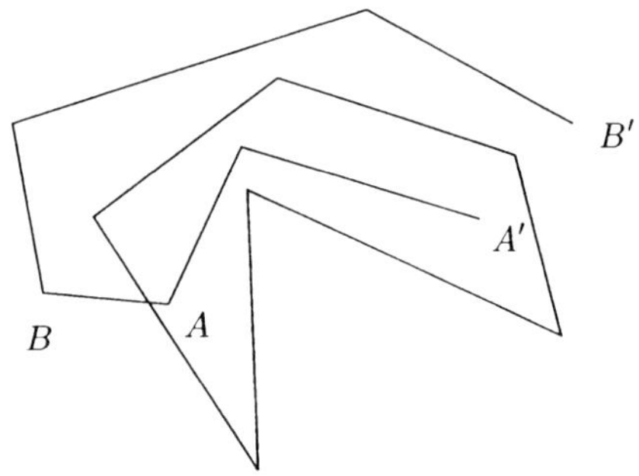

Fig. 7.

S a t z 7. Jede Ebene α trennt die übrigen Punkte des Raumes in zwei Gebiete von folgender Beschaffenheit: jeder Punkt A des einen Gebietes bestimmt mit jedem Punkt B des andern Gebietes eine Strecke AB, innerhalb derer ein Punkt von α liegt; dagegen bestimmen irgend zwei Punkte A und A' eines und desselben Gebietes stets eine Strecke AA', die keinen Punkt von α enthält.

E r k l ä r u n g. Indem wir die Bezeichnungen dieses Satzes 7 benutzen, sagen wir: die Punkte A, A' liegen im Raume *auf ein und derselben Seite von der Ebene* α und die Punkte A, B liegen im Raume *auf verschiedenen Seiten von der Ebene* α.

Der Satz 7 bringt die wichtigsten Thatsachen betreffs der Anordnung der Elemente im R a u m e zum Ausdruck; diese Thatsachen sind daher ledig-

8 Wenn] E r k l ä r u n g. Wenn

35 Punkte A und A'] Punkte von A und A'

lich Folgerungen aus den bisher behandelten Axiomen und es bedurfte in der Gruppe II keines neuen räumlichen Axioms.

§ 5.

Die Axiomgruppe III: Axiom der Parallelen (Euklidisches Axiom).

Die Einführung dieses Axioms vereinfacht die Grundlagen und erleichtert den Aufbau der Geometrie in erheblichem Masse; wir sprechen dasselbe wie folgt aus:

III. *In einer Ebene α lässt sich durch einen Punkt A ausserhalb einer Geraden a stets eine und nur eine Gerade ziehen, welche jene Gerade a nicht schneidet; dieselbe heisst die Parallele zu a durch den Punkt A.*

Diese Fassung des Parallelenaxioms enthält zwei Aussagen; nach der ersteren giebt es in der Ebene α durch A stets eine Gerade, die a nicht trifft, und zweitens wird ausgesprochen, dass keine andere solche Gerade möglich ist.

Die zweite Aussage unseres Axioms ist die wesentliche; sie nimmt auch folgende Fassung an:

Satz 8. Wenn zwei Geraden a, b in einer Ebene eine dritte Gerade c derselben Ebene nicht treffen, so treffen sie sich auch einander nicht.

In der That hätten a, b einen Punkt A gemein, so würden durch A in derselben Ebene die beiden Geraden a, b möglich sein, die c nicht treffen;

2–444.4 § 5. ⟨...⟩ *ebenes Axiom.*] ⟨In E2 the axiom groups III (§ 5 in E1) and IV (§§ 6, 7 in E1) are interchanged. Hence § 5 of E1 was moved to § 7 of E2 (cp. the synopsis on p. 420) and rephrased as follows:⟩

§ 7.

Die Axiomgruppe IV: Axiom der Parallelen

Aus den bisherigen Axiomen folgt in bekannter Weise der Euklidische Satz, daß der Außenwinkel eines Dreieckes stets größer ist, als jeder der beiden inneren Winkel.

Es sei nun α eine beliebige Ebene, a eine beliebige Gerade in α und A ein Punkt in α und außerhalb a. Ziehen wir dann in α eine Gerade c, die durch A geht und a schneidet, und sodann in α eine Gerade b durch A, so daß die Gerade c die Geraden a, b unter gleichen Gegenwinkeln schneidet, so folgt leicht aus dem erwähnten Satze vom Außenwinkel, daß die Geraden a, b keinen Punkt miteinander gemein haben, d. h. in einer Ebene α läßt sich durch einen Punkt A außer einer Geraden a stets eine Gerade ziehen, welche jene Gerade a nicht schneidet.

Das Parallelenaxiom lautet nun:

IV (Euklidisches Axiom). *Es sei a eine beliebige Gerade und A ein Punkt außerhalb a: dann gibt es in der durch a und A bestimmten Ebene α nur eine Gerade b, die durch A läuft und a nicht schneidet; dieselbe heißt die Parallele zu a durch A.*

Das Parallelenaxiom IV ist gleichbedeutend mit der folgenden Forderung:

Wenn zwei Geraden a, b in einer Ebene eine dritte Gerade c derselben Ebene nicht treffen, so treffen sie sich auch einander nicht.

In der Tat, hätten a, b einen Punkt A gemein, so würden durch A in derselben Ebene die beiden Geraden a, b möglich sein, die c nicht treffen; dieser Umstand widerspräche dem Parallelenaxiom IV. Ebenso leicht folgt umgekehrt das Parallelenaxiom IV aus der genannten Forderung.

Das Parallenaxiom IV ist ein *ebenes Axiom.*

Die Einführung des Parallelenaxioms vereinfacht die Grundlagen und erleichtert den Aufbau der Geometrie in erheblichem Maße.

dieser Umstand widerspräche der zweiten Aussage des Parallelenaxioms in unserer ursprünglichen Fassung. Auch folgt umgekehrt aus Satz 8 die zweite Aussage des Parallelenaxioms in unserer ursprünglichen Fassung.

Das Parallelenaxiom III ist ein *ebenes Axiom.*

§ 6.

Die Axiomgruppe IV: Axiome der Congruenz.

Die Axiome dieser Gruppe definieren den Begriff der Congruenz oder der Bewegung.

Erklärung. Die Strecken stehen in gewissen Beziehungen zu einander, zu deren Beschreibung uns insbesondere das Wort „*congruent*" dient.

IV 1. *Wenn A, B zwei Punkte auf einer Geraden a und ferner A' ein Punkt auf derselben oder einer anderen Geraden a' ist, so kann man auf einer gegebenen Seite der Geraden a' von A' stets einen und nur einen Punkt B' finden, so dass die Strecke AB (oder BA) der Strecke $A'B'$ congruent ist, in Zeichen:*

$$AB \equiv A'B'.$$

Jede Strecke ist sich selbst congruent, d. h. es ist stets:

$$AB \equiv AB.$$

Wir sagen auch kürzer, dass eine jede Strecke auf einer gegebenen Seite einer gegebenen Geraden von einem gegebenen Punkte in eindeutig bestimmter Weise *abgetragen* werden kann.

11 IV 2. *Wenn eine Strecke AB sowohl der Strecke $A'B'$ als auch der Strecke $A''B''$ congruent ist, so ist auch $A'B'$ der Strecke $A''B''$ congruent, d. h.: wenn $AB \equiv A'B'$ und $AB \equiv A''B''$, so ist auch $A'B' \equiv A''B''$.*

IV 3. *Es seien AB und BC zwei Strecken ohne gemeinsame Punkte auf der Geraden a und ferner $A'B'$ und $B'C'$ zwei Strecken auf derselben oder einer anderen Geraden a' ebenfalls ohne gemeinsame Punkte; wenn dann*

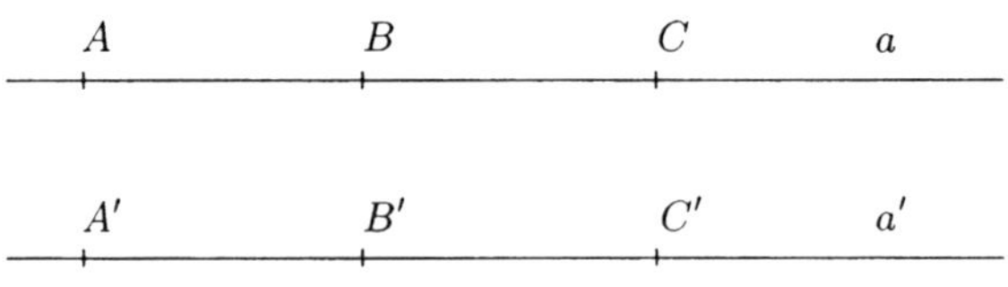

Fig. 8.

$AB \equiv A'B'$ und $BC \equiv B'C'$ ist, so ist auch stets $AC \equiv A'C'$.

---

7 oder] und damit auch den

10 insbesondere das Wort „*congruent*" dient] die Worte „*kongruent*" oder „*gleich*" dienen

14 *(oder BA)*] ⟨deleted⟩

14–15 *congruent*] *kongruent oder gleich*

18 $AB \equiv AB.$] $AB \equiv AB$ *und* $AB \equiv BA$.

19–21 kürzer ⟨...⟩ kann.] kürzer: eine jede Strecke kann auf einer gegebenen Seite einer gegebenen Geraden von einem gegebenen Punkte in eindeutig bestimmter Weise *abgetragen* werden.

24 *d. h.: wenn*] *d. h. wenn*

Definition. Es sei α eine beliebige Ebene und h, k seien irgend zwei verschiedene von einem Punkte O ausgehende Halbstrahlen in α, die verschiedenen Geraden angehören. Das System dieser beiden Halbstrahlen h, k nennen wir einen *Winkel* und bezeichnen denselben mit $\sphericalangle(h,k)$ oder $\sphericalangle(k,h)$. Aus den Axiomen II 1–5 kann leicht geschlossen werden, dass die Halbstrahlen h und k, zusammengenommen mit dem Punkt O die übrigen Punkte der Ebene α in zwei Gebiete von folgender Beschaffenheit teilen: Ist A ein Punkt des einen und B ein Punkt des anderen Gebietes, so geht jeder Streckenzug, der A mit B verbindet, entweder durch O oder hat mit h oder k wenigstens einen Punkt gemein; sind dagegen A, A' Punkte desselben Gebietes, so giebt es stets einen Streckenzug, der A mit A' verbindet und weder durch O, noch durch einen Punkt der Halbstrahlen h, k hindurchläuft. Eines dieser beiden Gebiete ist vor dem anderen ausgezeichnet, indem jede Strecke, die irgend zwei Punkte dieses ausgezeichneten Gebietes verbindet, stets ganz in demselben liegt; dieses ausgezeichnete Gebiet heisse das *Innere* des Winkels (h,k) zum Unterschiede von dem anderen Gebiete, welches das *Aeussere* des Winkels (h,k) genannt werden möge. Die Halbstrahlen h, k heissen *Schenkel* des Winkels und der Punkt O heisst der *Scheitel* des Winkels.

IV 4. *Es sei ein Winkel $\sphericalangle(h,k)$ in einer Ebene α und eine Gerade a' in einer Ebene α', sowie eine bestimmte Seite von a' auf α' gegeben. Es bedeute h' einen Halbstrahl der Geraden a', der vom Punkte O' ausgeht: dann giebt es in der Ebene α' einen und nur einen Halbstrahl k', so dass der Winkel (h,k) (oder (k,h)) congruent | dem Winkel (h',k') ist und zugleich alle* 12
inneren Punkte des Winkels (h',k') auf der gegebenen Seite von a' liegen, in Zeichen:

$$\sphericalangle(h,k) \equiv \sphericalangle(h',k').$$

Jeder Winkel ist sich selbst congruent, d. h. es ist stets

$$\sphericalangle(h,k) \equiv \sphericalangle(h,k).$$

Wir sagen auch kurz, dass ein jeder Winkel in einer gegebenen Ebene nach einer gegebenen Seite an einen gegebenen Halbstrahl auf eine eindeutig bestimmte Weise *abgetragen* werden kann.

IV 5. *Wenn ein Winkel (h,k) sowohl dem Winkel (h',k') als auch dem Winkel (h'',k'') congruent ist, so ist auch der Winkel (h',k') dem Winkel*

1 Definition.] Erklärung.

4 oder] oder mit

6 Punkt] Punkte

18 ⟨After this line the following paragraph is added:⟩
Erklärung. Die Winkel stehen in gewissen Beziehungen zueinander, zu deren Bezeichnung uns ebenfalls die Worte „*kongruent*“ oder „*gleich*“ dienen.

23 (*oder* (k,h))] ⟨deleted⟩

23 *congruent*] *kongruent oder gleich*

28 $\sphericalangle(h,k) \equiv \sphericalangle(h,k)$.] $\sphericalangle(h,k) \equiv \sphericalangle(h,k)$ und $\sphericalangle(h,k) \equiv \sphericalangle(k,h)$.

29–31 kurz, ⟨...⟩ kann.] kurz: ein jeder Winkel kann in einer gegebenen Ebene nach einer gegebenen Seite an einen gegebenen Halbstrahl auf eine eindeutig bestimmte Weise *abgetragen* werden.

(h'', k'') congruent, d. h. wenn $\sphericalangle(h,k) \equiv \sphericalangle(h',k')$ und $\sphericalangle(h,k) \equiv \sphericalangle(h'',k'')$, so ist auch stets $\sphericalangle(h',k') \equiv \sphericalangle(h'',k'')$.

E r k l ä r u n g. Es sei ein Dreieck ABC vorgelegt; wir bezeichnen die beiden von A ausgehenden durch B und C laufenden Halbstrahlen mit h bez. k. Der Winkel (h, k) heisst dann der von den Seiten AB und AC eingeschlossene oder der der Seite BC gegenüberliegende Winkel des Dreieckes ABC; er enthält in seinem Inneren sämtliche innere Punkte des Dreieckes ABC und wird mit $\sphericalangle BAC$ oder $\sphericalangle A$ bezeichnet.

IV 6. *Wenn für zwei Dreiecke ABC und $A'B'C'$ die Congruenzen*

$$AB \equiv A'B', \quad AC \equiv A'C', \quad \sphericalangle BAC \equiv \sphericalangle B'A'C'$$

gelten, so sind auch stets die Congruenzen

$$\sphericalangle ABC \equiv \sphericalangle A'B'C' \text{ und } \sphericalangle ACB \equiv \sphericalangle A'C'B'$$

erfüllt.

Die Axiome IV 1–3 enthalten nur Aussagen über die Congruenz von Strecken auf Geraden; sie mögen daher die *linearen* Axiome der Gruppe IV heissen. Die Axiome IV 4, 5 enthalten Aussagen über die Congruenz von Winkeln. Das Axiom IV 6 knüpft das Band zwischen den Begriffen der Congruenz von Strecken und von Winkeln. Die Axiome IV 3–6 enthalten Aussagen über die Elemente der ebenen Geometrie und mögen daher die *ebenen* Axiome der Gruppe IV heissen.

§ 7.

Folgerungen aus den Axiomen der Congruenz.

E r k l ä r u n g. Es sei die Strecke AB congruent der Strecke $A'B'$. Da
13 nach Axiom IV 1 auch die Strecke AB congruent AB | ist, so folgt aus Axiom IV 2 $A'B'$ congruent AB; wir sagen: die beiden Strecken AB und $A'B'$ sind *unter einander congruent.*

E r k l ä r u n g. Sind A, B, C, D, ..., K, L auf a und A', B', C', D', ..., K', L' auf a' zwei Reihen von Punkten, so dass die sämtlichen entsprechenden Strecken AB und $A'B'$, AC und $A'C'$, BC und $B'C'$, ..., KL und $K'L'$ bez. einander congruent sind, so heissen die beiden Reihen von Punkten *unter einander congruent*; A und A', B und B', ..., L und L' heissen die *entsprechenden Punkte* der congruenten Punktreihen.

Aus den linearen Axiomen IV 1–3 schliessen wir leicht folgende Sätze:

S a t z 9. Ist von zwei congruenten Punktreihen A, B, ..., K, L und A', B', ..., K', L' die erste so geordnet, dass B zwischen A einerseits und C, D, ..., K, L andererseits, C zwischen A, B einerseits und D, ..., K, L andererseits, u. s. w. liegt, so sind die Punkte A', B', ..., K', L' auf die gleiche Weise geordnet, d. h. B' liegt zwischen A' einerseits und C', D', ..., K', L' andererseits, C' zwischen A', B' einerseits und D', ..., K', L' andererseits u. s. w.

4 bez.] und

15 Strecken auf Geraden;] Strecken;

24–25 folgt aus Axiom IV 2] ist nach Axiom III 2 auch

Erklärung. Es sei Winkel (h, k) congruent dem Winkel (h', k'). Da nach Axiom IV 4 der Winkel (h, k) congruent $\sphericalangle(h, k)$ ist, so folgt aus Axiom IV 5, dass $\sphericalangle(h', k')$ congruent $\sphericalangle(h, k)$ ist; wir sagen: die beiden Winkel (h, k) und (h', k') sind *unter einander congruent.*

Definition. Zwei Winkel, die den Scheitel und einen Schenkel gemein haben und deren nicht gemeinsame Schenkel eine gerade Linie bilden, heissen *Nebenwinkel.* Zwei Winkel mit gemeinsamem Scheitel, deren Schenkel je eine Gerade bilden, heissen *Scheitelwinkel.* Ein Winkel, welcher einem seiner Nebenwinkel congruent ist, heisst ein *rechter Winkel.*

Erklärung. Zwei Dreiecke ABC und $A'B'C'$ heissen einander *congruent,* wenn sämtliche Congruenzen

$$AB \equiv A'B', \quad AC \equiv A'C', \quad BC \equiv B'C',$$
$$\sphericalangle A \equiv \sphericalangle A', \quad \sphericalangle B \equiv \sphericalangle B', \quad \sphericalangle C \equiv \sphericalangle C'$$

erfüllt sind.

Satz 10 (Erster Congruenzsatz für Dreiecke). Wenn für zwei Dreiecke ABC und $A'B'C'$ die Congruenzen

$$AB \equiv A'B', \quad AC \equiv A'C', \quad \sphericalangle A \equiv \sphericalangle A'$$

gelten, so sind die beiden Dreiecke einander congruent.

Beweis. Nach Axiom IV 6 sind die Congruenzen 14

$$\sphericalangle B \equiv \sphericalangle B' \text{ und } \sphericalangle C \equiv \sphericalangle C'$$

erfüllt und es bedarf somit nur des Nachweises, dass die Seiten BC und $B'C'$ einander congruent sind. Nehmen wir nun im Gegenteil an, es wäre etwa BC

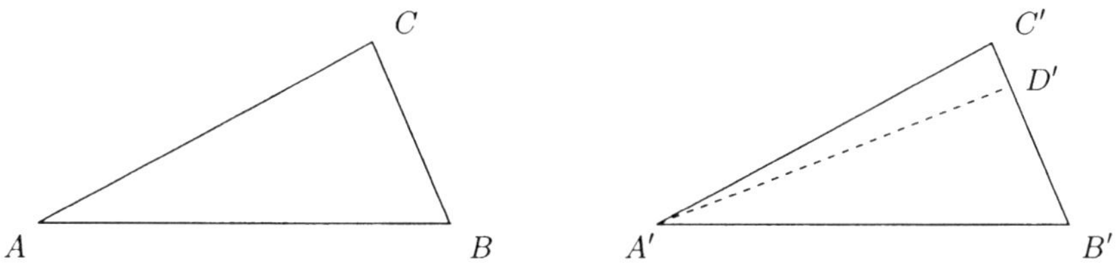

Fig. 9.

nicht congruent $B'C'$ und bestimmen auf $B'C'$ den Punkt D', so dass $BC \equiv B'D'$ wird, so stimmen die beiden Dreiecke ABC und $A'B'D'$ in zwei Seiten und dem von ihnen eingeschlossenen Winkel überein; nach Axiom IV 6 sind mithin insbesondere die beiden Winkel $\sphericalangle BAC$ und $\sphericalangle B'A'D'$ einander congruent. Nach Axiom IV 5 müssten mithin auch die beiden Winkel $\sphericalangle B'A'C'$ und $\sphericalangle B'A'D'$ einander congruent ausfallen; dies ist nicht möglich, da nach Axiom IV 4 ein jeder Winkel an einen gegebenen Halbstrahl nach einer gegebenen Seite in einer Ebene nur auf eine Weise abgetragen werden kann. Damit ist der Beweis für Satz 10 vollständig erbracht.

Ebenso leicht beweisen wir die weitere Thatsache:

5 Definition.] Erklärung.
10 Erklärung. Zwei] Zwei

Satz 11 (Zweiter Congruenzsatz für Dreiecke). Wenn in zwei Dreiecken je eine Seite und die beiden anliegenden Winkel congruent ausfallen, so sind die Dreiecke stets congruent.

Wir sind nunmehr im Stande, die folgenden wichtigen Thatsachen zu beweisen:

Satz 12. *Wenn zwei Winkel* $\sphericalangle ABC$ *und* $\sphericalangle A'B'C'$ *einander congruent sind, so sind auch ihre Nebenwinkel* $\sphericalangle CBD$ *und* $\sphericalangle C'B'D'$ *einander congruent.*

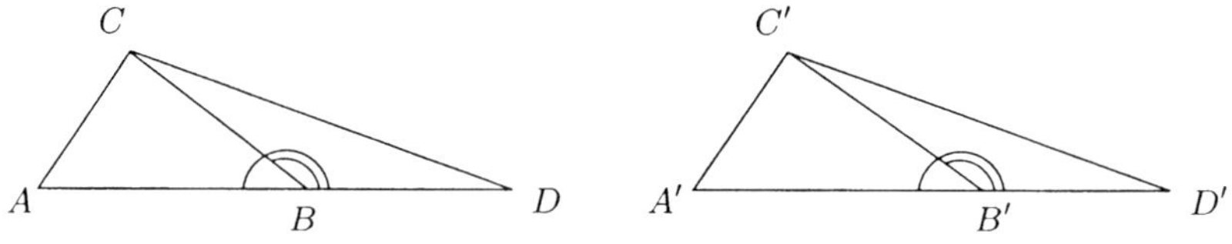

Fig. 10.

Beweis. Wir wählen die Punkte A', C', D' auf den durch B' gehenden Schenkeln derart, dass

15
$$A'B' \equiv AB, \quad C'B' \equiv CB, \quad D'B' \equiv DB$$

wird. In den beiden Dreiecken ABC und $A'B'C'$ sind dann die Seiten AB und CB bez. den Seiten $A'B'$ und $C'B'$ congruent und, da überdies die von diesen Seiten eingeschlossenen Winkel nach Voraussetzung einander congruent sein sollen, so folgt nach Satz 10 die Congruenz jener Dreiecke, d. h. es gelten die Congruenzen

$$AC \equiv A'C' \quad \text{und} \quad \sphericalangle BAC \equiv \sphericalangle B'A'C'.$$

Da andererseits nach Axiom IV 3 die Strecken AD und $A'D'$ einander congruent sind, so folgt wiederum aus Satz 10 die Congruenz der Dreiecke CAD und $C'A'D'$, d. h. es gelten die Congruenzen

$$CD \equiv C'D' \quad \text{und} \quad \sphericalangle ADC \equiv \sphericalangle A'D'C'$$

und hieraus folgt mittels Betrachtung der Dreiecke BCD und $B'C'D'$ nach Axiom IV 6 die Congruenz der Winkel $\sphericalangle CBD$ und $\sphericalangle C'B'D'$.

Eine unmittelbare Folgerung aus Satz 12 ist der Satz von der Congruenz der Scheitelwinkel.

Satz 13. Es sei der Winkel (h, k) in der Ebene α dem Winkel (h', k') in der Ebene α' congruent und ferner sei l ein Halbstrahl der Ebene α, der vom Scheitel des Winkels (h, k) ausgeht und im Inneren dieses Winkels verläuft: dann giebt es stets einen Halbstrahl l' in der Ebene α', der vom Scheitel des Winkels (h', k') ausgeht, und im Inneren dieses Winkels (h', k') verläuft, so dass

$$\sphericalangle(h, l) \equiv \sphericalangle(h', l') \quad \text{und} \quad \sphericalangle(k, l) \equiv \sphericalangle(k', l')$$

wird.

Beweis. Wir bezeichnen die Scheitel der Winkel (h, k) und (h', k') bez. mit O, O' und bestimmen dann auf den Schenkeln h, k, h', k' die Punkte A,

9 Schenkeln] Schenkel

13 einander congruent] kongruent

16 $\equiv \sphericalangle B'A'C'$.] und $\sphericalangle B'A'C'$.

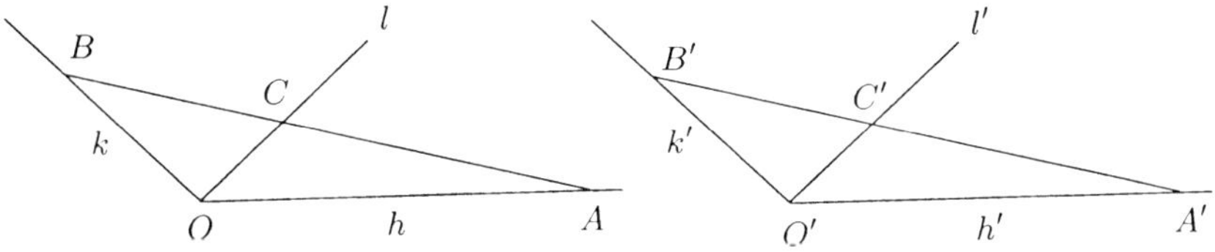

Fig. 11.

B, A', B' derart, dass die Congruenzen

$$OA \equiv O'A' \text{ und } OB \equiv O'B'$$

erfüllt sind. Wegen der Congruenz der Dreiecke OAB und $O'A'B'$ wird 16

$$AB \equiv A'B', \quad \sphericalangle OAB \equiv \sphericalangle O'A'B', \quad \sphericalangle OBA \equiv \sphericalangle O'B'A'.$$

Die Gerade AB schneide l in C; bestimmen wir dann auf der Strecke $A'B'$ den Punkt C', so dass $A'C' \equiv AC$ wird, so ist $O'C'$ der gesuchte Halbstrahl l'. In der That, aus $AC \equiv A'C'$ und $AB \equiv A'B'$ kann mittelst Axiom IV 3 leicht die Congruenz $BC \equiv B'C'$ geschlossen werden; nunmehr erweisen sich die Dreiecke OAC und $O'A'C'$, sowie ferner die Dreiecke OBC und $O'B'C'$ unter einander congruent; hieraus ergeben sich die Behauptungen des Satzes 13.

Auf ähnliche Art gelangen wir zu folgender Thatsache:

Satz 14. Es seien einerseits h, k, l und andererseits h', k', l' je drei von einem Punkte ausgehende und je in einer Ebene gelegene Halbstrahlen: wenn dann die Congruenzen

$$\sphericalangle(h,l) \equiv \sphericalangle(h',l') \text{ und } \sphericalangle(k,l) \equiv \sphericalangle(k',l')$$

erfüllt sind, so ist stets auch

$$\sphericalangle(h,k) \equiv \sphericalangle(h',k').$$

Auf Grund der Sätze 12 und 13 gelingt der Nachweis des folgenden einfachen Satzes, den *Euklid* – meiner Meinung nach mit Unrecht – unter die Axiome gestellt hat:

Satz 15. *Alle rechten Winkel sind einander congruent.*

Beweis: Der Winkel BAD sei seinem Nebenwinkel CAD congruent und desgleichen sei der Winkel $B'A'D'$ seinem Nebenwinkel $C'A'D'$ congruent;

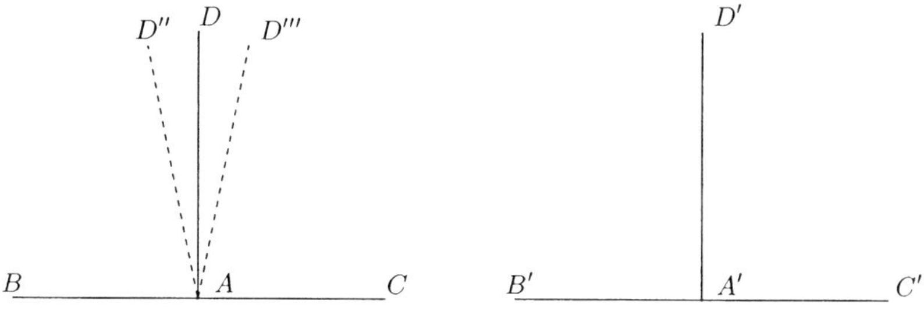

Fig. 12.

es sind dann $\sphericalangle BAD$, $\sphericalangle CAD$, $\sphericalangle B'A'D'$, $\sphericalangle C'A'D'$ sämtlich rechte Winkel. Wir nehmen im Gegensatz zu unserer Behauptung an, es wäre der rechte Winkel $B'A'D'$ nicht congruent dem rechten Winkel BAD und tragen dann $\sphericalangle B'A'D'$ an den Halbstrahl AB an, so dass der entstehende Schenkel AD''
17 entweder in das Innere des Winkels BAD oder des Winkels CAD | fällt; es treffe etwa die erstere Möglichkeit zu. Wegen der Congruenz der Winkel $B'A'D'$ und BAD'' folgt nach Satz 12, dass auch der Winkel $C'A'D'$ dem Winkel CAD'' congruent ist, und da die Winkel $B'A'D'$ und $C'A'D'$ einander congruent sein sollen, so lehrt Axiom IV 5, dass auch der Winkel BAD'' dem Winkel CAD'' congruent sein muss. Da ferner $\sphericalangle BAD$ congruent $\sphericalangle CAD$ ist, so können wir nach Satz 13 innerhalb des Winkels CAD einen von A ausgehenden Halbstrahl AD''' finden, so dass $\sphericalangle BAD''$ congruent $\sphericalangle CAD'''$ und zugleich $\sphericalangle DAD''$ congruent $\sphericalangle DAD'''$ wird. Nun war aber $\sphericalangle BAD''$ congruent $\sphericalangle CAD''$ und somit müsste nach Axiom IV 5 auch $\sphericalangle CAD''$ congruent $\sphericalangle CAD'''$ sein; das ist nicht möglich, weil nach Axiom IV 4 ein jeder Winkel an einen gegebenen Halbstrahl nach einer gegebenen Seite in einer Ebene nur auf e i n e Weise abgetragen werden kann; hiermit ist der Beweis für Satz 15 erbracht.

Wir können jetzt die Bezeichnungen „*spitzer Winkel*" und „*stumpfer Winkel*" in bekannter Weise einführen.

Der Satz von der Congruenz der Basiswinkel $\sphericalangle A$ und $\sphericalangle B$ im gleichschenkligen Dreiecke ABC folgt unmittelbar durch Anwendung des Axioms IV 6 auf Dreieck ABC und Dreieck BAC. Mit Hülfe dieses Satzes und unter Hinzuziehung des Satzes 14 beweisen wir dann leicht in bekannter Weise die folgende Thatsache:

S a t z 16 (D r i t t e r C o n g r u e n z s a t z f ü r D r e i e c k e). Wenn in zwei Dreiecken die drei Seiten entsprechend congruent ausfallen, so sind die Dreiecke congruent.

E r k l ä r u n g. Irgend eine endliche Anzahl von Punkten heisst eine *Figur*; liegen alle Punkte der Figur in einer Ebene, so heisst sie eine *ebene Figur*.

Zwei Figuren heissen *congruent*, wenn ihre Punkte sich paarweise einander so zuordnen lassen, dass die auf diese Weise einander zugeordneten Strecken und Winkel sämtlich einander congruent sind.

Congruente Figuren haben, wie man aus den Sätzen 12 und 9 erkennt, folgende Eigenschaften: Drei Punkte einer Geraden liegen auch in jeder congruenten Figur auf einer Geraden. Die Anordnung der Punkte in entsprechenden Ebenen in Bezug auf entsprechende Gerade ist in congruenten Figuren die nämliche; das Gleiche gilt von der Reihenfolge entsprechender Punkte in entsprechenden Geraden.

---

9 IV 5] ⟨Corresponding to III 5 in E2, where the reference to IV 5 is by mistake.⟩

14 congruent $\sphericalangle CAD'''$] kongruent CAD'''

21 Dreiecke] Dreieck

Der allgemeinste Congruenzsatz für die Ebene und für den Raum drückt sich, wie folgt, aus:

Satz 17. Wenn $(A, B, C, \ldots)$ und $(A', B', C', \ldots)$ congruente ebene 18
Figuren sind und P einen Punkt in der Ebene der ersten bedeutet, so lässt sich in der Ebene der zweiten Figur stets ein Punkt P' finden derart, dass $(A, B, C, \ldots, P)$ und $(A', B', C', \ldots, P')$ wieder congruente Figuren sind. Enthalten die beiden Figuren wenigstens drei nicht auf einer Geraden liegende Punkte, so ist die Construction von P' nur auf eine Weise möglich.

Satz 18. Wenn $(A, B, C, \ldots)$ und $(A', B', C', \ldots)$ congruente Figuren sind und P einen beliebigen Punkt bedeutet, so lässt sich stets ein Punkt P' finden, so dass die Figuren $(A, B, C, \ldots, P)$ und $(A', B', C', \ldots, P')$ congruent sind. Enthält die Figur $(A, B, C, \ldots)$ mindestens vier nicht in einer Ebene liegende Punkte, so ist die Construction von P' nur auf eine Weise möglich.

Dieser Satz enthält das wichtige Resultat, dass die sämtlichen räumlichen Thatsachen der Congruenz, d. h. der Bewegung im Raume – mit Hinzuziehung der Axiomgruppen I und II – lediglich Folgerungen aus den sechs oben aufgestellten linearen und ebenen Axiomen der Congruenz sind, also das Parallelenaxiom zu ihrer Feststellung nicht notwendig ist.

Nehmen wir zu den Congruenzaxiomen noch das Parallelenaxiom III hinzu, so gelangen wir leicht zu den bekannten Thatsachen:

Satz 19. Wenn zwei Parallelen von einer dritten Geraden geschnitten werden, so sind die Gegenwinkel und Wechselwinkel congruent, und umgekehrt: die Congruenz der Gegen- und Wechselwinkel hat zur Folge, dass die Geraden parallel sind.

Satz 20. Die Winkel eines Dreiecks machen zusammen zwei Rechte aus.

---

7 Enthalten die beiden Figuren] Enthält die Figur $(A, B, C, \ldots)$

15 Dieser Satz enthält das wichtige Resultat] Der Satz 18 spricht das wichtige Resultat aus

16 Congruenz, d. h.] Kongruenz und mithin

16–17 mit Hinzuziehung] unter Hinzuziehung

17 lediglich Folgerungen] Folgerungen

19 sind, ⟨...⟩ ist.] sind.

20 Nehmen wir] Nehmen wir nämlich

20 noch das] das

20 Parallelenaxiom III] Parallelenaxiom

27 aus.] ⟨After 'aus' a footnote is added in E2:⟩ Betreffs der Frage, inwieweit dieser Satz umgekehrt das Parallelenaxiom zu ersetzen vermag, vergleiche man die Bemerkungen am Schluß von Kap. II § 12.

D e f i n i t i o n. Wenn M ein beliebiger Punkt in einer Ebene α ist, so heisst die Gesamtheit aller Punkte A, für welche die Strecken MA einander congruent sind, ein *Kreis*; M heisst der *Mittelpunkt des Kreises.*

Auf Grund dieser Definition folgen mit Hülfe der Axiomgruppen III–IV leicht die bekannten Sätze über den Kreis, insbesondere die Möglichkeit der Konstruktion eines Kreises durch irgend drei nicht in einer Geraden gelegene Punkte sowie der Satz über die Congruenz aller Peripheriewinkel über der nämlichen Sehne und der Satz von den Winkeln im Kreisviereck.

19 § 8.

Die Axiomgruppe V: Axiom der Stetigkeit (Archimedisches Axiom).

Dieses Axiom ermöglicht die Einführung des Stetigkeitsbegriffes in die Geometrie; um dasselbe auszusprechen, müssen wir zuvor eine Festsetzung über die Gleichheit zweier Strecken auf einer Geraden treffen. Zu dem Zwecke können wir entweder die Axiome über Streckencongruenz zu Grunde legen und dementsprechend congruente Strecken als „g l e i c h e" bezeichnen oder auf Grund der Axiomgruppen I–II durch geeignete Constructionen (vgl. Kap. V § 24) festsetzen, wie eine Strecke von einem Punkte einer gegebenen Geraden abzutragen ist, so dass eine bestimmte neue ihr „g l e i c h e" Strecke entsteht. Nach einer solchen Festsetzung lautet das Archimedische Axiom, wie folgt:

V. *Es sei A_1 ein beliebiger Punkt auf einer Geraden zwischen den beliebig gegebenen Punkten A und B; man construire dann die Punkte A_2, A_3, $A_4, \ldots$, so dass A_1 zwischen A und A_2, ferner A_2 zwischen A_1 und A_3, ferner A_3 zwischen A_2 und A_4 u. s. w. liegt und überdies die Strecken*

$$AA_1,\ A_1A_2,\ A_2A_3,\ A_3A_4,\ \ldots$$

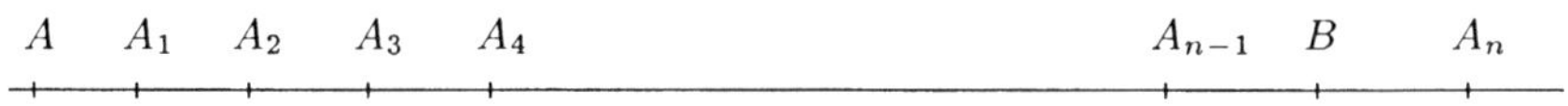

Fig. 13.

einander gleich sind: dann giebt es in der Reihe der Punkte A_2, A_3, $A_4, \ldots$ stets einen solchen Punkt A_n, dass B zwischen A und A liegt.

---

1 D e f i n i t i o n.] E r k l ä r u n g.

4 Definition] Erklärung

10 **Axiom der Stetigkeit (Archimedisches Axiom)] Axiome der Stetigkeit**

11–19 Dieses ⟨...⟩ folgt:] ⟨deleted⟩

20 V.] V 1 (A x i o m d e s M e s s e n s o d e r A r c h i m e d i s c h e s A x i o m).

1 *zwischen A und A*] *zwischen A und A_n*

Das Archimedische Axiom ist ein *lineares* Axiom.

---

1 ⟨Before this line the following paragraph is added:⟩

V 2 (Axiom der Vollständigkeit). *Die Elemente (Punkte, Geraden, Ebenen) der Geometrie bilden ein System von Dingen, welches bei Aufrechterhaltung sämtlicher genannten Axiome keiner Erweiterung mehr fähig ist,* d. h.: zu dem System der Punkte, Geraden, Ebenen ist es nicht möglich, ein anderes System von Dingen hinzuzufügen, so daß in dem durch Zusammensetzung entstehenden System sämtliche aufgeführten Axiome I – IV, V 1 erfüllt sind.

1 Archimedische Axiom] Archimedische Axiom V 1

1 ⟨The following paragraphs are added:⟩

Hinsichtlich des Axioms der Vollständigkeit V 2 füge ich hier folgende Bemerkungen hinzu.

Die Aufrechterhaltung sämtlicher Axiome, von der in diesem Axiom die Rede ist, hat man so zu verstehen, daß nach der Erweiterung sämtliche früheren Axiome in der früheren Weise gültig bleiben sollen, d.h. sofern man die vorhandenen Beziehungen der Elemente, nämlich die vorhandene Anordnung und Kongruenz der Strecken und Winkel nirgends stört, also z. B. ein Punkt, der vor der Erweiterung zwischen zwei Punkten liegt, dies auch nach der Erweiterung tut, Strecken und Winkel, die vorher einander kongruent sind, dies auch nach der Erweiterung bleiben.

Die Erfüllbarkeit des Vollständigkeitsaxioms ist wesentlich durch die Voranstellung des Archimedischen Axioms bedingt; in der Tat läßt sich zeigen, daß zu einem System von Punkten, Geraden und Ebenen, welche die Axiome I–IV erfüllen, stets noch auf mannigfache Weise solche Elemente hinzugefügt werden können, daß in dem durch Zusammensetzung entstehenden Systeme die Axiome I–IV ebenfalls sämtlich gültig sind; d. h. das Vollständigkeitsaxiom würde einen Widerspruch einschließen, wenn man den Axiomen I–IV nicht noch das Archimedische Axiom hinzufügt.

Das Vollständigkeitsaxiom ist nicht eine Folge des Archimedischen Axioms. In der Tat reicht das Archimedische Axiom allein nicht aus, um mit Benutzung der Axiome I–IV unsere Geometrie als identisch mit der gewöhnlichen analytischen „Cartesischen" Geometrie nachzuweisen (vgl. § 9 und § 12). Dagegen gelingt es unter Hinzunahme des Vollständigkeitsaxioms – obwohl dieses Axiom unmittelbar keine Aussage über den Begriff der Konvergenz enthält –, die Existenz der einem Dedekindschen Schnitte entsprechenden Grenze und den Bolzanoschen Satz vom Vorhandensein der Verdichtungsstellen nachzuweisen, womit dann unsere Geometrie sich als identisch mit der Cartesischen Geometrie erweist.

Durch die vorstehende Betrachtungsweise ist die Forderung der Stetigkeit in zwei wesentlich verschiedene Bestandteile zerlegt worden, nämlich in das Archimedische Axiom, dem zugleich die Rolle zukommt, die Forderung der Stetigkeit vorzubereiten, und in das Vollständigkeitsaxiom, das den Schlußstein des ganzen Axiomensystems bildet [1]).

In den nachfolgenden Untersuchungen stützen wir uns wesentlich nur auf das Archimedische Axiom und setzen im allgemeinen das Vollständigkeitsaxiom nicht voraus.

1) Man vergleiche auch die Bemerkungen am Schluß von § 17 sowie meinen Vortrag über den Zahlbegriff: Berichte der Deutschen Mathematiker-Vereinigung, 1900.[7] – Bei der Untersuchung des Satzes von der Gleichheit der Basiswinkel im gleichschenkligen Dreieck bin ich auf ein weiteres Stetigkeitsaxiom geführt worden, das ich Axiom der Nachbarschaft genannt habe; man sehe meine im Anhang abgedruckte Abhandlung „Über den Satz von der Gleichheit der Basiswinkel im gleichschenkligen Dreieck". Proceedings of the London Mathematical Society XXXV 1903.[8] Vgl. S. 92 und S. 107.

[7] *Hilbert 1900b.*

[8] *Hilbert 1902/03.*

Kapitel II.

Die Widerspruchslosigkeit und gegenseitige Unabhängigkeit der Axiome.

§ 9.

Die Widerspruchslosigkeit der Axiome.

Die Axiome der fünf in Kapitel I aufgestellten Axiomgruppen stehen mit einander nicht in Widerspruch, d. h. es ist nicht möglich, durch logische Schlüsse aus denselben eine Thatsache abzuleiten, welche einem der aufgestellten
20 Axiome widerspricht. Um | dies einzusehen, genügt es, eine Geometrie anzugeben, in der sämtliche Axiome der fünf Gruppen erfüllt sind.

Man betrachte den Bereich Ω aller derjenigen algebraischen Zahlen, welche hervorgehen, indem man von der Zahl 1 ausgeht und eine endliche Anzahl von Malen die vier Rechnungsoperationen: Addition, Subtraktion, Multiplikation, Division und die fünfte Operation $\sqrt{1+\omega^2}$ anwendet, wobei ω jedesmal eine Zahl bedeuten kann, die vermöge jener fünf Operationen bereits entstanden ist.

Wir denken uns ein Paar von Zahlen (x, y) des Bereiches Ω als einen Punkt und die Verhältnisse von irgend drei Zahlen $(u : v : w)$ aus Ω, falls u, v nicht beide Null sind, als eine Gerade; ferner möge das Bestehen der Gleichung

$$ux + vy + w = 0$$

ausdrücken, dass der Punkt (x, y) auf der Geraden $(u : v : w)$ liegt; damit sind, wie man leicht sieht, die Axiome I 1–2 und III erfüllt. Die Zahlen des Bereiches Ω sind sämtlich reell; indem wir berücksichtigen, dass dieselben sich ihrer Grösse nach anordnen lassen, können wir leicht solche Festsetzungen für unsere Punkte und Geraden treffen, dass auch die Axiome II der Anordnung sämtlich gültig sind. In der That, sind (x_1, y_1), (x_2, y_2), $(x_3, y_3), \ldots$ irgend welche Punkte auf einer Geraden, so möge dies ihre Reihenfolge auf der Geraden sein, wenn die Zahlen $x_1, x_2, x_3, \ldots$ oder $y_1, y_2, y_3, \ldots$ in dieser Reihenfolge entweder beständig abnehmen oder wachsen; um ferner die Forderung das Axioms II 5 zu erfüllen, haben wir nur nöthig festzusetzen, dass alle Punkte (x, y), für die $ux + vy + w$ kleiner oder grösser als 0 ausfällt, auf der einen bez. auf der anderen Seite der Geraden $(u : v : w)$ gelegen sein sollen. Man überzeugt sich leicht, dass diese Festsetzung sich mit der vorigen Festsetzung in Uebereinstimmung befindet, derzufolge ja die Reihenfolge der Punkte auf einer Geraden bereits bestimmt ist.

Das Abtragen von Strecken und Winkeln erfolgt nach den bekannten Methoden der analytischen Geometrie. Eine Transformation von der Gestalt

$$x' = x + a,$$
$$y' = y + b$$

vermittelt die Parallelverschiebung von Strecken und Winkeln. Wird ferner der Punkt $(0, 0)$ mit O, der Punkt $(1, 0)$ mit E und ein beliebiger Punkt (a, b)

11 betrachte] betrachte zunächst
14 $\sqrt{1+\omega^2}$] $|\sqrt{1+\omega^2}|$

mit C bezeichnet, so entsteht durch Drehung um den Winkel $\sphericalangle COE$, wenn O der feste Drehpunkt | ist, aus dem beliebigen Punkte (x, y) der Punkt (x', y'), 21
wobei

$$x' = \frac{a}{\sqrt{a^2+b^2}}x - \frac{b}{\sqrt{a^2+b^2}}y,$$
$$y' = \frac{b}{\sqrt{a^2+b^2}}x + \frac{a}{\sqrt{a^2+b^2}}y$$

zu setzen ist. Da die Zahl

$$\sqrt{a^2+b^2} = a\sqrt{1+\left(\tfrac{b}{a}\right)^2}$$

wiederum dem Bereiche Ω angehört, so gelten bei unseren Festsetzungen auch die Congruenzaxiome IV und offenbar ist auch das Archimedische Axiom V erfüllt.

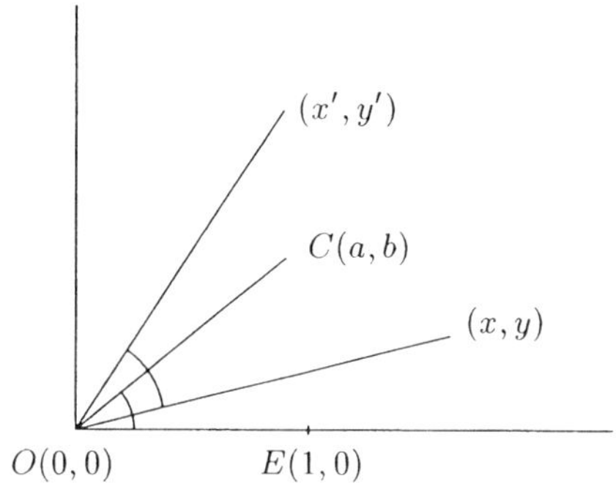

Fig. 14.

Wir schliessen hieraus, dass jeder Widerspruch in den Folgerungen aus unseren Axiomen auch in der Arithmetik des Bereiches Ω erkennbar sein müsste.

Die entsprechende Betrachtungsweise für die räumliche Geometrie bietet keine Schwierigkeit.

Wählen wir in der obigen Entwickelung statt des Bereiches Ω den Bereich a l l e r reellen Zahlen, so erhalten wir ebenfalls eine Geometrie, in der sämtliche Axiome I–V gültig sind. Für unseren Beweis genügte die Zuhülfenahme des Bereiches Ω, der nur eine abzählbare Menge von Elementen enthält.

---

7 $\sqrt{a^2+b^2} = a\sqrt{1+\left(\frac{b}{a}\right)^2}$] ⟨In E2 erroneously changed to:⟩ $\sqrt{a^2+b^2} = a\sqrt{1+\left(\frac{a}{b}\right)^2}$

12 erfüllt.] erfüllt. Das Axiom der Vollständigkeit V 2 ist nicht erfüllt.

13–14 Wir schliessen hieraus, dass jeder] Jeder

15 Axiomen] geometrischen Axiomen I–IV, V 1 müßte demnach

17 sein müsste.] sein[1]).

1) Betreffs der Frage nach der Widerspruchslosigkeit der arithmetischen Axiome vergleiche man meine Vorträge über den Zahlbegriff: Berichte der deutschen Mathematiker-Vereinigung, 1900, sowie Mathematische Probleme, gehalten auf dem internationalen Mathematikerkongreß 1900, Göttinger Nachr. 1900,[9] insbesondere Problem No. 2.

18–19 ⟨In E2, this sentence appears two paragraphs later.⟩

21 ebenfalls] ⟨deleted⟩

22 sind.] sind; diese Geometrie ist die gewöhnliche Cartesische Geometrie.

22–23 Für ⟨…⟩ enthält.] ⟨deleted⟩

23 ⟨After this line the following paragraph is added:⟩

Jeder Widerspruch in den Folgerungen aus den Axiomen I–V müßte demnach in der Arithmetik des Systems der reellen Zahlen erkennbar sein.

Die entsprechende Betrachtungsweise für die räumliche Geometrie bietet keine Schwierigkeit.

Wie man erkennt, gibt es unendlich viele Geometrien, die den Axiomen I–IV, V 1 genügen, dagegen nur e i n e, nämlich die Cartesische Geometrie, in der auch zugleich das Vollständigkeitsaxiom V 2 gültig ist.

[9] *Hilbert 1900b, 1900e*

§ 10.

Die Unabhängigkeit des Parallelenaxioms (Nicht-Euklidische Geometrie).

Nachdem wir die Widerspruchslosigkeit der Axiome erkannt haben, ist es von Interesse zu untersuchen, ob sie sämtlich von einander unabhängig sind. In der That zeigt sich, dass keines der Axiome durch logische Schlüsse aus den übrigen abgeleitet werden kann.

Was zunächst die einzelnen Axiome der Gruppen I, II und IV betrifft, so ist der Nachweis dafür leicht zu führen, dass die Axiome ein und derselben Gruppe je unter sich unabhängig sind[1]) .

22 Die Axiome der Gruppen I und II liegen bei unserer Darstellung den übrigen Axiomen zu Grunde, so dass es sich nur noch darum handelt, für jede der Gruppen III, IV und V die Unabhängigkeit von den übrigen nachzuweisen.

Die erstere Aussage des Parallelenaxioms kann aus den Axiomen der Gruppen I, II, IV bewiesen werden. Um dies einzusehen, verbinden wir den gegebenen Punkt A mit einem beliebigen Punkte B der Geraden a. Es sei ferner C irgend ein anderer Punkt der Geraden a; dann tragen wir $\sphericalangle ABC$ an AB im Punkte A nach derjenigen Seite in der nämlichen Ebene α an, auf der nicht der Punkt C liegt. Die so erhaltene Gerade durch A trifft die Gerade a nicht. In der That, schnitte sie a im Punkte D und nehmen wir etwa an, dass B zwischen C und D liege, so könnten wir auf a einen Punkt D' finden, so dass B zwischen D uud D' liegt und überdies

$$AD \equiv BD'$$

ausfiele. Wegen der Congruenz der Dreiecke ABD und BAD' würde die Congruenz

$$\sphericalangle ABD \equiv \sphericalangle BAD'$$

folgen und da die Winkel ABD' und ABD Nebenwinkel sind, so müssten sich dann mit Rücksicht auf Satz 12 auch die Winkel BAD und BAD' als Nebenwinkel erweisen; dies ist aber wegen Satz 1 nicht der Fall.

Die zweite Aussage des Parallelenaxioms III ist von den übrigen Axiomen unabhängig; dies zeigt man in bekannter Weise am einfachsten wie folgt. Man wähle die Punkte, Geraden und Ebenen der gewöhnlichen in § 9 construirten Geometrie, so weit sie innerhalb einer festen Kugel verlaufen, für sich

1) Vergl. meine Vorlesung über Euklidische Geometrie (Wintersemester 1898/99),[10] die nach einer Ausarbeitung des Herrn Dr. *von Schaper* für meine Zuhörer autographirt worden ist.

5 keines der Axiome] keine wesentlichen Bestandteile der genannten Axiomgruppen

6 übrigen] jedesmal voranstehenden Axiomgruppen

6 kann] können

9 ⟨Footnote deleted.⟩

13–28 Die ⟨...⟩ Fall.] ⟨deleted⟩

29 Die ⟨...⟩ Parallelenaxioms III] Das Parallelenaxiom IV

32 Geometrie,] (Cartesischen) Geometrie,

[10] *Hilbert 1899a**. See Chapter 4 of this Volume.

allein als Elemente einer räumlichen Geometrie und vermittle die Congruenzen dieser Geometrie durch solche lineare Transformationen der gewöhnlichen Geometrie, welche die feste Kugel in sich überführen. Bei geeigneten Festsetzungen erkennt man, dass in dieser *„Nicht-Euklidischen" Geometrie* sämtliche Axiome ausser dem Euklidischen Axiom III gültig sind und da die Möglichkeit der gewöhnlichen Geometrie in § 9 nachgewiesen worden ist, so folgt nunmehr auch die Möglichkeit der Nicht-Euklidischen Geometrie.

§ 11. 23

Die Unabhängigkeit der Congruenzaxiome.

Wir werden die Unabhängigkeit der Congruenzaxiome erkennen, indem wir den Nachweis führen, dass das Axiom IV 6 oder, was auf das nämliche hinausläuft, der erste Congruenzsatz für Dreiecke, d. i. Satz 10 durch logische Schlüsse nicht aus den übrigen Axiomen I, II, III, IV 1–5, V abgeleitet werden kann.

Wir wählen die Punkte, Geraden, Ebenen der gewöhnlichen Geometrie auch als Elemente der neuen räumlichen Geometrie und definiren das Abtragen der Winkel ebenfalls wie in der gewöhnlichen Geometrie, etwa in der Weise, wie in § 9 auseinandergesetzt worden ist; dagegen definiren wir das Abtragen der Strecken auf andere Art. Die zwei Punkte A_1, A_2 mögen in der gewöhnlichen Geometrie die Coordinaten x_1, y_1, z_1 bez. x_2, y_2, z_2 haben; dann bezeichnen wir den positiven Wert von

$$\sqrt{(x_1 - x_2 + y_1 - y_2)^2 + (y_1 - y_2)^2 + (z_1 - z_2)^2}$$

als die Länge der Strecke A_1A_2 und nun sollen zwei beliebige Strecken A_1A_2 und $A'_1A'_2$ einander congruent heissen, wenn sie im eben festgesetzten Sinne gleiche Längen haben.

Es leuchtet unmittelbar ein, dass in der so hergestellten räumlichen Geometrie die Axiome I, II, III, IV 1–2, 4–5, V gültig sind.

Um zu zeigen, dass auch das Axiom IV 3 erfüllt ist, wählen wir eine beliebige Gerade a und auf ihr drei Punkte A_1, A_2, A_3, sodass A_2 zwischen A_1 und A_3 liegt. Die Punkte x, y, z der Geraden a seien durch die Gleichungen

$$\begin{aligned} x &= \lambda t + \lambda', \\ y &= \mu t + \mu', \\ z &= \nu t + \nu' \end{aligned}$$

gegeben, worin λ, λ', μ, μ', ν, ν' gewisse Constante und t einen Parameter bedeutet. Sind t_1, $t_2(< t_1)$, $t_3(< t_2)$ die Parameterwerte, die den Punkten A_1, A_2, A_3 entsprechen, so finden wir für die Längen der drei Strecken A_1A_2, A_2A_3 und A_1A_3 bez. die Ausdrücke:

$$\begin{aligned} &(t_1 - t_2)\left|\sqrt{(\lambda + \mu)^2 + \mu^2 + \nu^2}\right|, \\ &(t_2 - t_3)\left|\sqrt{(\lambda + \mu)^2 + \mu^2 + \nu^2}\right|, \\ &(t_1 - t_3)\left|\sqrt{(\lambda + \mu)^2 + \mu^2 + \nu^2}\right| \end{aligned}$$

24 und mithin ist die Summe der Längen der Strecken A_1A_2 und | A_2A_3 gleich der Länge der Strecke A_1A_3; dieser Umstand bedingt die Gültigkeit des Axioms IV 3.

Das Axiom IV 6 oder vielmehr der erste Congruenzsatz für Dreiecke ist in unserer Geometrie nicht immer erfüllt. Betrachten wir nämlich in der Ebene $z = 0$ die vier Punkte

O	mit den	Coordinaten	$x = 0,$	$y = 0,$
A	"	"	$x = 1,$	$y = 0,$
B	"	"	$x = 0,$	$y = 1,$
C	"	"	$x = \frac{1}{2},$	$y = \frac{1}{2},$

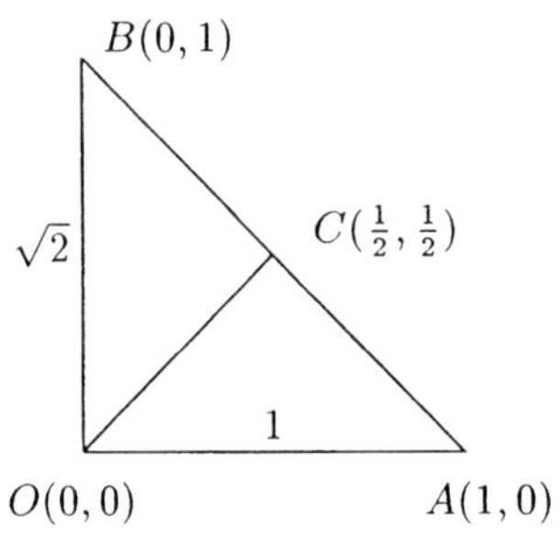

Fig. 15.

so sind in den beiden (rechtwinkligen) Dreiecken OAC und OBC die Winkel bei C und die anliegenden Seiten entsprechend congruent, da die Seite OC beiden Dreiecken gemeinsam ist und die Strecken AC und BC die gleiche Länge $\frac{1}{2}$ besitzen. Dagegen haben die dritten Seiten OA und OB die Länge 1, bez. $\sqrt{2}$ und sind daher nicht einander congruent.

Es ist auch nicht schwer, in dieser Geometrie zwei Dreiecke zu finden, für welche das Axiom IV 6 selbst nicht erfüllt ist.

§ 12.

Die Unabhängigkeit des Stetigkeitsaxioms V (Nicht-Archimedische Geometrie).

Um die Unabhängigkeit des Archimedischen Axioms V zu beweisen, müssen wir eine Geometrie herstellen, in der sämtliche Axiome mit Ausnahme des Archimedischen Axioms erfüllt sind[1]).

Zu dem Zwecke construiren wir den Bereich $\Omega(t)$ aller derjenigen algebraischen Funktionen von t, welche aus t durch die vier Rechnungsoperationen der Addition, Subtraktion, Multiplikation, Division und durch die fünfte Operation $\sqrt{1+\omega^2}$ hervorgehen; dabei soll ω irgend eine Funktion bedeuten, die

1) *G. Veronese* hat in seinem tiefsinnigen Werke, Grundzüge der Geometrie, deutsch von A. Schepp, Leipzig 1894[11] ebenfalls den Versuch gemacht, eine Geometrie aufzubauen, die von dem Archimedischen Axiom unabhängig ist.

---

5 nämlich] nämlich z. B.

27 Axioms V] Axioms V 1

28–29 des Archimedischen Axioms] der Axiome V, diese letzteren aber nicht

Note 1 Werke,] Werke:

31 vier] fünf

32 fünfte] ⟨deleted⟩

33 $\sqrt{1+\omega^2}$] $|\sqrt{1+\omega^2}|$

vermöge jener fünf Operationen bereits entstanden ist. Die Menge der Elemente von $\Omega(t)$ ist – ebenso wie von Ω – eine abzählbare. Die fünf Operationen sind sämtlich eindeutig und reell ausführbar; | der Bereich $\Omega(t)$ enthält daher 25
nur eindeutige und reelle Funktionen von t.

Es sei c irgend eine Funktion des Bereiches $\Omega(t)$; da die Funktion c eine algebraische Funktion von t ist, so kann sie jedenfalls nur für eine endliche Anzahl von Werten t verschwinden und es wird daher die Funktion c für genügend grosse positive Werte von t entweder stets positiv oder stets negativ ausfallen.

Wir sehen jetzt die Funktionen des Bereiches $\Omega(t)$ als eine Art complexer Zahlen an; offenbar sind in dem so definirten complexen Zahlensystem die gewöhnlichen Rechnungsregeln sämtlich gültig. Ferner möge, wenn a, b irgend zwei verschiedene Zahlen dieses complexen Zahlensystems sind, die Zahl a grösser oder kleiner als b, in Zeichen: $a > b$ oder $a < b$, heissen, je nachdem die Differenz $c = a - b$ als Funktion von t für genügend grosse positive Werte von t stets positiv oder stets negativ ausfällt. Bei dieser Festsetzung ist für die Zahlen unseres complexen Zahlensystems eine Anordnung ihrer Grösse nach möglich, die von der gewöhnlichen Art wie bei reellen Zahlen ist; auch gelten, wie man leicht erkennt, für unsere complexen Zahlen die Sätze, wonach Ungleichungen richtig bleiben, wenn man auf beiden Seiten die gleiche Zahl addirt oder beide Seiten mit der gleichen Zahl > 0 multiplicirt.

Bedeutet n eine beliebige positive ganze rationale Zahl, so gilt für die beiden Zahlen n und t des Bereiches $\Omega(t)$ gewiss die Ungleichung $n < t$, da die Differenz $n - t$, als Funktion von t betrachtet, für genügend grosse positive Werte von t offenbar stets negativ ausfällt. Wir sprechen diese Thatsache in folgender Weise aus: die beiden Zahlen 1 und t des Bereiches $\Omega(t)$, die beide > 0 sind, besitzen die Eigenschaft, dass ein beliebiges Vielfaches der ersteren stets kleiner als die letztere Zahl bleibt.

Wir bauen nun aus den complexen Zahlen des Bereiches $\Omega(t)$ eine Geometrie genau auf dieselbe Art auf, wie dies in § 9 unter Zugrundelegung des Bereiches Ω von algebraischen Zahlen geschehen ist: wir denken uns ein System von drei Zahlen (x, y, z) des Bereiches $\Omega(t)$ als einen Punkt und die Verhältnisse von irgend vier Zahlen $(u : v : w : r)$ aus $\Omega(t)$, falls u, v, w nicht sämtlich Null sind, als eine Ebene; ferner möge das Bestehen der Gleichung

$$ux + vy + wz + r = 0$$

ausdrücken, dass der Punkt (x, y, z) in der Ebene $(u : v : w : r)$ liegt und die Gerade sei die Gesamtheit aller in zwei Ebenen gelegenen Punkte. Treffen wir sodann die entsprechenden Fest|setzungen über die Anordnung der Ele- 26
mente und über Abtragen von Strecken und Winkeln, wie in § 9, so entsteht eine „*Nicht-Archimedische*“ *Geometrie*, in welcher, wie die zuvor erörterten Ei-

2 von Ω] von Ω in § 9

18 von der gewöhnlichen Art wie bei reellen Zahlen] derjenigen bei reellen Zahlen analog

37 Ebenen] Ebenen mit verschiedenen $u : v : w$

39 Abtragen] das Abtragen

11 Veronese 1894.

genschaften des complexen Zahlensystems $\Omega(t)$ zeigen, sämtliche Axiome mit Ausnahme des Archimedischen Axioms erfüllt sind. In der That können wir die Strecke 1 auf der Strecke t beliebig oft hinter einander abtragen, ohne dass der Endpunkt der Strecke t bedeckt wird; dies widerspricht der Forderung des Archimedischen Axioms.

---

2 des Archimedischen Axioms] der Stetigkeitsaxiome

4 bedeckt] überschritten

5 ⟨After this line the following paragraphs are added:⟩

Daß auch das Vollständigkeitsaxiom V 2 von allen voranstehenden Axiomen I–IV, V 1 unabhängig ist, zeigt die erste in § 9 aufgestellte Geometrie, da in dieser das Archimedische Axiom erfüllt ist.

Auch die Nicht-Archimedischen und zugleich Nicht-Euklidischen Geometrien sind von prinzipieller Bedeutung und insbesondere erschien mir die Frage nach der Winkelsumme im Dreiecke und nach der Abhängigkeit dieser Sätze vom Archimedischen Axiom von hohem Interesse. Die Untersuchung, die *M. Dehn* [1]) auf meine Anregung hin über diesen Gegenstand unternommen hat, führte zu einer vollen Aufklärung dieser Frage. Den Untersuchungen von M. Dehn liegen die Axiome I–III zu Grunde. Nur zum Schlusse der Dehnschen Arbeit – damit auch die Riemannsche (elliptische) Geometrie in den Bereich der Untersuchung hineinfällt – sind die Axiome II der Anordnung allgemeiner, als in der gegenwärtigen Abhandlung, nämlich etwa wie folgt zu fassen:

Vier Punkte A, B, C, D einer Geraden zerfallen stets in zwei Paare A, C und B, D, so daß A, C durch B, D getrennt sind und umgekehrt. Fünf Punkte auf einer Geraden können immer in der Weise mit A, B, C, D, E bezeichnet werden, daß A, C durch B, D und durch B, E, ferner daß A, D durch B, E und durch C, E u. s. f. getrennt sind.

Die hauptsächlichsten von M. Dehn auf Grund der Axiome I–III, also ohne Benutzung der Stetigkeit bewiesenen Sätze sind folgende:

Wenn in irgend *einem* Dreieck die Summe der Winkel größer bezüglich gleich oder kleiner als zwei Rechte ist, so ist sie es in jedem Dreieck.[2])

Aus der Annahme unendlich vieler Parallelen zu einer Geraden durch einen Punkt folgt, wenn man das Archimedische Axiom ausschließt, *nicht*, daß die Winkelsumme im Dreieck kleiner als zwei Rechte ist. Es gibt vielmehr sowohl eine Geometrie (die Nicht-Legendresche Geometrie), in der man durch einen Punkt zu einer Geraden unendlich viele Parallelen ziehen kann und in der trotzdem die Sätze der Riemannschen (elliptischen) Geometrie gültig sind. Andererseits gibt es eine Geometrie (die Semi-Euklidische Geometrie), in welcher es unendlich viele Parallelen durch einen Punkt zu einer Geraden gibt und in der dennoch die Sätze der Euklidischen Geometrie gelten.

Aus der Annahme, daß es keine Parallelen gibt, folgt stets, daß die Winkelsumme im Dreieck größer als zwei Rechte ist.

Ich bemerke endlich, daß, wenn man das Archimedische Axiom hinzunimmt, das Parallelenaxiom durch die Forderung ersetzt werden kann, es solle die Winkelsumme im Dreieck gleich zwei Rechten sein.

1) Die Legendreschen Sätze über die Winkelsumme im Dreieck. Math. Ann. Bd. 53. 1900.[12]

2) Einen Beweis für diesen Satz hat später auch *F. Schur* erbracht, Math. Ann. Bd. 55.[13]

[12] *Dehn 1900b.*

[13] *Schur 1902.*

Kapitel III.

Die Lehre von den Proportionen.

§ 13.

Complexe Zahlensysteme.

Am Anfang dieses Kapitels wollen wir einige kurze Auseinandersetzungen über complexe Zahlensysteme vorausschicken, die uns später insbesondere zur Erleichterung der Darstellung nützlich sein werden.

Die reellen Zahlen bilden in ihrer Gesamtheit ein System von Dingen mit folgenden Eigenschaften:

Sätze der Verknüpfung (1–12):

1. Aus der Zahl a und der Zahl b entsteht durch „Addition“ eine bestimmte Zahl c, in Zeichen

$$a + b = c \quad \text{oder} \quad c = a + b.$$

2. Es giebt eine bestimmte Zahl – sie heisse 0 –, so dass für jedes a zugleich

$$a + 0 = a \quad \text{und} \quad 0 + a = a$$

ist.

3. Wenn a und b gegebene Zahlen sind, so existirt stets eine und nur eine Zahl x und auch eine und nur eine Zahl y, so dass

$$a + x = b \quad \text{bez.} \quad y + a = b$$

wird.

4. Aus der Zahl a und der Zahl b entsteht noch auf eine andere Art durch „Multiplikation“ eine bestimmte Zahl c, in | Zeichen 27

$$ab = c \quad \text{oder} \quad c = ab.$$

5. Es giebt eine bestimmte Zahl – sie heisse 1 –, so dass für jedes a zugleich

$$a \cdot 1 = a \quad \text{und} \quad 1 \cdot a = a$$

ist.

6. Wenn a und b beliebig gegebene Zahlen sind und a nicht 0 ist, so existirt stets eine und nur eine Zahl x und auch eine und nur eine Zahl y, so dass

$$ax = b \quad \text{bez.} \quad ya = b$$

wird.

---

4 **Zahlensysteme]** ⟨After ‘Zahlensysteme’ a footnote is added:⟩ Vgl. meinen Vortrag: Über den Zahlbegriff, Jahresbericht der Deutschen Mathematiker-Vereinigung, Bd. 8. 1900.[14]

10 (1–12):] (1–6):

14–21 ⟨In E2 theorems 2 and 3 are interchanged.⟩

25–33 ⟨In E2 theorems 5 and 6 are interchanged.⟩

[14] *Hilbert 1900b.*

Wenn a, b, c beliebige Zahlen sind, so gelten stets folgende Rechnungsgesetze:

7.	$a + (b + c)$	$=$	$(a + b) + c$
8.	$a + b$	$=$	$b + a$
9.	$a(bc)$	$=$	$(ab)c$
10.	$a(b + c)$	$=$	$ab + ac$
11.	$(a + b)c$	$=$	$ac + bc$
12.	ab	$=$	$ba.$

Sätze der Anordnung (13–16).

13. Wenn a, b irgend zwei verschiedene Zahlen sind, so ist stets eine bestimmte von ihnen (etwa a) grösser ($>$) als die andere; die letztere heisst dann die kleinere, in Zeichen:

$$a > b \quad \text{und} \quad b < a.$$

14. Wenn $a > b$ und $b > c$, so ist auch $a > c$.

15. Wenn $a > b$ ist, so ist auch stets

$$a + c > b + c \quad \text{und} \quad c + a > c + b.$$

16. Wenn $a > b$ und $c > 0$ ist, so ist auch stets

$$ac > bc \quad \text{und} \quad ca > cb.$$

Archimedischer Satz (17).

17. Wenn $a > 0$ und $b > 0$ zwei beliebige Zahlen sind, so ist es stets möglich, a zu sich selbst so oft zu addiren, dass die entstehende Summe die Eigenschaft hat

$$a + a + \cdots + a > b.$$

Ein System von Dingen, das nur einen Teil der Eigenschaften 1–17 besitzt, heisse ein *complexes Zahlensystem* oder auch ein *Zahlensystem* schlechthin.
28 Ein Zahlensystem heisse ein *Archi|medisches* oder ein *Nicht-Archimedisches*, jenachdem dasselbe der Forderung 17 genügt oder nicht.

Von den aufgestellten Eigenschaften 1–17 sind einige Folgen der übrigen. Es entsteht die Aufgabe, die logische Abhängigkeit dieser Eigenschaften zu

---

1 ⟨Before this line the following heading is added in E2:⟩
Regeln der Rechnung (7–12):

9 (13–16).] (13–16):

19 Archimedischer Satz (17).] Sätze von der Stetigkeit (17–18):

20 17.] 17. (Archimedischer Satz.)

24 ⟨Before this line the following paragraph is added in E2:⟩
18. (Satz von der Vollständigkeit.) Es ist nicht möglich, dem System der Zahlen ein anderes System von Dingen hinzuzufügen, so daß auch in dem durch Zusammensetzung entstehenden Systeme die Sätze 1–17 sämtlich erfüllt sind; oder kurz: die Zahlen bilden ein System von Dingen, welches bei Aufrechterhaltung sämtlicher aufgeführten Sätze keiner Erweiterung mehr fähig ist.

24 1–17] 1–18

25 oder auch ein *Zahlensystem* schlechthin] ⟨deleted⟩

26 Zahlensystem] komplexes Zahlensystem

28 1–17] 1–18

untersuchen. Wir werden in Kapitel VI § 32 und § 33 zwei bestimmte Fragen der angedeuteten Art wegen ihrer geometrischen Bedeutung beantworten und wollen hier nur darauf hinweisen, dass jedenfalls die letzte Forderung 17 keine logische Folge der übrigen Eigenschaften ist, da ja beispielsweise das in § 12 betrachtete complexe Zahlensystem $\Omega(t)$ sämtliche Eigenschaften 1–16 besitzt, aber nicht die Forderung 17 erfüllt.

§ 14.

Beweis des Pascalschen Satzes.

In diesem und dem folgenden Kapitel legen wir unserer Untersuchung die ebenen Axiome sämtlicher Gruppen mit Ausnahme des Archimedischen Axioms, d. h. die Axiome I 1–2 und II–IV zu Grunde. In dem gegenwärtigen Kapitel III gedenken wir Euklids Lehre von den Proportionen mittelst der genannten Axiome, d. h. *in der Ebene und unabhängig vom Archimedischen Axiom* zu begründen.

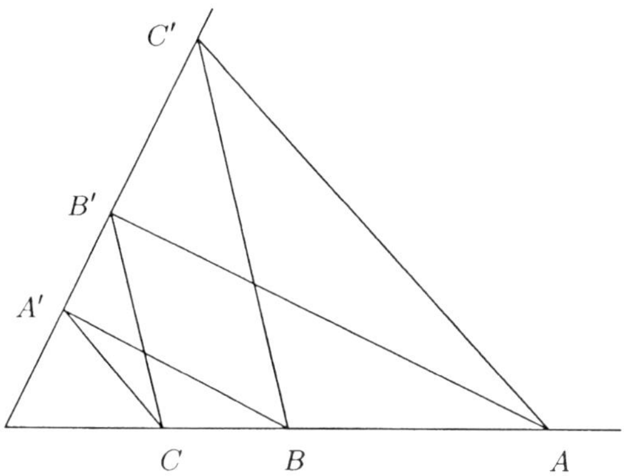

Fig. 16.

Zu dem Zwecke beweisen wir zunächst eine Thatsache, die ein besonderer Fall des bekannten Pascalschen Satzes aus der Lehre von den Kegelschnitten ist und die ich künftig kurz als den Pascalschen Satz bezeichnen will. Dieser Satz lautet:

Satz 21[1]) (Pascalscher Satz). *Es seien A, B, C bez. A', B', C' je drei Punkte auf zwei sich schneidenden Geraden, die vom Schnittpunkte der Geraden verschieden sind; ist dann CB' parallel BC' und CA' parallel AC', so ist auch BA' parallel AB'.*

1) *F. Schur* hat einen interessanten Beweis des Pascalschen Satzes auf Grund der sämtlichen Axiome I–II, IV in den Math. Ann. Bd. 51[15] veröffentlicht.

1 untersuchen.] ⟨After 'untersuchen' a footnote is added:⟩ Vgl. meinen bereits citierten Vortrag über den Zahlbegriff.

3 letzte] ⟨deleted⟩

4 übrigen] voranstehenden

6 ⟨After this line the following paragraph is added in E2:⟩

Im übrigen gelten betreffs der Sätze von der Stetigkeit (17–18) die entsprechenden Bemerkungen, wie sie in § 8 über die geometrischen Axiome der Stetigkeit gemacht worden sind.

10–11 des Archimedischen Axioms] der Stetigkeitsaxiome

[15] *Schur 1899.*

Um den Beweis für diesen Satz zu erbrin-
29 gen, führen wir | zunächst folgende Bezeichnungsweise ein. In einem rechtwinkligen Dreiecke ist offenbar die Kathete a durch die Hypotenuse c und den von a und c eingeschlossenen Basiswinkel α eindeutig bestimmt: wir setzen kurz

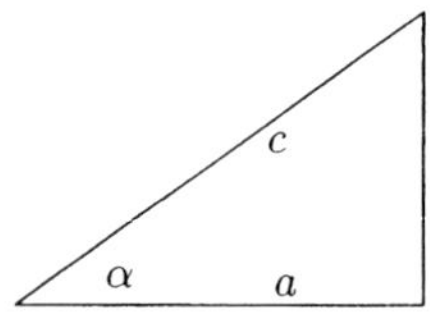

Fig. 17.

$$a = \alpha c,$$

sodass das Symbol αc stets eine bestimmte Strecke bedeutet, sobald c eine beliebig gegebene Strecke und α ein beliebig gegebener spitzer Winkel ist.

Nunmehr möge c eine beliebige Strecke und α, β mögen zwei beliebige spitze Winkel bedeuten; wir behaupten, dass allemal die Streckencongruenz

$$\alpha\beta c \equiv \beta\alpha c$$

besteht und somit die Symbole α, β stets mit einander vertauschbar sind.

Um diese Behauptung zu beweisen, nehmen wir die Strecke $c = AB$ und tragen an diese Strecke in A zu beiden Seiten die Winkel α und β an. Dann fällen wir von B aus auf die anderen Schenkel dieser Winkel die Lote BC und BD, verbinden C mit D und fällen schliesslich von A aus das Lot AE auf CD.

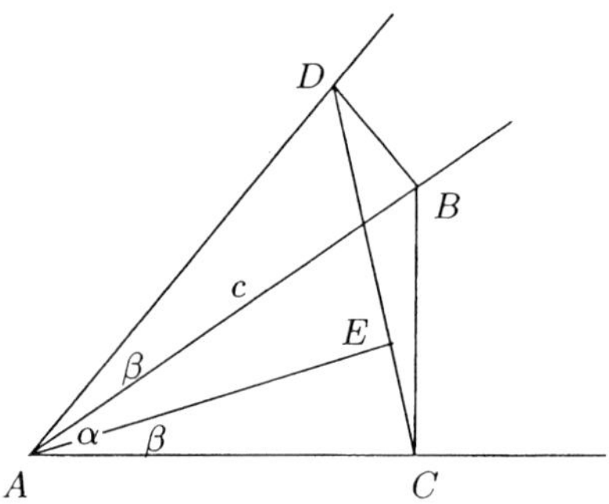

Fig. 18.

Da die Winkel $\sphericalangle ACB$ und $\sphericalangle ADB$ Rechte sind, so liegen die vier Punkte A, B, C, D auf einem Kreise und demnach sind die beiden Winkel $\sphericalangle ACD$ und $\sphericalangle ABD$ als Peripheriewinkel auf derselben Sehne AD einander congruent. Nun ist einerseits $\sphericalangle ACD$ zusammen mit dem $\sphericalangle CAE$ und andererseits $\sphericalangle ABD$ zusammen mit $\sphericalangle BAD$ je ein Rechter und folglich sind auch die Winkel $\sphericalangle CAE$ und $\sphericalangle BAD$ einander congruent, d. h. es ist

$$\sphericalangle CAE \equiv \beta,$$

und daher

$$\sphericalangle DAE \equiv \alpha.$$

Wir gewinnen nun unmittelbar die Streckencongruenzen

$$\begin{array}{r|l} \beta c \equiv AD & \alpha c \equiv AC \\ \alpha\beta c \equiv \alpha(AD) \equiv AE & \beta\alpha c \equiv \beta(AC) \equiv AE, \end{array}$$

30 und hieraus folgt die Richtigkeit der vorhin behaupteten Congruenz.

Wir kehren nun zur Figur des Pascalschen Satzes zurück und bezeichnen den Schnittpunkt der beiden Geraden mit O und die Strecken OA, OB, OC,

18 und β] bez. β

OA', OB', OC', CB', BC', CA', AC', BA', AB' bez. mit a, b, c, a', b', c', l, l^*, m, m^*, n, n^*. Sodann fällen wir von O Lote auf l, m, n; das Lot auf l schliesse mit den beiden Geraden OA, OA' die spitzen Winkel λ', λ ein und die Lote auf m bez. n mögen mit den Geraden OA und OA' die spitzen Winkel μ', μ bez. ν', ν bilden. Drücken wir nun diese drei Lote in der vorhin angegebenen Weise mit Hülfe der Hypotenusen und Basiswinkel in den betreffenden rechtwinkligen Dreiecken auf doppelte Weise aus, so erhalten wir folgende drei Streckencongruenzen

$$\lambda b' \equiv \lambda' c, \tag{1}$$

$$\mu a' \equiv \mu' c, \tag{2}$$

$$\nu a' \equiv \nu' b. \tag{3}$$

Da nach Voraussetzung l parallel l^* und m parallel m^* sein soll, so stimmen die von O auf l^* bez. m^* zu fällenden Lote mit den Loten auf l bez. m überein und wir erhalten somit

$$\lambda c' \equiv \lambda' b, \tag{4}$$

$$\mu c' \equiv \mu' a. \tag{5}$$

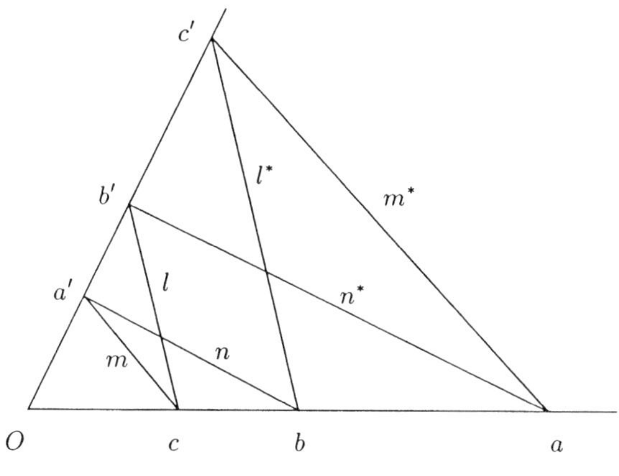

Fig. 19.

Wenn wir auf die Congruenz (3) links und rechts das Symbol $\lambda'\mu$ anwenden und bedenken, dass nach dem vorhin Bewiesenen die in Rede stehenden Symbole mit einander vertauschbar sind, so finden wir

$$\nu\lambda'\mu a' \equiv \nu'\mu\lambda' b.$$

In dieser Congruenz berücksichtigen wir links die Congruenz (2) und rechts (4); dann wird

$$\nu\lambda'\mu' c \equiv \nu'\mu\lambda c'$$

---

1 CA', AC'] AC', CA'

2 m, n;] m^*, n;

13 bez. m^*] bez. m

13 bez. m] bez. m^*

31 oder

$$\nu\mu'\lambda'c \equiv \nu'\lambda\mu c'.$$

Hierin berücksichtigen wir links die Congruenz (1) und rechts (5); dann wird

$$\nu\mu'\lambda b' \equiv \nu'\lambda\mu' a$$

oder

$$\lambda\mu'\nu b' \equiv \lambda\mu'\nu' a.$$

Wegen der Bedeutung unserer Symbole schliessen wir aus der letzten Congruenz sofort

$$\mu'\nu b' \equiv \mu'\nu' a$$

und hieraus

$$\nu b' \equiv \nu' a. \tag{6}$$

Fassen wir nun das von O auf n gefällte Lot in's Auge und fällen auf dasselbe Lote von A und B' aus, so zeigt die Congruenz (6), dass die Fusspunkte der letzteren beiden Lote zusammenfallen, d. h. die Gerade $n^* = AB'$ steht zu dem Lote auf n senkrecht und ist mithin zu n parallel. Damit ist der Beweis für den Pascalschen Satz erbracht.

---

16 ⟨After this line the following paragraphs are added in E2:⟩

Wir benutzen im folgenden zur Begründung der Geometrie lediglich denjenigen speziellen Fall des Pascalschen Satzes, in dem die Streckenkongruenz

$$OC \equiv OA'$$

und folglich auch

$$OA \equiv OC'$$

gilt. In diesem speziellen Fall gelingt der Beweis besonders einfach, nämlich folgendermaßen (Figur S. 30):

Wir tragen auf OA' von O aus die Strecke OB bis D' ab, so daß die Verbindungsgerade BD' parallel zu CA' und AC' wird. Wegen der Kongruenz der Dreiecke $OC'B$ und OAD' wird

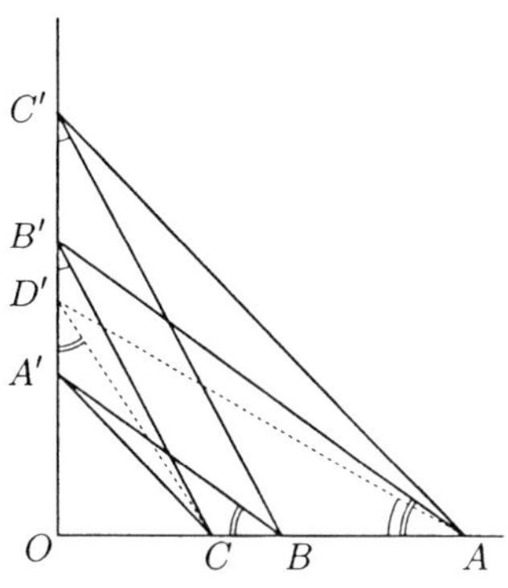

$$\sphericalangle OC'B \equiv \sphericalangle OAD'. \tag{1}$$

Da CB' und BC' nach Voraussetzung einander parallel sind, so ist

$$\sphericalangle OC'B \equiv \sphericalangle OBC; \tag{2}$$

aus (1) und (2) folgern wir

$$\sphericalangle OAD' \equiv \sphericalangle OB'C;$$

dann aber ist nach der Lehre vom Kreise $ACD'B'$ ein Kreisviereck und mithin gilt nach einem bekannten Satze von den Winkeln im Kreisviereck die Kongruenz

$$\sphericalangle OD'C \equiv \sphericalangle OAB'. \tag{3}$$

Andererseits ist wegen der Kongruenz der Dreiecke $OD'C$ und OBA' auch

$$\sphericalangle OD'C \equiv \sphericalangle OBA'; \tag{4}$$

aus (3) und (4) folgern wir

$$\sphericalangle OAB' \equiv \sphericalangle OBA'$$

und diese Kongruenz lehrt, daß AB' und BA' einander parallel sind, wie es der Pascalsche Satz verlangt.

Wenn irgend eine Gerade, ein Punkt ausserhalb derselben und irgend ein Winkel gegeben ist, so kann man offenbar durch Abtragen dieses Winkels und Ziehen einer Parallelen eine Gerade finden, die durch den gegebenen Punkt geht und die gegebene Gerade unter dem gegebenen Winkel schneidet. Im Hinblick auf diesen Umstand dürfen wir uns zum Beweise des Pascalschen Satzes folgenden einfachen Schlussverfahrens bedienen das ich einer Mittheilung von anderer Seite verdanke.

Man ziehe durch B eine Gerade, die OA' im Punkte D' unter dem Winkel OCA' trifft, so dass die Congruenz

$$\sphericalangle OCA' \equiv \sphericalangle OD'B \qquad (1^*)$$

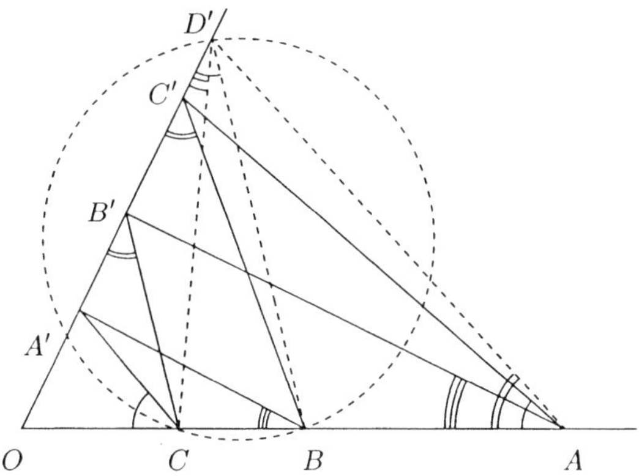

Fig. 20.

gilt; dann ist nach einem bekannten Satze aus der Lehre vom Kreise $CBD'A'$ 32
ein Kreisviereck und mithin gilt nach dem Satze von der Congruenz der Peripheriewinkel auf der nämlichen Sehne die Congruenz

$$\sphericalangle OBA' \equiv \sphericalangle OD'C. \qquad (2^*)$$

Da CA' und AC' nach Voraussetzung einander parallel sind, so ist

$$\sphericalangle OCA' \equiv \sphericalangle OAC'; \qquad (3^*)$$

aus (1^*) und (3^*) folgern wir die Congruenz

$$\sphericalangle OD'B \equiv \sphericalangle OAC';$$

dann aber ist auch $BAD'C'$ ein Kreisviereck und mithin gilt nach dem Satze von den Winkeln im Kreisviereck die Congruenz

$$\sphericalangle OAD' \equiv \sphericalangle OC'B. \qquad (4^*)$$

Da ferner nach Voraussetzung CB' parallel BC' ist, so haben wir auch

$$\sphericalangle OB'C \equiv \sphericalangle OC'B; \qquad (5^*)$$

aus (4^*) und (5^*) folgern wir die Congruenz

$$\sphericalangle OAD' \equiv \sphericalangle OB'C;$$

5–6 uns ⟨...⟩ bedienen] endlich zum Beweise des allgemeineren Pascalschen Satzes das folgende einfache Schlußverfahren anwenden

diese endlich lehrt, dass $CAD'B'$ ein Kreisviereck ist, und mithin gilt auch die Congruenz

$$\sphericalangle OAB' \equiv \sphericalangle OD'C. \qquad (6^*)$$

Aus (2^*) und (6^*) folgt

$$\sphericalangle OBA' \equiv \sphericalangle OAB'$$

und die Congruenz lehrt, dass BA' und AB' einander parallel sind, wie es der Pascalsche Satz verlangt.

Fällt D' mit einem der Punkte A', B', C' zusammen, so wird eine Abänderung dieses Schlussverfahrens nothwendig, die leicht ersichtlich ist.

§ 15.

Die Streckenrechnung auf Grund des Pascalschen Satzes.

Der im vorigen Paragraph bewiesene Pascalsche Satz setzt uns in den Stand, in die Geometrie eine Rechnung mit Strecken einzuführen, in der die Rechnungsregeln für reelle Zahlen sämtlich unverändert gültig sind.

33 Statt des Wortes „congruent" und des Zeichens $\equiv$ bedienen wir uns in der Streckenrechnung des Worte „gleich" und des Zeichens $=$.

Wenn A, B, C drei Punkte einer Geraden sind und B zwischen A und C liegt, so bezeichnen wir $c = AC$ als die *Summe* der beiden Strecken $a = AB$ und $b = BC$ und setzen

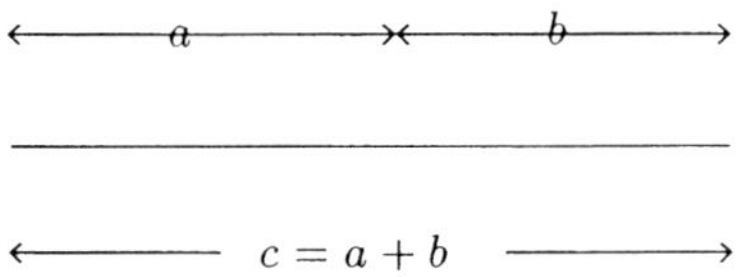

$$c = a + b.$$

Die Strecken a und b heissen kleiner als c, in Zeichen:

$$a < c, \quad b < c,$$

und c heisst grösser als a und b, in Zeichen:

$$c > a, \quad c > b.$$

Aus den linearen Congruenzaxiomen IV 1–3 entnehmen wir leicht, dass für die eben definirte Addition der Strecken das a s s o c i a t i v e Gesetz

$$a + (b + c) = (a + b) + c$$

sowie das c o m m u t a t i v e Gesetz

$$a + b = b + a$$

gültig ist.

Um das Produkt einer Strecke a in eine Strecke b geometrisch zu definiren, bedienen wir uns folgender Construktion. Wir wählen zunächst eine beliebige

---

6 die Congruenz] diese Congruenz

9 ist.] ⟨After 'ist' a footnote is added:⟩ Interesse verdient auch die Verwendung, die der Satz vom gemeinsamen Schnittpunkt der Höhen eines Dreieckes zur Begründung des Pascalschen Satzes bez. der Proportionenlehre findet; man vergleiche hierüber F. Schur, Math. Ann. Bd. 57 und J. Mollerup, Studier over den plane geometrics aksiomer, Kopenhagen 1903.[16]

[16] *Schur 1903*, *Mollerup 1903b*.

Strecke, die für die ganze Betrachtung die nämliche bleibt, und bezeichnen dieselbe mit 1. Nunmehr tragen wir auf dem einen Schenkel eines rechten Winkels vom Scheitel O aus die Strecke 1 und ferner ebenfalls vom Scheitel O aus die Strecke b ab; sodann tragen wir auf dem anderen Schenkel die Strecke a ab. Wir verbinden die Endpunkte der Strecken 1 und a durch eine Gerade und ziehen zu dieser Geraden durch den Endpunkt der Strecke b eine Parallele; dieselbe möge auf dem anderen Schenkel eine Strecke c abschneiden: dann nennen wir diese Strecke c das *Produkt* der Strecke a in die Strecke b und bezeichnen sie mit

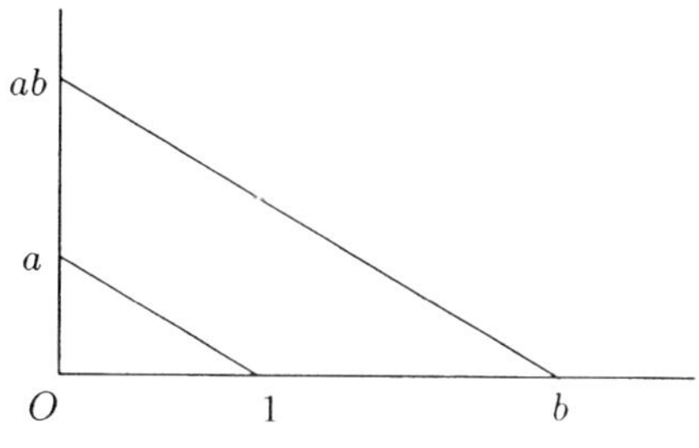

Fig. 21.

$$c = ab.$$

Wir wollen vor Allem beweisen, dass für die eben definirte Multiplikation 34
der Strecken das commutative Gesetz

$$ab = ba$$

gültig ist. Zu dem Zwecke construiren wir zuerst auf die oben festgesetzte Weise die Strecke ab. Ferner tragen wir auf dem ersten Schenkel des rechten Winkels die Strecke a und auf dem anderen Schenkel die Strecke b ab, verbinden den Endpunkt der Strecke 1 mit dem Endpunkt von b auf dem anderen Schenkel durch eine Gerade und ziehen zu dieser Geraden durch den Endpunkt von a auf dem ersten Schenkel eine Parallele: dieselbe schneidet auf dem anderen Schenkel die Strecke ba ab; in der That fällt diese Strecke ba, wie die Figur 22 zeigt, wegen der Parallelität der punktirten Hülfslinien nach dem Pascalschen Satze (Satz 21) mit der vorhin construirten Strecke ab zusammen.

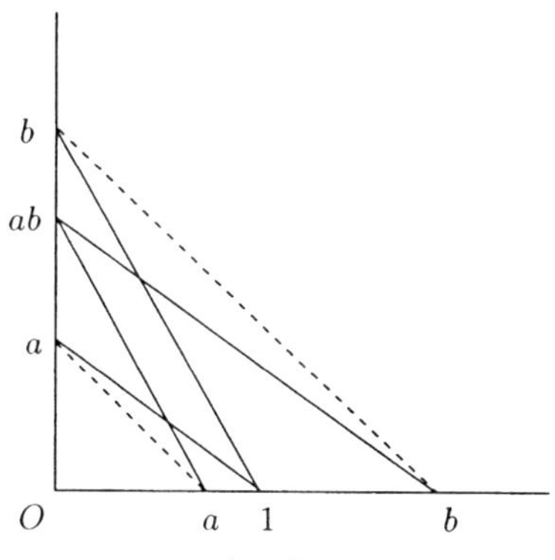

$ab = ba$

Fig. 22.

---

5 aus] ⟨deleted⟩

32–33 die Figur 22] untenstehende Figur

37 zusammen.] zusammen. Auch umgekehrt folgt, wie man sofort sieht, aus der Gültigkeit des kommutativen Gesetzes in unserer Streckenrechnung der Pascalsche Satz.

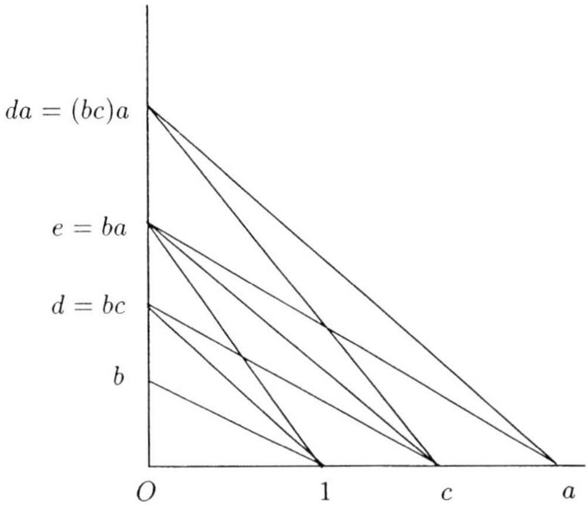

$a(bc) = (ab)c$

Fig. 23.

Um für unsere Multiplikation der Strecken das associative Gesetz

$$a(bc) = (ab)c.$$

zu beweisen, construiren wir erst die Strecke $d = bc$, dann da, ferner die Strecke $e = ba$ und dann ec. Dass die Endpunkte von da und ec zusammenfallen, ist wiederum auf Grund des Pascalschen 35 | Satzes aus der Figur 23 unmittelbar ersichtlich und mit Benutzung des bereits bewiesenen commutativen Gesetzes folgt hieraus die obige Formel für das associative Gesetz der Streckenmultiplikation.

5–19 construieren ⟨...⟩ Streckenmultiplikation.]

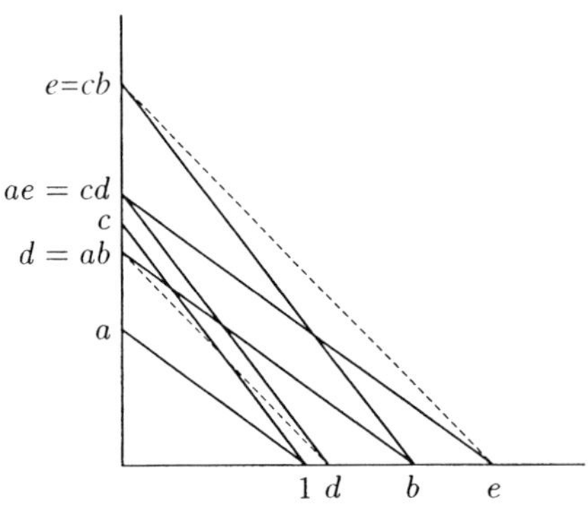

tragen wir auf dem einen Schenkel des rechten Winkels vom Scheitel O aus die Strecken 1 und b und auf dem anderen Schenkel ebenfalls von O aus die Strecken a und c ab. Sodann konstruieren wir die Strecken $d = ab$ und $e = cb$ und tragen diese Strecken d und e auf dem ersteren Schenkel von O aus ab. Konstruieren wir sodann ae und cd, so ist wiederum auf Grund des Pascalschen Satzes aus vorstehender Figur ersichtlich, daß die Endpunkte dieser Strecken zusammenfallen, d. h. es ist

$$ae = cd \quad \text{oder} \quad a(cb) = c(ab)$$

und hieraus folgt mit Zuhilfenahme des kommutativen Gesetzes auch

$$a(bc) = (ab)c^{1)}$$

1) Man vergleiche hierzu auch die Methoden zur Begründung der Proportionenlehre, die neuerdings von *A. Kneser*, Archiv für Math. und Phys., R. III, Bd. 2, und *J. Mollerup*, Math. Ann., Bd. 56 sowie Studier over den plane geometrics aksiomer, Kopenhagen 1903,[17] angegeben worden sind und bei denen die Proportionengleichung vorangestellt wird. *F. Schur*, Zur Proportionenlehre, Math. Ann. Bd. 57 bemerkt, daß bereits *Kupffer* (Sitzungsber. der Naturforscherges. zu Dorpat 1893) das kommutative Gesetz der Multiplikation richtig bewiesen hat.[18] Jedoch ist *Kupffers* weitere Begründung der Proportionenlehre als unzureichend anzusehen.

19 ⟨After this line the following paragraphs are added:⟩

[17] *Kneser 1902a, Mollerup 1903a, 1903b.*

[18] *Schur 1903, Kupffer 1894.*

Endlich gilt in unserer Streckenrechnung auch das d i s t r i b u t i v e Gesetz

$$a(b+c) = ab + ac.$$

Um dasselbe zu beweisen, construiren wir die Strecken ab, ac und $a(b+c)$ und ziehen dann durch den Endpunkt der Strecke c (s. Figur 24) eine Parallele zu dem anderen Schenkel des rechten Winkels. Die Congruenz der beiden rechtwinkli-

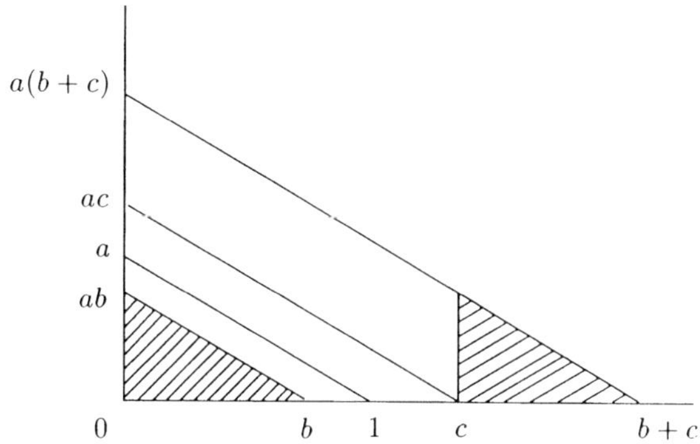

$a(b+c) = ab + ac$

Fig. 24.

Wie man sieht, haben wir im vorstehenden beim Nachweis sowohl des kommutativen wie des assoziativen Gesetzes der Multiplikation lediglich denjenigen speziellen Fall des Pascalschen Satzes benutzt, dessen Beweis auf S. 29–30 (§ 14) in besonders einfacher Weise durch einmalige Anwendung des Kreisvierecksatzes gelang.

Durch eine geringe Umordnung des obigen Schlußverfahrens erhält man folgende ebenfalls sehr kurze Begründung der Multiplikationsgesetze der Streckenrechnung.

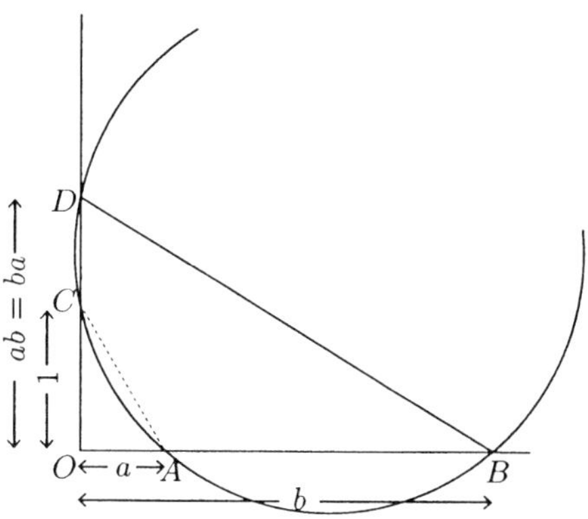

Auf dem einen Schenkel eines rechten Winkels trage man vom Scheitel O aus die Strecken $a = OA$ und $b = OB$ und außerdem auf dem anderen Schenkel die Einheitsstrecke $1 = OC$ ab. Der durch A, B, C gelegte Kreis schneide den letzteren Schenkel noch im Punkte D. Der Punkt D wird leicht in elementarer Weise gewonnen, indem man vom Mittelpunkt des Kreises das Lot auf OC fällt und an diesem den Punkt C spiegelt. Die Strecke OD heiße das Produkt der beiden Strecken a und b; es gilt dann offenbar das kommutative Gesetz der Multiplikation

$$ab = ba.$$

Wegen der Gleichheit der Winkel OCA und OBD stimmt diese Definition des Produktes mit der auf S. 32 angegebenen überein. Die Gleichung

$$ab = ba$$

beweist nunmehr nach S. 33 oben den speziellen Fall (S. 29) des Pascalschen Satzes und aus diesem wiederum folgt nach S. 33 das assoziative Gesetz der Multiplikation

$$a(bc) = (ab)c.$$

11 Figur 24] nebenstehende Figur

gen in der Figur 24 schraffirten Dreiecke und die Anwendung des Satzes von der Gleichheit der Gegenseiten im Parallelogramm liefert dann den gewünschten Nachweis.

Sind b und c zwei beliebige Strecken, so giebt es stets eine Strecke a, sodass $c = ab$ wird; diese Strecke a wird mit $\frac{c}{b}$ bezeichnet und der *Quotient* von c durch b genannt.

§ 16.

Die Proportionen und die Aehnlichkeitssätze.

Mit Hülfe der eben dargelegten Streckenrechnung lässt sich *Euklids* Lehre von den Proportionen einwandsfrei und ohne Archimedisches Axiom in folgender Weise begründen.

E r k l ä r u n g. Sind a, b, a', b' irgend vier Strecken, so soll die *Proportion*

$$a : b = a' : b'$$

nichts anderes bedeuten als die Streckengleichung

$$ab' = ba'.$$

D e f i n i t i o n. Zwei Dreiecke heissen *ähnlich*, wenn entsprechende Winkel in ihnen congruent sind.

36 S a t z 22. Wenn a, b und a', b' entsprechende Seiten in zwei ähnlichen Dreiecken sind, so gilt die Proportion

$$a : b = a' : b'.$$

Beweis. Wir betrachten zunächst den besonderen Fall, wo die von a, b und a', b' eingeschlossenen Winkel in beiden Dreiecken Rechte sind, und denken uns die beiden Dreiecke in ein und denselben rechten Winkel eingetragen. Wir tragen sodann vom Scheitel aus auf einem Schenkel die Strecke 1 ab und ziehen durch den Endpunkt dieser Strecke 1 die Parallele zu den beiden Hypotenusen; dieselbe schneide auf dem anderen Schenkel die Strecke e ab; dann ist nach unserer Definition des Streckenproduktes

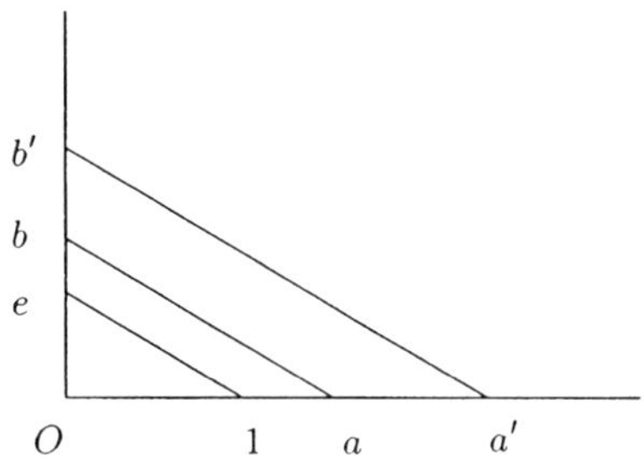

Fig. 25.

$$b = ea, \quad b' = ea';$$

mithin haben wir

$$ab' = ba'$$

d. h.

$$a : b = a' : b'.$$

Nunmehr kehren wir zu dem allgemeinen Falle zurück. Wir construiren in jedem der beiden ähnlichen Dreiecke den Schnittpunkt S bez. S' der

1 24] ⟨deleted⟩
16 D e f i n i t i o n.] E r k l ä r u n g.

drei Winkelhalbirenden und fällen von diesem die drei Lote r bez. r' auf die Dreiecksseiten; die auf diesen entstehenden Abschnitte bezeichnen wir mit

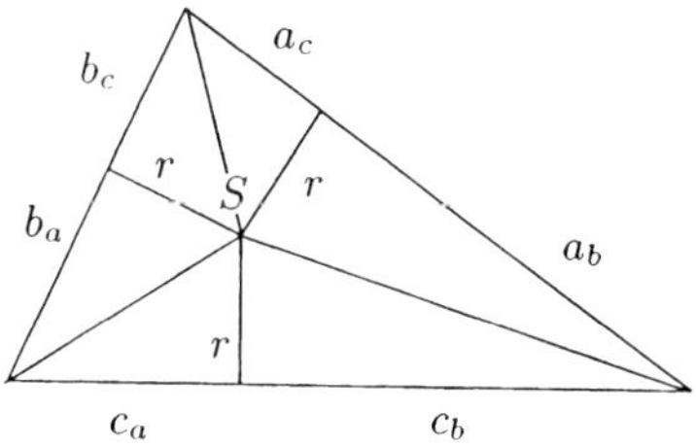

Fig. 26.

$$a_b,\ a_c,\ b_c,\ b_a,\ c_a,\ c_b$$

bez.

$$a'_b,\ a'_c,\ b'_c,\ b'_a,\ c'_a,\ c'_b.$$

Der vorhin bewiesene spezielle Fall unseres Satzes liefert dann die Proportionen

$$a_b : r = a'_b : r \qquad b_c : r = b'_c : r$$
$$a_c : r = a'_c : r \qquad b_a : r = b'_a : r$$

aus diesen schliessen wir mittelst des distributiven Gesetzes 37

$$a : r = a' : r', \qquad b : r = b' : r'$$

und folglich mit Rücksicht auf das commutative Gesetz der Multiplikation

$$a : b = a' : b'.$$

Aus dem eben bewiesenen Satze 22 entnehmen wir leicht den Fundamentalsatz in der Lehre von den Proportionen, der wie folgt lautet:

Satz 23. *Schneiden zwei Parallele auf den Schenkeln eines beliebigen Winkels die Strecken a, b bez. a', b' ab, so gilt die Proportion*

$$a : b = a' : b'.$$

Umgekehrt, wenn vier Strecken a, b, a', b' diese Proportion erfüllen und a, a' und b, b' je auf einem Schenkel eines beliebigen Winkels abgetragen werden, so sind die Verbindungsgeraden der Endpunkte von a, b bez. von a', b' einander parallel.

§ 17.

Die Gleichungen der Geraden und Ebenen.

Zu dem bisherigen System von Strecken fügen wir noch ein zweites ebensolches System von Strecken hinzu; die Strecken des neuen Systems denken wir uns durch ein Merkzeichen kenntlich gemacht und nennen sie dann „*negative*" Strecken zum Unterschiede von den bisher betrachteten „*positiven*" Strecken. Führen wir noch die durch einen einzigen Punkt bestimmte Strecke 0 ein, so gelten bei gehörigen Festsetzungen in dieser erweiterten Streckenrechnung sämtliche Rechnungsregeln für reelle Zahlen, die in § 13 zusammengestellt worden sind. Wir heben folgende specielle Thatsachen hervor:

Es ist stets $a \cdot 1 = 1 \cdot a = a$.

Wenn $ab = 0$, so ist entweder $a = 0$ oder $b = 0$.

Wenn $a > b$ und $c > 0$, so folgt stets $ac > bc$.

Wir nehmen nun in einer Ebene α durch einen Punkt O zwei zu einander senkrechte Gerade als festes rechtwinkliges Axenkreuz an und tragen dann

1 Winkelhalbirenden] Winkelhalbirenden, dessen Existenz aus dem Satze vom gleichschenkligen Dreiecke leicht abzuleiten ist,

1 diesem] diesen

die beliebigen Strecken x, y von O aus auf den beiden Geraden ab, und zwar nach der einen oder nach der anderen Seite hin, jenachdem die abzutragende
38 Strecke x bez. y positiv oder negativ ist; sodann errichten wir die Lote | in den Endpunkten der Strecken x, y und bestimmen den Schnittpunkt P dieser Lote: die Strecken x, y heissen die *Coordinaten* des Punktes P; jeder Punkt der Ebene α ist durch seine Coordinaten x, y, die positive oder negative Strecken oder 0 sein können, eindeutig bestimmt.

Es sei l irgend eine Gerade in der Ebene α, die durch O und durch einen Punkt C mit den Coordinaten a, b gehe. Sind dann x, y die Coordinaten irgend eines Punktes von l, so finden wir leicht aus Satz 22

$$a : b = x : y$$

oder

$$bx - ay = 0$$

als die Gleichung der Geraden l. Ist l' eine zu l parallele Gerade, die auf der x-Axe die Strecke c abschneidet, so gelangen wir zu der Gleichung der Geraden l', indem wir in der Gleichung der Geraden l die Strecke x durch die Strecke $x - c$ ersetzen; die gewünschte Gleichung lautet also

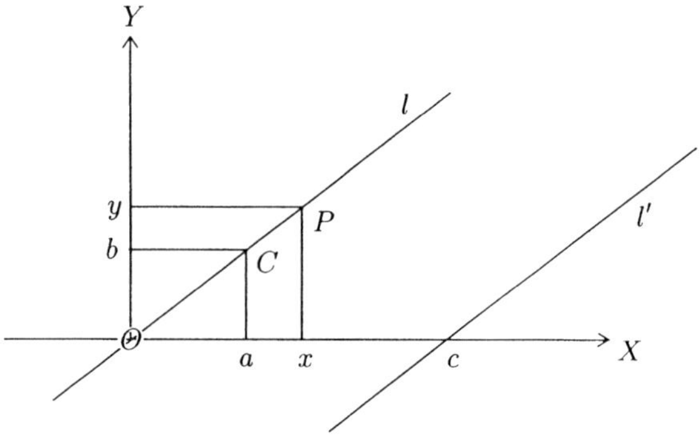

Fig. 27.

$$bx - ay - bc = 0.$$

Aus diesen Entwickelungen schliessen wir leicht auf eine Weise, die von dem Archimedischen Axiom unabhängig ist, dass jede Gerade in einer Ebene durch eine lineare Gleichung in den Coordinaten x, y dargestellt wird und umgekehrt jede solche lineare Gleichung eine Gerade darstellt, wenn die Coefficienten derselben in der betreffenden Geometrie vorkommende Strecken sind.

Die entsprechenden Resultate beweist man ebenso leicht in der räumlichen Geometrie.

Der weitere Aufbau der Geometrie kann von nun an nach den Methoden geschehen, die man in der analytischen Geometrie gemeinhin anwendet.

Wir haben bisher in diesem Kapitel III das Archimedische Axiom nirgends benutzt; setzen wir jetzt die Gültigkeit desselben voraus, so können wir den Punkten einer beliebigen Geraden im Raume reelle Zahlen zuordnen und zwar auf folgende Art.

39 Wir wählen auf der Geraden zwei beliebige Punkte aus und ordnen diesen die Zahlen 0 und 1 zu; sodann halbiren wir die durch sie bestimmte Strecke 01 und bezeichnen den entstehenden Mittelpunkt mit $\frac{1}{2}$, ferner den Mittelpunkt der Strecke $0\frac{1}{2}$ mit $\frac{1}{4}$ u. s. w.; nach n-maliger Ausführung dieses Verfahrens gelangen wir zu einem Punkte, dem die Zahl $\frac{1}{2^n}$ zuzuordnen ist. Nun tragen

29 wenn] wobei

wir die Strecke $0\frac{1}{2^n}$ an den Punkt 0 sowohl nach der Seite des Punktes 1 als auch nach der anderen Seite hin etwa m mal hintereinander ab und erteilen den so entstehenden Punkten die Zahlenwerte $\frac{m}{2^n}$ bez. $-\frac{m}{2^n}$. Aus dem Archimedischen Axiom kann leicht geschlossen werden, dass auf Grund dieser Zuordnung sich jedem beliebigem Punkte der Geraden in eindeutig bestimmter Weise eine reelle Zahl zuordnen lässt und zwar so dass dieser Zuordnung folgende Eigenschaft zukommt: wenn A, B, C irgend drei Punkte der Geraden und bez. α, β, γ die zugehörigen reellen Zahlen sind und B zwischen A und C liegt, so erfüllen dieselben stets entweder die Ungleichung $\alpha < \beta < \gamma$ oder $\alpha > \beta > \gamma$.

Aus den Entwicklungen in Kap. II § 9 leuchtet ein, dass dort für jede Zahl, die dem algebraischen Zahlkörper Ω angehört, notwendig ein Punkt der Geraden existiren muss, dem sie zugeordnet ist. Ob auch jeder anderen reellen Zahl ein Punkt entspricht, lässt sich im allgemeinen nicht entscheiden, sondern hängt von der vorgelegten Geometrie ab.

Dagegen ist es stets möglich, das ursprüngliche System von Punkten, Geraden und Ebenen so durch „ideale“ oder „irrationale“ Elemente zu erweitern, dass auf irgend einer Geraden der entstehenden Geometrie jedem System von drei reellen Zahlen ohne Ausnahme ein Punkt zugeordnet ist. Durch gehörige Festsetzung kann zugleich erreicht werden, dass in der erweiterten Geometrie sämtliche Axiome I–V gültig sind. Diese (durch Hinzufügung der irrationalen Elemente) erweiterte Geometrie ist keine andere als die gewöhnliche analytische Geometrie des Raumes.

Kapitel IV. 40

Die Lehre von den Flächeninhalten in der Ebene.

§ 18.

Die Flächengleichheit und Inhaltsgleichheit von Polygonen.

Wir legen den Untersuchungen des gegenwärtigen Kapitels IV dieselben

5 beliebigem] beliebigen

9 dieselben] diese Zahlen

14–15 lässt ⟨...⟩ Geometrie ab] hängt davon ab, ob in der vorgelegten Geometrie das Vollständigkeitsaxiom V 2 gilt oder nicht gilt

16 ist es] ist es, auch wenn man nur die Gültigkeit des Archimedischen Axioms annimmt,

16 ursprüngliche] ⟨deleted⟩

17 „ideale“ oder] ⟨deleted⟩

18 irgend einer] jeder

23 Geometrie] Cartesische Geometrie

23 Raumes.] Raumes, in welcher auch das Vollständigkeitsaxiom V 2 gilt[1]).

1) Vgl. die Bemerkungen am Schluß von § 8.

27 **Flächengleichheit] Zerlegungsgleichheit**

Axiome wie im Kapitel III zu Grunde, nämlich die ebenen Axiome sämtlicher Gruppen mit Ausnahme des Archimedischen Axioms, d. h. die Axiome I 1–2 und II–IV.

Die im Kapitel III erörterte Lehre von den Proportionen und die daselbst eingeführte Streckenrechnung setzt uns in den Stand, die Euklidische Lehre von den Flächeninhalten mittelst der genannten Axiome, d.h. *in der Ebene und unabhängig vom Archimedischen Axiom* zu begründen.

Da nach den Entwickelungen im Kapitel III die Lehre von den Proportionen wesentlich auf dem Pascalschen Satze (Satz 21) beruht, so gilt dies auch für die Lehre von den Flächeninhalten; diese Begründung der Lehre von den Flächeninhalten erscheint mir als eine der merkwürdigsten Anwendungen des Pascalschen Satzes in der Elementargeometrie.

Erklärung. Verbindet man zwei Punkte eines Polygons P durch irgend einen Streckenzug, der ganz im Inneren des Polygons verläuft, so entstehen zwei neue Polygone P_1 und P_2, deren innere Punkte alle im Inneren von P liegen; wir sagen: P zerfällt in P_1 und P_2, oder P_1 und P_2 setzen P zusammen.

Definition. Zwei Polygone heissen *flächengleich*, wenn sie in eine endliche Anzahl von Dreiecken zerlegt werden können, die paarweise einander congruent sind.

Definition. Zwei Polygone heißen *inhaltsgleich* oder *von gleichem Inhalte*, wenn es möglich ist, zu denselben flächengleiche Polygone hinzuzufügen, so dass die beiden zusammengesetzten Polygone einander flächengleich sind.

Aus diesen Definitionen folgt sofort: durch Zusammenfügung flächengleicher Polygone entstehen wieder flächengleiche Polygone, und wenn man flächengleiche Polygone von flächengleichen Polygonen wegnimmt, so sind die übrigbleibenden Polygone inhaltsgleich.

41 Ferner gelten folgende Sätze:

Satz 24. Sind zwei Polygone P_1 und P_2 mit einem dritten Polygon P_3 flächengleich, so sind sie auch unter einander flächengleich. Sind zwei Polygone mit einem dritten inhaltsgleich, so sind sie unter einander inhaltsgleich.

Beweis. Nach Voraussetzung lässt sich sowohl für P_1, als auch für P_2 eine Zerlegung in Dreiecke angeben, so dass einer jeden dieser beiden Zerlegungen je eine Zerlegung des Polygons P_3 in congruente Dreiecke entspricht. Indem wir diese Zerlegungen von P_3 gleichzeitig in Betracht ziehen, wird im

---

1 ebenen Axiome] linearen und ebenen Axiome

2 des Archimedischen Axioms] der Stetigkeitsaxiome

7 *vom Archimedischen Axiom*] *von den Stetigkeitsaxiomen*

16 und P_2,] und P_2,

16 und P_2] und P_2

18 Definition.] Erklärung.

18 ⟨In this paragraph 'flächengleich' is always replaced by 'zerlegungsgleich' in E2.⟩

21 Definition.] Erklärung.

24 diesen Definitionen] den letzteren Erklärungen

Allgemeinen jedes Dreieck der einen Zerlegung durch Strecken, welche der anderen Zerlegung angehören, in Polygone zerlegt. Wir fügen nun noch so viele Strecken hinzu, dass jedes dieser Polygone selbst wieder in Dreiecke zerfällt und bringen dann die zwei entsprechenden Zerlegungen in Dreiecke in P_1 und in P_2 an; dann zerfallen offenbar diese beiden Polygone P_1 und P_2 in

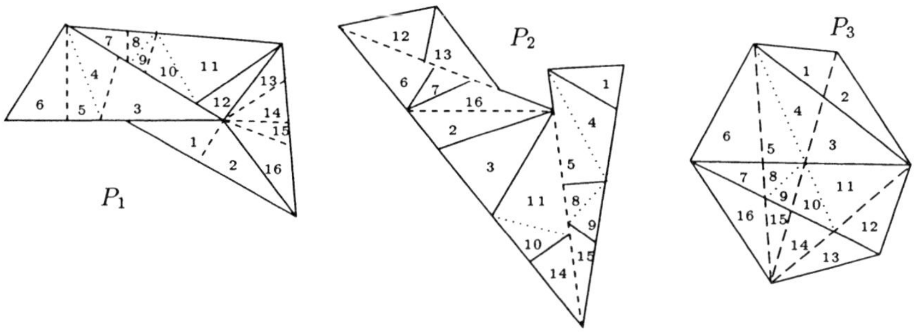

Fig. 28.

gleich viele paarweise einander congruente Dreiecke und sind somit nach der Definition einander flächengleich.

Der Beweis der zweiten Aussage des Satzes 24 ergiebt sich nunmehr ohne Schwierigkeit.

Wir definiren in der üblichen Weise die Begriffe: *Rechteck, Grundlinie* und *Höhe eines Parallelogrammes*, *Grundlinie* und *Höhe eines Dreiecks.*

§ 19.

Parallelogramme und Dreiecke mit gleicher Grundlinie und Höhe.

Die bekannte in den nebenstehenden Figuren illustrirte Schlussweise *Euklids* liefert den Satz:

Satz 25. Zwei Parallelogramme mit gleicher Grundlinie und Höhe sind 42
einander inhaltsgleich.

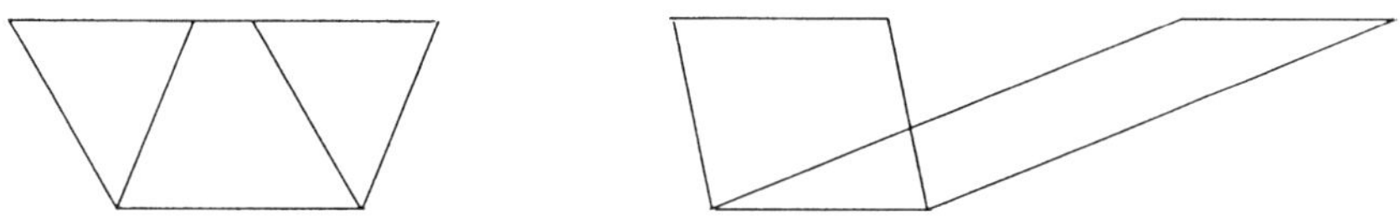

Fig. 29.

7 Definition] Erklärung
10 definiren] erklären
14 nebenstehenden] untenstehenden

Ferner gilt die bekannte Thatsache:

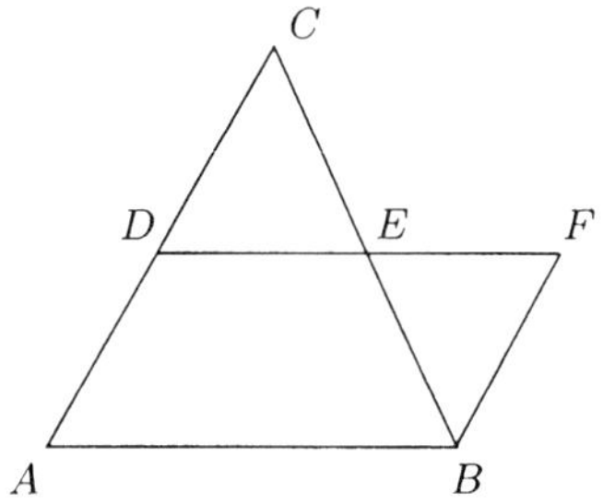

Fig. 30.

Satz 26. Ein jedes Dreieck ABC ist stets einem gewissen Parallelogramm mit gleicher Grundlinie und halber Höhe flächengleich.

Beweis: Halbirt man AC in D und BC in E und verlängert dann DE um sich selbst bis F, so sind die Dreiecke DEC und FBE einander congruent und folglich sind Dreieck ABC und Parallelogramm $ABFD$ einander inhaltsgleich.

Aus Satz 25 und Satz 26 folgt mit Hinzuziehung von Satz 24 unmittelbar:

Satz 27. Zwei Dreiecke mit gleicher Grundlinie und Höhe sind einander inhaltsgleich.

Bekanntlich zeigt man gewöhnlich, dass zwei Dreiecke mit gleicher Grundlinie und Höhe auch stets flächengleich sind. Wir bemerken jedoch, dass *dieser Nachweis ohne Benutzung des Archimedischen Axioms nicht möglich ist*; in der That lassen sich in unserer Nicht-Archimedischen Geometrie (vgl. Kap. II § 12) ohne Schwierigkeit solche zwei Dreiecke angeben, die gleiche Grundlinie und Höhe besitzen und folglich dem Satze 27 entsprechend inhaltsgleich, aber die dennoch nicht flächengleich sind. Als Beispiel mögen zwei Dreiecke ABC und ABD mit der gemeinsamen Grundlinie $AB = 1$ und der gleichen Höhe 1 dienen, wenn die Spitze C des ersteren Dreiecks senkrecht über A und im zweiten Dreiecke der Fusspunkt F der von der Spitze D gefällten Höhe so gelegen ist, dass $AF = t$ wird.

Die übrigen Sätze aus der elementaren Geometrie über die Inhaltsgleichheit von Polygonen, insbesondere der Pythagoräische Lehrsatz sind leichte Folgerungen der eben aufgestellten Sätze. Wir begegnen aber dennoch bei der weiteren Durchführung der Theorie der Flächeninhalte einer wesentlichen
43 Schwierigkeit. Ins|besondere lassen es unsere bisherigen Betrachtungen dahingestellt, ob nicht etwa alle Polygone stets einander inhaltsgleich sind. In diesem Falle wären die sämtlichen vorhin aufgestellten Sätze nichtssagend und ohne Bedeutung. Weiter entsteht die allgemeinere Frage, ob zwei inhaltsgleiche Rechtecke mit einer gemeinschaftlichen Seite auch notwendig in der an-

5 flächengleich] zerlegungsgleich

16 gewöhnlich] leicht

17 flächengleich] zerlegungsgleich

20 solche zwei] zwei solche

22 flächengleich] zerlegungsgleich

26 ⟨After this line the following paragraphs are added:⟩
Wir erwähnen noch den leicht zu beweisenden Satz:

Satz 28. Zu einem beliebigen Dreiecke und mithin auch zu einem beliebigen Polygon kann stets ein rechtwinkliges Dreieck konstruiert werden, das eine Kathete 1 besitzt und das mit dem Dreiecke bez. Polygon inhaltsgleich ist.

27 Die] Auch die

34 Weiter entsteht die allgemeinere Frage] Hiermit hängt die Frage zusammen

deren Seite übereinstimmen, d. h., ob ein Rechteck durch eine Seite und den Flächeninhalt eindeutig bestimmt ist.

Wie die nähere Ueberlegung zeigt, bedarf man zur Beantwortung der aufgeworfenen Fragen der Umkehrung des Satzes 27, die folgendermassen lautet:

Satz 28. *Wenn zwei inhaltsgleiche Dreiecke gleiche Grundlinie haben, so haben sie auch gleiche Höhe.*

Dieser fundamentale Satz 28 findet sich im ersten Buch der Elemente des *Euklid* als 39ster Satz; beim Beweise desselben beruft sich jedoch *Euklid* auf den allgemeinen Grössensatz: „*Καὶ τὸ ὅλον τοῦ μέρους μεῖζόν ἐστιν*“ – ein Verfahren, welches auf die Einführung eines neuen geometrischen Axioms über Flächeninhalte hinausläuft.

Es gelingt nun, den Satz 28 und damit die Lehre von den Flächeninhalten auf dem hier von uns in Ausssicht genommenen Wege, d. h. lediglich mit Hülfe der ebenen Axiome ohne Benutzung des Archimedischen Axioms zu begründen. Um dies einzusehen, haben wir den Begriff des Flächenmasses nöthig.

§ 20.

Das Flächenmass von Dreiecken und Polygonen.

Definition. Wenn wir in einem Dreieck ABC mit den Seiten a, b, c die beiden Höhen $h_a = AD$, $h_b = BE$ construiren, so folgt aus der Aehnlichkeit der Dreiecke BCE und ACD nach Satz 22 die Proportion

$$a : h_b = b : h_a,$$

d. h.

$$ah_a = bh_b;$$

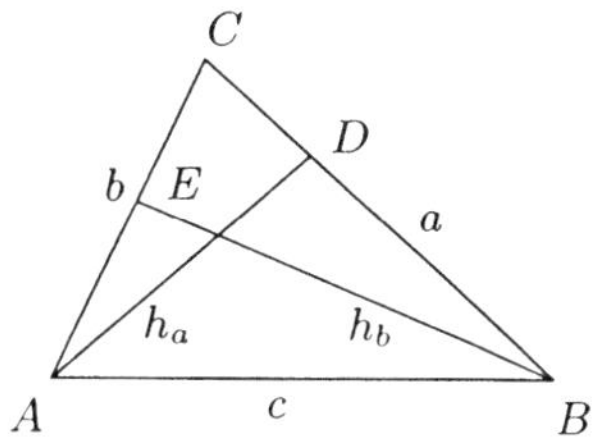

Fig. 31.

mithin ist in jedem Dreiecke das Produkt aus einer Grundlinie und der zu ihr gehörigen Höhe davon unabhängig, welche Seite des Dreiecks man als Grundlinie wählt. Das halbe Produkt aus Grundlinie und Höhe eines Dreiecks Δ heisse das *Flächenmass des Dreiecks* Δ und werde mit $F(\Delta)$ bezeichnet.

Erklärung. Eine Strecke, welche eine Ecke eines Dreiecks mit einem 44
Punkte der gegenüberliegenden Seite verbindet, heisst *Transversale*; dieselbe zerlegt das Dreieck in zwei Dreiecke mit gemeinsamer Höhe, deren Grundlinien in dieselbe Gerade fallen; eine solche Zerlegung heisse eine *transversale Zerlegung des Dreiecks.*

---

15 Flächenmasses] Inhaltsmaßes

18 ⟨In this paragraph the expressions 'Flächenmass' and 'Flächenmasse' are replaced by 'Inhaltsmaß' and 'Inhaltsmaße' respectively in E2.⟩

19 Definition.] Erklärung:

30 heisse] ist also eine für das Dreieck charakteristische Strecke; sie heiße

31 ⟨In this paragraph the notation 'F' is always replaced by 'J' in E2.⟩

33 *Transversale*;] *Transversale des Dreiecks*;

S a t z 29. Wenn ein Dreieck Δ durch beliebige Gerade irgendwie in eine gewisse endliche Anzahl von Dreiecken Δ_k zerlegt wird, so ist stets das Flächenmass des Dreiecks Δ gleich der Summe der Flächenmasse der sämtlichen Dreiecke Δ_k.

Beweis. Aus dem distributiven Gesetze in unserer Streckenrechnung folgt unmittelbar, dass das Flächenmass eines beliebigen Dreiecks gleich der Summe der Flächenmasse zweier solcher Dreiecke ist, die durch irgendwelche transversale Zerlegung aus jenem Dreieck hervorgehen. Die wiederholte Anwendung dieser Thatsache zeigt, dass das Flächenmass eines beliebigen Dreiecks auch gleich der Summe der Flächenmasse der sämtlichen Dreiecke ist, die aus dem vorgelegten Dreiekke entstehen, wenn man nach einander beliebig viele transversale Zerlegungen ausführt.

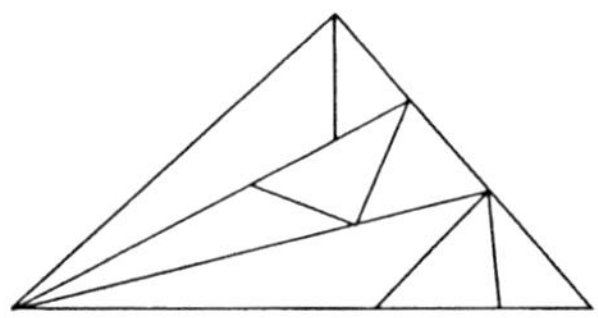

Fig. 32.

Um nun den entsprechenden Nachweis für die beliebige Zerlegung des Dreiecks Δ in Dreiecke Δ_k zu erbringen, ziehen wir von der einen Ecke A des Dreieckes Δ durch jeden Teilpunkt der Zerlegung, d. h. durch jeden Eckpunkt der Dreiecke Δ_k eine Transversale; durch diese Transversalen zerfällt das Dreieck Δ in gewisse Dreiecke Δ_t. Jedes dieser Dreiekke Δ_t zerfällt durch die Teilstrecken der gegebenen Zerlegung in gewisse Dreiecke und Vierecke. Wenn wir in den Vierecken noch je eine Diagonale construiren, so zerfällt jedes Dreieck Δ_t in gewisse Dreiecke Δ_{ts}. Wir wollen nun zeigen, dass die Zerlegung in Dreiecke Δ_{ts} sowohl für die Dreiecke Δ_t als auch für die Dreiecke Δ_k nichts anderes als eine Kette von transversalen Zerlegungen bedeutet.

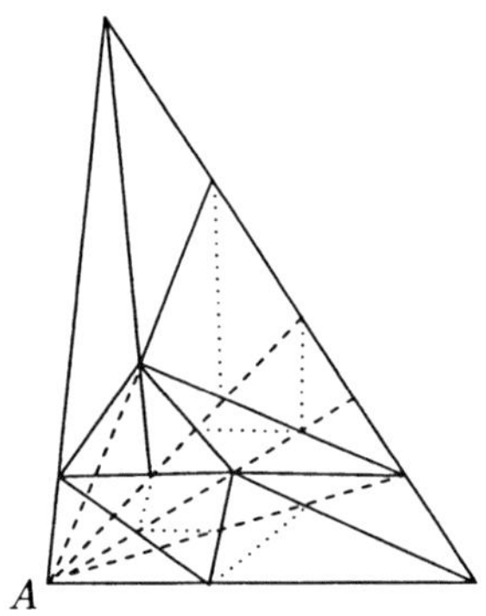

Fig. 33.

In der That, zunächst ist klar, dass jede Zerlegung eines Dreiecks in
45 Teildreiecke stets durch eine Reihe von transversalen | Zerlegungen bewirkt werden kann, wenn bei der Zerlegung im Inneren des Dreiecks keine Teilpunkte liegen und überdies wenigstens eine Seite des Dreiecks von Teilpunkten frei bleibt.

Nun ist für die Dreiecke Δ_t unsere Behauptung aus dem Umstande ersichtlich, dass für jedes derselben das Innere sowie eine Seite, nämlich die dem Punkte A gegenüberliegende Seite von weiteren Teilpunkten frei ist.

Aber auch für jedes Δ_k ist die Zerlegung in Δ_{ts} auf transversale Zerlegungen zurückführbar. Betrachten wir nämlich ein Dreieck Δ_k, so wird es unter den von A ausgehenden Transversalen im Dreieck Δ eine bestimmte Transversale geben, in welche entweder eine Seite von Δ_k hineinfällt oder welche selbst das Dreieck Δ_k in zwei Dreiecke zerlegt. Im ersten Fall bleibt die betreffende Seite des Dreiecks Δ_k überhaupt frei von weiteren Teilpunkten bei der Zerle-

44 Fall] Falle

gung in Dreiecke Δ_{ts}; im letzteren Falle ist die im Inneren des Dreiecks Δ_k verlaufende Strecke jener Transversale in den beiden entstehenden Dreiecken eine Seite, die bei der Teilung in Dreiecke Δ_{ts} von weiteren Teilpunkten gewiss frei bleibt.

Nach der am Anfang dieses Beweises angestellten Betrachtung ist das Flächenmass $F(\Delta)$ des Dreiecks Δ gleich der Summe aller Flächenmasse $F(\Delta_t)$ der Dreiecke Δ_t, und diese Summe ist gleich der Summe aller Flächenmasse $F(\Delta_{ts})$. Andererseits ist auch die Summe über die Flächenmasse $F(\Delta_k)$ aller Dreiecke Δ_k gleich der Summe aller Flächenmasse $F(\Delta_{ts})$, und hieraus folgt endlich, dass das Flächenmass $F(\Delta)$ auch gleich der Summe aller Flächenmasse $F(\Delta_k)$ ist. Damit ist der Beweis des Satzes 29 vollständig erbracht.

Definition. Definiren wir das Flächenmass $F(P)$ eines Polygons als die Summe der Flächenmasse aller Dreiecke, in die dasselbe bei einer bestimmten Zerlegung zerfällt, so erkennen wir auf Grund des Satzes 29 durch eine ähnliche Schlussweise, wie wir sie in § 18 beim Beweise des Satzes 24 angewandt haben, dass das Flächenmass eines Polygons von der Art der Zerlegung in Dreiecke unabhängig ist und mithin allein durch das Polygon sich eindeutig bestimmt. Aus dieser Definition entnehmen wir vermittels des Satzes 29 die Thatsache, dass flächengleiche Polygone gleiches Flächenmass haben.

Sind ferner P und Q zwei inhaltsgleiche Polygone, so muss es nach der Definition zwei flächengleiche Polygone P' und Q' geben, so dass das aus P und P' zusammengesetzte Polygon mit | dem aus Q und Q' zusammengesetz- 46
ten Polygon flächengleich ausfällt. Aus den beiden Gleichungen:

$$F(P+P') = F(Q+Q'),$$
$$F(P') = F(Q')$$

schliessen wir leicht

$$F(P) = F(Q),$$

d. h. inhaltsgleiche Polygone haben gleiches Flächenmass.

Aus der letzteren Thatsache entnehmen wir unmittelbar den Beweis des Satzes 28. Bezeichnen wir nämlich die gleiche Grundlinie der beiden Dreiecke mit g, die zugehörigen Höhen mit h und h', so schliessen wir aus der angenommenen Inhaltsgleichheit der beiden Dreiecke, dass dieselben auch gleiches

2 Transversale] Transversalen
12 Definition.] Erklärung.
18 Definition] Erklärung
19 flächengleiche] zerlegungsgleiche
22 Definition] Erklärung
22 flächengleiche] zerlegungsgleiche
23 Polygon] Polygon $(P+P')$
24 Polygon] Polygon $(Q+Q')$
24 flächengleich] zerlegungsgleich
29 haben] haben
30 der letzteren] dieser
30–482.5 ⟨In E2 this paragraph is moved to the beginning of the next section (§ 21).⟩

Flächenmass haben müssen, d. h. es folgt:

$$\tfrac{1}{2}gh = \tfrac{1}{2}gh'$$

und mithin nach Division durch $\frac{1}{2}g$

$$h = h';$$

dies ist die Aussage des Satzes 28.

§ 21.

Die Inhaltsgleichheit und das Flächenmass.

In § 20 haben wir gefunden, dass inhaltsgleiche Polygone stets gleiches Flächenmass haben. Diese Aussage lässt sich auch umkehren.

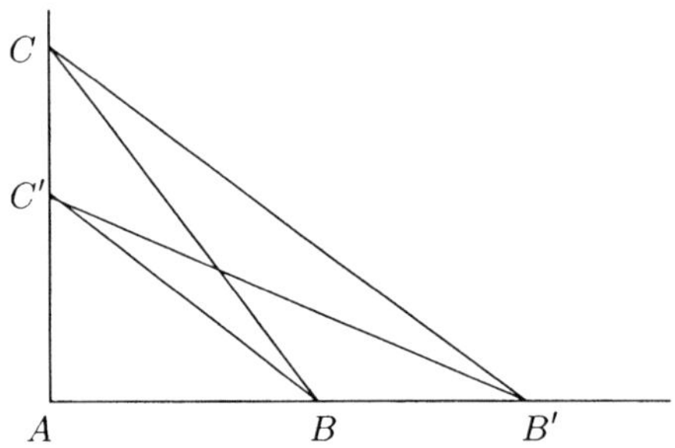

Fig. 34.

Um diese Umkehrung zu beweisen, betrachten wir zunächst zwei Dreiecke ABC und $AB'C'$ mit gemeinsamem rechten Winkel bei A. Die Flächenmasse dieser beiden Dreiecke drücken sich durch die Formeln

$$F(ABC) = \tfrac{1}{2}AB \cdot AC,$$
$$F(AB'C') = \tfrac{1}{2}AB' \cdot AC'$$

47 aus. Nehmen wir an, dass diese beiden Flächenmasse einander gleich sind, so folgt:

$$AB \cdot AC = AB' \cdot AC'$$

oder

$$AB : AB' = AC' : AC$$

und hieraus ergiebt sich nach Satz 23, dass die beiden Geraden BC' und $B'C$ einander parallel sind und sodann erweisen sich nach Satz 27 die beiden Dreiecke $BC'B'$ und $BC'C$ als inhaltsgleich. Durch Hinzufügen des Dreiecks ABC' folgt, dass die beiden Dreiecke ABC und $AB'C'$ einander inhaltsgleich sind. Wir haben damit erkannt, dass zwei rechtwinklige Dreiecke mit gleichem Flächenmass auch stets einander inhaltsgleich sind.

Nehmen wir jetzt ein beliebiges Dreieck mit der Grundlinie g und der Höhe h, so ist dasselbe nach Satz 27 inhaltsgleich einem rechtwinkligen Dreieck mit den beiden Katheten g und h; und da das ursprüngliche Dreieck offenbar dasselbe Flächenmass wie das rechtwinklige Dreieck besitzt, so folgt, dass in

---

7 **Flächenmass] Inhaltsmaß**

9 Flächenmass] Inhaltsmaß

9 ⟨In E2 after 'haben.' the last paragraph of the previous section (§ 20) is added.⟩

9–11 Diese Aussage lässt sich auch umkehren.] Auch läßt sich nunmehr die am Schluß von § 20 gemachte Aussage umkehren.

12–483.3 Um ⟨...⟩ inhaltsgleich sind.] ⟨deleted⟩

der letzten Betrachtung die Einschränkung auf rechtwinklige Dreiecke nicht nötig war; damit haben wir erkannt, dass zwei beliebige Dreiecke mit gleichem Flächenmass auch stets einander inhaltsgleich sind.

Es sei nunmehr ein beliebiges Polygon P mit dem Flächenmass g vorgelegt; P zerfalle in n Dreiecke mit den Flächenmassen bez. $g_1, g_2, \ldots, g_n$; dann ist

$$g = g_1 + g_2 + \cdots + g_n.$$

Wir construiren nun ein Dreieck ABC mit der Grundlinie $AB = g$ und der

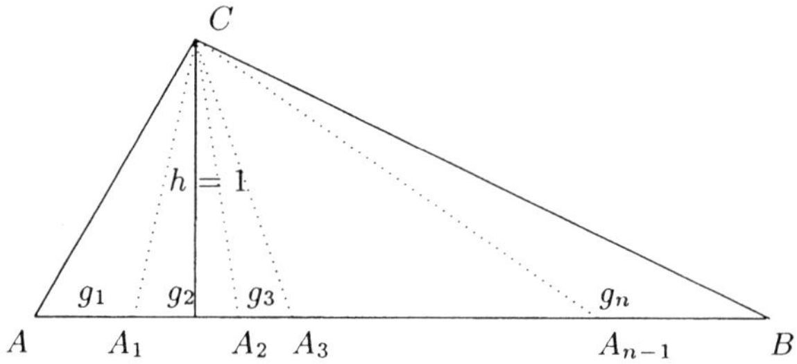

Fig. 35.

Höhe $h = 1$ und markiren auf der Grundlinie die Punkte $A_1, A_2, \ldots, A_{n-1}$, sodass

$$g_1 = AA_1, \quad g_2 = A_1A_2, \ldots, \quad g_{n-1} = A_{n-2}A_{n-1}, \quad g_n = A_{n-1}B$$

ausfällt. Da die Dreiecke innerhalb des Polygons P bez. die | gleichen Flächen- 48
masse besitzen, wie die Dreiecke AA_1C, A_1A_2C, ..., $A_{n-2}A_{n-1}C$, $A_{n-1}BC$, so sind sie nach dem vorhin Bewiesenen diesen auch inhaltsgleich; folglich ist das Polygon P einem Dreiecke mit der Grundlinie g und der Höhe $h = 1$ inhaltsgleich. Hieraus folgt mit Hülfe von Satz 24, dass zwei Polygone mit gleichem Flächenmass einander stets inhaltsgleich sind.

Die beiden in diesem und dem vorigen Paragraph gefundenen Thatsachen fassen wir in den folgenden Satz zusammen:

Satz 30. *Zwei inhaltsgleiche Polygone haben stets das gleiche Flächenmass*; und umgekehrt: *Zwei Polygone mit gleichem Flächenmass sind stets einander inhaltsgleich.*

4–17 Es sei ⟨...⟩ inhaltsgleich sind.] In der Tat, seien P und Q zwei Polygone mit gleichem Inhaltsmaß, so konstruieren wir gemäß Satz 28 zwei rechtwinklige Dreiecke Δ und E, von denen jedes eine Kathete 1 besitzt und so daß das Dreieck Δ mit dem Polygon P und das Dreieck E mit dem Polygon Q inhaltsgleich ist. Aus dem am Schluß von § 20 bewiesenen Satze folgt dann, daß Δ mit P und ebenso E mit Q gleiche Inhaltsmaße haben. Wegen des gleichen Inhaltsmaßes von P und Q folgt hieraus, daß auch Δ und E gleiches Inhaltsmaß haben. Da nun diese beiden rechtwinkligen Dreiecke in der Kathete 1 übereinstimmen, so stimmen notwendig auch ihre anderen Katheten überein, d. h. die beiden Dreiecke Δ und E sind einander kongruent und mithin sind nach Satz 24 die beiden Polygone P und Q einander inhaltsgleich.

20–21 *Flächenmass*; und umgekehrt: *Zwei*] *Inhaltsmaß und zwei*

21 *Flächenmass*] *Inhaltsmaß*

Insbesondere müssen zwei inhaltsgleiche Rechtecke mit einer gemeinsamen Seite auch in den anderen Seiten übereinstimmen. Auch folgt der Satz:

S a t z 31. Zerlegt man ein Rechteck durch Gerade in mehrere Dreiecke und lässt auch nur eines dieser Dreiecke fort, so kann man mit den übrigen Dreiecken das Rechteck nicht mehr ausfüllen.

Dieser Satz 31 ist von *F. Schur* 1) und *W. Killing* 2) mit Hülfe des Archimedischen Axioms bewiesen und von *O. Stolz* 3) als Axiom hingestellt worden. Im Vorstehenden ist gezeigt, dass derselbe völlig unabhängig von dem Archimedischen Axiom gilt. Der Satz 31 reicht übrigens zur Begründung des Euklidischen Satzes von der Gleichheit der Höhen in inhaltsgleichen Dreiekken auf gleicher Grundlinie (Satz 28) an sich nicht aus, wenn man von der Anwendung des Archimedischen Axioms absieht.

Zum Beweise der Sätze 28, 29, 30 haben wir wesentlich die in Kapitel III § 15 eingeführte Streckenrechnung benutzt, und da diese im Wesentlichen auf dem Pascalschen Satze (Satz 21) beruht, so erweist sich für die Lehre von den Flächeninhalten der Pascalsche Satz als der wichtigste Baustein. Wir erkennen leicht, dass auch umgekehrt aus den Sätzen 27 und 28 der Pascalsche Satz wieder gewonnen werden kann.

---

1) Sitzungsberichte der Dorpater Naturf. Ges. 1892.[19]

2) Grundlagen der Geometrie, Bd. 2, Absch. 5, § 5, 1898.[20]

3) Monatshefte für Math. und Phys., Jahrgang 5, 1894.[21]

---

5 ausfüllen.] ⟨After 'ausfüllen' a footnote is added:⟩ Hinsichtlich der merkwürdigen Rolle, die dieser Satz 32 ⟨31⟩ behufs Ergänzung der sogenannten „Kongruenzsätze im engeren Sinne" spielt, vergleiche man den Schluß von Anhang II S. 105–106.

6 Dieser ⟨...⟩ worden.] Dieser Satz ist von *de Zolt* 1) und von *O. Stolz* 2) als Axiom hingestellt und von *F. Schur* 3) und *W. Killing* 4) mit Hilfe des Archimedischen Axioms bewiesen worden.

1) Principii della eguaglianza di polizoni preceduti da alcuni critici sulla teoria della equivalenza geometrica. Milano, Briola 1881. Vgl. auch Principii della eguaglianza di poliedri e di polizoni sferici. Milano, Briola 1883.[22]

2) Monatshefte für Math. und Phys., Jahrgang 5, 1894.[23]

3) Sitzungsberichte der Dorpater Naturf. Ges. 1892.[24]

4) Grundlagen der Geometrie, Bd. 2, Abschnitt 5, § 5, 1898.[25]

9–12 Der ⟨...⟩ absieht.] ⟨deleted⟩

16 ⟨In E2, a new paragraph begins after 'Baustein'.⟩

18 ⟨After this line the following paragraph is added in E2:⟩

[19] *Schur 1893.*

[20] *Killing 1898.*

[21] *Stolz 1894.*

[22] *de Zolt 1881, 1883.*

[23] *Stolz 1894.*

[24] *Schur 1893.*

[25] *Killing 1898.*

Von zwei Polygonen P und Q nennen wir P *inhaltskleiner* bez. *inhaltsgrösser* als Q, je nachdem das Flächenmass $F(P)$ kleiner oder grösser als $F(Q)$ ausfällt. Es ist nach dem Vorstehenden klar, dass die Begriffe inhaltsgleich, inhaltskleiner und inhaltsgrösser sich gegenseitig ausschliessen. Ferner erkennen wir leicht, | dass ein Polygon, welches ganz im Inneren eines anderen Polygons 49
liegt, stets inhaltskleiner als dieses sein muss.

Hiermit haben wir die wesentlichen Sätze der Lehre von den Flächeninhalten begründet.

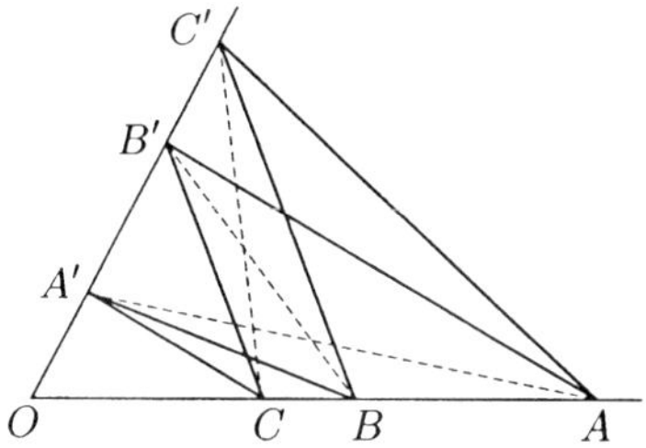

In der Tat, aus der Parallelität der Geraden CB' und $C'B$ folgt nach Satz 27 die Inhaltsgleichheit der Dreiecke OBB' und OCC'; ebenso folgt aus der Parallelität der Geraden CA' und AC' die Inhaltsgleichheit der Dreiecke OAA' und OCC' Da hiernach auch die Dreiecke OAA' und OBB' einander inhaltsgleich sind, so ergibt Satz 29, daß auch BA' zu AB' parallel sein muß.

1 Polygonen] nicht inhaltsgleichen Polygonen

2 Flächenmass $F(P)$] Inhaltsmaß $J(P)$

2–3 $F(Q)$ ausfällt.] $J(Q)$ ausfällt.

7–8 Flächeninhalten] Flächeninhalten in der Ebene

8 ⟨After this line the following paragraphs are added in E2:⟩

Bereits Gauß hat die Aufmerksamkeit der Mathematiker auf die entsprechende Frage für den Raum gelenkt. Ich habe die Vermutung der Unmöglichkeit einer analogen Begründung der Lehre von den Inhalten im Raume ausgesprochen und die bestimmte Aufgabe 1) gestellt, zwei Tetraeder mit gleicher Grundfläche und von gleicher Höhe anzugeben, die sich auf keine Weise in kongruente Tetraeder zerlegen lassen, und die sich auch durch Hinzufügung kongruenter Tetraeder nicht zu solchen Polyedern ergänzen lassen, für die ihrerseits eine Zerlegung in kongruente Tetraeder möglich ist.

M. Dehn 2) ist dieser Nachweis in der Tat gelungen; er hat damit in strenger Weise die Unmöglichkeit dargetan, die Lehre von den räumlichen Inhalten so zu begründen, wie dies im vorstehenden für die ebenen Inhalte geschehen ist.

Hiernach wären zur Behandlung der analogen Fragen für den Raum andere Hilfsmittel, etwa das Cavalierische Prinzip heranzuziehen.

1) Vgl. meinen Vortrag „Mathematische Probleme“[26] Nr. 3.

2) Über raumgleiche Polyeder, Göttinger Nachr. 1900, sowie Über den Rauminhalt, Math. Ann. Bd. 55. 1902. Vgl. ferner Kagan, Math. Ann. Bd. 57.[27]

[26] *Hilbert 1900e.*

[27] *Dehn 1900c, 1902, Kagan 1903.*

Kapitel V.

Der Desarguessche Satz.

§ 22.

Der Desarguessche Satz und der Beweis desselben in der Ebene mit Hülfe der Congruenzaxiome.

Von den im Kapitel I aufgestellten Axiomen sind diejenigen der Gruppen II–V sämtlich teils lineare, teils ebene Axiome; die Axiome 3–7 der Gruppe I sind die einzigen räumlichen Axiome. Um die Bedeutung dieser räumlichen Axiome klar zu erkennen, denken wir uns irgend eine e b e n e Geometrie vorgelegt und untersuchen allgemein die Bedingungen dafür, dass diese ebene Geometrie sich als Teil einer räumlichen Geometrie auffassen lässt, in welcher wenigstens die Axiome der Gruppen I–III sämtlich erfüllt sind.

Auf Grund der Axiome der Gruppen I–III ist es bekanntlich leicht möglich, den sogenannten Desarguesschen Satz zu beweisen; derselbe ist ein ebener Schnittpunktsatz. Wir nehmen insbesondere die Gerade, auf der die Schnittpunkte entsprechender Seiten der beiden Dreiecke liegen sollen, zur „Unendlichfernen", wie man sagt, und bezeichnen den so entstehenden Satz nebst seiner Umkehrung schlechthin als Desarguesschen Satz; dieser Satz lautet, wie folgt:

S a t z 32. (D e s a r g u e s s c h e r S a t z.) Wenn zwei Dreiecke in einer Ebene so gelegen sind, dass je zwei entsprechende Seiten einander parallel sind, so laufen die Verbindungslinien der entsprechenden Ecken durch ein und denselben Punkt oder sind einander parallel, und umgekehrt:

Wenn zwei Dreiecke in einer Ebene so gelegen sind, dass die Verbindungs-
50 linien der entsprechenden Ecken durch einen Punkt | laufen oder einander

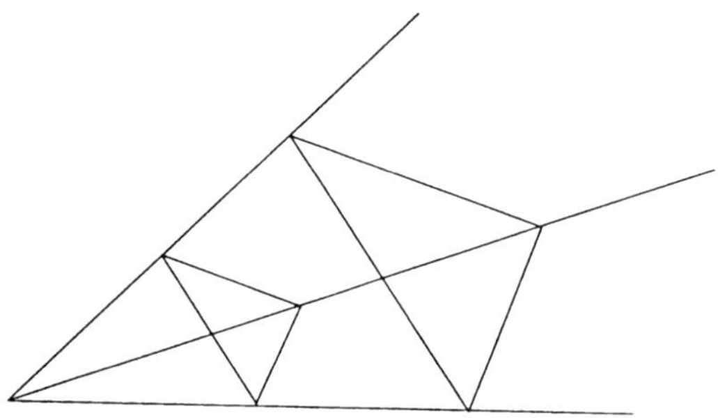

parallel sind, und wenn ferner zwei Paare entsprechender Seiten in den Dreiecken einander parallel sind, so sind auch die dritten Seiten der beiden Dreiecke einander parallel.

12 wenigstens] ⟨deleted⟩

Wie bereits erwähnt, ist der Satz 32 eine Folge der Axiome I–III; dieser Thatsache gemäss ist die Gültigkeit des Desarguesschen Satzes in der Ebene jedenfalls eine notwendige Bedingung dafür, dass die Geometrie dieser Ebene sich als Teil einer räumlichen Geometrie auffassen lässt, in welcher die Axiome der Gruppen I–III sämtlich erfüllt sind.

Wir nehmen nun wie in den Kapiteln III und IV eine ebene Geometrie an, in welcher die Axiome I 1–2 und II–IV gelten, und denken uns in derselben nach § 15 eine Streckenrechnung eingeführt: dann lässt sich, wie in § 17 dargelegt worden ist, jedem Punkte der Ebene ein Paar von Strecken (x, y) und jeder Geraden ein Verhältnis von drei Strecken $(u : v : w)$ zuordnen derart, dass die lineare Gleichung

$$ux + vy + w = 0$$

die Bedingung für die vereinigte Lage von Punkt und Gerade darstellt. Das System aller Strecken in unserer Geometrie bildet nach § 17 einen Zahlenbereich, für welchen die in § 13 aufgezählten Eigenschaften 1–16 bestehen, und wir können daher mittelst dieses Zahlenbereiches, ähnlich wie es in § 9 oder § 12 mittelst des Zahlensystems Ω bez. $\Omega(t)$ geschehen ist, eine räumliche Geometrie construiren: wir setzen zu dem Zwecke fest, dass ein System von drei Strecken (x, y, z) einen Punkt, die Verhältnisse von vier Strecken $(u : v : w : r)$ eine Ebene darstellen möge, während die Geraden als Schnitte zweier Ebenen definirt sind; dabei drückt die lineare Gleichung

$$ux + vy + wz + r = 0$$

aus, dass der Punkt (x, y, z) auf der Ebene $(u : v : w : r)$ liegt. Was endlich die Anordnung der Punkte auf einer Geraden oder | der Punkte einer Ebene in 51
Bezug auf eine Gerade in ihr oder endlich die Anordnung der Punkte in Bezug auf eine Ebene im Raume anbetrifft, so wird diese in analoger Weise durch Ungleichungen zwischen Strecken bestimmt, wie dies in § 9 für die Ebene geschehen ist.

Da wir durch das Einsetzen des Wertes $z = 0$ die ursprüngliche ebene Geometrie wieder gewinnen, so erkennen wir, dass unsere ebene Geometrie als Teil einer räumlichen Geometrie betrachtet werden kann. Nun ist hierfür die Gültigkeit des Desarguesschen Satzes nach den obigen Ausführungen eine notwendige Bedingung, und daher folgt, dass in der angenommenen ebenen Geometrie auch der Desarguessche Satz gelten muss.

Wir bemerken, dass die eben gefundene Thatsache sich auch direkt aus dem Satze 23 in der Lehre von den Proportionen ohne Mühe ableiten lässt.

§ 23.

Die Nichtbeweisbarkeit des Desarguesschen Satzes in der Ebene ohne Hülfe der Congruenzaxiome.

Wir untersuchen nun die Frage, ob in der ebenen Geometrie auch ohne Hülfe der Congruenzaxiome der Desarguessche Satz bewiesen werden kann, und gelangen dabei zu folgendem Resultate:

---

10 $(u : v : w)$] $(u : v : w)$, wobei u, v nicht beide Null sind,

Satz 33. *Es giebt eine ebene Geometrie, in welcher die Axiome* I 1–2, II–III, IV 1–5, V, *d. h. sämtliche linearen und ebenen Axiome mit Ausnahme des Congruenzaxioms* IV 6 *erfüllt sind, während der Desarguessche Satz* (Satz 32) *nicht gilt. Der Desarguessche Satz kann mithin aus den genannten Axiomen allein nicht gefolgert werden: es bedarf zum Beweise desselben notwendig entweder der räumlichen Axiome oder der sämtlichen Congruenzaxiome.*

Beweis. Wir wählen in der gewöhnlichen ebenen Geometrie, deren Möglichkeit bereits im Kapitel II § 9 erkannt worden ist, irgend zwei zu einander senkrechte Gerade als Coordinatenaxen X, Y und denken uns um den Nullpunkt O dieses Coordinatensystems als Mittelpunkt eine Ellipse mit den Halbaxen 1 und $\frac{1}{2}$ construirt; endlich bezeichnen wir mit F den Punkt, welcher in der Entfernung $\frac{3}{2}$ von O auf der positiven X-Axe liegt.

Wir fassen nun die Gesamtheit aller Kreise ins Auge, welche die Ellipse in vier reellen – getrennten oder beliebig zusammenfallenden – Punkten schneiden, und suchen unter allen auf diesen Kreisen gelegenen Punkten denjenigen Punkt zu bestimmen, der auf der positiven X-Axe am weitesten vom
52 Nullpunkt entfernt | ist. Zu dem Zwecke gehen wir von einem beliebigen Kreise aus, der die Ellipse in vier Punkten schneidet und die positive X-Axe im Punkte C treffen möge. Diesen Kreis denken wir uns dann um den Punkt C derart gedreht, dass zwei von den vier Schnittpunkten oder mehr in einen einzigen Punkt A zusammenfallen, während die übrigen reell bleiben. Der so entstehende Berührungskreis werde alsdann vergrössert derart, dass stets der Punkt A Berührungspunkt mit der Ellipse bleibt; hierdurch gelangen wir notwendig zu einem Kreise, der die Ellipse entweder noch in einem anderen Punkte B berührt oder in A mit der Ellipse eine vierpunktige Berührung aufweist und der überdies die positive X-Axe in einem entfernteren Punkte als C trifft. Der gesuchte entfernteste Punkt wird sich demnach unter denjenigen Punkten befinden, die von den doppeltberührenden ausserhalb der Ellipse verlaufenden Kreisen auf der positiven X-Axe ausgeschnitten werden. Die doppeltberührenden ausserhalb der Ellipse verlaufenden Kreise liegen nun, wie man leicht sieht, sämtlich zur Y-Axe symmetrisch. Es seien a, b die Coordinaten irgend eines Punktes der Ellipse: dann lehrt eine leichte Rechnung, dass der in diesem Punkte berührende zur Y-Axe symmetrische Kreis auf der positiven X-Axe die Strecke

$$x = \left|\sqrt{1+3b^2}\right|$$

abschneidet. Der grösstmögliche Wert dieses Ausdrucks tritt für $b = \frac{1}{2}$ ein und wird somit gleich $\frac{1}{2}|\sqrt{7}|$. Da der vorhin mit F bezeichnete Punkt auf der X-Axe die Abscisse $\frac{3}{2} > \frac{1}{2}|\sqrt{7}|$ aufweist, so folgt, dass unter den die Ellipse viermal treffenden Kreisen sich gewiss keiner befindet, der durch den Punkt F läuft.

---

4 (Satz 32)] ⟨This should be '(Satz 33)' in E2 but was erroneously left unchanged.⟩

6–7 *der ⟨...⟩ Congruenzaxiome*] *des Axioms III 6 über die Kongruenz der Dreiecke*

11 Ellipse] Ellipse z. B.

Nunmehr stellen wir uns eine neue ebene Geometrie in folgender Weise her. Als Punkte der neuen Geometrie nehmen wir die Punkte der XY-Ebene; als Gerade der neuen Geometrie nehmen wir diejenigen Geraden der XY-Ebene unverändert, welche die feste Ellipse berühren oder gar nicht treffen; ist dagegen g eine Gerade der XY-Ebene, die die Ellipse in zwei Punkten P und Q trifft, so construiren wir durch P, Q und den festen Punkt F einen Kreis; dieser Kreis hat, wie aus dem oben Bewiesenen hervorgeht, keinen weiteren Punkt mit der Ellipse gemein. Wir denken uns nun an Stelle des zwischen P und Q innerhalb der Ellipse gelegenen Stückes der Geraden g denjenigen Bogen des eben construirten Kreises gesetzt, der zwischen P und Q innerhalb der Ellipse verläuft. Den Linienzug, welcher aus den beiden | von P und Q 53
ausgehenden unendlichen Stücken der Geraden g und dem eben construirten Kreisbogen PQ besteht, nehmen wir als Gerade der neu herzustellenden Geometrie. Denken wir uns für alle Geraden der XY-Ebene die entsprechenden Linienzüge construirt, so entsteht ein System von Linienzügen, welche, als Gerade einer Geometrie aufgefasst, offenbar den Axiomen I 1–2 und III genügen. Bei Festsetzung der natürlichen Anordnung für die Punkte und Geraden in unserer neuen Geometrie erkennen wir unmittelbar, dass auch die Axiome II gültig sind.

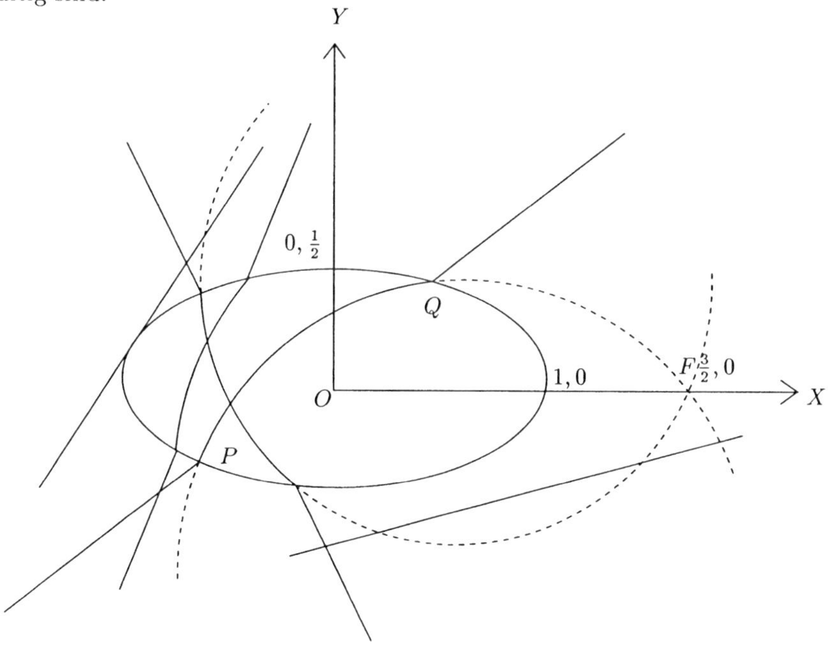

Fig. 36.

Ferner nennen wir zwei Strecken AB und $A'B'$ in unserer neuen Geometrie congruent, wenn der zwischen A und B verlaufende Linienzug die gleiche natürliche Länge hat, wie der zwischen A' und B' verlaufende Linienzug.

16 Axiomen I 1–2 und III] ⟨Wrongly this way in E2, instead of 'Axiomen I 1–3 und IV'.⟩

Endlich bedürfen wir einer Festsetzung betreffs der Congruenz der Winkel. Sobald keiner von den Scheiteln der zu vergleichenden Winkel auf der Ellipse liegt, nennen wir zwei Winkel einander congruent, wenn sie im gewöhnlichen Sinne einander gleich sind. Im anderen Falle treffen wir folgende
54 Festsetzung. Es | mögen die Punkte A, B, C in dieser Reihenfolge und die Punkte A', $B' C'$ in dieser Reihenfolge je auf einer Geraden unserer neuen Geometrie liegen; D sei ein Punkt ausserhalb der Geraden ABC und D' ein Punkt ausserhalb der Geraden $A'B'C'$: dann mögen für die Winkel zwischen diesen Geraden in unserer neuen Geometrie die Congruenzen

$$\sphericalangle ABD \equiv \sphericalangle A'B'D' \quad \text{und} \quad \sphericalangle CBD \equiv \sphericalangle C'B'D'$$

gelten, sobald für die natürlichen Winkel zwischen den entsprechenden Linienzügen in der gewöhnlichen Geometrie die Proportion

$$\sphericalangle ABD : \sphericalangle CBD = \sphericalangle A'B'D' : \sphericalangle C'B'D'$$

erfüllt ist. Bei diesen Festsetzungen sind auch die Axiome IV 1–5 und V gültig.

Um einzusehen, dass in unserer neu hergestellten Geometrie der Desarguessche Satz nicht gilt, betrachten wir folgende drei gewöhnliche gerade Linien in der XY-Ebene: die X-Axe, die Y-Axe und die Gerade, welche die

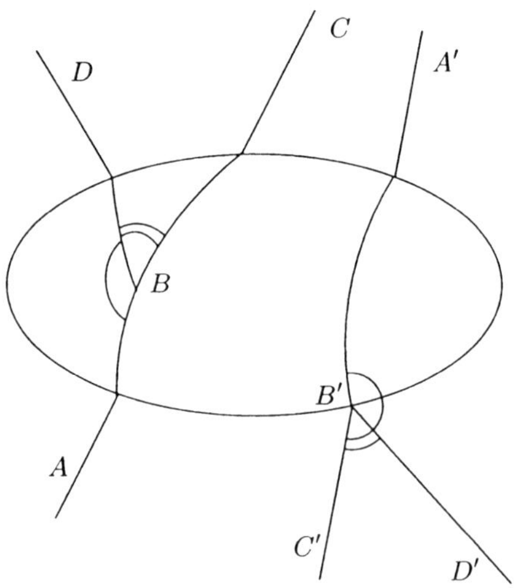

Fig. 37.

beiden Ellipsenpunkte $x = \frac{3}{5}$, $y = \frac{2}{5}$ und $x = -\frac{3}{5}$, $y = -\frac{2}{5}$ mit einander verbindet. Da diese drei gewöhnlichen geraden Linien durch den Nullpunkt O laufen, so können wir leicht zwei solche Dreiecke angeben, deren Ecken bez. auf jenen drei Geraden liegen, deren entsprechende Seiten paarweise einander parallel laufen und die sämtlich ausserhalb der Ellipse gelegen sind. Da die Linienzüge,

6 A', $B' C'$] A', B', C'
10 $\sphericalangle ABD$ $\langle \dots \rangle$ $\sphericalangle C'B'D'$] $\sphericalangle ABD = \sphericalangle A'B'D'$ und $\sphericalangle CBD = \sphericalangle C'B'D'$

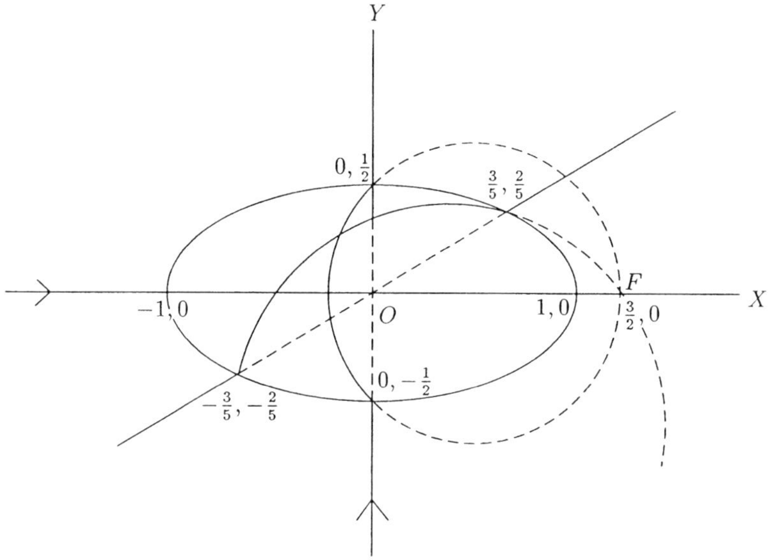

Fig. 38.

welche aus den genannten drei geraden Linien entspringen, sich, wie Figur 38 | zeigt und wie man leicht durch Rechnung bestätigt, nicht in einem Punkte 55 treffen, so folgt, dass der Desarguessche Satz in der neuen ebenen Geometrie für die beiden vorhin construirten Dreiecke gewiss nicht gilt.

Die von uns hergestellte ebene Geometrie dient zugleich als Beispiel einer ebenen Geometrie, in welcher die Axiome I 1–2, II–III, IV 1–5, V gültig sind, und die sich dennoch n i c h t als Teil einer räumlichen Geometrie auffassen lässt.

§ 24.

Einführung einer Streckenrechnung ohne Hülfe der Congruenzaxiome auf Grund des Desarguesschen Satzes.

Um die Bedeutung des Desarguesschen Satzes (Satz 32) vollständig zu erkennen, legen wir eine ebene Geometrie zu Grunde, in welcher die Axiome I 1–2, II–III, d. h. die sämtlichen ebenen Axiome der ersten drei Axiomgruppen gültig sind, und führen in diese Geometrie u n a b h ä n g i g v o n d e n C o n g r u e n z a x i o m e n auf folgende Weise eine neue Streckenrechnung ein.

Wir nehmen in der Ebene zwei feste Geraden an, die sich in dem Punkte O schneiden mögen, und rechnen im Folgenden nur mit solchen Strecken, deren Anfangspunkt O ist und deren Endpunkte auf einer dieser beiden festen

1 Figur 38] die Figur auf S. 53

7 dennoch] doch

14 ebenen] linearen und ebenen

14 der ersten drei Axiomgruppen] außer den Kongruenz- und Stetigkeitsaxiomen

56 Geraden liegen. Auch | den Punkt O allein bezeichnen wir als Strecke und nennen ihn dann die Strecke 0, in Zeichen

$$OO = 0 \quad \text{oder} \quad 0 = OO.$$

Es seien E und E' je ein bestimmter Punkt auf den festen Geraden durch O; dann bezeichnen wir die beiden Strecken OE und OE' als die Strekken 1, in Zeichen:

$$OE = OE' = 1 \quad \text{oder} \quad 1 = OE = OE'.$$

Die Gerade EE' nennen wir kurz die Einheitsgerade. Sind ferner A und A' Punkte auf den Geraden OE bez. OE' und läuft die Verbindungsgerade AA' parallel zu EE', so nennen wir die Strecken OA und OA' einander gleich, in Zeichen:

$$OA = OA' \quad \text{oder} \quad OA' = OA.$$

Um zunächst die Summe der Strecken $a = OA$ und $b = OB$ zu definiren, construire man AA' parallel zur Einheitsgeraden EE' und ziehe sodann durch A' eine Parallele zu OE und durch B eine Parallele zu OE'. Diese beiden Parallelen mögen sich in A'' schneiden. Endlich ziehe man durch A'' zur Einheitsgeraden EE' eine Parallele; dieselbe treffe die festen Geraden OE und OE' in C und C': dann heisse $c = OC = OC'$ die *Summe* der Strecke $a = OA$ mit der Strecke $b = OB$, in Zeichen:

$$c = a + b \quad \text{oder} \quad a + b = c.$$

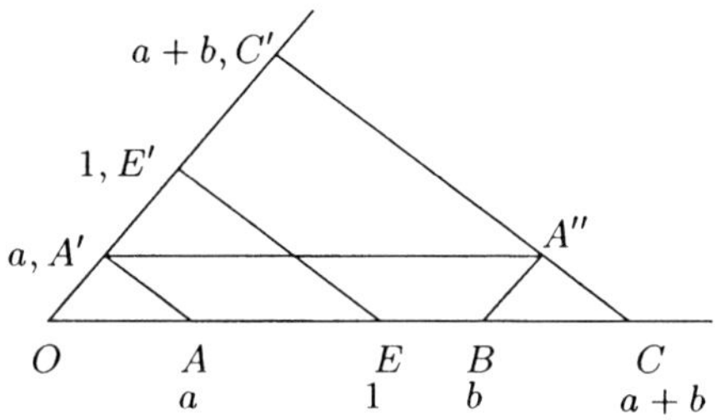

Fig. 39.

Um das Produkt einer Strecke $a = OA$ in eine Strecke $b = OB$ zu definiren, bedienen wir uns genau der in § 15 angegebenen Konstruktion, nur dass an Stelle der Schenkel des rechten Winkels hier die beiden festen Geraden OE und OE' treten. Die Konstruktion ist demnach folgende. Man bestimme auf OE' den Punkt A', sodass AA' parallel der Einheitsgeraden EE' wird,
57 ver|binde E mit A' und ziehe durch B eine Parallele zu EA'; trifft diese Parallele die feste Gerade OE' im Punkte C', so heisst $c = OC'$ das Produkt der Strecke $a = OA$ in die Strecke $b = OB$, in Zeichen:

$$c = ab \quad \text{oder} \quad ab = c.$$

---

1 liegen] beliebig liegen

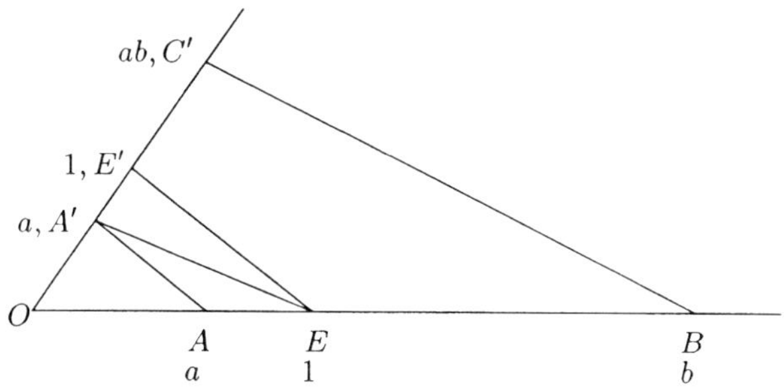

Fig. 40.

§ 25.

Das commutative und associative Gesetz der Addition in der neuen Streckenrechnung.

Wir untersuchen jetzt, welche von den in § 13 aufgestellten Rechnungsgesetzen für unsere neue Streckenrechnung gültig sind, wenn wir eine ebene Geometrie zu Grunde legen, in der die Axiome I 1–2, II–III erfüllt sind und überdies der Desarguessche Satz gilt.

Vor Allem wollen wir beweisen, dass für die in § 24 definirte Addition der Strecken das commutative Gesetz

$$a + b = b + a$$

gilt. Es sei

$$a = OA = OA',$$
$$b = OB = OB',$$

wobei unserer Festsetzung entsprechend AA' und BB' der Einheitsgeraden parallel sind. Nun construiren wir die Punkte A'' und B'', indem wir $A'A''$ sowie $B'B''$ parallel OA und ferner AB'' und BA'' parallel OA' ziehen; wie man sofort sieht, sagt dann unsere Behauptung aus, dass die Verbindungslinie $A''B''$ parallel mit AA' läuft. Die Richtigkeit dieser Behauptung erkennen wir auf Grund des Desarguesschen Satzes (Satz 32) wie folgt. | Wir bezeichnen 58
den Schnittpunkt von AB'' und $A'A''$ mit F und den Schnittpunkt von BA'' und $B'B''$ mit D; dann sind in den Dreiecken $AA'F$ und $BB'D$ die entsprechenden Seiten einander parallel. Mittelst des Desarguesschen Satzes schlies-

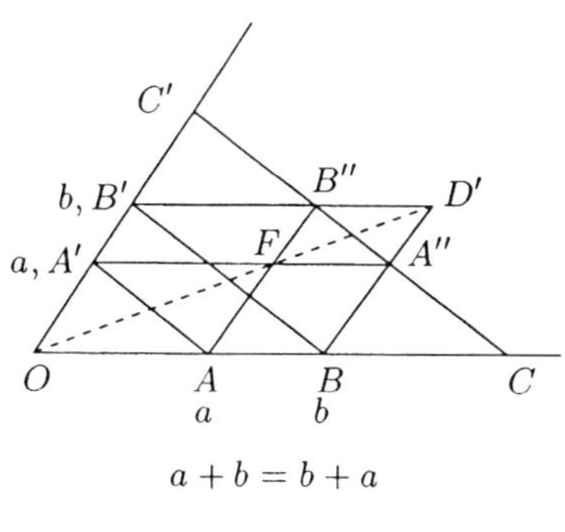

$a + b = b + a$

Fig. 41.

6 II–III] II–IV ⟨Wrongly, instead of 'II, IV'.⟩

7 überdies der Desarguessche Satz gilt] überdies der Desarguessche Satz gilt

22 (Satz 32)] ⟨Wrongly, instead of '(Satz 33)'.⟩

sen wir hieraus, dass die drei Punkte O, F, D in einer Geraden liegen. In Folge dieses Umstandes liegen die beiden Dreiecke OAA' und $DB''A''$ derart, dass die Verbindungslinien entsprechender Ecken durch den nämlichen Punkt F laufen und da überdies zwei Paare entsprechender Seiten, nämlich OA und DB'' sowie OA' und DA'' einander parallel sind, so laufen nach der zweiten Aussage des Desarguesschen Satzes (Satz 32) auch die dritten Seiten AA' und $B''A''$ einander parallel.

Zum Beweise des associativen Gesetzes der Addition

$$a + (b + c) = (a + b) + c$$

dient die Figur 42. Mit Berücksichtigung des eben bewiesenen commutativen Gesetzes der Addition spricht sich die obige Behauptung, wie man sieht, darin aus, dass die Gerade $A''B''$ parallel der Einheitsgeraden verlaufen muss. Die

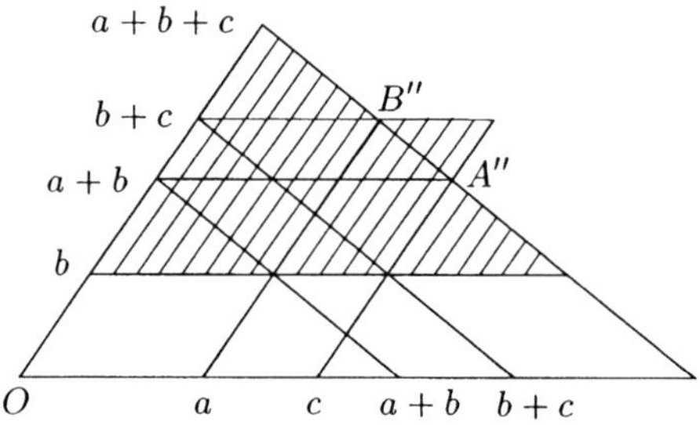

$$a + (b + c) = (a + b) + c$$

Fig. 42.

Richtigkeit dieser Behauptung ist offenbar, da der schraffirte Teil der Figur 42 mit der Figur 41 genau übereinstimmt.

59 § 26.

Das associative Gesetz der Multiplikation und die beiden distributiven Gesetze in der neuen Streckenrechnung.

Bei unseren Annahmen gilt auch für die Multiplikation der Strecken das associative Gesetz:

$$a(bc) = (ab)c.$$

Es seien auf der ersteren der beiden festen Geraden durch O die Strecken

$$1 = OA, \quad b = OC, \quad c = OA'$$

---

6 (Satz 32)] ⟨Erroneously for '(Satz 33)'.⟩

10 Figur 42] Figur auf S. 56

13 der Figur 42] dieser Figur

14 Figur 41] vorhergehenden Figur

und auf der anderen Geraden die Strecken

$$a = OG \quad \text{und} \quad b = OB$$

gegeben. Um gemäss der Vorschrift in § 24 der Reihe nach die Strecken

$$bc = OB' \text{ und } bc = OC',$$
$$ab = OD,$$
$$(ab)c = OD'$$

zu construiren, ziehen wir $A'B'$ parallel AB, $B'C'$ parallel BC, CD parallel AG sowie $A'D'$ parallel AD; wie wir sofort erkennen, läuft dann unsere

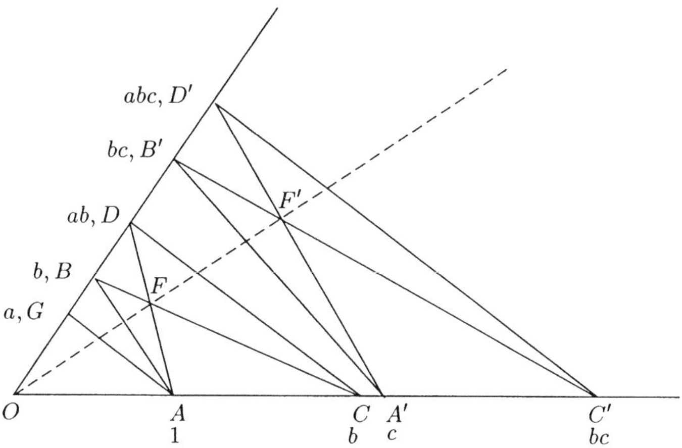

$a(bc) = (ab)c$

Fig. 43.

Behauptung darauf hinaus, dass auch CD parallel $C'D'$ sein muss. Bezeichnen wir nun den Schnittpunkt der Geraden AD und BC mit F und den Schnittpunkt der | Geraden $A'D'$ und $B'C'$ mit F', so sind in den Dreiecken ABF 60 und $A'B'F'$ die entsprechenden Seiten einander parallel; nach dem Desarguesschen Satze liegen daher die drei Punkte O, F, F' auf einer Geraden. Wegen dieses Umstandes können wir die zweite Aussage des Desarguesschen Satzes auf die beiden Dreiecke CDF und $C'D'F'$ anwenden und erkennen hieraus, dass in der That CD parallel $C'D'$ ist.

Wir beweisen endlich in unserer Streckenrechnung auf Grund des Desarguesschen Satzes die beiden distributiven Gesetze:

$$a(b + c) = ab + ac$$

und

$$(a + b)c = ac + bc.$$

Zum Beweise des e r s t e r e n Gesetzes dient die Figur 44[1]). In derselben ist

$$b = OA', \qquad c = OC'$$
$$ab = OB', \qquad ab = OA'', \qquad ac = OC'' \quad \text{u. s. f.}$$

und es läuft

$B''D_2$ parallel $C''D_1$ parallel zur festen Geraden OA',
$B'D_1$ " $C'D_2$ " " " " OA'';

ferner ist

$$A'A'' \text{ parallel } C'C''$$

und

$$A'B'' \text{ parallel } B'A'' \text{ parallel } F'D_2 \text{ parallel } F''D_1.$$

Die Behauptung läuft darauf hinaus, dass dann auch

$$F'F'' \text{ parallel } A'A'' \text{ und } C'C''$$

sein muss.

Wir construiren folgende Hülfslinien:

$F''J$ parallel der festen Geraden OA',
$F'J$ " " " " OA'';

die Schnittpunkte der Geraden $C''D_1$ und $C'D_2$, $C''D_1$ und $F'J$, $C'D_2$ und $F''J$ heissen G, H_1, H_2; endlich erhalten wir noch die weiteren in der Figur 44 punktirten Hülfslinien durch Verbindung bereits construirter Punkte.

In den beiden Dreiecken $A'B''C''$ und $F'D_2G$ laufen die Verbindungsge-
61 raden entsprechender Ecken einander parallel; nach | der zweiten Aussage des Desarguesschen Satzes folgt daher, dass

$$A'C'' \text{ parallel } F'G$$

sein muss. In den beiden Dreiecken $A'C''F''$ und $F'GH_2$ laufen ebenfalls die Verbindungsgeraden entsprechender Ecken einander parallel; wegen der vorhin gefundenen Thatsache folgt nach der zweiten Aussage des Desarguesschen Satzes, dass

$$A'F'' \text{ parallel } F'H_2$$

sein muss. Da somit in den beiden wagerecht schraffirten Dreiecken $OA'F''$ und JH_2F' die entsprechenden Seiten parallel sind, so lehrt der Desarguessche Satz, dass die drei Verbindungsgeraden

$$OJ, \qquad A'H_2, \qquad F''F'$$

1) Die Figuren 44, 45 und 47 hat Herr Dr. *von Schaper* entworfen[28] und auch die zugehörigen Beweise ausgeführt.

4 $bc = OC'$,] $bc = OC'$

1 Figur 44] Figur S. 58

1 ⟨In the footnote, 'Figuren 44, 45 und 47' is changed in E2 to 'Figuren S. 58, 60 und 62'.⟩

19 Figur 44] untenstehende Figur

[28] Cf. *Hilbert 1899a**, 152, 153, 156.

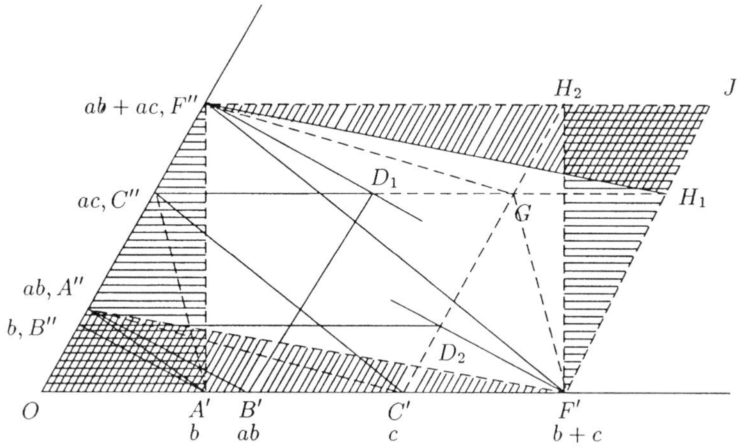

$$a(b+c) = ab + ac$$

Fig. 44.

sich in einem und demselben Punkte, etwa in P, treffen.

Auf dieselbe Weise finden wir, dass auch

$$A''F' \text{ parallel } F''H_1$$

sein muss und da somit in den beiden schräg schraffierten Dreiecken $OA''F'$ und JH_1F'' die entsprechenden Seiten parallel laufen, so treffen sich dem Desarguesschen Satze zufolge die drei Verbindungsgeraden

$$OJ, \qquad A''H_1, \qquad F'F''$$

ebenfalls in einem Punkte – dem Punkte P.

Nunmehr laufen für die Dreiecke $OA'A''$ und JH_2H_1 die Verbindungsgeraden entsprechender Ecken durch den nämlichen Punkt | P, und mithin 62
folgt, dass

$$H_1H_2 \text{ parallel } A'A''$$

sein muss; mithin ist auch

$$H_1H_2 \text{ parallel } C'C''.$$

Endlich betrachten wir die Figur $F''H_2C'C''H_1F'F''$. Da in derselben

$$\begin{array}{lllll} F''H_2 & \text{parallel} & C'F' & \text{parallel} & C''H_1 \\ H_2C' & \text{"} & F''C'' & \text{"} & H_1F' \\ C'C'' & \text{"} & H_1H_2 & & \end{array}$$

ausfällt, so erkennen wir hierin die Figur 41 wieder, die wir in § 25 zum Beweise für das commutative Gesetz der Addition benutzt haben. Die entsprechenden Schlüsse wie dort zeigen dann, dass

$$F'F'' \text{ parallel } H_1H_2$$

---

19 Figur 41] Figur S. 55 ($B'B''AA'A''BB'$)

sein muss und da mithin auch

$$F'F'' \text{ parallel } A'A''$$

ausfällt, so ist der Beweis unserer Behauptung vollständig erbracht.

Zum Beweise der zweiten Formel des distributiven Gesetzes dient die völlig verschiedene Figur 45. In derselben ist

$$1 = OD, \quad a = OA, \quad a = OB, \quad b = OG, \quad c = OD'$$
$$ac = OA', \quad ac = OB', \quad bc = OG' \text{ u. s. f.}$$

und es läuft

GH parallel $G'H'$ parallel zur festen Geraden OA,
AH " $A'H'$ " " " " OB

und ferner

AB parallel $A'B'$,
BD " $B'D'$,
DG " $D'G'$,
HJ " $H'J'$.

Die Behauptung läuft darauf hinaus, dass dann auch

$$DJ \text{ parallel } D'J'$$

sein muss.

Wir bezeichnen die Punkte, in denen BD und GD die Gerade AH treffen, bez. mit C und F und ferner die Punkte, in denen $B'D'$ und $G'D'$ die
63 Gerade $A'H'$ treffen, bez. mit C' und F'; | endlich ziehen wir noch die in der Figur 45 punktirten Hülfslinien FJ und $F'J'$.

In den Dreiecken ABC und $A'B'C'$ laufen die entsprechenden Seiten parallel; mithin liegen nach dem Desarguesschen Satze die drei Punkte O, C, C' auf einer Geraden. Ebenso folgt dann aus der Betrachtung der Dreiecke CDF und $C'D'F'$, dass O, F, F' auf einer Geraden liegen, und die Betrachtung der Dreiecke FGH und $F'G'H'$ lehrt, dass O, H, H' Punkte einer Geraden sind. Nun laufen in den Dreiecken FHJ und $F'H'J'$ die Verbindungsgeraden entsprechender Ecken durch den nämlichen Punkt O, und mithin sind zufolge der zweiten Aussage des Desarguesschen Satzes die Geraden FJ und $F'J'$ einander parallel. Endlich zeigt dann die Betrachtung der Dreiecke DFJ und $D'F'J'$, dass die Geraden DJ und $D'J'$ einander parallel sind, und damit ist der Beweis unserer Behauptung vollständig erbracht.

§ 27.

Die Gleichung der Geraden auf Grund der neuen Streckenrechnung.

Wir haben in § 24 bis § 26 mittelst der in § 24 angeführten Axiome und unter Voraussetzung der Gültigkeit des Desarguesschen Satzes in der Ebene eine Streckenrechnung eingeführt, in welcher das commutative Gesetz der Addition, die associativen Gesetze der Addition und Multiplikation, sowie die

5 Figur 45] Figur S. 60
22 Figur 45] Figur S. 60

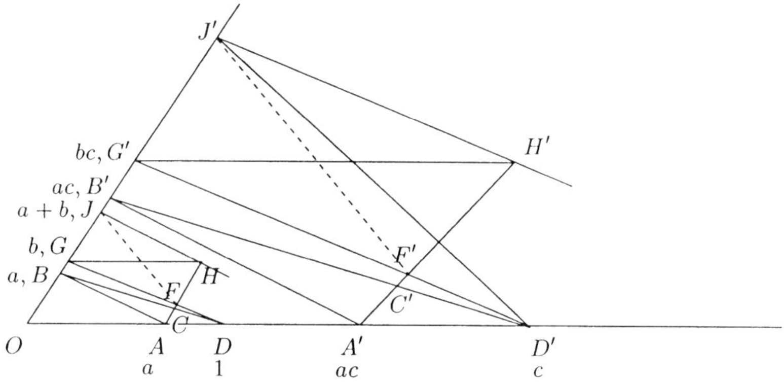

$$(a+b)c = ac + bc$$

Fig. 45.

beiden distributiven Gesetze gültig sind. Wir wollen in diesem Paragraphen | zeigen, in welcher Weise auf Grund dieser Streckenrechnung eine analytische 64
Darstellung der Punkte und Geraden in der Ebene möglich ist.

D e f i n i t i o n. Wir bezeichnen in der Ebene die beiden angenommenen festen Geraden durch den Punkt O als die X- und Y-Axe und denken uns irgend einen Punkt P der Ebene durch die Strecken x, y bestimmt, die man auf der X- bez. Y-Axe erhält, wenn man durch P zu diesen Axen Parallelen zieht. Diese Strecken x, y heissen *Coordinaten* des Punktes P. Auf Grund der neuen Streckenrechnung und mit Hülfe des Desarguesschen Satzes gelangen wir zu der folgenden Thatsache:

S a t z 34. *Die Coordinaten x, y der Punkte auf einer beliebigen Geraden erfüllen stets eine Streckengleichung von der Gestalt:*

$$ax + by + c = 0;$$

in dieser Gleichung stehen die Strecken a, b notwendig l i n k s s e i t i g von den Coordinaten x, y; die Strecken a, b sind niemals beide Null und c ist eine beliebige Strecke.

Umgekehrt: jede Streckengleichung der beschriebenen Art stellt stets eine Gerade in der zu Grunde gelegten ebenen Geometrie dar.

Beweis. Wir nehmen zunächst an, die Gerade l gehe durch O. Ferner sei C ein bestimmter von O verschiedener Punkt auf l und P ein beliebiger Punkt auf l; C habe die Coordinaten OA, OB und P habe die Coordinaten x, y; wir bezeichnen die Verbindungsgerade der Endpunkte von x, y mit g.

Endlich ziehen wir durch den Endpunkt der Strecke 1 auf der X-Axe eine Parallele h zu AB; diese Parallele schneide auf der Y-Axe die Strecke e | ab. Aus 65
der zweiten Aussage des Desarguesschen Satzes folgt leicht, dass die Gerade g stets parallel zu AB läuft. Da somit auch g stets zu h parallel ist, so folgt für

4 D e f i n i t i o n.] E r k l ä r u n g.

19 durch O.] durch O und sei von den Achsen verschieden.

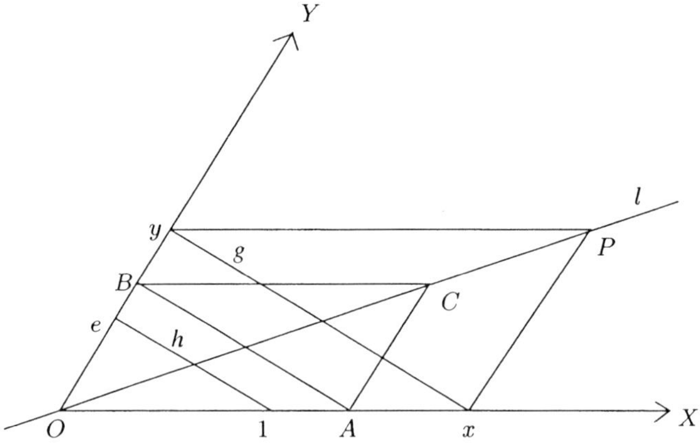

Fig. 46.

die Coordinaten x, y des beliebigen Punktes P auf l die Streckengleichung

$$ex = y.$$

Nunmehr sei l' eine beliebige Gerade in unserer Ebene; dieselbe schneide auf der X-Axe die Strecke $c = OO'$ ab. Wir ziehen ferner die Gerade l durch O parallel zu l'. Es sei P' ein beliebiger Punkt auf l'; die Parallele durch P' zur X-Axe treffe die Gerade l in P und schneide auf der Y-Axe die Strecke $y = OB$ ab; ferner mögen die Parallelen durch P und P' zur Y-Axe auf der X-Axe die Strecken $x = OA$ und $x' = OA'$ abschneiden.

Wir wollen nun beweisen, dass die Streckengleichung

$$x' = x + c$$

besteht. Zu diesem Zwecke ziehen wir $O'C$ parallel zur Einheitsgeraden, ferner CD parallel zur X-Axe und AD parallel zur Y-Axe; dann läuft unsere Behauptung darauf hinaus, dass

$$A'D \text{ parallel } O'C$$

sein muss. Wir construiren noch D' als Schnittpunkt der Geraden CD und $A'P'$ und ziehen $O'C'$ parallel zur Y-Axe.

Da in den Dreiecken OCP und $O'C'P'$ die Verbindungsgeraden entsprechender Ecken parallel laufen, so folgt mittelst der zweiten Aussage des Desarguesschen Satzes, dass

$$CP \text{ parallel } C'P'$$

66 sein muss; auf gleiche Weise lehrt die Betrachtung der Dreiecke | ACP und $A'C'P'$, dass

$$AC \text{ parallel } A'C'$$

ist. Da somit in den Dreiecken ACD und $C'A'O'$ die entsprechenden Seiten einander parallel laufen, so treffen sich die Geraden AC', CA', DO' in einem

3 Gerade] nicht zu den Achsen parallele Gerade

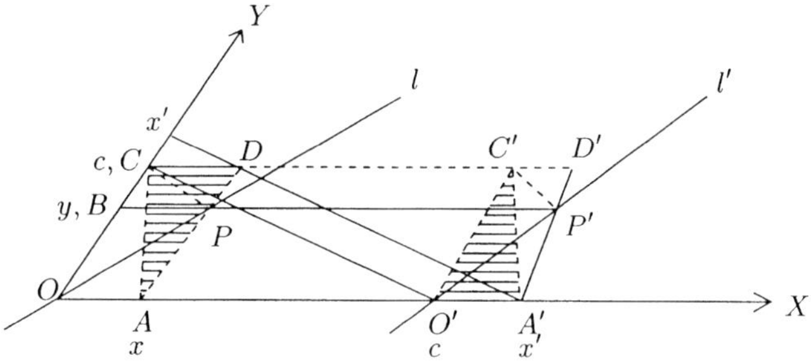

Fig. 47.

Punkte und die Betrachtung der beiden Dreiecke $C'A'D$ und ACO' zeigt dann, dass $A'D$ und CO' einander parallel sind.

Aus den beiden bisher gefundenen Streckengleichungen

$$ex = y \quad \text{und} \quad x' = x + c$$

folgt sofort die weitere Gleichung

$$ex' = y + ec.$$

Bezeichnen wir schliesslich mit n die Strecke, die zur Strecke 1 addirt die Strecke 0 liefert, so folgt, wie man leicht beweist, aus der letzten Gleichung

$$ex' + ny + nec = 0$$

und diese Gleichung ist von der Gestalt, wie der Satz 34 behauptet.

Die zweite Aussage des Satzes 34 erkennen wir nun ohne Mühe als richtig; denn eine jede vorgelegte Streckengleichung

$$ax + by + c = 0$$

lässt sich offenbar durch linksseitige Multiplikation mit einer geeigneten Strecke in die vorhin gefundene Gestalt

$$ex + ny + nec = 0$$

bringen.

Es sei noch ausdrücklich bemerkt, dass bei unseren Annahmen eine Streckengleichung von der Gestalt

$$xa + yb + c = 0,$$

in der die Strecken a, b **rechtsseitig** von den Coordinaten x, y stehen, im Allgemeinen **nicht** eine Gerade darstellt.

12–14 Streckengleichung $\langle \ldots \rangle$ lässt sich] Streckengleichung

$$ax + by + c = 0,$$

wo $b \neq 0$ ist, lässt sich

Wir werden in § 30 eine wichtige Anwendung von dem Satze 34 machen.

§ 28.

Der Inbegriff der Strecken aufgefasst als complexes Zahlensystem.

Wir sehen unmittelbar ein, dass für unsere neue in § 24 begründete Streckenrechnung die Sätze 1–6 in § 13 erfüllt sind.

67 Ferner haben wir in § 25 und § 26 mit Hülfe des Desarguesschen Satzes erkannt, dass für diese Streckenrechnung die Rechnungsgesetze 7–11 in § 13 gültig sind; es bestehen somit sämtliche Sätze der Verknüpfung, abgesehen vom commutativen Gesetze der Multiplikation.

Um endlich eine Anordnung der Strecken zu ermöglichen, treffen wir folgende Festsetzung. Es seien A, B irgend zwei verschiedene Punkte der Geraden OE; dann bringen wir gemäss Axiom II 4 die vier Punkte O, E, A, B in eine Reihenfolge. Ist dies auf eine der folgenden sechs Arten

$$ABOE,\ AOBE,\ AOEB,\ OABE,\ OAEB,\ OEAB$$

möglich, so nennen wir die Strecke $a = OA$ *kleiner* als die Strecke $b = OB$, in Zeichen:

$$a < b.$$

Findet dagegen eine der sechs Reihenfolgen

$$BAOE,\ BOAE,\ BOEA,\ OBAE,\ OBEA,\ OEBA$$

statt, so nennen wir die Strecke $a = OA$ *grösser* als die Strecke $b = OA$, in Zeichen

$$a > b.$$

Diese Festsetzung bleibt auch in Kraft, wenn A oder B mit O oder E zusammenfallen, nur dass dann die zusammenfallenden Punkte als ein einziger Punkt anzusehen sind und somit lediglich die Anordnung dreier Punkte in Frage kommt.

Wir erkennen leicht, dass nunmehr in unserer Streckenrechnung auf Grund der Axiome II die Rechnungsgesetze 13–16 in § 13 erfüllt sind; somit bildet die Gesamtheit aller verschiedenen Strecken ein complexes Zahlensystem, für welches die Gesetze 1–11, 13–16 in § 13, d. h. die sämtlichen Vorschriften ausser dem commutativen Gesetze der Multiplikation und dem Archimedischen Satze gewiss gültig sind; wir bezeichnen ein solches Zahlensystem im Folgenden kurz als ein *Desarguessches Zahlensystem.*

§ 29.

Aufbau einer räumlichen Geometrie mit Hülfe eines Desarguesschen Zahlensystems.

Es sei nun irgend ein Desarguessches Zahlensystem D vorgelegt; dasselbe ermöglicht uns den Aufbau einer räum|lichen Geometrie,
68 in der die Axiome I, II, III sämtlich erfüllt sind.

32 dem Archimedischen Satze] den Sätzen von der Stetigkeit

Um dies einzusehen, denken wir uns das System von irgend drei Zahlen (x, y, z) des Desarguesschen Zahlensystems D als einen Punkt und das System von irgend vier Zahlen $(u : v : w : r)$ in D, von denen die ersten drei Zahlen nicht zugleich 0 sind, als eine Ebene; doch sollen die Systeme $(u : v : w : r)$ und $(au : av : aw : ar)$, wo a irgend eine von 0 verschiedene Zahl in D bedeutet, die nämliche Ebene darstellen. Das Bestehen der Gleichung

$$ux + vy + wz + r = 0$$

möge ausdrücken, dass der Punkt (x, y, z) auf der Ebene $(u : v : w : r)$ liegt. Die Gerade endlich definiren wir mit Hülfe eines Systems zweier Ebenen $(u' : v' : w' : r')$ und $(u'' : v'' : w'' : r'')$, wenn es nicht möglich ist, zwei von 0 verschiedene Zahlen a', a'' in D zu finden, sodass gleichzeitig

$$a'u' = a''u'', \qquad a'v' = a''v'', \qquad a'w' = a''w''$$

wird. Ein Punkt (x, y, z) heisst auf dieser Geraden $[(u' : v' : w' : r'), (u'' : v'' : w'' : r'')]$ gelegen, wenn er den beiden Ebenen $(u' : v' : w' : r')$ und (u'', v'', w'', r'') gemeinsam ist. Zwei Gerade, welche dieselben Punkte enthalten, gelten als nicht verschieden.

Indem wir die Rechnungsgesetze 1–11 in § 13 anwenden, die nach Voraussetzung für die Zahlen in D gelten sollen, gelangen wir ohne Schwierigkeit zu dem Resultate, dass in der soeben aufgestellten räumlichen Geometrie die Axiome I und III sämtlich erfüllt sind.

Damit auch den Axiomen II der Anordnung Genüge geschehe, treffen wir folgende Festsetzungen. Es seien

$$(x_1, y_1, z_1), \qquad (x_2, y_2, z_2), \qquad (x_3, y_3, z_3)$$

irgend drei Punkte einer Geraden

$$[(u' : v' : w' : r'), \qquad (u'' : v'' : w'' : r'')];$$

dann heisse der Punkt (x_2, y_2, z_2) zwischen den beiden anderen gelegen, wenn wenigstens eine der sechs Doppelungleichungen

$$x_1 < x_2 < x_3, \qquad x_1 > x_2 > x_3 \tag{1}$$
$$y_1 < y_2 < y_3, \qquad y_1 > y_2 > y_3 \tag{2}$$
$$z_1 < z_2 < z_3, \qquad z_1 > z_2 > z_3 \tag{3}$$

erfüllt ist. Besteht nun etwa eine der beiden Doppelungleichungen (1), so schliessen wir leicht, dass entweder $y_1 = y_2 = y_3$ oder | notwendig eine der 69
beiden Doppelungleichungen (2) und ebenso dass entweder $z_1 = z_2 = z_3$ oder eine der Doppelungleichungen (3) gelten muss. In der That, aus den Gleichungen

$$u'x_i + v'y_i + w'z_i + r' = 0,$$
$$u''x_i + v''y_i + w''z_i + r'' = 0,$$
$$(i = 1, 2, 3)$$

leiten wir durch linksseitige Multiplikation derselben mit geeigneten Zahlen aus D und durch nachherige Addition der entstehenden Gleichungen ein Glei-

15 $(u'', v'', w'', r'')]$ $(u'' : v'' : w'' : r'')$

27 Doppelungleichungen] Paare von Ungleichungen

40 aus D] aus D, die $\neq 0$ sind,

chungssystem von der Gestalt

$$u'''x_i + v'''y_i + r''' = 0 \qquad (4)$$
$$(i = 1,\, 2,\, 3)$$

ab. Hierin ist der Coefficient v''' sicher nicht 0, da sonst die Gleichheit der drei Zahlen x_1, x_2, x_3 folgen würde. Aus

$$x_1 \lesseqgtr x_2 \lesseqgtr x_3$$

schliessen wir

$$u'''x_1 \overset{<}{\underset{>}{=}} u'''x_2 \overset{<}{\underset{>}{=}} u'''x_3$$

und mithin wegen (4)

$$v'''y_1 + r''' \overset{<}{\underset{>}{=}} v'''y_2 + r''' \overset{<}{\underset{>}{=}} v'''y_3 + r'''$$

und daher

$$v'''y_1 \overset{<}{\underset{>}{=}} v'''y_2 \overset{<}{\underset{>}{=}} v'''y_3,$$

und da v''' nicht 0 ist, so haben wir

$$y_1 \overset{<}{\underset{>}{=}} y_2 \overset{<}{\underset{>}{=}} y_3;$$

in jeder dieser Doppelungleichungen soll stets entweder durchweg das obere oder durchweg das mittlere oder durchweg das untere Zeichen gelten.

Die angestellten Ueberlegungen lassen erkennen, dass in unserer Geometrie die linearen Axiome II 1–4 der Anordnung zutreffen. Es bleibt noch zu zeigen übrig, dass in unserer Geometrie auch das ebene Axiom II 5 gültig ist.

Zu dem Zwecke sei eine Ebene $(u : v : w : r)$ und in ihr eine Gerade $[(u : v : w : r),\ (u' : v' : w' : r')]$ gegeben. Wir setzen fest, dass alle in der Ebene $(u : v : w : r)$ gelegenen Punkte (x, y, z), für die der Ausdruck $u'x + v'y + w'z + r'$ kleiner oder grösser als 0 ausfällt, auf der einen bez. auf der anderen Seite von jener Geraden gelegen sein sollen, und haben dann zu
70 beweisen, dass diese Fest|setzung sich mit der vorigen in Uebereinstimmung befindet, was leicht geschehen kann.

Damit haben wir erkannt, dass die sämtlichen Axiome I, II, III in derjenigen räumlichen Geometrie erfüllt sind, die in der oben geschilderten Weise aus dem Desarguesschen Zahlensystem D entspringt. Bedenken wir, dass der Desarguessche Satz eine Folge der Axiome I, II, III ist, so sehen wir, dass die eben gefundene Thatsache die genaue Umkehrung desjenigen Ergebnisses darstellt, zu dem wir in § 28 gelangt sind.

§ 30.

Die Bedeutung des Desarguesschen Satzes.

Wenn in einer ebenen Geometrie die Axiome I 1–2, II, III erfüllt sind und überdies der Desarguessche Satz gilt, so ist es nach § 24–§ 28 in dieser Geometrie stets möglich, eine Streckenrechnung einzuführen, für die die Regeln 1–11, 13–16 in § 13 anwendbar sind. Wir betrachten nun weiter den Inbegriff dieser Strecken als ein complexes Zahlensystem und bauen aus denselben nach den Entwickelungen in § 29 eine räumliche Geometrie auf, in der sämtliche Axiome I, II, III gültig sind.

37 für die] für welche

Fassen wir in dieser räumlichen Geometrie lediglich die Punkte $(x, y, 0)$ und diejenigen Geraden ins Auge, auf denen nur solche Punkte liegen, so entsteht eine ebene Geometrie, und wenn wir die in § 27 abgeleitete Thatsache berücksichtigen, so leuchtet ein, dass diese ebene Geometrie genau mit der zu Anfang vorgelegten ebenen Geometrie übereinstimmen muss. Damit gewinnen wir folgenden Satz, der als das Endziel der gesamten Entwickelungen dieses Kapitels V anzusehen ist:

Satz 35. *Es seien in einer ebenen Geometrie die Axiome* I 1–2, II, III *erfüllt: dann ist die Gültigkeit des Desarguesschen Satzes die notwendige und hinreichende Bedingung dafür, dass diese ebene Geometrie sich auffassen lässt als ein Teil einer räumlichen Geometrie, in welcher die sämtlichen Axiome* I, II, III *erfüllt sind.*

Der Desarguessche Satz kennzeichnet sich so gewissermassen für die ebene Geometrie als das Resultat der Elimination der räumlichen Axiome.

Die gefundenen Resultate setzen uns auch in den Stand zu erkennen, dass jede räumliche Geometrie, in der die Axiome I, II, III sämtlich erfüllt sind, sich stets als ein Teil einer „Geometrie von beliebig vielen Dimensionen“ auffassen lässt; dabei ist | unter einer Geometrie von beliebig vielen Dimensionen eine 71
Gesamtheit von Punkten, Geraden, Ebenen und noch weiteren linearen Elementen zu verstehen, für welche die entsprechenden Axiome der Verknüpfung und Anordnung sowie das Parallelenaxiom erfüllt sind.

Kapitel VI.

Der Pascalsche Satz.

§ 31.

Zwei Sätze über die Beweisbarkeit des Pascalschen Satzes.

Der Desarguessche Satz (Satz 32) lässt sich bekanntlich aus den Axiomen I, II, III, d. h. unter wesentlicher Benutzung der räumlichen Axiome beweisen; in § 23 habe ich gezeigt, dass der Beweis desselben ohne die räumlichen Axiome der Gruppe I und ohne die Congruenzaxiome IV nicht möglich ist, selbst wenn die Benutzung des Archimedischen Axioms gestattet wird.

In § 14 ist der Pascalsche Satz (Satz 21) und damit nach § 22 auch der Desarguessche Satz aus den Axiomen I 1–2, II–IV, also mit Ausschluss der räumlichen Axiome und unter wesentlicher Benutzung der Congruenzaxiome abgeleitet worden. Es entsteht die Frage, ob auch der Pascalsche Satz ohne Hinzuziehung der Congruenzaxiome bewiesen werden kann. Unsere Untersuchung wird zeigen, dass in dieser Hinsicht der Pascalsche Satz sich völlig anders als der Desarguessche Satz verhält, indem bei dem

8 Satz 35.] ⟨deleted⟩

19 linearen] ⟨deleted⟩

29 IV] II ⟨Wrongly, instead of 'III'.⟩

30 des Archimedischen Axioms] der Stetigkeitsaxiome

Beweise des Pascalschen Satzes die Zulassung oder Ausschliessung des Archimedischen Axioms von entscheidendem Einflusse für seine Gültigkeit ist. Die wesentlichen Ergebnisse unserer Untersuchung fassen wir in den folgenden zwei Sätzen zusammen:

Satz 36. *Der Pascalsche Satz* (Satz 21) *ist beweisbar auf Grund der Axiome* I, II, III, V, *d. h. unter Ausschliessung der Congruenzaxiome mit Zuhülfenahme des Archimedischen Axioms.*

Satz 37. *Der Pascalsche Satz* (Satz 21) *ist nicht beweisbar auf Grund der Axiome* I, II, III, *d. h. unter Ausschliessung der Congruenzaxiome sowie des Archimedischen Axioms.*

72 In der Fassung dieser beiden Sätze können nach dem allge|meinen Satze 35 die räumlichen Axiome I 3–7 auch durch die ebene Forderung ersetzt werden, dass der Desarguessche Satz (Satz 32) gelten soll.

§ 32.

Das commutative Gesetz der Multiplikation im Archimedischen Zahlensystem.

Die Beweise der Sätze 36 und 37 beruhen wesentlich auf gewissen gegenseitigen Beziehungen, welche für die Rechnungsregeln und Grundthatsachen der Arithmetik bestehen und deren Kenntnis auch an sich von Interesse erscheint. Wir stellen die folgenden zwei Sätze auf:

Satz 38. *Für ein Archimedisches Zahlensystem ist das commutative Gesetz der Multiplikation eine notwendige Folge der übrigen Rechnungsgesetze; d. h., wenn ein Zahlensystem die in* § 13 *aufgezählten Eigenschaften* 1–11, 13–17 *besitzt, so folgt notwendig, dass dasselbe auch der Formel* 12 *genügt.*

Beweis. Zunächst bemerken wir: wenn a eine beliebige Zahl des Zahlensystems und

$$n = 1 + 1 + \cdots + 1$$

eine ganze rationale positive Zahl ist, so gilt für a und n stets das commutative Gesetz der Multiplikation; es ist nämlich

$$an = a(1 + 1 + \cdots + 1) = a \cdot 1 + a \cdot 1 + \cdots + a \cdot 1 = a + a + \cdots + a$$

und ebenso

$$na = (1 + 1 + \cdots + 1)a = 1 \cdot a + 1 \cdot a + \cdots + 1 \cdot a = a + a + \cdots + a$$

Es seien nun im Gegensatz zu unserer Behauptung a, b solche zwei Zahlen des Zahlensystems, für welche das commutative Gesetz der Multiplikation nicht gültig ist. Wir dürfen dann, wie leicht ersichtlich, die Annahmen

$$a > 0, \quad b > 0 \qquad ab - ba > 0$$

machen. Wegen der Forderung 6 in § 13 giebt es eine Zahl $c (> 0)$, so dass

$$(a + b + 1)c = ab - ba$$

12 ebene Forderung] Forderung der ebenen Geometrie

ist. Endlich wählen wir eine Zahl d, die zugleich den Ungleichungen

$$d > 0, \quad d < 1, \quad d < c$$

genügt, und bezeichnen mit m und n zwei solche ganze rationale | Zahlen $\geqq 0$, 73
für die

$$md < a \leqq (m+1)d$$

bez.

$$nd < b \leqq (n+1)d$$

wird. Das Vorhandensein solcher Zahlen m, n ist eine unmittelbare Folgerung des Archimedischen Satzes (Satz 17 in § 13). Mit Rücksicht auf die Bemerkung zu Anfang dieses Beweises erhalten wir aus den letzteren Ungleichungen durch Multiplikation

$$ab \leqq mnd^2 + (m+n+1)d^2,$$
$$ba > mnd^2,$$

also durch Subtraktion

$$ab - ba \leqq (m+n+1)d^2.$$

Nun ist

$$md < a, \quad nd < b, \quad d < 1$$

und folglich

$$(m+n+1)d < a+b+1,$$

d. h.

$$ab - ba < (a+b+1)d$$

oder wegen $d < c$

$$ab - ba < (a+b+1)c.$$

Diese Ungleichung widerspricht der Bestimmung der Zahl c, und damit ist der Beweis für den Satz 38 erbracht.

§ 33.

Das commutative Gesetz der Multiplikation im Nicht-Archimedischen Zahlensystem.

Satz 39. *Für ein Nicht-Archimedisches Zahlensystem ist das commutative Gesetz der Multiplikation* nicht *eine notwendige Folge der übrigen Rechnungsgesetze; d. h. es giebt ein Zahlensystem, das die in* § 13 *aufgezählten Eigenschaften* 1–11, 13–16 *besitzt – ein Desarguessches Zahlensystem nach* § 28 –, *in welchem* nicht *das commutative Gesetz* (12) *der Multiplikation besteht.*

Beweis. Es sei t ein Parameter und T irgend ein Ausdruck mit einer endlichen oder unendlichen Gliederzahl von der Gestalt

$$T = r_0 t^n + r_1 t^{n+1} + r_2 t^{n+2} + r_3 t^{n+3} + \cdots;$$

darin mögen $r_0 (\neq 0)$, r_1, r_2, ... beliebige rationale Zahlen bedeuten und n sei eine beliebige ganze rationale Zahl $\gtreqless 0$. Ferner sei s ein anderer Parameter

und S irgend ein Ausdruck mit einer | endlichen oder unendlichen Gliederzahl 74
von der Gestalt

$$S = s^m T_0 + s^{m+1} T_1 + s^{m+2} T_2 + \cdots;$$

darin mögen $T_0(\neq 0)$, T_1, $T_2, \ldots$ beliebige Ausdrücke von der Gestalt T bezeichnen und m sei wiederum eine beliebige ganze rationale Zahl $\gtreqless 0$. Die Gesamtheit aller Ausdrücke von der Gestalt S sehen wir als ein complexes Zahlensystem $\Omega(s,t)$ an, in dem wir folgende Rechnungsregeln festsetzen: man rechne mit s und t, wie mit Parametern nach den Regeln 7–11 in § 13, während man an Stelle der Regel 12 stets die Formel

$$ts = -st \tag{1}$$

anwende.

Sind nun S', S'' irgend zwei Ausdrücke von der Gestalt S:

$$S' = s^{m'} T'_0 + s^{m'+1} T'_1 + s^{m'+2} T'_2 + \ldots,$$
$$S'' = s^{m''} T''_0 + s^{m''+1} T''_1 + s^{m''+2} T''_2 + \ldots,$$

so kann man offenbar durch Zusammenfügung einen neuen Ausdruck $S' + S''$ bilden, der wiederum von der Gestalt S und zugleich eindeutig bestimmt ist; dieser Ausdruck $S' + S''$ heisst die Summe der durch S', S'' dargestellten Zahlen.

Durch gliedweise Multiplikation der beiden Ausdrücke S', S'' gelangen wir zunächst zu einem Ausdrucke von der Gestalt

$$S'S'' = s^{m'} T'_0 s^{m''} T''_0 + (s^{m'} T'_0 s^{m''+1} T''_1 + s^{m'+1} T'_1 s^{m''} T''_0)$$
$$+ (s^{m'} T'_0 s^{m''+2} T''_2 + s^{m'+1} T'_1 s^{m''+1} T''_1 + s^{m'+2} T'_2 s^{m''} T''_0) + \cdots.$$

Dieser Ausdruck wird bei Benutzung der Formel (1) offenbar ein eindeutig bestimmter Ausdruck von der Gestalt S; der letztere heisse das Produkt der durch S' dargestellten Zahl in die durch S'' dargestellte Zahl.

Bei der so festgesetzten Rechnungsweise leuchtet die Gültigkeit der Rechnungsregeln 1–5 in § 13 unmittelbar ein. Auch die Gültigkeit der Vorschrift 6 in § 13 ist nicht schwer einzusehen. Zu dem Zwecke nehmen wir an, es seien etwa

$$S' = s^{m'} T'_0 + s^{m'+1} T'_1 + s^{m'+2} T'_2 + \cdots$$

und

$$S''' = s^{m'''} T'''_0 + s^{m'''+1} T'''_1 + s^{m'''+2} T'''_2 + \cdots$$

gegebene Ausdrücke von der Gestalt S, und bedenken, dass unseren Festsetzungen entsprechend der erste Coefficient r'_0 aus T'_0 von 0 verschieden sein
75 muss. Indem wir nun die nämlichen Po|tenzen von s auf beiden Seiten einer Gleichung

$$S'S'' = S''' \tag{2}$$

vergleichen, finden wir in eindeutig bestimmter Weise zunächst eine ganze Zahl m'' als Exponenten und sodann der Reihe nach gewisse Ausdrücke

$$T''_0, T''_1, T''_2, \ldots$$

---

10 $-st$] $2st$

derart, dass der Ausdruck

$$S'' = s^{m''}T_0'' + s^{m''+1}T_1'' + s^{m''+2}T_2'' + \cdots$$

bei Benutzung der Formel (1) der Gleichung (2) genügt; hiermit ist der gewünschte Nachweis erbracht.

Um endlich die Anordnung der Zahlen unseres Zahlensystems $\Omega(s,t)$ zu ermöglichen, treffen wir folgende Festsetzungen. Eine Zahl des Systems heisse $<$ oder $>$ 0, jenachdem in dem Ausdrucke S, der sie darstellt, der erste Coefficient r_0 von T_0 $<$ oder $>$ 0 ausfällt. Sind irgend zwei Zahlen a und b des complexen Zahlensystems vorgelegt, so heisse $a < b$ bez. $a > b$, jenachdem $a - b < 0$ oder > 0 wird. Es leuchtet unmittelbar ein, dass bei diesen Festsetzungen die Regeln 13–16 in § 13 gültig sind, d. h. $\Omega(s,t)$ ist ein Desarguessches Zahlensystem (vgl. § 28).

Die Vorschrift 12 in § 13 ist, wie Gleichung (1) zeigt, für unser complexes Zahlensystem $\Omega(s,t)$ n i c h t erfüllt und damit ist die Richtigkeit des Satzes 39 vollständig erkannt.

In Uebereinstimmung mit Satz 38 gilt der Archimedische Satz (Satz 17 in § 13) für das soeben aufgestellte Zahlensystem $\Omega(s,t)$ nicht.

Es werde noch hervorgehoben, dass das Zahlensystem $\Omega(s,t)$ – ebenso wie die in § 9 und § 12 benutzten Zahlensysteme Ω und $\Omega(t)$ – nur eine abzählbare Menge von Zahlen enthält.

§ 34.

Beweis der beiden Sätze über den Pascalschen Satz. (Nicht-Pascalsche Geometrie.)

Wenn in einer räumlichen Geometrie die sämtlichen Axiome I, II, III erfüllt sind, so gilt auch der Desarguessche Satz (Satz 32) und mithin ist nach Kapitel V § 24 bis § 26 in dieser Geometrie die Einführung einer Streckenrechnung möglich, für welche die Vorschriften 1–11, 13–16 in § 13 gültig sind. Setzen wir nun das Archimedische Axiom V in unserer Geometrie voraus, so gilt | offenbar für die Streckenrechnung der Archimedische Satz (Satz 17 76
in § 13) und mithin nach Satz 38 auch das commutative Gesetz der Multiplikation. Da aber die hier in Rede stehende in § 24 (Fig. 40) eingeführte Definition des Streckenproduktes mit der in § 15 (Fig. 21) angewandten Definition übereinstimmt, so bedeutet gemäss der in § 15 ausgeführten Construktion das commutative Gesetz der Multiplikation zweier Strecken auch hier nichts anderes als den Pascalschen Satz. Damit ist die Richtigkeit des Satzes 36 erkannt.

Um den Satz 37 zu beweisen, fassen wir das in § 33 aufgestellte Desarguessche Zahlensystem $\Omega(s,t)$ ins Auge und construiren mit Hülfe desselben auf die in § 29 beschriebene Art eine räumliche Geometrie, in der die sämtlichen Axiome I, II, III erfüllt sind. Trotzdem gilt der Pascalsche Satz in dieser Geometrie nicht, da das commutative Gesetz der Multiplikation in dem Desarguesschen Zahlensystem $\Omega(s,t)$ nicht besteht. Die so aufgebaute „*Nicht-*

31 (Fig. 40)] (Fig. S. 54)
32 (Fig. 21)] (Fig. S. 32)

Pascalsche" *Geometrie* ist in Uebereinstimmung mit dem vorhin bewiesenen Satz 36 notwendig zugleich auch eine „*Nicht-Archimedische*" *Geometrie.*

Es ist offenbar, dass der Pascalsche Satz sich bei unseren Annahmen auch dann nicht beweisen lässt, wenn man die räumliche Geometrie als einen Teil einer Geometrie von beliebig vielen Dimensionen auffasst, in welcher neben den Punkten, Geraden und Ebenen noch weitere lineare Elemente vorhanden sind und für diese ein entsprechendes System von Axiomen der Verknüpfung und Anordnung, sowie das Parallelenaxiom zu Grunde gelegt wird.

§ 35.

Beweis eines beliebigen Schnittpunktsatzes mittelst des Desarguesschen und des Pascalschen Satzes.

Ein jeder ebener Schnittpunktsatz hat notwendig diese Form: Man wähle zunächst ein System von Punkten und Geraden willkürlich, bez. mit der Bedingung, dass für gewisse von diesen Punkten und Geraden die vereinigte Lage vorgeschrieben ist; wenn man dann in bekannter Weise Verbindungsgerade und Schnittpunkte construirt, so gelangt man schliesslich zu einem bestimmten System von drei Geraden, von denen der Satz aussagt, dass sie durch den nämlichen Punkt hindurchlaufen.

Es sei nun eine ebene Geometrie vorgelegt, in der sämtliche Axiome I 1–2, II–V gültig sind; nach Kap. III § 17 können wir dann vermittelst eines rechtwinkligen Axenkreuzes jedem Punkte ein Zahlenpaar (x, y) und jeder Geraden
77 ein Verhältnis | von drei Zahlen $(u : v : w)$ entsprechen lassen; hierbei sind x, y, u, v, w jedenfalls r e e l l e Zahlen, von denen u, v nicht beide verschwinden, und die Bedingung für die vereinigte Lage von Punkt und Geraden

$$ux + vy + w = 0$$

bedeutet eine Gleichung im gewöhnlichen Sinne. Umgekehrt dürfen wir, falls x, y, u, v, w Zahlen des in § 9 construirten algebraischen Bereiches Ω sind und u, v nicht beide verschwinden, gewiss annehmen, dass das Zahlenpaar (x, y) und das Zahlentripel $(u : v : w)$ einen Punkt bez. eine Gerade in der vorgelegten Geometrie liefert.

Führen wir für alle Punkte und Geraden, die in einem beliebigen ebenen Schnittpunktsatze auftreten, die betreffenden Zahlenpaare und Zahlentripel ein, so wird dieser Schnittpunktsatz aussagen, dass ein bestimmter, von gewissen Parametern $p_1, \ldots, p_r$ rational abhängiger Ausdruck $A(p_1, \ldots, p_r)$ mit reellen Coefficienten stets verschwindet, sobald wir für jene Parameter irgend welche Zahlen des in § 9 betrachteten Bereiches Ω einsetzen. Wir schlies-

6 lineare] ⟨deleted⟩
8 wird.] ist.
17 drei] ⟨deleted⟩
26 Umgekehrt] Andererseits
27 Zahlen] insbesondere Zahlen
28 gewiss] ⟨deleted⟩
28 annehmen, dass] annehmen, dass umgekehrt
35–36 irgend welche] insbesondere irgend welche

sen hieraus, dass der Ausdruck $A(p_1, \ldots, p_r)$ auch identisch auf Grund der Rechnungsgesetze 7–12 in § 13 verschwinden muss.

Da in der vorgelegten Geometrie nach § 22 der Desarguessche Satz gilt, so können wir gewiss auch die in § 24 eingeführte Streckenrechnung benutzen, und wegen der Gültigkeit des Pascalschen Satzes trifft für diese Streckenrechnung auch das commutative Gesetz der Multiplikation zu, sodass in dieser Streckenrechnung sämtliche Rechnungsgesetze 7–12 in § 13 gültig sind.

Indem wir die Axen des bisher benutzten Axenkreuzes auch als Axen dieser neuen Streckenrechnung gewählt und die Einheitspunkte E und E' geeignet festgesetzt denken, erkennen wir die Uebereinstimmung der neuen Streckenrechnung mit der früheren Coordinatenrechnung.

Um in der neuen Streckenrechnung das identische Verschwinden des Ausdruckes $A(p_1, \ldots, p_r)$ nachzuweisen, genügt die Anwendung des Desarguesschen und Pascalschen Satzes und damit erkennen wir, dass jeder in der vorgelegten Geometrie geltende Schnittpunktsatz durch Konstruktion geeigneter Hülfspunkte und Hülfsgeraden sich stets als eine Kombination des Desarguesschen und Pascalschen Satzes herausstellen muss. Zum Nachweise der Richtigkeit des Schnittpunktsatzes brauchen wir also nicht auf die Congruenzsätze zurückzugreifen.

Kapitel VII. 78

Die geometrischen Kontruktionen auf Grund der Axiome I–V.

§ 36.

Die geometrischen Konstruktionen mittelst Lineals und Streckenübertragers.

Es sei eine räumliche Geometrie vorgelegt, in der die sämtlichen Axiome I–V gelten; wir fassen der Einfachheit wegen in diesem Kapitel eine ebene Geometrie ins Auge, die in dieser räumlichen Geometrie enthalten ist, und untersuchen dann die Frage, welche elementaren Konstruktionsaufgaben in einer solchen Geometrie notwendig ausführbar sind.

Auf Grund der Axiome I ist die Ausführung der folgenden Aufgabe stets möglich:

Aufgabe 1. Zwei Punkte durch eine Gerade zu verbinden und den Schnittpunkt zweier Geraden zu finden, falls die Geraden nicht parallel sind.

Das Axiom III ermöglicht die Ausführung der folgenden Aufgabe:

---

26 Streckenübertragers] Eichmaßes

36–512.12 Das ⟨...⟩ lösbar;] ⟨In E2, following the new arrangement of the axioms, the parts of this sections to do with the Congruence and Parallel Axioms are exchanged, and the *Aufgaben* correspondingly renumbered. The last sentence 'Auf Grund ⟨...⟩ somit;', now merely referring to Axiom II, is truncated, and moved to the beginning.⟩

Aufgabe 2. Durch einen gegebenen Punkt zu einer Geraden eine Parallele zu ziehen.

Auf Grund der Congruenzaxiome IV ist das Abtragen von Strecken und Winkeln möglich, d. h. es lassen sich in der vorgelegten Geometrie folgende Aufgaben lösen:

Aufgabe 3. Eine gegebene Strecke auf einer gegebenen Geraden von einem Punkte aus abzutragen.

Aufgabe 4. Einen gegebenen Winkel an eine gegebene Gerade anzutragen oder eine Gerade zu konstruiren, die eine gegebene Gerade unter einem gegebenen Winkel schneidet.

Auf Grund der Axiome der Gruppen II und V werden keine neuen Aufgaben lösbar;

wir sehen somit, dass unter ausschliesslicher Benutzung der Axiome I–V alle und nur diejenigen Konstruktionsaufgaben lösbar sind, die sich auf die eben genannten Aufgaben 1–4 zurückführen lassen.

Wir fügen den fundamentalen Aufgaben 1–4 noch die folgende hinzu:

Aufgabe 5. Zu einer gegebenen Geraden eine Senkrechte zu ziehen.

79 Wir erkennen unmittelbar, dass diese Aufgabe 5 auf verschiedene Arten durch die Aufgaben 1–4 gelöst werden kann.

Zur Ausführung der Aufgabe 1 bedürfen wir des Lineals. Ein Instrument, welches zur Ausführung der Aufgabe 3 dient d. h. das Abtragen einer Strecke auf einer gegebenen Geraden ermöglicht, nennen wir einen Streckenübertrager. Wir wollen nunmehr zeigen, dass die Aufgaben 2, 4 und 5 auf die Lösung der Aufgaben 1 und 3 zurückgeführt werden können, und mithin die sämtlichen Aufgaben 1–5 lediglich mittelst Lineals und Streckenübertragers lösbar sind. Wir finden damit folgendes Resultat:

Satz 40. *Diejenigen geometrischen Konstruktionsaufgaben, die unter ausschliesslicher Benutzung der Axiome* I–V *lösbar sind, lassen sich notwendig mittelst Lineals und Streckenübertragers ausführen.*

---

12 Gruppen II und V] Gruppe II

13 ausschliesslicher Benutzung] Zugrundelegung

20–26 Ein ⟨...⟩ Resultat:] Um die Aufgaben 2–5 auszuführen, genügt es, wie im Folgenden gezeigt wird, neben dem Lineal das Eichmaß anzuwenden, ein Instrument, welches das Abtragen einer einzigen bestimmten Strecke, etwa der Einheitsstrecke ermöglicht[1]). Wir gelangen damit zu folgendem Resultat:

1) Diese Bemerkung ist von *J. Kürschák* gemacht worden; vgl. dessen Note „Das Streckenabtragen" Math. Ann. Bd. 55. 1902.[29]

28 *ausschliesslicher Benutzung*] *Zugrundelegung*

1 *Streckenübertragers*] *Eichmaßes*

[29] *Kürschák 1902.*

Beweis. Um die Aufgabe 2 auf die Aufgaben 1 und 3 zurückzuführen, verbinden wir den gegebenen Punkt P mit irgend einem Punkte A der gegebenen Geraden und verlängern PA über A hinaus um sich selbst bis C. Sodann verbinden wir C mit irgend einem andern Punkte B der gegebenen Geraden und verlängern CB über B hinaus um sich selbst bis Q; die Gerade PQ ist die gesuchte Parallele.

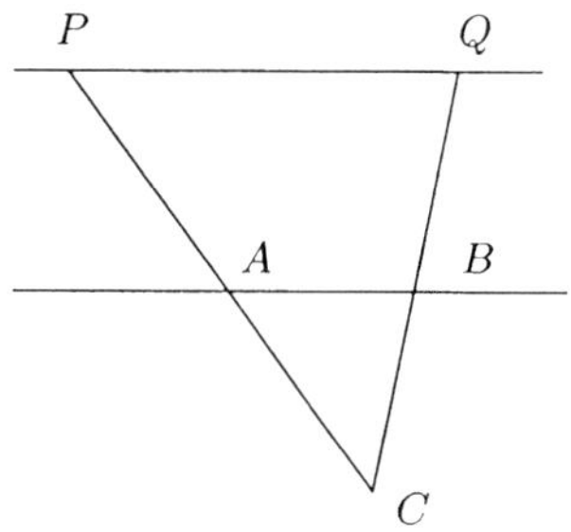

Fig. 48.

Die Aufgabe 5 lösen wir auf folgende Weise. Es sei A ein beliebiger Punkt der gegebenen Geraden; dann tragen wir von A aus auf dieser Geraden nach beiden Seiten hin zwei gleiche

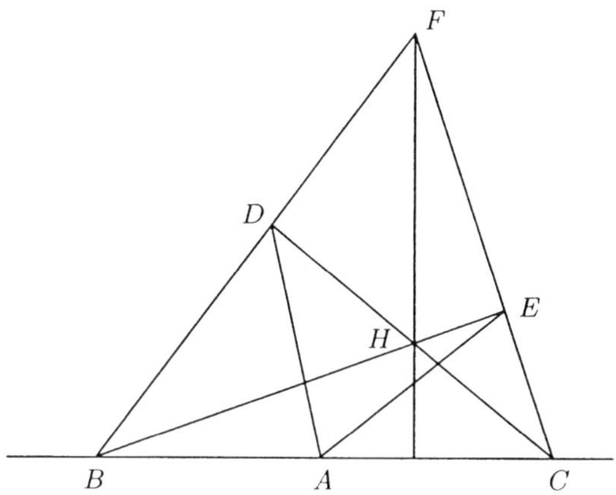

Fig. 49.

Strecken AB und AC ab und bestimmen dann auf zwei beliebigen anderen
| durch A gehenden Geraden die Punkte E und D, so dass auch die Strek- 80
ken AD und AE den Strecken AB und AC gleich werden. Die Geraden
BD und CE mögen sich in F, die Geraden BE und CD in H schneiden:
dann ist FH die gesuchte Senkrechte. In der That: die Winkel $\sphericalangle BDC$ und

1–10 Um ⟨...⟩ Parallele.]

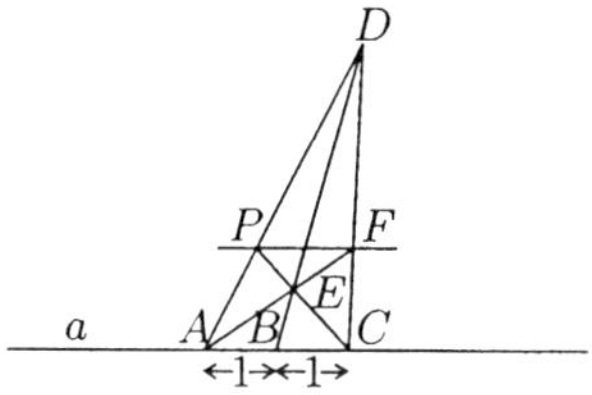

Um die Aufgabe 4 auszuführen, verbinden wir den gegebenen Punkt P mit irgend einem Punkte A der gegebenen Geraden a und tragen von A aus auf a zweimal hintereinander mittels des Eichmaßes die Einheitsstrecke ab, etwa bis B und C. Es sei nun D irgend ein Punkt auf AP, ferner E der Treffpunkt von CP und BD und endlich F der Treffpunkt von AE und CD: dann ist nach *Steiner*[30] PF die gesuchte Parallele zu a.

13–14 zwei gleiche Strecken] mittels des Eichmaßes die Einheitsstrecken

16 den Strecken AB und AC gleich] gleich der Einheitsstrecke

[30] *Steiner 1833*, 15.

$\sphericalangle BEC$ sind als Winkel im Halbkreise über BC Rechte und daher steht nach dem Satze vom Höhenschnittpunkt eines Dreieckes, auf das Dreieck BCF angewandt, auch FH auf BC senkrecht.

Wir können nunmehr leicht auch die Aufgabe 4 allein mittelst Ziehens von Geraden und Abtragens von Strecken lösen; wir schlagen etwa folgendes Verfahren ein, welches nur das Ziehen von Parallelen und Fällen von Loten erfordert. Es sei β der abzutragende Winkel und A der Scheitel dieses Winkels. Wir ziehen die Gerade l durch A parallel zu der gegebenen Geraden, an welche der gegebene Winkel β angetragen werden soll. Von einem beliebigen Punkte B eines Schenkels von β fällen wir Lote auf den anderen Schenkel des Winkels β und auf l. Die Fusspunkte dieser Lote seien D und C. Das Fällen von Loten geschieht vermöge der Aufgaben 2 und 5. Sodann fällen wir von A eine Senkrechte auf CD, ihr Fusspunkt sei E. Nach dem in § 14 ausgeführten Beweise ist $\sphericalangle CAE = \beta$; die Aufgabe 4 ist somit ebenfalls auf die Aufgaben 1 und 3 zurückgeführt und damit der Satz 40 vollständig bewiesen.

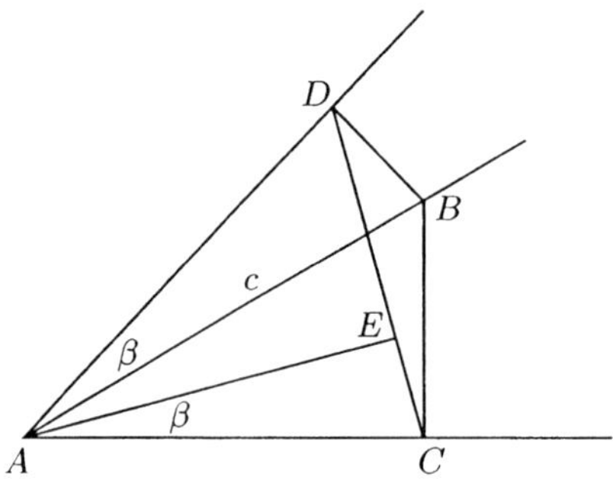

Fig. 50.

2–3 eines ⟨...⟩ angewandt] eines Dreieckes, den wir auf das Dreieck BCF anwenden
4–5 Ziehens ⟨...⟩ Strecken] Lineals und Eichmaßes
6 Fällen] das Fällen
8 Gerade l] Gerade l (AC)
13 Aufgaben 2 und 5] ⟨Wrongly so instead of 'Aufgaben 4 und 5'⟩
17 in § 14] in § 14 S. 27
19–21 ebenfalls ⟨...⟩ zurückgeführt] gelöst.

Um endlich die Aufgabe 2 auszuführen, benutzen wir die einfache von J. Kürschák angegebene Konstruktion:[31] es sei AB die abzutragende Strecke und P der gegebene Punkt auf der gegebenen Geraden l. Man ziehe durch P die Parallele zu AB und trage auf derselben mittels des Eichmaßes von P aus die Einheitsstrecke ab etwa bis C; ferner trage man auf l von P aus die Einheitsstrecke bis D ab. Die zu AP durch B gezogene Parallele treffe PC in Q und die durch Q zu CD gezogene Parallele treffe l in E: dann ist $PE = AB$. Damit ist gezeigt, daß die Aufgaben 1–5 sämtlich durch Lineal und Eichmaß lösbar sind,

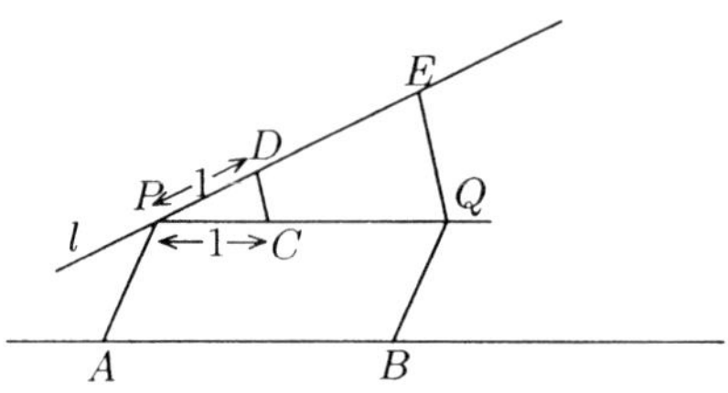

21 damit] folglich

[31] *Kürschák 1902.*

§ 37.

Analytische Darstellung der Coordinaten konstruirbarer Punkte.

Ausser den in § 36 behandelten elementargeometrischen Aufgaben giebt es noch eine grosse Reihe weiterer Aufgaben, zu deren Lösung man lediglich das Ziehen von Geraden und das Abtragen von Strecken nötig hat. Um den Bereich aller auf diese Weise lösbaren Aufgaben überblicken zu können, legen wir bei der weiteren Betrachtung ein rechtwinkliges Coordinatensystem zu Grunde und denken uns die Coordinaten der Punkte in der üblichen Weise als reelle Zahlen oder Funktionen von gewissen | willkürlichen Parametern. Um 81
die Frage nach der Gesamtheit aller konstruirbaren Punkte zu beantworten, stellen wir folgende Betrachtung an:

Es sei ein System von bestimmten Punkten gegeben; wir setzen aus den Coordinaten dieser Punkte einen Bereich R zusammen; derselbe enthält gewisse reelle Zahlen und gewisse willkürliche Parameter p. Nunmehr denken wir uns die Gesamtheit aller derjenigen Punkte, die durch Ziehen von Geraden und Abtragen von Strecken aus dem vorgelegten System von Punkten konstruirbar sind. Der Bereich, der von den Coordinaten dieser Punkte gebildet wird, heisse $\Omega(R)$; derselbe enthält gewisse reelle Zahlen und Funktionen der willkürlichen Parameter p.

Unsere Betrachtungen in § 17 zeigen, dass das Ziehen von Geraden und Parallelen analytisch auf die Anwendung der Addition, Multiplikation, Subtraktion, Division von Strecken hinausläuft; ferner lehrt die bekannte in § 9 aufgestellte Formel für die Drehung, dass das Abtragen von Strecken auf einer beliebigen Geraden keine andere analytische Operation erfordert, als die Quadratwurzel zu ziehen aus einer Summe von zwei Quadraten, deren Basen man bereits konstruirt hat. Umgekehrt kann man zufolge des Pythagoräischen Lehrsatzes vermöge eines rechtwinkligen Dreiecks die Quadratwurzel aus der Summe zweier Streckenquadrate durch Abtragen von Strecken stets konstruiren.

Aus diesen Betrachtungen geht hervor, dass der Bereich $\Omega(R)$ alle diejenigen und nur solche reellen Zahlen und Funktionen der Parameter p enthält, die aus den Zahlen und Parametern in R vermöge einer endlichen Anzahl von Anwendungen von fünf Rechnungsoperationen, nämlich der vier elementaren Rechnungsoperationen hervorgehen, wenn man noch das Ziehen der Quadratwurzel aus einer Summe zweier Quadrate als fünfte Rechnungsoperation zulässt. Wir sprechen dieses Resultat wie folgt aus:

Satz 41. Eine geometrische Konstruktionsaufgabe ist dann und nur dann durch Ziehen von Geraden und Abtragen von Strecken, d. h. mittels Lineals und Streckenübertragers lösbar, wenn bei der analytischen Behandlung der Aufgabe die Coordinaten der gesuchten Punkte solche Funktionen

---

11 Betrachtung] Überlegung

33–36 Rechnungsoperationen, ⟨...⟩ zulässt.] Rechnungsoperationen hervorgehen, nämlich der vier elementaren Rechnungsoperationenen und einer fünften Operation, als die man das Ziehen der Quadratwurzel aus einer Summe zweier Quadrate betrachtet.

39 Streckenübertragers] Eichmaßes

der Coordinaten der gegebenen Punkte sind, deren Herstellung nur rationale Operationen und die Operation des Ziehens der Quadratwurzel aus der Summe zweier Quadrate erfordert.

Wir können aus diesem Satze sofort erkennen, dass nicht jede mittelst Zirkels lösbare Aufgabe auch allein mittelst Lineals und Streckenübertragers
82 gelöst werden kann. Zu dem Zwecke | legen wir diejenige Geometrie zu Grunde, die in § 9 mit Hilfe des algebraischen Zahlenbereichs Ω aufgebaut worden ist; in dieser Geometrie giebt es lediglich solche Strecken, die mittelst Lineals und Streckenübertragers konstruirbar sind, nämlich die durch Zahlen des Bereichs Ω bestimmten Strecken.

Ist nun ω irgend eine Zahl in Ω, so erkennen wir aus der Definition des Bereichs Ω leicht, dass auch jede zu ω conjugirte algebraische Zahl in Ω liegen muss, und da die Zahlen des Bereichs Ω offenbar sämtlich reell sind, so folgt hieraus, dass der Bereich Ω nur solche reelle algebraische Zahlen enthalten kann, deren Conjugirte ebenfalls reell sind.

Wir stellen jetzt die Aufgabe, ein rechtwinkliges Dreieck mit der Hypotenuse 1 und einer Kathete $|\sqrt{2}|-1$ zu konstruiren. Nun kommt die algebraische Zahl $\sqrt{2|\sqrt{2}|-2}$, die den Zahlenwert der anderen Kathete ausdrückt, im Zahlenbereich Ω nicht vor, da die zu ihr konjugirte Zahl $\sqrt{-2|\sqrt{2}|-2}$ imaginär ausfällt. Die gestellte Aufgabe ist mithin in der zu Grunde gelegten Geometrie nicht lösbar, und kann daher überhaupt nicht mittelst Lineals und Streckenübertragers lösbar sein, obwohl die Konstruktion mittelst des Zirkels sofort ausführbar ist.

§ 38.

Die Darstellung algebraischer Zahlen und ganzer rationaler Funktionen als Summe von Quadraten.

Die Frage nach der Ausführbarkeit geometrischer Konstruktionen mittelst Lineals und Streckenübertragers erfordert zu ihrer weiteren Behandlung einige Sätze zahlentheoretischen und algebraischen Charakters, die, wie mir scheint, auch an sich von Interesse sind.

Nach *Fermat* ist bekanntlich jede ganze rationale positive Zahl als Summe von vier Quadratzahlen darstellbar. Dieser Fermatsche Satz gestattet eine merkwürdige Verallgemeinerung von folgender Art:

D e f i n i t i o n. Es sei k ein beliebiger Zahlkörper; der Grad dieses Körpers k heisse m, und die $m-1$ zu k konjugirten Zahlkörper mögen mit k', k'', ..., $k^{(m-1)}$ bezeichnet werden. Trifft es sich, dass unter den m Kör-

3 Quadrate] Quadrate und diese fünf Operationen in endlicher Anzahl
5 Streckenübertragers] Eichmaßes
9 Streckenübertragers] Eichmaßes
13 Bereichs] Bereiches
21–22 Streckenübertragers] Eichmaßes
28 Streckenübertragers] Eichmaßes
34 D e f i n i t i o n.] E r k l ä r u n g.

pern $k, k', \ldots, k^{(m-1)}$ einer oder mehrere aus lauter reellen Zahlen gebildet sind, so nennen wir diese | Körper selbst reell; es seien diese Körper etwa 83 $k, k', \ldots, k^{(s-1)}$. Eine Zahl α des Körpers k heisst in diesem Falle *total positiv in* k, falls die s zu α konjugirten bez. in $k, k', \ldots, k^{(s-1)}$ gelegenen Zahlen sämtlich positiv sind. Kommen dagegen in jedem der m Körper $k, k', \ldots, k^{(m-1)}$ auch imaginäre Zahlen vor, so heisst eine jede Zahl α in k stets *total positiv.*

S a t z 42. *Jede total positive Zahl in k lässt sich als Summe von vier Quadraten darstellen, deren Basen ganze oder gebrochene Zahlen des Körpers k sind.*

Der Beweis dieses Satzes bietet erhebliche Schwierigkeiten dar; er beruht wesentlich auf der Theorie der relativquadratischen Zahlkörper, die ich unlängst in mehreren Arbeiten[1]) entwickelt habe. Es sei hier nur auf denjenigen Satz aus dieser Theorie hingewiesen, der die Bedingungen für die Lösbarkeit einer ternären Diophantischen Gleichung von der Gestalt

$$\alpha\xi^2 + \beta\eta^2 + \gamma\zeta^2 = 0$$

angiebt, worin die Coefficienten α, β, γ gegebene Zahlen in k und ξ, η, ζ gesuchte Zahlen in k bedeuten. Der Beweis des Satzes 42 wird durch wiederholte Anwendung des eben genannten Satzes erbracht.

Aus dem Satze 42 folgen eine Reihe von Sätzen über die Darstellung solcher rationaler Funktionen einer Veränderlichen mit rationalen Coefficienten, die niemals negative Werte haben; ich hebe nur den folgenden Satz hervor, der uns im nächsten Paragraphen von Nutzen sein wird.

S a t z 43. Es bedeute $f(x)$ eine solche ganze rationale Funktion von x mit rationalen Zahlencoefficienten, die niemals negative Werte annimmt, wenn man für x beliebige reelle Werte einsetzt: dann lässt sich $f(x)$ stets als Quotient zweier Summen von Quadraten darstellen, so dass die sämtlichen Basen

1) Ueber die Theorie der relativquadratischen Zahlkörper, Jahresber. d. Deutschen Math.-Vereinigung Bd. 6, 1899 und Math. Ann. Bd. 51;[32] ferner: Ueber die Theorie der relativ-Abelschen Zahlkörper, Nachr. d. K. Ges. d. Wiss. zu Göttingen 1898.[33]

6 heisst] heiße

13 ⟨In E2 is added to the end of the footnote:⟩ und Acta mathematica Bd. 26'[34]

22 haben; ⟨...⟩ hervor] haben.
Erwähnt sei noch folgender Satz

26–27 als Quotient zweier Summen] als Summe

[32] *Hilbert 1898a, 1899d.*

[33] *Hilbert 1898d.*

[34] *Hilbert 1902a.*

dieser Quadrate ganze rationale Funktionen von x mit rationalen Coefficienten sind.

Beweis. Den Grad der vorgelegten Funktion $f(x)$ wollen wir mit m bezeichnen; offenbar muss derselbe jedenfalls gerade sein. Für den Fall $m = 0$, d. h. wenn $f(x)$ eine rationale Zahl ist, folgt die Richtigkeit des Satzes 43 unmittelbar aus dem Fermatschen Satze von der Darstellung einer positiven Zahl als Summe von vier Quadratzahlen. Wir nehmen nun an, der Satz sei
84 | bereits für die Funktionen vom Grade 2, 4, 6, ..., $m-2$ bewiesen, und zeigen dann seine Richtigkeit für den vorliegenden Fall einer Funktion m^{ten} Grades auf folgende Weise.

Zunächst behandeln wir kurz die Annahme, dass $f(x)$ in das Produkt von zwei oder mehreren ganzen Funktionen von x mit rationalen Coefficienten zerfällt. Ist $p(x)$ eine solche in $f(x)$ aufgehende Funktion, die selbst nicht mehr in ein Produkt von ganzen Funktionen mit rationalen Coefficienten zerlegt werden kann, so folgt aus dem vorausgesetzten definiten Charakter der Funktion $f(x)$ leicht, dass der Faktor $p(x)$ entweder in $f(x)$ zu einer geraden Potenz erhoben vorkommen muss oder dass $p(x)$ selbst definit, d. h. eine solche Funktion ist, die für reelle Werthe von x niemals negativ ausfällt. Im ersteren Falle ist der Quotient $\dfrac{f(x)}{\{p(x)\}^2}$, im letzteren Falle sind sowohl $p(x)$ als auch $\dfrac{f(x)}{p(x)}$ wiederum definite Funktionen und diese Funktionen besitzen einen geraden Grad $< m$. Zufolge unserer Annahme sind daher im ersteren Falle $\dfrac{f(x)}{\{p(x)\}^2}$, im letzteren Falle sowohl $p(x)$ wie auch $\dfrac{f(x)}{p(x)}$ als Quotienten von Quadratsummen von der im Satze 43 angegebenen Art darstellbar, und daher gestattet notwendig in beiden Fällen auch die Funktion $f(x)$ die verlangte Darstellung.

Wir behandeln nunmehr die Annahme, dass $f(x)$ nicht in das Produkt von zwei ganzen Funktionen mit rationalen Coefficienten zerlegt werden kann. Die Gleichung $f(\vartheta) = 0$ definirt dann einen algebraischen Zahlkörper $k(\vartheta)$ vom m-ten Grade, der nebst seinen sämtlichen conjugirten Körpern imaginär ausfällt. Da somit nach der Definition, die wir dem Satze 42 vorangestellt haben, jede in $k(\vartheta)$ gelegene Zahl, also auch insbesondere die Zahl -1 total positiv in $k(\vartheta)$ ist, so giebt es nach diesem Satz 42 eine Darstellung der Zahl -1 als Summe von 4 Quadraten gewisser Zahlen in $k(\vartheta)$; es sei

$$-1 = \alpha^2 + \beta^2 + \gamma^2 + \delta^2, \tag{1}$$

wobei α, β, γ, δ ganze oder gebrochene Zahlen in $k(\vartheta)$ sind. Wir setzen

2 sind.] ⟨After 'sind' a footnote is added in E2:⟩ Der Beweis für die Darstellbarkeit von $f(x)$ als Quotient zweier Quadratsummen ist von mir auf Grund des Satzes 42 in der ersten Auflage ausgeführt worden. Inzwischen ist es *E. Landau* gelungen, den Beweis für die Darstellbarkeit von $f(x)$ direkt als Quadratsumme, wie oben behauptet, zu erbringen und zwar lediglich mit Benutzung sehr einfacher und elementarer Hilfsmittel. Math. Ann. Bd. 57 1903.[35]

3–519.22 Beweis. ⟨...⟩ vollständig erbracht.] ⟨deleted⟩

[35] *Landau 1903.*

$$\begin{aligned}\alpha &= a_1\vartheta^{m-1} + a_2\vartheta^{m-2} + \cdots + a_m = \varphi(\vartheta),\\ \beta &= b_1\vartheta^{m-1} + b_2\vartheta^{m-2} + \cdots + b_m = \psi(\vartheta),\\ \gamma &= c_1\vartheta^{m-1} + c_2\vartheta^{m-2} + \cdots + c_m = \chi(\vartheta),\\ \delta &= d_1\vartheta^{m-1} + d_2\vartheta^{m-2} + \cdots + d_m = \varrho(\vartheta);\end{aligned}$$

hierin bedeuten $a_1, a_2, \ldots, a_m, \ldots, d_1, d_2, \ldots, d_m$ rationale Zahlencoefficien- 85
ten und $\varphi(\vartheta)$, $\psi(\vartheta)$, $\chi(\vartheta)$, $\varrho(\vartheta)$ die betreffenden ganzen rationalen Funktionen vom $(m-1)$-ten Grade in ϑ.

Wegen (1) ist

$$1 + \{\varphi(\vartheta)\}^2 + \{\psi(\vartheta)\}^2 + \{\chi(\vartheta)\}^2 + \{\varrho(\vartheta)\}^2 = 0,$$

und mit Rücksicht auf die Irreducibilität der Gleichung $f(x) = 0$ stellt daher der Ausdruck

$$F(x) = 1 + \{\varphi(x)\}^2 + \{\psi(x)\}^2 + \{\chi(x)\}^2 + \{\varrho(x)\}^2$$

notwendig eine solche ganze rationale Funktion von x dar, die durch $f(x)$ teilbar ist. $F(x)$ ist offenbar eine definite Funktion vom $(2m-2)$-ten oder von niederem Grade und daher wird der Quotient $\dfrac{F(x)}{f(x)}$ eine definite Funktion vom $(m-2)$-ten oder von niederem Grade in x mit rationalen Coefficienten. Infolgedessen lässt sich im Hinblick auf unsere Annahme $\dfrac{F(x)}{f(x)}$ als Quotient zweier Summen von Quadraten von der im Satze 43 angegebenen Art darstellen, und da $F(x)$ selbst eine solche Summe von Quadraten ist, so folgt, dass auch $f(x)$ ein Quotient zweier Summen von Quadraten von der im Satze 43 angegebenen Art sein muss. Damit ist der Beweis des Satzes 43 vollständig erbracht.

Es dürfte sehr schwierig sein, die entsprechenden Thatsachen für ganze rationale Funktionen von zwei oder mehr Veränderlichen aufzustellen und zu beweisen, doch sei hier darauf hingewiesen, dass die Darstellbarkeit einer beliebigen definiten ganzen rationalen Funktion zweier Veränderlicher als Quotient von Quadratsummen ganzer Funktionen auf einem völlig anderen Wege von mir bewiesen worden ist – unter der Voraussetzung, dass für die darstellenden Funktionen nicht blos rationale, sondern beliebige reelle Coefficienten zulässig sind[1]).

§ 39.

Kriterium für die Ausführbarkeit geometrischer Konstruktionen mittelst Lineals und Streckenübertragers.

Es sei eine geometrische Konstruktionsaufgabe vorgelegt, die mittelst des Zirkels ausführbar ist; wir wollen dann ein Kriterium aufzustellen versuchen, welches unmittelbar aus der analy|tischen Natur der Aufgabe und ihrer Lö- 86

1) Ueber ternäre definite Formen, Acta Mathematica Bd. 17.[36]

26 definiten] definitiven

33 **Streckenübertragers] Eichmaßes**

[36] *Hilbert 1893a.*

sungen beurteilen lässt, ob die Konstruktion auch allein mittelst Lineals und Streckenübertragers ausführbar ist. Wir werden bei dieser Untersuchung auf den folgenden Satz geführt:

Satz 44. *Es sei eine geometrische Konstruktionsaufgabe vorgelegt von der Art, dass man bei analytischer Behandlung derselben die Coordinaten der gesuchten Punkte aus den Coordinaten der gegebenen Punkte lediglich durch rationale Operationen und durch Ziehen von Quadratwurzeln finden kann; es sei n die kleinste Anzahl der Quadratwurzeln, die hierbei zur Berechnung der Coordinaten der Punkte ausreichen: soll dann die vorgelegte Konstruktionsaufgabe sich auch allein durch Ziehen von Geraden und Abtragen von Strecken ausführen lassen, so ist dafür notwendig und hinreichend, dass die geometrische Aufgabe genau 2^n reelle Lösungen besitzt und zwar für alle Lagen der gegebenen Punkte, d. h. für alle Werte der in den Coordinaten der gegebenen Punkte auftretenden willkürlichen Parameter.*

Beweis. Wir beweisen diesen Satz 44 ausschliesslich für den Fall, dass die Coordinaten der gegebenen Punkte rationale Funktionen eines Parameters p mit rationalen Coefficienten sind.

Die Notwendigkeit des aufgestellten Kriteriums leuchtet ein. Um zu zeigen, dass dasselbe auch hinreicht, setzen wir dieses Kriterium als erfüllt voraus und betrachten zunächst eine solche von jenen n Quadratwurzeln, die bei der Berechnung der Coordinaten der gesuchten Punkte zuerst zu ziehen ist. Der Ausdruck unter dieser Quadratwurzel ist eine rationale Funktion $f_1(p)$ des Parameters p mit rationalen Coefficienten; diese rationale Funktion darf für beliebige reelle Parameterwerte p niemals negative Werte annehmen, da sonst entgegen der Voraussetzung die vorgelegte Aufgabe für gewisse Werte p imaginäre Lösungen haben müsste. Aus Satz 43 schliessen wir daher, dass $f_1(p)$ als Quotient von Summen von Quadraten ganzer rationaler Funktionen darstellbar ist.

Nunmehr zeigen die Formeln

$$\sqrt{a^2+b^2+c^2} = \sqrt{\left(\sqrt{a^2+b^2}\right)^2+c^2},$$
$$\sqrt{a^2+b^2+c^2+d^2} = \sqrt{\left(\sqrt{a^2+b^2+c^2}\right)^2+d^2},$$
$$\cdots\cdots\cdots\cdots\cdots\cdots\cdots\cdots\cdots\cdots\cdots$$

dass allgemein das Ziehen der Quadratwurzel aus einer Summe von beliebig
87 vielen Quadraten sich stets zurückführen lässt auf | wiederholtes Ziehen der Quadratwurzel aus der Summe zweier Quadrate.

Nehmen wir diese Bemerkung mit dem vorigen Ergebnisse zusammen, so erkennen wir, dass der Ausdruck $\sqrt{f_1(p)}$ gewiss mittelst Lineals und Streckenübertragers construirt werden kann.

Wir betrachten ferner eine solche von den n Quadratwurzeln, die bei der Berechnung der Coordinaten der gesuchten Punkte an zweiter Stelle zu

2 Streckenübertragers] Eichmaßes

18 leuchtet] leuchtet aus § 37

21 ziehen] suchen

37–38 Streckenübertragers] Eichmaßes

ziehen ist. Der Ausdruck unter dieser Quadratwurzel ist eine rationale Funktion $f_2\left(p, \sqrt{f_1}\right)$ des Parameters p und der zuerst betrachteten Quadratwurzel; auch diese Funktion f_2 ist bei beliebigen reellen Parameterwerten p und für jedes Vorzeichen von $\sqrt{f_1}$ niemals negativer Werte fähig, da sonst entgegen der Voraussetzung die vorgelegte Aufgabe unter ihren 2^n Lösungen für gewisse Werte p auch imaginäre Lösungen haben müsste. Aus diesem Umstande folgt, dass f_2 einer quadratischen Gleichung von der Gestalt

$$f_2^2 - \varphi_1(p) f_2 + \psi_1(p) = 0$$

genügen muss, worin $\varphi_1(p)$ und $\psi_1(p)$ notwendig solche rationale Funktionen von p mit rationalen Coefficienten sind, die für reelle Werte von p niemals negative Werte besitzen. Aus der letzteren quadratischen Gleichung entnehmen wir

$$f_2 = \frac{f_2^2 + \psi_1(p)}{\varphi_1(p)}.$$

Nun müssen wiederum nach Satz 43 die Funktionen $\varphi_1(p)$ und $\psi_1(p)$ Quotienten von Summen von Quadraten rationaler Funktionen sein und andererseits ist nach dem Vorigen der Ausdruck f_2 mittelst Lineals und Streckenübertragers construirbar; der gefundene Ausdruck für f_2 zeigt somit, dass f_2 ein Quotient von Summen von Quadraten construirbarer Funktionen ist. Also lässt sich auch der Ausdruck $\sqrt{f_2}$ mittelst Lineals und Streckenübertragers construiren.

Ebenso wie der Ausdruck f_2 erweist sich auch jede andere rationale Funktion $\varphi_2\left(p, \sqrt{f_1}\right)$ von p und $\sqrt{f_1}$ als Quotient zweier Summen von Quadraten construirbarer Funktionen, sobald diese rationale Funktion φ_2 die Eigenschaft besitzt, niemals negative Werte anzunehmen bei reellem Parameter p und für beiderlei Vorzeichen von $\sqrt{f_1}$.

Diese Bemerkung gestattet uns, das eben begonnene Schlussverfahren in folgender Weise fortzusetzen.

Es sei $f_3\left(p, \sqrt{f_1}, \sqrt{f_2}\right)$ ein solcher Ausdruck, der von den drei | Argumen- 88
ten $p, \sqrt{f_1}, \sqrt{f_2}$ in rationaler Weise abhängt und aus dem bei der analytischen Berechnung der Coordinaten der gesuchten Punkte an dritter Stelle die Quadratwurzel zu ziehen ist. Wie vorhin schliessen wir, dass f_3 bei beliebigen reellen Werten p und für beiderlei Vorzeichen von $\sqrt{f_1}$ und $\sqrt{f_2}$ niemals negative Werte annehmen darf; dieser Umstand wiederum zeigt, dass f_3 einer quadratischen Gleichung von der Gestalt

$$f_3^2 - \varphi_2\left(p, \sqrt{f_1}\right) f_3 + \psi_2\left(p, \sqrt{f_1}\right) = 0$$

genügen muss, worin φ_2 und ψ_2 solche rationale Funktionen von p und $\sqrt{f_1}$ bedeuten, die für reelle Werte p und beiderlei Vorzeichen von $\sqrt{f_1}$ negativer Werte nicht fähig sind. Da mithin φ_2 und ψ_2 nach der vorigen Bemerkung Quotienten zweier Summen von Quadraten construirbarer Ausdrücke sind, so

---

16–17 Streckenübertragers] Eichmaßes

19 Streckenübertragers] Eichmaßes

folgt das gleiche auch für den Ausdruck

$$f_3 = \frac{f_3^2 + \psi_2\left(p, \sqrt{f_1}\right)}{\varphi_2\left(p, \sqrt{f_1}\right)}$$

und mithin ist auch $\sqrt{f_3}$ mittelst Lineals und Streckenübertragers construirbar.

Die Fortsetzung dieser Schlussweise führt zum Beweise des Satzes 44 in dem betrachteten Falle eines Parameters p.

Die allgemeine Richtigkeit des Satzes 44 hängt davon ab, ob der Satz 43 in entsprechender Weise sich auf den Fall mehrerer Veränderlicher verallgemeinern lässt.

Als Beispiel für die Anwendung des Satzes 44 mögen die regulären mittelst Zirkels construirbaren Polygone dienen; in diesem Falle kommt ein willkürlicher Parameter p nicht vor, sondern die zu construirenden Ausdrücke stellen sämtlich algebraische Zahlen dar. Man sieht leicht, dass das Kriterium des Satzes 44 erfüllt ist, und somit ergiebt sich, dass man jene regulären Polygone auch allein mittelst Ziehens von Geraden und Abtragens von Strecken construiren kann – ein Resultat, welches sich auch aus der Theorie der Kreisteilung direkt entnehmen lässt.

Was weitere aus der Elementargeometrie bekannte Konstruktionsaufgaben anbetrifft, so sei hier nur erwähnt, dass das Malfattische Problem, nicht aber die Apollonische Berührungsaufgabe allein mittelst Lineals und Streckenübertragers gelöst werden kann.

---

89 ## Schlusswort.

Die vorstehende Abhandlung ist eine kritische Untersuchung der Principien der Geometrie; in dieser Untersuchung leitete uns der Grundsatz, eine jede sich darbietende Frage in der Weise zu erörtern, dass wir zugleich prüften, ob ihre Beantwortung auf einem vorgeschriebenen Wege mit gewissen eingeschränkten Hilfsmitteln möglich oder nicht möglich ist. Dieser Grundsatz scheint mir eine allgemeine und naturgemässe Vorschrift zu enthalten; in der That wird, wenn wir bei unseren mathematischen Betrachtungen einem Probleme begegnen oder einen Satz vermuten, unser Erkenntnistrieb erst dann befriedigt, wenn uns entweder die völlige Lösung jenes Problems und der strenge Beweis dieses Satzes gelingt oder wenn der Grund für die Unmöglichkeit des Gelingens und damit zugleich die Notwendigkeit des Misslingens von uns klar erkannt worden ist.

---

3 Streckenübertragers] Eichmaßes

20–21 Streckenübertragers] Eichmaßes

21 kann.] ⟨After 'kann' a footnote is added in E2:⟩ Betreffs weiterer geometrischer Konstruktionen mittels Lineals und Eichmaßes vgl. *M. Feldblum*, Über elementargeometrische Konstruktionen, Inauguraldissertation, Göttingen 1899.[37]

[37] *Feldblum 1899*.

So spielt denn in der neueren Mathematik die Frage nach der Unmöglichkeit gewisser Lösungen oder Aufgaben eine hervorragende Rolle und das Bestreben, eine Frage solcher Art zu beantworten, war oftmals der Anlass zur Entdeckung neuer und fruchtbarer Forschungsgebiete. Wir erinnern nur an *Abel*'s Beweis für die Unmöglichkeit der Auflösung der Gleichungen fünften Grades durch Wurzelziehen, ferner an die Erkenntnis der Unbeweisbarkeit des Parallelenaxioms und an *Hermite*'s und *Lindemann*'s Sätze von der Unmöglichkeit, die Zahlen e und π auf algebraischem Wege zu construiren.

Der Grundsatz, demzufolge man überall die Principien der Möglichkeit der Beweise erläutern soll, hängt auch aufs Engste mit der Forderung der „Reinheit" der Beweismethoden zusammen, die von mehreren Mathematikern der neueren Zeit mit Nachdruck erhoben worden ist. Diese Forderung ist im Grunde nichts Anderes als eine subjektive Fassung des hier befolgten Grundsatzes. In der That sucht die vorstehende geometrische Untersuchung allgemein darüber Aufschluss zu geben, welche Axiome, Voraussetzungen oder Hilfsmittel zum Beweise einer elementar-geo|metrischen Wahrheit nötig sind, 90
und es bleibt dann dem jedesmaligen Ermessen anheim gestellt, welche Beweismethode von dem gerade eingenommenen Standpunkte aus zu bevorzugen ist.

Bei der Anfertigung der Figuren sowie bei der Durchsicht der Correcturbogen habe ich mich der Hülfe des Herrn Dr. *Hans von Schaper* erfreut; ich spreche ihm hierfür meinen Dank aus. Desgleichen danke ich meinem Freunde *Hermann Minkowski* und Herrn Dr. *Julius Sommer* für ihre Unterstützung beim Lesen der Correctur.

20–24 Bei ⟨...⟩ Correctur.] Zum Schlusse spreche ich meinem Freunde Hermann Minkowski und Dr. Max Dehn für ihre Unterstützung beim Lesen der Korrektur und die empfangenen mannigfachen und wertvollen Ratschläge meinen herzlichsten Dank aus.

Göttingen, August 1903. ⟨In E2, this paragraph is moved to the conclusion after *Anhang* V.⟩

Inhalt.

---

12 Axiom der Stetigkeit (Archimedisches Axiom)] Axiome der Stetigkeit

25 Flächengleichheit] Zerlegungsgleichheit

27 Flächenmass] Inhaltsmaß

28 Flächenmass] Inhaltsmaß

14 Streckenübertragers] Eichmaßes
19 Streckenübertragers] Eichmaßes

Appendix: Extracts from 'Les principes fondamentaux de la géométrie' (1900), the French translation of the *Festschrift*

⟨*Extract 1, Hilbert 1900c*, pp. 123–124:⟩

Note (1). – Remarquons qu'aux cinq précédents groupes d'axiomes l'on peut encore ajouter l'axiome suivant qui n'est pas d'une nature purement géométrique et qui, au point de vue des principes, mérite une attention particulière.

AXIOME D'INTÉGRITÉ (*Vollständigkeit*) (2)

Au système des points, droites et plans, il est impossible d'adjoindre d'autres êtres de manière que le système ainsi généralisé forme une nouvelle géométrie où les axiomes des cinq groupes I–V *soint tous vérifiés ; en d'autres termes : les éléments de la Géométrie forment un système d'êtres qui, si l'on conserve tous les axiomes, n'est susceptible d'aucune extension*

Cet axiome ne nous dit rien sur l'existence de points limites ni sur la notion de convergence ; néanmoins l'on peut en l'invoquant démonstrer ce théorème de Bolzano en vertu duquel, pour tout ensemble de points situés sur une droite entre deux points de celle-ci, il doit toujours nécessairement exister un point de condensation. La valeur de cet axiome au point de vue des principes tient donc à ce que l'existence de tous les points limites en est une
124 consequénce et que, par | suite, cet axiome rend possible la correspondance univoque et reversible des points d'une droite et de tous les nombres réels. D'ailleurs, dans le cours des présentes recherches, nous ne nous sommes servi nulle part de cet « axiome d'intégrité ».

(1) M. Hilbert a bien voulu écrire cette Note inédite pour la traduction de son Mémoire, ainsi qu'une longue addition à la conclusion. Le traducteur saisit avec empressement cette occasion pour présenter ici ses très cordiaux remerciements à M. L. Gérard, professeur au Lycée Charlemagne, et à M. P. Stäckel, professeur à l'Université de Kiel, pour leurs précieux conseils et leur aide dans la correction des épreuves. Il ne saurait oublier non plus de remercier encore une fois M. Hilbert et M. Teubner d'avoir autorisé la publication de ce Mémoire. ⟨L. Laugel⟩

(2) Comparer ma Communication à la réunion de savants tenue à Munich en 1899 : *Ueber den Zahlbegriff* (*Berichte der Deutschen Mathematiker-Vereinigung,* 1900).

D. HILBERT

⟨*Extract 2*, *Hilbert 1900c*, pp. 204–208:⟩

Conclusion (¹).

Le précédent Travail ne traite essentiellement que les problèmes de la Géométrie *euclidienne*, c'est-à-dire qu'il n'y est discuté que les questions qui se présentent quand on admet l'exactitude d l'axiome | des parallèles. Il n'en est 205
pas moins important de discuter les principes et les théorèmes fondamentaux de la Géométrie quand on fait abstraction de l'axiome des parallèles. Nous avons aussi exclu de notre étude la question importante de savoir s'il est possible, sans la notion du plan ni de la droite, au seul moyen des points comme éléments et en employant la notion des groupes des déplacements, ou à l'aide de la notion de distance, d'édifier la Géométrie d'une manière logique. Cette dernière question a fait récemment des progrès considérables, grâce aux travaux fondamentaux et féconds de Sophus Lie. Néanmoins, pour éclaircir complètement la question, il serait bon de subdiviser en plusieurs l'axiome de Lie que l'espace est une multiplicité numérique ; et avant tout il me semblerait désirable que l'on fit une discussion approfondie de l'hypothèse de Lie que les fonctions qui donnent les déplacements sont non seulement continues, mais encore susceptibles de différentiation. Quant à moi, il ne me semble pas probable que les axiomes géométriques renfermés dans la condition de la possibilité de la différentiation soient tous nécessaires.

Dans le traitement de toutes les questions de ce genre, je crois que les méthodes et les principes développés dans le précédent Mémoire seront utiles. Comme exemple je renverrai à une étude entreprise à mon instigation par M. Dehn et qui vient de paraitre (²). Dans cette étude sont discutés les théorèmes connus de Legendre sur la somme des angles d'un triangle, que ce géomètre a démonstrés au moyen de la continuité.

Les considérations de M. Dehn reposent sur les axiomes de l'association, de la distribution et de la congruence, c'est-à-dire les groupes d'axiomes I, II, IV ; au contraire, l'axiome des parallèles et l'axiome d'Archimède sont exclus. D'autre part, les axiomes de distributions sont énoncés d'une manière plus générale que dans le travail actuel, à peu près comme il suit : Parmi quatre points A, B, C, D d'une droite, il y en a toujours deux, A, C, par exemple, qui sont séparés par les deux autres B et D, et réciproquement. Cinq points A, B, C, D, E sur une droite peuvent toujours être distribués de telle sorte
que A, C | soient séparés par B, D et par B, E, ensuite que A, D soient 206
séparés par B, E et par C, E et ainsi de suite. De cette façon, ce qui n'a pas lieu dans mon présent Mémoire, la Géométrie riemannienne (elliptique) n'est pas exclue *a priori*.

En se basant sur les axiomes d'association, de distribution et de congruence, c'est-à-dire sur les axiomes I, II, IV, on peut introduire de la manière

(¹) A partir d'ici jusqu'au Tableau de la page 208, le texte est entièrement nouveau et n'existe pas dans l'édition allemande. (*Le Traducteur.*)

(²) *Math. Annalen,* t. LIII (1900).

connue les éléments dits *idéaux* (points, droites, plans idéaux). Cela fait, M. Dehn démontre la théorème suivant :

Si l'on regarde toutes les droites et tous les points (idéaux et réels) du plan, à l'exception d'une droite unique t et des points situés sur t, comme éléments d'une nouvelle Géométrie, on peut pour cette nouvelle Géométrie définir un nouveau genre de congruence de telle sorte que cette Géométrie vérifie tous les axiomes d'association, de distribution, de congruence, ainsi que l'axiome d'Euclide, *la droite t dans cette Géométrie jouant le rôle de la* droite de l'infini.

Cette Géométrie *euclidienne* imposée pour ainsi dire au plan *non euclidien* sera dite une *pseudo-Géométrie* et le nouveau genre de congruence une *pseudo-congruence.*

En invoquant le théorème qui précède, on peut alors introduire un calcul segmentaire relatif au plan, en s'appuyant sur les développements du Chap. III, § 15. Ce calcul segmentaire permet de démontrer l'important théorème suivant :

Si dans un *triangle quelconque la somme des angles est*

$$\left.\begin{array}{l}\text{plus grande que} \ldots\ldots\ldots\ldots\ldots \\ \text{égale à} \ldots\ldots\ldots\ldots\ldots\ldots\ldots\ldots \\ \text{plus petite que} \ldots\ldots\ldots\ldots\ldots\end{array}\right\} \text{2 droits}$$

il en sera de même dans tout *triangle.*

Le cas où la somme des angles est égale à deux droits donne le théorème bien connu de Legendre ; mais, pour le démontrer, Legendre s'est servi de la continuité.

M. Dehn discute alors la connexion entre les trois différentes hypothèses relatives à la somme des angles et les trois différentes hypothèses relatives aux parallèles.

207 Il arrive ainsi aux remarquables propositions suivantes :

De l'hypothèse que par un point donné l'on peut mener à une droite une infinité de parallèles il s'ensuit, si l'on exclut l'axiome d'Archimède, NON PAS que la somme des angles d'un triangle est plus petite que deux droits, mais au contraire que cette somme peut être

(*a*) plus grande que 2 droits
(*b*) égale à 2 droits.

Pour démontrer le cas (*a*) de ce théorème, M. Dehn édifie une Géométrie où l'on peut mener par un point une infinité de parallèles à une droite et où, d'ailleurs, sont aussi vérifiés tous les théorèmes de la Géométrie riemanienne (elliptique). A cette Géométrie convient le nom de Géométrie *non legendrienne*, car elle est en contradiction avec le théorème de Legendre en vertu duquel la somme des angles d'un triangle n'est jamais plus grande que 2 droits. De l'existence de cette Géométrie non legendrienne il résulte immédiatement qu'il est impossible de démontrer le précédent théorème de Legendre sans employer l'axiome d'Archimède ; et, en effet, Legendre se sert de la continuité pour démontrer son théorème.

Pour démontrer le cas (b) du théorème précité, on édifie une Géométrie *sans* axiome des parallèles et où sont néanmoins vérifiés tous les théorèmes de la Géométrie euclidienne : la somme des angles d'un triangle est égale à deux droits, il y a des triangles semblables, les extrémités de perpendiculaires de même longueur menées à une droite sont toutes situées sur la même droite, etc. De l'existence de cette Géométrie s'ensuit que, si l'on fait abstraction de l'axiome d'Archimède, l'axiome des parallèles ne peut être remplacé par aucune des propositions que l'on regarde d'habitude comme lui étant équivalentes.

Cette nouvelle Géométrie peut être dite une *Géométrie semi-euclidienne.* De même que la Géométrie non legendrienne, il est clair que la Géométrie semi-euclidienne est en même temps une Géométrie non archimédienne.

M. Dehn arrive finalement à ce théorème surprenant :

De l'hypothèse qu'il n'existe aucune parallèle, il s'ensuit que la somme des angles d'un triangle est plus grande que deux droits.

Ce théorème montre que les deux hypothèses non euclidienne sur les par- 208
allèles se comportent d'une manière absolument différente vis-à-vis de l'axiome d'Archimède.

On peut réunir les résultats énoncés par les théorèmes précédents dans le Tableau suivant :

La somme des angles d'un triangle est	Par un point donné l'on put mener à une droit		
	aucune parallèle.	une parallèle.	une infinité de parallèles.
$>$ 2 droits	Géométrie de Riemann (elliptique)	Cas impossible	Géométrie non legendrienne
$<$ 2 droits	Cas impossible	Géométrie euclidienne (parabolique)	Géométrie semi-euclidienne
$=$ 2 droits	Cas impossible	Cas impossible	Géométrie de Lobatschewski (hyperbolique)

Maintenant mon présent Travail, comme je l'ai dit, est plutôt une recherche critique sur les principes de la Géométrie *euclidienne.* $\langle \dots \rangle$

Chapter 6

Lectures on the Foundations of Geometry (1902)

Introduction

This Chapter presents Hilbert's lectures on the foundations of geometry from the summer semester of 1902 in an *Ausarbeitung* by August Adler (*Hilbert 1902d**), and presumably written in his hand. It exists in the *Lesesaal* of the Göttingen Mathematical Institute. Unusually, Adler was neither an *Assistent* nor a student of Hilbert's, nor was he even a student in Göttingen. On his application for admission as *Gasthörer*, Adler himself stated that he was a Professor at the German *Oberrealschule* in Prague-Karolinenthal. Hence, this is very likely the August Adler to whose paper (*Adler 1890*) Hilbert refers in the sixth of the remarks written on his copy of the *Ausarbeitung* of the 1898/1899 lectures, and which form the Appendix to Chapter 4. This Adler (born in 1863, died in 1923) is described in Poggendorf as a teacher at the German *Realschule* in Prague (1895–1906), and from 1901–1906 as also a *Privatdozent* at the Prague *Technische Hochschule*. From 1907 on he was Director of the *Staatsrealschule* in Vienna, and also *Privatdozent* at the *Technische Hochschule* there. Hilbert already refers to his work as early as the 1897/1898 lectures on 'Zahlbegriff und Quadratur des Kreises' (*Hilbert 1897b**, 48), and again in the sketch for the lectures on the same subject for 1901/1902 (*Hilbert 1897b**, 52–54). If this identification is correct, then Adler was certainly known to Hilbert (at least through his papers) before he undertook the *Ausarbeitung* which is the subject of this Chapter.

These lectures are rather different, both from those of 1898/1899 (Chapter 4) and from the *Festschrift* (Chapter 5). Certainly the Introduction exhibits many familiar themes: (i) Geometry is the oldest natural science, with a reference again, if only indirect, to Hertz's *Bildtheorie* characterization of natural science. (ii) The presentation of a theory according to the axiomatic method furnishes a 'Fachwerk von Begriffen' abstracted from 'Tatsachen', with the axioms corresponding to 'Grundtatsachen'. (iii) This *Fachwerk* must be examined for certain properties, completeness, independence (of the axioms from one another), and freedom from contradiction, the conditions given in the *Festschrift* and in *Hilbert 1900b*. (iv) The difference between analytic geometry (reliance on number) and synthetic geometry (reliance on intuition) is stressed, as is the difference of Hilbert's project from both of these, two of the aims of Hilbert's treatment being to investigate the theoretical basis in synthetic geometry for the 'Einführung der Zahl', thus to exhibit the bridge between this and analytic geometry, and at the same time to carry out a 'logische Analyse unseres Anschauungsvermögens'. (v) As in the 1898/1899 lectures, this exercise relies on the 'Gesetze der reinen Logik und Zahlentheorie' as given. In addition, some of the important theoretical material of the earlier lectures and/or the *Festschrift* is presented, for example, the theory of proportions is developed, with an appropriate *Streckenrechnung* (pp. 91–100); the theory of surface content is quickly presented (pp. 100–107); the same model of Bolyai-Lobatchevsky geometry is given as in 1898/1899 (pp. 59–72); non-Archimedean geometries are specified (pp. 49–59); and the Legendre Theorems are set out and proved (pp. 72–75). There is, too, the same connection drawn (p. 47) between mathematical existence and consistency which we find in the 1898/1899 lectures and in the correspondence with Frege, although there is a strong reiteration of one part of Hilbert's position heavily criticised by Frege, namely that:

> Die Dinge, mit denen die Mathematik sich beschäftigt, werden durch Axiome definiert, *in's Leben gerufen* (p. 47).

What, then, are the differences? Firstly, the treatments here are, on the whole, shorter and tighter, although correspondingly sometimes less well motivated and sometimes also not as clear. Secondly, there are some structural differences in the axiomatic framework which point the way to the revisions incorporated in the second edition of the *Festschrift* in 1903. To give one small example, Axiom II 4 of the original *Festschrift* is dropped, and proved instead (see pp. 14–15). But it would be quite wrong to think that the lectures are a preparation for the new edition of the *Festschrift*. For one thing, there is important *Festschrift* material that is not explicitly treated at all, e.g., the Desargues and Pascal Theorems, so central in the original development of the segment calculi. For another, there is more concentration on incidence and order structures than there is in the *Festschrift*, and much more attention is paid to the congruence axioms, specifically the Triangle Congruence Axiom. But more importantly, the lectures present developments in Hilbert's analysis of geometry which came after the original *Festschrift*, and which were never incorporated into the body of that work, though the resultant papers did find their way into the second and subsequent editions as appendices. These concern the deductive weight of the elementary theorem on the base angles in isosceles triangles, the development of Bolyai-Lobatchevsky geometry, and the Helmholtz-Lie approach to geometry. There is also a brief treatment of semi-Euclidean geometry. A few words on each of these developments is in order.

In the brief treatment of semi-Euclidean geometry, which derives from Dehn's work (see the Introduction to Chapter 4, Specific Note 5(a) and the second extract from the French translation of the *Festschrift* in the Appendix to Chapter 5), Hilbert constructs a simple non-Archimedean model in which the central Euclidean theorem (*Elements*, I, 32) holds that the angles of a triangle are two right-angles, but in which the Parallel Axiom itself fails. The theorem on the angle-sum is a straightforward consequence of the Parallel Axiom, and was standardly taken to be a possible replacement for it. Later in the lectures, Hilbert proves that, if the Archimedean Axiom is assumed, then the angle-sum theorem is indeed equivalent to the Parallel Axiom. (See pp. 75–83.) He concludes, therefore, that this theorem can *only* serve as a replacement for the Parallel Axiom in the presence of the Archimedean Axiom (p. 83).

The work on the theorem that the base angles in an isosceles triangle are equal (call this proposition P) is the reason for Hilbert's reexamination of the Triangle Congruence Axiom earlier in the lectures. P follows easily from the usual form of this, or from its immediate consequence, the First Triangle

Congruence Theorem, which legitimizes the side-angle-side criterion of congruence. Hilbert states a weakening of the axiom on pp. 22–25, and then devotes a special section (section A, pp. 26–31) to showing that certain central consequences still follow from this weaker version. This is important for the later consideration of P (pp. 107–125a), for P no longer follows from the weakening. The weakened form of triangle congruence proposed by Hilbert governs only pairs of triangles which are in the same orientation in the plane, which is natural if one takes seriously the view that congruence is intuitively based on displacement and superposition in the *plane*, where 'sliding' can be appealed to, but no use can be made of a spatial dimension in which figures can be 'flipped' over. The usual axiom allows 'flipping', and the usual proof of P exploits this, which is why P no longer follows from the weaker axiom alone. One of the central questions in the examination of P, therefore, centres on what has to be added to the normal system, with both the Parallel Axiom and the Archimedean Axiom but only the weaker triangle congruence axiom, to enable P to be proved. Hilbert's first answer is that one can prove P if one adopts a second continuity axiom which he calls the *Axiom der Nachbarschaft.* This latter axiom (p. 84) states that, given any segment AB, there exists a triangle ('oder Quadrat etc.') in whose interior there is no segment congruent to AB. The bulk of Hilbert's investigation is devoted to showing that this axiom and the Archimedean Axiom are both essential if P is to be proved in this way. In short, geometries can be constructed in which all the plane axioms hold (with the weaker version of triangle congruence), and where *Nachbarschaft* holds, but where the Archimedean Axiom and P both fail, or similarly where the Archimedean Axiom holds, but where *Nachbarschaft* and P fail. Hilbert points out that, in the model where the Archimedean Axiom is not fulfilled, the Pythagorean Theorem holds (in the form where one shows the *Inhaltsgleichheit* of the square figure constructed on the hypotenuse of a right-angled triangle with the combination of the square figures constructed on the other two sides), but then goes on to explain (p. 125a):

> ... man kann aber *nicht mehr hier aus der Gleichheit der Quadrate auf die Gleichheit ihrer Seiten schließen.*

In other words, the usual formula for expressing the Pythagorean theorem ($a^2 + b^2 = c^2$) fails. For this reason, the geometry constructed is called by Hilbert *non-Pythagorean geometry.* Hilbert then draws the conclusion (ibid.) that:

> *Die Lehre vom Flächeninhalt hängt wesentlich von dem Satze über das gleichschenkliche Dreieck ab; ist also nicht Folge der Lehre von den Proportionen allein.*

In the 1902 lectures, there is no elucidation, either of these models, or of this latter remark. Much more information, however, is contained in the paper *Hilbert 1902/03*, which is marked as having been received by the *Proceedings of the London Mathematical Society* on 22 August, 1902, thus a little after the 1902 lecture course would have concluded.

The remark cited is elucidated there as follows. Euclid proved the fundamental theorem that triangles on the same base and with the same area must have the same height (*Elements*, I, 39). In establishing his version of this theorem (that two triangles which are *inhaltsgleich* and on the same base have the same height), Hilbert, in the *Festschrift* (p. 43, or the *Ausarbeitung* of the 1898/1899 lectures, p. 131), defines the notion of the *Flächenmass* of a triangle as the product of half the base and the height. But in order for this definition to be a proper one, it has to be shown that this quantity is independent of the choice of the base. However, this apparently simple fact depends on the recognition that two right-angled triangles constructed within the given triangle are similar, thus on the theory of proportions and on the theory of triangle congruence. But the triangles in question are not in the same orientation, so this must ultimately depend on the unrestricted Triangle Congruence Axiom—the weaker version is not sufficient. (This version, however, *is* sufficient for developing the theory of proportion, i.e., P is not required.) Indeed, in the model he gives in which the Archimedean Axiom fails, but *Nachbarschaft* holds, Hilbert constructs a right-angled triangle with one leg of the right-angle on the x-axis, and the other perpendicular to it, and considers its reflection in the x-axis, obtained by rotations. In this reflection, the two legs of the right-angle are of the same length, but the length of the hypotenuse is different. From this follows, with the help of the Pythagorean theorem, the existence of two triangles which are *inhaltsgleich* and on the same base but with different height. The model is too complex to be described here, but at its basis is a complex field built over a non-Archimedean field. The geometry results from the Cartesian coordinate system constructed from pairs of numbers in this field, with angles and segments taken to be congruent when they can be mapped to each other by parallel transport and rotation. The central results derive from the formula Hilbert introduces for rotation of a straight line about an angle α; this contains a positive factor depending on α, and it is this factor which is responsible for the infinitesimal 'distortion' produced by rotations in different directions.

Hilbert's paper (*Hilbert 1902/03*) was reproduced as Appendix II to the various editions of the *Festschrift* from the second on. But it was never, however, a straight reprinting. For instance, in the 1902 lectures (p. 32), Hilbert raises a second question, namely of what must be added to the weaker triangle congruence axiom in the presence of just the other plane axioms so that full triangle congruence can be regained. The answer he gives in the lectures (ibid.) is just the proposition P itself (this is shown by p. 34), a claim repeated in the paper (*Hilbert 1902/03*, 54). This claim, however, is not unproblematic, for insufficient attention is paid to those congruence principles which, although they are easy consequences of the full Triangle Congruence Axiom, are no longer consequences of the restricted axiom. Using P itself to bridge the gap between the usual Triangle Congruence Axiom and its weaker cousin is a case in point. Hilbert adds an important note (pp. 259–262) to the fifth edition version of Appendix II (i.e., that in *Hilbert 1922*, 119–143) which points out that, in addition to P, it is also necessary to add a guarantee that angle addition is commutative. For this purpose he adds a new congruence axiom (III 7) stating that if the same angle is added to either leg of a given angle, the two total angles will be congruent. Another example of significant alteration is provided by the second edition version of the Appendix (*Hilbert 1903a*, 88–107). To this, Hilbert adds a section (pp. 104–106) giving another answer to the question mentioned above by showing that, instead of adding P as a new axiom, one can recapture the full Triangle Congruence Axiom from the weaker by adding what, from the third edition of the *Festschrift* on, Hilbert calls the *Axiom der Einlagerung*. (See *Hilbert 1909*, 152.) This says roughly that a polygon is never equi-decomposable (*Zerlegungsgleich*) with a polygon which is embedded in it.[1] More radically, the Appendix (thus, the original paper) was completely rewritten and expanded for the seventh edition (*Hilbert 1930a*, Appendix II, pp. 133–158), dropping the discussion of the *Axiom der Einlagerung*, and of the fifth edition III 7, and adding new axioms III 6 and III 7, both easy consequences of the usual Triangle Congruence Axiom (III 5 in the seventh edition, not III 6 as earlier), but not consequences of the weaker version. In this revision, the explanation of the construction of the relevant models is clearer; in particular, much more information is given about the role of the proposition that, in any triangle, any two sides taken together must be greater than the third (*Elements* I, 20), and its violation in the models. According to Hilbert, this revision is entirely due to Arnold Schmidt.[2] Schmidt was the *Ausarbeiter* of Hilbert's lectures on the foundations of geometry in 1927 (*Hilbert 1927**, not included in this Volume), and this *Ausarbeitung* was clearly an important step in the development of the radically revised seventh edition of the *Festschrift*. Hilbert makes it clear that, besides writing the new version of Appendix II, Schmidt was also responsible, in whole or in part, for many of the other revisions executed for the edition. Supplement III added by Bernays to the eight edition of the *Festschrift* (i.e., *Hilbert 1956*) restores the discussion of *Einlagerung*. (It is revised as Supplement V 2 in the ninth and subsequent editions.) Note that Bernays's Supplement V I (ninth and subsequent editions) adds further elucidation and commentary on the subject matter of Appendix II.

Two remarks about the *Axiom der Nachbarschaft*. (i) In the paper *Hilbert 1902/03*, 54–55, Hilbert states that the axiom is a consequence of P, indeed of the Euclidean proposition I, 20 just mentioned which follows from P, and which fails in the models Hilbert constructs. (ii) As we have seen, Hilbert designates the axiom a 'continuity axiom' (the 1902 lectures, p. 84, and p. 54 of the paper). With respect to this aspect of the axiom, Freudenthal offers the following comment:

---

[1]The axiom is equivalent to Theorem 32 in the second edition of the *Festschrift*, p. 46, and Theorem 31 in the first, p. 48. See also the 1898/1899 lectures (Chapter 4 in this Volume), e.g., the *Ausarbeitung*, pp. 130 and 137.

[2]Hilbert writes:

> ... ins besondere ist die neue Fassung des Anhanges II von ihm [i.e., Schmidt] selbständig hergestellt worden. (*Hilbert 1930a*, *Vorwort*.)

Man braucht ein zweites [Stetigkeitsaxiom], ein ganz überraschendes, das wir heute so formulieren würden: Durch die Anordnungsaxiome wird eine Topologie festgelegt und durch die Kongruenzaxiome eine Art von Metrik (allerdings nicht mit Zahlwerten und noch ohne Dreiecksaxiom). Die Metrik soll nun mit der Topologie in gewisser Weise verträglich sein; es soll nämlich der Abstand zweier Punkte x, y beliebig klein werden, wenn diese Punkte sich nur in einer genügend kleinen Umgebung eines festen Punktes a befinden. (*Freudenthal 1957*, 136–137.)

In the lectures (p. 114), in the course of developing the work just described), Hilbert refers to the 'forthcoming' paper *Hilbert 1903c* for some results on groups. (The paper is also referred to in *Hilbert 1902/03*, 56. It is dated 10 May, 1902.) This is the paper which investigates the Helmholtz-Lie conception of geometry, an investigation touched on at the very end of the 1902 lectures.[3] Lie's investigation of the work of Riemann and Helmholtz on the foundations of geometry was based on the concept of transformation group; a central assumption is that the functions which make up the groups are differentiable. Hilbert's work seeks to show that this latter assumption is not essential. His approach is based on the definition of a *motion* (*Bewegung*) as a certain kind of one-to-one transformation of the plane in itself, and then on three axioms: I. The motions form a group. II. If A and M are two different points of the plane, then A can be transformed into infinitely many different positions by rotations of the plane about M. III. The motions form a closed system, an axiom which expresses a continuity condition on the group of motions. Only the axioms are stated in the 1902 lectures (right at the end, p. 136, although the statement is slightly different); no theoretical development is given, and there is no motivation, nor is it clear what the theoretical purpose of the work is. This, however, *is* clear from the paper. After setting out the axioms, Hilbert states:

Ich beweise nun folgende Behauptung:
Eine ebene Geometrie, in welcher die Axiome I–III *erfüllt sind, ist entweder die Euklidische oder die Bolyai-Lobatschefskysche ebene Geometrie.* (*Hilbert 1903c*, 385–386.)

This work demonstrates the catholicity of Hilbert's approach to geometry. As Freudenthal says:

Hilberts Methode ist natürlich topologisch. Es versteht sich, dass wir heute den ganzen Beweis anders einrichten und darstellen würden. Die Idee aber ist klassisch geblieben und hat sich – reichlich spät – als fruchtbar erwiesen. Es ist eine Idee, die der Tendenz der „Grundlagen" diametral widerspricht. Die „Grundlagen" hatten das Ziel einer Geometrie ohne Stetigkeitsaxiome. Aber dies Ziel war kein Dogma. Hier, in dieser neuen Untersuchung, lässt Hilbert gerade Stetigkeitsaxiome vom Anfang an mit voller Gewalt eingreifen, ja alle Forderungen sind topologischer Natur. (*Freudenthal 1957*, 140.)

[3]The introductory part of this paper already appears as *Hilbert 1902f*, dated 8 November, 1901; this was reproduced in English translation in *Hilbert 1902c*. See the Introduction to Chapter 5.

Hilbert's work on the Helmholtz-Lie approach to geometry makes essential use of the third main piece of new work represented in the 1902 lectures, Hilbert's attempt to give a new foundation for plane hyperbolic geometry, to which we now briefly turn.

Hilbert points out that the investigations of Bolyai and Lobatchevsky into the hyperbolic plane assume a three-dimensional (analytic) space, and also a 'limit sphere', on the surface of which Euclidean geometry is assumed to hold. In addition, Klein's projective foundation of the same geometry uses both a spatial assumption, and a significant continuity principle. Hilbert's work is designed to establish a more elementary approach. What he presents is an axiomatisation of plane Bolyai-Lobatchevsky geometry which uses only the familiar plane incidence, order and congruence axioms, and a suitable non-Euclidean axiom governing parallels. Continuity axioms are expressly avoided. The replacement for the parallel axiom is roughly this. Suppose a line a given, and a point A outside that line. Then the axiom says that there are always two half-lines a_1, a_2 through A, which are not part of the same straight line, but which do not meet a. These half-lines are said to be *parallel* to a. The masterstroke of Hilbert's construction is his algebra of *ends*. Every half-line is said to determine an *end* (every full straight line determines two ends), and parallel half-lines determine the same end. Hilbert introduces addition and multiplication functions for ends, and shows that one gets thereby an ordered field in which addition and multiplication are associative and commutative, and multiplication is also distributive over addition. The field is also Euclidean, since each element has a square root. One can, therefore, 'calculate' with these ends 'as with ordinary numbers' (these lectures, p. 133). This is as far as Hilbert goes here in his justification of the construction. The paper which follows (*Hilbert 1903b*, which bears no submission date) is more explicit in its claim. In fact, from the coordinatisation which the theory of ends allows, one can, says Hilbert (*Hilbert 1903b*, 150), reproduce the standard theorems of hyperbolic geometry, and moreover all the basic theorems of projective geometry, thus reproducing both the Bolyai-Lobatchevsky and the Klein insights without any use of spatial or continuity assumptions. (In this coordinatisation, it is convenient to think of the straight lines as determined by a pair of coordinates, the two ends, whereas points are given by linear equations.) Hilbert does not actually carry out any of this development, not even in the paper. But one can find a very detailed account of how the essentials of hyperbolic geometry can be developed using Hilbert's field of ends in *Hartshorne 2000*, §§ 41–43. Hilbert's paper appears as Appendix III to all editions of the *Festschrift* from the second on. The third edition reprinting (*Hilbert 1909*, 156–176) contains a new, long footnote (pp. 156–157) making reference to more recent, related work, and to the fifth edition version (*Hilbert 1922*, 144–162) there is appended a note (pp. 263-264) containing an additional proof. The bulk of the paper, though, remained basically unchanged in all editions.

The text of the *Ausarbeitung* is essentially very clean, except for various remarks which have been added in Hilbert's hand. For the most part, these are given in a first series of footnotes. There is no way of dating these remarks, or indeed the changes Hilbert made to Adler's manuscript. Some changes were certainly executed in connection with the 1927 lectures (*Hilbert 1927**); for example, the revision to the bibliography which is carried out on p. 4 is certainly from later (given the dates of some of the publications), yielding exactly the same list as that given on p. 3 of *Hilbert 1927**, except that *Reidemeister 1930* has been added to this by hand. (*Hilbert 1927** is largely typed; the addition is not in Hilbert's hand.) Therefore some caution is to be exercised in any interpretation of these remarks or additions

The editorial notes, which are by Michael Hallett and Albert Krayer, have been kept to the minimum. In particular, there is no systematic reference to, or comparison with, other texts of Hilbert in this Volume.

Michael Hallett

Grundlagen der Geometrie

Einleitung.[A]

Jede Wissenschaft nimmt ihren Anfang dann, wenn ein genügendes zusammenhängendes Tatsachenmaterial vorliegt. Sie entsteht aber erst durch *Ordnen* dieses Materials. Die Ordnung erfolgt durch die *axiomatische Methode*[B], d. h.[C] man construiert ein *logisches Fachwerk von Begriffen* derart, dass den Beziehungen zwischen diesen Begriffen auch Beziehungen der zu ordnenden Thatsachen entsprechen.

In der Aufstellung eines solchen Fachwerks von Begriffen liegt eine Willkürlichkeit[D]; wir *verlangen* aber von ihnen:

1.) *Vollständigkeit*, 2.) *Unabhängigkeit*, 3.) *Widerspruchslosigkeit.*[4]

Wir wollen die Methode anwenden auf die einfachste und vollkommenste Naturwissenschaft: Die Geometrie.

[A] Als Beispiel dafür, dass das Abtragen von Strecken und Winkeln erfüllt sein kann, ohne dass $AB = BA$ ist, wähle eine Gebirgslandschaft und als Länge die Zeit, die zur Durchwanderung der Strecke nöthig ist. Gerade sollen dabei die Kurven sein, deren Projektionen auf die Horizontalebene wirkliche gerade Linien sind und Winkel sollen die wirklichen Winkel sein, die diese Projektionen einschliessen.[1]

Beispiel einer Nicht-Desarguesschen Geometrie vgl. Fortschr. d. Math. 1902, S. 497.[2] ⟨Added by Hi^s on the inside of the cover, opposite the title page.⟩

[B] das ist heute bei der hohen Ausbildung dieser Methode der allgemeine Weg, wenn man die Grundlagen einer Wissenschaft untersuchen will. ⟨Added by Hi^g.⟩

[C] Ueber das Wesen der axiomatischen Methode werde ich ⟨mich⟩ am Schlusse meiner Vorlesung ausführlicher aussprechen;[3] hier nur soviel ... ⟨Added by Hi^g in the upper right-hand margin and directed to this position by an arrow.⟩

[D] z. B. Stellung des Begriffes der Stetigkeit, ob zum Schluss oder gleich am Anfang des Axiomensystems – liefert ganz verschiedene Zugänge. ⟨Added by Hi^g.⟩

---

[1] See Specific Note 6(a) in the Introduction to Chapter 4.

[2] The reference is to *Steinitz 1904*, which is a review of *Moulton 1902*.

[3] Hilbert does not return to this, at least not according to the record provided by these notes. Neither is there any further discussion in the 1927 lectures, nor is it announced there.

[4] Hi^g has changed this line to '1.) Gesichtspunkt der *Vollständigkeit*, 2.) der *Unabhängigkeit*, 3.) der *Widerspruchsfreiheit*'.

Die Gesetze der reinen Logik und Zahlenlehre setzen wir dabei voraus und fragen uns dann: Welche Sätze müssen wir diesem Bereiche adjungieren, damit wir die Geometrie aufbauen können?

Unsere *Hauptfrage* wird also sein: Welches sind die nothwendigen und un- 2
ter einander unabhängigen Bedingungen, denen wir ein System von Dingen unterwerfen müssen, damit jeder Eigenschaft dieser Dinge eine geometrische Tatsache entspricht und umgekehrt; wie müssen wir es ferner einrichten, damit diese Dinge ein vollständiges und unabhängiges „*Bild*" der geometrischen Wirklichkeit sind?

Wir können unsere Aufgabe auch als eine *logische Analyse unseres Anschauungsvermögens* bezeichnen. Ein näheres Eingehen auf die erkenntnistheoretische Seite der Untersuchung soll auf den Schluß aufgespart bleiben.[5]

Durch das Gesagte bestimmt sich auch das Verhältnis dieser Vorlesung zu denen über *analytische und synthetische Geometrie*. In diesen beiden Disciplinen werden die principiellen Fragen nicht behandelt: In der analytischen Geometrie beginnt man ja mit der *Einführung der Zahlen*; wir dagegen werden die Berechtigung dazu genau zu untersuchen haben und erst am Schluße zeigen, dass sie erlaubt ist. In der synthetischen Geometrie appelirt man an die *Anschauung*, d. h. | man nimmt die Erscheinungsbilder möglichst so, wie 3
sie sich darbieten und sucht von da aus neue Erscheinungen zu gewinnen; wir dagegen werden die Anschauung analysieren.

Wir werden in unserem „Fachwerk" die Begriffe mit ihren üblichen Namen bezeichnen, obwohl dies nicht nothwendig wäre.

Man könnte auch von *Figuren* absehen, wir werden dies aber nicht thun, sondern häufig Figuren gebrauchen, uns aber *niemals auf sie verlassen*. In dem Gebrauche der Figuren ist besondere Vorsicht nötig; wir werden stets dafür sorgen, dass die an einer Figur vorgenommenen Operationen auch rein logisch richtig bleiben. Der Gebrauch von Figuren schließt oft Gefahren in sich, als besonders drastisches Beispiel mag folgender „Beweis" des absurden Satzes gelten:

Jedes Dreieck ist gleichschenklig (gleichseitig).

ABC sei ein beliebiges Dreieck, CM: die Halbierende des Winkels C, DM: die Mittelsenkrechte der Seite AB; außerdem sei: $ME \perp AC$ und $MF \perp BC$.

Es ist dann:

$$\left.\begin{array}{rll} & \triangle CEM \cong CFM, & \text{also: } EC = CF \\ \text{weiter: } & \triangle ADM \cong BDM & \\ \text{daher: } & \triangle AME \cong BMF, & \text{also: } \underline{EA = FB} \end{array}\right\} + \qquad 4$$

$$\text{daraus: } AC = BC. \text{ q. e. d.}$$

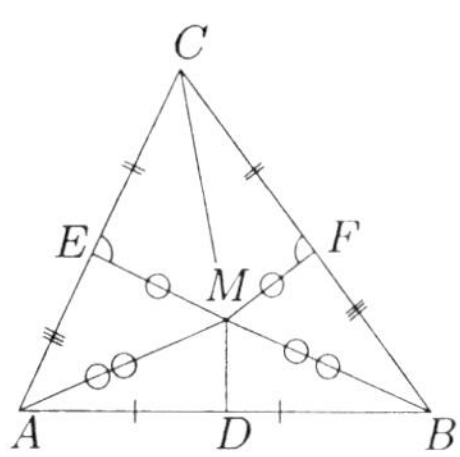

Der Fehler liegt lediglich in der falsch gezeichneten Figur.

Litteratur. (Wir nennen dabei nur einige der Hauptwerke, welche zusammenfassende Darstellungen unseres Stoffes darbieten):

1.) *Euklid*: Elemente.
2.) *Frischauf*: Absolute Geometrie nach J. Bólyai, 1872.

[5]Hilbert does not return to this.

" : Elemente der absoluten Geometrie, 1876.
3.) *Killing*: Grundlagen der Geometrie, 2 Bd. 1893, 1898 (eigentlich das einzige vollständige Werk, aber nicht überall leicht lesbar).
4.) *Hilbert*: Grundlagen der Geometrie, Festschrift zur Feier der Enthüllung des Gauss-Weberdenkmals, 1899.
5.) *Baltzer*: Elemente der Mathematik. 2 Bd. (gegenwärtig wohl die beste Darstellung der Elementarmathematik).
6.) *Stäckel und Engel*: Die Theorie der Parallellinien von Euklid bis auf Gauss. 2 Bd. 1895.
7.) *Hölder*: Anschauung und Denken in der Geometrie, 1899 (darin kann man sich am besten über die philosophische Seite der Frage unterrichten).[6]

Nun wollen wir die Grundthatsachen aufzählen; *wir nennen diese Grundthatsachen Axiome**. Wir beschränken uns dabei vorläufig nur mit der *ebenen* Geometrie und werden erst später hinzufügen, was nöthig ist, um die räumliche Geometrie aufzubauen.

Erklärung: Wir denken zwei verschiedene Systeme von Dingen: die Dinge des ersten Systems nennen wir *Punkte* und bezeichnen sie mit A, B, C, …; die Dinge des zweiten Systems nennen wir *Gerade* und bezeichnen sie mit a, b, c, … .

Punkte und Gerade heißen die *Elemente* der ebenen Geometrie.

Die Beziehungen zwischen diesen Elementen werden ausgedrückt und vollständig beschrieben durch fünf Gruppen von Axiomen; wir benennen die fünf Gruppen von Axiomen in folgender Weise:

*vgl. Meine Grundlagen, S. 2[7]

[6] In this list, 1.) and 4.) are obvious; the others are respectively: 2.) *Frischauf 1872* and *Frischauf 1876*; 3.) *Killing 1893*, *Killing 1898*; 5.) *Baltzer 1879* and *Baltzer 1883*; 6.) *Stäckel and Engel 1895*, of which, incidentally, there is only one volume; and 7.) *Hölder 1900*. Hilbert subsequently revised this list by deleting some items, adding more and renumbering, yielding the following list, much of the punctuation in which has been supplied by the editors:
'1.) *Euklid*: Elemente.
2.) *Hilbert*: Grundlagen der Geometrie, Festschrift zur Feier der Enthüllung des Gauss-Weberdenkmals, 1899.
3.) *Stäckel und Engel*: Die Theorie der Parallellinien von Euklid bis auf Gauss. 2 Bd. 1895.
4.) Schur, Grundlagen der Geometrie.
5.) Th. Vahlen, Abstrakte Geometrie, 1905.
6.) *Bonola*-Liebmann, Die Nicht-Euklidische Geometrie, Teubner 1908.
7.) Enriques, Prinzipien der Geometrie, Enziklopädie-Art. 1907.
8.) Pasch-Dehn, Vorlesungen über neuere Geometrie, 1926.
9.) Weyl, Mathematische Analyse des Raumproblems, 1923. Modernstes Werk mit den neuesten gruppentheoretischen Methoden.'

Again 1.), 2.) are obvious, and 3.) is the same as 6.) previously. The others are: 4.) (probably) *Schur 1909*; 5.) *Vahlen 1905*; 6.) *Bonola 1908*; 7.) *Enriques 1907*; 8.) *Pasch 1926*; 9.) *Weyl 1923*. The addition of '*Bonola* etc.' was in black ink; the other changes were executed by Hilbert in pencil. 'Enziklopädie' in 7.) is Hilbert's spelling.

[7] The reference is to the *Festschrift*. There is no p. 2; however, the first page of the text proper (p. 4) states the view that each group of axioms expresses 'gewisse zusammenghörige Grundthatsachen'. See Chapter 5 in this Volume.

I. Axiome der Verknüpfung,
II. " " Anordnung,
III. " " Kongruenz,
IV. Parallelenaxiom,
V. Axiome der Stetigkeit.

I. 6

Axiome der Verknüpfung.

Sie geben die einfachsten und nothwendigsten Verknüpfungen zwischen Punkten und Geraden an; wir haben in der ebenen Geometrie nur 3 Verknüpfungsaxiome:

I 1. *Zwei von einander verschiedene Punkte A und B bestimmen stets eine Gerade.*

Wir schreiben $AB = a$ oder auch $BA = a$. Statt „*bestimmen*" werden wir auch andere Wendungen brauchen, z. B. A „*liegt auf*" a, A „*ist ein Punkt von*" a, a „*geht durch*" A, a „*verbindet*" A „*und*" oder „*mit*" B. u. s. w.

Wenn A auf a und außerdem auf einer anderen Geraden b liegt, so gebrauchen wir auch die Wendung: „*die Geraden*" a „*und*" b „*haben den Punkt* A *gemein*".

I 2. *Irgend zwei von einander verschiedene Punkte einer Geraden bestimmen diese Gerade.*

Wir wollen schreiben: Wenn $AB = a$ und $AC = a$, so ist auch $BC = a$, wenn $B \neq C$.

Diese beiden Axiome erfüllen die einleitend gestellten Anforderungen: Sie widersprechen einander nicht und sind | auch von einander *unabhängig*. Letz- 7
teres wollen wir beweisen, indem wir zeigen, dass I 2 keine Folge von I 1 ist. *Beweis*: Wir nehmen als Punkte die ganzen rationalen positiven Zahlen und als Gerade die ganzen rationalen negativen Zahlen. Wir wollen nun zwei Punkten p_1 und p_2 jene Gerade g zuordnen, welche aus der Gleichung:

$$-\left[\frac{p_1 \cdot p_2}{2}\right] = g$$

folgt, wo $[x]$ die größte in x enthaltene ganze Zahl ist.

Es ist dann I 1 erfüllt, I 2 aber nicht, wie man an einem Beispiel sofort ersieht.

Es seien 3 Punkte: A $(p_1 = 1)$, B $(p_2 = 2)$, C $(p_3 = 3)$ gegeben; dann bestimmen, nach den eben gemachten Festsetzungen, die Punkte A u. B die Gerade $AB = -1$, die Punkte A und C die Gerade $AC = -1$; die Punkte B u. C aber die Gerade $BC = -\left[\frac{2 \cdot 3}{2}\right] = -3$ und nicht -1, q. e. d.

Da durch unsere Annahme I 1 erfüllt ist, I 2 aber nicht, so folgt daraus die Unabhängigkeit dieser beiden Axiome.

I 3. *Auf jeder Geraden giebt es wenigstens zwei* [8]*Punkte. Es giebt*[8] *wenigstens drei Punkte, welche nicht in einer Geraden liegen.*

Damit sind die ebenen Verknüpfungsaxiome abgeschloßen. 8

[8–8] Substituted by *Hi*[s] for: *Punkte; in jeder Ebene giebt es*

1.) Wir wollen noch folgenden *Satz* erwähnen:
Zwei Gerade einer Ebene haben entweder einen oder keinen Punkt gemein; denn zwei Punkte können zwei nicht zusammenfallende Gerade nicht gemein haben (I 1).

II.

Axiome der Anordnung.

Diese Axiomsgruppe wird den Begriff „*zwischen*" definieren:

II 1. *Wenn A, B, C Punkte einer Gerade sind und B zwischen A und C liegt, so liegt auch B zwischen C und A.*

II 2. *Wenn A u. B zwei Punkte einer Geraden sind, so giebt es wenigstens einen Punkt C, der zwischen A u. B liegt und wenigstens einen Punkt D, so dass B zwischen A und D liegt.*

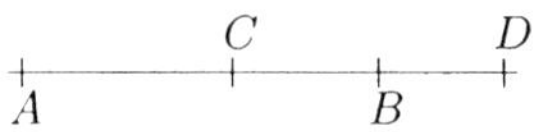

II 3. *Unter irgend 3 Punkten einer Geraden giebt es stets einen und nur einen, der zwischen den beiden andern liegt.*

Diese 3 Axiome beziehen sich alle auf die Lage von Punkten in einer Geraden; wir wollen sie auch *lineare Axiome* nennen.

9 2.) *Definition*: Das System zweier Punkte A u. B, die auf einer Geraden a liegen, nennen wir eine *Strecke*; wir bezeichnen dieselbe mit AB oder BA; die Punkte zwischen A u. B heißen *Punkte der Strecke* oder auch *innerhalb* der Strecke AB gelegen; alle übrigen Punkte der Geraden a heißen *außerhalb* der Strecke AB gelegen.

II 4. *Es seien A, B, C drei nicht in gerader Linie gelegene Punkte und a eine Gerade in der Ebene ABC, die keinen der Punkte A, B, C trifft; wenn dann diese Gerade durch einen Punkt innerhalb der Strecke AB geht, so geht sie gewiss auch durch einen Punkt der Strecke BC oder der Strecke AC.*

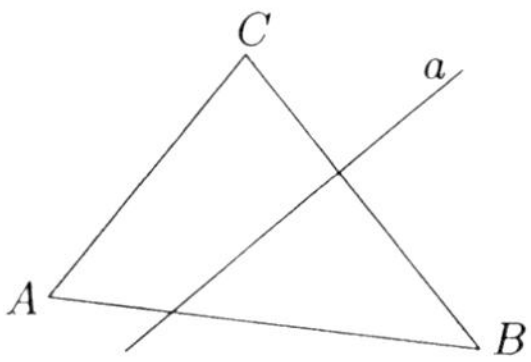

3.) Wir müssen nun wieder eine *Definition* voranschicken:

Eine Gerade a liegt zwischen den Punkten A u. B, wenn sie einen Punkt enthält, der zwischen den A u. B liegt; man sagt dann auch: *die Punkte A u. B werden durch a von einander getrennt.*

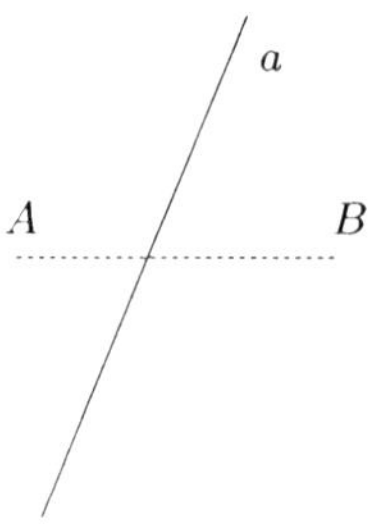

4.) Wir wollen nun folgenden *Hülfssatz* beweisen:

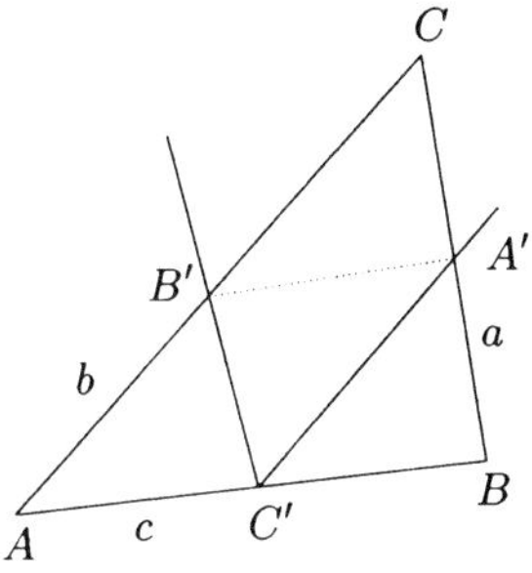

Wenn die Punkte A, B, C nicht auf einer Geraden liegen, so können auch 3 Punkte A', B', C' der Strecken BC resp. CA u. AB (siehe 2.)) nicht auf einer Geraden liegen.

Beweis: Wir nehmen an, die drei Punkte A', B', C' liegen auf einer Geraden und A' möge dabei zwischen B' und C' liegen (II 3). 10

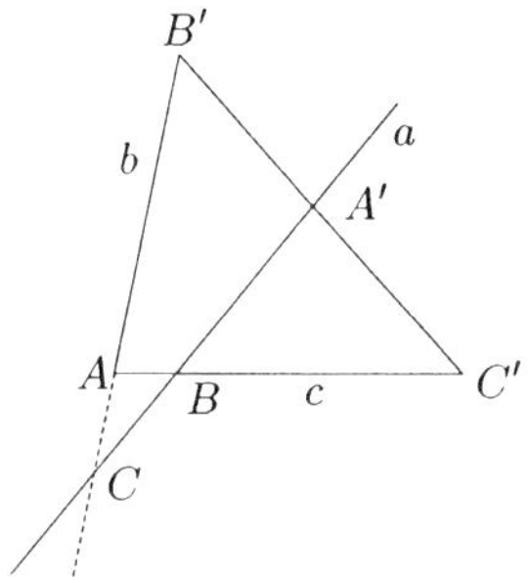

Durch A' geht nun die Gerade a, welche mit den Geraden b und c die Punkte C resp. B gemein hat.

Nach II 4 muss nun wenigstens einer dieser Punkte (z. B. B) zwischen die Ecken des Dreieckes $AB'C'$ fallen; dies widerspricht aber Axiom II 3, da C' zwischen A und B der Voraussetzung nach liegt.

5.) Wir können nun folgenden *Hauptsatz* beweisen:
Jede Gerade a trennt die Punkte[9], *die nicht in ihr liegen, in zwei Gebiete von folgender Beschaffenheit: Punkte, welche verschiedenen Gebieten angehören, werden durch a getrennt; Punkte, welche demselben Gebiete angehören, werden durch a nicht getrennt.*

Der Beweis dieses Hauptsatzes besteht aus zwei Theilen:

1.) Wir nehmen zwei Punkte A und B an, die nicht auf a liegen und durch a getrennt werden.
Ist dann C ein beliebiger Punkt, welcher aber nicht auf AB liegt, so wird nach II 4 und 4.) eine und nur eine der Verbindungsstrecken AC u. BC von a getroffen. Wird z. B. AB von a getroffen also BC nicht, so sagen wir: C liegt mit B auf derselben Seite von a, mit A dagegen auf verschiedenen Seiten von a.

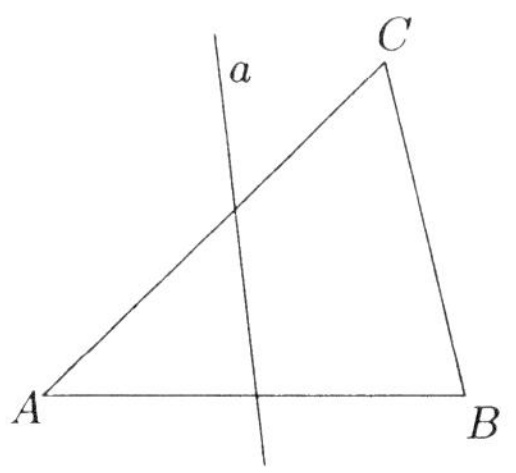

Liegt C auf der Geraden AB, so nehmen wir einen Punkt D außerhalb der Geraden AB an und verbinden ihn mit A, B und C. (D möge von A nicht getrennt sein). Es können nun zwei Fälle eintreten: 11

[9]Deleted by *Hi*^{g}: *der Ebene*

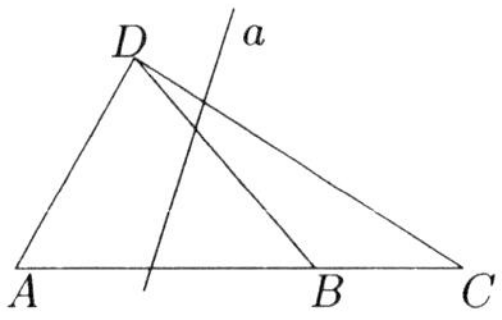

α.) D und C sind durch a getrennt; dann sind auch A und C durch a getrennt ($\triangle ACD$, siehe II 4 und 4.)) ebenso B und D ($\triangle ABD$) dagegen B und C nicht ($\triangle BCD$, II 4 und 4.)).

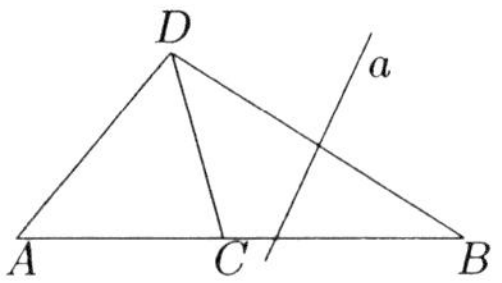

β.) D und C sind durch a nicht getrennt; dann sind auch C und A nicht getrennt durch a ($\triangle ACD$), jedoch sind getrennt D und B ($\triangle ABD$) und C von B (siehe $\triangle CDB$, II 4 und 4.)).

2.) Wir haben in 1.) bewiesen, dass sämmtliche Punkte der Ebene in zwei Gebiete getrennt werden, wir müssen noch beweisen, dass irgend zwei Punkte verschiedener Gebiete durch a getrennt sind, während dies bei Punkten desselben Gebietes nicht eintritt:

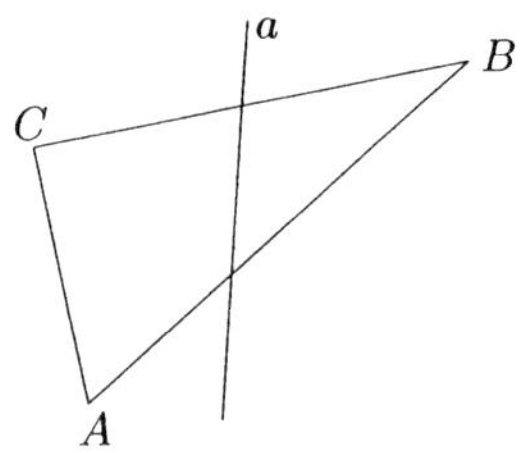

Sind C und D Punkte verschiedenen Gebietes, so muss einer, z. B. C, mit A auf derselben Seite von a liegen, der andere, D, ist dann von A getrennt; nach II 4 und 4.) muss dann die Gerade a nicht nur die Strecke AD sondern auch CD treffen. q. e. d.

12 Sind C und D Punkte desselben Gebietes, so müssen beide mit einem unserer früheren Punkte A, B auf verschiedenen Seiten von a liegen, z. B. mit A. Verbinden wir nun A, D, C mit einander, so müssen daher AC und AD von a getroffen werden; die Strecke DC also nach II 4 und 4.) nicht. q. e. d.

Unser Hauptsatz 5.) ist damit vollständig bewiesen.

6.) Wir wollen noch folgenden *Satz* hinzufügen:

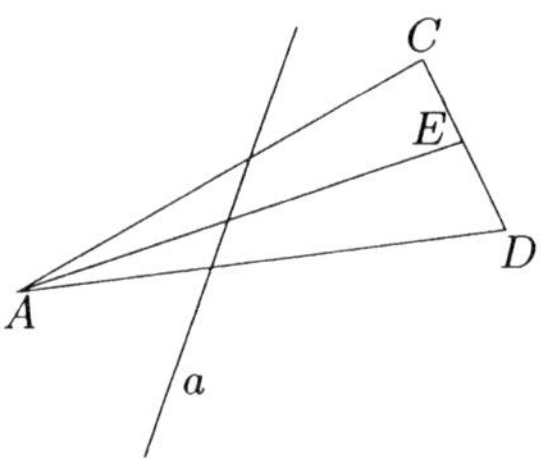

Gehören die Punkte C und D beide einem der beiden Gebiete an, in welchem die Gerade a die Punkte der Ebene trennt, so gehören auch sämmtliche Punkte der Strecke CD demselben Gebiete an.

Beweis: Da C und D auf derselben Seite von a liegen, so müssen sie beide mit einem der beiden Punkte A und B auf verschiedenen Seiten von a liegen, z. B. mit A, daher müssen die Strecken AC und AD von a getroffen werden, DC aber nicht (nach II 4 und 4.)).

Ist nun E ein beliebiger Punkt von CD, so verbindet man E mit A und erhält das $\triangle ACE$. a kann die Strecke CD und daher auch die Strecke CE nicht treffen; a muss daher die Strecke AE treffen. q. e. d.

13 7.) Wir können nun folgenden Satz beweisen:

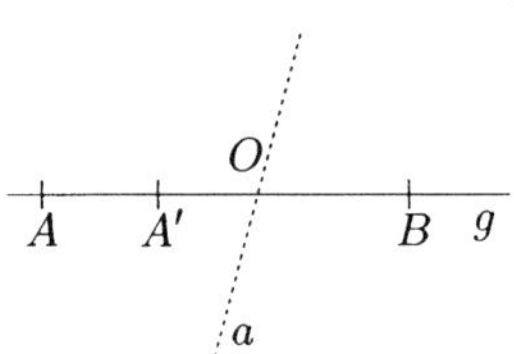

Jeder Punkt O, welcher auf einer Geraden g liegt, trennt die übrigen Punkte der Geraden in zwei Abtheilungen von folgender Beschaffenheit: Ist A ein Punkt der einen, B ein Punkt der anderen Abtheilung, so liegt O

zwischen A und B; sind dagegen A u. A′ Punkte derselben Abtheilung, so liegt O nicht zwischen A und A′.

Bew.: Zieht man durch *O* die Hilfsgerade *a*, so sind (nach 5.)) die Punkte der ganzen Ebene und damit auch die Punkte von *g* in der gewünschten Weise abgetheilt. Man sagt auch: *A und B liegen auf verschiedenen Seiten von O.* Die sämmtlichen auf ein und derselben Seite von *O* gelegenen Punkte der Geraden *g* heißen auch ein *von O ausgehender Halbstrahl.*

8.) Wir müssen nun folgenden Hilfssatz anführen:

Sind A, B, C, D vier Punkte einer Geraden a und liegt C zwischen A u. D, ferner B zwischen A u. C, so liegt auch stets B zwischen A u. D aber nicht zwischen C u. D.

Der Beweis besteht aus 2 Theilen:

α.) Nach 7.) werden die Punkte von *a* durch *B* in zwei Abtheilungen getrennt.

Würden A und D auf derselben Seite von B liegen, so | müssten (nach 6.)) 14
sämmtliche Punkte der Strecke *AD* auf derselben Seite von *B* liegen; nach Voraussetzung ist *C* ein Punkt der Strecke *AD* (siehe 2.)); er muss aber nach dem 2. Theile der Voraussetzung mit *A* auf verschiedener Seite von *B* liegen. *A* u. *D* können also nicht auf derselben Seite von *B* liegen; damit ist der erste Theil unseres Satzes bewiesen.

β.) *Würde B zwischen C u. D liegen*, dann müssten entweder *A* u. *C* oder *A* und *D* auf derselben Seite von *B* sich befinden; ersteres ist aber nach Voraussetzung unmöglich u. die Unmöglichkeit des zweiten ist in *α.*) nachgewiesen.

B kann also nicht zwischen *C* u. *D* liegen.

9.) *Erklärung:* Sind *A*, *B* zwei Punkte eines von *O* ausgehenden Halbstrahls, so wollen wir sagen: *B liegt hinter A*, wenn *A* zwischen *O* und *B* liegt.

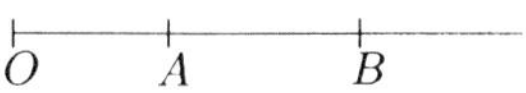

Wir können nun folgenden Satz beweisen:

10.) *Sind A, B, C drei Punkte eines Halbstrahles O und liegt B hinter A, C hinter B, so liegt auch C hinter A.*

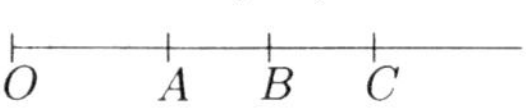

Der Beweis folgt aus 8.) ohne weiteres.

Damit ist ein Satz bewiesen, der früher von mir als *Axiom* aufgestellt wurde nämlich der *Anordnungssatz*[10]:

11.) *Irgend 4 Punkte A, B, C, D einer Geraden können stets so angeordnet werden, dass B zwischen | A u. C und* 15
auch zwischen A u. D und ferner C zwischen A u. D und auch zwischen B u. D liegt.

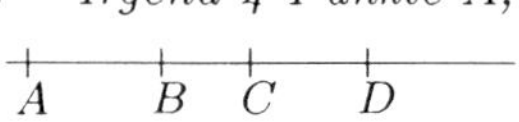

12.) Man kann diesen Satz auf beliebig viele Punkte der Ebene verallgemeinern und sagen:

[10]In *Hilbert 1902/03*, 51, footnote, and then again in the second edition of the *Festschrift* (*Hilbert 1903a*, 5, n. 1), Hilbert acknowledges *Moore 1902* as the source of this as a theorem, not an axiom. (It is Axiom II 4 in the first edition of the *Festschrift*, p. 6; see Chapter 5 in this Volume.)

Sind A, B, C, D, ..., K eine endliche Anzahl von Punkten einer Geraden, so lassen sich dieselben stets in eine Reihe bringen, so dass B zwischen A einerseits und C, D, ..., K andererseits liegt, ferner C zwischen A, B einerseits u. D, E, ..., K andererseits liegt und so weiter. Der Beweis gelingt leicht durch Schluß von n auf $\overline{n+1}$ Punkte.

Wir wollen nun beweisen, dass II 4 keine Folge von II 1, 2, 3 ist; früher wollen wir aber noch zeigen, dass man auch die Punkte der gewöhnlichen Cartesischen Zahlenebene in eine Reihe bringen kann: Wir brauchen zu diesem Zweck nur jeden Punkt[A] durch rechtwinklige Coordinaten x, y zu bestimmen und dann festzusetzen, dass jeder Punkt hinter jedem Punkte mit kleinerem x stehen soll und dass bei Punkten mit gleichem x das y den Ausschlag giebt.

16 Nun wollen wir eine *Geometrie construieren*, in welcher außer den Axiomen *I 1, 2* noch die *II 1, 2, 3 gelten* aber der *Anordnungssatz 11.) nicht*; damit ist auch die Unabhängigkeit des Axioms II 4 von den vorhergehenden erwiesen:

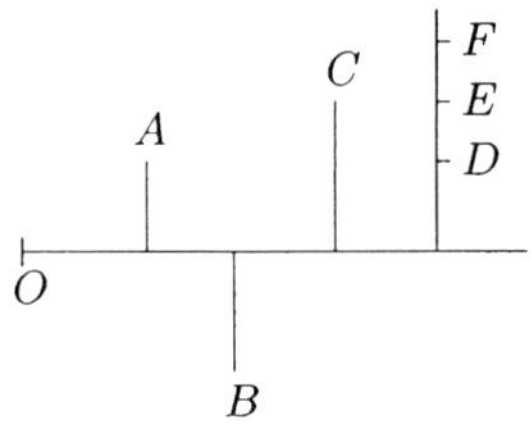

Wir betrachten zu diesem Zwecke wieder die Punkte der Descartes'schen Zahlenebene und machen folgende Festsetzungen:

1.) Haben drei Punkte A, B, C verschiedene x, so soll ihre Reihenfolge durch ihre Coordinaten x bestimmt werden.

2.) Haben von 3 Punkten A, E, D zwei, E und D, dasselbe x, so soll zunächst das x und für E, D ihr y den Ausschlag geben.

3.) Haben aber 3 Punkte D, E, F *dasselbe* x, so soll der Punkt F mit dem *größten* y als zwischen den beiden anderen bezeichnet werden.

Diese Festsetzungen erfüllen II 1, 2, 3 aber nicht den Anordnungssatz 11.); denn die 4 Punkte C, D, E, F lassen sich nicht in eine Reihe bringen.

Denn nach 2.)

$\underbrace{\text{liegt } E \text{ zwischen } F \text{ und } C}_{\alpha}$ und $\underbrace{D, C \text{ auf derselben Seite von } E,}_{\beta}$

17 nach 3.) liegt | F zwischen E und D also mit $\underbrace{D \text{ auf derselben Seite}}_{\gamma}$ von E.

Aus $\underbrace{\quad}_{\beta}$ und $\underbrace{\quad}_{\gamma}$ folgt, dass F, C auf derselben Seite von E liegen, was $\underbrace{\quad}_{\alpha}$ widerspricht. Eine Anordnung dieser 4 Punkte ist also nicht möglich.

Wir wollen noch einen *Unabhängigkeitsbeweis* führen: *Wir wollen nämlich eine Geometrie construieren, in welcher alle bisher genannten Axiome erfüllt sind, auch der Anordnungssatz 11.), in welcher aber II 4* **nicht** *erfüllt ist.*

Dies gelingt unschwer auf 2 Arten:

[A] Es genügt, sich auf rationale Punkte zu beschränken! ⟨Added by *Hi*[g] in the right-hand margin alongside this paragraph, although not assigned to any particular place.⟩

1.) Nehmen wir als Dinge die Punkte und Geraden des ganzen Euklidischen Raumes, so genügen sie den Axiomen I, II 1, 2, 3 und sie lassen sich auch vermittelst ihrer Coordinaten x, y, z in eine Reihe bringen. II 4 ist aber nicht erfüllt, denn eine Gerade, welche eine Seite eines Dreieckes trifft, braucht nicht in der Ebene des Dreieckes zu liegen und damit auch nicht eine andere Seite desselben Dreieckes zu treffen.

2.) Wir nehmen wieder die Punkte und Geraden | der Cartesischen Zahlenebene und 18
lassen die Anordnungsgesetze überall gelten bis auf die Gerade $x = 0$; die Punkte dieser Geraden wollen wir so anordnen, *dass jeder Punkt mit irrationalem y* ***hinter*** *jedem Punkte mit rationalem y zu stehen kommt*; die Punkte jedes dieser beiden Bereiche wollen wir nach der Größe der y ordnen.

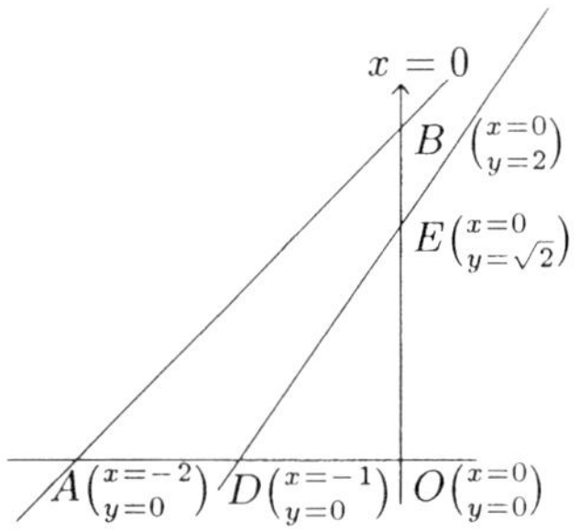

Der Punkt E $(x = 0, y = \sqrt{2})$ kann dann nicht zwischen O u. B liegen, denn E kommt hinter beiden Punkten.

Ist nun D ein Punkt der Strecke AO, so wird infolge unserer Festsetzungen die Gerade g (DE) eine und nur eine der 3 begrenzten Seiten des Dreieckes schneiden.

II 4 ist also hier nicht erfüllt.

13.) Wir wollen noch einen Satz hinzufügen, den wir schon früher hätten beweisen können:

„*Auf jeder Strecke giebt es unendlich viele Punkte.*"

Beweis ergiebt sich aus II 2 sofort.

III. 19

Axiome der Congruenz.

Die Axiome dieser Gruppe definieren den Begriff der *Congruenz* oder der *Bewegung*.

III 1. *Wenn A, B zwei Punkte der Geraden a sind und A' ein Punkt derselben oder einer anderen Geraden a' ist, so kann man stets auf einer gegebenen Seite der Geraden a' von A' aus stets einen und nur einen Punkt B' finden, so dass die Strecke AB (oder BA) der Strecke $A'B'$ congruent ist*;

in Zeichen $AB \equiv A'B'$ oder $BA \equiv A'B'$.

Jede Strecke ist sich selbst congruent.

$$AB \equiv AB \quad \text{oder} \quad BA \equiv AB.$$

Dieses Axiom bestimmt das *Abtragen der Strecken*.

III 2. *Wenn die Strecke AB sowohl der Strecke $A'B'$ als auch der Strecke $A''B''$ congruent ist, so ist auch $A'B'$ der Strecke $A''B''$ congruent*, d. h.

Wenn

$$AB \equiv A'B'$$

und

$$AB \equiv A''B'',$$

so ist

$$A'B' \equiv A''B''.$$

20 Man kann diesen Satz als die *Gruppeneigenschaft des Congruenzbegriffes* bezeichnen.

III 3. *Es seien AB und BC zwei Strecken ohne gemeinsame Punkte auf a und ferner $A'B'$ und $B'C'$ zwei Strecken ohne gemeinsame Punkte auf derselben oder einer anderen Geraden a'; wenn dann $AB \equiv A'B'$ und $BC \equiv B'C'$, so ist auch*

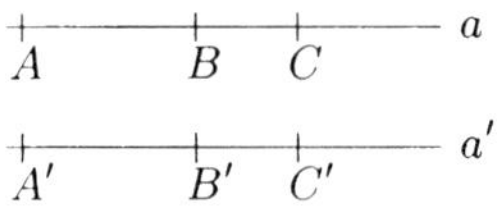

$$AC \equiv A'C'.$$

III 3 ist keine Folge von III 1, 2.

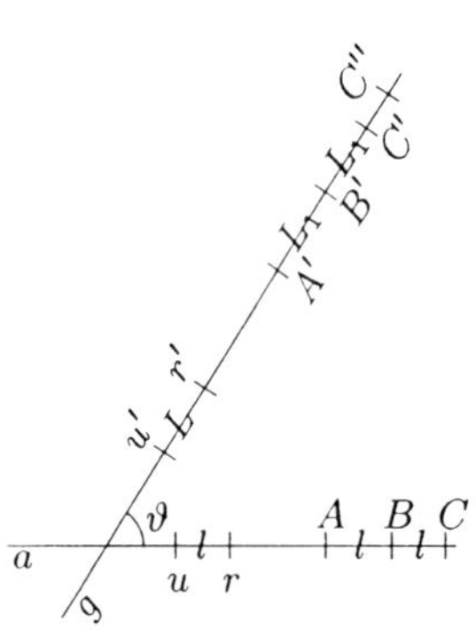

Beweis: Wir nehmen die Punkte und Geraden der Euklidischen Geometrie, definieren aber die Congruenz der Strecken auf folgende Weise:

a sei eine feste, g eine beliebige Gerade der Ebene. Wir wollen nun zwei Strecken UV und $U'V'$ von a resp. g congruent nennen, wenn die Gleichung

$$L = l + l^2 \sin^2 \vartheta$$

besteht, wobei l und L die Euklidischen Längen von UV resp. $U'V'$ sind.

Haben nun die Strecken AB, BC auf a die Länge $l = 1$, so haben congruente Strecken $A'B'$ und $B'C'$ von g jede die
21 Länge: $L_1 = 1 + 1 \sin^2 \vartheta$, | die zu AC congruente Strecke $A'C''$ auf g hat aber die Länge

$$L_2 = 2 + 4 \sin^2 \vartheta,$$

d. h. *C'' fällt nicht mit C' zusammen.*

III 3 gilt also hier nicht.

14.) *Definition*: *Zwei von einem Punkte ausgehende Strahlen nennen wir einen Winkel und bezeichnen ihn entweder mit:*

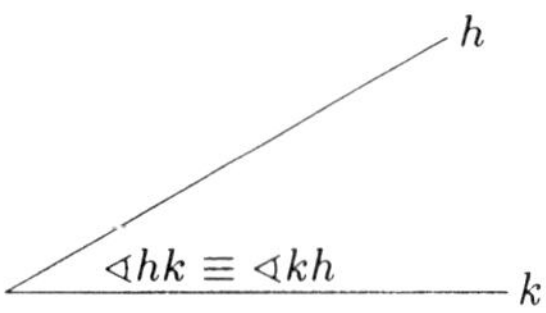

$$\sphericalangle(hk) \quad \text{oder} \quad \sphericalangle(kh).$$

Gehören diese beiden Halbstrahlen nicht derselben Geraden an, so theilen sie die Ebene in *zwei Gebiete*: *ein äußeres*, in welchem die *Verlängerungen* der Halbstrahlen liegen und in ein *inneres.*

III 4. *Es sei ein Winkel (hk), eine Gerade a' u. eine bestimmte Seite von a' gegeben. Es bedeute h' einen Halbstrahl der Geraden a', der vom Punkte O' ausgeht:*

dann giebt es einen und nur einen Halbstrahl k', so dass

$$\sphericalangle(hk) \equiv \sphericalangle(h'k')$$

oder

$$\sphericalangle(hk) \equiv \sphericalangle(k'h')$$

ist und | *zugleich alle inneren Punkte des Winkels* $(h'k')$ *auf der gegebenen Seite von* a' *liegen.* 22

Jeder Winkel ist sich selbst congruent, d. h.

$$\sphericalangle(hk) \equiv (hk).$$

Damit ist das *Abtragen der Winkel* definiert.

III 5. *Wenn ein Winkel* (hk) *sowohl dem* $\sphericalangle(h'k')$ *als auch dem* $\sphericalangle(h''k'')$ *congruent ist, so ist auch* $\sphericalangle(h'k')$ *congruent dem* $\sphericalangle(h''k'')$*, d. h.:*

Wenn

$$\sphericalangle(hk) \equiv \sphericalangle(h'k')$$

und

$$\sphericalangle(hk) \equiv \sphericalangle(h''k''),$$

so ist auch

$$\sphericalangle(h'k') \equiv \sphericalangle(h''k'').$$

Wir brauchen noch ein Axiom, welches die *3 linearen* Axiome III 1, 2, 3 mit den Axiomen III 4, 5 verknüpft.

III 6_1. *Wenn für zwei Dreiecke* ABC *und* $A'B'C'$ *gilt:* $AB \equiv A'B'$, $AC \equiv A'C'$ *und* $\sphericalangle BAC \equiv \sphericalangle B'A'C'$*, so gilt auch stets:*

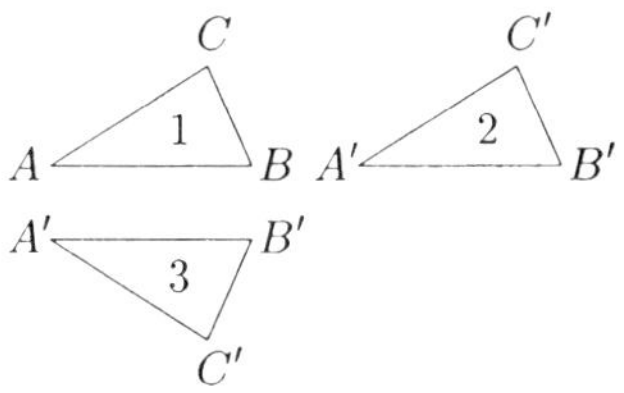

$$\sphericalangle ABC \equiv \sphericalangle A'B'C' \text{ und } \sphericalangle ACB \equiv A'C'B'.$$

In dieser Fassung des Axioms können die Dreiecke entweder wie 1 und 2 oder wie 1 u. 3 zu einander liegen.

Es ist aber *wichtig*, das Axiom so zu *fassen*, dass es nur für *Dreiecke gilt*, die in der Ebene durch *bloße Bewegung* ohne Umklappung zur Deckung gebracht werden können. 23

Wir wollen nun den *Zusatz* entwickeln, der zur ***weiteren Fassung*** des Axioms hinzugefügt werden muß, um die ***engere Fassung*** desselben zu erhalten:

Erklärung: Auf jeder Geraden r ist durch[A] 2 Punkte A, B und deren Reihenfolge ein *Sinn (Pfeilrichtung)* bestimmt.

r theilt die Ebene in zwei Theile;[B] mit Bezug auf die Pfeilrichtung wollen wir die *eine Seite die rechte*, die *andere die linke Seite von* r *nennen.*

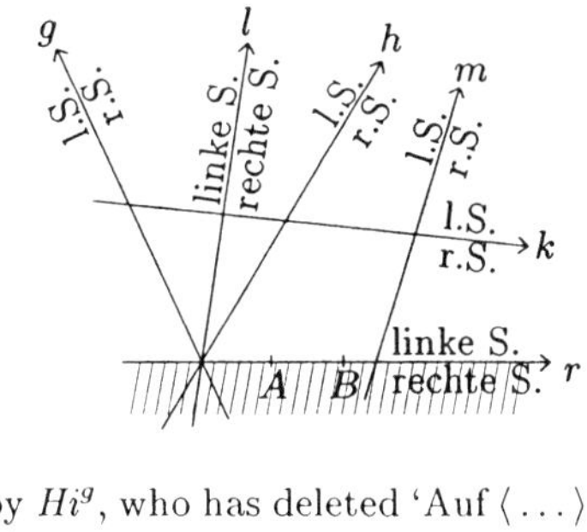

Ist nun l eine beliebige Gerade sammt Pfeilrichtung, so setzen wir fest:

[A] Wir wählen einen Halbstrahl r. ⟨Interlineated by Hi^g, who has deleted 'Auf ⟨…⟩ durch', and also added 'auf r' after 'Punkte A, B'.⟩

[B] Die Punkte, die nicht auf r und nicht auf der zu r gehörigen Grade liegen nennen wir die eine Klasse rechts, die andere Klasse links vom Halbstrahl r. ⟨Interlineated by Hi^g, who has deleted 'r ⟨…⟩ Theile;'.⟩

Wenn der Pfeil l in die linke Seite von r hineinzeigt, so soll r in die rechte Seite von l zeigen; dadurch sind die Seiten von l, g, h *eindeutig* bestimmt.

24 Schneidet die Gerade (z. B. k) die r *nicht*, | so bestimmt man zunächst die Seiten einer Geraden, welche r und k schneidet und daraus die Seiten von k.

Man kommt zu *gleichem Resultate*, ob man von r oder l, g, ... ausgeht, wenn nur deren Seiten früher richtig bestimmt wurden.

Es ist dem Anfänger sehr zu empfehlen, sich dies klar zu machen ohne jede Benützung der Anschauung, streng aus der Definition allein.

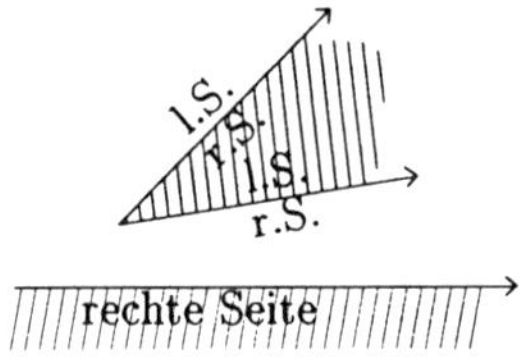

Ist ein Winkel (hk) gegeben, so kann man jetzt von einem rechten und linken Schenkel desselben sprechen bei Zugrundelegung einer Pfeilrichtung r; wir wollen dabei jenen Schenkel den rechten nennen, für welchen der Winkel auf der linken Seite liegt.

Damit haben wir die Worte gefunden, welche dem Axiom III 6 angehängt werden müssen, damit die *weitere* Fassung des Axioms in die *engere* übergeht.

Dieser Zusatz muss lauten:

25 III 6_2. ... *„vorausgesetzt, dass AB und $A'B'$ | die rechten Schenkel, AC und $A'C'$ die linken Schenkel der Winkel BAC resp. $B'A'C'$ darstellen“.*[11]

Folgerungen aus den Congruenzaxiomen.

1.) *Satz:* Es sei $AB \equiv A'B'$, dann ist auch (III 1)

$$A'B' \equiv AB.$$

Man kann daher sagen: *zwei Strecken* (ebenso zwei Winkel) *sind einander congruent.*

16.) 2.) *Definition:* Zwei Winkel, die den Scheitel und einen Schenkel gemein haben und deren nicht gemeinsame Schenkel eine gerade Linie bilden, heißen *Nebenwinkel.*

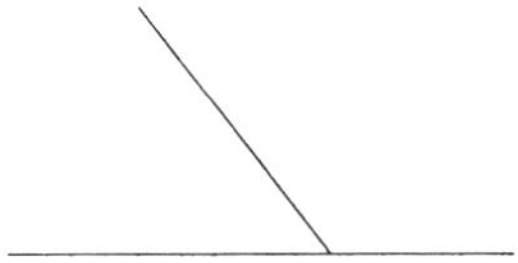

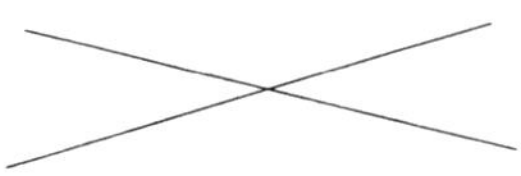

Zwei Winkel, deren Schenkel je eine Gerade bilden, heißen *Scheitelwinkel.*

Ein Winkel, welcher seinem Neben-Winkel gleich ist heißt *rechter Winkel.*

Zwei Dreiecke ABC, $A'B'C'$ sind einander congruent, wenn

$$AB \equiv A'B', \quad AC \equiv A'C', \quad BC \equiv B'C'$$

und

$$\sphericalangle A \equiv A', \quad \sphericalangle B \equiv B', \quad \sphericalangle C \equiv C'$$

ist.

[11] This 'engere Fassung' of III 6 is used in Hilbert's analysis of the theorem that the base angles in an isoceles triangle are equal; see pp. 107–125 of this *Ausarbeitung*, and the Introduction to this Chapter. The section marked 'A' below (pp. 26–32) gives the proofs of some central theorems usually proved with the standard version of the triangle congruence axiom, but using now only this weaker version.

A. 26

In diesem Abschnitte wollen wir nur III 6_2 voraussetzen; die vorkommenden congruenten Dreiecke dürfen nur so liegen, wie es III 6_2 verlangt.

17.) *Erster Congruenzsatz.* Wenn für 2 Dreiecke ABC und $A'B'C'$ die Congruenzen bestehen

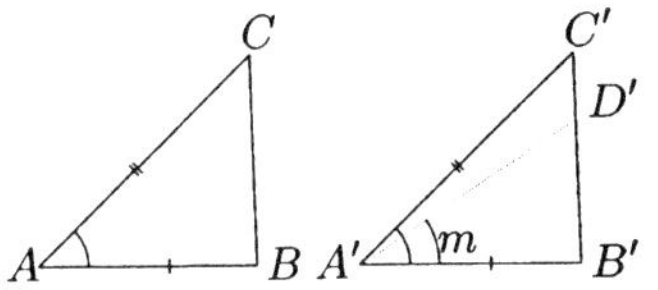

$$AB \equiv A'B'$$
$$AC \equiv A'C'$$
$$\sphericalangle A \equiv A',$$

dann sind die beiden Dreiecke congruent.

Nach III 6_2 ist ja $\sphericalangle C' \equiv C$, $\sphericalangle B' \equiv B$; wäre $BC \not\equiv B'C'$, so bestimme man D' derart, dass $B'D' \equiv BC$; dann ist

$$\sphericalangle ACB \equiv A'D'B', \quad (\text{III } 6_2)$$

also

$$\sphericalangle m \equiv A \equiv A',$$

was nicht möglich ist (III 4).

Ganz analog beweist man die Thatsache:

18.) *Zweiter Congruenzsatz:* Wenn in zwei Dreiecken je eine Seite und die beiden anliegenden Winkel congruent sind, dann sind die Dreiecke congruent.

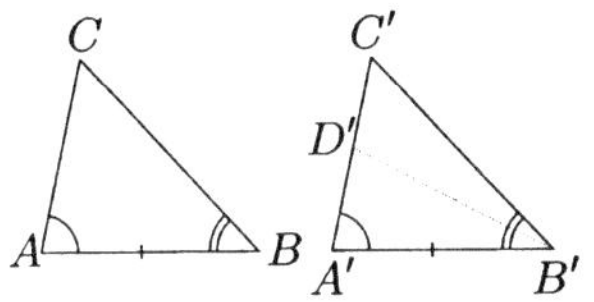

Jetzt sind wir imstande folgende wichtige Thatsache nachzuweisen: 27

19.) *Wenn zwei Winkel ABC und A'B'C' congruent sind, so sind auch deren Nebenwinkel congruent.* Diese Nebenwinkel müssen aber beidemale durch Verlängerung gleicher Schenkel (l oder r) entstehen.

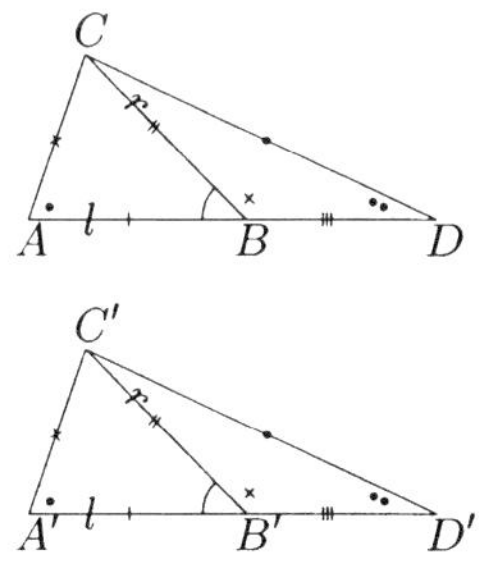

Zum Beweise mache man:

$$A'B' \equiv AB, \quad C'B' \equiv CB, \quad D'B' \equiv DB,$$

dann ist

$$\triangle ABC \equiv A'B'C', \quad \text{also } \sphericalangle A \equiv A', \ AC \equiv A'C',$$

demnach

$$\triangle ACD \equiv A'C'D'$$

und

$$\triangle CBD \equiv C'B'D',$$

also

$$\sphericalangle CBD \equiv C'B'D'. \quad \text{q. e. d.}$$

20.) *Es sei* $\sphericalangle(hk) \equiv (h'k')$, *$l$ ein Halbstrahl, der vom Scheitel des Winkels* (hk) *ausgeht und im Innern desselben verläuft; dann gibt es stets einen Halbstrahl l', der vom Scheitel des Winkels* $(h'k')$ *ausgeht und im Innern desselben verläuft, so dass*

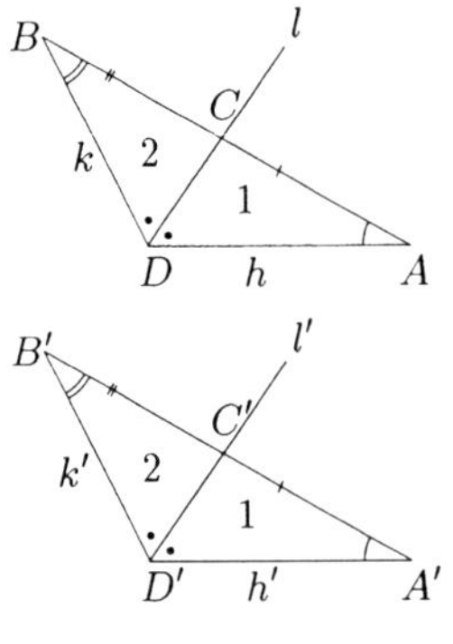

$$\sphericalangle(hl) \equiv (h'l'), \quad \sphericalangle(kl) \equiv (k'l')$$

ist. Dabei müssen wieder k, k' z. B. *linke* Schenkel sein.

28 *Bew.:* Mache

$$OA \equiv OA', OB \equiv O'B',$$

also

$$\triangle AOB \equiv A'O'B'.$$

l muss AB in C schneiden.
Bestimmt man C' auf $A'B'$ so, dass

$$A'C' \equiv AC,$$

so muss auch

$$C'B' \equiv CB$$

sein (III 3), daher

$$\triangle 1 \equiv \triangle 1'$$
$$\triangle 2 \equiv \triangle 2',$$

so dass $C'O'$ der gewünschte Halbstrahl l' ist.

21.) Mit Hilfe dieser beiden letzten Sätze kann man die *Gleichheit der Scheitelwinkel* aus III 6_2 allein folgern.

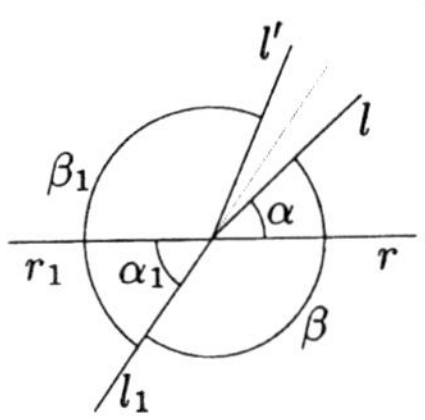

Gegeben sei $\sphericalangle\alpha$, r sei dessen rechter Schenkel.

Nun construieren wir $\sphericalangle\alpha_1 \equiv \alpha$ und es soll gezeigt werden, dass l_1 und l in eine Gerade fallen müssen.

Ist dies nicht der Fall, so machen wir

$$\sphericalangle\beta_1 \equiv \beta$$

und suchen in β_1 nach 20.) den zu r entsprechenden Strahl, welcher mit r_1 zusammenfallen muss.

29 Nach 19.) muß aber auch $\sphericalangle(r_1 l) \equiv (r_1 l')^A$ sein, also l mit l' zusammenfallen; dies kann aber nur in der Verlängerung von l_1 eintreten, denn nach 14.) *lassen wir nur Winkel zu, die kleiner als ein gestreckter sind.*

22.) Auch die *Halbierung einer Strecke* gelingt mit III 6_2 allein.

$^A r l_1 = r_1 l$ ⟨Added by Hi^g over this formula.⟩

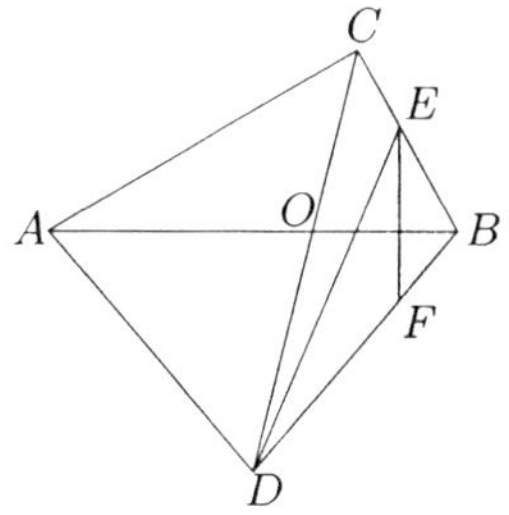

Zuerst müssen wir aber folgenden *Hilfssatz vorausschicken*:

Ist AB eine Strecke, O ein Punkt derselben, sind C und D Punkte, die auf einer Geraden durch O, aber getrennt durch O liegen, zeichnet man endlich die Linienzüge ACB und ADB, so wird jede Strecke (DE und EF z. B.), welche einen Punkt des einen Zuges mit einem Punkt des anderen Zuges verbindet, von der Strecke AB getroffen.

Bew.: E liegt zwischen C, B; O, C liegt auf derselben Seite von DE, also liegen O und B auf verschiedenen Seiten von DE (5.), 6.)). Analog für EF.

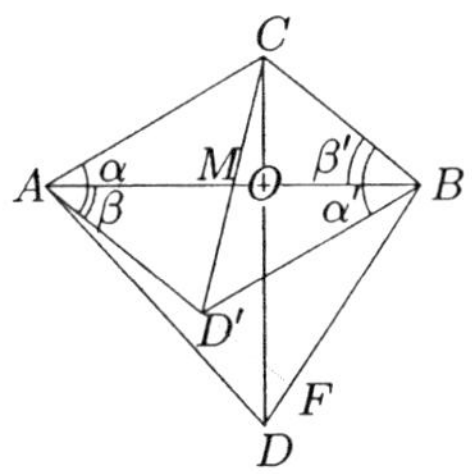

Wir construieren nun, um unsere Strecke AB zu *halbieren*, $\sphericalangle\alpha' \equiv \alpha$, $\sphericalangle\beta' \equiv \beta$ und nehmen dabei an, dass D' innerhalb ABD zu liegen kommt. 30

Es muss dann CD' die Gerade AB in einem Punkte M schneiden, wie aus dem Vierecke $AFBC$ unter Benutzung des eben bewiesenen Hilfssatzes folgt.

Damit ist die Hauptschwierigkeit überwunden.

Nun gilt

$$\triangle ABC \equiv ABD'$$

daraus

$$\triangle ACD' \equiv BD'C$$

″

$$\underline{\triangle AMD' \equiv CBM}$$

also

$$AM \equiv MB. \quad \text{q. e. d.}$$

23.) Mit Hilfe von III 6_2 können wir auch ganz *nach Euklid* folgenden Satz beweisen:

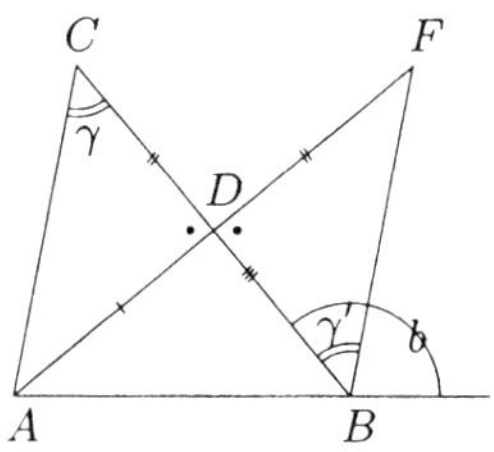

In jedem Dreiecke ist ein Außenwinkel stets größer als ein nicht anliegender Innenwinkel.

Bew.: Ist D der Mittelpunkt der Seite BC, | und 31

$$FD \equiv DA,$$

so ist

$$\triangle ADC \equiv FDB,$$

also

$$\sphericalangle\gamma' \equiv \gamma.$$

Da außerdem F ins Innere des Winkels b zu liegen kommt, so ist

$$b > \gamma \quad \text{q. e. d.}$$

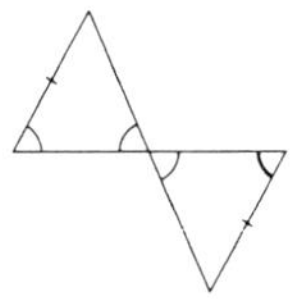

Wir können jetzt auch den *zweiten Congruenzsatz erweitern.*

Wir können sagen:

24.) *Wenn in zwei Dreiecken* ABC u. $A'B'C'$ *eine Seite* ($AB \equiv A'B'$) *und zwei beliebige Winkel* ($\sphericalangle A \equiv A'$, $\sphericalangle C \equiv C'$) *congruent sind, dann sind auch die Dreiecke congruent.*

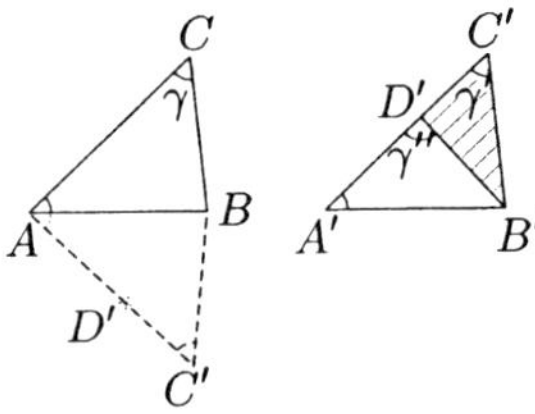

Wäre nämlich $AC \not\equiv A'C'$, so construiere man $A'D' \equiv AC$, wobei D' innerhalb $A'C'$ fallen möge; dann ist $\triangle A'B'D' \equiv ABC$ (17.)), also $\sphericalangle\gamma'' \equiv \gamma' \equiv \gamma$; γ'' muß aber nach 23.) größer als γ' sein.

32 B.

Wir wollen nun III 6_1 voraussetzen.

Zunächst fragen wir: *Was muß man zu III 6_2 hinzufügen, damit III 6_1 entsteht?*

Die Antwort lautet:

25.) *Man muss den Satz von der Gleichheit der Basiswinkel im gleichschenkeligen Dreiecke hinzufügen.*

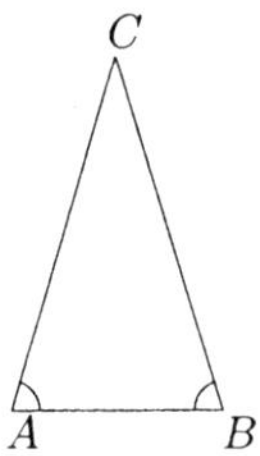

Der Satz vom gleichschenkeligen Dreieck folgt aus III 6_1 ohne weiteres, denn ist

$$AC \equiv BC,$$

so ist

$$\triangle ABC \equiv BCA,$$

also

$$\sphericalangle A \equiv B.$$

26.) *Sind daher in einem Dreiecke zwei Seiten einander gleich, so sind auch die denselben gegenüberliegenden Winkel einander gleich.*

27.) Die *Umkehrung* dieses Satzes vom gleichschenkligen Dreieck lässt sich auch leicht beweisen:

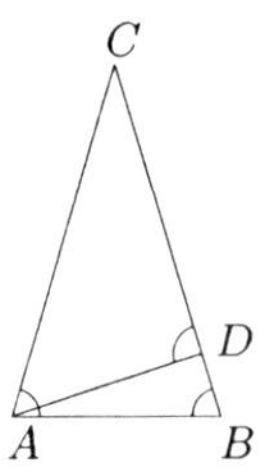

Wäre nämlich

$$\sphericalangle CAB \equiv CBA$$

und

$$AC \not\equiv BC,$$

33 so construiere man $DC \equiv AC$, wobei D zwi|schen C u. B fallen möge; dann wird $\sphericalangle CAB$ größer als $\sphericalangle CAD$ sein, da D innerhalb $\sphericalangle CAB$ fällt.

Nach 26.) ist aber $\sphericalangle CDA \equiv CAD$ also auch kleiner als $\sphericalangle CAB$ und demnach auch kleiner als $\sphericalangle CBA$; letzteres ist aber nach 23.) unmöglich.

28.) Wir können jetzt den *III. Congruenzsatz* beweisen; derselbe lautet:

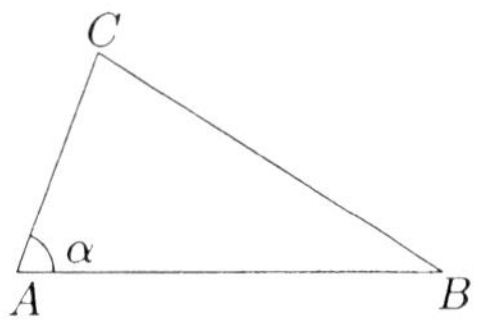

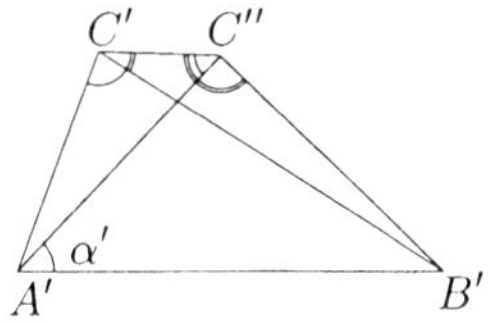

Wenn in zwei Dreiecken ABC und $A'B'C'$ die Seiten übereinstimmen, so sind die Dreiecke congruent.

Beim Beweise müssen wir *zwei Fälle* unterscheiden:

1.) *Beide Dreiecke seien gleichliegend.*

Ist dann z. B. $\sphericalangle A > \sphericalangle A'$, so construiere man

$$\sphericalangle \alpha' \equiv \alpha \quad \text{und} \quad A'C'' \equiv AC;$$

nach 26.) sind dann:

$$\sphericalangle A'C'C'' \equiv A'C''C'$$

und

$$\sphericalangle B'C'C'' \equiv B'C''C'.$$ [12]

Dies ist unmöglich; denn bei zwei gleichen Winkeln (1 u. 2) kann nicht ein Theil des 1. Winkels einem 3. Winkel gleich sein, der größer als der 2. Winkel ist.

2.) *$\triangle ABC$ und $A'B'C'$ hätten verschiedenen Umlaufssinn.* 34

Man construiere dann:

$$\sphericalangle BAC'' \equiv B'A'C'$$

u.

$$\underline{AC'' \equiv AC,}$$

so ist

$$\triangle AC''B \equiv A'B'C' (\text{nach } 17.))$$

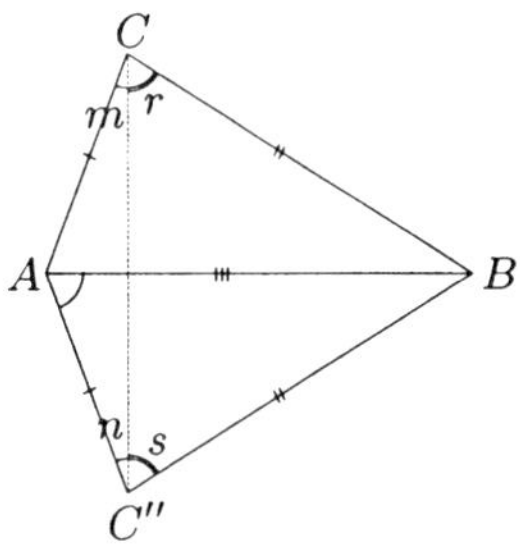

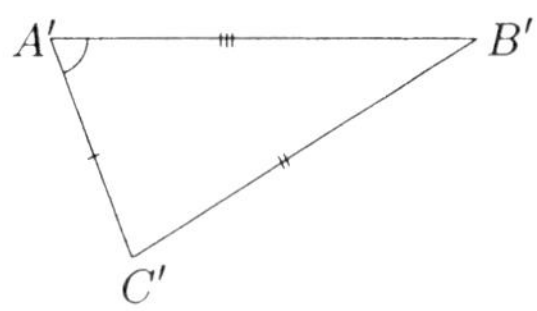

Zieht man CC'', so entstehen zwei gleichschenkelige Dreiecke, daher ist

$$\left.\begin{array}{r}\sphericalangle m = n\\ \sphericalangle r = \sphericalangle s\end{array}\right\} +$$

also

$$\sphericalangle C \equiv \sphericalangle C'' \equiv \sphericalangle C'.$$

Auf analoge Weise zeigt man die Congruenz der Winkel A, A' resp. B, B'; damit ist die Congruenz der beiden Dreiecke ABC und $A'B'C'$ erwiesen.[13]

[12] This follows from the restricted version of the First Congruence Theorem, 17.), which shows that $\triangle A'C''B' \equiv \triangle ABC$ (since the triangles are *gleichliegend*). Therefore $BC \equiv B'C''$, and so $B'C' \equiv B'C''$, and $\triangle C'B'C''$ is iscoceles. But 17.) is proved on the basis of the restricted III 6_2 alone.

[13] Thus, the Third Triangle Congruence Theorem is proved on the basis of the restricted III 6_2 alone together with 26.), the theorem showing the equality of the base angles in an isoceles triangle. The unrestricted Axiom III 6_1 follows easily (see p. 38 below). Thus, 25.) on p. 32 is established, namely that it is sufficient to add 26.) to III 6_2 to get III 6_1. See also the further investigation of 25.), pp. 108–125a, as well as the Introduction to this Chapter.

29.) Jetzt gelingt es uns auch, *die Existenz des rechten Winkels zu beweisen.*

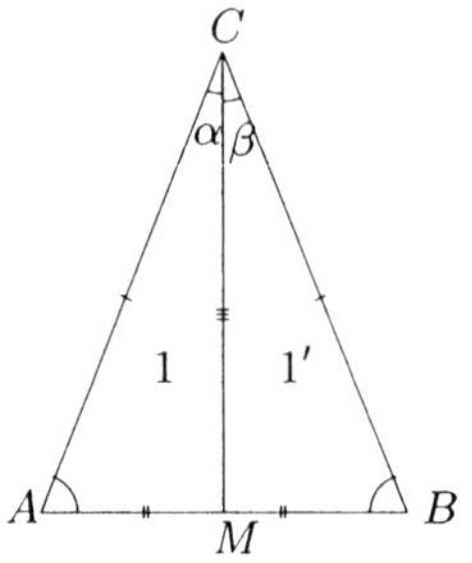

Trägt man nämlich auf den Schenkeln irgend eines Winkels gleiche Strecken auf, verbindet die Endpunkte B und A und halbiert diese Strecke BA in M, so schließt CM mit AB
35 rechte Winkel | ein; nach der Construction ist ja (28.))

$$\triangle 1' \equiv \triangle 1,$$

also

$$\sphericalangle AMC \equiv \sphericalangle CMB = R \quad (16.)).$$

Es folgt aber auch

$$\sphericalangle \alpha \equiv \sphericalangle \beta;$$

man ersieht daraus *die Möglichkeit jeden Winkel zu halbieren und zwar nur auf eine einzige Weise.*

30.) *Alle rechten Winkel sind einander congruent;*
der Beweis ergiebt sich unschwer, indem man zeigt, dass zwei rechte Winkel nicht ungleich sein können. Was die Ausführung des Beweises anlangt, so verweisen wir auf unser Buch „Grundlagen der Geometrie, pag. 16“.[14]

Bemerkenswert ist, dass *Euklid* diesen Satz 30.) unter die Axiome aufgenommen hat; er musste dies thun, da er unsere Axiomgruppe II, welche den fundamentalen Begriff „zwischen“ definiert, nicht angiebt.[15]

Wir müssen noch *zwei Congruenzsätze über das rechtwinklige Dreieck* angeben:

36 31.) 1.) *Wenn in zwei rechtwinkligen Dreiecken ABC und $A'B'C'$ je eine Kathete und die Basiswinkel einander gleich sind, dann sind die Dreiecke congruent.*

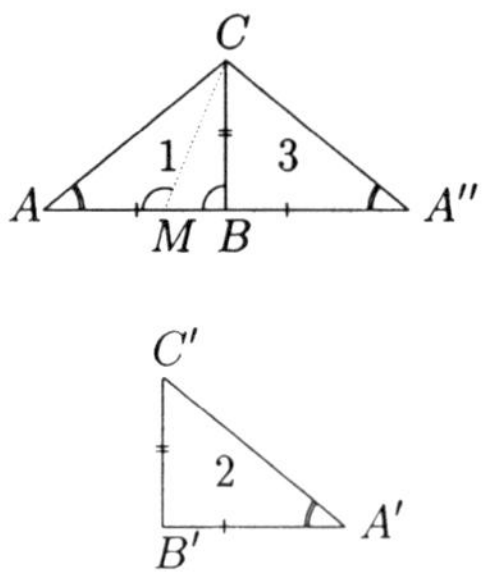

Bew.: Haben die Dreiecke gleichen Umlaufssinn, so folgt der Beweis aus 24.) sofort; wir nehmen daher an, dass ABC und $A'B'C'$ *verschiedenen Umlaufssinn* hätten.

Wir construieren dann:

$$\triangle 3 \equiv 2,$$

also

$$\sphericalangle A'' \equiv A' \equiv A,$$

daher ist

$$AC \equiv A''C, \qquad (27.))$$

folglich

$$AB \equiv A''B.$$

Wäre nämlich $AB \not\equiv A''B$ und M der Halbierungsmittelpunkt der Strecke AA'', so müsste nach 29.) die Gerade MC senkrecht auf AA'' stehen, was 23.) widerspricht, sobald M von B verschieden.

2.) *Wenn in zwei rechtwinkligen Dreiecken die Katheten übereinstimmen, dann sind die Dreiecke congruent.*

[14] Hilbert means the *Festschrift* (*Hilbert 1899c*, Chapter 5 in this Volume), Theorem 15, pp. 16–17.

[15] See the Introduction to Chapter 4, Specific Note (3)(*a*).

Zum *Beweise* construiere man 37

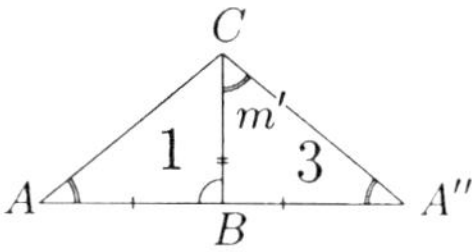

$$\sphericalangle m' \equiv \sphericalangle m,$$

dann ist

$$\triangle 3 \equiv \triangle 2, \qquad (31.))$$

also auch

$$A''B \equiv A'B' \equiv AB$$

und

$$A''C \equiv AC;$$

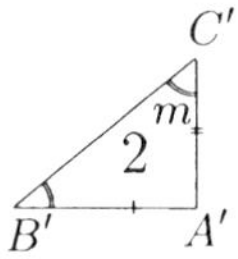

daher

$$\triangle 3 \equiv \triangle 1, \qquad (28.))$$

folglich

$$\triangle 1 \equiv \triangle 2 \qquad \text{q. e. d.}$$

Jetzt endlich können wir III 6₁ beweisen und damit die Antwort 25.) erledigen. Wir nehmen dabei gleich die beiden Dreiecke ABC und ABC' so an, dass sie eine Seite AB gemein haben und symmetrisch dazu liegen.

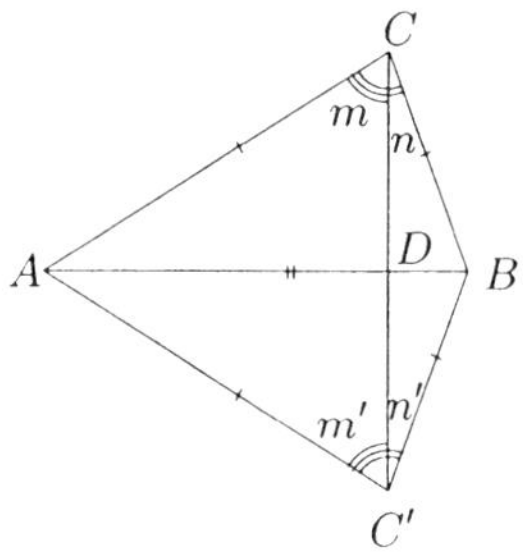

Der Voraussetzung nach ist nun:

$$AC \equiv AC'$$

und

$$\sphericalangle CAB \equiv C'AB,$$

also $\triangle ACC'$ gleichschenkelig, demnach

$$\sphericalangle m \equiv m'.$$

Nach 24.) ist nun

$$\triangle ADC \equiv ADC',$$

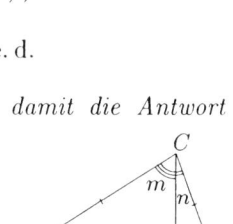

also sind die Winkel bei D lauter rechte Winkel, | und 38

$$CD \equiv C'D;$$

daraus folgt

$$\triangle CDB \equiv C'DB,$$

also

$$\sphericalangle n \equiv \sphericalangle n'$$

$$CB \equiv C'B,$$

endlich

$$\sphericalangle C \equiv \sphericalangle C'$$

und

$$\sphericalangle \beta \equiv \sphericalangle \beta',$$

womit die Congruenz der beiden Dreiecke vollständig erwiesen ist.

Unabhängigkeit des Congruenzaxioms.

Wir wollen zeigen, dass III $6_{1 \text{ und } 2}$ nicht durch logische Schlüsse aus I, II, III 1, 2, 3, 4, 5 abgeleitet werden können.

Zu diesem Zwecke werden wir eine Geometrie, *ein System von Punkten und Geraden*, construieren, in welchem alle bisherigen Axiome außer III $6_{1,\,2}$ gelten.

Wir wählen die Punkte und Geraden der *gewöhnlichen Euklidischen Geometrie* als Elemente, und legen ein rechtwinkliges Coordinatensystem zu Grun-
39 de, so dass jeder | Punkt durch zwei Zahlen x, y festgelegt ist. *Wir wollen nun festsetzen:*

„*Zwei Winkel sollen in unserer Geometrie gleich heißen, wenn sie es in der gewöhnlichen Geometrie sind.*

Die Länge der Verbindungslinie der Punkte $1\,(x_1\,y_1)$ *mit* $2\,(x_2\,y_2)$ *soll dagegen von der gewöhnlichen Geometrie verschieden sein und gegeben sein durch*

$$\overline{12} = \left|\sqrt{(x_1 - x_2 + y_1 - y_2)^2 + (y_1 - y_2)^2}\right|.\text{“}$$

Man sieht sofort ein, dass *I, II, III 1, 2, 4, 5* erfüllt sind.

Aber auch dem *Axiom III 3 wird* in unserer Geometrie Genüge geleistet: *Bew.:* Die Gleichung einer Geraden kann immer dargestellt ⟨werden⟩ durch

$$\begin{aligned} x &= \alpha t + \alpha' \\ y &= \beta t + \beta', \end{aligned}$$

wobei α, α', β, β' Constante sind und t einen Parameter bedeutet und x, y die laufenden Coordinaten euklidisch gemessen.

40 Sind nun 1, 2, 3 drei Punkte einer Geraden, t_1, t_2, t_3 ihre Parameter, so gilt

$$\begin{array}{ll} \text{für Punkt 1:} & \begin{aligned} x_1 &= \alpha t_1 + \alpha' \\ y_1 &= \beta t_1 + \beta' \end{aligned} \\ \text{" \quad " \quad 2:} & \begin{aligned} x_2 &= \alpha t_2 + \alpha' \\ y_2 &= \beta t_2 + \beta' \end{aligned} \\ \text{" \quad " \quad 3:} & \begin{aligned} x_3 &= \alpha t_3 + \alpha' \\ y_3 &= \beta t_3 + \beta' \end{aligned} \end{array}$$

Nach unseren Festsetzungen ist nun, wie man durch einfache Rechnung zeigt:

$$\begin{aligned} \overline{12} &= |t_1 - t_2| \cdot \sqrt{(\alpha + \beta)^2 + \beta^2} \\ \overline{23} &= |t_2 - t_3| \cdot \sqrt{(\alpha + \beta)^2 + \beta^2} \\ \overline{13} &= |t_1 - t_3| \cdot \sqrt{(\alpha + \beta)^2 + \beta^2}. \end{aligned}$$

Liegt nun 2 zwischen 1 und 3, so ist t_2 zwischen t_1 und t_3, daher

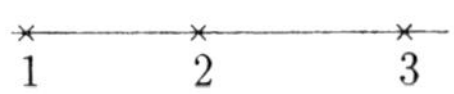

$$|t_1 - t_2| + |t_2 - t_3| = |t_1 - t_3|,$$

also dann:

$$\overline{12} + \overline{23} = \overline{13}.$$

Daraus folgt sofort die Richtigkeit von III 3.

41 Sind nämlich $1'$, $2'$, $3'$ drei Punkte einer anderen Geraden und ist

$$\overline{1'2'} \equiv \overline{12}, \ \overline{2'3'} \equiv \overline{23},$$

so ist auch

$$\overline{1'2'} + \overline{2'3'} \equiv \overline{12} + \overline{23},$$

also nach eben Bewiesenem

$$\overline{1'3'} \equiv \overline{13}.$$

In unserer Geometrie sind also I, II, III 1–5 erfüllt, nun wollen wir zeigen, dass III 6 nicht erfüllt wird.

Nehmen wir die Punkte $A\,(1,0)$, $B\,(-1,0)$ und $C\,(0, \frac{1}{\sqrt{2}})$, so sind nach unseren Festsetzungen

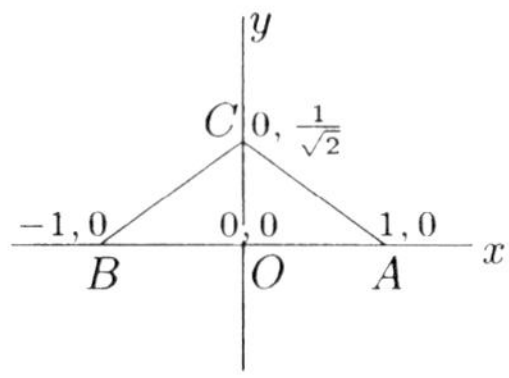

$$\overline{OA} = 1$$
$$\overline{OC} = 1$$
$$\overline{OB} = 1.$$

Da nun die Winkel bei O rechte sind in der euklidischen Geometrie und daher auch in unserer, so sollten nach 17.) die Dreiecke AOC und COB also insbesondere die Strecken AC und CB congruent sein, falls III 6_2 gilt.

Nun findet man aber in unserer Geometrie 42

$$\overline{AC} \equiv \sqrt{2 - \frac{2}{\sqrt{2}}}$$

und

$$\overline{BC} \equiv \sqrt{2 + \frac{2}{\sqrt{2}}}.$$

17.) und damit *auch III 6_2 gelten in unserer Geometrie also nicht.*[A]

Beide Dreiecke hatten *gleichen Umlaufssinn*; wir wollen nun noch zwei Dreiecke von *verschiedenem Umlaufssinn* betrachten und damit auf III 6_1 schließen.

In unserer Geometrie nehmen wir die Punkte $A\,(1,0)$, $B\,(0,1)$, $C\,(\frac{1}{2}, \frac{1}{2})$ an; dann ist nach unseren Festsetzungen:

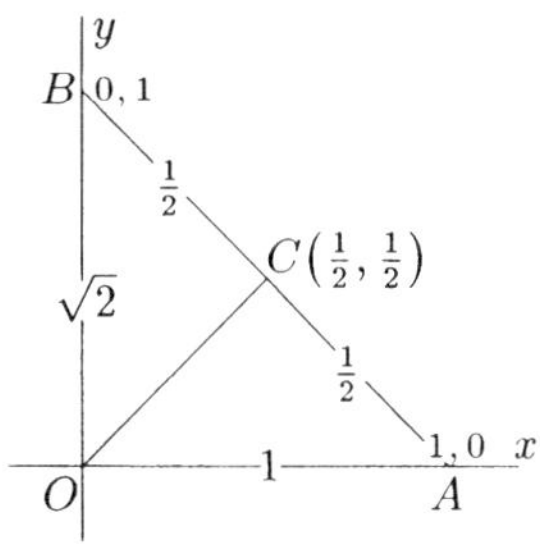

$$\overline{AC} = \tfrac{1}{2},$$
$$\overline{BC} = \tfrac{1}{2}.$$

Da außerdem $OC \equiv OC$ und die Winkel bei C rechte sind in unserer Geometrie geradeso wie in der Euklidischen, so müssten die Dreiecke ACO und OCB congruent sein, falls III 6_1 gelten würde, und damit OA und OB gleich sein.

Letzteres ist aber nicht der Fall; in unserer Geometrie ist nämlich 43

$$OA = 1, \quad OB = \sqrt{2}.$$

Damit ist streng nachgewiesen, dass III $6_{1 \text{ und } 2}$ keine Folge der früheren Axiome sind.

[A] Uebrigens sind unsere Dreiecke beide rechtwinklig u. gleichschenklich und es müssten, wenn das Dreiecksaxiom gültig wäre, auch die Basiswinkel in ihnen übereinstimmen, was nicht der Fall ist. Man kann mit Zuhilfenahme des Raumes unsere Konstruktion offenbar sehr einfach deuten: Die Winkel werden in einer Horizontalebene gleich den wirklichen Winkeln genommen, die Strecken aber so dass als Längen die wirklichen Längen der Projektionen genommen werden auf eine unter 45° geneigte schiefe Ebene. ⟨Added by *Hi*^{g} in the left-hand margin and continuing at the bottom of the page. The remark is not assigned to a particular line.⟩

IV.

Axiom des Euklid, Parallelenaxiom.

Dieses Axiom wollen wir so aussprechen:

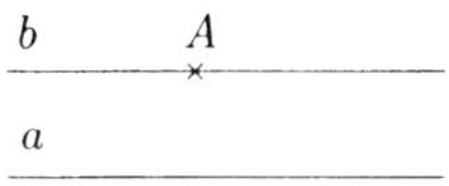

Durch den Punkt A außerhalb einer Geraden a lässt sich [16] *nur eine Gerade b ziehen, welche a nicht schneidet.*

Dieses Axiom enthält *zwei Aussagen:*

1.) Es giebt durch A eine Gerade, welche a nicht schneidet.

2.) Es giebt außer dieser Gerade keine weitere Gerade von derselben Eigenschaft.

44 [17]Man kann das Parallelenaxiom auch so aussprechen:[17]

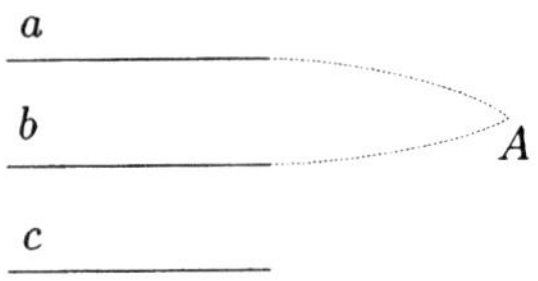

Wenn zwei Gerade a und b eine dritte c nicht treffen, so treffen sie auch einander nicht.

Diese Aussage ist mit unserer Aussage vollkommen äquivalent, wie man wohl sofort erkennt.

Wir wollen nun einige Folgerungen aus dem Parallelenaxiom geben. Euklid hat dieselben in einer Einfachheit und Schönheit gegeben, welche man oft in Büchern namentlich Schulbüchern vermisst, so dass in diesem Sinne ein *wirklicher Rückschritt* zu verzeichnen ist.

34.)[18] *Satz von der Gleichheit der Wechselwinkel:*

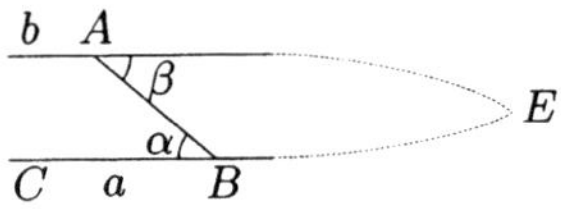

45 Ist $\sphericalangle BAD \equiv \sphericalangle CBA$, so können sich die Geraden a und b nicht treffen; | denn in diesem Falle würde ein Dreieck ABE entstehen, und der Satz über die Außenwinkel desselben nicht erfüllt sein (vergl. 23.)). [19]Es giebt also durch einen Punkt A ausserhalb a stets eine und nur eine Parallele zu a.[19]

35.) *Die Umkehrung des Satzes 34.) gilt auch* und man beweist sie auf das einfachste:

Ist nämlich $a \parallel b$ und $\sphericalangle\beta \neq \alpha$, so construire man in A

$$\sphericalangle\beta' = \alpha,$$

wodurch man eine zweite Gerade b' parallel zu a erhält, b' muss mit b zusammenfallen.

---

[16]Deleted by Hi^s: *stets eine und*

[17–17]This, written in Hi^s, replaces: *Euklid* hat sein Axiom auch in zwei Theile zerlegt und sie so ausgesprochen:
1.) Durch A giebt es immer eine Gerade, welche a nicht schneidet.
2.)

[18]There are no propositions marked 32.) and 33.).

[19–19]Added by Hi^s

36.) *Die Winkelsumme in jedem Dreiecke beträgt 2R.*
Bew.: Man zieht durch den Scheitel des Dreieckes eine Parallele zur Grundlinie, so sind die am Scheitel entstehenden Winkel den Dreieckswinkeln congruent und letztere daher zusammen $2R$.

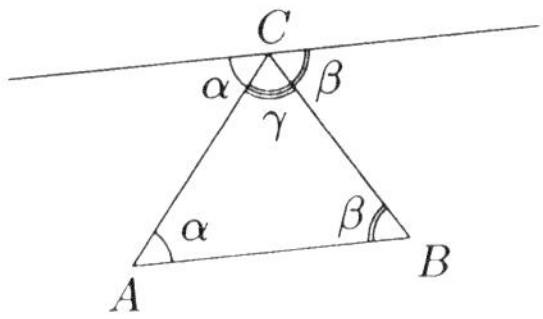

Aus derselben Figur folgt auch der Satz: 46

37.) In jedem Dreiecke ist ein Außenwinkel gleich der Summe der beiden nicht anliegenden Innenwinkel.

V.

Archimedisches Axiom.

Wir wollen dies Axiom so aussprechen:

> Es sei A_1 ein beliebiger Punkt auf einer Geraden zwischen den beliebigen Punkten A und B derselben Geraden; man construiere nun die Punkte $A_2, A_3, A_4, \ldots$ so, dass A_1 zwischen A und A_2, ferner A_2 zwischen A_1 und A_3, ferner A_3 zwischen A_2 und A_4 u. s. w. liegt und überdies die Strecken
>
> $$A\,A_1, A_1\,A_2, A_2\,A_3, A_3\,A_4, \ldots$$
>
> einander gleich sind; dann gibt es in der Reihe der Punkte A_2, A_3, A_4, ... stets einen solchen Punkt A_n, dass B zwischen A u. A_n liegt.

Damit sind die rein geometrischen Axiome gegeben.

Wir müssen nun die *Widerspruchslosigkeit dieser gesammten Axiome* zei- 47
gen; außerdem die *Unabhängigkeit der letzten zwei* von den übrigen.

Um aber das *Verständnis* für diese Beweise zu *erleichtern*, wollen wir *eine Bemerkung* vorausschicken:

Die Dinge, mit denen die Mathematik sich beschäftigt, werden durch Axiome definiert, *in's Leben gerufen.*

Diese Axiome selbst können ganz *willkürlich* angenommen werden. Wenn aber diese Axiome einander widersprechen, so lassen sich aus ihnen keine logischen Schlüsse ziehen; das definierte System von Dingen existiert dann für uns Mathematiker nicht.

Aus diesem Grunde müssen wir die Widerspruchslosigkeit unserer geometrischen Axiome untersuchen.

Wir werden uns dabei des *Systems der reellen Zahlen* bedienen, deren *Existenz wir voraussetzen.*

Für das *System der reellen Zahlen lassen sich auch Axiome aufstellen*, 48
welche die Eigenschaften der reellen Zahlen definieren.

Diese Axiome, welche in unserem *Buche „Die Grundlagen der Geometrie“* auf Seite 26 näher auseinandergesetzt wurden[20], lassen sich analog unseren geometrischen Axiomen, in Axiome der Verknüpfung und der Anordnung eintheilen; außerdem gilt auch für reelle Zahlen das Archimedische Axiom. Dieses lautet hier so:

---

[20]I.e., the *Festschrift* (*Hilbert 1899c*, here Chapter 5), pp. 26–28. See also *Hilbert 1900b*.

„*Wenn $a > 0$ und $b > 0$ zwei beliebige Zahlen sind, so ist es stets möglich, a so oft zu sich zu addieren, dass die entstehende Summe die Eigenschaft hat*

$$a + a + a + \ldots + a > b.\text{“}$$

Der Inbegriff der reellen Zahlen ist dann das System von Dingen, deren gegenseitige Beziehungen durch diese aufgestellten Axiome geregelt wurden.

Ist die Widerspruchslosigkeit dieser Axiome bewiesen, so ist damit die Existenz der reellen Zahlen evident. *Wir nehmen dies im folgenden an.*

49 Das gewöhnliche complexe Zahlensystem $a + ib$ erfüllt nicht sämmtliche Axiome der reellen Zahlen, da die complexen Zahlen sich nicht anordnen lassen. („Wenn $a > b$ u. $c > 0$, so ist bei reellen Zahlen $ac > bc$, bei complexen nicht.“)

Analog wollen wir jedes System von Dingen, welches nicht sämmtliche Axiome der reellen Zahlen erfüllt, ein complexes Zahlensystem nennen; die reellen Zahlen (auch 0) gehören offenbar jedem complexen Zahlensystem an.

Wir wollen nun ein solches complexes Zahlensystem folgendermaßen aufstellen:

Sei t irgend ein Parameter, ein *Symbol* (analog i in obigem Beispiele), so wollen wir *das Ding*

$$\alpha = a_0 t^n + a_1 t^{n+1} + a_2 t^{n+2} + \ldots$$

eine Zahl unseres neuen Systems heißen, wenn $a_0, a_1, a_2, \ldots$ u. n gewöhnliche reelle positive oder negative Zahlen sind; das $+$ Zeichen hat offenbar nur symbolische Bedeutung.

50 *Wir wollen nun festsetzen:*

1.) Mit unseren Zahlen sollen die 4 Rechnungsoperationen geradeso ausgeführt werden, wie wenn die rechte Seite der Zahlen wirkliche Potenzzeichen wären.

Daraus folgt z. B. für α^2

$$\alpha^2 = a_0 t^{2n} + 2a_0 a_1 t^{2n+1} + \ldots;$$

außerdem dass die Axiome *der Verknüpfung* reeller Zahlen auch hier gelten.

2.) Ist für α die reelle Zahl a_0 *größer* als 0, dann soll auch α *größer als Null heißen*, ist dagegen $a_0 < 0$, dann soll auch $\alpha < 0$ heißen.

3.) Ist gemäß 2.) die Zahl

$$\alpha - \alpha' > 0, \textit{ so soll } \alpha > \alpha' \textit{ heißen},$$

ist

$$\alpha - \alpha' < 0, \quad \text{"} \quad \text{"} \quad \alpha < \alpha' \quad \text{"} \quad .$$

Aus 2.) und 3.) folgt, dass für unser neues Zahlensystem auch die Axiome *der Anordnung* erfüllt sind.

51 *Das Archimedische Axiom ist aber nicht | erfüllt*:

Nehmen wir nämlich t, welches ja eine Zahl unseres Systems ist ($t = 1 \cdot t^0$)[21] und sind außerdem a und m beliebige reelle Zahlen; so ist $mt - a$ auch eine Zahl unseres Systems.

[21]Hilbert (in Hi^s) has added a question mark here, and written '$t^0 = 1$' in the right-hand margin. From the development which follows, it is clear that Adler should have written here '$t = 1 \cdot t^1$'.

Nach 2.) ist nun $mt - a$ kleiner als Null, denn

$$mt - a = -a \cdot t^0 + m \cdot t^1,$$

also $mt - a < 0$ und daraus folgt nach 3.)

$$mt < a,$$

wie groß auch immer a und m sein mögen, d. h. das Archimedische Axiom gilt für unser Zahlensystem nicht, denn sonst müsste ja für gegebene t und a sich m immer so bestimmen lassen, dass $mt > a$ wird.

Unser Zahlensystem können wir also ein nichtarchimedisches Zahlensystem nennen.

Es existiert, da das reelle Zahlensystem existiert.

Mit Hilfe dieser Zahlensysteme (reelles und complexes) können wir nun 52
unschwer die Widerspruchslosigkeit unserer geometrischen Axiome I–V und die Unabhängigkeit einer Anzahl derselben beweisen.

Nach dem *Principe der analytischen Geometrie definieren wir*:

„Ein (endliches) *Zahlenpaar x, y* heiße *Punkt*, das *Verhältnis dreier Zahlen* $(u : v : w)$ heiße *Gerade*. Der *Punkt (xy) soll auf der Geraden $(u : v : w)$ liegen*, wenn

$$ux + vy + w = 0$$

ist.

Die Punkte einer Geraden sollen entsprechend ihrem x geordnet werden (oder y)."

Dann *erfüllen* unsere so definierten Dinge die *Axiome I, II, IV.*

Um auch die *Congruenzaxiome* zu erhalten, setzen wir fest:

Zwei Strecken oder Winkel (welche wie früher definiert werden) sollen *con-* 53
gruent heißen, wenn sie durch *Parallelverschiebung, Drehung und Umklappung um die x Axe* auseinander entstehen.

Die *Parallelverschiebung* ist dabei gegeben durch:

$$x' = x + \alpha$$
$$y' = y + \beta.$$

Die *Drehung um einen Winkel* $[x, 0, \overline{ab}]$ durch:

$$x' = \frac{a}{\sqrt{a^2+b^2}}x - \frac{b}{\sqrt{a^2+b^2}}y$$
$$y' = \frac{b}{\sqrt{a^2+b^2}}x + \frac{a}{\sqrt{a^2+b^2}}y.$$

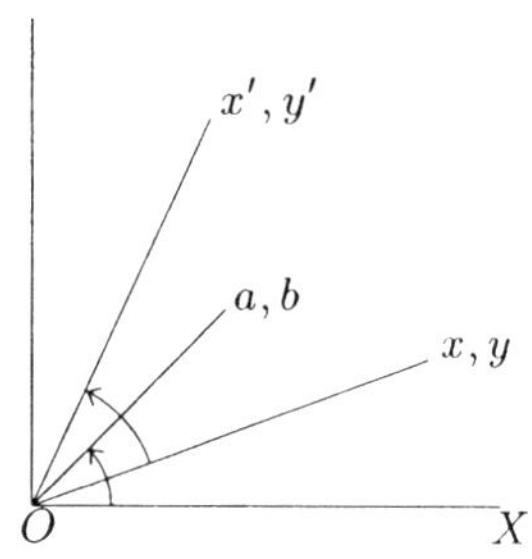

Die *Umklappung* durch:

$$x' = x$$
$$y' = -y.$$

Unsere *so* definierten Punkte und Geraden *erfüllen* jetzt *alle* früher aufgestellten *Axiome I–V*.

Diese Axiome I–V können also einander nicht widersprechen; denn | sie 54
erfordern nur Operationen mit reellen Zahlen; ein Widerspruch zwischen den Axiomen würde einen Widerspruch zwischen den Operationen herbeiführen, welcher nicht existiert.

Unabhängigkeit des Axioms V von den übrigen.

Ein *System von Dingen*, welche wir *Punkte* und *Geraden* nennen und denen wir gewisse gegenseitige Beziehungen auferlegen, wollen wir eine *Geometrie* nennen.

Mit Hilfe der complexen Zahlen (pag. 49, 50) können wir nun leicht eine *Geometrie construieren*, in welcher sämmtliche Axiome *I–IV gelten, V aber nicht.*

Damit ist dann die Unabhängigkeit von V nachgewiesen.

Zunächst *bemerken* wir noch Folgendes: Ist α eine complexe Zahl, so ist
55 nicht nur α^2, wie früher gezeigt, eine derartige Zahl | sondern auch $\sqrt{\alpha}$; denn

$$\begin{aligned}\sqrt{\alpha} &= \sqrt{a_0 t^n + a_1 t^{n+1} + \ldots} = \\ &= t^{\frac{n}{2}} \sqrt{a_0 + a_1 t^1 + a_2 t^2 + \ldots} \\ &= A_0 t^m + A_1 t^{m+1} + \ldots\end{aligned}$$

Sind also a, b Zahlen unseres Systems, so gehört auch $\sqrt{a^2 + b^2}$ dem Systeme an.

Nun construieren wir unsere neue Geometrie mit Hilfe der complexen Zahlen:

Wir definieren ganz analog pag. 52, 53:

Ein Zahlenpaar xy heiße Punkt, das Verhältnis dreier Zahlen $(u : v : w)$ *Gerade*, $ux + vy + w = 0$ *gebe die incidente Lage von Punkt und Gerade an;* ferner *Parallelverschiebung, Drehung, Umklappung* mittelst der früheren Formeln aus unseren Zahlen.

Wir erhalten dadurch eine Geometrie, in welcher *sämmtliche Axiome I–IV erfüllt sind, V ist aber nicht erfüllt; denn unser Zahlensystem erfüllt dieses Axiom nicht.*

[22]Man könnte in Analogie mit den Zahlensystemen solche Geometrien, in denen nicht alle Axiome als gültig angenommen werden „complexe Geometrien" nennen:[22] [23]

56

Semi-euklidische Geometrie.

Wir wollen eine merkwürdige Geometrie aufbauen, wieder mit Hilfe unserer complexen Zahlen.

Wir setzen fest:

Ist

$$\alpha = a_0 t^n + a_1 t^{n+1} + \ldots$$

eine Zahl unseres Systems und ist

$$n \geqq 0,$$

so soll α eine *ganze Zahl heißen.*

[22–22] Added by *Hi*^{s}

[23] See also Hilbert's Remark [17] in the Appendix to Chapter 4.

Ist dann α eine ganze Zahl, so ist es auch α^2 und auch $\sqrt{\alpha}$; sind a, b ganze Zahlen in unserem Sinne, so sind es auch

$$\frac{a}{\sqrt{a^2+b^2}} \quad \text{und} \quad \frac{b}{\sqrt{a^2+b^2}},$$

wie man leicht einsieht.

Mit *Hilfe dieser ganzen (complexen) Zahlen* allein definieren wir genau wie früher nach dem Princip der analytischen Geometrie ein System von Punkten und Geraden, [24]so jedoch dass ein Punkt ein Paar *ganzer* Zahlen x, y ist, die die Coordinaten des Punktes heissen und die Geraden ein System $(u:v:w)$ sind, wenn die lineare Gleichung $ux+vy+w=0$ durch mindestens 2 Paare ganzer x, y befriedigt wird.[24]

In dieser Geometrie gilt *Axiom V nicht*, da es für complexe Zahlen nicht 57
gilt.

Die Axiome *I, II und III gelten aber sämmtlich.* Wir wollen nun beweisen, dass *Axiom IV nicht gilt.*

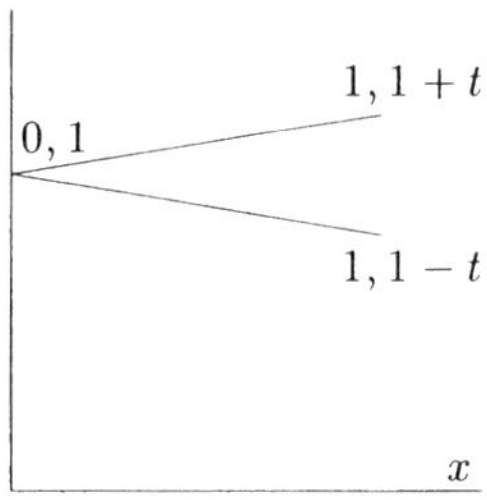

Wir nehmen zu diesem Zwecke die Gleichungen

$$tx+y=1$$

und

$$-tx+y=1.$$

Beide stellen Geraden dar, die erste die Verbindungslinie der Punkte $(0,1)$ u. $(1,1-t)$, die zweite die Verbindungslinie von $(0,1)$ mit $(1,1+t)$.

Setzen wir in unseren Gleichungen $y=0$, (bringen also die Geraden mit der x Axe zum Schnitte), so erhalten wir $x=\frac{1}{t}$ resp. $-\frac{1}{t}$.

Punkte mit den Coordinaten $\frac{1}{t}$, $-\frac{1}{t}$ kennt unsere Geometrie nicht; durch den Punkt $(0,1)$ gehen also zwei die x Axe nicht schneidende Gerade; *IV gilt hier nicht.*

Dagegen ist die Winkelsumme eines jeden Dreieckes unserer Geometrie $2R$.

1. Beweis: Wir haben pag. 54, 55 mit Hilfe der complexen Zahlen eine Geo- 58
metrie construiert, in welcher I, II, III, IV sämmtlich galten, also auch der Satz über die Winkelsumme im Dreiecke.

Unsere jetzige Geometrie, welche nur die ganzen complexen Zahlen benützt, ist aber in der früheren Geometrie enthalten.

2. Beweis: Die Winkel unserer Geometrie sind ja *definiert* durch die Drehungsformel pag. 53. *Wir wollen nun für den Augenblick 2 Geraden parallel nennen*, welche durch die Transformation

$$x' = x + \alpha$$
$$y' = y + \beta$$

aus einander entstehen.

Dann sind offenbar die *Wechselwinkel* einander gleich, welche beim *Schnitte* von *2 unserer „Parallelen" mit einer dritten* Geraden entstehen; es gilt also auch der | Euklidische *Beweis für die Winkelsumme* im Dreiecke (pag. ⟨45⟩). 59

[24–24]Added by *Hi*^{s}

Der Satz von der Winkelsumme im Dreiecke ist also nicht äquivalent dem Parallelaxiom; aus letzterem folgt der erste Satz, nicht aber umgekehrt, wenn man nicht V hinzunimmt.

Unabhängigkeit des Axioms IV von I, II, III, V.
(Nichteuklidische o. Bólyai-Lobatschefsky'sche Geometrie).

Wir construieren ein System von Dingen, welche wir Punkte und Gerade nennen, und welche die Axiome I, II, III, V erfüllen (IV aber nicht) in folgender Weise:

60 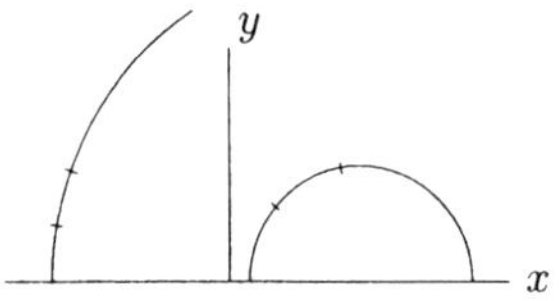

Wir nehmen die *gewöhnliche* Euklidische Zahlenebene. *Als Punkte unserer Geometrie nehmen wir die Punkte der Halbebene oberhalb der x Axe mit Ausschluss der Punkte der x Axe selbst.*

Als Gerade nehmen wir die zur x Axe orthogonalen Halbkreise der Halbebene.

In dieser Geometrie gelten dann die Axiomsgruppen I, II.

Um auch III zu erfüllen, definieren wir die *nichteuklidische Entfernung* der Punkte 1, 2 durch das Integral

$$\int_{t_1}^{t_2} \left| \sqrt{\left(\frac{dx}{dt}\right)^2 + \left(\frac{dy}{dt}\right)^2} \right| \frac{dt}{y} \text{ [25]}$$

Dabei denken wir uns den Kreis durch 1, 2 dargestellt durch

$$x = x(t)$$
$$y = y(t).$$

61 Setzt man

$$z = x + iy,$$

also

$$dz = dx + idy,$$

so ist

$$\left|\frac{dz}{dt}\right| = \sqrt{\left|\left(\frac{dx}{dt}\right)^2 + \left(\frac{dy}{dt}\right)^2\right|},$$

⟪???⟫ unser Integral $= \int_1^2 \frac{|dz|}{y}$.

Daraus ersieht man sofort, dass die so definierte Entfernung $\overline{12}$ wächst, wenn bei festem 1 sich 2 davon entfernt, denn dz wird ja nur absolut genommen.

---

[25] '$\frac{dt}{y}$' has been substituted for '$\frac{dt}{dy}$' in pencil, certainly not in Hilbert's hand, and probably not Adler's either. The next occurence of '$\frac{dt}{y}$' (in the integral inequality) is a similar replacement, this time in Adler's hand.

Wenn sich 2 der x Axe unendlich nähert, so wächst seine Entfernung von 1 über alle Grenzen; denn es ist jedenfalls

$$\int_{t_1}^{t_2} \left| \sqrt{\left(\frac{dx}{dt}\right)^2 + \left(\frac{dy}{dt}\right)^2} \right| \frac{dt}{y} > \int_{t_1}^{t_2} \left| \sqrt{0 + \left(\frac{dy}{dt}\right)^2} \right| \frac{dt}{y},$$

also

$$> \int_{t_1}^{t_2} \frac{|dy|}{y},$$

daher

$$> |\mathrm{l}\, y_2 - \mathrm{l}\, y_1| .$$

Damit ist auch der nothwendige Nachweis erbracht, dass unsere Gerade unendlich lang ist. 62

Es sei auch vorweg bemerkt, *dass V in unserer Geometrie erfüllt ist*, da man in dieser Geometrie jeden Punkt durch Auftragen einer Strecke überschreiten kann.

Unter Winkel wollen wir in unserer Geometrie die wirklichen Winkel verstehen und mit ihnen in gewöhnlicher Weise operieren.

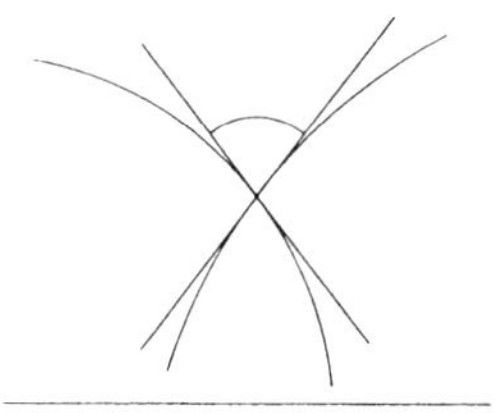

Man beweist nun unschwer, *dass in unserer Geometrie die Axiome III 1, 2, 3, 4, 5 erfüllt sind.* Um aber zu zeigen, dass auch III 6 gilt, wollen wir zunächst einen Excurs über

lineare Transformationen der ganzen Ebene 63

vorausschicken.

Ein Punkt dieser Ebene ist gegeben durch

$$z = x + iy,$$

die *lineare Transformation* selbst durch

$$z' = \frac{az + b}{cz + d},$$

wobei a, b, c, d complex sein dürfen, aber $ad - bc$ von Null verschieden sein muss; wir wollen annehmen

$$ad - bc = 1.$$

Unter dem *Doppelverhältnis* von 4 beliebigen Punkten z_1, z_2, z_3, z_4 versteht man:

$$(z_1\, z_2\, z_3\, z_4) = \frac{z_1 - z_2}{z_3 - z_2} : \frac{z_1 - z_4}{z_3 - z_4},$$

wobei diese 4 Punkte nicht in einer Geraden zu liegen brauchen.

64 Durch lineare Transformation entstehen aus z_1, z_2, z_3, z_4 die Punkte z_1', z_2', z_3', z_4', deren Doppelverhältnis ist:

$$(z_1'\,z_2'\,z_3'\,z_4') = \frac{z_1'-z_2'}{z_3'-z_2'} : \frac{z_1'-z_4'}{z_3'-z_4'}.\langle A\rangle$$

Durch einfaches Nachrechnen findet man nun, dass

$$(z_1'\,z_2'\,z_3'\,z_4') = (z_1\,z_2\,z_3\,z_4),$$

d. h.: *Das Doppelverhältnis von 4 Punkten bleibt bei einer linearen complexen Transformation erhalten.*

Wir wollen nun eine *geometrische Deutung* dieses Satzes angeben.

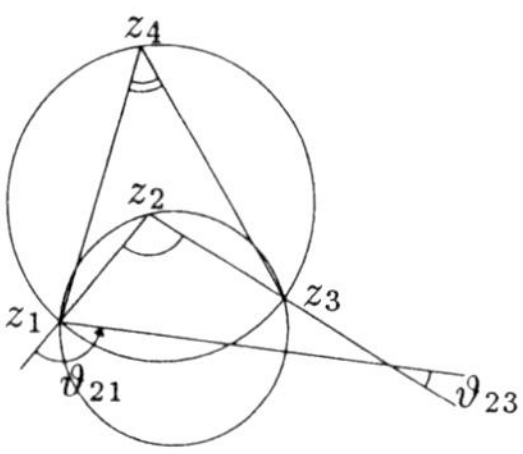

Es sei

$$z_1 = x_1 + iy_1$$
$$z_2 = x_2 + iy_2,$$

also

$$z_1 - z_2 = (x_1 - x_2) + i(y_1 - y_2) = \varrho_{12}e^{i\vartheta_{21}}.$$

(Die Strecke $z_1 - z_2$ hat die Richtung von z_2 nach z_1.)

Ebenso ist

$$z_3 - z_2 = \varrho_{32}e^{i\vartheta_{23}}.$$

65 Daraus

$$\frac{z_1-z_2}{z_3-z_2} = \frac{\varrho_{12}}{\varrho_{32}}e^{i(\vartheta_{21}-\vartheta_{23})}.$$

Aus der Figur folgt, dass

$$\sphericalangle\vartheta_{21} - \vartheta_{23} = \sphericalangle(z_1\,z_2\,z_3)$$

ist.

Analog wird $\frac{z_1-z_4}{z_3-z_4}$ eine complexe Zahl, deren Argument der Winkel $z_1\,z_4\,z_2$ sein wird.

Schließlich wird

$$\frac{z_1-z_2}{z_3-z_2} : \frac{z_1-z_4}{z_3-z_4} = \varrho e^{i\vartheta}$$

ebenfalls eine complexe Zahl, deren Argument die Differenz der Winkel bei z_2 und z_4 ist.

Nach bekannten Sätzen über Peripheriewinkel folgt nun, dass $\sphericalangle(z_2 - z_4)$ gleich dem Winkel ist, unter welchem sich die beiden Kreise unserer Figur schneiden.

Das Doppelverhältnis $(z_1\,z_2\,z_3\,z_4)$ ist im Allgemeinen complex und reell
66 nur, wenn | $\vartheta = 0, \pi$ ist, also wenn die Punkte z_1, z_2, z_3, z_4 auf einem Kreise liegen.

Daraus folgt:

1.) Vier Punkte, deren Doppelverhältnis reell ist, liegen auf einem Kreise und umgekehrt.

[A]Denn:

$$z_2' - z_1' = \frac{z_2-z_1}{(cz_1+d)(cz_2+d)}$$

etc. ⟨Added by *Hi*[s] in the left-hand margin and not assigned to any particular line.⟩

2.) Jeder Kreis geht also durch eine lineare complexe Transformation wieder in einen Kreis über, der auch eine Gerade sein kann.[B]

3.) Die Winkel, unter welchen sich zwei Curven schneiden, bleiben bei der linearen complexen Transformation erhalten; die Transformation ist winkeltreu.

4.) Wir wollen zeigen, dass auch die sogenannten nicht euklidischen Längen erhalten bleiben bei unserer linearen complexen Transformation.

Es ist 67

$$z' = \frac{az+b}{cz+d} \quad (ad - bc = 1),$$

daraus folgt:

$$\frac{|dz'|}{|dz|} = \frac{1}{|cz+d|^2}. \qquad (\alpha$$

Bezeichnet man die zu z conjugierte complexe Zahl mit $\overline{z}$, so ergiebt sich:

$$z' - \overline{z}' = 2iy' = \frac{az+b}{cz+d} - \frac{a\overline{z}+b}{c\overline{z}+d}$$
$$= \frac{z-\overline{z}}{|cz+d|^2} = \frac{2iy}{|cz+d|^2},$$

also

$$y' = \frac{y}{|cz+d|^2}. \qquad (\beta$$

Nun ist die nicht euklidische Länge von $\overline{1'2'}$:

$$\int\limits_{1'}^{2'} \frac{|dz'|}{y'} = \int\limits_{1}^{2} \frac{dz}{|cz+d|^2} : \frac{y}{|cz+d|^2} = \int\limits_{1}^{2} \frac{|dz|}{y},$$

d. h. *die nichteuklidische Länge wird durch die lineare Transformation nicht geändert.*[26]

Nun können wir leicht nachweisen, dass in unserer Geometrie das Axiom $III\,6_1$ *gilt:*

In den beiden Dreiecken 1 und 2 sei 68

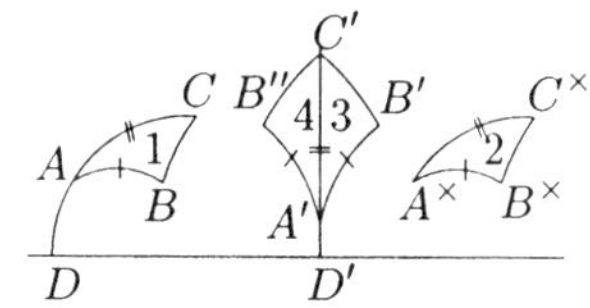

$$\sphericalangle A^\times = A$$
$$A^\times B^\times = AB$$
$$A^\times C^\times = AC.$$

Machen wir nun:

$$D'A' = DA \quad \langle A \rangle$$
$$A'C' = AC$$
$$\sphericalangle A' = A,$$

endlich

$$A'B'(A'B'') = AB.$$

[B] Gl⟨eichung⟩ des Kreises ist $A(x^2+y^2)+Bx+Cy+D=0$. $\mathrm{A}z\overline{z}+\mathrm{B}z+\Gamma\overline{z}+\Delta=0$ geht bei der Transformation wieder in eine solche über. ⟨Added by Hi^g in the upper margin of p. 67, and not assigned to any particular line.⟩

[A] hat $\int \frac{dz}{y}$ Sinn? ⟨Added by Hi^g to the right of this equation.⟩

[26] Throughout this proof, the absolute value signs have been added (or replace parentheses) by Hi^g except in equation (β, where the replacement has been executed by the editors.

Nun stellen wir eine lineare Transformation so her, dass A in A', D in D' übergeht.

Sind die Coordinaten von A $(z = x + iy)$, von D $(\delta + i \cdot 0)$, die von A' $(z' = x' + iy')$, so gelten die Gleichungen:

$$(cz + d)z' = az + b$$

und

$$(c\delta + d) \cdot 0 = a\delta + b.$$

Aus der ersten folgen 2 neue, indem man die reellen Theile und die imaginären für sich einander gleich setzt.

69 Wir haben also zur Bestimmung der Verhältnisse

$$a : b : c : d$$

3 Gleichungen, was genügend ist.

Wir müssen noch zeigen, dass

$$ad - bc \neq 0$$

ist.

Wäre dies der Fall, so wäre

$$a : b = c : d,$$

also

$$az + b : cz + d = a : c = z' : 1.$$

Die Proportion $a : c = z' : 1$ kann nicht existieren, denn a, c, 1 sind reell, z' aber complex.

Bei unserer so gefundenen linearen Transformation geht das Dreieck 1 in 3 über (oder in 4). Mit dem Dreiecke 2 kann man analog verfahren, folglich ist

$$\text{Dreieck } 1 \equiv \text{Dreieck } 2 \quad \text{q. e. d.}$$

Nun wollen wir zeigen, dass in unserer Geometrie das Parallelenaxiom nicht gilt:

Nehmen wir zu dem Zwecke irgend eine unserer „Geraden“ a an und einen
70 außer|halb liegenden Punkt A, so gehen durch A zwei Geraden g_1 und g_2 (Grenzgeraden), welche a berühren und die Ebene in 4 Theile I, II, III, IV zerlegen.

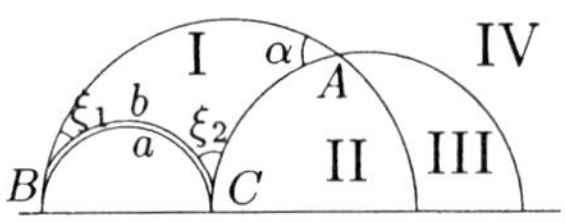

Jede Gerade durch A in I oder III schneidet a; jede Gerade durch A, welche in II oder IV verläuft, schneidet a nicht.

Also gehen durch A sehr viele a nicht schneidende Gerade.

Wir wollen nun zeigen, dass in unserer Geometrie auch die Euklidische Winkelsumme im Dreiecke nicht gilt:

Zu diesem Zwecke ziehen wir nahe an a eine Gerade b, so nahe, dass sowohl $\sphericalangle\xi_1$ als auch $\sphericalangle\xi_2$ kleiner als $\frac{2R-\alpha}{2}$ sind; wobei α den Winkel bei A bedeutet.

In dem Dreiecke ABC ist dann die Winkelsumme kleiner als $2R$.

71 Hat überhaupt ein Dreieck unserer Geometrie die Winkel α, β, γ, so gilt allgemein:

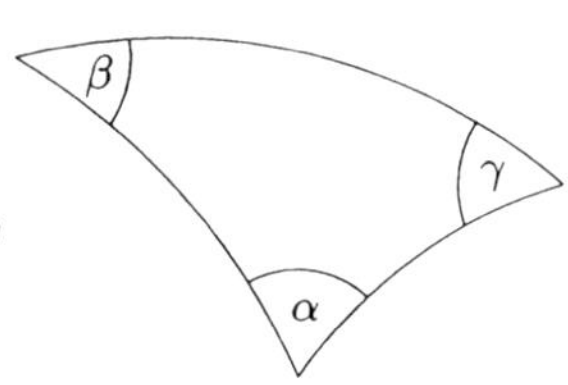

$$\alpha + \beta + \gamma = \pi - \frac{I}{4},$$

wobei I eine positive Größe ist, gegeben durch

$$I = \iint \frac{dx\,dy}{y^2}, \text{ genommen über das ganze Dreieck hin.}$$

I heißt der *nicht euklidische Inhalt des Dreieckes.*

Man kann auch leicht nachweisen, dass bei einer beliebigen linearen Transformation I ungeändert bleibt.

Aus der Formel

$$\alpha + \beta + \gamma = \pi - \frac{I}{4}$$

folgt, dass *die Winkelsumme im Dreiecke kleiner als $2R$ ist*; man kann auch Dreiecke construieren, in denen die *Winkelsumme Null ist*

In unserer Geometrie gelten manche Sätze der Euklidischen Geometrie nicht, | z. B. schneidet nicht jede Gerade durch einen Punkt innerhalb eines 72
Winkels die Schenkel des Winkels.

Bemerkung: Im Vorhergehenden haben wir nicht nur bewiesen, dass IV nicht aus den übrigen Axiomen gefolgert werden kann, sondern wir haben damit auch ein Mittel erhalten, um die Fehler in derartigen häufig vorkommenden, oft sehr scharfsinnigen Beweisen aufzudecken.[27]

Legendre'sche Sätze

(Dissertation v. Dehn, Göttingen 1900)[28]

1.) *In einem Dreiecke kann die Winkelsumme niemals größer als $2R$ sein.*

2.) *Wenn in einem einzigen Dreiecke die Winkelsumme $2R$ ist, dann ist sie es in jedem Dreiecke.*[A]

ad 1.) 73

Es sei

$$\alpha \leqq \beta, \quad \alpha \leqq \gamma, \quad AC \leqq AB.$$

Wir halbieren BC in D, machen $DE \equiv AD$, dann ist

$$\triangle ADC \equiv EDB,$$

also

$$\sphericalangle \delta' \equiv \delta.$$

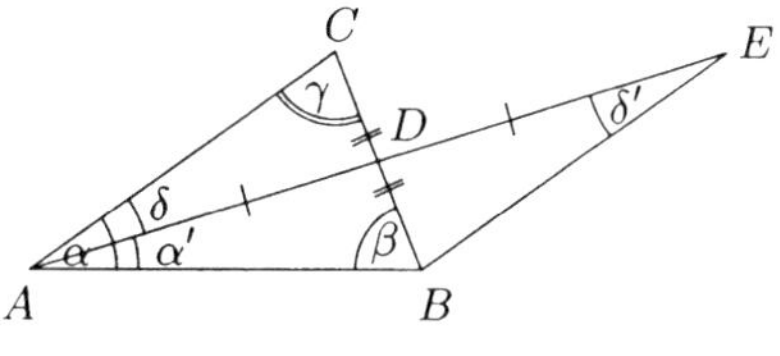

[A]Satz 2.) kann nach Dehn auch ohne Stetigkeitsaxiome bewiesen werden.

[27]The material in this section is treated in more expansive form in the 1898/1899 lectures; see, for instance, von Schaper's *Ausarbeitung* (*Hilbert 1899a**, in Chapter 4 of this Volume), pp. 81–94.

[28]The dissertation referred to is *Dehn 1900a*. The work is also published in *Dehn 1900b*.

Aus $\triangle ABE$ folgt ferner

$$\sphericalangle \alpha' \leqq \delta', \quad \text{also } \alpha = \alpha' + \delta' \geqq 2\alpha'$$
$$\alpha' \leqq \frac{\alpha}{2}.$$

$\triangle ABE$ und ABC haben gleiche Winkelsumme [29]und es ist auch wieder der Winkel bei A nämlich $\alpha' \leqq$ den beiden anderen Winkeln bei B und bei E nämlich δ'[29].

Aus $\triangle ABE$ können wir wieder ein Dreieck von derselben Winkelsumme construieren, bei welchem aber

$$\sphericalangle \alpha'' \leqq \frac{\alpha'}{2}, \quad \text{also} \quad \leqq \frac{\alpha}{4}.$$

Durch n malige Wiederholung erhalten wir ein Dreieck ABX mit der ursprünglichen Winkelsumme, wo aber der Winkel bei A [30]$\leqq$[30] $\frac{\alpha}{2^n}$ ist.

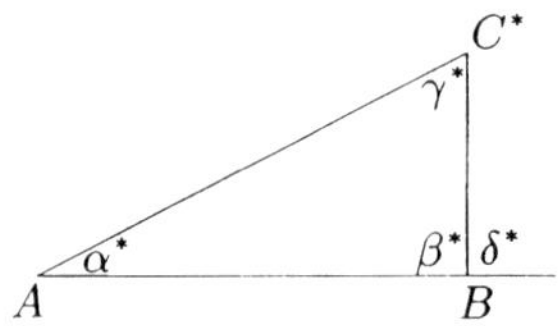

Wäre die ursprüngliche Dreieckswinkelsumme größer als $2R$ also gleich $2R + \delta$, so können wir ja n so groß nehmen, [31]dass $\frac{\alpha}{2^n} < \delta$ ist.

$$2R + \delta = \alpha^* + \beta^* + \gamma^*$$
$$\alpha^* \leqq \frac{\alpha}{2^n}$$
$$\frac{\alpha}{2^n} < \delta$$
$$\gamma^* < \delta^*$$
$$\overline{} $$
$$2R < \beta^* + \delta^*,$$

was nicht der Fall ist.[31] [32]

[29-29]Added by Hi^9

[30-30]Substituted by Hi^9 for: nur

[31-31]Substituted by Hi^9 for: dass $\frac{\alpha}{2^n} < \delta$ ist.

Es müsste dann $b + x \geqq 2R$ also umsomehr $b + w > 2R$; was nicht der Fall ist.

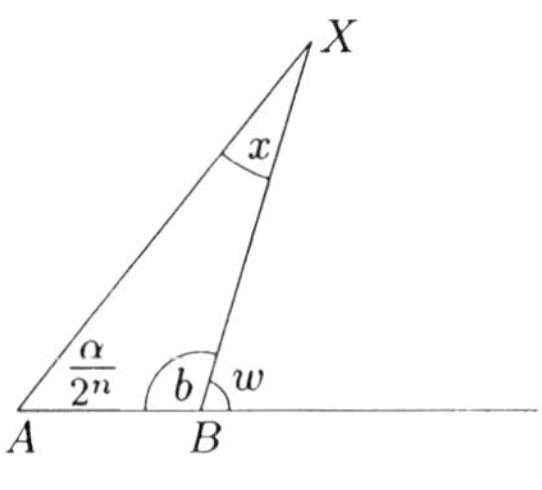

[32]The second part of this proof is in Hilbert's hand, and replaces what Adler had written, which took up the first four lines of p. 74. Indeed, Hilbert made two attempts, the first of which has been scribbled out and is unreadable; the second is what is given here. Part of the change involved replacing Adler's diagram, reproduced in the previous note, with the one now in the text, which accidentally appears as a right-angled triangle, and which should presumably be labelled with A^*, B^*, C^* and not with A, B, C^*. Adler's diagram is what explains the reference to $\triangle ABX$ in the early part of the proof not altered by Hilbert.

ad 2.) 74

Ist in dem Dreiecke ABC die Winkelsumme $2R$, so muss auch für jedes innerhalb ABC liegende Dreieck $A'B'C'$ die Winkelsumme $2R$ betragen.[A]

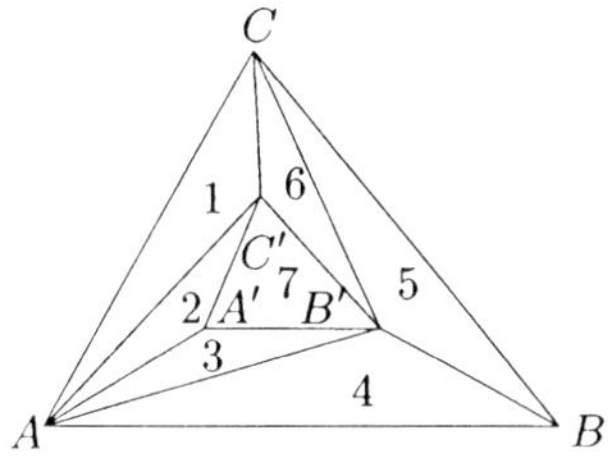

Bezeichnen nämlich $w_1, \ldots, w_7$[33] die Winkelsummen der Dreiecke $1, \ldots, 7$, so ist

$$w_1 + w_2 + w_3 + \ldots + w_7 = 2R + 3 \cdot 4R = 14R.$$

Nun folgt aus Legendre 1, dass $w_1 \leqq 2R \ \ldots \ w_7 \leqq 2R$, also $\sum_1^7 w_i \leqq 14R$, daraus folgt

$$w_1 = w_2 = w_3 = \ldots = w_7 = 2R \qquad \text{q. e. d.}$$

Aus Dreieck $A'B'C'$ können wir nun ein Rechteck construieren, indem wir die Normale $| \ C'B'$ fällen und $\triangle 2 \equiv \triangle 1$ machen. 75

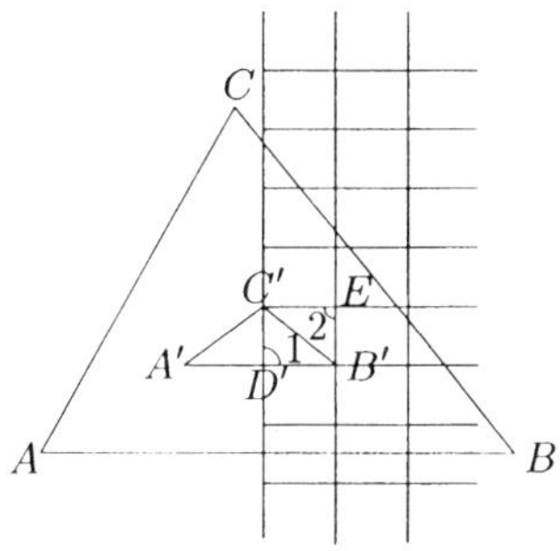

Anschließend an dieses erste Rechteck können wir die ganze Ebene mit congruenten Rechtecken bedecken und dadurch immer ein so großes Rechteck erhalten, dass ein beliebig gegebenes Dreieck innerhalb dieses Rechteckes zu liegen kommt; und damit analog dem früheren nachweisen, dass die Winkelsumme unseres Dreieckes $2R$ beträgt.[B]

Bei beiden Legendre'schen Sätzen wurde aber das Archimedische Axiom V benutzt.

[A]Ist in irgend einem Viereck V Winkelsumme $4R$, so ist in jedem innerhalb V gelegenen Dreiecke die Winkelsumme gleich $2R$.

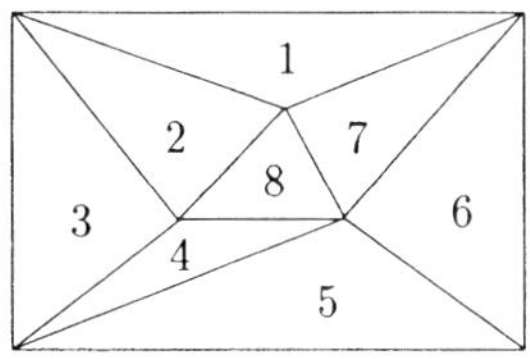

⟨Added by Hi^g in the left-hand margin.⟩

[B]Nun machen wir nur aus dem geg. △ ein Viereck mit 4 Rechten, nehmen darin ein kleines rechtwinkl⟨iges⟩ △ mit 2 Rechten, daraus ein Rechteck ⟪mi⟫⟨t⟩ lauter rechten Winkeln, bedecken damit die Ebene, bis wir ein beliebig vorgelegt⟨es⟩ △ bedeckt haben, u. in diesem muss also auch die Winkelsumme = 2 Rechte sein. ⟨Added by Hi^g in the lower left-hand and lower margins of p. 74.⟩

[33]In the light of his marginal addition concerning the quadrilateral, Hilbert (in Hi^g) has adjusted the following proof by replacing 7, $2R$ and $14R$ by 8, $4R$ and $16R$ respectively.

[34]Nun stelle ich noch folgende Behauptung ohne Beweis auf.[34]

Das Parallelenaxiom kann durch die Forderung ersetzt werden, dass die Winkelsumme im [35]d. h. in einem[35] Dreiecke gleich $2R$ ist, wenn man das Archimedische Axiom hinzunimmt.

Beweis:

1.) In den beiden Punkten A, B der Geraden a, errichte man gleichlange Normale BC und AD. Wir wollen nun nachweisen, dass die Gerade CD eine äquidistante zu a ist:

76 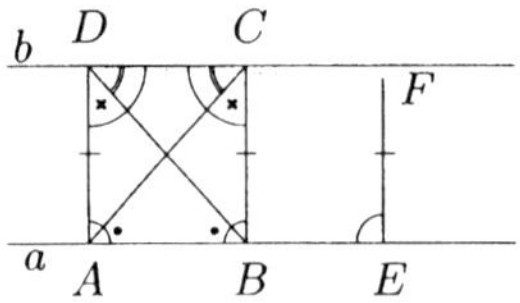

Es ist

$$\triangle ABD \equiv ABC,$$

folglich

$$AC = BD$$
$$\sphericalangle ACB = ADB$$
$$\sphericalangle DBA = CAB.$$

Ferner ist

$$\triangle ADC \equiv BCD,$$

folglich

$$\sphericalangle BDC = ACD$$

und

$$\sphericalangle ADC = BCD = R,$$

weil die Winkelsumme im Viereck $4R$ beträgt.

Ferner ist

$$\triangle ADC \equiv ADB,$$

also

$$CD = AB.$$

Ist nun E ein beliebiger Punkt von a und $EF \perp a$, außerdem $EF = AD$, so folgt durch ganz analoge Schlüsse wie oben, dass $AEFD$ ein Rechteck ist, also F auf DC liegen muss.

CD ist also der Ort gleicher Entfernungen von a.

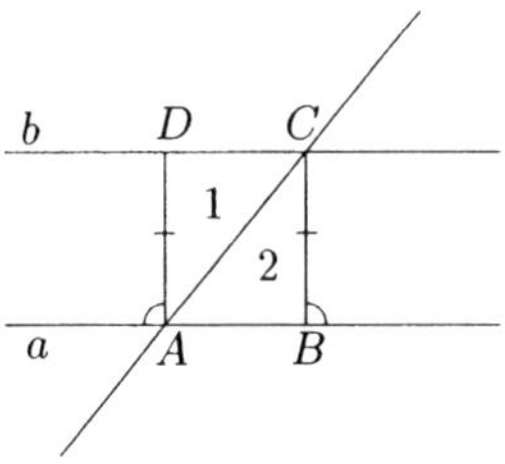

Werden zwei äquidistante Gerade a, b von einer dritten geschnitten, so sind die entstehenden Gegenwinkel und Wechselwinkel einander gleich.

Bew.: Ist AC die Schneidende, so construiere man AD und BC normal zu a, dann sind
77 | die Dreiecke $1 \equiv 2$ und daraus folgt sofort die Gleichheit der Gegenwinkel und Wechselwinkel.

Es gelten also auch die Sätze über das Parallelogramm.

2.) *Wir wollen nun zeigen, dass es außer dieser äquidistanten Geraden b keine weiteren Geraden giebt, welche a nicht schneiden.*

Zu diesem Zwecke müssen wir zunächst mit Hilfe von Axiom V die Zahl einführen und hierauf mit Hilfe desselben Axioms die Lehre von den Proportionen begründen.

[34-34]Interlineated by Hi^g

[35-35]Added by Hi^g

α.) *Einführung der Zahl.*

Gegeben sei eine beliebige Gerade g; zwei Punkte derselben seien mit 0 und 1 bezeichnet, halbieren wir diese Strecke, sowie jede entstehende und tragen alle diese Strecken hintereinander auf, so erhalten wir die Punkte

$$\frac{1}{2}, \frac{1}{4}, \frac{3}{4}, 2, 3, \ldots, \frac{3}{2}, \frac{5}{4}, \ldots,$$

kurz alle Punkte $\frac{A}{2^n}$, wobei A eine beliebige ganze Zahl darstellt, gerade so wie n.

Wenn man n genügend groß nimmt, so erhält man dadurch innerhalb jeder noch so kleinen Strecke Punkte $\frac{A}{2^n}$.

Beweis: 78

Sei T_0T_1 die gegebene Strecke, so wollen wir annehmen, dass noch keiner unserer Punkte $\frac{A}{2^n}$ innerhalb derselben liegt, und zeigen, dass man durch wiederholtes Halbieren und Streckenauftragen sicher einen unserer Punkte in diese Strecke hineinbringen kann.

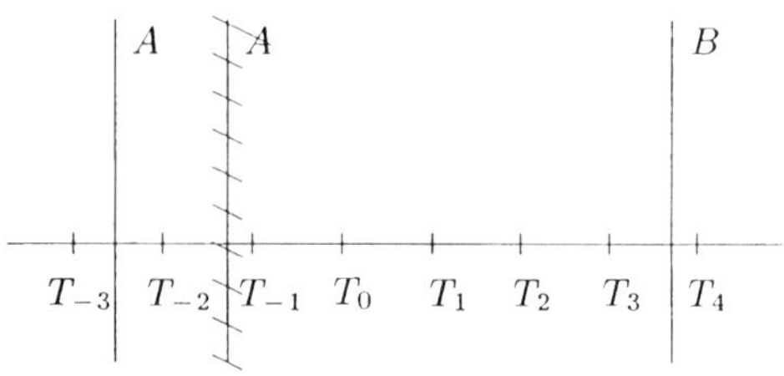

Wir tragen zu dem Zwekke T_0T_1 nach rechts und links sooft auf, bis wir schon vorhandene Punkte A, B unserer Form überschreiten; dies muss nach V durch endliche Operationen erreicht werden können.

Wir wollen $T_{-3}T_{-2}$ unserer Figur den ersten, T_3T_4 den letzten Theil zwischen AB nennnen; die Anzahl dieser Theile zwischen AB ist jedenfalls endlich und mindestens gleich 3.[36]

Halbiert man nun die Strecke AB, so kann der Halbierungspunkt nicht in den ersten Theil ($T_{-3}T_{-2}$) fallen, weil ja dann die rechte Hälfte größer als ein Theil, die linke kleiner als ein Theil wäre.

Der Halbierungspunkt C kann also nur innerhalb | eines Theiles t zu liegen 79
kommen, in welchem A oder B nicht liegen.

Wenn wir nur die Strecken AC und BC halbieren und so fort, so bringen wir jedesmal einen Punkt von der Form $\frac{A}{2^n}$ in einen der vorhandenen, noch keinen derartigen Punkt enthaltenden Theile.

Da die Anzahl der vorhandenen Theile endlich ist, so werden wir nach einer endlichen Anzahl von Halbierungen sicher einen Punkt in den Theil T_0T_1 hineinbringen.

Nimmt man nun irgend einen Punkt der Geraden g an, so kann man sämmtliche Zahlen durch diesen Punkt in zwei Kategorien theilen: in die kleineren und in die größeren; dieser Punkt macht einen *Schnitt in diese Zahlen.* Zu jedem Punkte gehört also eine Zahl α in einen *Dualbruch* entwickelt.

[36]Hilbert has corrected the diagram by crosssing out the line marking A between T_{-2} and T_{-1} and setting it instead between T_{-3} and T_{-2}. This correction restores the consistency between text and diagram, as is indicated by a line drawn by Hilbert from the '$T_{-3}T_{-2}$' at the beginning of the paragraph to the new line marking A in the diagram.

Auch durch die Zahl ist der Punkt bestimmt. Gäbe es nämlich außer α
80 noch einen anderen Punkt α', welcher denselben Schnitt ⟨macht⟩, | dem also dieselbe Zahl entspricht, so müsste ja innerhalb der Strecke $\alpha\alpha'$ keiner unserer Punkte liegen, was nach früherem unmöglich ist.

Also die Zahl bestimmt den Punkt eindeutig; wir werden später mit Hilfe des Vollständigkeitsaxioms zeigen, dass auch jeder Zahl ein Punkt entspricht.

β.) Lehre von den Proportionen.

Sei ACB ein beliebiger Winkel, ferner A' die Mitte von AC, endlich $A'B'$ äquidistant zu AB, dann ist auch B' die Mitte von CB; und umgekehrt.

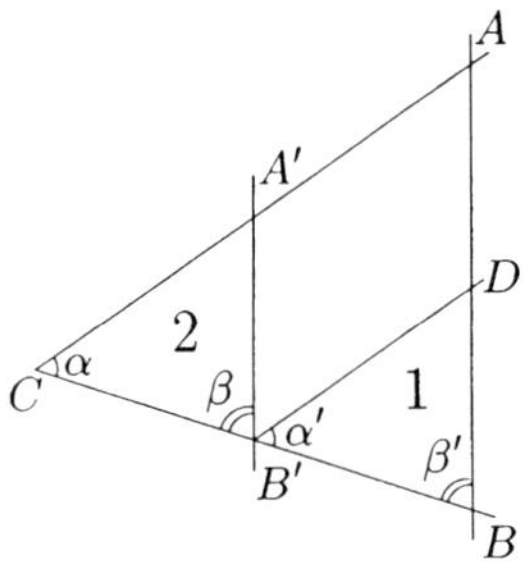

Macht man nämlich

$$\sphericalangle \alpha' = \alpha,$$

so ist

$$B'D = A'A = A'C$$

und

$$\sphericalangle \beta' = \beta,$$

also

$$\triangle 1 \equiv 2,$$

daher

$$CB' = B'B. \qquad \text{q. e. d.}$$

81 Umgekehrt findet man auch, dass wenn $CA' = A'A$ und $CB' = B'B$ ist, auch $A'B'$ und AB äquidistant sein müssen.

Durch wiederholtes Halbieren und Ziehen von Äquidistanzen findet man:

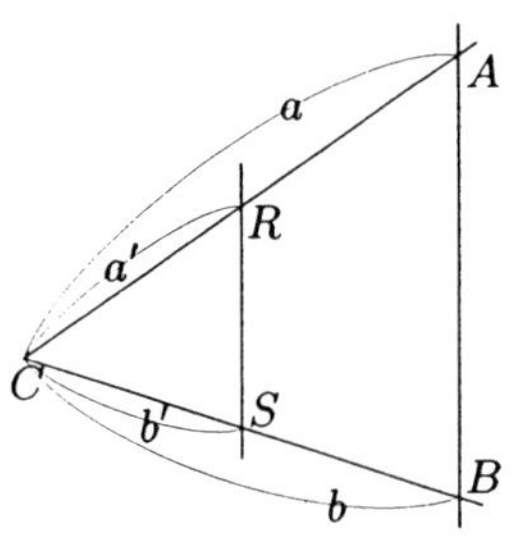

Entsprechen den Punkten A und B die Zahlen a und b, den durch Halbieren gefundenen Punkten R und S die Zahlen a' und b', so müssen ja $\frac{a}{a'}$ und $\frac{b}{b'}$ zwei Zahlen von der Form $\frac{A}{2^n}$ sein.

Sind nun $\frac{a}{a'} = \frac{b}{b'}$, so muss die Verbindungslinie von R und S äquidistant zu *AB sein und umgekehrt.*

Satz: Sind AB und $A'B'$ äquidistant aber a, a', b, b' beliebige Zahlen, so muss auch $\frac{a}{a'} = \frac{b}{b'}$ sein.

82 *Beweis:* Es sei $\frac{a}{a'} > \frac{b}{b'}$, dann giebt es zwischen $\frac{a}{a'}$ und $\frac{b}{b'}$ jedenfalls | eine Zahl

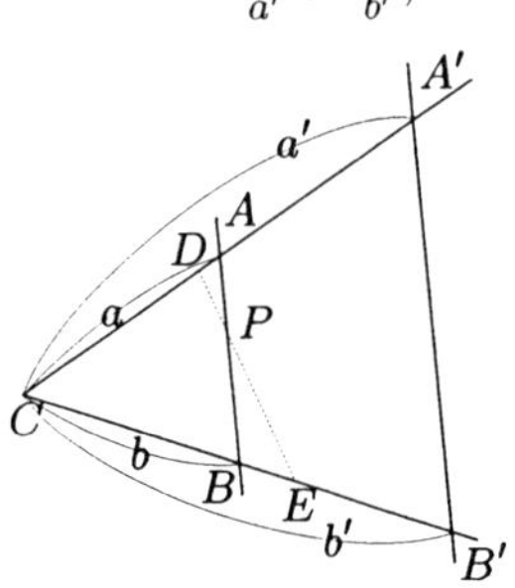

$\frac{A}{2^n}$, also

$$\frac{a}{a'} > \frac{A}{2^n} > \frac{b}{b'}.$$

Wir konstruieren nun den Punkt D so, dass

$$\frac{CD}{a'} = \frac{A}{2^n}$$

und außerdem einen Punkt E so, dass

$$\frac{CE}{b'} = \frac{A}{2^n} \quad \text{ist;}$$

dann müssen

$$CD < a$$

und

$$CE > b$$

sein. Außerdem muss aber jetzt die Proportion gelten

$$CD : a' = CE : b',$$

d. h. die Verbindungslinie DE müsste äquidistant zu $A'B'$ sein.

Die Linien AB und DE müssen einander in einem Punkte P schneiden; durch diesen gingen dann 2 äquidistante Geraden zu $A'B'$, was unmöglich ist.

Daher kann nicht $\frac{a}{a'}$ größer als $\frac{b}{b'}$ sein, ebensowenig kann es kleiner sein 83
und demnach nur gleich sein.

Nun endlich können wir den Satz von pag. 75 beweisen, den Satz nämlich, dass das Parallelenaxiom durch die Forderung, die Winkelsumme im Dreiecke sei 2R, nur mit Zuhilfenahme von V. ersetzt werden kann.

Man kann nämlich auf Grund der Entwickelungen jetzt leicht zeigen, dass jede durch P gehende Gerade h die Gerade g schneiden muss außer der äquidistanten a.

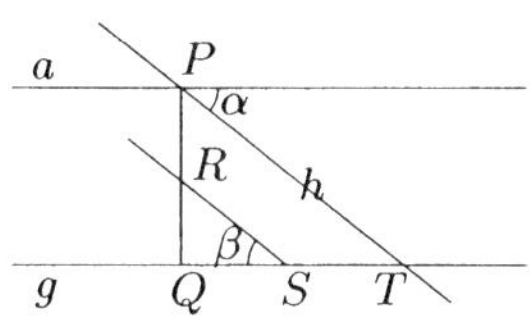

Fällt man nämlich das Loth PQ, macht $\sphericalangle\beta = \alpha$, also h äquidistant zu RS, und bestimmt T aus folgender Proportion

$$QT : QS = QP : QR,$$

so muss T in endlicher Ferne vorhanden sein.
T mit P verbinden liefert eine äquidistante zu RS; da aber durch P nur eine äquidistante zu RS geht, so muss h durch T laufen.

Wir wollen uns nun näher mit den 84

V. *Axiomen der Stetigkeit*

beschäftigen und wollen drei derartige Axiome geben:

1.) *Das Archimedische Axiom,* welches uns schon bekannt ist (pag. 46).

2.) *Das Axiom der Nachbarschaft*, welches wir so aussprechen wollen:

Liegt irgend eine Strecke AB vor, so gibt es gewiss ein Dreieck (oder Quadrat etc.), in dessen Innern keine zu AB gleich lange Strecke vorhanden ist.

Die Bedeutung dieses Axioms wird später hervortreten.

3.) *Das Axiom der Vollständigkeit.* Dieses wollen wir so aussprechen:

Es ist nicht möglich, dem System der Punkte und Geraden ein anderes System von Dingen hinzuzufügen, so dass auch in dieser verallgemeinerten Geometrie sämmtliche Axiome erfüllt sind.

Man kann auch sagen: 85

Die Elemente der Geometrie bilden ein System von Dingen, welches bei Aufrechterhaltung sämmtlicher Axiome keiner Erweiterung mehr fähig ist.

Der Inhalt dieses Vollständigkeitsaxioms wird uns am besten klar, wenn wir an die reellen Zahlen denken:

Dieselben sind definiert als ein System von Dingen, welche die Axiome der Rechnung, Verknüpfung und Anordnung erfüllen, außerdem noch das Archimedische Axiom.

Wenn wir nun das System der rationalen Zahlen nehmen, so erfüllt dasselbe sämmtliche ebengenannten Axiome; aber es erfüllt nicht das Vollständigkeitsaxiom.

Adjungieren wir nämlich den rationalen Zahlen die Zahl $\sqrt{2}$ und bilden den Rationalitätsbereich $R(\sqrt{2})$, so wird dieses erweiterte System auch alle Axio-
86 me der rationalen Zahlen erfüllen; das Vollständigkeitsaxiom | aber ebenfalls nicht, denn man könnte noch $\sqrt[3]{2}$ oder π, e etc. adjungieren; oder das System der algebraischen Zahlen betrachten und wieder Systeme erhalten, welche alle Axiome der rationalen Zahlen erfüllen.

Erst das System sämmtlicher reeller Zahlen erfüllt alle Axiome und außerdem das Vollständigkeitsaxiom.

Bew.:

Wäre nämlich a_1 ein Ding, welches sich dem System der sämmtlichen reellen Zahlen noch hinzufügen liesse so, dass in diesem erweiterten System alle früheren Axiome erhalten bleiben, so müsste ja a_1, verglichen mit jeder der reellen Zahlen, entweder größer oder kleiner sein als dieselbe.

a_1 würde also die reellen Zahlen in zwei Theile zerlegen: in solche, welche größer sind als a_1 und solche, welche kleiner als a_1 sind und beide Theile
87 zusammen | müssten die Gesammtheit der reellen Zahlen ausmachen.

Der Zahlbegriff lehrt aber, dass wenn man eine solche Theilung der reellen Zahlen hat, es dann eine reelle Zahl giebt, welche zwischen beiden Theilen liegt, Schnitt macht und keinem der beiden Theile angehört.

a_1 kann also nicht existieren.

Für das nichtarchimedische Zahlensystem gilt das Vollständigkeitsaxiom nicht:

Beweis: Das nichtarchimedische Zahlensystem (pag. ⟨49⟩) sei gegeben durch

$$f(t) = a_0 t^n + a_1 t^{n+1} + \dots$$

und die Zahlen desselben seien τ_1, τ_2, ...

Nun betrachten wir das erweiterte nichtarchimedische Zahlensystem, welches aus einer Potenzreihe nach s hervorgeht und deren Coefficienten die Zahlen $\tau_1, \tau_2, \dots$ sind.

88 Dann gelten für dieses erweiterte Zahlensystem sämmtliche Rechnungsregeln, Verknüpfungsaxiome und Anordnungssätze wie für das gewöhnliche nichtarchimedische Zahlensystem.

Man sieht also, dass das Vollständigkeitsaxiom dem nichtarchimedischen Zahlensystem nicht mehr hinzugefügt werden kann.

Das Archimedische Axiom sagt aus, dass wir jede Strecke durch eine Zahl messen können; *aus dem Vollständigkeitsaxiom folgt aber unmittelbar, dass auch jeder Zahl eine Strecke entspricht.*

Das Vollständigkeitsaxiom ist von den übrigen verschieden; dies wollen wir beweisen:

Wir konstruieren einen Zahlenbereich dadurch, dass wir alle Zahlen nehmen, welche aus der Zahl 1 durch die vier Species und die Operation $\sqrt{1+\omega^2}$ hervorgehen, wobei ω eine Zahl unseres Systems ist.

Mit Hilfe dieses Zahlenbereiches construieren wir nun eine Geometrie nach 89
der Methode der analytischen Geometrie, indem wir (wie pag. 52) einen Punkt durch ein Zahlenpaar x, y, eine Gerade durch das Verhältnis $u:v:w$ definieren, weiter ganz wie früher die Parallelverschiebung, Drehung festsetzen. In dieser Geometrie gelten dann die Axiome I–V; das Vollständigkeitsaxiom gilt aber nicht, denn man kann ja die übrigen reellen Zahlen hinzufügen und bekommt dann die gewöhnliche Cartesische Geometrie, von welcher unsere obige Geometrie nur ein Theil ist.

Die eben construierte Geometrie ist besonders interessant; *sie enthält nur Punkte und Geraden, welche bloß auf Grund des Abtragens von Strecken und Winkeln gefunden werden können.*

Wie wir in unseren „Grundlagen", pag. ⟨81f.⟩ gezeigt haben[37], kann nicht jede Strecke durch | bloßes Abtragen von Strecken construiert werden; z. B. nicht 90
die Strecke $\sqrt{2\sqrt{2}-2}$.

Aufbau der Geometrie.

Wir wollen im folgenden uns mit der Euklidischen Geometrie beschäftigen, und sämtliche Axiome gelten lassen, nur das archimedische nicht.

Die Hauptfragen, welche wir unter dieser Beschränkung behandeln wollen, werden sein:

1.) *Die Proportionenlehre,*
2.) *Die Lehre vom Flächeninhalt,*
3.) *Der Satz von der Gleichheit der Basiswinkel im gleichschenkeligen Dreieck.*

Der Kürze halber werden wir nicht alle Beweise durchführen, sondern wiederholt auf unser Buch „Grundlagen der Geometrie" verweisen.

I. 91

Die Lehre von den Proportionen[A].

Wir beweisen zuerst einen Schnittpunktsatz, welchen wir den *Pascal'schen Satz* nennen wollen:

[A]Zur Proportionenlehre vgl. meine Vorles. Prinz. d. Math. u. Logik W.S. 17/18,[38] S. 37 – S. 61. ⟨Added by *Hi*⁹ in the upper margin, and not assigned to any particular line.⟩

[37]I.e., the *Festschrift* (*Hilbert 1899c*, Chapter 5 in this Volume), pp. 81f. In his text, Adler has left a space for the page number which the editors have filled.

[38]The reference is to *Hilbert 1917** which will be included in Volume 3 of this Edition.

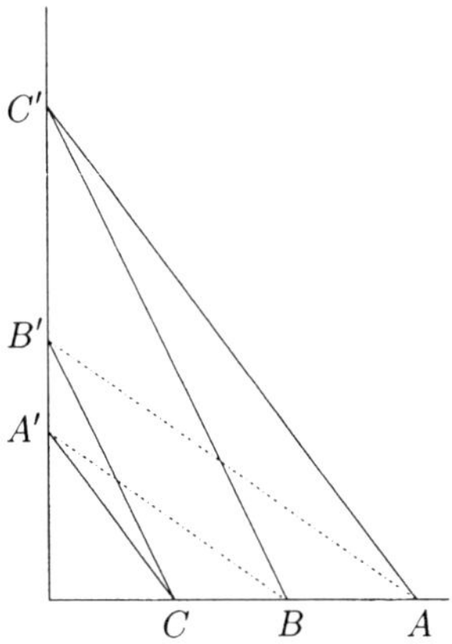

Sind A, B, C und A', B', C' je 3 Punkte auf den Schenkeln eines Winkels, und ist

$$CB' \parallel BC'$$
$$CA' \parallel AC'$$

so ist auch:

$$B'A \parallel BA'.$$

Was den *Beweis des Satzes* anlangt, so verweisen wir auf unsere „Grundlagen" und geben hier nur den *Gedankengang* desselben an:

Aus I, II, III[39], IV ohne V können wir einige Sätze über den Kreis nachweisen, so insbesondere den Satz über das Sehnenviereck.

Von diesem Satze brauchen wir folgenden Fall: Wenn A, B, C, D vier Punkte sind und $\sphericalangle ACB = \sphericalangle ADB$ ist, so ist auch stets

$$\sphericalangle BDC = \sphericalangle BAC.$$

92 Aus diesem Satz folgt leicht unser Pascal'scher Satz (Grundlagen pag. 31). – [40]

Nun sind wir im Stande eine *Streckenrechnung* zu begründen:

Mit Hilfe der *linearen* Congruenzaxiome ergeben sich die *Regeln über das Addieren* von Strecken a, b, $c, \ldots$, nämlich

$$a + b = b + a$$
$$a + (b + c) = (a + b) + c.$$

Aus den Anordnungsaxiomen folgen: der Begriff *„größer"* und *„kleiner"* von Strecken.

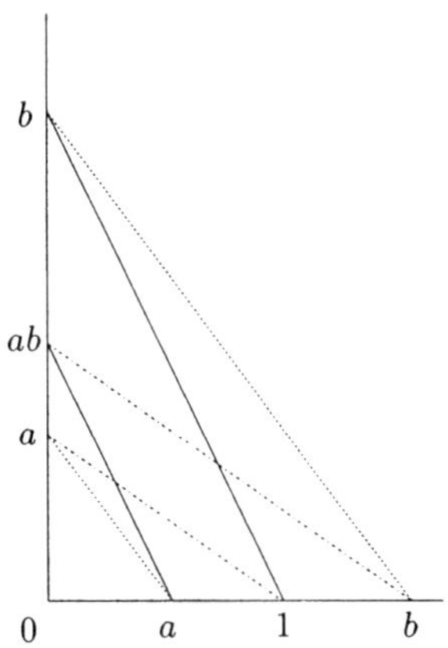

Mit Hilfe des Pascal'schen Satzes gelingt es nun, das *Produkt* (und damit den Quotienten) von zwei Strecken zu *definieren*:

Man nehme einen rechten Winkel. Einen Punkt des einen Schenkels bezeichne man mit 1 und nenne die Strecke 01 1. –

93 Außerdem trage man die Strecken a, b auf | wie in der Nebenfigur gezeigt.

Um nun $a \cdot b$ zu erhalten, verbinde man 1 mit b und ziehe durch a eine Parallele zu dieser Linie.

[39] Above 'III', *Hi⁹* has interlineated '($III6_1$)'

[40] I.e., the *Festschrift* (*Hilbert 1899c*, Chapter 5 in this Volume), p. 31. The theorem is stated on p. 28, and the proof ends on p. 31.

Will man $b \cdot a$ construieren, so muss man zunächst die Strecken b, a richtig auftragen, und hierauf nach der festegesetzten Regel construieren.
Aus dem Pascal'schen Satz folgt dann:

$$b \cdot a = a \cdot b,$$

womit das commutative Gesetz der Multiplikation bewiesen ist.

Auch das assoziative Gesetz

$$a(bc) = (ab)c$$

folgt aus dem Pascal'schen Satze, wenn man diese Ausdrücke nach unserer Regel construiert.

Das distributive Gesetz

$$a(b + c) = ab + ac$$

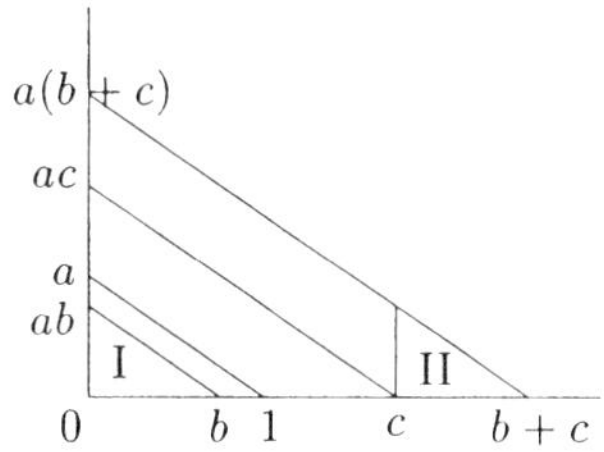

ergibt sich, wenn man zuerst $b + c$, hierauf $a(b+c)$, endlich $a \cdot b$ und $a \cdot c$ construiert und dabei beachtet, dass $\triangle I \equiv \triangle II$ ist. 94

Die Definition der Quotienten ergibt sich sofort: Wenn $\frac{a}{b} = c$ ist, so muss $a = b \cdot c$ sein.

Wir können nun den *Hauptsatz der Proportionenlehre begründen*; müssen aber zunächst ähnliche Dreiecke definieren:

Wir nennen zwei Dreiecke ähnlich, wenn sie gleiche Winkel haben.

Nun nehmen wir zuerst zwei rechtwinklige ähnliche Dreiecke und legen sie mit dem rechten Winkel aufeinander; ihre Hypotenusen sind dann parallel. Fügen wir noch die Strecken 1 und c hinzu, wobei $1c$ parallel den Hypotenusen, so ist nach Obigem:

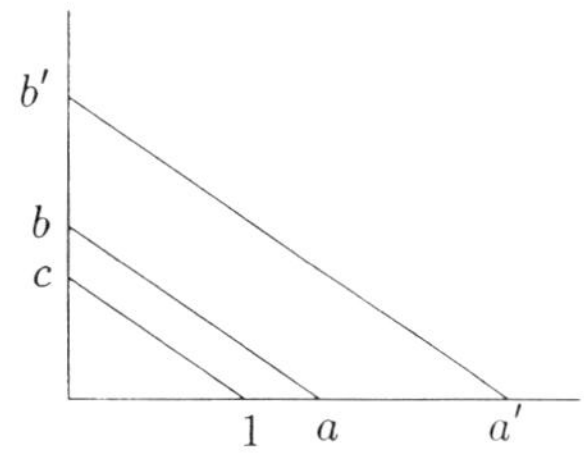

$$b = c \cdot a$$

und 95

$$b' = ca'$$

also

$$ab' = a'b$$

oder

$$a : b = a' : b'.$$

Wir wollen nun diesen Hauptsatz der Proportionenlehre auf beliebige ähnliche Dreiecke ausdehnen:

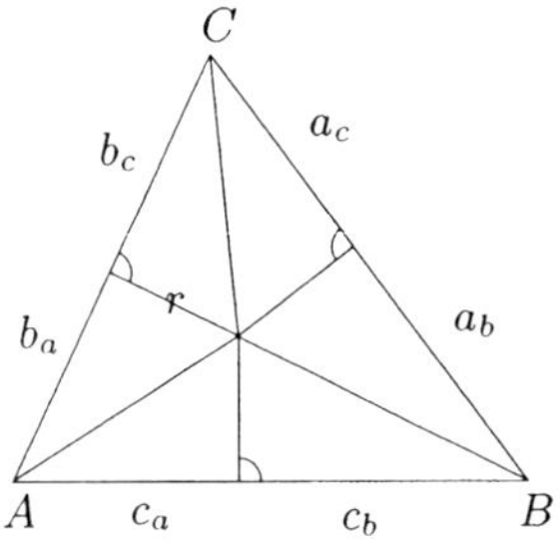

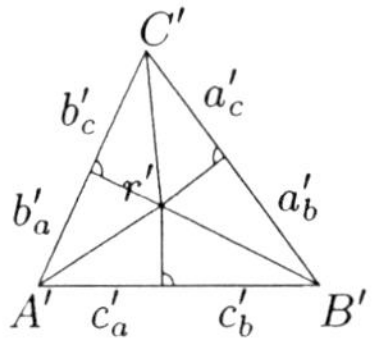

Wir ziehen in diesen beiden Dreiecken die Winkelhalbierenden, welche sich in einem Punkte schneiden müssen, ⟨fällen von diesem Schnittpunkt die Lote auf die Seiten⟩ und erhalten dadurch eine Reihe ähnlicher [41]rechtwinkliger[41] Dreiecke; für diese gelten die Proportionen:

$$+\begin{cases} a_b : r = a'_b : r' \\ a_c : r = a'_c : r' \end{cases}$$
$$b_c : r = b'_c : r'$$

und so weiter.

Durch Addition erhält man daraus:

$$a : r = a' : r'$$

ebenso

$$b : r = b' : r'$$

und

$$c : r = c' : r'$$

oder

$$a : b : c = a' : b' : c'$$

96 Daraus folgt endlich unser Hauptsatz der Proportionenlehre:

Schneiden zwei Parallele auf den Schenkeln eines beliebigen Winkels die Strecken a, b beziehungsweise a', b' ab, so gilt:

$$a : b = a' : b'$$

und umgekehrt.

Damit ist der Satz bewiesen ohne V; dabei wurde aber das *Dreieckscongruenzaxiom im weiteren Sinne* benutzt.

(*Andere Beweise* desselben Hauptsatzes geben *Kneser* in den Sitzungsberichten der Berliner math. Gesellschaft 1901 und *Mollerup* in den Math. Annalen 1902.)[42]

Wir sind nun im Stande die richtige Gleichung der Geraden zu beweisen; müssen aber erst noch bemerken dass nach unseren Festsetzungen sofort folgt, dass

$$a \cdot 1 = a$$

97 ist, | und dass $a \cdot b$ nur dann Null sein kann, wenn wenigstens einer der Faktoren Null ist.

[41-41] Added by *Ad*ˢ

[42] The works referred to here are *Kneser 1902b*, and *Mollerup 1903a*. In the latter paper, marked as having been submitted in January, 1902, Mollerup refers to a suggestion of Hilbert's, presumably in a personal communication. See the note to Hilbert's Remark [12] in the Appendix to Chapter 4.

Geht die Gerade l durch den Ursprung, so gilt

$$a : b = x : y$$

also

$$bx - ay = 0$$

Für die zu l parallele Gerade l' gilt ebenso:

$$b(x - c) - (a - c)y = 0 \text{ [43]}$$

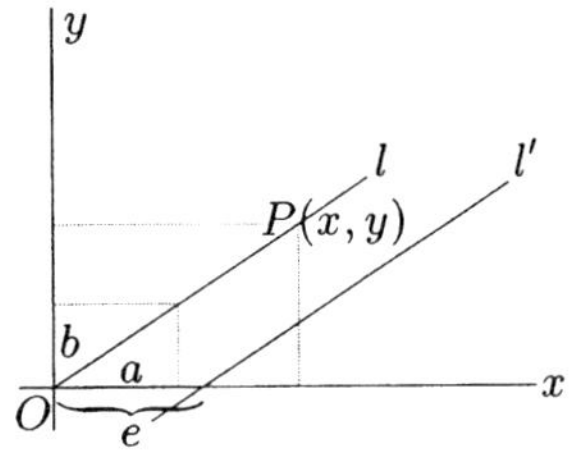

Damit ist bewiesen, dass *die Gleichung der Geraden eine lineare ist* und umgekehrt jede lineare Gleichung zwischen Strecken eine Gerade darstellt.

Um nun die analytische Geometrie ganz auf unserer Streckenrechnung aufzubauen, ist es noch nöthig, die *Parallelverschiebungsformeln* und die Formeln *für die Drehung* zu geben.

Erstere sind ohneweiteres einzusehen; letzere wollen wir jetzt entwickeln:

Wir wollen die Formeln für die Drehung 98
um den Winkel ϑ aufstellen; die Coordinaten des ursprünglichen Punktes seien x, y; die des gedrehten x', y'; das Coordinatensystem sei schiefwinklig. Es ist dann:

$$SX' + SP' = y'$$

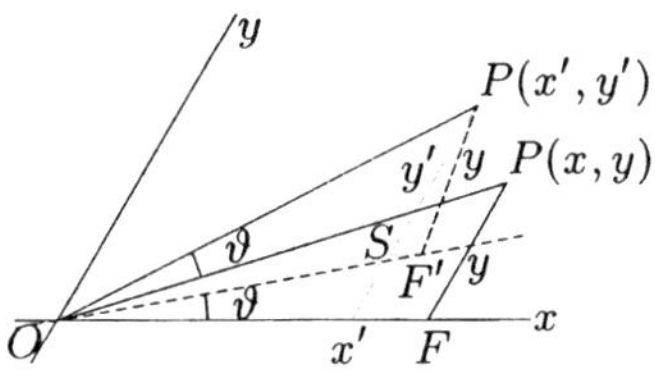

$$\sphericalangle SP'F' = \vartheta \text{ [44]}$$
$$\sphericalangle F'SX' = \sphericalangle SF'P' + \sphericalangle SP'F'$$
$$\sphericalangle P'F'S + \vartheta = \sphericalangle PFO = \sphericalangle X'SF'$$

daraus folgt:

$$\sphericalangle SP'F' = \vartheta.$$

Daher ist $SP' = \delta \cdot y$, wobei δ nur von ϑ und der Wahl des Coordinatensystems abhängt. Ebenso ist $SX' = \gamma \cdot x$, wobei γ ebenfalls nur von ϑ und dem Coordinatenwinkel abhängt.
Es ist also

$$y' = \gamma x + \delta y$$

ebenso

$$x' = \alpha x + \beta y$$

d.h. *Die Drehung wird durch eine lineare homogene Transformation vermittelt. α, β, γ, δ sind dabei Strecken.*

Wenn wir ein rechtwinkliges Coordinatensystem einführen, so fällt es leicht 99
α, β, γ, δ zu bestimmen.

---

[43]This equation is not corrrect, since the gradients of the two lines must be the same. It should be: $b(x - c) - ay = 0$.

[44]The proof of this follows.

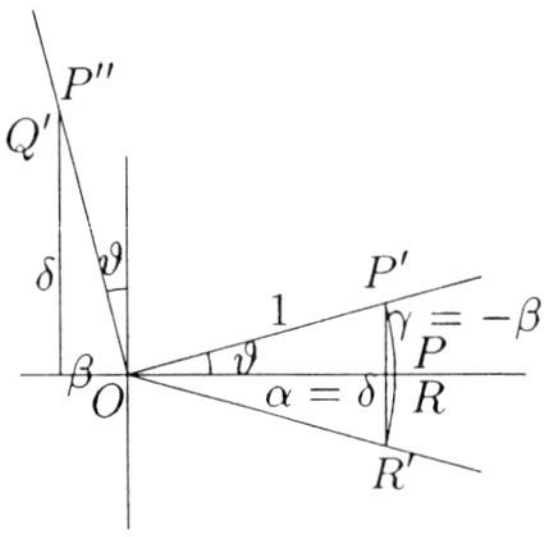

Setzt man nämlich

$$x = 1, \quad y = 0$$

so erhält man $x' = \alpha$, $y' = \gamma$ und damit ergeben sich die Bedeutung von α und γ.

Setzt man

$$y = 1, \quad x = 0$$

so folgt weiter

$$\delta = \alpha, \quad \beta = -\gamma$$

Die Drehungsformel vereinfacht sich also auf:

$$x' = \alpha x - \beta y$$
$$y' = \beta x + \alpha y$$

Wir müssen noch zeigen, dass

$$\alpha^2 + \beta^2 = 1.$$

100 Zu dem Zwecke drehen wir den Punkt R' (mit den Coordinaten $x = \alpha$, $y = -\beta$) um den Winkel ϑ. Wir erhalten dann

$$x' = \alpha^2 + \beta^2$$
$$y' = 0.$$

Unter Benutzung des Satzes vom gleichschenkeligen Dreieck ersieht man aber dass $OR' = OP'$ ist und demnach

$$\alpha^2 + \beta^2 = 1.$$

II.
Lehre vom Flächeninhalt

1.) *Zwei Polygone heißen flächengleich,*[45] wenn sie in eine endliche Anzahl von Dreiecken zerlegt werden können, die paarweise congruent sind.

2.) *Zwei Polygone heißen inhaltsgleich*, wenn es möglich ist, zu ihnen flächen-
101 gleiche Polygone | hinzuzufügen (wegzunehmen), so dass die beiden zusammengesetzten Polygone dann flächengleich sind.

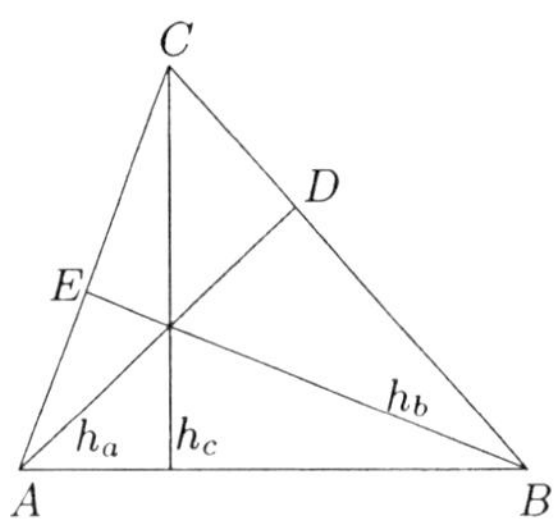

3.) *Wir wollen noch einen Begriff einführen, nämlich das Inhaltsmaß.*

Zieht man in einem Dreiecke die Höhen h_a, h_b, h_c, so entstehen zwei ähnliche Dreiecke ADC und ECB und es ist daher

$$a : b = h_b : h_a$$

also

$$a\frac{h_a}{2} = b\frac{h_b}{2} = c\frac{h_c}{2} = I(\triangle)$$

$I(\triangle)$ ist also eine für das Dreieck *charakteristische Strecke*; wir wollen sie das *Inhaltsmaß des Dreieckes* nennen.

[45] Hi^{9} changed 'flächengleich' to 'zerlegungsgleich'. This alteration of terminology is carried out only sporadically here, but is done systematically in the second edition of the *Festschrift*: see Chapter 5 in this Volume.

Wird ein Dreieck durch eine Ecktransversale in zwei Dreiecke zerlegt, so ist offenbar das Inhaltsmaß des ganzen Dreieckes gleich der Summe der Inhaltsmaße der einzelnen Dreiecke.

Auch bei einem Polygon kann man von einem Inhaltsmaß sprechen, indem man das | Polygon in Dreiecke zerlegt und die Summe der Inhaltsmaße 102
derselben nimmt.

Diese Summe ist dabei unabhängig von der Eintheilung des Polygons. Wir wollen dies in einem speziellen Falle zeigen:

Das nebenstehende Fünfeck sei eingetheilt in je 3 Dreiecke, einmal durch die Diagonalen aus A, dann aus E. Ziehen wir noch eine Hilfslinie, welche die Dreiecke 5, 6 liefert und bezeichnen wir mit den Ziffern die Inhaltsmaße der Dreiecke, so sind die Inhaltsmaße des Polygons

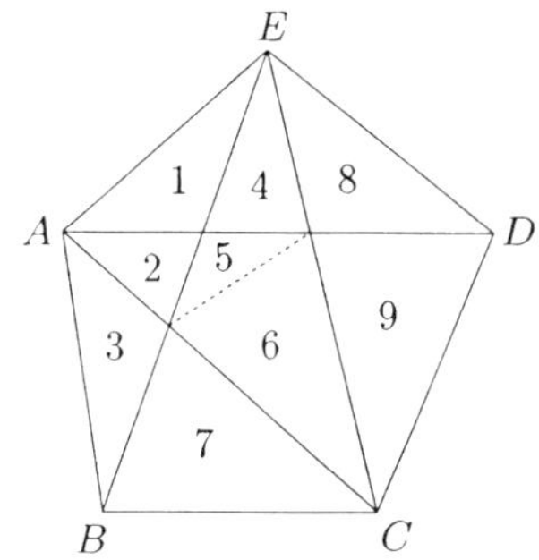

bei der Eintheilung durch Diagonalen aus A:

$$I_1(ABCDE) = (1+4+8)+(2+5+6+9)+(3+7);$$

bei der zweiten Eintheilung ist:

$$I_2(ABCDE) = (1+2+3)+(4+5+6+7)+(8+9)$$

also

$$I_1 = I_2 \qquad \text{q. e. d.}$$

Das Inhaltsmaß ist also auch für Polygone eine charakteristische Strecke.

Zwischen den Begriffen: Flächengleichheit, Inhaltsgleichheit, Inhaltsmaß 103
bestehen einige Beziehungen, welche wir jetzt aufdecken wollen:

1.) *Zwei flächengleiche Polygone haben gleiches Inhaltsmaß*, weil sie ja in congruente Dreiecke zerlegt werden können.

2.) *Zwei inhaltsgleiche Polygone P, P' haben auch gleiches Flächenmaß.* P, P' werden ja durch 2 neue flächengleiche Polygone Q, Q' zu flächengleichen Polygonen $P+Q$, $P'+Q'$ ergänzt. Nun ist $I(P+Q) = I(P'+Q')$ und auch $I(Q) = I(Q')$, also auch $I(P) = I(P')$, da ja $I(P+Q) = I(P) + I(Q)$ ist.

Jetzt können wir in aller Strenge den Euklidischen Hauptsatz beweisen:

Wenn zwei Dreiecke gleiche Grundlinie haben und inhaltsgleich sind, so haben sie auch gleiche Höhen.

Da sie inhaltsgleich sind, so muss 104

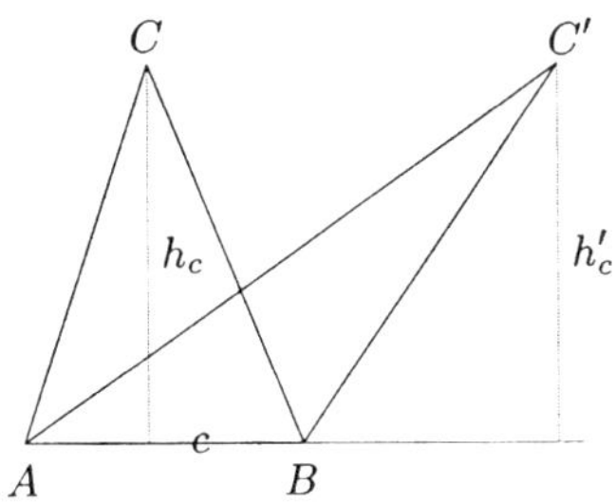

$$c \cdot \frac{h_c}{2} = c \cdot \frac{h'_c}{2}$$

sein, also

$$h_c = h'_c.$$

Wir bemerken, dass wir bei den bisherigen Beweisen über das Flächenmaß von der Stetigkeit keine Anwendung machten.

Satz: Wenn zwei Dreiecke gleiches Inhaltsmaß besitzen, so sind sie auch inhaltsgleich.

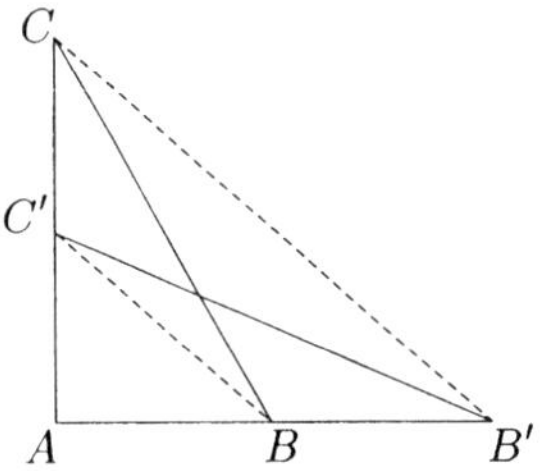

Beweis: 1.) Für zwei rechtwinklige Dreiecke.

Wir legen die Dreiecke ABC und $AB'C'$ so aufeinander, dass die rechten Winkel zusammenfallen. Da sie gleiches Inhaltsmaß besitzen, so muss

$$AC \cdot \frac{AB}{2} = AC' \cdot \frac{AB'}{2}$$

also

$$AB : AB' = AC' : AC$$

daher

$$BC' \parallel B'C$$

sein.

Folglich ist

$$\left.\begin{array}{rl} & \triangle C'BC = \triangle C'BB' \\ \text{und} & \underline{\triangle C'BA = \triangle C'BA} \end{array}\right\} + \\ \text{also} \quad \triangle ABC = AB'C' \qquad \text{q. e. d.}$$

105 *2.) Sind die beiden Dreiecke beliebig*, so verwandle man jedes von ihnen in ein rechtwinkliges.

Das Inhaltsmaß ist also wirklich ein Maß für den Inhalt.

Wir haben 3 wichtige Begriffe kennengelernt: *Flächengleichheit, Inhaltsgleichheit, Inhaltsmaß.*

Letztere 2 Begriffe bedingen sich gegenseitig.

Mit Zuhilfenahme des Axiom V stimmt auch der erste Begriff mit den beiden anderen überein.

Beweis. α.) Wir zeigen zunächst, dass zwei Parallelogramme von gleicher Grundlinie und Höhe (also vom selben Inhaltsmaß) auch [46]flächengleich[46] sind.

106 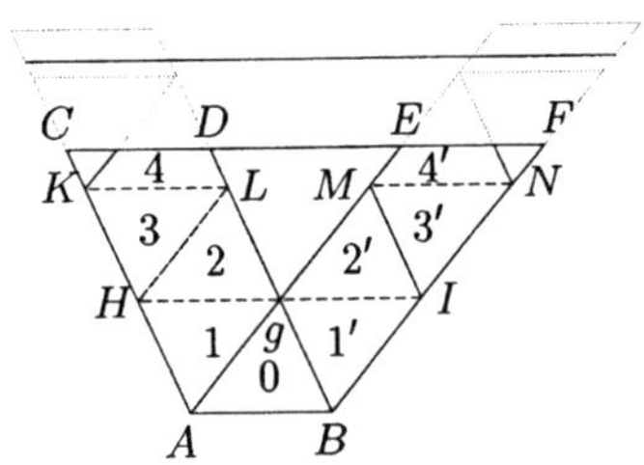

Man ziehe die Hilfslinie $HI \parallel$ zur Grundlinie und hierauf weitere Parallele, wie Figur lehrt; dann sind offenbar die Flächentheile

$$1' \equiv 1, \quad 2' \equiv 2 \quad \text{u. s. w.}$$

und die Parallelogramme in congruente Flächentheile zerlegt.

Dabei haben wir aber V benutzt; denn auf AC wurde die Strecke AH wiederholt aufgetragen und angenommen, dass einmal C überschritten wird.

β.) *Wir wollen nun zeigen, dass in der nichtarchimedischen Geometrie inhaltsgleiche Dreiecke nicht auch flächengleich* [47] *zu sein brauchen.*[47]

[46–46]By Hi^9 changed to: zerlegungsgleich

[47–47]Substituted by Hi^s for: *sein müssen*

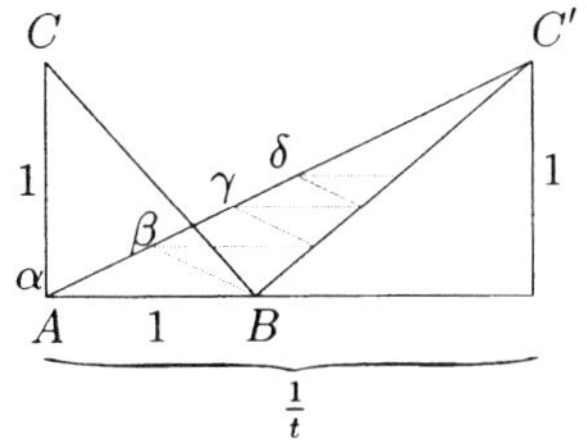

Nehmen wir die inhaltsgleichen Dreiecke ABC und ABC' der Figur und erinnern wir uns, dass

$$nt < 1 \quad \text{also} \quad \frac{1}{t} > n$$

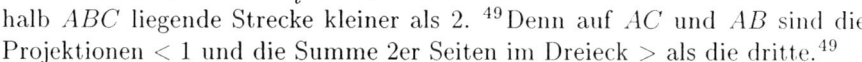

ist[48], wobei n jede beliebige ganze Zahl bedeutet.
AC' ist dann größer als $\frac{1}{t}$ und jede innerhalb ABC liegende Strecke kleiner als 2. [49]Denn auf AC und AB sind die Projektionen < 1 und die Summe 2er Seiten im Dreieck $>$ als die dritte.[49]

Lässt sich nun eine Zerlegung von ABC und ABC' in congruente Dreiecke ausführen, so muss jede der Strecken $\alpha\beta$, $\beta\gamma$ etc. auch in ABC vorkommen, also kleiner 2 sein. 107
Die Anzahl der Theildreiecke von ABC' ist endlich, z. B. die Theile auf AC' seien n. Die Seite AC' müßte demnach kleiner als $n \cdot 2$ sein, was dem widerspricht dass $AC' > \frac{1}{t}$ und $\frac{1}{t} >$ als jede ganze Zahl ist.

III.

Das Dreieckscongruenzaxiom im engeren und weiteren Sinne (die Bedeutung des Satzes von der Gleichheit der Basiswinkel im gleichschenkligen Dreiecke).[50]

Wir wollen im folgenden die *Giltigkeit sämtlicher Axiomgruppen* annehmen, | das *Dreieckscongruenzaxiom* aber nur im *engeren* Sinne anwenden, *außerdem das archimedische Axiom, das Axiom der Nachbarschaft; das Vollständigkeitsaxiom aber nicht.* 108

Wir wollen zeigen, dass wir dann den Satz vom gleichschenkligen Dreiecke beweisen und die ganze Geometrie aufbauen können.

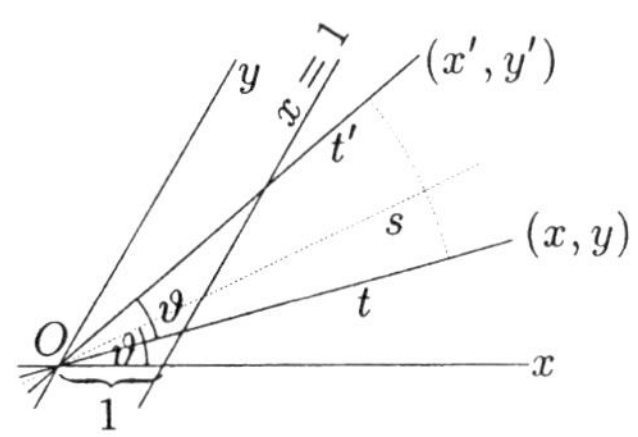

Wir führen zunächst *zwei* sich schneidende Gerade als schiefwinklige Coordinatenaxen ein; ordnen jedem Punkte dieser Axen mit ⟪?⟫ eine Zahl zu, also jedem Punkte der Ebene ein Zahlenpaar. Wir begründen wie früher die Lehre von den Proportionen und stellen mit ihrer Hilfe wie früher die Drehungsformel auf

$$\begin{aligned} x' &= \alpha x + \beta y \\ y' &= \gamma x + \delta y \end{aligned} \qquad 1.)$$

[48] *Hi*^{g} has written 'statt t schreibe $\frac{1}{t}$' above 'nt' and added '$1 + \ldots + 1 < t$' and '$n < t$' in the left margin. Furthermore, he has enclosed in parentheses the fraction '$\frac{1}{t}$' in the diagram and added 't' instead. In the first new sentence following the formula, he has enclosed the fraction '$\frac{1}{t}$' in parentheses and added t.

[49–49] Added by *Hi*^{g}

[50] See the Introduction to this Chapter.

wobei α, β, γ, δ nur vom Drehungswinkel ϑ abhängen.

109 Vor allem suchen wir eine lineare Beziehung

$$\kappa x' + \rho y' = \lambda(\kappa x + \rho y) \ldots \qquad 2.)$$

d. h. die Gerade $\kappa x + \rho y = 0$ soll bei der Drehung unverändert bleiben.

Führen wir die Werthe von x' und y' aus 1.) in 2.) ein, so folgt

$$\begin{aligned} \kappa\alpha + \rho\gamma &= \lambda\kappa \\ \kappa\beta + \rho\delta &= \kappa\rho \end{aligned} \qquad 3.)$$

d. h.

$$\begin{vmatrix} \alpha - \lambda & \gamma \\ \beta & \delta - \lambda \end{vmatrix} = 0 \qquad 4.)$$

diese Gleichung hat complexe Wurzeln λ, $\overline{\lambda}$; setzen wir dieselben in 3.) ein, so erhalten wir für $\frac{\kappa}{\rho}$ auch complexe Werthe.

Die Form

$$Q(xy) = (\kappa x + \rho y)(\overline{\kappa} x + \overline{\rho} y)$$

ist eine quadratische definite Form mit reellen Coeffizienten.

110 Führen wir in sie die $x'y'$ ein, so wird

$$Q(x'y') = \lambda\overline{\lambda} Q(xy) = (\alpha\delta - \beta\gamma) Q(xy) \qquad 5.)$$

Satz:

$$\alpha\delta - \beta\gamma = \Delta_\vartheta = 1.$$

Beweis: Wäre $\Delta_\vartheta < 1$, so wäre nach n maliger Drehung

$$Q(x^{(n)} y^{(n)}) = (\alpha\delta - \beta\gamma)^n Q(xy)$$

Für genügend großes n könnte dann $Q(x^{(n)} y^{(n)})$ beliebig klein gemacht werden und da $Q(xy) = l^2(xy) + m^2(xy) = \text{const.}$ eine Ellipse ist, so könnte dieselbe durch wiederholtes Drehen in ein beliebiges Parallelogramm um O hineingebracht werden; dies widerspricht aber dem vorausgesetzen Axiom der Nachbarschaft.

$\alpha\delta - \beta\gamma$ kann aber auch nicht größer als 1 sein, man hätte dann nur entgegengesetzt zu drehen, da es auch nicht negativ sein kann, weil λ complex sein muß, so folgt $\alpha\delta - \beta\gamma = 1$. q. e. d.

111 Alle Drehungen um O bilden eine Gruppe, wie aus dem Congruenzaxiom folgt.

Wir ordnen nun eine gruppentheoretische Betrachtung ein:

Wir ziehen die Gerade $x = 1$ parallel y Axe und schneiden sie mit den Schenkeln des Drehungswinkel ϑ und mit den Radienvectoren der Punkte xy, $x'y'$. Wir erhalten dadurch auf dieser Geraden von der x Axe gerechnet die Strecken s, t, t'.[51]

Nach Figur ist:

$$\begin{aligned} x : y &= 1 : t \\ x' : y' &= 1 : t'. \end{aligned}$$

Da außerdem

$$\begin{aligned} x' &= \alpha x + \beta y \\ y' &= \gamma x + \delta y, \end{aligned}$$

[51] See the figure on the previous page.

so folgt

$$t' = \frac{y'}{x'} = \frac{\gamma + \delta t}{\alpha + \beta t}.$$

Auf der Geraden $x = 1$ haben wir nun eine ∞ lineare Gruppe von folgender 112
Eigenschaft:

1.) *Die Gruppe ist einparametrig (s)*

2.) *Bei rationalem s kann ich sicherlich entsprechende Punkte auf der Geraden $x = s$ finden.*

3.) *Die Gruppe ist stetig.*

Bew. zu 3.: Jedem s entspricht eine bestimmte Drehung und damit bestimmte α, β, γ, δ.

Wenn die Zahlen $s_{i_1}, s_{i_2}, s_{i_3}, \ldots$ wachsende Zahlen sind, ergibt sich für jedes s ein

$$t'(s_i) = \frac{\gamma_{s_i} + \delta_{s_i} t}{\alpha_{s_i} + \beta_{s_i} t}$$

und aus den Anordnungsaxiomen ersieht man sofort, dass auch

$$t'(s_{i_1}), t'(s_{i_2}), t'(s_{i_3}), \ldots$$

eine wachsende Reihe bilden müssen.

Nehmen wir $s_{k_1}, s_{k_2}, \ldots$ als fallende Reihe an, so werden auch die entsprechenden | $t'(s_{k_1}), t'(s_{k_2}), t'(s_{k_3}), \ldots$ eine fallende Reihe bilden. 113

Wenn die beiden s Reihen gegen dieselbe Grenze s convergieren, so müssen auch die beiden t' Reihen gegen eine Grenze convergieren.

Es sei dies nicht der Fall, also alle $t'(s_i) \leqq \tau$ und alle $t'(s_k) \geqq \tau'$, also

$$t'(s_i) \leqq \tau < \tau' \leqq t'(s_k).$$

Es müsste dann aber auch

$$s_i \leqq \sigma < \sigma' \leqq s_k$$

sein, wobei σ und σ' ganze Werthe von s sind, welche τ und τ' entsprechen. Letzteres darf nicht eintreten, da beide s Reihen gegen eine bestimmte Grenze s convergieren, also muss auch $t'(s_i)$ gegen einen bestimmten Grenzwerth hin convergieren.

t' ist also eine überall stetige, eindeutig umkehrbare Function von t und s; 114
daher sind auch α, β, γ, δ stetige Functionen von s.

In einer Arbeit in den Math. Annalen[52], *auf die wir hier verweisen, werden wir zeigen:*

Eine Gruppe, welche umkehrbar eindeutig ist und bei welcher die einzelnen Substitutionen durch einen Parameter s bestimmt werden, ist holoëdrisch-isomorph den Drehungen eines Kreises.

Man kann dann statt t und s neue Veränderliche τ und σ einführen und erhält dann

$$\tau' = \frac{\sigma - \tau}{s + \sigma\tau}$$

wie in unserer Arbeit bewiesen wird.

[52] I.e., *Hilbert 1903c*.

Die Drehungsformeln gehen dann über in

$$X' = (\sigma X - Y)\Delta_\sigma$$
$$Y' = (X + \sigma Y)\Delta_\sigma$$

115 Δ_σ kann leicht bestimmt werden.

Wir wissen, dass bei jeder Drehungsformel $\alpha\delta - \beta\gamma = 1$ sein muss; daraus folgt:

$$\Delta_\sigma = \frac{1}{\sqrt{1+\sigma^2}}$$

und unsere Drehungsformeln werden:

$$X' = \frac{\sigma}{\sqrt{s+\sigma^2}}X - \frac{1}{\sqrt{s+\sigma^2}}Y$$
$$Y' = \frac{1}{\sqrt{1+\sigma^2}}X + \frac{\sigma}{\sqrt{s+\sigma^2}}Y.$$

Wenn man in diesen Formeln $\sigma = 0$ setzt, so bemerkt man, dass die Linien $-Y$ in X' und X in Y' übergehen d. h. es gibt ein Linienpaar X, Y, das durch Drehung in einander übergeht.

X und Y müssen also aufeinander senkrecht stehen.

Die Existenz des rechten Winkels ist damit bewiesen.

116 X, Y, können auch als die rechtwinkligen Coordinaten eines Punktes bezogen auf diese beiden rechtwinkligen Axen betrachtet werden.

Damit ist die gewöhnliche Drehungsformel abgeleitet; aus derselben folgen die Congruenzsätze; die ganze Geometrie ist damit aufgebaut.

Wir können mit diesem Coordinatensystem jetzt leicht die *Giltigkeit des pythagoräischen Lehrsatzes nachweisen:*
Zu dem Zwecke schreiben wir die Drehungsformeln in der Form:

$$X' = \alpha X - \gamma Y$$
$$Y' = \gamma X + \alpha Y$$

117 Wir wissen, dass dann $\alpha^2 + \gamma^2 = 1$ sein muss. | Setzt man nun

$$X = \gamma, \quad Y = \alpha,$$

so wird

$$X' = 0, \quad Y' = 1,$$

d. h. Punkt P geht in P' über.
Da $OP' = Y' = 1$ ist, so ist auch $OP = 1 = \alpha^2 + \gamma^2$ *und damit der pythagoräische Lehrsatz für das Dreieck OBP bewiesen.*
Setzt man $X = \gamma$, $Y = -\alpha$, so folgt $X' = 1$, $Y' = 0$; Q geht in Q' über; also auch $OQ = 1$.

Aus den beiden congruenten Dreiecken OBP und OBQ folgt nun der Satz vom gleichschenkeligen Dreieck.

Damit ist die ganze Geometrie aus unseren Annahmen begründet.

118 *Den Satz vom gleichschenkeligen Dreieck* wollen wir noch näher beweisen: Sei ABC dasselbe und $AC = BC$, so suche man zu A in bezug auf BC das Spiegelbild A'; dann ist nach dem Vorhergehenden

$$\triangle A'BC \equiv \triangle BCA$$

also

$$\sphericalangle A' \equiv \sphericalangle B \equiv \sphericalangle A.$$

Wir wollen im folgenden die Axiome I–IV gelten lassen, das Dreieckscongruenzaxiom jedoch nur im engeren Sinne; außerdem wollen wir noch folgende drei Sätze als giltig voraussetzen:

1.) Es gibt rechte Winkel.

2.) Wenn in zwei symmetrisch liegenden rechtwinkligen Dreiecken die beiden Katheten übereinstimmen, so sind auch die Winkel an der Hypotenuse einander gleich.[A]

3.) Wenn in einem Vierecke $ABCD$ die $\sphericalangle ADB$ und ACB rechte sind, so sind die Winkel ACD und ABD einander gleich.

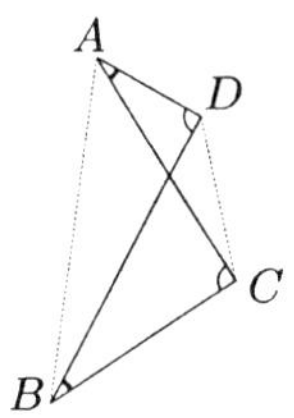

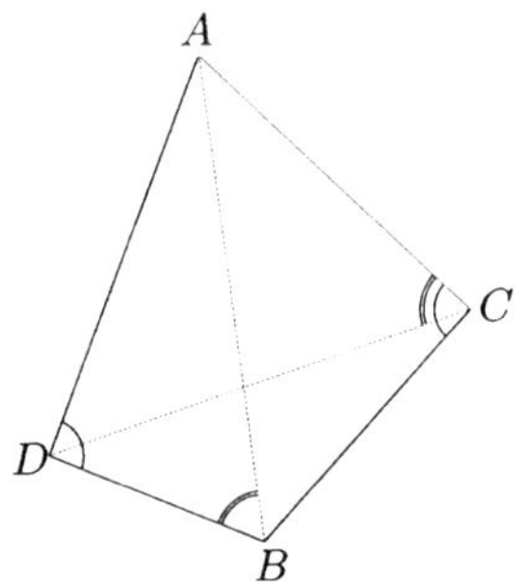

Wir wollen zuerst die Bedeutung dieser drei Sätze untersuchen: 119

ad 1.) Auf Grund des rechten Winkels können wir die *Spiegelung eines Punktes A an einer Geraden a definieren.*

Wir zeigten früher (pag. ⟨98⟩), dass die Drehung, welche auf den Congruenzsätzen im engeren Sinne beruht, zu folgenden Drehungsformeln führten (schiefwinkliges Coordinatensystem):

$$x' = \alpha_\vartheta x + \beta_\vartheta y$$
$$y' = \gamma_\vartheta x + \delta_\vartheta y$$

dabei bedeutet α_ϑ eine symbolische Operation, die vom Winkel ϑ abhängt; ist r die Hypotenuse eines rechtwinkligen Dreieckes ABC, $\sphericalangle CAB = \vartheta$, so ist

$$AB = \alpha_\vartheta r.$$

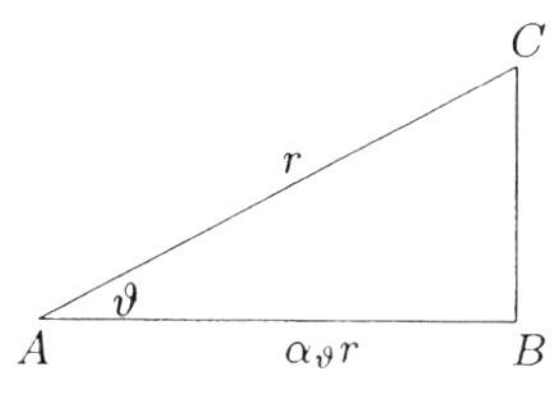

Ist $r = 1$, so ist $AB = \alpha_\vartheta$.

Beim *rechtwinkligen Coordinatensystem*, das wir jetzt einführen können, be|merken wir bei den Annahmen: 120

[A] 2.) ist überflüssig zur Begründung der Proportionenlehre, wahrscheinlich auch 1.), so dass nur der Satz vom Kreisviereck übrig bliebe als nothwendig zur Begründung der Proportionenlehre. ⟨Added by *Hi*[s] in the margin to the left of this paragraph.⟩

1.) $y = 0$, dass $x' = \alpha_\vartheta x$, $y' = \gamma_\vartheta x$ wird und
2.) wenn $x = 0$, " $x' = \beta_\vartheta y$, $y' = \delta_\vartheta y$ werden.
Für den rechten Winkel werden also die Drehungsformeln:

$$x' = \alpha_\vartheta x - \gamma_\vartheta y$$
$$y' = \gamma_\vartheta x + \alpha_\vartheta y.$$

ad 2.) *Im Spiegelbilde einer Figur stimmen die Winkel stets mit den Winkeln der ursprünglichen Figur überein; die Spiegelung ist winkeltreu, conform.*

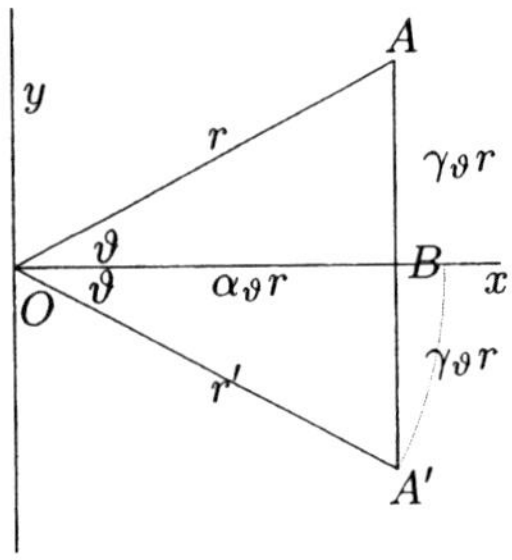

Hat Punkt A die Coordinaten $\alpha_\vartheta r$, $\gamma_\vartheta r$ so hat sein Spiegelbild A' die Coordinaten

$$x = \alpha_\vartheta r$$
$$y = -\gamma_\vartheta r$$

Drehen wir A' um den Winkel ϑ, so wird A' in die X Axe fallen nach B und es wird sein

121 $$x' = \alpha_\vartheta(\alpha_\vartheta r) - \gamma_\vartheta(-\gamma_\vartheta r) = OA' \qquad (1.$$
$$y' = \gamma_\vartheta(\alpha_\vartheta r) + \alpha_\vartheta(-\gamma_\vartheta r) = 0 \qquad (2.$$

Aus der letzten Gleichung folgt

$$\alpha_\vartheta\gamma_\vartheta = \gamma_\vartheta\alpha_\vartheta \qquad \text{I.)}$$

d. h. *Die Reihenfolge der Anwendung der Symbole α_ϑ und γ_ϑ ist gleichgiltig.*

Eine Spiegelung der Strecke r mittels des Winkels ϑ wollen wir mit $\Delta_\vartheta r$ bezeichnen; aus 1.) folgt:

$$r' = \Delta_\vartheta r = (\alpha_\vartheta^2 + \gamma_\vartheta^2) r \qquad \text{II.)}$$

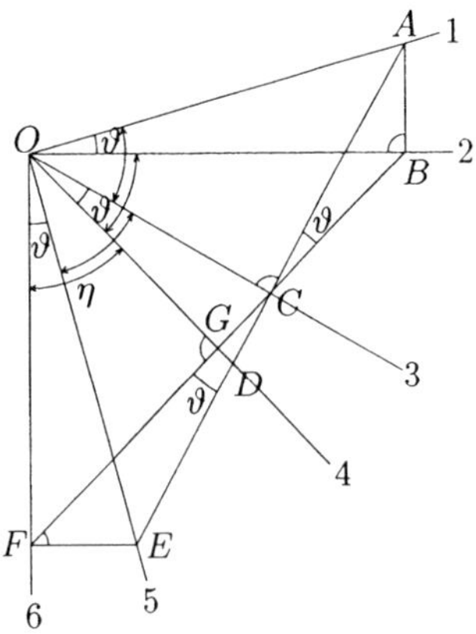

ad 3.) Ist

$$\sphericalangle 1O2 = \sphericalangle 3O4 = \sphericalangle 5O6 = \vartheta$$
$$\sphericalangle 1O3 = \sphericalangle 2O4 = \sphericalangle 3O5$$
$$= \sphericalangle 4O6 = \eta$$

ferner

$$AB \perp O2, \quad AC \perp O3,$$

und

$$EF \perp O6,$$

so sind nach Satz 3, pag. 118:

$$\sphericalangle ACB = \vartheta, \quad \sphericalangle FCE = \vartheta,$$

122 also | liegen B, C, F in einer geraden Linie. Außerdem ist:

$$AC \equiv EC,$$
$$FG \equiv BG.$$

Es sei

$$OA \equiv r.$$

Daher

$$OB = \alpha_\vartheta r$$

und

$$OF = \Delta_\eta \cdot \alpha_\vartheta r.$$

Andererseits

$$OE = \Delta_\eta r$$

und

$$OF = \alpha_\vartheta \Delta_\eta r.$$

Daraus folgt

$$\Delta_\eta \cdot \alpha_\vartheta r = \alpha_\vartheta \cdot \Delta_\eta r,$$

d. h.: *Die Operationen Δ_η und α_ϑ sind ebenfalls vertauschbar.*

Jetzt wollen wir sehen, welche Schlüsse sich aus unseren Voraussetzungen (pag. 118) ziehen lassen:

Unsere Drehungsformel lautete:

$$\begin{aligned} x' &= \alpha_\vartheta x - \gamma_\vartheta y \\ y' &= \gamma_\vartheta x + \alpha_\vartheta y. \end{aligned}$$

Wir haben bewiesen, dass aus den Sätzen 1, 2, 3 (pag. 118) folgt:

123

$$\left.\begin{aligned} &\alpha_\vartheta \cdot \gamma_\vartheta = \gamma_\vartheta \cdot \alpha_\vartheta \\ &\text{ferner} \\ &\Delta_\vartheta = \alpha_\vartheta^2 + \gamma_\vartheta^2 \\ &\text{endlich} \\ &\alpha_\vartheta \cdot \Delta_\eta = \Delta_\eta \cdot \alpha_\vartheta. \end{aligned}\right\}$$

Es sei noch bemerkt, dass man mit Zuhilfenahme unserer Voraussetzungen (pag. 118) leicht *folgende 2 Sätze nachweist:*

1.) *Drei Spiegelungen* an drei von einem Punkte ausgehenden Geraden können immer durch eine Spiegelung ersetzt werden.

2.) *Die Mittelsenkrechten* der Seiten eines Dreieckes schneiden sich in einem Punkte.

Um nun die [53] Gültigkeit der[53] Proportionenlehre zu begründen, müssen wir zunächst zeigen, [54] dass:

$$\alpha_\vartheta \cdot \alpha_\eta = \alpha_\eta \alpha_\vartheta$$

und

$$\alpha_\vartheta \gamma_\eta = \gamma_\eta \alpha_\vartheta.$$

Der Beweis gelingt auf folgendem Wege[A]:

[A] Es scheint mir, dass man auch direkt an der Figur die Richtigkeit dieser Formeln beweisen kann. ⟨Added by Hi^g to the right of the above formulas.⟩

[53-53] Substituted by Hi^g for: *Nothwendige*

[54] Deleted by Hi^g: dass zwei beliebige Drehungen commutativ sind; also

Wir drehen zunächst um den Winkel ϑ und hierauf um $\sphericalangle\eta$ und erhalten die Formel:

124
$$\begin{aligned} x' &= \alpha_\vartheta x - \gamma_\vartheta y \\ y' &= \gamma_\vartheta x + \alpha_\vartheta y \end{aligned} \qquad 1.)$$

$$\begin{aligned} x'' &= \alpha_\eta(\alpha_\vartheta x - \gamma_\vartheta y) - \gamma_\eta(\gamma_\vartheta x + \alpha_\vartheta y) = \alpha_{\vartheta+\eta} x - \gamma_{\vartheta+\eta} y \\ y'' &= \gamma_\eta(\alpha_\vartheta x - \gamma_\vartheta y) + \alpha_\eta(\gamma_\vartheta x + \alpha_\vartheta y) = \gamma_{\vartheta+\eta} x + \alpha_{\vartheta+\eta} y \end{aligned} \qquad 2.)$$

oder:

$$\begin{aligned} x'' &= (\alpha_\eta \cdot \alpha_\vartheta - \gamma_\eta\gamma_\vartheta)x - (\alpha_\eta\gamma_\vartheta + \gamma_\eta\alpha_\vartheta)y \\ y'' &= (\gamma_\eta\alpha_\vartheta + \alpha_\eta\gamma_\vartheta)x + (\alpha_\eta\alpha_\vartheta - \gamma_\eta\gamma_\vartheta)y. \end{aligned} \qquad 3.)$$

Nach 1.) folgt jetzt daraus:

$$\begin{aligned} \alpha_\eta \cdot \alpha_\vartheta - \gamma_\eta\gamma_\vartheta &= \alpha_\eta\alpha_\vartheta - \gamma_\eta\gamma_\vartheta \\ \alpha_\eta\gamma_\vartheta + \gamma_\eta\alpha_\vartheta &= \gamma_\eta\alpha_\vartheta + \alpha_\eta\gamma_\vartheta \end{aligned} \qquad 4.)$$

Wir drehen nun um den Winkel $-\eta$, dann müssen wir die Drehungsgleichungen

$$\begin{aligned} x' &= \alpha_\eta x - \gamma_\eta y \\ y' &= \gamma_\eta x + \alpha_\eta y \end{aligned} \qquad 5.)$$

nach x, y auflösen; wir erhalten $(\alpha_\eta^2 + \gamma_\eta^2 = \Delta_\eta)$

$$\begin{aligned} \Delta_\eta x &= \alpha_\eta x' + \gamma_\eta y' \\ \Delta_\eta y &= -\gamma_\eta x' + \alpha_\eta y' \end{aligned} \qquad 6.)$$

125 Wir wollen nun zusammensetzen: 1.) eine Drehung um ϑ und eine um $-\eta$, 2.) eine Drehung um $-\eta$ und hierauf um $\sphericalangle\vartheta$. Wegen der Vertauschbarkeit der α und Δ, können wir mit den Δ so operieren, als wenn sie Zahlen wären:

Wir erhalten:

bei der 1. Drehung:

$$\begin{aligned} \Delta_\eta x'' &= \alpha_\eta(\alpha_\vartheta x - \gamma_\vartheta y) + \gamma_\eta(\gamma_\vartheta x + \alpha_\vartheta y) \\ \Delta_\eta y'' &= -\gamma_\eta(\alpha_\vartheta x - \gamma_\vartheta y) + \alpha_\eta(\gamma_\vartheta x + \alpha_\vartheta y) \end{aligned} \qquad 7.)$$

bei der 2. Drehung

$$\begin{aligned} \Delta_\eta x'' &= \alpha_\vartheta(\alpha_\eta x + \gamma_\eta y) - \gamma_\vartheta(-\gamma_\eta x + \alpha_\eta y) \\ \Delta_\eta y'' &= \gamma_\vartheta(\alpha_\eta x + \gamma_\eta y) + \alpha_\vartheta(-\gamma_\eta x + \alpha_\eta y) \end{aligned} \qquad 8.)$$

Vergleichen wir die Coeffizienten von x bei beiden Drehungen ebenso die von y, so erhalten wir:

$$\begin{aligned} \alpha_\eta\alpha_\vartheta + \gamma_\eta\gamma_\vartheta &= \alpha_\vartheta\alpha_\eta + \gamma_\vartheta\gamma_\eta \\ \alpha_\eta\gamma_\vartheta - \gamma_\eta\alpha_\vartheta &= \gamma_\vartheta\alpha_\eta - \alpha_\vartheta\gamma_\eta \end{aligned} \qquad 9.)$$

Wenn wir System 4) und 9) addieren, so erhalten wir:

$$\left.\begin{aligned} \alpha_\eta\alpha_\vartheta &= \alpha_\vartheta\alpha_\eta \\ \gamma_\eta\gamma_\vartheta &= \gamma_\vartheta\gamma_\eta \\ \alpha_\eta\gamma_\vartheta &= \gamma_\vartheta\alpha_\eta \end{aligned}\right\} \quad \text{q. e. d.}$$

124a *Damit sind die verlangten Formeln nachgewiesen, welche zur Einführung der Proportionenlehre genügen*; daraus folgt, dass die Gleichung der geraden Linie linear ist, ebenso die Schnittpunktsätze.

Dies Alles ohne den Satz vom gleichschenkeligen Dreiecke zu benutzen.

Nichtpythagoräische Geometrie.

Wir wollen eine Geometrie construieren, in welcher alle unsere Voraussetzungen (pag. 118) erfüllt sind, aber Δ_ϑ nicht 1 ist.

Diese Geometrie wird entweder eine nichtarchimedische sein oder eine solche, in welcher das 2. Stetigkeitsaxiom nicht gilt; denn mit Hilfe dieser beiden Axiome kann man ja den Satz vom gleichschenkeligen Dreieck beweisen, wie wir wissen (pag. ⟨107–118⟩).

Gelingt es eine solche Geometrie unter Außerachtlassung des archimedischen Axioms zu construieren, so würde daraus auch Folgendes sich ergeben:

Der pythagoräische Lehrsatz gilt wohl in | unserer Geometrie, denn er 125a
basiert ja nur auf der Congruenz gleichstimmiger Dreiecke; man kann aber *nicht mehr hier aus der Gleichheit der Quadrate auf die Gleichheit ihrer Seiten schließen.*

Daraus ergibt sich:

1.) *Die Lehre vom Flächeninhalt hängt wesentlich von dem Satze über das gleichschenkelige Dreieck ab; ist also nicht Folge der Lehre von den Proportionen allein.*

2.) Aus dem Satze vom gleichschenkeligen Dreiecke zieht auch Euklid die Folgerung, dass die Summe zweier Seiten eines Dreieckes größer als die dritte ist. Auch dieser Satz wird in unserer Geometrie nicht mehr gelten.

Begründung der nicht-euklidischen Geometrie.

Bolyai und Lobatschefsky benützen zum Aufbau derselben den dreidimensionalen Raum; beide gehen auf dieselbe Weise vor, indem | sie mit Hilfe der 126
Stetigkeit ein Gebilde „*die Grenzkugel*" einzuführen, auf welcher die gewöhnliche Euklidische Geometrie gilt.

Klein geht auf dem Wege der projektiven Geometrie vor, benutzt auch den Raum und die Stetigkeit.

Wir wollen hier die nichteuklidische Geometrie auf einem neuen Wege begründen.

Wir setzen die Axiome I, II, III (das Dreiecksaxiom im weiteren Sinne) *voraus; statt Axiom IV nehmen wir folgendes an:*

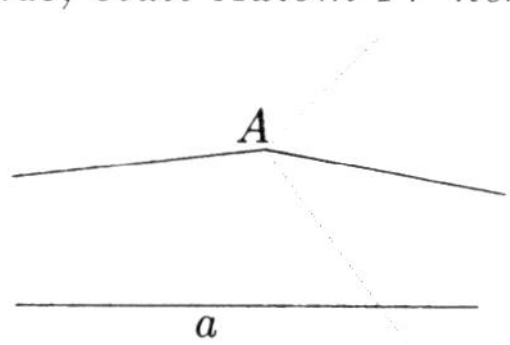

Es gibt durch jeden Punkt A zwei Halbgerade, die nicht ein und dieselbe Gerade ausmachen und welche eine gegebene Gerade a nicht schneiden; während jede in dem durch sie gebildeten Winkelraum gelegene von A ausgehende Gerade a schneidet;

und umgekehrt:

Zu irgend zwei von dem Punkte A ausgehenden Halbgeraden, die nicht ein und | dieselbe Gerade ausmachen, gibt es stets einen bestimmte Gerade a, 127
die jene beiden nicht schneidet, dagegen von jeder anderen Halbgeraden ge-

troffen wird, die von A ausgeht und in dem Winkelraume zwischen beiden Halbgeraden verläuft.

Die zwei Halbgeraden durch A wollen wir Grenzgerade oder Parallele zu a nennen.

Wir wollen nun zwei Sätze beweisen:

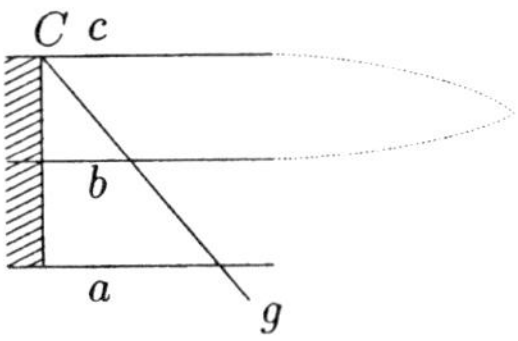

1.) *Ist $b \parallel a$ und $c \parallel a$, dann ist auch $b \parallel c$.*
Bew.: Wenn $b \parallel a$ und $c \parallel a$ beide nach derselben Seite hin, und $b \nparallel c$ wäre, so hätten wir durch den Schnittpunkt von b und c zwei Parallele zu a nach derselben Seite, was unmöglich ist.

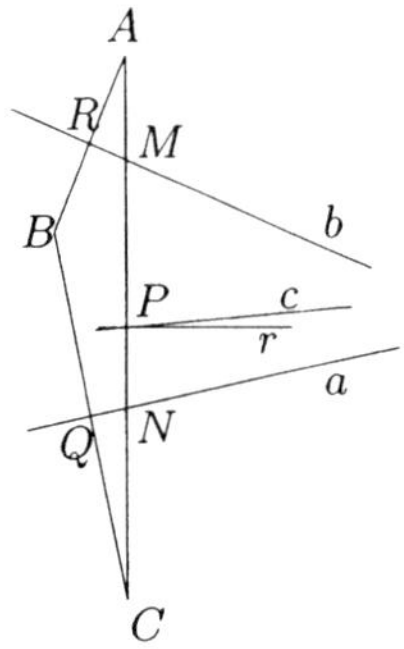

2.) *Wenn ein Dreieck ABC so beschaffen ist, dass die Mittelsenkrechten auf den zwei Seiten AB und BC einander parallel sind, dann ist auch die Mittelsenkrechte von AC zu den beiden anderen $\parallel$.*

128 Es ist

$$AM = MB,$$
$$BN = CN$$
$$\sphericalangle NBA > \sphericalangle BAN$$

daher

$$AN > BN$$

also

$$AN > NC$$

daher muss N zwischen P und C liegen; ebenso ergibt sich, dass M zwischen A und P liegt; P liegt also zwischen M und N.

Wenn die 3. Mittelsenkrechte aus P die obere im Punkte S schneiden würde, dann wird

$$SA = SC \quad \langle \text{und} \quad SA = SB \rangle$$

daraus

$$SB = SC$$

d. h. S muss auch auf der Mittelsenkrechten aus Q liegen; dann würden aber die beiden Mittelsenkrechten aus Q und R nicht parallel sein, wie die Voraussetzung verlangt.

Jetzt ist noch zu zeigen, dass die Mittelsenkrechte c zu a und b parallel ist.
Wäre dies nicht der Fall, so liesse sich durch P eine Parallele zu a nach rechts
129 ziehen und c | müsste also entweder a oder b schneiden, was nach Obigem nicht erlaubt ist.

Mit Hilfe dieser Sätze können wir nun eine Rechnung mit Enden begründen.

Wir setzen fest: Jede Gerade besitzt zwei Enden α, β; je eines derselben hat sie mit sämtlichen parallelen, nach derselben Seite gehenden, Geraden gemein.

Nach Axiom IV, pag. 126, bestimmen zwei Enden eine Gerade eindeutig.

Addition von 2 Enden

Wir zeichnen eine kreisförmige Figur; auf derselben mögen alle Ende liegen.

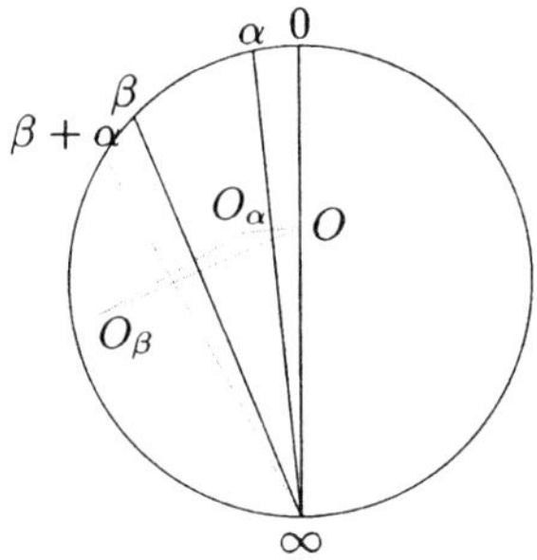

Wir nehmen eine Gerade 0∞ und auf derselben den Punkt O als fest an.

Sind jetzt $\alpha\infty$ und $\beta\infty$ zwei andere Geraden, so spiegeln wir O an diesen beiden; erhalten so O_α und O_β; die Mittelsenkrechte von $O_\alpha O_\beta$ soll dann das Ende $\beta+\alpha$ liefern.

Daraus ergibt sich sofort 130

$$\alpha+\beta=\beta+\alpha.$$

Wir müssen jetzt zeigen, dass

$$\alpha+(\beta+\gamma)=(\alpha+\beta)+\gamma$$

ist. Zu diesem Zwecke bedeute S_ξ die Spiegelung an der Gerade $\xi\infty$.

Dann ist $S_\alpha\cdot S_{\alpha+\beta}S_\beta S_0=1$, weil bei der resultierenden Congruenz der Punkt ∞ und der Punkt O fest bleiben.

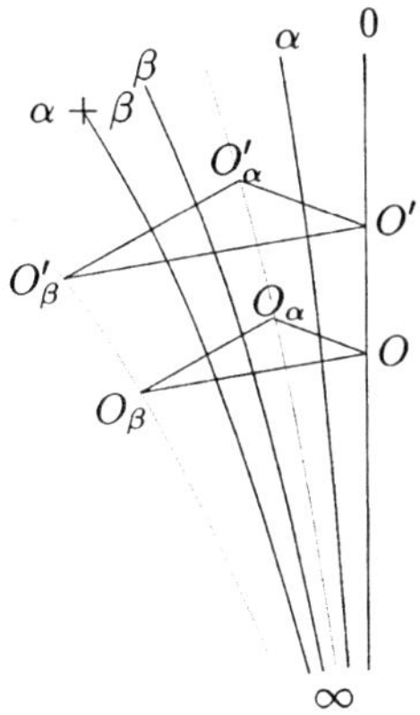

Da auch $S_\alpha^2=1$ etc. so folgt,

$$S_{\alpha+\beta}=S_\alpha S_0 S_\beta.$$

Demnach ist

$$S_{\alpha+(\beta+\gamma)}=S_\alpha S_0 S_{\beta+\gamma}=S_\alpha S_0 S_\beta S_0 S_\gamma$$

und

$$S_{(\alpha+\beta)+\gamma}=S_{\alpha+\beta}S_0 S_\gamma=S_\alpha S_0 S_\beta S_0 S_\gamma$$

also *das associative Gesetz der Addition gilt.*

Die Construction von $\alpha+\beta$ aus α und β ist | von der Wahl des Punktes O unabhängig, wie leicht aus der Figur zu 131
ersehen ist; man hat nur die Geraden $O'_\alpha O_\alpha$ und $O'_\beta O_\beta$ zu ziehen.

Nebenbei sei bemerkt:

Spiegelt man das Ende α an der Geraden 0∞, so erhält man das Ende $-\alpha$.

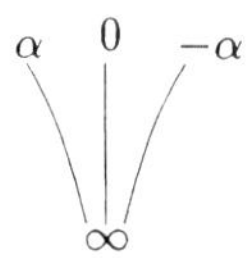

Definition der Multiplication.

Man zeichne die Gerade 0∞ und durch den Punkt O eine Normale; deren Enden wollen wir mit $+1$ und -1 bezeichnen. Ist nun $\alpha\cdot\beta$ zu bilden, so ziehe man von diesen Enden die Normalen auf 0∞, erhält dadurch die Punkte A und B; nun mache man

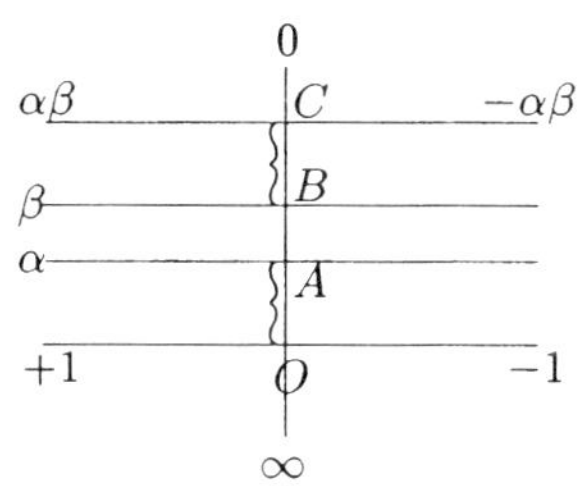

$$BC=OA,$$

das Lot durch den Punkt C hat die Enden $\alpha\cdot\beta$ und $-\alpha\beta$.

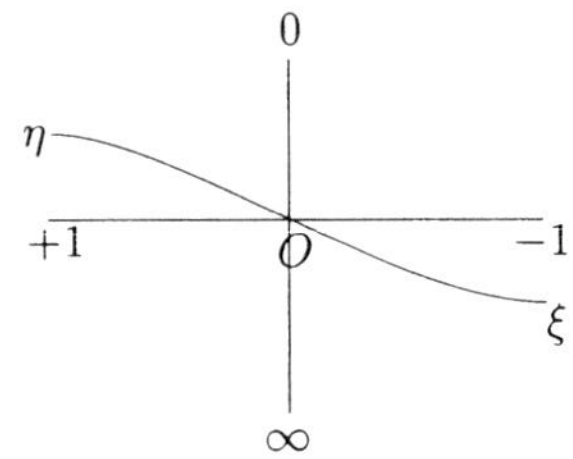

Aus dieser Construction ersieht man auch, dass für jede durch O gehende Gerade
132 *mit den | Enden η, ξ die Gleichung bestehen muss*

$$\eta \cdot \xi = -1.$$

Aus der Construction von $\alpha\beta$ ergiebt sich auch sofort, dass

$$\beta\alpha = \alpha\beta$$

ist.

Es ist noch das distributive Gesetz

$$\alpha(\beta + \gamma) = \alpha\beta + \alpha\gamma$$

zu beweisen.

Man construiere zu dem Zwecke $\beta + \gamma$; dann handelt es sich bei der Construction von $\alpha(\beta + \gamma)$ nur um eine Verschiebung nach dem Ende 0 um den Betrag OA.

Bestimmt man nach früherem $\alpha\beta$ und $\alpha\gamma$, so handelt es sich auch um eine Verschiebung der Enden β und γ um einen Betrag.

Nun führe man die Construction $\alpha\beta + \alpha\gamma$ von dem Punkte A aus; man ersieht dann leicht die Richtigkeit des Gesetzes

$$\alpha(\beta + \gamma) = \alpha\beta + \alpha\gamma.$$

133 Aus der Figur erkennt man unschwer, dass

$$1 \cdot \alpha = \alpha \quad \text{und} \quad 0 \cdot \alpha = 0$$

sein muss.

Damit ist gezeigt, dass man mit den Enden gerade so rechnen kann wie mit gewöhnlichen Zahlen; unsere Geometrie ist also begründet.

Noch eines wollten wir hinzufügen:

Man kann jetzt einen Punkt durch Liniencoordinaten bestimmen.

Wir setzen

$$u = \xi \cdot \eta$$
$$v = \frac{\xi + \eta}{2}, \qquad (\xi \text{ und } \eta \text{ sind Enden})$$

Ist dann $4ac - b^2 > 0$, so repräsentiert die Gleichung

$$au + bv + c = 0$$

einen Punkt; d. h. alle Geraden mit den Enden ξ, η gehen durch einen Punkt, wenn sie dieser linearen Gleichung genügen.

Um dies zu beweisen, setzen wir:

$$\kappa = 1, \quad \lambda = \frac{b}{2a}$$

134

$$\mu = 0 \quad \nu = \frac{\sqrt{4ac - b^2}}{2a}$$

dann ist

$$\kappa^2 + \mu^2 = 1$$
$$\kappa\lambda + \mu\nu = \frac{b}{2a}$$
$$\lambda^2 + \nu^2 = \frac{c}{a}$$

und unsere lineare Punktgleichung wird:

$$(\kappa^2 + \mu^2)\xi\eta + (\kappa\lambda + \mu\nu)(\xi + \eta) + \lambda^2 + \nu^2 = 0$$

oder anders geschrieben:

$$\frac{\kappa\xi + \lambda}{\mu\xi + \nu} \cdot \frac{\kappa\eta + \lambda}{\mu\eta + \nu} = -1.$$

Diese Gleichung hat die Form

$$\xi' \cdot \eta' = -1$$

und zeigt, dass die Gerade $\xi'\eta'$ durch O geht. ξ', η' gehen aus ξ, η durch dieselbe lineare Transformation hervor, zeigen also eine Verschiebung der Ebene an; die Geraden $\xi\eta$ müssen sich daher auch in einem Punkte schneiden.

Wir wollen zum Schlusse noch einige Litteratur angeben: 135

1.) Dass *der Pascal'sche Satz vom Desargues*'schen unabhängig ist, bei Ausschluss von V, wurde in der Festschrift gezeigt.[55]

2.) *Dr. Hamel* hat in seiner Doctordissertation die Stellung der geraden Linie als kürzeste zwischen zwei Punkten gezeigt.[56]

3.) *Gauss zeigt* (Bd. 8, pag. 193–195), dass die Ebene mit Hilfe des Congruenzaxioms und einer anderen Annahme definiert werden kann.[57]

4.) *Dehn* handelt in den Math. Annalen über den Rauminhalt (Raumgleichheit) der Polyeder im Sinne unserer Flächengleichheit.[58]

5.) Wenn nur die *Oberfläche der Kugel* | *gegeben* ist, so müsste sich die sphäri- 136
sche Trigonometrie mit Hilfe der Congruenzaxiome allein *ohne Zuhilfenahme des Raumes* begründen lassen (elliptische Geometrie).[59]

6.) In einer Arbeit, die demnächst in den Math. Annalen erscheint, zeigen wir, dass die von *Riemann, Helmholtz, Lie* unter Zugrundelegung der Stetigkeit begründete nichteuklidische Geometrie sich durch eine geringe Anzahl von Axiomen begründen lässt; diese Axiome sind:

I.) *Die Bewegungen bilden eine Gruppe.*

---

[55]Hilbert is apparently referring to Theorem 37 of the *Festschrift* (*Hilbert 1899c*), p. 71. See Chapter 5 of this Volume.

[56]Hamel's work is presented in *Hamel 1901* and *1903*. See the Introduction to Chapter 4 of this Volume, Specific Note 6(*c*).

[57]The reference is to *Gauß 1900*, and most likely to the pieces marked as [2] and [3], pp. 194–195. See the note to Hilbert's remark [4] in the Appendix to Chapter 4.

[58]Dehn's paper is *Dehn 1902*. See the Introduction to Chapter 4, Specific Note 5(*c*).

[59]There is a note about this (n. 2) added to the reprinting of *Hilbert 1903b* (as Anhang III) in the third edition of the *Festschrift*, i.e., *Hilbert 1909*. (The same Anhang appears in *Hilbert 1903a*, but this note is new.) Hilbert there refers to three papers, by Dehn (*Dehn 1905*), Hessenberg (*Hessenberg 1905a*) and Hjelmslev (*Hjelmslev 1907*).

II.) *Bei Festhaltung eines Punktes, sind noch unendlich viele Drehungen möglich.*

III.) *Die Bewegungen bilden ein abgeschlossenes System.*[60]

[60]The paper Hilbert refers to is *Hilbert 1903c*. See the Introduction to this Chapter.

Textual Notes

540.3–4: zusammenhängendes] ↑zusammenhängendes↓ ⟨added by Hi^g⟩

540.13: Naturwissenschaft] Naturwissenschaft↑, an der sie auch zuerst ausgebildet worden ist.↓ ⟨added by Hi^g⟩

540n: wirkliche] ↑wirkliche↓ ⟨Added by Hi^s⟩

540n: wirklichen] ↑wirkliche↓ ⟨Added by Hi^s⟩

540n: Nicht-Desarguesschen] Nicht-Desargueschen

540n: der] die

540n: Wissenschaft] Wiss.

540n: Vorlesung] Vorl.

540n: z. B.] ↑z. B↓ ⟨added by Hi^g⟩

541.8: und unabhängiges] ⟨Enclosed in parentheses by Hi^g.⟩

541.11–12: Ein näheres ... bleiben.] ⟨Enclosed in parentheses by Hi^g.⟩

541.14: *analytische und synthetische*] ⟨Hi^g has reversed the original order by writing '1.' above 'analytische' and '2.' above 'synthetische'.⟩

541.26: ist ⟨...⟩ nötig] ⟦liegt eine Hauptschwierigkeit⟧⊣↑ist besondere Vorsicht nötig↓ ⟨substituted by Hi^g⟩

541.29: besonders drastisches] ↑besonders drastisches↓ ⟨added by Hi^g⟩

541.38: BMF] ⟦D⟧⊣↑F↓ ⟨substituted by Hi^g⟩

542.13: *] ⟨Footnote added by Hilbert in black ink, with the asterisk in blue pencil.⟩

542n: Meine Grundlagen, S. 2] ⟦Beiblatt⟧⊣↑Meine Grundlagen S. 2↓ ⟨substituted by Hi^g⟩

542n: 5.) ⟨...⟩ 1905.] ⟨Added by Hi^g in the left-hand margin, presumably after the titles which follow, whose numbers have correspondingly been increased by one.⟩

542n: Pasch-Dehn,] ⟨Added by Hi^g in the left-hand margin 'Pasch, Vorlesungen üb Teubner 1882'; this was then crossed out and replaced by a reference to the second edition.⟩

542n: gruppentheoretischen] gruppentheor.

543.12: *Gerade.*] *Gerade;*

543.18: *gemein.*] *gemein.* ⟦Wir müssen noch hinzufügen: *Auf jeder Geraden giebt es wenigstens zwei Punkte.*⟧ ⟨deleted by Hi^g⟩

543.30: in x enthaltene] ⟦unterhalb x liegende⟧⊣↑in x enthaltene↓ ⟨substituted by Hi^g⟩

543.33: A ⟨...⟩ = 3)] ⟨Throughout this paragraph, Hi^g has replaced A, B, C by C, A, B respectively and added the reference 'vgl. Beiblatt' and a marking in the margin. The 'Beiblatt' seems not to have been preserved.⟩

545.34: also BC nicht] ↑also BC nicht↓ ⟨added by Ad^s⟩

548.30: F zwischen E und D] $\underbrace{F \text{ zwischen } E \text{ und } D}_{[\![\gamma]\!]\langle\text{deleted by } Ad^s\rangle}$

548n: sich ⟨...⟩ beschränken!] ⟦die rationalen Punkte ins Auge zu fa⟧⊣↑sich auf rationale Punkte zu beschränken!↓ ⟨substituted by Hi^g⟩

549.1: Geraden] Gerade

549.13: *rationalem*] *rationalen*

549.30–31: *derselben oder*] ⟨Enclosed in parentheses by Hi^g.⟩

549.31: *anderen*] ⟨Enclosed in parentheses by Hi^g.⟩

550.12: $AC \equiv A'C'$.] ⟨*Hi*[g] has added after this formula the first of a number of markings which look like '×'. In the following text over the next three pages, these markings appear in pairs of upper and lower ones and seem to have been used to select ˣcertain passages ₓ in the indicated way. In this first case, however, the opening ˣ is missing.⟩

550.32–35: *Definition:* ⟨...⟩ $\triangleleft(kh)$.] ⟨Enclosed in 'ˣ ₓ' by *Hi*[g].⟩

550.32: *ausgehende*] *ausgehenden*

550.40–551.6: Es sei ⟨...⟩ $(hk) \equiv (hk)$.] ⟨Enclosed in 'ˣ ₓ' by *Hi*[g].⟩

551.8–9: Wenn ⟨...⟩ $\triangleleft(h''k'')$,] ⟨Enclosed in 'ˣ ₓ' by *Hi*[g].⟩

551.18–21: Wenn ⟨...⟩ $ACB \equiv A'C'B'$.] ⟨Enclosed in 'ˣ ₓ' by *Hi*[g].⟩

551n: einen Halbstrahl r.] eine ⟦bestimmte Gerade r und⟧↑Halbstrahl r.↓ ⟨substituted by *Hi*[g]⟩

551n: Punkte] Pkte

551n: gehörigen] gehörg

551n: die eine Klasse rechts] ⟦rechts⟧↑die eine Klasse rechts↓ ⟨substituted by *Hi*[g]⟩

552.14: linken] rechten

552.19: *darstellen*"] *darstellen*

554.24: Gegeben] ⟨Indentation omitted by editor.⟩

555.15: ABD] ⟦ABC⟧↑ABD↓ ⟨substituted by *Hi*[g]⟩

556.2: ⟨This figure was added by *Hi*[g].⟩

556.5–10: ⟨This figure has been extended (the part with the dotted lines) by *Hi*[g].⟩

556.39: letzteres] letztere

557.7: z. B.] ⟨Added by *Ad*[s].⟩

558.3: Trägt] ⟨Indentation omitted by editor.⟩

558.5: A] ⟦C⟧↑A↓ ⟨Replaced by *Hi*[g] ⟩

558.6: BA] ⟦BC⟧↑BA↓ ⟨Replaced by *Hi*[g] ⟩

558.28: *Bew.:*] ⟨Indentation omitted by editor.⟩

562.14: Aussage ⟨...⟩ Aussage] ⟦Aussagen sind mit unseren Aussagen⟧↑Aussage ist mit unserer Aussage↓ ⟨substituted by *Hi*[s]⟩

563.2: *Bew.:*] ⟨Indentation omitted by editor.⟩

564.10–11: („Wenn ⟨...⟩ nicht.")] ⟨Added by *Ad*[s].⟩

564.15–16: *aufstellen:*] ⟦*definieren:*⟧↑*aufstellen:*↓ ⟨substituted by *Ad*[s]⟩

564.30: 0] ⟦0⟧ ⟨deleted by *Ad*[s]⟩

567.13: *gelten aber sämmtlich.*] *gelten* ↑*aber*↓ ⟨added by *Ad*[s]⟩ *sämmtlich.*

568.12: *Ausschluss*] *Auschluss*

568.19: $\left|\sqrt{\left(\frac{dx}{dt}\right)^2 + \left(\frac{dy}{dt}\right)^2}\right|$] $\left|\sqrt{\left(\frac{dx}{dt}\right)^2 + \left(\frac{dy}{dt}\right)^2}\right.$

569.2: es] est

570.27: Argument] Argumment

570.30: welchem] welchen

572.3: $(z = x + iy)$] (↑$z =$ ↓⟨added by *Hi*[g]⟩ ⟦x, y⟧↑$x + iy$↓⟨substituted by *Ad*[s]⟩)

572.3: $(\delta + i \cdot 0)$] (⟦$0, \delta$⟧↑$\delta + i \cdot 0$↓ ⟨substituted by *Hi*[g]⟩)

572.7: $c\delta$] c⟦z⟧↑δ↓ ⟨substituted by *Hi*[g]⟩

574.2: $\alpha = \alpha' + \delta' \geqq 2\alpha'$] ↑$\alpha = \alpha' + \delta'$⟦ $< 2\delta'$⟧↑ $\geqq 2\alpha'$↓↓⟨added by *Hi*[g]⟩

575n: bedeckt] bedekt

576.27: Schlüsse] Schlüße

577.4: Strecken] Strecke↑n↓ ⟨added by *Hi*[g]⟩

581.33: den] ↑den↓ ⟨added by Hi^s⟩
582.36: erhalten] erhalte
583.11: a] ⟦b⟧⊣↑a↓ ⟨substituted by Hi^g⟩
583.24: Strecken] ⟦Punkte⟧⊣↑Strecken↓ ⟨substituted by Ad^s⟩
583.25: Obigem] Obigen
584.1: Dreiecken] Dreiecke
586.14–15: Zu dem ⟨...⟩ Winkel ϑ.] ⟦Dies folgt leicht aus dem Secantensatz für den Kreis K, welcher Satz ja aus dem Hauptsatz der Proportionenlehre folgt.⟧⊣↑Zu dem ⟨...⟩ Winkel ϑ.↓ ⟨substituted by Ad^s⟩
586.14: Punkt R'] ⟨According to the diagram substituted for:⟩ Punkt R
587.27: *inhaltsgleiche*] *Inhaltsgleiche*
588.10: AB'] $A'B'$
588.27: flächengleich] Flächengleich
589.8: AB] BC
589.11: etc.] ⟪??⟫
590.11: Wurzeln] Wurzel
590.33: Axe] axe
591.30: daher] ⟦ebenso⟧⊣↑daher↓ ⟨substituted by Ad^s⟩
591.34–35: holoëdrisch-isomorph] ⟦⟪???⟫ ⟪???⟫ auf⟧⊣↑holoëdrisch-isomorph↓ ⟨substituted by Ad^s⟩
592.8: Drehungsformeln] Drehungsformel
592.25: $\alpha^2 + \gamma^2 = 1$] $\alpha^2 + \beta^2 = 1$
592.30: $\alpha^2 + \gamma^2$] $\alpha^2 + \beta^2$
593.11–12: rechte ⟨...⟩ gleich.] ⟦einander gleich sind, so sind es auch die Winkel CAD und CBD.⟧⊣↑rechte ⟨...⟩ gleich↓ ⟨substituted by Ad^s⟩
593.13: *zuerst*] ⟦*nun*⟧⊣↑*zuerst*↓ ⟨substituted by Ad^s⟩
593.17: Drehungsformeln] Drehungsformel
594.6: *Winkeln*] *Winkel*
594.43: Andererseits] Anderseits
594.44: $OE = \Delta_\eta r$] ⟦$OC = \alpha_\eta r$⟧⊣↑$OE = \Delta_\eta r$↓ ⟨substituted by Ad^s⟩
595.15: $\alpha_\vartheta x - \gamma_\vartheta y$] ⟦$\alpha_\vartheta x + \gamma_\vartheta y$⟧⊣↑$\alpha_\vartheta x - \gamma_\vartheta y$↓ ⟨substituted by Ad^s⟩
595.20: γ_ϑ^2] ⟦β_ϑ^2⟧⊣↑γ_ϑ^2↓ ⟨substituted by Ad^s⟩
595.22: Δ_η] $\Delta_{[\![\vartheta]\!]\dashv\uparrow\eta\downarrow}$⟨substituted twice by Ad^s⟩
596.6: α_η] $\alpha_{[\![\vartheta]\!]\dashv\uparrow\eta\downarrow}$ ⟨substituted by Ad^s⟩
596.8: $\alpha_\eta\gamma_\vartheta$] $\alpha_{[\![\vartheta]\!]\dashv\uparrow\eta\downarrow}$⟨substituted by Ad^s⟩γ_ϑ
596.14: $\alpha_\eta x - \gamma_\eta y$] ⟨In this, as well as in the four following lines, the manuscript originally read 'x''' and 'y''' for 'x' and 'y' respectively, the primes having been erased by Ad^s.⟩
597.30: das] Das
597.32–34: Halbgerade, ⟨...⟩ ausmachen] ⟦Gerade, die nicht zusammenfallen⟧⊣↑HalbGerade, ⟨...⟩ ausmachen↓ ⟨substituted by Ad^s⟩
597.39: ausgehenden] ausgehende
598.42: *eines derselben*] *eine derselbe*
600.3: *Gleichung*] *Gleichungen*
600.27: *einen*] ⟦*den*⟧⊣↑*einen*↓ ⟨substituted by Hi^g⟩
601.8: $\frac{\kappa\xi+\lambda}{\mu\xi+\nu}\cdot\frac{\kappa\eta+\lambda}{\mu\eta+\nu}$] $\frac{\kappa\xi+\lambda}{\mu\xi+\mu}\cdot\frac{\kappa\mu+\lambda}{\mu\eta+\nu}$
601.15–16: Ausschluss] Auschluss

Description of the Text

Collection: Georg-August-Universität Göttingen, Mathematisches Institut, Inv. Nr. 16205h.

Size: Cover size 22.7 × 29.5 cm; page size approx. 22 × 29 cm.

Cover Annotations: On the spine, in gold lettering, is the notation, 'D. Hilbert // Grundlagen der Geometrie // S.S. 1902'. On the inside front cover is the library numbering '45', corresponding to the numbering in the catalogue list. (This is in contrast to the other bound texts in the Mathematisches Institut, where the number appears on the outside.)

Composition: Seven signatures of five or six double pages each; in all, 78 sheets, inclusive of front- and end-papers.

Pagination: The pages are continuously numbered from 1 to 136; the numbers 124f. appear twice. Twenty blank sheets appear at the end of the volume. The recto side of the front end paper and of the title page (i.e., the first page of the first signature) were subsequently numbered by Hilbert as −1 and 0, in red pencil.

Original Title: On the title page (numbered 0 by Hilbert in red pencil): 'Grundlagen der Geometrie // Vorlesungen von // D. Hilbert // gehalten im Sommersemester 1902. // 2st. // ausgearbeitet von // August Adler.'. The '2st.' was added by Hilbert in black ink, and underlined three times.

Text: Handwritten text by August Adler, in ocher ink. A text area of 15.5 × 22 cm is stamped on the page, inside of which is to be found the lecture text. Page numbers, drawings, and formula numbers sometimes occur outside of this text area.

Hilbert's Lecture Courses 1886–1934

The following table gives an overview of the lecture courses held by Hilbert throughout his teaching career. It also presents an overview of the accompanying documentary record, at least to the extent that these documents exist in the Hilbert *Nachlaß* held in the Staats- und Universitätsbibliothek of Göttingen University or in the library of the Mathematical Institute in Göttingen. For instance, listed are Hilbert's own lecture manuscripts or approved notes taken by collaborators or notes taken by members of Hilbert's lecture audience. A few comparable documents known to the Editors and kept in others archives or libraries are briefly listed at the end of the table.

The main sources for the list of lectures (given in the central column) are the official lecture listings of the Universities of Königsberg and Göttingen. All Hilbert's lectures listed there have been cited. Use was also made of two additional sources. The first is the *Kuratorialakten* in the archives of the University of Göttingen; the call-signs for these are 'UAG Kur 4.I.81' and 'UAG Kur 4.I.104 (II)', abbreviated here as 'UAG Kur'. These extend up the 'Herbstzwischensemester' of 1919, and list both the lecture courses announced and those actually held, and also give, where appropriate, the number of participants. The second source is a list contained in the Hilbert *Nachlaß* (*Hilbert 1934**) which bears the title, in Hilbert's hand, 'Verzeichnis meiner Vorlesungen (1886-1932)'. This list is abbreviated here as 'HVV', for 'Hilbert's own *Vorlesungsverzeichnis*'.[1] When lecture courses given in UAG Kur or HVV could not be correlated with the titles in the official lecture listings, they have been entered in the list in *Sperrschrift* (i.e., s p a c e d   l e t t e r s).[2] If the source indicates a reason for doubting whether a course of lectures announced was

[1]This list records the lectures in chronological order. The entries up to the WS 1917/18 are in Käthe Hilbert's hand, and she clearly relied mainly on those manuscripts which were available in Hilbert's personal library. Consequently, a great many lectures listed in the first two sources specified are missing in this third source for the period up to 1917/18. The list was continued up to the WS 1926/27 by Paul Bernays; there is a further handwritten entry for the following summer semester added by Arnold Schmidt. The remainder of the list, which, *pace* Hilbert's title, extends to the WS 1933/34, is typed.

[2]The 'Verzeichnis der von Hilbert gehaltenen Vorlesungen' published in the third volume of Hilbert's *Gesammelte Abhandlungen* (*Hilbert 1935*, 430) is merely a list organized by subject matter of titles of some of Hilbert's lecture courses, and takes no account of how many times he lectured on a particular subject or under what title. The list was therefore of little use for our purposes.

actually held, the title is written in *italics*, and the situation is explained in a footnote.

In order not to overburden this table with an excesss of detail, Hilbert's 'exercise classes' ('Übungen') and seminars have been omitted. Of these, only a few have been recorded in surviving documents, and these often present problems of dating. On the other hand, the third column in the table lists all the texts for lectures courses which survive in Göttingen; to one and the same lecture course there may be correlated several documents, for example, Hilbert's own manuscript, an official protocol, and/or notes by a participant. If a document is not a manuscript by Hilbert, the name of the author is given in parentheses after its title, and when the author is unknown, this is marked by 'N.N.'. The titles of the various documents sometimes differ significantly from the title announced for the lectures. If so, the new title is given; if not, then the entry just records 'Title as announced'. Where there is no document extant, as far as is known, the third column lists a horizontal line.

The following abbreviations are used. In the first column (and in the second list of those documents not kept in Göttingen), 'WS' and 'SS' are short for 'Wintersemester' and 'Sommersemester' respectively. ('HZS 1919' denotes the exceptional *Herbstzwischensemester* in 1919.) In the central column, the indication 'nst.' following a title is short for 'n-stündig', and means that n academic 'hours' were planned each week for the lecture course, i.e., n periods of forty-five minutes each. The third column also contains the abbreviations 'SUB' and 'MI'; as usual, these stand respectively for the Handschriftenabteilung (Manuscript Division) of the Staats- und Universitätsbibliothek of the Georg-August Universität Göttingen, and the Mathematisches Institut in Göttingen. More particularly, 'SUB xyz' here is really short for '*Cod. Ms. D. Hilbert xyz*', the call-sign under which the Manuscript Division keeps the document in question.

Hilbert's Lecture Courses in Königsberg und Göttingen, 1886–1934

Semester	*Title Announced*	*Corresponding Documents*
Privatdozent in Königsberg		
WS 1886/87[3]	Invariantentheorie	SUB 521: Invariantentheorie mit Uebungsstunde
SS 1887	Theorie der Determinanten, verbunden mit geometrischen Übungen, 2st.	SUB 523: Determinantentheorie und Uebungen
	Theoretische Hydrodynamik, 3st.	SUB 522: Hydrodynamik
WS 1887/88	Theorie der numerischen Gleichungen, 3st.	SUB 525: Numerische Gleichungen
	Theorie der Kugelfunctionen, 1st.	SUB 524: Kugelfunctionen
SS 1888	Darstellende Geometrie, verbunden mit Übungen, 2st.[4]	———
	Theorie der höheren algebraischen Gleichungen, 2st.	SUB 526: Höhere algebraische Gleichungen
WS 1888/89	Über Linien- und Kugelgeometrie, 3st.	SUB 527: Kugel- und Liniengeometrie[5]
	Theorie der Determinanten, 1st.	———
SS 1889	Zahlentheorie, 2st.	SUB 528: Zahlentheorie und Uebungen
WS 1889/90	Allgemeine Theorie der algebraischen Gebilde, 2st.	SUB 529: title as announced
	Einleitung in die höhere Analysis, 2st.	SUB 530: Einführung in das Studium der Mathematik[6]
SS 1890	Bestimmte Integrale, 2st.	SUB 531: title as announced
	Theorie der krummen Linien und Flächen, 2st.	SUB 532: Krumme Linien und Flächen
WS 1890/91	Theorie der ebenen algebraischen Curven, 2st.	SUB 534: title as announced
	Theorie der algebraischen Zahlen und Functionen, 2st.	SUB 533: title as announced
SS 1891	Geometrie der Lage, 2st.	SUB 535: Projektive Geometrie
WS 1891/92	Theorie der linearen Differentialgleichungen, 2st.	SUB 536: Lineare Differentialgleichungen und Uebungen

[3] Not announced in the official *Vorlesungsverzeichnis*. The title follows HVV.
[4] Not in HVV.
[5] Added in parentheses: 'enthält die Theorie der quadratischen Formen'.
[6] Added in parentheses: 'Zahlbegriff, Höhere Analysis, Analytische Geometrie, Differential- und Integralrechnung'.

Hilbert's Lecture Courses (cont.)

Semester	*Title Announced*	*Corresponding Documents*
SS 1892	Über die eindeutigen Functionen mit linearen Transformationen in sich, 2st.	SUB 537: Eindeutige Functionen mit linearen Transformationen in sich
WS 1892/93	Einleitung in das Studium der höheren Mathematik, 2st.[7]	———
	Integralrechnung, 4st.[8]	SUB 538: Integralrechnung und Uebungen
	Theorie der partiellen Differentialgleichungen, 2st.	SUB 539: Partielle Differentialgleichungen
Promotion to Außerordentlicher Professor in Königsberg		
SS 1893	Theorie der bestimmten Integrale, 4st.	SUB 540, pp. 67–149: Bestimmte Integrale
	Die Grundlagen der Geometrie, 1st.[9]	SUB 541: Die Grundlagen der Geometrie.[10]
	Ausgewählte Capitel aus der Invariantentheorie, 1st.	SUB 521, pp. 353–369: title as announced
WS 1893/94	Theorie der Kettenbrüche, 2st.[11]	———
	Theorie der elliptischen Functionen, 4st.[12]	———
	Analytische Geometrie des Raumes, 2st.[13]	SUB 543, pp. 77ff.: Analytische Geometrie des Raumes
	Einleitung in die Functionentheorie, 1st.[14]	SUB 540, pp. 1–31: Einleitung in die Functionentheorie

[7] Not in HVV.
[8] According to HVV.
[9] Not read, according to the title page of the manuscript.
[10] The manuscript was revised for the following semester; see the Introduction to Chapter 2.
[11] Not in HVV.
[12] Not in HVV.
[13] According to HVV.
[14] According to HVV.

Hilbert's Lecture Courses (cont.)

Semester	*Title Announced*	*Corresponding Documents*
Promotion to Ordentlicher Professor		
SS 1894	Theorie der Functionen einer complexen Veränderlichen, 4st.[15]	———
	Über die Axiome der Geometrie, 2st.	SUB 541: Die Grundlagen der Geometrie
WS 1894/95	Analytische Geometrie der Ebene und des Raumes, 2st.	SUB 543: title as announced
	Das Problem der Quadratur des Kreises, 1st.	SUB 542: Die Quadratur des Kreises.
	Die Dichtigkeit und Vertheilung der Primzahlen, 1–2st.	———
Appointment in Göttingen		
SS 1895	Determinanten, 2st.[16]	———
	Elliptische Functionen, 4st.	SUB 544: Doppelperiodische Functionen MI: Vorlesung über die Theorie der elliptischen Functionen
WS 1895/96	Integralrechnung, 4st.[17]	———
	Theorie der partiellen Differentialgleichungen, 4st.	SUB 545: Partielle Differentialgleichungen MI: Partielle Differentialgleichungen
SS 1896	Theorie der gewöhnlichen Differentialgleichungen, 4st.	SUB 546: Gewöhnliche Differentialgleichungen
	Über die Quadratur des Kreises, 2st.[18]	———

[15] Not in HVV.
[16] Not in HVV.
[17] Not in HVV.
[18] Not in HVV.

Hilbert's Lecture Courses (cont.)

Semester	*Title Announced*	*Corresponding Documents*
WS 1896/97	Theorie der algebraischen Gleichungen, 2st.	SUB 547: Algebraische Gleichungen
	Theorie der Functionen einer complexen Veränderlichen, 4st.	MI: Theorie der Functionen einer complexen Variablen (Dörrie, 2 copies)
SS 1897	Theorie der algebraischen Invarianten nebst Anwendungen auf Geometrie, 4st.	MI: Theorie der algebraischen Invarianten (Marxen, 2 copies)
	Theorie der Functionen einer complexen Veränderlichen II. Teil, 2st.[19]	—
WS 1897/98	Über die Focaleigenschaften der Curven und Flächen zweiter Ordnung, 2st.	SUB 548: Focaleigenschaften der Curven und Flächen zweiter Ordnung
	Zahlentheorie, 4st.	MI: title as announced (N.N.)[20]
	Über den Begriff der Irrationalzahl und das Problem der Quadratur des Kreises, 2st.	SUB 549: Zahlbegriff und Quadratur des Kreises
SS 1898	Einleitung in die Theorie der Differentialgleichungen, 4st.[21]	
	Bestimmte Integrale und Fouriersche Reihen, 2st.	SUB 550: title as announced
	Ausgewählte Capitel aus der Zahlentheorie, 2st.[22]	MI: Ausgewählte Kapitel der Zahlentheorie (N.N.)[23]
WS 1898/99	Determinantentheorie, 2st.[24]	—
	Elemente der Euklidischen Geometrie, 2st.	SUB 551: Grundlagen der Euklidischen Geometrie SUB 552: title as announced (v. Schaper) MI: title as announced (v. Schaper)
	Mechanik, 4st.	SUB 553: title as announced

[19] Not in HVV.
[20] Published as *Hilbert 1990b*.
[21] Not in HVV.
[22] Not in HVV.
[23] Published as *Hilbert 1990a*.
[24] Not in HVV.

Hilbert's Lecture Courses (cont.)

Semester	*Title Announced*	*Corresponding Documents*
SS 1899	Differentialrechnung, 4st.	SUB 554: title as announced
	Ausgewählte Capitel aus der Gruppentheorie, 2st.	SUB 556: title as announced
	Einführung in die Variationsrechnung, 2st.	SUB 555: title as announced
WS 1899/1900	Integralrechnung, 4st.[25]	———
	Zahlbegriff und Quadratur des Kreises, 2st.[26]	[27]
	Theorie der Flächenkrümmung, 2st.	SUB 557: Einleitung in die Flächentheorie
SS 1900	Theorie der Differentialgleichungen, 4st.[28]	———
	Ausgewählte Capitel aus der Flächentheorie, 2st.	SUB 557: title as announced
WS 1900/01	Theorie der analytischen Functionen, 4st.[29]	———
	Theorie der partiellen Differentialgleichungen, 4st.	SUB 558: title as announced
SS 1901	Algebra, 4st.[30]	———
	Anwendungen der Theorie der partiellen Differentialgleichungen, 2st.	MI: Vorlesung über lineare partielle Differentialgleichungen (Andrae)
WS 1901/02	Zahlbegriff und Quadratur des Kreises, 2st.[31]	SUB 549, pp. 52–54: only a 'Disposition' of the course
	Potentialtheorie, 4st.	MI: Vorlesung über Potentialtheorie (Andrae)
SS 1902	Differential- und Integralrechnung I. Teil, 4st.	Inserted in SUB 554 (SS 1899): title as announced
	Grundlagen der Geometrie, 2st.	MI: title as announced (Adler)
	Ausgewählte Capitel der Potentialtheorie, 2st.	MI: title as announced (Andrae)

[25] Not in HVV.
[26] Not in HVV.
[27] The manuscript of the lecture course from WS 1897/98 was probably used in this course, too. It contains many signs of revision, and in particular on pp. 38–51, it contains additions which, at least in part, stem from a period later than 1897–1898.
[28] Not in HVV.
[29] Not in HVV.
[30] Not in HVV.
[31] Not in HVV.

Hilbert's Lecture Courses (cont.)

Semester	*Title Announced*	*Corresponding Documents*
WS 1902/03	Differential- und Integralrechnung II. Teil, 4st.	Inserted in SUB 554: title as announced
	Mechanik der Continua, 4st.	MI: Mechanik der Continua. Teil I (Berkowski)
SS 1903	Theorie der Differentialgleichungen, 4st.[32]	——
	Ausgewählte Capitel aus der Mechanik der Continua, 2st.	MI: Mechanik der Continua. Teil II (Berkowski)
WS 1903/04	*Zahlbegriff und Quadratur des Kreises*, 2st.[33]	——
	Theorie der partiellen Differentialgleichungen, 4st.	MI: Partielle Differentialgleichungen (Prinz) MI: title as announced (Tieffenbach)
SS 1904	Funktionentheorie, 4st.[34]	
	Zahlbegriff und Quadratur des Kreises, 2st.	MI: title as announced (Born)
WS 1904/05	Variationsrechnung, 4st.	MI: title as announced (Hellinger, 2 different documents, of different length)
	Bestimmte Integrale, 2st.	MI: Bestimmte Integrale und Fouriersche Reihen (Born)
SS 1905	*Zahlentheorie*, 4st.[35]	——
	Einführung in die Theorie der Integralgleichungen, 2st.[36]	MI: Einführung in die Theorie der Integralgleichungen (Hellinger) SUB 558a: title as announced (Born)[37]
	Logische Prinzipien des mathematischen Denkens, 2st.	MI: title as announced (Hellinger)

[32]Not in HVV.
[33]Not in HVV, or in UAG Kur.
[34]Not in HVV.
[35]Not in HVV, or in UAG Kur.
[36]According to HVV; UAG Kur has 'Theorie der Integralgleichungen'.
[37]This *Ausarbeitung* was acquired by the *Nachlaß* only in 1984.

Hilbert's Lecture Courses (cont.)

Semester	*Title Announced*	*Corresponding Documents*
WS 1905/06	Einleitung in die Theorie der partiellen Differentialgleichungen, 2st.	MI: title as announced (Hellinger)[38]
	Mechanik, 4st.	MI: title as announced (Hellinger, 2 copies)
SS 1906	Differential- und Integralrechnung erster Teil mit Übungen, mit Carathéodory, 4st.[39]	———
	Mechanik der Continua, 4st.[40]	———
	T h e o r i e d e r I n t e g r a l g l e i c h u n g e n, 4st.[41]	MI: Integralgleichungen (Hellinger)
WS 1906/07	Differential- und Integralrechnung II. Teil, 4st.	MI: title as announced (Ewald)
	Mechanik der Continua, 4st.	MI: title as announced (Hellinger)
		MI: Vorlesung über die Mechanik der Continua (Marshall/Crathorne)[42]
SS 1907	Theorie der Differentialgleichungen einer unabhängigen Variabeln, 4st.	MI: Theorie der Differentialgleichungen (Haar)
WS 1907/08	Theorie der partiellen Differentialgleichungen, 4st.	MI: title as announced (Haar)
	Einführung in die Theorie der Funktionen unendlich vieler Variabler (Integralgleichungen), 2st.	MI: title as announced (Haar)

[38] For this lecture course, there exist in addition two partial *Ausarbeitungen* with the titles 'Quadratische Formen mit unendlich vielen Variablen. Aus der Vorlesung über partielle Differentialgleichungen' and 'Theorie der quadratischen Formen mit unendlich vielen Variablen. Auszug aus der Vorlesung über partielle Differentialgleichungen im WS 1905/06'.

[39] Not in HVV.

[40] Not in HVV, or in UAG Kur.

[41] According to UAG Kur; HVV has 'Integralgleichungen'.

[42] Not in HVV.

Hilbert's Lecture Courses (cont.)

Semester	*Title Announced*	*Corresponding Documents*
SS 1908	Prinzipien der Mathematik, 4st.	MI: title as announced (N.N.)
WS 1908/09	Funktionentheorie, 4st.	MI: Funktionentheorie (Courant)[43]
		MI: Funktionentheorie (N.N.)
	Prinzipien der Mathematik, 2st.[44]	—
SS 1909	Zahlentheorie, 4st.	MI: Vorlesung über Zahlentheorie (Courant)
	Ausgewählte Fragen aus der Funktionentheorie, 2st.	MI: Ausg. Kapitel der Funktionentheorie (Konforme Abbildungen) (Courant)
WS 1909/10	Allgemeine Theorie der partiellen Differentialgleichungen, 4st.	MI: Theorie der partiellen Differentialgleichungen (Courant)
SS 1910	Elemente und Prinzipienfragen der Mathematik, 4st.	MI: title as announced (Courant)
	Ausgewählte Kapitel aus der Theorie der partiellen Differentialgleichungen, 2st.	MI: title as announced (Courant)
WS 1910/11	Mechanik, 4st.	MI: title as announced (Behrens, 2 copies)
SS 1911	Mechanik der Kontinua, 4st.	MI: title as announced (Hecke)
WS 1911/12	Logische Grundlagen der Mathematik, 1st.[45]	—
	Mechanik der Kontinua, 4st.	MI: Kinetische Gastheorie (Hecke)
SS 1912	Gewöhnliche Differentialgleichungen, 4st.	MI: title as announced (N.N.)
	Mathematische Grundlagen der Physik, 4st.	MI: Strahlungstheorie (Hecke)
WS 1912/13	Einführung in die Theorie der partiellen Differentialgleichungen, 2st.	MI: Partielle Differentialgleichungen (Baule)
	Mathematische Grundlagen der Physik, 2st.	MI: Molekulartheorie der Materie (Hecke, 2 copies)

[43] 2 copies, one in Courant's hand, entitled 'Vorlesung über Funktionentheorie' with the addition 'Ausgearbeitet von R. Courant (Abschrift)'

[44] Not in HVV, or in UAG Kur.

[45] Not in HVV. UAG Kur has 'Grundlagen der Analysis und Geometrie', assigned 2 (academic) hours weekly and with 87 participants.

Hilbert's Lecture Courses (cont.)

Semester	*Title Announced*	*Corresponding Documents*
SS 1913	Elemente und Prinzipien der Mathematik, 4st.	SUB 559: contains draft as well as 'Einige Abschnitte aus der Vorlesung über die Grundlagen der Mathematik und Physik' (Baule)
	Theorie der Elektronenbewegung, 2st.	MI: Elektronentheorie (Hecke)
WS 1913/14	Analytische Mechanik, 4st.[46]	MI: Entwurf in Mechanik-Ausarbeitung WS 1910/11
	Elektromagnetische Schwingungen, 2st.	MI: title as announced (Hecke)
SS 1914	Differentialgleichungen, 4st.	Inserted in the document for the WS 1909/10: 'Theorie der partiellen Differentialgl. (1914)' (N.N.)
	Ausgewählte Kapitel der statistischen Mechanik, 2st.	MI: Statistische Mechanik (Lange)
WS 1914/15	Probleme und Prinzipienfragen der Mathematik, 4st.	SUB 559 enthält Entwurf
SS 1915	Variationsrechnung, 4st.	———
	Ausgewählte Kapitel über Struktur der Materie, 2st.	MI: Vorlesung über die Struktur der Materie
WS 1915/16	Differentialgleichungen, 4st.	MI: title as announced (N.N.)
SS 1916	Partielle Differentialgleichungen, 2st.	MI: title as announced (Bär)
	Einleitung in die Prinzipien der Physik, 2st.	MI: Die Grundlagen der Physik (N.N.)
WS 1916/17	*Mengenlehre*, 4st.[47]	———
	Die Grundlagen der Physik, 4st.[48]	MI: Die Grundlagen der Physik II (Bär)
SS 1917	Mengenlehre, 4st.	MI: title as announced (Goeb, 2 copies)

[46] There is a reference to this lecture course in the entry 'Mechanik mit Übungen: Prof. *Hilbert* und Dr. *Weyl*' under 'Theoretische, experimentelle und angewandte Physik'.

[47] Not in HVV. In UAG Kur the lecture course is listed, although in the column listing the number of participants a '*' appears, probably indicating that the course was not held.

[48] Not announced in the *Vorlesungsverzeichnis*. The title comes from HVV, and the number of hours assigned from UAG Kur.

Hilbert's Lecture Courses (cont.)

Semester	*Title Announced*	*Corresponding Documents*
WS 1917/18	Prinzipien der Mathematik, 2st.[49]	MI: title as announced (Bernays, 2 copies)
	Elektronentheorie, 2st.	SUB 560: title as announced (Humm)
		MI: = SUB 560
SS 1918	Differentialgleichungen, 4st.	———
WS 1918/19	Partielle Integral- und Differentialgleichungen, 4st.[50]	———
	Mengenlehre, 2st.[51]	———
	Über Raum und Zeit (allgemeinverständlich), 1st.	SUB 561: Raum und Zeit (Bernays)
		MI: = SUB 561
SS 1919	Zahlbegriff und Quadratur des Kreises, 4st.	———
	Denkmethoden in den exakten Wissenschaften, 1st.	———[52]
HZS 1919	Natur und mathematisches Erkennen, 2st.	MI: title as announced (Bernays)[53]
WS 1920	Formale Logik und ihr erkenntnistheoretischer Wert, 2st.	MI: Logik-Kalkül (Bernays)
	Mechanik, 4st.	MI: Mechanik und neue Gravitationstheorie (Kratzer)
SS 1920	Höhere Mechanik und neue Gravitationstheorie, 4st.	SUB 562: title as announced (Kratzer)
		MI: = SUB 562
	Probleme der mathematischen Logik, 1st.	MI: title as announced (Schönfinkel, Bernays)
WS 1920/21	Anschauliche Geometrie, 4st.	SUB 563: title as announced (Rosemann)
		MI: = SUB 563 (2 copies)
SS 1921	Einsteinsche Gravitationstheorie, 4st.	———
	Grundgedanken der Relativitätstheorie (für Hörer aller Fakultäten), 1st.	MI: title as announced (Bernays)

[49] In HVV, the title is 'Prinzipien der Mathematik und Logik'.

[50] In HVV, the title is 'Partielle Differentialgleichungen'.

[51] Not in HVV, or in UAG Kur.

[52] *Verzeichnis 1943** lists notes for this (a 'Nachschrift') which are no longer traceable. According to HVV, Bernays was the *Ausarbeiter*.

[53] Published as *Hilbert 1992*.

Hilbert's Lecture Courses (cont.)

Semester	*Title Announced*	*Corresponding Documents*
WS 1921/22	Grundlegung der Mathematik, 4st.	MI: Grundlagen der Mathematik (Bernays)
SS 1922	Statistische Methoden insbesondere der Physik, 4st.	SUB 565 (Nordheim)
		MI: = SUB 565
WS 1922/23	Wissen und mathematisches Denken (für Hörer aller Fakultäten), 1st.	MI: title as announced (Ackermann)
	Grundlagen der Arithmetik, 2st.[54]	SUB 567: Logische Grundlagen der Mathematik (Bernays)[55]
	Mathematische Grundlagen der Quantentheorie, 2st.	SUB 566: title as announced (Nordheim, Heckhausen)
		MI: = SUB 566
SS 1923	Anschauliche Geometrie, 4st.	———[56]
WS 1923/24	*Mengenlehre*, 4st.[57]	———
	Logische Grundlagen der Mathematik, 4st.	[58]
	Unsere Vorstellungen von Gravitation und Elektrizität (allgemeinverständlich), 1st.	SUB 568: Über die Einheit in der Naturerkenntnis
		SUB 569: title as announced[59] (Diestel)
		MI: = Typescript of SUB 568 (N.N.)
SS 1924	Mechanik, 4st.	SUB 570: Einleitung in die Vorlesung über Mechanik
		MI: Relativitätstheorie; Ergänzung zur Vorlesung über allgemeine Mechanik im SS 24 (Nordheim)[60]

[54] HVV gives the title of the extant document.

[55] In part, a typescript prepared by Bernays, in part a manuscript in Hilbert's hand

[56] Addition in HVV, probably in Hilbert's hand: 's. 1920/21'.

[57] Not in HVV. See also footnote 68, below.

[58] Addition in HVV, probably in Hilbert's hand: 's. 1922/23'. Indeed, the handwritten part of *Bernays 1922** contains passages which could not have been composed before the end of 1923. This is clear form the paper on which they are written.

[59] Added in parentheses on the title page of SUB 569: '(Einheit in der Naturerkenntnis)'

[60] HVV notes: 'teilweise ausgearb., Nordheim'

Hilbert's Lecture Courses (cont.)

Semester	*Title Announced*	*Corresponding Documents*
WS 1924/25	Quadratur des Kreises (Elementarmathematik nach höheren Methoden), 4st.[61]	MI: Notizen siehe Nachschrift SS 1904
	Über das Unendliche (allgemeinverständlich), 1st.	MI: title as announced (Nordheim, 2 copies)
SS 1925	Anschauliche Geometrie, 4st.	———
WS 1925/26	Über Raum und Zeit (allgemeinverständlich), 1st.	———
	Zahlentheorie, 4st.	SUB 570a: title as announced (Struik)
SS 1926	Theorie der algebraischen Zahlen, 4st.[62]	MI: Vorlesung über die Theorie der algebraischen Zahlen (N.N.)
WS 1926/27	Mathematische Methoden der Quantentheorie, 4st.	MI: title as announced (Nordheim)
SS 1927	Grundlagen der Geometrie, 4st.	MI title as announced (Schmidt)
SS 1927/28	Grundlagen der Mathematik, 4st.	———
SS 1928	Anschauliche Geometrie, 4st.	———
WS 1928/29	Zahlbegriff und Quadratur des Kreises, 4st.	MI: Notizen siehe Nachschrift SS 1904
SS 1929	Mengenlehre, 4st.	———
WS 1929/30	Theorie der algebraischen Invarianten, 4st.	———
SS 1930	Mathematische Methoden der neuen Physik, 2st.	———
WS 1930/31	Natur und Denken, 1st.	———[63]
SS 1931	Grundlagen der Mathematik, 1st.[64]	———
WS 1931/32	Einleitung in die Philosophie auf Grund moderner Naturwissenschaft (für Hörer aller Fakultäten), 1st.	SUB 607/1: 'Einlage zu Vorl. Winter 31/32'[65]

[61] HVV gives only the subtitle in parentheses.
[62] HVV gives the title as 'Algebraische Zahlkörper'
[63] *Verzeichnis 1943** lists notes for this course, but these could not be found.
[64] According to HVV, this was a seminar
[65] *Verzeichnis 1943** lists notes for this course, but these could not be found.

Hilbert's Lecture Courses (cont.)

Semester	*Title Announced*	*Corresponding Documents*
SS 1932	Grundlagenfragen der Geometrie, 1st.	———
WS 1932/33	Grundlagen der Logik, allgemeinverständlich (für Hörer aller Fakultäten), 1st.	SUB 607/6: 'Einleitung zu Kolleg WS 32/33'
SS 1933	Wissen und Denken (für Hörer aller Fakultäten), 1st.	———
WS 1933/34	Grundlagen der Geometrie, mit Schmidt 2st.	———

The following documents (listed in chronological order) relating to Hilbert's lecture courses are not in either of the two main official repositories of Hilbert's work. In most cases, the Editors are not directly acquainted with the documents themselves.

SS 1891: Transcription or reworking by Julius Hurwitz of Hilbert's lectures on projective geometry with the title 'Geometrie der Lage', held in the library of the Eidgenössische Technische Hochschule in Zürich, Handschriftenabteilung, *Nachlaß* of Adolf Hurwitz, call sign 'Hs. 582: 158'.

SS 1895: Manuscript of notes on the lectures on elliptic functions bearing the title 'Theorie der elliptischen Functionen'; pp. 314–336 appear to have been added (in pencil) in 1942. Kept in the Rare Book and Special Collections section of the Library of the University of Illinois at Urbana-Champaign under the call-sign '515.983 H54T'.

WS 1895/96: Manuscript of notes on the lectures on partial differential equations bearing the title 'Partielle Differentialgleichungen'. Kept in the Rare Book and Special Collections section of the Library of the University of Illinois at Urbana-Champaign under the call-sign '515.353 H54P'.

WS 1896/97: Manuscript of notes on the lectures on the theory of functions of a complex variable bearing the title 'Theorie der Functionen einer complexen Variablen', apparently in two parts of 271 and 105 pages respectively. (Pp. 241–271 and pp. 104–105 appear to have been added in pencil in 1942.) Kept in the Rare Book and Special Collections section of the Library of the University of Illinois at Urbana-Champaign under the call-sign '515.93 H54T'.[66]

[66] The two parts might well correspond to the two courses 'Theorie der Funktionen einer complexen Veränderlichen' in WS 1896/97 and 'Theorie der Funktionen einer complexen Veränderlichen, II Teil' in SS 1897.

WS 1898/99: 3 copies of the autograph of the *Ausarbeitung* prepared by Hans von Schaper of Hilbert's lectures on the foundations of Euclidean geometry from 1898/99. These copies exist in university libraries in Berlin, Bremen and Oslo. See the Introduction to Chapter 4. In addition, the Widener Library at Harvard University in Cambridge, Massachussetts, possesses a copy of this autograph made in 1902.

WS 1904/05: Manuscript of notes on the lectures on the calculus of variations bearing the title 'Variations-rechnung'. The fly-leaf bears the autograph 'Arthur R. Crathorne', and there is a note on the fly-leaf which reads: 'Up to page 68 where the red lines are drawn across the pages, these notes are as I took them during the lectures, with additions in pencil on the opposite page. From about page 69 the notes are copied from the Hefts in the Lesezimmer'. Kept in the Mathematics Section of the Library of the University of Illinois at Urbana-Champaign under the call-sign '515.64 H54V'.[67]

SS 1905: Notes and an incomplete *Ausarbeitung* by Otto Birck for the lecture course 'Logische Prinzipien des mathematischen Denkens'. Kept under this title in the *Universitäsarchiv* of the University of Bonn.

WS 1905/06: A copy of Hellinger's *Ausarbeitung* for the lecture course 'Mechanik', in the library of the Mechanics Centre of the Technische Universität Braunschweig.

WS 1906/07: 6 copies of Hellinger's *Ausarbeitung* for the lecture course 'Mechanik der Continua', in the library of the Mechanics Centre of the Technische Universität Braunschweig.

WS 1910/11: A copy of Frankfurter's *Ausarbeitung* for the lecture course 'Mechanik', in the library of the Mechanics Centre of the Technische Universität Braunschweig.

SS 1913: A copy of Hecke's *Ausarbeitung* for the lecture course 'Elektronentheorie', in the library of the Fakkultät für Mathematik und Informatik of the University of Heidelberg.

WS 1913/14: 2 copies of Hecke's *Ausarbeitung* for the lecture course 'Elektromagnetische Schwingungen', in the library of the Fakkultät für Mathematik und Informatik of the University of Heidelberg.

WS 1914/15: *Ausarbeitung* by an unknown hand of the lecture course 'Probleme und Prinzipienfragen der Mathematik', held under this title in the library of the Institute for Advanced Study in Princeton.

WS 1917/18: Notes by Paul Bernays in (Gabelsberger shorthand) relating to the lecture course 'Prinzipien der Mathematik', in the Archives of the Eidgenössiche Technische Hochschule in Zürich (call-sign: Hs. 973.184).

---

[67]The Library of the University of Illinois at Urbana-Champaign also contains two documents listed under Hilbert which are possibly of relevance here, bearing the titles 'Mechanik' (531 H54M), dated as 1905, and 'Mechanik der Continua' (532.5 H54M), no date listed. Cf. the second document listed above in connection with the lecture course 'Mechanik der Continua', namely 'Vorlesung über die Mechanik der Continua' (Marshall/Crathorne).

WS 1921/22: Notes taken by Hellmuth Kneser for the lecture course 'Grundlegung der Mathematik' with the abbreviated title 'Grundl. d. Math.'; in the possession of Martin Kneser.

Incomplete notes for the same lecture course by an unknown hand, in the Archiv of the University of Bonn under the title 'Grundlegung der Mathematik'.

WS 1922/23: Notes taken by Hellmuth Kneser for the lecture course announced as 'Grundlagen der Arithmetik'; in the possession of Martin Kneser.

WS 1923/24: Notes by Hellmuth Kneser relating to lectures by Hilbert on the 21st, 25th and 28th February, 1924, probably from the course 'Logische Grundlagen der Mathematik';[68] in the possession of Martin Kneser.

[68]The dates named fall on Thursday, Monday and again Thursday, the weekdays on which Hilbert in these years ordinarily held his four hour lectures. These were the days for which the lecture course 'Mengenlehre' was also announced, which makes it likely that this latter course did not take place.

Bibliography

In the bibliographical details which follow, the author's name or parts of the title enclosed in (round brackets) signify that the information does not appear on the title page of the book or essay, but elsewhere in the work, and its enclosure in [square brackets] signifies that it has been supplied by the Editors. ⟨Angled brackets⟩ are used for parentheses which are part of the original title. Entries marked with an asterisk refer to unpublished items, either from Hilbert's *Nachlaß*, or from the Mathematisches Institut of Göttingen University.

Adler 1890

August Adler: *Über die zur Ausführung geometrischer Constructionsaufgaben zweiten Grades nothwendigen Hilfsmittel.* ⟨Vorgelegt in der Sitzung vom 9. October 1890.⟩ – Sitzungsberichte der Mathematisch-Naturwissenschaftlichen Classe der Kaiserlichen Akademie der Wissenschaften. Abtheilung II. a 99 (1890), 846–859, 1 table.

d' Alembert 1759

(Jean Le Rond d'Alembert): *Mélanges de Littérature, d'Histoire, et de Philosophie. Nouvelle édition, revue, corrigée & augmentée très-considérablement. Tome premier; Tome second; Tome troisieme; Tome quatrieme.* – Amsterdam: Chatelain 1759.

d' Alembert 1767

(Jean Le Rond d'Alembert): *Mélanges de Littérature, d'Histoire, et de Philosophie. Tome cinquieme.* – Amsterdam: Chatelain 1767.

Artin 1957

E. Artin: *Geometric algebra.* – New York, London: Interscience (1957). x, 214 pages.

Bachmann 1872

Paul Bachmann: *Die Lehre von der Kreistheilung und ihre Beziehungen zur Zahlentheorie. Academische Vorlesungen.* [Zahlentheorie. Versuch einer Gesammtdarstellung dieser Wissenschaft in ihren Haupttheilen. Dritter Theil.] – Leipzig: Teubner 1872. XII, 299 pages, 1 table.

Baldus 1928

Richard Baldus: *Zur Axiomatik der Geometrie. I. Über Hilberts Vollständigkeitsaxiom.* – Mathematische Annalen 100 (1928), 321–333.

Baltzer 1868

Richard Baltzer: *Die Elemente der Mathematik. Erster Band. Gemeine Arithmetik, Allgemeine Arithmetik, Algebra. Dritte verbesserte Auflage.* – Leipzig: Hirzel 1868. VI, 289 pages.

Baltzer 1870

Richard Baltzer: *Die Elemente der Mathematik. Zweiter Band. Planimetrie, Stereometrie, Trigonometrie. Dritte verbesserte Auflage.* – Leipzig: Hirzel 1870. VIII, 385 pages.

Baltzer 1871

R. Baltzer: *Ueber die Hypothese der Parallelentheorie.* – Journal für die reine und angewandte Mathematik 73 (1871), 372f.

Baltzer 1879

Richard Baltzer: *Die Elemente der Mathematik. Erster Band. Gemeine Arithmetik, Allgemeine Arithmetik, Algebra. Sechste verbesserte Auflage.* – Leipzig: Hirzel 1879. VI, 387 pages.

Baltzer 1883

Richard Baltzer: *Die Elemente der Mathematik. Zweiter Band. Planimetrie, Stereometrie, Trigonometrie. Sechste verbesserte Auflage.* – Leipzig: Hirzel 1883. VIII, 388 pages.

Beltrami 1868

E. Beltrami: *Saggio di interpetrazione della geometria non-Euclidea.* – Giornale di matematiche ad uso degli studenti delle università italiane 6 (1868) [printed: 1869], 285–315. Reprinted in *Opere Matematiche* 1 (1902), 374–405.

Beltrami 1869

E. Beltrami: *Essai d'interprétation de la géométrie non euclidienne.* Partial Translation by J. Hoüel of *Beltrami 1868*. – Annales scientifique de l'École Normale Supérieure 6 (1869), 251–288.

Bernays 1922*

Paul Bernays: *Logische Grundlagen der Mathematik.* – Niedersächsische Staats- und Universitätsbibliothek Göttingen, Handschriftenabteilung, *Cod. Ms. D. Hilbert* 567.

Bernays 1955

Paul Bernays: *Betrachtungen über das Vollständigkeitsaxiom und verwandte Axiome.* – Mathematische Zeitschrift 63 (1955), 217–229.

Blumenthal 1922

Otto Blumenthal: *David Hilbert.* – Die Naturwissenschaften, Wochenschrift für die Fortschritte der Naturwissenschaft, der Medizin und der Technik 10 (1922), 67–72.

Blumenthal 1935

Otto Blumenthal: *Lebensgeschichte [David Hilberts].* – In *Hilbert 1935*, 388–429.

du Bois-Reymond 1882

Paul du Bois-Reymond: *Die allgemeine Functionentheorie. Erster Theil. Metaphysik und Theorie der mathematischen Grundbegriffe: Grösse, Grenze, Argument und Function.* – Tübingen: Laupp 1882. XIV, 292 pages.

Boltianskii 1978

Vladimir G. Boltianskii: *Hilbert's third problem. Translated by Richard A. Silverman and introduced by Albert B. Novikoff.* – New York [and others]: Wiley (1978). x, 228 pages.

Boltzmann 1892

L. Boltzmann: *Über die Methoden der theoretischen Physik.* – In *Dyck 1893*, 89–98.

Bolyai 1832a

Johann Bolyai: *Appendix, Scientiam spatii absolute veram exhibens: a veritate aut falsitate axiomatis XI Euclidei (a priori haud unquam decidenda) Independentem; adjecta ad casum falsitatis, quadratura circuli geometrica.* Auctore Johanne Bolyai in eadem, Geometrarum in Exercitu Caesareo Regio Austriaco Castrensium Capitaneo.. – In *Bolyai 1832b*, Appendix.

Bolyai 1832b

Wolfgang Bolyai: *Tentamen juventutem studiosam in elementa matheseos purae, elementaris ac sublimioris, methodo intuitiva, evidentiaque huic propria introducendi. Tomus Primus.* Cum Appendice Triplici. – : 1832. 2 uncounted, LVIII, 502 pages, [Appendix:] 2 uncounted, 26, 2 uncounted pages, 4 tables.

Bonola 1908

Roberto Bonola: *Die nichteuklidische Geometrie. Historisch-kritische Darstellung ihrer Entwicklung.* German translation of the Italian original. – Leipzig, Berlin: Teubner 1908. (Wissenschaft und Hypothese; IV.) VIII, 245 pages.

Bonola 1912

Roberto Bonola: *Non-Euclidean Geometry. A Critical and Historical Study of its Development.* Authorised English Translation with Additional Appendices by H.S. Carslaw. With an Introduction by Frederigo Enriques. – Chicago: Open Court 1912. XII, 268 pages.

Busemann 1976

Herbert Busemann: *Problem IV: Desarguesian spaces.* – In *Developments 1976*, 131–141.

Cantor 1872

Georg Cantor: *Über die Ausdehnung eines Satzes aus der Theorie der trigonometrischen Reihen.* – Mathematische Annalen 5 (1872), 123–132.

Cantor 1883

Georg Cantor: *Grundlagen einer allgemeinen Mannichfaltigkeitslehre.* Ein mathematisch-philosophischer Versuch in der Lehre des Unendlichen. – Leipzig: 1883.

Cantor 1895

Georg Cantor: *Beiträge zur Begründung der transfiniten Mengenlehre. ⟨Erster Artikel.⟩* – Mathematische Annalen 46 (1895), 481–512.

Cantor 1897

Georg Cantor: *Beiträge zur Begründung der transfiniten Mengenlehre. ⟨Zweiter Artikel.⟩* – Mathematische Annalen 49 (1897), 207–246.

Carnap and Bachmann 1936

Rudolf Carnap and Friedrich Bachmann: *Über Extremalaxiome.* – Erkenntnis 6 (1936), 166–188.

Cayley 1859

Arthur Cayley: *A sixth memoir on quantics.* – Philosophical Transactions of the Royal Society of London CXLIX (1859), 61–90. Reprinted in *Cayley 1889*, 561–606.

Cayley 1889

Arthur Cayley: *Collected mathematical papers. Vol. II.* – Cambridge: University Press 1889.

Clavius 1574

Christophorus Clavius: *Euclidis Elementorum Libri XV. Accessit XVI. De Solidorum Regularium comparatione.* – Romae: 1574.

Clebsch 1891

Alfred Clebsch: *Vorlesungen über Geometrie unter besonderer Benutzung der Vorträge von Alfred Clebsch bearbeitet von Ferdinand Lindemann. Zweiten Bandes erster Theil. Die Flächen erster und zweiter Ordnung oder Klasse und der lineare Complex.* – Leipzig: Teubner 1891. VIII, 650 pages.

Congresso 1928

Congresso internazionale dei matematici. Bologna, 3–10 Settembre 1928 ⟨VI⟩. Tomo I. Rendiconto del congresso conferenze. – Bologna: Nicola Zanichelli (1929).

Congrès 1900

Compte rendu du deuxième congrès international des mathématiciens tenu à Paris du 6 au 12 août 1900. Procès-verbaux et communications. – Paris: Gauthier-Villars 1902. 4 uncounted, 455 pages.

Coolidge 1916

Julian Lowell Coolidge: *A treatise on the circle and the sphere.* – Oxford: Clarendon 1916. 603 pages.

Coxeter 1993

H. S. M. Coxeter: *The real projective plane.* – New York [and others]: Springer (1993). xiv, 222 pages.

Dedekind 1872

Richard Dedekind: *Stetigkeit und irrationale Zahlen.* – Braunschweig: Vieweg 1872. 31 pages. Reprinted in *Dedekind 1932*, 315–332.

Dedekind 1888

Richard Dedekind: *Was sind und was sollen die Zahlen?* – Braunschweig: Vieweg 1888. XVIII, 58 pages. Reprinted in *Dedekind 1932*, 335–341, 344–390.

Dedekind 1932

Richard Dedekind: *Gesammelte mathematische Werke. Herausgegeben von Robert Fricke in Braunschweig, Emmy Noether in Göttingen, Öystein Ore in New Haven. Dritter Band.* – Braunschweig: Vieweg 1932. IV, 508 pages.

Dehn 1900a

Max Dehn: *Die Legendre'schen Sätze über die Winkelsumme im Dreieck.* – Georg-August-Universität zu Göttingen, Phil. Diss. 1900.

Dehn 1900b

Max Dehn: *Die Legendre'schen Sätze über die Winkelsumme im Dreieck.* – Mathematische Annalen 53 (1900), 404–439.

Dehn 1900c

M Dehn: *Ueber raumgleiche Polyeder.* – Nachrichten von der Königlichen Gesellschaft der Wissenschaften zu Göttingen. Mathematisch-physikalische Klasse (1900), 345–354.

Dehn 1902

M. Dehn: *Ueber den Rauminhalt.* – Mathematische Annalen 55 (1902), 465–478.

Dehn 1905

M. Dehn: *Über den Inhalt sphärischer Dreiecke.* – Mathematische Annalen 60 (1905), 166–174.

Descartes 1637

René Descartes: *Discours de la methode pour bien conduire sa raison, & chercher la verité dans les sciences. Plus la dioptrique. Les meteores. Et la geometrie. Qui sont des essais de cete methode.* – Leyde: Ian Maire 1637. 413, 34 uncounted pages.

Developments 1976

Felix Browder [ed.]: *Mathematical developments arising from Hilbert problems.* – Providence: American Mathematical Society 1976. (Proceedings of Symposia in Pure Mathematics; 28). xii, 628 pages.

Dirichlet 1894

P. G. Lejeune Dirichlet: *Vorlesungen über Zahlentheorie. Herausgegeben und mit Zusätzen versehen von R. Dedekind, Professor an der technischen Hochschule Carolo-Wilhelmina zu Braunschweig. Vierte umgearbeitete und vermehrte Auflage.* – Braunschweig: Vieweg 1894. XVIII, 657 pages.

Dyck 1892

Walter Dyck [ed.]: *Katalog mathematischer und mathematisch-physikalischer Modelle, Apparate und Instrumente.* – München: Wolf & Sohn 1892. XVI, 430 pages.

Dyck 1893

Walter Dyck [ed.]: *Katalog mathematischer und mathematisch-physikalischer Modelle, Apparate und Instrumente. Nachtrag.* – München: Wolf & Sohn 1893. X, 135 pages.

Encyklopädie 1899–1916

Encyklopädie der mathematischen Wissenschaften mit Einschluss ihrer Anwendungen. Zweiter Band in drei Teilen. Analysis. Erster Teil. Erste Hälfte. – Leipzig: Teubner 1899–1916. XXII, 694 pages.

Encyklopädie 1907–10

Encyklopädie der mathematischen Wissenschaften mit Einschluss ihrer Anwendungen. Dritter Band in drei Teilen. Geometrie. Erster Teil. Erste Hälfte. – Leipzig: Teubner 1907–10. XXII, 770 pages.

Enriques 1907

F. Enriques: *Prinzipien der Geometrie.* – In *Encyklopädie 1907–10*, 1–129.

Erdmann 1877

Benno Erdmann: *Die Axiome der Geometrie. Eine philosophische Untersuchung der Riemann-Helmholtz'schen Raumtheorie.* – Leipzig: Voss 1877. X, 174 pages.

Euclid 1883

Euclid: *Elementa. Edidit et latine interpretatus est I. L. Heiberg. Uol. I. Libros I–IV continens. (Euclidis Opera omnia. Ediderunt I. L. Heiberg et H. Menge. [I].)* – Lipsiae: Teubner 1883. X, 333 pages.

Feldblum 1899

Michael Feldblum: *Ueber elementar-geometrische Constructionen.* – Georg-August-Universität zu Göttingen, Phil. Diss. 1899.

Festschrift 1899

Festschrift zur Feier der Enthüllung des Gauss-Weber-Denkmals in Göttingen. Herausgegeben von dem Fest-Comitee. – Leipzig: Teubner 1899. 2 uncounted, 92, 112 pages.

Festschrift 1901

Festschrift zur Feier des hundertfünzigjährigen Bestehens der Königlichen Gesellschaft der Wissenschaften zu Göttingen. Beiträge zur Gelehrtengeschichte Göttingens. – Berlin: Weidmannsche Buchhandlung 1901. 1 Titelbild, 4 uncounted, 688 pages, 13 tables.

Fraenkel 1928

Adolf Fraenkel: *Einleitung in die Mengenlehre.* Dritte umgearbeitete und stark erweiterte Auflage. – Berlin: Springer 1928. (Die Grundlehren der mathematischen Wissenschaften in Einzeldarstellungen mit besonderer Berücksichtigung der Anwendungsgebiete.) XIV, 424 pages.

Frege 1903a

G. Frege: *Über die Grundlagen der Geometrie.* – Jahresbericht der Deutschen Mathematiker Vereinigung 12 (1903), 319–324.

Frege 1903b

G. Frege: *Über die Grundlagen der Geometrie. II.* – Jahresbericht der Deutschen Mathematiker Vereinigung 12 (1903), 368–375.

Frege 1906

G. Frege: *Über die Grundlagen der Geometrie.* – Jahresbericht der Deutschen Mathematiker-Vereinigung 15 (1906), 293–309; 377–403; 423–430.

Frege 1976

Gottlob Frege: *Wissenschaftlicher Briefwechsel. Herausgegeben, bearbeitet, eingeleitet und mit Anmerkungen versehen von Gottfried Gabriel, Hans Hermes, Friedrich Kambartel, Christian Thiel, Albert Veraart.* (Nachgelassene Schriften und Wissenschaftlicher Briefwechsel. Zweiter Band.) – Hamburg: Meiner 1976. XXVIII, 310 pages.

Frege 1980

Gottlob Frege: *Philosophical and Mathematical Correspondence.* Edited by Gottfried Gabriel [and others]. Abridged for the English edition by Brian McGuinness and translated by Hans Kaal. – Oxford: Blackwell (1980). xviii, 2 uncounted, 214 pages.

Freudenthal 1957

Hans Freudenthal: *Zur Geschichte der Grundlagen der Geometrie. Zugleich eine Besprechung der 8. Aufl. von Hilberts „Grundlagen der Geometrie".* – Nieuw Archief voor Wiskunde. Derde Serie 5 (1957), 105–142.

Frewer 1979

Magdalene Frewer: *Das mathematische Lesezimmer der Universität Göttingen unter der Leitung von Felix Klein ⟨1886–1922⟩.* Hausarbeit zur Prüfung für den höheren Bibliotheksdienst. Bibliothekar-Lehrinstitut des Landes Nordrhein-Westfalen, Köln 1979.

Fricke and Klein 1897

Robert Fricke and Felix Klein: *Vorlesungen über die Theorie der automorphen Functionen. Erster Band: Die gruppentheoretischen Grundlagen.* – Leipzig: Teubner 1897. XIV, 634 pages.

Frischauf 1872

J. Frischauf: *Absolute Geometrie nach Johann Bolyai.* – Leipzig: Teubner 1872. XII, 96 pages.

Frischauf 1876

J. Frischauf: *Elemente der absoluten Geometrie.* – Leipzig: Teubner 1876. VI, 142 pages.

Gauß 1801

Carolus Fridericus Gauss: *Disquisitiones arithmeticae.* – Lipsiae: Fleischer 1801. XVIII, 668 pages. Reprinted in *Gauß 1863.*

Gauß 1808

C. F. Gauß [announcement]: *Summatio quarumdam serierum singularium.* – Göttingische gelehrte Anzeigen (1808), 151. Stück, 1505–1509.

Gauß 1811

Carolus Fridericus Gauss: *Summatio quarumdam serierum singularium.* – Commentationes societatis regiae scientiarum Gottingensis, Classis mathematicae 1 (MDCCCVIII–XI) [printed: MDCCCXI]. 40 pages.

Gauß 1816

[Carl Friedrich Gauß] [review]: *[Review of Schwab 1814, Metternich 1815].* – Göttingische gelehrte Anzeigen (1816), 63. Stück, 617–622. Reprinted in *Gauß 1873*, 364–368; *Gauß 1900*, 170–174.

Gauß et al. 1860–1865

C. F. Gauss and H. C. Schumacher: *Briefwechsel zwischen C. F. Gauss und H. C. Schumacher. Hrsg. von C. A. F. Peters. 6 Bände.* – Altona: Esch 1860–1865.

Gauß 1863

Carl Friedrich Gauss: *Werke. Erster Band. Herausgegeben von der Königlichen Gesellschaft der Wissenschaften zu Göttingen.* – (Göttingen): (Dieterich) 1863. 4 uncounted, 478 pages.

Gauß 1873

Carl Friedrich Gauss: *Werke. Vierter Band. Herausgegeben von der Königlichen Gesellschaft der Wissenschaften zu Göttingen.* – (Göttingen): (Dieterich) 1873. 2 uncounted, 492 pages.

Gauß et al. 1880

C. F. Gauss and F. W. Bessel: *Briefwechsel. Herausgegeben auf Veranlassung der königlich Preussischen Akademie der Wissenschaften.* – Leipzig: Engelmann 1880. XXVI, 598 pages.

Gauß et al. 1899

Carl Friedrich Gauss and Wolfgang Bolyai: *Briefwechsel. Mit Unterstützung der ungarischen Akademie der Wissenschaften herausgegeben von Franz Schmidt und Paul Stäckel.* – Leipzig: Teubner 1899. XVI, 208 pages.

Gauß 1900

Carl Friedrich Gauss: *Werke. Achter Band. (Arithmetik und Algebra. Nachträge zu Band I–III.)* – Leipzig: Teubner 1900. 2 uncounted, 458 pages.

Gauß 1901

Carl Friedrich Gauss: *Gauss' wissenschaftliches Tagebuch 1796–1814. Mit Anmerkungen herausgegeben von Felix Klein. Hierzu ein Portrait von Gauss aus dem Jahre 1803 und ein Facsimile.* – In *Festschrift 1901*, 1–44.

Gauß 1917

Carl Friedrich Gauss: *Werke. Zehnten Bandes erste Abteilung. Herausgegeben von der Königlichen Gesellschaft der Wissenschaften zu Göttingen.* – Göttingen: Teubner 1917. 2 uncounted, 586 pages.

Gauß 1922–1933

Carl Friedrich Gauss: *Werke. Zehnten Bandes zweite Abteilung. Herausgegeben von der Gesellschaft der Wissenschaften zu Göttingen.* – Berlin: Springer 1922–1933.

Gérard 1892

L. Gérard: *Sur la géométrie non euclidienne.* – Paris: Gauthier-Villars 1892. (Thèses de la Faculté des sciences de Paris; 768.) 111 pages.

Gérard 1895a

L. Gérard: *Sur le postulat relatif à l'équivalence des polygones, considéré comme corrolaire du théorème de Varignon.* – Bulletin de la Société mathématique de France 23 (1895), 268–269.

Gérard 1895b

L. Gérard: *Volume d'un polyèdre.* – Bulletin de mathématiques spéciales 1 (1894/95), 127–128.

Gérard 1896

L. Gérard: *Sur la mesure des polygones.* – Bulletin de mathématiques élémentaires 1 (1895/96), 100–102.

Gérard 1897

L. Gérard: *Sur l'équivalence.* – Bulletin de mathématiques élémentaires 2 (1896/97), 273–276.

Gérard 1898

L. Gérard: *Sur la notion d'aire et de volume.* – Bulletin de Sciences Mathématiques et Physiques élémentaires 3 (1897/98), 257–258.

Grünwald 1900

Josef Grünwald: *Lineare Lösung der Aufgaben über das Verbinden und Schneiden imaginärer Punkte, Geraden und Ebenen.* – Zeitschrift für Mathematik und Physik 45 (1900), 10–22.

Halsted 1902a

George Bruce Halsted [review]: *The Foundations of Geometry. By D. Hilbert.* – Science 16 (1902), 307–308.

Halsted 1902b

George Bruce Halsted [review]: *[Hilbert:] The foundations of geometry.* – The Open Court 16 (1902), 513–521.

Hamel 1901

Georg Hamel: *Ueber die Geometrieen.* – Georg-August-Universität zu Göttingen, Phil. Diss. 1901.

Hamel 1903

Georg Hamel: *Über die Geometrieen, in denen die Geraden die Kürzesten sind.* – Mathematische Annalen 57 (1903), 231–264.

Hankel 1875

Hermann Hankel: *Die Elemente der projectivischen Geometrie in synthetischer Behandlung. Vorlesungen.* – Leipzig: Teubner 1875. VIII, 256 pages.

Hartshorne 2000

Robin Hartshorne: *Geometry: Euclid and Beyond.* – New York, Berlin, Heidelberg: Springer (2000). (Undergraduate Texts in Mathematics). xii, 526 pages.

Heath 1926a

Thomas Heath: *The Thirteen Books of Euclid's Elements. Translated from the text of Heiberg. With introduction and commentary. Volume I: Introduction and Books I, II.* – Cambridge: University Press 1926. xii, 432 pages.

Heath 1926b

Thomas L. Heath: *The Thirteen Books of Euclid's Elements. Translated from the text of Heiberg. With introduction and commentary. Volume II: Books III–IX.* – Cambridge: University Press 1926. 6 uncounted, 436 pages.

Heath 1926c

Thomas L. Heath: *The Thirteen Books of Euclid's Elements. Translated from the text of Heiberg. With introduction and commentary. Volume III: Books X–XIII and Appendix.* – Cambridge: University Press 1926. 6 uncounted, 546 pages.

Hedrick 1903

(E. R. Hedrick) [review]: *The English and French translations of Hilbert's Grundlagen der Geometrie.* – Bulletin of the American Mathematical Society 9 (1903), 158–165.

Helmholtz 1868

H. Helmholtz: *Ueber die Thatsachen, die der Geometrie zum Grunde liegen.* – Nachrichten von der K. Gesellschaft der Wissenschaften und der Georg-Augusts-UniversitätVerhandlungen des Naturhistorisch medizinischen Vereins zu Heidelberg (1868), 193–221. Reprinted in H. von Helmholtz: *Wissenschaftliche Abhandlungen.* Leipzig: J. Barth 1883, 618–639.

Helmholtz 1876a

H. Helmholtz: *Über den Ursprung und die Bedeutung der geometrischen Axiome. Vortrag, gehalten im Docentenverein zu Heidelberg im Jahre 1870.* – In *Helmholtz 1876b*, 21–54.

Helmholtz 1876b

H. Helmholtz: *Populäre wissenschaftliche Vorträge. Drittes Heft.* – Braunschweig: Vieweg 1876. XII, 139 pages.

von Helmholtz 1896

Hermann von Helmholtz: *Vorträge und Reden. Zweiter Band.* – Braunschweig: Vieweg 1896. XII, 434 pages.

Hertz 1894

Heinrich Hertz: *Die Prinzipien der Mechanik. (Gesammelte Werke. Band III.)* In neuem Zusammenhange dargestellt. Mit einem Vorworte von H. von Helmholtz. – Leipzig: Barth ⟨Meiner⟩ 1894. XXX, 312 pages.

Hessenberg 1905a

Gerhard Hessenberg: *Begründung der elliptischen Geometrie.* – Mathematische Annalen 61 (1905), 173–184.

Hessenberg 1905b

Gerhard Hessenberg: *Beweis des Desarguesschen Satzes aus dem Pascalschen.* – Mathematische Annalen 61 (1905), 161–172.

Hilbert 1890*

David Hilbert: *Krumme Linien und Flächen.* – Niedersächsische Staats- und Universitätsbibliothek Göttingen, Handschriftenabteilung, *Cod. Ms. D. Hilbert* 532.

Hilbert 1891a*

David Hilbert: *Projektive Geometrie.* – Niedersächsische Staats- und Universitätsbibliothek Göttingen, Handschriftenabteilung, *Cod. Ms. D. Hilbert* 535.

Hilbert 1891b

David Hilbert: *Ueber die stetige Abbildung einer Linie auf ein Flächenstück.* – Mathematische Annalen 38 (1891), 459–460.

Hilbert 1892*

David Hilbert: *Eindeutige Funktionen mit linearen Transformationen in sich.* – Niedersächsische Staats- und Universitätsbibliothek Göttingen, Handschriftenabteilung, *Cod. Ms. D. Hilbert* 537.

Hilbert 1893a

David Hilbert: *Über ternäre definite Formen.* – Acta Mathematica 17 (1893), 169–197.

Hilbert 1893b

David Hilbert: *Ueber die Transcendenz der Zahlen e und π.* – Mathematische Annalen 43 (1893), 216–219.

Hilbert 1894a*

David Hilbert: *Grundlagen der Geometrie.* – Niedersächsische Staats- und Universitätsbibliothek Göttingen, Handschriftenabteilung, *Cod. Ms. D. Hilbert* 541.

Hilbert 1894b*

David Hilbert: *Quadratur des Kreises.* – Niedersächsische Staats- und Universitätsbibliothek Göttingen, Handschriftenabteilung, *Cod. Ms. D. Hilbert* 542.

Hilbert 1895a*

David Hilbert: *Rede über meinen Bericht, gehalten in Lübeck 1895.* – Niedersächsische Staats- und Universitätsbibliothek Göttingen, Handschriftenabteilung, *Cod. Ms. D. Hilbert* 598.

Hilbert 1895b

David Hilbert: *Ueber die gerade Linie als kürzeste Verbindung zweier Punkte. ⟨Aus einem an Herrn F. Klein gerichteten Briefe.⟩* – Mathematische Annalen 46 (1895), 91–96.

Hilbert 1896*

David Hilbert: *Feriencursus Ostern 1896.* – Niedersächsische Staats- und Universitätsbibliothek Göttingen, Handschriftenabteilung, *Cod. Ms. D. Hilbert* 597, f. 0r–9r.

Hilbert 1897a

David Hilbert: *Die Theorie der algebraischen Zahlkörper.* Bericht, erstattet der Deutschen Mathematiker-Vereinigung. – Jahresbericht der Deutschen Mathematiker-Vereinigung 4 (1894/95) [printed: 1897], 175–546. Reprinted in *Hilbert 1932*, 63–363.

Hilbert 1897b*

David Hilbert: *Zahlbegriff und Quadratur des Kreises.* – Niedersächsische Staats- und Universitätsbibliothek Göttingen, Handschriftenabteilung, *Cod. Ms. D. Hilbert* 549.

Hilbert 1897c

David Hilbert: *Zum Gedächtnis an Karl Weierstraß.* – Nachrichten von der Königlichen Gesellschaft der Wissenschaften zu Göttingen. Geschäftliche Mitteilungen (1897), 60–69.

Hilbert 1898a

David Hilbert: *Über die Theorie der relativquadratischen Zahlkörper.* – Jahresbericht der Deutschen Mathematiker-Vereinigung 6 (1897) [printed: 1898], Erstes Heft, 88–94. Reprinted in *Hilbert 1932*, 364–369.

Hilbert 1898b*

David Hilbert: *Fereiencursus: Uber den Begriff des Unendlichen. Ostern 1898.* – Niedersächsische Staats- und Universitätsbibliothek Göttingen, Handschriftenabteilung, *Cod. Ms. D. Hilbert* 597, f. 15r–29r.

Hilbert 1898c*

David Hilbert: *Grundlagen der Euklidischen Geometrie.* – Niedersächsische Staats- und Universitätsbibliothek Göttingen, Handschriftenabteilung, *Cod. Ms. D. Hilbert* 551.

Hilbert 1898d

David Hilbert: *Ueber die Theorie der relativ-Abel'schen Zahlkörper.* Vorgelegt in der Sitzung am 10. Dezember 1898. – Nachrichten von der Königl. Gesellschaft der Wissenschaften zu Göttingen. Mathematisch-physikalische Klasse (1898), 370–399. Reprinted in *Hilbert 1932*, 483–509.

Hilbert 1899a*

David Hilbert: *Elemente der Euklidischen Geometrie.* – Niedersächsische Staats- und Universitätsbibliothek Göttingen, Handschriftenabteilung, *Cod. Ms. D. Hilbert* 552.

Hilbert 1899b*

David Hilbert: *Elemente der euklidischen Geometrie.* – Georg-August-Universität Göttingen, Mathematisches Institut, Lesesaal Inv. Nr. 6808.

Hilbert 1899c

David Hilbert: *Grundlagen der Geometrie.* – In *Festschrift 1899*. 92 pages.

Hilbert 1899d

David Hilbert: *Ueber die Theorie des relativquadratischen Zahlkörpers.* – Mathematische Annalen 51 (1899), 1–127. Reprinted in *Hilbert 1932*, 370–482.

Hilbert 1900a

David Hilbert: *Über das Dirichlet'sche Princip.* – Jahresbericht der Deutschen Mathematiker-Vereinigung 8 (1900), 184–188.

Hilbert 1900b

David Hilbert: *Über den Zahlbegriff.* – Jahresbericht der Deutschen Mathematiker-Vereinigung 8 (1900), 180–184.

Hilbert 1900c

D. Hilbert: *Les principes fondamentaux de la géométrie. Festschrift publiée à l'occasion des fêtes pour l'inauguration du monument Gauss-Weber à Göttingen. Publiée par les soins du Comité des fêtes.* Leipzig: Teubner 1899. Authorised translation of *Hilbert 1899c* by Léonce Laugel. – Annales scientifiques de l'école normale supérieure. Troisième série 17 (1900), 103–209.

Hilbert 1900d

D. Hilbert: *Les principes fondamentaux de la géométrie. Traduit par L. Laugel.* Unaltered Reprint of *Hilbert 1900c*. – Paris: Gauthier-Villars 1900. 114 pages.

Hilbert 1900e

David Hilbert: *Mathematische Probleme.* Vortrag, gehalten auf dem internationalen Mathematiker-Kongreß zu Paris 1900. – Nachrichten von der Königl. Gesellschaft der Wissenschaften zu Göttingen. Mathematisch-physikalische Klasse (1900), 253–297.

Hilbert 1900f

David Hilbert: *Sur le principe de Dirichlet.* – Nouvelles annales de mathématiques. Journal des candidats aux écoles polytechniques et normales (série 3) 19 (1900), 337–344. Translation of *Hilbert 1900a.*

Hilbert 1901

David Hilbert: *Ueber Flächen von constanter Gaussscher Krümmung.* – Transactions of the American Mathematical Society 2 (1901), 87–99.

Hilbert 1902a

David Hilbert: *Über die Theorie der relativ-Abel'schen Zahlkörper.* – Acta Mathematica 26 (1902), 99–131.

Hilbert 1902b*

David Hilbert: *Feriencurs. Ostern 1902.*. – Niedersächsische Staats- und Universitätsbibliothek Göttingen, Handschriftenabteilung, *Cod. Ms. D. Hilbert* 597, f. 9v–11v.

Hilbert 1902c

David Hilbert: *The Foundations of Geometry.* Authorised translation by E. J. Townsend of *Hilbert 1899c.* – Chicago: The Open Court Publishing Company 1902. viii, 143 pages.

Hilbert 1902d*

David Hilbert: *Grundlagen der Geometrie.* – Georg-August-Universität Göttingen, Mathematisches Institut, Lesesaal.

Hilbert 1902e

David Hilbert: *Sur les problèmes futurs des mathématiques.* Traduit par L. Laugel. – In *Congrès 1900* [printed: 1902], 58–114.

Hilbert 1902f

David Hilbert: *Ueber die Grundlagen der Geometrie.* Vorgelegt in der Sitzung vom 8. November 1901. – Nachrichten von der Königl. Gesellschaft der Wissenschaften zu Göttingen. Mathematisch-physikalische Klasse (1902), 233–241.

Hilbert 1902/03

David Hilbert: *Über den Satz von der Gleichheit der Basiswinkel im gleichschenkligen Dreieck.* – Proceedings of the London Mathematical Society 35 (1902/03), 50–67.

Hilbert 1903a

David Hilbert: *Grundlagen der Geometrie.* Zweite, durch Zusätze vermehrte und mit fünf Anhängen versehene Auflage. – Leipzig: Teubner 1903. VI, 175 pages.

Hilbert 1903b

David Hilbert: *Neue Begründung der Bolyai-Lobatschefskyschen Geometrie.* – Mathematische Annalen 57 (1903), 137–150.

Hilbert 1903c

David Hilbert: *Ueber die Grundlagen der Geometrie.* – Mathematische Annalen 56 (1903), 381–422.

Hilbert 1904a

David Hilbert: *Über das Dirichletsche Prinzip.* Abdruck aus der Festschrift zur Feier des 150jährigen Bestehens der Königl. Gesellschaft der Wissenschaften zu Göttingen 1901. – Mathematische Annalen 59 (1904), 161–186.

Hilbert 1904b

D. Hilbert: *Über die Grundlagen der Logik und der Arithmetik.* – In *Verhandlungen 1905*, 174–185.

Hilbert 1904c*

David Hilbert: *Bestimmte Integrale und Fouriersche Reihen.* – Georg-August-Universität Göttingen, Mathematisches Institut, Lesesaal.

Hilbert 1905a*

David Hilbert: *Logische Principien des mathematischen Denkens.* – Georg-August-Universität Göttingen, Mathematisches Institut, Lesesaal.

Hilbert 1905b*

David Hilbert: *Mechanik.* – Georg-August-Universität Göttingen, Mathematisches Institut, Lesesaal.

Hilbert 1909

David Hilbert: *Grundlagen der Geometrie.* Dritte, durch Zusätze und Literaturhinweise von Neuem vermehrte und durch sieben Anhänge erweiterte Auflage. – Leipzig, Berlin: Teubner 1909. (Wissenschaft und Hypothese; 7.) VI, 279 pages.

Hilbert 1910*

David Hilbert: *Mechanik.* – Georg-August-Universität Göttingen, Mathematisches Institut, Lesesaal.

Hilbert 1911*

David Hilbert: *Kinetische Gastheorie.* – Georg-August-Universität Göttingen, Mathematisches Institut, Lesesaal.

Hilbert 1912*

David Hilbert: *Statistische Mechanik (gehalten im Oberlehrerkurs 26.-27. April).* – Niedersächsische Staats- und Universitätsbibliothek Göttingen, Handschriftenabteilung, *Cod. Ms. D. Hilbert* 594.

Hilbert 1913

David Hilbert: *Grundlagen der Geometrie.* Vierte, durch Zusätze und Literaturhinweise von neuem vermehrte und mit sieben Anhängen versehene Auflage. – Leipzig, Berlin: Teubner 1913. (Wissenschaft und Hypothese; VII.) VI, 258 pages.

Hilbert 1917*

David Hilbert: *Prinzipien der Mathematik.* – Georg-August-Universität Göttingen, Mathematisches Institut, Lesesaal.

Hilbert 1920*

David Hilbert: *Anschauliche Geometrie.* – Georg-August-Universität Göttingen, Mathematisches Institut, Lesesaal.

Hilbert 1922

David Hilbert: *Grundlagen der Geometrie.* Fünfte, durch Zusätze und Literaturhinweise von Neuem vermehrte und mit sieben Anhängen versehene Auflage. – Leipzig, Berlin: Teubner 1922. (Wissenschaft und Hypothese; 7.) V, 265 pages.

Hilbert 1923

David Hilbert: *Grundlagen der Geometrie.* 6., unveränderte Auflage, anastatischer Nachdruck. – Leipzig, Berlin: Teubner 1923. (Wissenschaft und Hypothese; 7.) V, 265 pages.

Hilbert 1926

David Hilbert: *Über das Unendliche.* – Mathematische Annalen 95 (1926), 161–190.

Hilbert 1927*

David Hilbert: *Grundlagen der Geometrie.* – Georg-August-Universität Göttingen, Mathematisches Institut, Lesesaal.

Hilbert 1928a

David Hilbert: *Die Grundlagen der Mathematik.* – Abhandlungen aus dem Mathematischen Seminar der Hamburgischen Universität 6 (1928), 65–85.

Hilbert and Ackermann 1928

D. Hilbert and W. Ackermann: *Grundzüge der theoretischen Logik*. – Berlin: Springer 1928. (Die Grundlehren der mathematischen Wissenschaften in Einzeldarstellungen mit besonderer Berücksichtigung der Anwendungsgebiete; 27.) VIII, 120 pages.

Hilbert 1928b

D. Hilbert: *Probleme der Grundlegung der Mathematik*. – In *Congresso 1928* [printed: 1929], 135–141.

Hilbert 1930a

David Hilbert: *Grundlagen der Geometrie*. Siebente umgearbeitete und vermehrte Auflage. – Leipzig, Berlin: Teubner 1930. (Wissenschaft und Hypothese; 7.) VII, 326 pages.

Hilbert 1930b

David Hilbert: *Probleme der Grundlegung der Mathematik*. – Mathematische Annalen 102 (1930), 1–9.

Hilbert and Cohn-Vossen 1932

D. Hilbert and S. Cohn-Vossen: *Anschauliche Geometrie*. – Berlin: Springer 1932. (Die Grundlehren der mathematischen Wissenschaften in Einzeldarstellungen mit besonderer Berücksichtigung der Anwendungsgebiete; 37.) VIII, 310 pages.

Hilbert 1932

David Hilbert: *Gesammelte Abhandlungen. Erster Band. Zahlentheorie*. – Berlin: Springer 1932. XIV, 539 pages.

Hilbert 1933

David Hilbert: *Gesammelte Abhandlungen. Zweiter Band. Algebra, Invariantentheorie, Geometrie*. – Berlin: Springer 1933. VIII, 453 pages.

Hilbert 1934*

David Hilbert: *Verzeichniss meiner Vorlesungen*. – Niedersächsische Staats- und Universitätsbibliothek Göttingen, Handschriftenabteilung, *Cod. Ms. D. Hilbert* 520.

Hilbert 1935

David Hilbert: *Gesammelte Abhandlungen. Dritter Band. Analysis, Grundlagen der Mathematik, Physik, Verschiedenes. Nebst einer Lebensgeschichte*. – Berlin: Springer 1935. VIII, 435 pages.

Hilbert and Cohn-Vossen 1952

D. Hilbert and S. Cohn-Vossen: *Geometry and the imagination*. Translated by P. Nemenyi. – New York: Chelsea 1952. X, 357 pages.

Hilbert 1956

David Hilbert: *Grundlagen der Geometrie.* Achte Auflage, mit Revisionen und Ergänzungen von Dr. Paul Bernays. – Stuttgart: Teubner 1956. VII, 251 pages.

Hilbert 1962

David Hilbert: *Grundlagen der Geometrie.* Neunte Auflage, revidiert und ergänzt von Dr. Paul Bernays. – Stuttgart: Teubner 1962. VII, 271 pages.

Hilbert 1968

David Hilbert: *Grundlagen der Geometrie.* Mit Supplementen von Paul Bernays. 10. Auflage. – Stuttgart: Teubner 1968. VII, 271 pages.

Hilbert 1971a

D. Hilbert: *Foundations of geometry.*. Revised and enlarged by P. Bernays. – La Salle: Open Court 1971. IX, 226 pages.

Hilbert 1971b

David Hilbert: *Les fondements de la géométrie.* Édition critique avec introduction et compléments préparée par Paul Rossier. – Paris: Dunod 1971. XV, 310 pages.

Hilbert 1972

David Hilbert: *Grundlagen der Geometrie.* Mit Supplementen von Dr. Paul Bernays. 11. Auflage. – Stuttgart: Teubner 1972. VII, 271 pages.

Hilbert 1977

D. Hilbert: *Grundlagen der Geometrie.* Mit Ergänzungen von Paul Bernays. 12. Auflage. – Stuttgart: Teubner 1977. VII, 271 pages.

Hilbert and Klein 1985

David Hilbert and Felix Klein: *Der Briefwechsel David Hilbert – Felix Klein ⟨1886–1918⟩.* Mit Anmerkungen herausgegeben von Günther Frei. – Göttingen: Vandenhoeck & Ruprecht 1985. (Arbeiten aus der Niedersächsischen Staats- und Universitätsbibliothek Göttingen; 19.) XII, 153 pages.

Hilbert 1987

D. Hilbert: *Grundlagen der Geometrie.* Mit Ergänzungen von Paul Bernays. 13. Auflage. – Stuttgart: Teubner 1987. VII, 271 pages.

Hilbert 1990a

Hilbert: *Ausgewählte Kapitel der Zahlentheorie.* Sommersemester 1898. – Göttingen: (1990). 2 uncounted, 77 pages.

Hilbert 1990b

Hilbert: *Zahlentheorie.* Wintersemester 1897/98. – Göttingen: (1990). 2 uncounted, II, 204 pages.

Hilbert 1992

David Hilbert: *Natur und mathematisches Erkennen.* Vorlesung, gehalten 1919–1920 in Göttingen. Nach der Ausarbeitung von Paul Bernays. Hrsg. von David E. Rowe. – Basel, Boston, Berlin: Birkhäuser 1992. xxiv, 101 pages.

Hilbert 1999a

David Hilbert: *Grundlagen der Geometrie.* Mit Supplementen von Paul Bernays. 14. Auflage. Herausgegeben und mit Anhängen versehen von Michael Toepell. – Leipzig: Teubner 1999. (Teubner-Archiv zur Mathematik; Supplement 6.) xvi, VIII, 408 pages.

Hilbert 1999b

David Hilbert: *Schulheft zur Neueren Geometrie. Wilhelmsgymnasium Königsberg i. Pr. 1879/80.* Kommentiert und mit einer Einleitung versehen von Michael Toepell. [*Cod. Ms. D. Hilbert 758*]. – In *Hilbert 1999a*, 327–345.

Hjelmslev 1907

J. Hjelmslev: *Neue Begründung der ebenen Geometrie.* – Mathematische Annalen 64 (1907), 449–474.

Hölder 1900

Otto Hölder: *Anschauung und Denken in der Geometrie.* Akademische Antrittsvorlesung gehalten am 22. Juli 1899. Mit Zusätzen, Anmerkungen und einem Register. – Leipzig: Teubner 1900.

Hoüel 1876

J. Hoüel: *Ueber die Rolle der Erfahrung in den exacten Wissenschaften.* Mit Bewilligung des Verfassers übersetzt von Felix Müller. – Archiv der Mathematik und Physik mit besonderer Rücksicht auf die Bedürfnisse der Lehrer an höheren Unterrichtsanstalten 59 (1876), 65–75.

Kagan 1903

B. Kagan: *Über die Transformation der Polyeder.* – Mathematische Annalen 57 (1903), 521–524.

Karzel and Kroll 1988

Helmut Karzel and Hans-Joachim Kroll: *Geschichte der Geometrie seit Hilbert.* – Darmstadt: Wissenschaftliche Buchgesellschaft (1988).

Kiechle et al. 1999

Hubert Kiechle et al.: *Das Forschungsgebiet „Grundlagen der Geometrie" seit Hilbert.* – In *Hilbert 1999a*, 365–384.

Killing 1878

Wilhelm Killing: *Ueber zwei Raumformen mit constanter positiver Krümmung.* Mit Rücksicht auf die Abhandlung des Herrn Newcomb im 83. Bande dieses Journals. – Journal für die reine und angewandte Mathematik 86 (1878), 72–83.

Killing 1885

Wilhelm Killing: *Die Nicht-Euklidischen Raumformen in analytischer Behandlung.* – Leipzig: Teubner 1885. XII, 264 pages.

Killing 1892

Wilhelm Killing: *Ueber die Grundlagen der Geometrie.* – Journal für die reine und angewandte Mathematik 109 (1892), 121–186.

Killing 1893

Wilhelm Killing: *Einführung in die Grundlagen der Geometrie.* Erster Band. – Paderborn: Schöningh 1893. X, 357 pages.

Killing 1898

Wilhelm Killing: *Einführung in die Grundlagen der Geometrie.* Zweiter Band. – Paderborn: Schöningh 1898. VI, 361 pages.

Kirchhoff 1877

Gustav Kirchhoff: *Vorlesungen über Mathematische Physik. Mechanik.*. Zweite Auflage. – Leipzig: Teubner 1877. VII, 466 pages.

Klein 1871

Felix Klein: *Ueber die sogenannte Nicht-Euklidische Geometrie.* – Mathematische Annalen 4 (1871), 573–625.

Klein 1872

Felix Klein: *Vergleichende Betrachtungen über neuere geometrische Forschungen. Programm zum Eintritt in die philosophische Facultät und den Senat der k. Friedrich-Alexanders-Universität zu Erlangen.* – Erlangen: Deichert 1872. 48 pages.

Klein 1873

Felix Klein: *Ueber die sogenannte Nicht-Euklidische Geometrie. ⟨Zweiter Aufsatz⟩.* – Mathematische Annalen 6 (1873), 112–145.

Klein 1890

F. Klein: *Nicht-Euklidische Geometrie. II. Vorlesung, gehalten während des Sommersemesters 1890.* – Göttingen: [no publisher] 1890. 2 uncounted, 257, III pages.

Klein 1892

F. Klein: *Nicht-Euklidische Geometrie.* I. Vorlesung, gehalten während des Wintersemesters 1889–90. – Göttingen: [no publisher] 1892. 2 uncounted, 364, IV pages.

Klein 1893a

F. Klein: *Einleitung in die höhere Geometrie.* II. Vorlesung, gehalten im Sommersemester 1893. Ausgearbeitet von Fr. Schilling. – Göttingen: Autographie und Druck von Baumann & Cie., Cassel 1893. 6 uncounted, 388 pages.

Klein 1893b

F. Klein: *Einleitung in die höhere Geometrie.* I. Vorlesung, gehalten im Wintersemester 1892–93. Ausgearbeitet von Fr. Schilling. – Göttingen: [no publisher] 1893. VIII, 566 pages.

Klein 1894

F. Klein: *Riemann und seine Bedeutung für die Entwicklung der modernen Mathematik.* – Verhandlungen der Gesellschaft Deutscher Naturforscher und Ärzte 66 (1894), Erster Theil. Die allgemeinen Sitzungen, 57–72.

Klein 1895

F. Klein: *Vorträge über ausgewählte Fragen der Elementargeometrie.* Ausgearbeitet von F. Tägert. Eine Festschrift zu der Pfingsten 1895 in Göttingen stattfindenden dritten Versammlung des Vereins zur Förderung des mathematischen und naturwissenschaftlichen Unterrichts. – Leipzig: Teubner 1895. VI, 66 pages, 2 tables.

Klein 1898

Felix Klein: *Gutachten, betreffend den dritten Band der Theorie der Transformationsgruppen von S. Lie anlässlich der ersten Vertheilung des Lobatschewsky-Preises.* – Mathematische Annalen 50 (1898), 583–600.

Klein 1923

Felix Klein: *Gesammelte mathematische Abhandlungen. Dritter Band. Elliptische Funktionen, insbesondere Modulfunktionen. Hyperelliptische und Abelsche Funktionen. Riemannsche Funktionentheorie und automorphe Funktionen.* – Berlin: Springer 1923. X, 774, 36 pages.

Klein 1926

Felix Klein: *Vorlesungen über höhere Geometrie.* – Berlin: Springer 1926. (Die Grundlehren der mathematischen Wissenschaften in Einzeldarstellungen mit besonderer Berücksichtigung der Anwendungsgebiete; 22.) VIII, 405 pages.

Kline 1972

Morris Kline: *Mathematical Thought from Ancient to Modern Times.* – New York: Oxford University Press 1972. xviii, 1238 pages.

Kneser 1902a

A. Kneser: *Neue Begründung der Proportions- und Ähnlichkeitslehre unabhängig vom Archimedischen Axiom und dem Begriff des Inkommensurabeln.* – Archiv der Mathematik und Physik mit besonderer Rücksicht auf die Bedürfnisse der Lehrer an höheren Unterrichtsanstalten. 3. Reihe 2 (1902).

Kneser 1902b

A. Kneser: *Neue Begründung der Proportions- und Ähnlichkeitslehre unabhängig vom Archimedischen Axiom und dem Begriff des Inkommensurabeln.* – Sitzungsberichte der Berliner Mathematischen Gesellschaft 1 (1902), 4–9.

Knopp 1947

Konrad Knopp: *Theorie und Anwendung der unendlichen Reihen.* – Berlin, Heidelberg: Springer 1947. (Die Grundlehren der mathematischen Wissenschaften in Einzeldarstellungen mit besonderer Berücksichtigung der Anwendungsgebiete; 2.) XII, 582 pages.

Kürschák 1902

Josef Kürschák: *Das Streckenabtragen.* – Mathematische Annalen 55 (1902), 597–598.

Kupffer 1894

Karl Kupffer: *Die Darstellung einiger Kapitel der Elementarmathematik.* – Sitzungsberichte der Naturforscher-Gesellschaft bei der Universität Jurjew ⟨Dorpat⟩ 10 (1894), Zweites Heft, 359–385.

Kutta 1897

W. M. Kutta: *Zur Geschichte der Geometrie mit constanter Zirkelöffnung.* – Nova Acta, Abhandlungen der Kaiserlichen Leopoldinisch-Carolinischen Deutschen Akademie der Naturforscher 71 (1897), 69–102.

Lambert 1768

Johann Heinrich Lambert: *Mémoire sur quelques propriétés remarquables des quantités transcendentes circulaires et logarithmiques.* – In *Lambert 1948*, 112–159.

Lambert 1770a

Johann Heinrich Lambert: *Beyträge zum Gebrauche der Mathematik und deren Anwendung. Zweyter Theil. Erster Abschnitt.* – Berlin: Buchhandlung der Realschule 1770. 24 uncounted, 816 pages, Appendix, 11 tables.

Lambert 1770b

Johann Heinrich Lambert: *Vorläufige Kenntnisse für die, so die Quadratur und Rectification des Circuls suchen.* – In *Lambert 1770a*, 140–169.

Lambert 1786

Johann Heinrich Lambert: *Theorie der Parallellinien.* – Leipziger Magazin für reine und angewandte Mathematik (1786), 137–164, 325–358.

Lambert 1948

Iohannes Henricus Lambert: *Opera Mathematica. Volumen secundum. Commentationes arithmeticae algebraicae et analyticae. Pars altera. Edidit Andreas Speiser.* – Turici: Orell Füssli 1948. XXX, 324 pages.

Landau 1903

Edmund Landau: *Über die Darstellung definiter binärer Formen durch Quadrate.* – Mathematische Annalen 57 (1903), 53–64.

Legendre 1794

Adrien Marie Legendre: *Eléments de géométrie.* – Paris: 1794.

Legendre 1800

Adrien Marie Legendre: *Éléments de Géométrie, avec des notes.* – Paris: Didot An IX [1800].

Legendre 1833

Adrien Marie Legendre: *Réflexions sur différentes manières de démontrer la théorie des parallèles ou le théorème sur la somme des trois angles du triangle.* – Mémoires de l'Académie Royale des Sciences de l'Institut de France 12 (1833), 367–410.

Leibniz 1858

G. W. Leibniz: *Mathematische Schriften herausgegeben von C. I. Gerhardt. (Gesammelte Werke herausgegeben von Georg Heinrich Pertz. Dritte Folge Mathematik. Fünfter Band.)* Zweite Abtheilung. Die mathematischen Abhandlungen. Band I. – Halle: Schmidt 1858. X, 418 pages.

Lie 1890a

Sophus Lie: *Ueber die Grundlagen der Geometrie. (Erste Abhandlung).* – Berichte über die Verhandlungen der königlich sächsischen Gesellschaft der Wissenschaften zu Leipzig. Mathematisch-physische Classe 42 (1890), 284–321.

Lie 1890b

Sophus Lie: *Ueber die Grundlagen der Geometrie. (Zweite Abhandlung).* – Berichte über die Verhandlungen der königlich sächsischen Gesellschaft der Wissenschaften zu Leipzig. Mathematisch-physische Classe 42 (1890), 355–418.

Lie 1893

Sophus Lie: *Theorie der Transformationsgruppen.* Dritter und letzter Abschnitt. Unter Mitwirkung von Friedrich Engel. – Leipzig: Teubner 1893. XXVIII, 831 pages.

Liebmann 1898

(Heinrich Liebmann): *Noch einmal der Göttinger Ferienkursus.* – Unterrichtsblätter für Mathematik und Naturwissenschaften. Organ des Vereins zur Förderung des Unterrichts in der Mathematik und den Naturwissenschaften 4 (1898), 71–72.

Lindemann 1891

Ferdinand Lindemann: *Die Hypothesen der Geometrie.* – Schriften der physikalisch-ökonomischen Gesellschaft zu Königsberg in Pr. Bericht über die in den Sitzungen [...] gehaltenen Vorträge 32 (1891), 20–23.

Lipschitz 1877

Rudolf Lipschitz: *Lehrbuch der Analysis. Erster Band. Grundlagen der Analysis.* – Bonn: Cohen 1877. XVI, 594 pages.

Lipschitz 1880

Rudolf Lipschitz: *Lehrbuch der Analysis. Zweiter Band. Differential- und Integralrechnung.* – Bonn: Cohen 1880. XIV, 734 pages.

Lobachevsky 1837

N. I. Lobachevsky: *Géométrie imaginaire.* – Journal für die reine und angewandte Mathematik 17 (1837), 295–320.

Lobachevsky 1840

N. I. Lobachevsky: *Geometrische Untersuchungen zur Theorie der Parallellinien.* – Berlin: Fincke 1840. 2 tables, 2 uncounted, 61 pages.

Lobachevsky 1886

N. I. Lobachevsky: *Collection complète des œuvres géométriques. Édition de l'université Impériale de Kasan. Tome second. Ouvrages en langues française et allemande.* – Kazan: Tipografia Imperatorskago Universiteta 1886. 2 uncounted, 8 pages, pages 551–680, XX pages.

Lorey 1916

Wilhelm Lorey: *Das Studium der Mathematik an den deutschen Universitäten seit Anfang des 19. Jahrhunderts.* – Leipzig, Berlin: Teubner 1916. (Abhandlungen über den mathematischen Unterricht in Deutschland, veranlasst durch die Internationale Mathematische Unterrichtskommission, herausgegeben von F. Klein; III, Heft 9.) XII, 431 pages.

Loria 1888

Gino Loria: *Die hauptsächlichsten Theorien der Geometrie in ihrer früheren und heutigen Entwickelung. Historische Monographie. Ins Deutsche übertragen von Fritz Schütte.* – Leipzig: Teubner 1888. VI, 132 pages.

Lotze 1879

Hermann Lotze: *Metaphysik. Drei Bücher der Ontologie, Kosmologie und Psychologie. (System der Philosophie. Zweiter Theil.)* – Leipzig: Hirzel 1879. 8 uncounted, 604 pages.

Lüroth 1875

J. Lüroth: *Das Imaginäre in der Geometrie und das Rechnen mit Würfen. Darstellung und Erweiterung der v. Staudt'schen Theorie.* – Mathematische Annalen 8 (1875), 145–214.

Möbius 1827

August Ferdinand Möbius: *Der barycentrische Calcul[,] ein neues Hülfsmittel zur analytischen Behandlung der Geometrie dargestellt und insbesondere auf die Bildung neuer Classen von Aufgaben und die Entwickelung mehrerer Eigenschaften der Kegelschnitte angewendet.* – Leipzig: Barth 1827. XXIV, 455 pages. Reprinted in *Möbius 1885*, 1–388.

Möbius 1885

August Ferdinand Möbius: *Gesammelte Werke. Erster Band. Mit einem Bildnisse von Möbius. Herausgegeben von R. Baltzer.* – Leipzig: Hirzel 1885. XX, 633 pages.

Metternich 1815

Matthias Metternich: *Vollständige Theorie der Parallellinien.* – Mainz: Kupferberg 1815. XVI, 44 pages, 1 table.

Minkowski 1896/1910

Hermann Minkowski: *Geometrie der Zahlen. In zwei Lieferungen. Erste Lieferung; zweite Lieferung.* – Leipzig, Berlin: Teubner 1896; 1910. X pages, pages 1–240; pages 241–256.

Minkowski 1905

Hermann Minkowski: *Peter Gustav Lejeune Dirichlet und seine Bedeutung für die heutige Mathematik. Rede gehalten in der Festsitzung der Göttinger Mathematischen Gesellschaft am 13. Februar 1905.* – Jahresbericht der Deutschen Mathematiker-Vereinigung 14 (1905), 149–163. Reprinted in *Minkowski 1911*, 445–461.

Minkowski 1911

Hermann Minkowski: *Gesammelte Abhandlungen. Zweiter Band.* Unter Mitwirkung von Andreas Speiser und Hermann Weyl herausgegeben von David Hilbert. – Leipzig, Berlin: Teubner 1911. IV, 466 pages, double table.

Minkowski 1973

Hermann Minkowski: *Briefe an David Hilbert. Mit Beiträgen und herausgegeben von L. Rüdenberg und H. Zassenhaus.* – Berlin, Heidelberg, New York: Springer 1973.

Mitteilungen 1900

Mitteilungen der Verlagsbuchhandlung B. G. Teubner in Leipzig. 33. Jahrgang. – Leipzig: Teubner 1900. 192 pages.

Mollerup 1903a

J. Mollerup: *Die Lehre von den geometrischen Proportionen.* – Mathematische Annalen 56 (1903) [printed: 1902], 277–280.

Mollerup 1903b

Johannes Mollerup: *Studier over den plane Geometris Aksiomer.* – København: Høst 1903. 4 uncounted, 88 pages.

Mollerup 1904

J. Mollerup: *Die Beweise der ebenen Geometrie ohne Benutzung der Gleichheit und Ungleichheit der Winkel.* – Mathematische Annalen 58 (1904), 479–496.

Monge 1799

Gaspard Monge: *Géométrie descriptive. Leçons données aux Écoles Normales, l'an 3 de la République.* – Paris: Baudouin VII [1799]. viii, 132 pages, Planche I–XXV.

Monge 1811

G. Monge: *Géométrie descriptive. Nouvelle édition, Avec un Supplément, par M. Hachette.* – Paris: Klostermann M. DCCC. XI. xii, 162, 24 tables, viii, 118, 2 uncounted pages, 11 tables.

Monna 1975

A. F. Monna: *Dirichlet's principle. A mathematical comedy of errors and its influence on the development of analysis.* – Utrecht: Oosthoek, Scheltema & Holkema 1975. VIII, 138 pages.

Moore 1902

Eliakim Hastings Moore: *On the projective axioms of geometry.* – Transactions of the American Mathematical Society 3 (1902), 142–158.

Moulton 1902

Forest Ray Moulton: *A simple non-desarguesian plane geometry.* – Transactions of the American Mathematical Society 3 (1902), 192–195.

Neugebauer 1928

O. Neugebauer: *Über die Einrichtungen des Mathematischen Institutes der Universität Göttingen.* – Minerva-Zeitschrift 4 (1928), 5 und 6, 107–111.

Niewenglowski and Gérard 1898

B. Niewenglowski and L. Gérard: *Cours de géométrie élémentaire à l'usage des élèves de mathématiques élémentiares, de mathématiques spéciales, ⟨...⟩. I. Géométrie plane.* – Paris: Carré & Naud 1898. XX, 362 pages.

Noth 1882

Hermann Noth: *Die Arithmetik der Lage. Ein neues Hilfsmittel zur analytischen Behandlung der Raumlehre. Mit Berücksichtigung ebener geometrischer Gebilde erster und zweiter Ordnung.* – Leipzig: Barth 1882. VI, 89 pages.

Olbers 1900

Wilhelm Olbers: *Sein Leben und seine Werke. Im Auftrage der Nachkommen herausgegeben von C. Schilling. Zweiter Band. Briefwechsel zwischen Olbers und Gauss. Erste Abtheilung.* – Berlin: Springer 1900. VIII, 767 pages.

Pasch 1882

Moritz Pasch: *Vorlesungen über neuere Geometrie.* – Leipzig: Teubner 1882. IV, 202 pages.

Pasch 1926

Moritz Pasch: *Vorlesungen über neuere Geometrie.* Zweite Auflage. Mit einem Anhang: Die Grundlegung der Geometrie in historischer Entwicklung von Max Dehn. – Berlin: Springer 1926. (Die Grundlehren der mathematischen Wissenschaften in Einzeldarstellungen mit besonderer Berücksichtigung der Anwendungsgebiete; 23.) X, 275 pages.

Peano 1891

G. Peano: *Die Grundzüge des geometrischen Calculs.* Autorisirte deutsche Ausgabe von Adolf Schepp. – Leipzig: Teubner 1891. 4 uncounted, 38 pages.

Playfair 1795

John Playfair: *Elements of Geometry; containing the First Six Books of Euclid, with two Books on the Geometry of Solids. To which are added, Elements of Plane and Spherical Trigonometry.* – Edinburgh; London: Bell & Bradfute; Robinson MDCCXCV. xvi, 400 pages.

Poincaré 1902

(H. Poincaré) [review]: *Hilbert. Les Fondements de la Géométrie.* – Bulletin des sciences mathématiques. Deuxième série 26 (1902), 249–272.

Pringsheim 1899

Alfred Pringsheim: *Grundlagen der allgemeinen Funktionenlehre.* – In *Encyklopädie 1899–1916*, 1–53.

Probleme 1971

Die Hilbertschen Probleme. Vortrag „Mathematische Probleme" von D. Hilbert, gehalten auf dem 2. Internationalen Mathematikerkongreß Paris 1900, erläutert von einem Autorenkollektiv unter der Redaktion von P. S. Alexandrov. – Leipzig: Akademische Verlagsgesellschaft Geest & Portig 1971. 302 pages.

Rédei 1965

László Rédei: *Begründung der Euklidischen und Nichteuklidischen Geometrie.* – Leipzig: Teubner 1965.

Reid 1970

Constance Reid: *Hilbert.* – Berlin, New York: Springer 1970.

Reidemeister 1930

Kurt Reidemeister: *Grundlagen der Geometrie.* – Berlin: Springer 1930. (Die Grundlehren der mathematischen Wissenschaften in Einzeldarstellungen mit besonderer Berücksichtigung der Anwendungsgebiete; 32.) X, 147 pages.

Reye 1886

Theodor Reye: *Die Geometrie der Lage. Vorträge. Erste Abtheilung. Dritte vermehrte Auflage.* – Leipzig: Baumgärtner 1886. XIV, 248 pages.

Reye 1892a

Theodor Reye: *Die Geometrie der Lage. Vorträge. Dritte Abtheilung der dritten vermehrten Auflage.* – Leipzig: Baumgärtner 1892. XIV, 224 pages.

Reye 1892b

Theodor Reye: *Die Geometrie der Lage. Vorträge. Zweite Abtheilung. Dritte vermehrte Auflage.* – Leipzig: Baumgärtner 1892. XIV, 330 pages.

Riemann 1851

B. Riemann: *Grundlagen für eine allgemeine Theorie der Functionen einer veränderlichen complexen Grösse.* – Göttingen, Phil. Diss. 1851. 2 uncounted, 32 pages.

Riemann 1857

B. Riemann: *Theorie der* Abel*'schen Functionen.* – Journal für die reine und angewandte Mathematik 54 (1857), 115–155.

Riemann 1866/67

B. Riemann: *Ueber die Hypothesen, welche der Geometrie zu Grunde liegen.* Aus dem Nachlass des Verfassers mitgetheilt durch R. Dedekind. – Abhandlungen der mathematischen Classe der Königlichen Gesellschaft der Wissenschaften zu Göttingen 13 (1866/67) [printed: 1868], 133–152.

Riemann 1876

Bernhard Riemann: *Gesammelte mathematische Werke und wissenschaftlicher Nachlass. Herausgegeben unter Mitwirkung von R. Dedekind von H. Weber.* – Leipzig: Teuber 1876. VIII, 526 pages.

Réthy 1891

Moritz Réthy: *Endlich-gleiche Flächen.* – Mathematische Annalen 38 (1891), 405–428.

Réthy 1893

Moritz Réthy: *Ueber endlich-gleiche Flaechen.* – Mathematische Annalen 42 (1893), 297–307.

Réthy 1894

Moritz Réthy: *Zum Beweis des Hauptsatzes über die Endlichgleichheit zweier ebener Systeme.* – Mathematische Annalen 45 (1894), 471–472.

Süß 1921

Wilhelm Süß: *Begründung der Lehre vom Polyederinhalt.* Mathematische Annalen 82 (1921), 297–305.

Saccheri 1733

Hieronymus Saccherius: *Euclides ab omni naevo vindicatus: sive conatus geometricus quo stabiliuntur prima ipsa universae Geometriae Principia.* – Mediolani: Montani 1733. XVI, 141 pages, 6 tables.

Samuel 1988

Pierre Samuel: *Projective Geometry.* Translated from *Géométrie projective* (Presses Universitaires de France) by Silvio Levy. – New York [and others]: Springer 1988. (Undergraduate texts in mathematics. Readings in mathematics.) x, 156 pages.

Savile 1621

(Henricus Savilius): *Praelectiones tresdecim in principium elementorum Euclidis, Oxonii habitae. M. DC. XX.* – Oxonii: Iohannes Lichfield, & Iacobus Short 1621. 4 uncounted, 260 pages.

Schlegel 1872

Victor Schlegel: *System der Raumlehre.* Nach den Prinzipien der Grassmann'schen Ausdehnungslehre und als Einleitung in dieselbe dargestellt. Erster Theil: Geometrie. Die Gebiete des Punktes, der Geraden, der Ebene. – Leipzig: Teubner 1872. XVI, 156 pages.

Schlegel 1896

V. Schlegel: *Die Grassmann'sche Ausdehnungslehre. Ein Beitrag zur Geschichte der Mathematik in den letzten fünfzig Jahren.* – Zeitschrift für Mathematik und Physik 41 (1896), 1–21, 41–59.

Schmidt 1933

Arnold Schmidt: *Zu Hilberts Grundlegung der Geometrie.* – In *Hilbert 1933*, 404–414.

Schoenflies 1888

A. Schoenflies: *Ueber die regelmässigen Configurationen n_3.* – Mathematische Annalen 31 (1888), 43–69.

Schoenflies 1891

A. Schoenflies: *Ueber Configurationen, welche sich aus gegebenen Raumelementen durch blosses Schneiden und Verbinden ableiten lassen.* – In *Verhandlungen 1891*, Zweiter Theil. Abtheilungs-Sitzungen, 12f.

Schubert 1898

Hermann Schubert: *Mathematische Mussestunden.* Eine Sammlung von Geduldspielen, Kunststücken und Unterhaltungsaufgaben mathematischer Natur. – Leipzig: Göschen 1898. VI, 286 pages.

Schur 1891

Friedrich Schur: *Ueber die Einführung der sogenannten idealen Elemente in die projective Geometrie.* – Mathematische Annalen 39 (1891), 113–124.

Schur 1893

Friedrich Schur: *Ueber den Flächeninhalt geradlinig begrenzter ebener Figuren.* – Sitzungsberichte der Naturforscher-Gesellschaft bei der Universität Dorpat 10 (1893), Erstes Heft, 2–6.

Schur 1898

Friedrich Schur: *Lehrbuch der analytischen Geometrie.* – Leipzig: Veit & Comp. 1898. X, 216 pages.

Schur 1899

Friedrich Schur: *Ueber den Fundamentalsatz der projectiven Geometrie.* – Mathematische Annalen 51 (1899), 401–409.

Schur 1902

Friedrich Schur: *Ueber die Grundlagen der Geometrie.* – Mathematische Annalen 55 (1902), 265–292.

Schur 1903

Friedrich Schur: *Zur Proportionslehre.* – Mathematische Annalen 57 (1903), 205–208.

Schur 1909

Friedrich Schur: *Grundlagen der Geometrie.* – Leipzig, Berlin: Teubner 1909. X, 192 pages.

Schwab 1814

J. C. Schwab: *Commentatio in primum Elementorum Euclidis Librum, qua veritatem Geometriae principiis ontologicis niti evincitur, omnesque propositiones, axiomatum geometricorum loco habitae, demonstrantur.* – Stuttgartiae: Steinkopf 1814. 67 pages, 1 table.

Serret 1878

J. A. Serret: *Handbuch der höheren Algebra. Deutsche Übersetzung von G. Wertheim. Erster Band.* – Leipzig: Teubner 1878. VIII, 528 pages.

Serret 1879

J. A. Serret: *Handbuch der höheren Algebra. Deutsche Übersetzung von G. Wertheim. Zweiter Band.* – Leipzig: Teubner 1879. VIII, 574 pages.

Simon 1890

Max Simon: *Die Elemente der Geometrie mit Rücksicht auf die absolute Geometrie.* – Strassburger Druckerei und Verlagsanstalt 1890. IV, 74 pages.

Sommer 1899/1900

(J. Sommer) [review]: *Hilbert's foundations of geometry.* – Bulletin of the American Mathematical Society 6 (1899/1900), 287–299.

von Staudt 1847

Georg Karl Christian v. Staudt: *Geometrie der Lage.* – Nürnberg: Friedr. Korn'sche Buchhandlung 1847. VI, 216 pages.

von Staudt 1856

Karl Georg Christian v. Staudt: *Beiträge zur Geometrie der Lage. Erstes Heft.* – Nürnberg: Bauer und Raspe 1856. VI, 1–129.

von Staudt 1857

Karl Georg Christian v. Staudt: *Beiträge zur Geometrie der Lage. Zweites Heft.* – Nürnberg: Bauer und Raspe 1857. IV, 131–286.

von Staudt 1860

Karl Georg Christian v. Staudt: *Beiträge zur Geometrie der Lage. Drittes Heft. ⟨Schlussheft⟩.* – Nürnberg: Bauer und Raspe 1860. VI, 285–396.

Stäckel and Engel 1895

Paul Stäckel and Friedrich Engel: *Die Theorie der Parallellinien von Euklid bis auf Gauss, eine Urkundensammlung zur Vorgeschichte der nichteuklidischen Geometrie.* – Leipzig: Teubner 1895. X, 325 pages, 1 table.

Stäckel 1923

Paul Stäckel: *Gauss als Geometer.* – In *Gauß 1922–1933.* 123 pages.

Steck 1981

Max Steck: *Bibliographia Euclideana.* – Hildesheim: Gerstenberg 1981. 444 pages.

Steiner 1826

Jacob Steiner: *Einige geometrische Betrachtungen.* – Journal für die reine und angewandte Mathematik 1 (1826), 161–184/252–288.

Steiner 1832

Jacob Steiner: *Systematische Entwickelung der Abhängigkeit geometrischer Gestalten von einander, mit Berücksichtigung der Arbeiten alter und neuer Geometer über Porismen, Projections-Methoden, Geometrie der Lage, Transversalen, Dualität und Reciprocität, etc.* Erster Theil. – Berlin: Fincke 1832. XVI, 322 pages.

Steiner 1833

Jacob Steiner: *Die geometrischen Konstructionen, ausgeführt mittelst der geraden Linie und eines festen Kreises, als Lehrgegenstand auf höheren Unterrichts-Anstalten und zur praktischen Benutzung.* – Berlin: Dümmler 1833. 4 uncounted, 110 pages, II tables.

Steiner 1931

Jakob Steiner: *Allgemeine Theorie über das Berühren und Schneiden der Kreise und der Kugeln, worunter eine grosse Anzahl neuer Untersuchungen und Sätze vorkommen, in einem systematischen Entwicklungsgange dargestellt.* Von Dr. Rud. Fueter aus Steiners Nachlass herausgegeben unter Mitwirkung von Dr. F. Gonseth. – Zürich, Leipzig: Orell Füssli 1931. (Veröffentlichungen der Schweizer Mathematischen Gesellschaft / Publications de la Société Mathématique Suisse; 5.) XVIII, 345 pages.

Steiner 1867

Jacob Steiner: *Die Theorie der Kegelschnitte, gestützt auf projektivische Eigenschaften. Auf Grund von Universitätsvorträgen und mit Benutzung hinterlassener Manuscripte bearbeitet von Heinrich Schröter. (Vorlesungen über synthetische Geometrie. Zweiter Theil.)* – Leipzig: Teubner 1867. XX, 566 pages.

Steinitz 1904

(E. Steinitz) [review]: F. R. Moulton: *A simple non-Desarguesian plane geometry.* – Jahrbuch über die Fortschritte der Mathematik 33 (1902), 497–498.

Stolz 1885

Otto Stolz: *Vorlesungen über allgemeine Arithmetik. Nach den neueren Ansichten. Erster Theil: Allgemeines und Arithmetik der reellen Zahlen.* – Leipzig: Teubner 1885. VI, 344 pages.

Stolz 1886

Otto Stolz: *Vorlesungen über allgemeine Arithmetik. Nach den neueren Ansichten. Nach den neueren Ansichten. Zweiter Theil: Arithmetik der complexen Zahlen mit geometrischen Anwendungen.* – Leipzig: Teubner 1886. VIII, 326 pages.

Stolz 1894

O. Stolz: *Die ebenen Vielecke und die Winkel mit Einschluss der Berührungs-Winkel als Systeme von absoluten Grössen.* – Monatshefte für Mathematik und Physik 5 (1894), 233–254.

Sturm, Schröder and Sohnke 1878

(R. Sturm, E. Schröder and L. Sohnke): *Hermann Grassmann. Sein Leben und seine mathematisch-physikalischen Arbeiten.* ⟨Mit einem Verzeichnisse sämmtlicher Schriften Grassmann's.⟩ – Mathematische Annalen 14 (1878), 1–45.

Tannery 1895

Jules Tannery: *Introduction à l'étude de la Théorie des Nombres et de l'Algèbre Supérieure par Émile Borel et Jules Drach. D'après des Conférences faites à l'École Normale Supérieure.* – Paris: Libraire Nony 1895. IX, 350 pages.

Thibaut 1809

Bernhard Friedrich Thibaut: *Grundriß der reinen Mathematik zum Gebrauch bey academischen Vorlesungen.* – Göttingen: Vandenhoek und Ruprecht 1809. X, 2 uncounted, 448 pages, 5 tables.

Thiele 1878

Günther Thiele: *Grundriss der Logik und Metaphysik, dargestellt als Entwicklung des endlichen Geistes.* – Halle: Niemeyer 1878. XII, 214 pages.

Thomae 1873

Johannes Thomae: *Ebene geometrische Gebilde erster und zweiter Ordnung vom Standpunkte der Geometrie der Lage betrachtet.* – Halle a/S.: Nebert 1873. IV, 44 pages.

Thomae 1880

Johannes Thomae: *Elementare Theorie der analytischen Functionen einer complexen Veränderlichen.* – Halle a/S.: Nebert 1880. VIII, 132 pages.

Tijdeman 1976

R. Tijdeman: *Hilbert's seventh problem: on the Gel'fond Baker method and its applications.* – In *Developments 1976*, 241–268.

Toepell 1986

Michael-Markus Toepell: *Über die Entstehung von David Hilberts „Grundlagen der Geometrie“.* – Göttingen: Vandenhoeck & Ruprecht (1986). (Studien zur Wissenschafts-, Sozial- und Bildungsgeschichte der Mathematik; 2.) XIV, 293 pages.

Toepell 1999a

Michael Toepell: *Die projektive Geometrie als Forschungsgrundlage David Hilberts.* – In *Hilbert 1999a*, 347–361.

Toepell 1999b

Michael Toepell: *Zur Entstehung und Weiterentwicklung von David Hilberts „Grundlagen der Geometrie“.* – In *Hilbert 1999a*, 283–324.

Vahlen 1905

Karl Theodor Vahlen: *Abstrakte Geometrie. Untersuchungen über die Grundlagen der Euklidischen und nicht-Euklidischen Geometrie.* – Leipzig: Teubner 1905. XII, 302 pages.

Verhandlungen 1891

Verhandlungen der Gesellschaft deutscher Naturforscher und Ärzte. 64. Versammlung zu Halle a. S. 21.–25. September 1891. – Leipzig: Vogel 1892.

Verhandlungen 1905

Verhandlungen des Dritten Internationalen Mathematiker-Kongresses in Heidelberg vom 8. bis 13. August 1904. – Leipzig: Teubner 1905. X, 756 pages.

Veronese 1894

Guiseppe Veronese: *Grundzüge der Geometrie von mehreren Dimensionen und mehreren Arten gradliniger Einheiten in elementarer Form entwickelt.* Mit Genehmigung des Verfassers nach einer neuen Bearbeitung des Originals übersetzt von Adolf Schepp. – Leipzig: Teubner 1894. XLVIII, 710 pages.

Verzeichnis 1943*

Verzeichnis der von Geheimrat David Hilbert hinterlassenen Handschriften, Bücher, Sonderdrucke, Zeitschriften, [sic]. – Georg-August-Universität Göttingen, Mathematisches Institut, Lesesaal.

Wallis 1663

Johannis Wallis: *De Postulato Quinto et Definitione Quinta. Lib. 6. Euclidis; Disceptatio Geometrica.* – In *Wallis 1693*, 665–678.

Wallis 1693

Johannes Wallis: *De Algebra Tractatus; Historicus & Practicus. Cum variis Appendicibus; Partim prius editis anglice, partim nunc primum editis. Operum Mathematicorum Volumen alterum.* – Oxoniae: Theatrum Sheldonianum 1693. 16 uncounted, 879, 1 uncounted page.

Weber 1895

Heinrich Weber: *Lehrbuch der Algebra. In zwei Bänden. Erster Band.* – Braunschweig: Vieweg 1895. XVI, 653 pages.

Weierstraß 1870

Karl Weierstrass: *Über das sogenannte Dirichlet'sche Princip.* ⟨Gelesen in der Königl. Akademie der Wissenschaften am 14. Juli 1870.⟩ – In *Weierstraß 1895*, 49–54.

Weierstraß 1895

Karl Weierstrass: *Mathematische Werke. Zweiter Band. Abhandlungen II.* – Berlin: Mayer & Müller 1895. VI, 362 pages, 1 uncounted page.

Weyl 1923

Hermann Weyl: *Mathematische Analyse des Raumproblems.* Vorlesungen, gehalten in Barcelona und Madrid. – Berlin: Springer 1923. VIII, 117 pages.

Wiechert 1899

Emil Wiechert: *Grundlagen der Elektrodynamik.* – In *Festschrift 1899*. 112 pages.

Wiener 1891

H. Wiener: *Ueber Grundlagen und Aufbau der Geometrie.* – In *Verhandlungen 1891*, Zweiter Theil. Abtheilungs–Sitzungen, 8f.

Worpitzky 1873

Worpitzky: *Ueber die Grundbegriffe der Geometrie.* – Archiv der Mathematik und Physik mit besonderer Rücksicht auf die Bedürfnisse der Lehrer an höheren Unterrichtsanstalten 55 (1873), 405–421.

Wundt 1893

Wilhelm Wundt: *Erkenntnisslehre. (Logik. Eine Untersuchung der Principien der Erkenntniss und der Methoden wissenschaftlicher Forschung. Zwei Bände. Erster Band.) Zweite umgearbeitete Auflage.* – Stuttgart: Enke 1893. XIV, 651 pages.

Wundt 1894

Wilhelm Wundt: *Methodenlehre. (Logik. Eine Untersuchung der Principien der Erkenntniss und der Methoden wissenschaftlicher Forschung. Zwei Bände. Zweiter Band.) Erste Abtheilung. Allgemeine Methodenlehre. Logik der Mathematik und der Naturwissenschaften.* Zweite umgearbeitete Auflage. – Stuttgart: Enke 1894. XI, 590 pages.

Wundt 1895

Wilhelm Wundt: *Methodenlehre. (Logik. Eine Untersuchung der Principien der Erkenntniss und der Methoden wissenschaftlicher Forschung. Zwei Bände. Zweiter Band.) Zweite Abtheilung. Logik der Geisteswissenschaften.* Zweite umgearbeitete Auflage. – Stuttgart: Enke 1895. VII, 643 pages.

Wundt 1902

Wilhelm Wundt: *Methodenlehre. (Logik. Eine Untersuchung der Principien der Erkenntniss und der Methoden wissenschaftlicher Forschung. Zwei Bände. Zweiter Band.) Namensverzeichnis und Sachregister. Von Hans Lindau.* – Stuttgart: Enke 1902. 74 pages.

Yaglom 1971

I. M. Jaglom: *Zum vierten Hilbertschen Problem.* – In *Probleme 1971*, 118–125.

de Zolt 1881

Antonio de Zolt: *Principii della eguaglianza di poligoni (equivalenza di poligoni) preceduti da alcuni cenni critici sulla teoria della equivalenza geometrica.* – Milano: Briola 1881.

de Zolt 1883

Antonio de Zolt: *Principii della eguaglianza di poliedri e di poligoni sferici.* – Milano: Briola 1883.

Printing: Strauss GmbH, Mörlenbach
Binding: Schäffer, Grünstadt

LaVergne, TN USA
21 August 2009
155561LV00015B/1/A

9 783540 643739